See Right Through Me

Savvas Andronikou
Editor

See Right Through Me

An Imaging Anatomy Atlas

Second Edition

Springer

Editor
Savvas Andronikou
Department of Radiology
Faculty of Health Sciences
University of the Witwatersrand
Schnetler Corbett and Partners
Cape Town
South Africa

ISBN 978-3-642-23892-5 ISBN 978-3-642-23893-2 (eBook)
DOI 10.1007/978-3-642-23893-2
Springer Heidelberg New York Dordrecht London

Library of Congress Control Number: 2012952997

Printed on acid-free paper

Springer is part of Springer Science+Business Media (www.springer.com)

Preface

If one can point at a structure of the human body on a medical image, there is a name for that structure. *See Right Through Me* is an anatomical atlas that attempts to provide this information in every plane and with every imaging modality, in significant detail from head to toe.

This atlas differs from any predecessors, as it offers not only organ-specific anatomy but also clinically relevant modality-specific information, such as the expected density or signal intensity of normal structures. It also provides information that is grouped for clinical relevance, such as a pediatric anatomy, endocrine anatomy, genital anatomy according to gender, and cranial nerve pathway imaging.

We have made a clear choice at the outset to provide an anatomy atlas with more images than text for an enhanced visual experience. The scattered "pearl" boxes, however, offer insight from experts "in the know," as well as summaries of normal numerical values that are difficult to remember.

See Right Through Me is ideal for radiologists, surgeons, physicians, and even patients, and it has detail enough to satisfy both the expert and student for learning or quick reference purposes. We hope you enjoy it.

Acknowledgements

The chief editor wishes to acknowledge the massive effort of the subeditors who helped to produce the original version of this atlas and those who contributed to this improved product.

Mrs. Annalie Rich has my personal gratitude for her constant administrative support throughout preparation of the material.

Miss Isabella Rangel spent hours of her personal time to help me with this atlas, and I thank her for this.

Lee Klein of Springer who spent hours patiently replacing images, lines, and labels has my eternal gratitude.

Lastly the chief editor acknowledges the Radiological Society of South Africa, which supported production of the original version of this atlas, and set the stage for the educational material for the benefit of their members and trainees.

Savvas Andronikou

Contents

List of Editors and Contributors

Editor

Savvas Andronikou Department of Radiology, Faculty of Health Sciences, University of the Witwatersrand, Schnetler Corbett and Partners, Cape Town, South Africa

Subeditors

Owen Terblanche Department of Radiology, Faculty of Health Sciences, University of the Witwatersrand, Johannesburg, South Africa

Pieter Ernst Boshoff Department of Radiology, Faculty of Health Sciences, University of the Witwatersrand, Johannesburg, South Africa

Matthew Goodier Department of Radiology, Faculty of Health Sciences, University of the Witwatersrand, Johannesburg, South Africa

Tracy Kilborn Department of Paediatric Radiology, Red Cross Children's Hospital, University of Cape Town, Cape Town, South Africa

Matthys van Wyk Department of Radiology, Faculty of Health Sciences, University of the Witwatersrand, Johannesburg, South Africa

Hofmeyr Viljoen Department of Radiology, University of Stellenbosch, Cape Town, South Africa

Nicky Wieselthaler Department of Paediatric Radiology, Red Cross Children's Hospital, University of Cape Town, Cape Town, South Africa

Contributors

Christelle Ackerman Department of Radiology, University of Stellenbosch, Stellenbosch, South Africa

Amarjit Kaur Bajwa Department of Radiology, Faculty of Health Sciences, University of the Witwatersrand, Johannesburg, South Africa

Steve Beningfield Department of Radiology, University of Cape Town, Cape Town, South Africa

Manitha Bindapersad Department of Radiology, Faculty of Health Sciences, University of the Witwatersrand, Johannesburg, South Africa

Andrew Brandt Department of Radiology, University of Stellenbosch, Stellenbosch, South Africa

B.J. Brauer Department of Radiology, Faculty of Health Sciences, University of the Witwatersrand, Johannesburg, South Africa

Anthony Briede Department of Radiology, Faculty of Health Sciences, University of the Witwatersrand, Johannesburg, South Africa

Humza Carim Department of Radiology, Faculty of Health Sciences, University of the Witwatersrand, Johannesburg, South Africa

Dilesh Chhiba Department of Radiology, University of Cape Town, Cape Town, South Africa

Glen Cilliers Department of Radiology, University of Stellenbosch, Stellenbosch, South Africa

Neil Collins Interventional Radiology, North Bristol NHS Trust, Bristol, UK

Vijay Dahya Department of Radiology, University of Stellenbosch, Stellenbosch, South Africa

Richard de Villiers Department of Radiology, University of Stellenbosch, Stellenbosch, South Africa

Coert de Vries Department of Radiology, University of the Free State, Bloemfontein, South Africa

Gerard Deib Mercy Radiology Group, Auckland, New Zealand

Salome Dlangamandla Department of Radiology, Faculty of Health Sciences, University of the Witwatersrand, Johannesburg, South Africa

Jaco du Plessis Department of Radiology, University of Stellenbosch, Stellenbosch, South Africa

Zaeem I. Ebrahim Department of Radiology, University of Pretoria, Pretoria, South Africa

Jenny Edwards Department of Radiology, Faculty of Health Sciences, University of the Witwatersrand, Johannesburg, South Africa

Zelda Erasmus Department of Radiology, University of Pretoria, Pretoria, South Africa

Albert Eshun Department of Radiology, University of Cape Town, Cape Town, South Africa

Gisella Furlan Department of Radiology, University of Cape Town, Cape Town, South Africa

Wolfgang Gaisbichler Department of Radiology, University of Stellenbosch, Stellenbosch, South Africa

Fekade Gebramariam Department of Radiology, University of the Free State, Bloemfontein, South Africa

Alan Gouveia Department of Radiology, Faculty of Health Sciences, University of the Witwatersrand, Johannesburg, South Africa

Nishentha Govender Department of Radiology, Faculty of Health Sciences, University of the Witwatersrand, Johannesburg, South Africa

Patrick Govender Department of Radiology, Faculty of Health Sciences, University of the Witwatersrand, Johannesburg, South Africa

Murray Hayes Department of Radiology, University of Stellenbosch, Stellenbosch, South Africa

Linda Tebogo Hlabangana Department of Radiology, Faculty of Health Sciences, University of the Witwatersrand, Johannesburg, South Africa

Dipesh Himatlal Jogi Department of Radiology, Faculty of Health Sciences, University of the Witwatersrand, Johannesburg, South Africa

Glenda Jonathan Department of Radiology, University of Cape Town, Cape Town, South Africa

Budresh Joshi Department of Radiology, University of Stellenbosch, Stellenbosch, South Africa

Nausheen Khan Department of Radiology, University of Pretoria, Pretoria, South Africa

Paul Kitsa Department of Radiology, University of Cape Town, Cape Town, South Africa

Zarina Lockhat Department of Radiology, University of Pretoria, Pretoria, South Africa

Jan Lotz Department of Radiology, University of Stellenbosch, Stellenbosch, South Africa

H. Brand Louw Department of Radiology, University of Stellenbosch, Stellenbosch, South Africa

J. Jeanne Louw Department of Radiology, University of Pretoria, Pretoria, South Africa

Arthur T. Maydell Department of Radiology, University of Stellenbosch, Stellenbosch, South Africa

Bruce McIvor Department of Radiology, University of Cape Town, Cape Town, South Africa

Carey McKenzie Department of Radiology, University of Cape Town, Cape Town, South Africa

Sandile Mdunge Department of Radiology, University of Cape Town, Cape Town, South Africa

Debra Meerkotter Department of Radiology, Faculty of Health Sciences, University of the Witwatersrand, Johannesburg, South Africa

Joan Masetumo Mbengo Department of Radiology, Faculty of Health Sciences, University of the Witwatersrand, Johannesburg, South Africa

Victor Mngomezulu Department of Radiology, Faculty of Health Sciences, University of the Witwatersrand, Johannesburg, South Africa

Kuben Naidu Department of Radiology, University of Stellenbosch, Stellenbosch, South Africa

Maya Nathu Patel Department of Radiology, Faculty of Health Sciences, University of the Witwatersrand, Johannesburg, South Africa

Jaco Parsons Department of Radiology, University of Stellenbosch, Stellenbosch, South Africa

Linda Potgieter Department of Radiology, Faculty of Health Sciences, University of the Witwatersrand, Johannesburg, South Africa

Sanjay Prabhu Pediatric Radiology, Children's Hospital Boston, Boston, USA

Hendrik Prinsloo Department of Radiology, Faculty of Health Sciences, University of the Witwatersrand, Johannesburg, South Africa

Stefan Jerzy Przybojewski Department of Radiology, University of Stellenbosch, Stellenbosch, South Africa

Jerzy Zbigniew Przybojewski Christian Barnard Memorial Hospital, Cardiology Unit, Cape Town, South Africa

Amer Rauf Department of Radiology, Faculty of Health Sciences, University of the Witwatersrand, Johannesburg, South Africa

Jenny Rolands Department of Radiology, Shrewsbury, UK

Sugania Reddy Department of Radiology, Faculty of Health Sciences, University of the Witwatersrand, Johannesburg, South Africa

Leonie Scholtz Department of Radiology, University of Pretoria, Pretoria, South Africa

Russel Seider Mammography Unit, Wits Donald Gordon Medical Centre, Johannesburg, South Africa

Melanie Strachan Department of Radiology, University of Stellenbosch, Stellenbosch, South Africa

Rachel Such Department of Radiology, Faculty of Health Sciences, University of the Witwatersrand, Johannesburg, South Africa

Liora Sweidan Department of Radiology, Faculty of Health Sciences, University of the Witwatersrand, Johannesburg, South Africa

Catharine Elizabeth Tagg Department of Radiology, Aneurin Bevan Health Board, Wales, UK

Salomine Theron Radiology Department, Schnetler Corbett and Partners Radiology, Cape Town, South Africa

Jaques Janse Van Rensburg Department of Radiology, University of the Free State, Bloemfontein, South Africa

Bretton Woods Department of Radiology, Faculty of Health Sciences, University of the Witwatersrand, Johannesburg, South Africa

Yousuf Vadachia Department of Radiology, University of Stellenbosch, Stellenbosch, South Africa

Andrew van den Heever Department of Radiology, University of Cape Town, Cape Town, South Africa

Helga von Bezing Department of Radiology, University of Stellenbosch, Stellenbosch, South Africa

Part I

Central Nervous System

1 The Brain

The central nervous system (CNS) comprises the brain and the spinal cord and includes the supplying vessels, meningeal coverings, and cerebrospinal fluid (CSF). Any single cross-sectional image contains all of these. It is useful to appreciate each of these components for a fixed number of cross-sectional positions. Each of these positions should be named according to some characteristic finding on that slice. More detailed anatomy is required for the midline sagittal cut. The suprasellar region, the pineal region, the posterior fossa, and the region of the major deep grey nuclei have many detailed anatomical labels and have been highlighted individually. This chapter provides a tour in the axial, coronal, and sagittal planes, and then draws attention to details of specific regions. There are also special sections on the hippocampus, cerebral vessels, cranial nerves, and spinal cord. Surface anatomy, functional areas, and white matter tracts are presented in detail as they represent emerging tools representing anatomy not well-visualised using traditional methods. The globe and pituitary gland will be dealt with in more detail in other sections of this book.

S. Andronikou (ed.), *See Right Through Me*,
DOI 10.1007/978-3-642-23893-2_1, © Springer-Verlag Berlin Heidelberg 2012

Brain Cross-Sectional Tour

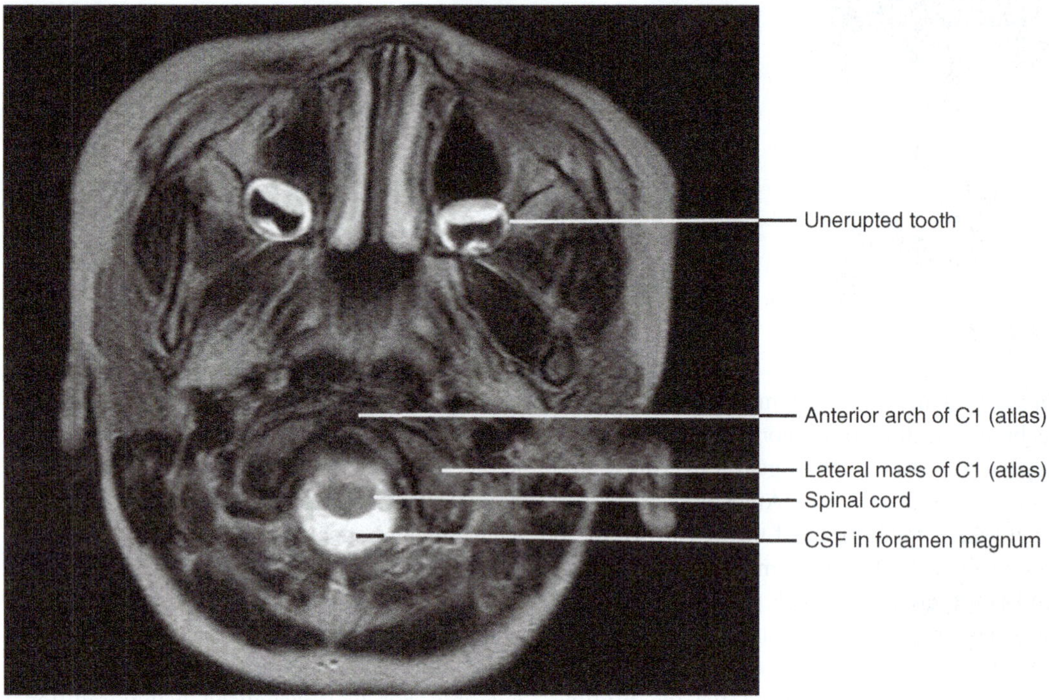

Fig. 1.1 Axial T2 at anterior arch of atlas

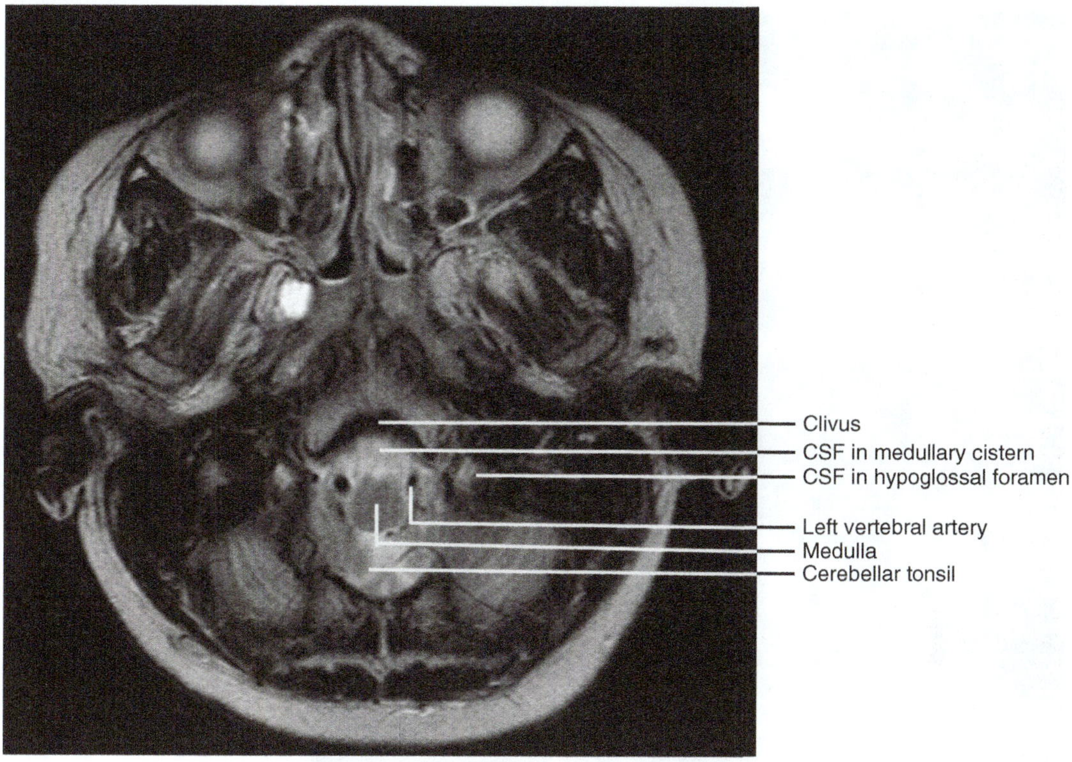

Fig. 1.2 Axial T2 at foramen magnum

Pearl: The foramen magnum should contain the cord, CSF, and vessels

Absence of CSF is abnormal and may represent descent of the cerebellar tonsils or vermis.

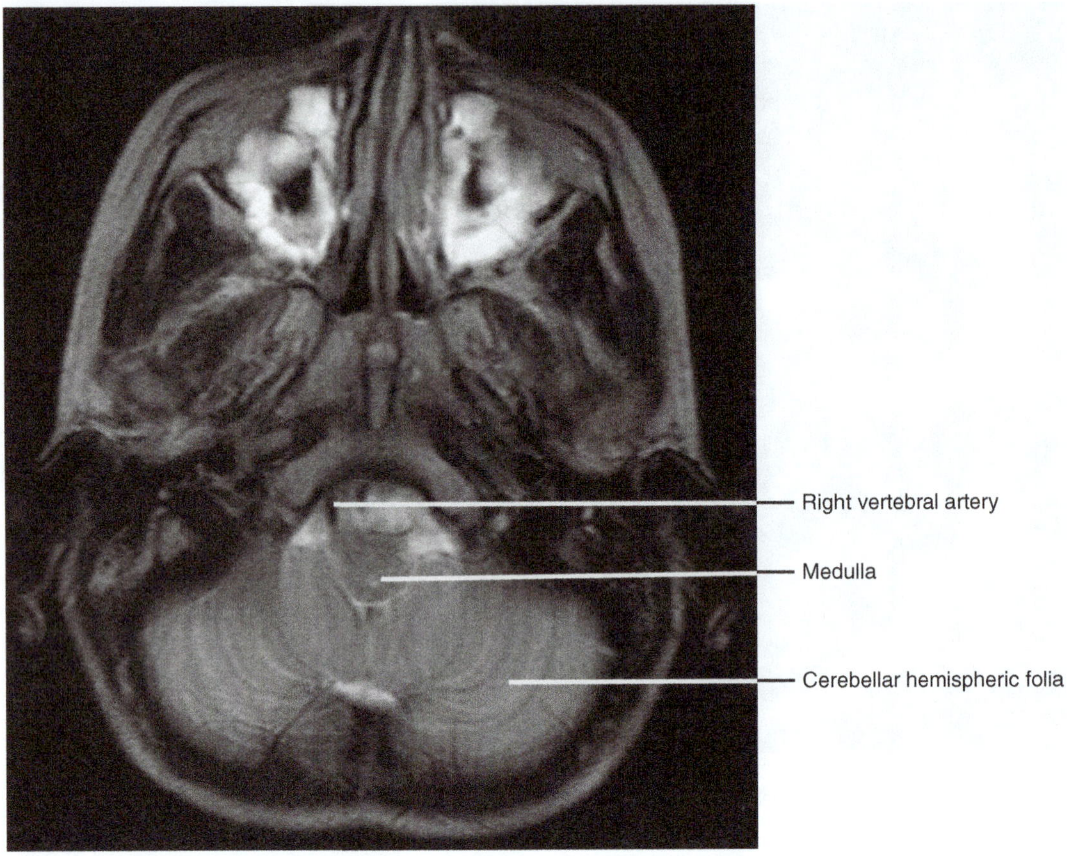

Fig. 1.3 Axial T2 at the level of the medulla

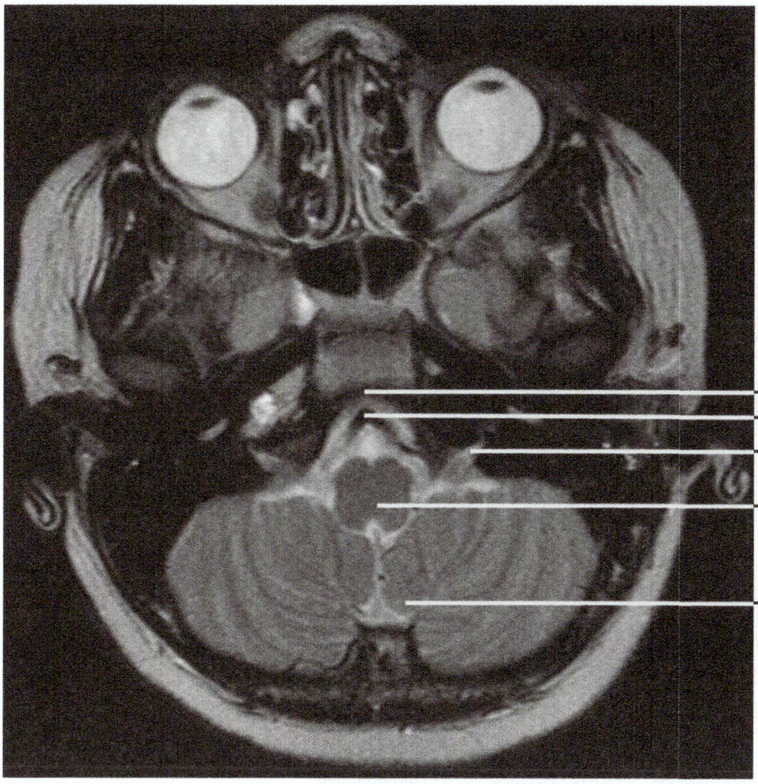

Fig. 1.4 Axial T2 at the level of the medulla and cerebellar tonsils

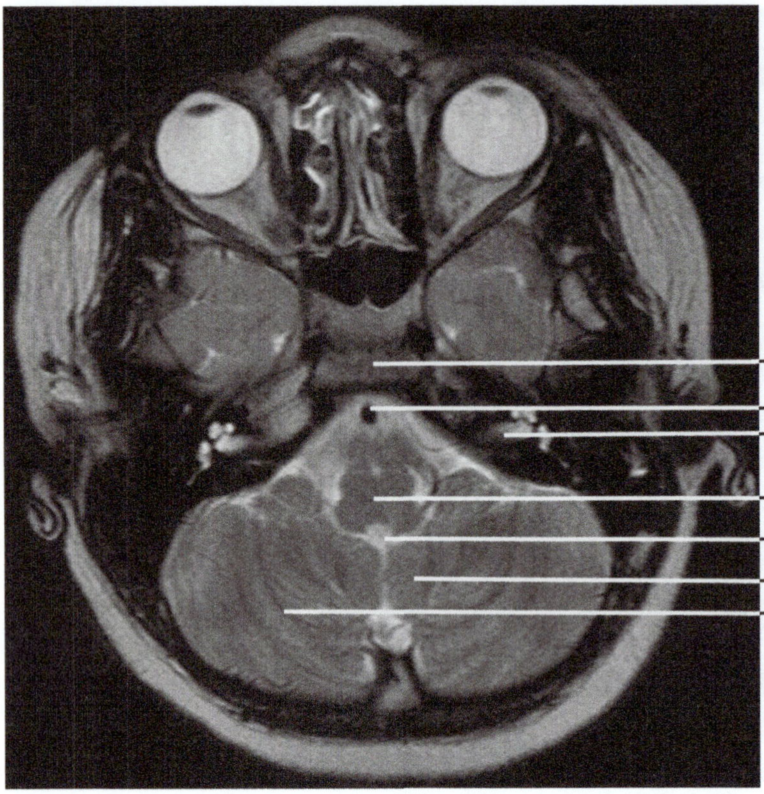

Fig. 1.5 Axial T2 at the level of the obex (lower portion of V4)

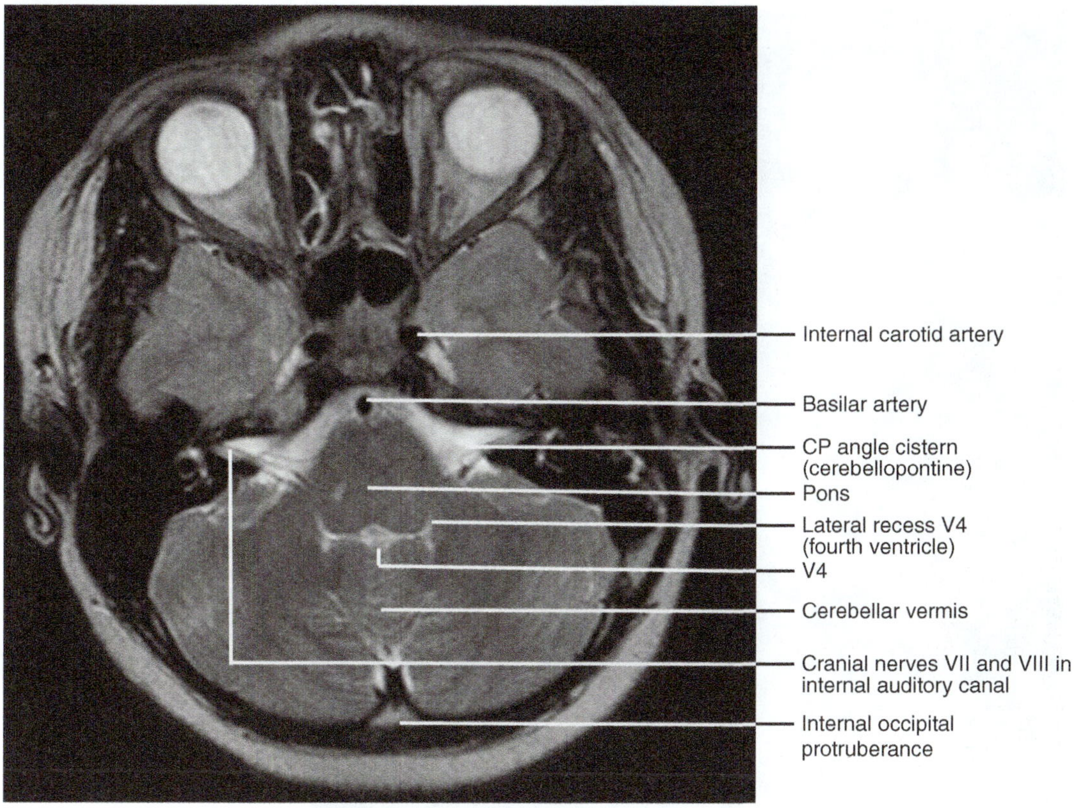

Fig. 1.6 Axial T2 at the level of the pons and vestibulo-cochlear nerves

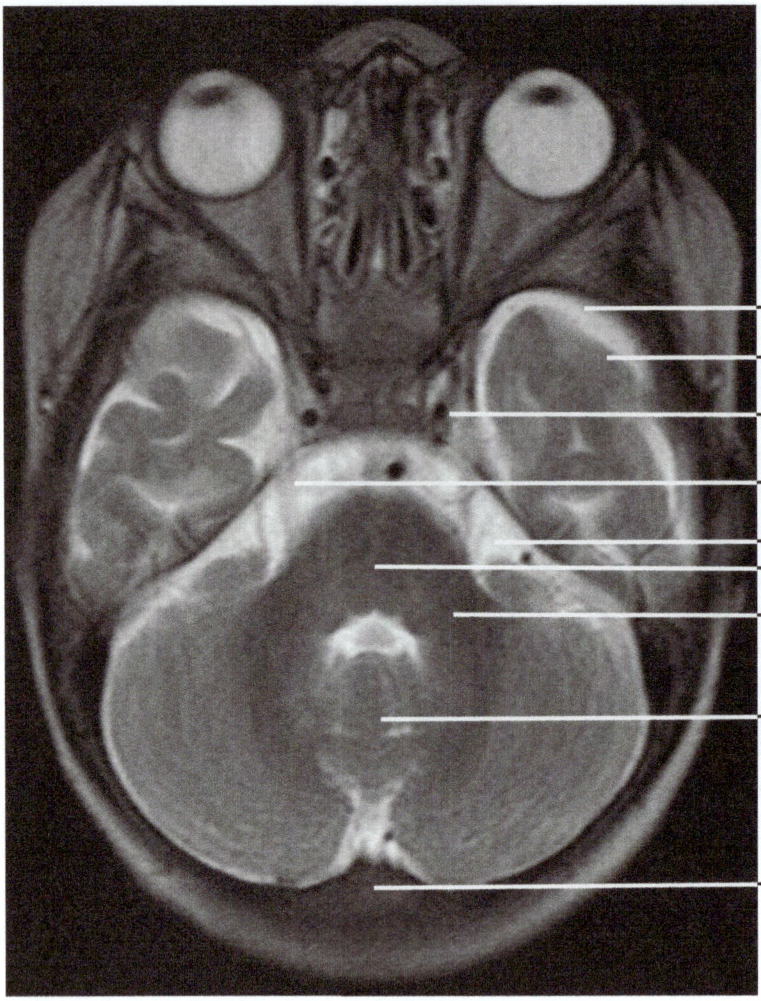

Fig. 1.7 Axial T2 at the level of pons and brachium pontis

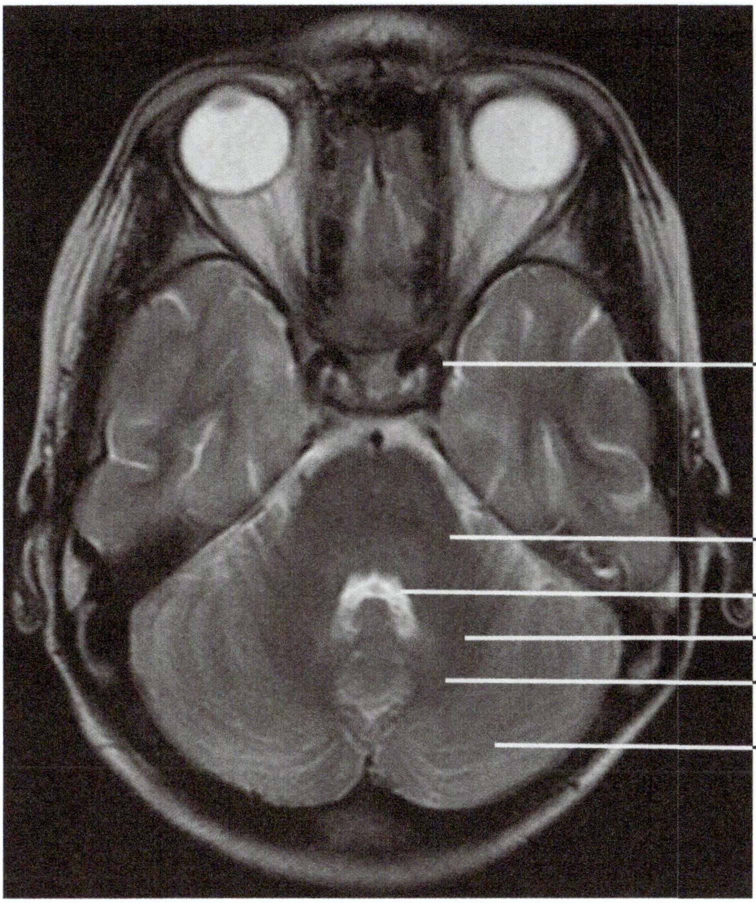

Fig. 1.8 Axial T2 at the level of the pons and V4 proper

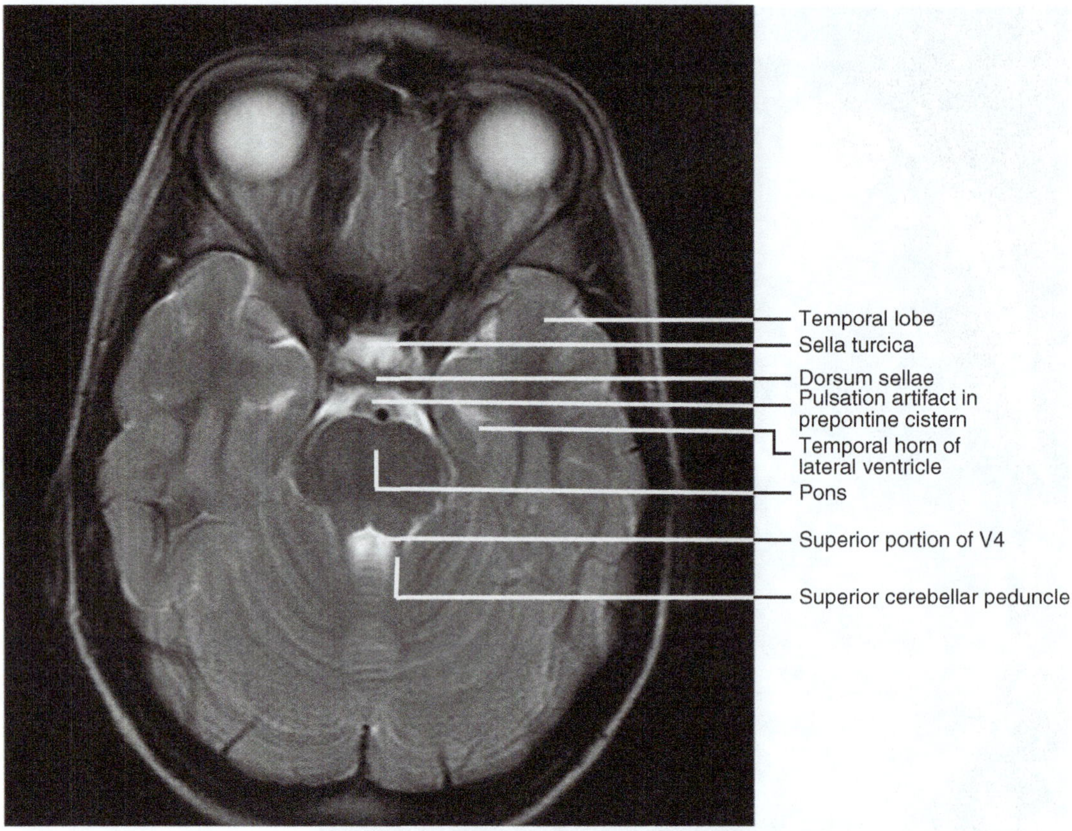

Fig. 1.9 Axial T2 at the level of the pituitary fossa and junction pons/midbrain

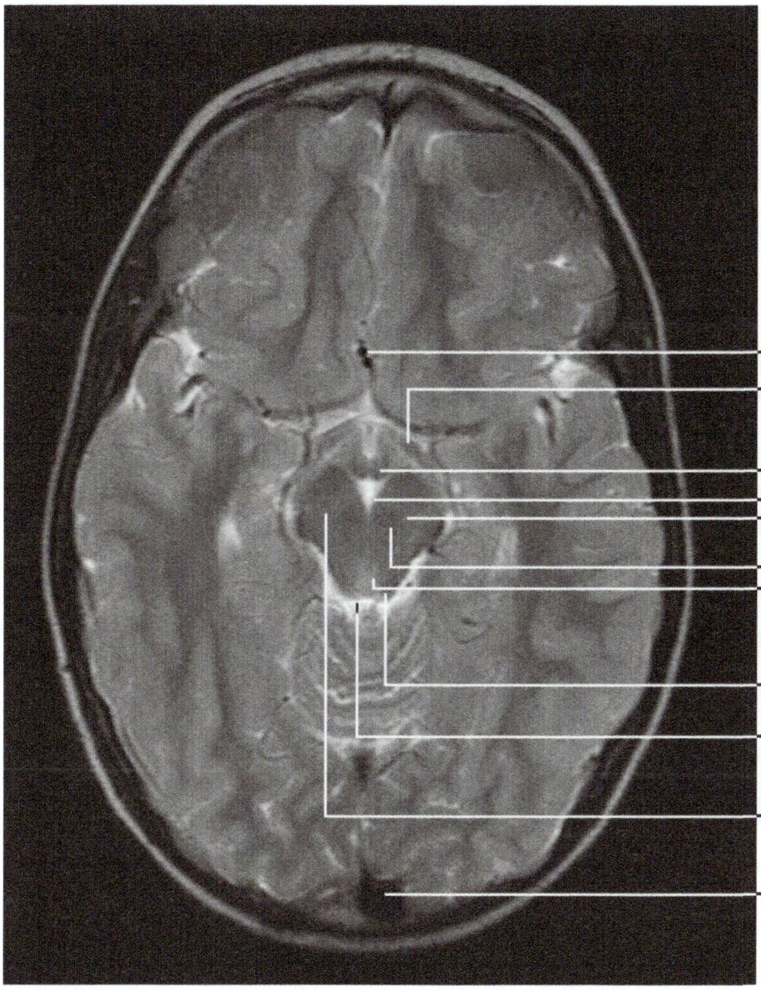

Fig. 1.10 Axial T2 at the level of the cerebral peduncles and mammillary bodies

Pearl

The Tectum = quadrigeminal plate = colliculi.

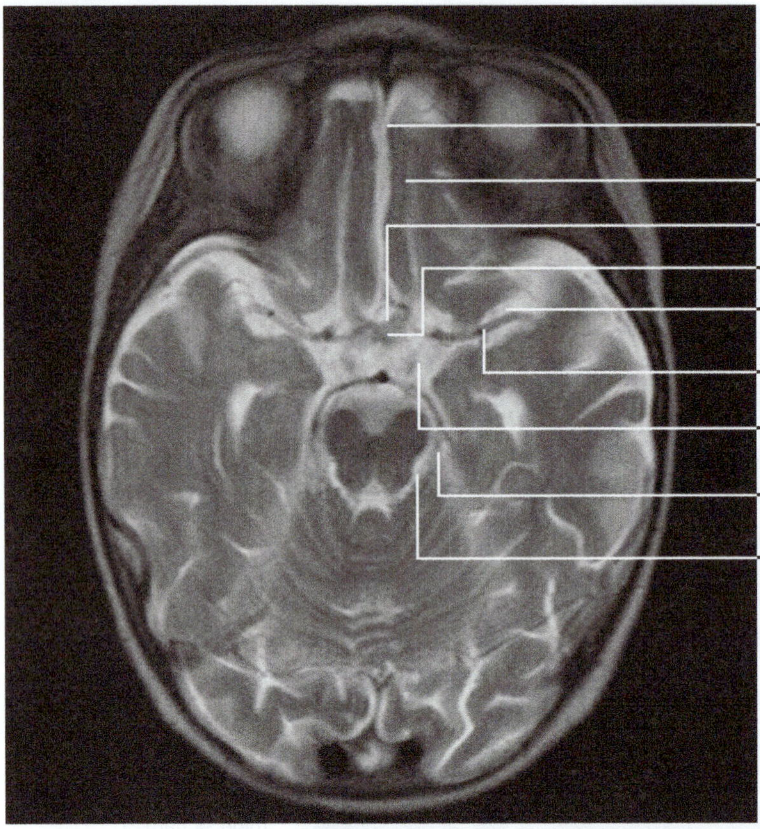

Fig. 1.11 Axial T2 at the level of the cerebral peduncles and circle of Willis

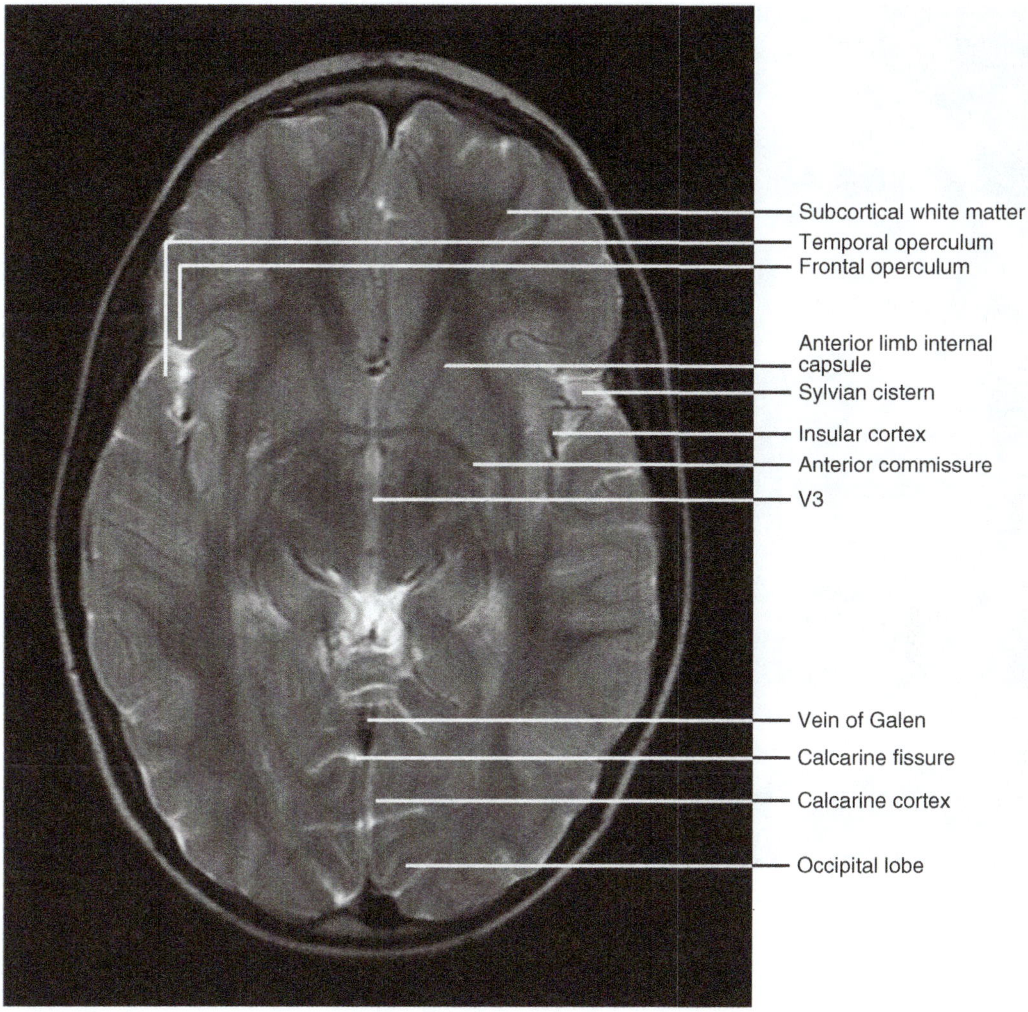

Fig. 1.12 Axial T2 at the level of the anterior commissure and V3

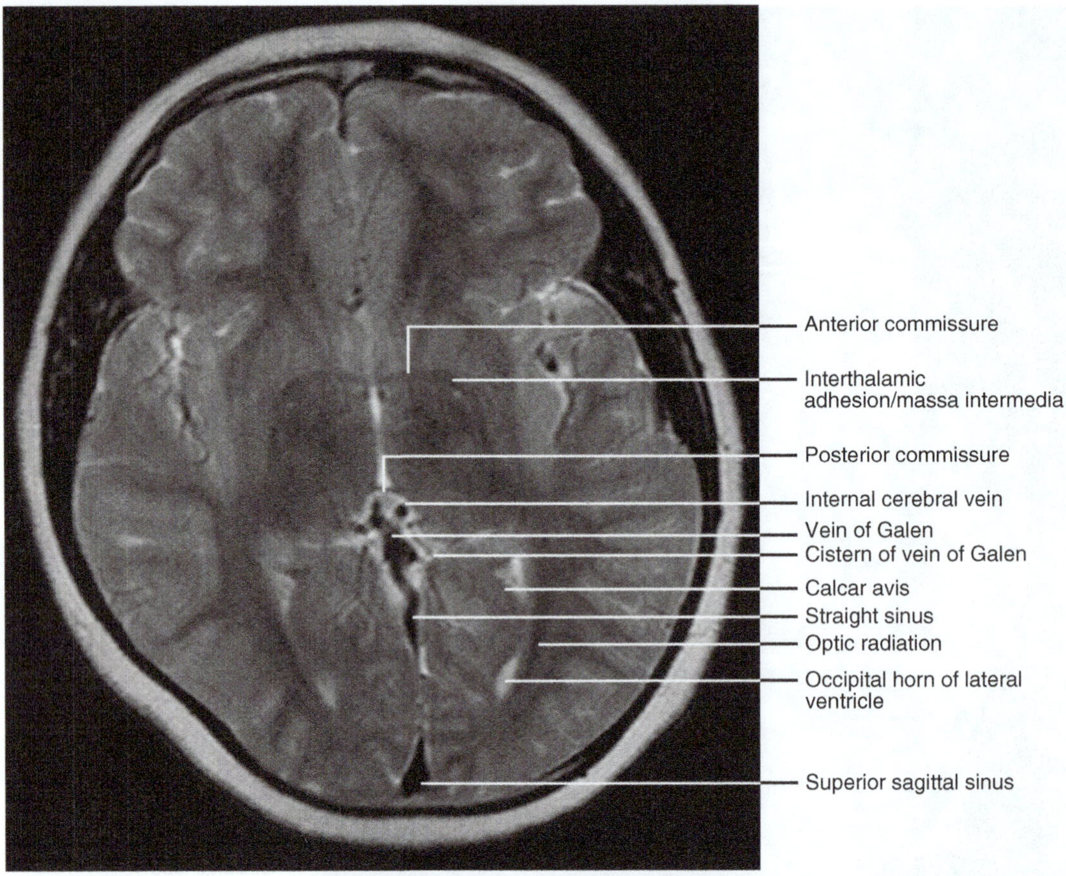

Fig. 1.13 Axial T2 at the level of V3 and the vein of Galen

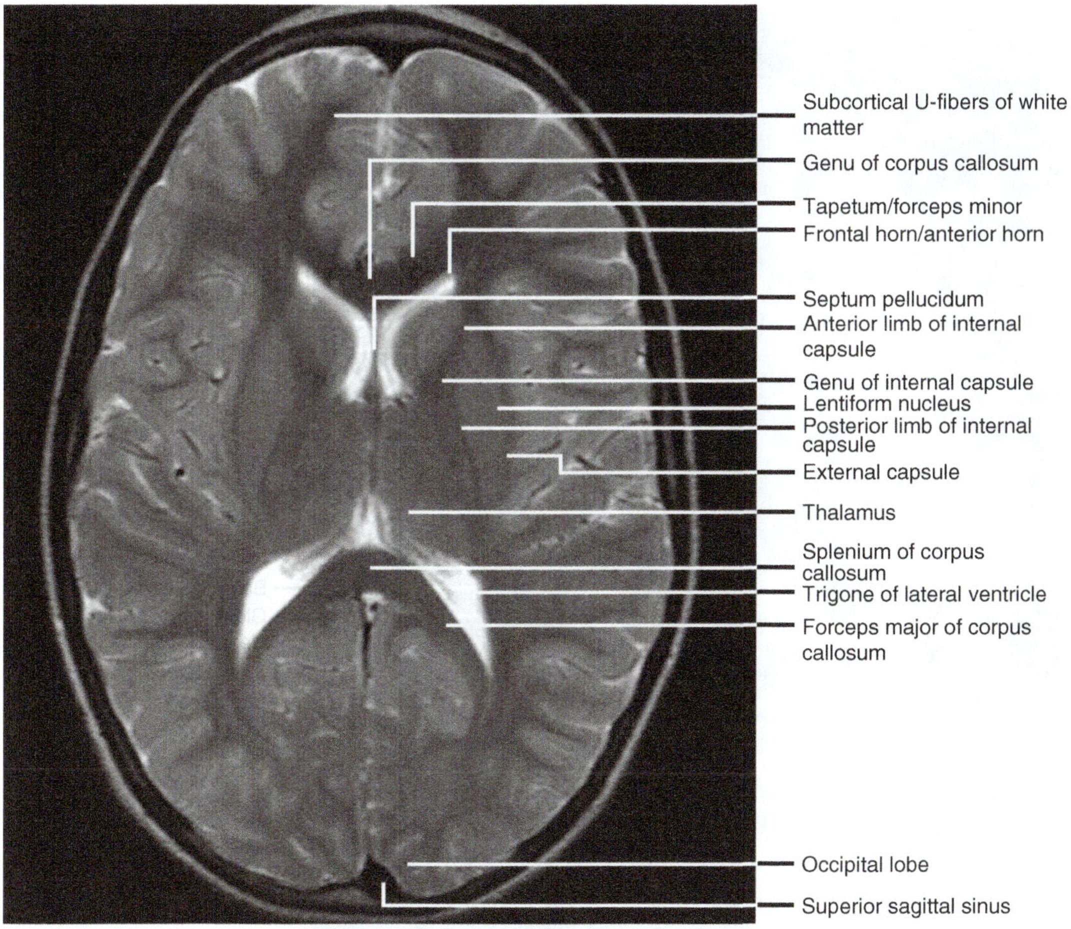

Fig. 1.14 Axial T2 at the level of the basal ganglia and trigone

Pearl

The corpus striatum = caudate nucleus + putamen.

Lentiform nucleus = globus pallidus medial and putamen lateral.

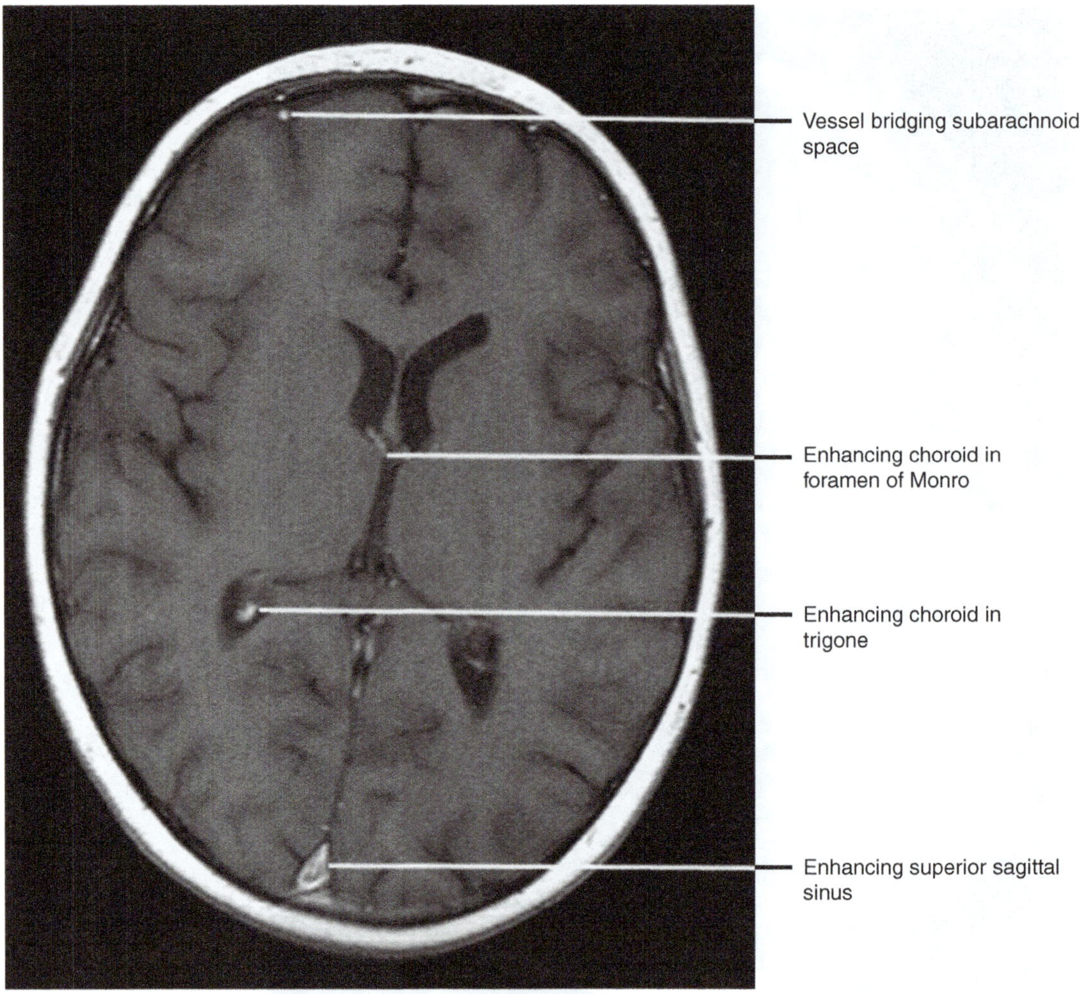

Fig. 1.17 Axial T1 post gadolinium at the level of the foramina of Monro

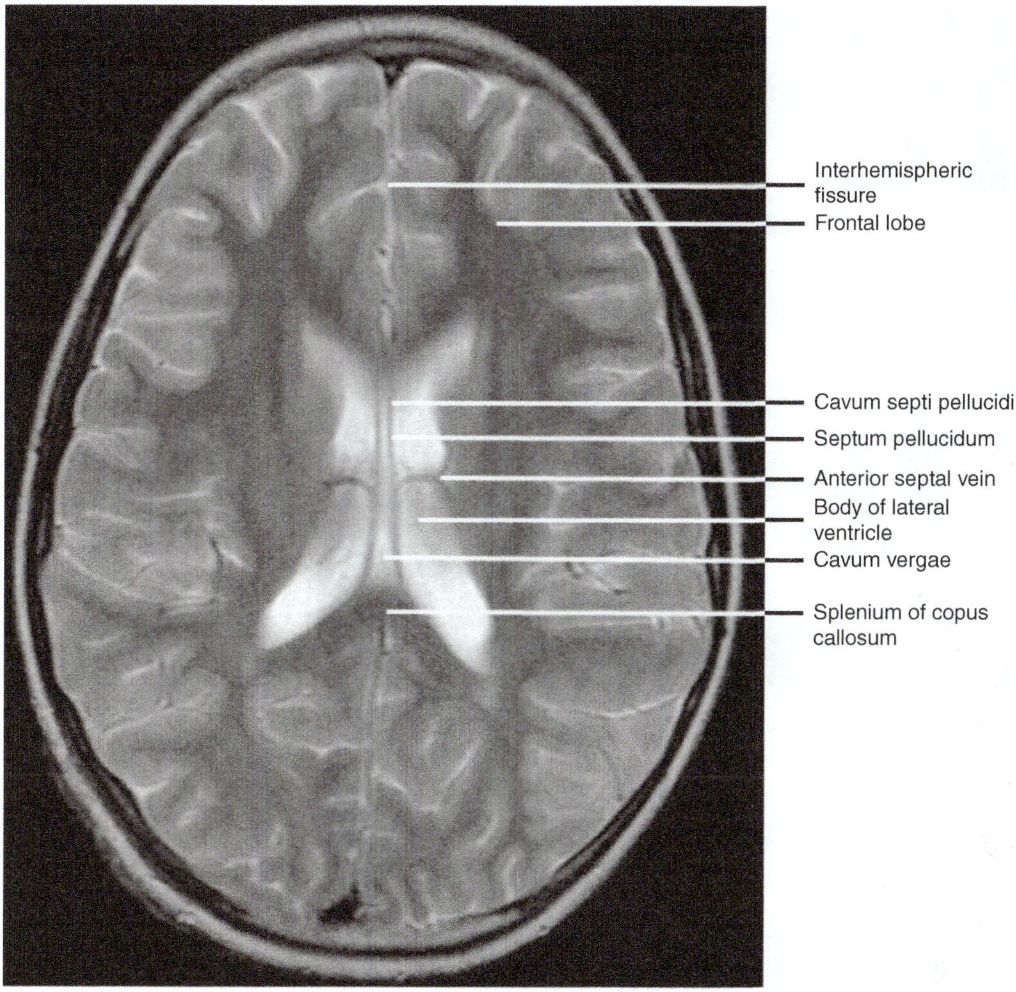

Fig. 1.18 Axial T2 at the level of the bodies of the lateral ventricles

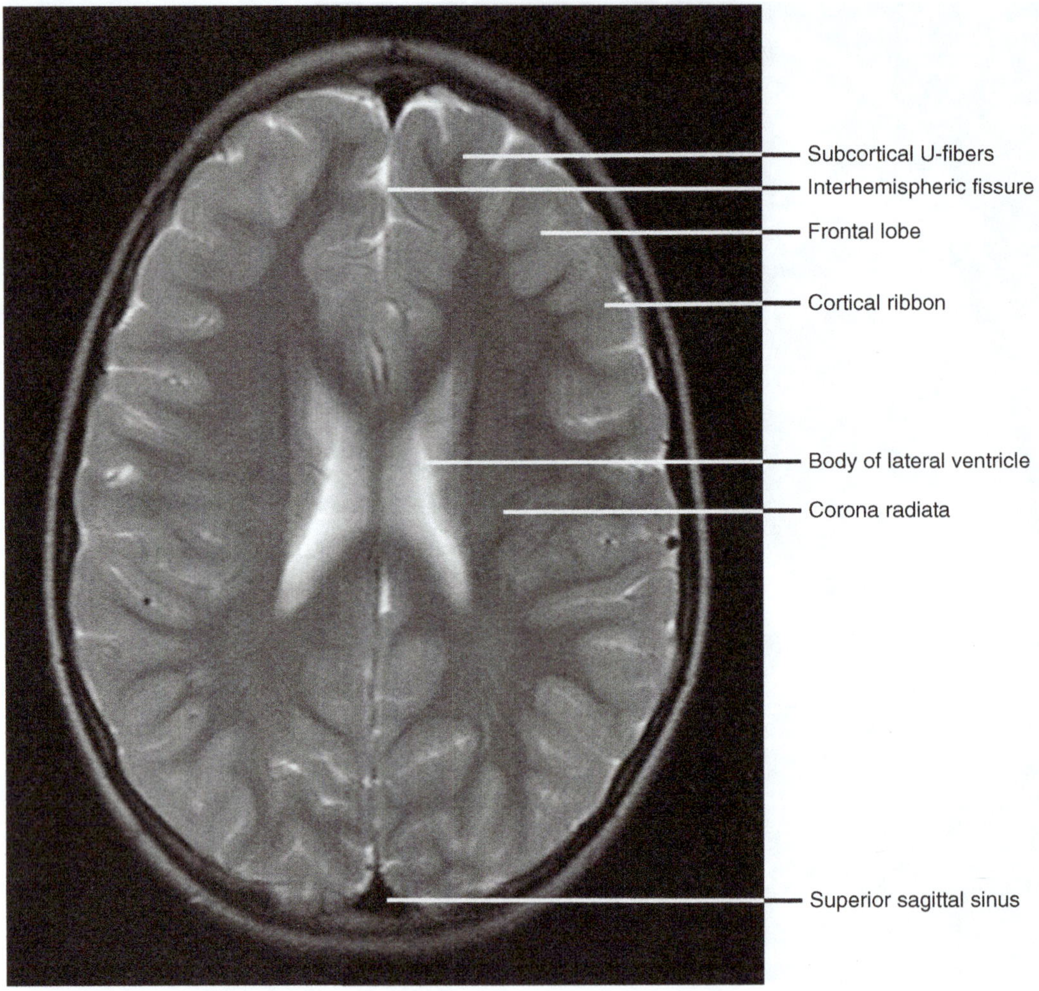

Fig. 1.19 Axial T2 at the level of the corona radiata

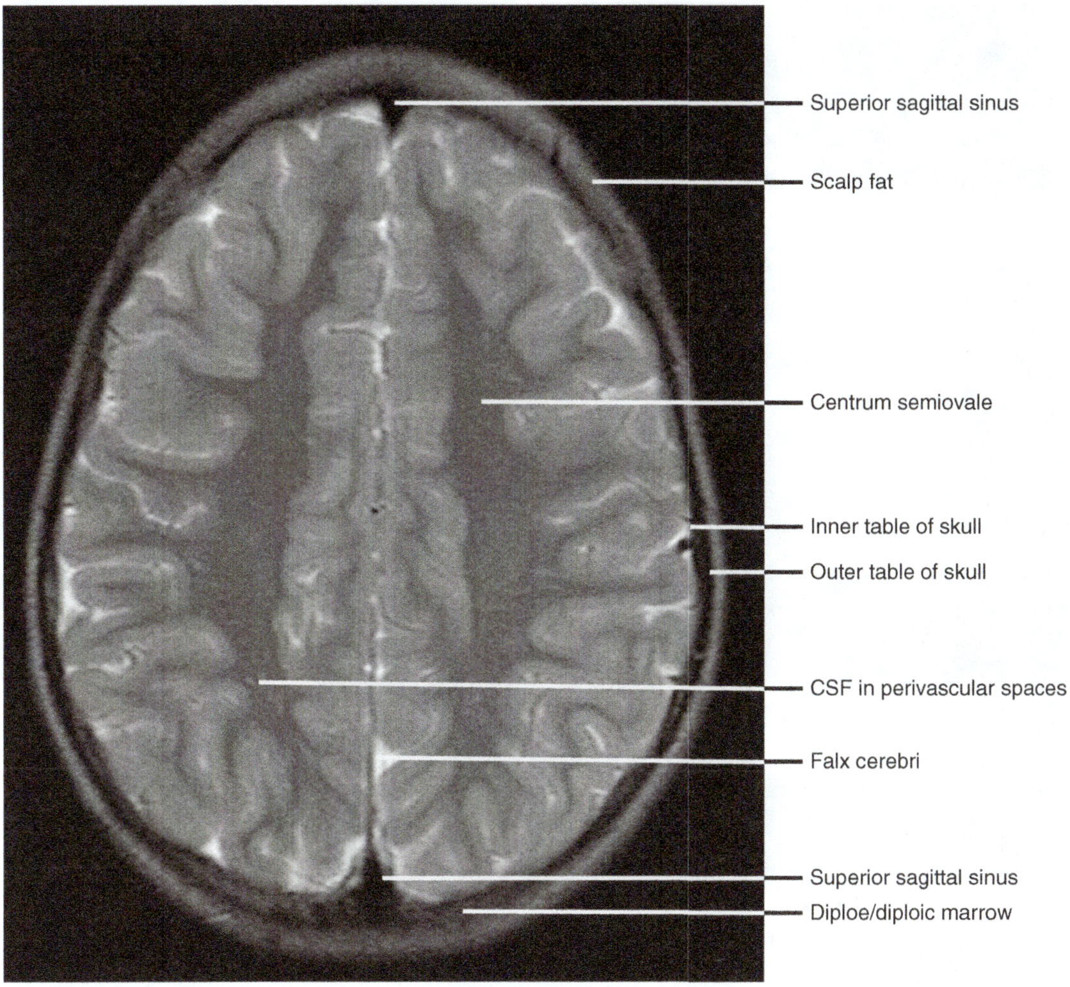

Fig. 1.20 Axial T2 at the level of the centrum semiovale

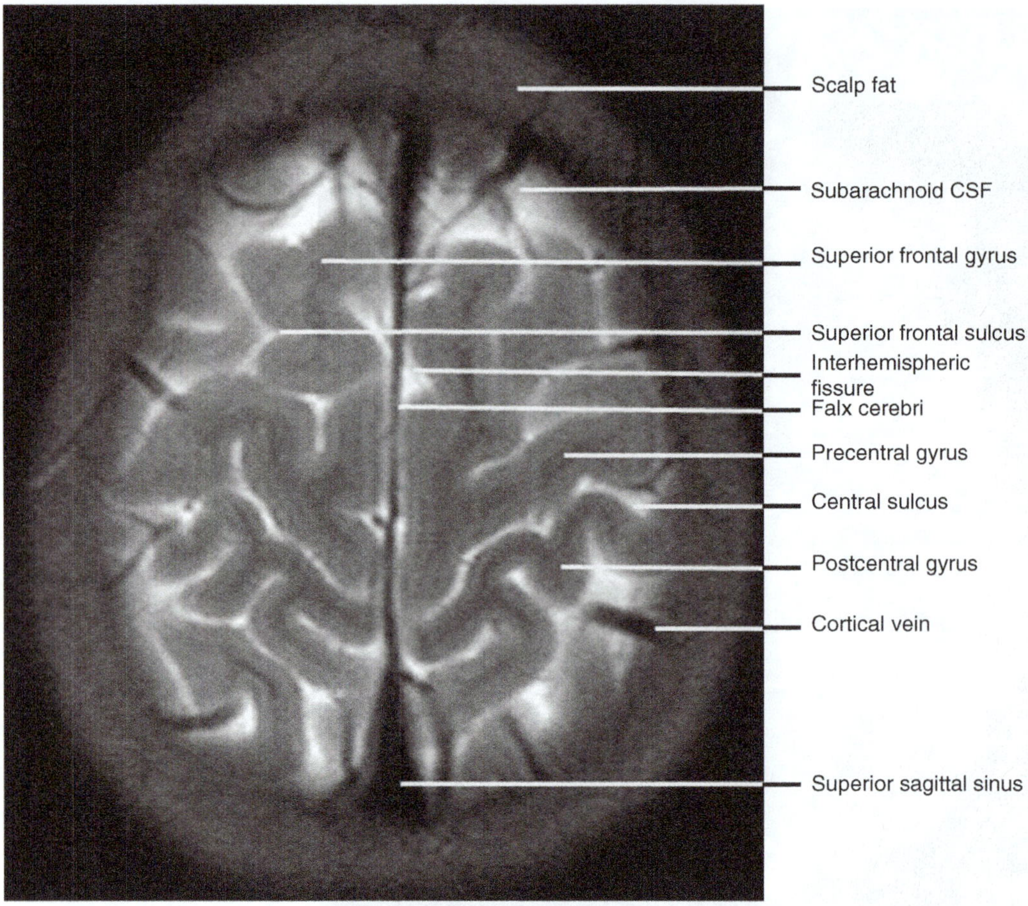

Fig. 1.21 Axial T2 at the level of the pre- and postcentral gyri

Coronal Anatomy of the Brain

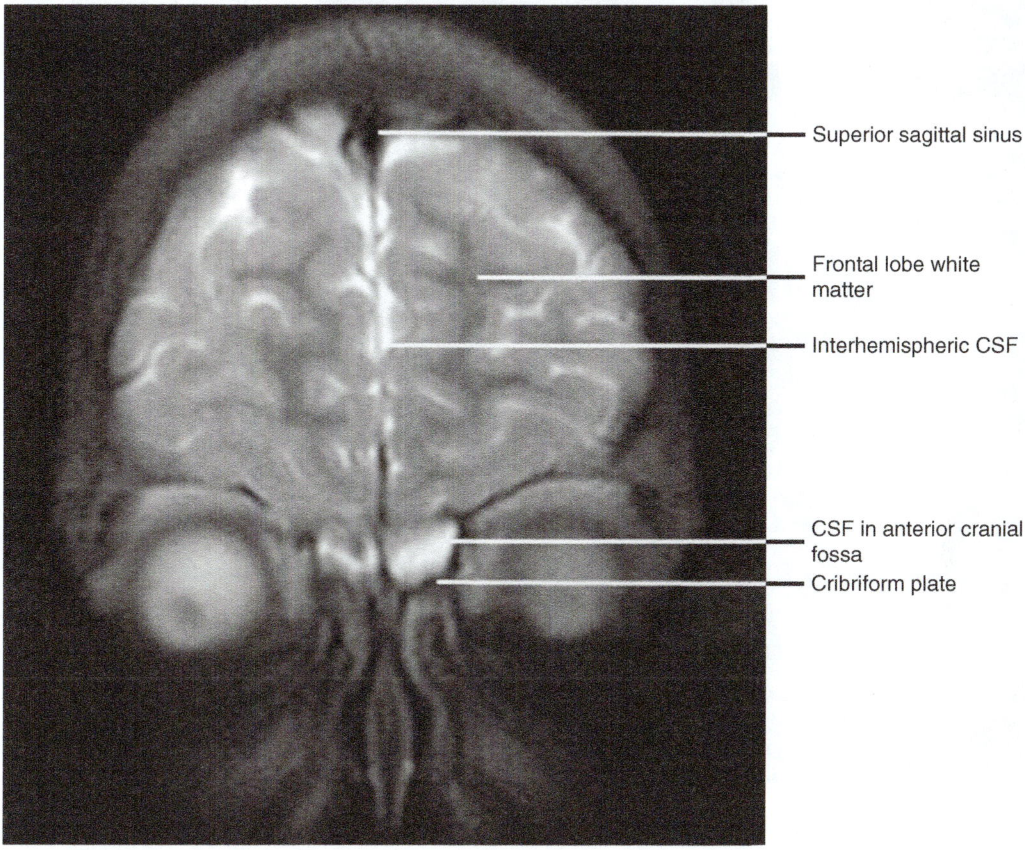

Fig. 1.22 Coronal T2 at the level of the frontal lobes

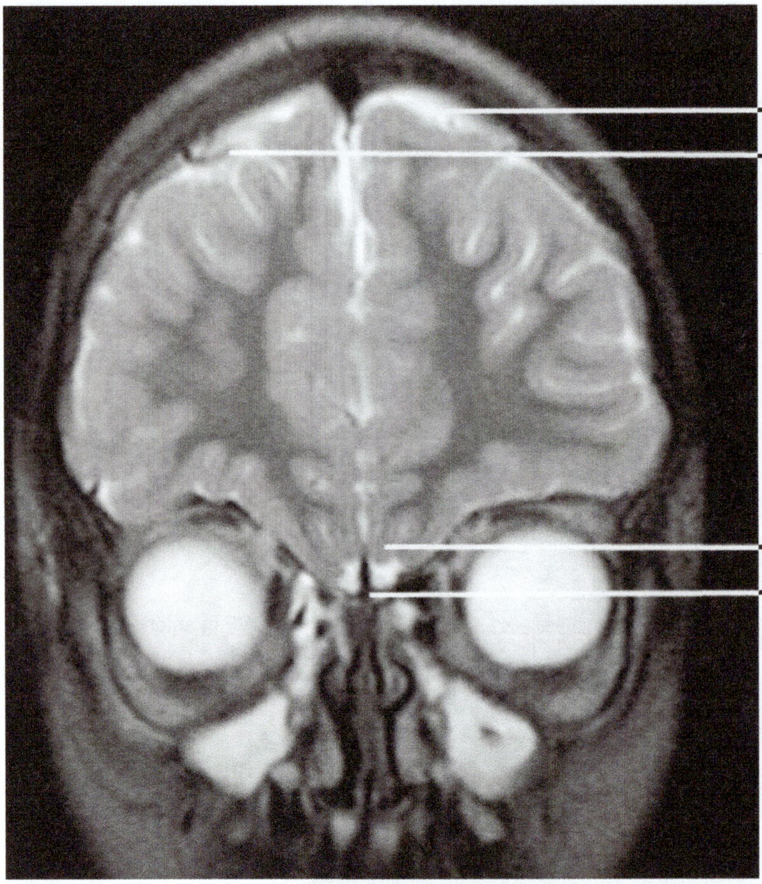

Fig. 1.23 Coronal T2 at the level of the frontal lobes and gyri recti

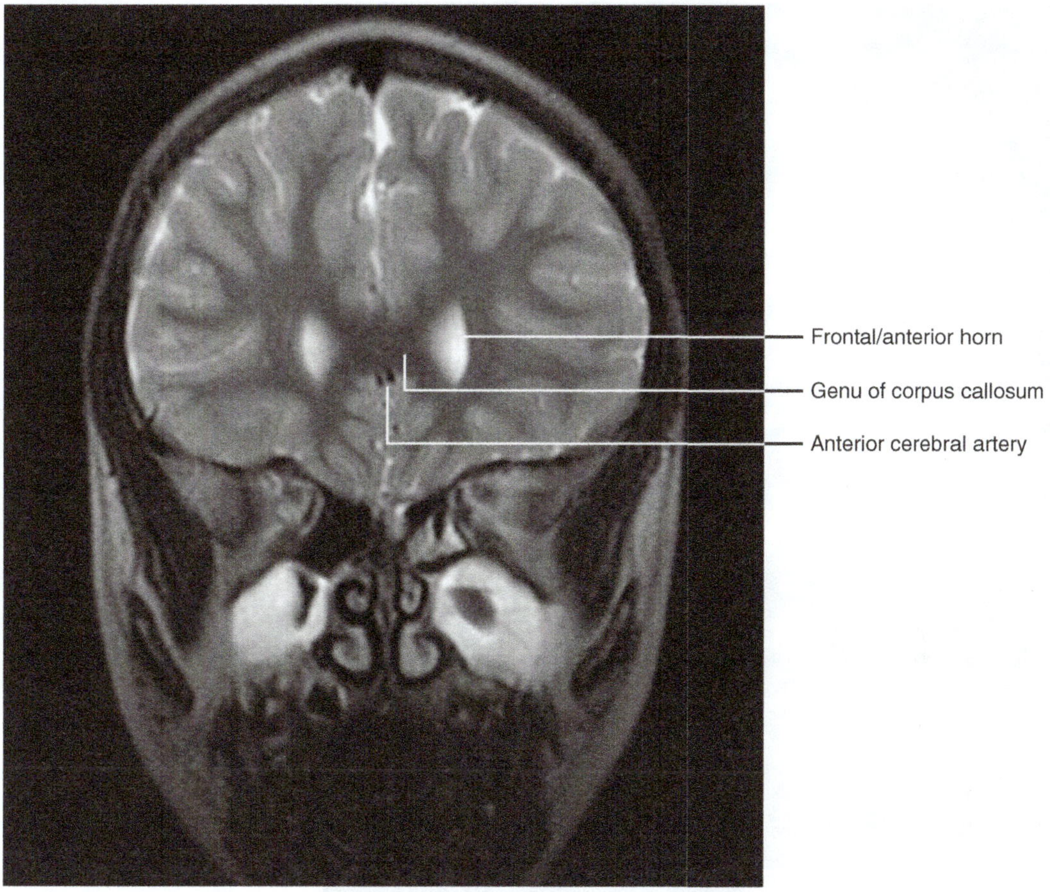

Fig. 1.24 Coronal T2 at the level of the frontal horns and genu of the corpus callosum

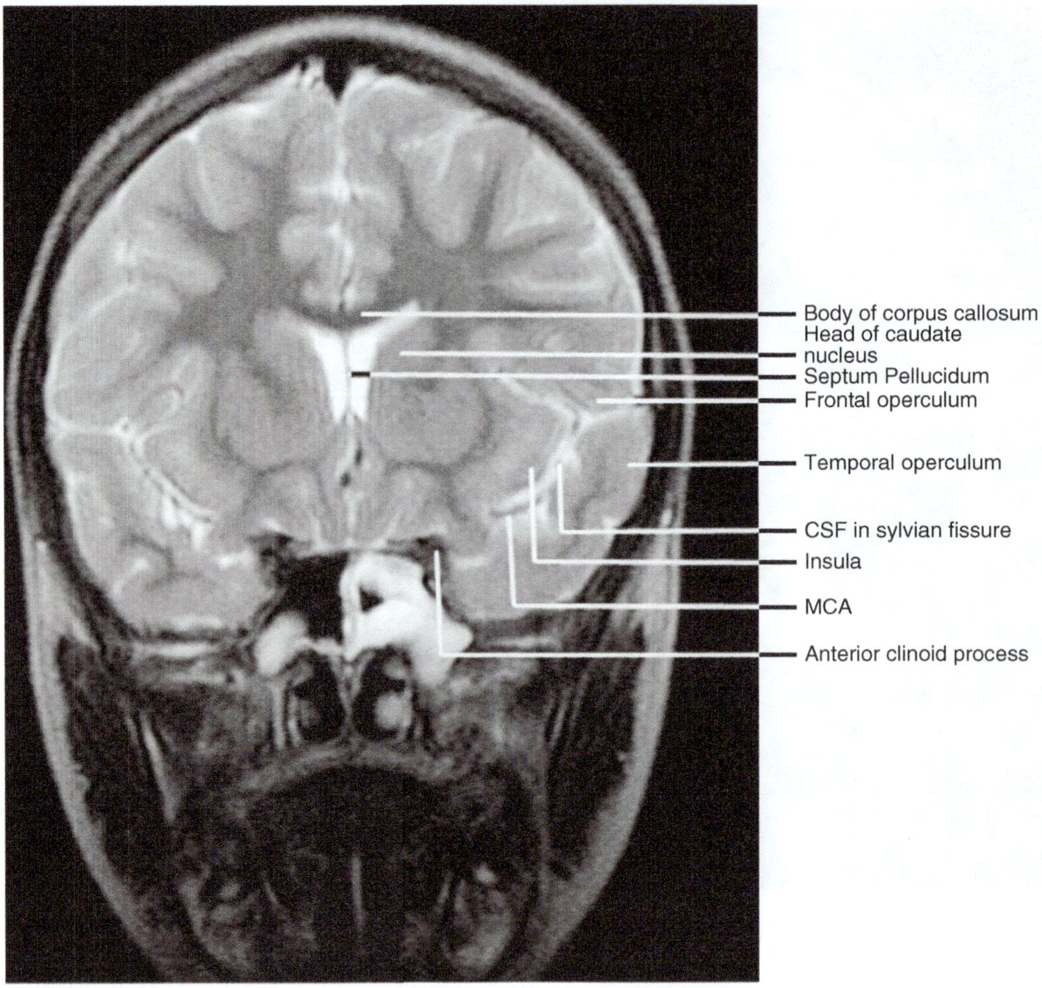

Fig. 1.25 Coronal T2 at the level of the septum pellucidum and Sylvian fissures

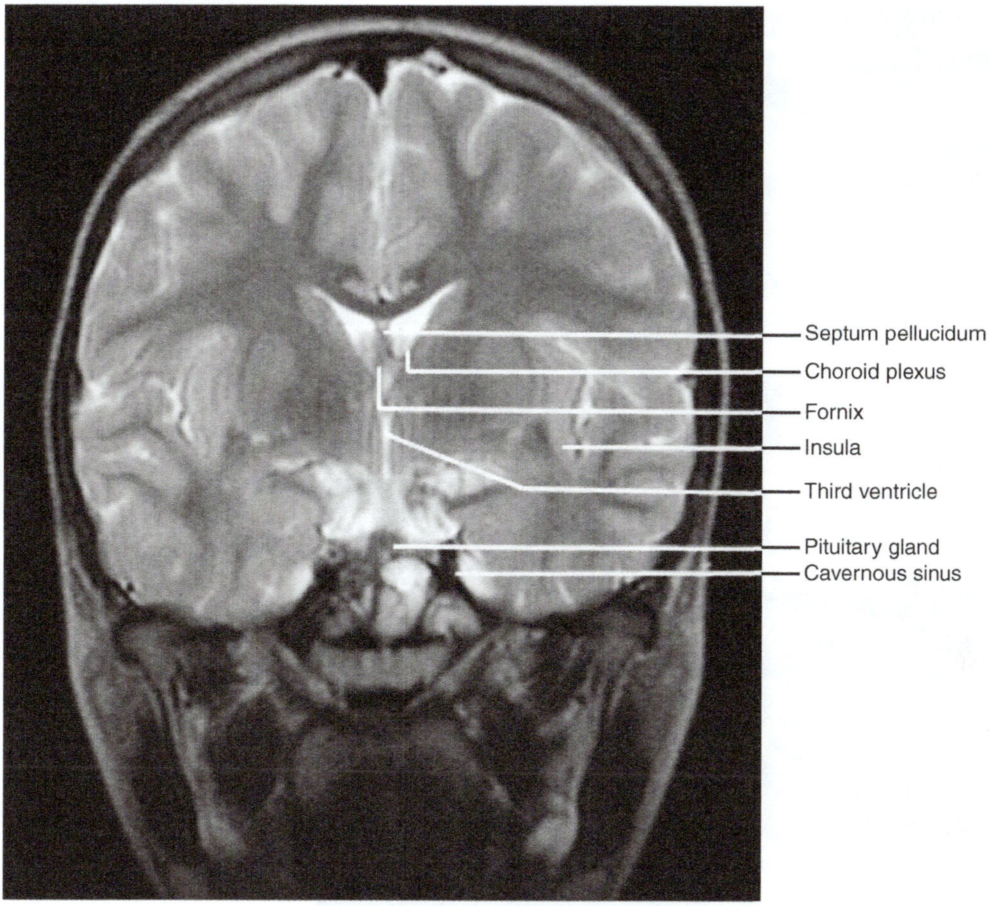

Fig. 1.26 Coronal T2 at the level of the foramina of Monro/third ventricle and insula

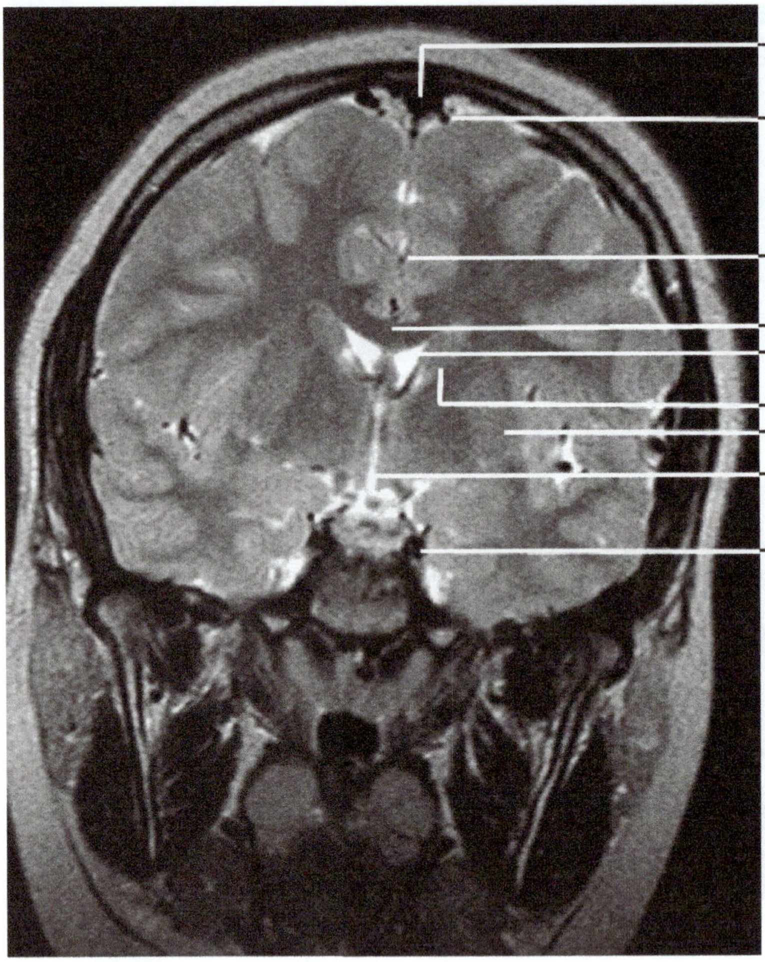

Fig. 1.27 Coronal T2 at the level of the third ventricle

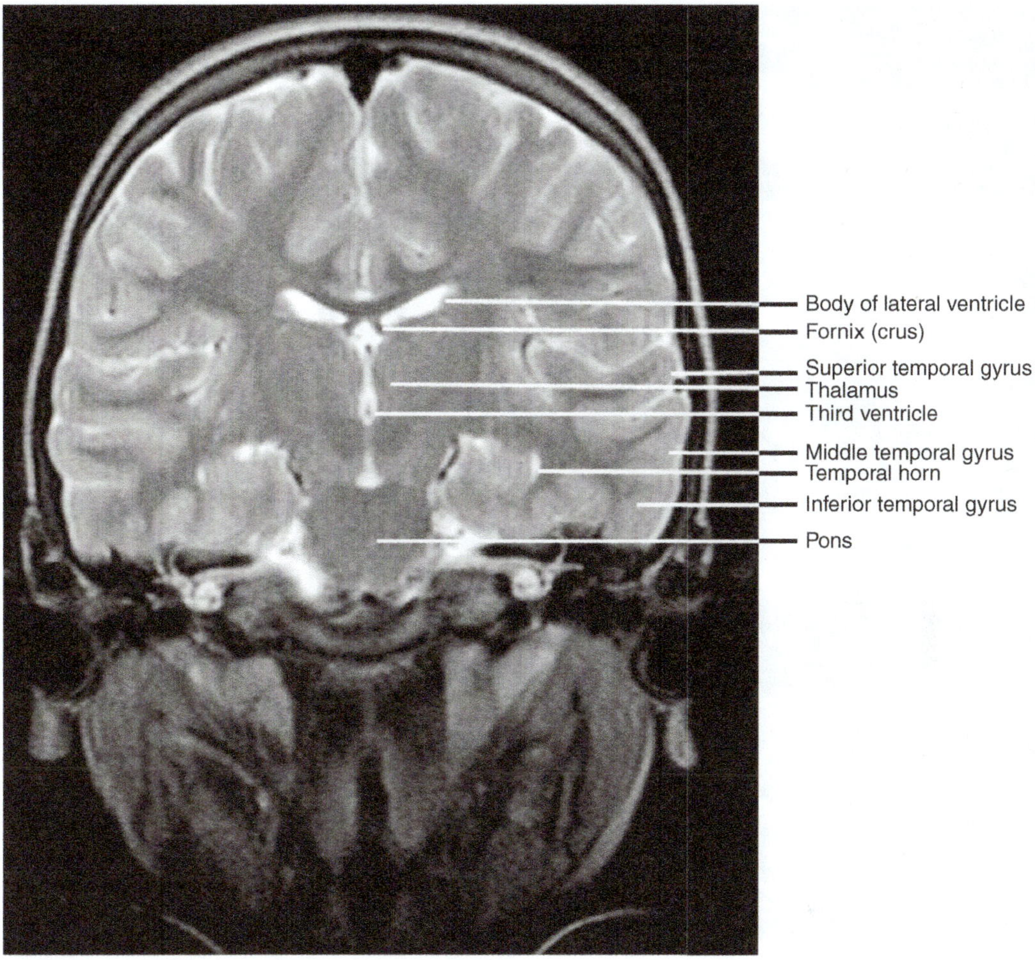

Fig. 1.28 Coronal T2 at the level of the thalami and pons

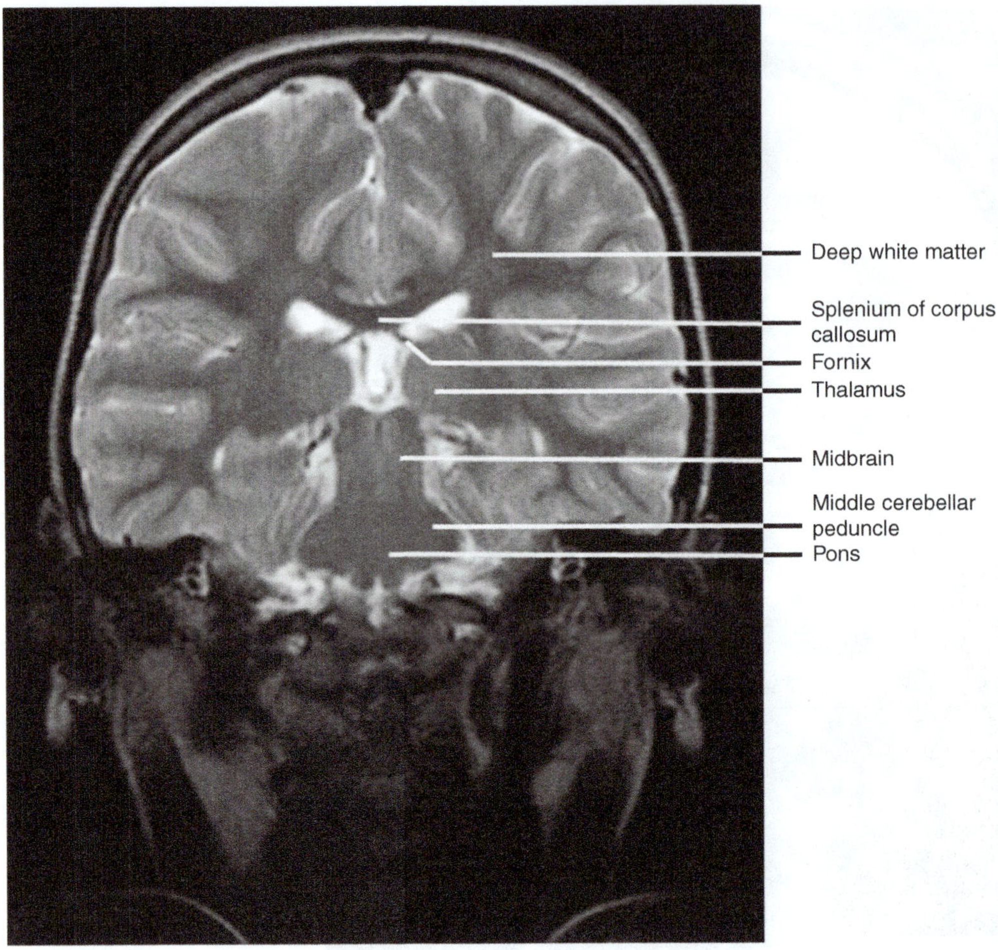

Fig. 1.29 Coronal T2 at the level of the brachium pontis

Pearl
Brachium pontis = middle cerebellar peduncle.

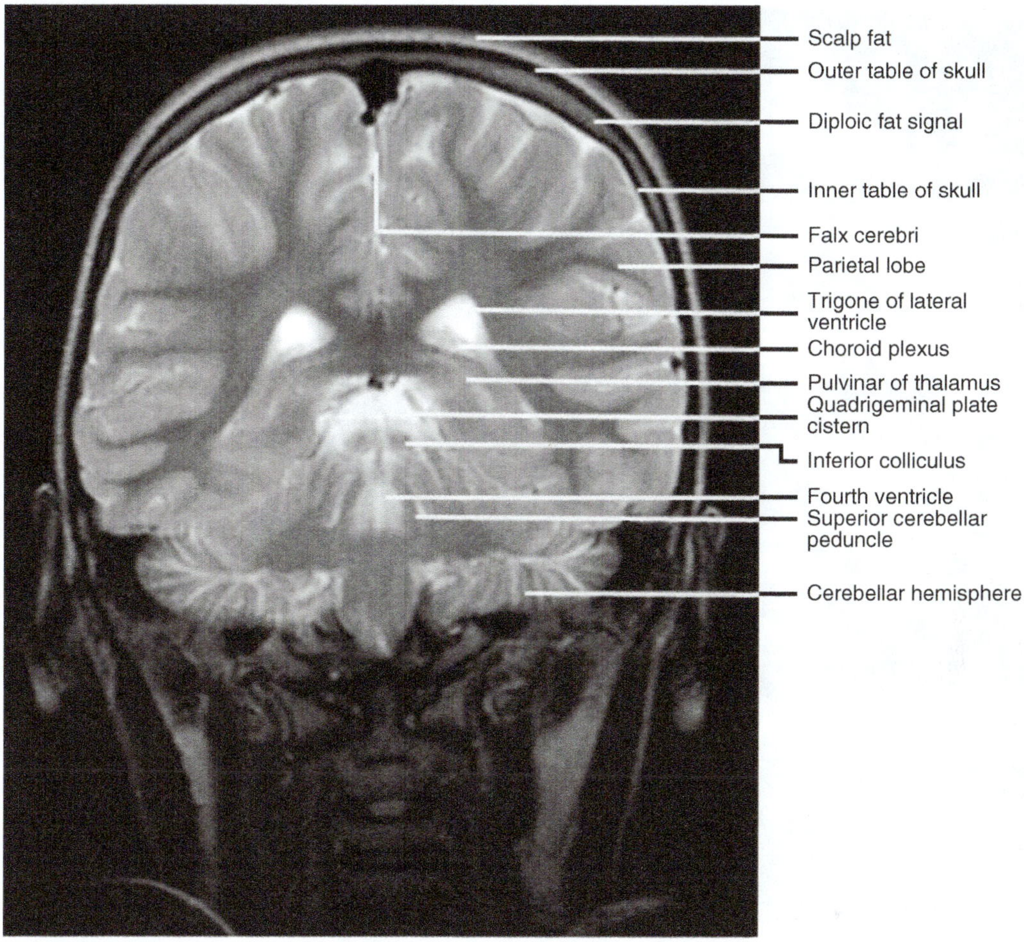

Fig. 1.30 Coronal T2 at the level of the bodies of the lateral ventricles and colliculi

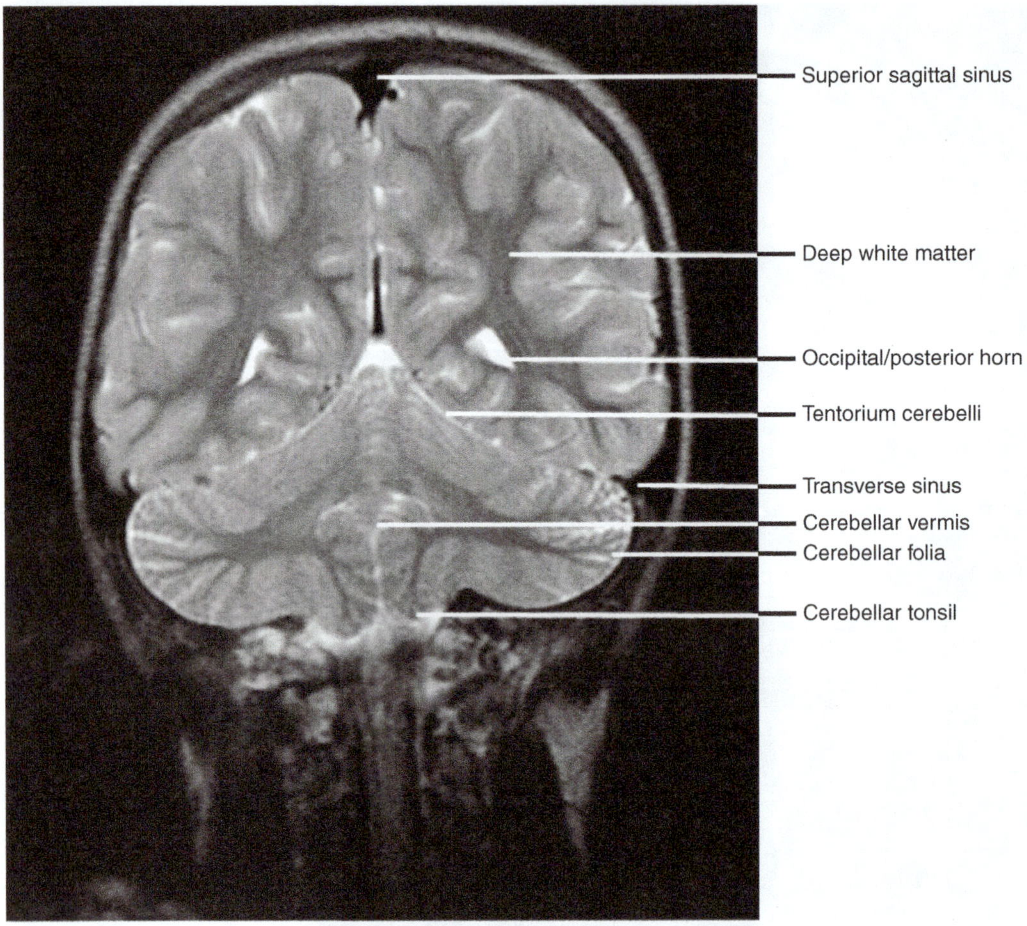

Fig. 1.31 Coronal T2 at the level of the cerebellar vermis and straight sinus

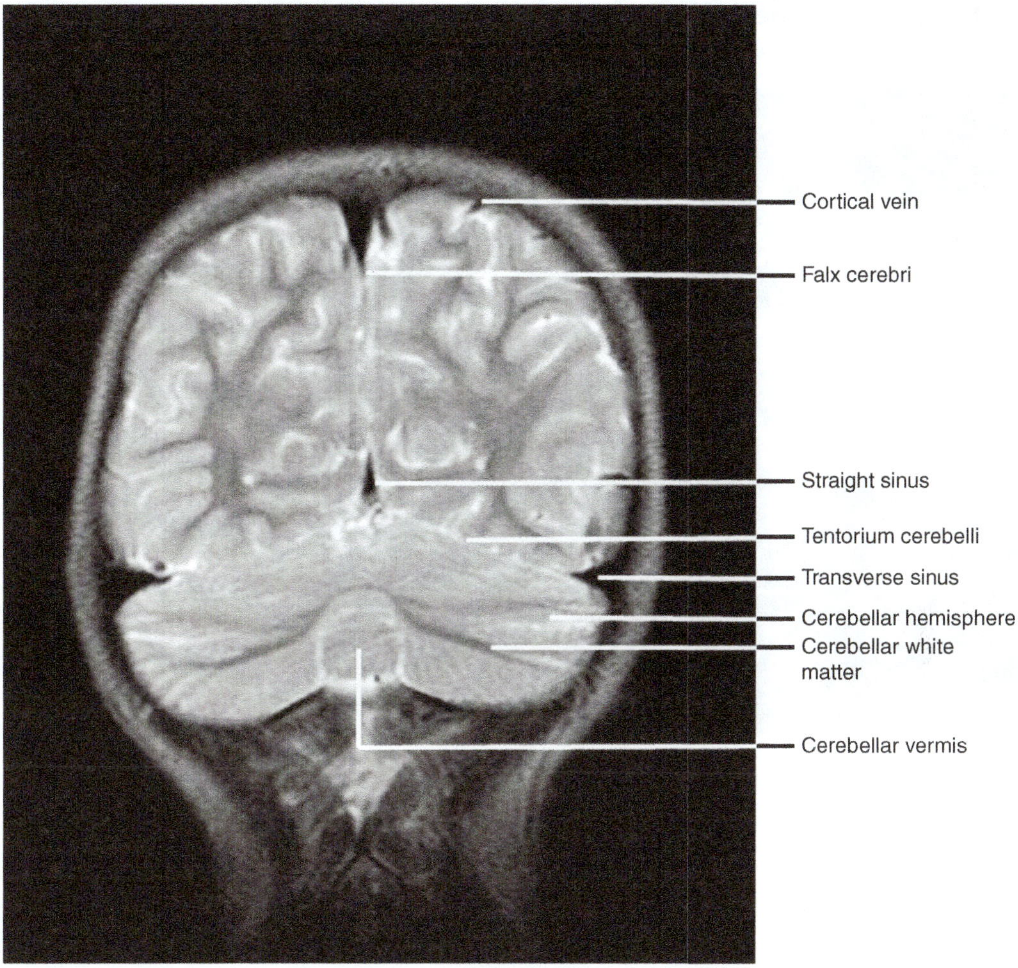

Fig. 1.32 Coronal T2 at the level of the posterior aspect of the vermis

Sagittal Anatomy of the Brain

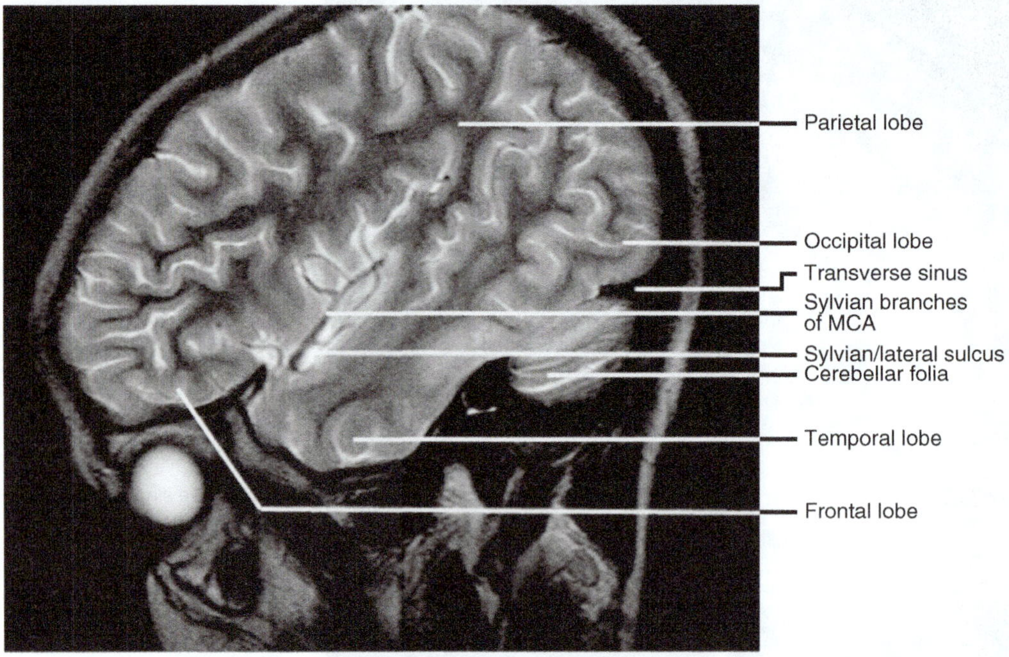

Fig. 1.33 Sagittal T2 at the level of the Sylvian fissure

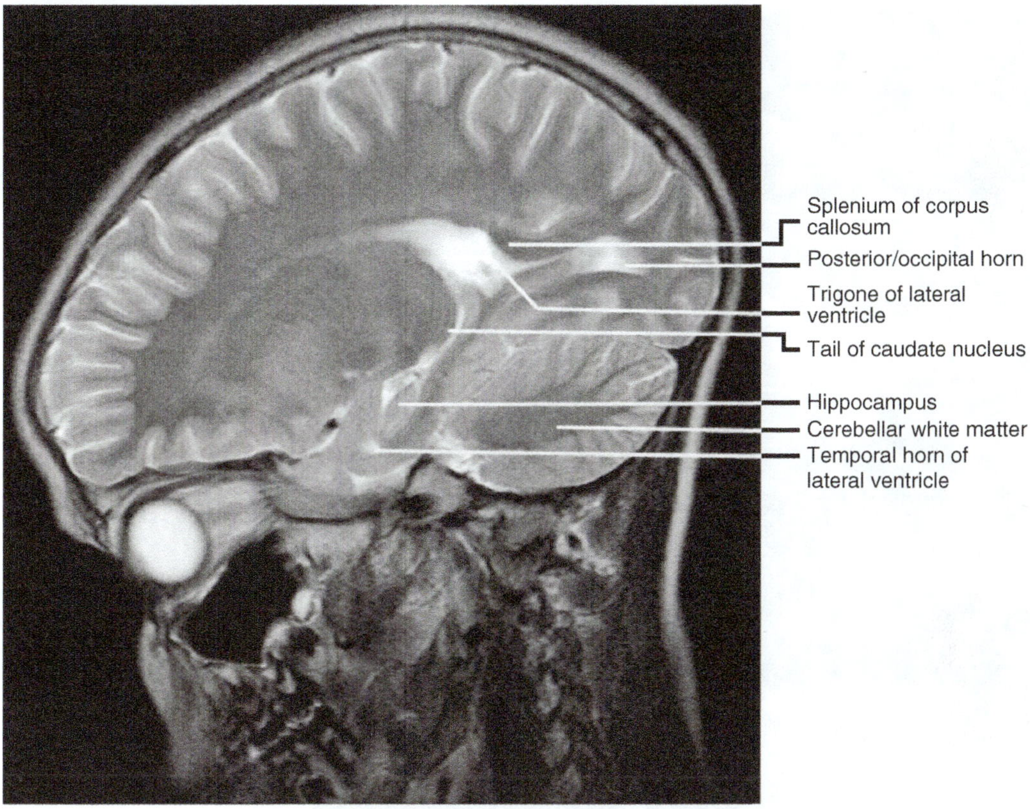

Fig. 1.34 Sagittal T2 at the level of the basal ganglia, hippocampus, and trigone

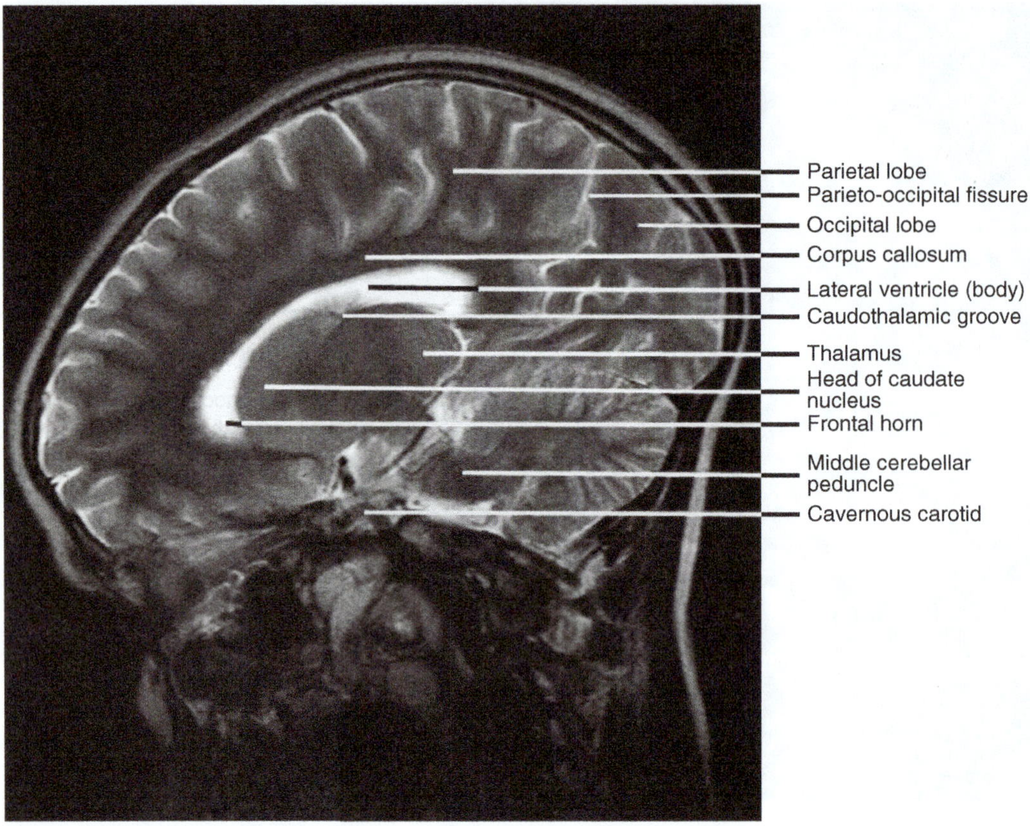

Fig. 1.35 Sagittal T2 at the level of the basal ganglia

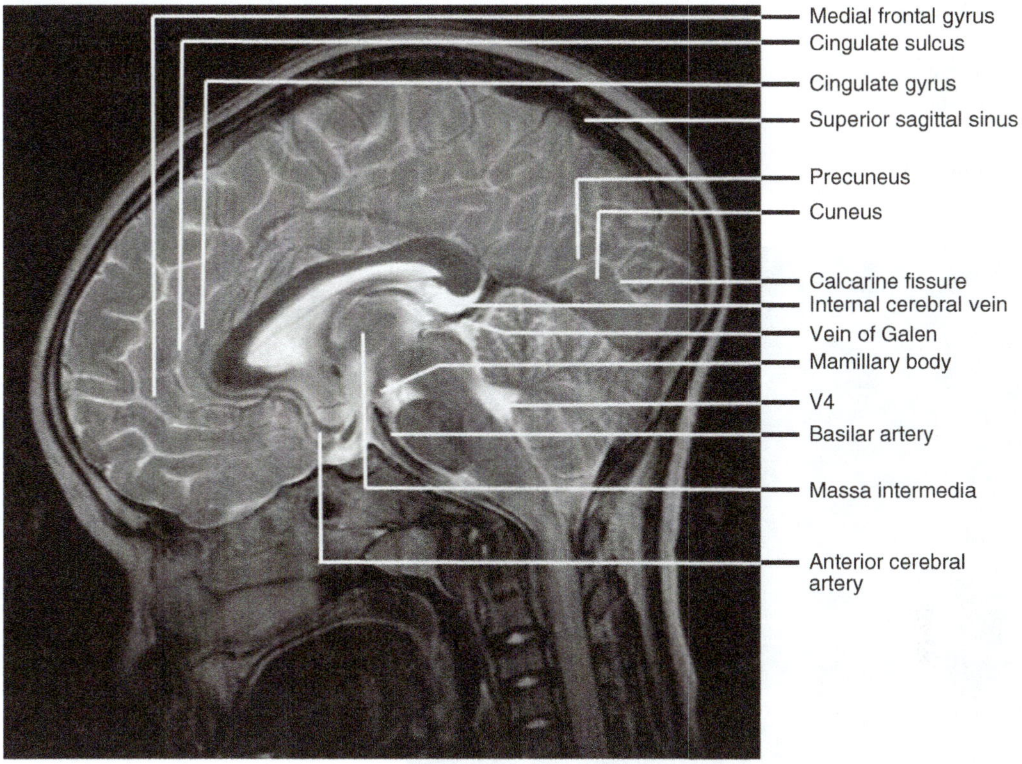

Fig. 1.36 Sagittal T2 at the midline

Special Areas

Midline sagittal
Corpus callosum
Cerebellar vermis
Coronal brainstem

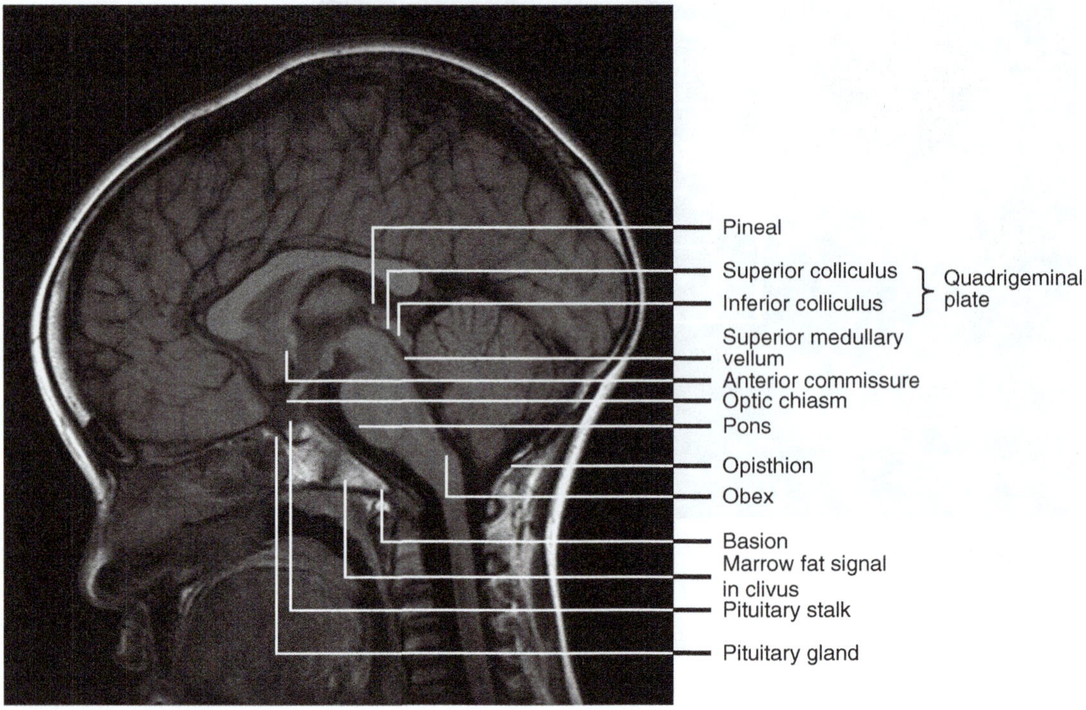

Fig. 1.37 Sagittal T1 midline

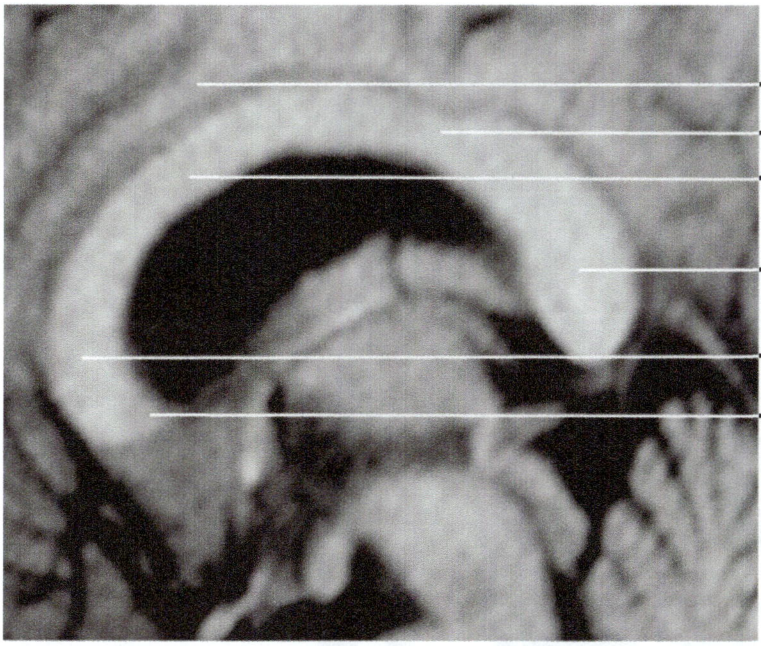

Fig. 1.38 Zoomed sagittal T1 of the corpus callosum (midline)

Pearl
The presence of the CC can be inferred by visualizing a horizontal cingulate gyrus.

Pearl
CC is a white matter tract. On T2 it is hypointense and on T1 it is hyperintense when myelinated.

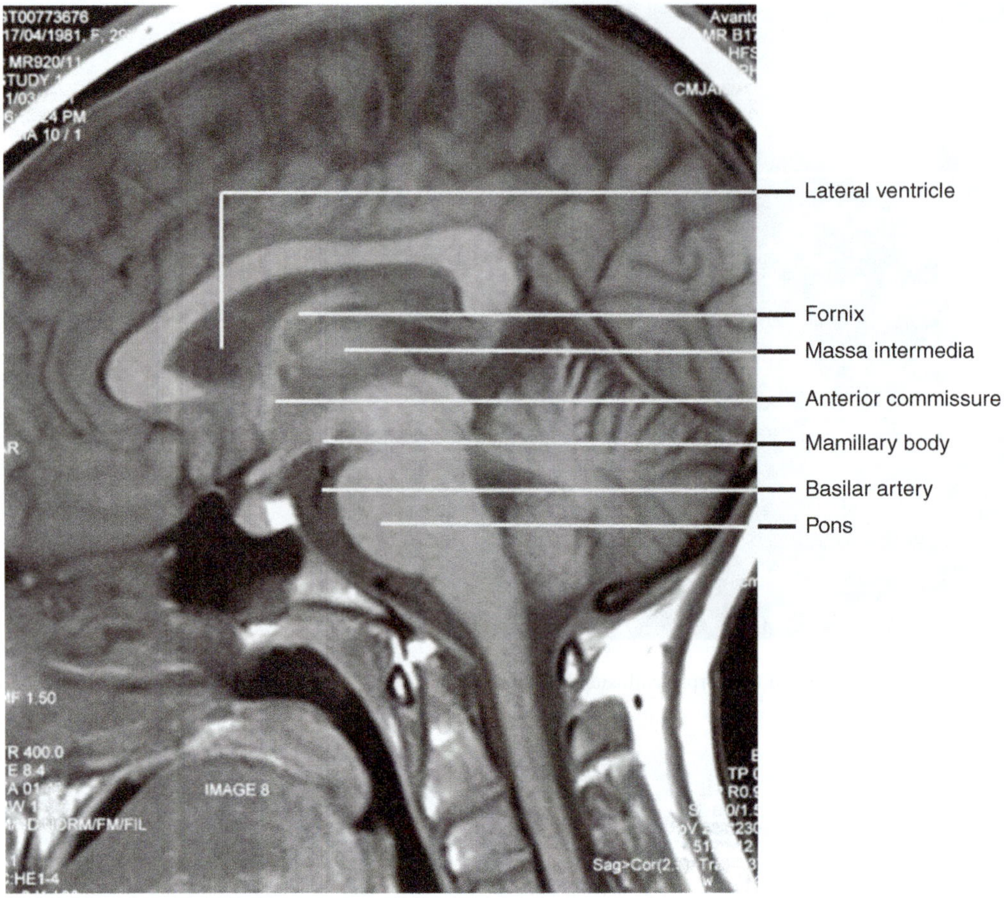

Fig. 1.39 Brain sagittal midline T1 MRI

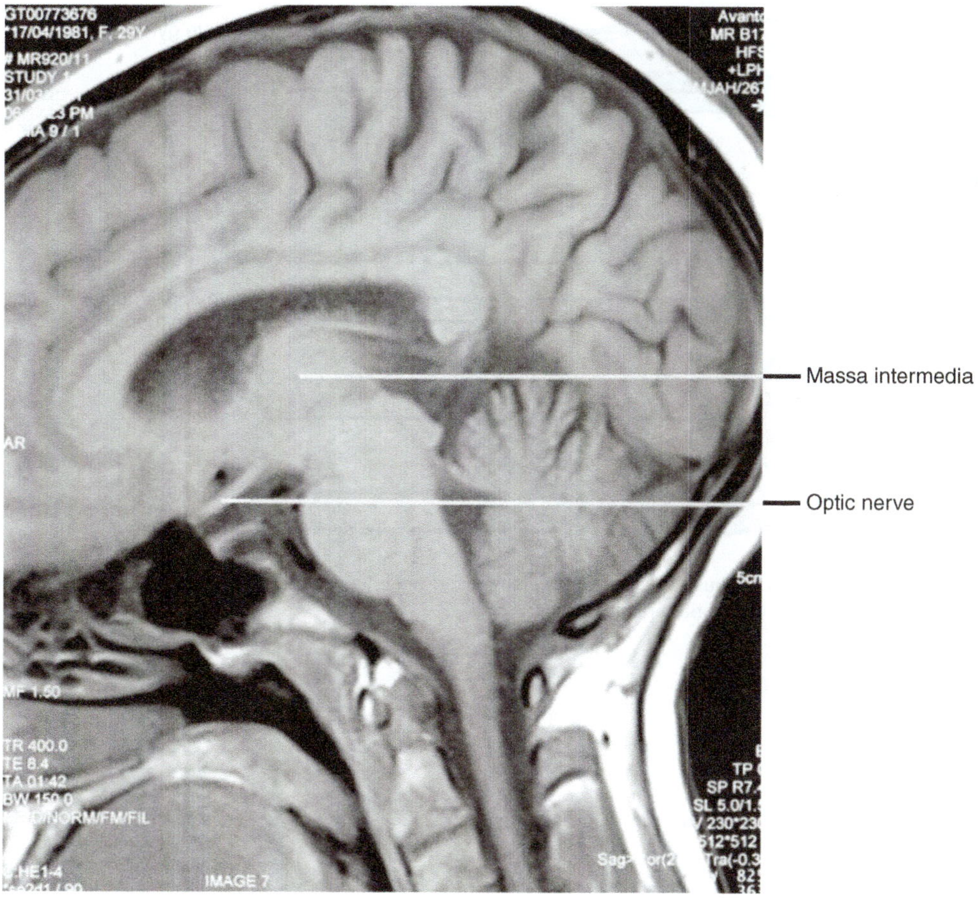

Fig. 1.40 Brain parasagittal T1 MRI

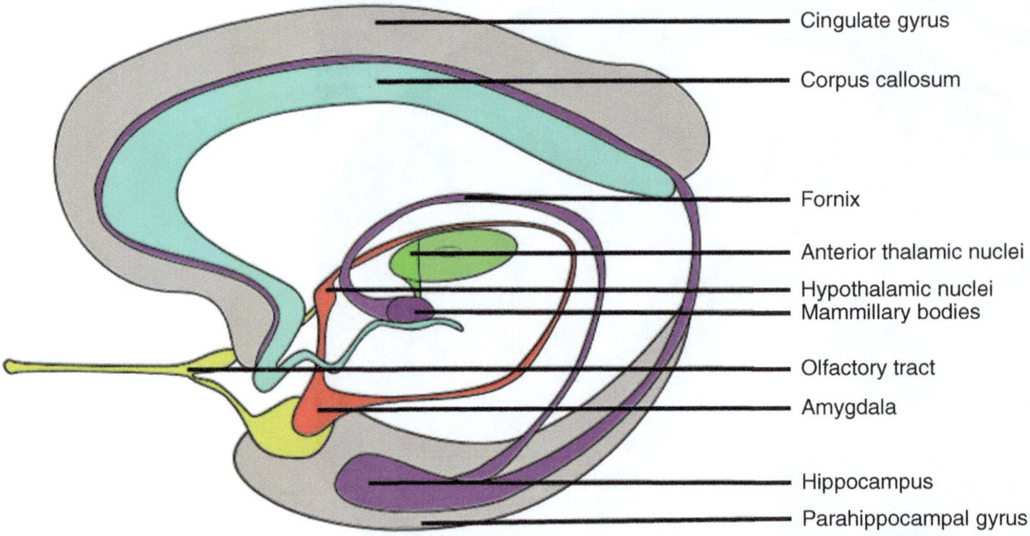

Fig. 1.41 Arcs of cingulate gyrus, corpus callosum, fornix, amygdala, and hippocampus. Concentric layers around the corpus callosum are flattened into one sagittal plane

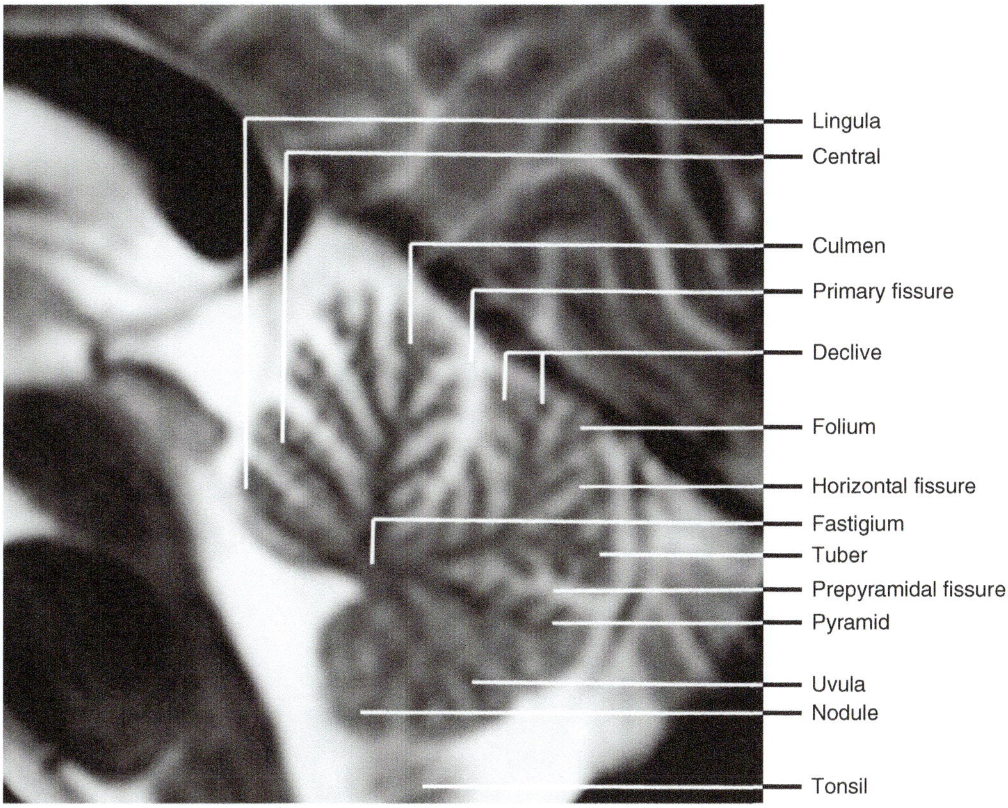

Fig. 1.42 Zoomed sagittal T2 of the cerebellar vermis (midline)

Pearl

Fastigium is the central white matter. From there, the individual lobes are identified by the white matter leading to them.

2 Temporal Lobe and Hippocampus

Anatomy

The temporal lobe comprises the lateral, inferior, and medial aspects of the cerebral hemispheres. Four sulci divide the temporal lobe into five gyri, one of which is the parahippocampal gyrus (T5). Included in the temporal lobe is the **hippocampus**, which is also part of the limbic system.

The hippocampus is a cortical fold that bulges into the floor of the temporal horn of the lateral ventricles. This fold is formed by two V-shaped interlocking cortical laminae, the **cornu ammonis** (Ammon's horn) and the **dentate gyrus**. These together constitute the archicortex (phylogenically old, three-layered cortex) of the temporal lobe.

The subiculum is a transitional cortex, which links the archicortex with the modified six layered neocortex of the parahippocampal gyrus. The fibres of the **perforant band** demarcate the subiculum from the parahippocampal cortex.

The endoventricular aspect of the cornu ammonis is covered by a thin layer of white matter, the **alveus**, which condenses medially to form a thickened band, the **fimbria**, which extends posteriorly to become the fornix.

Macroscopically and viewed axially, the hippocampus appears as an arc, medially concave, composed of three segments:

1. Anteriorly, a **head** which is transversely orientated. The digitationes hippocampi or **pes hippocampus**, which are folds of the cornu ammonis, are seen here.
2. Middle part or **body**, which is sagittally orientated. This is the most typical aspect and is composed of the cornu ammonis and gyrus dentatus. The cornu ammonis is covered by the alveus.
3. Posteriorly, a **tail**, again transversely orientated and situated beneath the splenium. This is a thin part of the hippocampal arc. Each of the hippocampal constituents can be found in the tail but with different names.

Imaging

For imaging purposes, the anatomy of the medial temporal lobe can be simplified and thought of as comprising three fibre arcs extending in a C-shape from anterior to posterior, which then course superiorly in reverse from posterior to anterior. They are:

1. The alveus and fimbria extending posteriorly to become the fornix, which ultimately terminates in the mammillary bodies anteriorly.
2. The hippocampus extending posteriorly to become the indusium griseum on the dorsal surface of the corpus callosum, terminating in the septal nuclei anteriorly.
3. The parahippocampal gyrus and subiculum becoming continuous with the cingulate gyrus to terminate in the association areas of the frontal lobe.

S. Andronikou (ed.), *See Right Through Me*,
DOI 10.1007/978-3-642-23893-2_2, © Springer-Verlag Berlin Heidelberg 2012

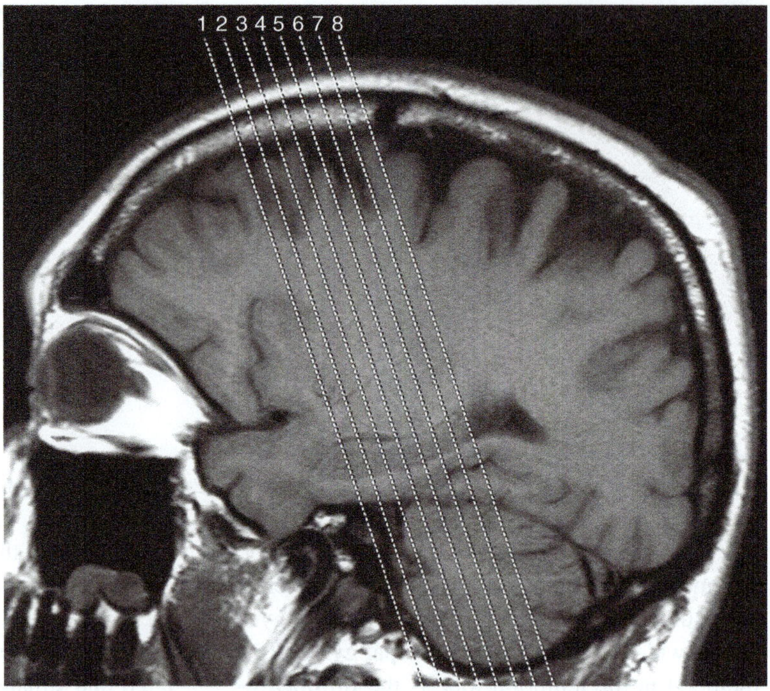

Fig. 2.1 Scan angle hippocampus. Sagittal T1 scout view shows cut lines perpendicular to the long axis of the hippocampus

Pearl

Coronal magnetic resonance imaging (MRI) is performed perpendicular to the hippocampus as planned on a sagittal image. Computed tomography (CT) is not useful.

Pearl

Best sequence is a coronal short tau inversion-recovery (STIR), usually displayed as an inverted greyscale.

Hippocampal component	MRI signal: inverted STIR	MRI signal: T2
Alveus/fimbria	White	Black
Cornu ammonis	Grey	Grey
Gyrus dentatus		
Subiculum	Grey	Grey
Hippocampal sulcus	Black	White
Perforant band	White	Black

Pearl

Potential pitfall: A "buried" or entrapped hippocampal sulcus should not be misinterpreted as a hippocampal cyst.

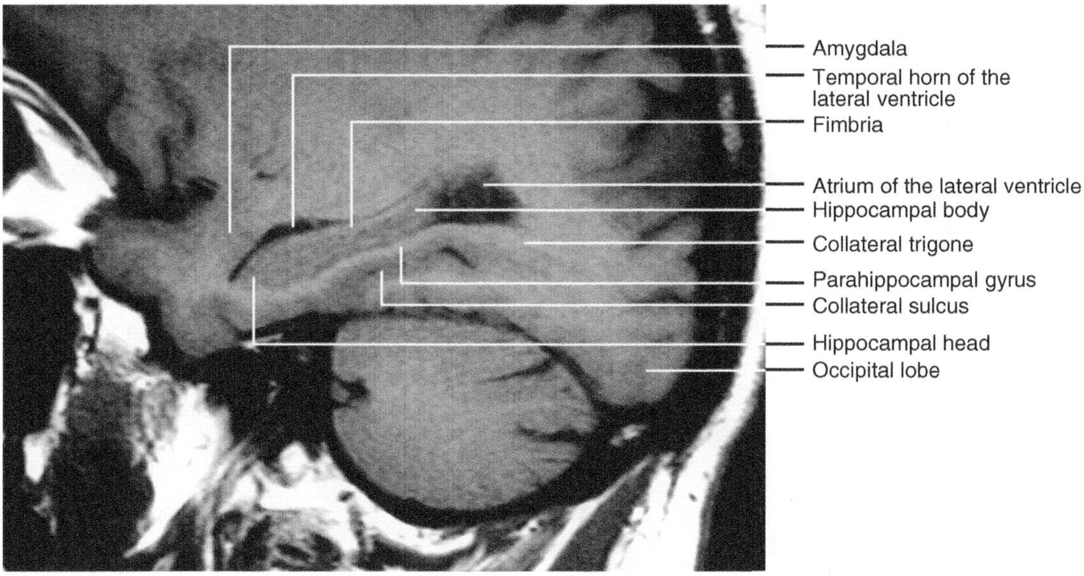

Fig. 2.2 Sagittal T1 of the hippocampus

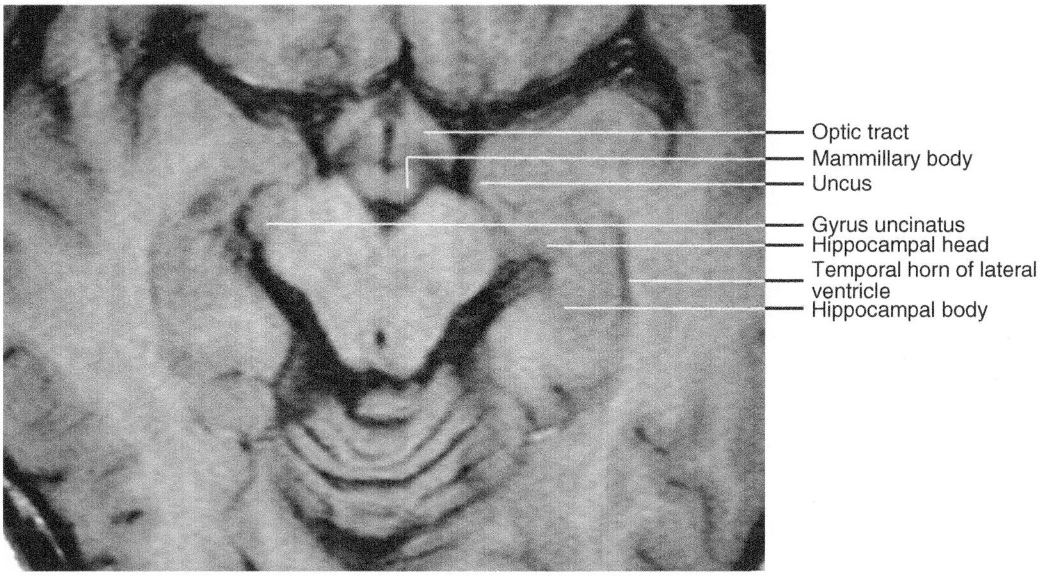

Fig. 2.3 Axial T1 of the hippocampus. Level at the mammillary bodies

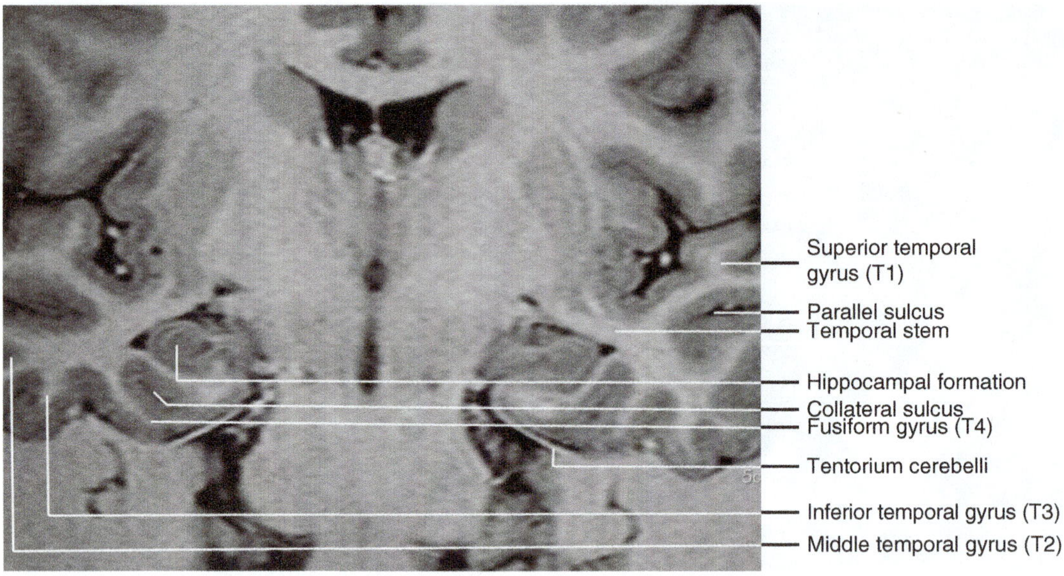

Fig. 2.4 General relations of the hippocampus to adjacent structures. Coronal inverted STIR image

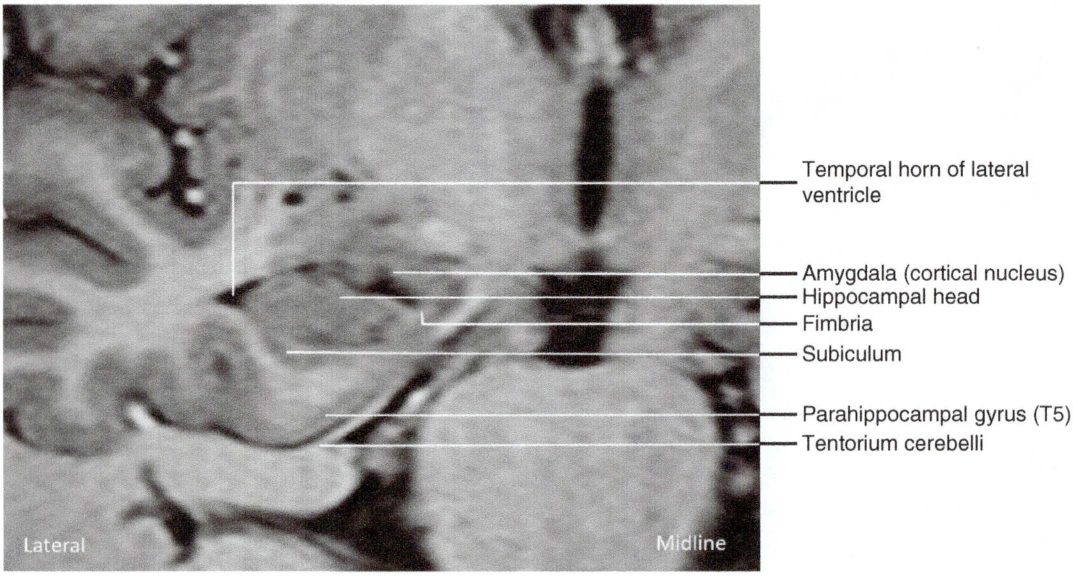

Fig. 2.5 Detail of the right hippocampal formation. Coronal inverted STIR of the hippocampal head

Pearl

Hippocampal head = digitationes hippocampi = pes hippocampus.

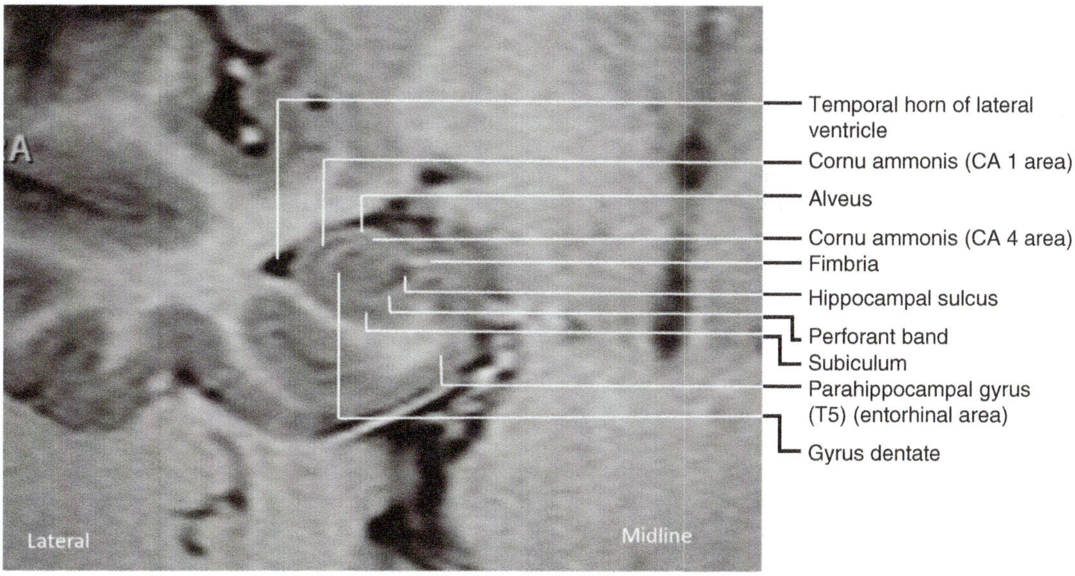

Fig. 2.6 Detail of the right hippocampal formation. Coronal inverted STIR of the hippocampal body immediately behind head

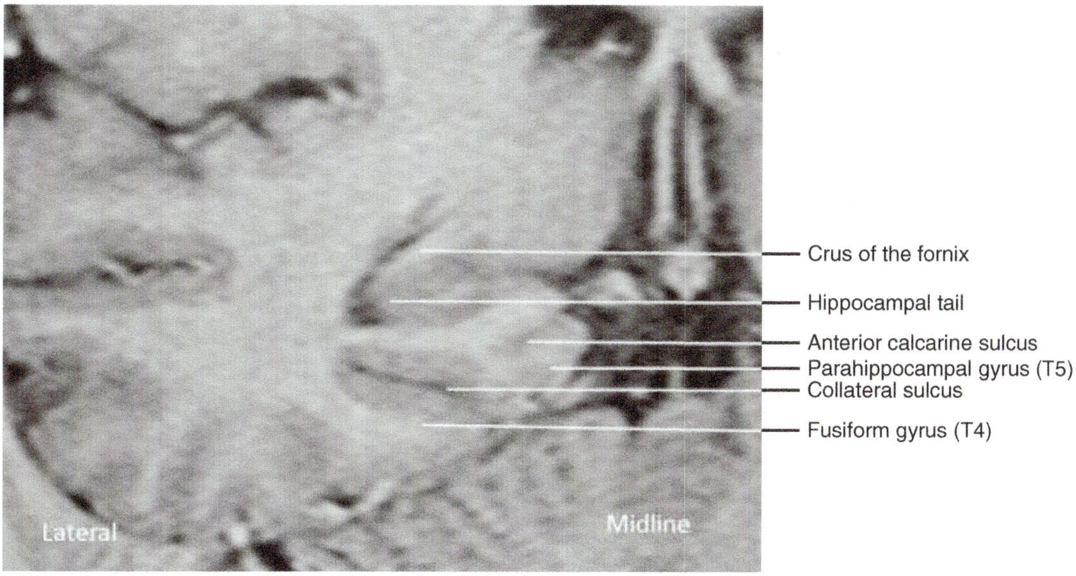

Fig. 2.7 Detail of right hippocampal formation. Coronal inverted STIR of the hippocampal tail

Acknowledgement Professor R. Hewlett for advice.

Sources

1. Duvernoy HM (1995) In: Kuzniecky RI, Jackson GD (eds) Magnetic resonance in epilepsy. Raven Press, New York
2. Felten DL, Jozefowicz RF (2003) Netter's atlas of human neuroscience. Icon Learning System, Teterboro
3. FitzGerald MJT, Folan-Curran J (2002) Clinical neuroanatomy and related neuroscience, 4th edn. WB Saunders, Philadelphia
4. Williams PL, Warwick R (1980) Grey's anatomy, 26th edn. Churchill Livingstone, London

3 Cerebral Vessels

Fig. 3.1 Vascular territories of the brain

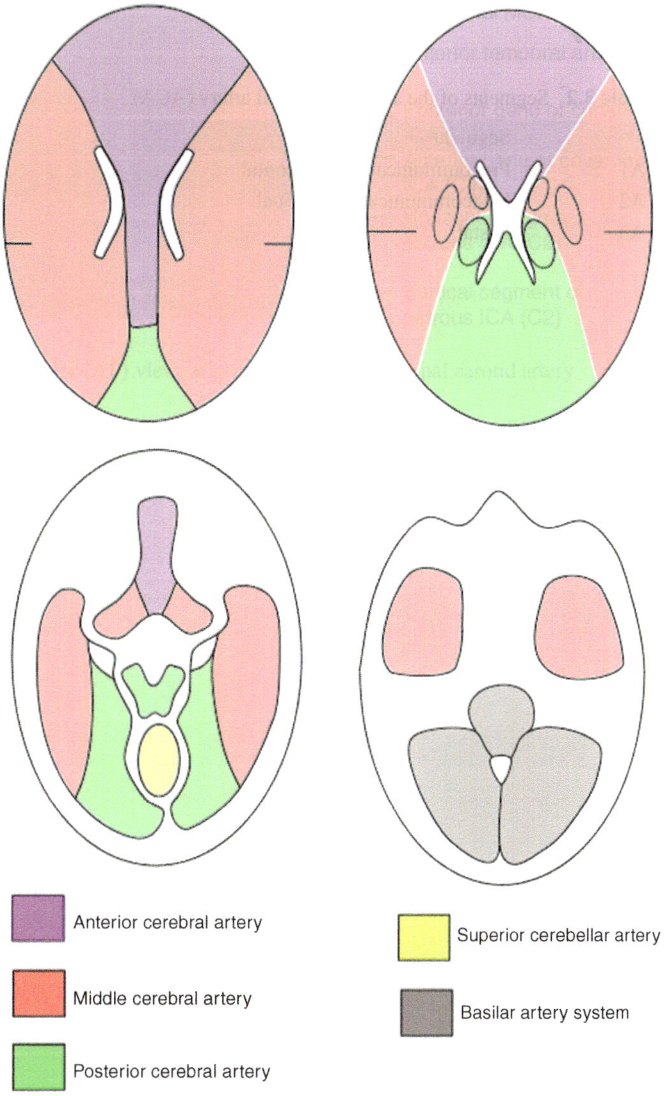

S. Andronikou (ed.), *See Right Through Me*,
DOI 10.1007/978-3-642-23893-2_3, © Springer-Verlag Berlin Heidelberg 2012

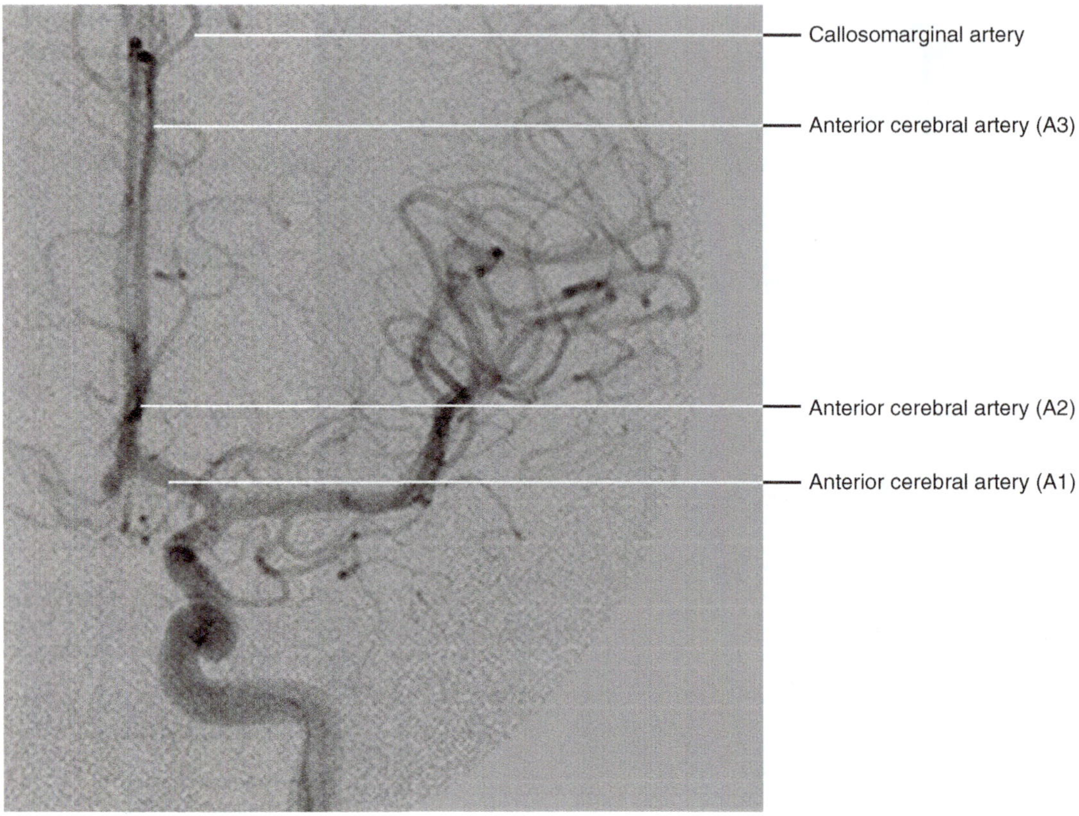

Fig. 3.3 Conventional angiography. AP view: arterial phase of left internal carotid artery

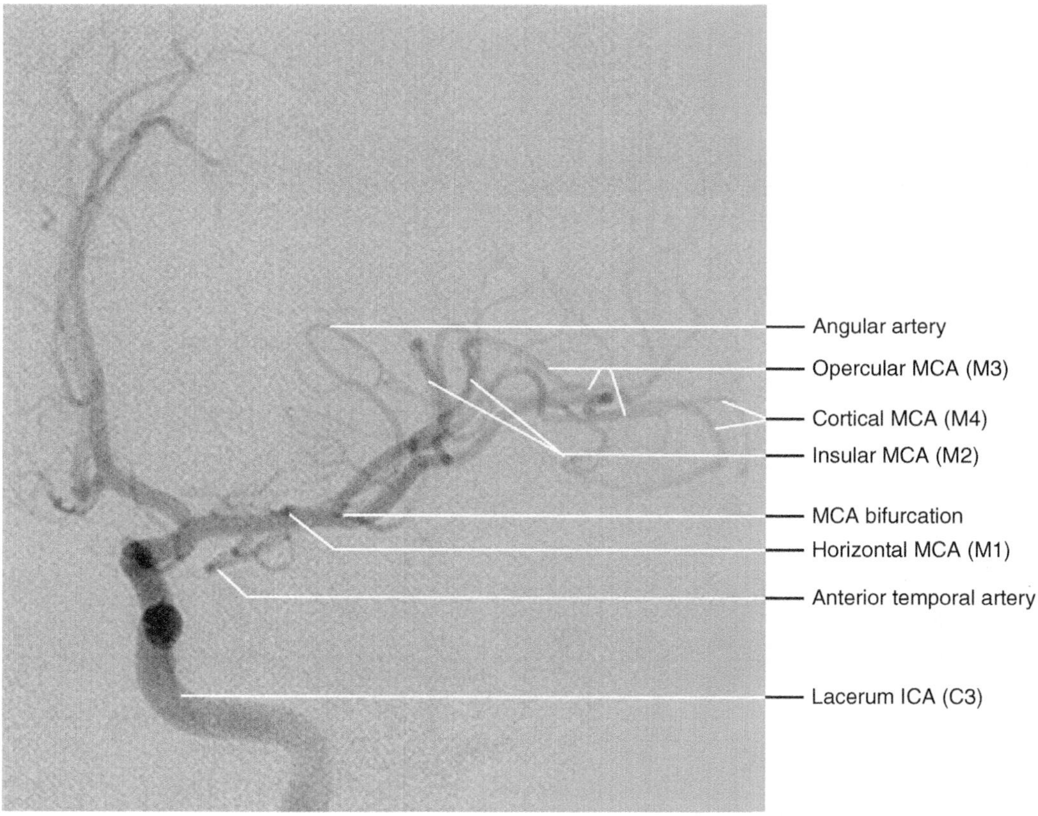

Fig. 3.4 Conventional angiography. Oblique view: arterial phase of left internal carotid artery

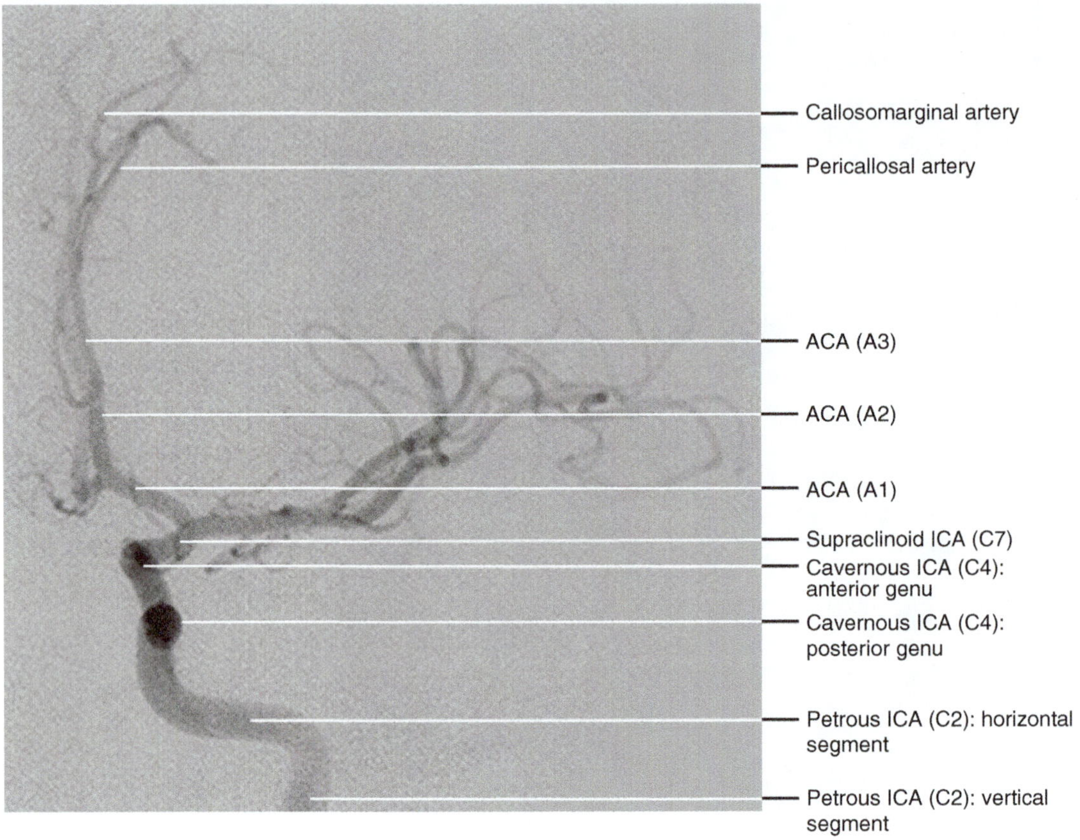

Fig. 3.5 Conventional angiography. Oblique view: arterial phase of left internal carotid artery

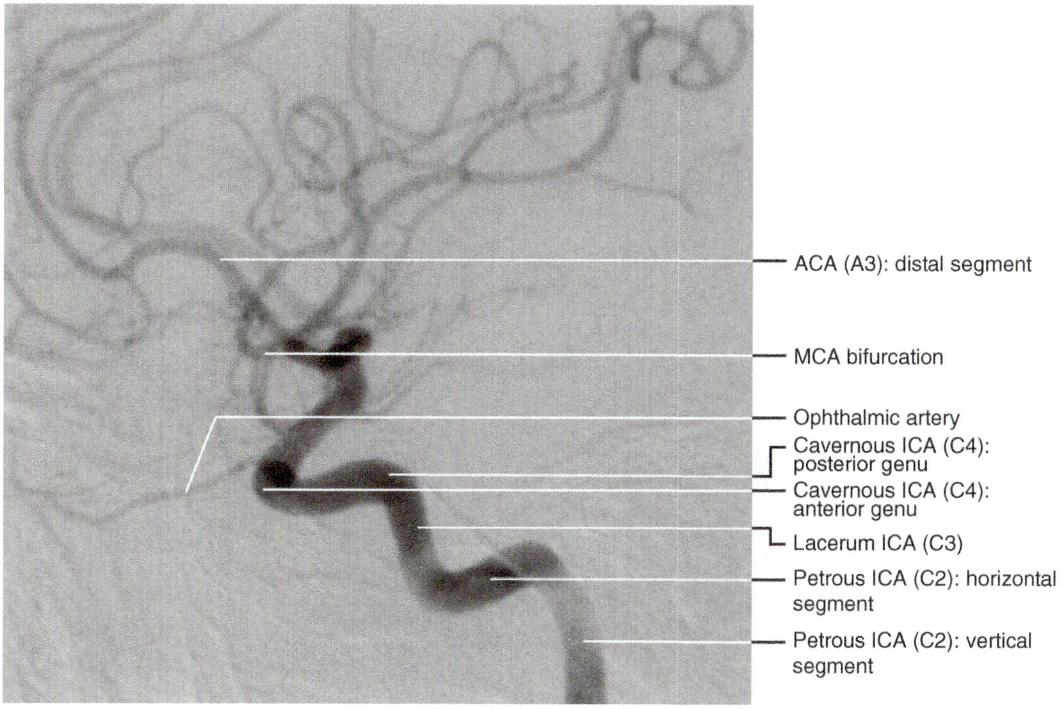

Fig. 3.6 Conventional angiography. Lateral view: arterial phase of left internal carotid artery

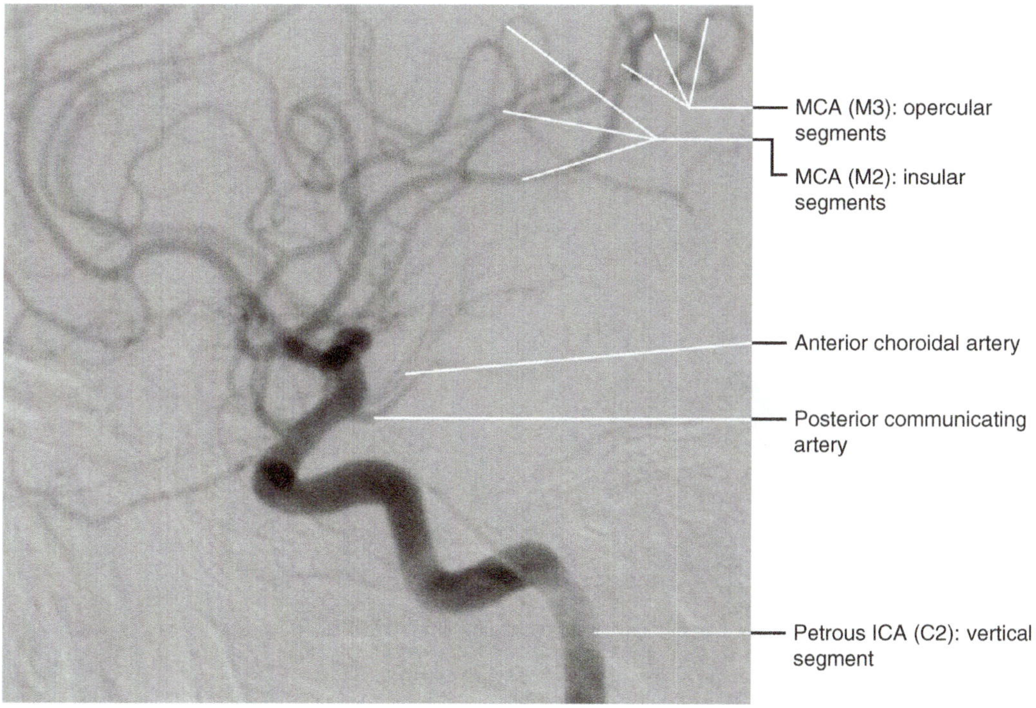

Fig. 3.7 Conventional angiography. Lateral view: arterial phase of left internal carotid artery

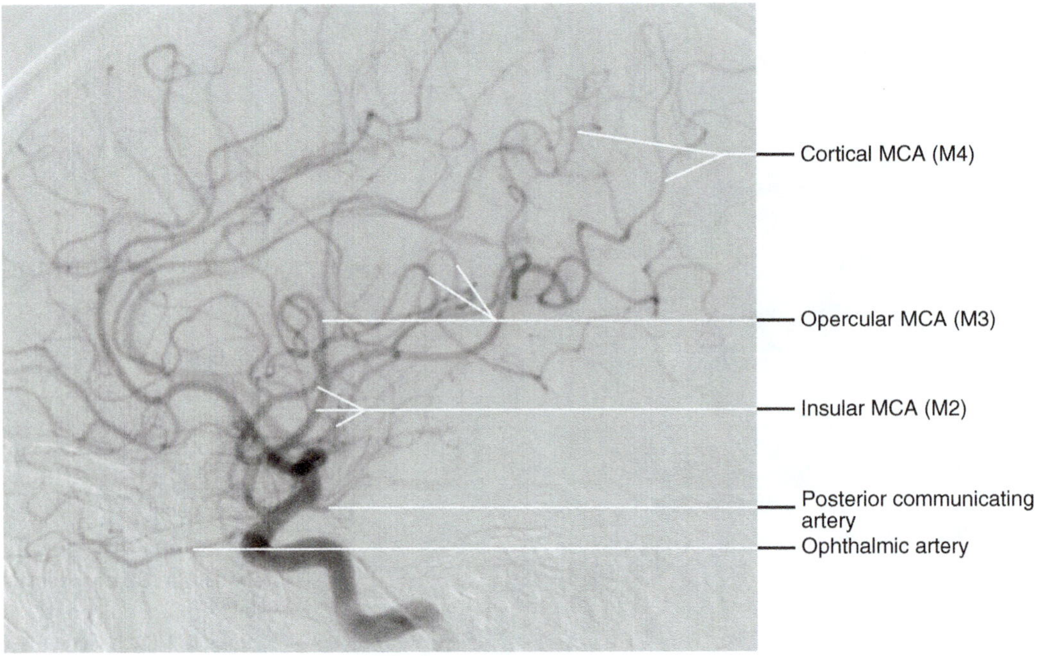

Fig. 3.8 Conventional angiography. Lateral view: arterial phase of left internal carotid artery

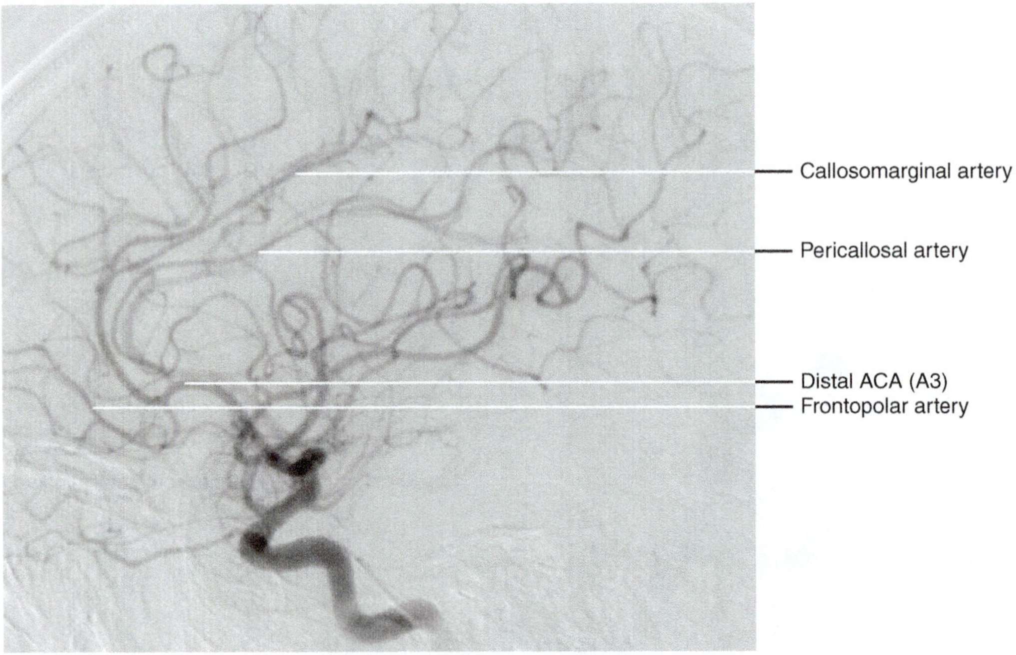

Fig. 3.9 Conventional angiography. Lateral view: arterial phase of left internal carotid artery

Table 3.4 Segments of the posterior cerebral artery (PCA)

	Segments
P1	Precommunicating
P2	Ambient
P3	Quadrigeminal
P4	Calcarine

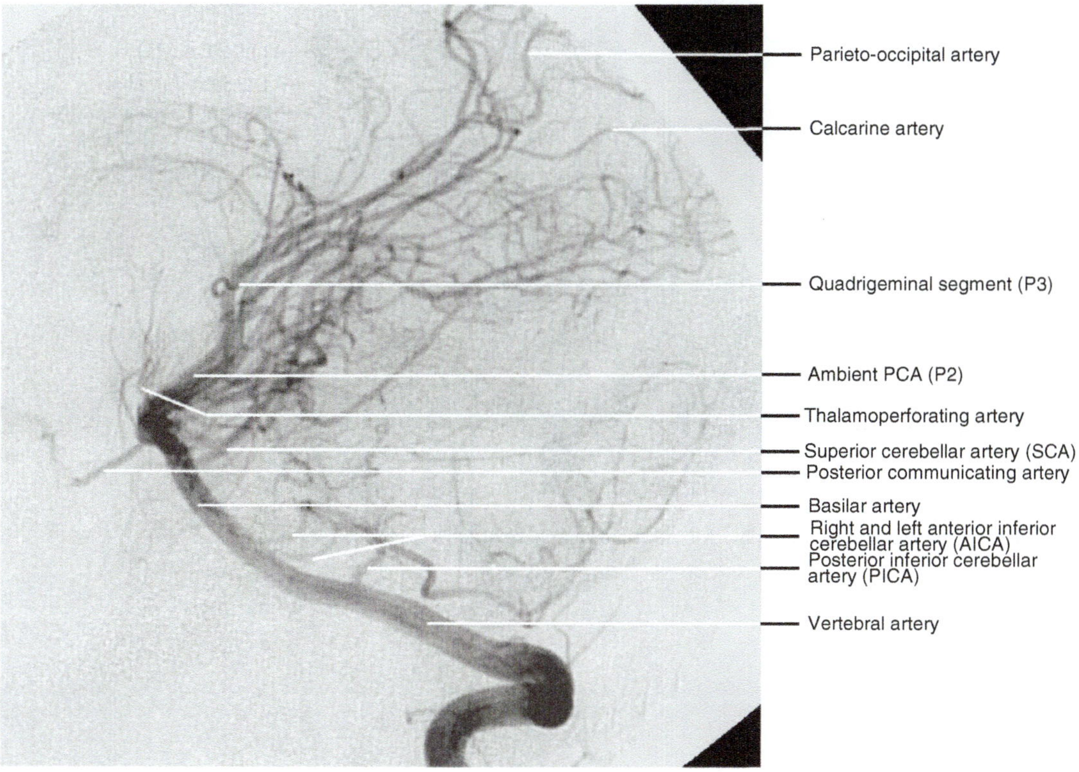

Fig. 3.10 Conventional angiography. Lateral view: arterial phase of left vertebral artery

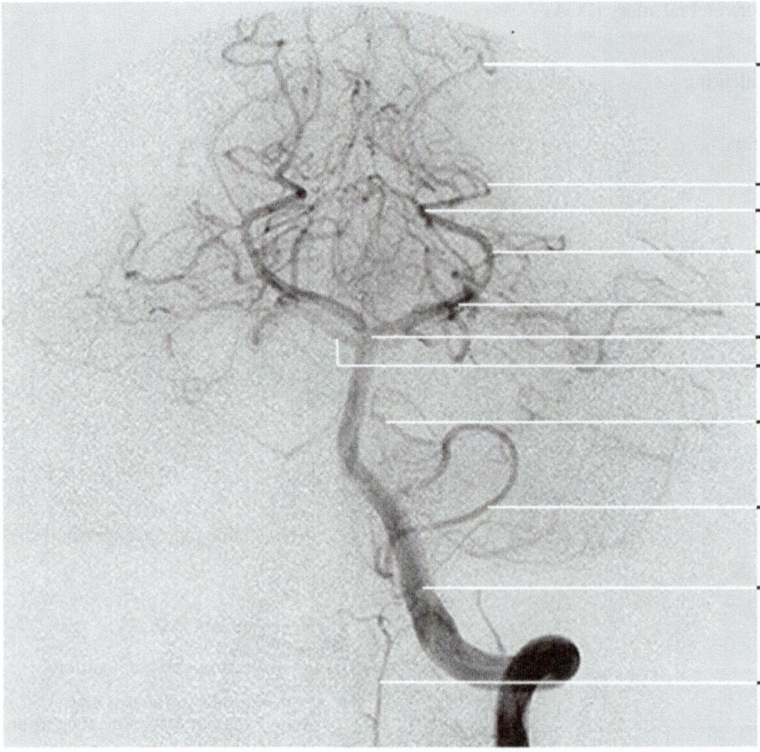

Fig. 3.11 Conventional angiography. AP view: arterial phase of left vertebral artery

External Carotid Artery

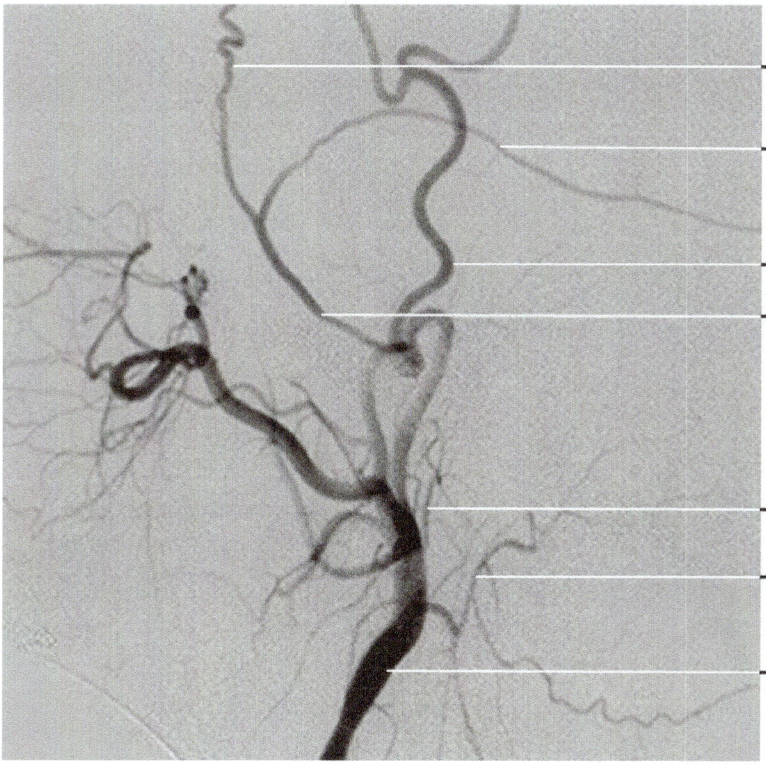

Fig. 3.12 Conventional angiography. Lateral view: arterial phase of left external carotid artery

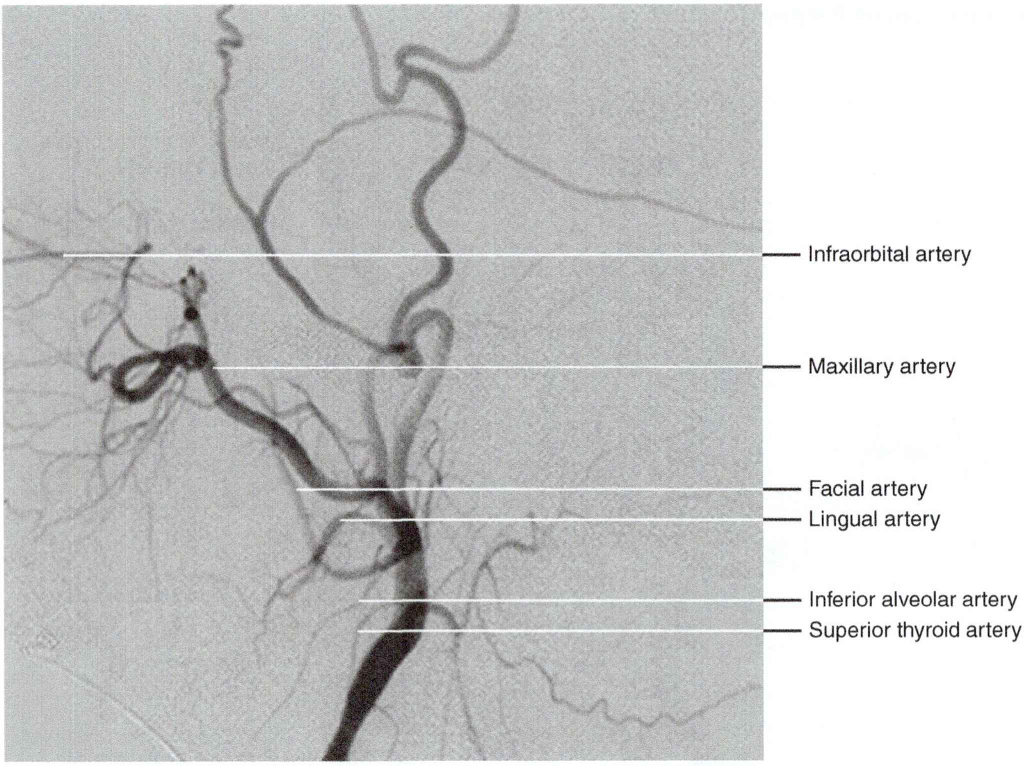

Fig. 3.13 Conventional angiography. Lateral view: arterial phase of left external carotid artery

Table 3.5 Segments of the vertebral artery

	Segments
V1	Extraosseous (from aortic arch to C6)
V2	Foraminal (from C6 to C1)
V3	Extraspinal VA (from C1 to foramen magnum)
V4	Intradural

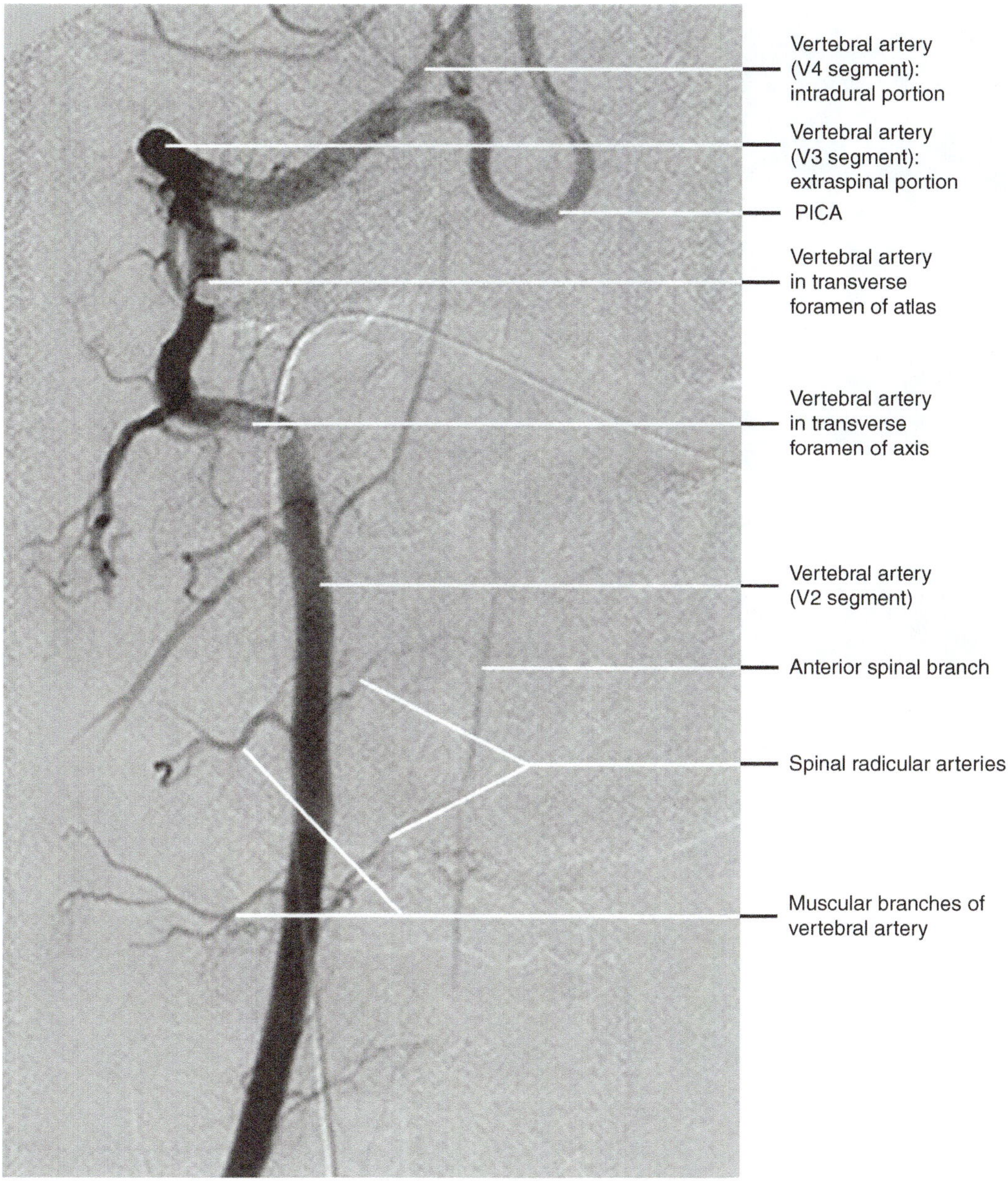

Fig. 3.14 Conventional angiography. Lateral view: arterial phase of right vertebral artery

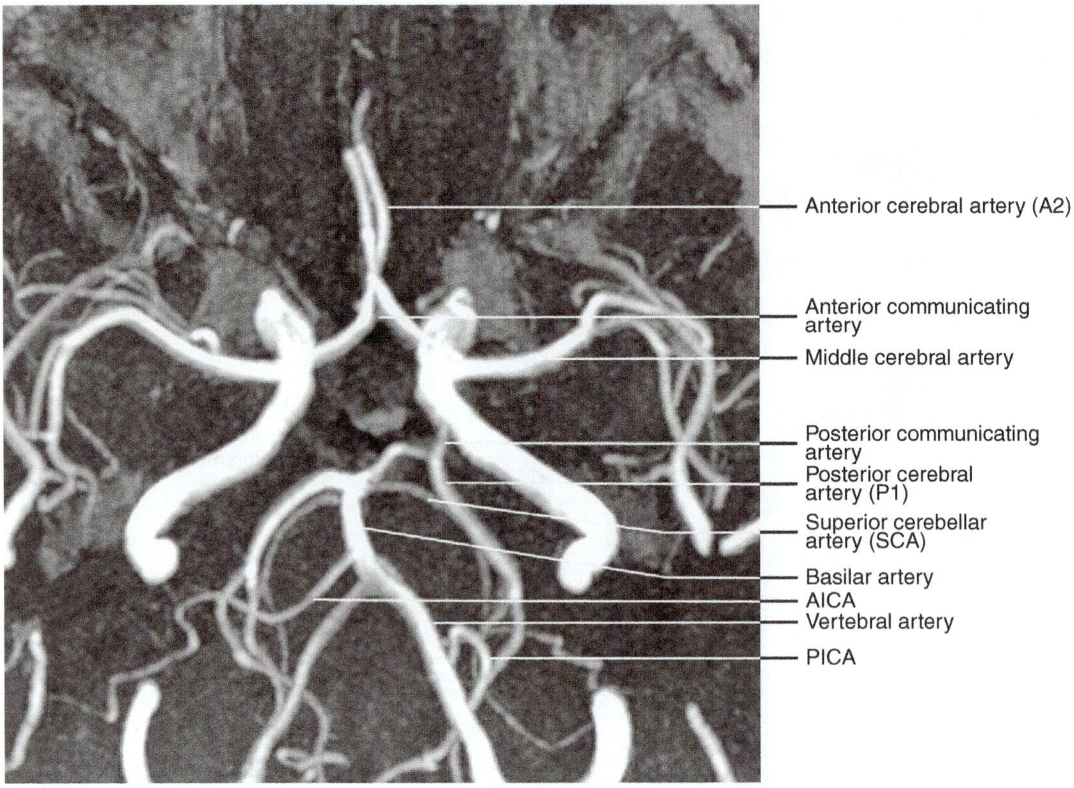

Fig. 3.15 Magnetic resonance angiography. Circle of Willis: AP view

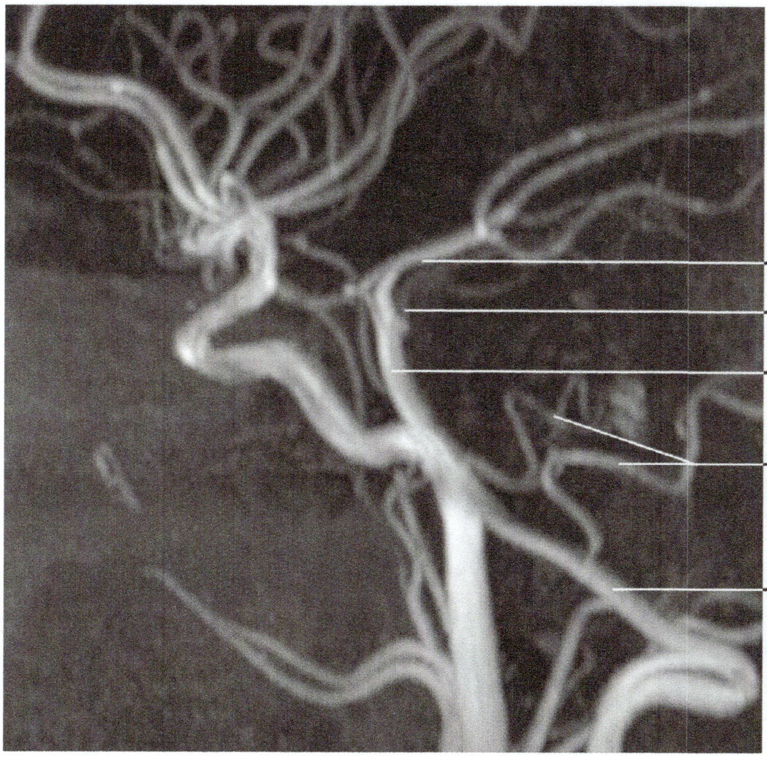

Fig. 3.16 Magnetic resonance angiography. Circle of Willis: lateral view

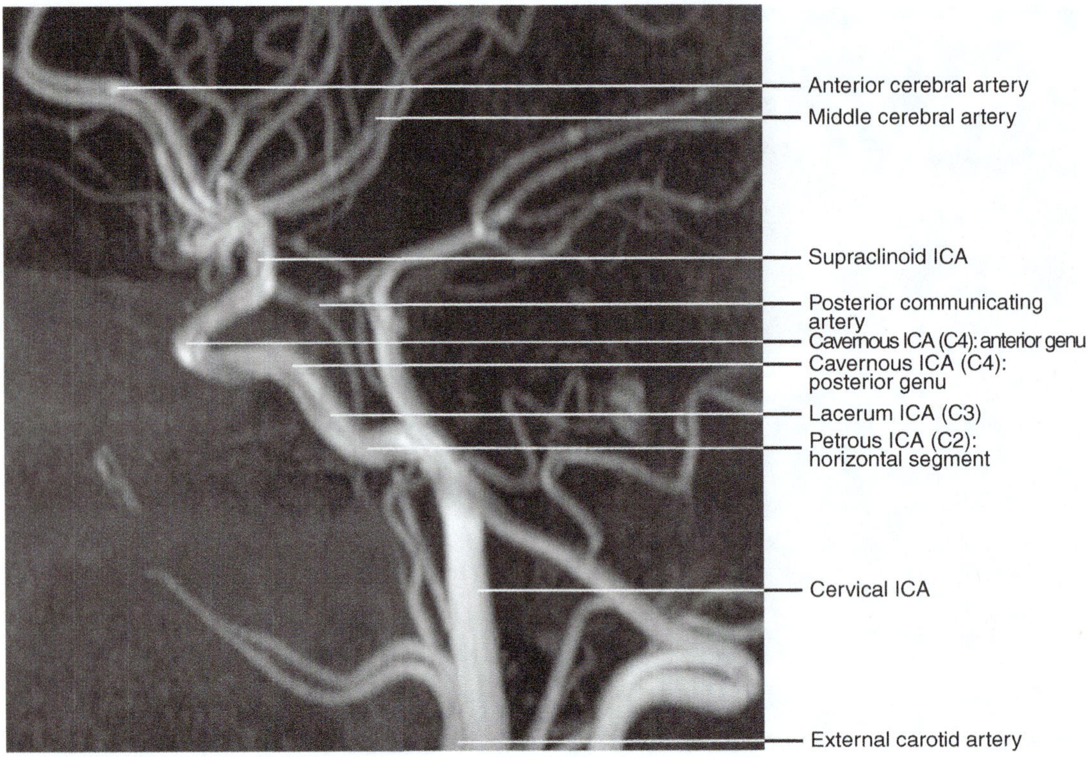

Fig. 3.17 Magnetic resonance angiography. Circle of Willis: lateral view

Cerebral Veins

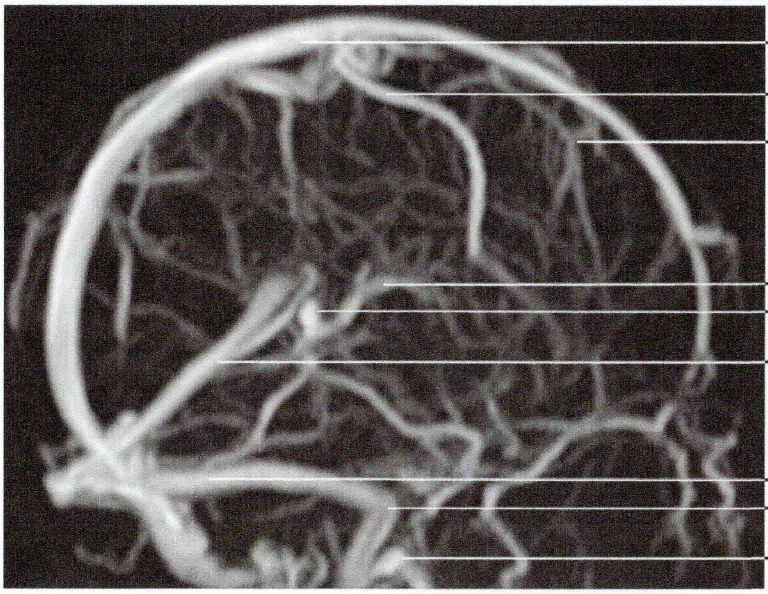

Fig. 3.18 Magnetic resonance venography lateral oblique view

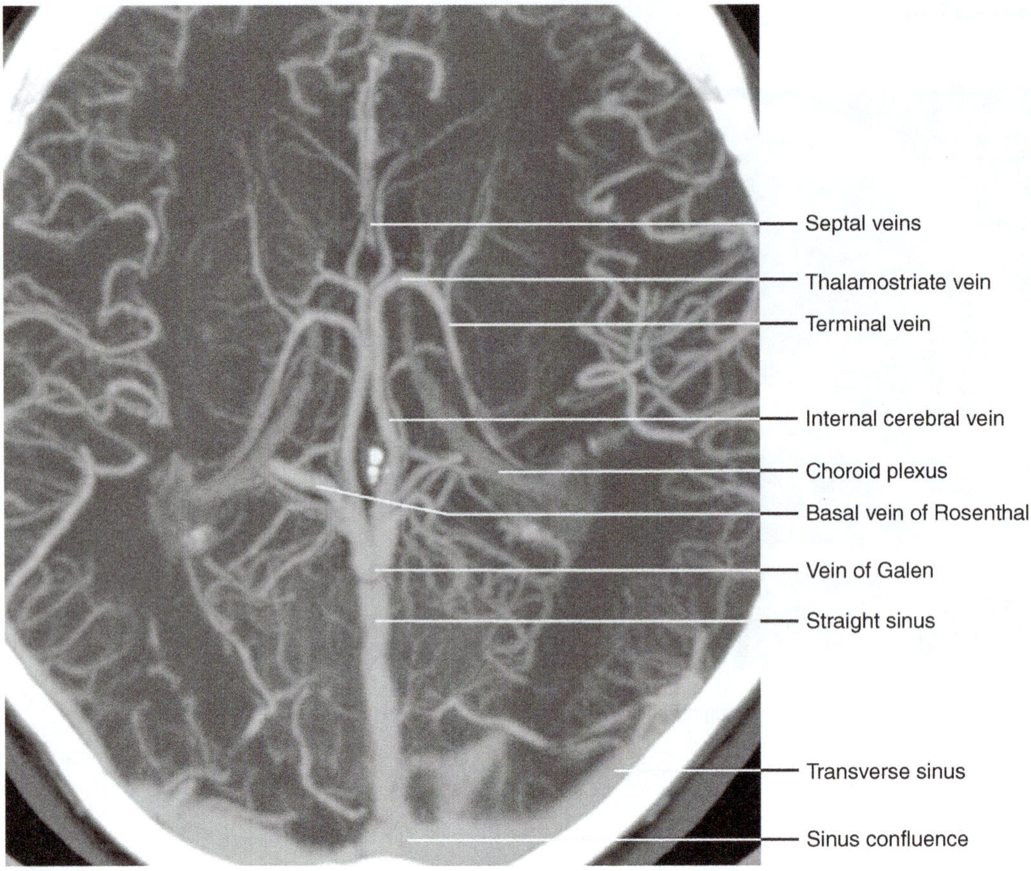

Fig. 3.19 Computed tomography axial maximum intensity projection through the cerebral veins

Sources

1. Harnsberger HR, Osborn A, Macdonald AJ et al (2006) Diagnostic and surgical imaging anatomy, 1st edn. Amirsys, Salt Lake City
2. Weir J, Abrahams PH (2003) Imaging atlas of human anatomy, 3rd edn. Mosby, London

4 White Matter Tracts

Pearls

There are three types of white matter tracts:

- Association pathways are intrahemispheric tracts joining different cortical areas in the same hemisphere. Short fibres make up the subcortical U-fibres. Long tracts are located in the deep white matter.
- Commissural pathways are interhemispheric tracts joining different cortical areas in both hemispheres, and are either homotopic (similar anatomic area) or heterotopic (different anatomic area).
- Projection tracts either originate (motor) or terminate (sensory) in the cortex connecting the cortex to subcortical structures like the basal ganglia, cerebellum or spine.

Fig. 4.1 Standardized colour-coding indicating the direction of the white matter tract: *Red* indicates left to right or vice-versa. *Green* indicates anterior to posterior or vice-versa. *Blue* indicates superior to inferior or vice-versa. *A mix of colours* indicates oblique directions or various combinations of tract directions in that area

S. Andronikou (ed.), *See Right Through Me*,
DOI 10.1007/978-3-642-23893-2_4, © Springer-Verlag Berlin Heidelberg 2012

Colour-Coded Maps of White Matter Tract Direction

Axial

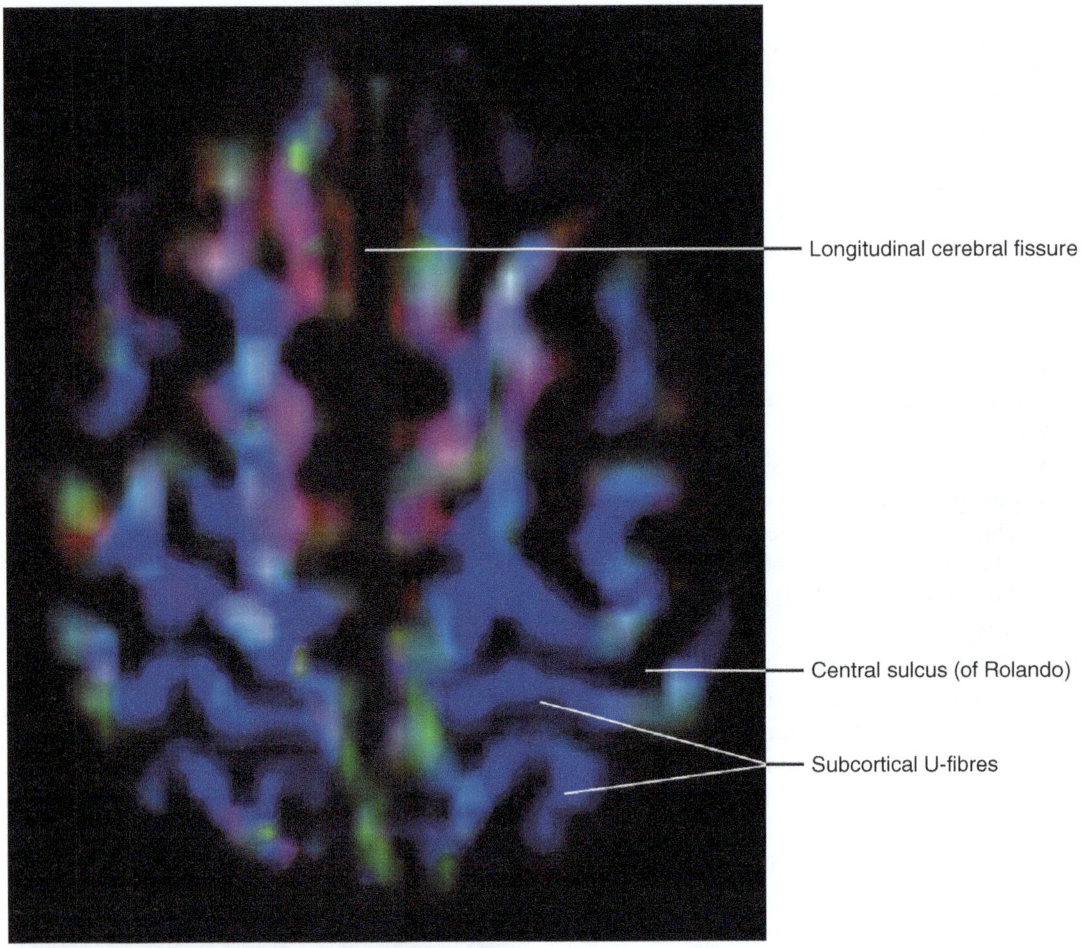

Fig. 4.2 The level of the cerebral convexities

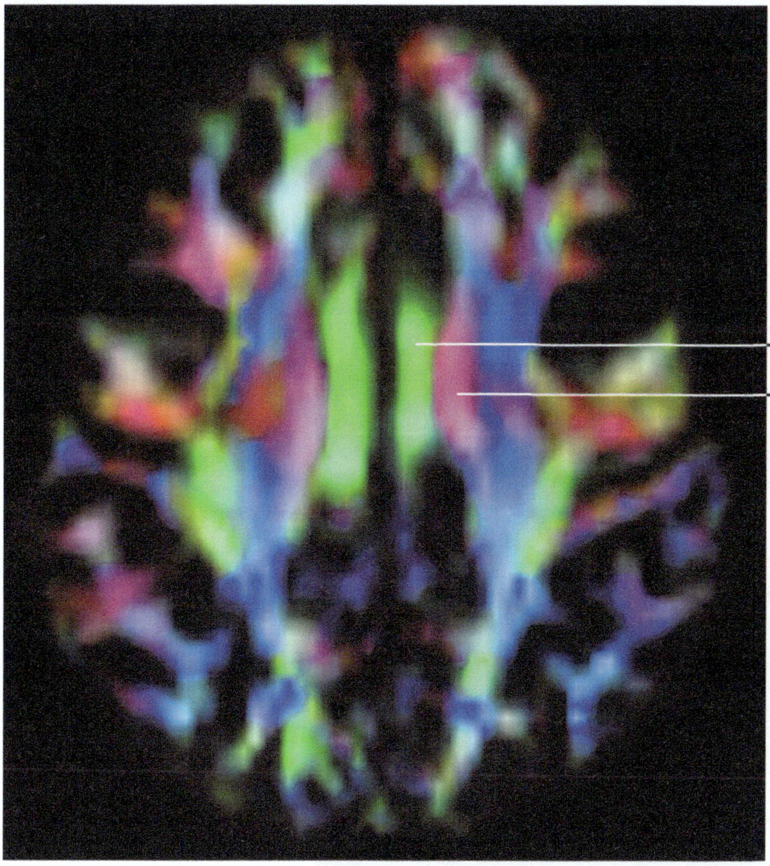

Fig. 4.3 The level of the centrum semi-ovale

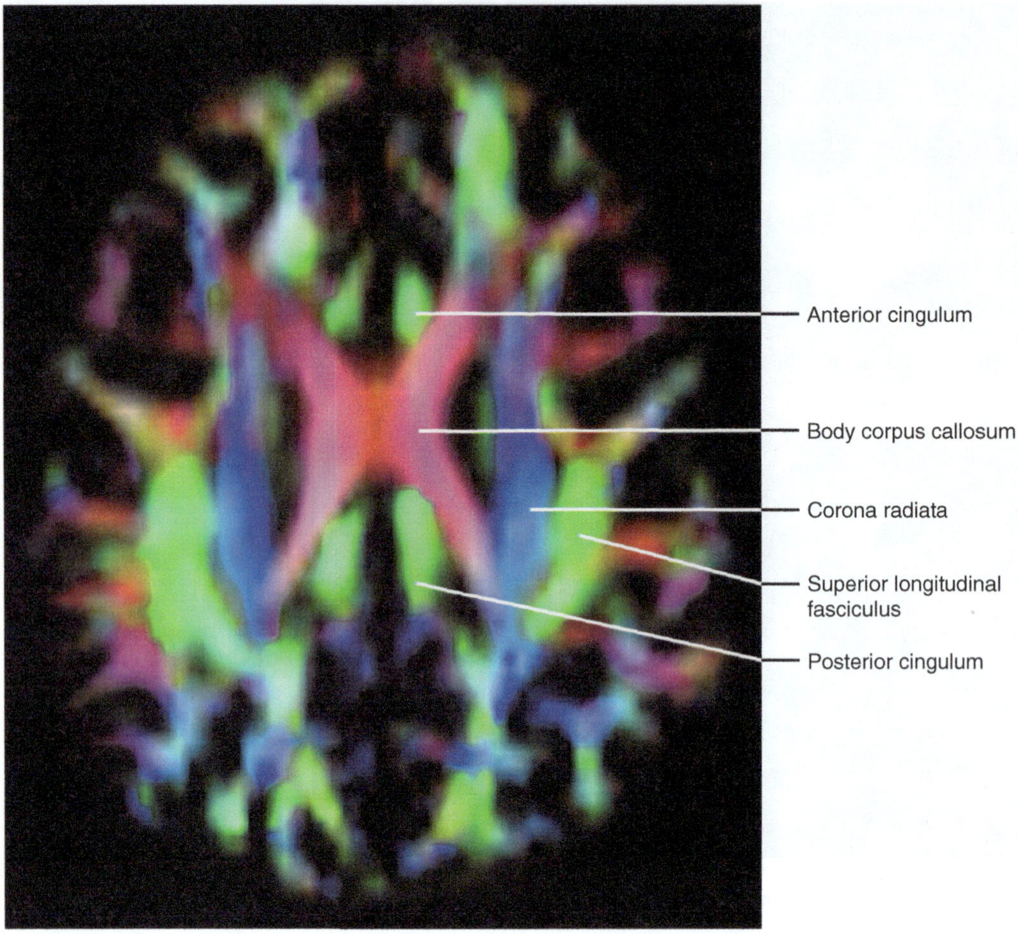

Fig. 4.4 The level of the corona radiata and superior part of the corpus callosum

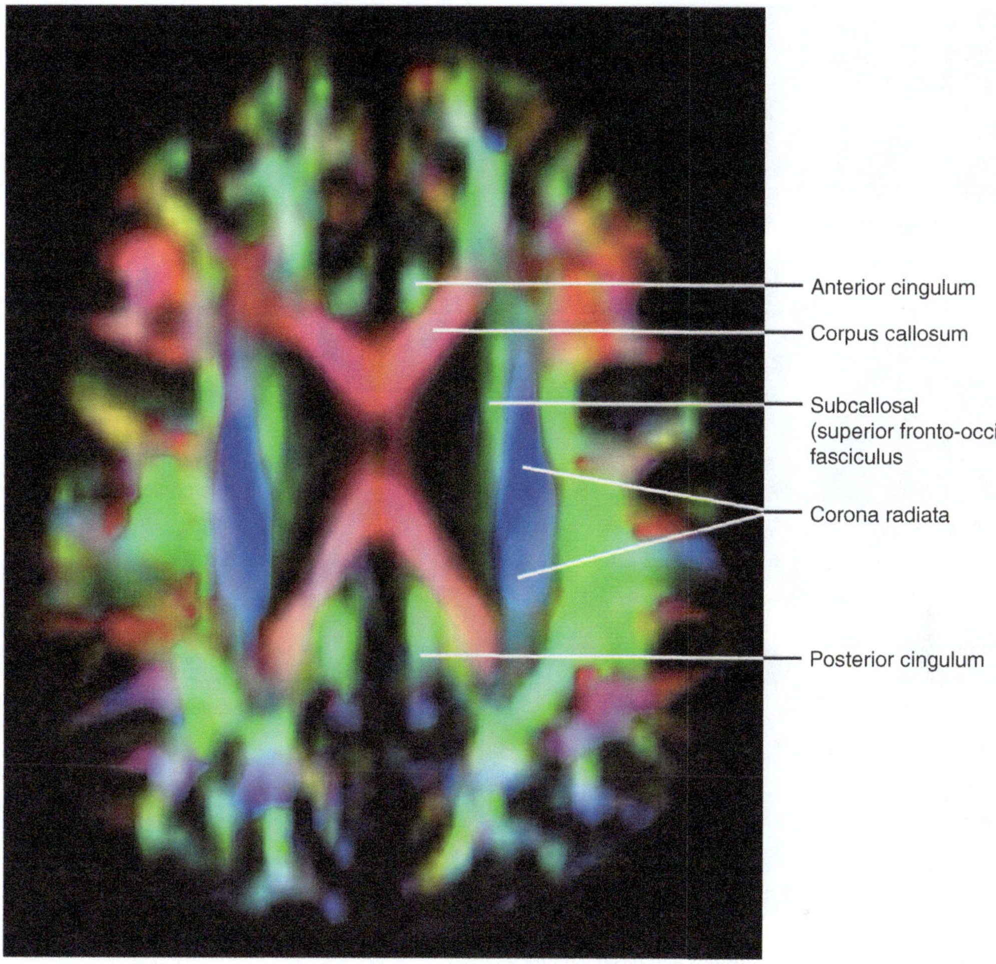

Fig. 4.5 The level of the corona radiata and genu of the corpus callosum

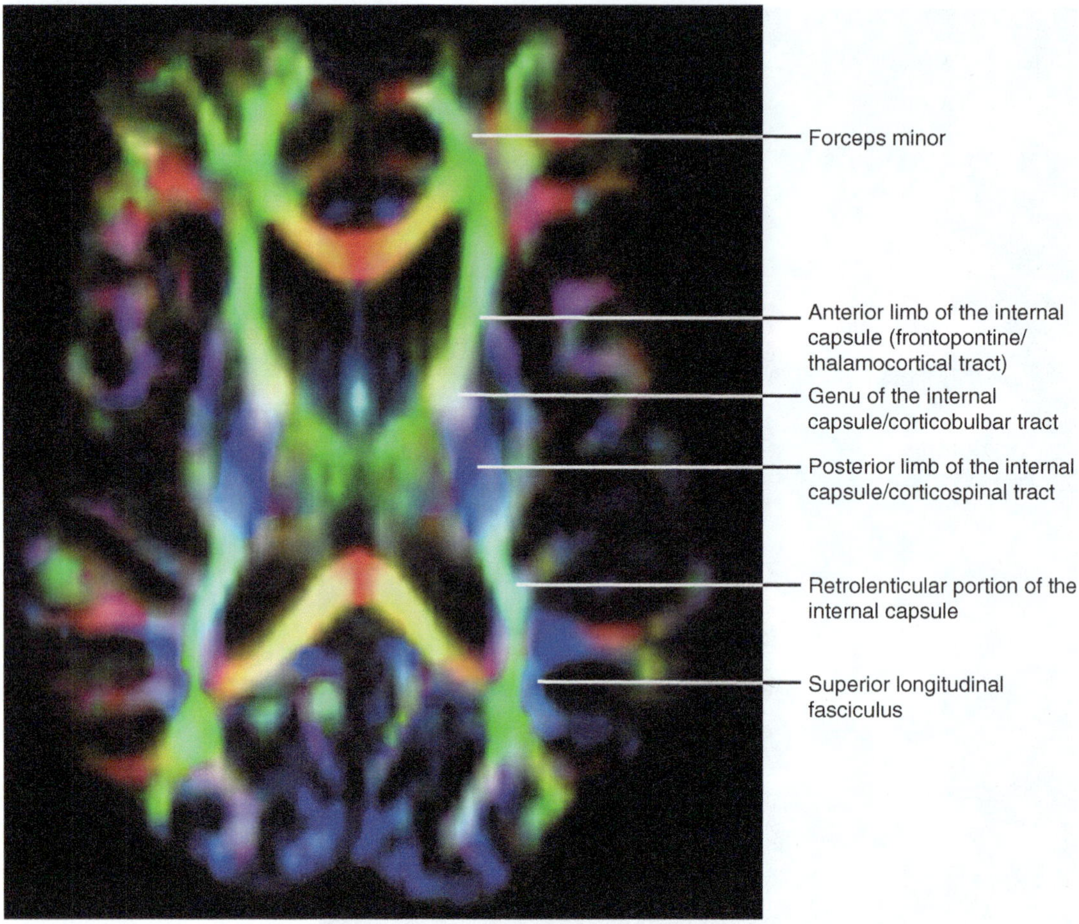

Fig. 4.6 The level of the internal capsules and basal ganglia

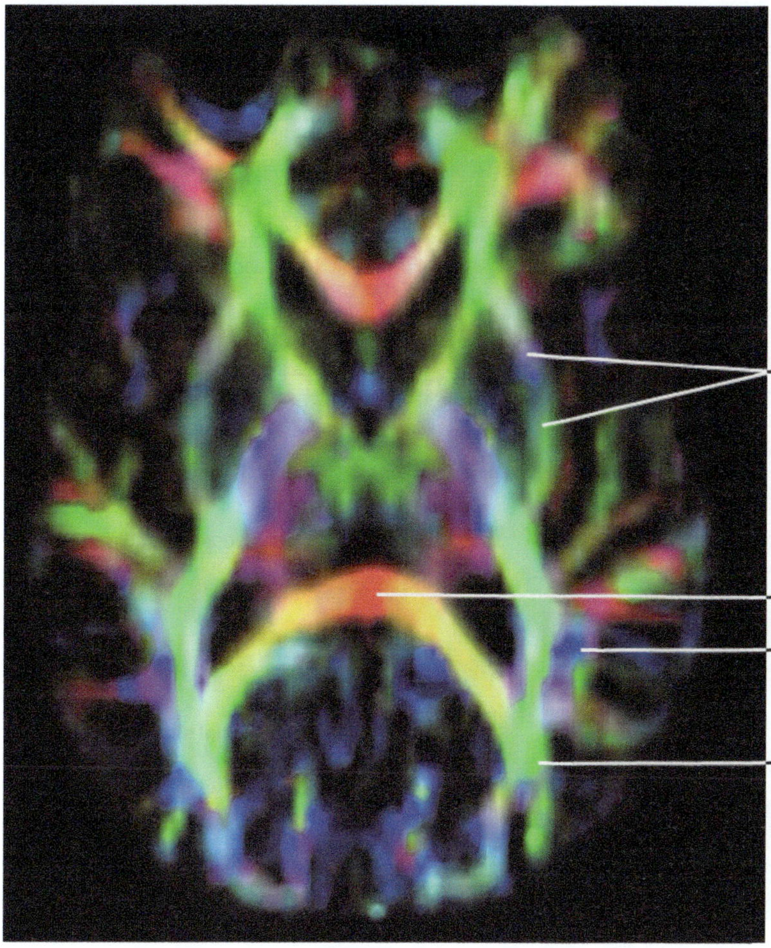

Fig. 4.7 The level of the internal capsules and splenium of the corpus callosum

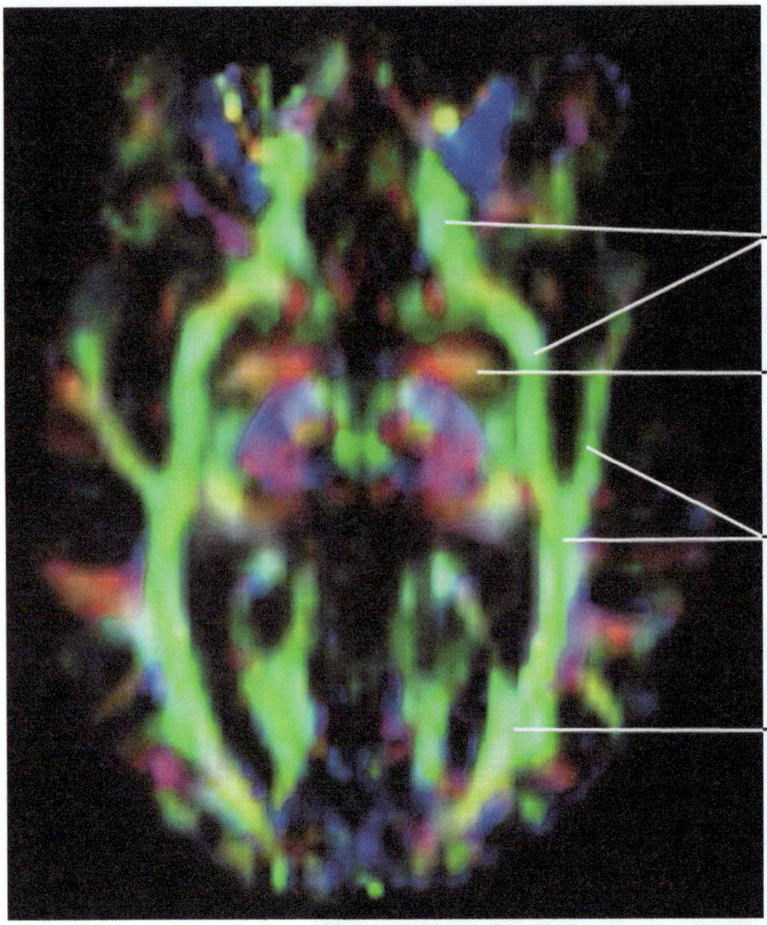

Fig. 4.8 The level of the occipital horns of the lateral ventricles

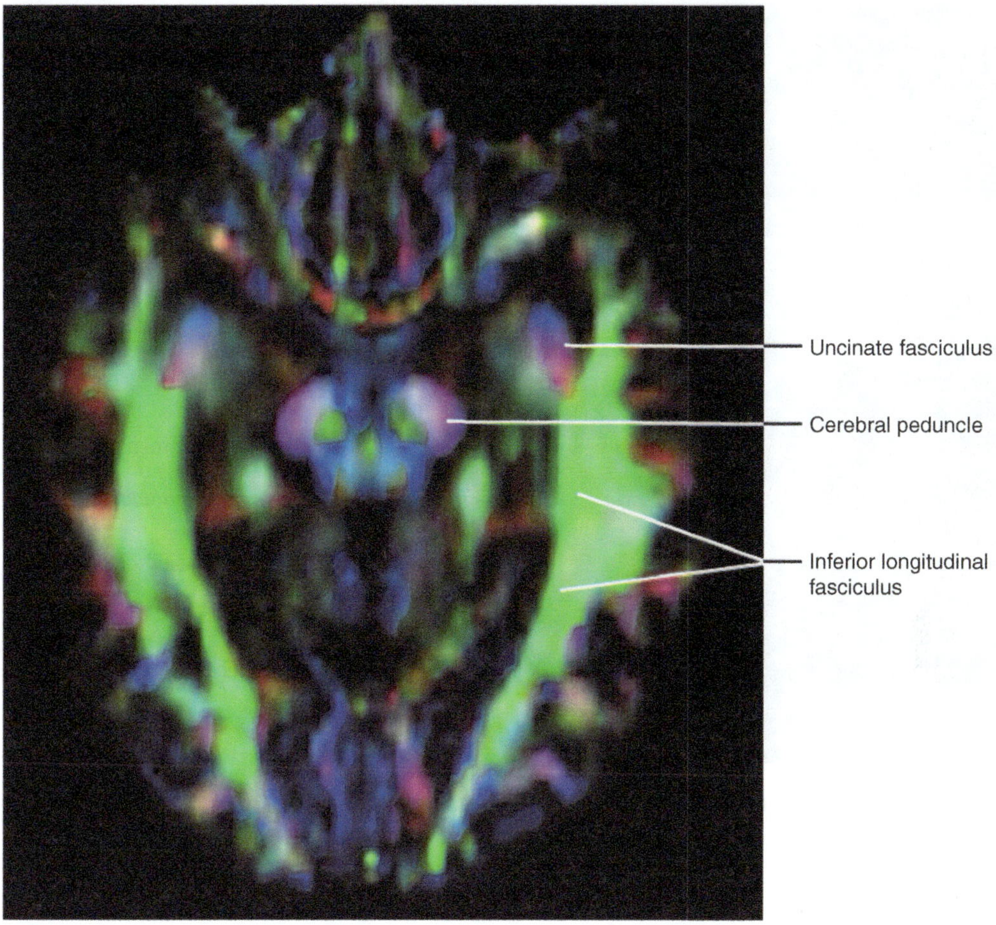

Fig. 4.9 The level of the cerebral peduncles

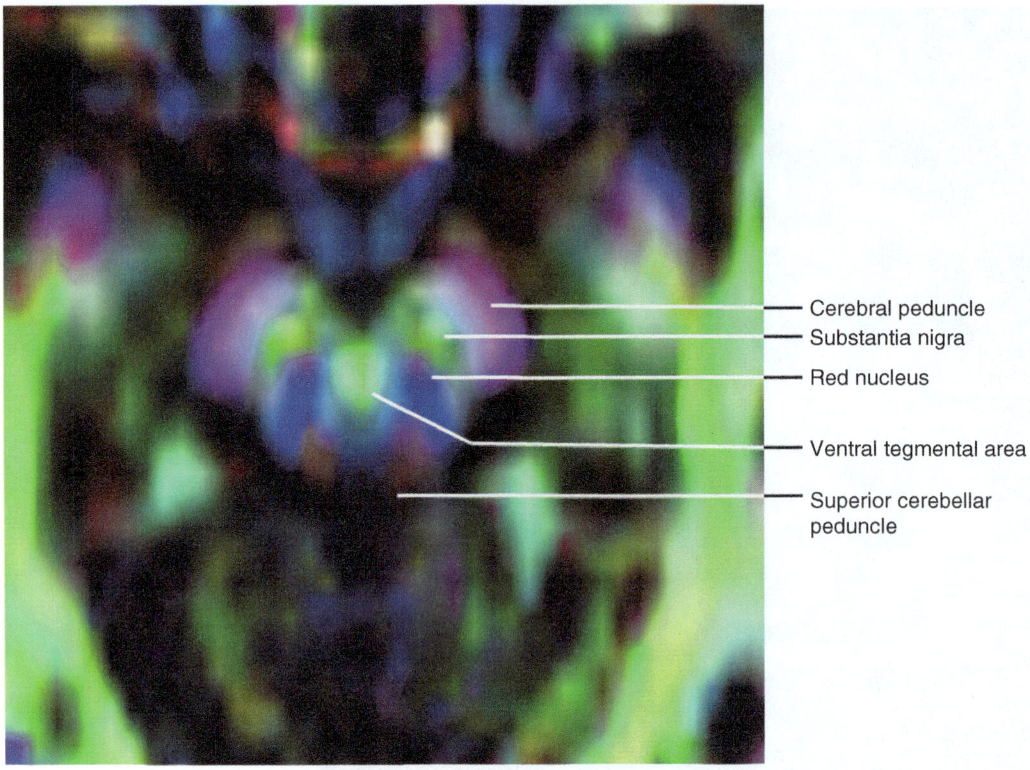

Fig. 4.10 Zoomed at the level of the cerebral peduncles

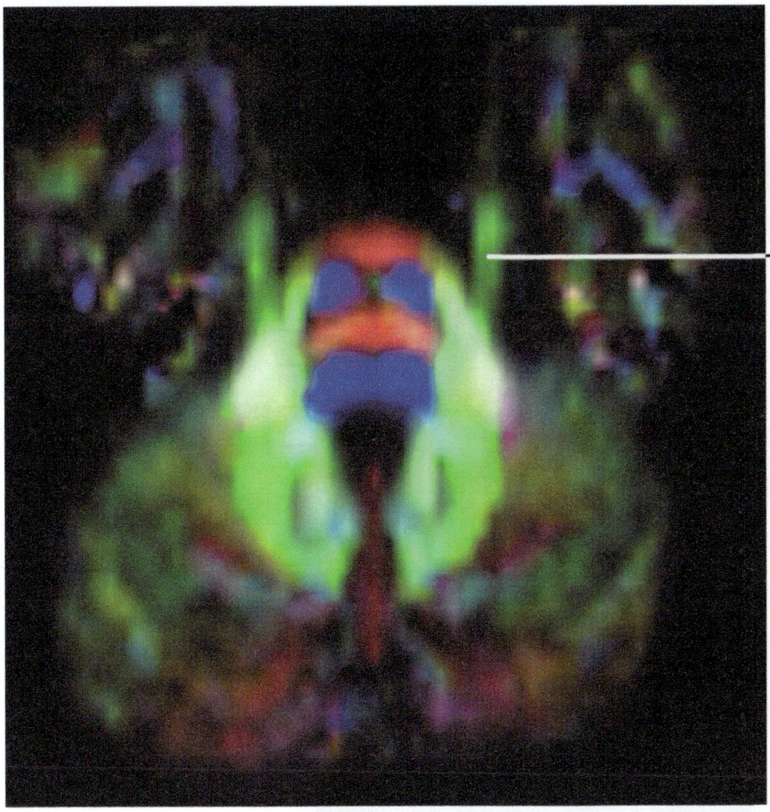

Fig. 4.11 The level of the pons and exiting trigeminal nerve

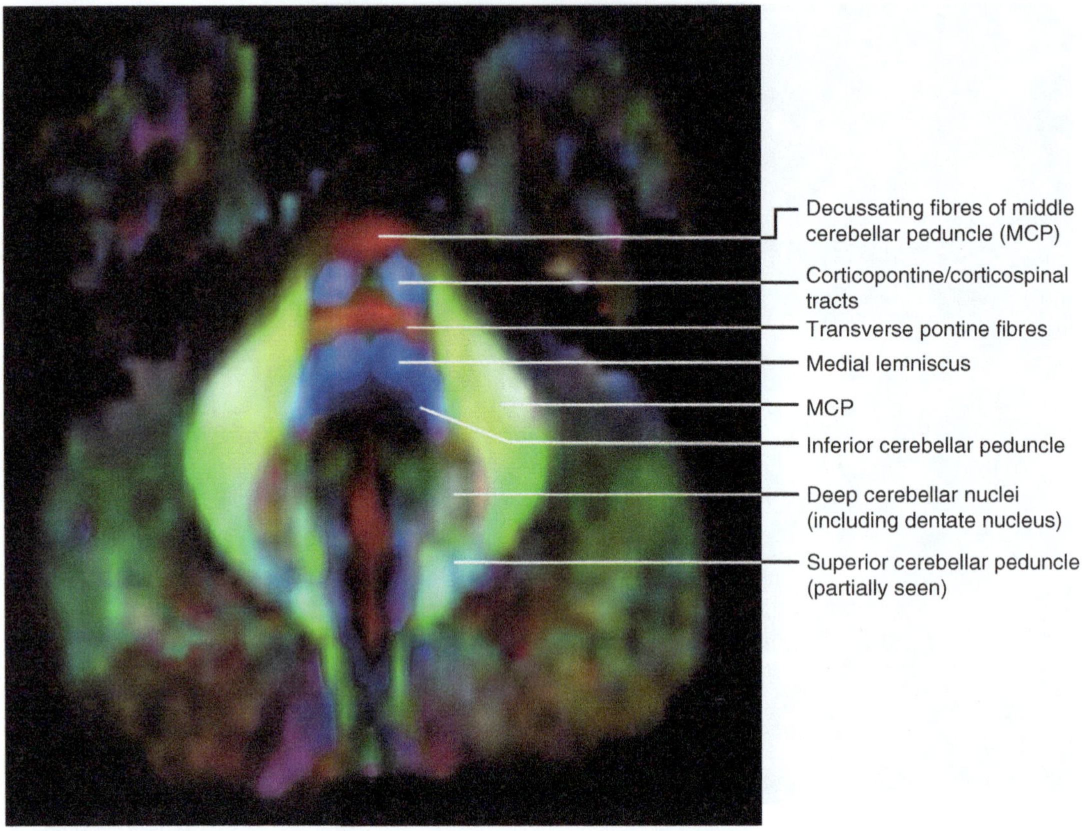

Fig. 4.12 The level of the brachium pontis (MCP)

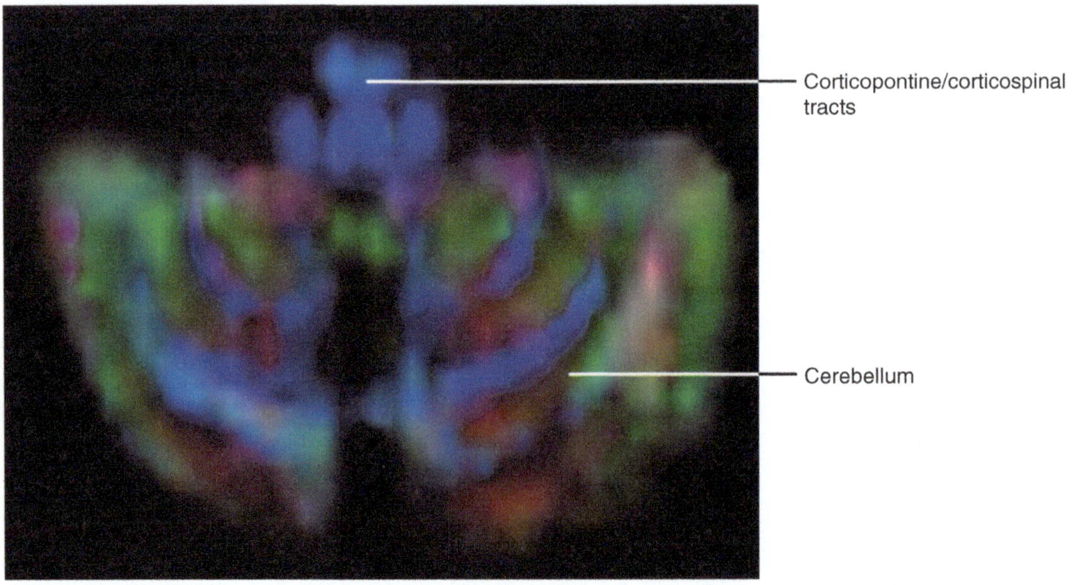

Fig. 4.13 The level of the medulla/craniocervical junction

Coronal

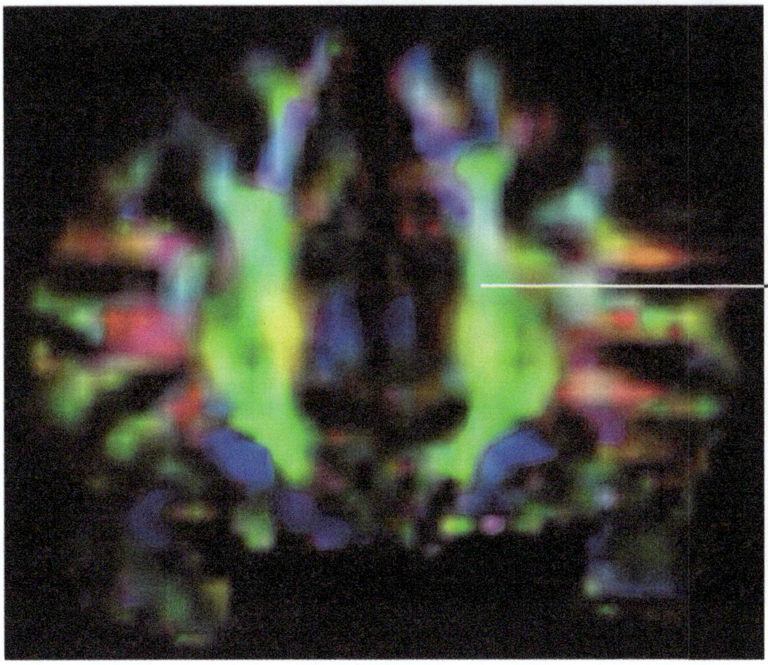

Fig. 4.14 The level of the frontal lobes

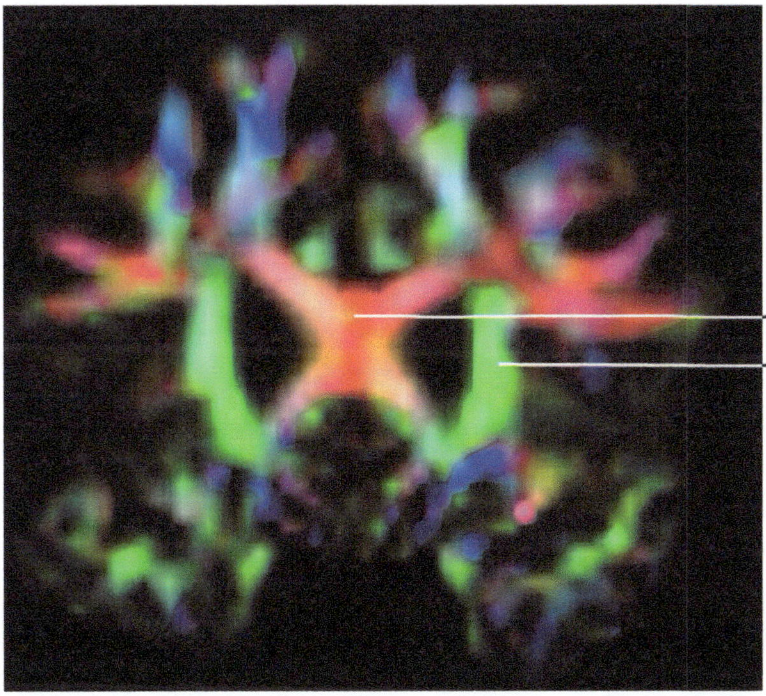

Fig. 4.15 The level of the genu of the corpus callosum

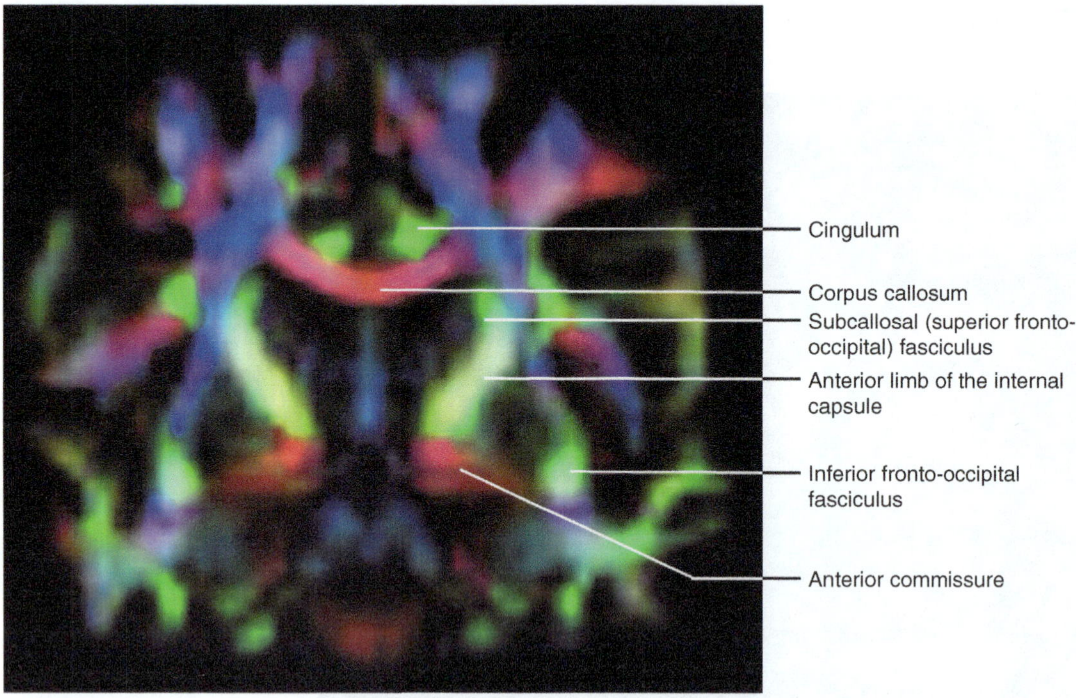

Fig. 4.16 The level of the caudate heads and anterior limbs of the internal capsules

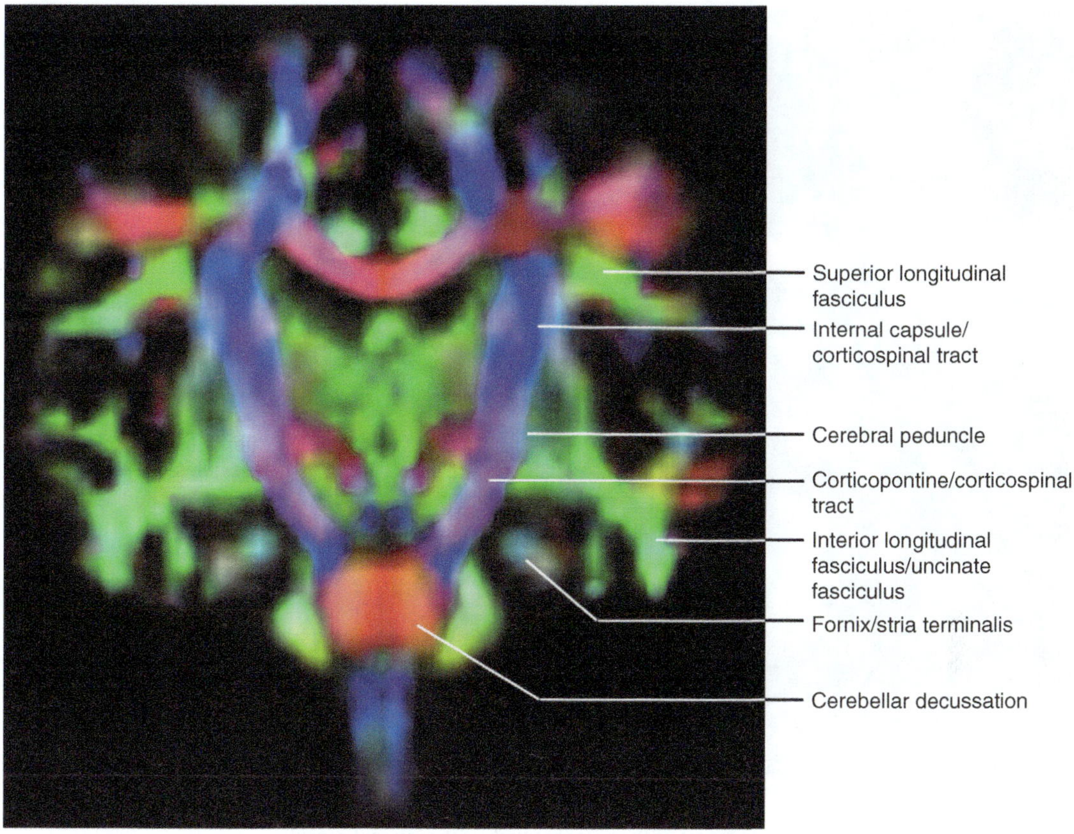

Fig. 4.17 The level of the posterior limbs of the internal capsules

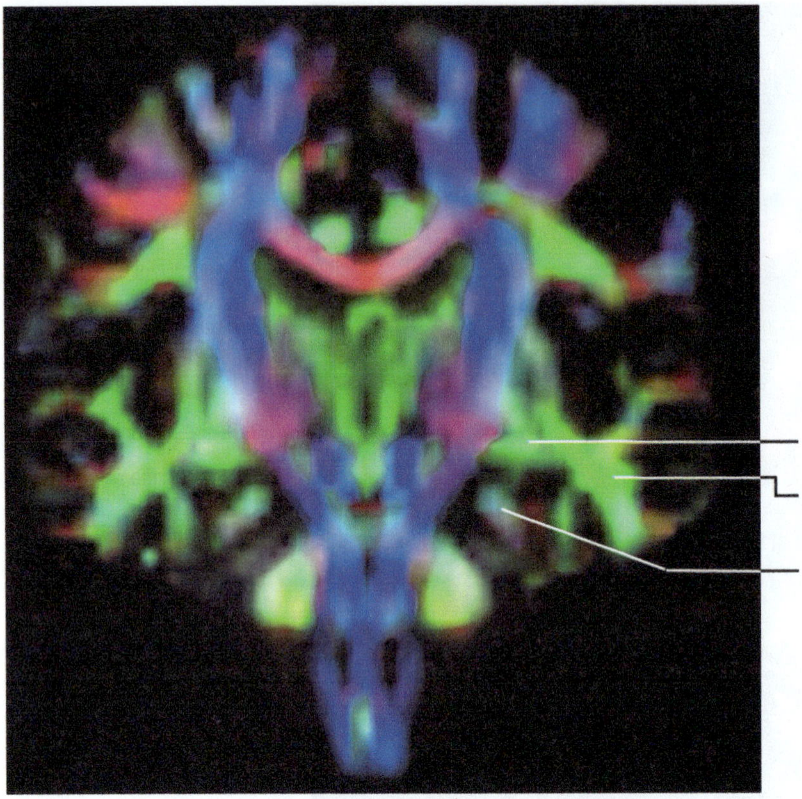

Fig. 4.18 The level of the posterior aspect of the posterior limb of the internal capsule

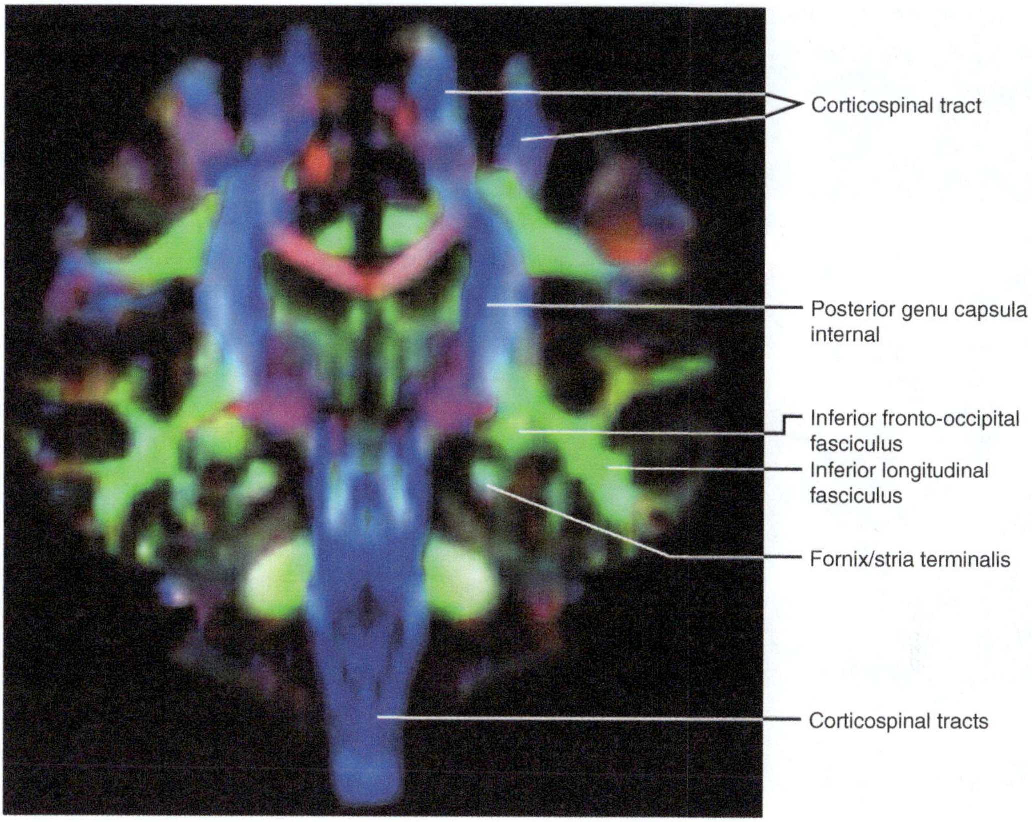

Fig. 4.19 The level of the posterior aspect of the posterior limb of the internal capsule

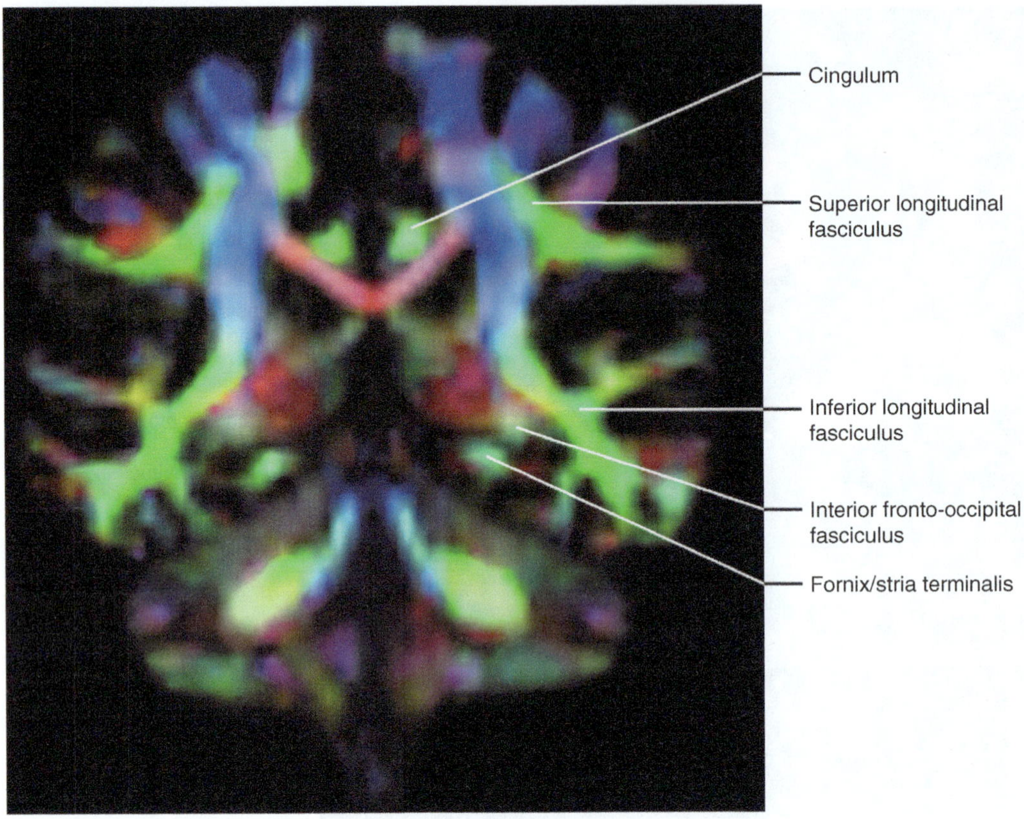

Fig. 4.20 The level of the isthmus of the corpus callosum

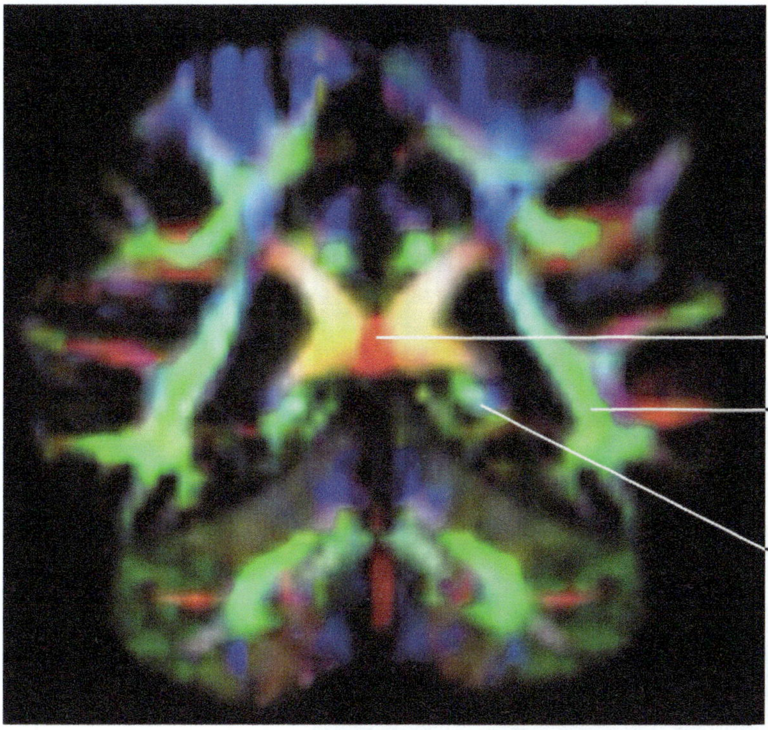

Fig. 4.21 The level of the splenium of the corpus callosum

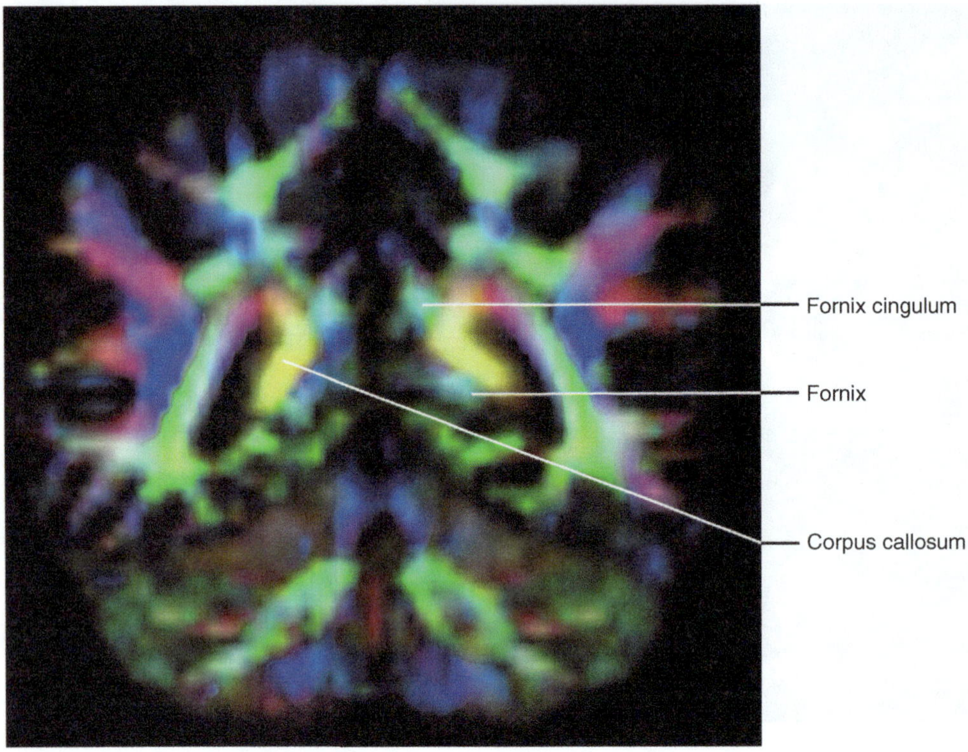

Fig. 4.22 The level of the trigone of the lateral ventricle

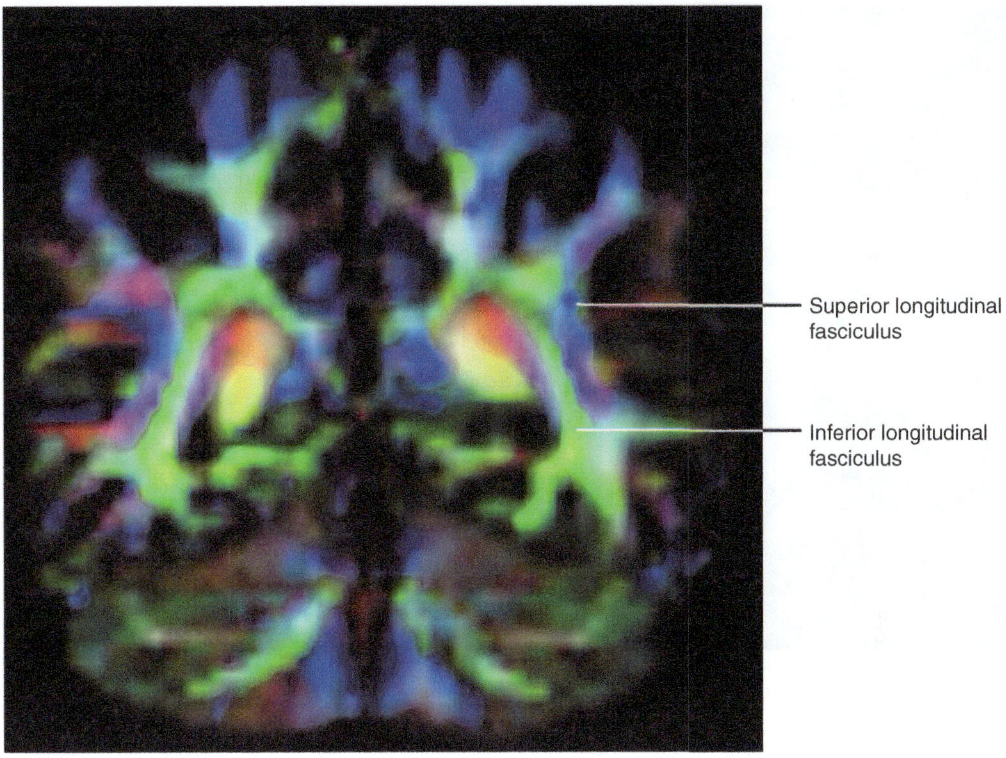

Fig. 4.23 The level of the occipital horn of the lateral ventricles

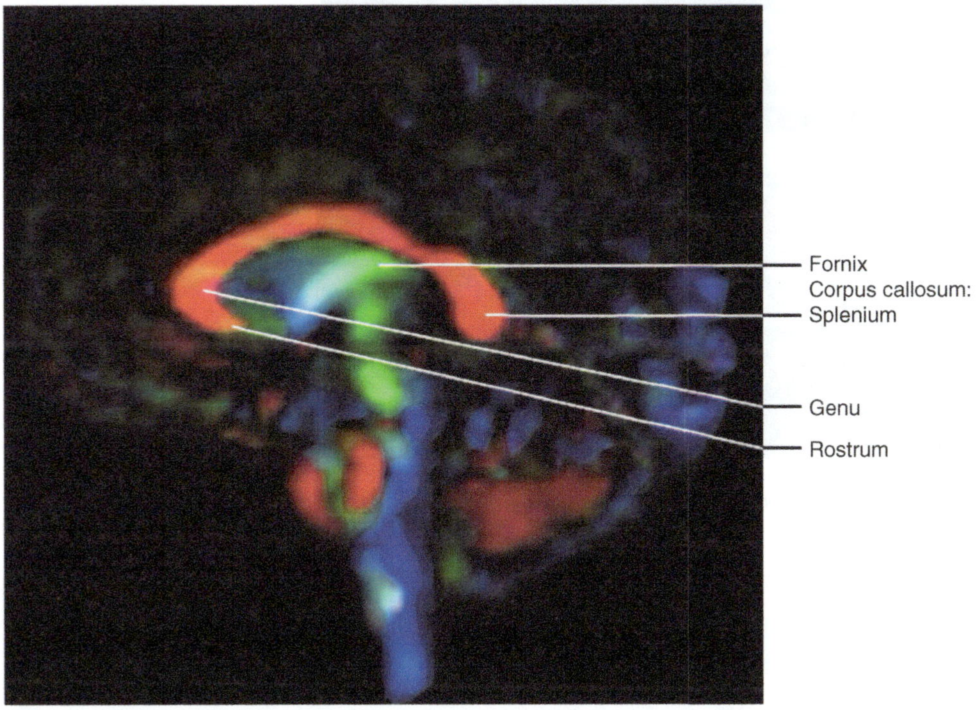

Fig. 4.24 Midline at the corpus callosum

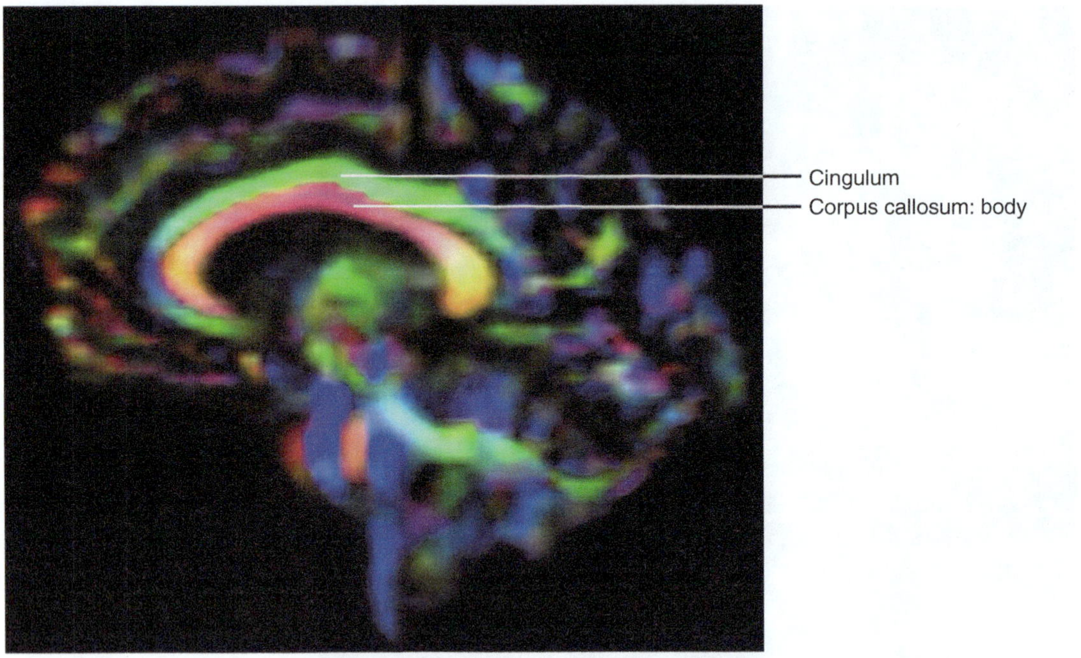

Fig. 4.25 Parasagittal at the corpus callosum and cingulum

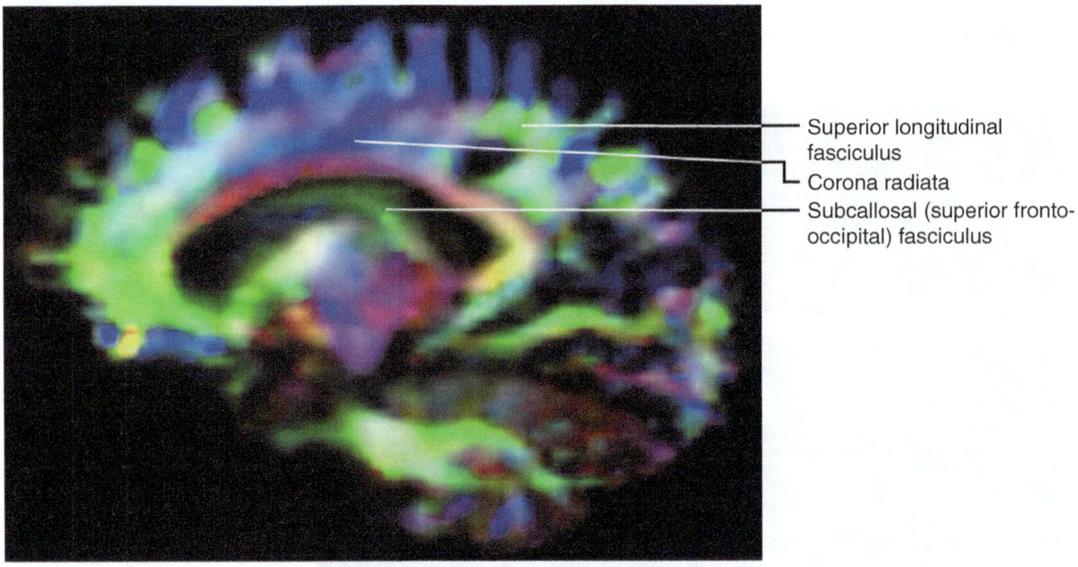

Fig. 4.26 Parasagittal at the corona radiata

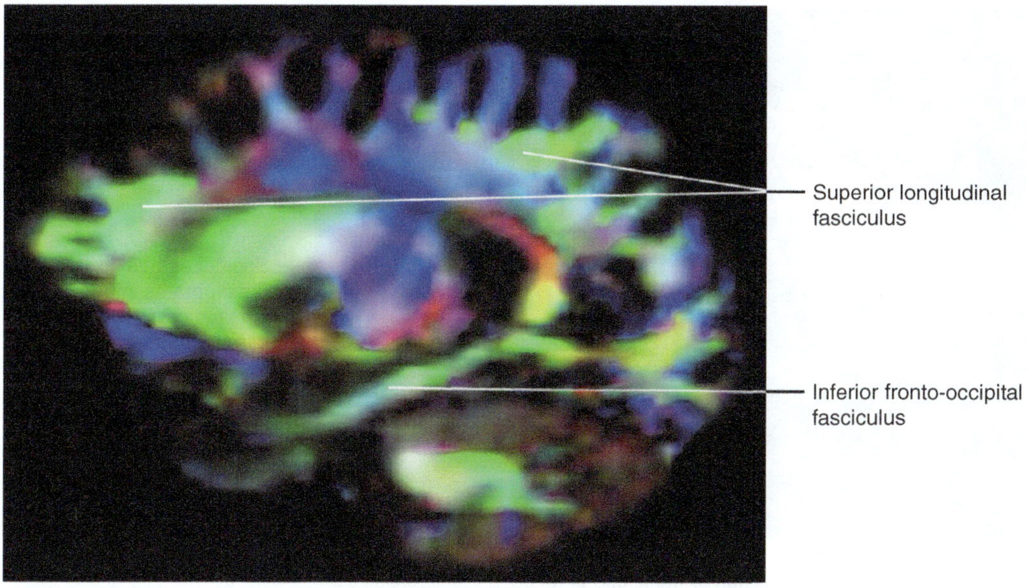

Fig. 4.27 Parasagittal at the level of the thalamus

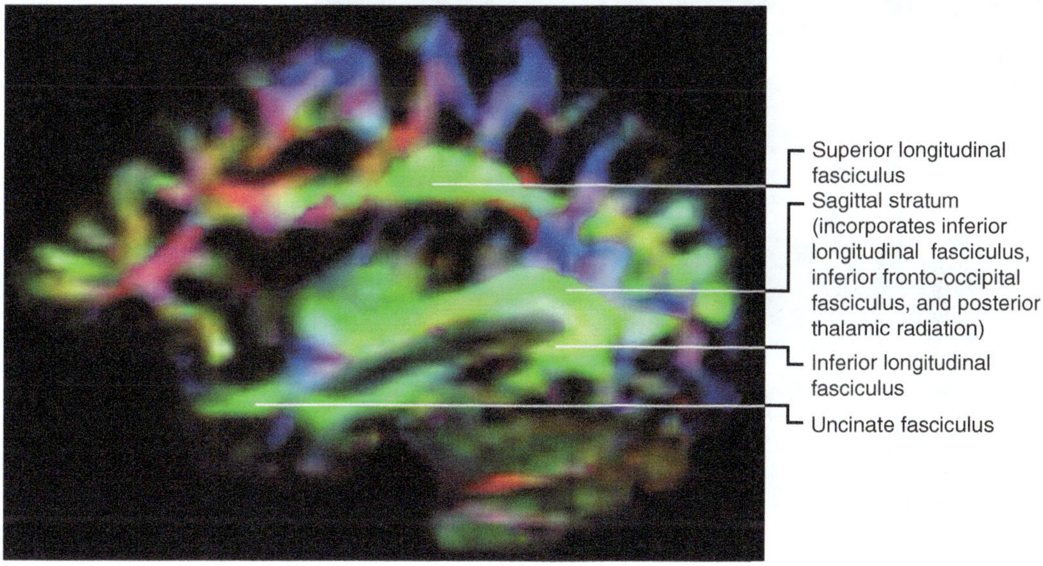

Fig. 4.28 The medial aspect of the temporal lobe

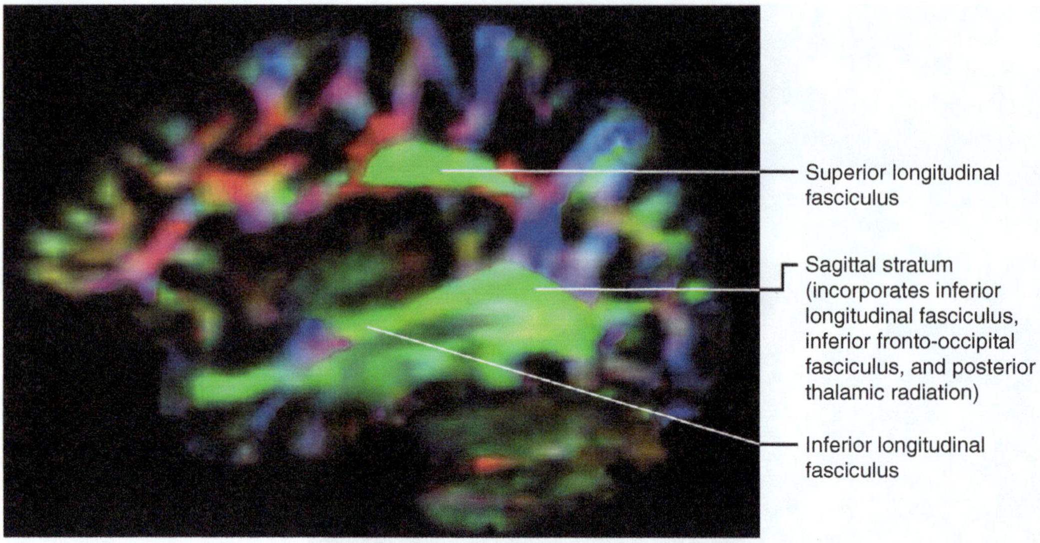

Fig. 4.29 The level of the mid-portion of the temporal lobe

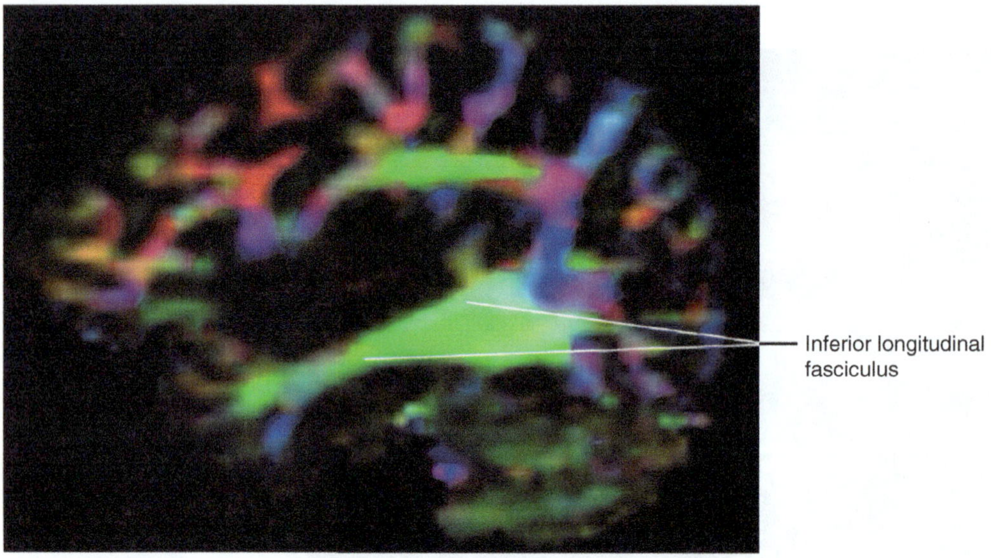

Fig. 4.30 The level of the most lateral extent of the temporal lobe

Fibre Tracking of White Matter

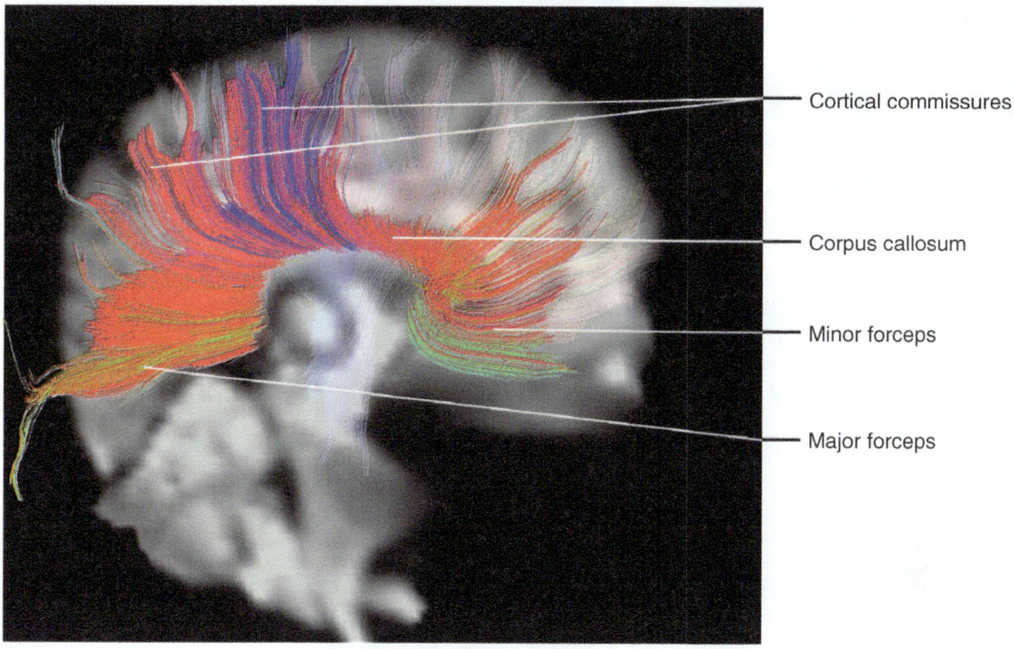

Fig. 4.31 Fibres crossing to the corpus callosum: lateral oblique view

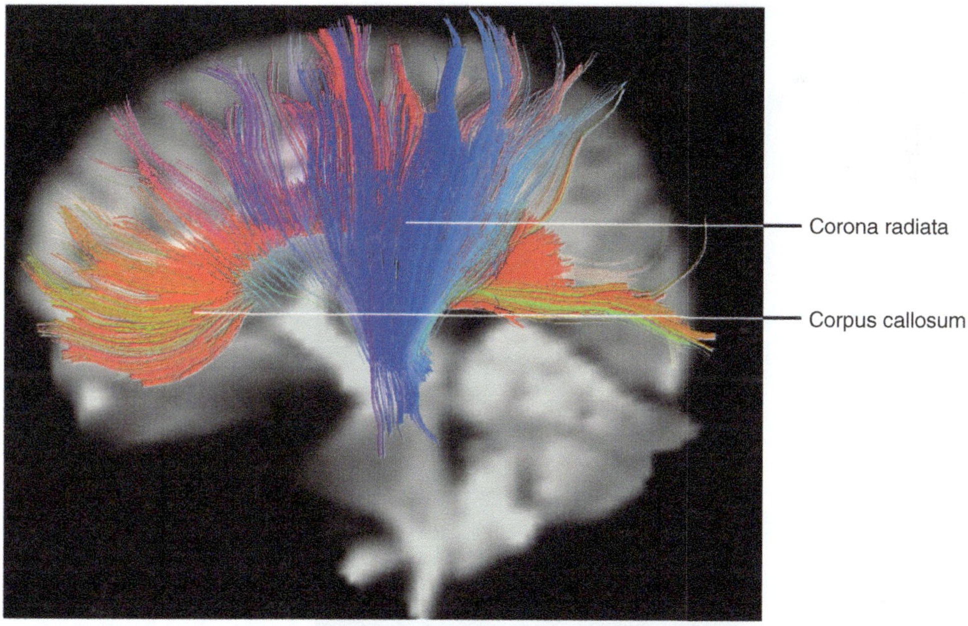

Fig. 4.32 Coronal radiata and corpus callosum fibres: lateral oblique view

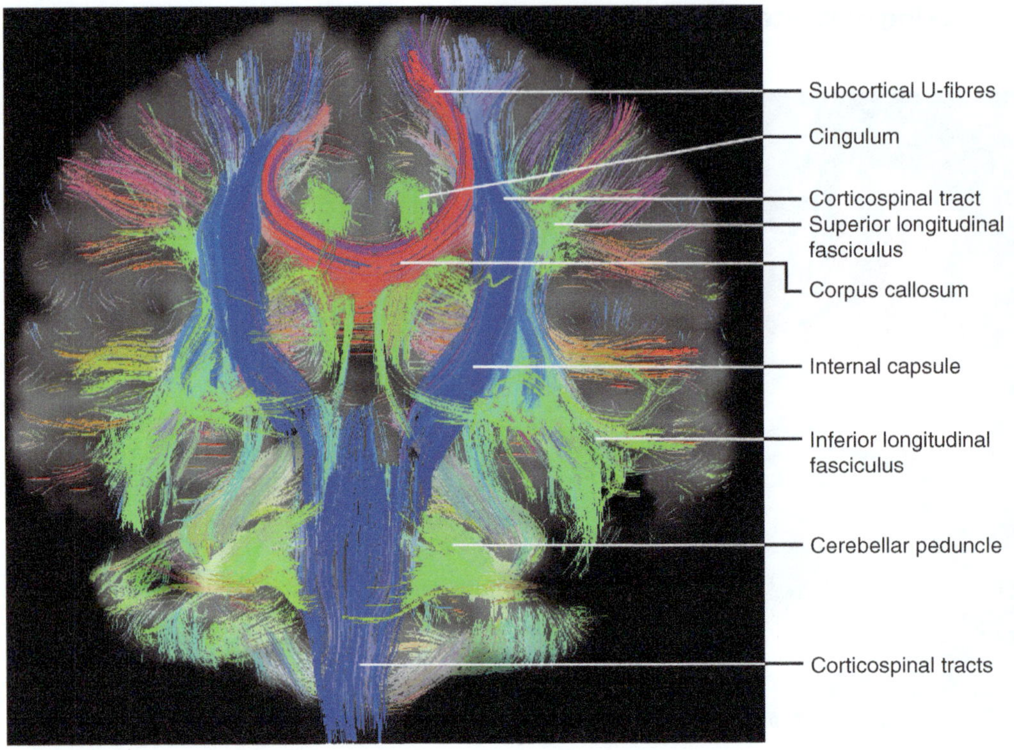

Fig. 4.33 Pathway of the corona radiata to the corticospinal tracts and relationship to other fibre tracts: coronal view

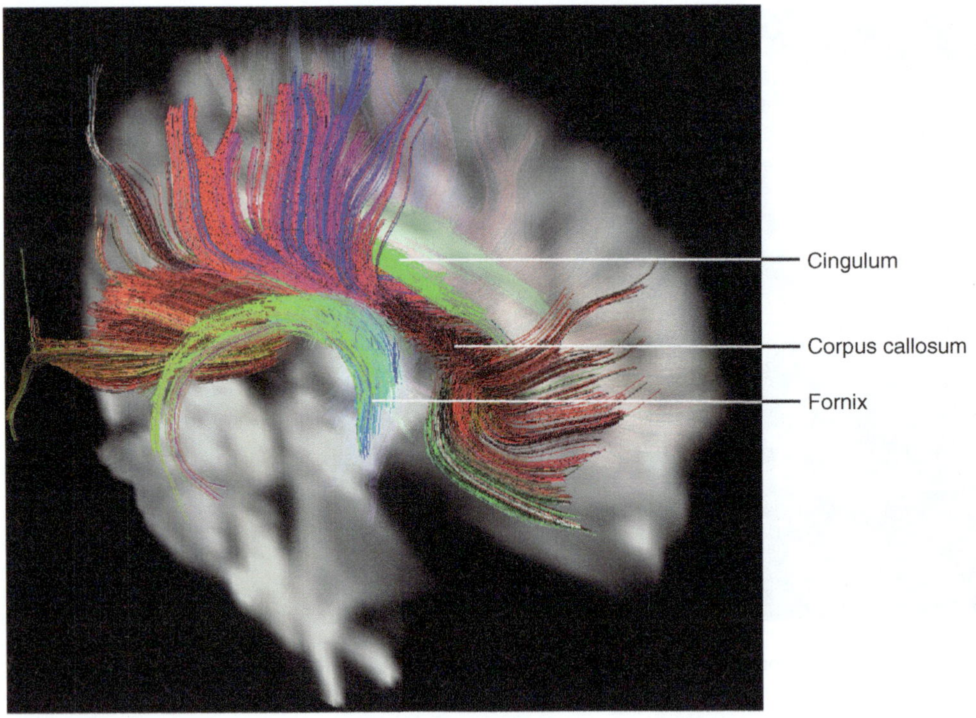

Fig. 4.34 Curvilinear relationship of the cingulum, corpus callosum and fornix: lateral oblique view

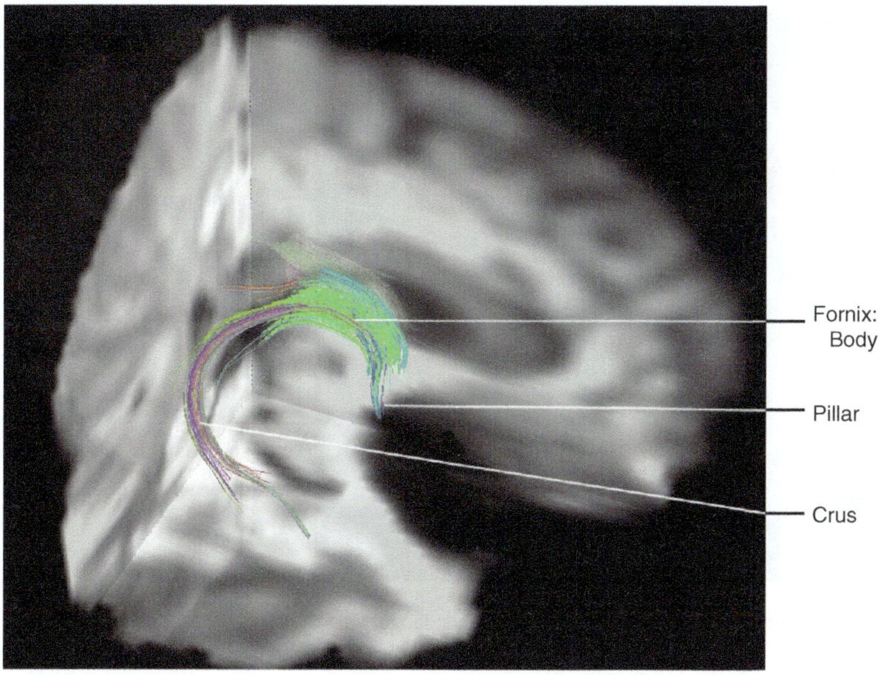

Fig. 4.35 Isolation of the fibres making up the fornix

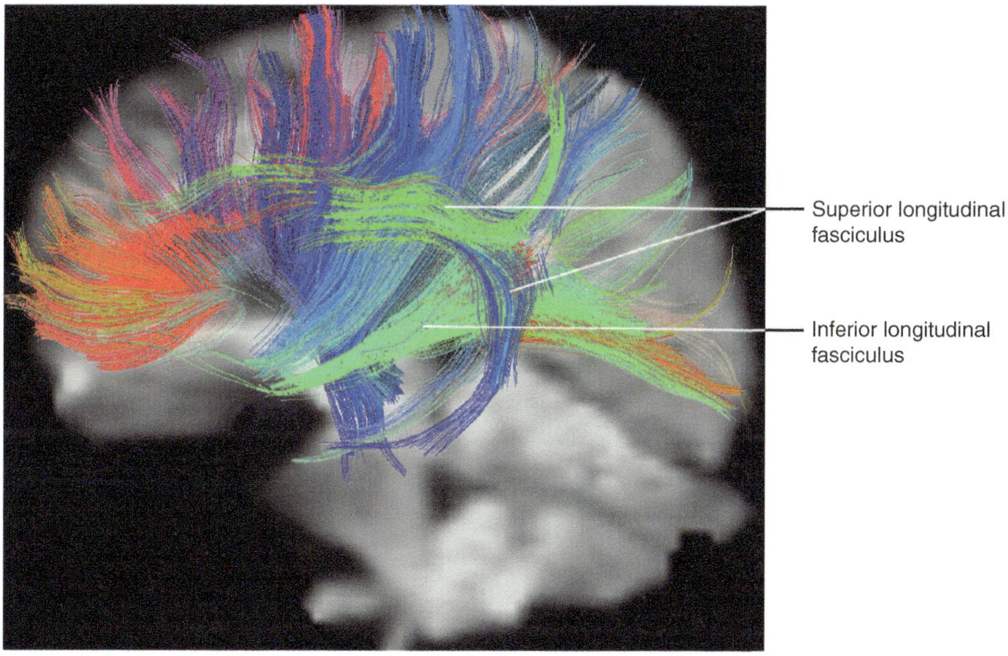

Fig. 4.36 Relations of the superior and inferior longitudinal fasciculus: lateral oblique view

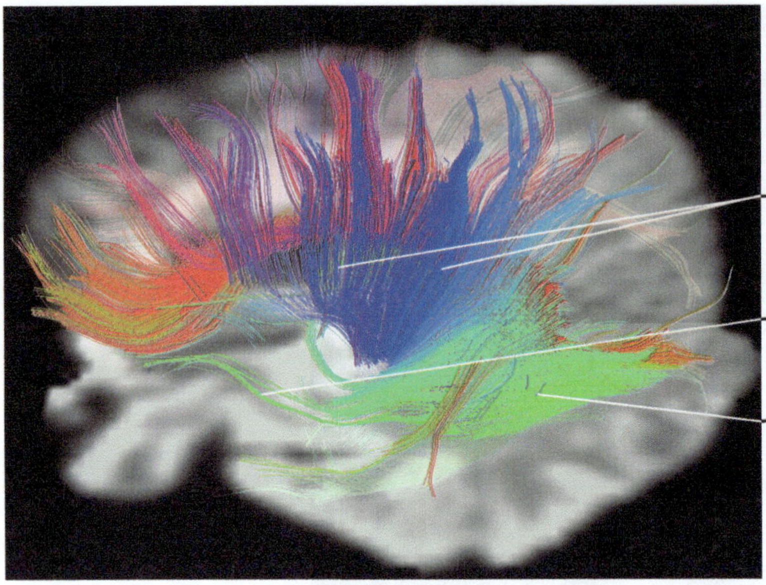

Fig. 4.37 Relations of the inferior fasciculus with the corona radiata: lateral view

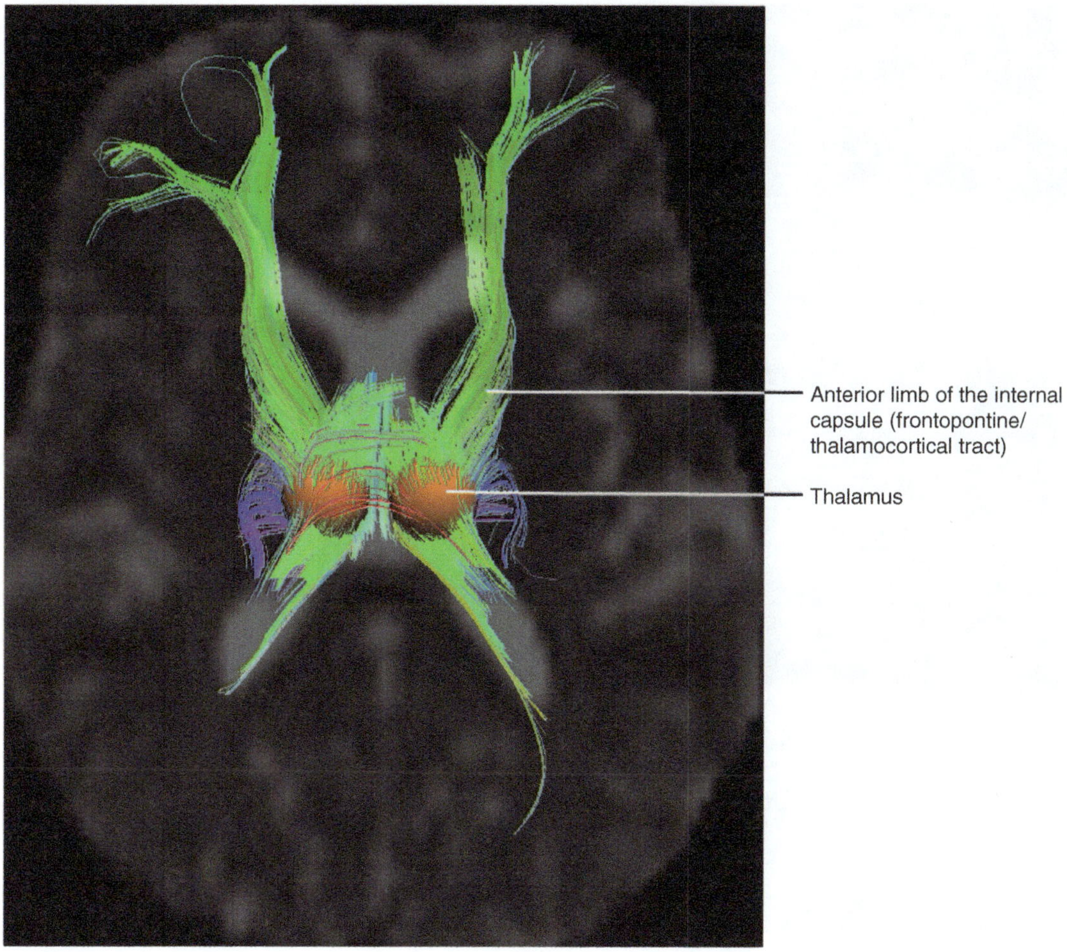

Fig. 4.38 Anterior limbs of the internal capsules and relationship to thalamus: axial view

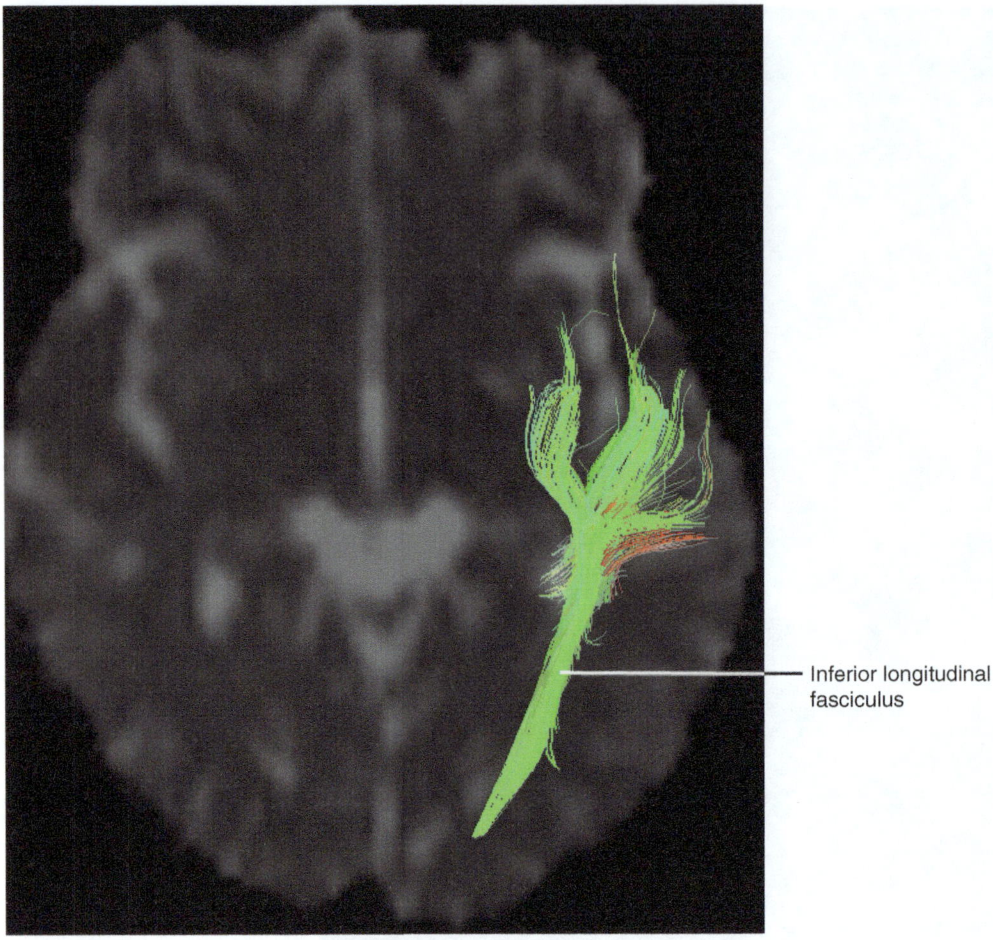

Fig. 4.39 Inferior longitudinal fasciculus: axial view

Brain Surface Anatomy and Functional Regions

5

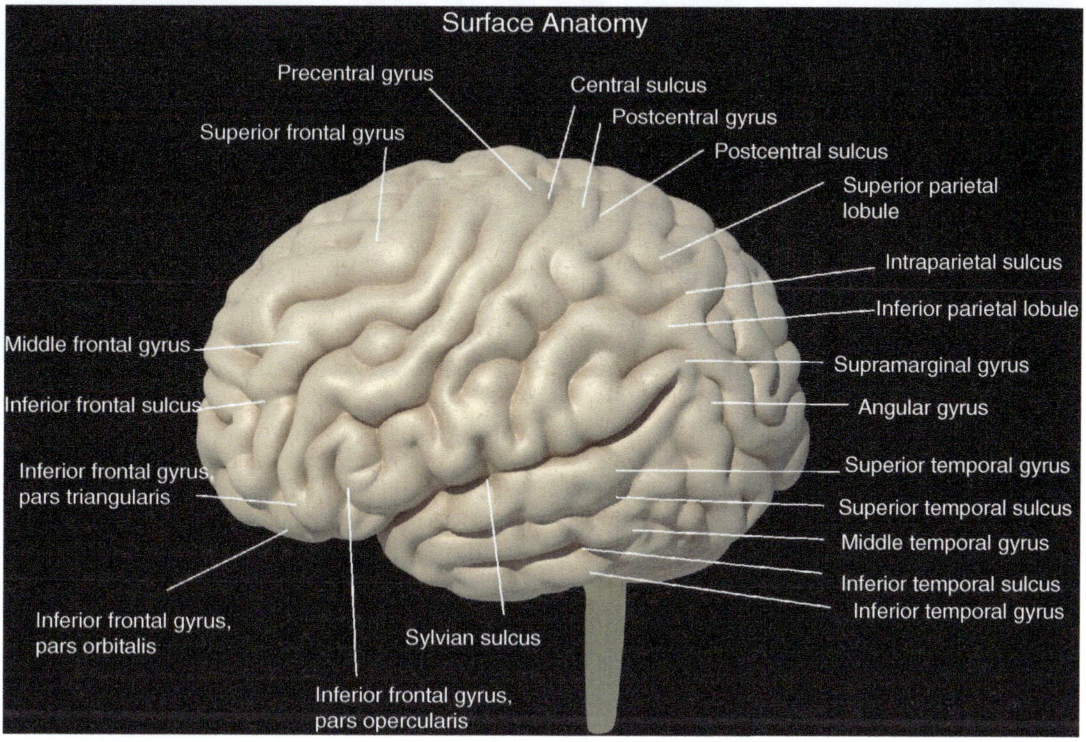

Fig. 5.1 Surface anatomy

S. Andronikou (ed.), *See Right Through Me*,
DOI 10.1007/978-3-642-23893-2_5, © Springer-Verlag Berlin Heidelberg 2012

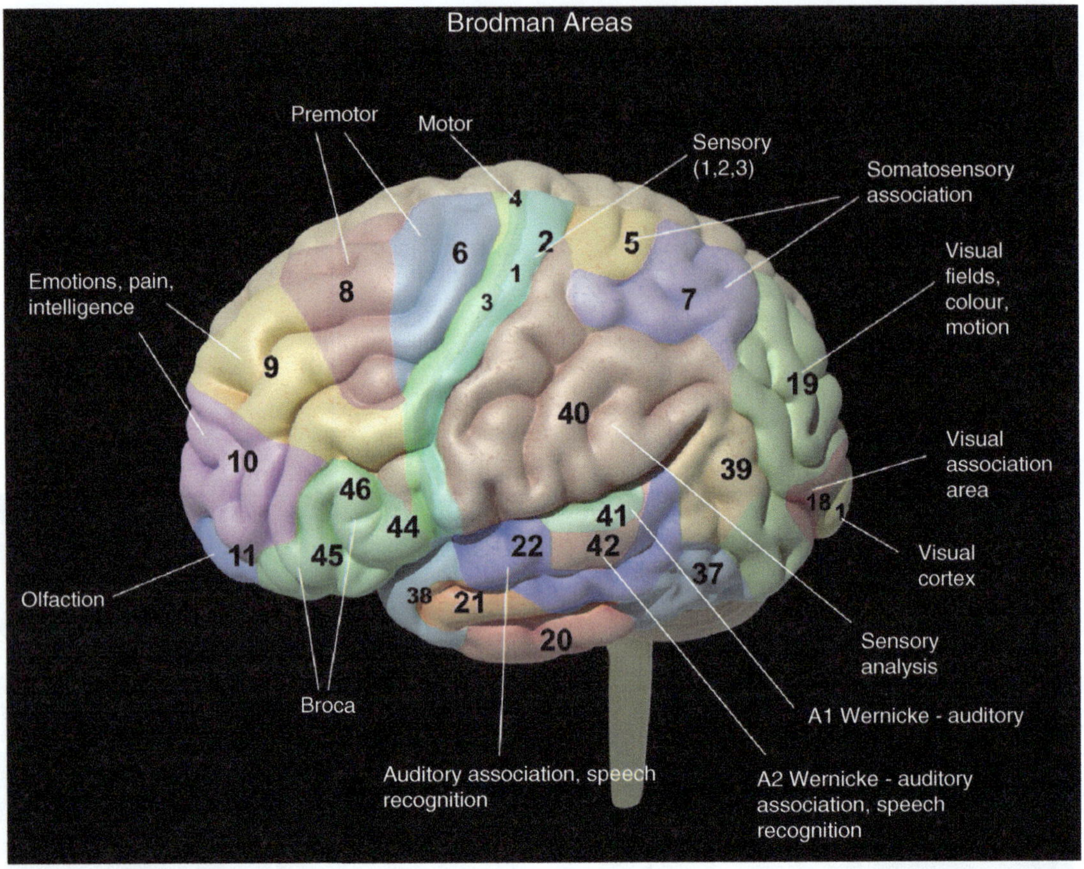

Fig. 5.2 Brodman areas

Table 5.1 Why do functional magnetic resonance imaging (fMRI)?

Presurgical mapping (most pertinent to clinical radiology)
Functional assessment (predominantly in psychiatric and neuropsychology research)
Functional connectivity MRI; "resting state fMRI"

Table 5.2 Tasks and paradigms

fMRI detects the difference in brain activity between several conditions (rest vs. task or task vs. control task).
During image acquisition, the patient is asked to alternatively perform several tasks or is stimulated to trigger several processes or emotions.
Each of these conditions is repeated several times and can be separated by rest periods.
The combination of these conditions is called a paradigm.
The stimuli are usually audiovisual but can involve more complex systems such as taste or smell.

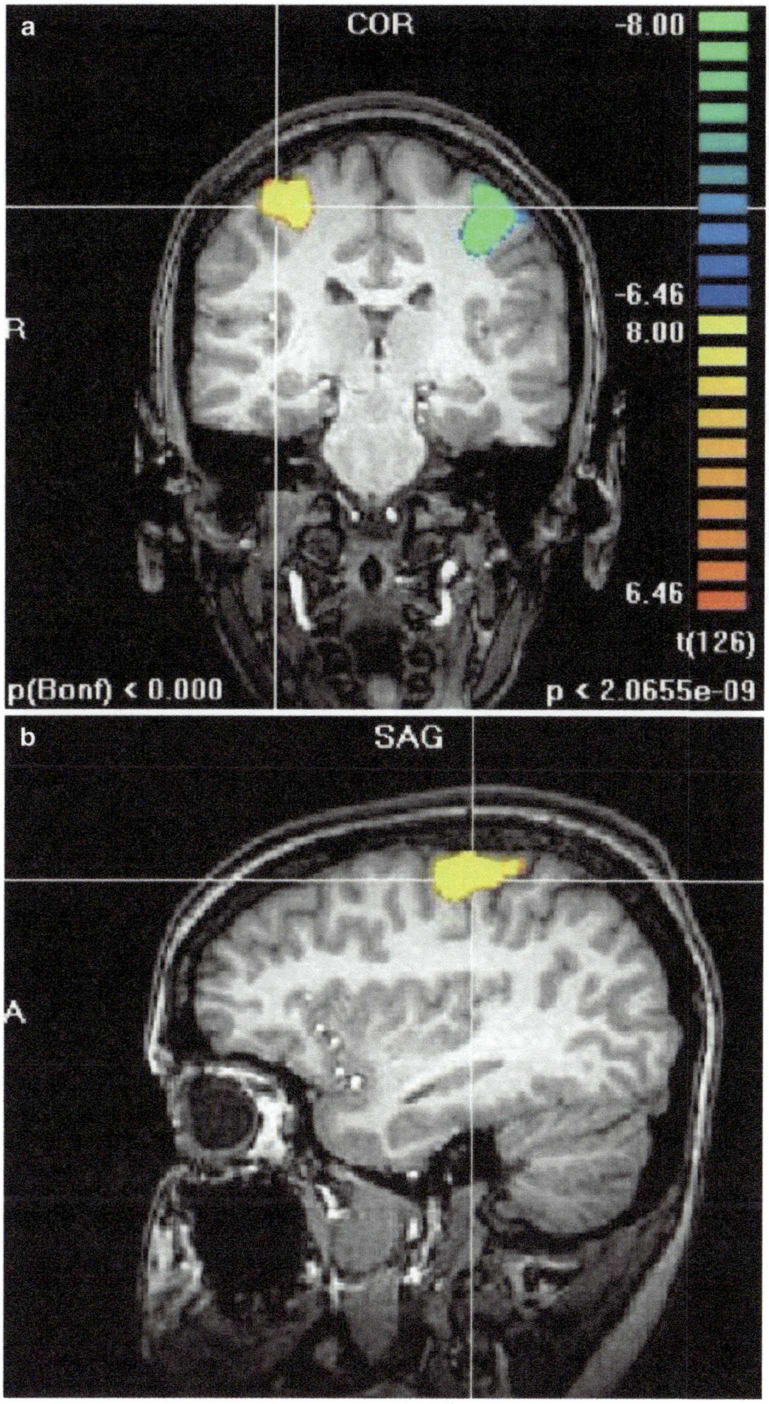

Fig. 5.3 Motor areas. (**a**–**c**) Sequential left and right finger tap

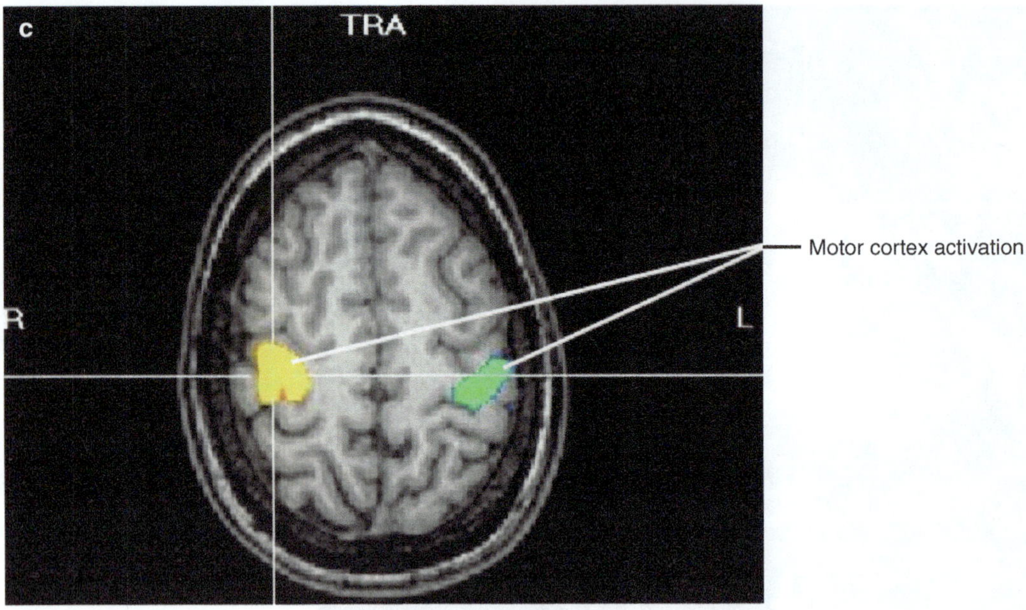

Fig. 5.3 (continued)

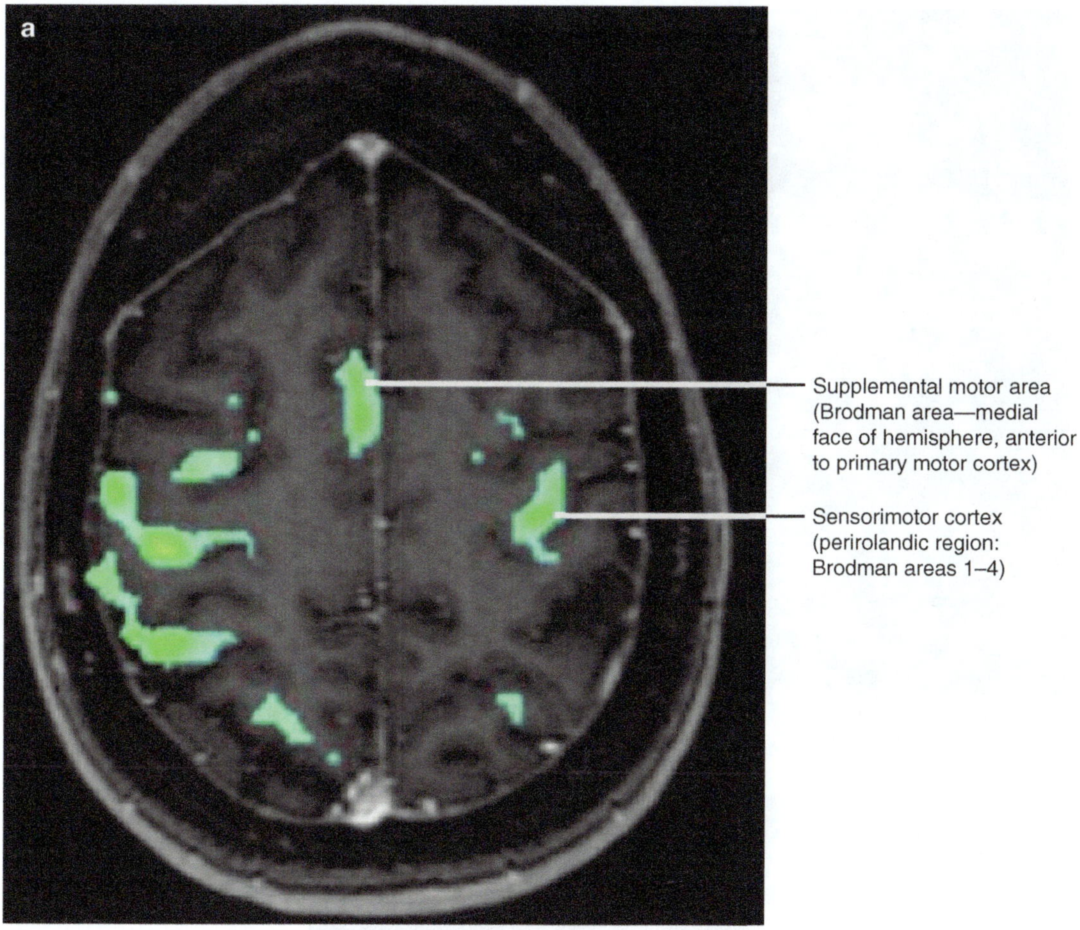

Fig. 5.4 Motor and premotor. (**a**, **b**) Bilateral complex finger tapping. Bilateral sensorimotor cortex around superior lateral central sulcus in expected somatotopic area of activation for body part. Hand motor portion of the homunculus. The "omega" of the precentral gyrus perirolandic region: Brodman areas 1–4 (Images courtesy of JJ Pillai, M.D., Neuroradiology Division, Johns Hopkins Hospital.)

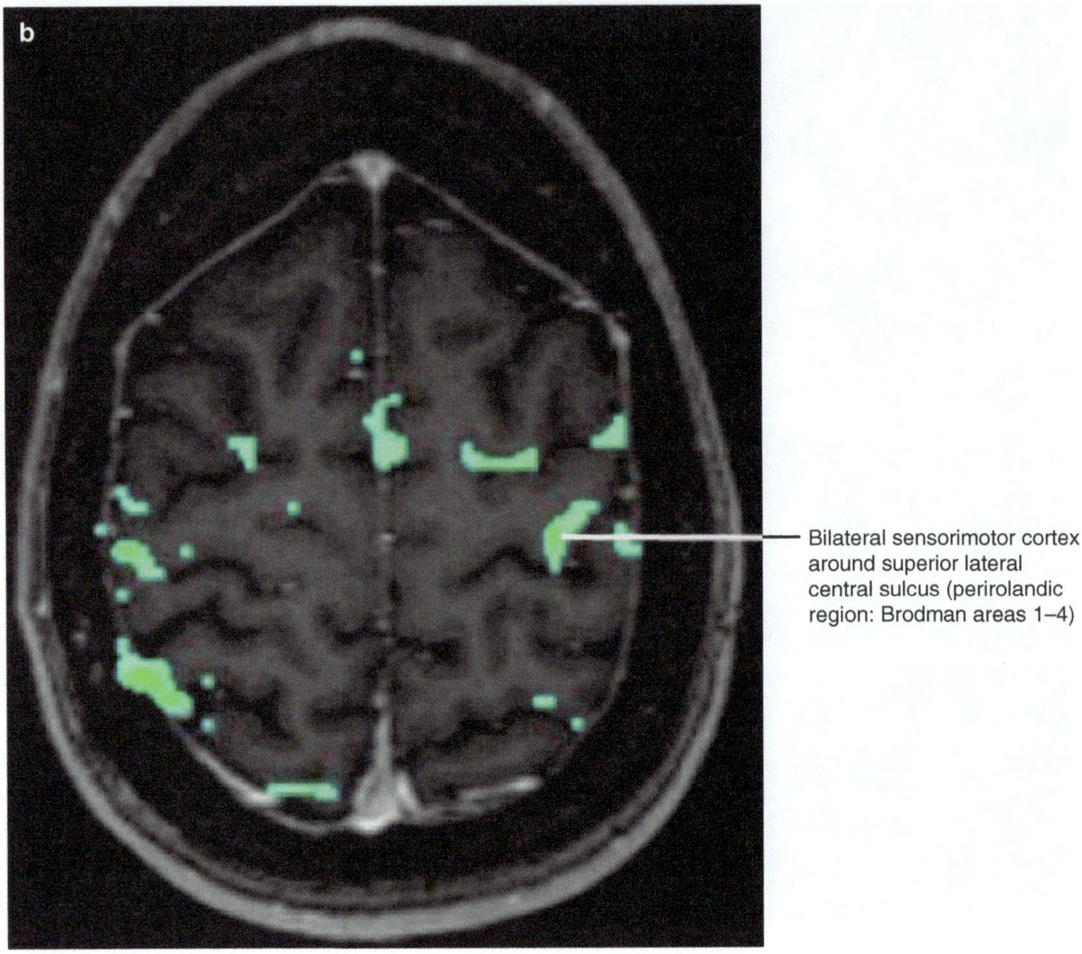

Fig. 5.4 (continued)

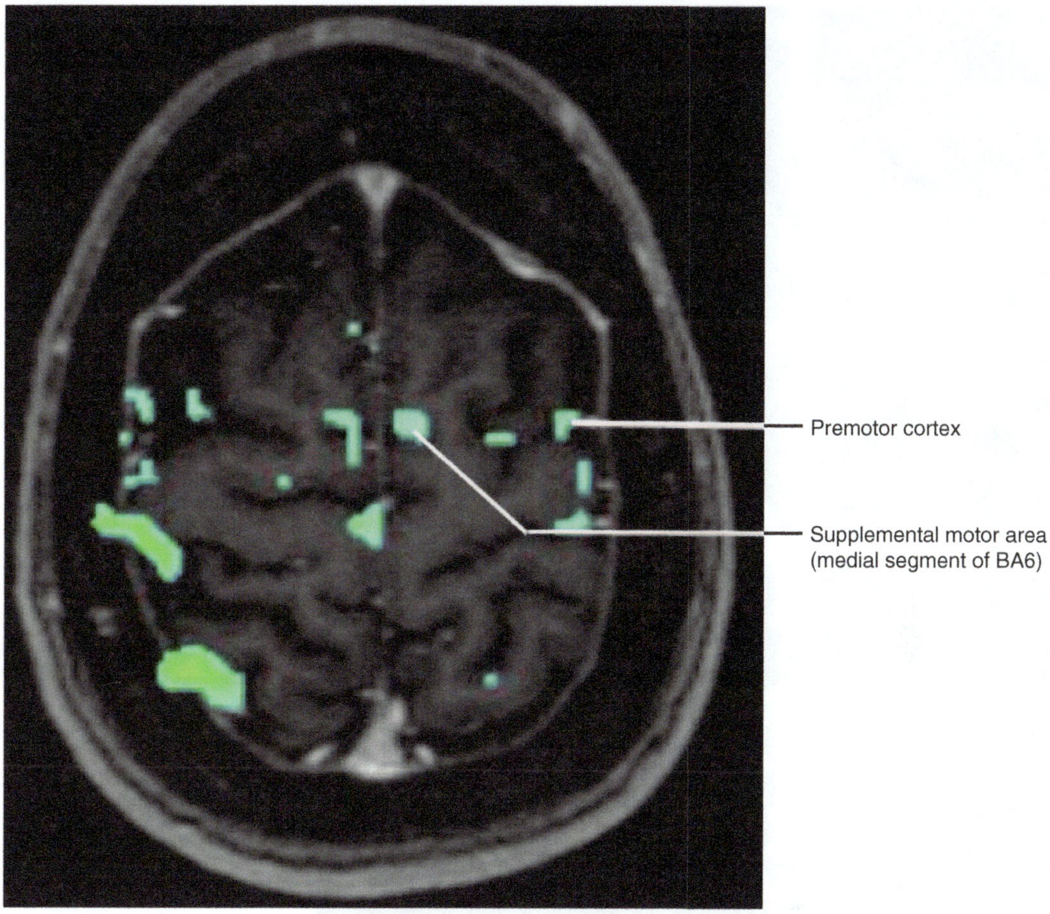

Fig. 5.5 Motor and premotor. Unilateral sequential (simple) finger tapping. Premotor cortex extends 3 mm anterior to primary motor cortex and corresponds to Brodman Area 6, with the exception that the medial surface of this is the site of the accessory motor cortex. Contralateral dominance is seen in the bilateral supplemental motor area (medial segment of BA6) when task is carried out on dominant hand (Image courtesy of JJ Pillai, M.D., Neuroradiology Division, Johns Hopkins Hospital.)

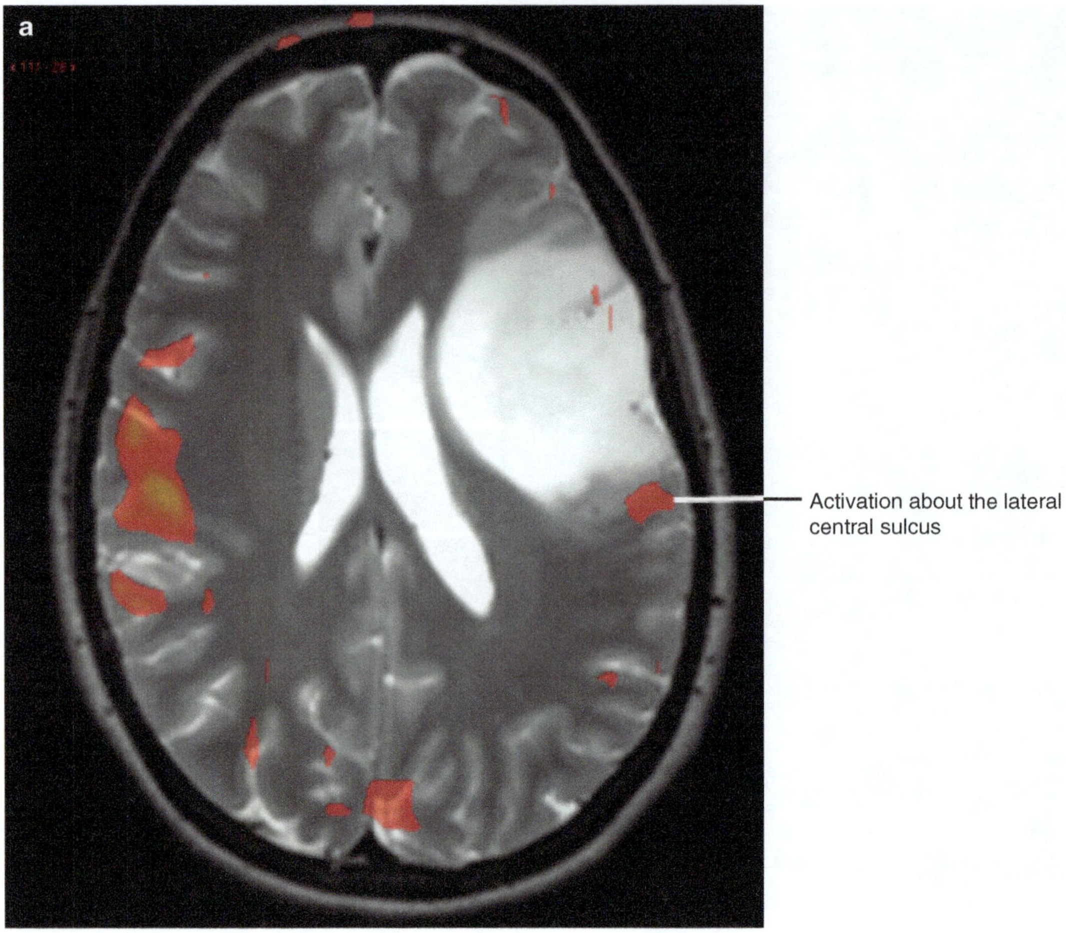

Fig. 5.6 Motor and premotor. (**a**, **b**) Tongue movement. Activation about the lateral central sulcus in the expected somatotopic location for lower face sensori-motor cortex. Sensori-motor cortex located along the homunculus lower than, and overlapping, that produced by lip-puckering (Images courtesy of Neuroradiology division, University of Virginia.)

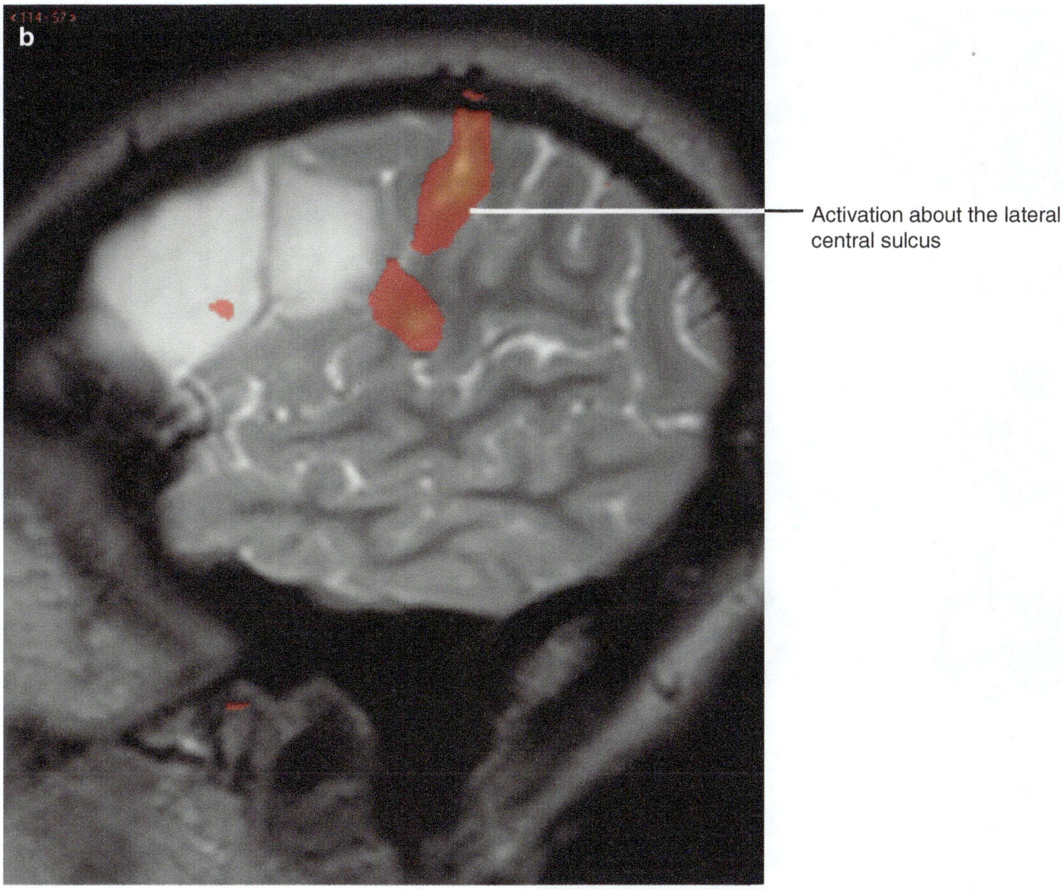

Fig. 5.6 (continued)

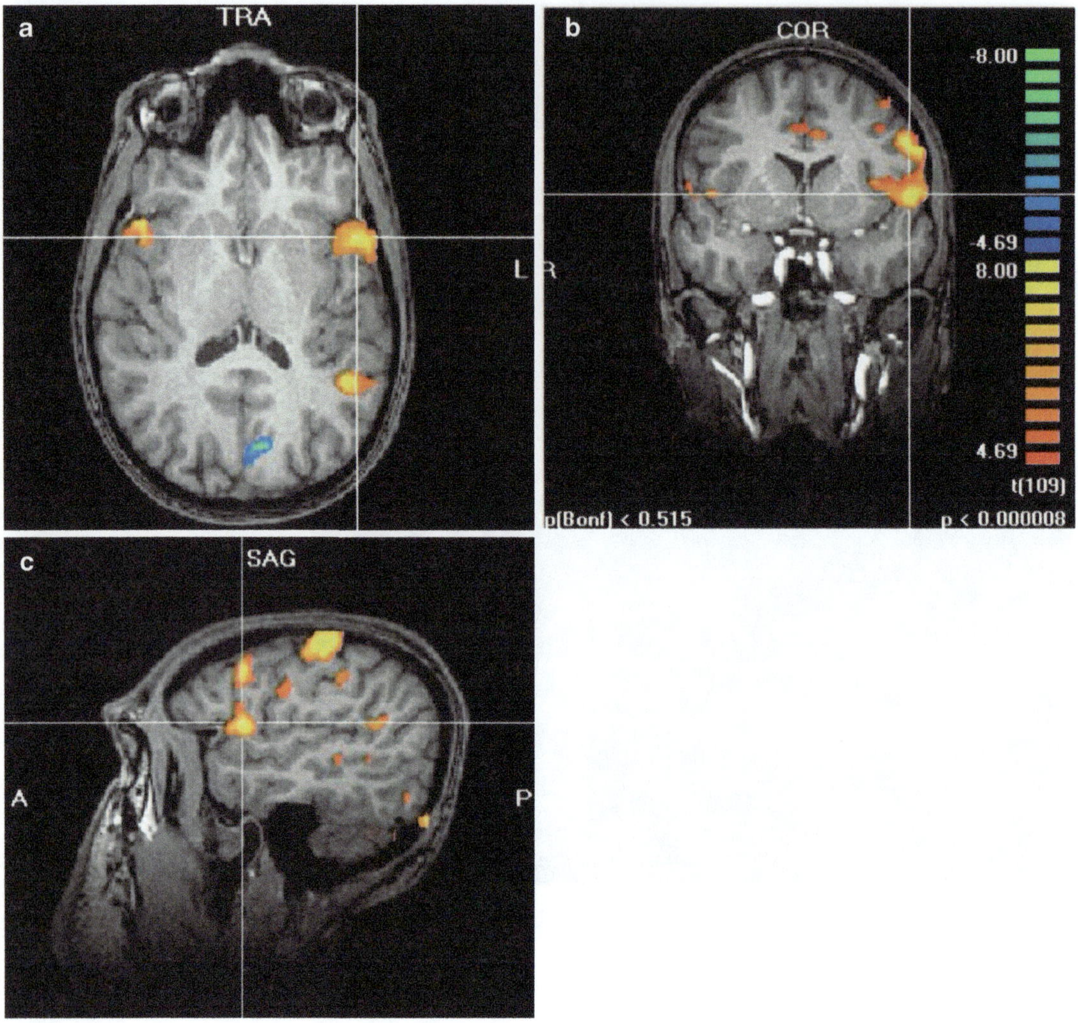

Fig. 5.7 Speech and language. (**a**–**c**) Word generation

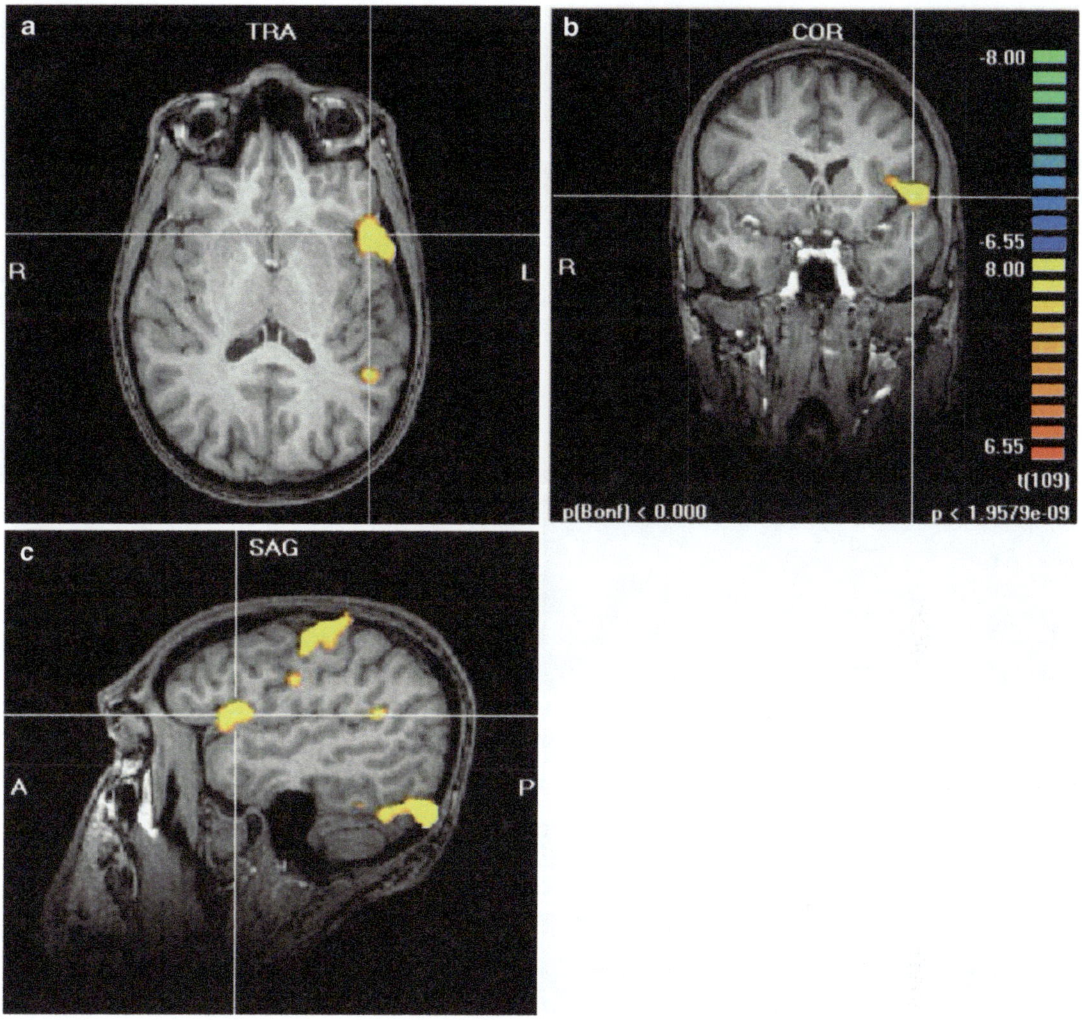

Fig. 5.8 Speech and language. (**a**–**c**) Reading comprehension

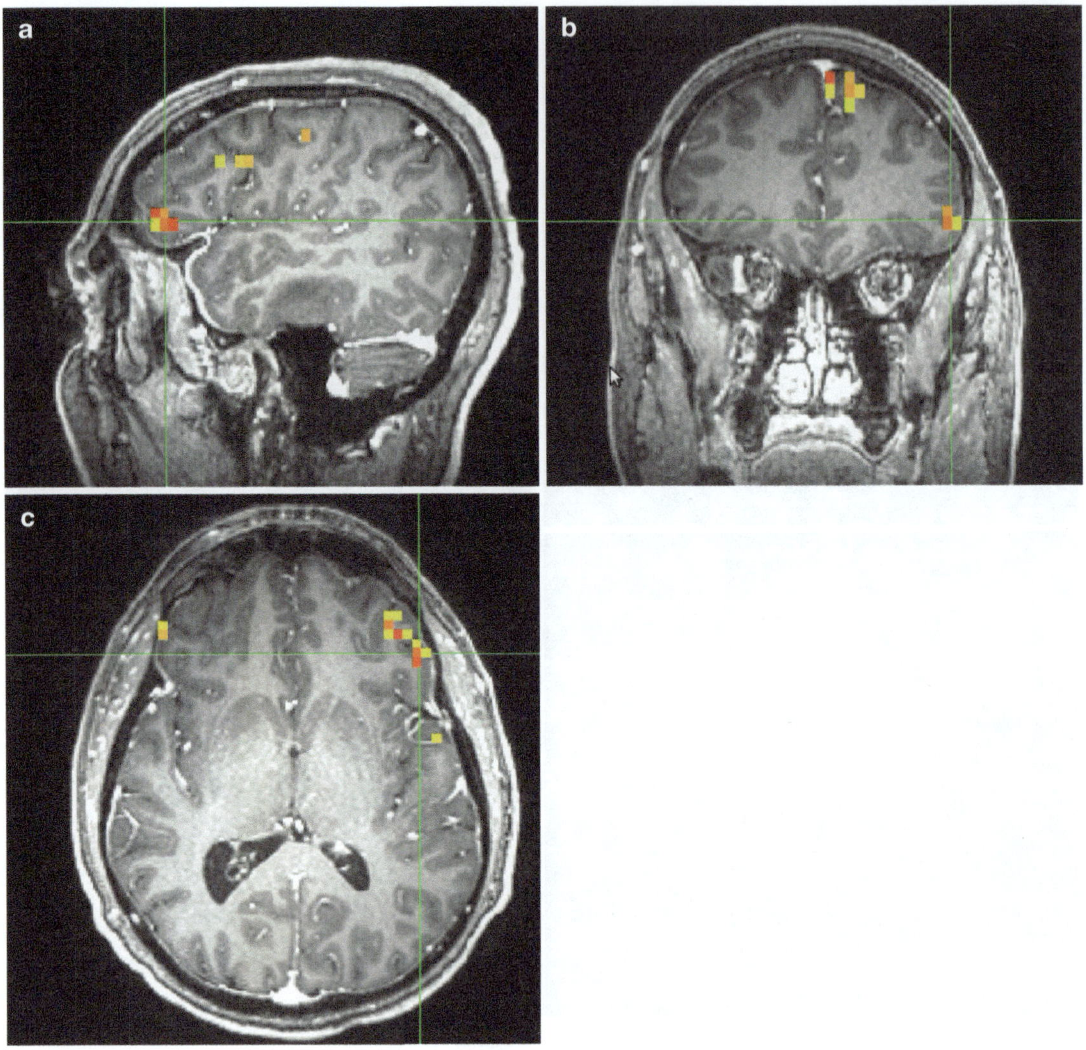

Fig. 5.9 (**a**–**c**) Broca

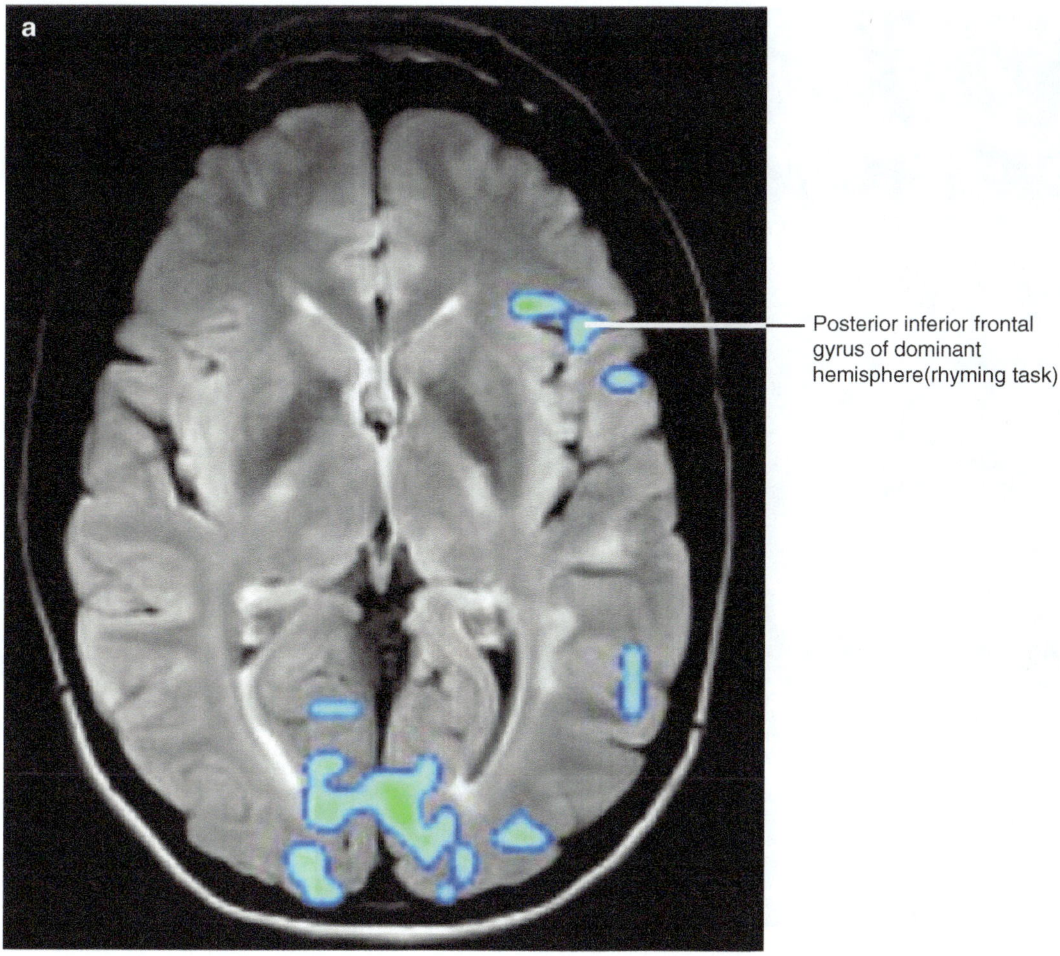

Fig. 5.10 Speech and language. (**a**) Pars opercularis and pars triangularis constitute Broca's area (Broadman area 44 and 45). (**b**) Primary auditory cortex, transverse temporal gyrus (Brodman areas 41 and 42) (Images courtesy of JJ Pillai, M.D., Neuroradiology Division, Johns Hopkins Hospital.)

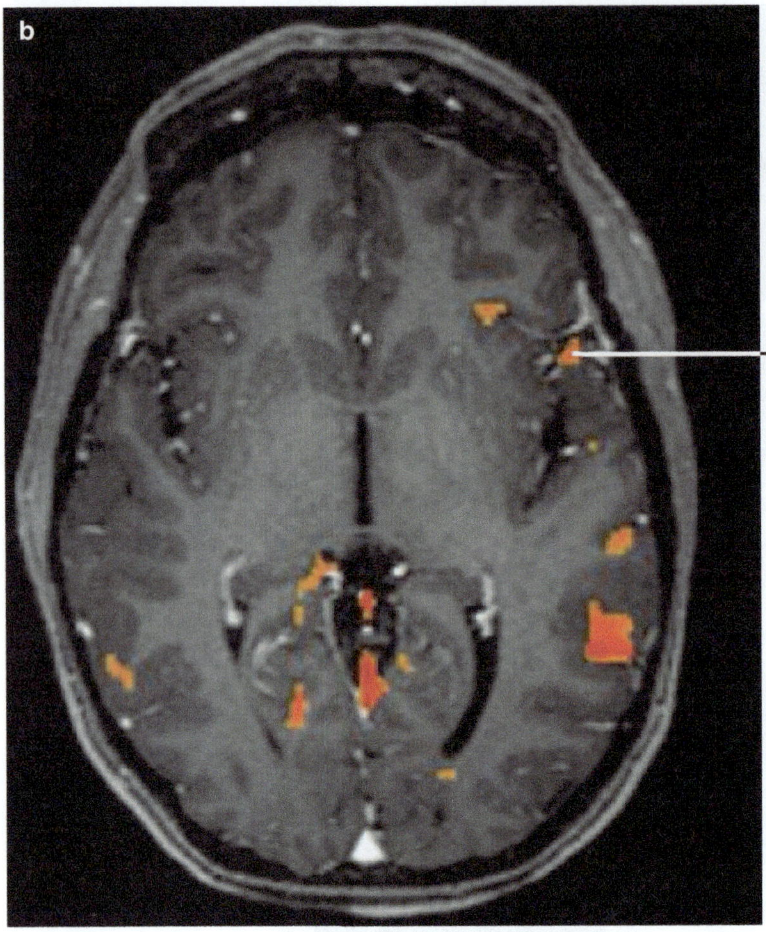

Fig. 5.10 (continued)

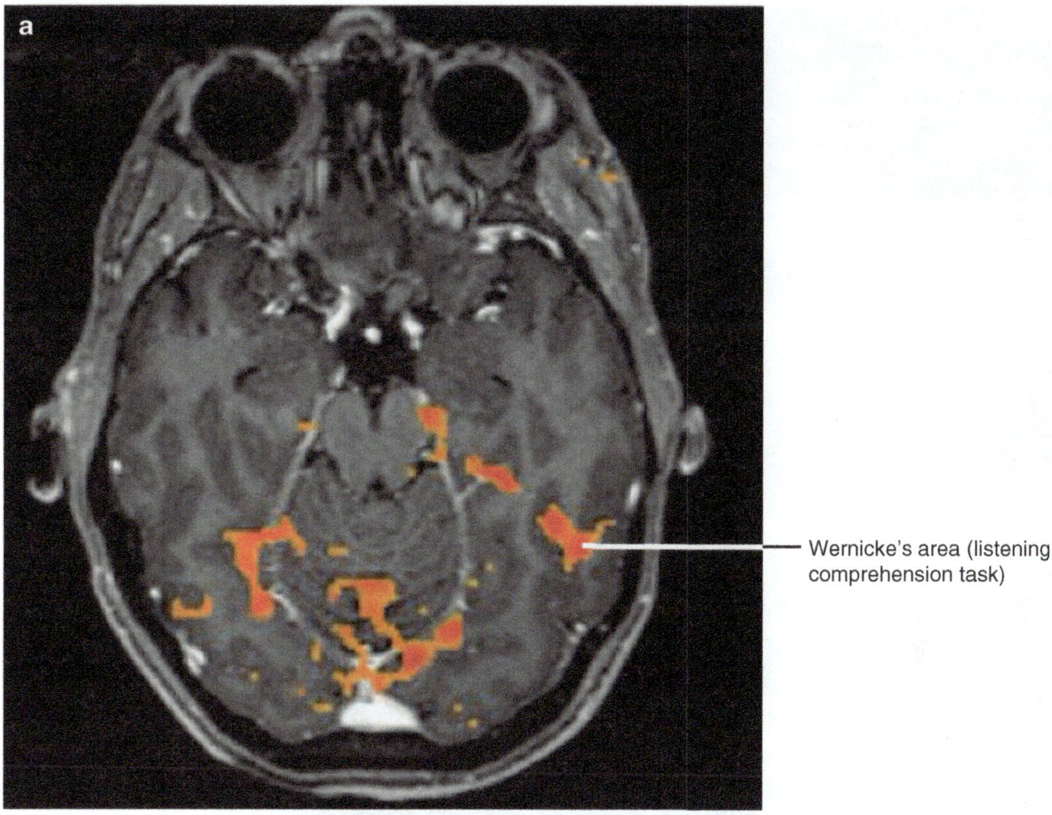

Fig. 5.11 Speech and language. (**a**) Posterior superior temporal gyrus of dominant hemisphere. (**b**) Posterior superior temporal gyrus of dominant hemisphere (Images courtesy of JJ Pillai, M.D., Neuroradiology Division, Johns Hopkins Hospital.)

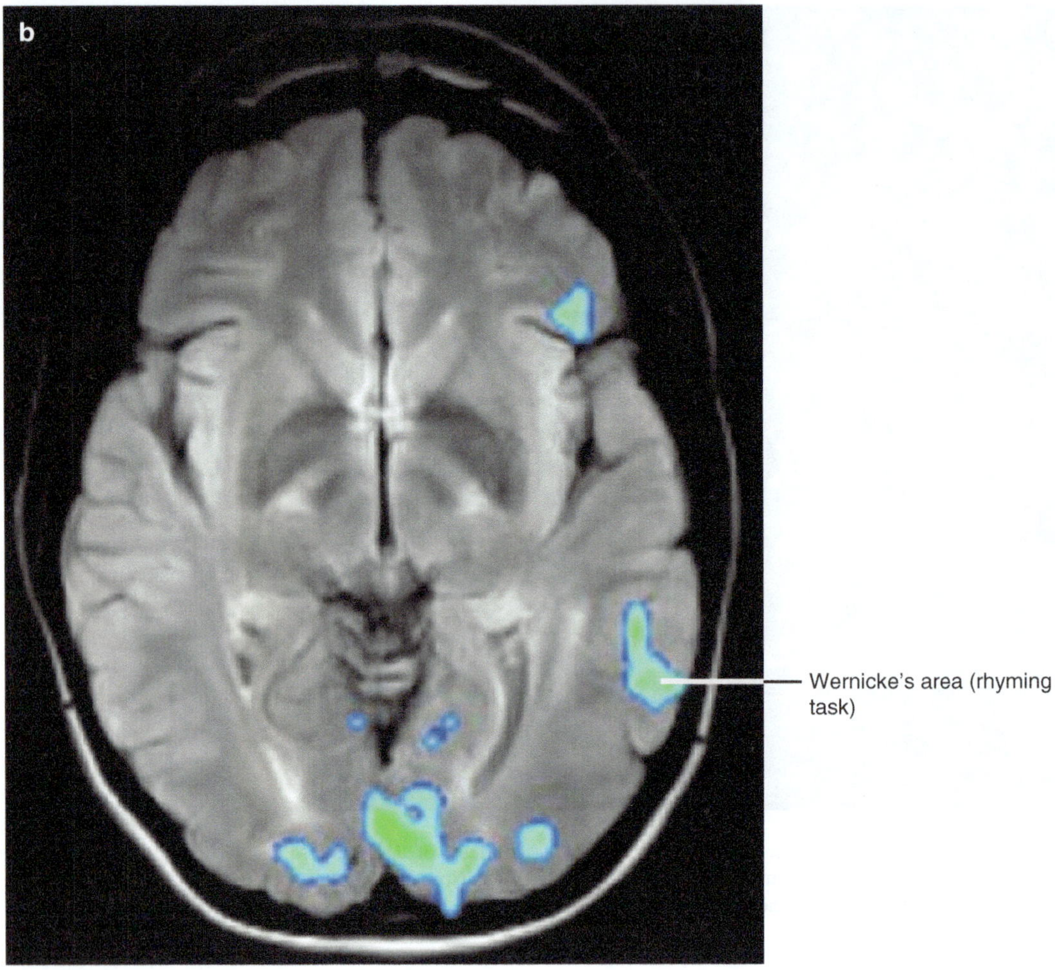

Fig. 5.11 (continued)

6 Cranial Nerves

This chapter will illustrate the imaging of the 12 cranial nerves (CNs). The pathway of each nerve will be demonstrated by the modality best suited to each portion. Where the nerve itself may not actually be seen, a landmark structure is indicated in its path instead.

Each nerve will be illustrated in the following format:

1. End organ
2. Course
3. Exit from brain
4. Origin in brain

S. Andronikou (ed.), *See Right Through Me*,
DOI 10.1007/978-3-642-23893-2_6, © Springer-Verlag Berlin Heidelberg 2012

CN I Olfactory

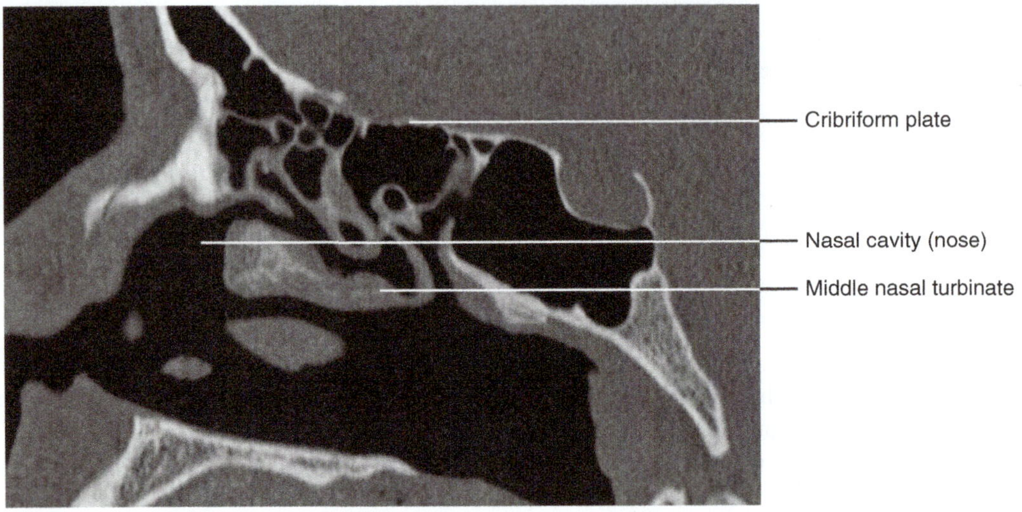

Fig. 6.1 Sagittal CT end organ: nose

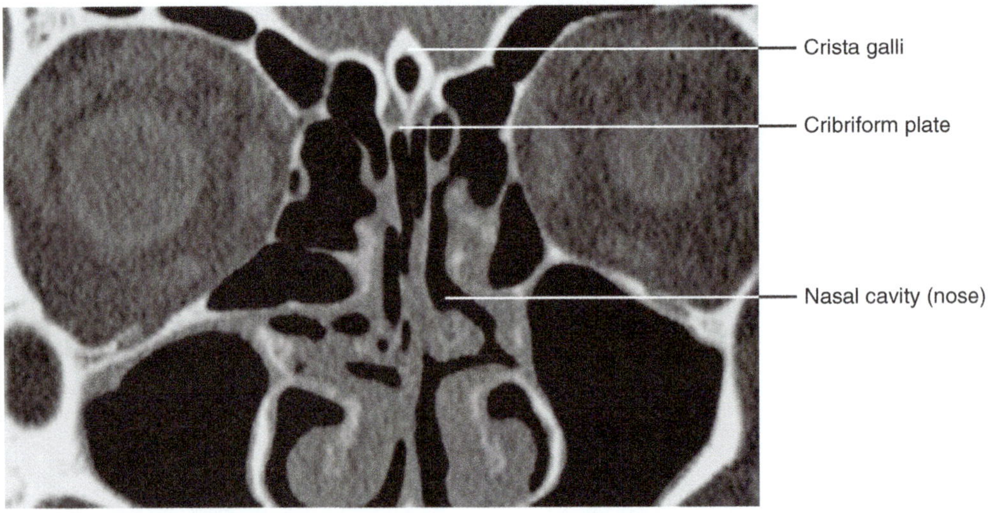

Fig. 6.2 Coronal CT course: cribriform plate

Fig. 6.3 Coronal magnetic resonance imaging (MRI) exit from brain: olfactory bulb/white matter adjacent to gyrus rectus of frontal lobe

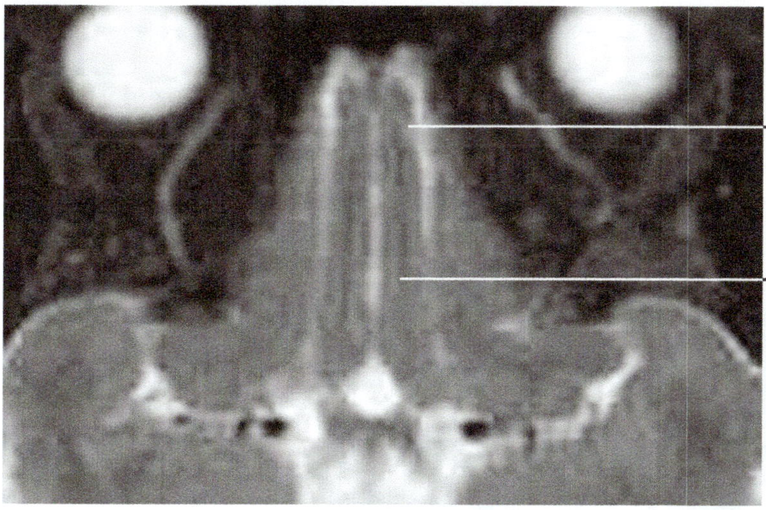

Fig. 6.4 Axial MRI exit from brain: olfactory bulb/white matter adjacent to gyrus rectus of frontal lobe

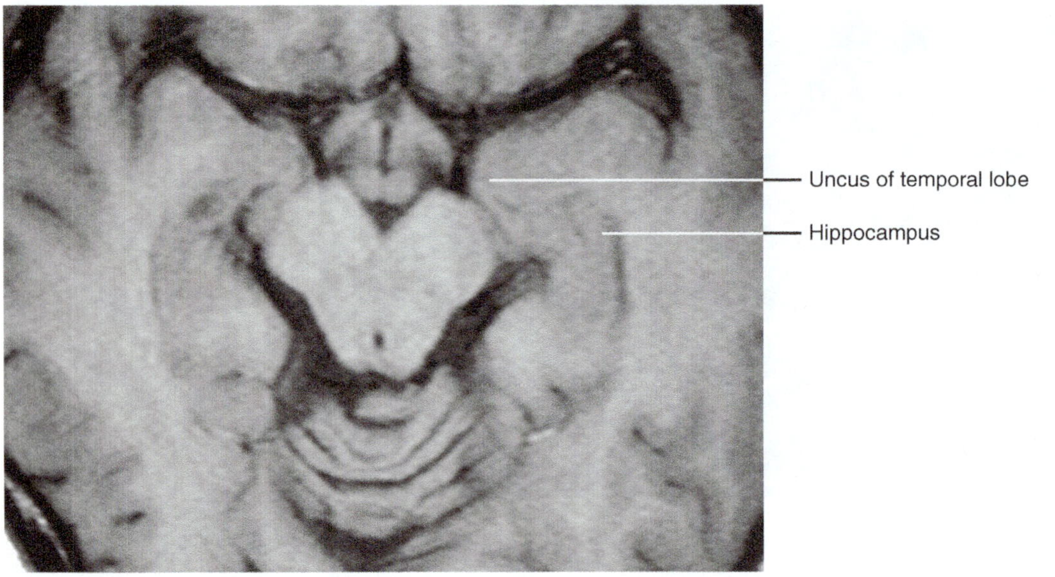

Fig. 6.5 Axial MRI course: posterior to uncus and adjacent temporal cortex

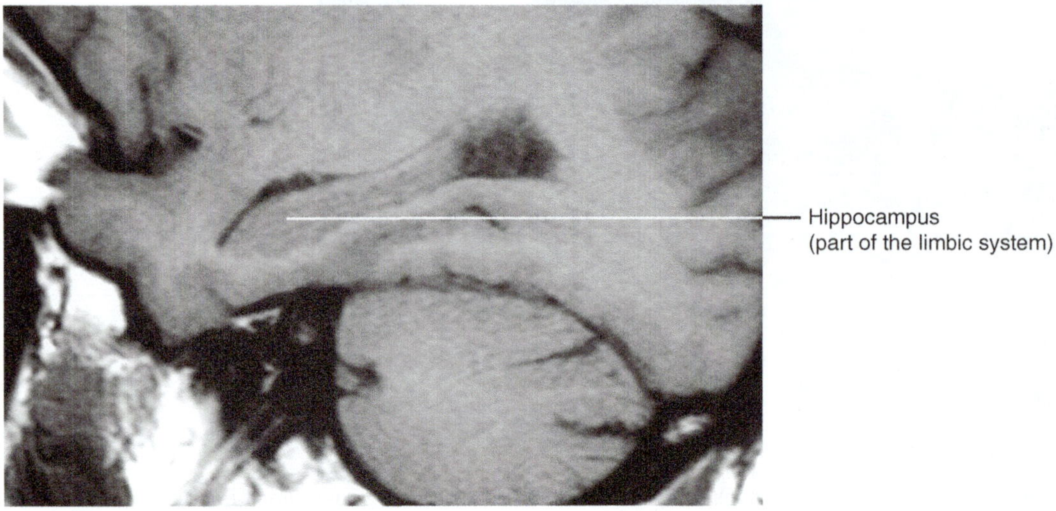

Fig. 6.6 Sagittal MRI origin in brain: limbic system

CN II Optic

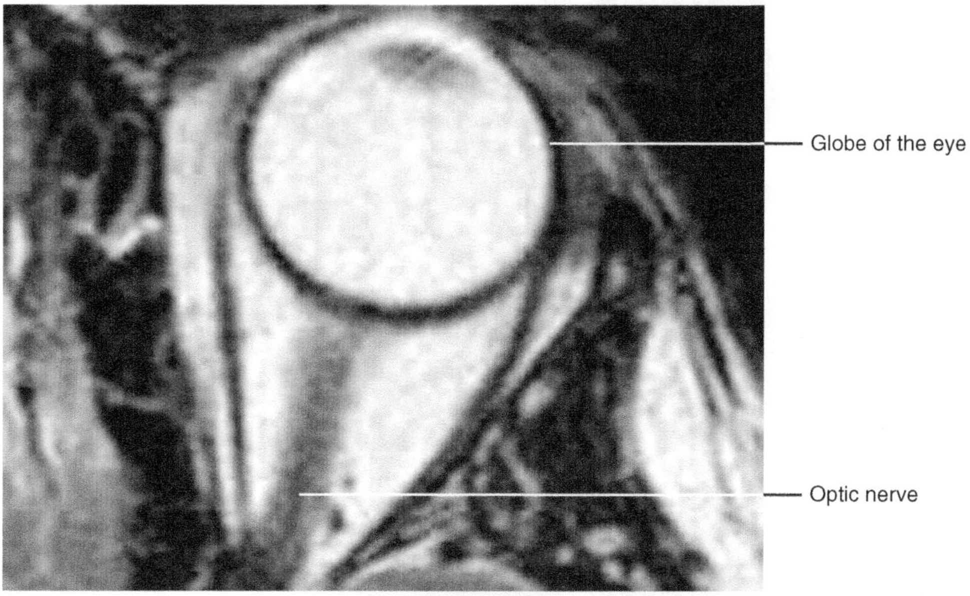

Fig. 6.7 Axial MRI end organ: eye

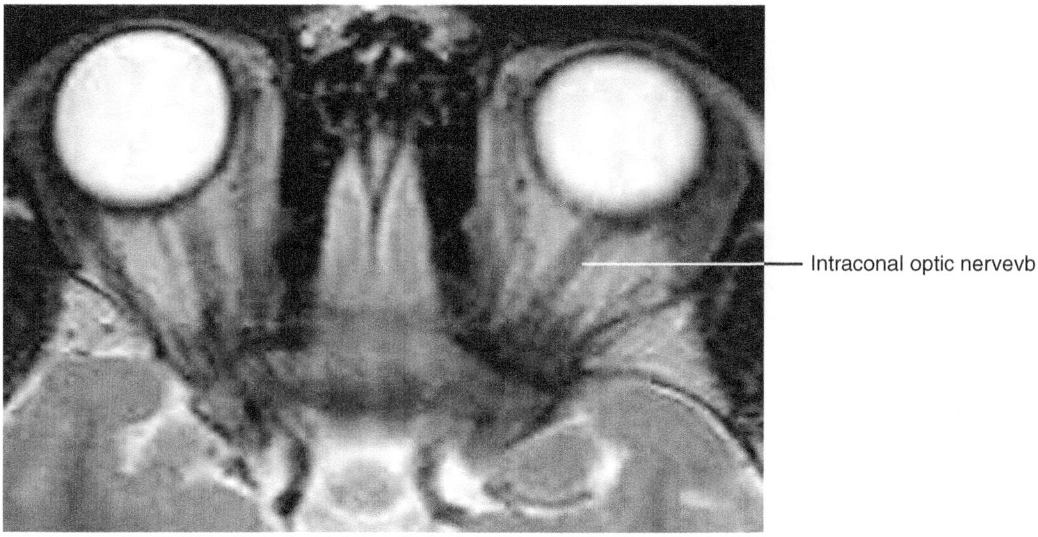

Fig. 6.8 Axial MRI course: optic nerve

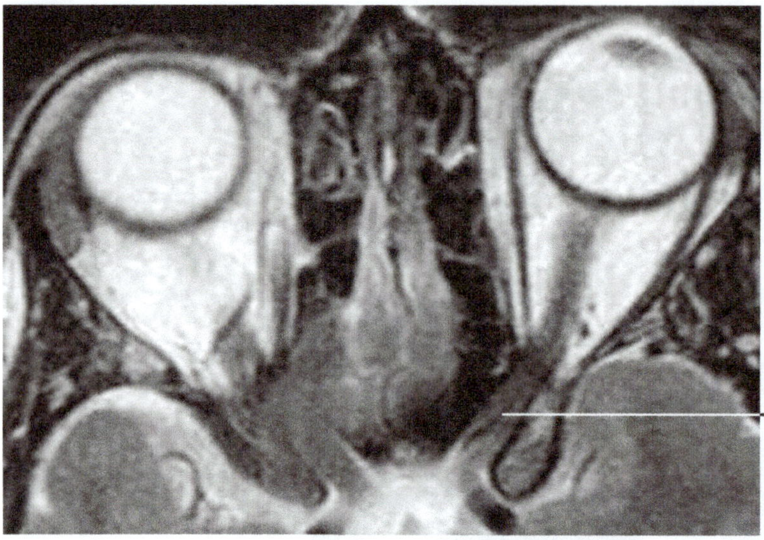

Fig. 6.9 Axial MRI course: optic canal

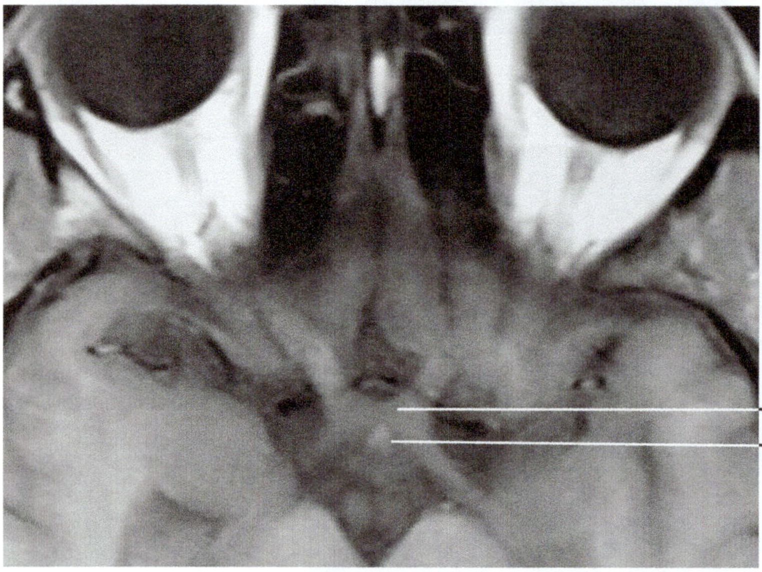

Fig. 6.10 Axial MRI course: optic chiasm

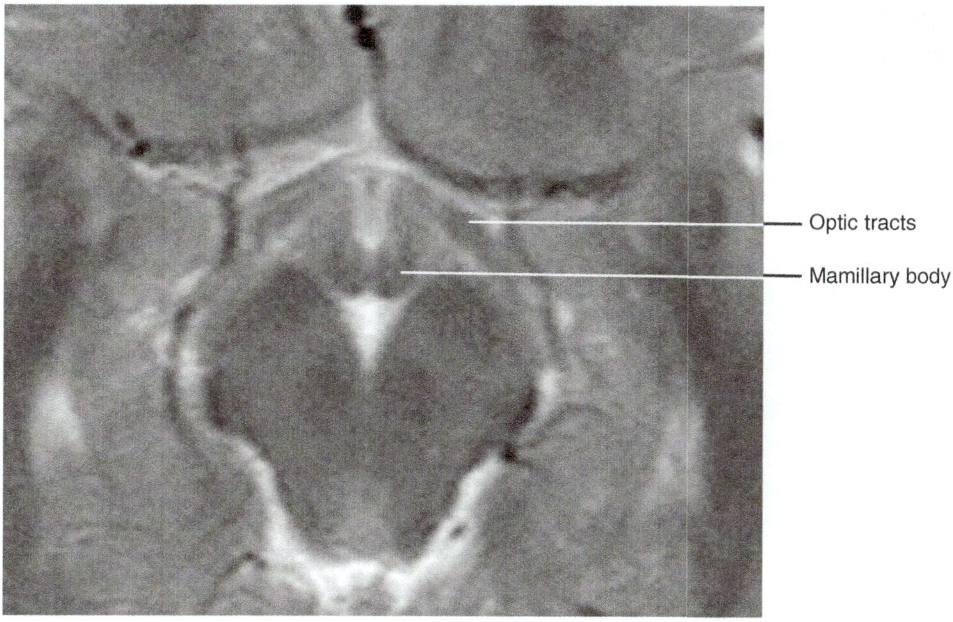

Fig. 6.11 Axial MRI exit from brain: optic tracts

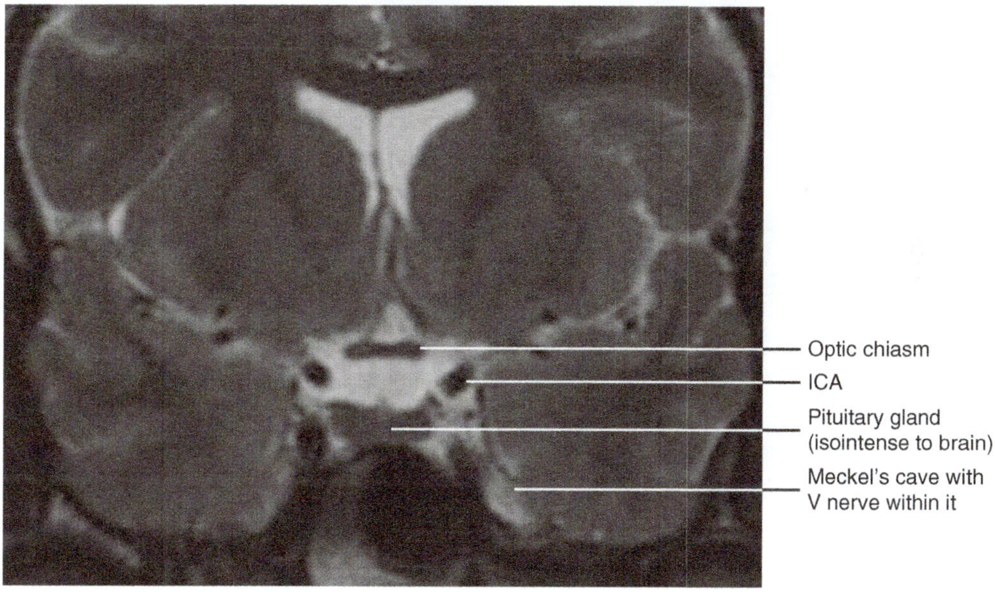

Fig. 6.12 Coronal MRI course: optic chiasm

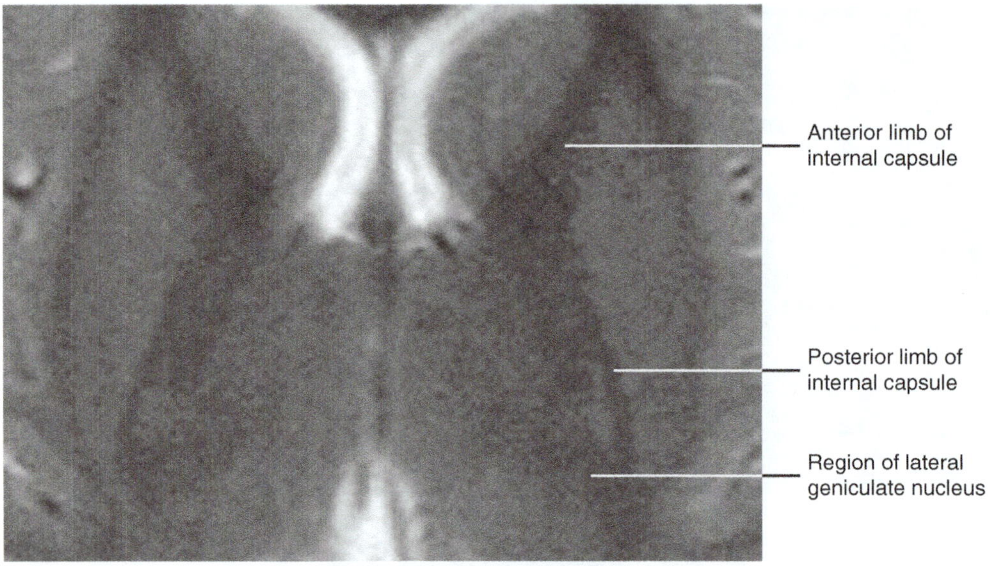

Fig. 6.13 Axial MRI course: superior fibres—internal capsule

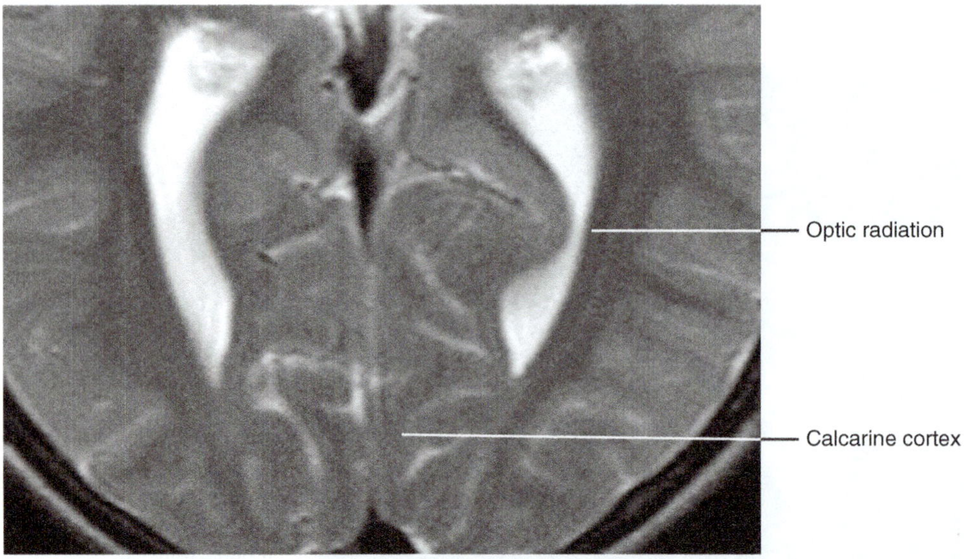

Fig. 6.14 Axial MRI course: inferior fibres—around inferior horn of ventricle

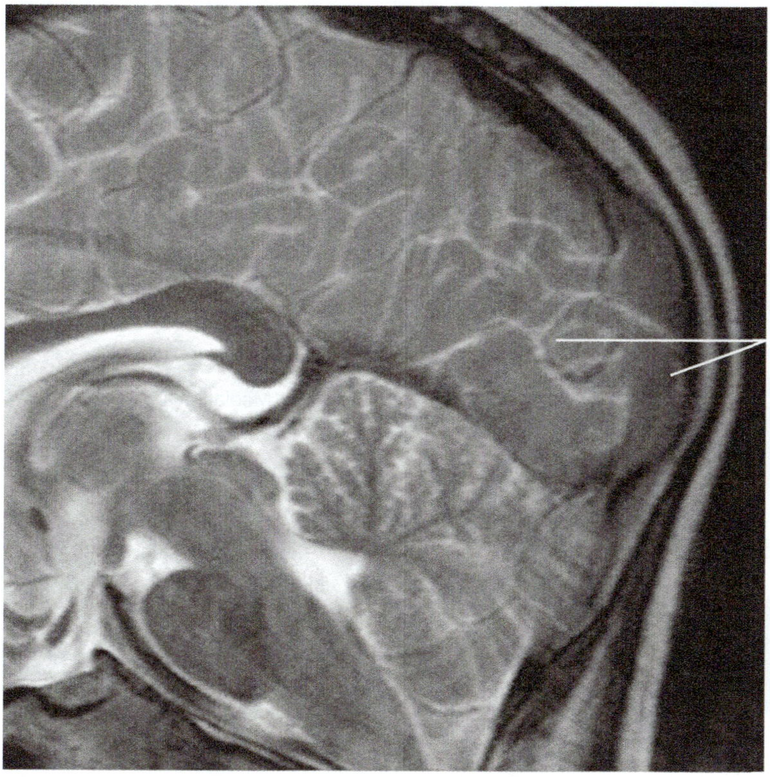

Fig. 6.15 Sagittal MRI origin: calcarine/visual cortex

CN III Occulomotor

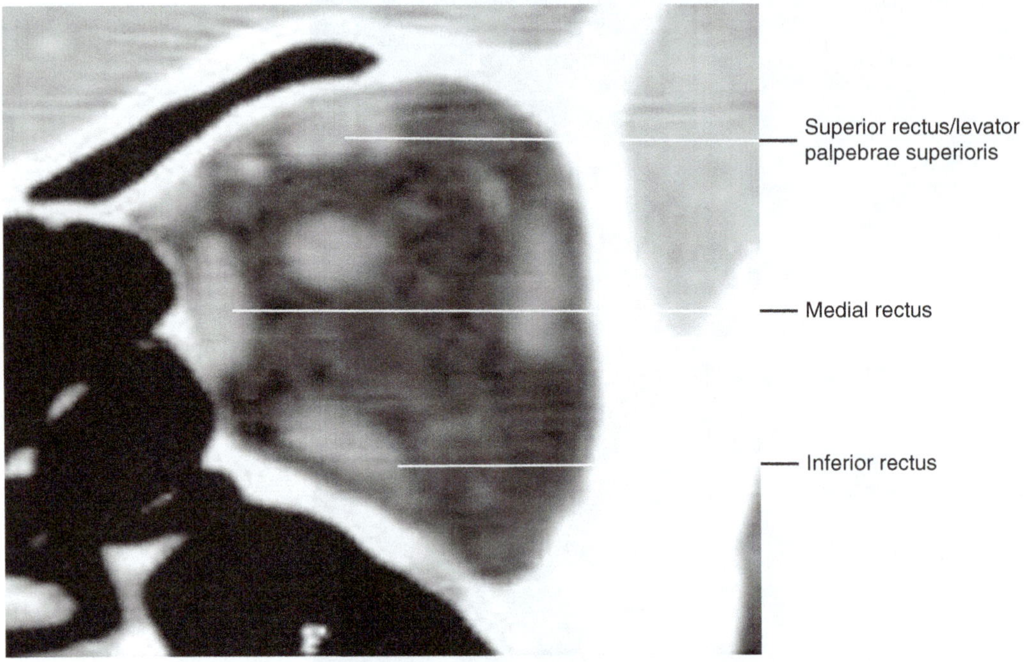

Fig. 6.16 Coronal CT end organ. Superior ramus: superior rectus/levator palpebrae superioris. Inferior ramus: medial rectus/inferior rectus/inferior oblique

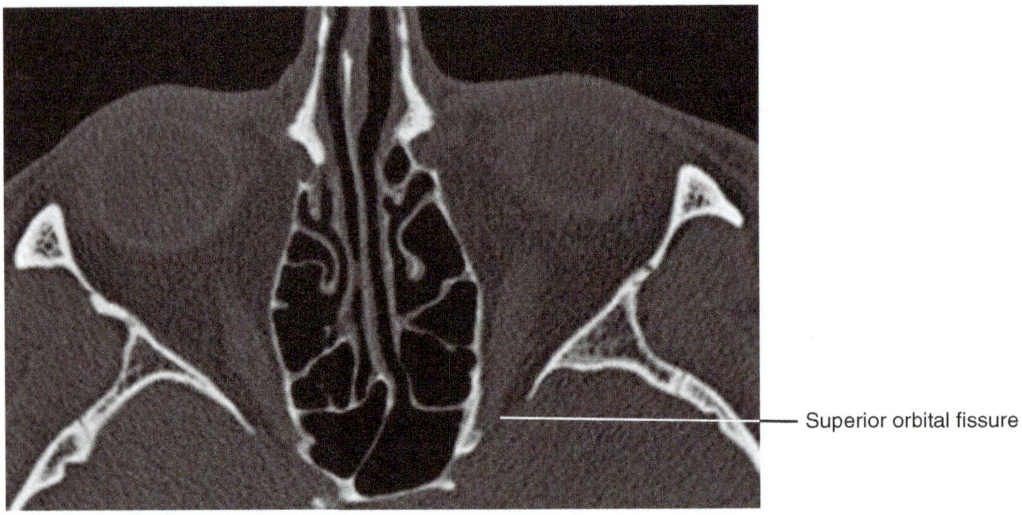

Fig. 6.17 Axial CT course: superior orbital fissure

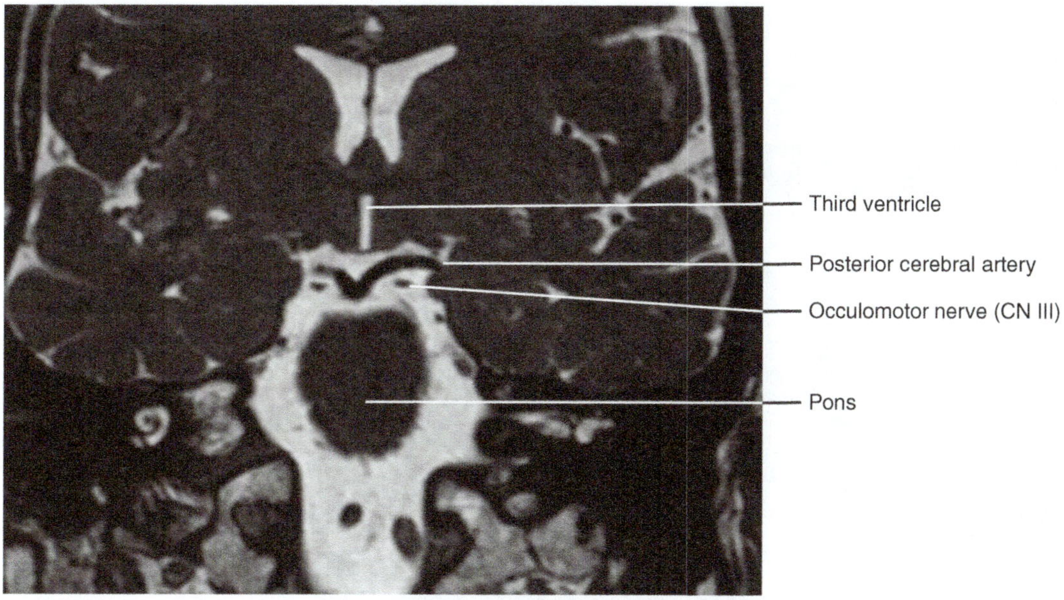

Fig. 6.18 Course: relationship between posterior cerebral artery and occulomotor nerve

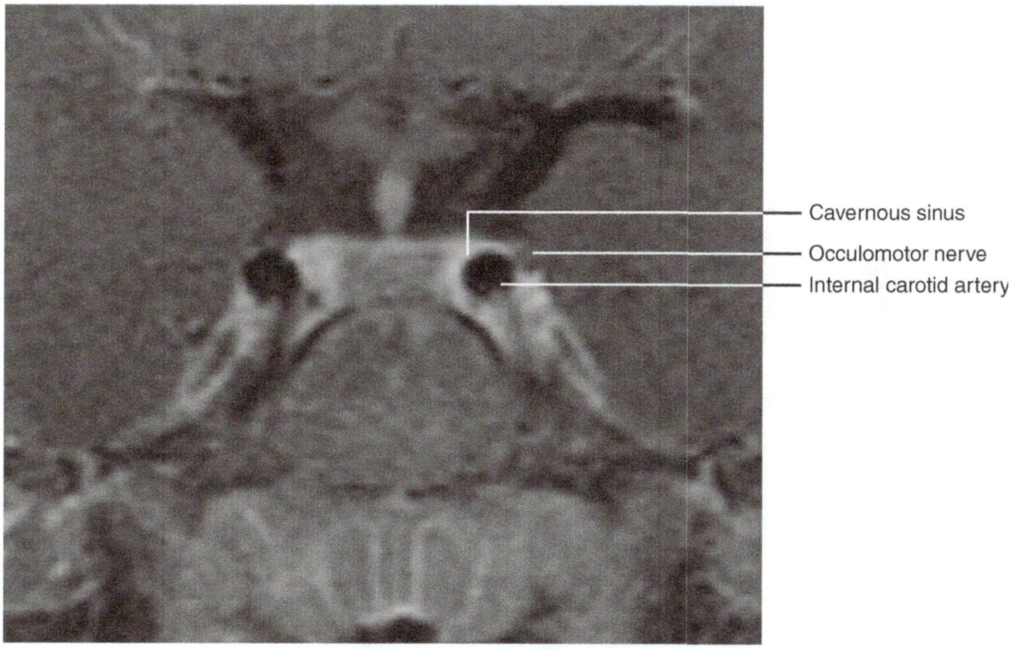

Fig. 6.19 Coronal MRI course: cavernous sinus (lateral most superior nerve)

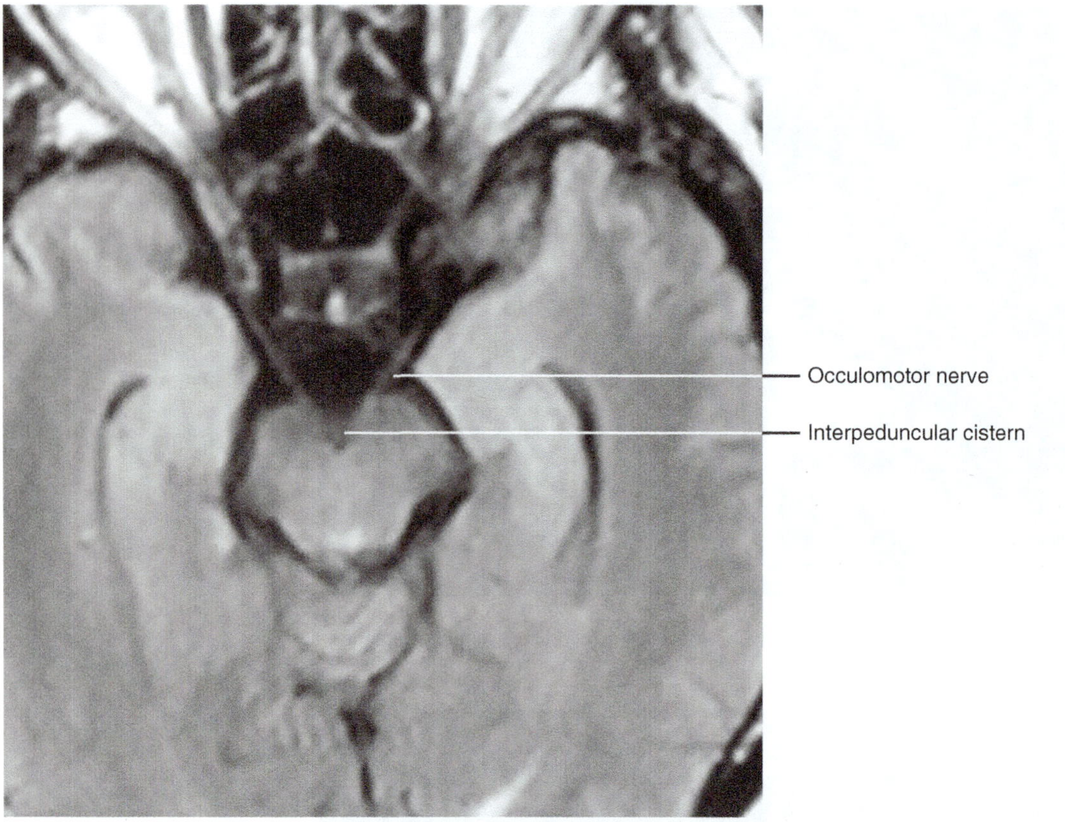

Fig. 6.20 Axial MRI course: crosses interpeduncular cistern (between PCA/SCA)

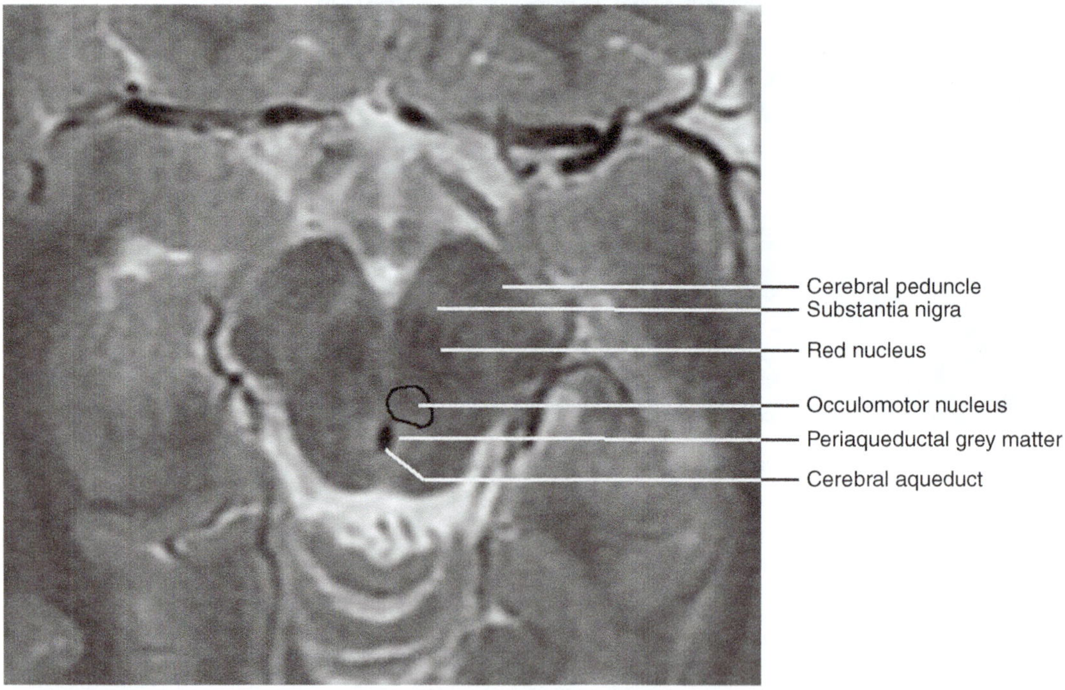

Fig. 6.21 Axial MRI course tegmentum: red nucleus and medial part substantia nigra

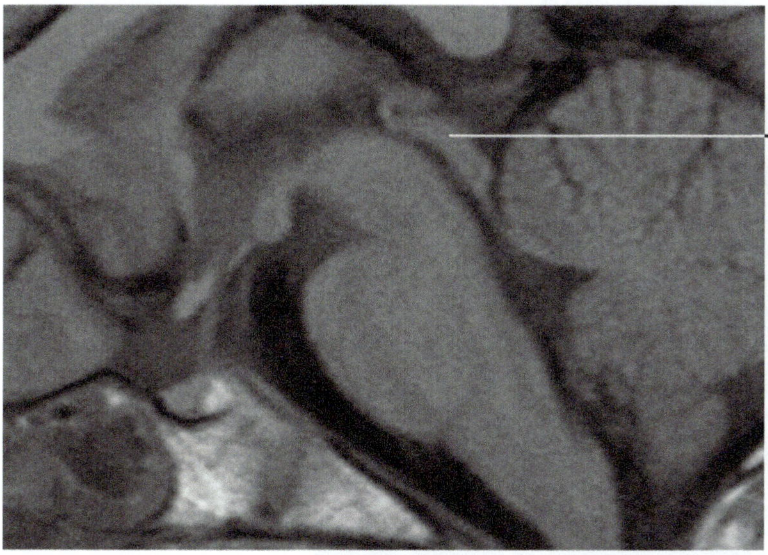

Fig. 6.22 Sagittal MRI origin. Superior colliculus: occulomotor nucleus, Edinger Westphal nucleus

CN IV Trochlear

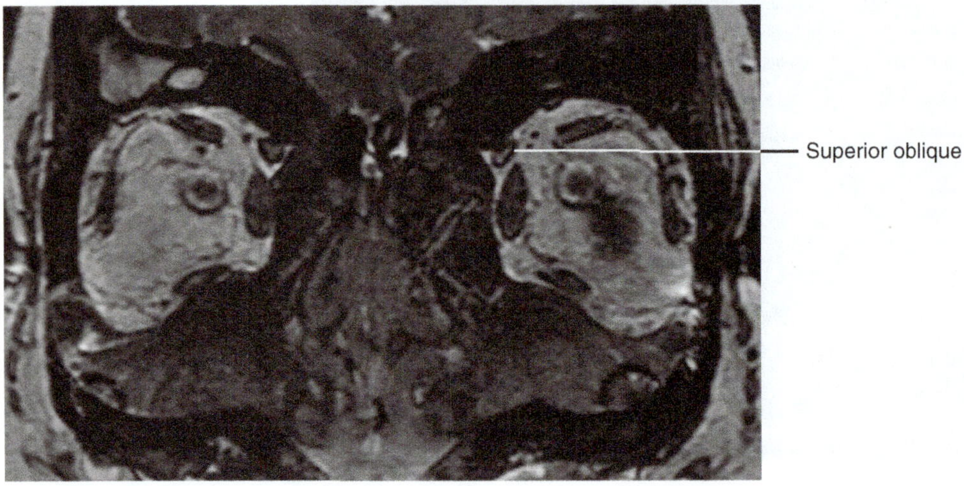

Fig. 6.23 Coronal MRI end organ: superior oblique

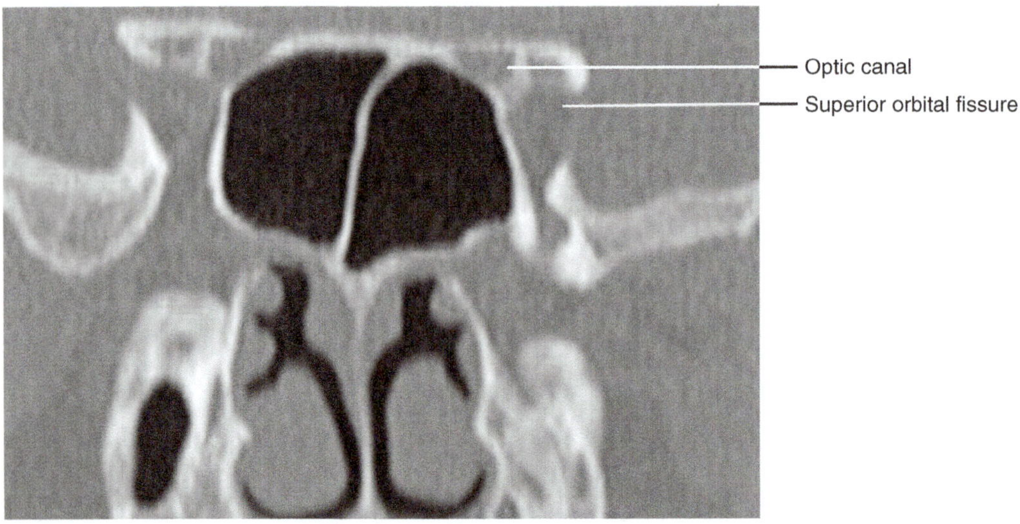

Fig. 6.24 Coronal CT course: superior orbital fissure

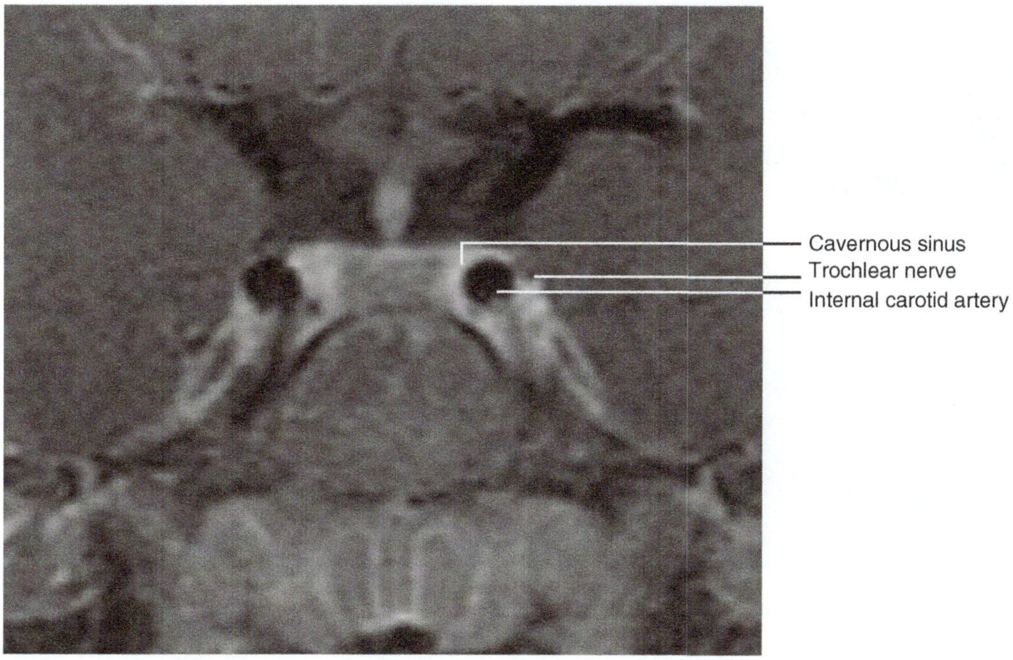

Fig. 6.25 Coronal MRI course: cavernous sinus

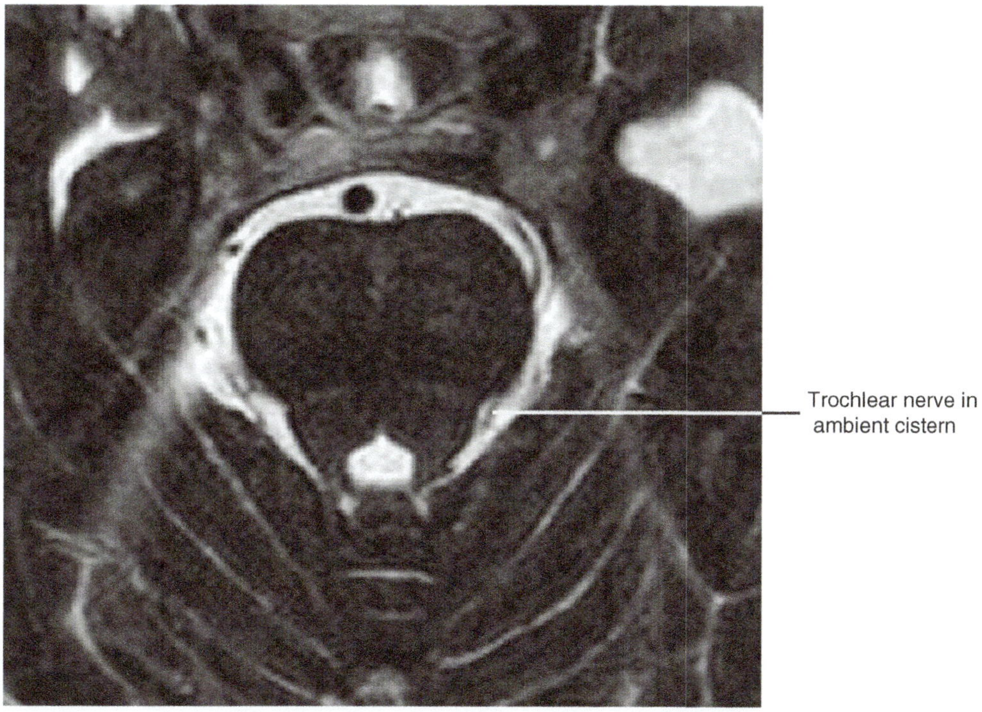

Fig. 6.26 Axial MRI exit from brain: dorsal midbrain inferior to inferior colliculus

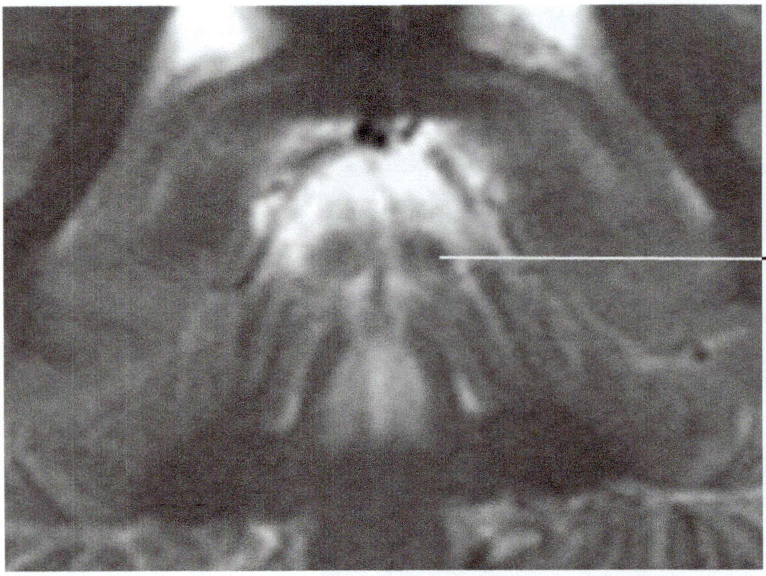

Fig. 6.27 Coronal MRI origin: inferior colliculus trochlear nucleus

CN V Trigeminal

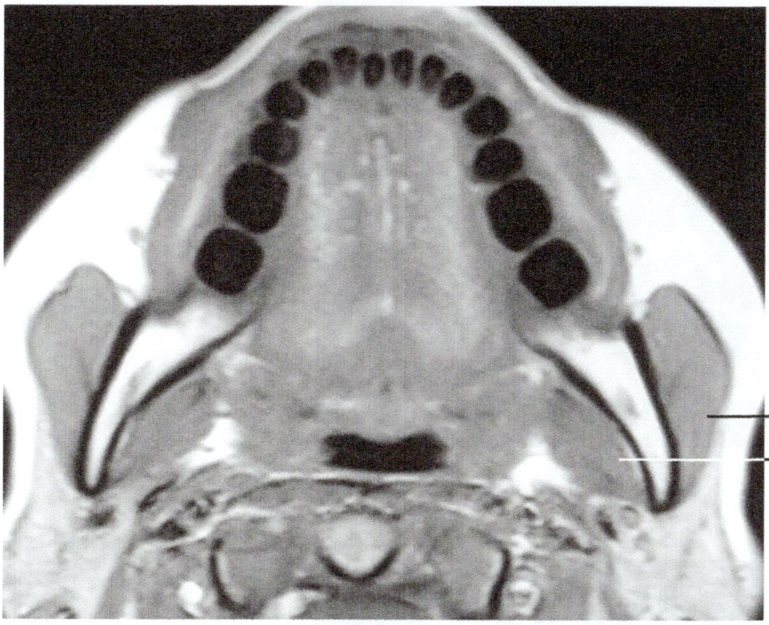

Fig. 6.28 Axial MRI end organ: muscles of mastication/floor of mouth/tensor tympani

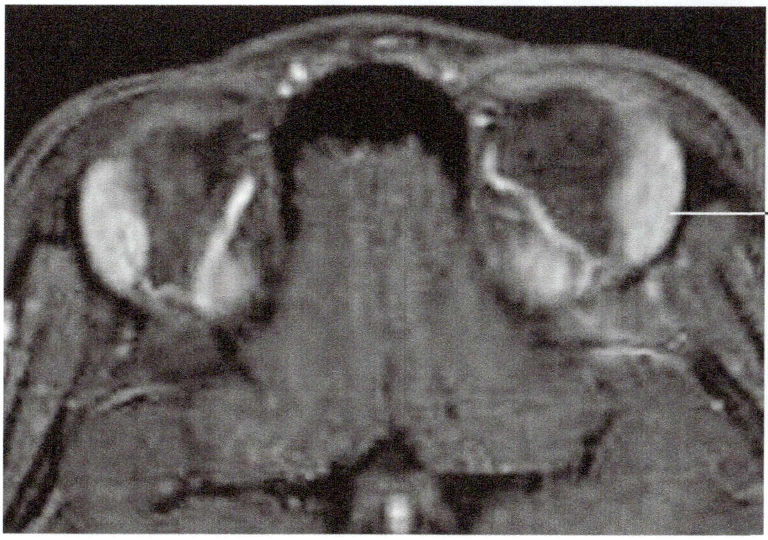

Fig. 6.29 Axial MRI end organ: nasolacrimal system

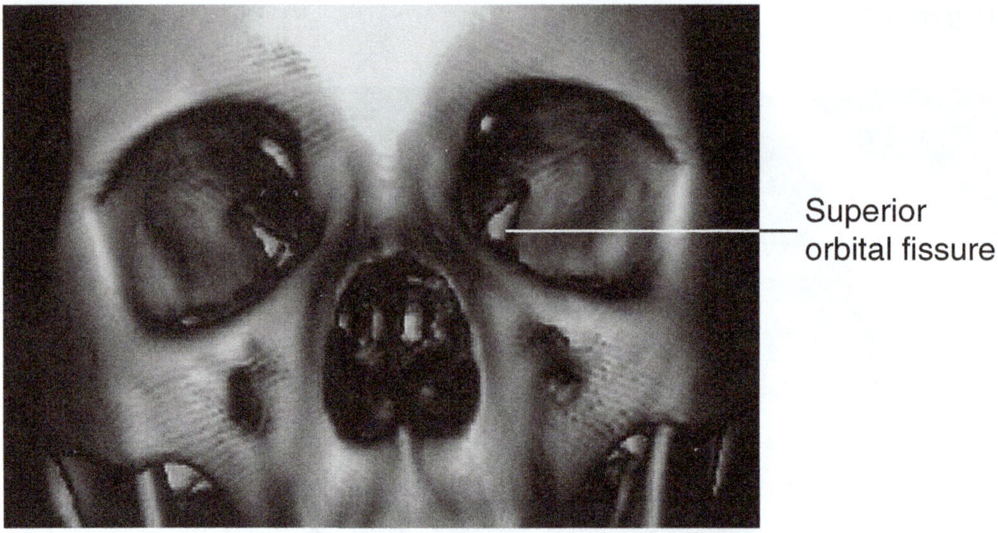

Fig. 6.30 Three-dimensional (3D) CT course. Ophthalmic division VI: superior orbital fissure

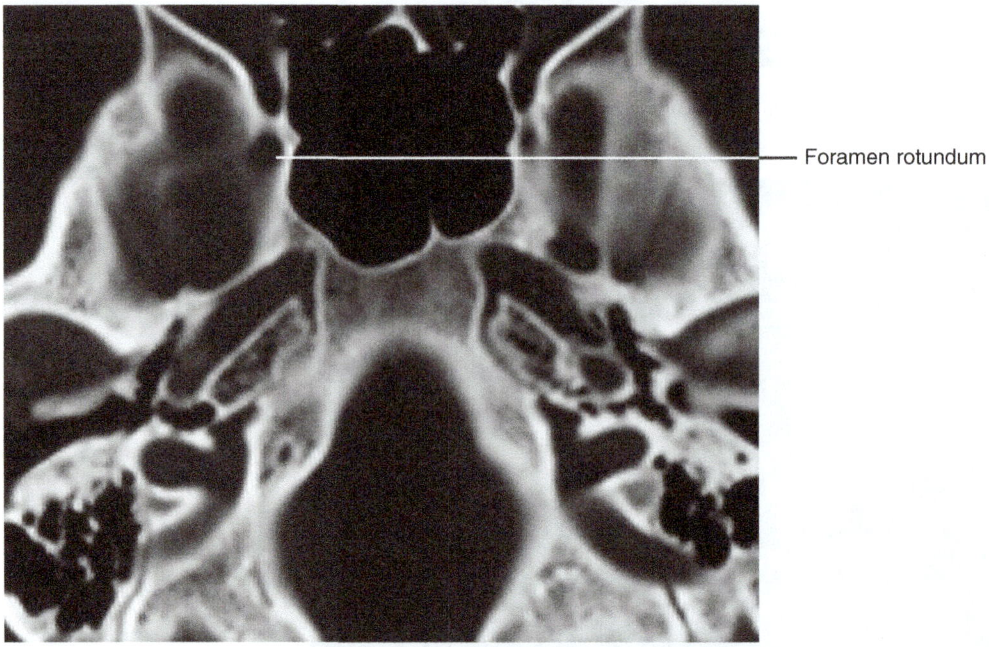

Fig. 6.31 Axial CT course. Maxillary division V2: foramen rotundum

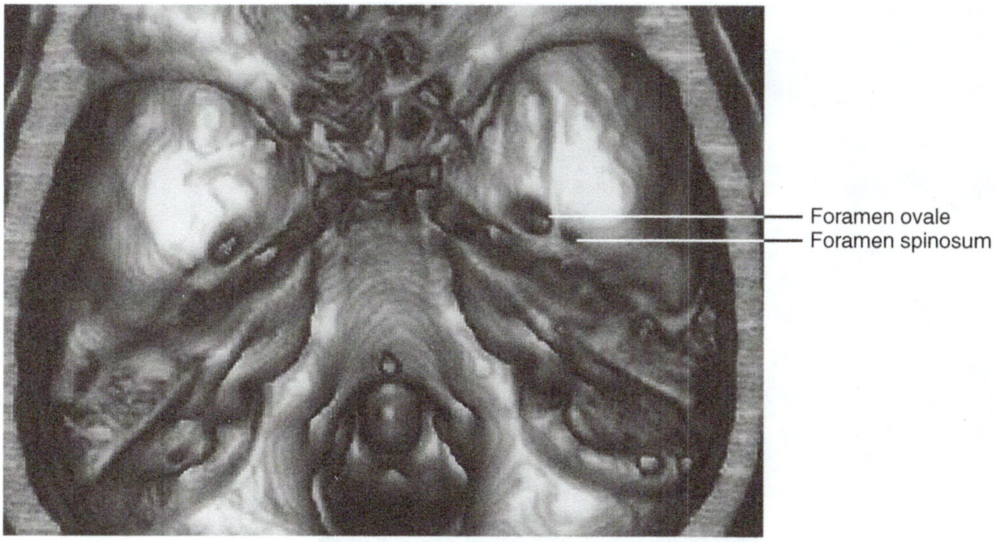

Fig. 6.32 3D CT course. Mandibular division V3: foramen ovale

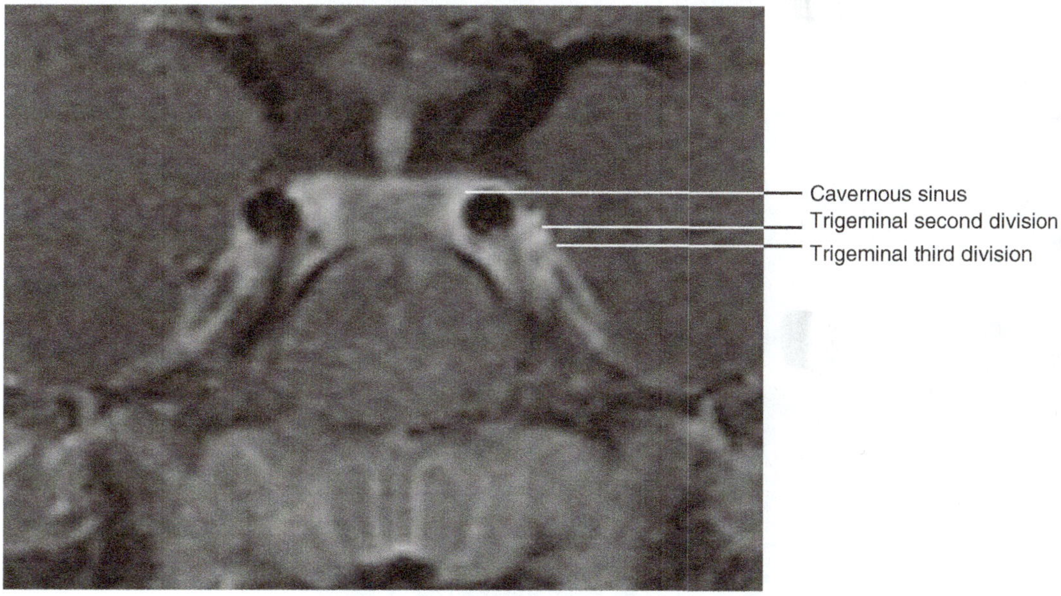

Fig. 6.33 Course. Cavernous sinus

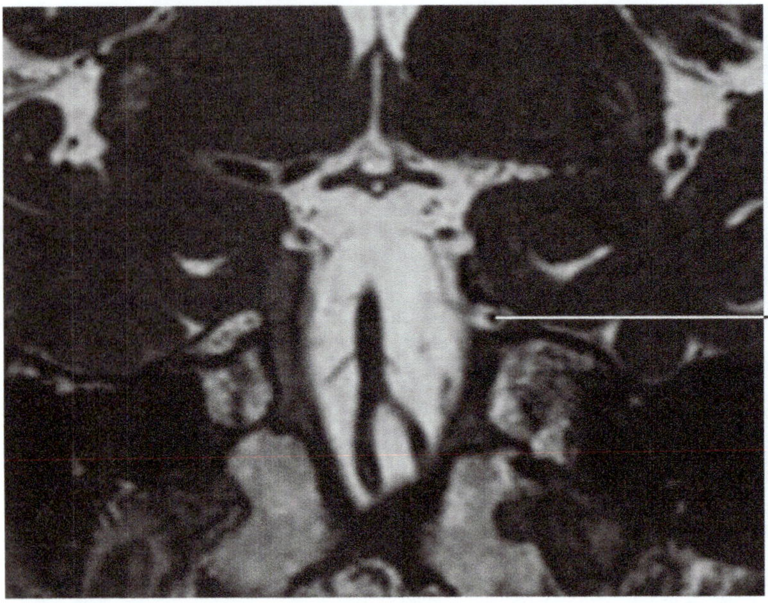

Fig. 6.34 Coronal MRI course: Meckel's cave

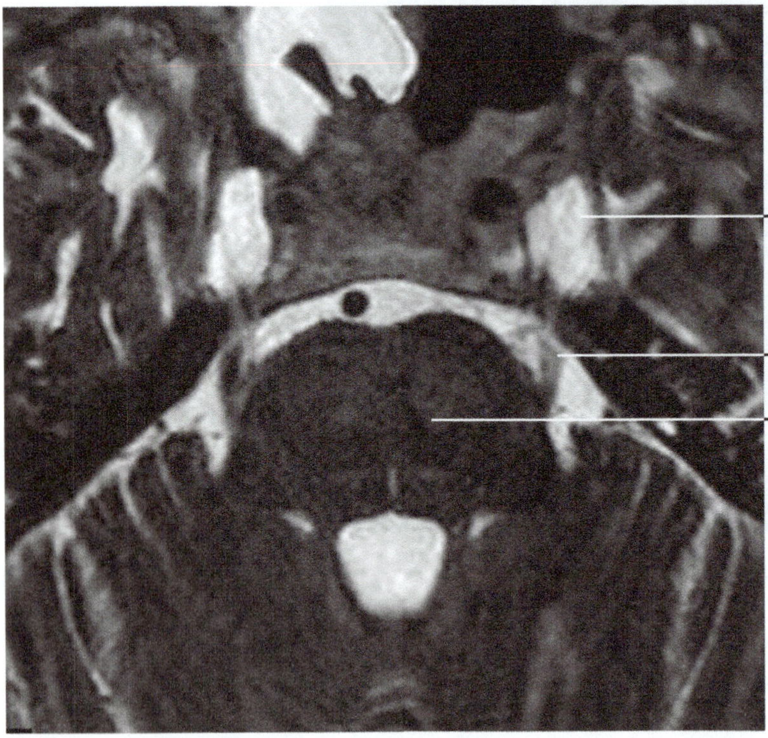

Fig. 6.35 Axial MRI course: prepontine cistern

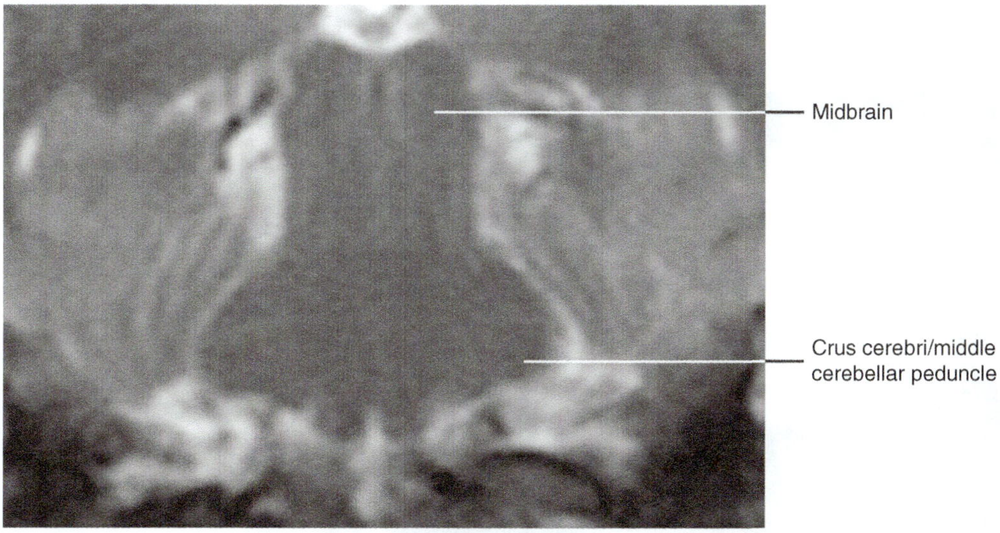

Fig. 6.36 Coronal MRI exit: junction of pons and middle cerebellar peduncle

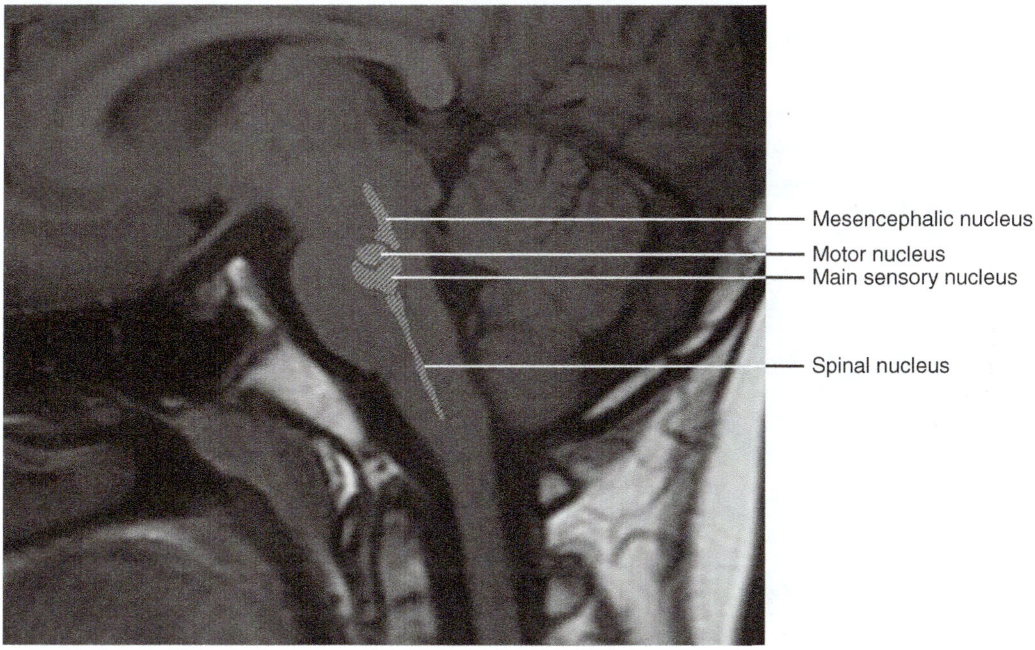

Fig. 6.37 Origin. Trigeminal nuclei. Mesencephalic nucleus: brainstem. Pontine nuclei (motor and sensory): pons. Spinal nucleus: medulla

CN VI Abducent

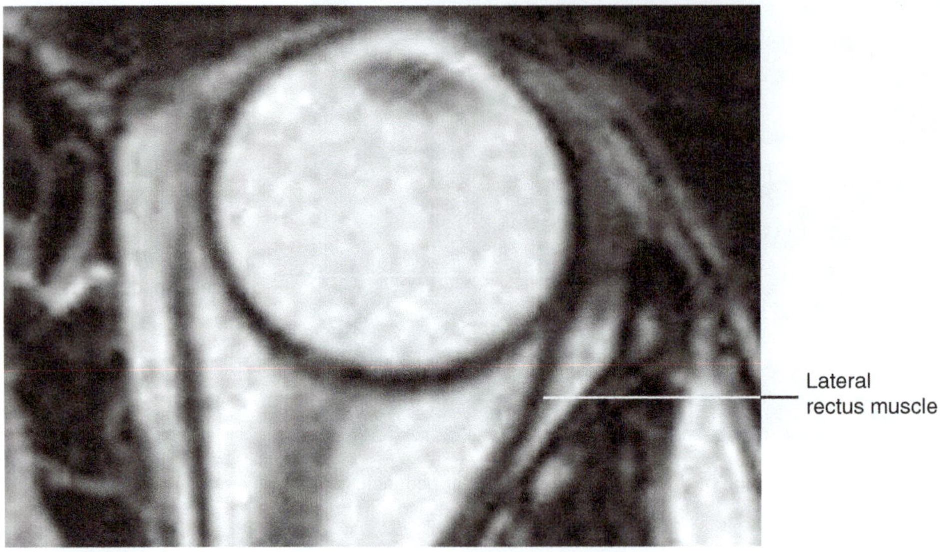

Fig. 6.38 Axial MRI end organ: lateral rectus

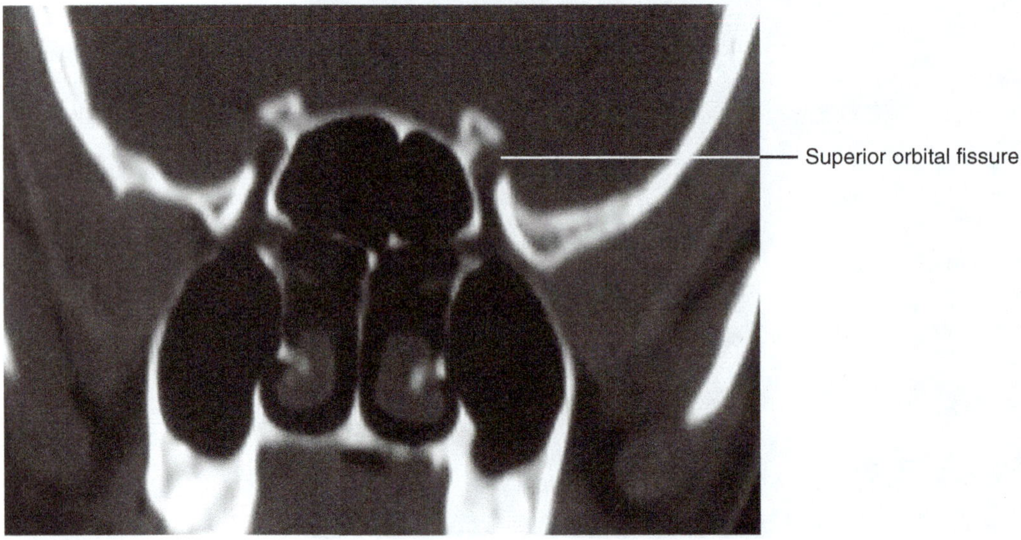

Fig. 6.39 Coronal CT course: superior orbital fissure

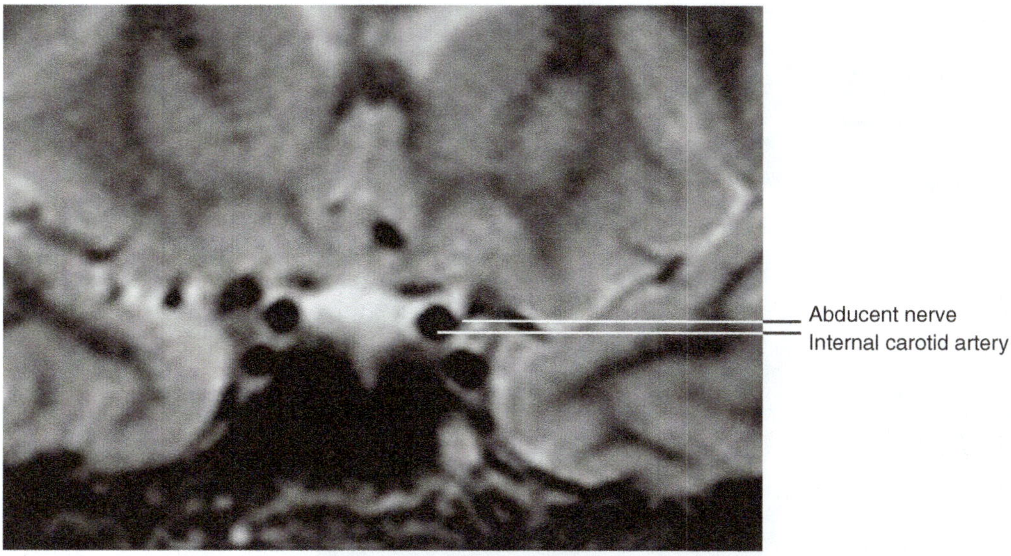

Fig. 6.40 Coronal MRI course. Cavernous sinus: medial to CN V and lateral to carotid artery

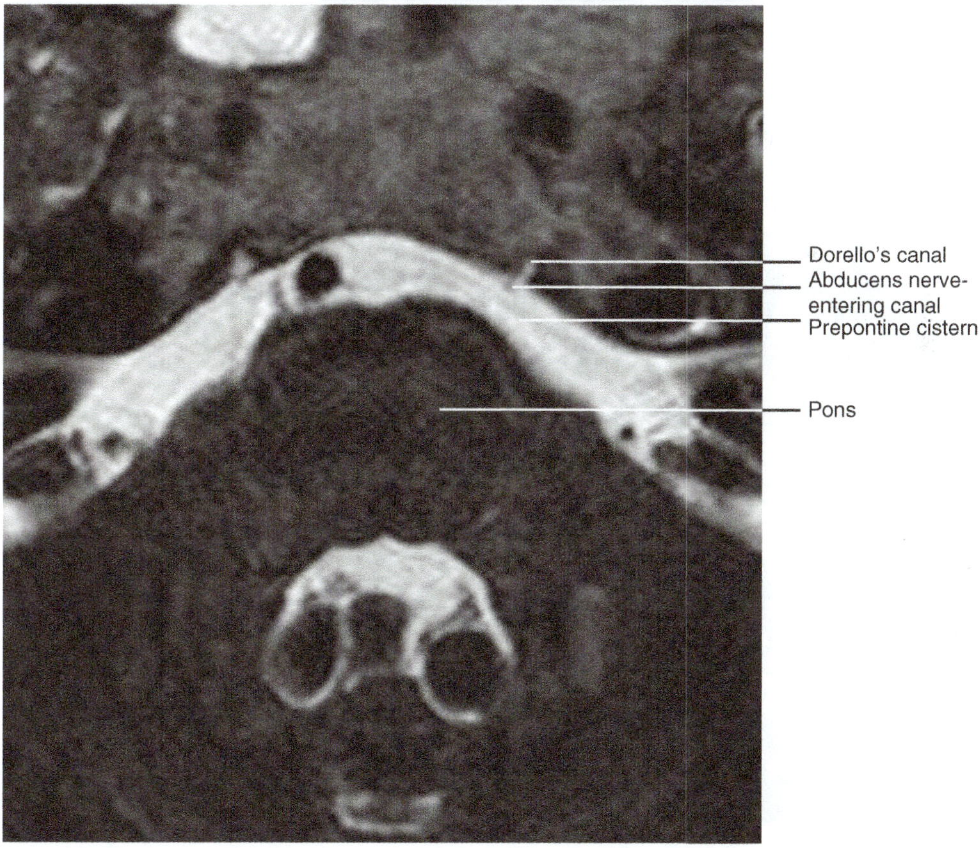

Fig. 6.41 Axial MRI course: prepontine cistern and Dorello's canal

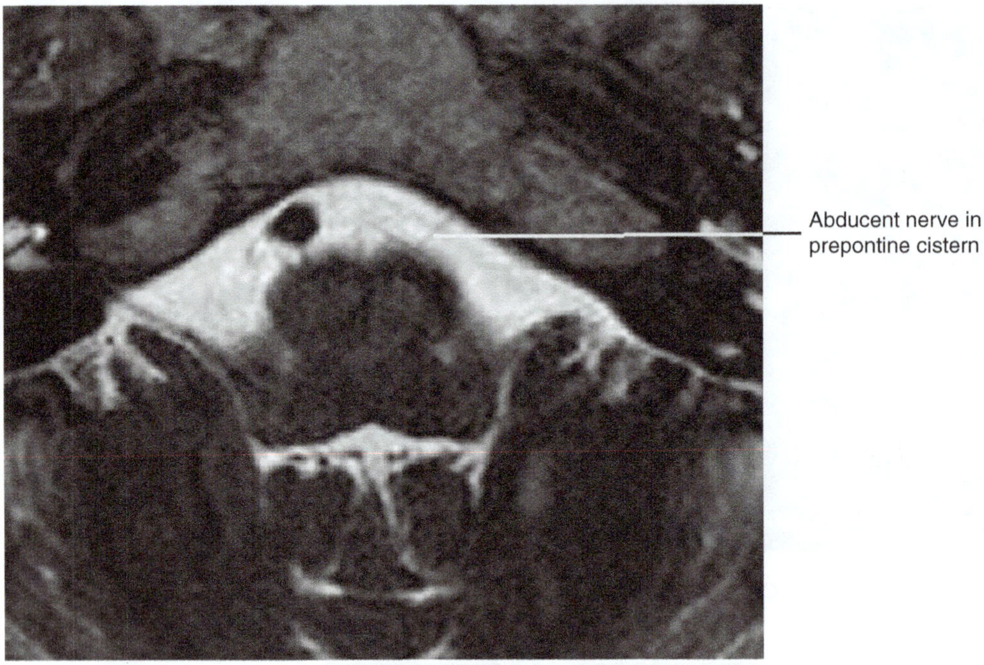

Fig. 6.42 Axial MRI: exit of junction of pons and medulla (bulbopontine sulcus)

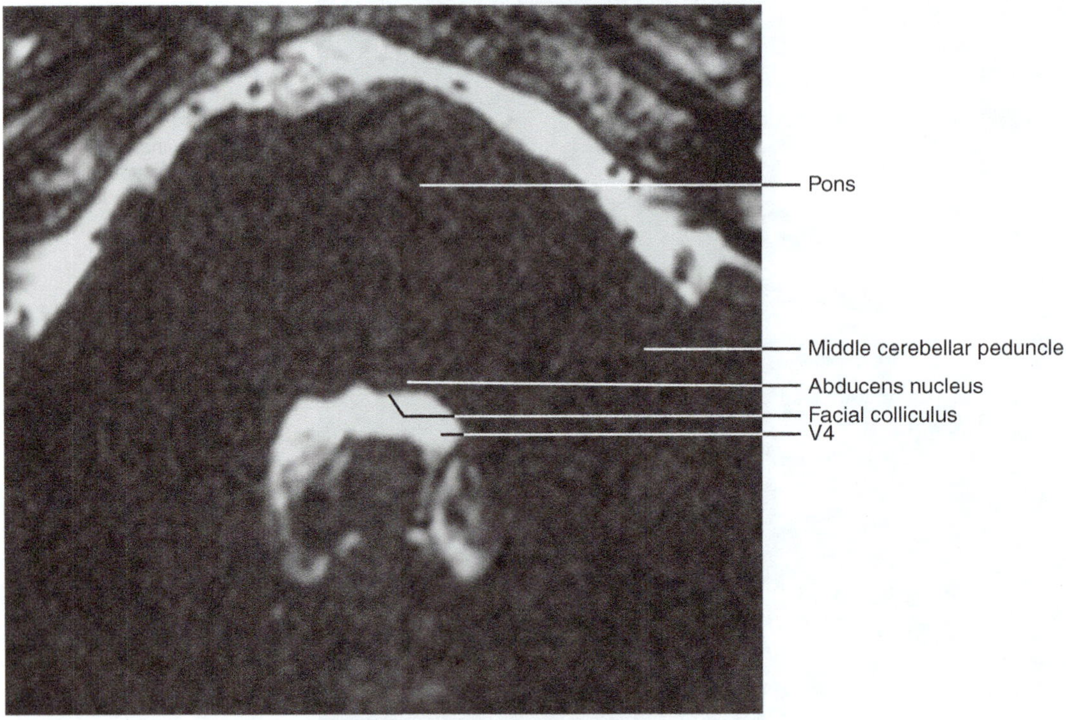

Fig. 6.43 Origin: ventral pons, midline of fourth ventricle

CN VII Facial

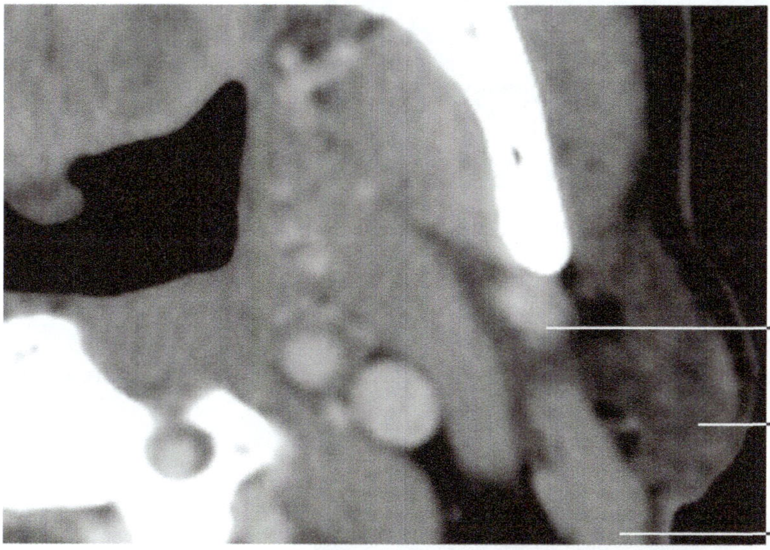

Fig. 6.44 End organ: parotid gland

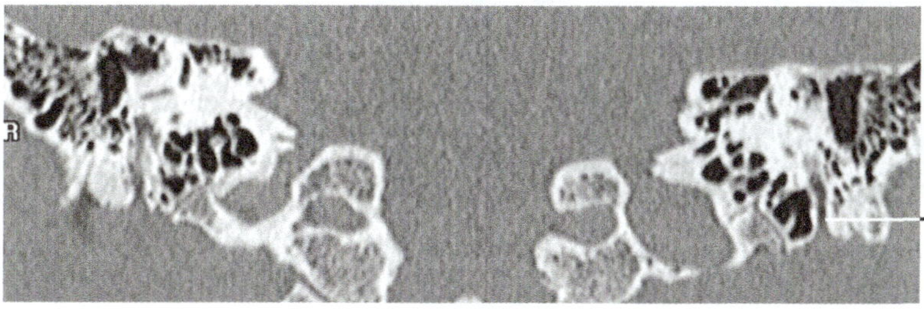

Fig. 6.45 Coronal CT course: exits bony facial canal at stylomastoid foramen

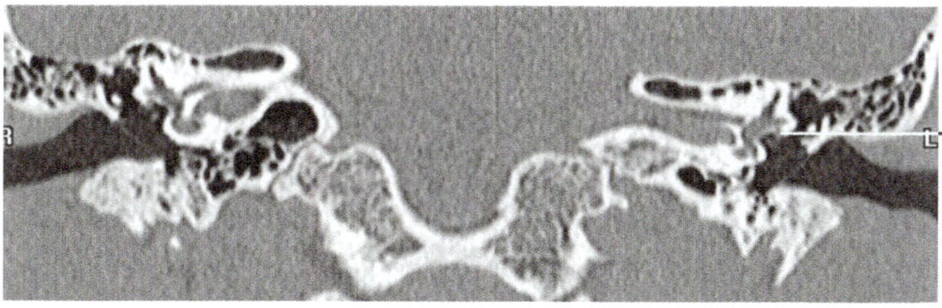

Fig. 6.46 Coronal CT course: horizontal portion of facial nerve canal

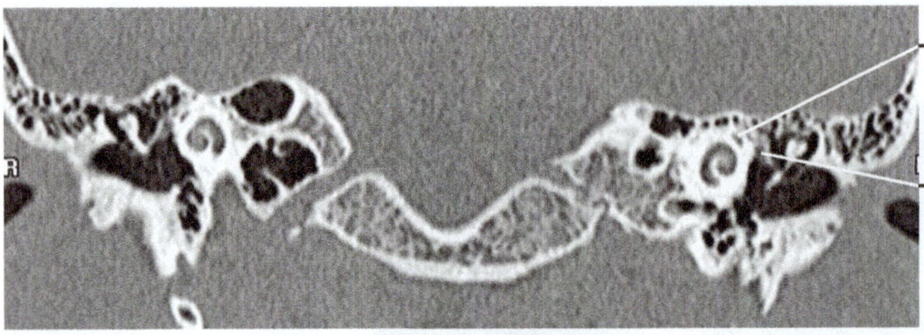

Fig. 6.47 Coronal CT course: genu of facial nerve canal (joined by sensory component from geniculate ganglion)

Pearl

Coronal through tympanic and labyrinthine segments of the facial nerve at the first genu resemble "snake eyes."

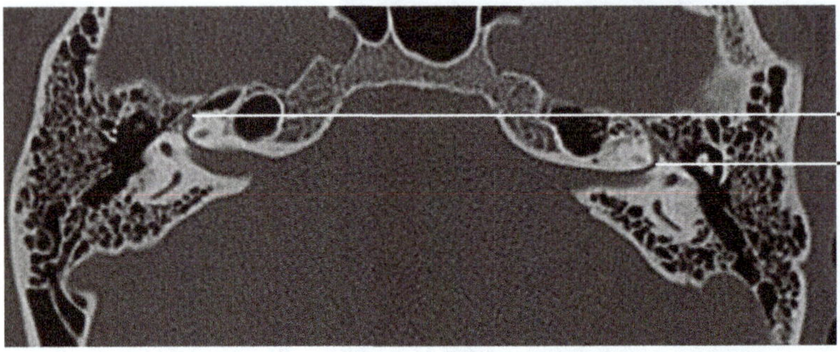

Fig. 6.48 Axial CT course: first part of facial nerve canal above vestibule

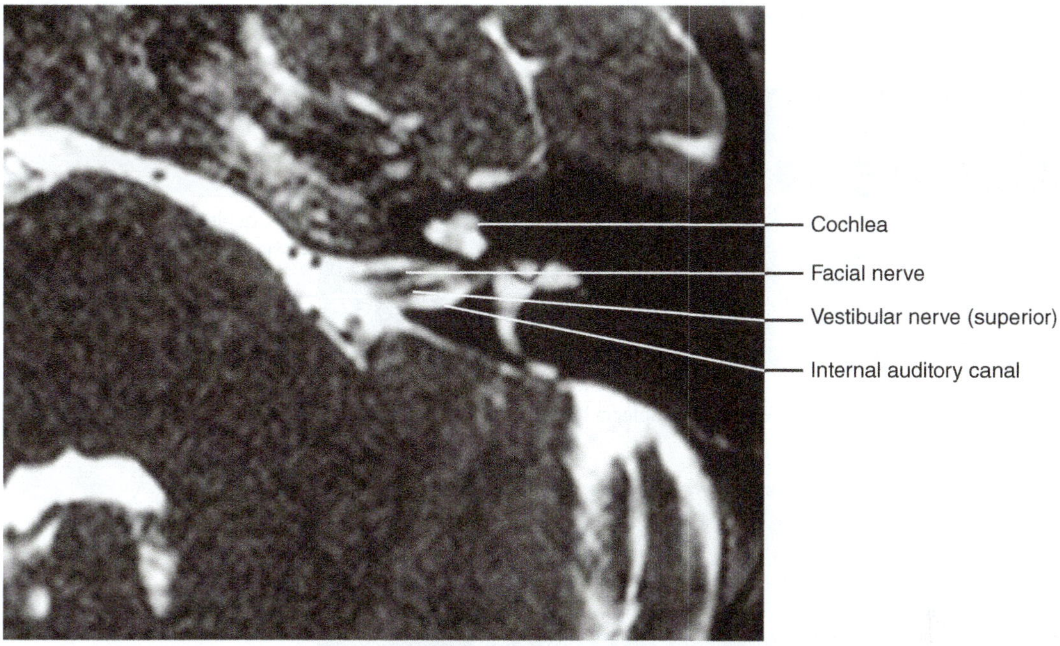

Fig. 6.49 Axial MRI course: internal auditory canal—superior-anterior compartment

Pearl

The upper group of nerves in the internal auditory canal (facial and superior division of vestibular) are distinguished from the lower group (cochlear and inferior division of vestibular) by showing a course toward each other in the axial plane.

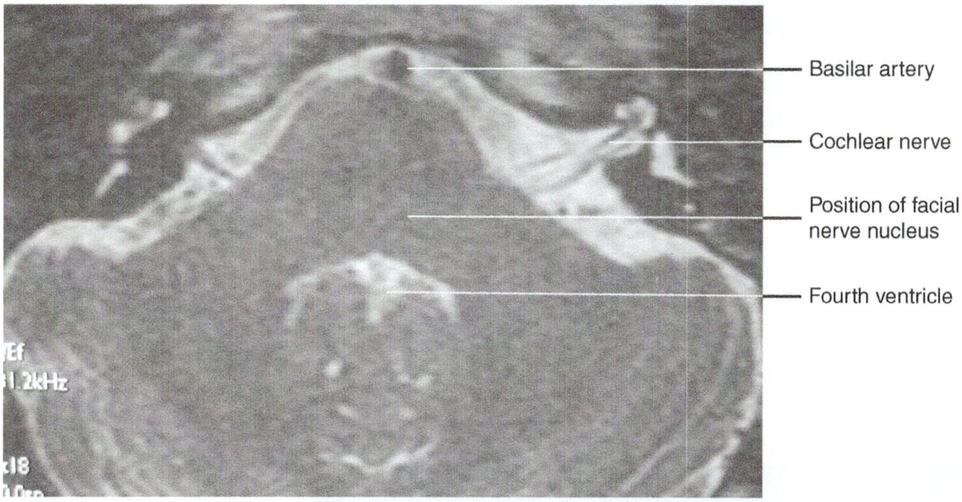

Fig. 6.50 Axial MRI origin: junction lower pons and medulla at internal auditory canals

CN VIII Vestibulocochlear

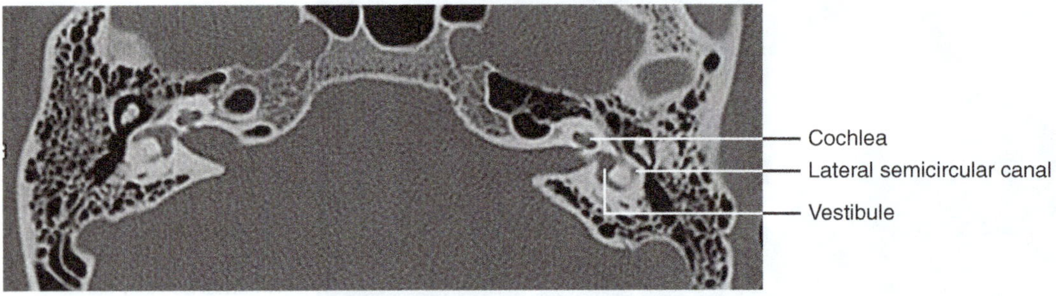

Fig. 6.51 Axial CT end organ. Vestibular: utricle and semicircular ducts

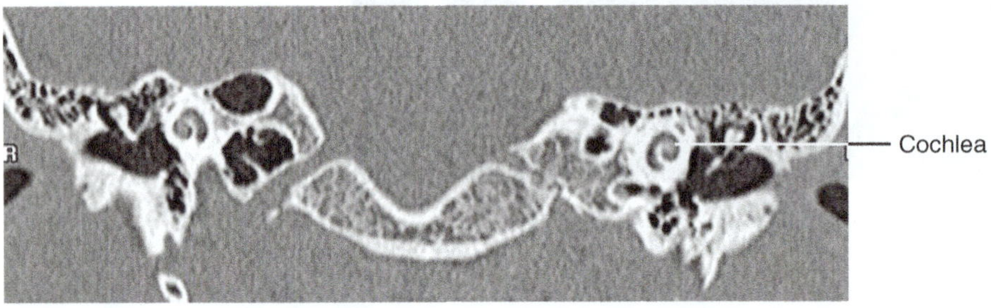

Fig. 6.52 Coronal CT end organ. Cochlear: cochlea and organ of corti

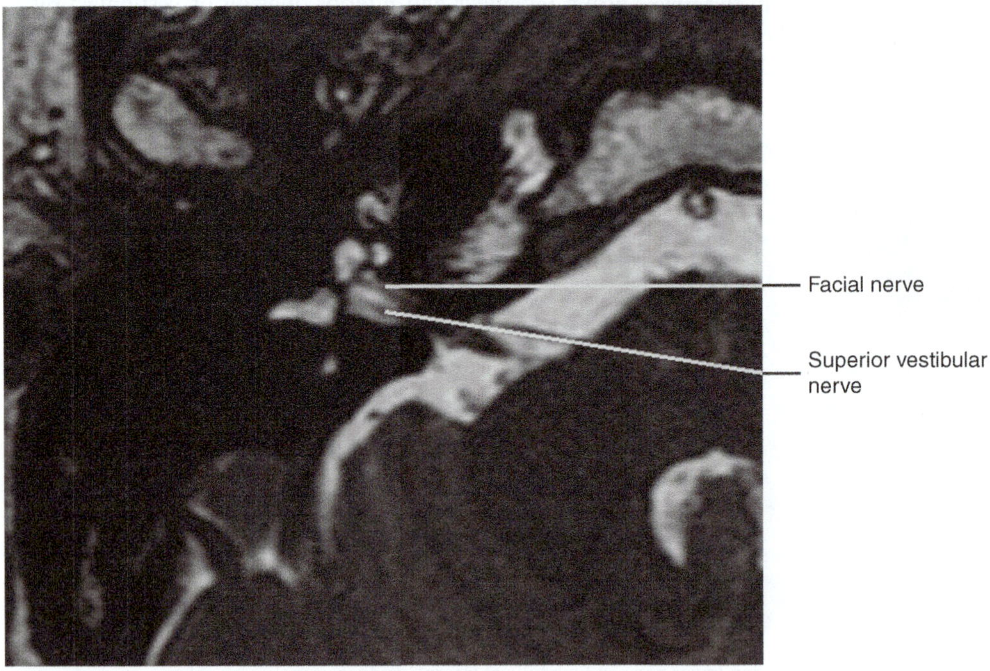

Fig. 6.53 Axial MRI course: crosses cerebellopontine angle cistern to enter internal auditory meatus; superior aspect

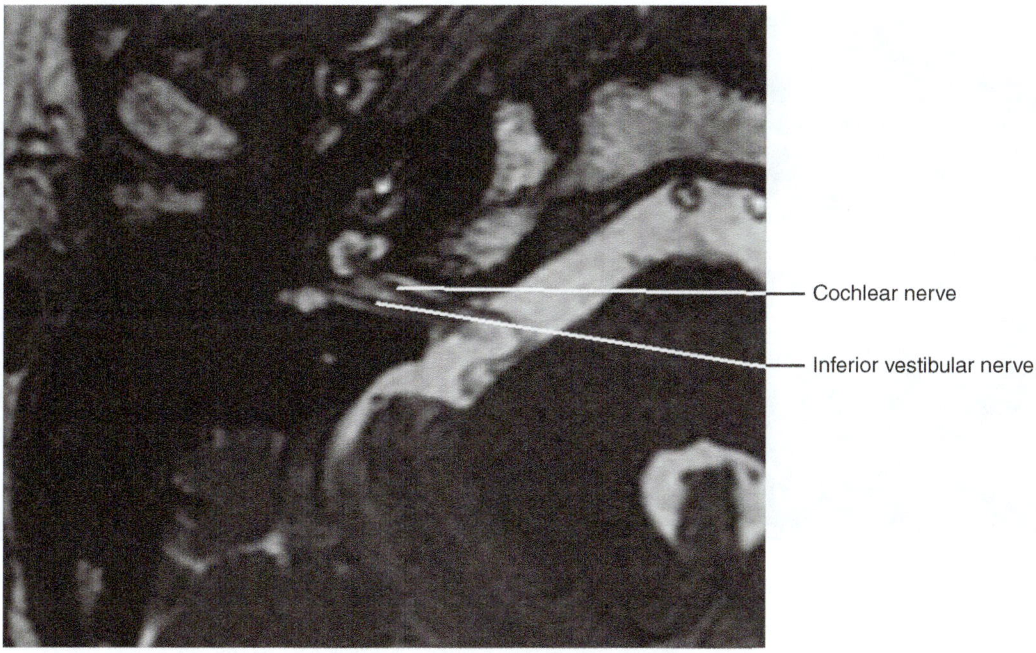

Fig. 6.54 Axial MRI course: inferior aspect of internal auditory canal

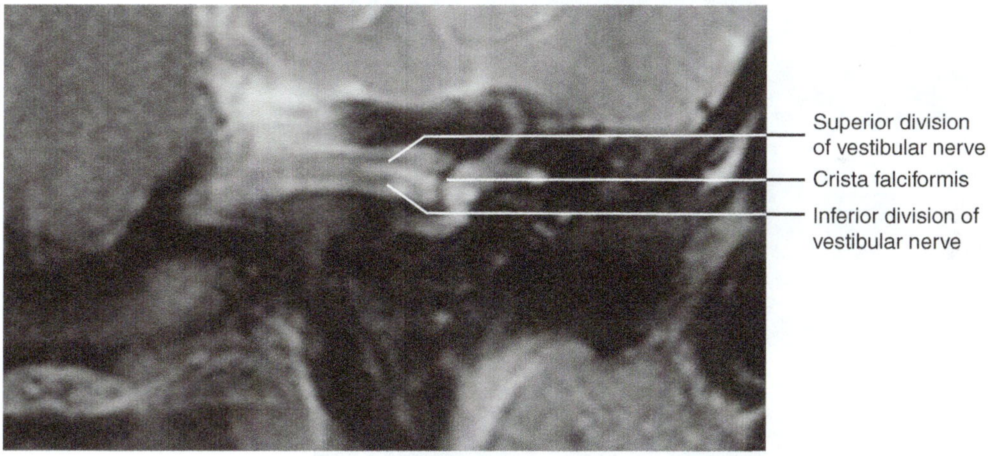

Fig. 6.55 Coronal MRI demonstrating the superior and inferior divisions of the vestibular nerve within the internal auditory canal

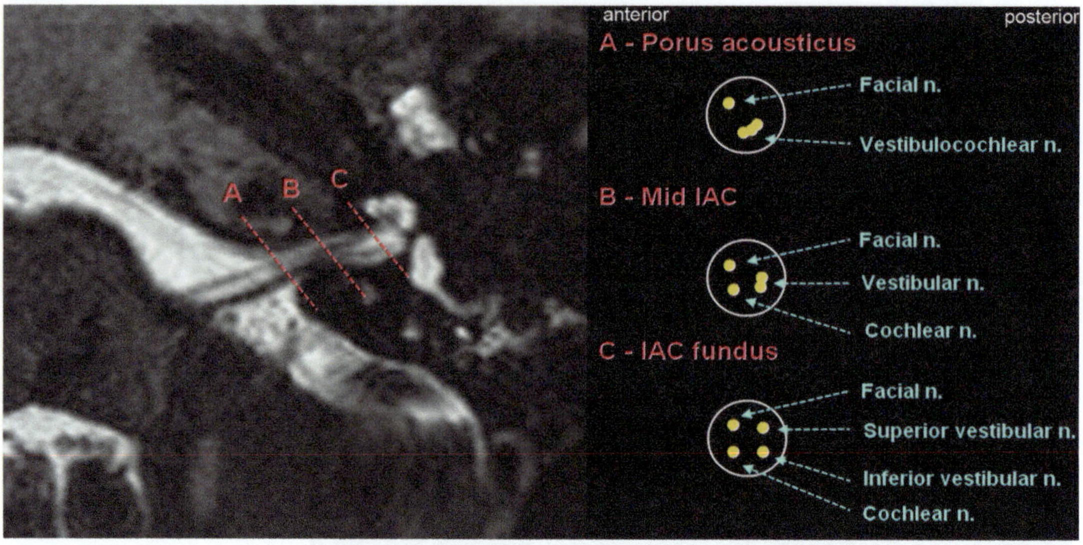

Fig. 6.56 Axial MRI and corresponding sagittal representation of the position of the nerve components in the sagittal oblique. *n* nerve

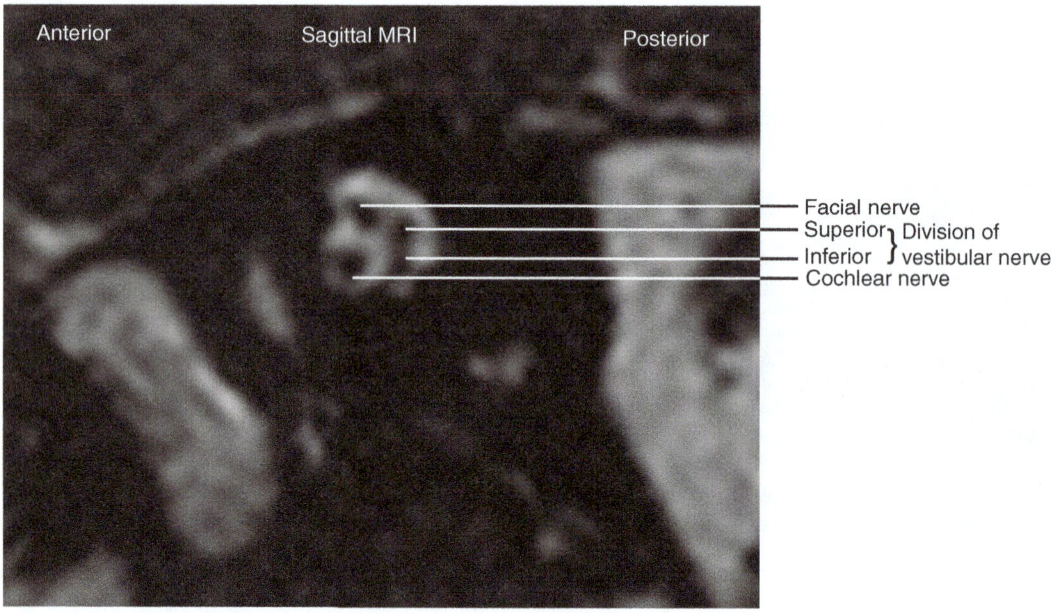

Fig. 6.57 Course. Sagittal MRI internal auditory canal

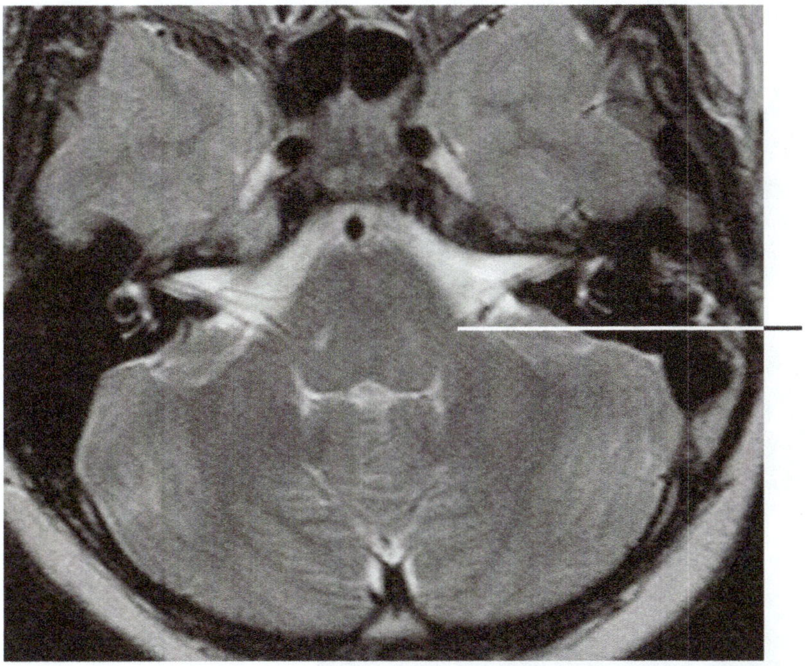

Fig. 6.58 Axial MRI origin. Medulla (next to inferior cerebellar peduncle): cochlear nuclei

CN IX Glossopharyngeal, CN X Vagus, CN XI Accessory

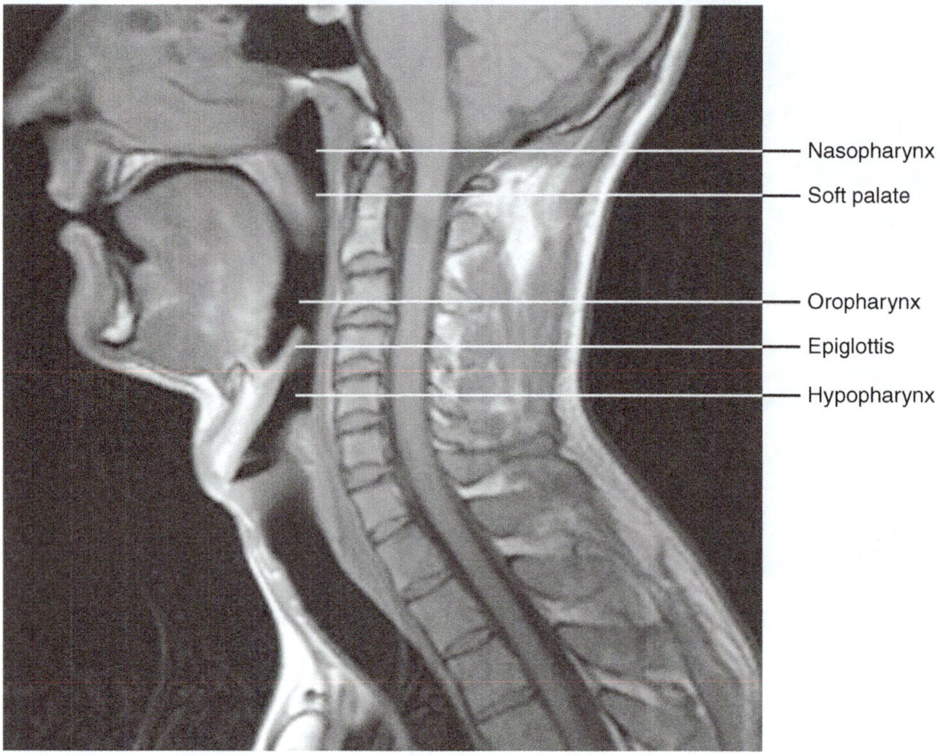

Fig. 6.59 Sagittal MRI end organ IX: oropharynx

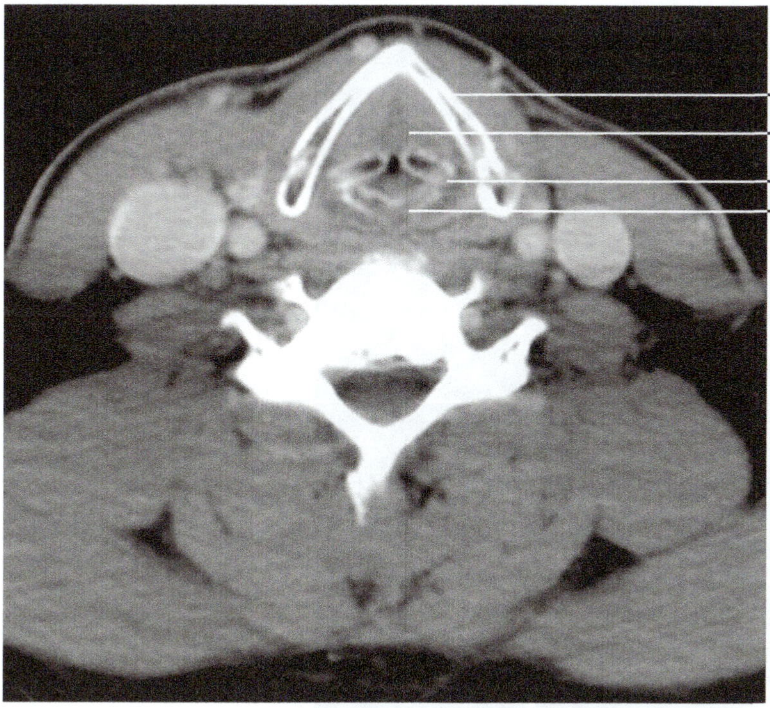

Fig. 6.60 Axial CT end organ X. Right vagus: larynx, pulmonary, oesophageal, cardiac plexus

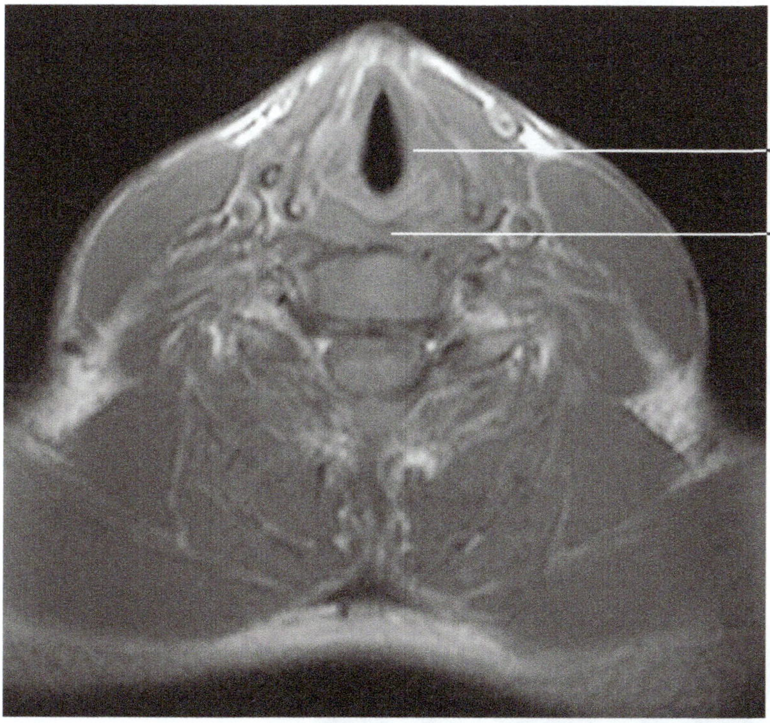

Fig. 6.61 Axial MRI end organ X. Left vagus: larynx, oesophageal plexus, gastrointestinal tract to left colon flexure

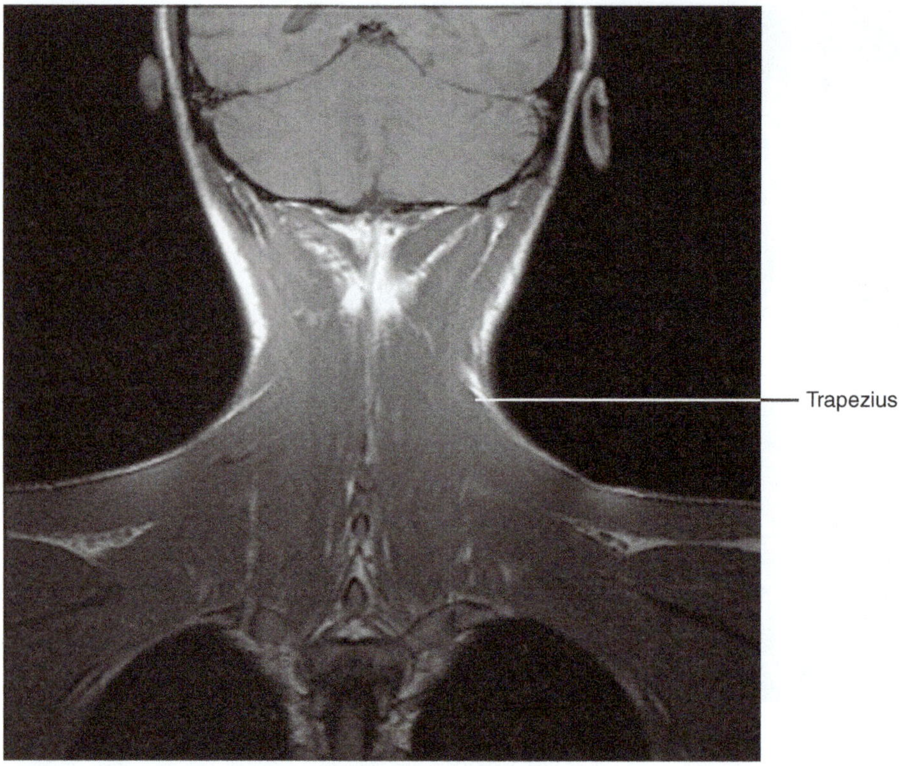

Fig. 6.62 Coronal MRI end organ XI: soft palate, pharynx, trapezius

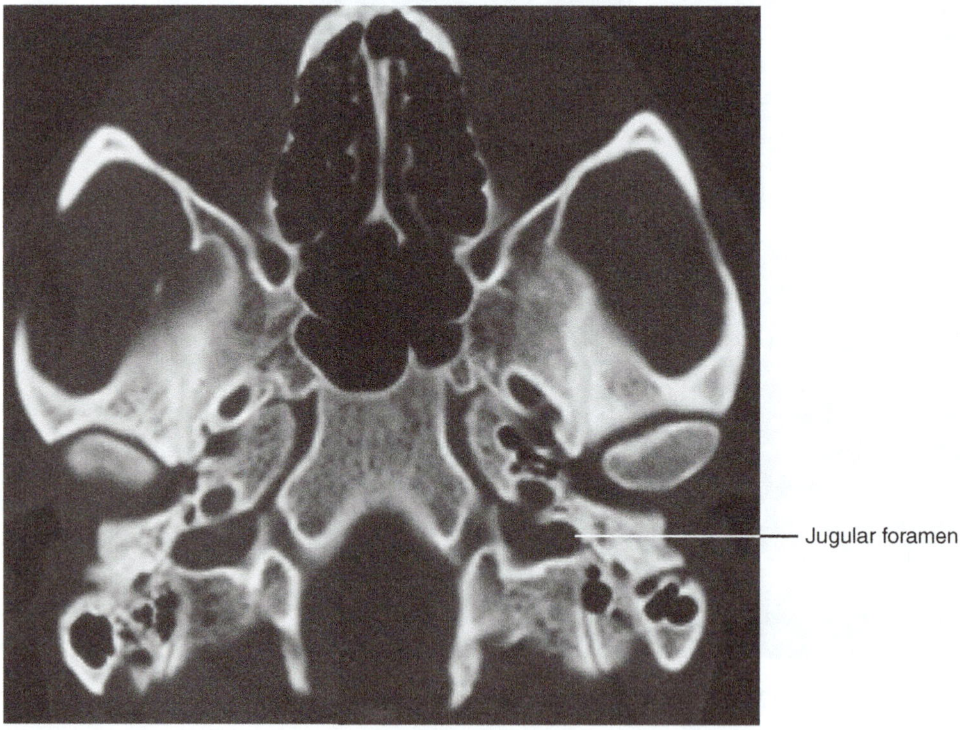

Fig. 6.63 Axial CT course: jugular foramen IX, X, XI

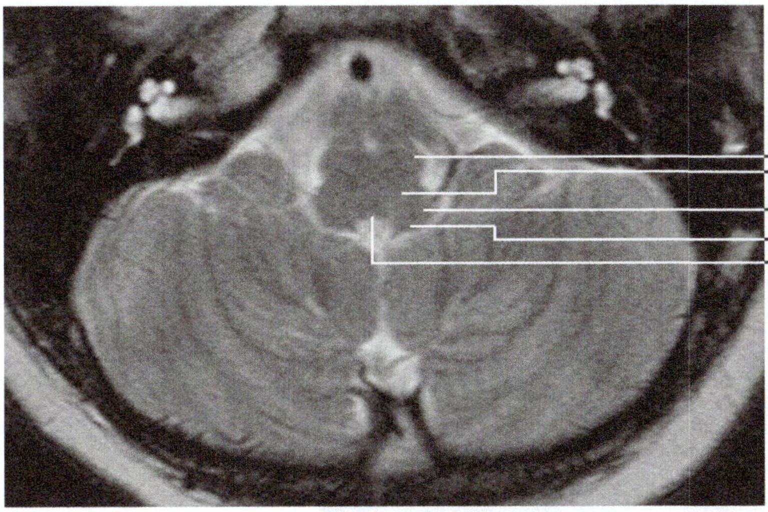

Fig. 6.64 Axial MRI origin: medulla CN IX, X, XI. Nucleus ambiguous, inferior olivary nucleus, solitary nucleus, spinal nucleus

CN XII Hypoglossal

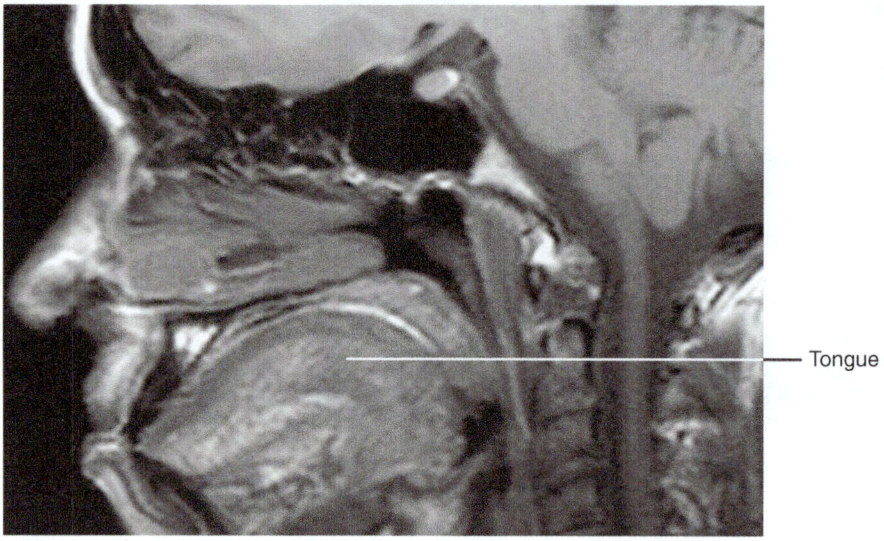

Fig. 6.65 Sagittal MRI end organ: tongue

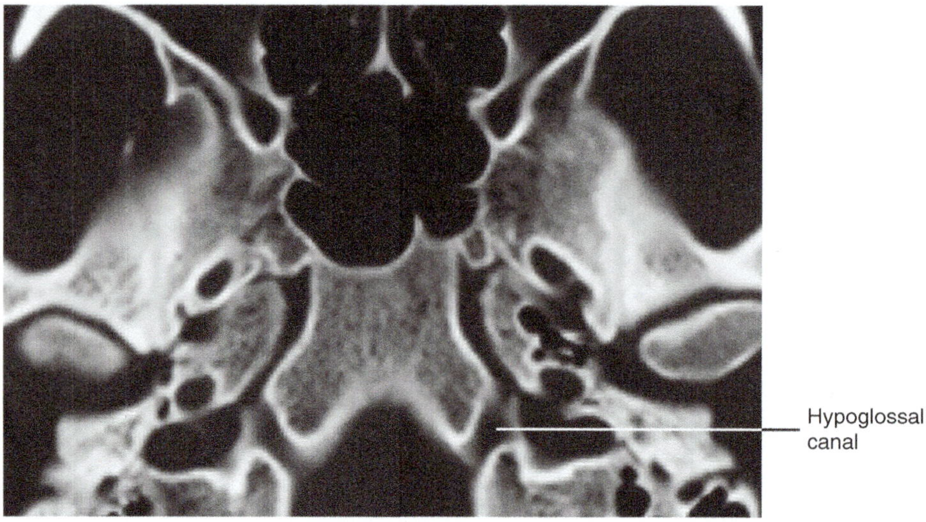

Fig. 6.66 Axial CT course. Hypoglossal canal

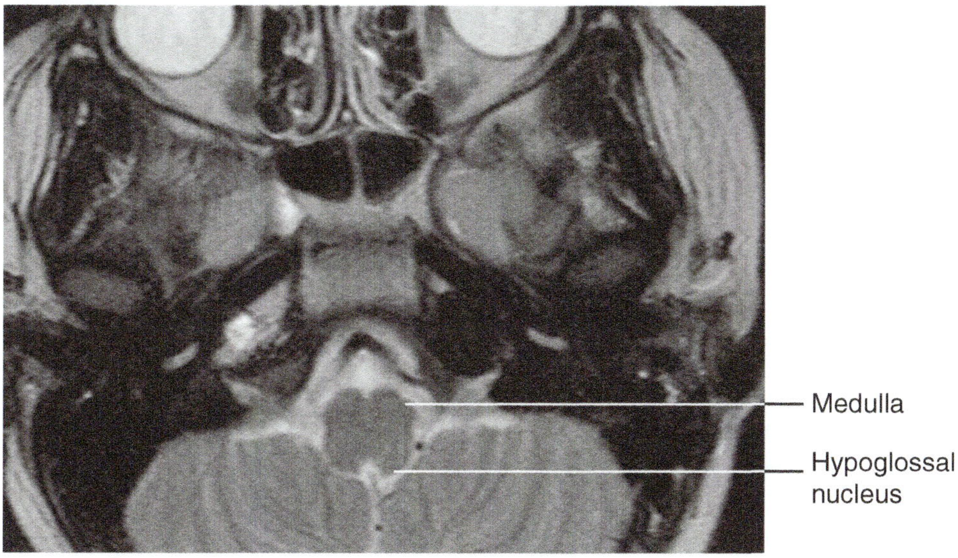

Fig. 6.67 Axial MRI origin: medulla

Acknowledgement Simon McGurk (Edinburgh) and Jan Lotz for contributing images.

7 Spinal Cord

The various levels are demonstrated using multiplanar magnetic resonance imaging (MRI).

Magnetic Resonance Imaging

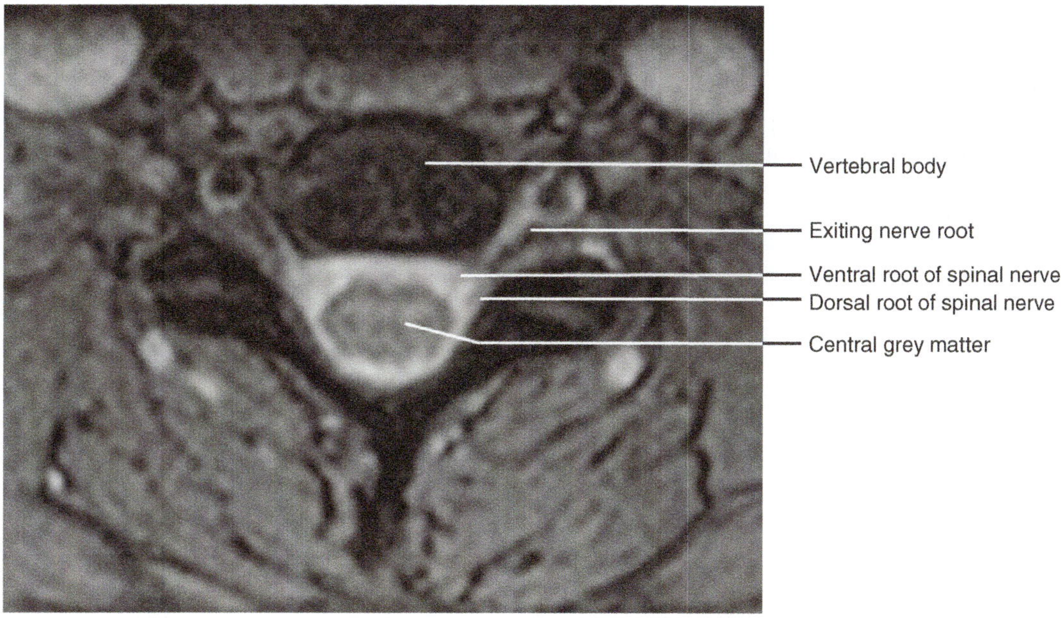

Fig. 7.1 Cervical cord: T2 axial

S. Andronikou (ed.), *See Right Through Me*,
DOI 10.1007/978-3-642-23893-2_7, © Springer-Verlag Berlin Heidelberg 2012

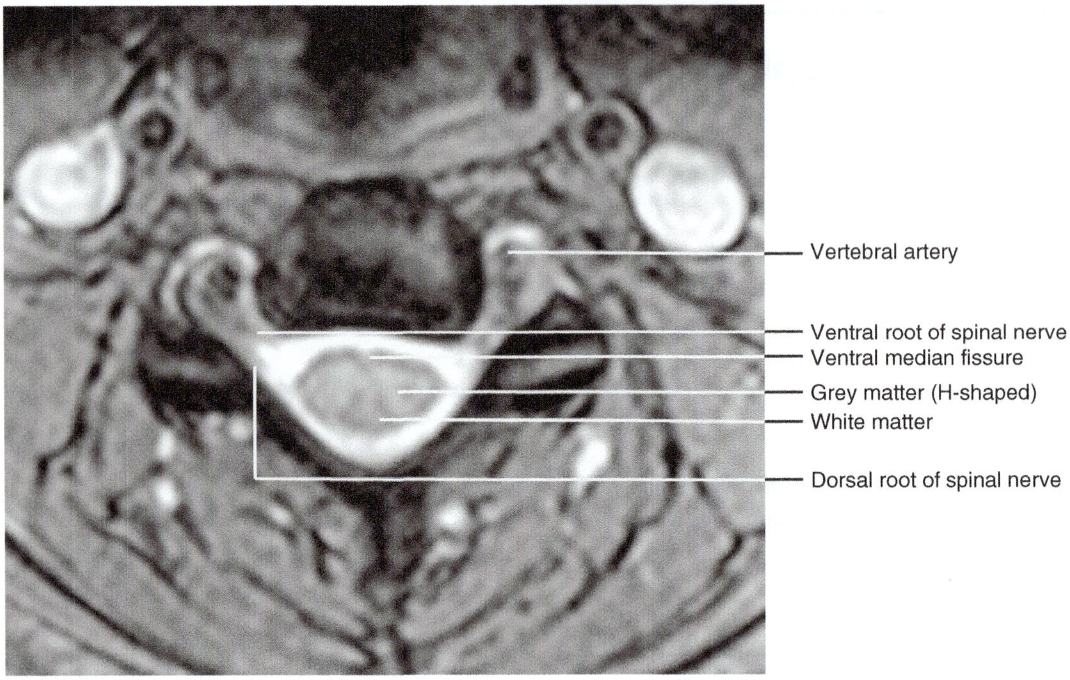

Fig. 7.2 Cervical cord: T2

Pearl

Grey matter of the cord is hyperintense on T2 compared with white matter. The grey matter of the cord is located centrally, with the white matter at the periphery.

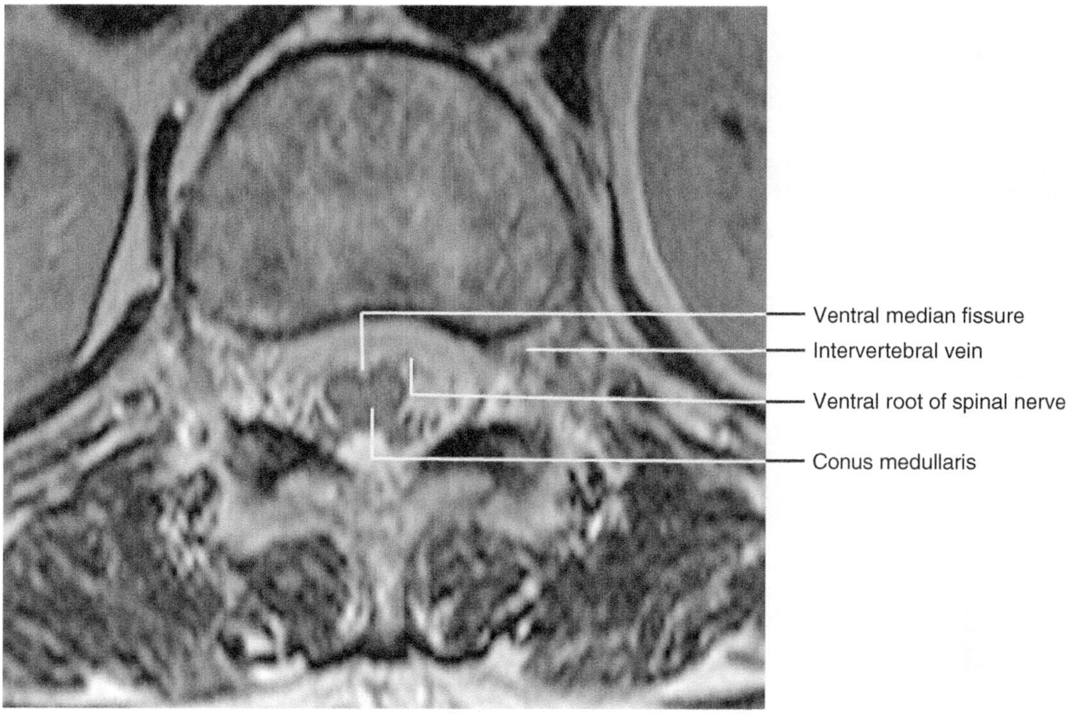

Fig. 7.3 Spinal cord at conus medullaris: T2 axial

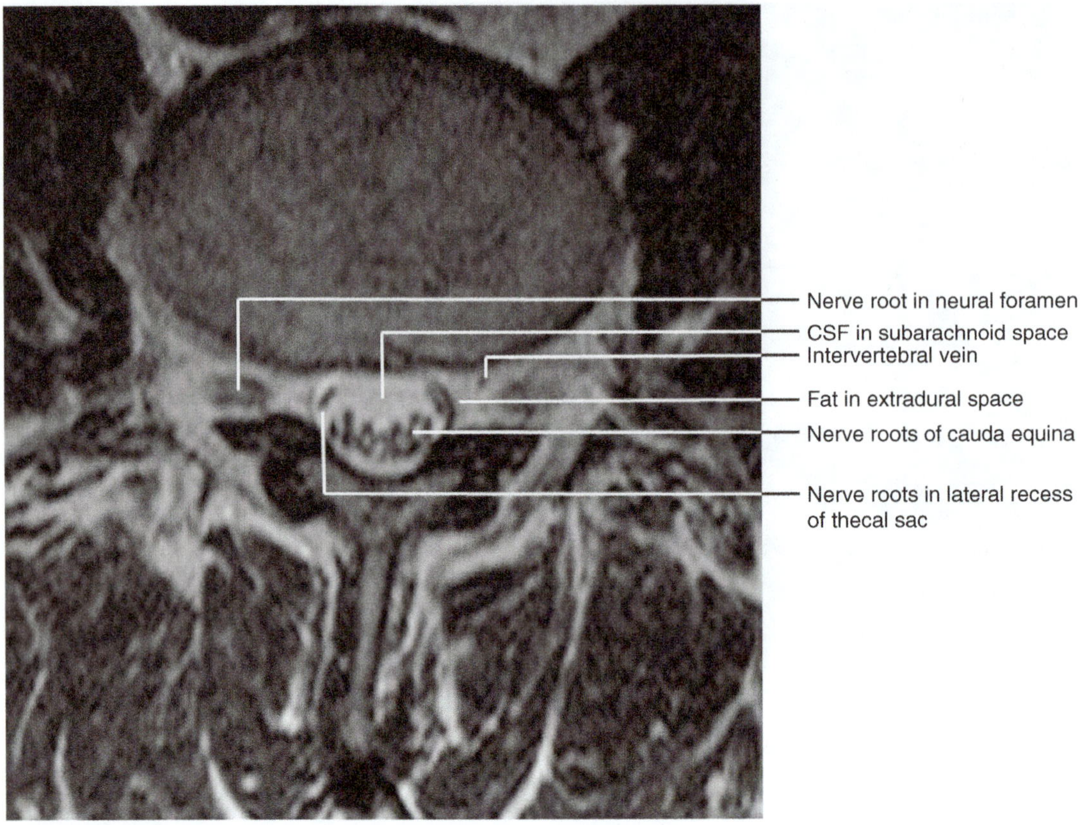

Fig. 7.4 Cauda equina (lumbar level): T2 axial

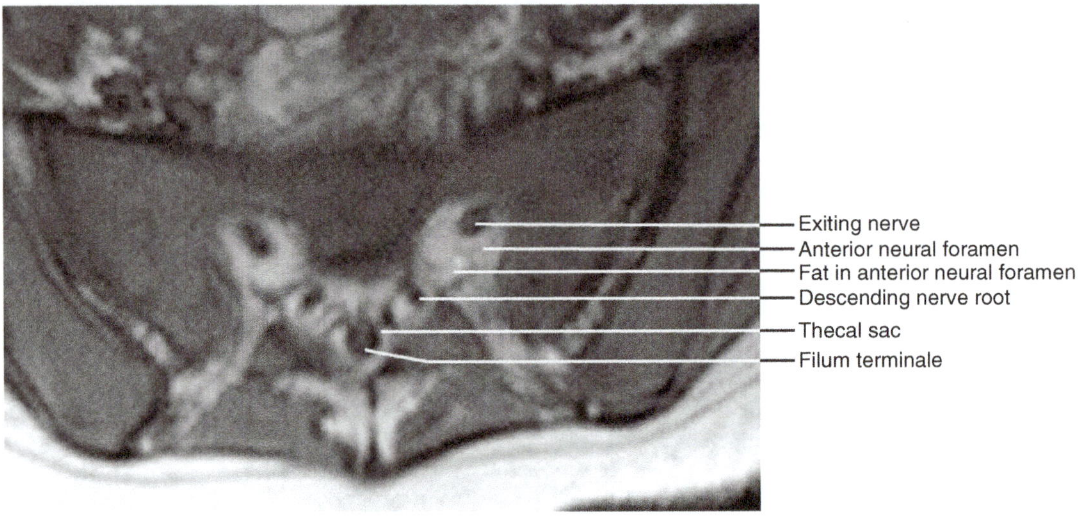

Fig. 7.5 Sacral level: T2 axial

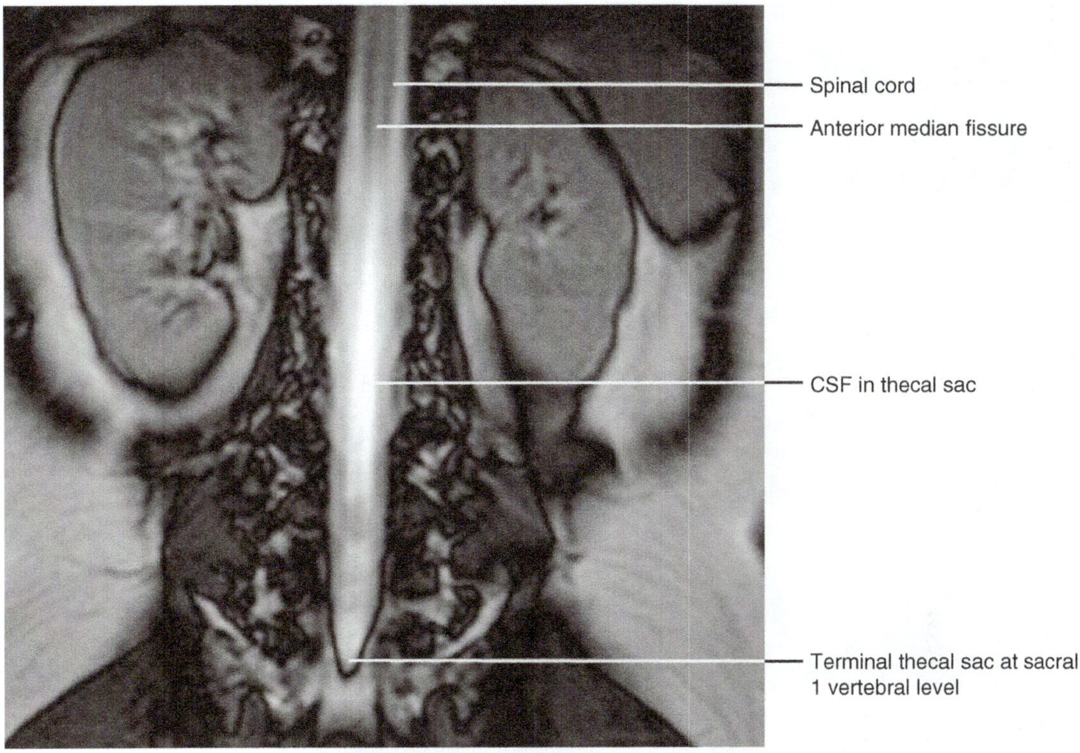

Fig. 7.6 Thoracolumbar level: T2 coronal

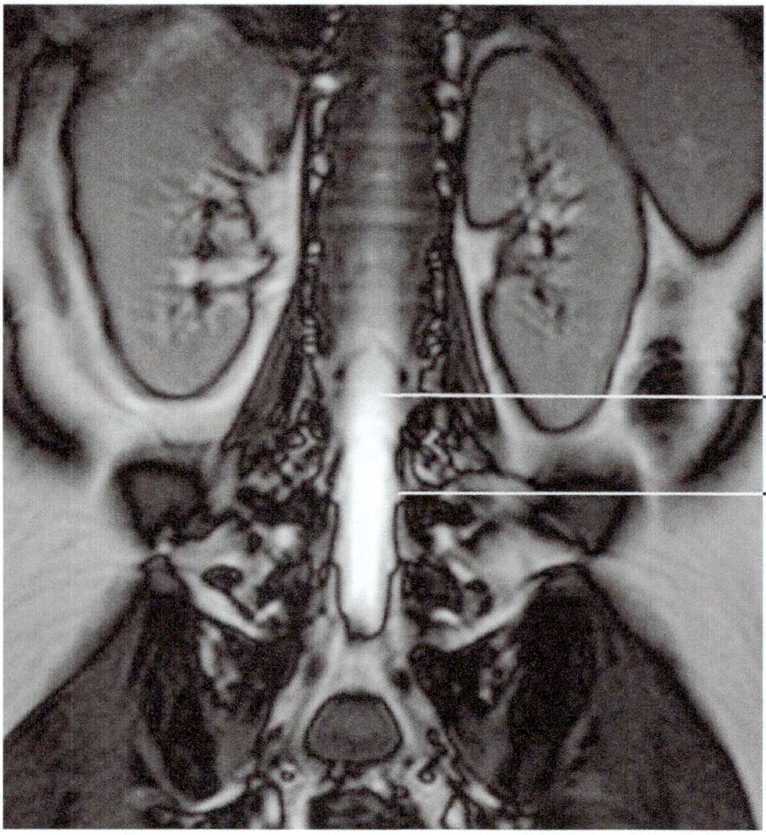

Fig. 7.7 Thoracolumbar level: T2 coronal

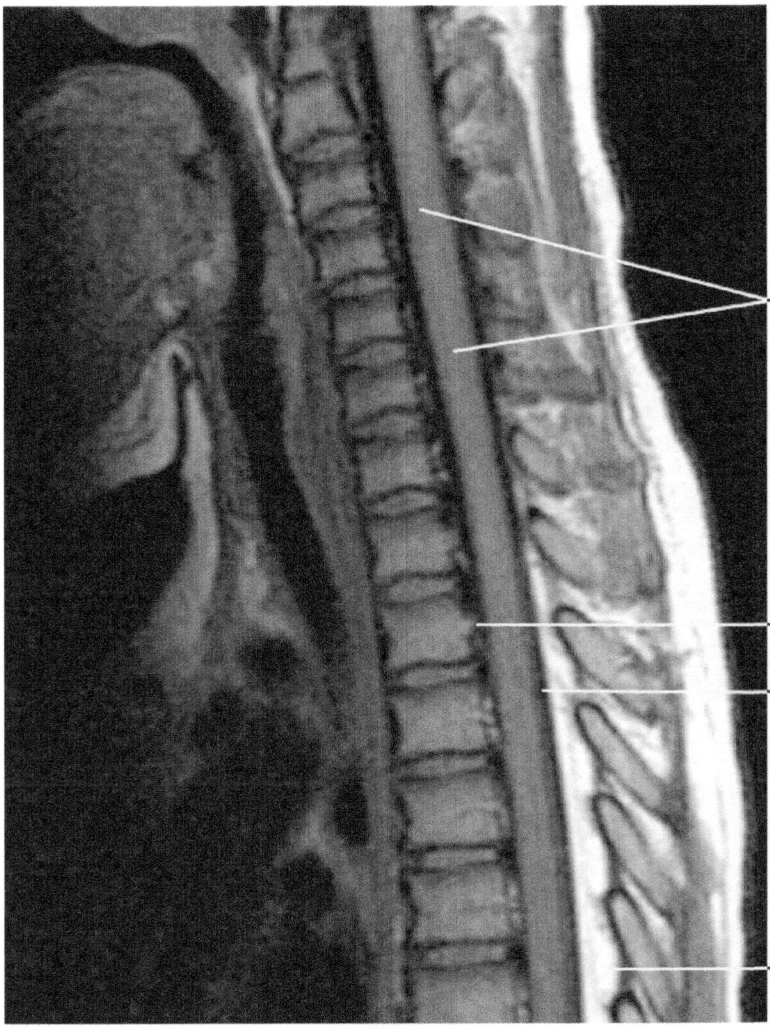

Fig. 7.8 Cervicothoracic cord: T1 sagittal

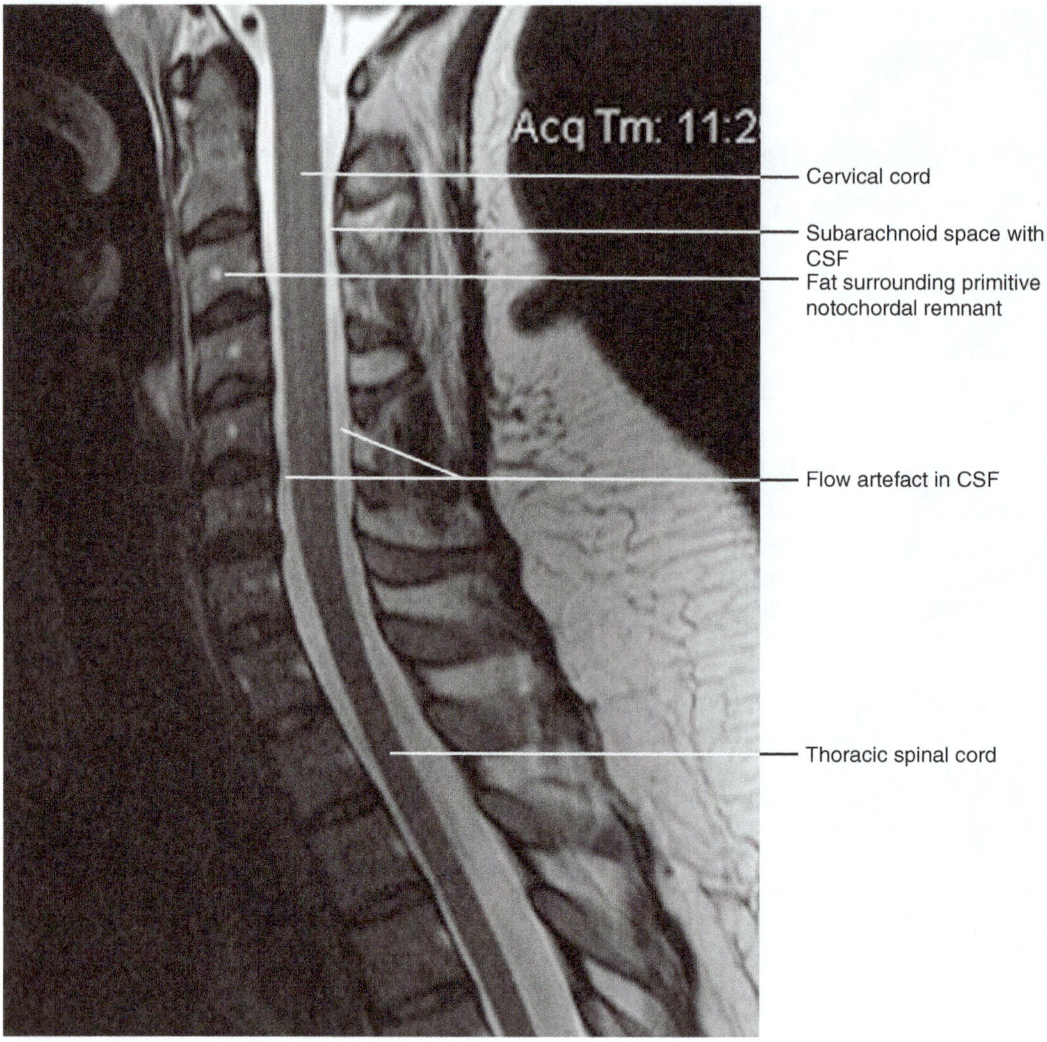

Fig. 7.9 Cervicothoracic cord: T2 sagittal

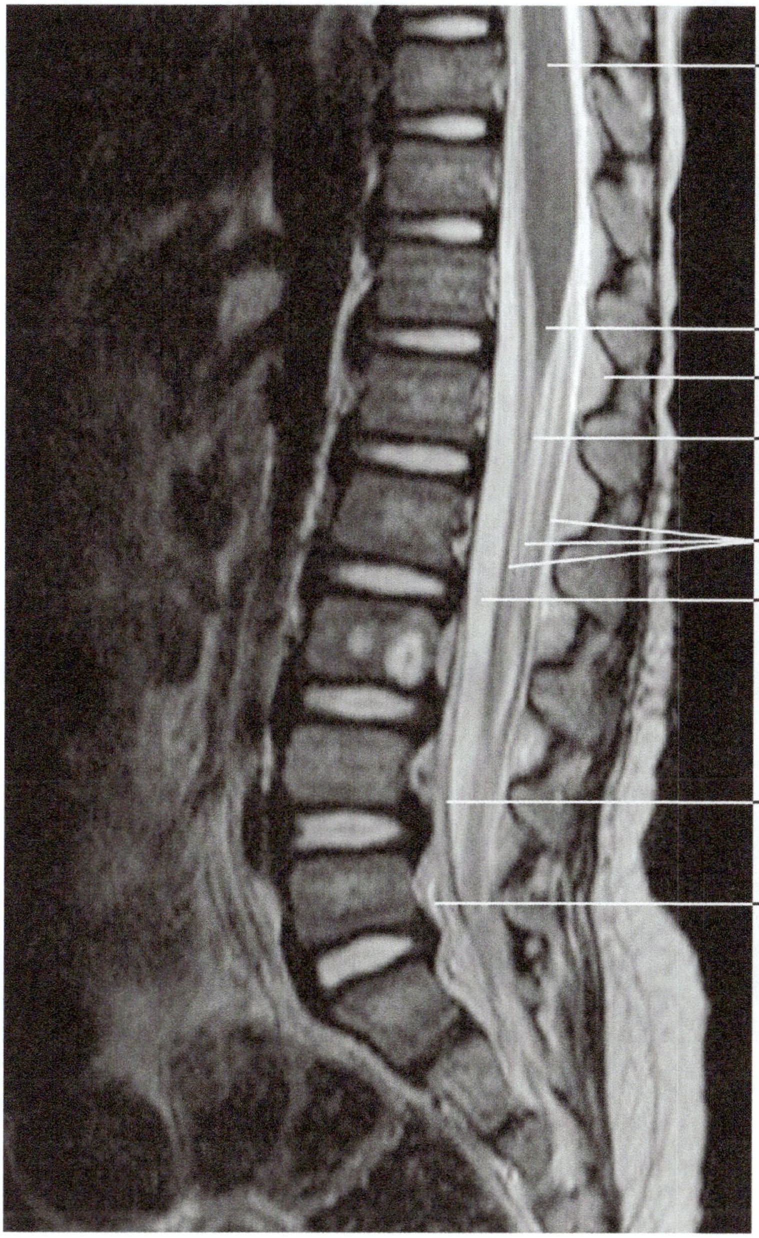

Fig. 7.10 Lumbarsacral level: T2 sagittal

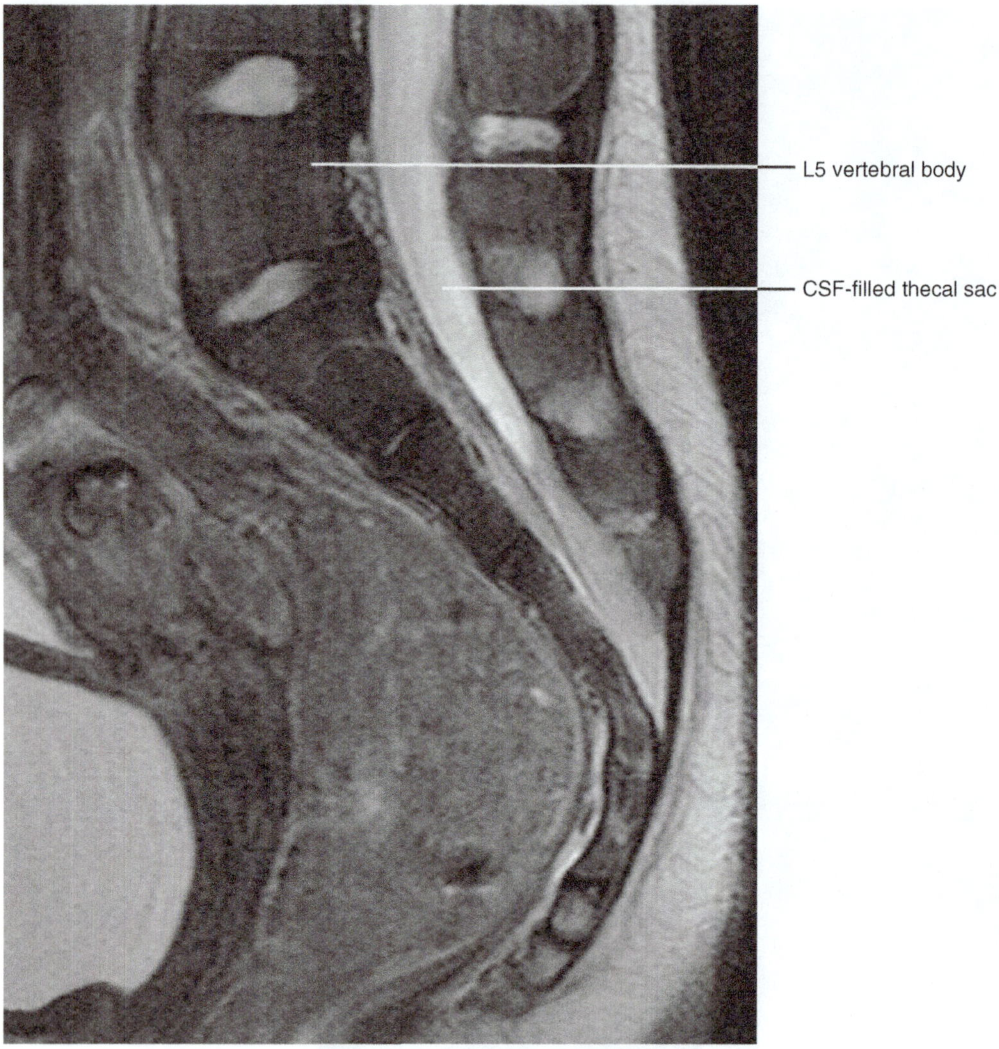

Fig. 7.11 Sacral level: T2 sagittal

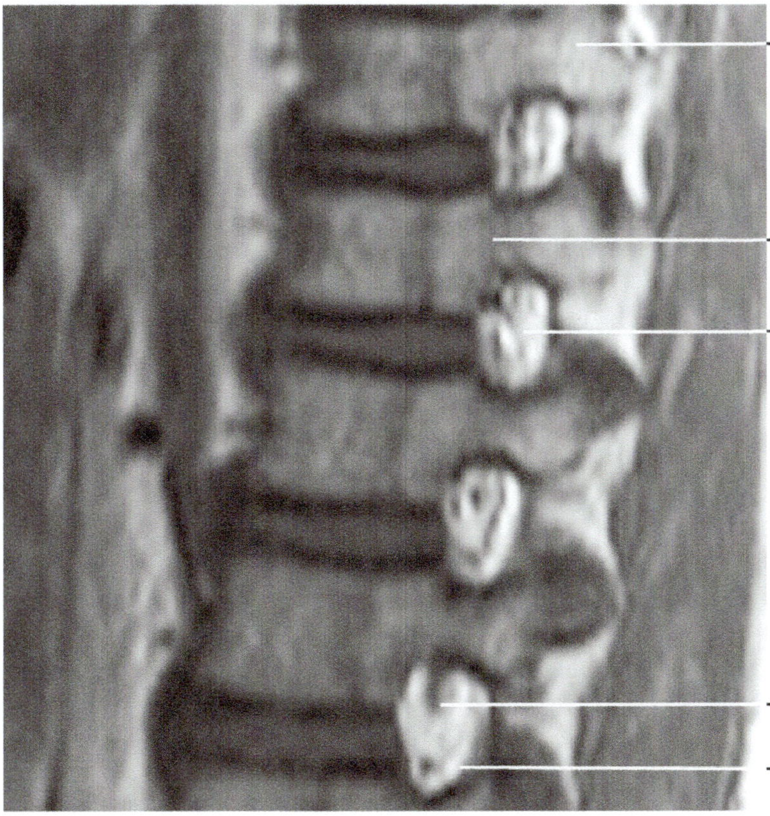

Fig. 7.12 Thoracolumbar level: parasagittal T2

Part II

Head and Neck

This section ecnompasses the head and neck regions seperately from the brain. The skull is included here rather than with the axial skeletal structure because it has significant foramina and fissures which relate to the exit and entry of nerves and vessels but also relate to specific spaces of the neck that are important with regard to disease spread.

The eyes and ears are detailed areas that require specialised knowledge and imaging techniques. In particular the ear structures are intricate and have complex anatomy which mus be known before attempting to identify any pathology.

This chapter also presents a novel endocrine section. The thyroid and parathyroid glands give us liberty to present this section under 'head and neck' but the pituitary gland is seen when performing CNS imaging while the adrenal glands could possibly be adressed with the urogenital and abdominal imaging.

8 Paranasal Sinuses

The paranasal sinuses are air cavities situated within the cranial and facial bones. The apertures of these cavities communicate with the nasal cavity. There are four paired paranasal sinuses:

- **Maxillary antra**, draining via the maxillary ostia into the middle meatus at the ostiomeatal complex.
- **Ethmoidal air cells**, consisting of:
  - Anterior and middle groups, draining into the middle meatus.
  - Posterior group, draining into the superior meatus.
- **Sphenoidal sinus**, draining into the sphenoethmoidal recess situated in the superior meatus.
- **Frontal sinus**, draining into the middle meatus posteriorly.

The nose consists of the cartilaginous septum, the vomer and perpendicular plate of the ethmoid, and the central nasal cavity. There are paired superior, middle, and inferior turbinates. The nasolacrimal duct drains at the inferior meatus. The paranasal sinuses and nasal cavity can be imaged using plain film, computed tomography (CT), and magnetic resonance imaging (MRI).

Plain Film Imaging

The sinuses can be visualised using all standard radiographic views of the skull but dedicated imaging usually entails a posteroanterior (PA), lateral and occipitomental (OM) 45° (Waters') view.

Computed Tomography Imaging

With multidetector CT, imaging is usually performed in the axial plane with reconstructions in coronal and sagittal planes. This is the best method of demonstrating a bony anatomical "roadmap" for clinical applications.

Magnetic Resonance Imaging

This is the optimal imaging technique for demonstrating mucosal abnormalities and adjacent soft tissue anatomy and pathology.

S. Andronikou (ed.), *See Right Through Me*,
DOI 10.1007/978-3-642-23893-2_8,

Plain Film

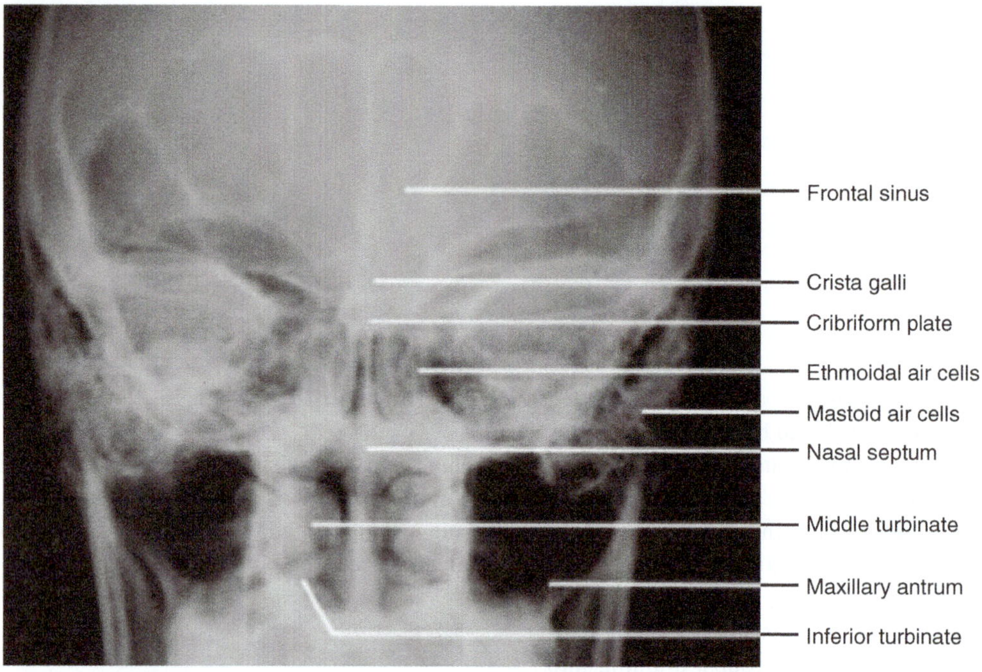

Fig. 8.1 Posteroanterior skull radiograph

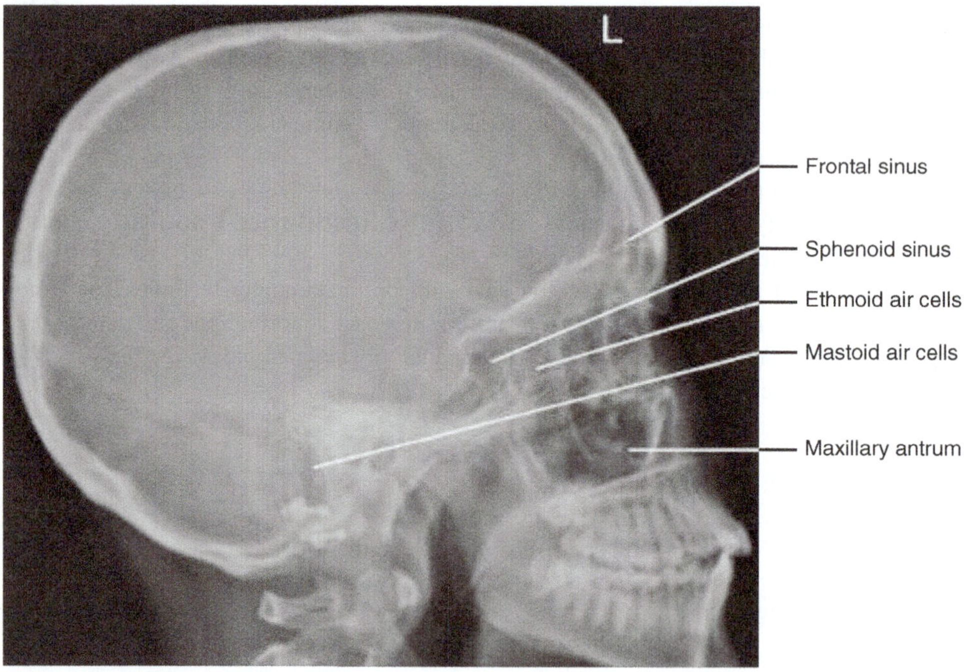

Fig. 8.2 Lateral skull radiograph

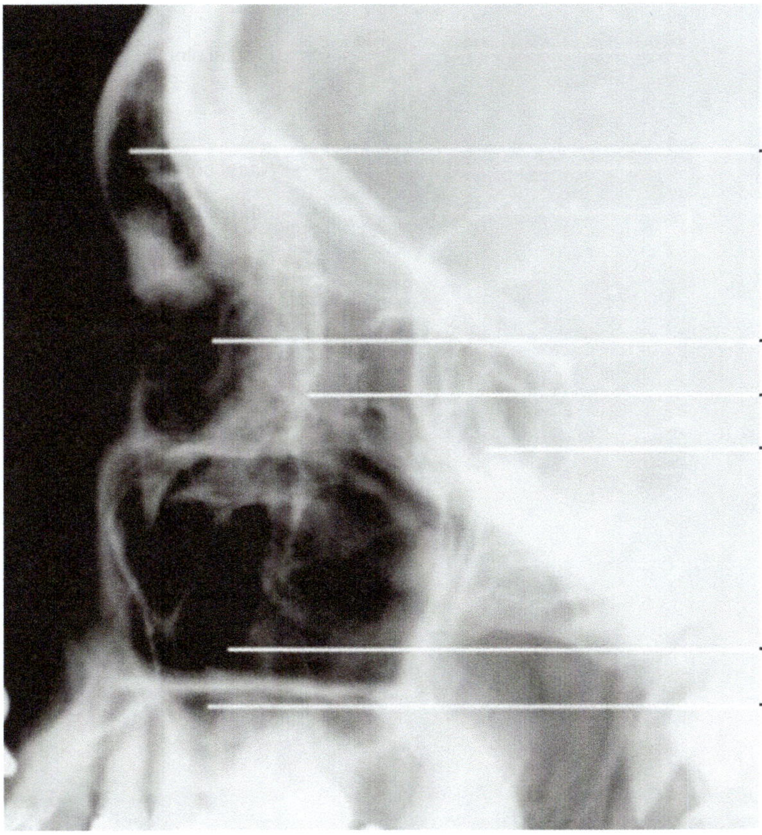

Fig. 8.3 Lateral skull radiograph

Pearl

Composite shadowing of the coronoid process of the mandible can mimic a nasopharyngeal mass lesion on a lateral skull radiograph.

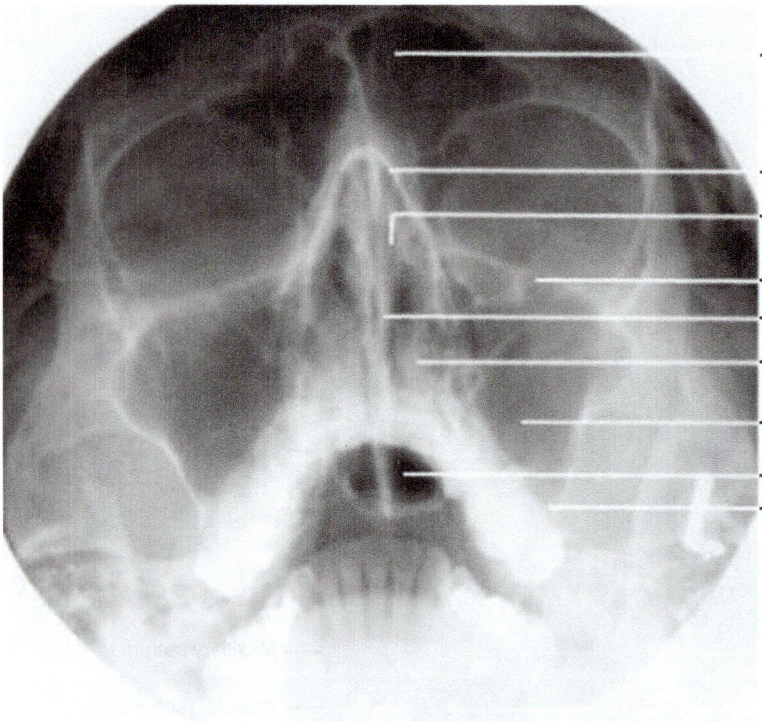

Fig. 8.4 Water's view (OM 45°)

Pearl
Sphenoidal sinus is optimally visualized through the open mouth on a Waters' view.

Computed Tomography

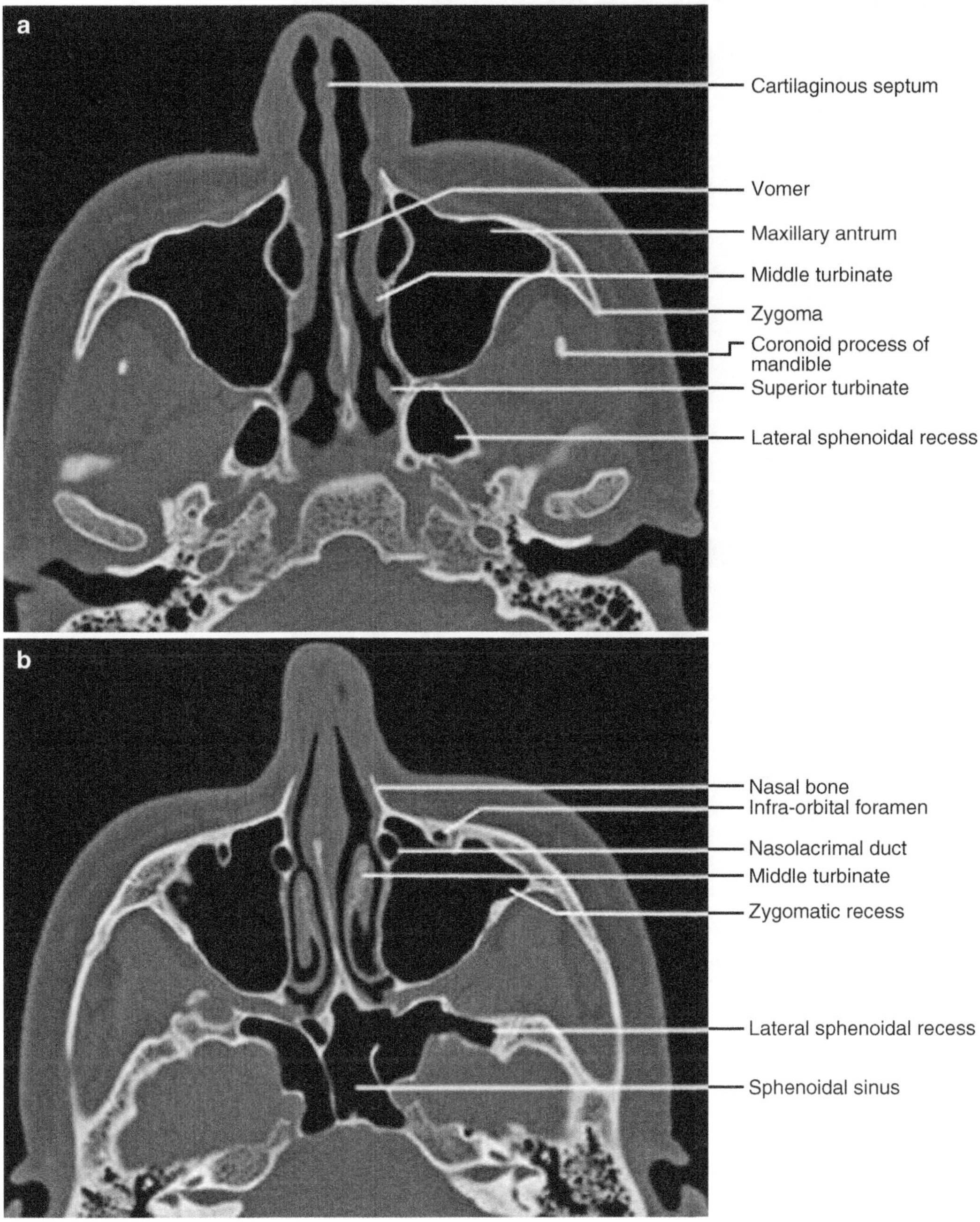

Fig. 8.5 Axial CT (bony window). (**a**–**d**) Sequential images from inferior to superior

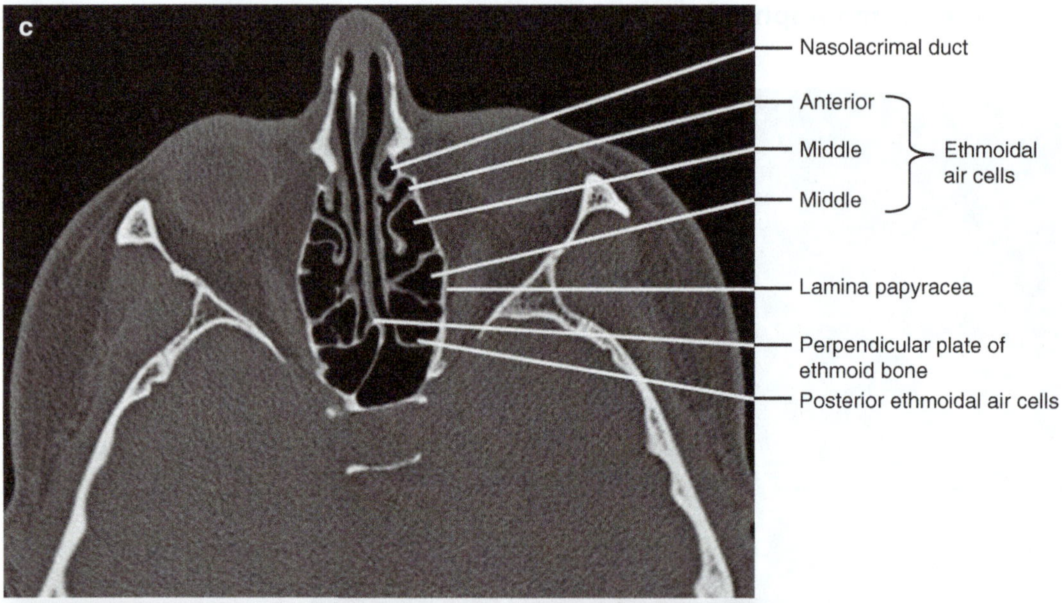

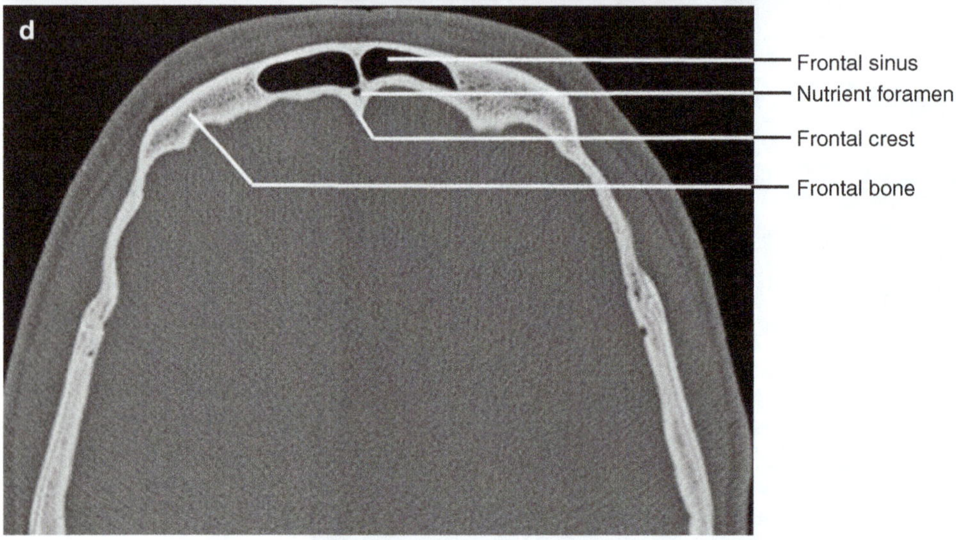

Fig. 8.5 (continued)

Pearl

Coronal bony CT images are vital for planning of endoscopic sinus surgery.

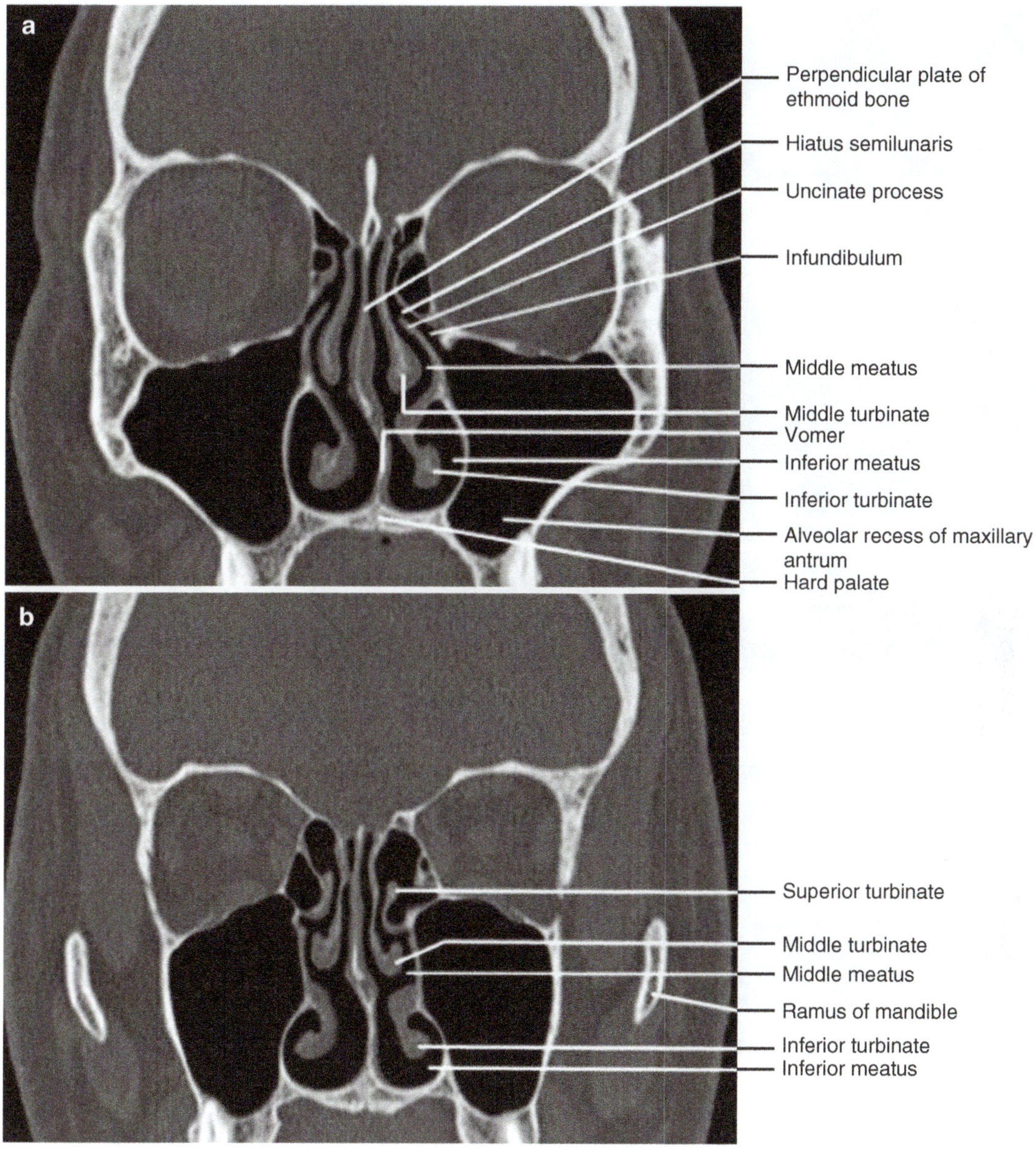

Fig. 8.6 Coronal CT (bony window). (**a**–**e**) Sequential images from anterior to posterior

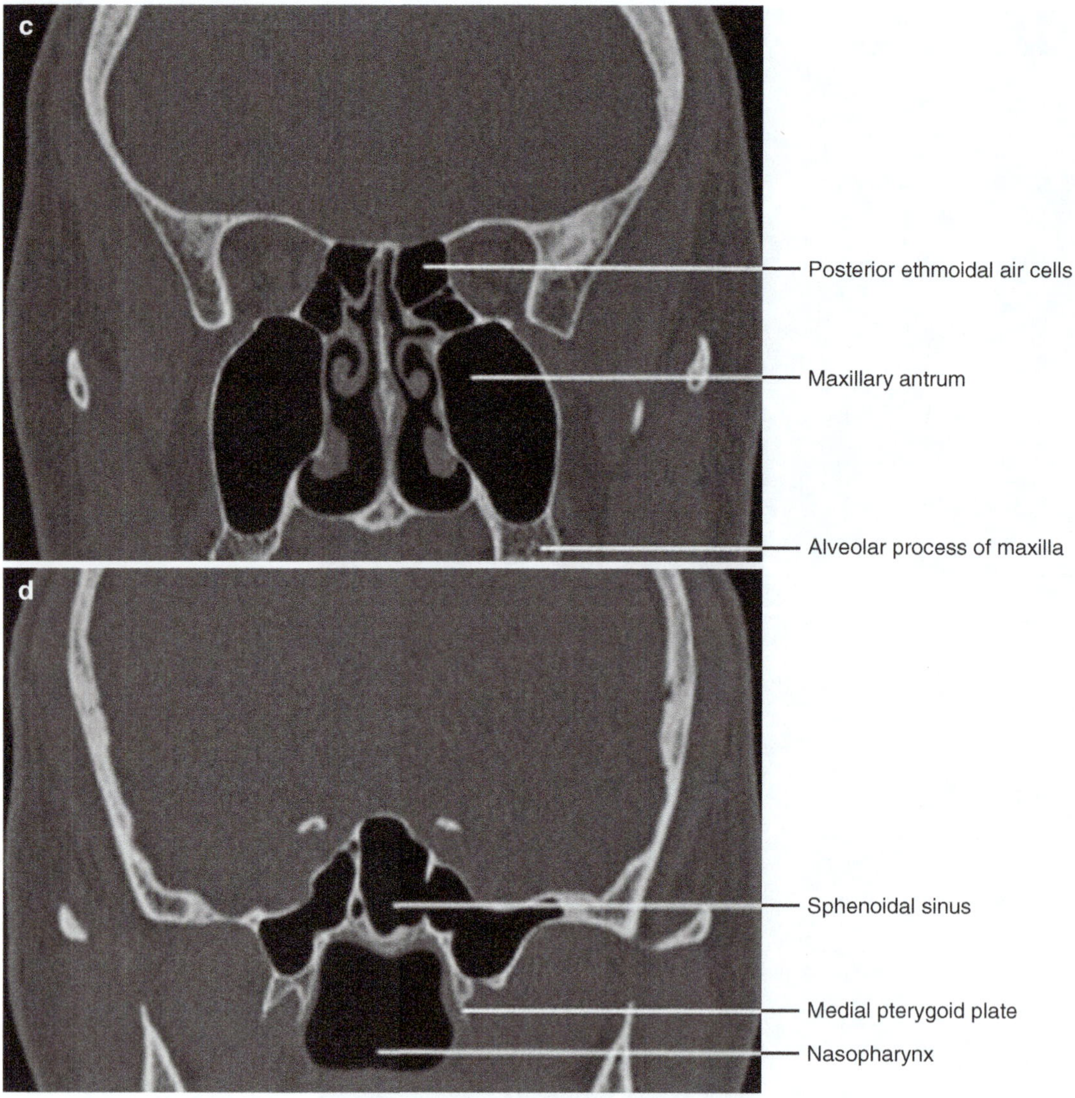

Fig. 8.6 (continued)

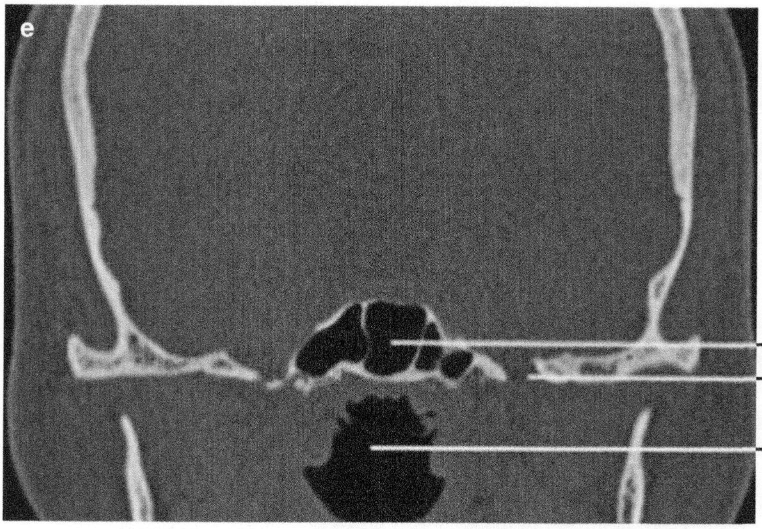

Fig. 8.6 (continued)

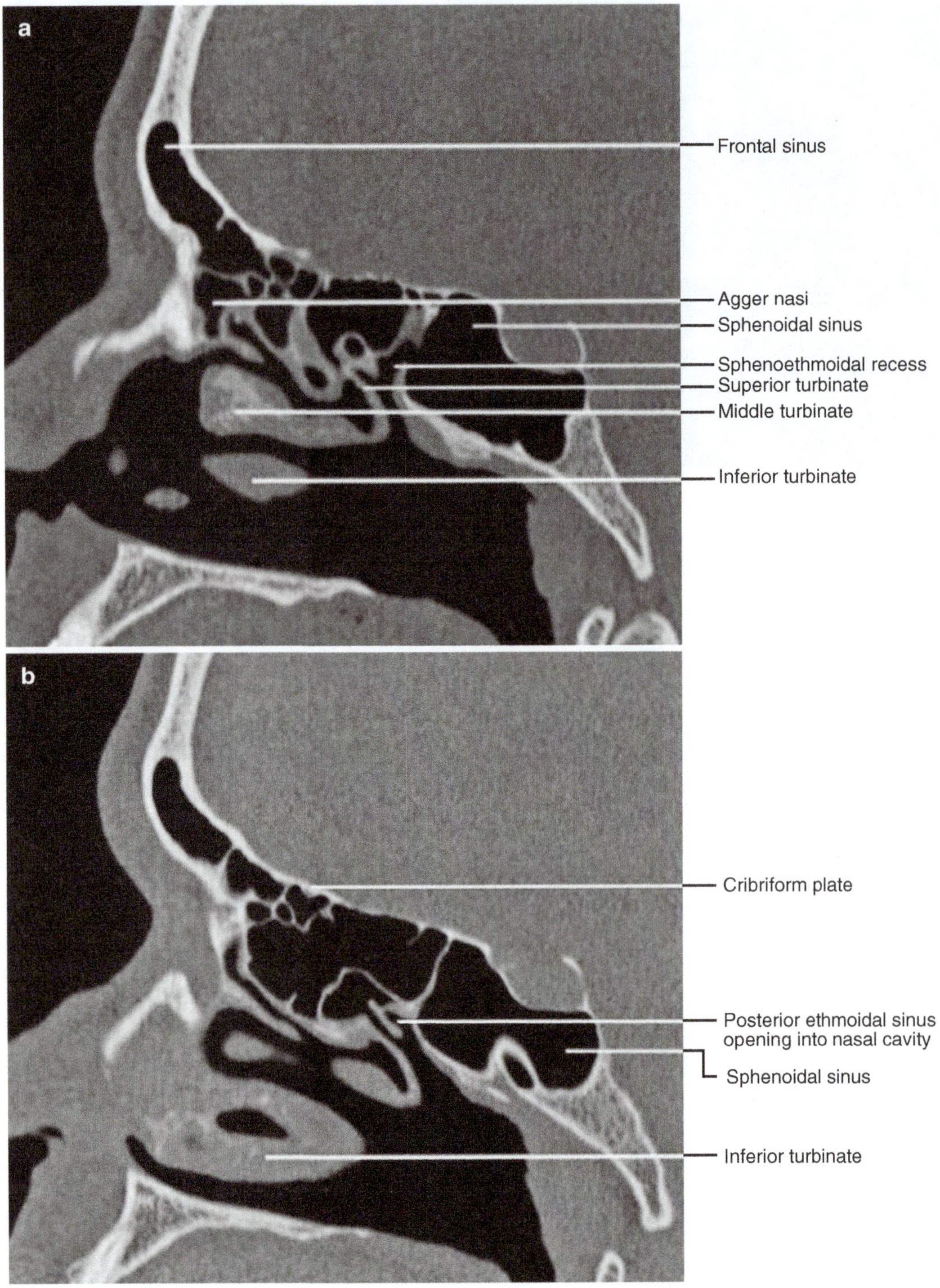

Fig. 8.7 Sagittal CT (bony window). (**a**, **b**) Sequential imaging from the midline laterally

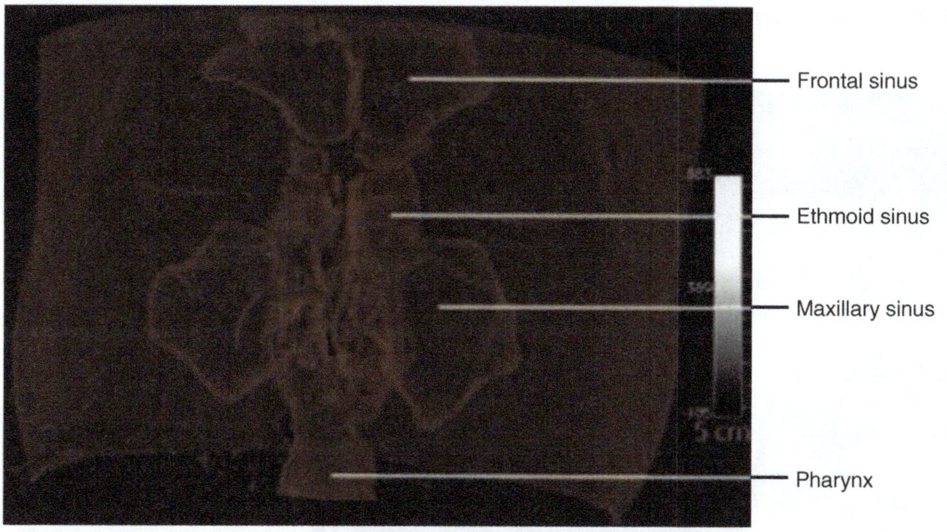

Fig. 8.8 Coronal three-dimensional (3D) reconstruction of paranasal sinuses

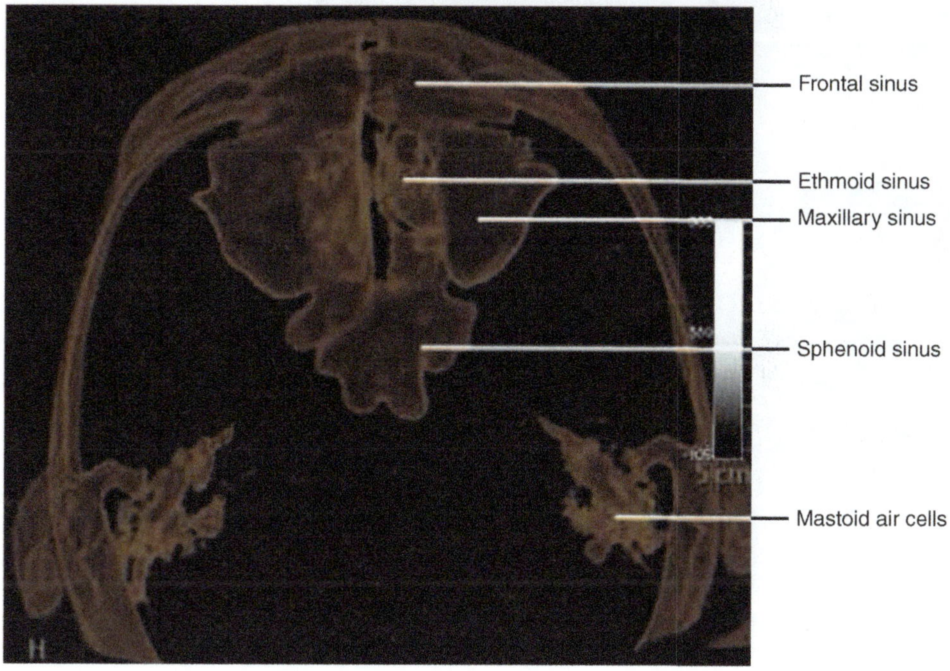

Fig. 8.9 Axial 3D reconstruction of paranasal sinuses

Magnetic Resonance Imaging

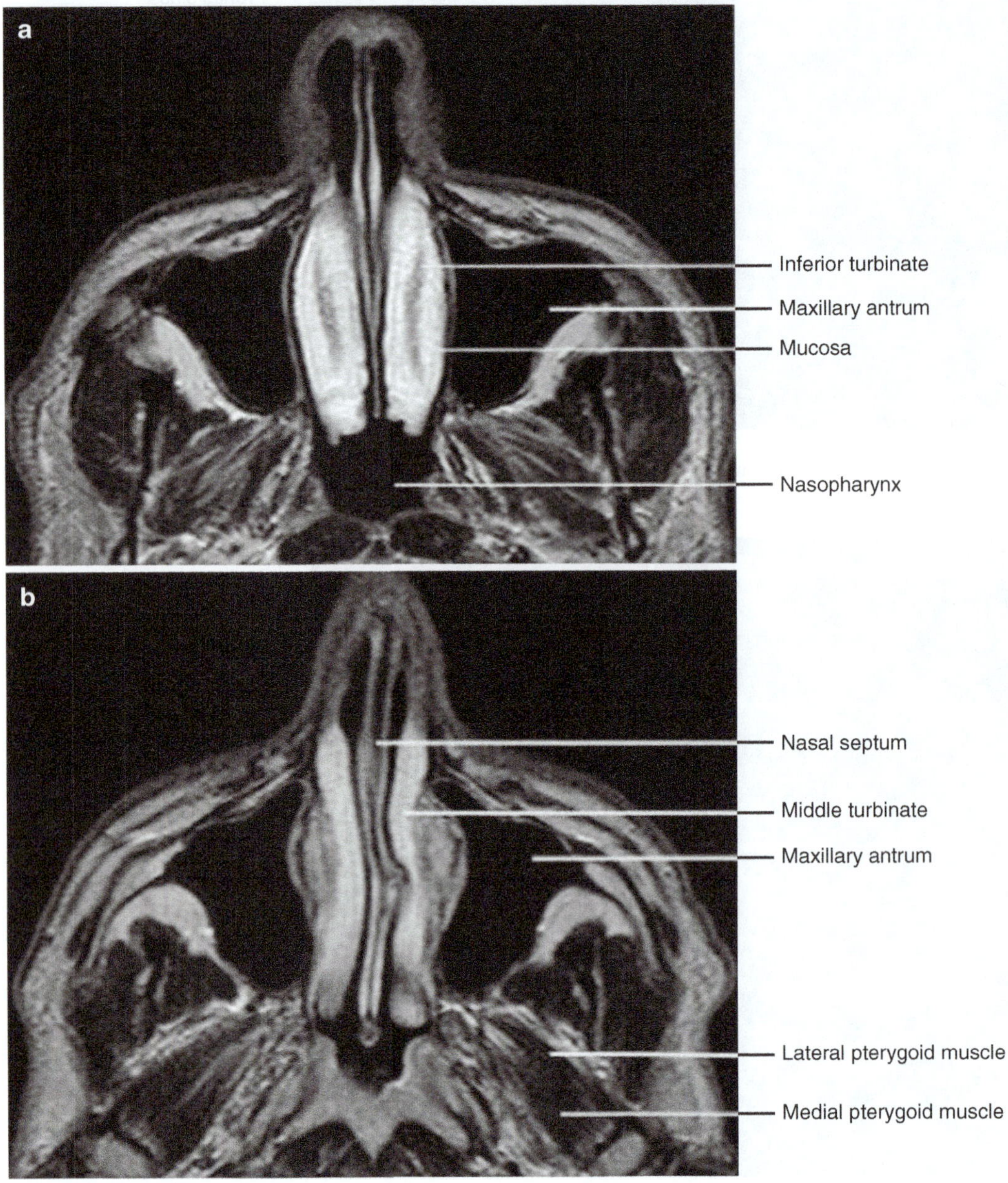

Fig. 8.10 Axial T2. (**a**–**c**) Sequential images from inferior to superior

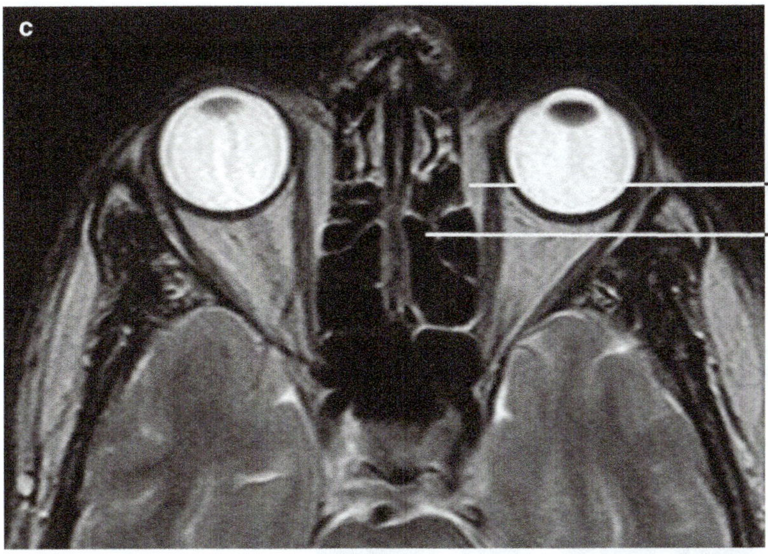

Fig. 8.10 (continued)

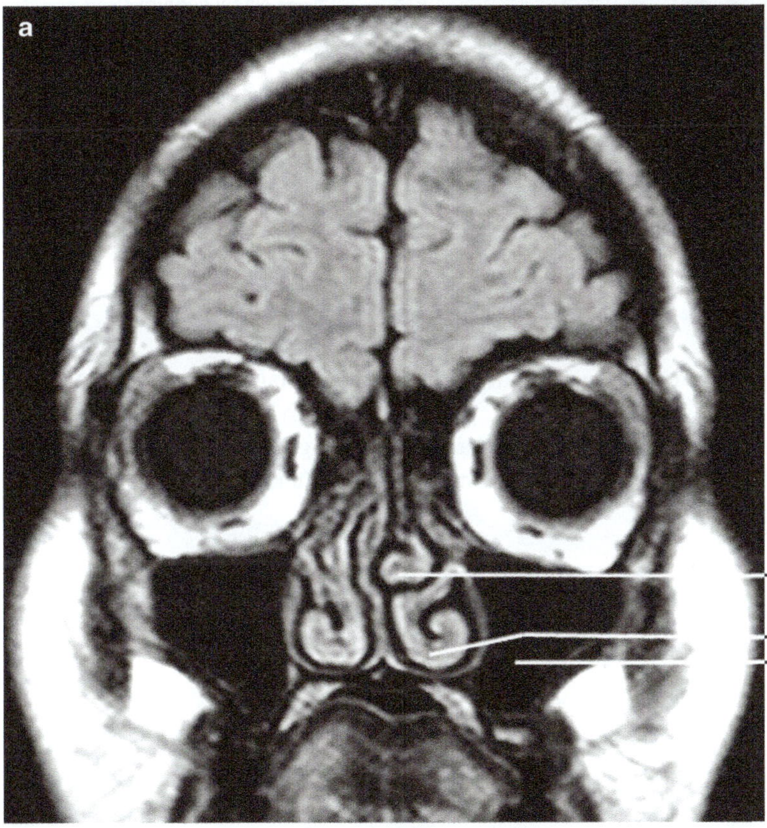

Fig. 8.11 Coronal T1. (**a**, **b**) Sequential images from anterior to posterior

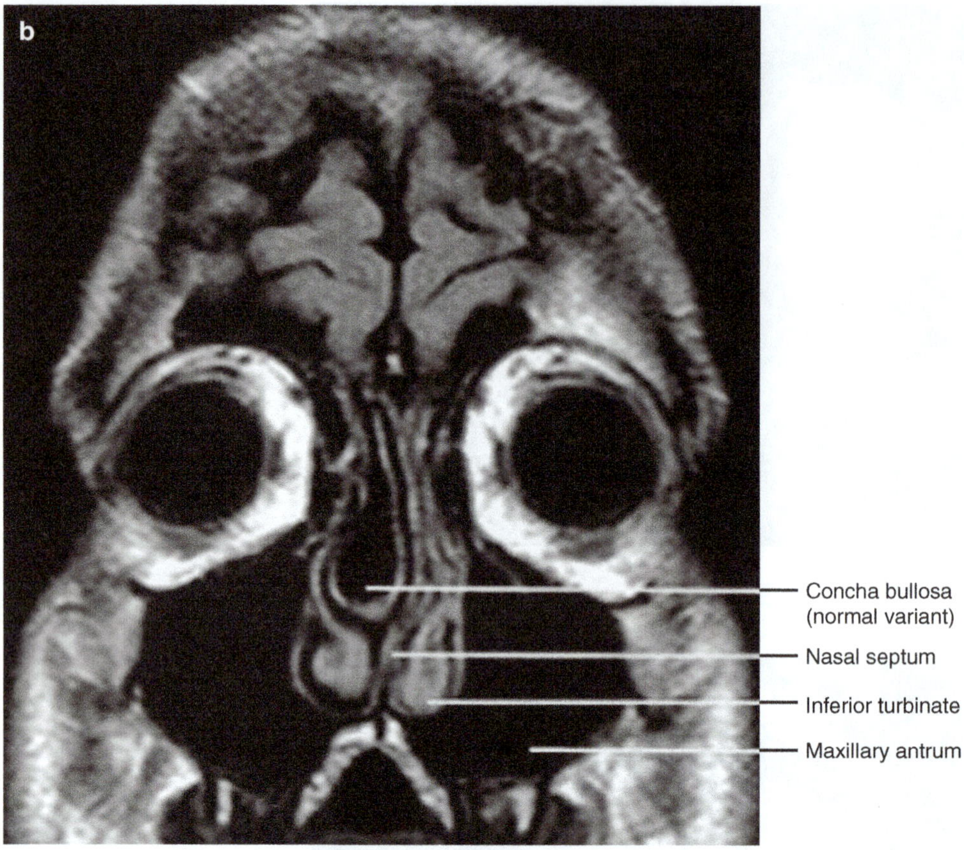

Fig. 8.11 (continued)

Pearl

Sinus disease is most conspicuous on T2 weighted, fluid-attenuated inversion recovery (FLAIR), and T1 post-gadolinium.

Intense enhancement of the nasopharyngeal mucosa is noted post-gadolinium and this serves as a single feature to differentiate a post-contrast image from an unenhanced T1-weighted image.

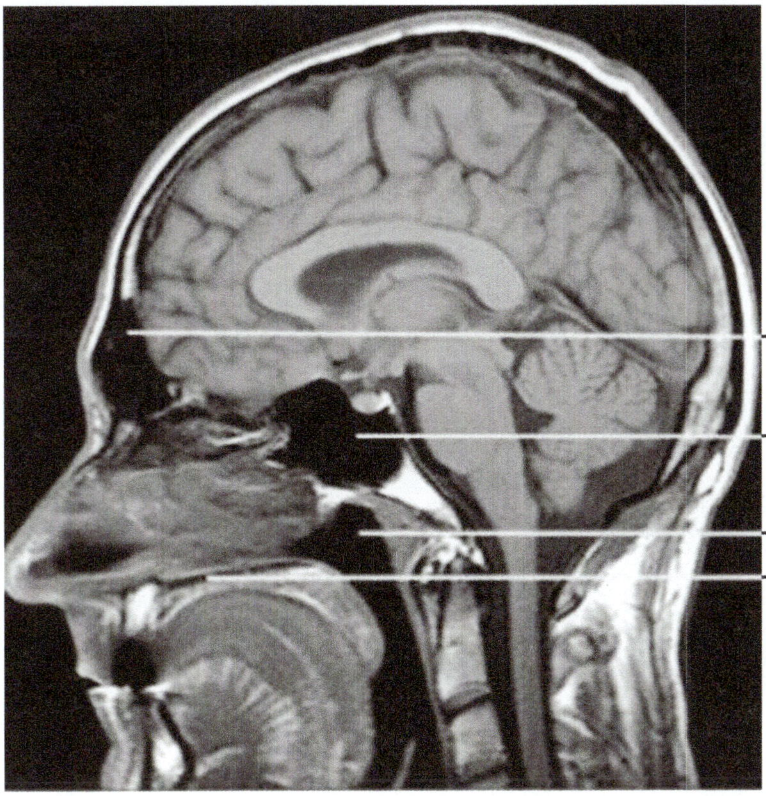

Fig. 8.12 Midline sagittal T1

Sources

1. Butler P, Mitchell AWM, Ellis H (1999) Applied radiological anatomy. Cambridge University Press, Cambridge, pp 17–120
2. Ellis H, Logan BM, Dixon AK (1991) Human cross-sectional anatomy: atlas of body sections and CT images. Butterworth-Heinemann, Oxford, pp 33–48
3. Grainger RG, Allison DJ, Adam A, Dixon AK (2001) Grainger and Allison's diagnostic radiology: a textbook of medical imaging, 4th edn. Churchill Livingstone, Philadelphia; Chap 107
4. Weir B, Abrahams PH (2001) Imaging atlas of human anatomy. Harcourt, London, pp 1–47

9 Skull and Jaw

Plain Film

Plain Film Imaging of the Skull

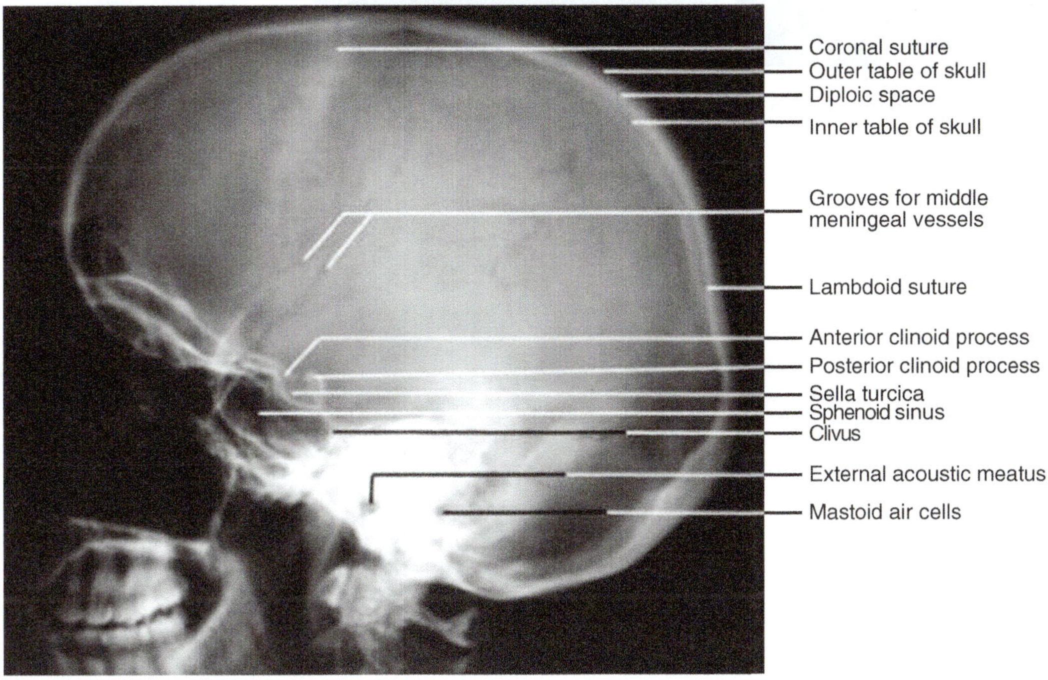

Fig. 9.1 Skull: lateral projection

S. Andronikou (ed.), *See Right Through Me*,
DOI 10.1007/978-3-642-23893-2_9, © Springer-Verlag Berlin Heidelberg 2012

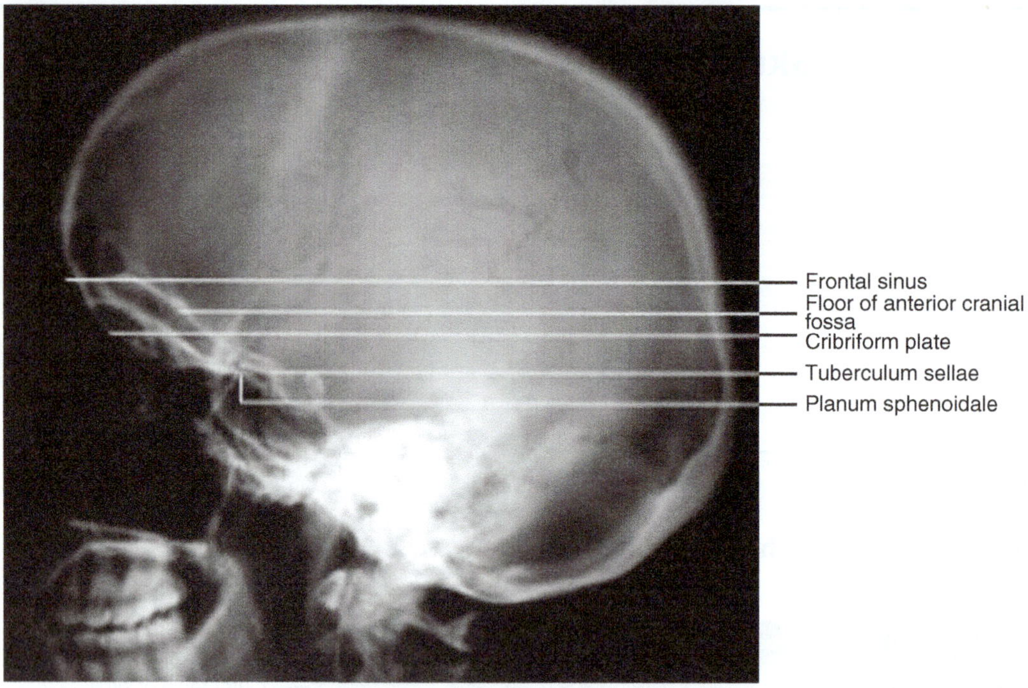

Fig. 9.2 Skull: lateral projection

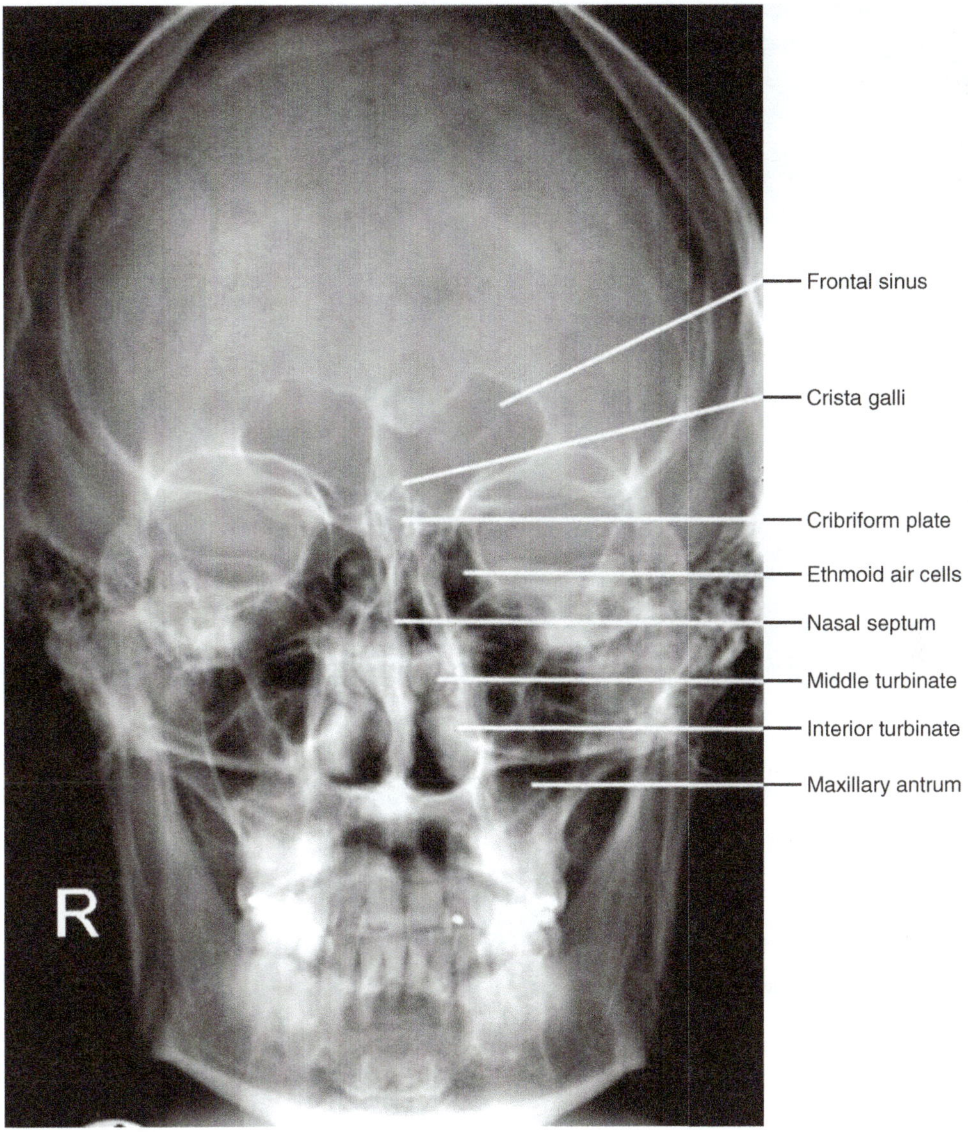

Fig. 9.3 Skull: posteroanterior radiograph

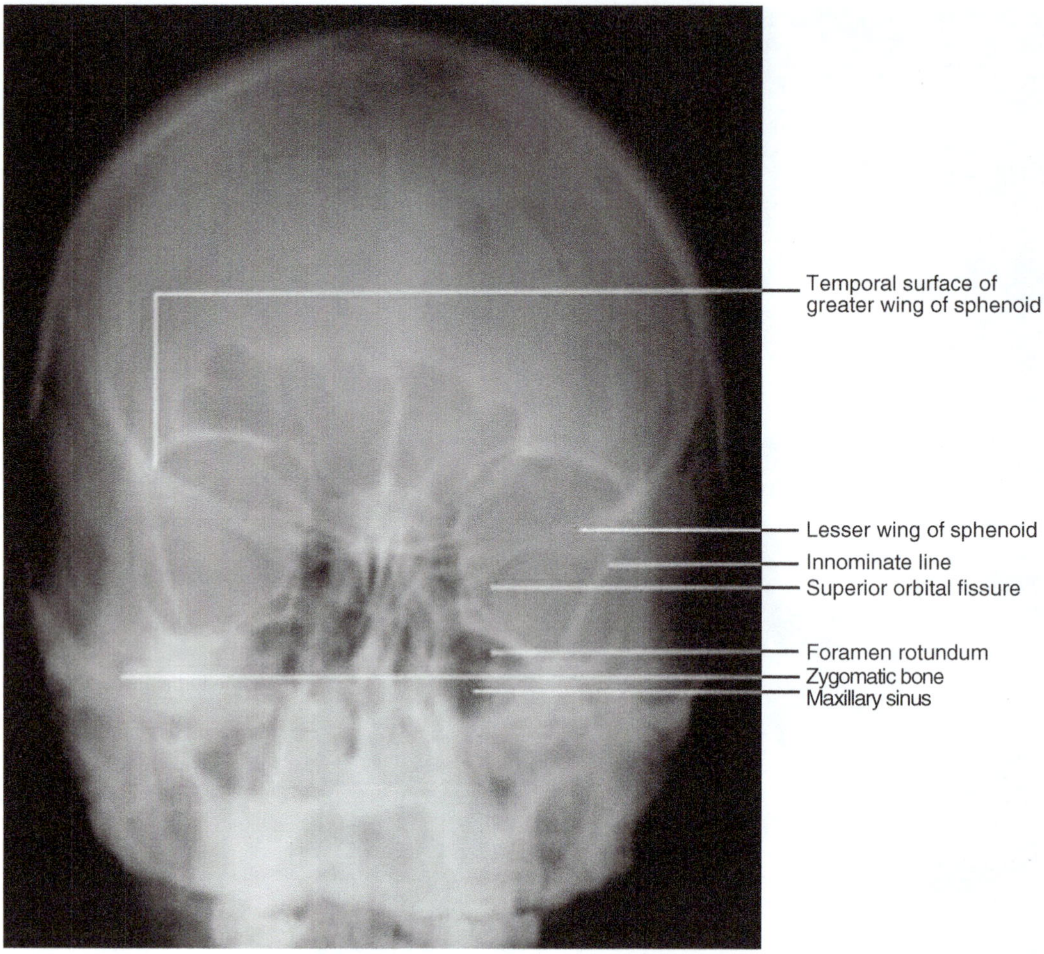

Fig. 9.4 Skull: 23° occipital projection (Caldwell view)

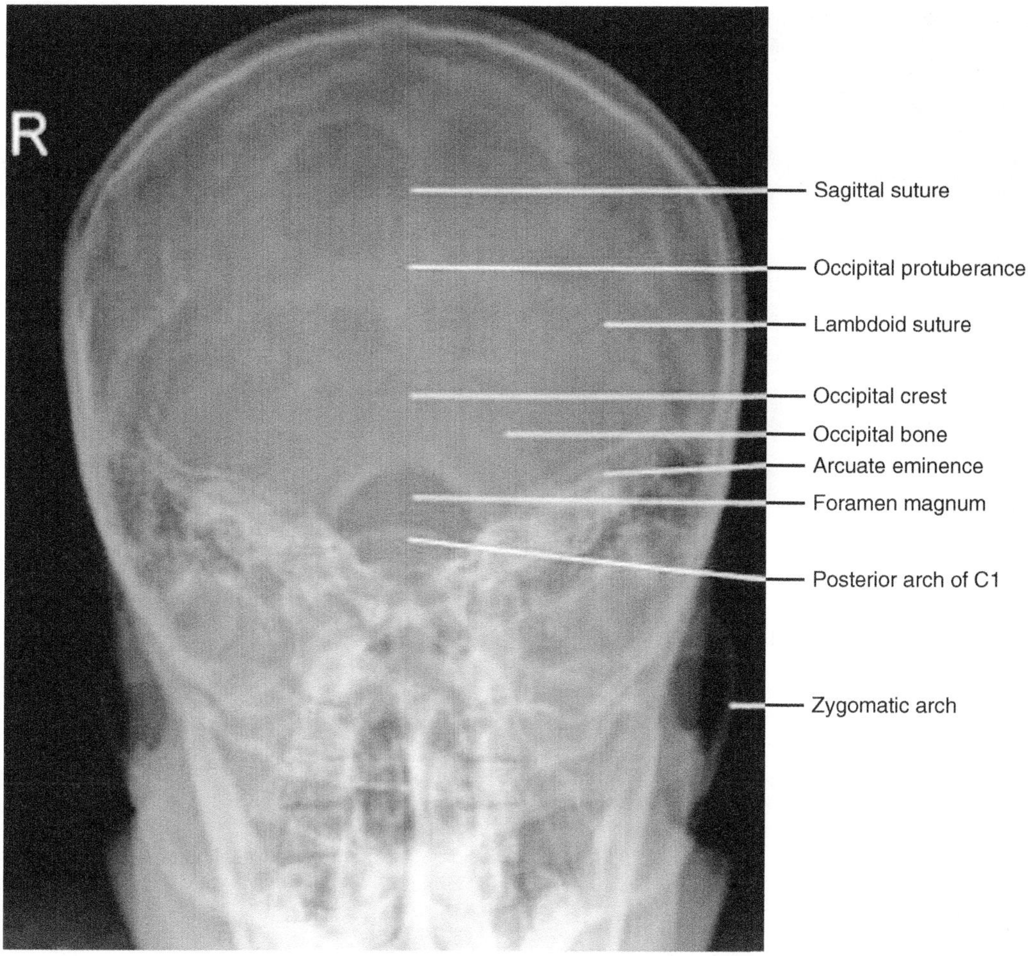

Fig. 9.5 Skull: 23° fronto-occipital projection (Townes' view)

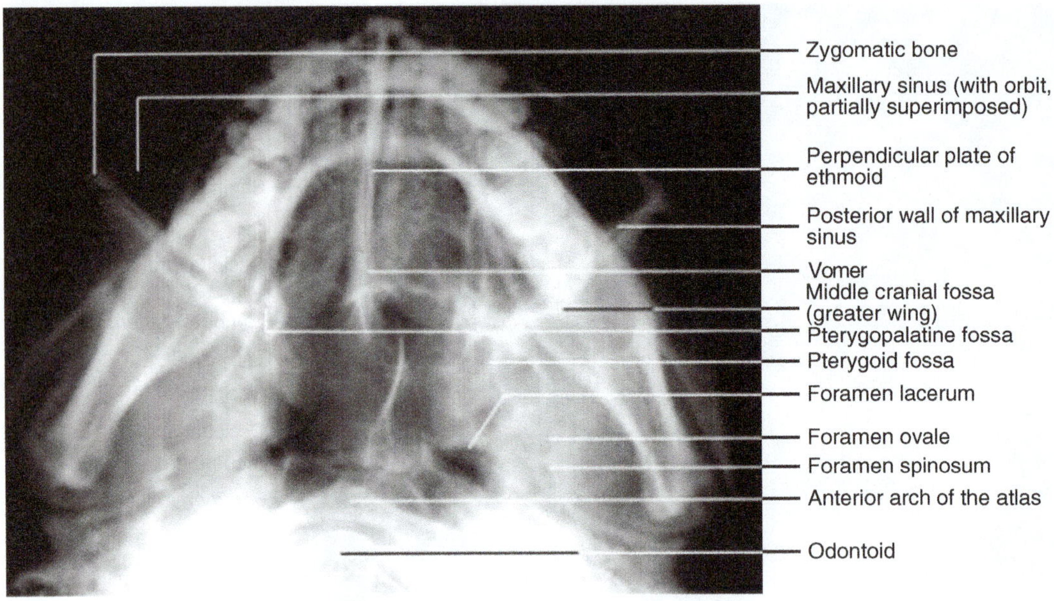

Fig. 9.6 Skull: submentovertical projection

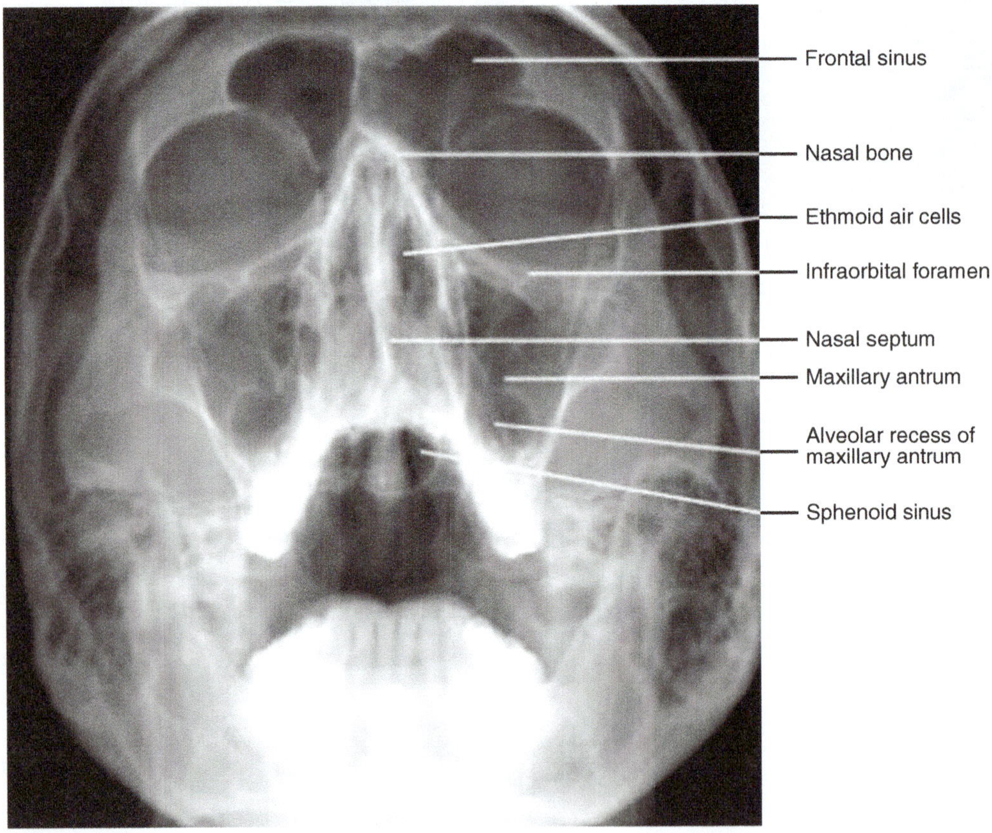

Fig. 9.7 Skull: OM projection (Waters' view)

Plain Film Imaging of the Nasal Bone

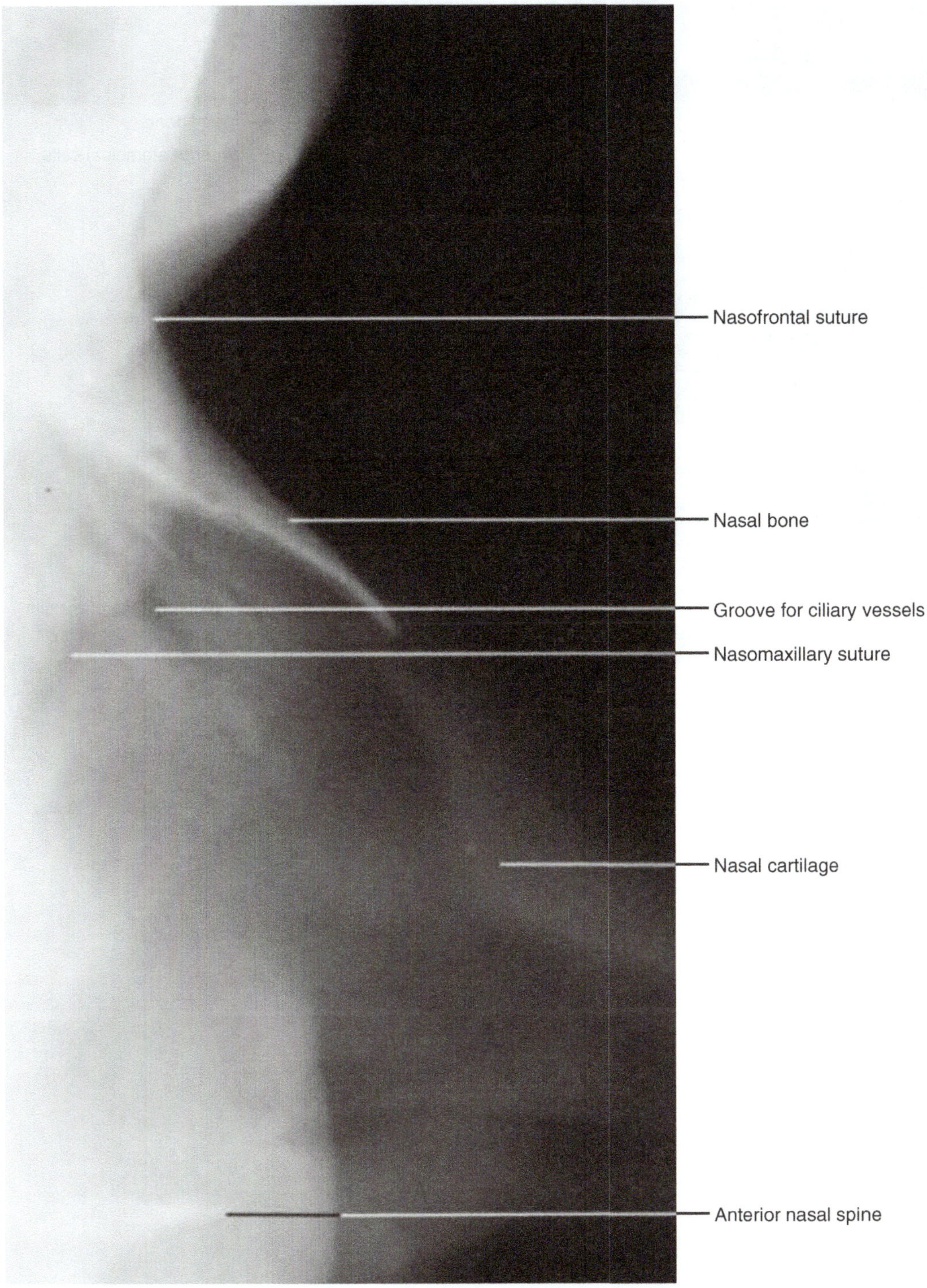

Fig. 9.8 Nasal bone: lateral projection

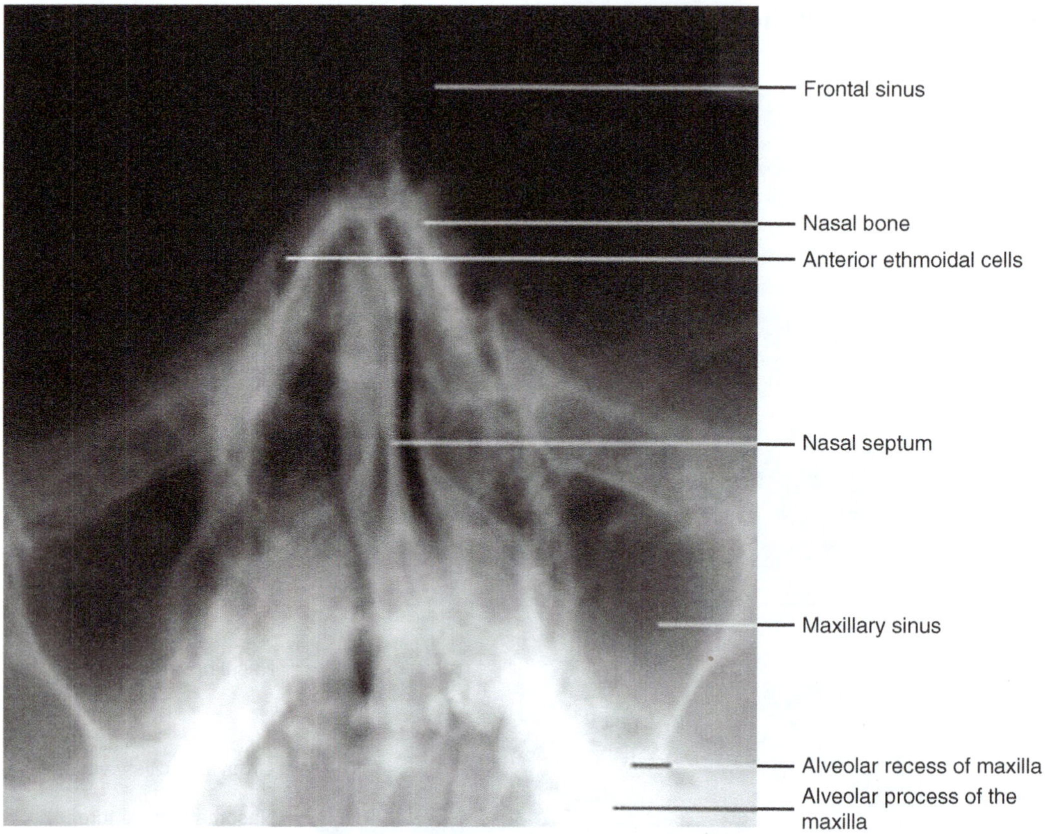

Fig. 9.9 Nasal bone: superoinferior projection

Plain Film Imaging of the Mandible

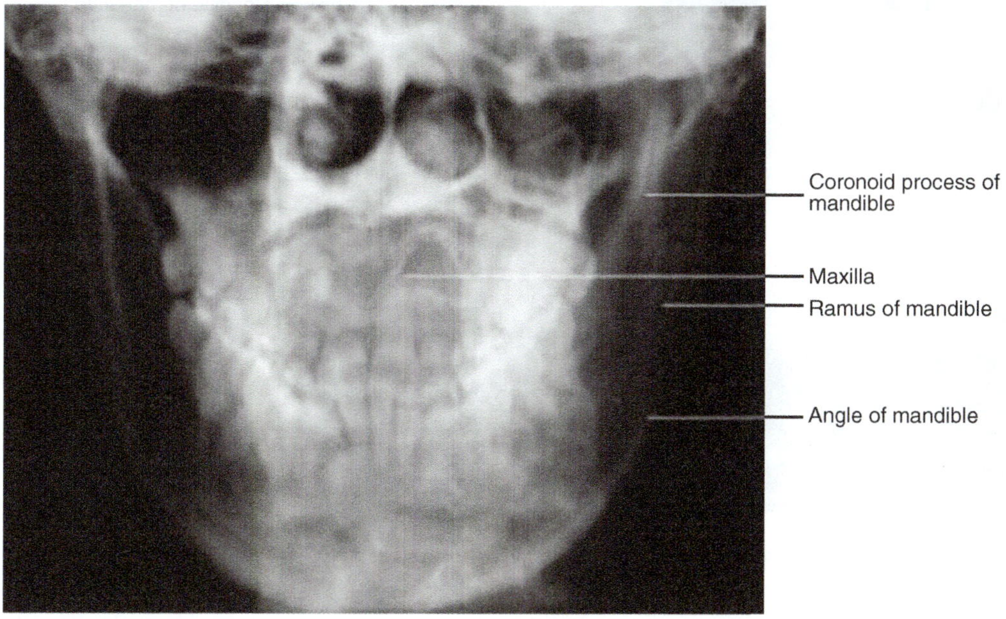

Fig. 9.10 Mandible: occipitofrontal projection

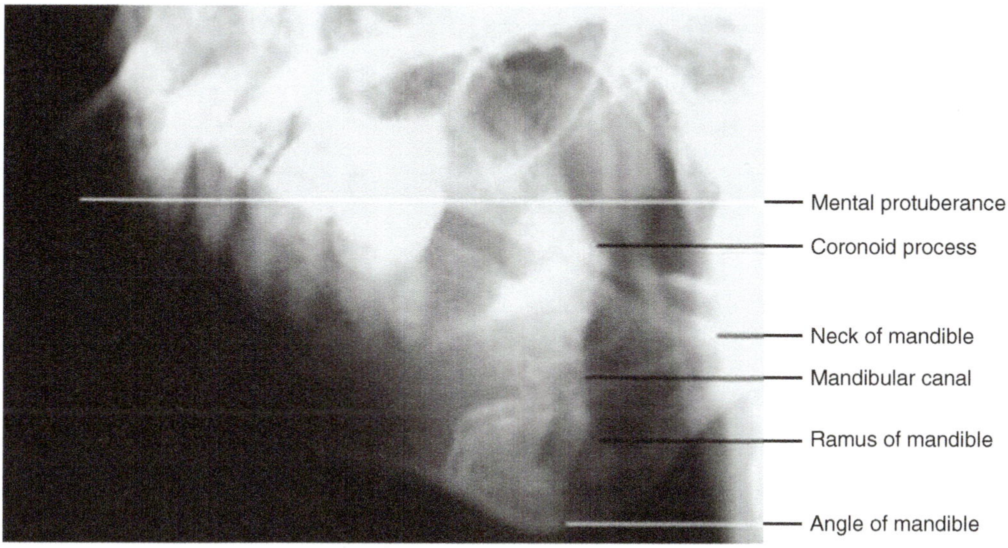

Fig. 9.11 Mandible: lateral oblique projection

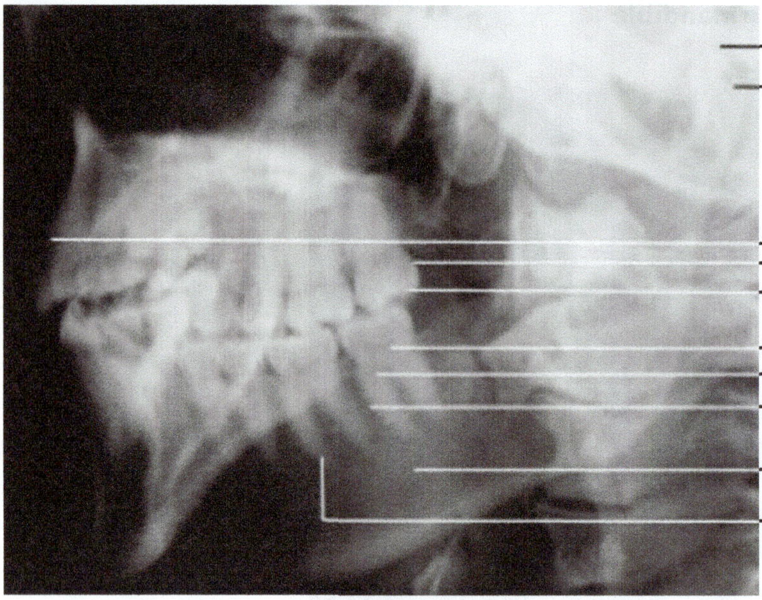

Fig. 9.12 Mandible: posteroanterior oblique projection

Arthrogram of the Temporomandibular Joint

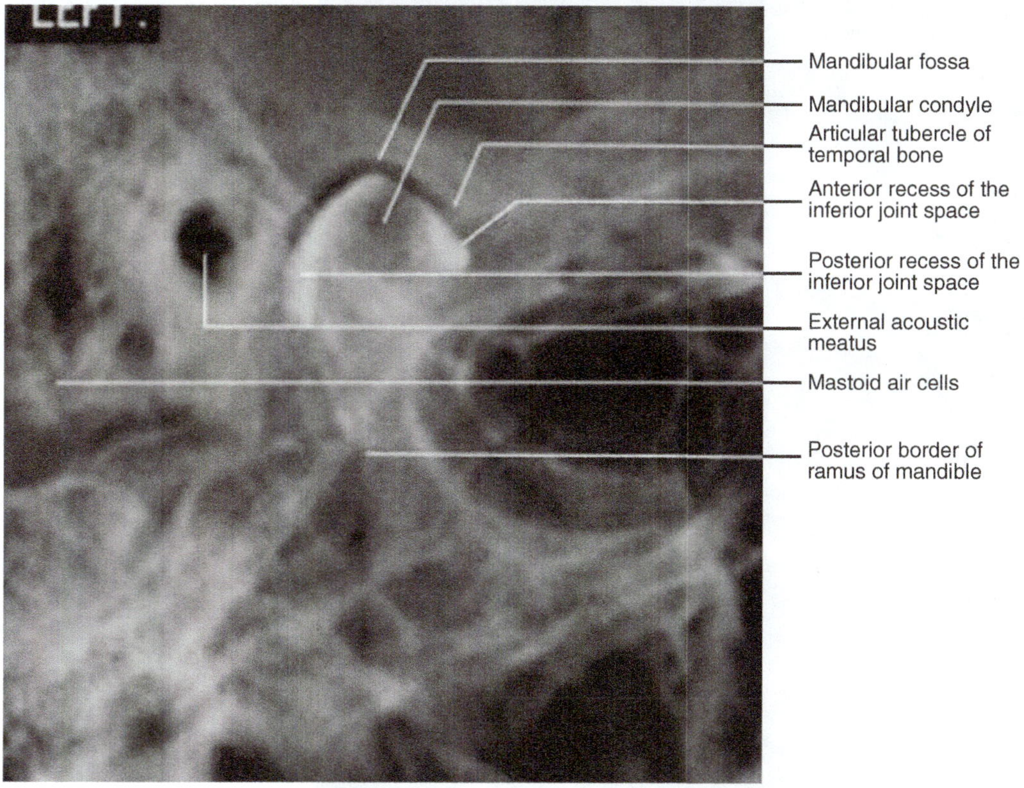

Fig. 9.13 Lateral oblique view (mouth closed)

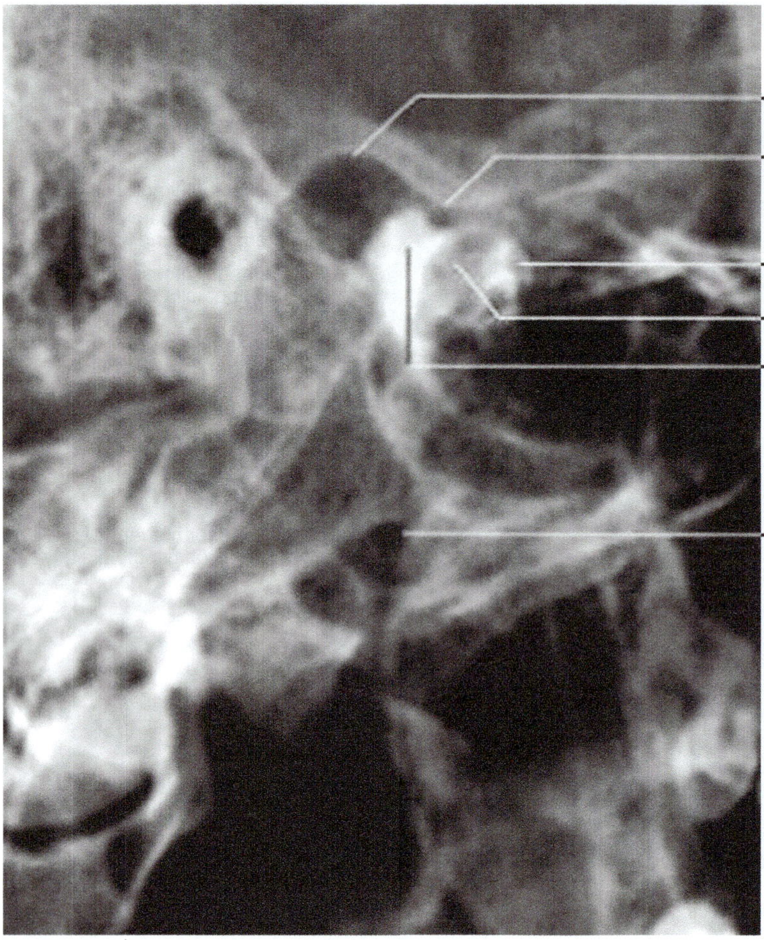

Fig. 9.14 Lateral oblique view (mouth open)

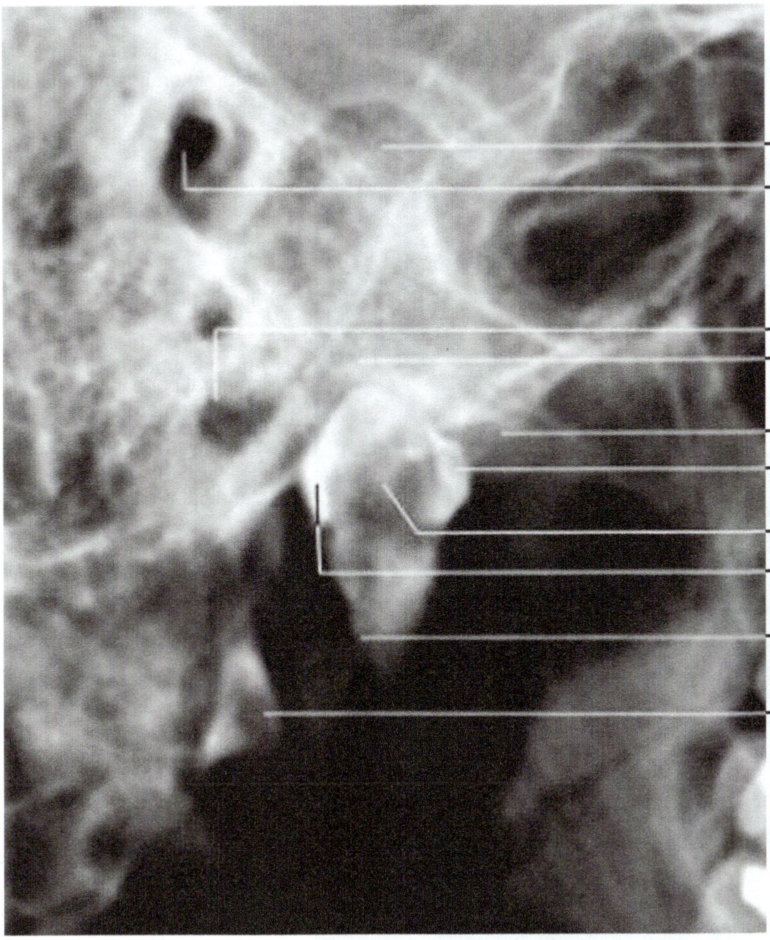

Fig. 9.15 Reverse lateral oblique view (mouth closed)

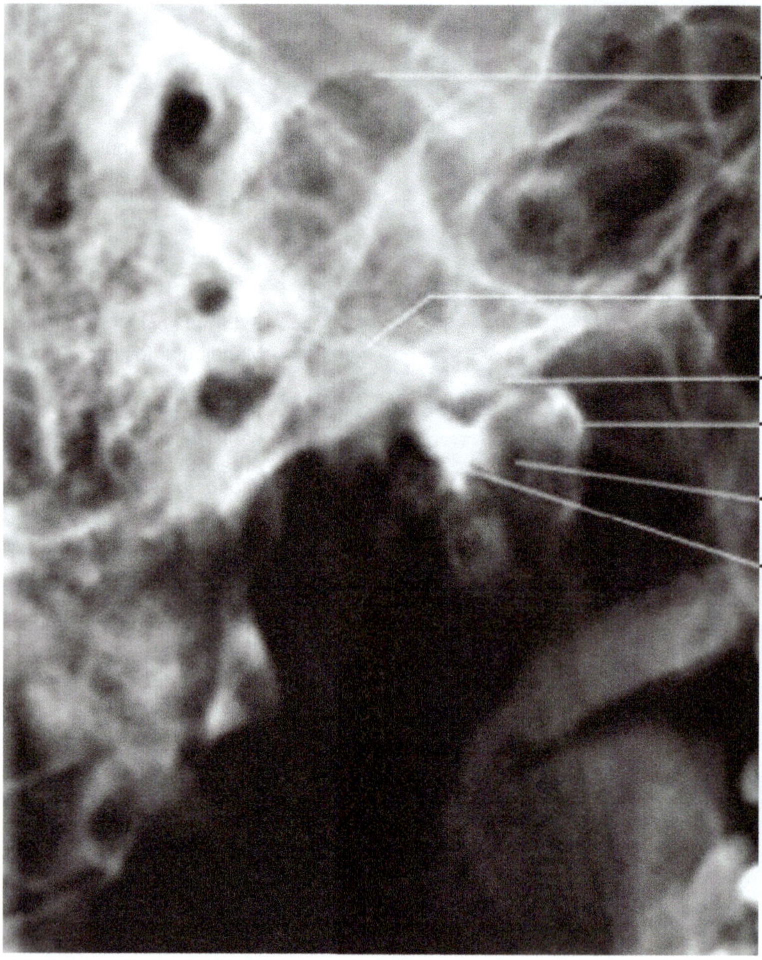

Fig. 9.16 Reverse lateral oblique view (mouth open)

Pearl

Routinely, only the inferior joint space is accessed for the arthrogram, but the superior joint space may be accessed separately.

The normal lateral oblique view demonstrates the lateral aspect of the inferior joint space well. The reverse view demonstrates the medial aspect of the inferior joint space well.

Orthopantomogram

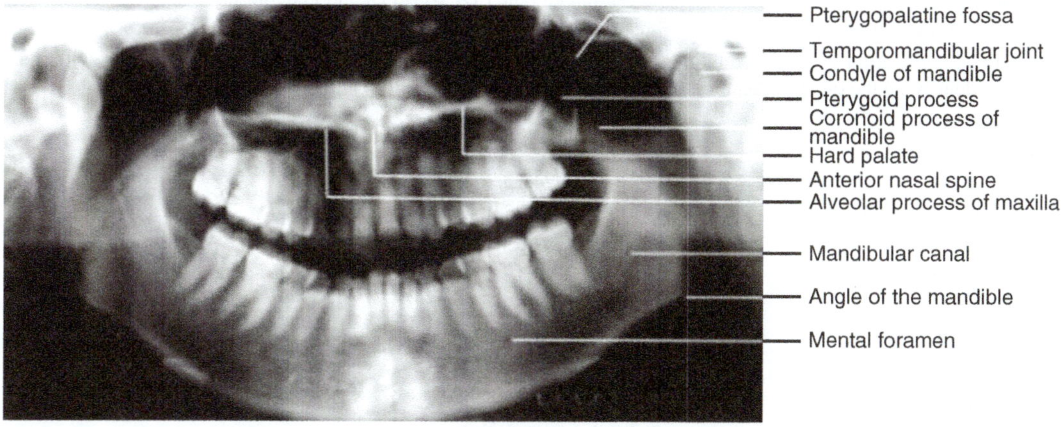

Fig. 9.17 Orthopantomogram

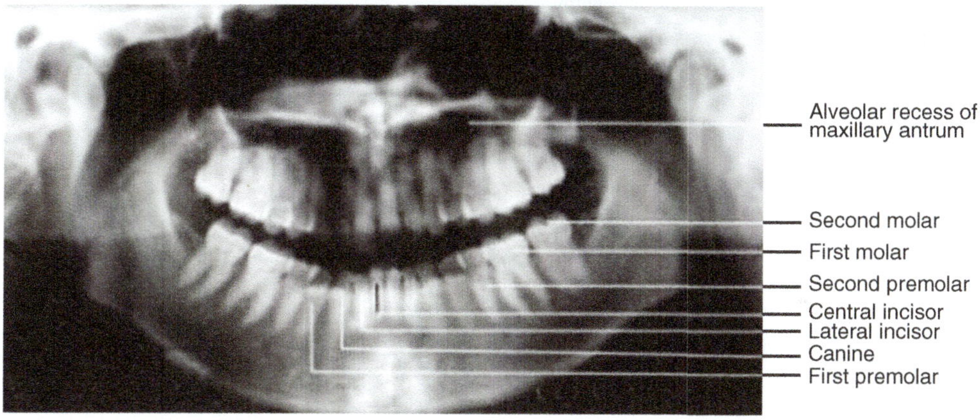

Fig. 9.18 Orthopantomogram

Table 9.1 Tooth identification

Tooth identification	
(According to the International Standards Organisation and the FDI World Dental Federation)	
FDI World Dental Federation Two-Digit Notification	
International notification system used to identify a specific tooth. The mouth is divided into four quadrants with the first digit identifying the quadrant and the second digit identifying the tooth within that quadrant.	
Permanent teeth	
Upper right (1)	**Upper left (2)**
18 17 16 15 14 13 12 11	21 22 23 24 25 26 27 28
48 47 46 45 44 43 42 41	31 32 33 34 35 36 37 38
Lower right (4)	**Lower left (3)**
Deciduous teeth	
Upper right (5)	**Upper left (6)**
55 54 53 52 51	61 62 63 64 65
85 84 83 82 81	71 72 73 74 75
Lower right (8)	**Lower left (7)**
Quadrants (first digit)	**Teeth (second digit)**
1. Upper right permanent	1. Central incisors
2. Upper left permanent	2. Lateral incisors
3. Lower left permanent	3. Canines
4. Lower right permanent	4. First premolar (permanent)/first molar (deciduous)
5. Upper right deciduous	5. Second premolar (permanent)/second molar (deciduous)
6. Upper left deciduous	6. First molar
7. Lower left deciduous	7. Second molar
8. Lower right deciduous	8. Third molar

Computed Tomography

Axial Slices: Superior to Inferior

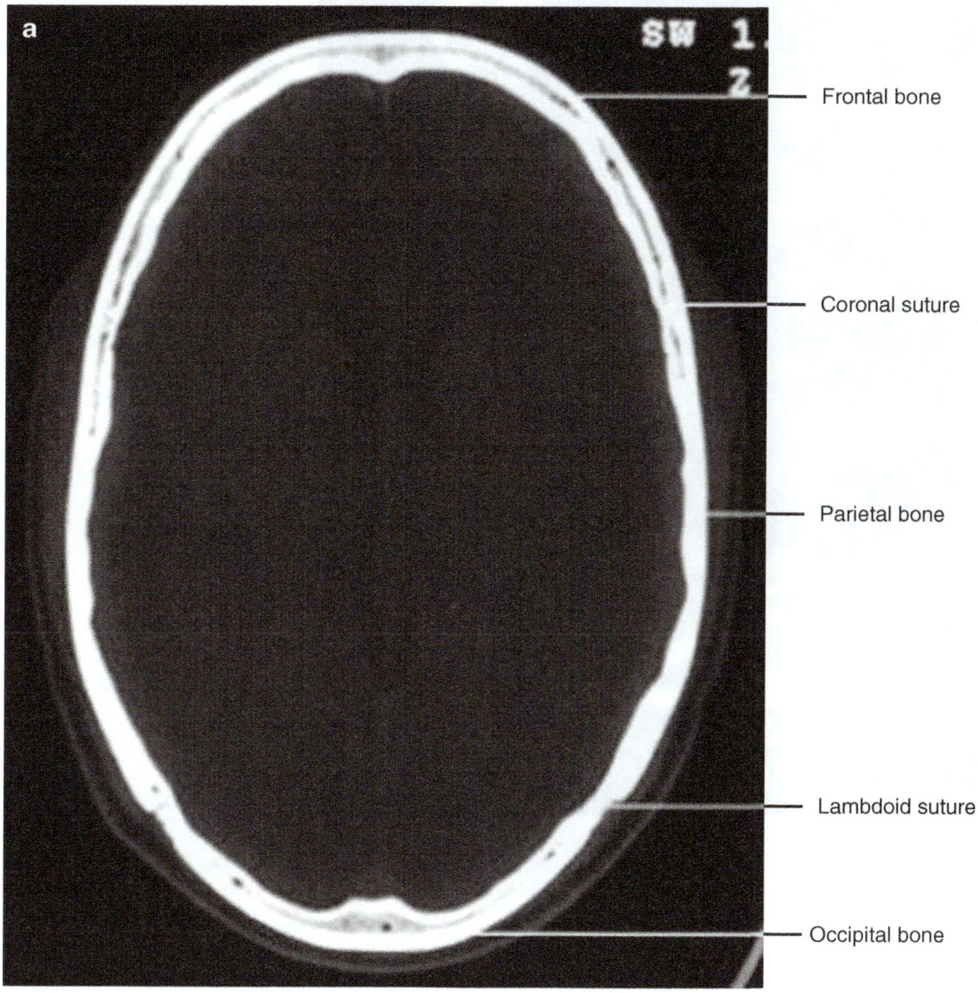

Fig. 9.19 (**a**–**l**) Axial slices: superior to inferior

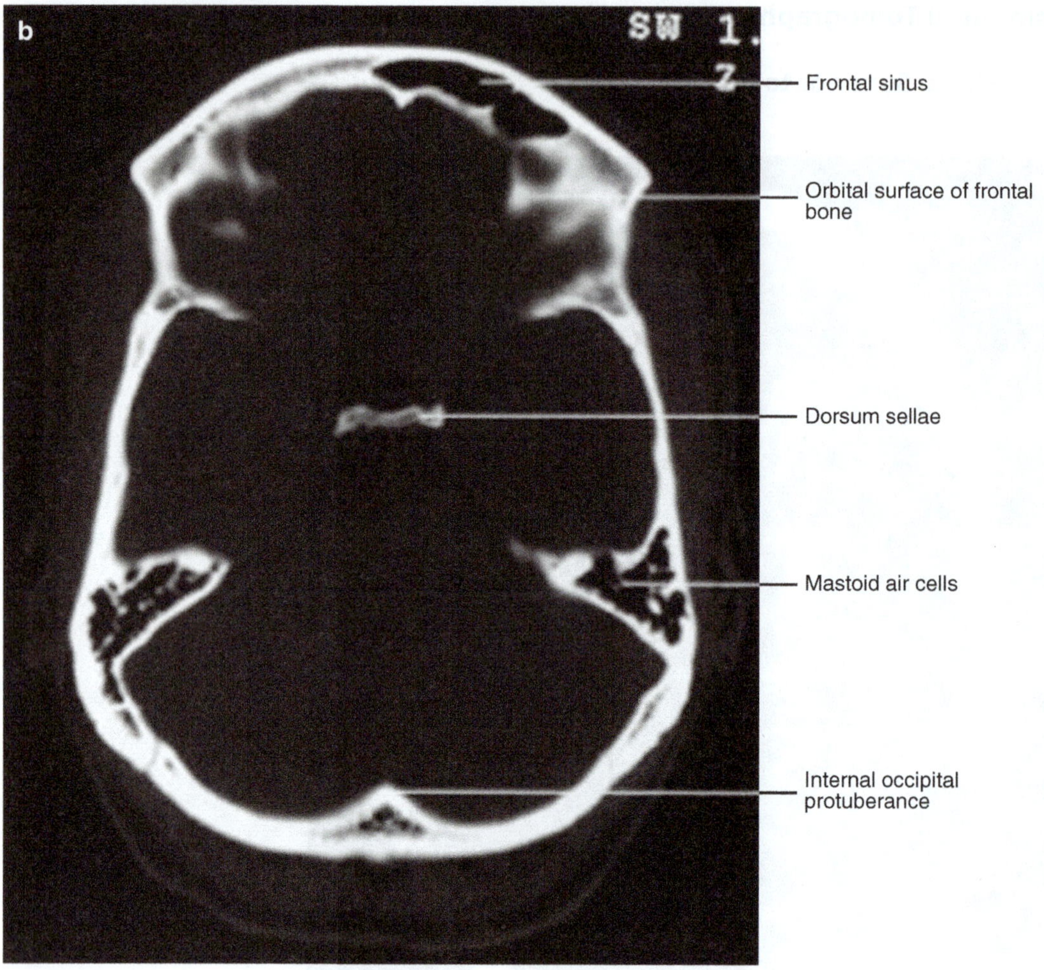

Fig. 9.19 (continued)

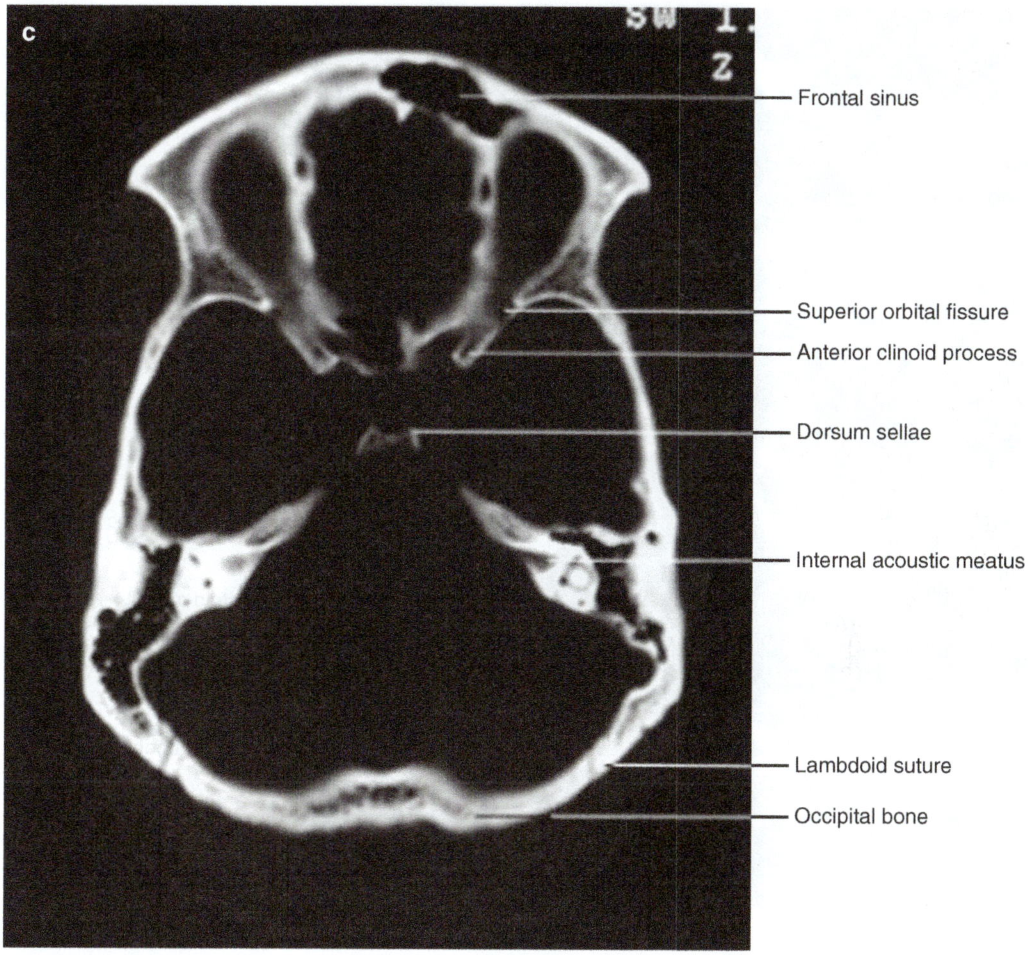

Fig. 9.19 (continued)

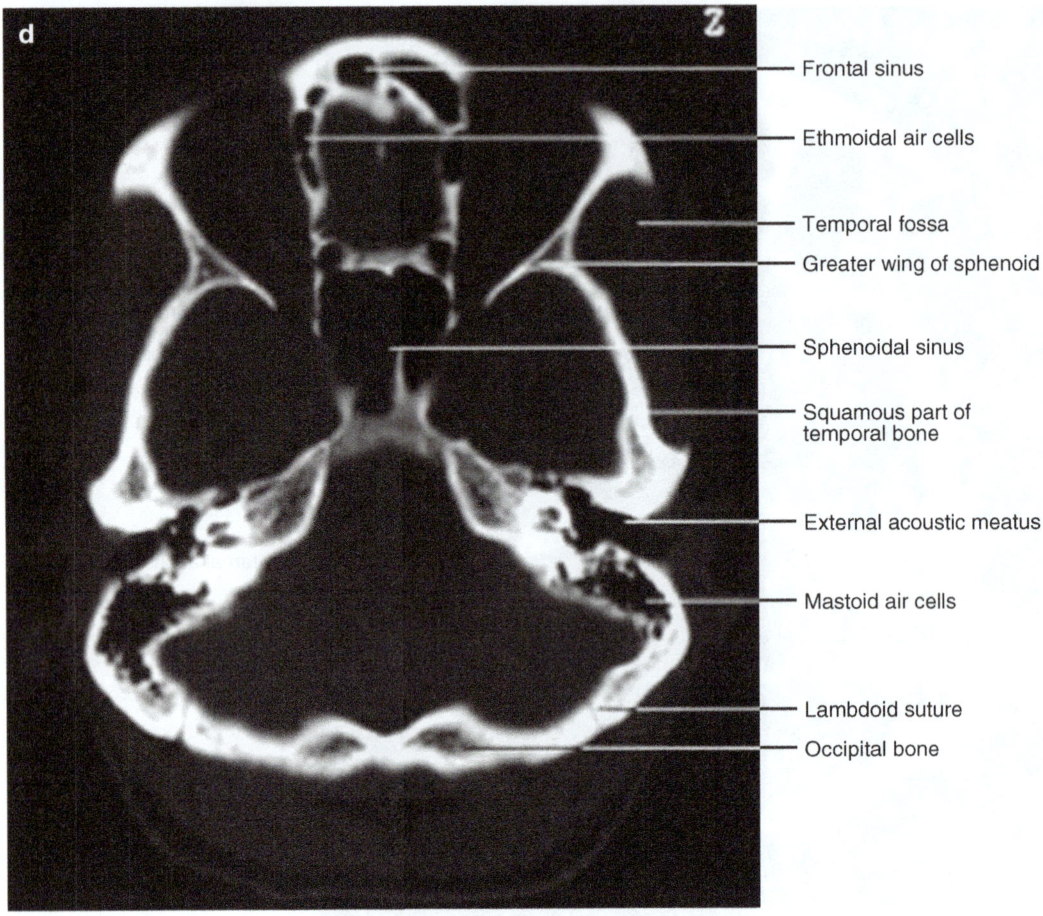

Fig. 9.19 (continued)

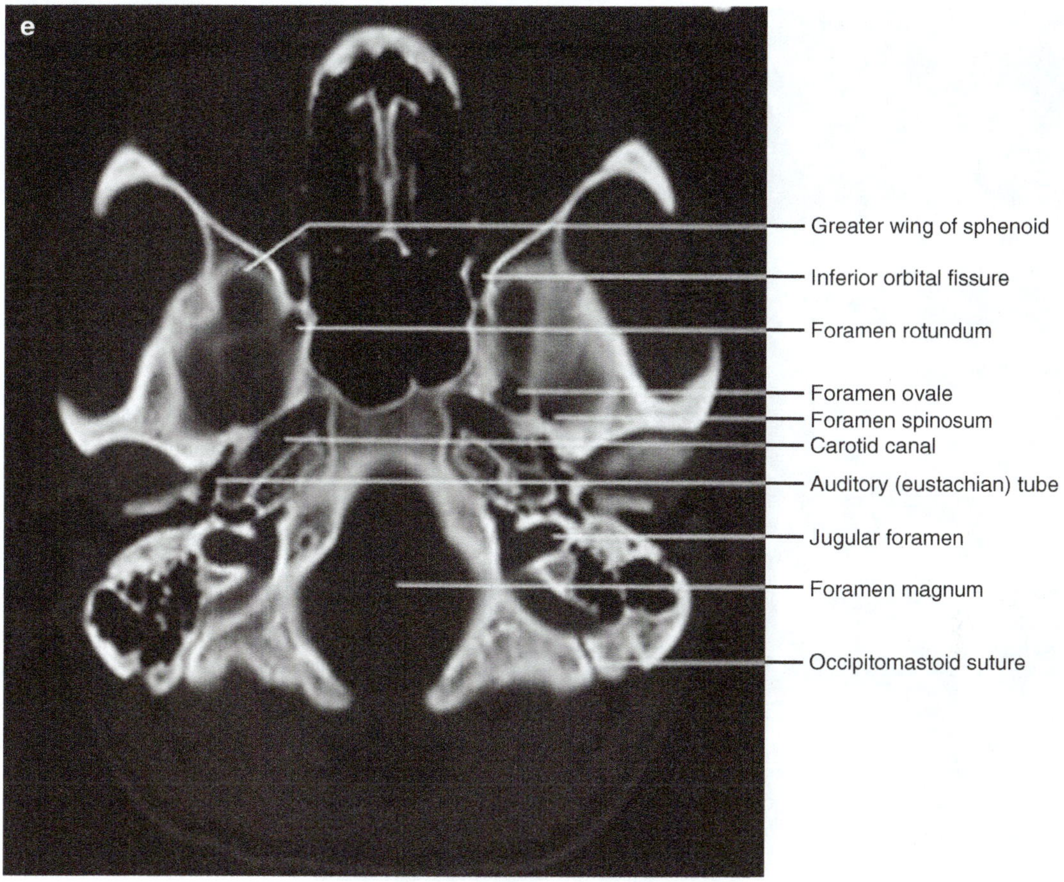

Fig. 9.19 (continued)

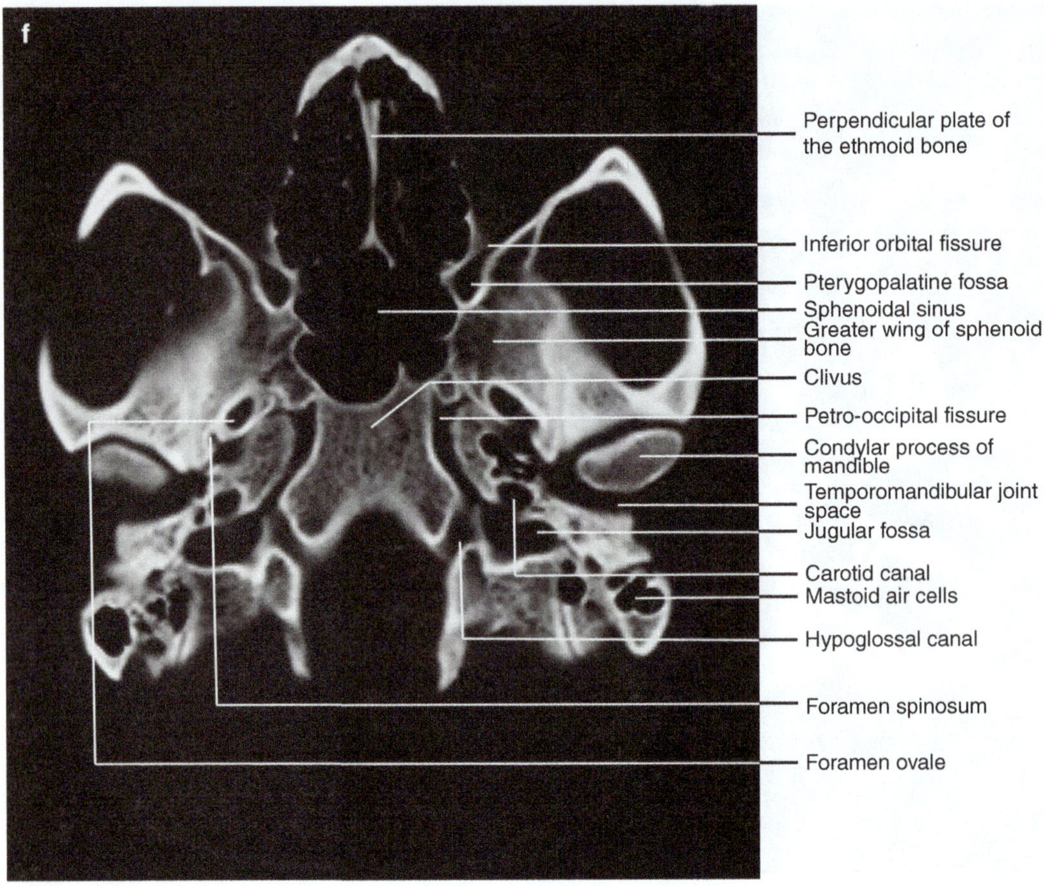

Fig. 9.19 (continued)

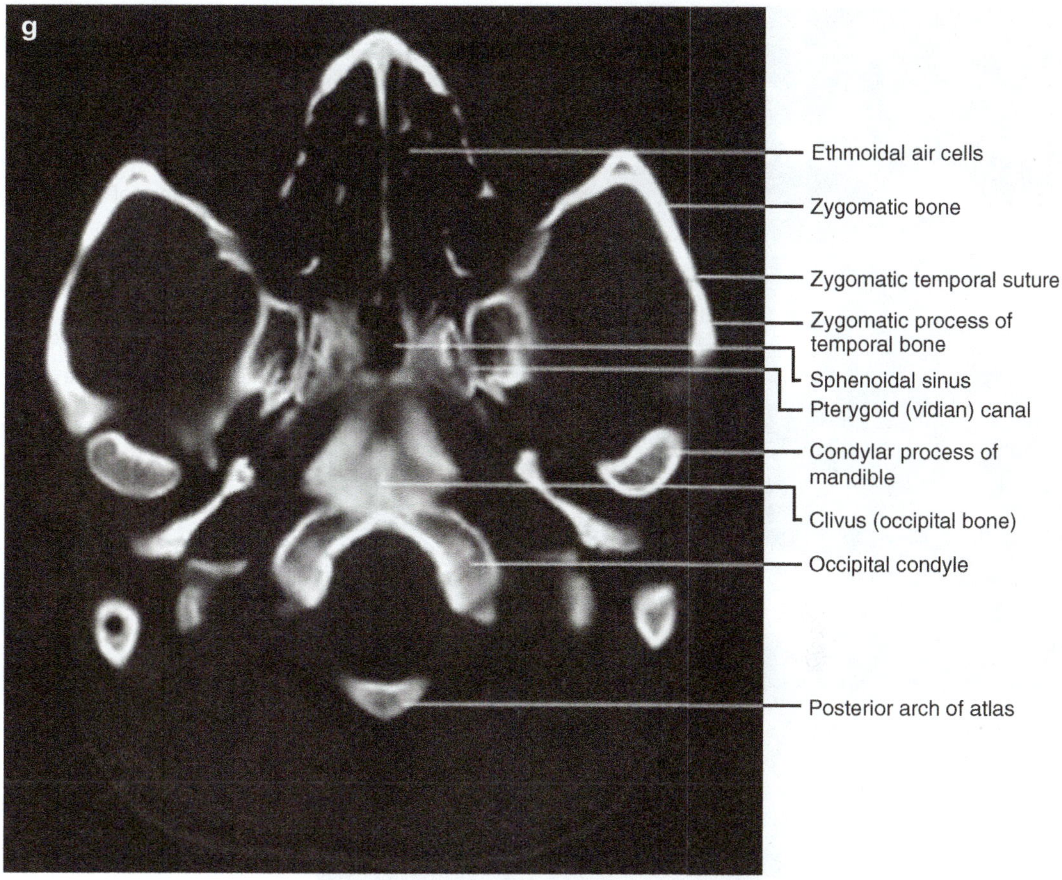

Fig. 9.19 (continued)

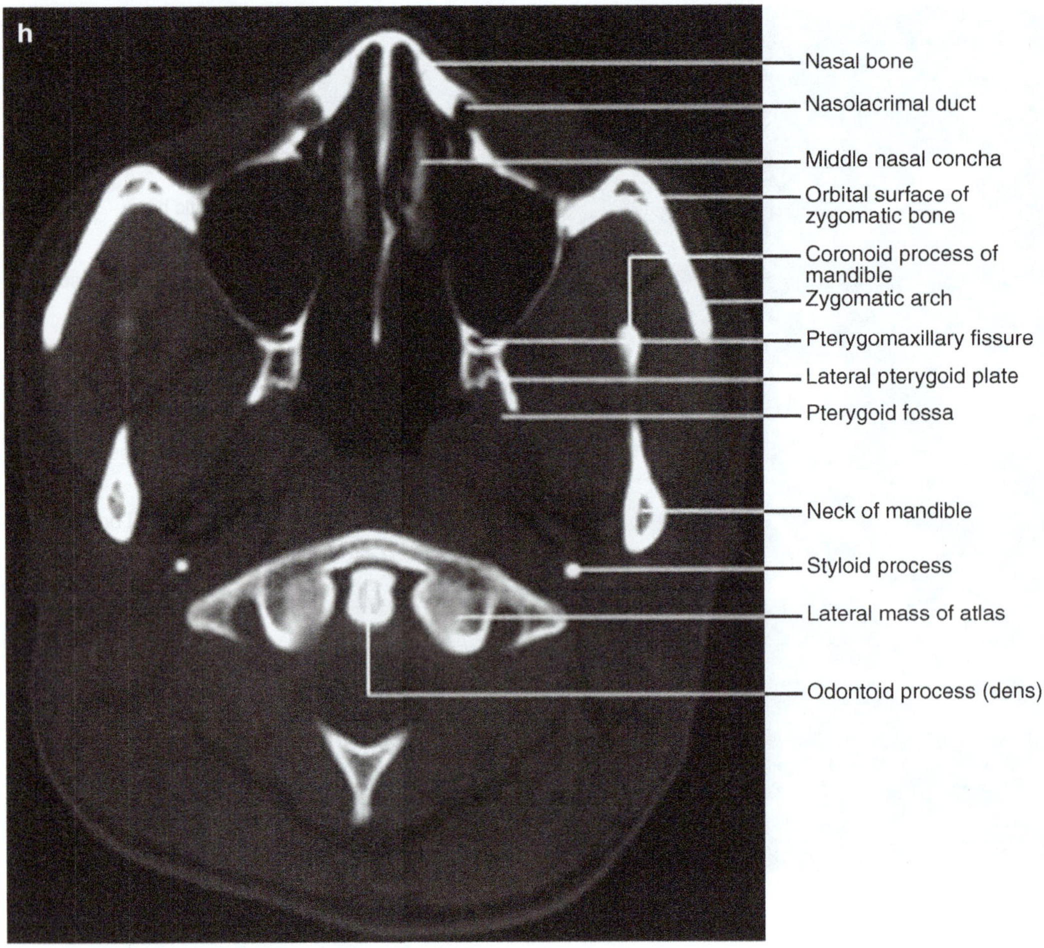

Fig. 9.19 (continued)

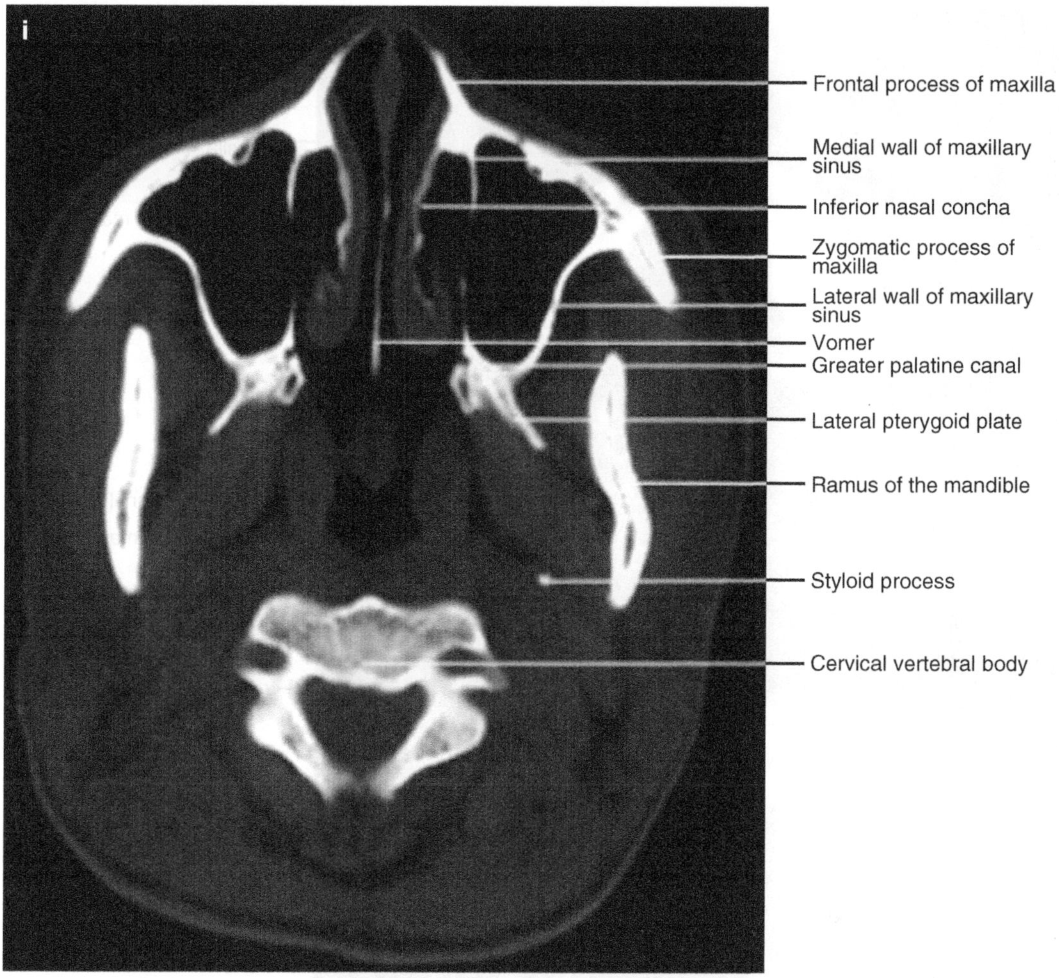

Fig. 9.19 (continued)

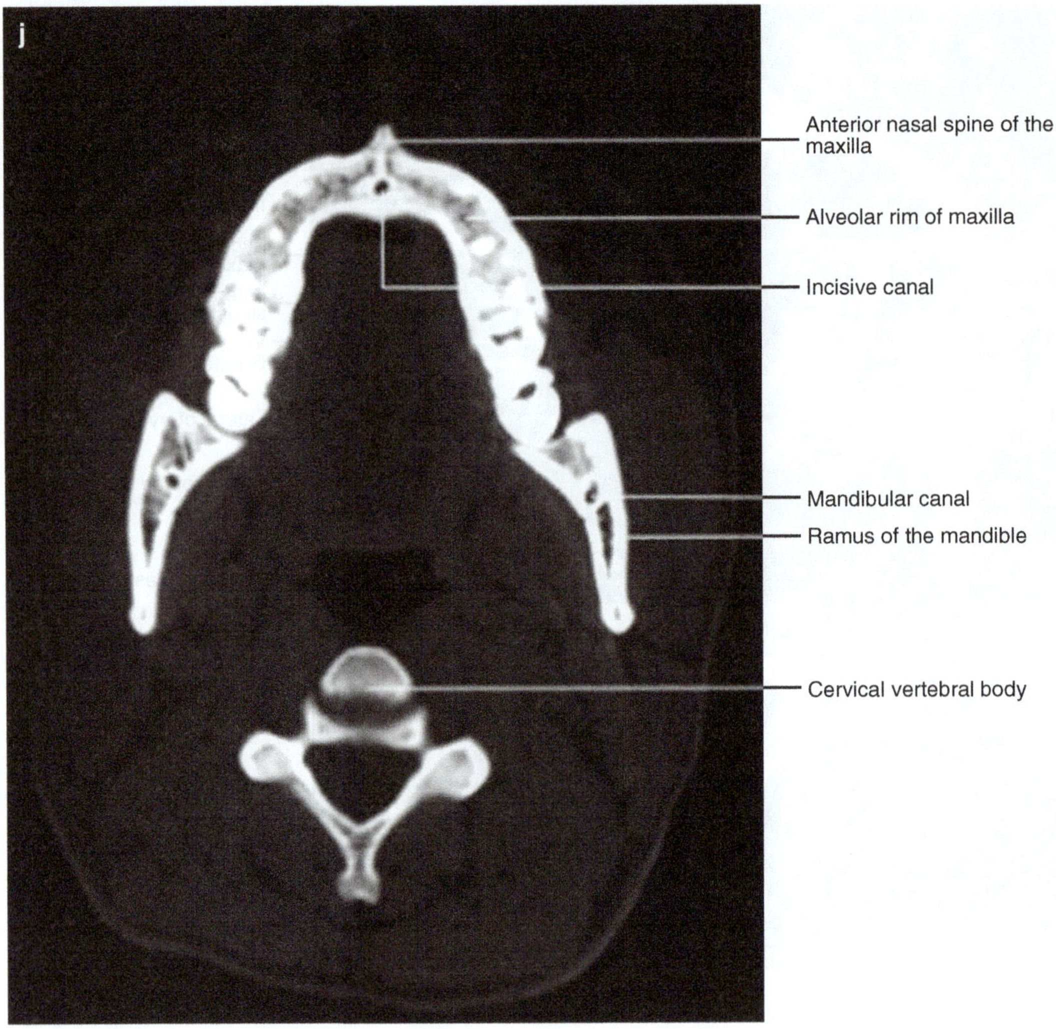

Fig. 9.19 (continued)

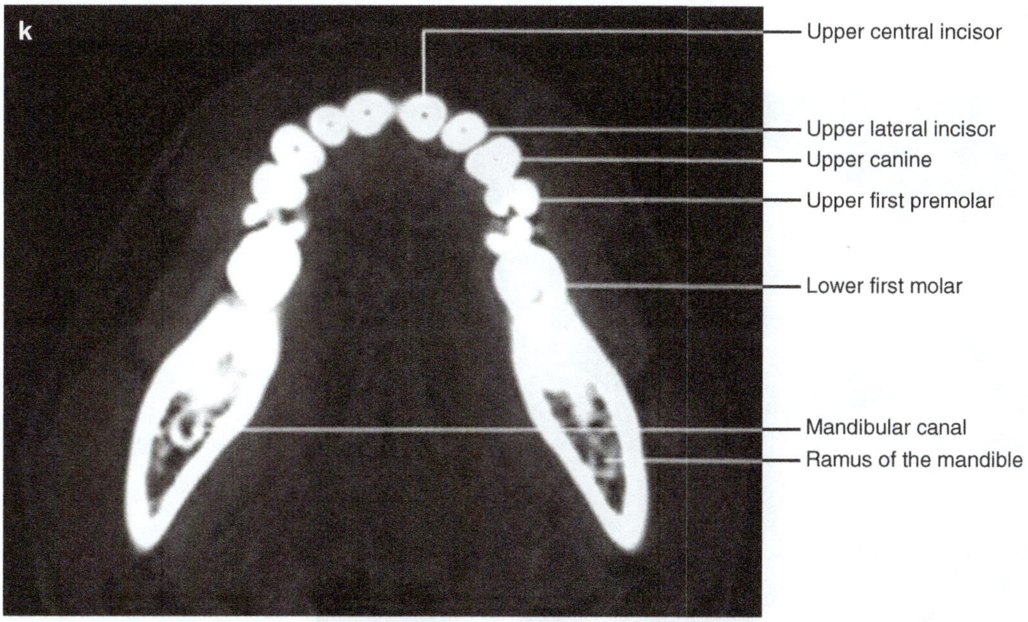

Fig. 9.19 (continued)

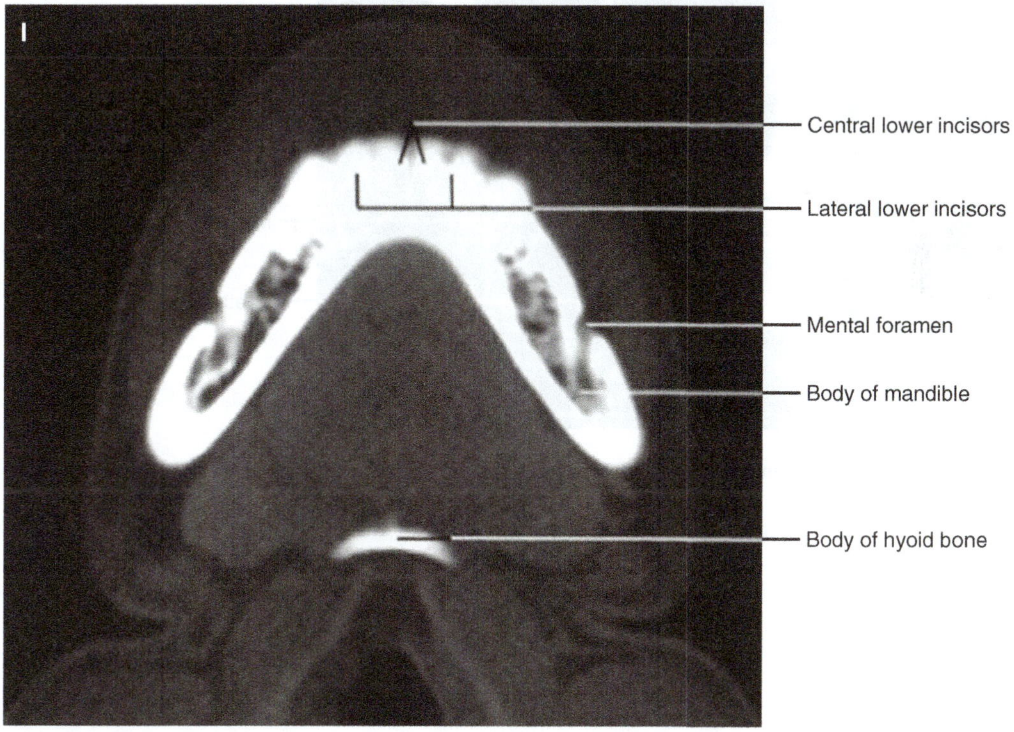

Fig. 9.19 (continued)

Coronal Slices: Anterior to Posterior

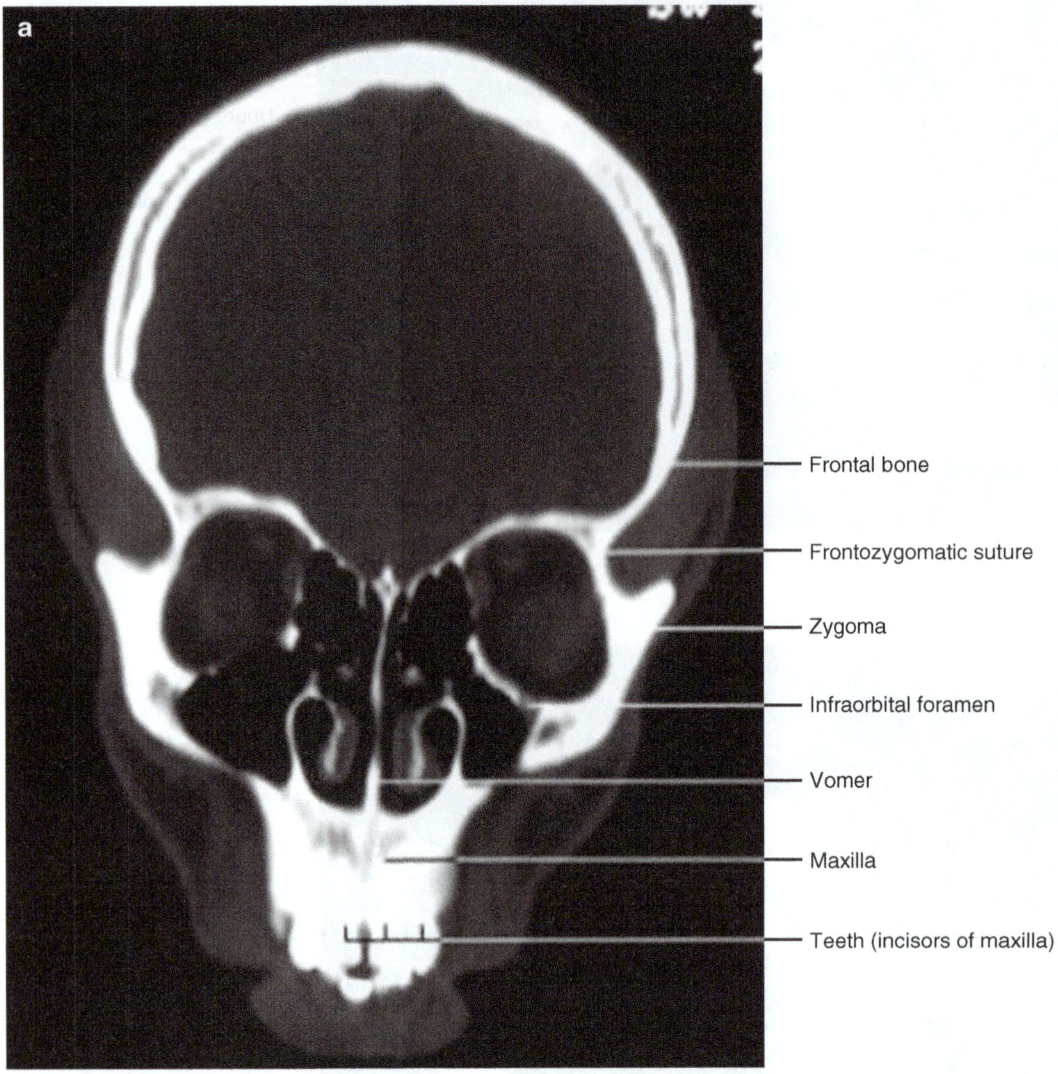

Fig. 9.20 (**a–j**) Coronal slices: anterior to posterior

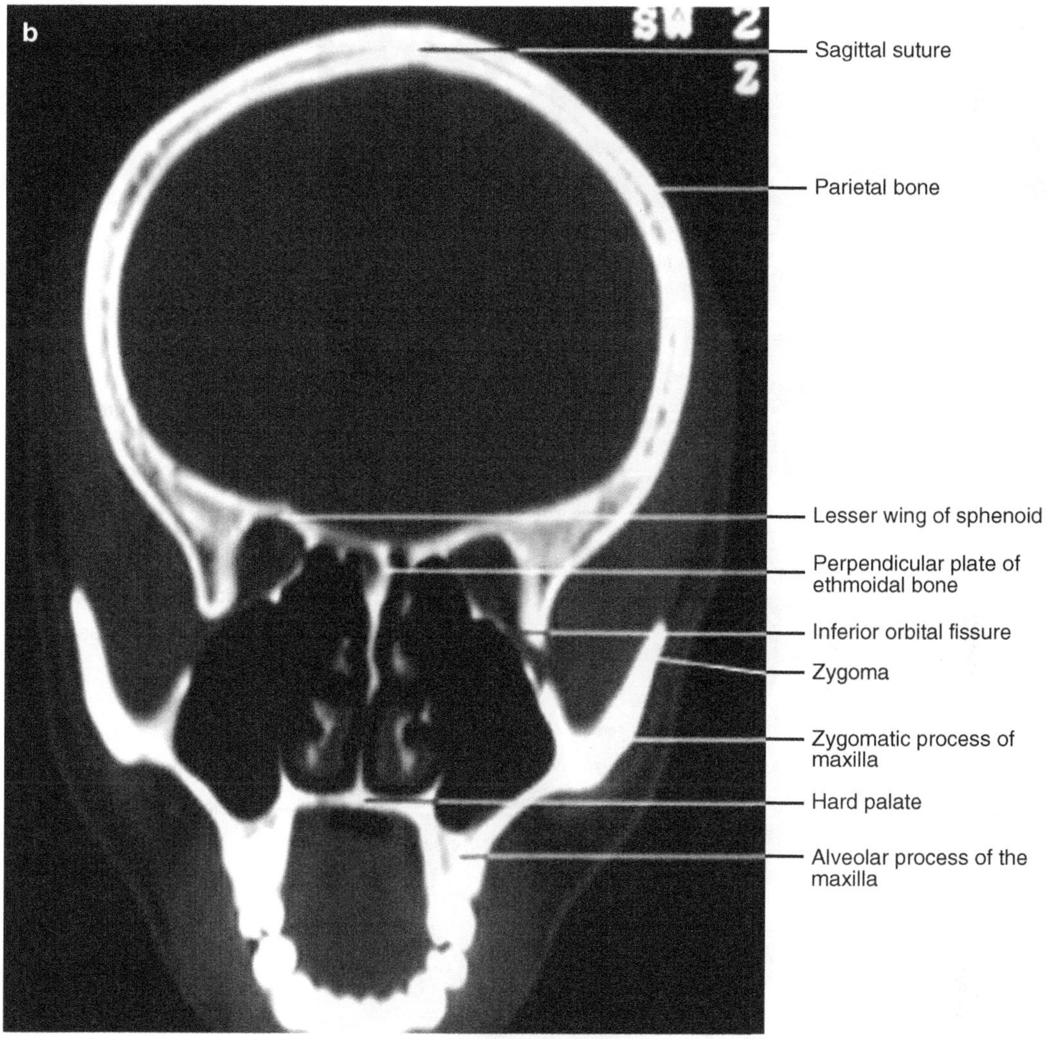

Fig. 9.20 (continued)

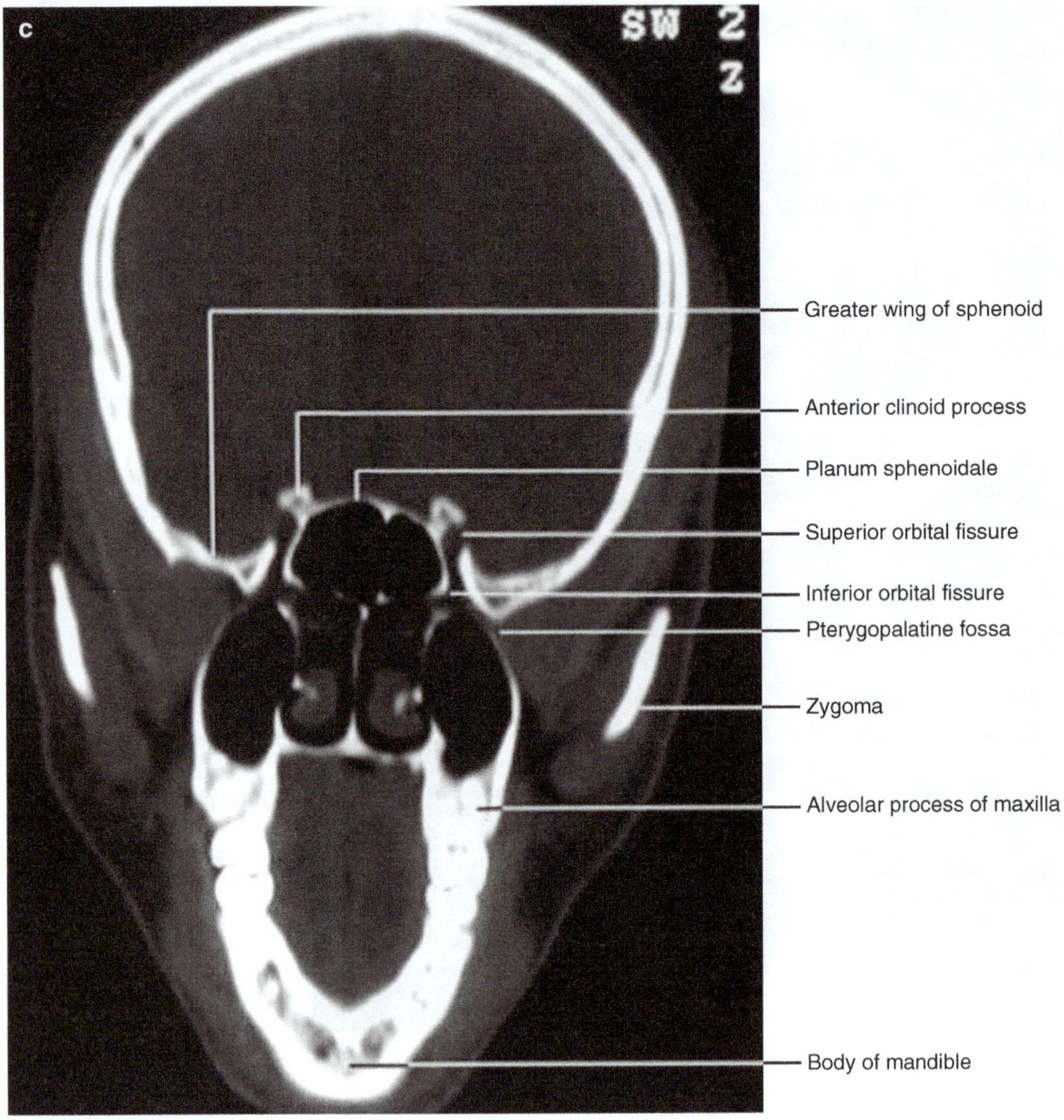

Fig. 9.20 (continued)

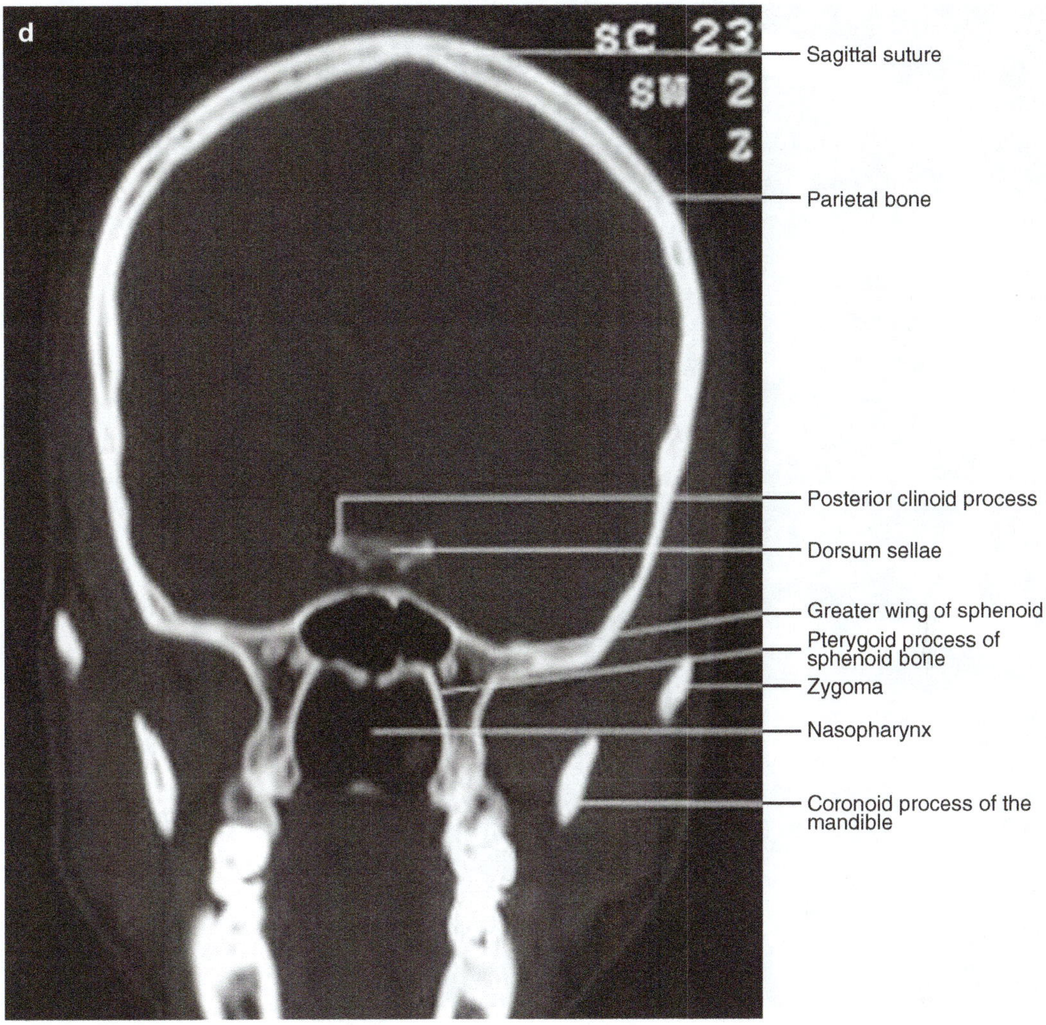

Fig. 9.20 (continued)

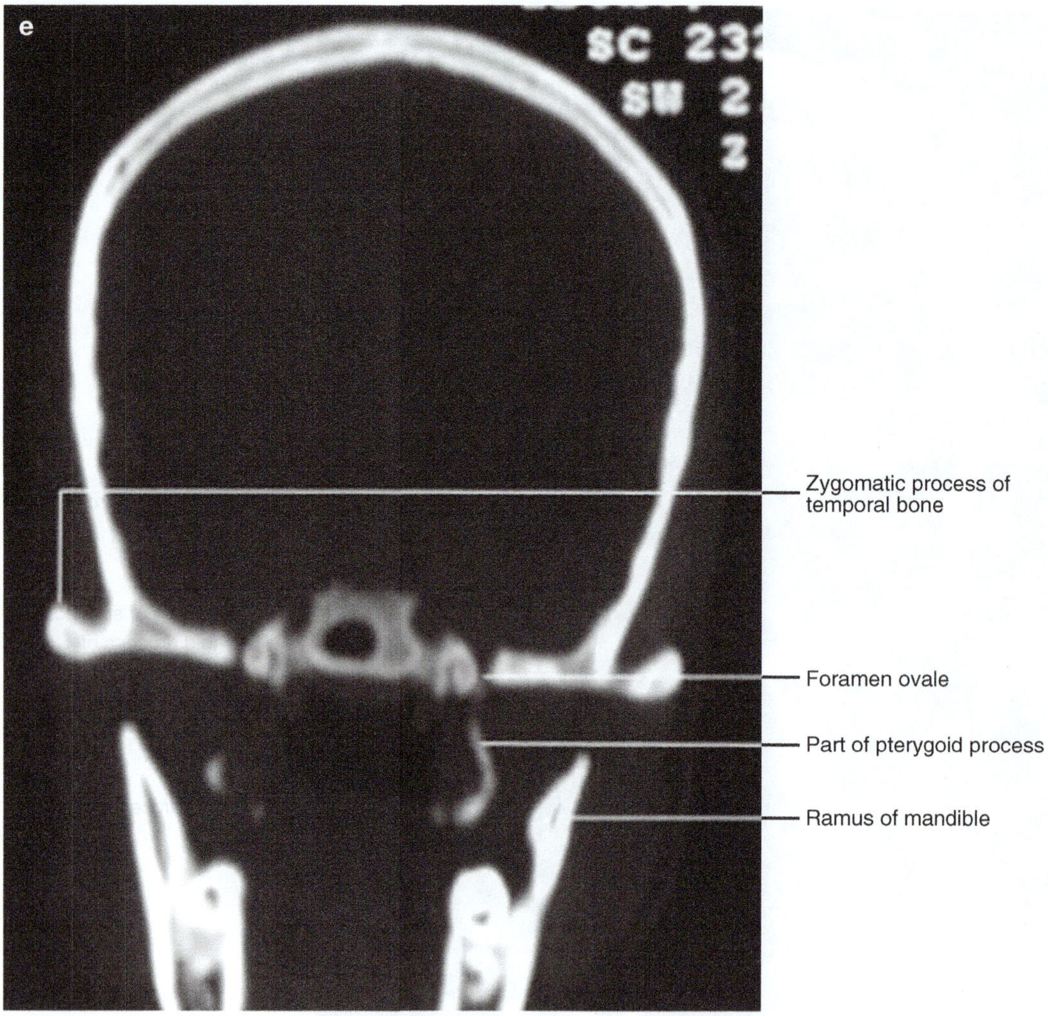

Fig. 9.20 (continued)

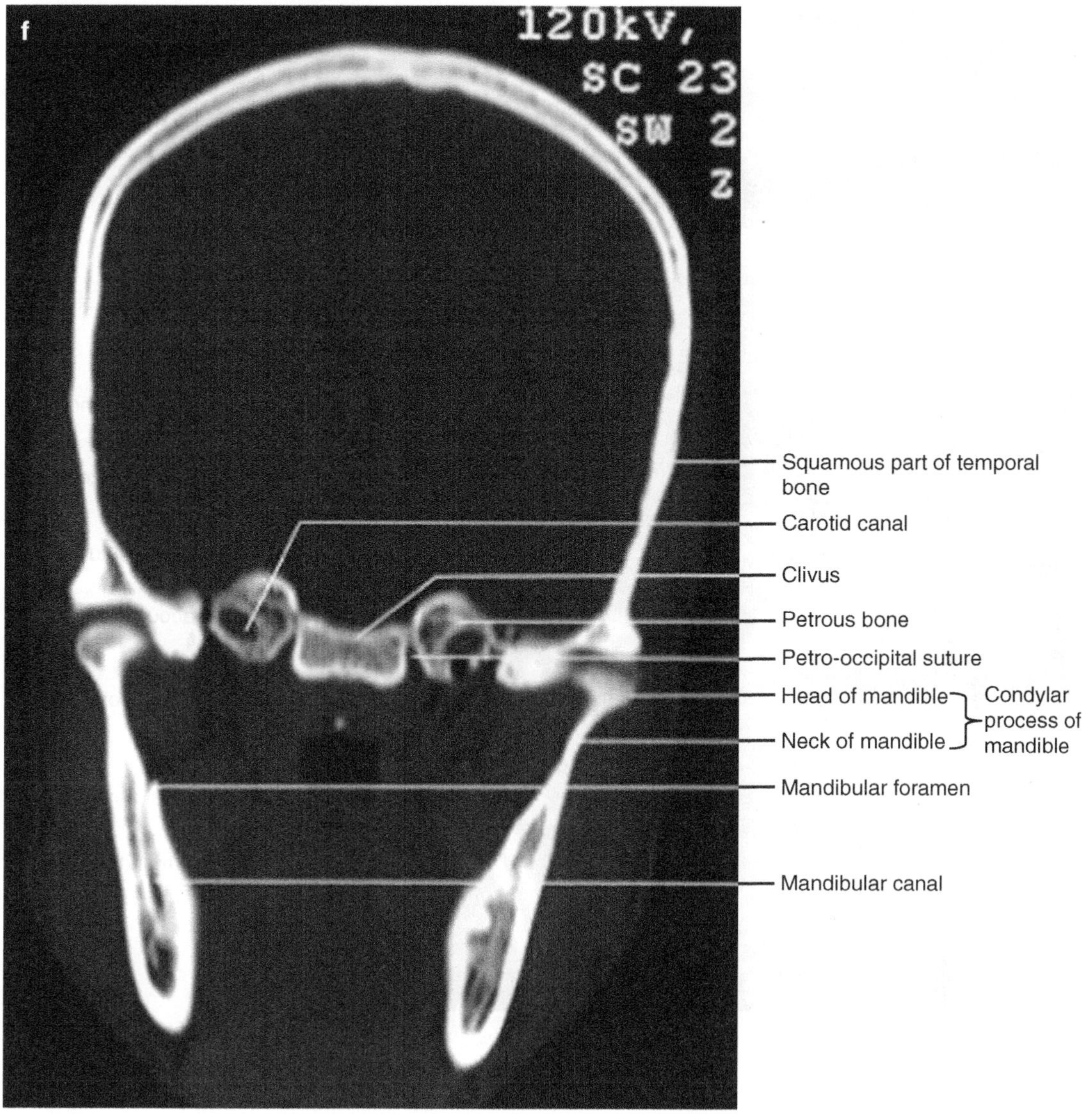

Fig. 9.20 (continued)

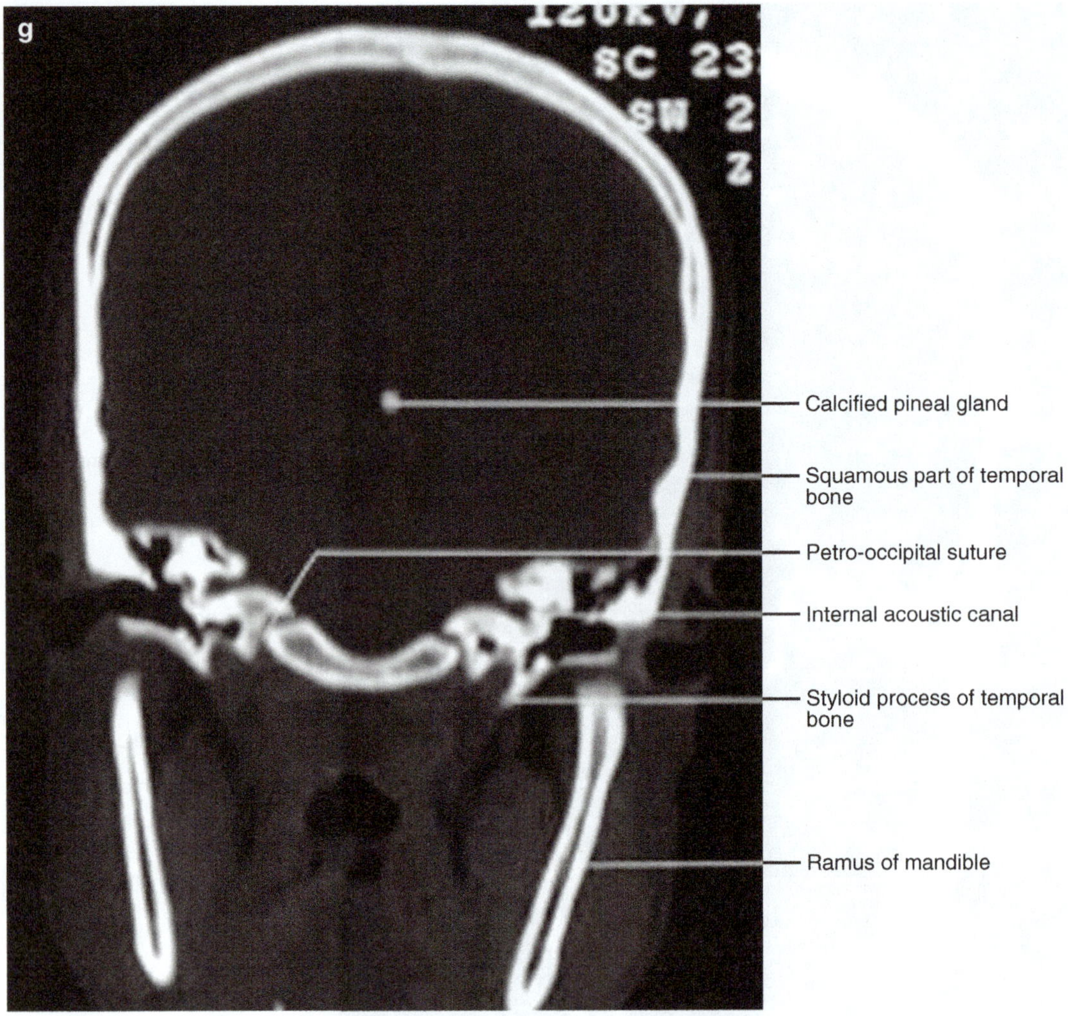

Fig. 9.20 (continued)

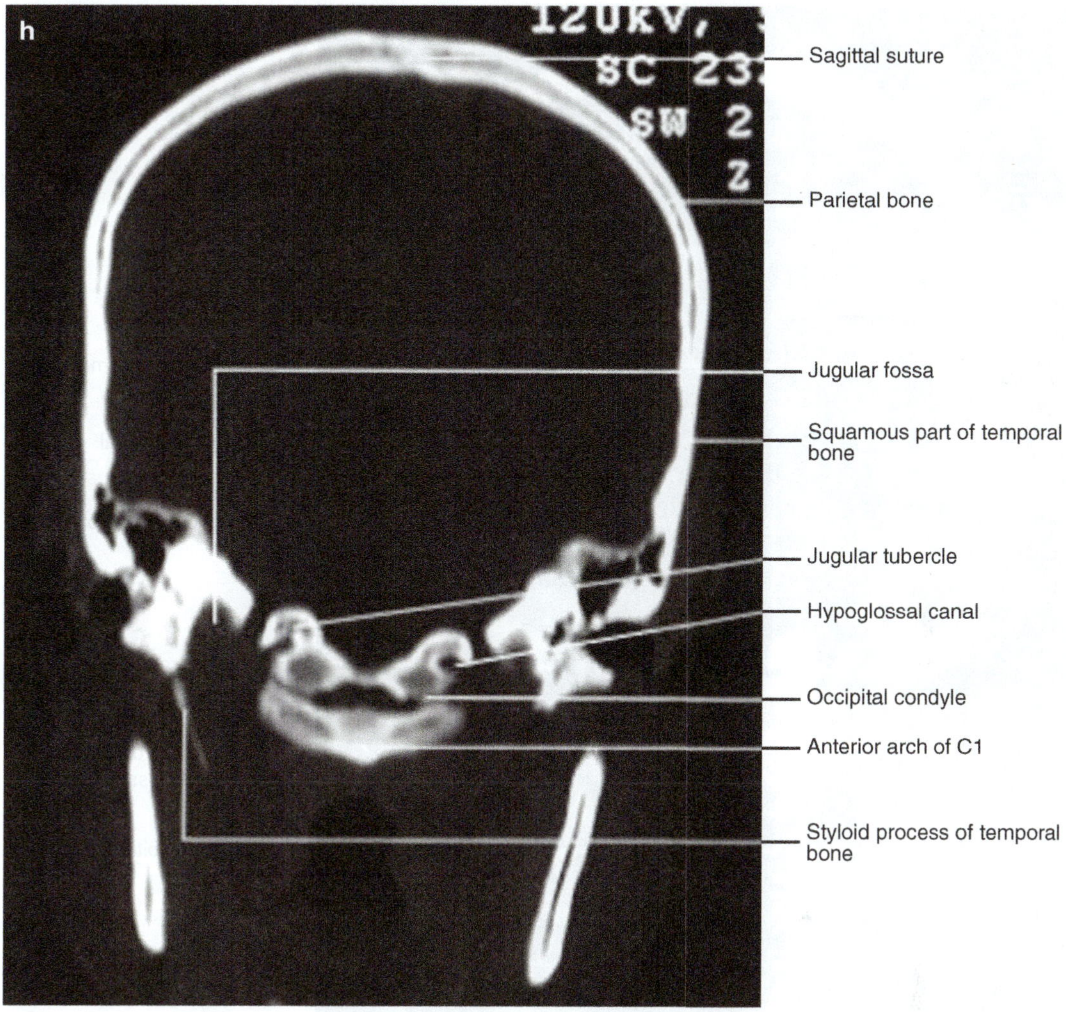

Fig. 9.20 (continued)

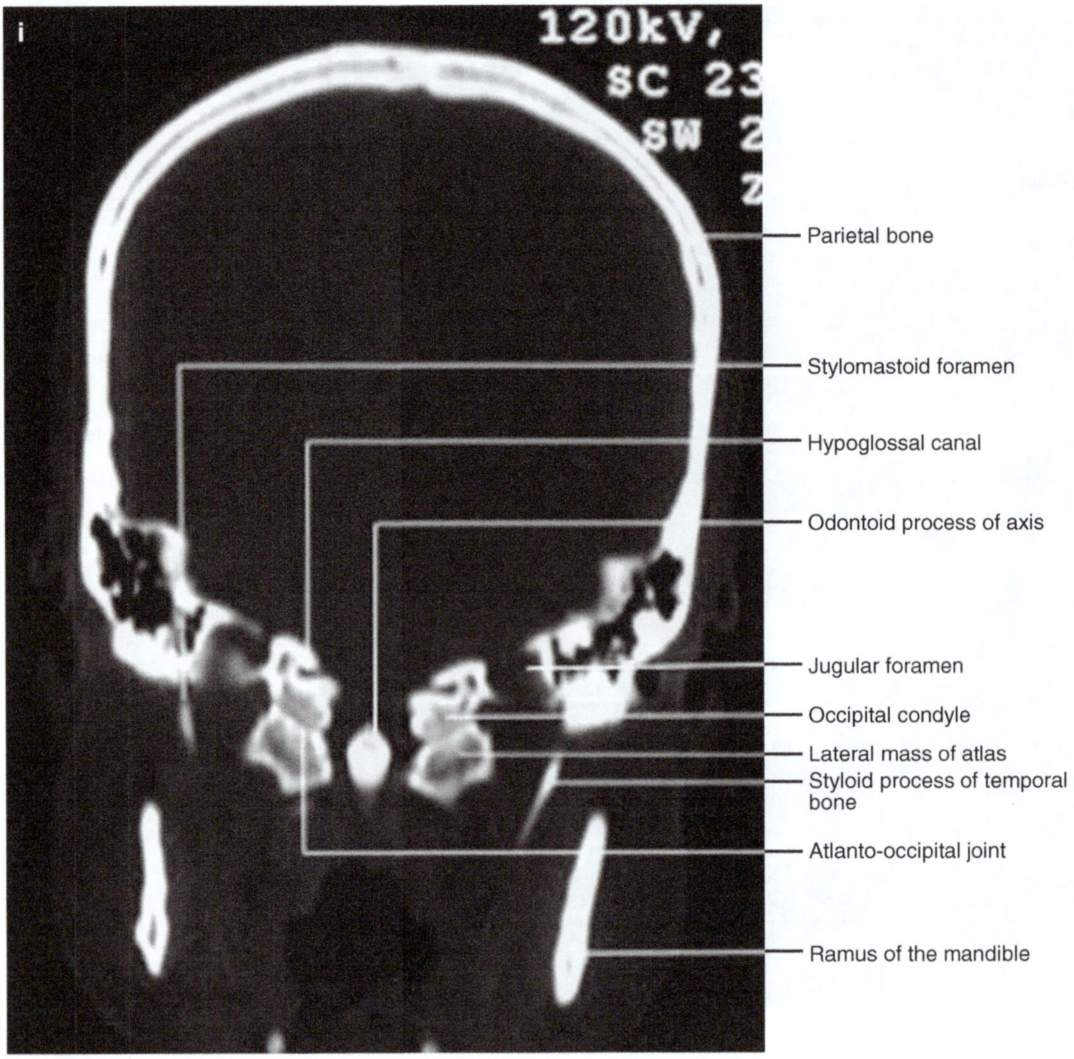

Fig. 9.20 (continued)

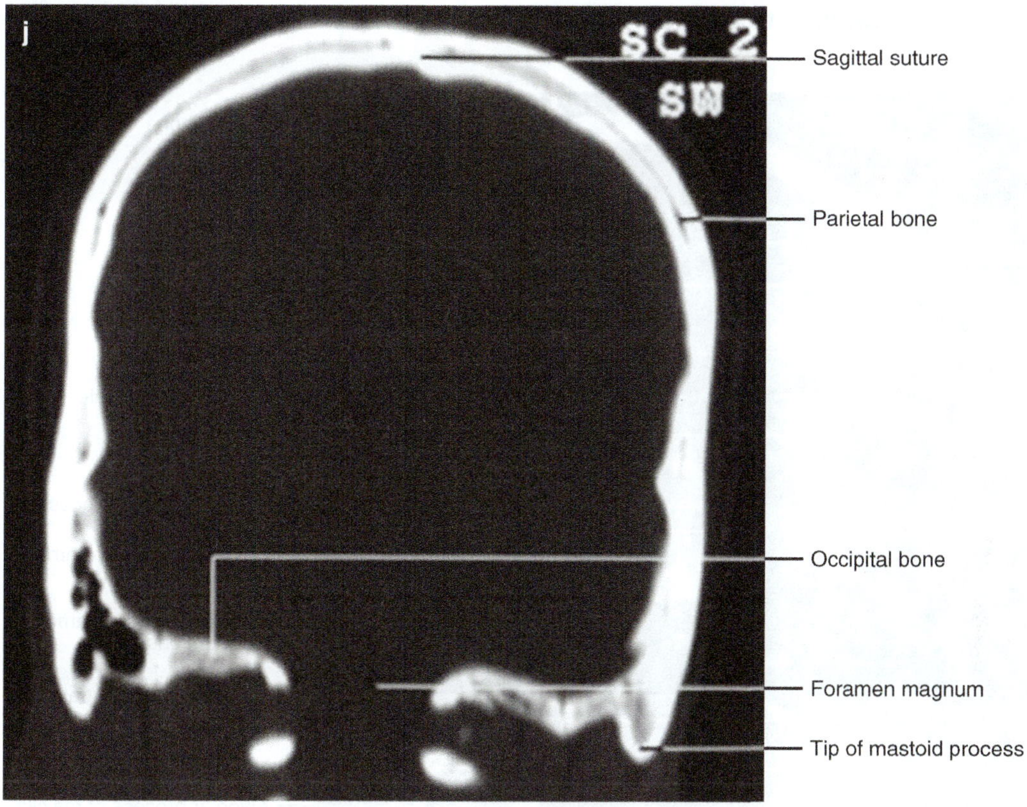

Fig. 9.20 (continued)

Sagittal Slices: Midline to Lateral

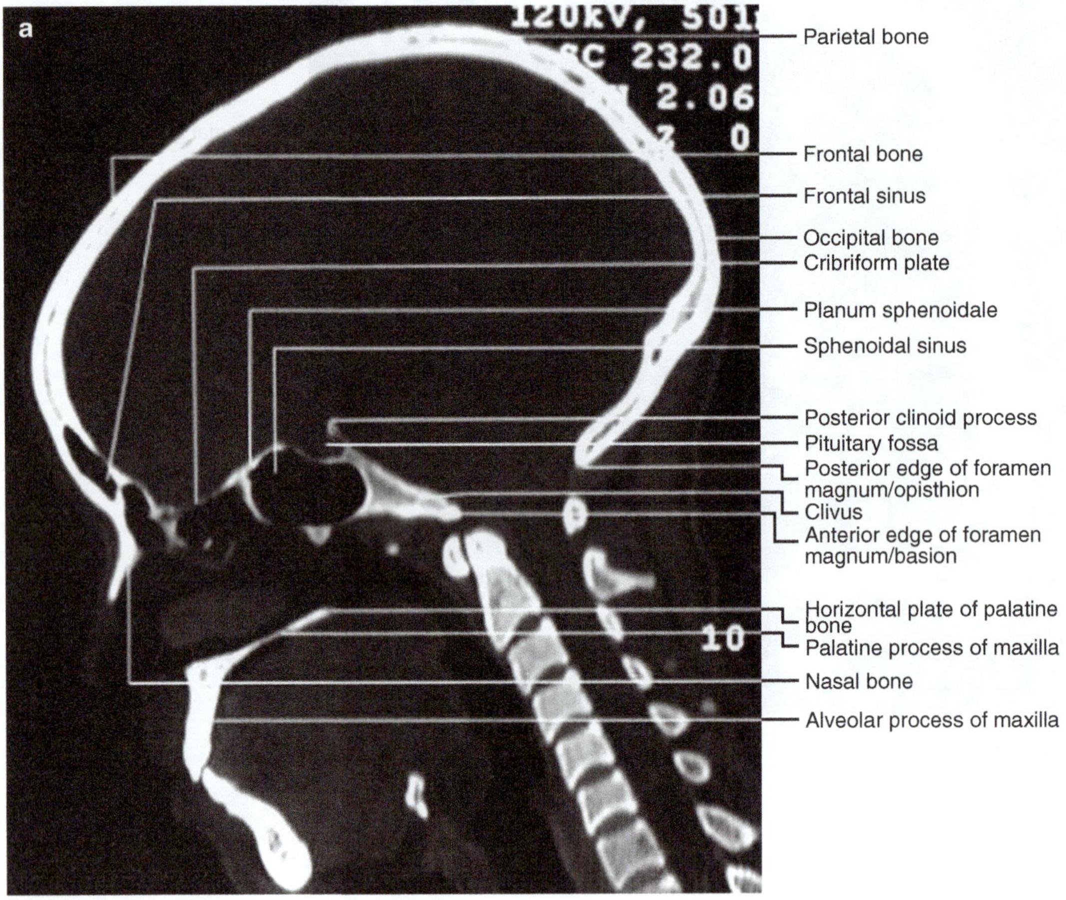

Fig. 9.21 (**a–d**) Sagittal slices: midline to lateral

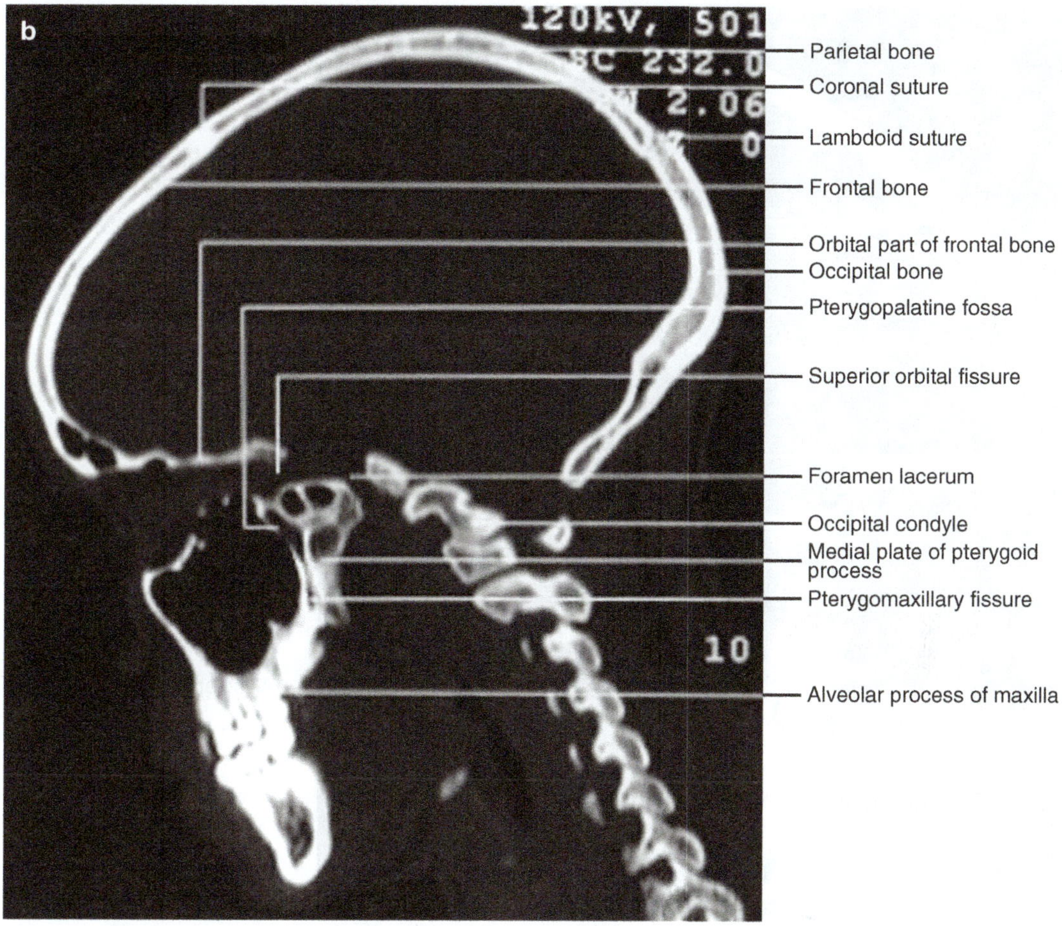

Fig. 9.21 (continued)

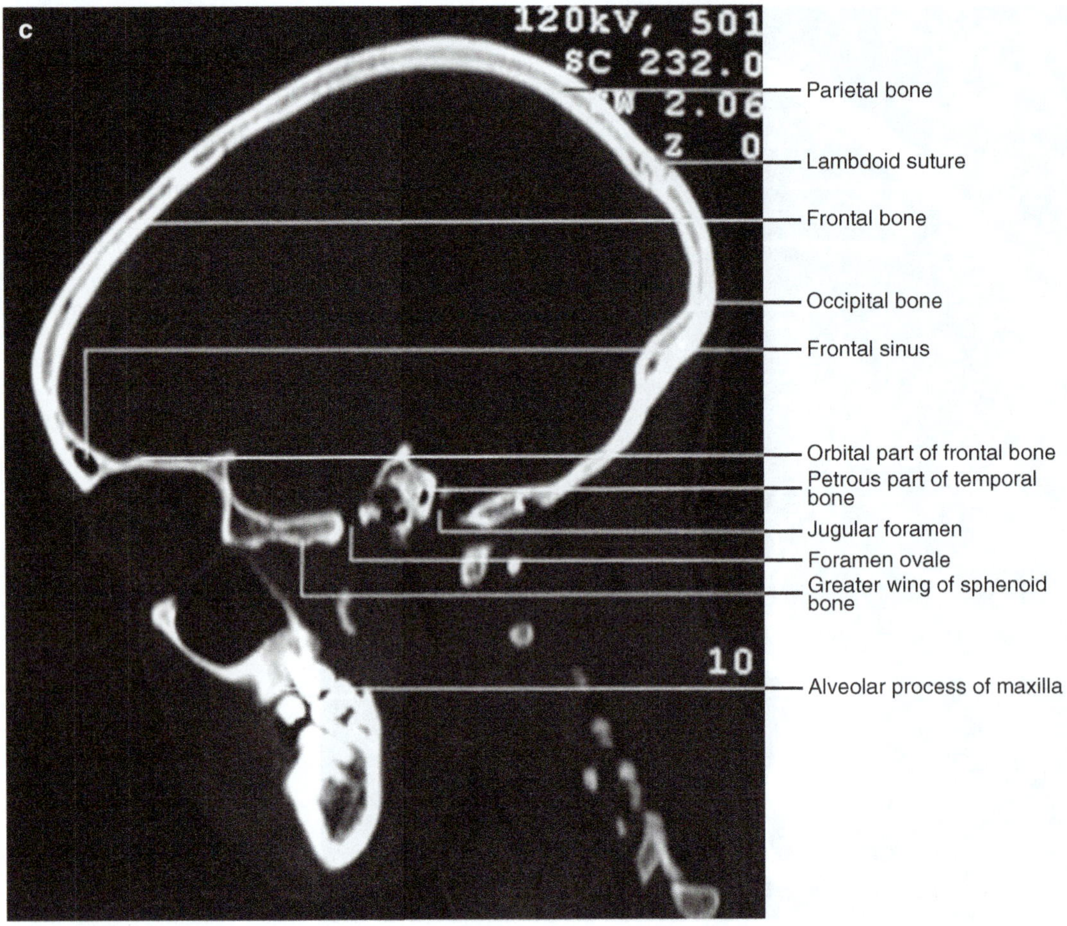

Fig. 9.21 (continued)

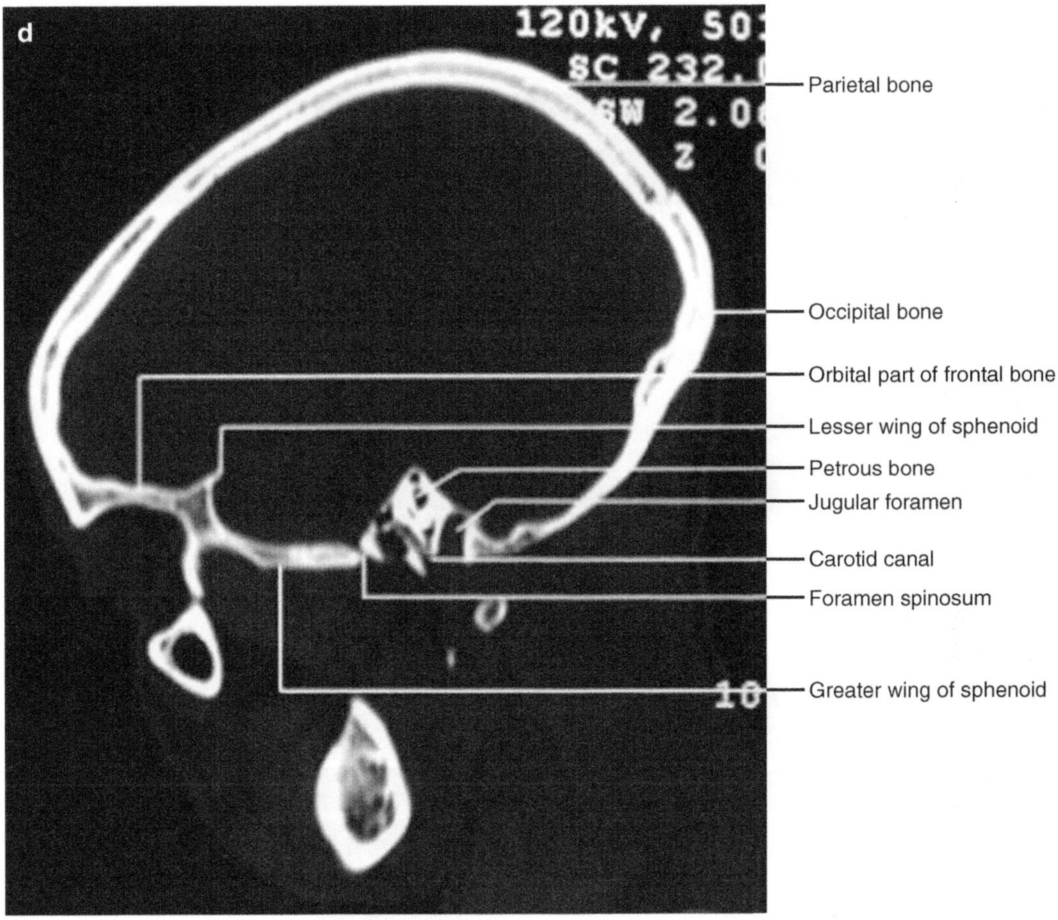

Fig. 9.21 (continued)

3D Reconstructions

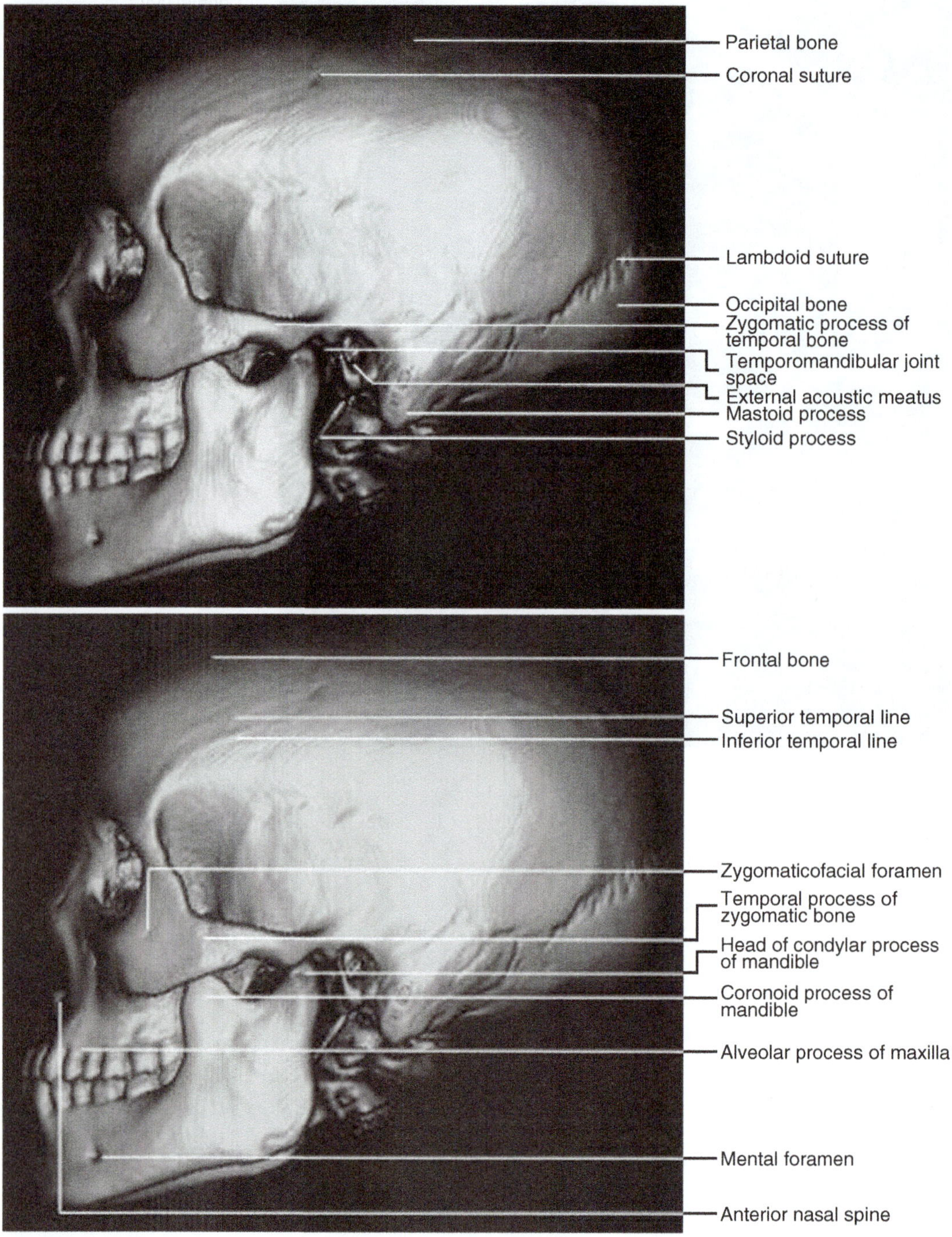

Fig. 9.22 Lateral view 3D Volume rendered reconstructions of the skull

Fig. 9.23 Anterior view 3D Volume rendered reconstructions of the skull

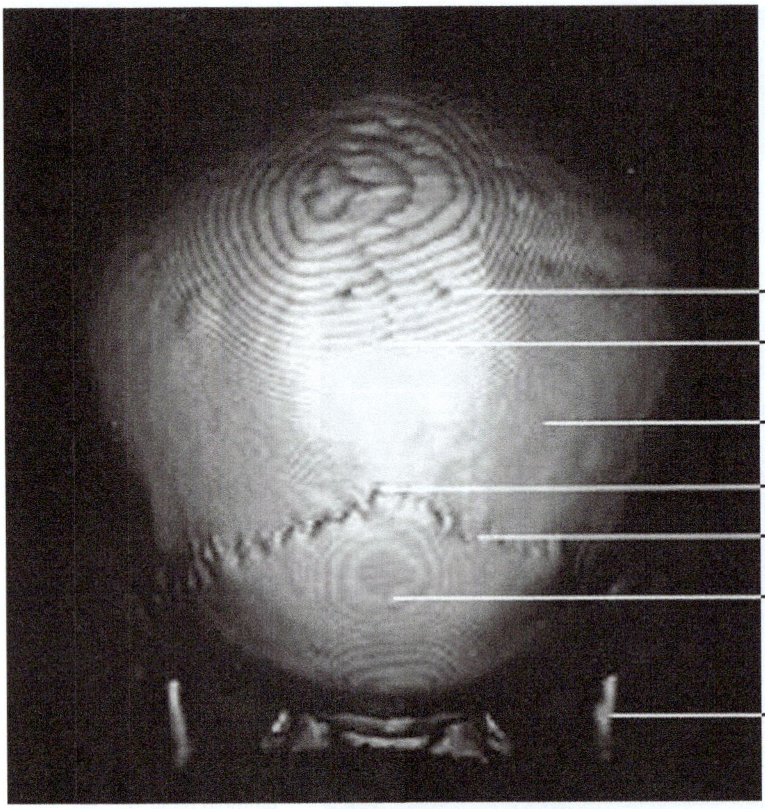

Fig. 9.24 Posterior view

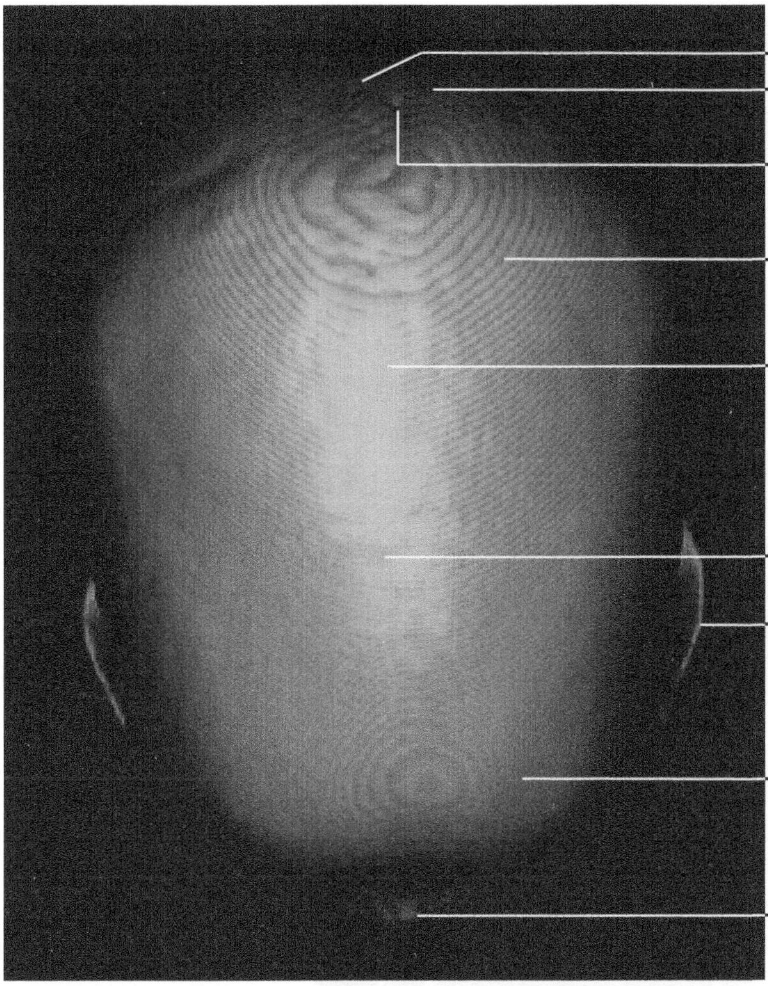

Fig. 9.25 Superior view

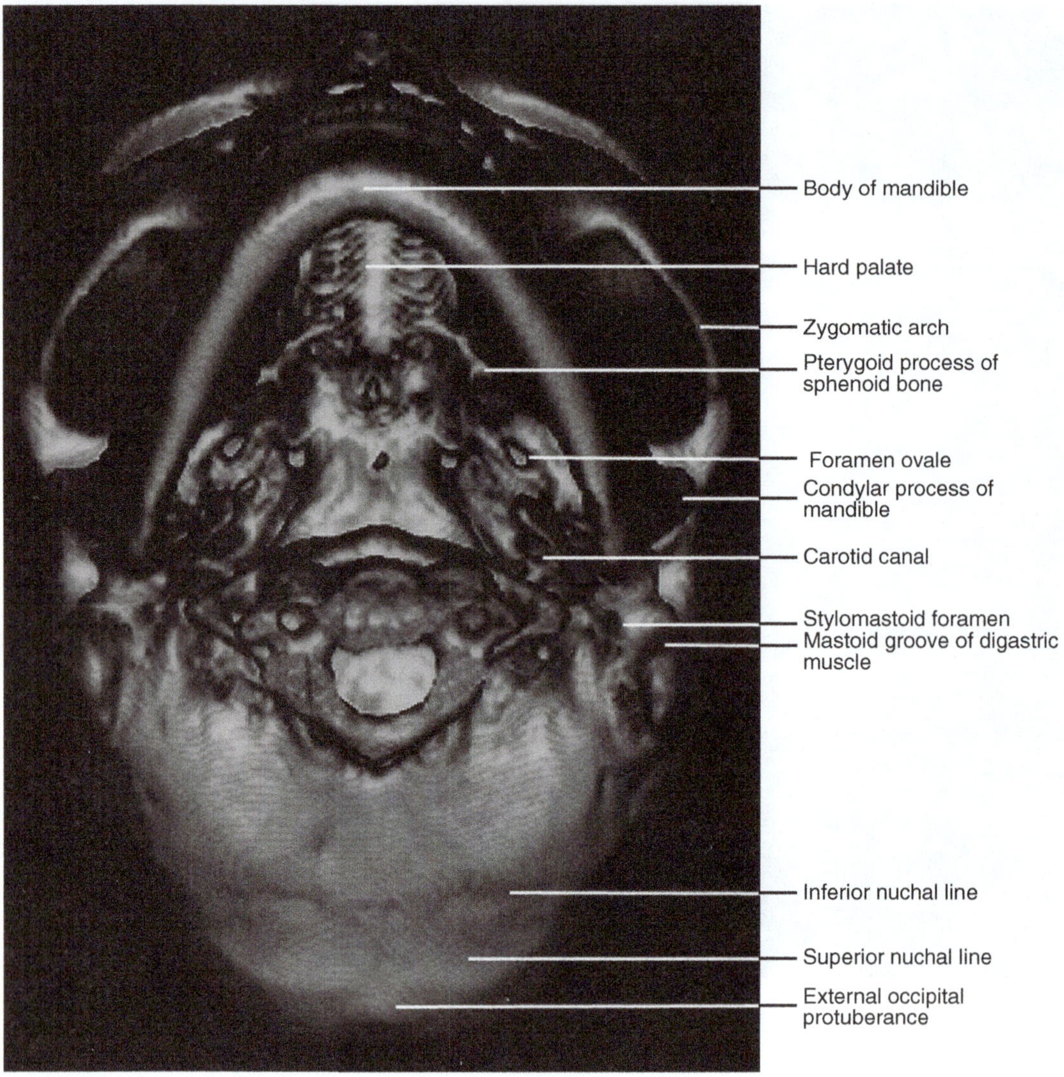

Fig. 9.26 Inferior view

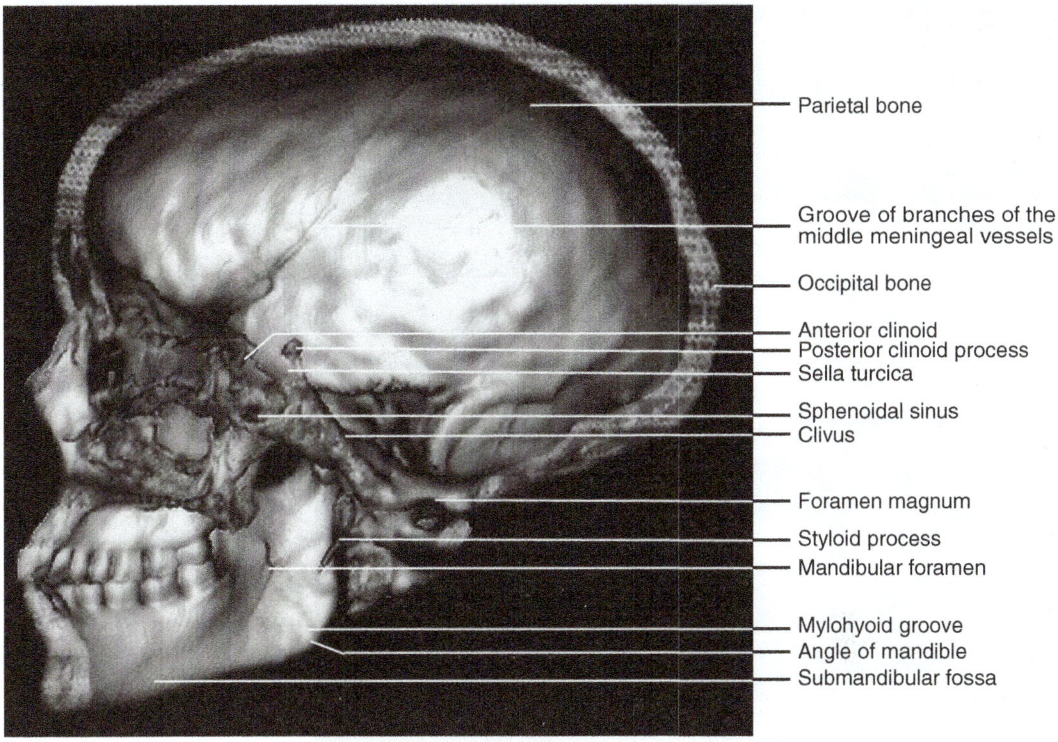

Fig. 9.27 Internal lateral view

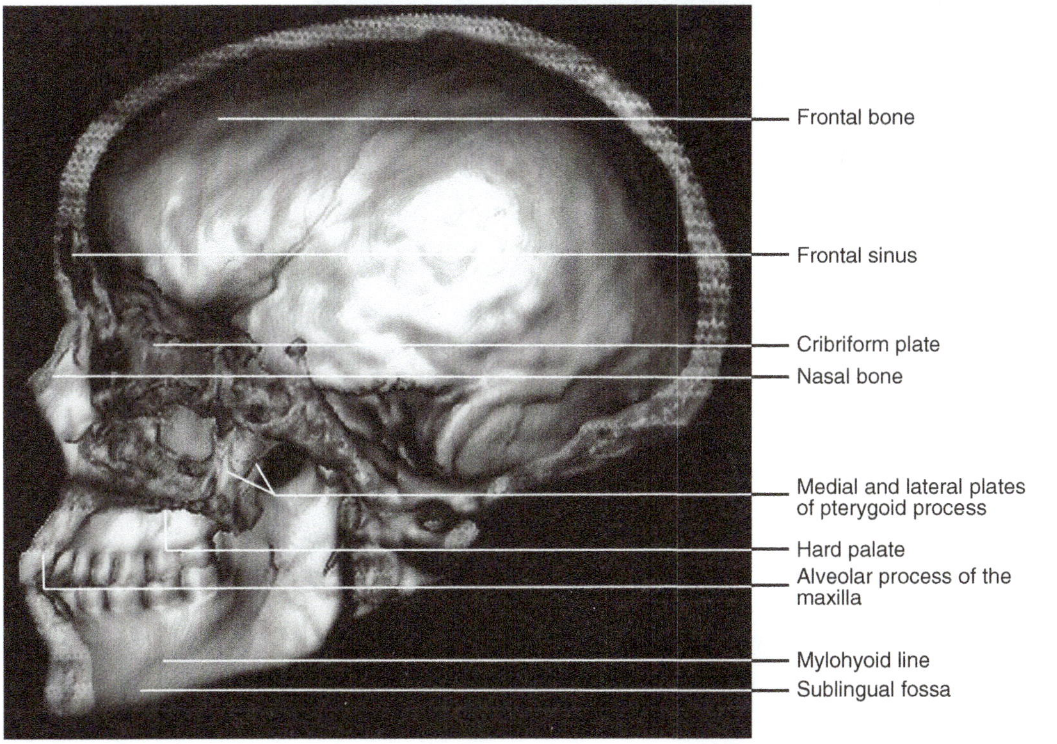

Fig. 9.28 Internal lateral view

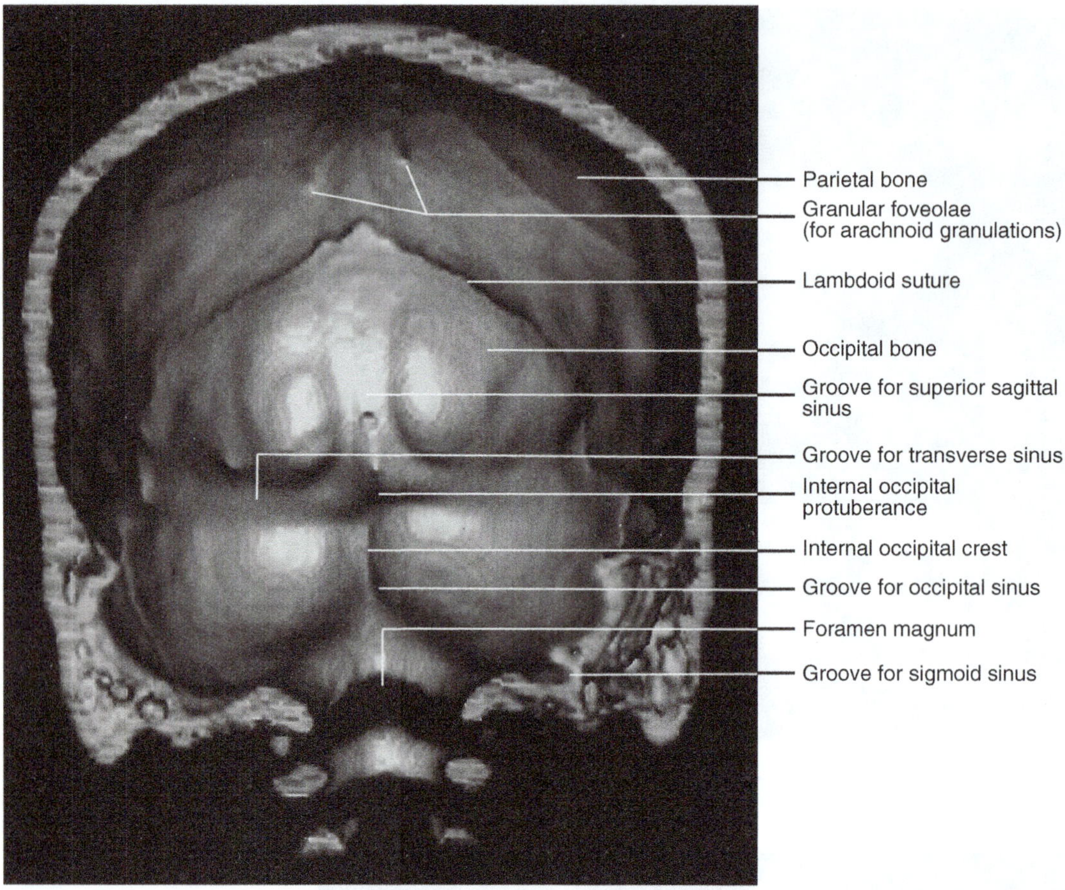

Fig. 9.29 Internal anteroposterior view

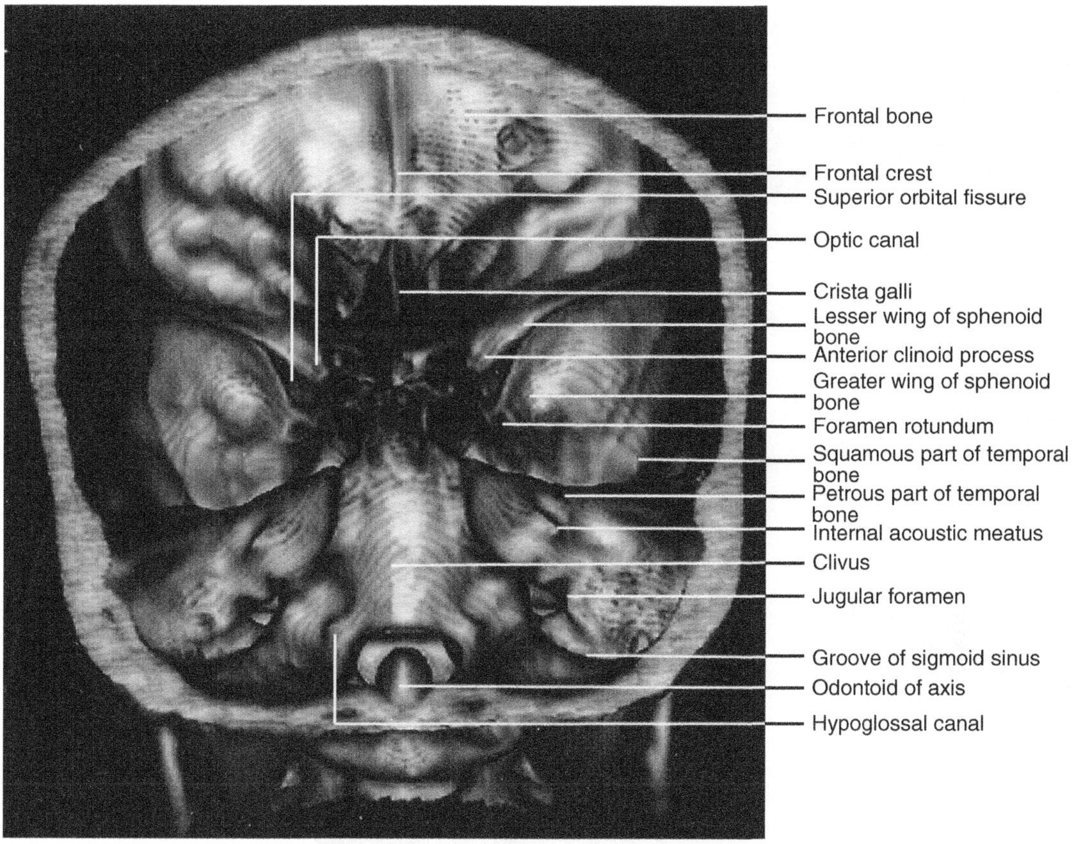

Fig. 9.30 Lateral posteroanterior view

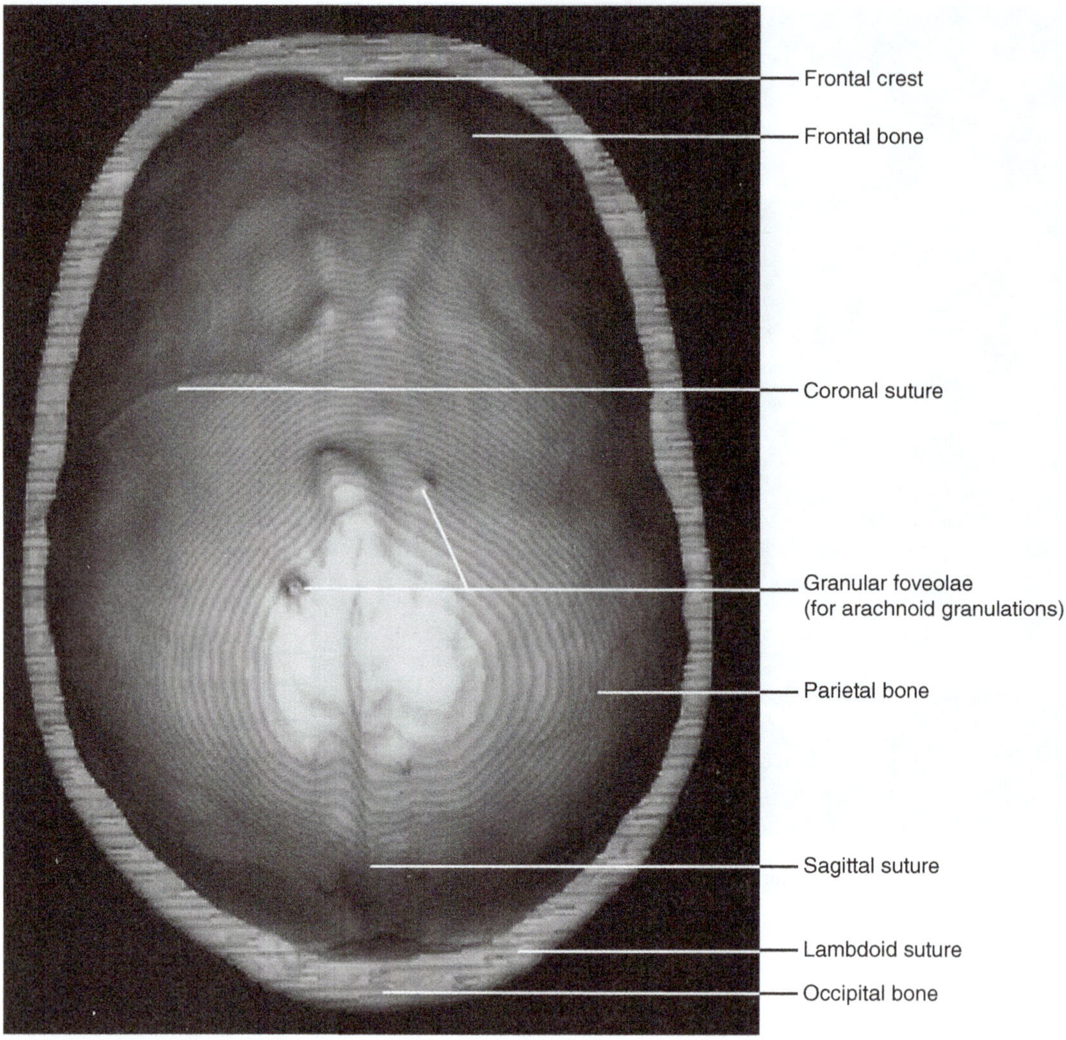

Fig. 9.31 Internal inferosuperior view

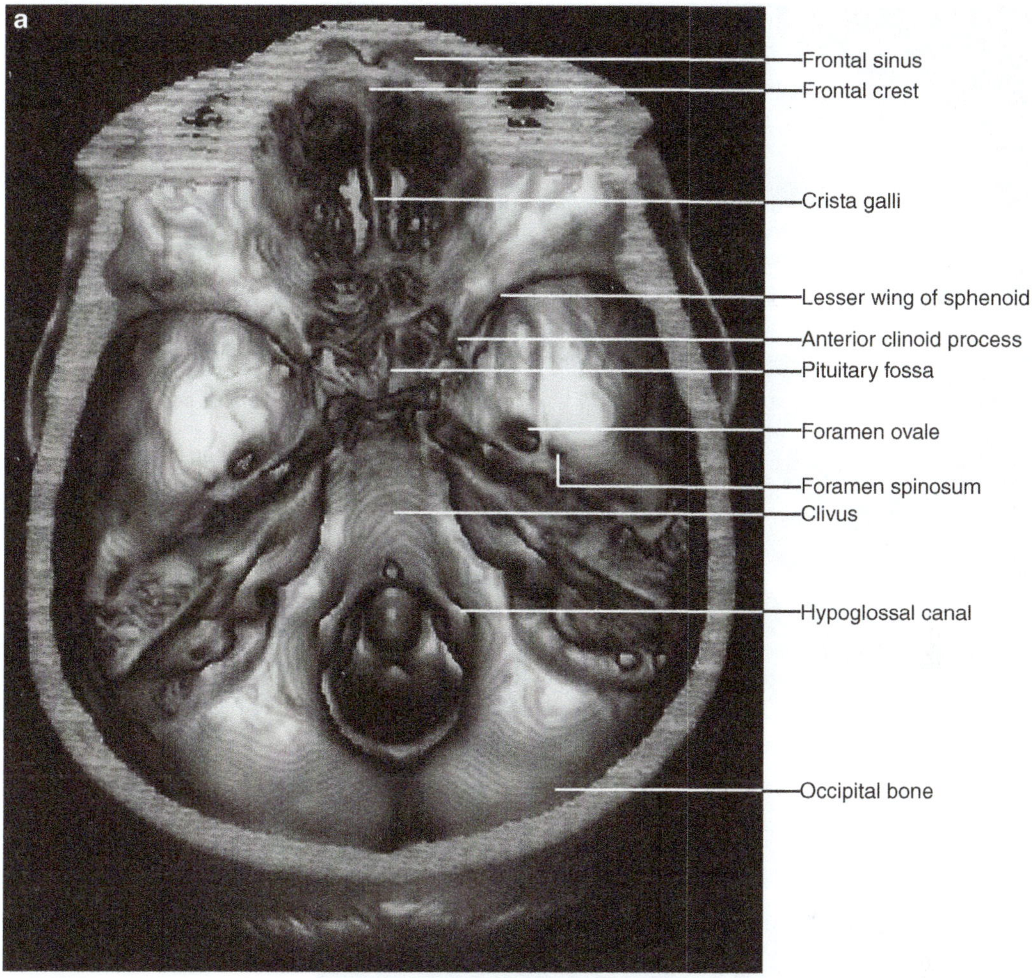

Fig. 9.32 Internal superoinferior view 3D Volume rendered reconstructions of the skull

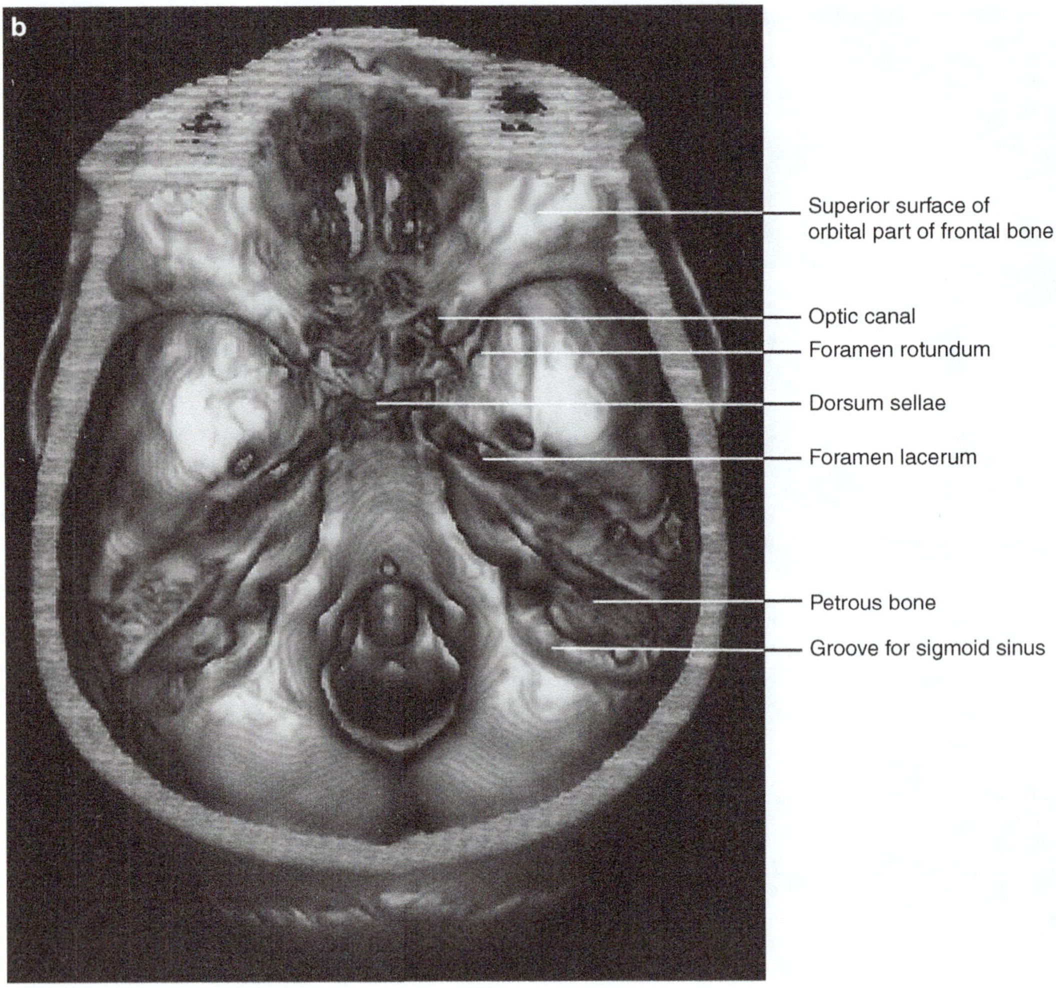

Fig. 9.32 (continued)

Magnetic Resonance Imaging

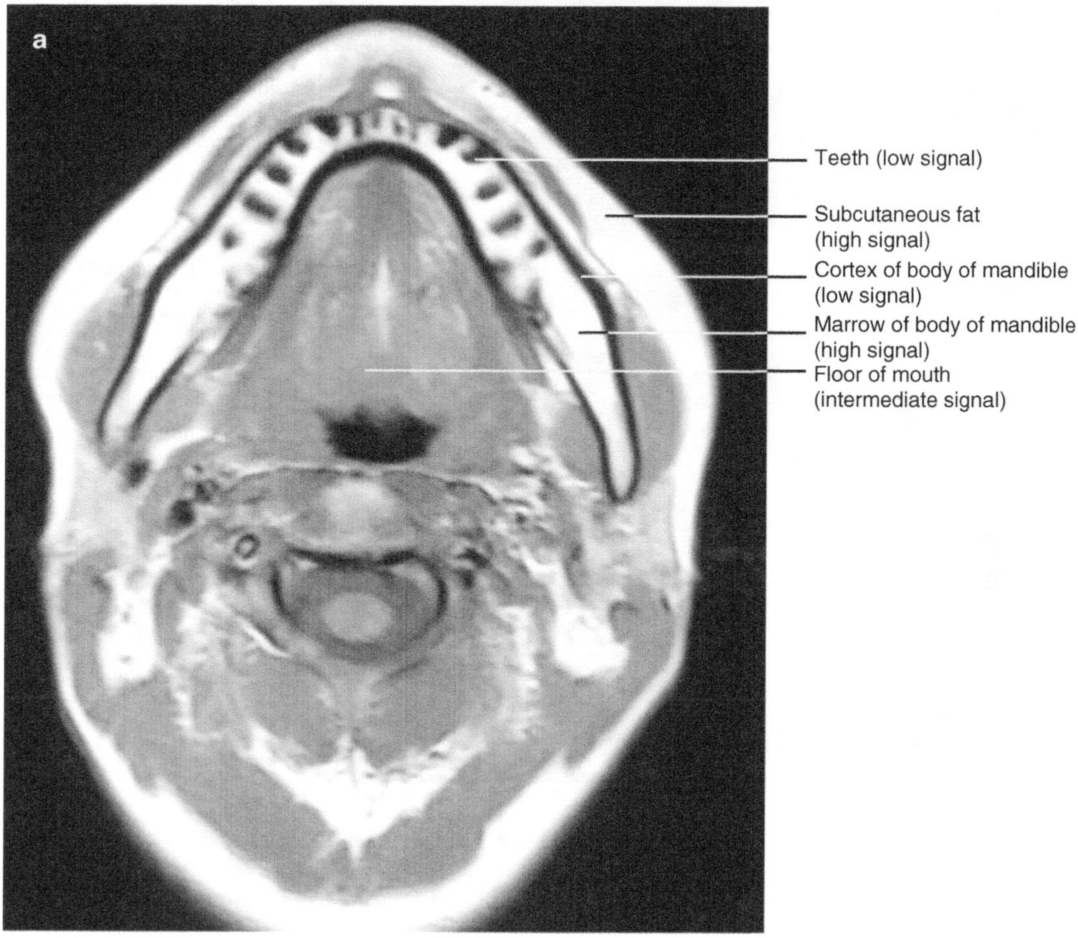

Fig. 9.33 T1 Axial 3D Volume rendered reconstructions of the skull

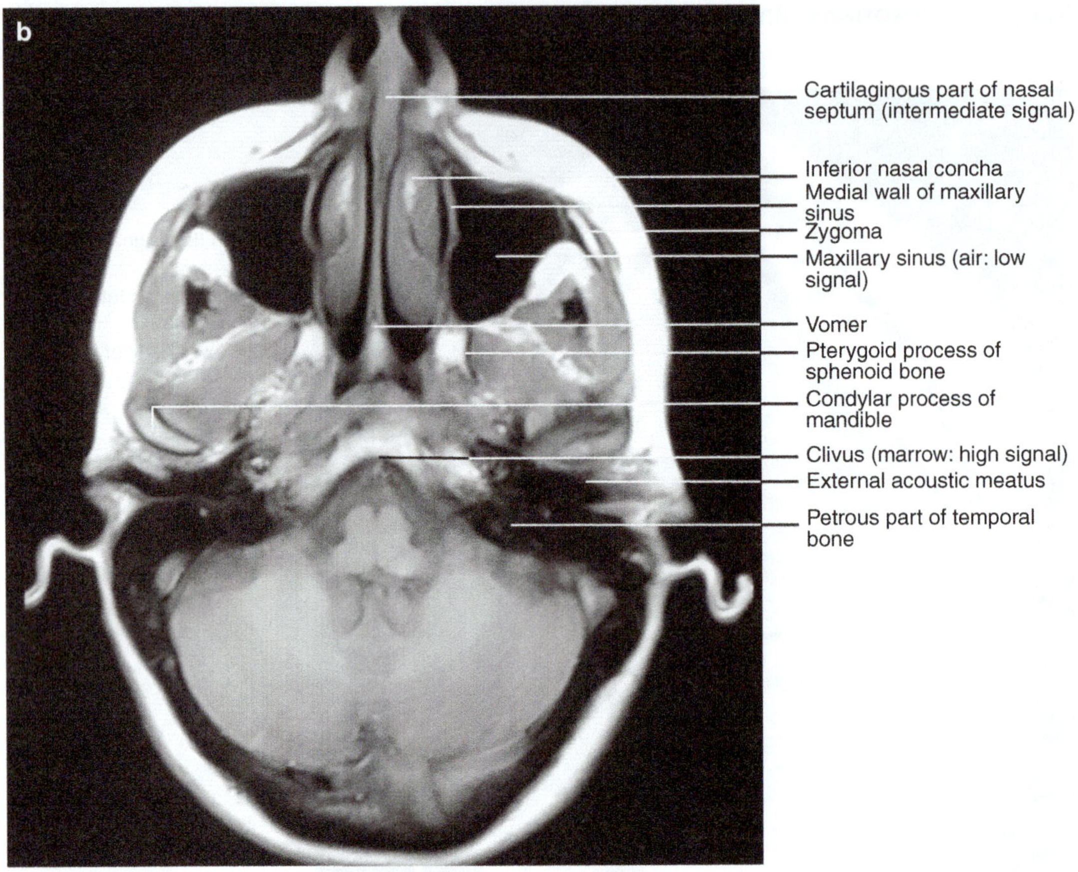

Fig. 9.33 (continued)

a

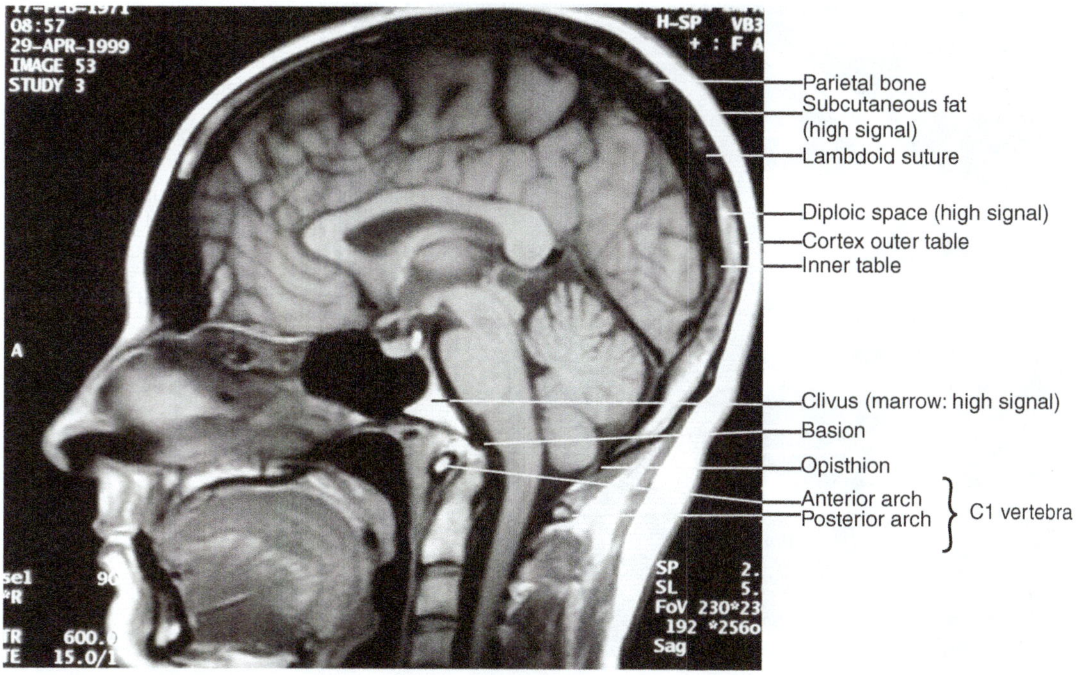

b

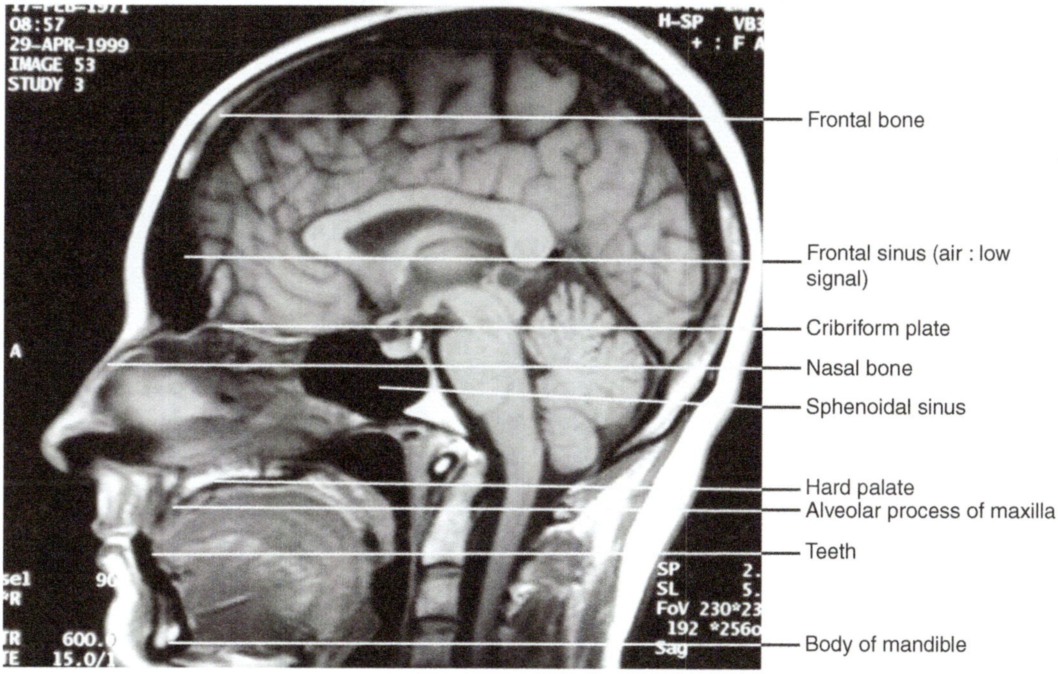

Fig. 9.34 a–c: T1 sagital of the skull

c

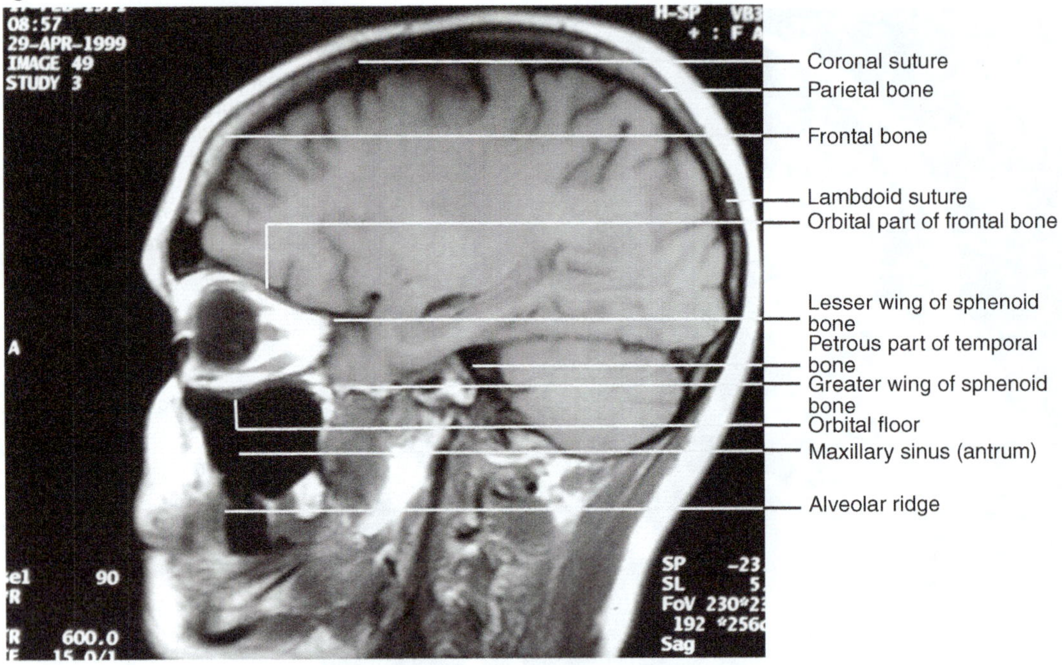

Fig. 9.34 (continued)

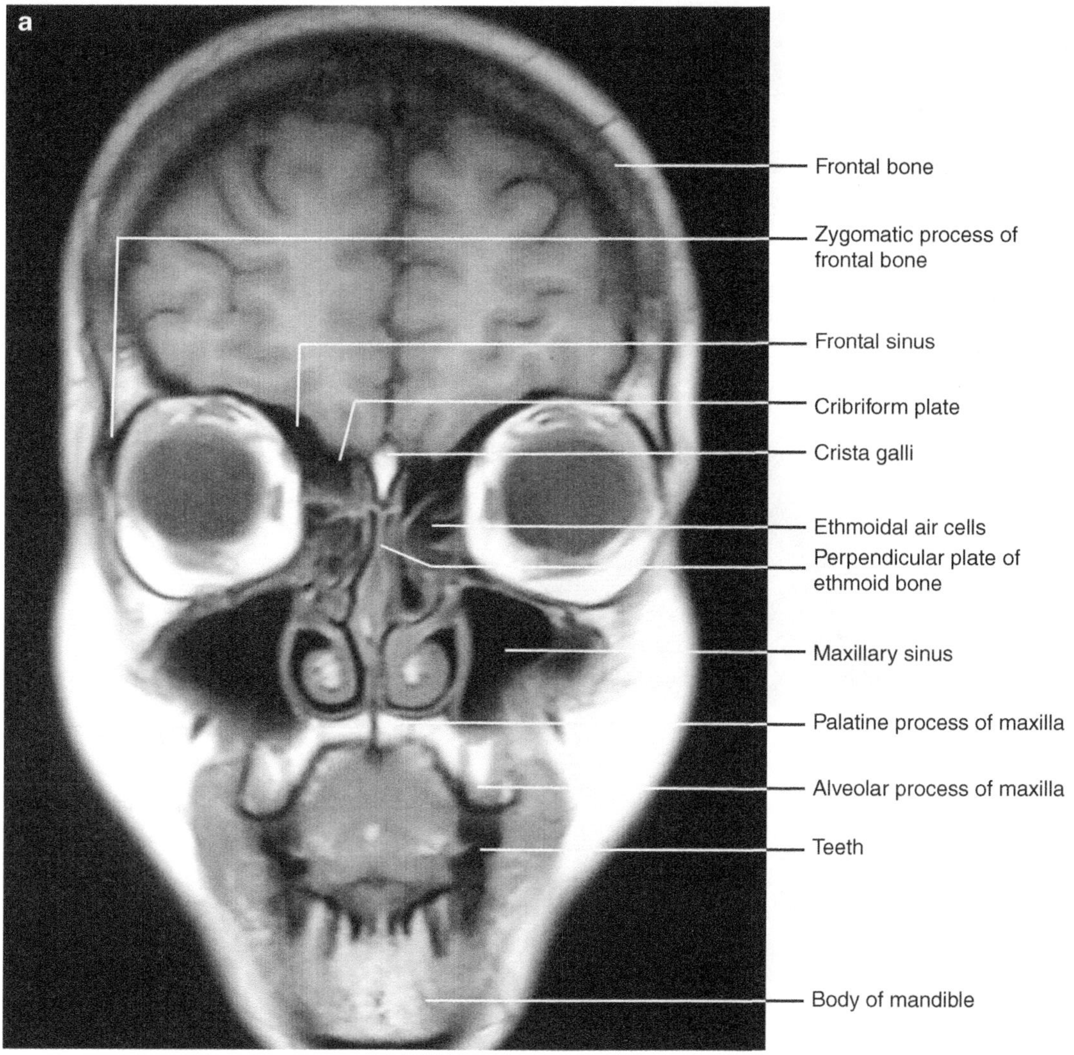

Fig. 9.35 T1 Coronal MRI of the temporomandibular joints

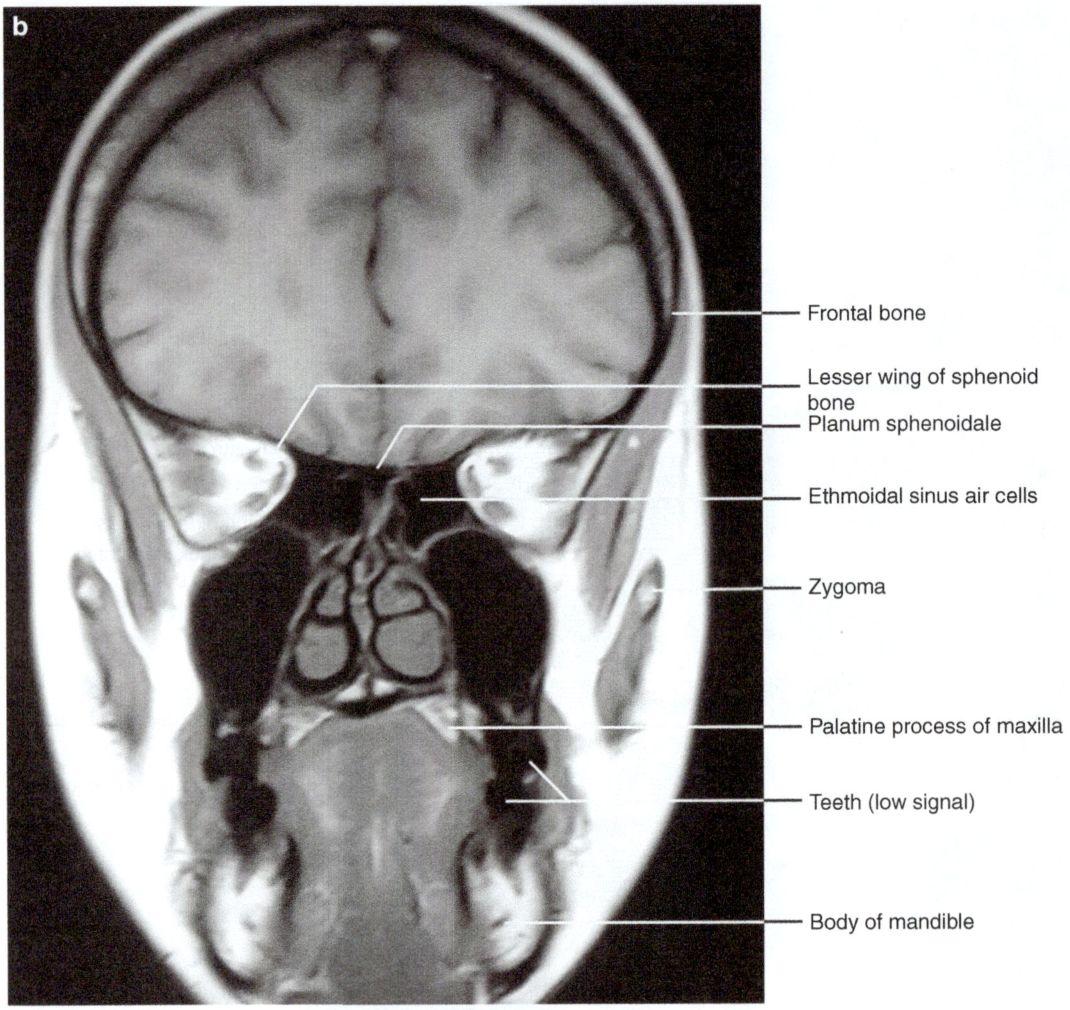

Fig. 9.35 (continued)

Temporomandibular Joint

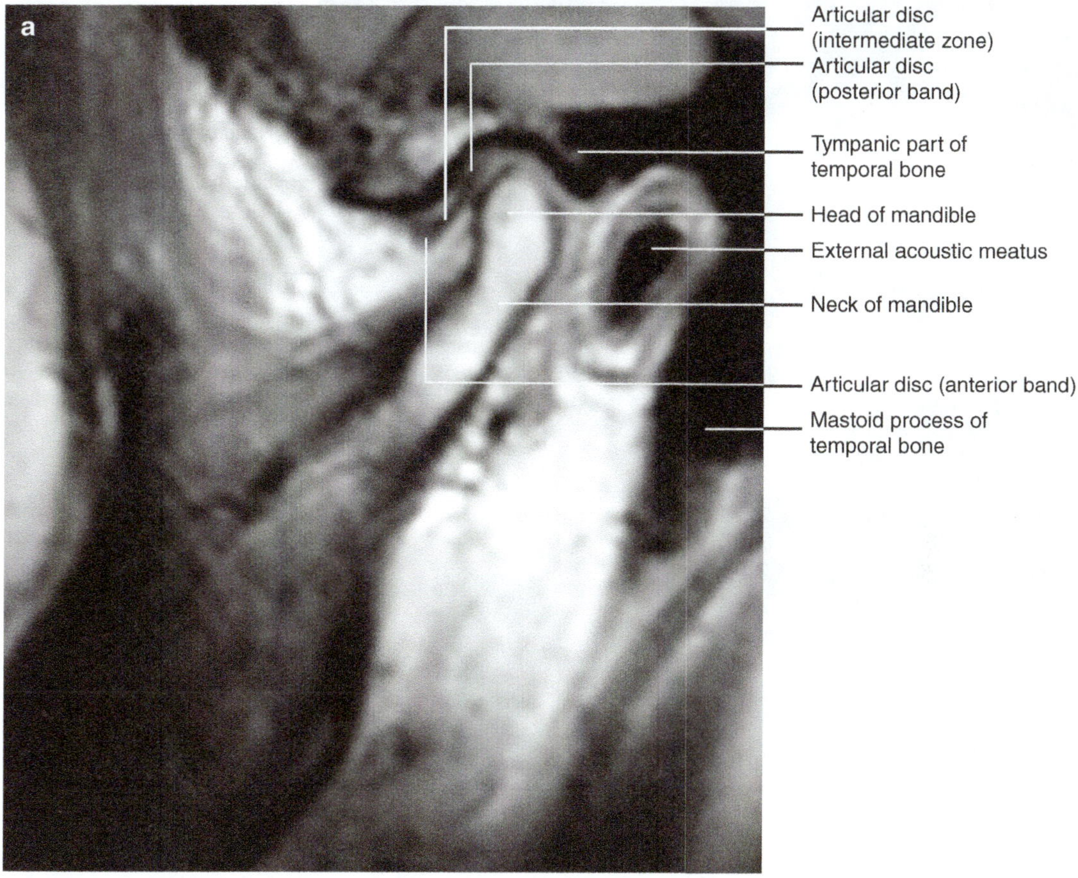

Fig. 9.36 Mouth closed MRI of the temporomandibular joints

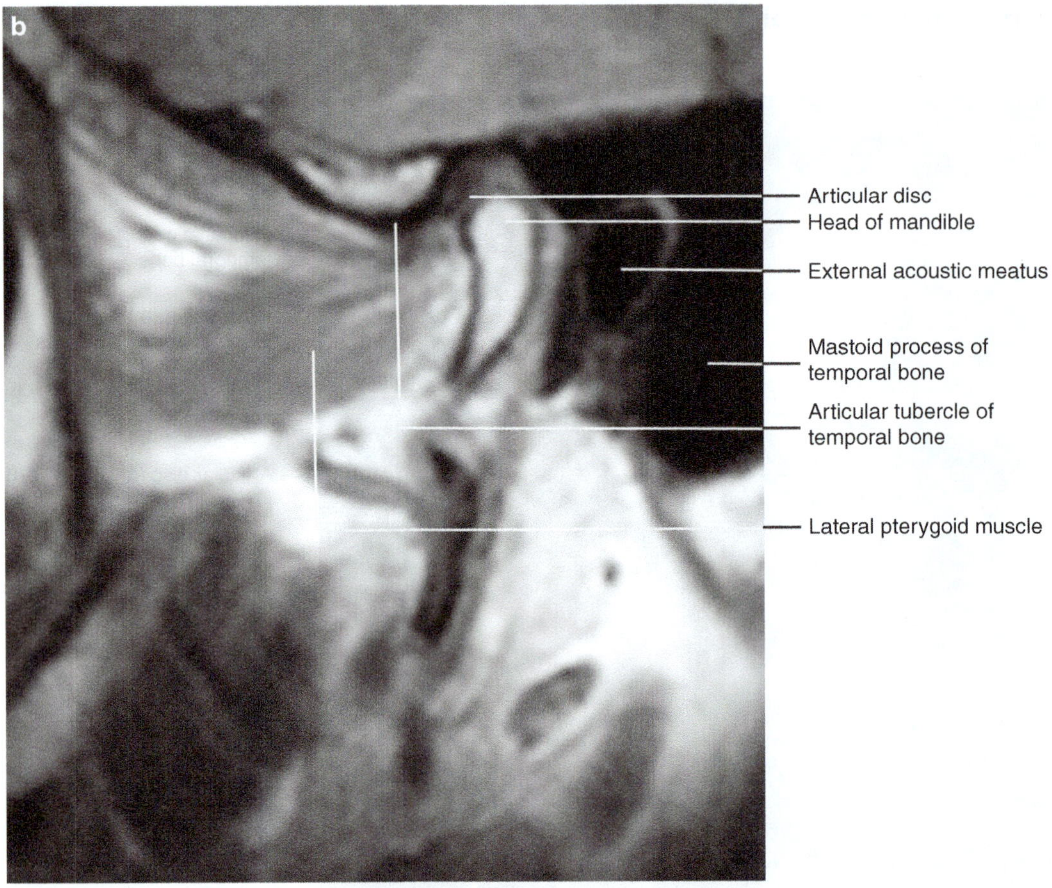

Fig. 9.36 (continued)

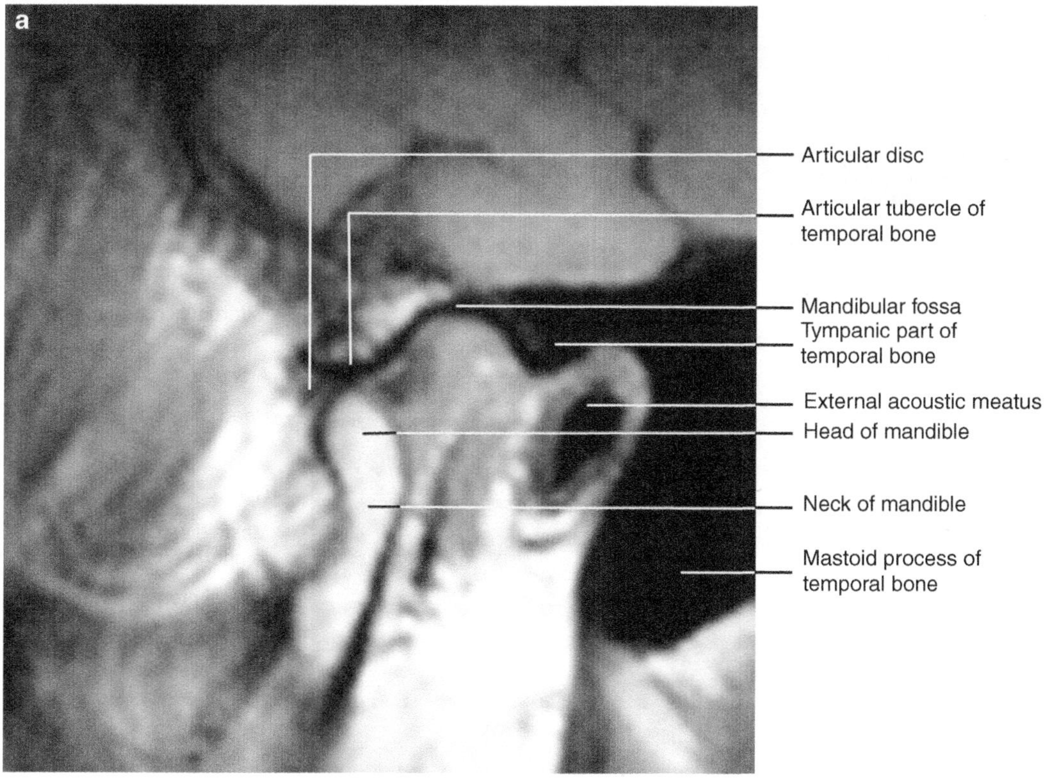

Fig. 9.37 Mouth open MRI of the temporomandibular joints

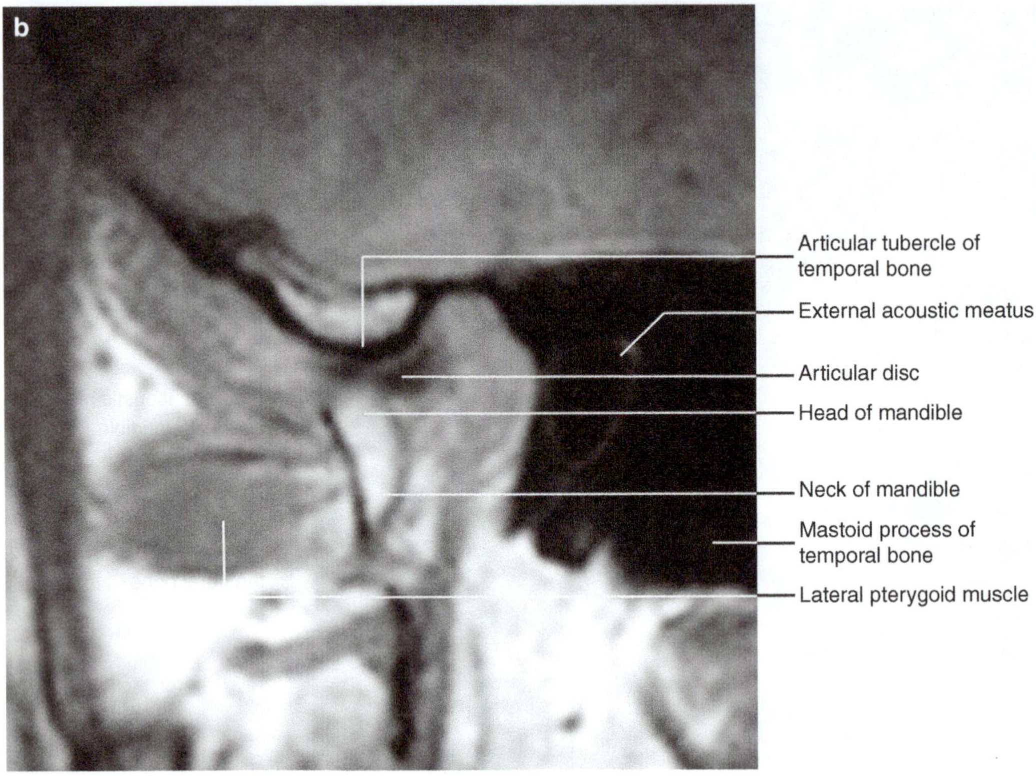

Fig. 9.37 (continued)

Acknowledgements With thanks to Dr. P. Evan from Parklane Clinic for the arthrogram images and information.

Sources

1. Moller TB, Reif E, Stark P (1993) Pocket atlas of radiographic anatomy. George Thieme Verlag, Stuttgart, pp 1–39
2. Netter FH (1991) Atlas of human anatomy. Ciba-Geigy, Basel, pp 1–63
3. Weir J, Abrahams PH (2000) Imaging atlas of human anatomy, 2nd edn. Mosby, London, pp 1–49

10 Brachial Plexus

This complicated area is best imaged with MRI in the coronal plane. The brachial plexus is found by using the subclavian artery as a landmark on the coronal images.

S. Andronikou (ed.), *See Right Through Me*,
DOI 10.1007/978-3-642-23893-2_10, © Springer-Verlag Berlin Heidelberg 2012

Magnetic Resonance Imaging

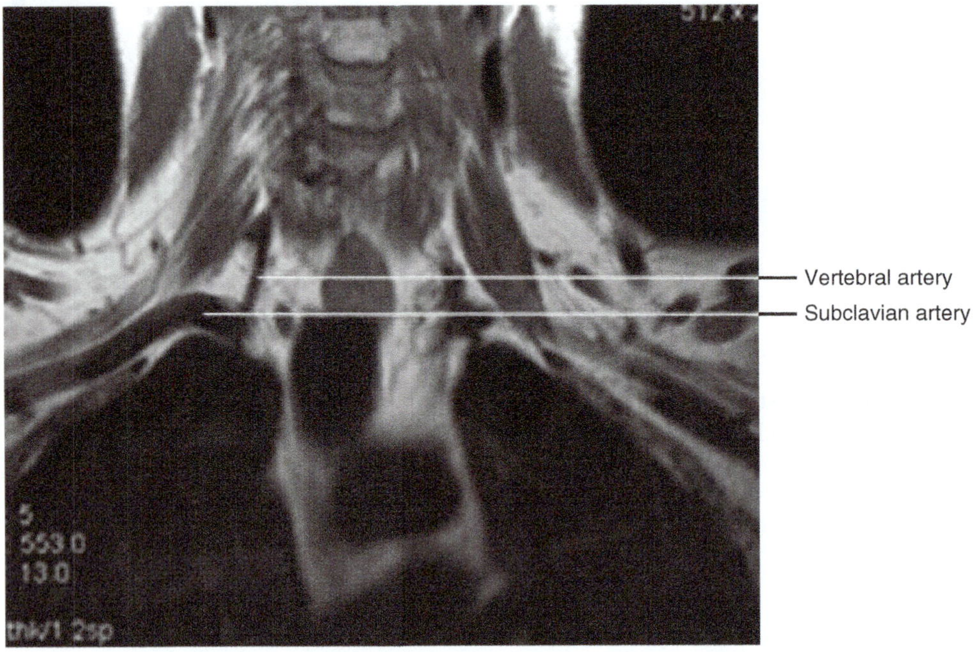

Fig. 10.1 T1 coronal (anterior to posterior)

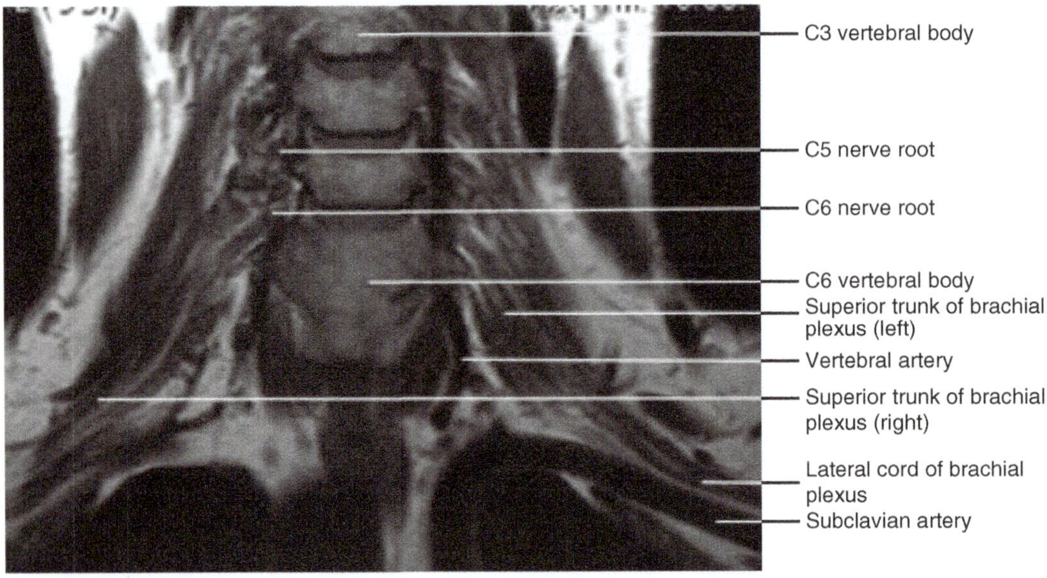

Fig. 10.2 T1 coronal

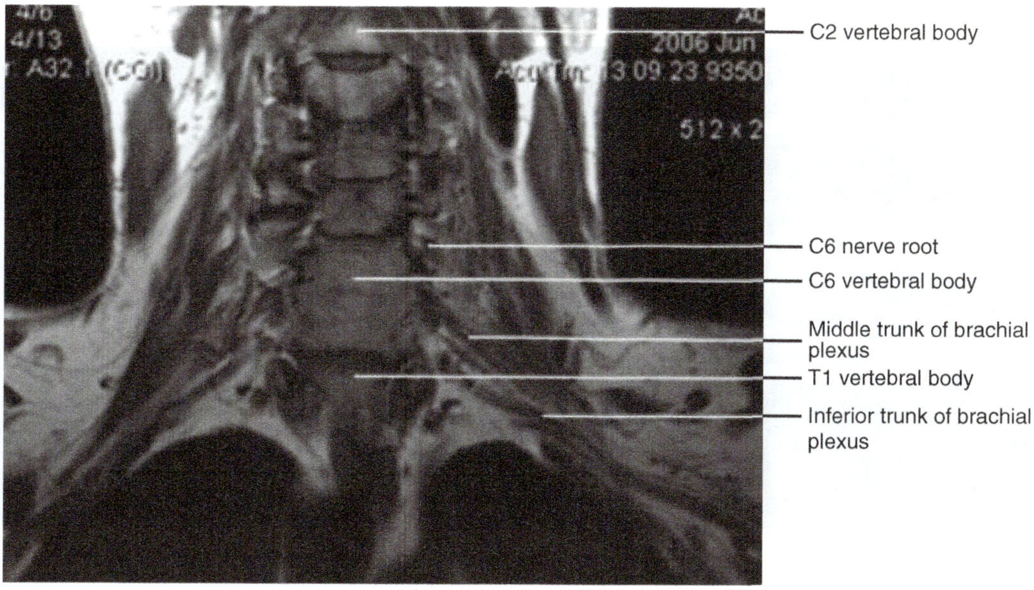

Fig. 10.3 T1 coronal

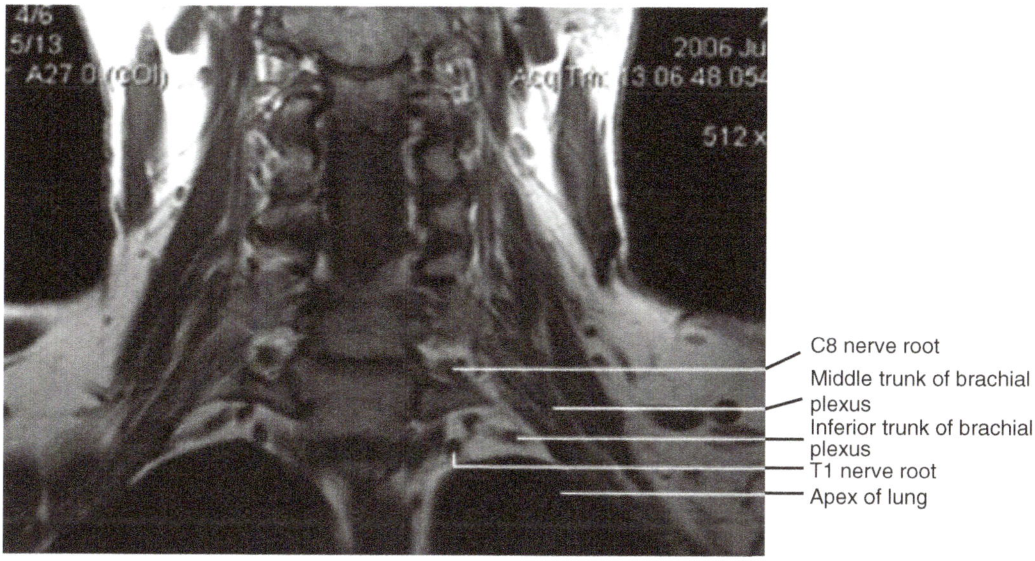

Fig. 10.4 T1 coronal

External Ear

The external ear is composed of a soft tissue canal and a bony canal. It terminates medially at the tympanic membrane.

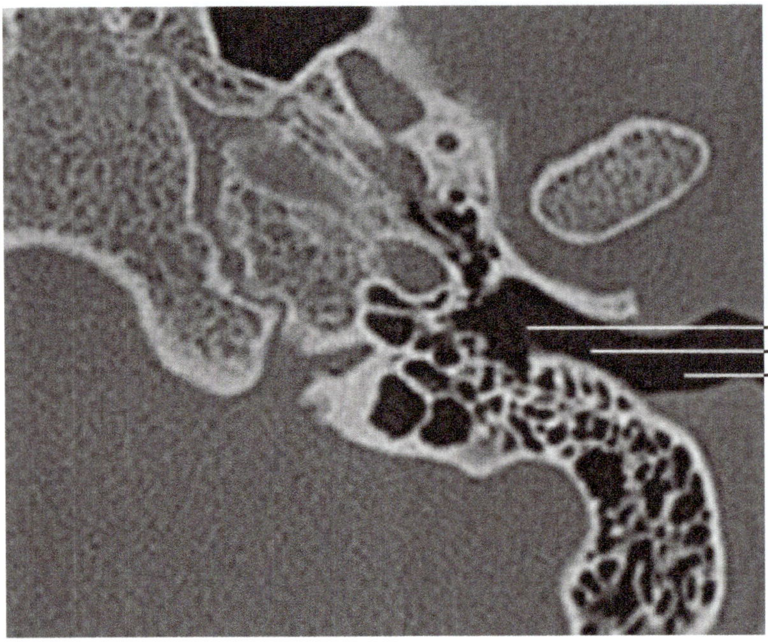

Fig. 11.1 Axial CT (bone window)

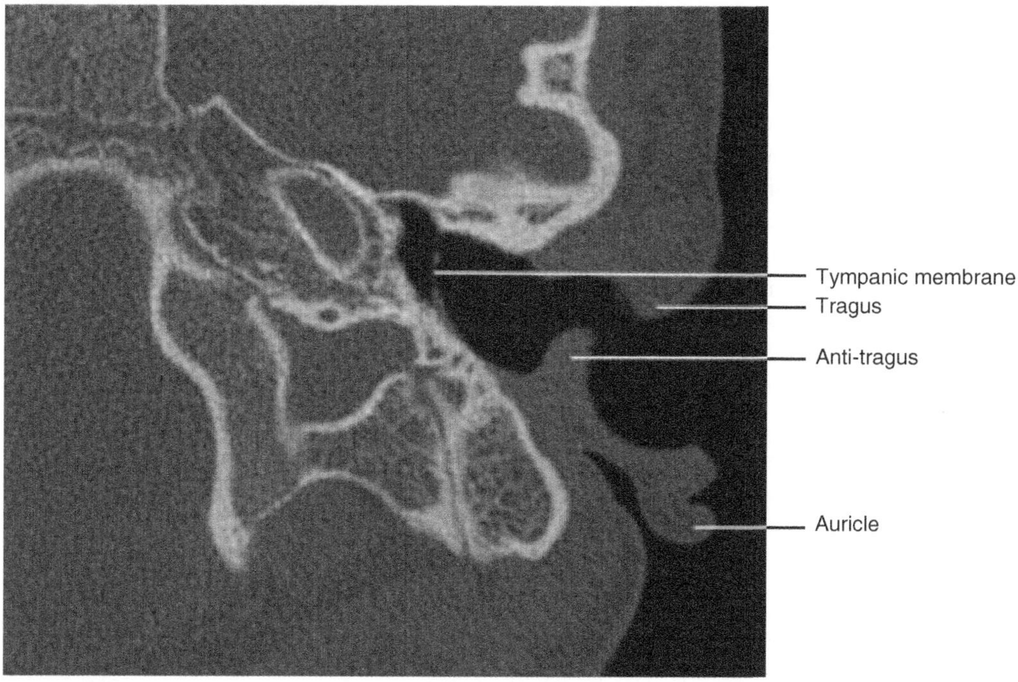

Fig. 11.2 Axial CT (bone window)

Middle Ear

Compartments in the coronal plane:

Tympanic membrane laterally.

Epitympanum: above the line between the inferior tip of the scutum and the tympanic portion of the facial nerve.

Mesotympanum: between the above line and the inferior aspect of the bony external canal.

Hypotympanum: the shallow cavity at the floor.

Inner Ear

Components include the cochlea, vestibule, semicircular canals, cochlear aqueduct, and vestibular aqueduct.

Imaging of the Middle and Inner Ear

Computed Tomography

Axial CT (from Superior to Inferior)

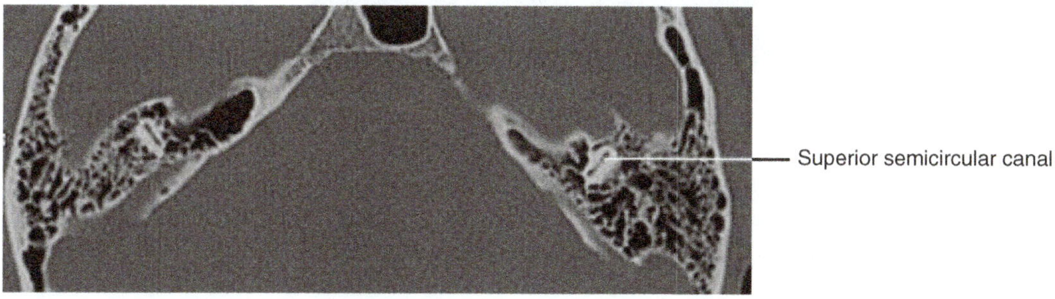

Fig. 11.3 Superior semicircular canal

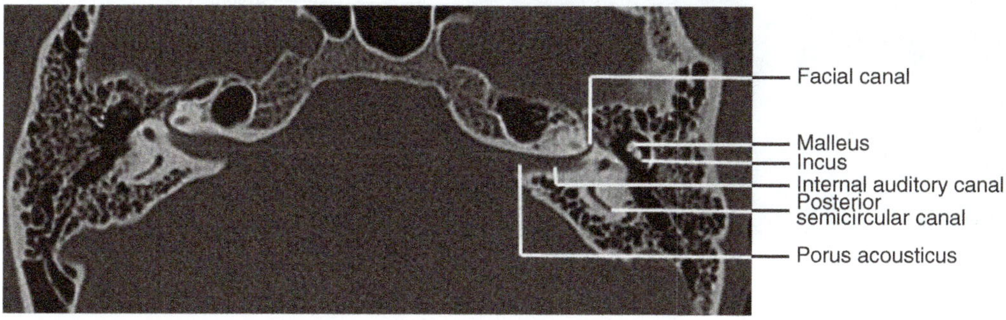

Fig. 11.4 Level of the porus acousticus

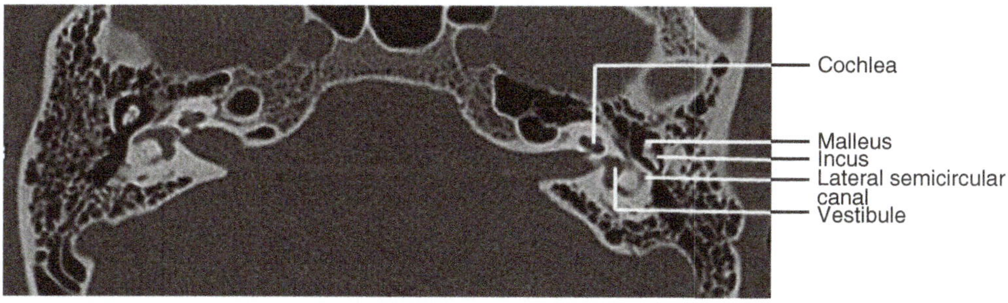

Fig. 11.5 Cochlea, vestibule, and internal auditory canal

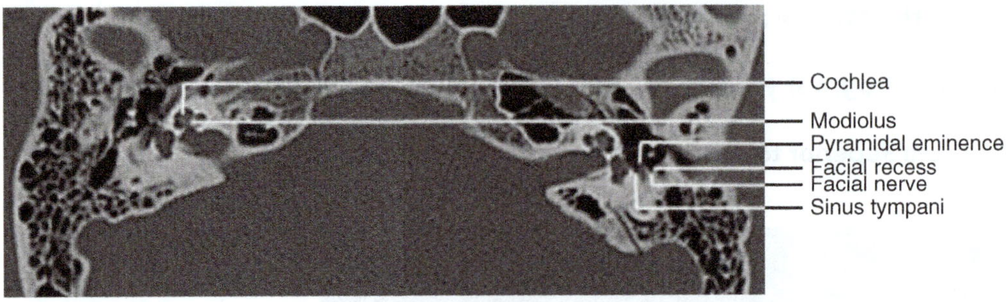

Fig. 11.6 Modiolus

Pearl

The modiolus is likened to the central spire in a snail shell.

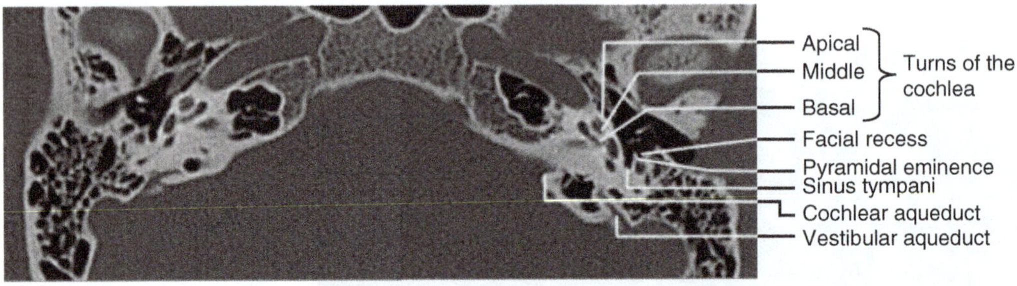

Fig. 11.7 Cochlear turns

Coronal CT (from Anterior to Posterior)

Pearl

The labyrinthine segments of the facial nerve on coronal section have been likened to "snake eyes."

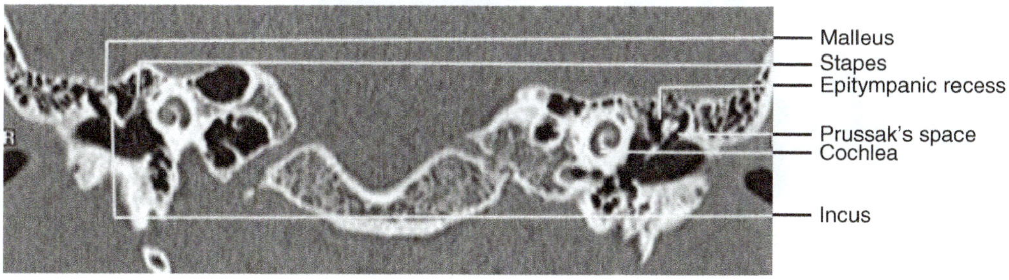

Fig. 11.8 Cochlear turns

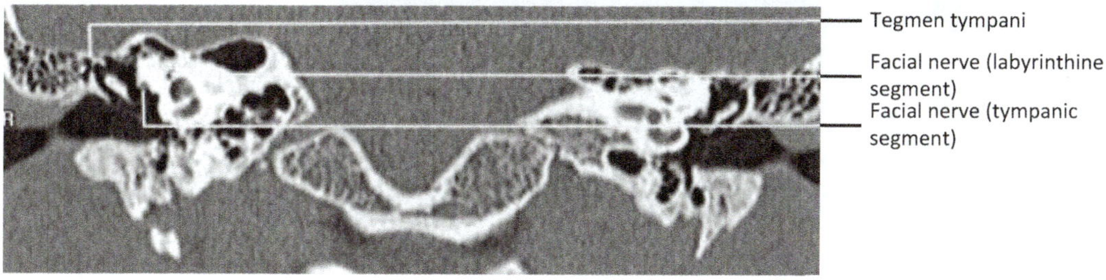

Fig. 11.9 Scutum and ossicles

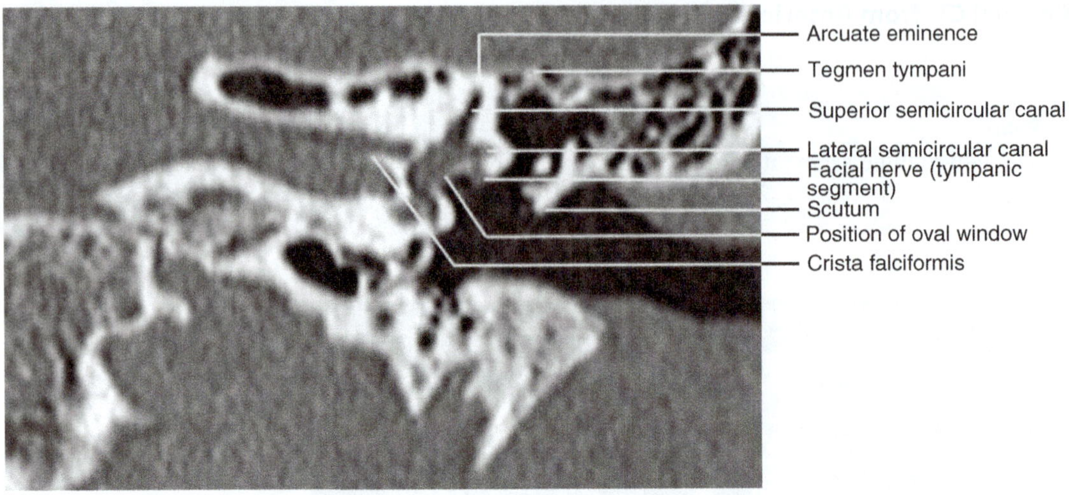

Fig. 11.10 Crista falciformis and "duck's beak" of lateral and superior semicircular canals

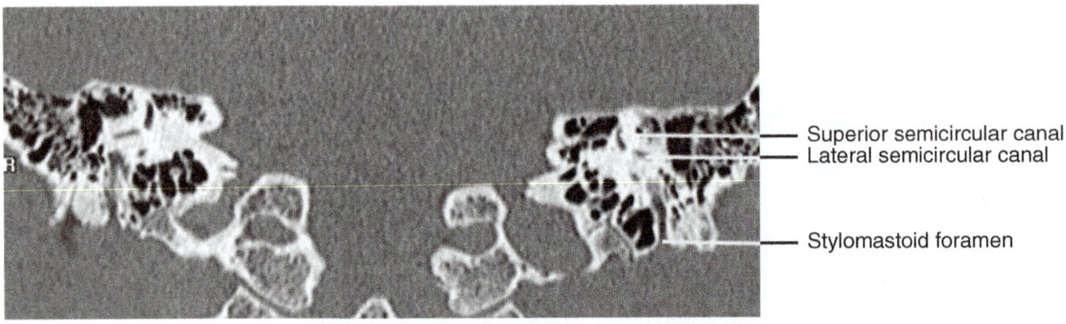

Fig. 11.11 Stylomastoid foramen

Magnetic Resonance Imaging

T2 Coronal (Anterior to Posterior)

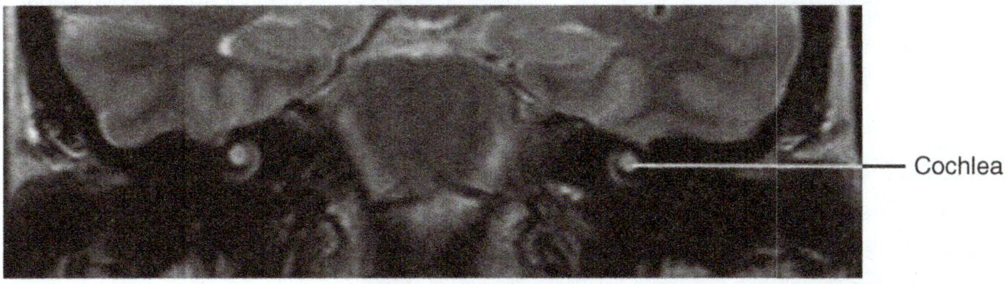

Fig. 11.12 Cochlea

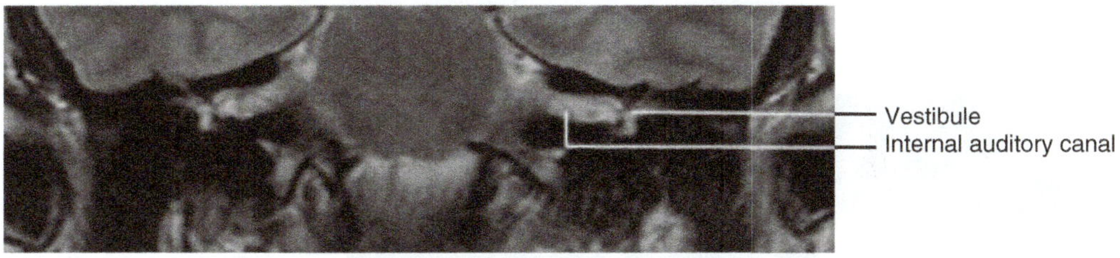

Fig. 11.13 Internal auditory canal

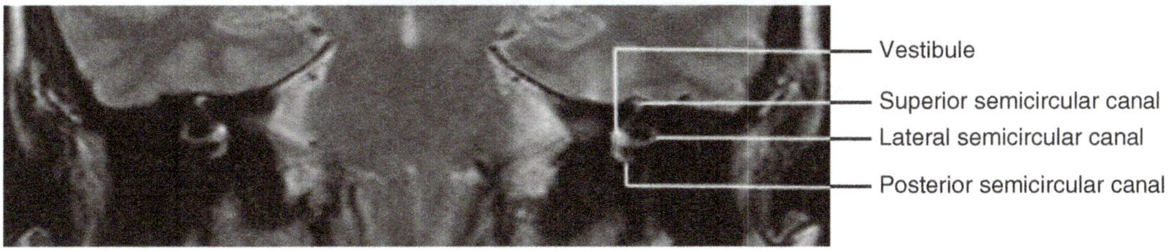

Fig. 11.14 Vestibule and semicircular canals

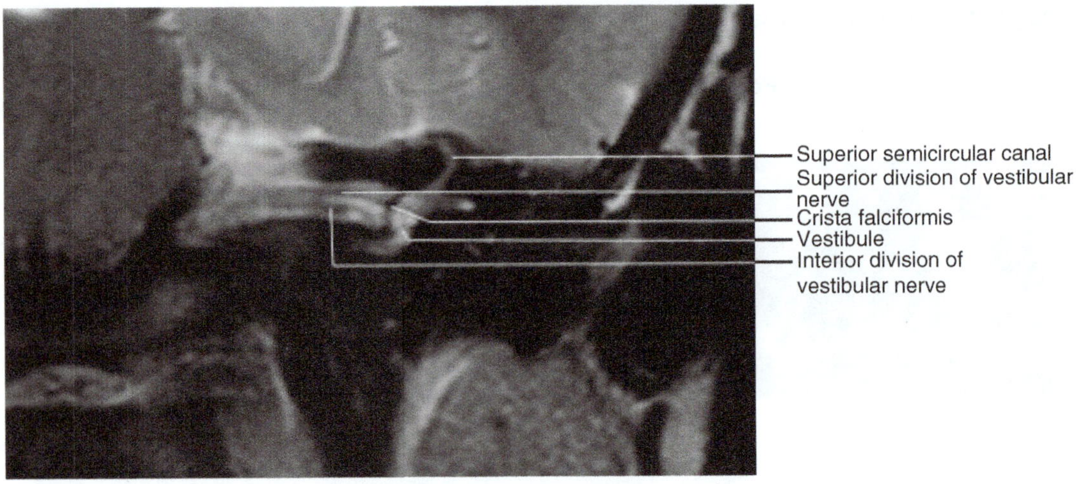

Fig. 11.15 T2 coronal (posterior aspect of IAM)

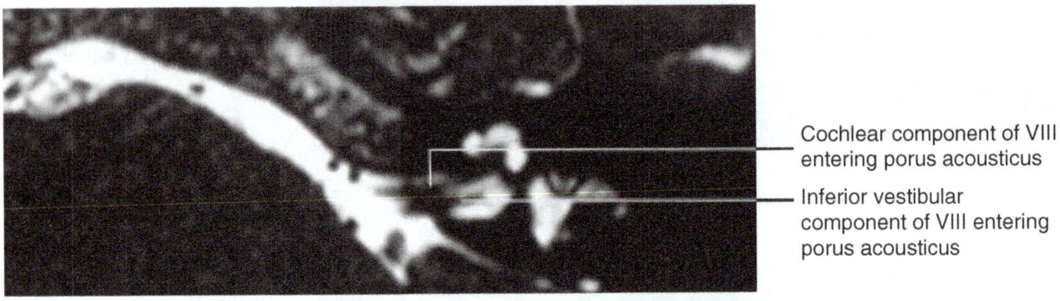

Fig. 11.16 T2 axial (inferior aspect of IAM)

12 The Orbit

Table 12.1 Components of the orbit

A. The bony orbit	B. Orbital contents	C. The lacrimal system
Plain film	Ultrasound	Contrast study—dacrocystogram
CT	CT MRI Angiography	CT

S. Andronikou (ed.), *See Right Through Me*,
DOI 10.1007/978-3-642-23893-2_12, © Springer-Verlag Berlin Heidelberg 2012

The Bony Orbit

Plain Film

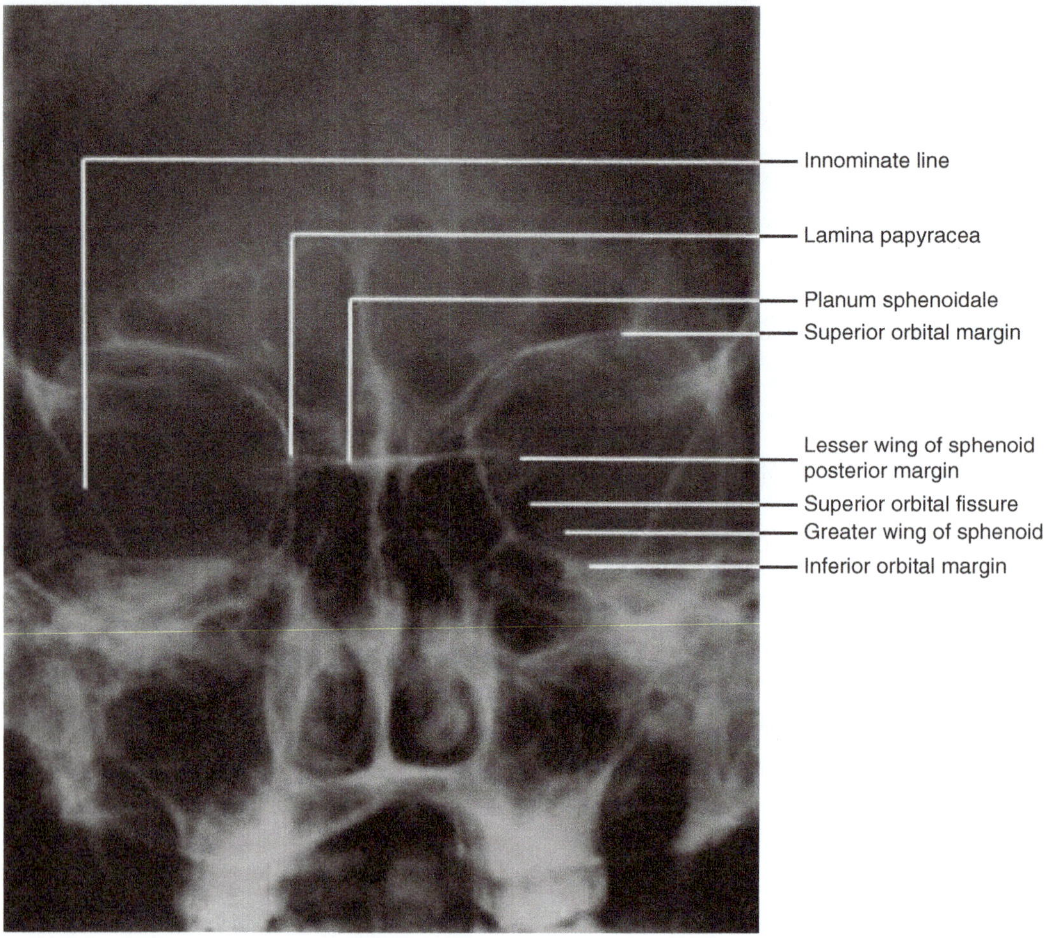

Fig. 12.1 Frontal projection of the skull

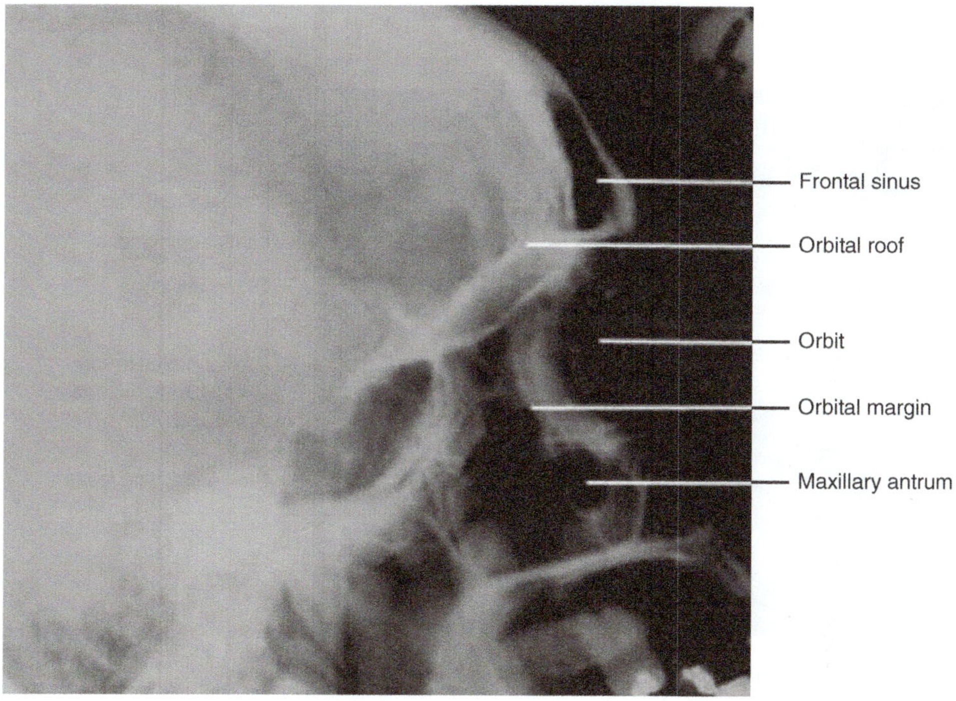

Fig. 12.2 Lateral projection of the skull

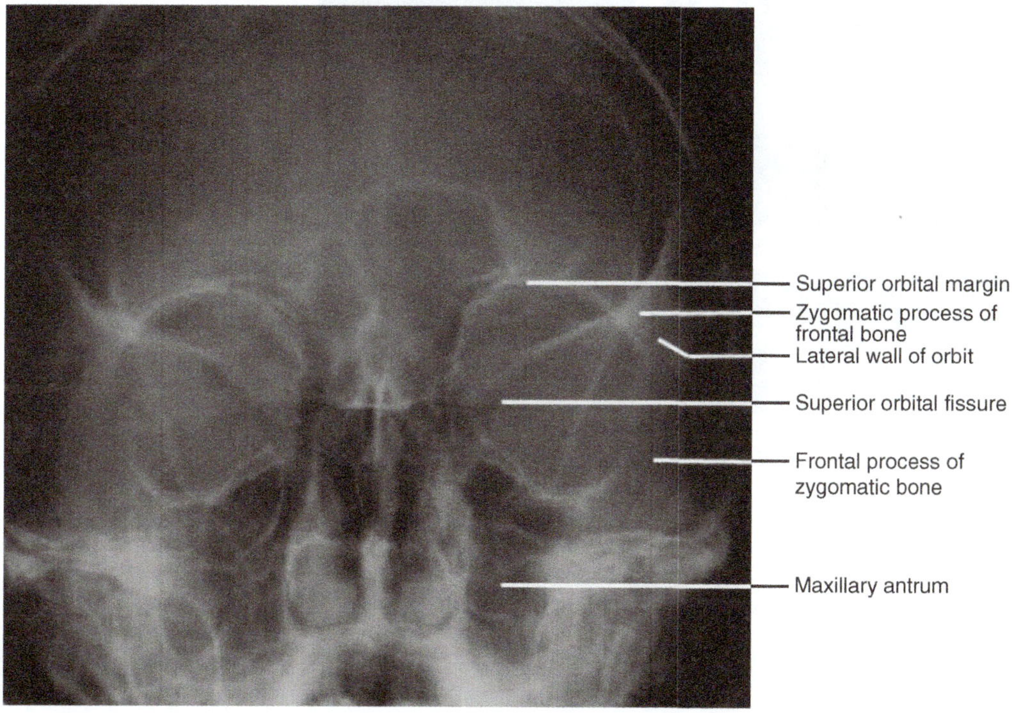

Fig. 12.3 Occipitofrontal view of the skull

Computed Tomography

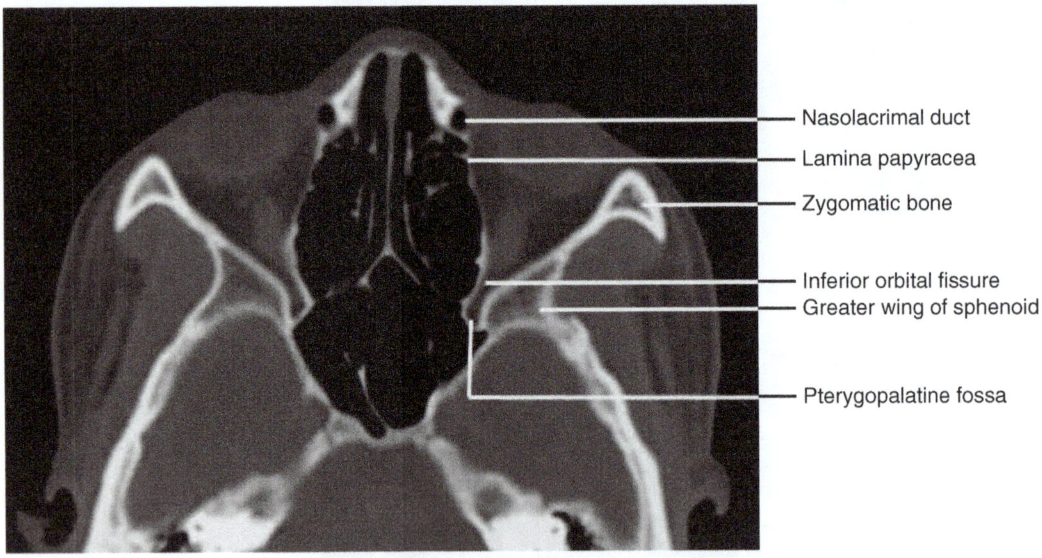

Fig. 12.4 Axial CT of bony orbit at level of inferior orbital fissure

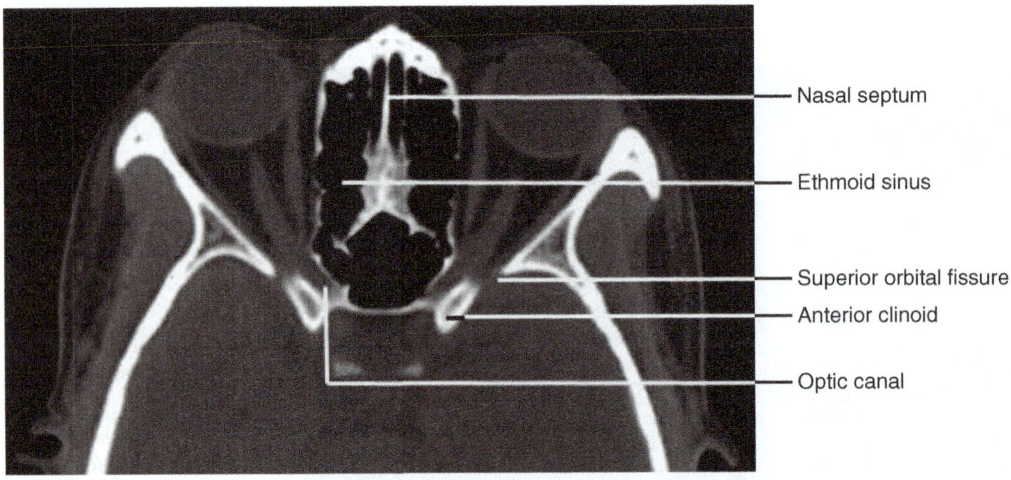

Fig. 12.5 Axial CT of bony orbit at level of superior orbital fissure

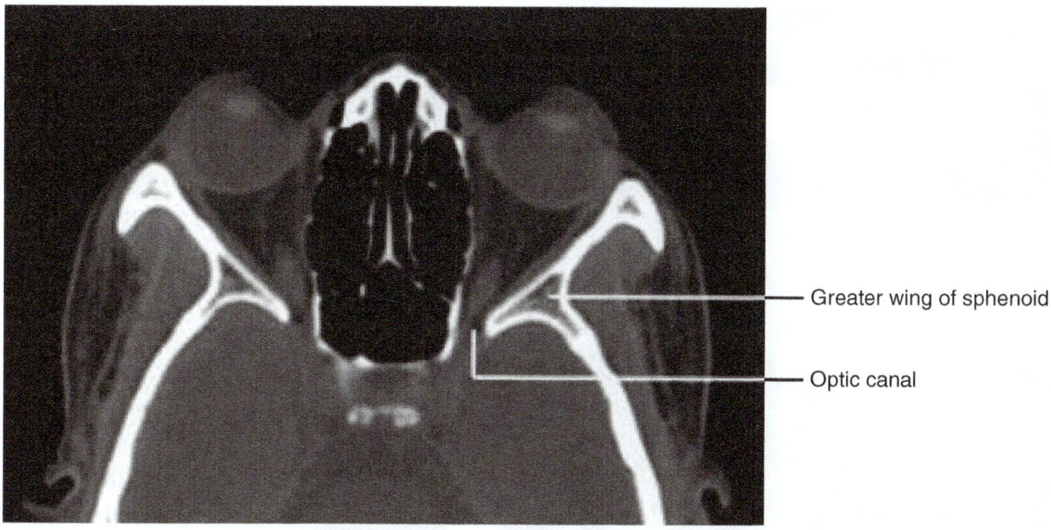

Fig. 12.6 Axial CT of bony orbit at level of the optic canal

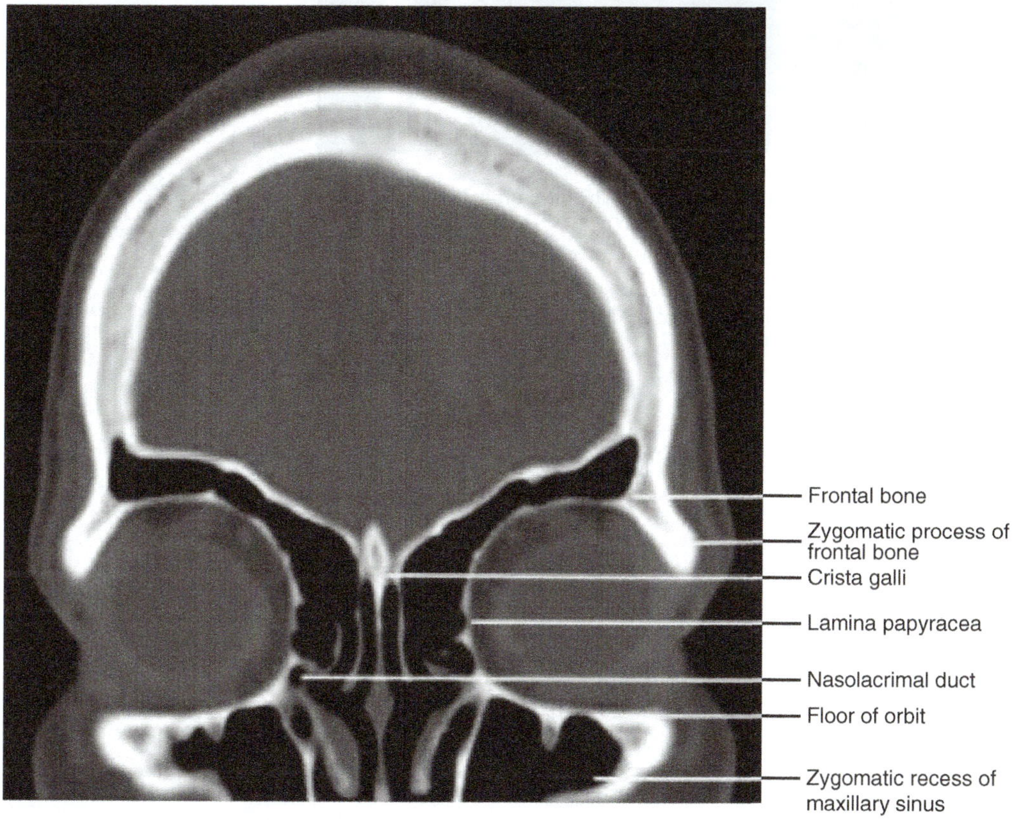

Fig. 12.7 Coronal CT of bony orbit at the level of the lacrimal gland

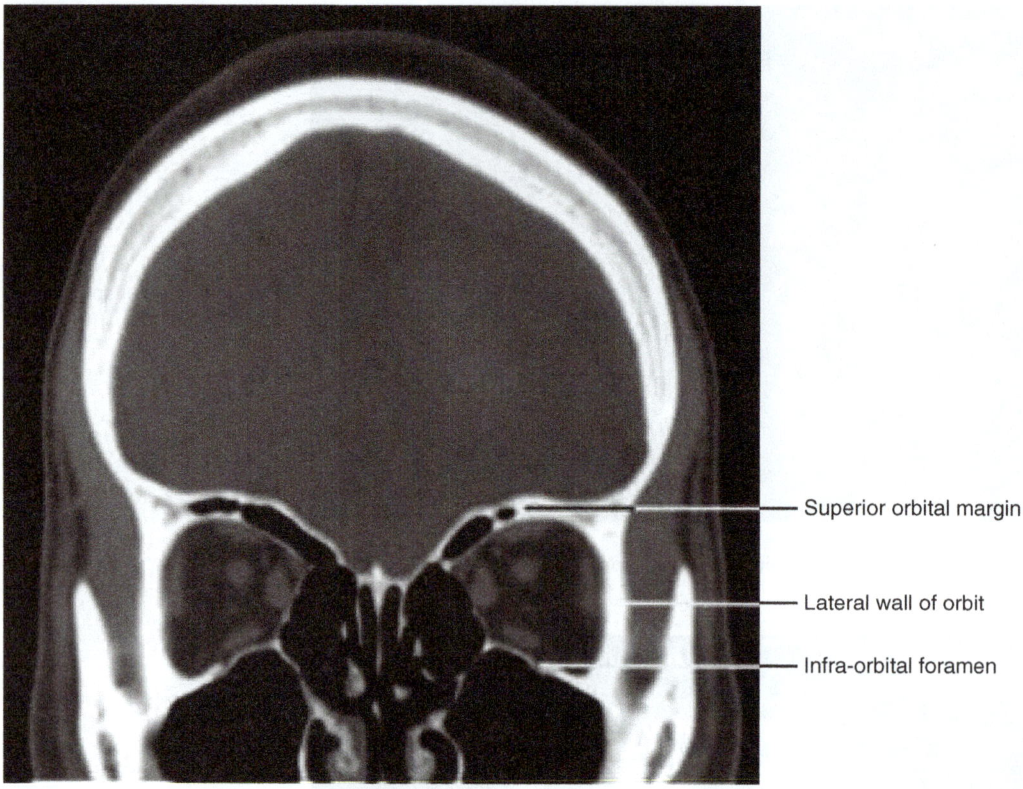

Fig. 12.8 Coronal CT of bony orbit at the level of the infra-orbital canal

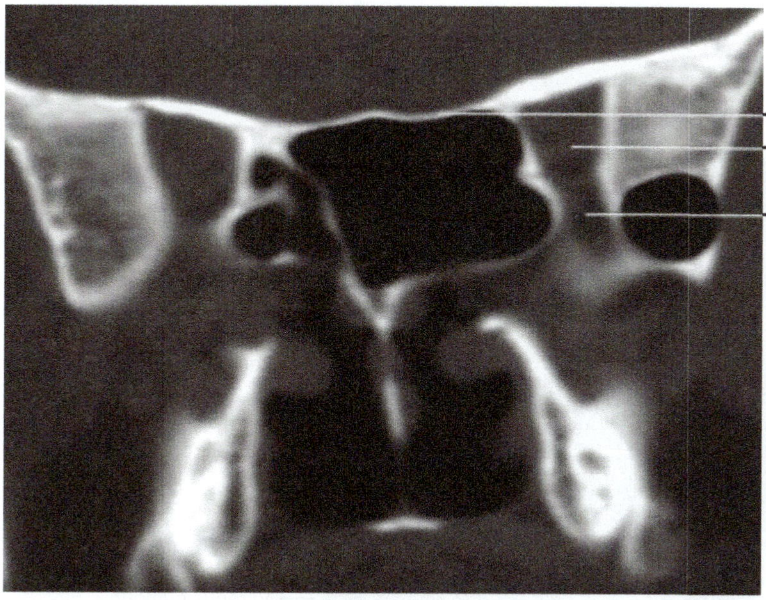

Fig. 12.9 Coronal CT of bony orbit at the level of the inferior orbital fissure

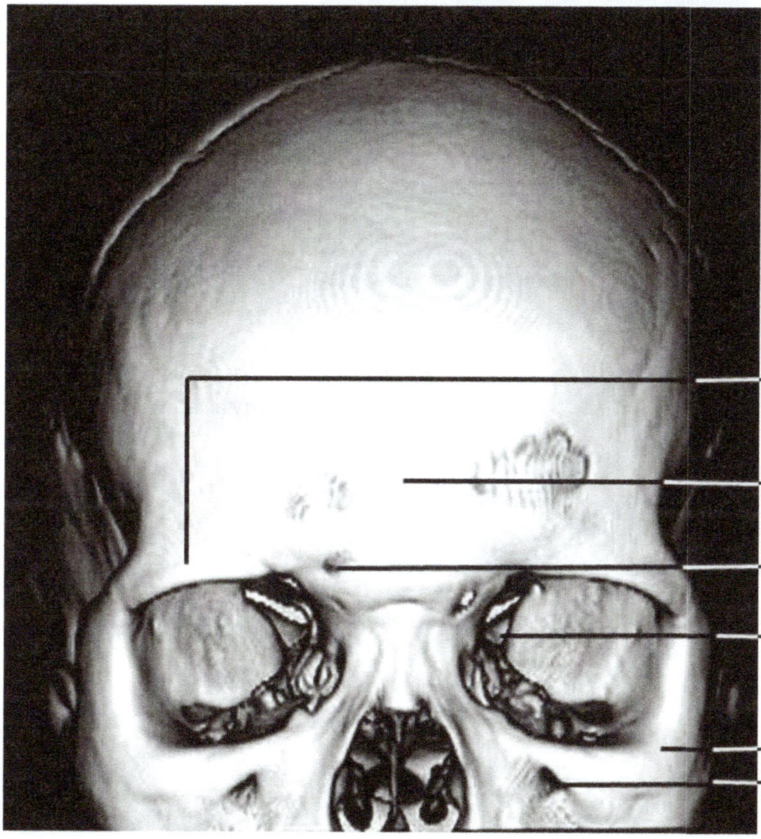

Fig. 12.10 Three-dimensional surface-shaded coronal reconstruction of the bony orbit

Orbital Contents

Ultrasound

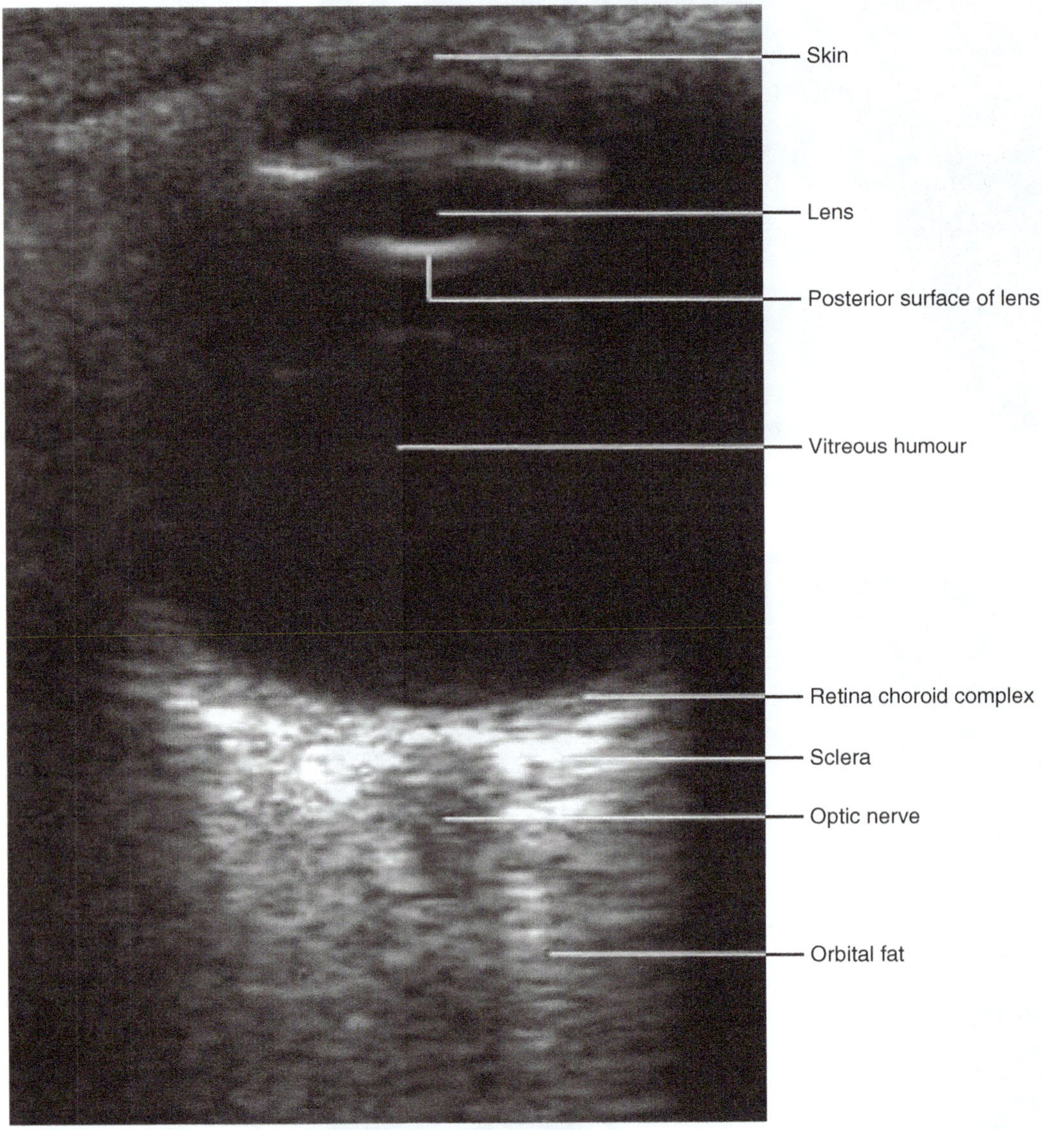

Fig. 12.11 Horizontal (axial) ultrasound scan demonstrating the globe

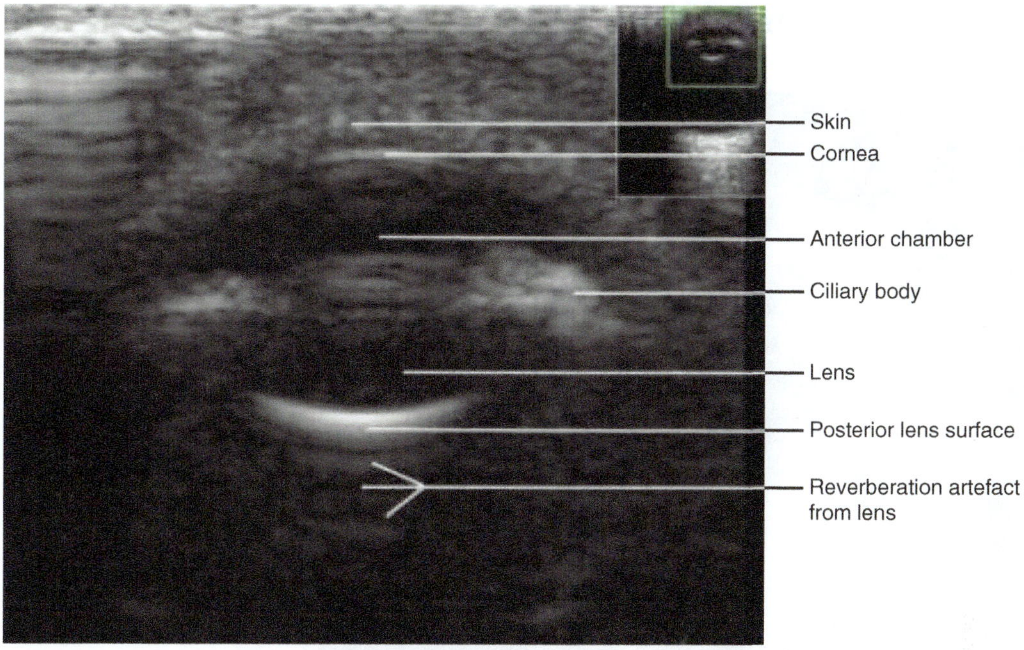

Fig. 12.12 Horizontal ultrasound scan: zoomed on the anterior globe

Computed Tomography

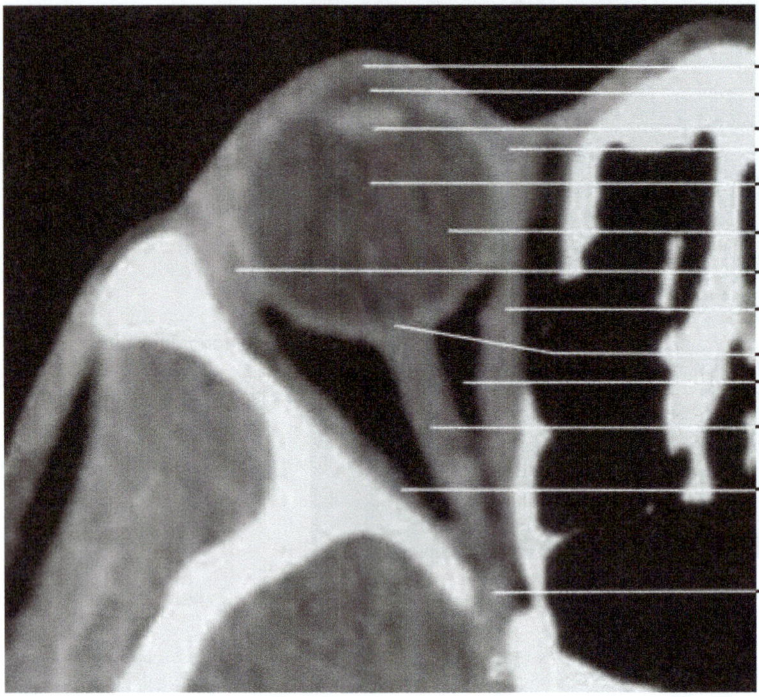

Fig. 12.13 Axial CT of the orbit (soft tissue window)

Pearl

The conus comprises the rectus muscles, retrobulbar fat, optic nerve and vessels that course through it.

The optic canal points medially to enter intracranially and one must see the optic nerve passing through. The superior orbital fissure opens laterally.

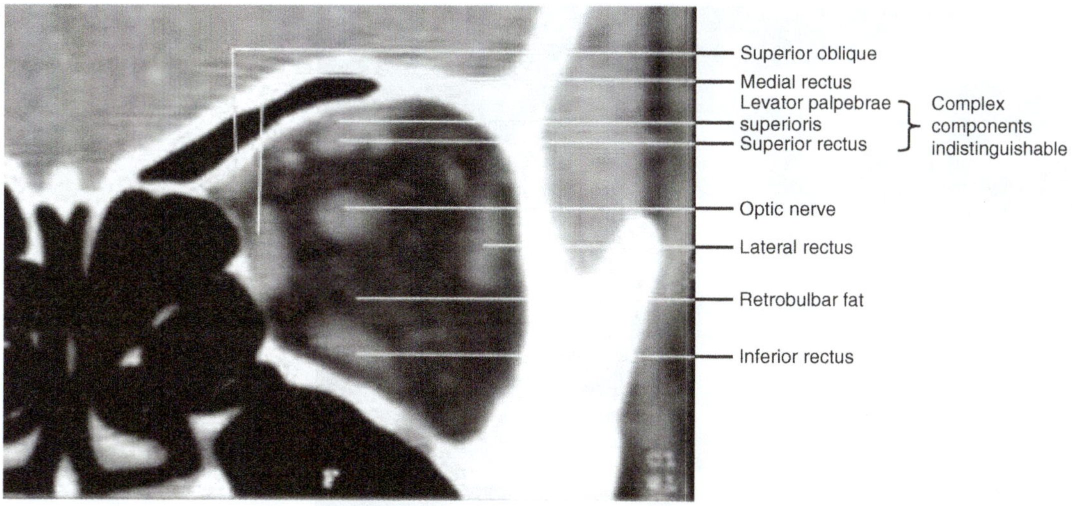

Fig. 12.14 Coronal CT of the orbit (soft tissue window: posterior to globe)

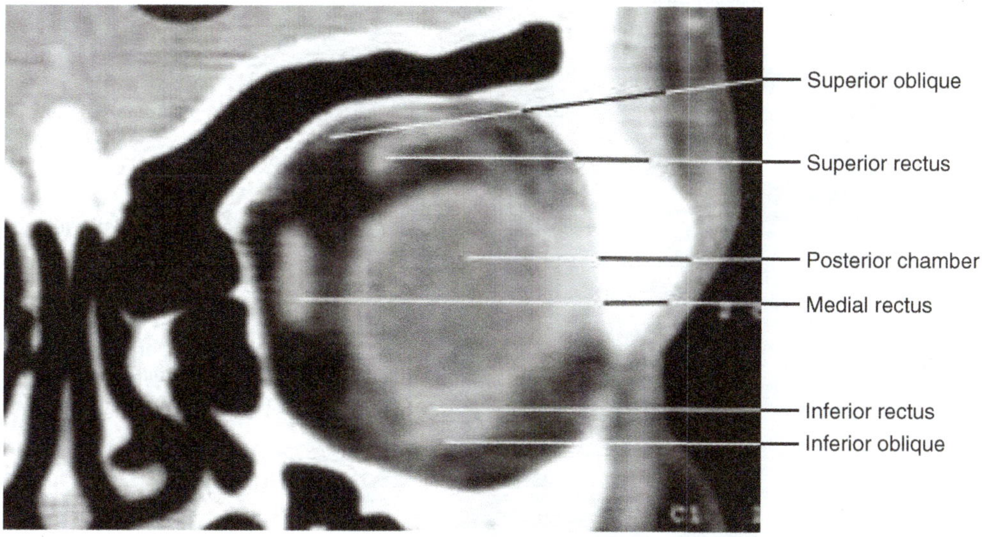

Fig. 12.15 Coronal CT of the orbit post contrast (soft tissue window through the globe)

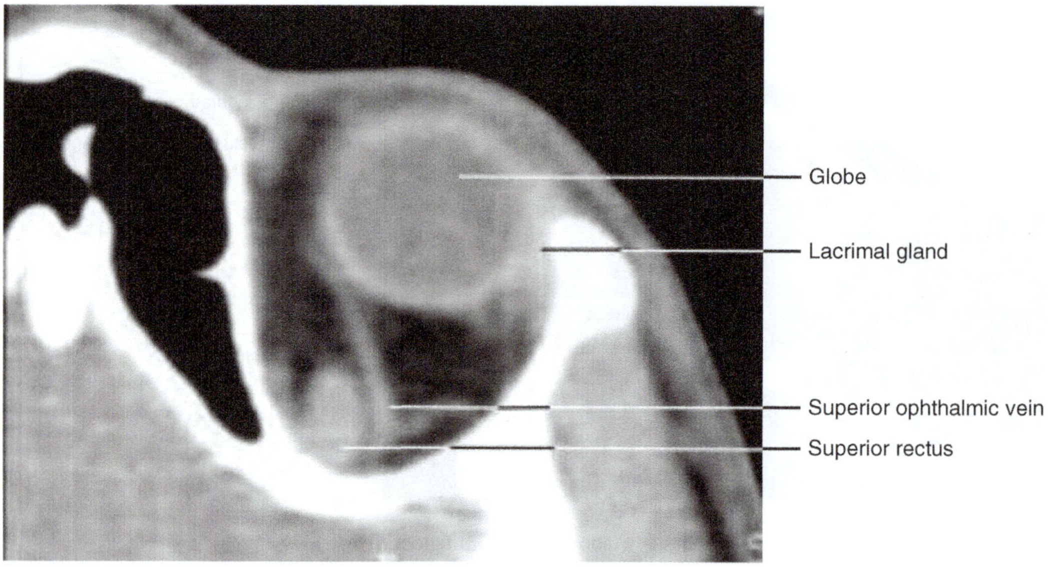

Fig. 12.16 Axial CT of the orbit post contrast (soft tissue window)

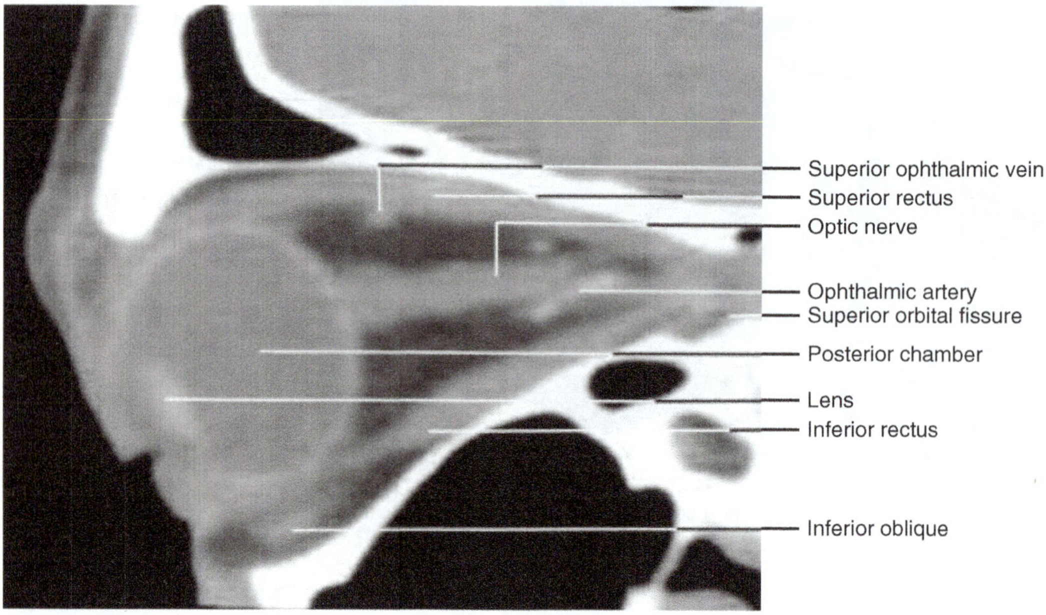

Fig. 12.17 Sagittal CT of the orbit post contrast (soft tissue window)

Magnetic Resonance Imaging

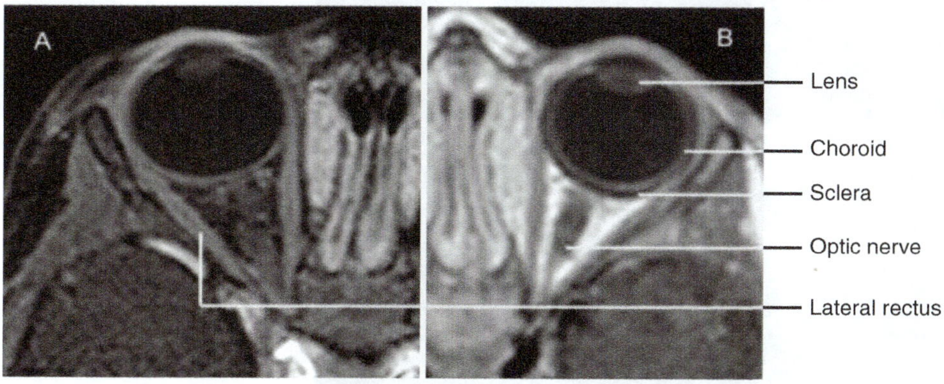

Fig. 12.18 Axial T1 fat-saturated post-gadolinium (**a**) and T1 post-gadolinium without fat saturation (**b**)

Pearl

High water content of vitreous body produces hypointense signal on T1-weighted image.
The choroid can be demonstrated on T1 with gadolinium as an enhancing layer deep to the sclera.
Extraoccular muscles of the conus enhance with gadolinium in-normal patients.

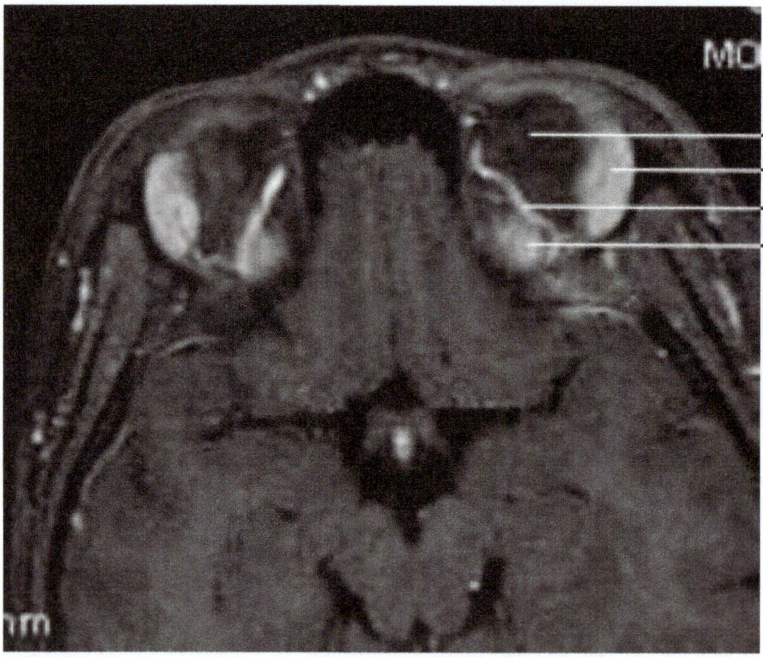

Fig. 12.19 Axial T1 fat-saturated post-gadolinium

Pearl
There is intense enhancement of the lacrimal gland post-gadolinium.

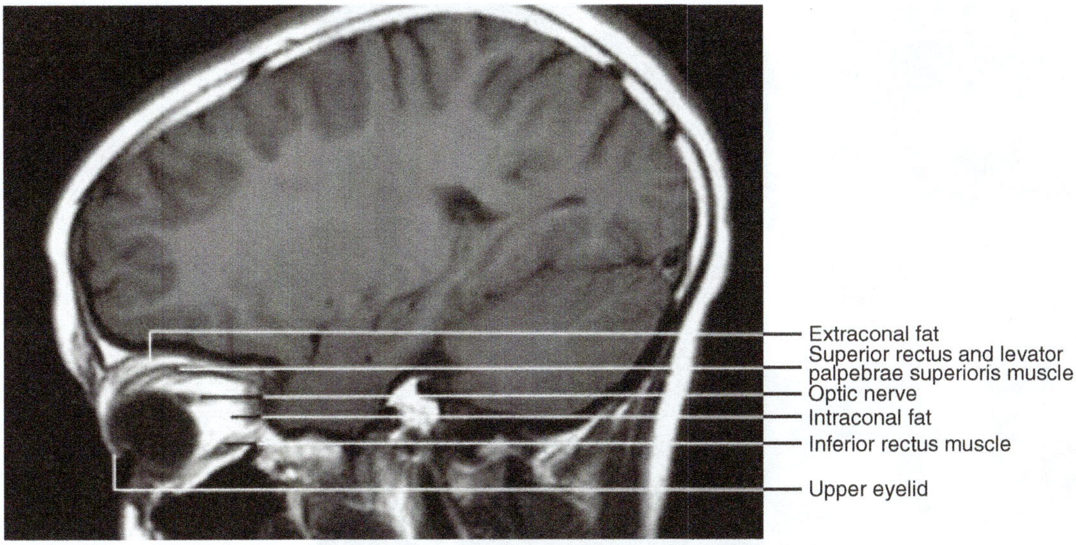

Fig. 12.20 Sagittal T1

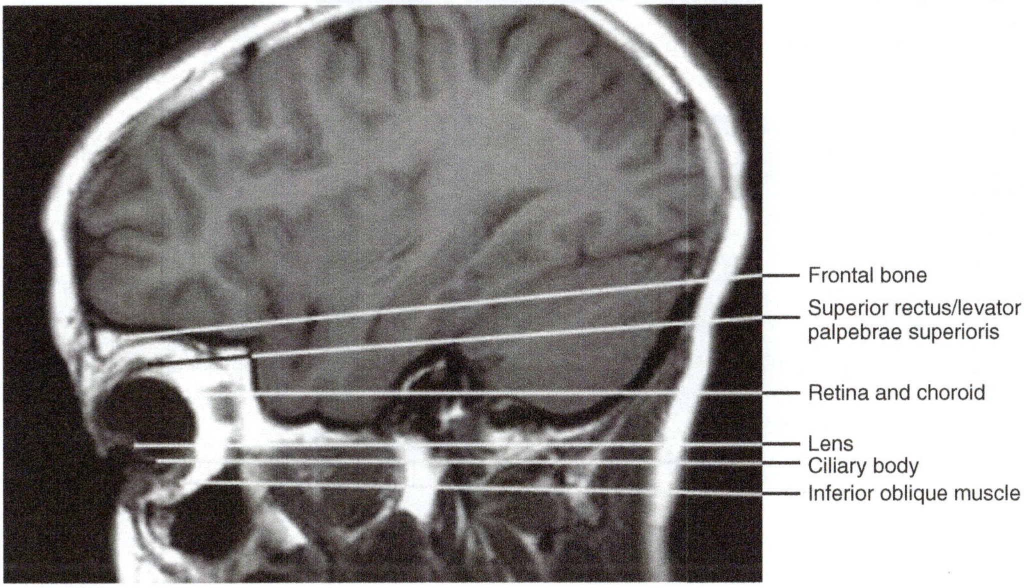

Fig. 12.21 Sagittal T1

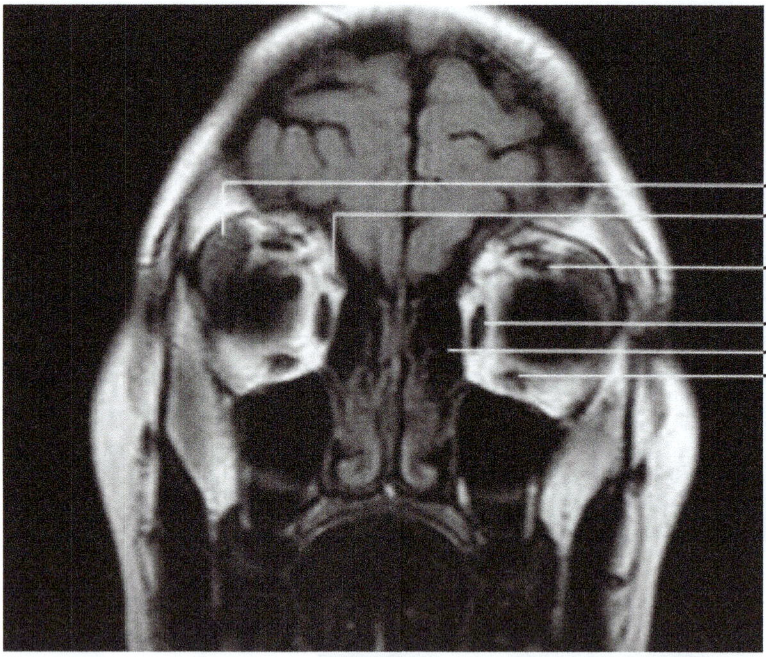

Fig. 12.22 Coronal FLAIR at level of globe

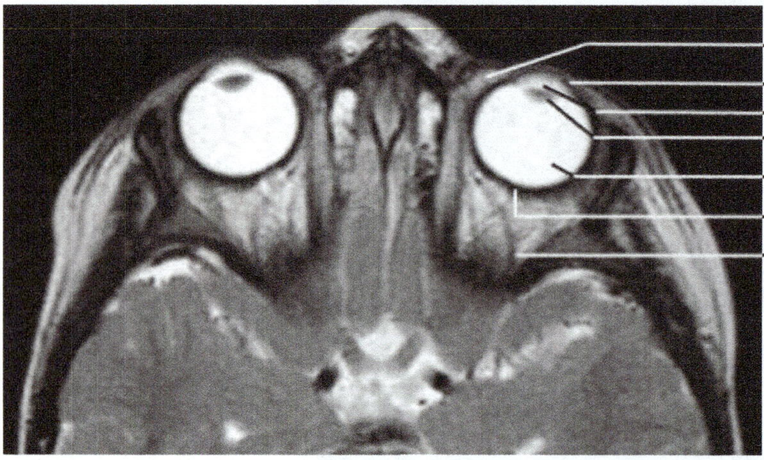

Fig. 12.23 Axial T2 at level of optic chiasm

Pearl

Aqueous and vitreous humour due to high water content appear bright/hyperintense on T2-weighted images.

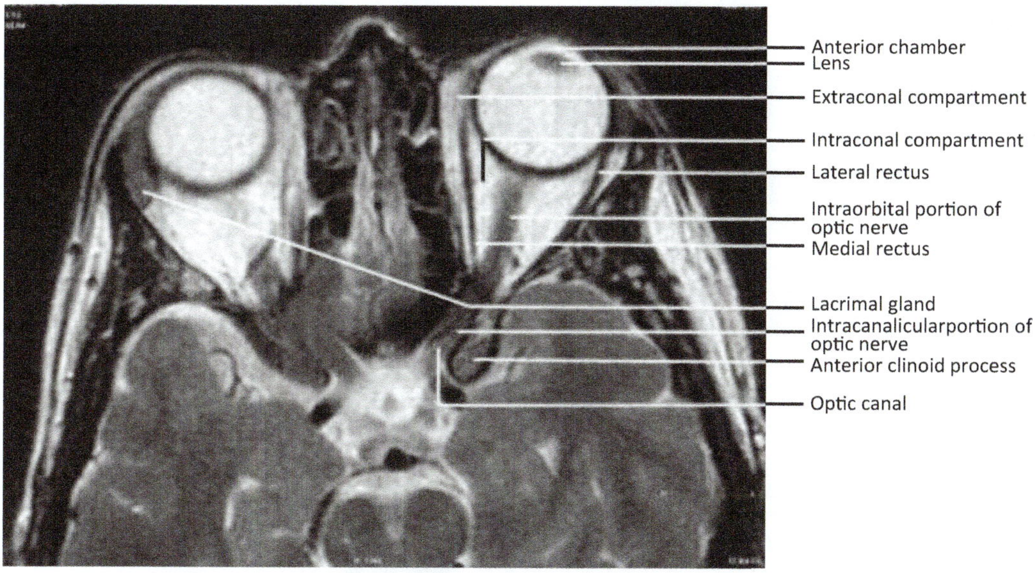

Fig. 12.24 Axial T2 at level of optic nerve

Pearl

Axial MR imaging demonstrates all three parts (namely intraorbital, intracanalicular, and intracranial) of the optic nerve with correct angulation.

Table 12.2 MRI signals of orbital contents

Anatomical part	T1	T2	Other sequence
Vitreous and aqueous humour	Low	High	FLAIR low
Lens	Higher	Lower	–
Lacrimal gland	Intermediate	Intermediate	High post-gadolinium
Intraorbital fat	High	High	Short T1-inversion recovery (STIR) low

Angiography

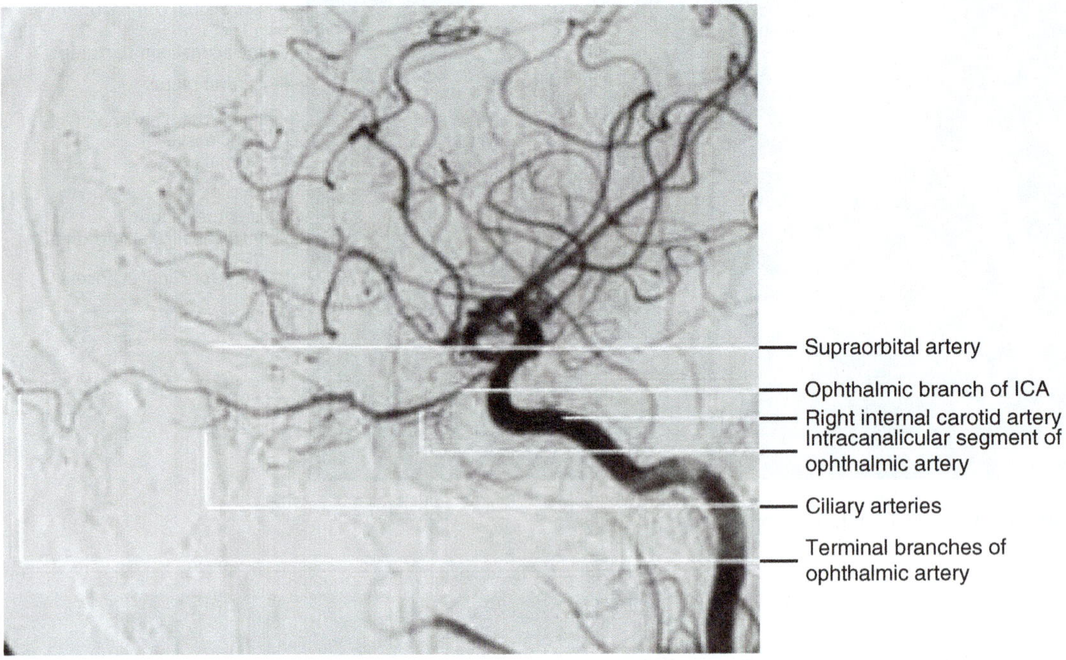

Fig. 12.25 Digital subtraction angiography of internal carotid artery and its branches—lateral projection

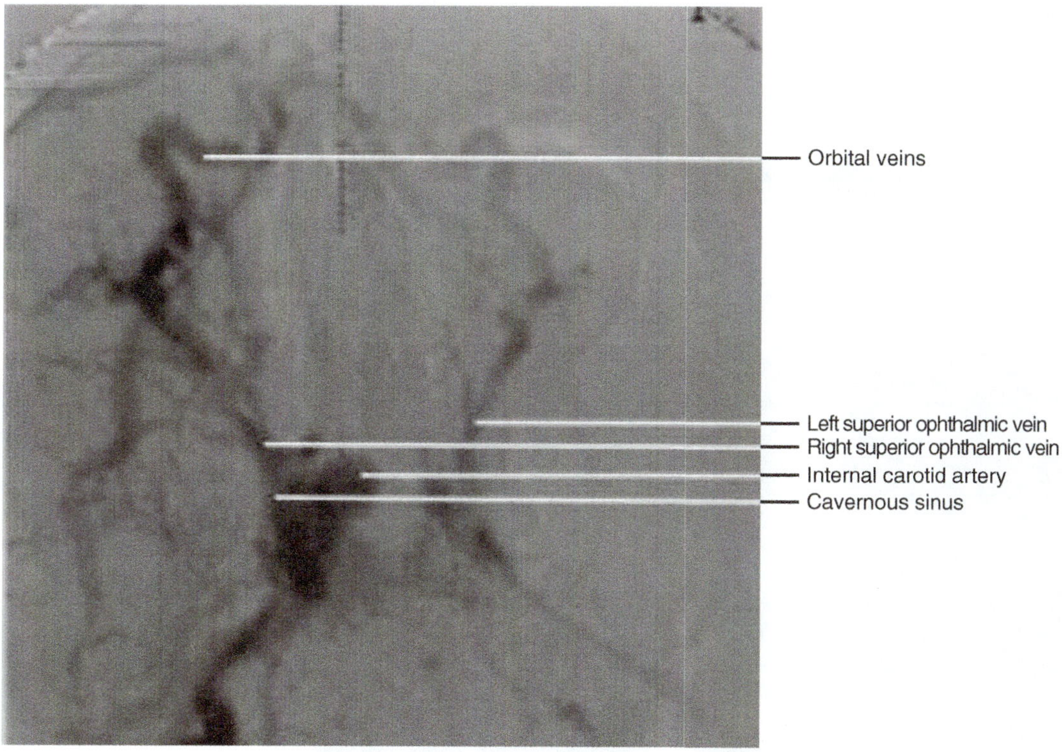

Fig. 12.26 Digital subtraction angiography of carotid-cavernous fistula demonstrating the venous drainage to good advantage

The Lacrimal System

Contrast Study

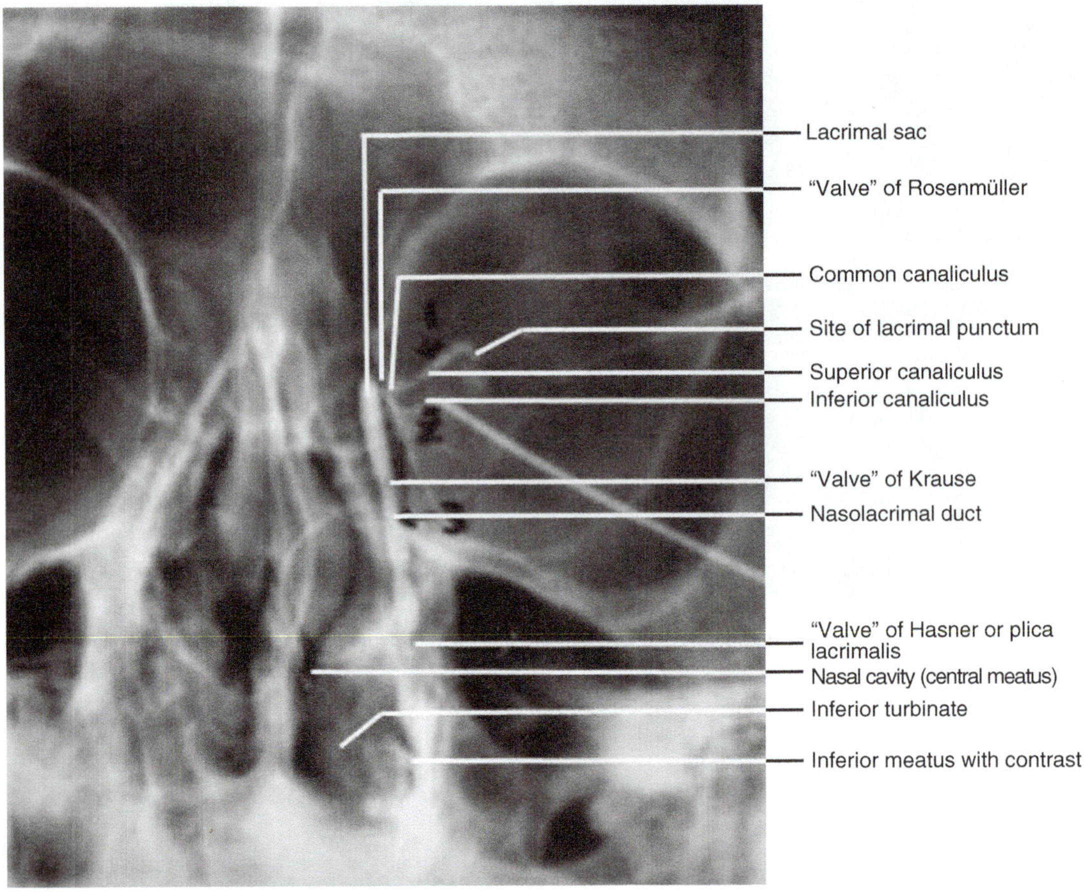

Fig. 12.27 Dacrocystogram: anteroposterior view

Computed Tomography

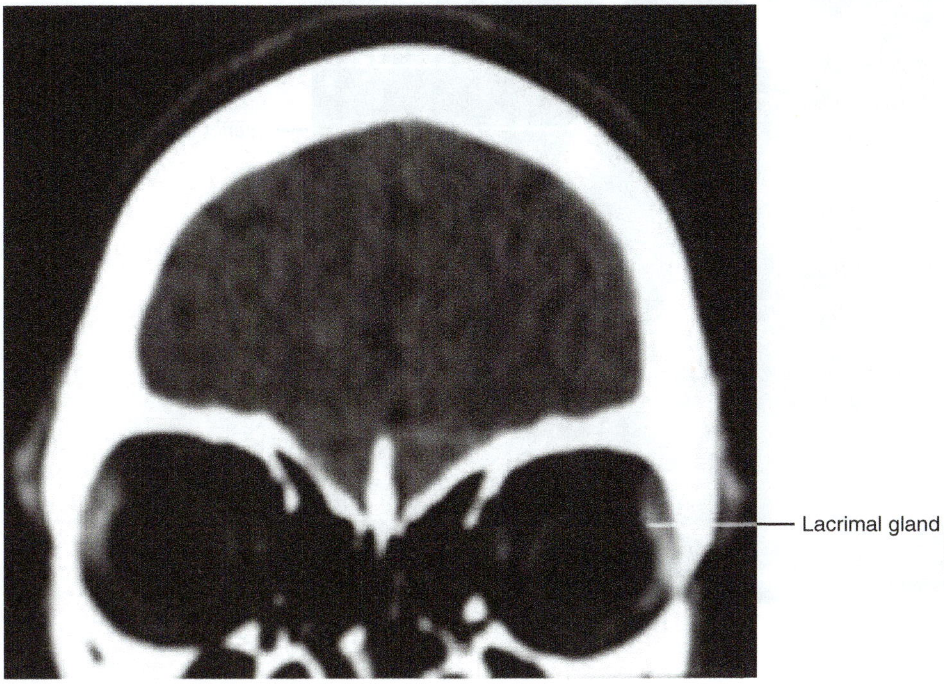

Fig. 12.28 Coronal CT

Pearl
The nasolacrimal duct is the only structure that opens at the inferior meatus (inferior to inferior turbinate).

Pearl
Lacrimal gland = lateral
Lacrimal sac = medial

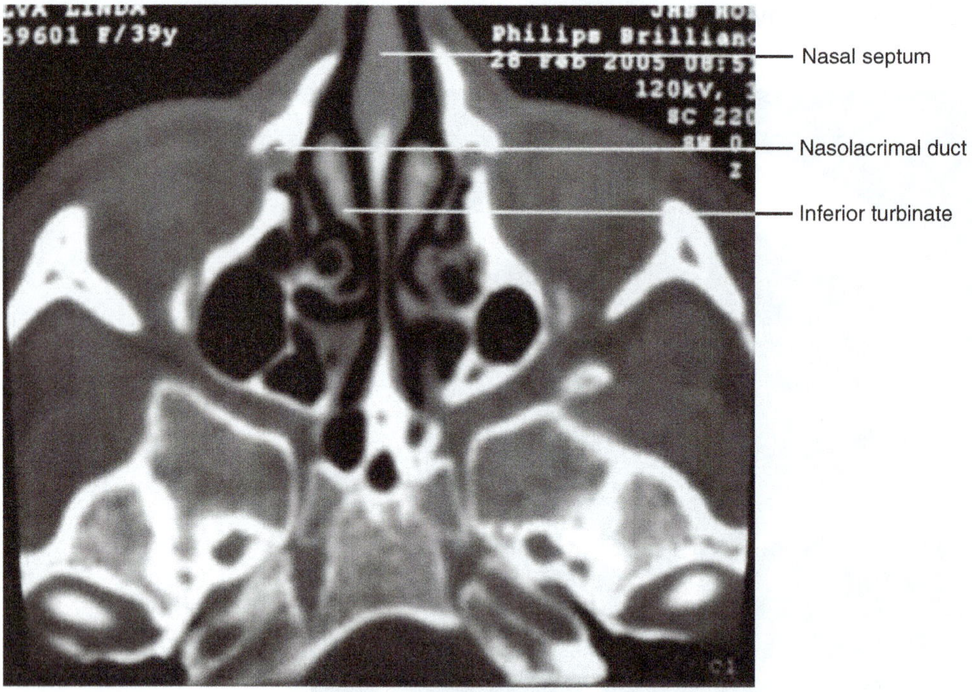

Fig. 12.29 Axial CT of nasolacrimal duct

Sources

1. Butler P, Mitchell AWH, Ellis H (eds) (1999) Applied radiology anatomy. Cambridge University Press, Cambridge
2. Csillag A (1999) Anatomy of the living human. Könemann, Cologne
3. Weir J, Abrahams PH (2003) Imaging atlas of human anatomy, 3rd edn. Elsevier, St. Louis

The Neck

13

When imaging the soft tissues of the neck, plain radiography is of limited value compared to the newer modalities. Sialography is a useful diagnostic tool for assessment of the anatomy of the salivary ducts. Soft tissue views of the larynx are sometimes used to search for radio-opaque foreign bodies and for evaluation of laryngeal cartilaginous structures.

Plain Film

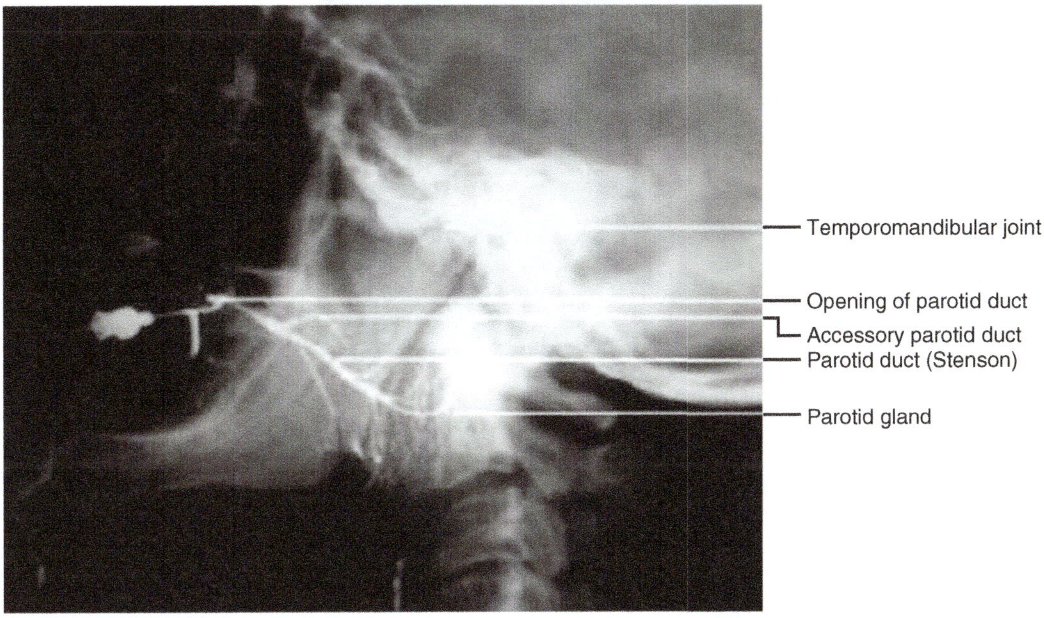

Fig. 13.1 Parotid sialogram: lateral view

> **Pearl**
> Parotid (Stenson's) duct opens opposite the second molar of the upper jaw.

S. Andronikou (ed.), *See Right Through Me*,
DOI 10.1007/978-3-642-23893-2_13, © Springer-Verlag Berlin Heidelberg 2012

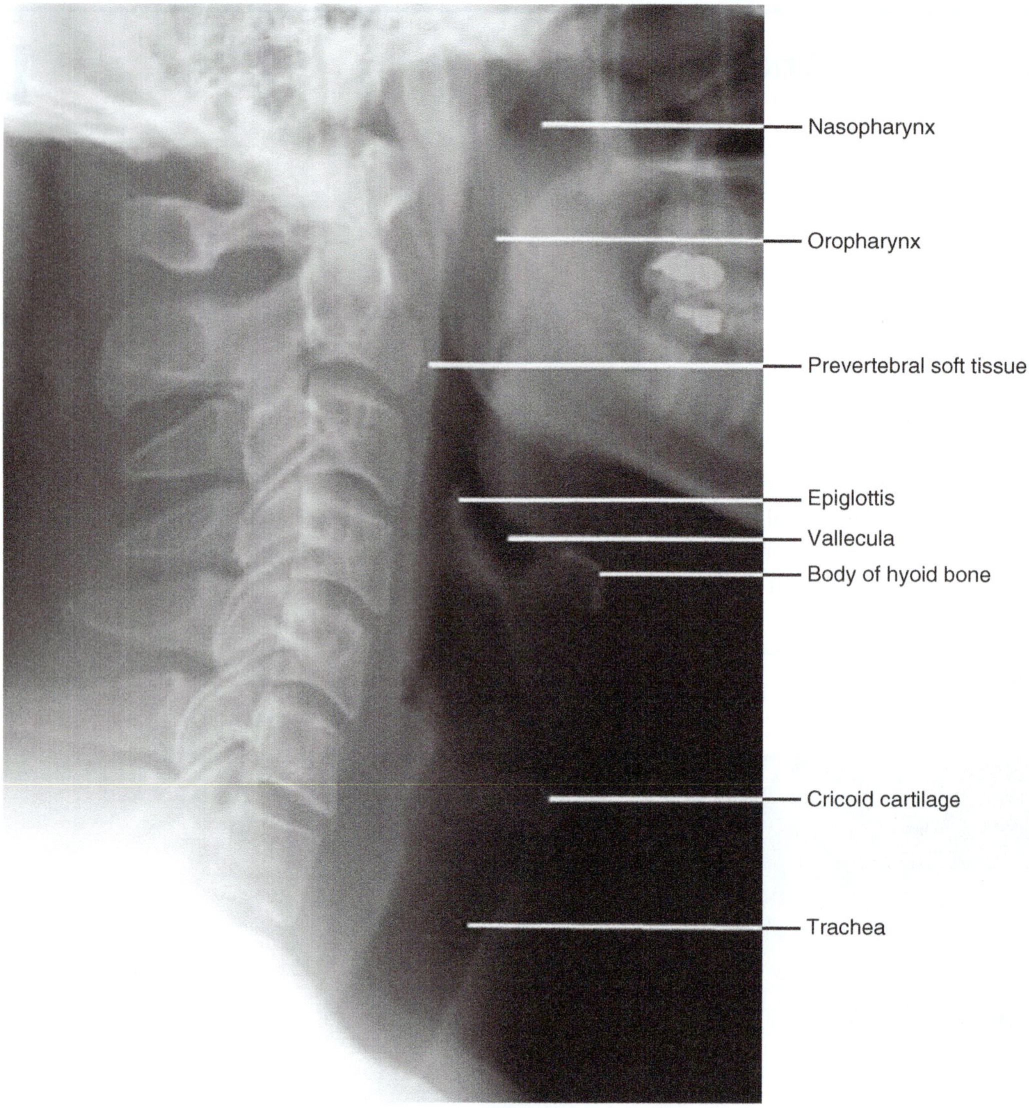

Fig. 13.2 Lateral soft tissue view of the larynx

Computed Tomography

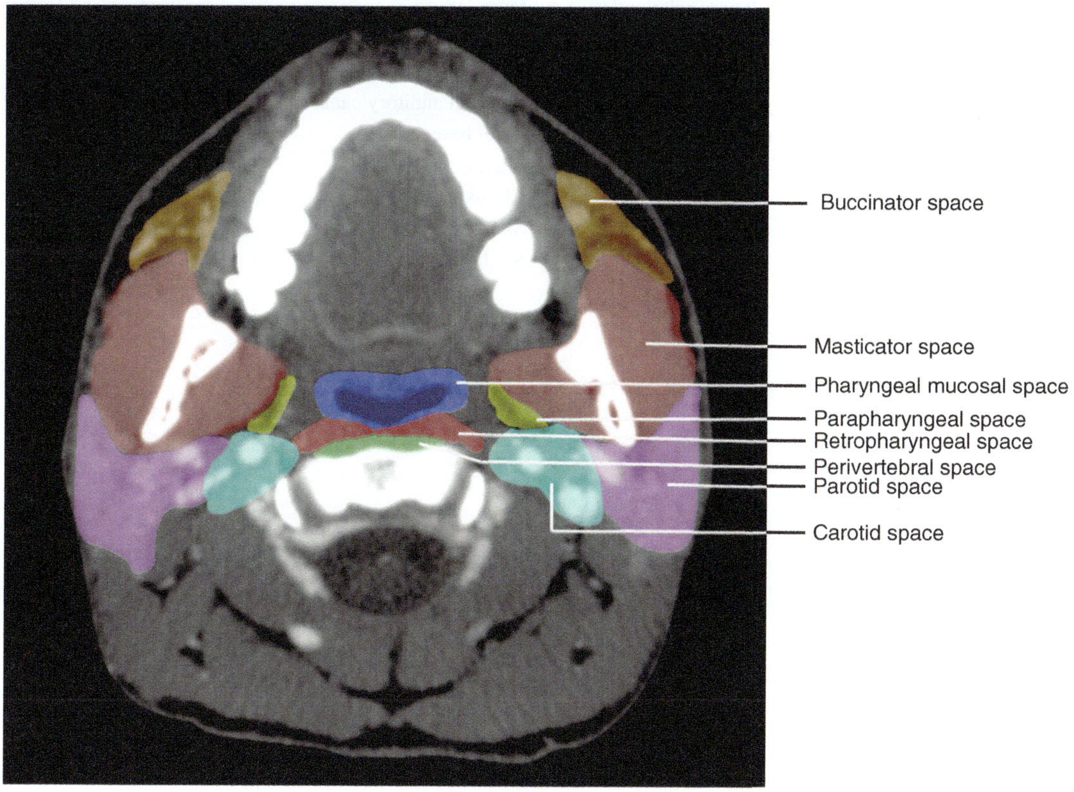

Fig. 13.3 Suprahyoid spaces

Table 13.1 Supraphyoid spaces and contents

Masticator space (MS)	Mandible, masticator muscles, CN 5 (V3)
Parotid space (PS)	Parotid gland, parotid nodes, external carotid artery, retro-mandibular vein, CN 7
Carotid space (CS)	Internal carotid artery, internal jugular vein, CN 9–12, associated deep cervical nodes
Parapharyngeal space (PPS)	Fat, ± minor salivary glands
Pharyngeal mucosal space (PMS)	Mucosa, minor salivary glands, lymphoid tissue
Perivertebral space (PVS; two components: prevertebral and paraspinal)	**Prevertebral:** prevertebral muscles, scalene muscles, brachial plexus, vertebral body, artery and vein
	Paraspinal: paraspinal muscles, posterior elements
Retropharyngeal space (RPS)	Fat, nodes

Table 13.2 Anatomical relations of suprahyoid spaces

		Extent	
Suprahyoid space	Skull base relation	Superior	Inferior
Masticator space	Foramen ovale Foramen spinosum	Temporalis muscle attachment	Mandible
Parotid space	Stylomastoid foramen	External auditory canal	Angle of mandible
Carotid space	Carotid canal and jugular	Skull base	Hyoid
Pharyngeal mucosal space	Foramen lacerum	Skull base	Hyoid
Retropharyngeal space		Clivus	T4
Perivertebral space		Skull base	Sacrum
Parapharyngeal space		Skull base	Hyoid

Axial Images

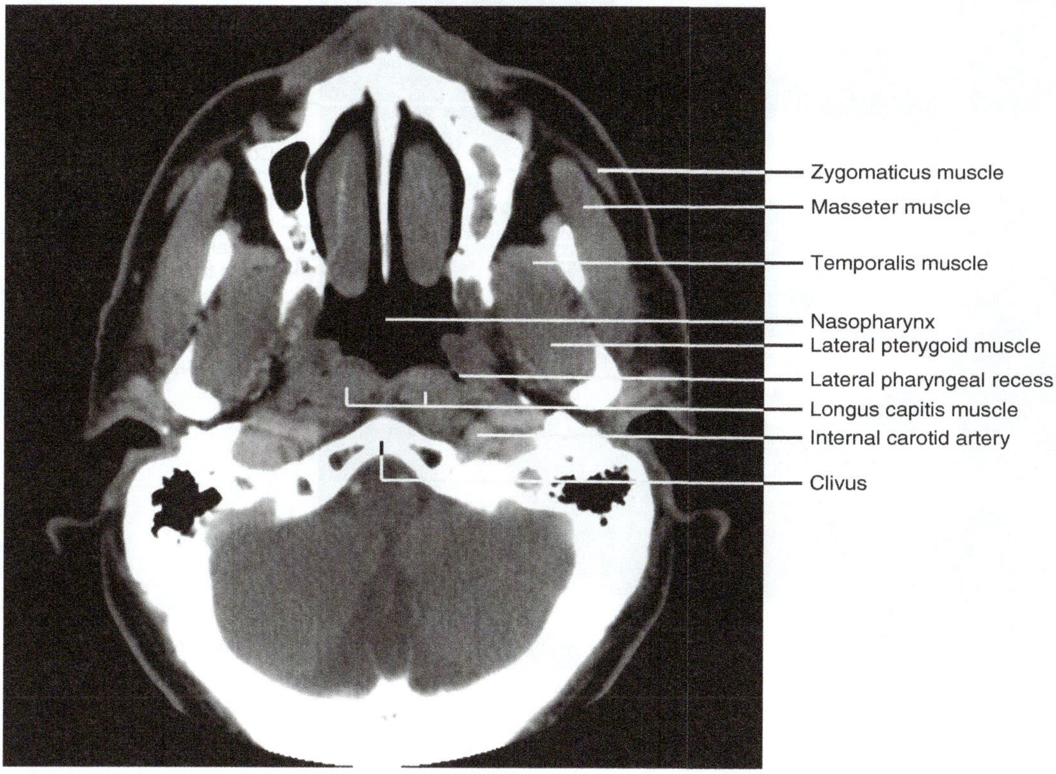

Fig. 13.4 Level of the external acoustic meatus

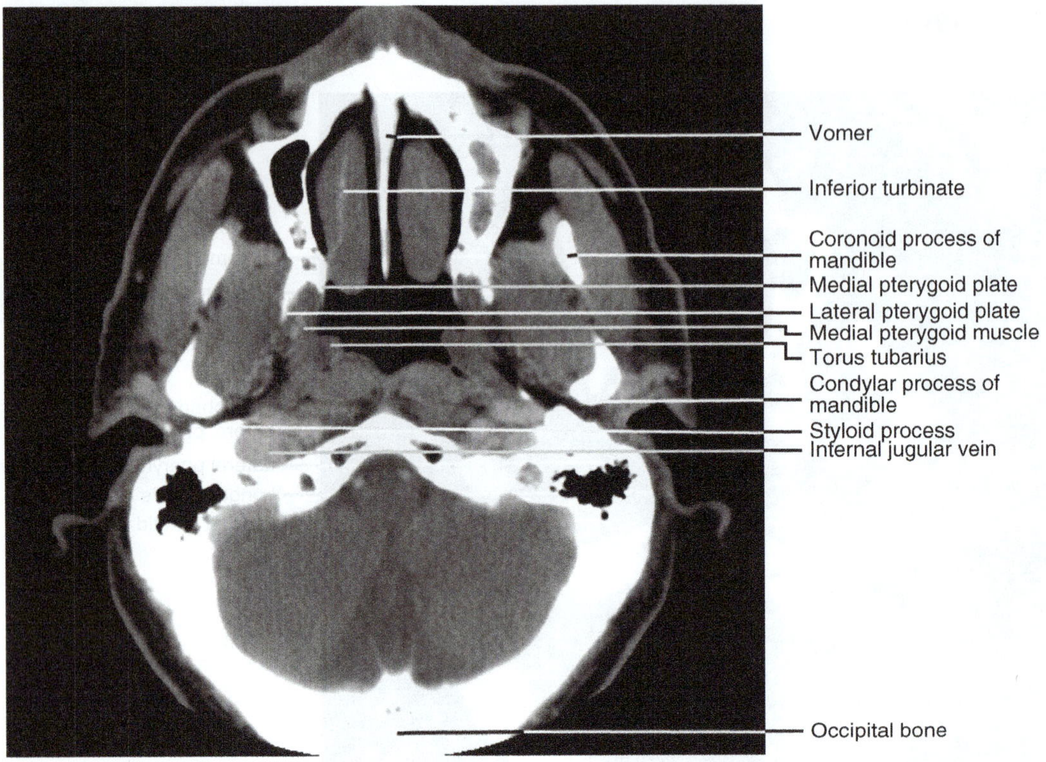

Fig. 13.5 Level of the external acoustic meatus

Pearl

Torus tubarius = pharyngeal cartilaginous end of the auditory tube. This raised elevation lies between the opening of the Eustachian tube laterally and the fossa of Rosenmüller medially.

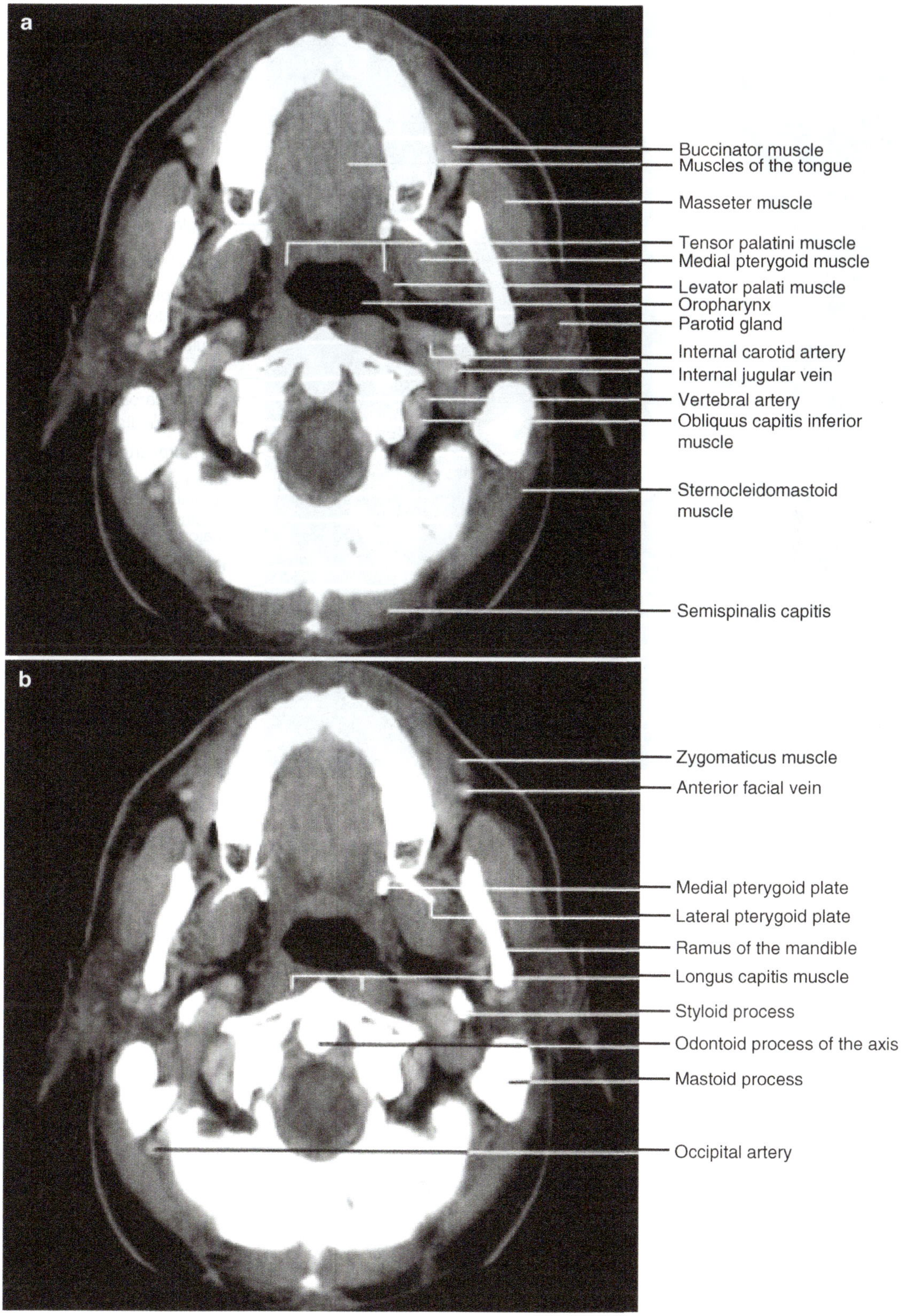

Fig. 13.6 (**a**, **b**) Level of the atlas

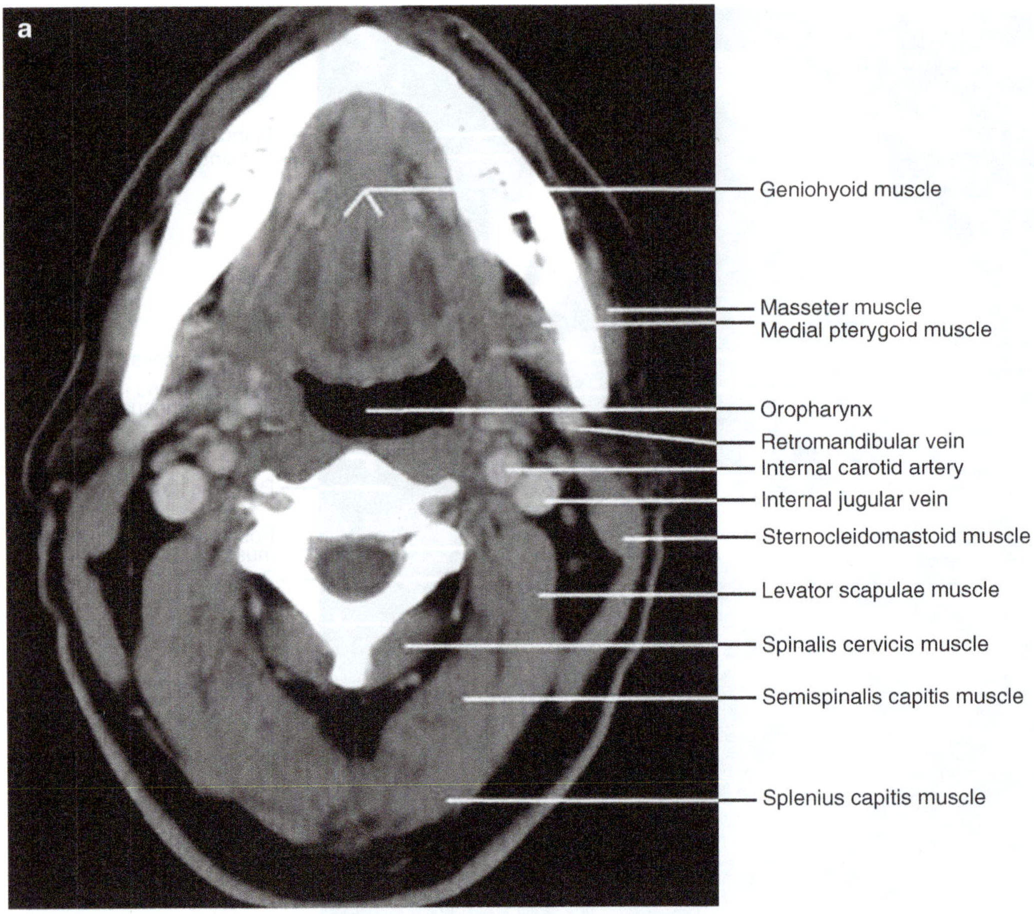

Fig. 13.8 Level of the floor of the mouth

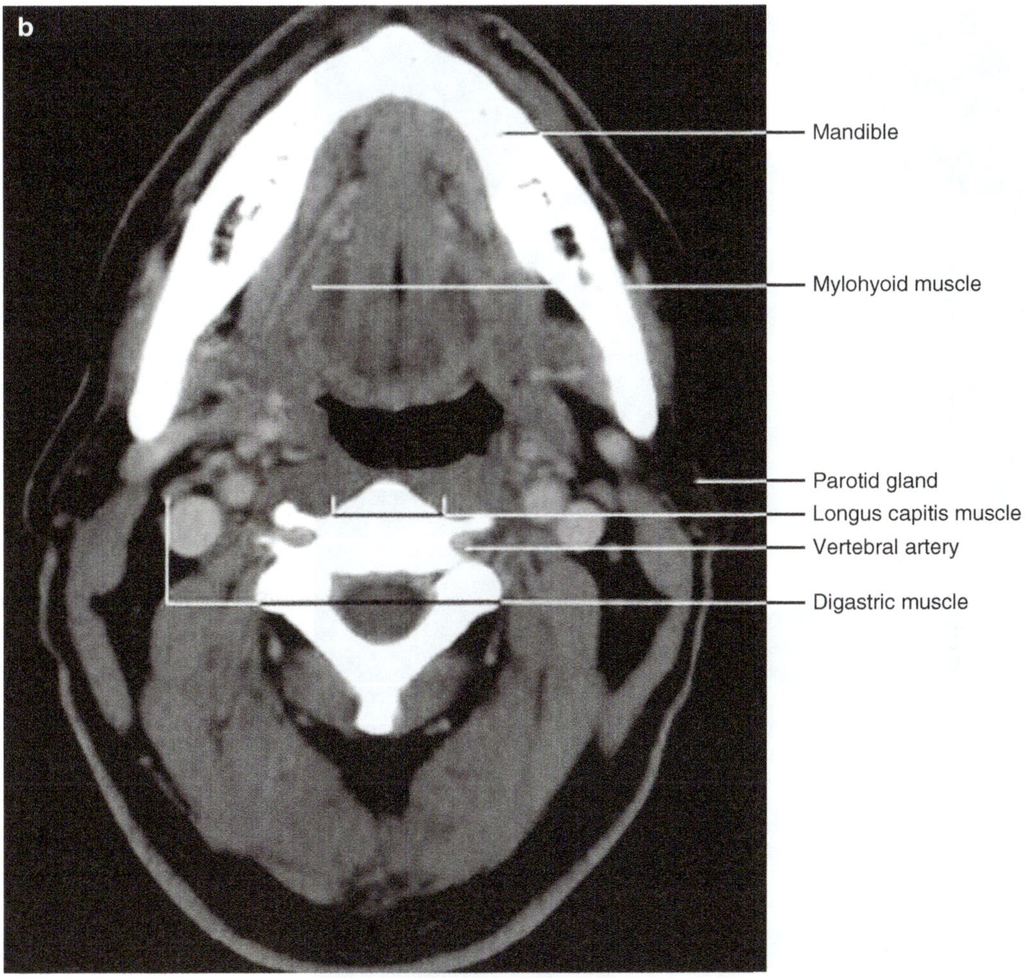

Fig. 13.8 (continued)

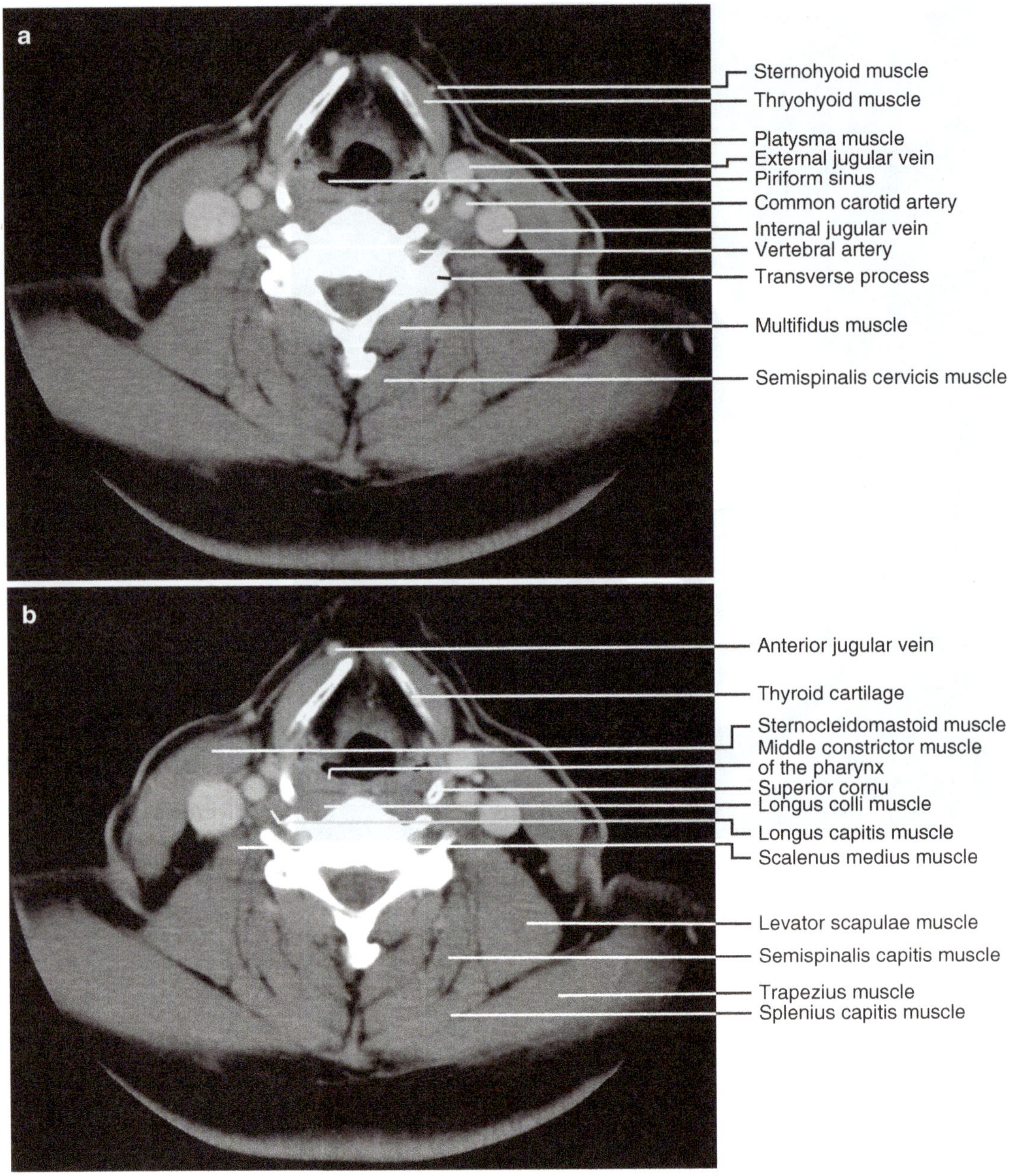

Fig. 13.11 Level of the superior cornu of the thyroid cartilage

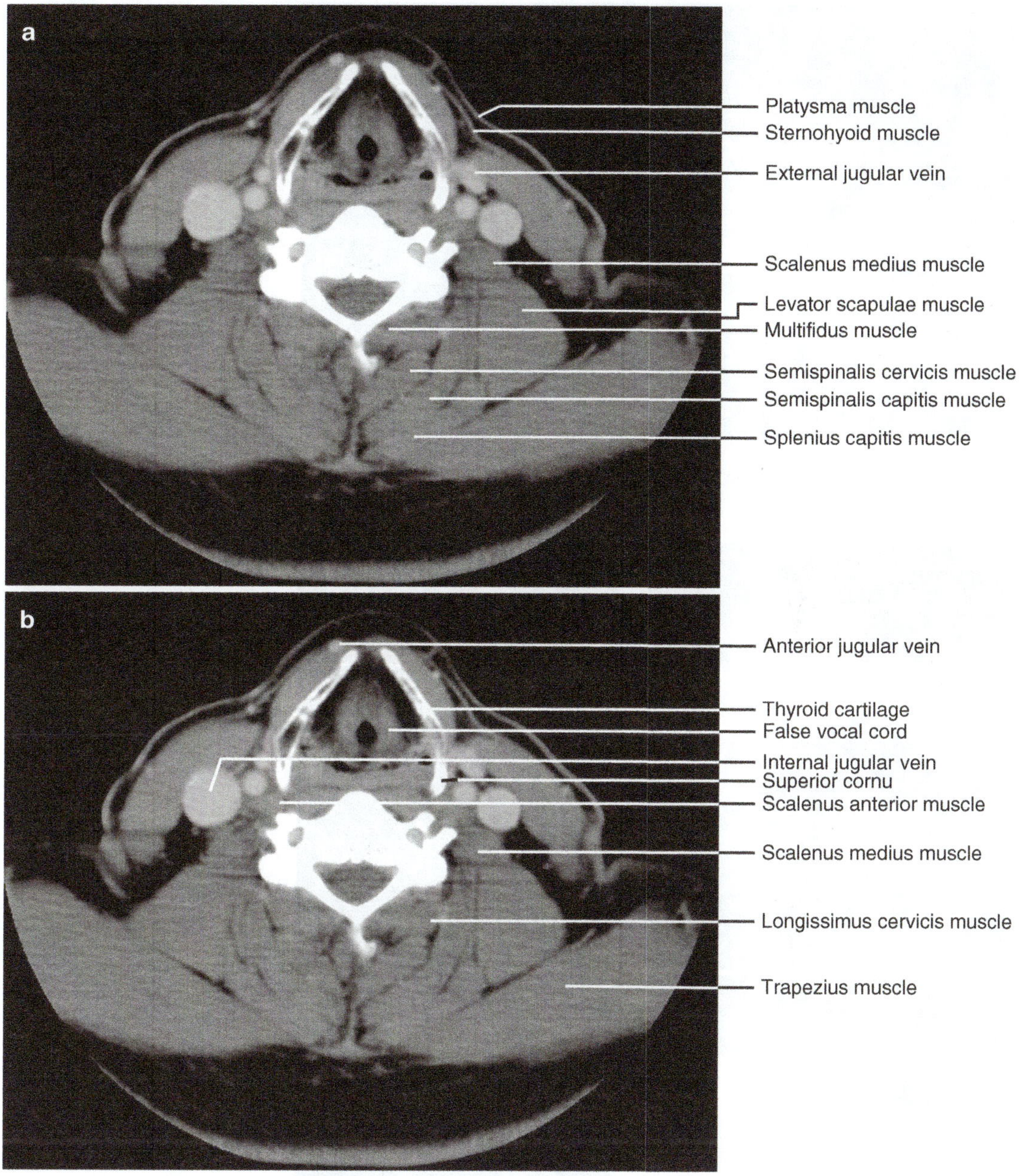

Fig. 13.12 Level of the false cords

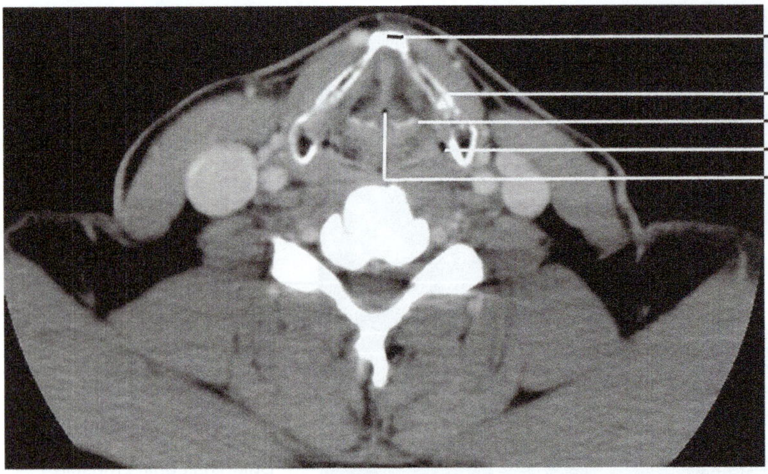

Fig. 13.13 Level superior to true vocal cords

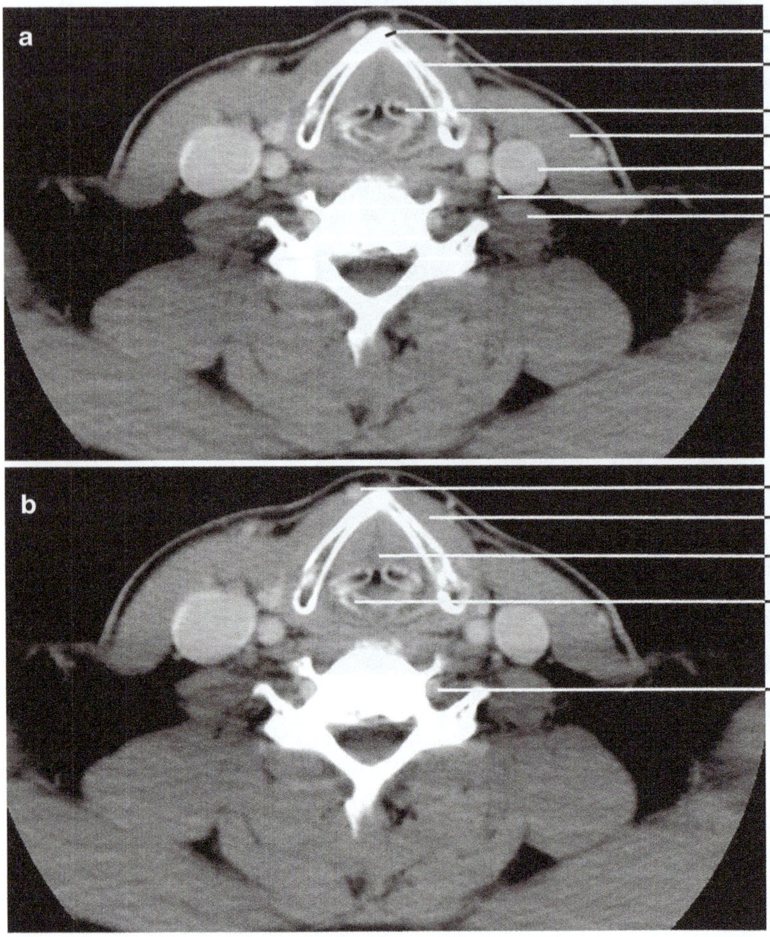

Fig. 13.14 Level at the true vocal cords

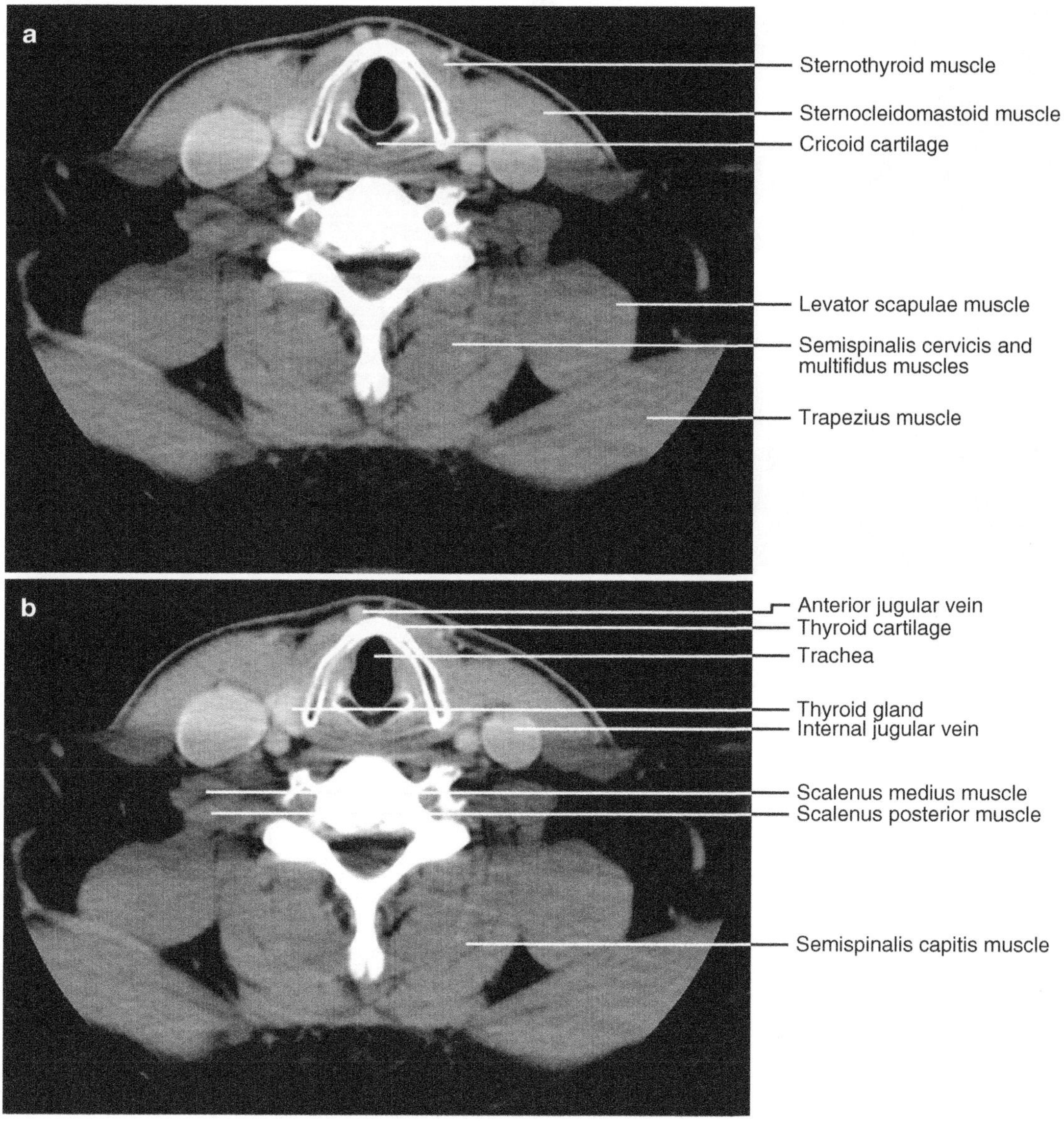

Fig. 13.15 Level of the lower thyroid cartilage

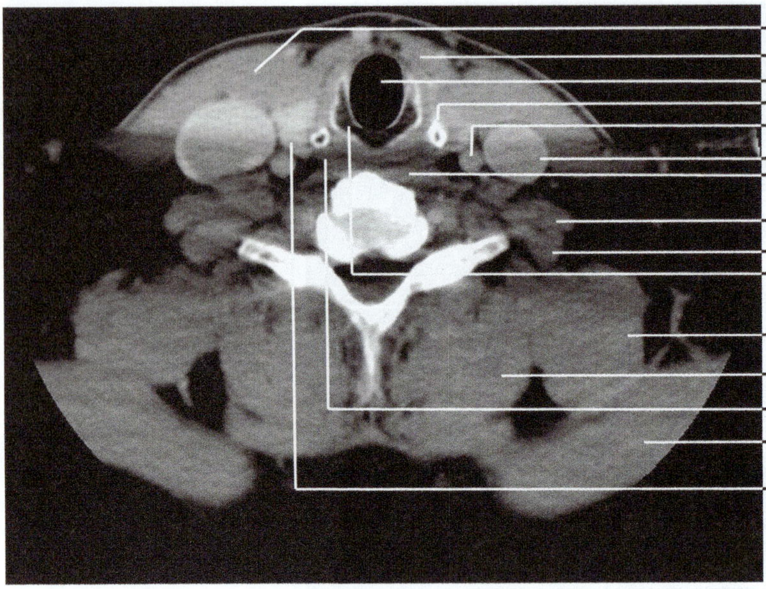

Fig. 13.16 Level at the cricoid cartilage

Coronal Images

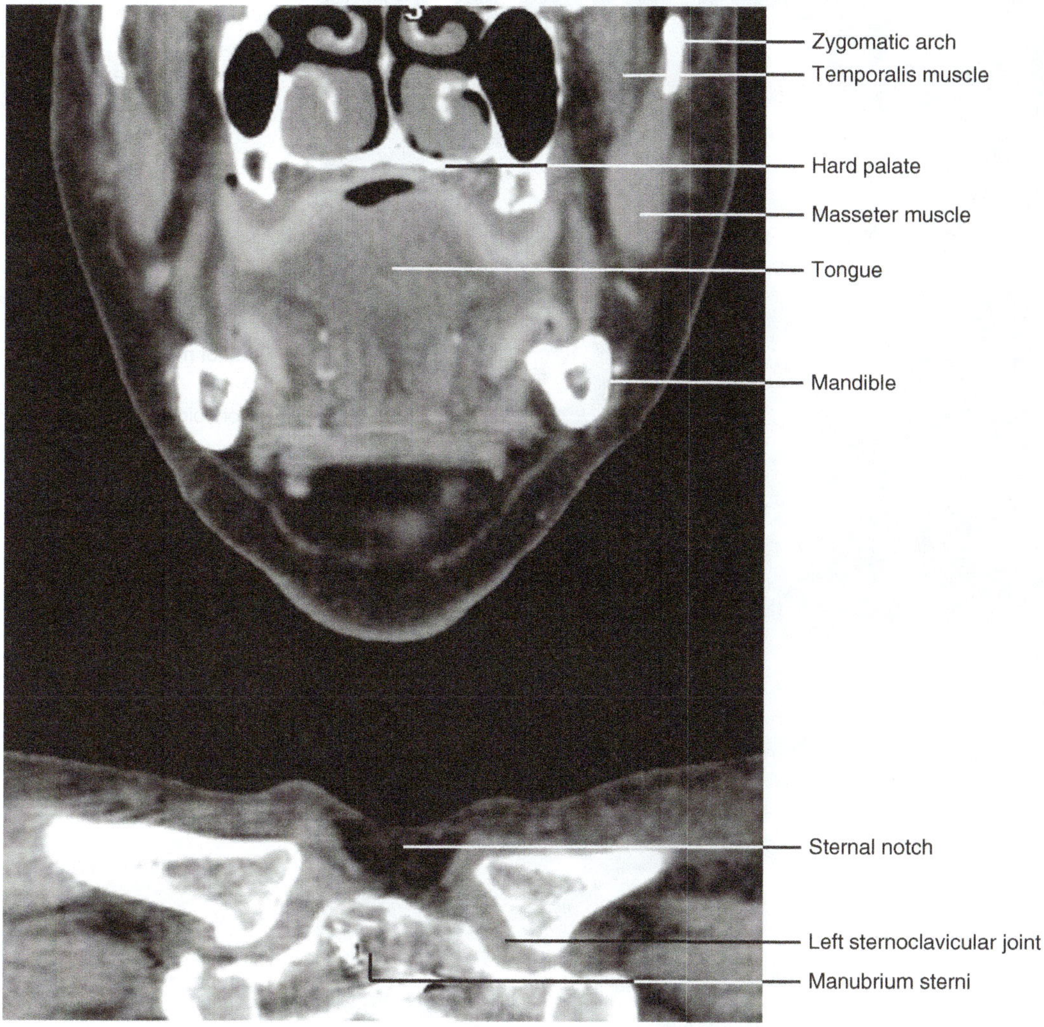

Fig. 13.17 Level of the sternal notch

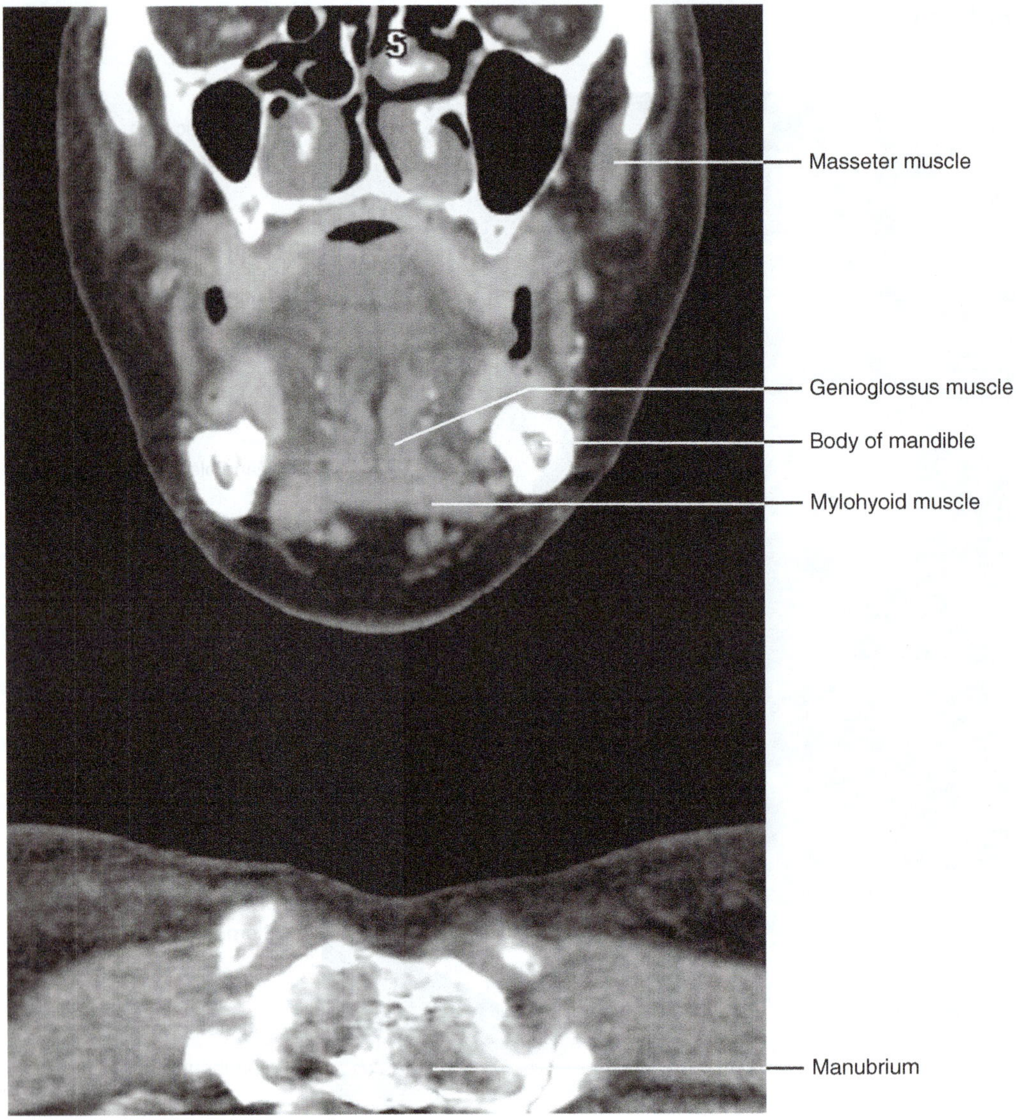

Fig. 13.18 Midway between sternum and larynx

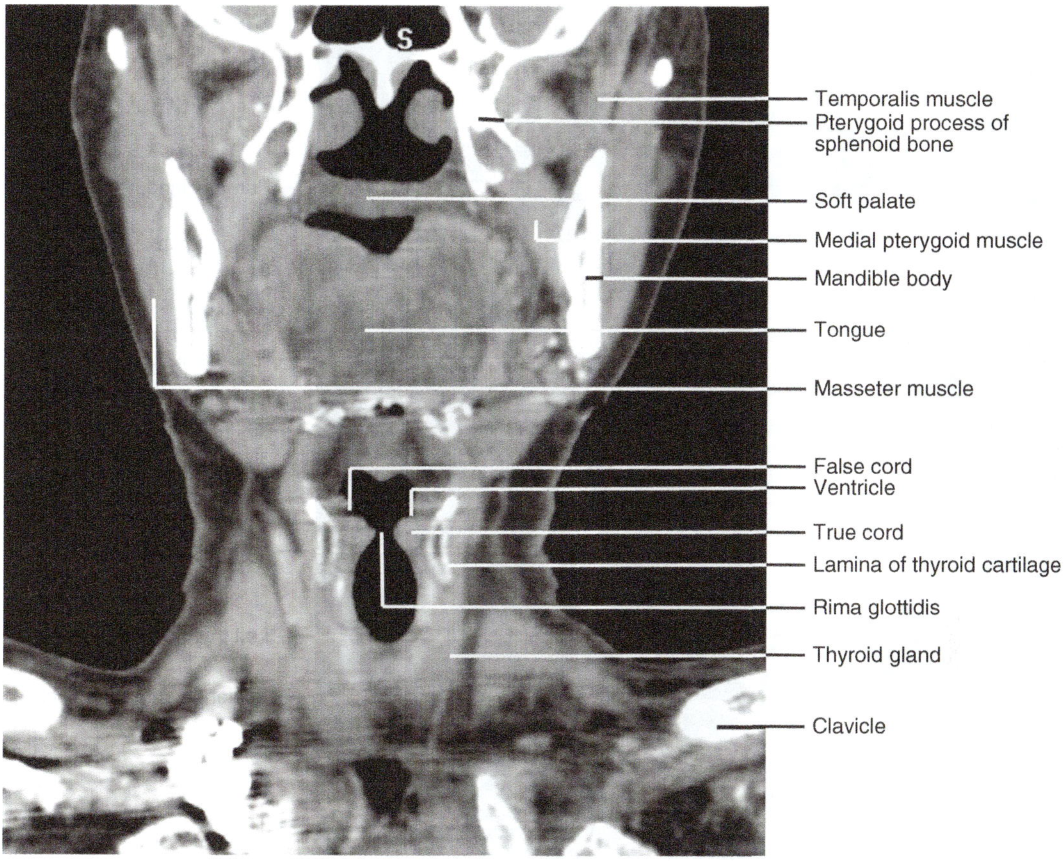

Fig. 13.19 Level at the larynx

T1 Axial

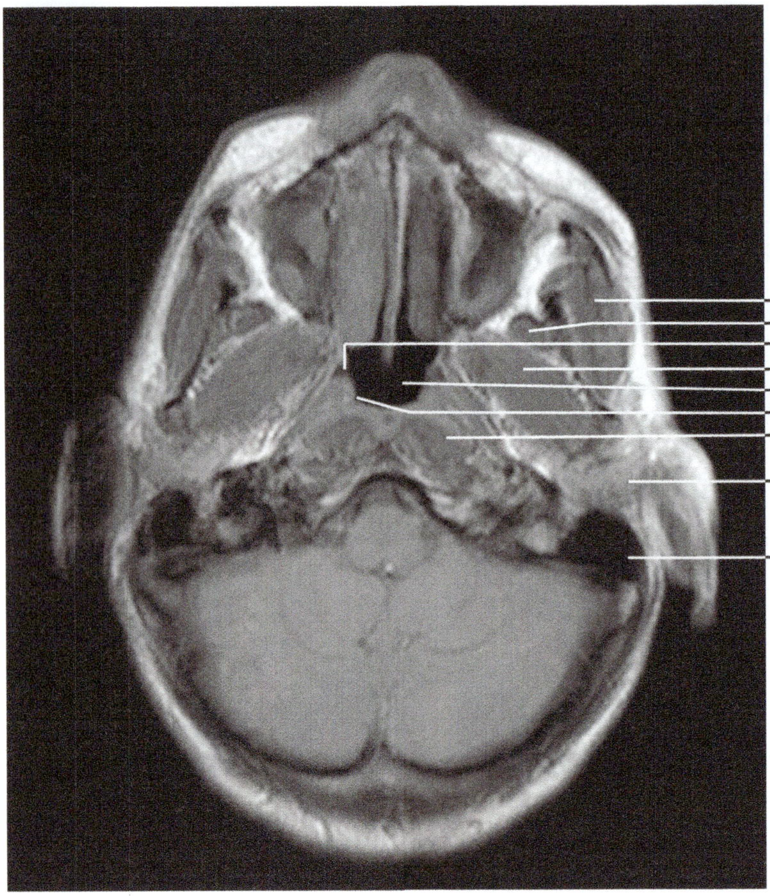

Fig. 13.21 Level of mastoids

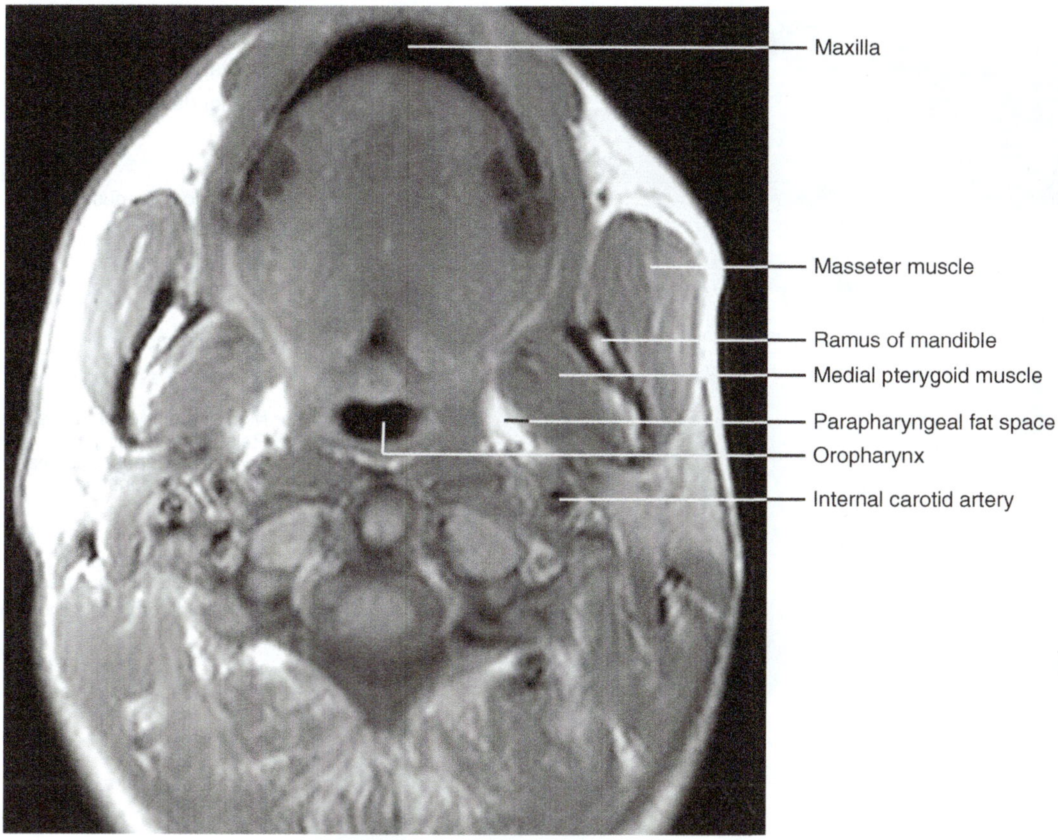

Fig. 13.22 Level of C1

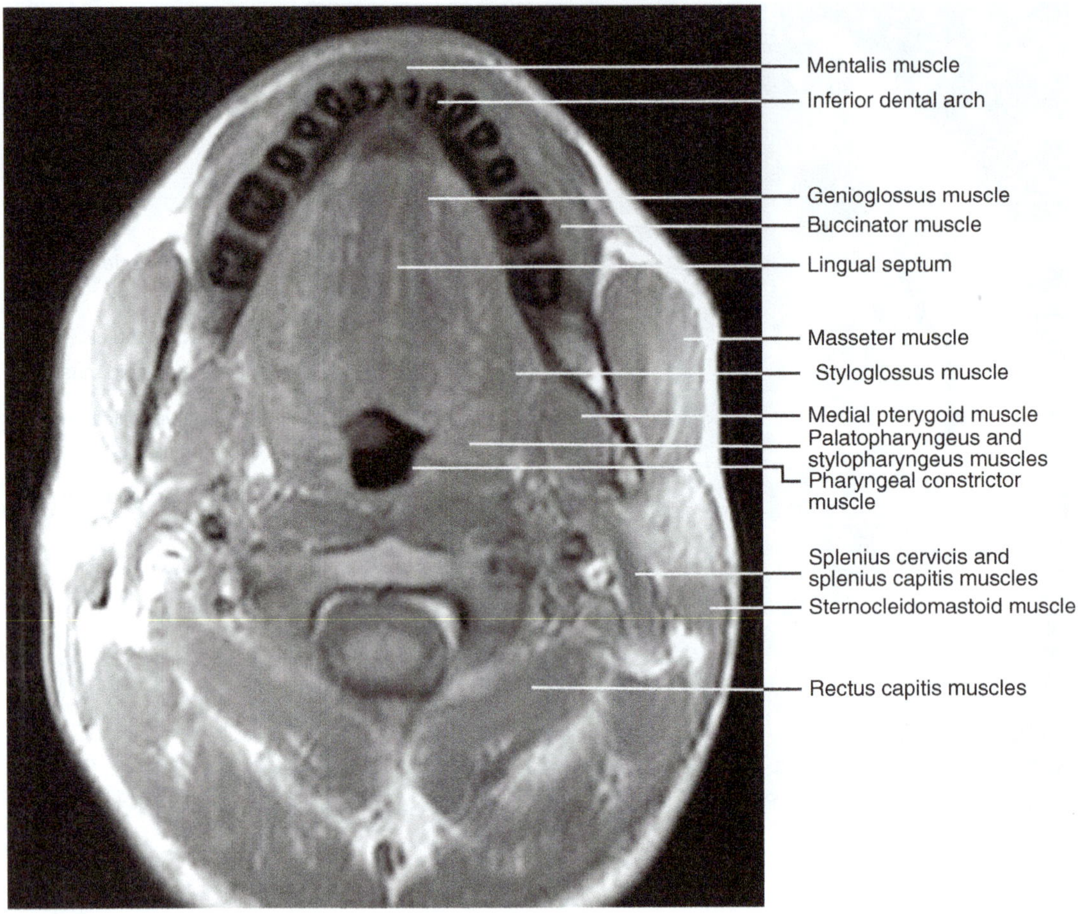

Fig. 13.23 Level of C2

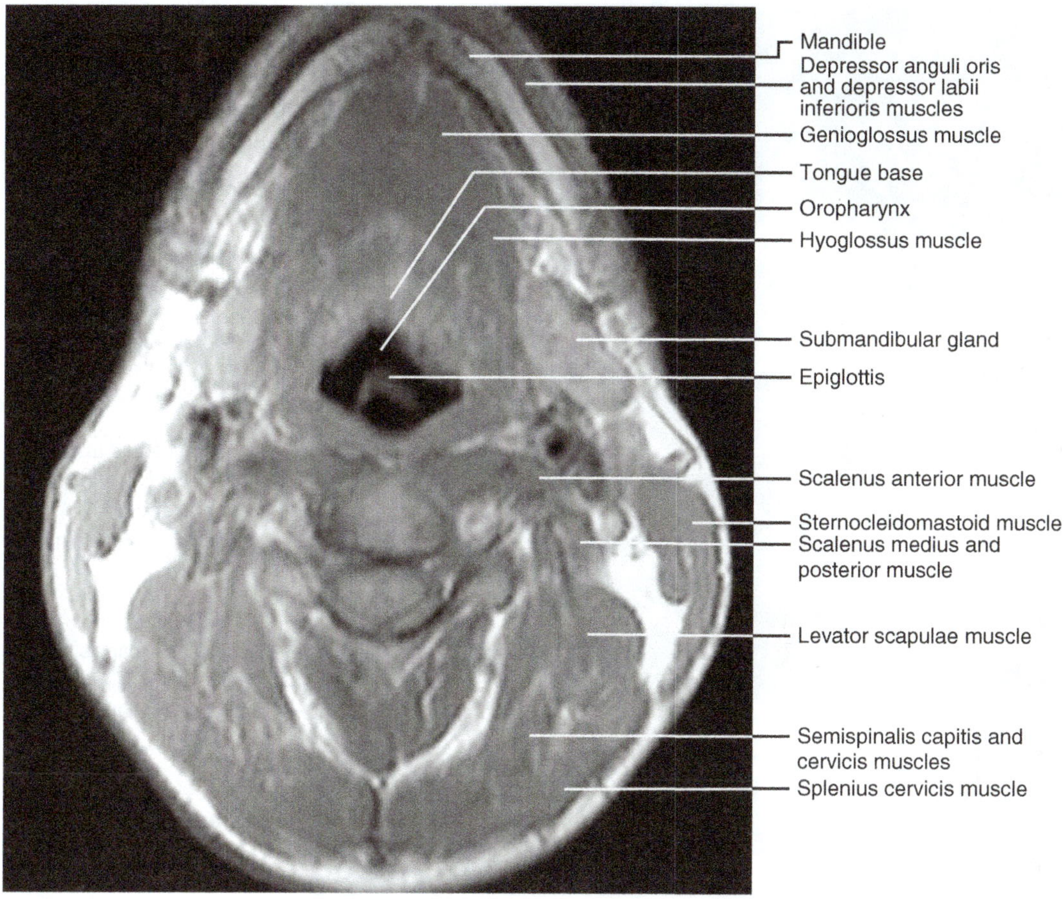

Fig. 13.24 Level of C4

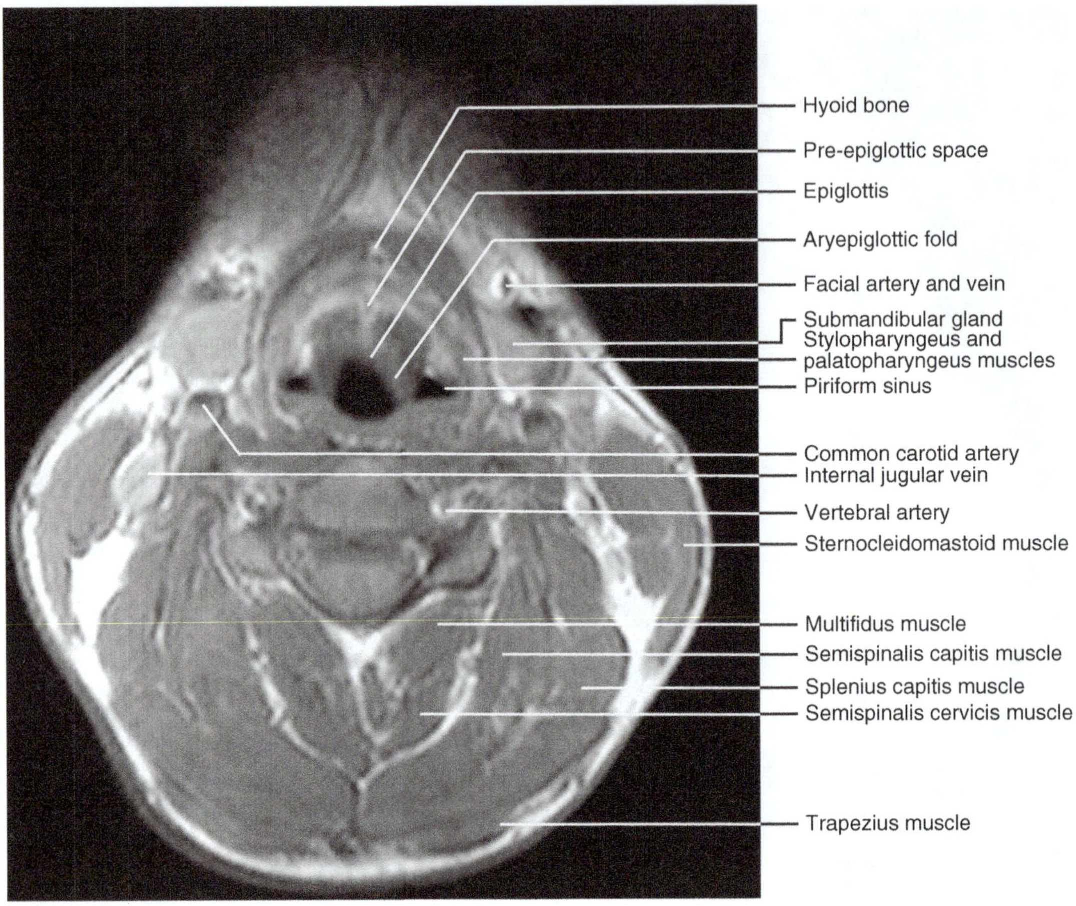

Fig. 13.25 Level of hyoid

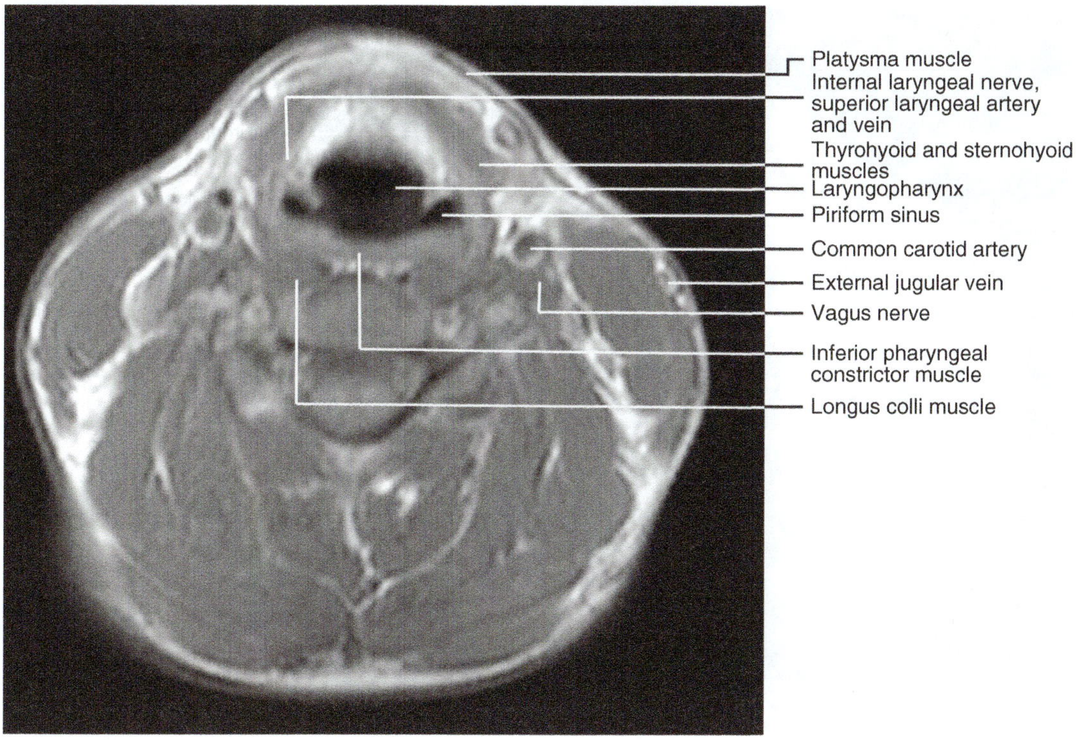

Fig. 13.26 Level of superior extent of laryngeal cartilage

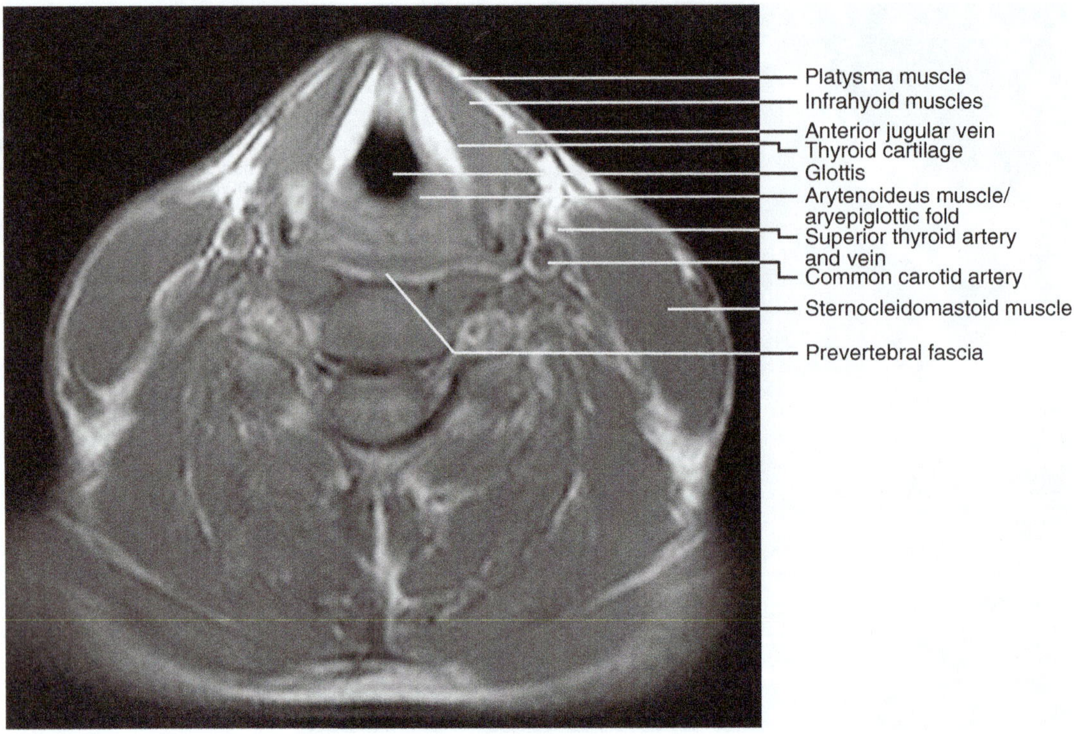

Fig. 13.27 Supraglottic level

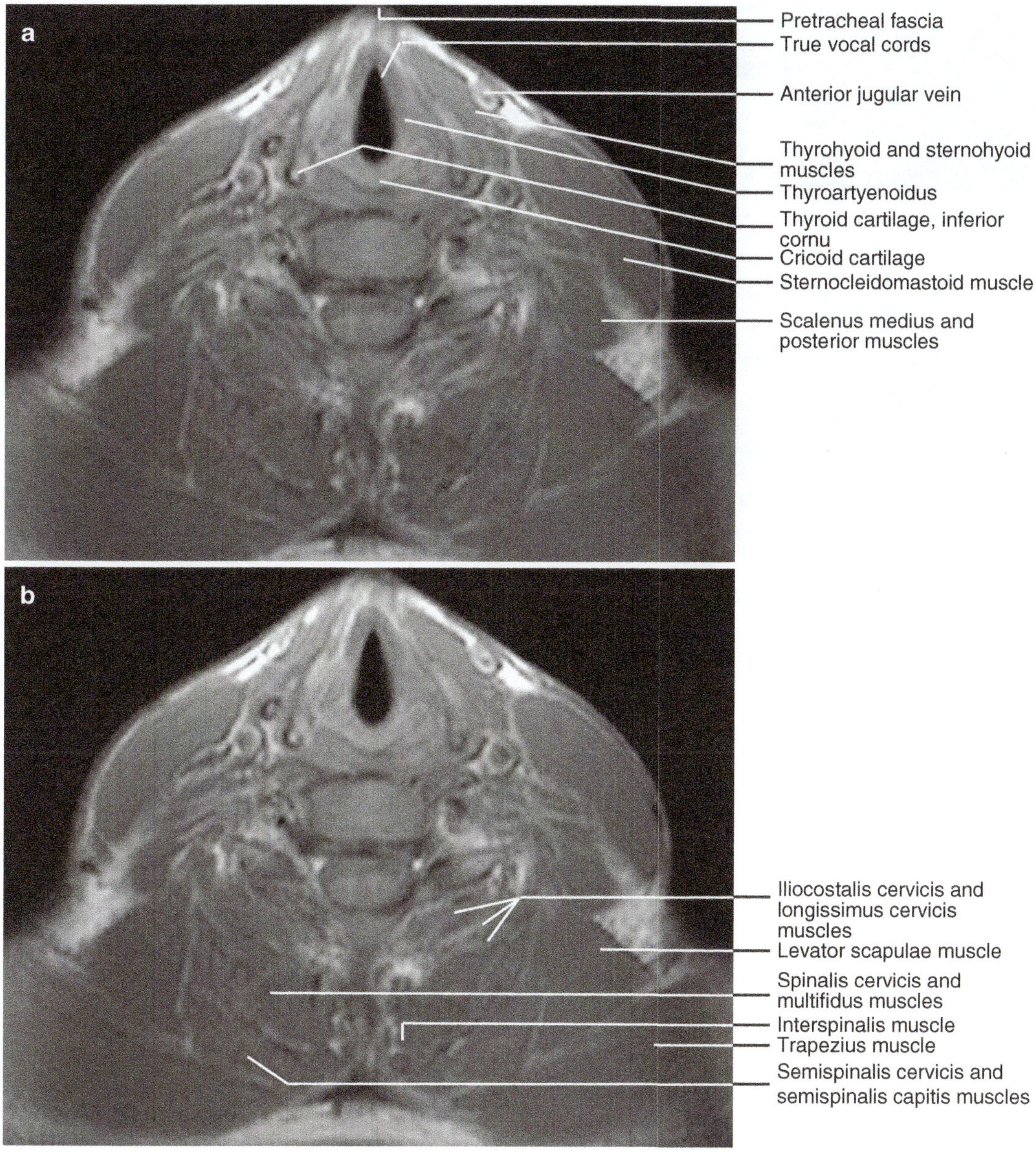

Fig. 13.28 Level of glottis

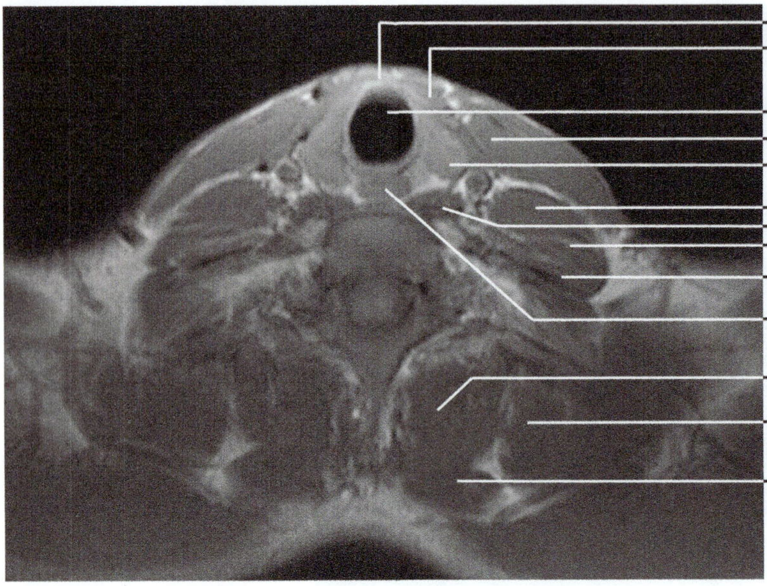

Fig. 13.29 Level of thyroid

T1 Sagittal

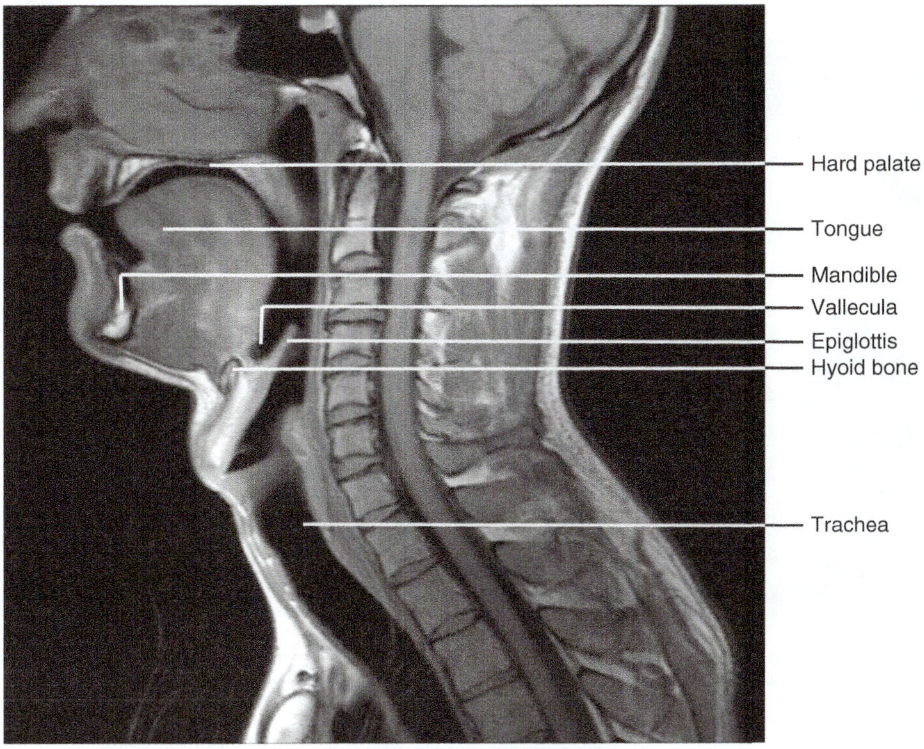

Fig. 13.30 Midline T1

T1 Coronal

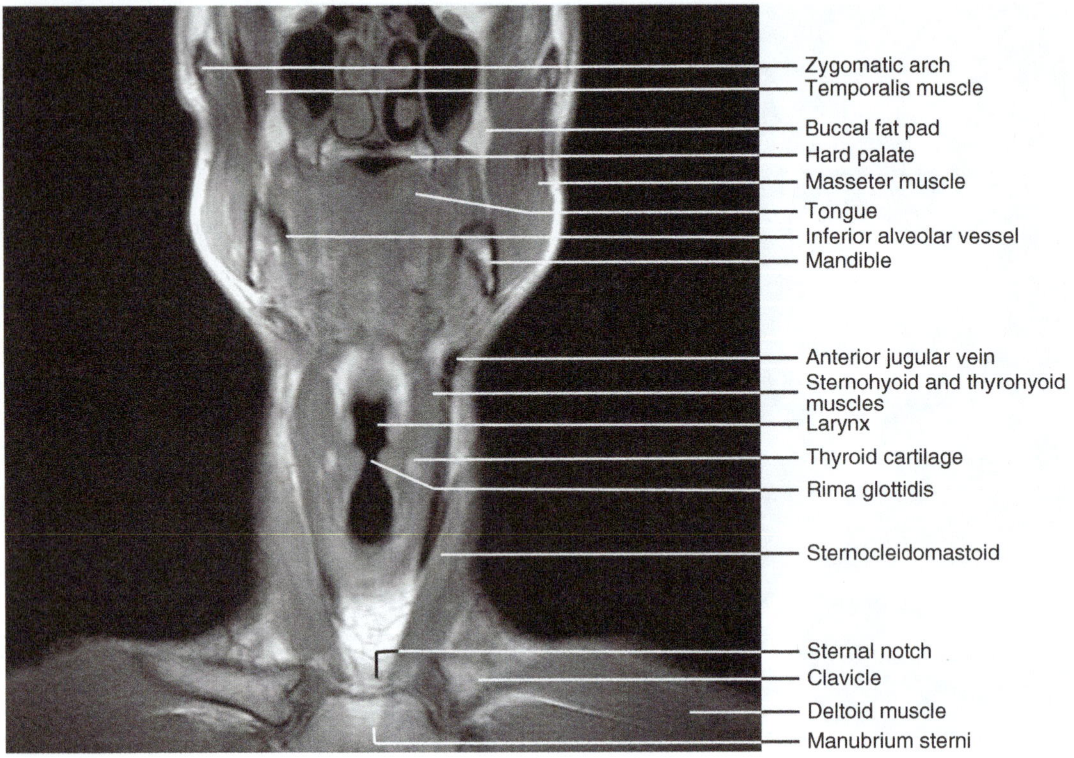

Fig. 13.31 Level of suprasternal notch

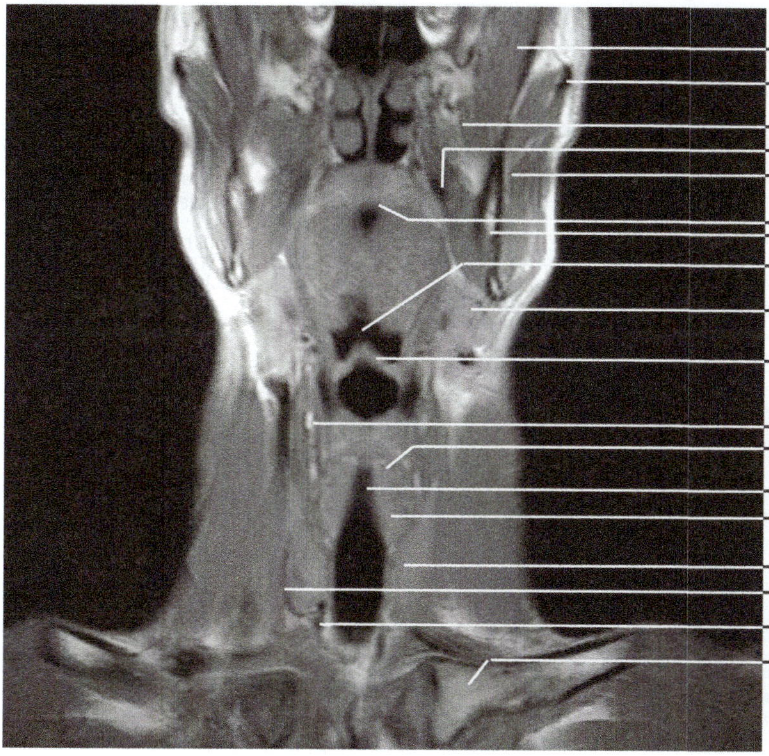

Fig. 13.32 Level at base of tongue

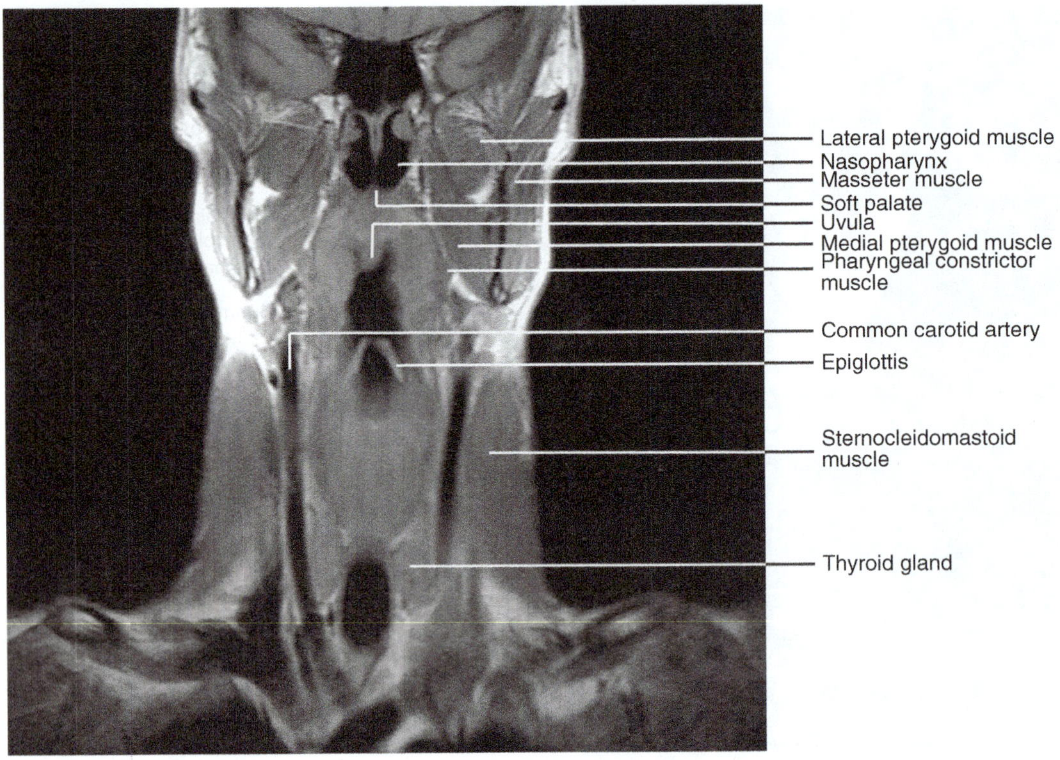

Fig. 13.33 Level of uvula

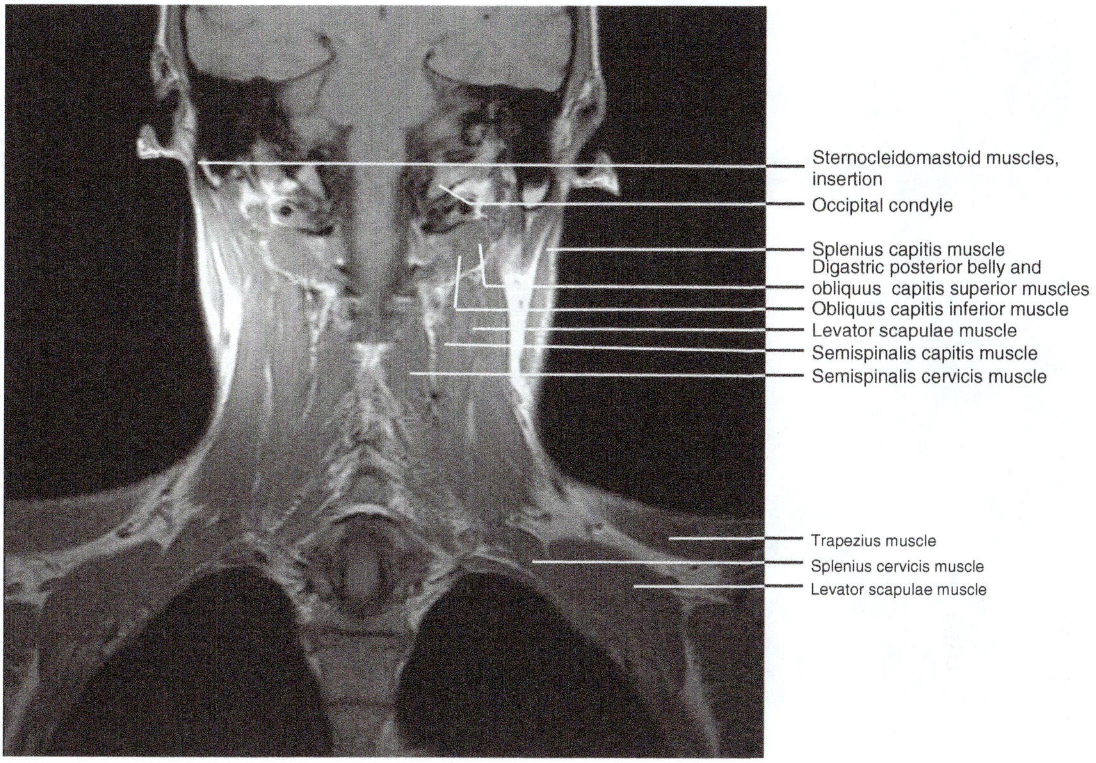

Fig. 13.34 Level of mastoids

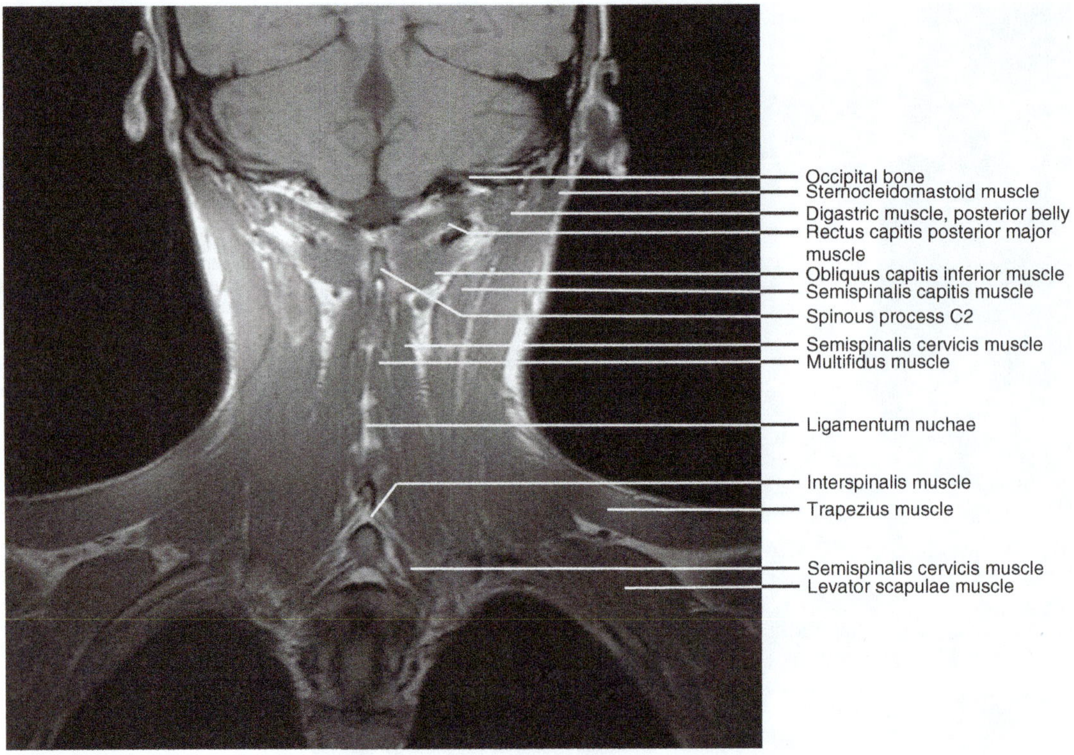

Fig. 13.35 Level through fourth ventricle

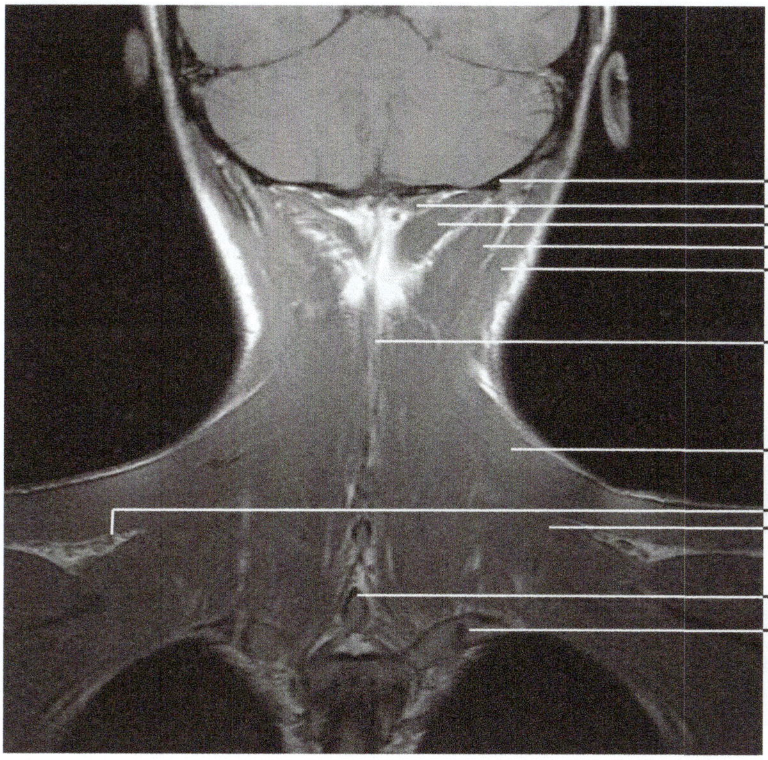

Fig. 13.36 Level through mid cerebellum

Vascular Imaging

Doppler

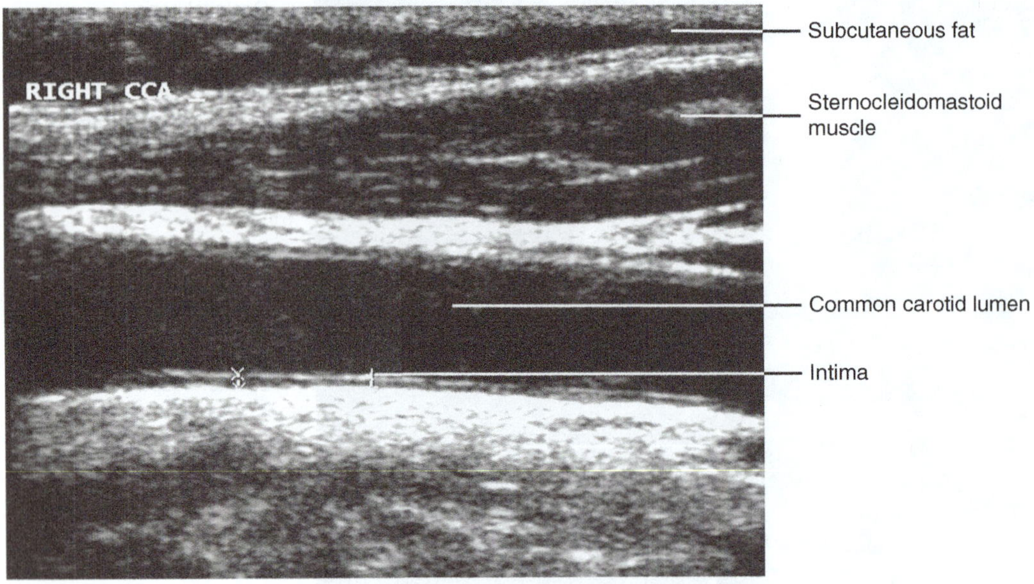

Fig. 13.37 Common carotid artery sagittal with intima measurement

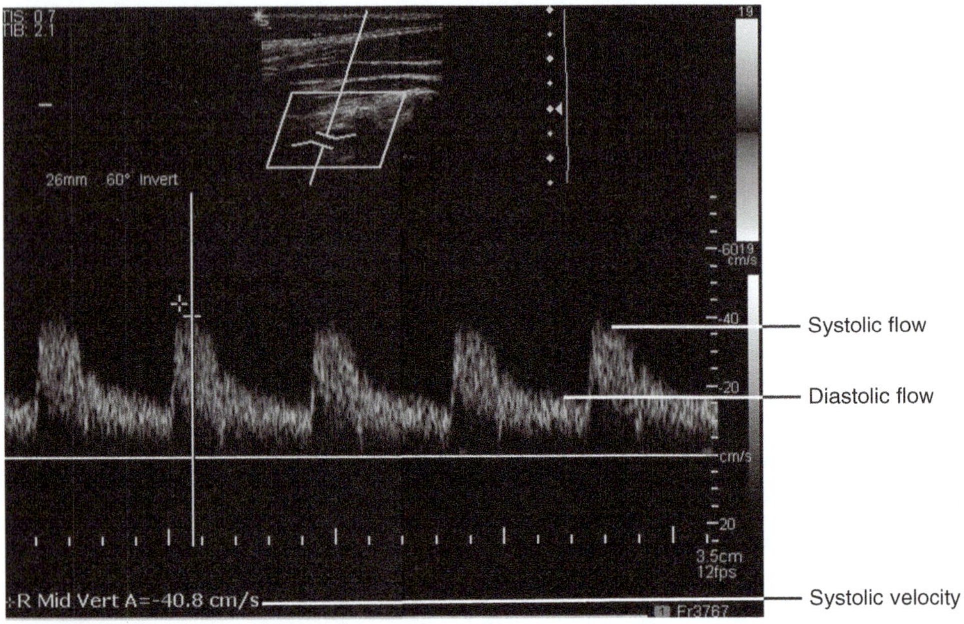

Fig. 13.38 Carotid bulb waveform

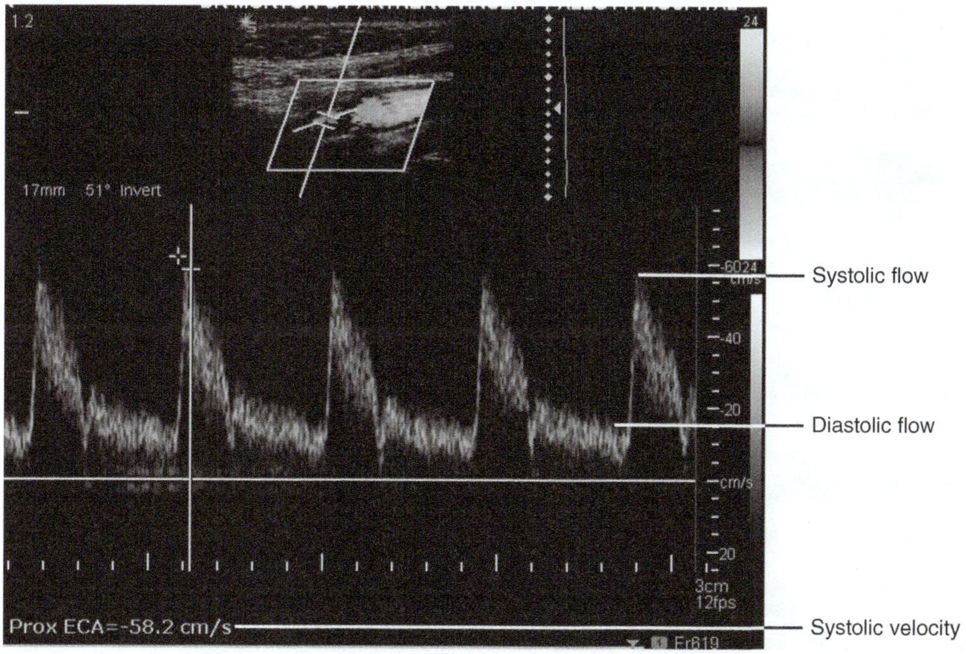

Fig. 13.39 Middle common carotid artery waveform

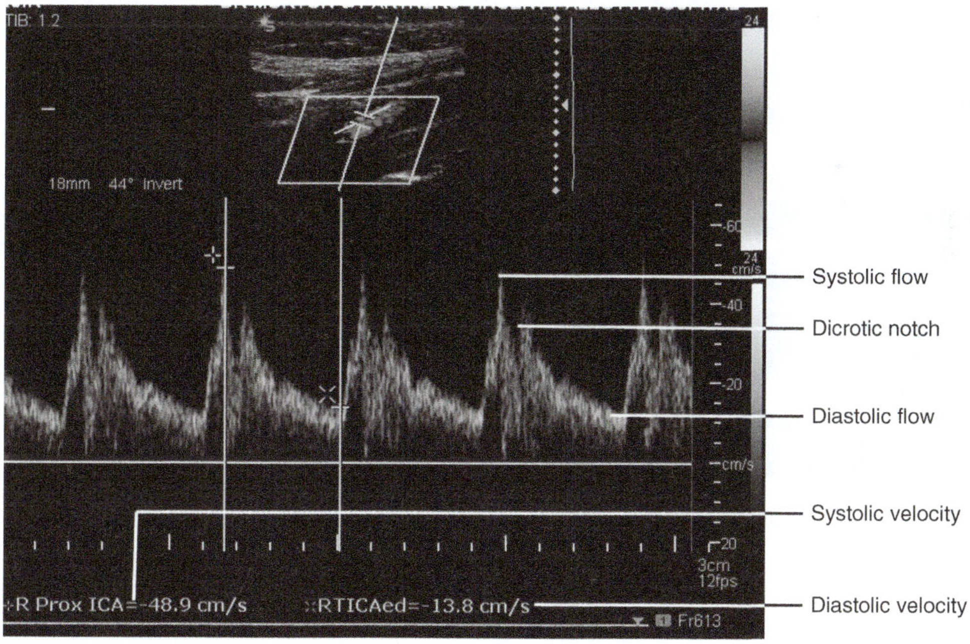

Fig. 13.40 Proximal internal carotid artery biphasic waveform

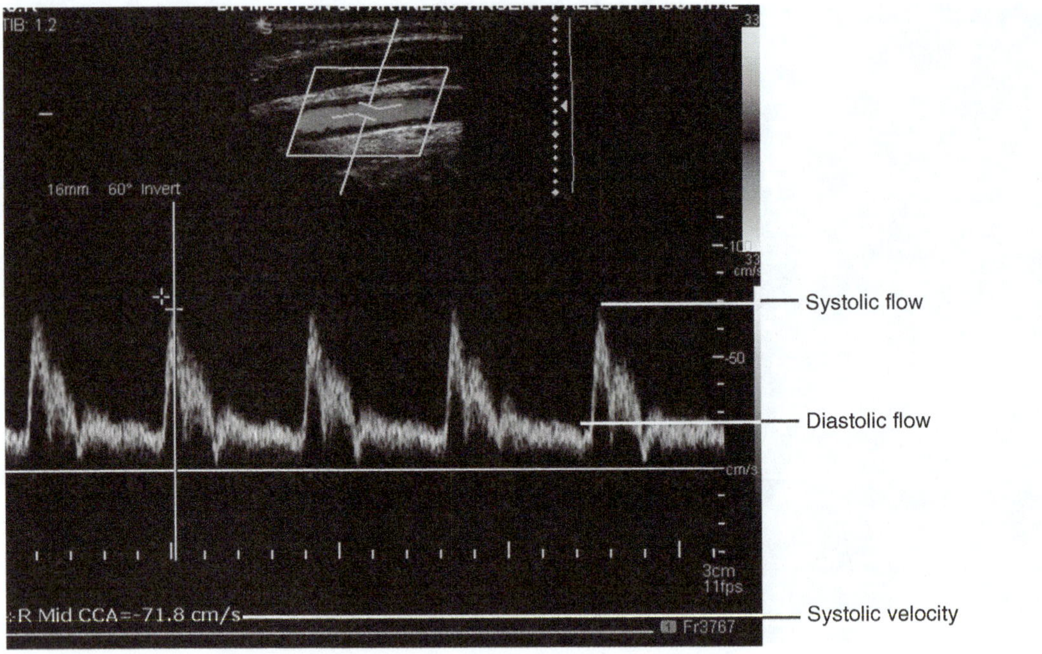

Fig. 13.41 Proximal external carotid artery waveform

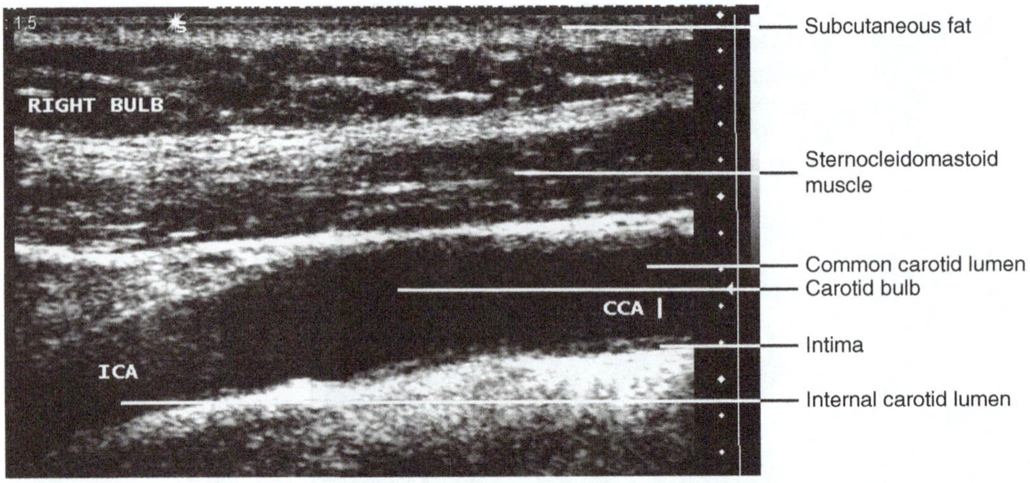

Fig. 13.42 Longitudinal ultrasound of carotid artery

Computer Tomography Angiography (CTA)

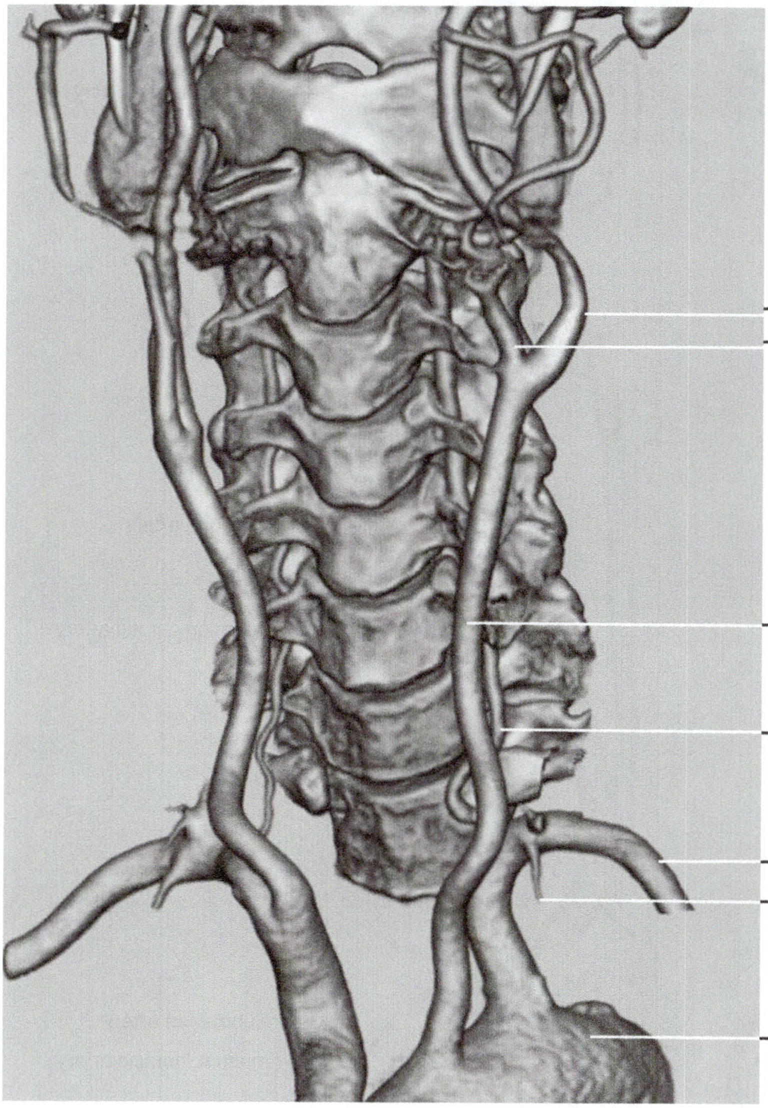

Fig. 13.43 Anteroposterior with bony detail

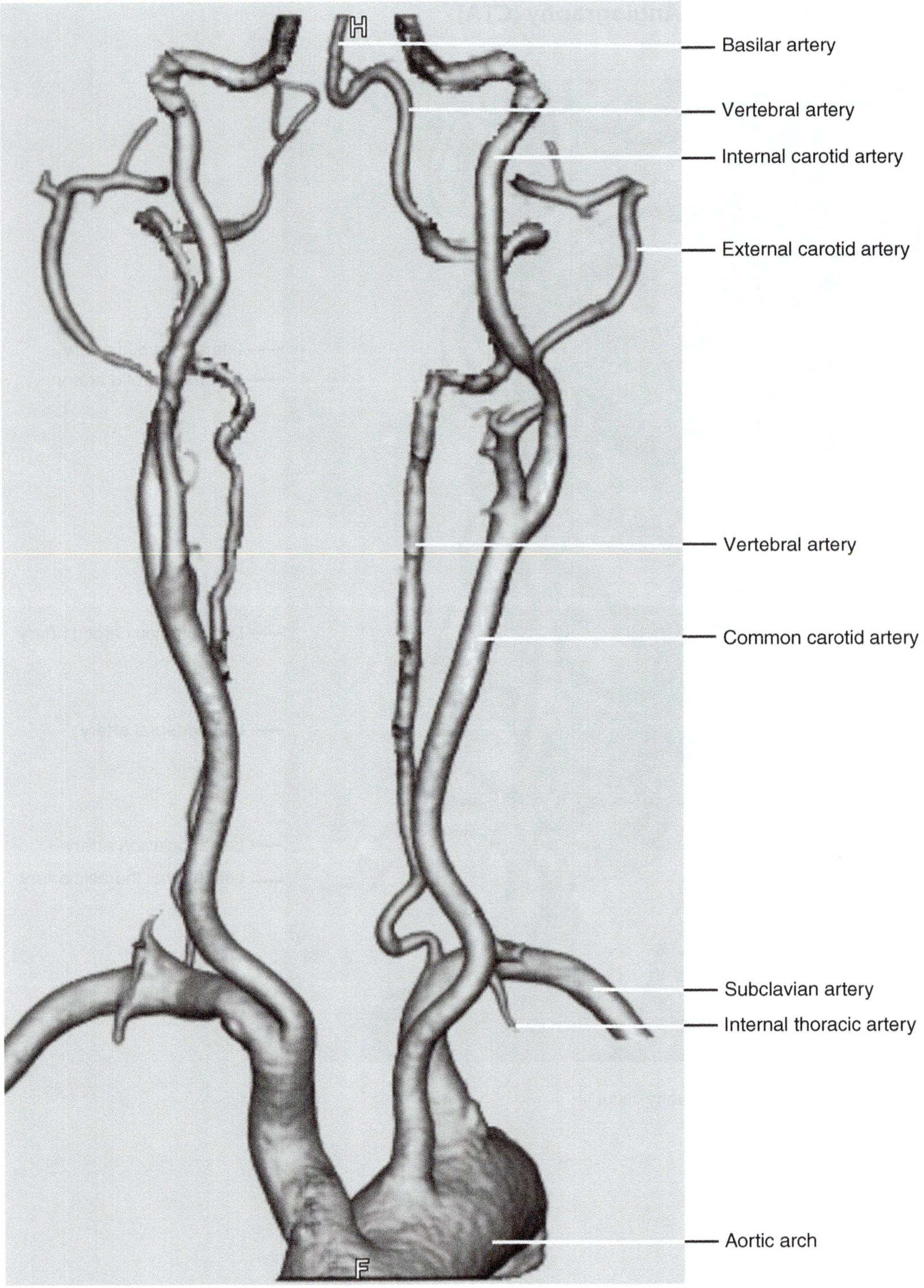

Fig. 13.44 Anteroposterior view of neck vessels

Digital Subtraction Angiography

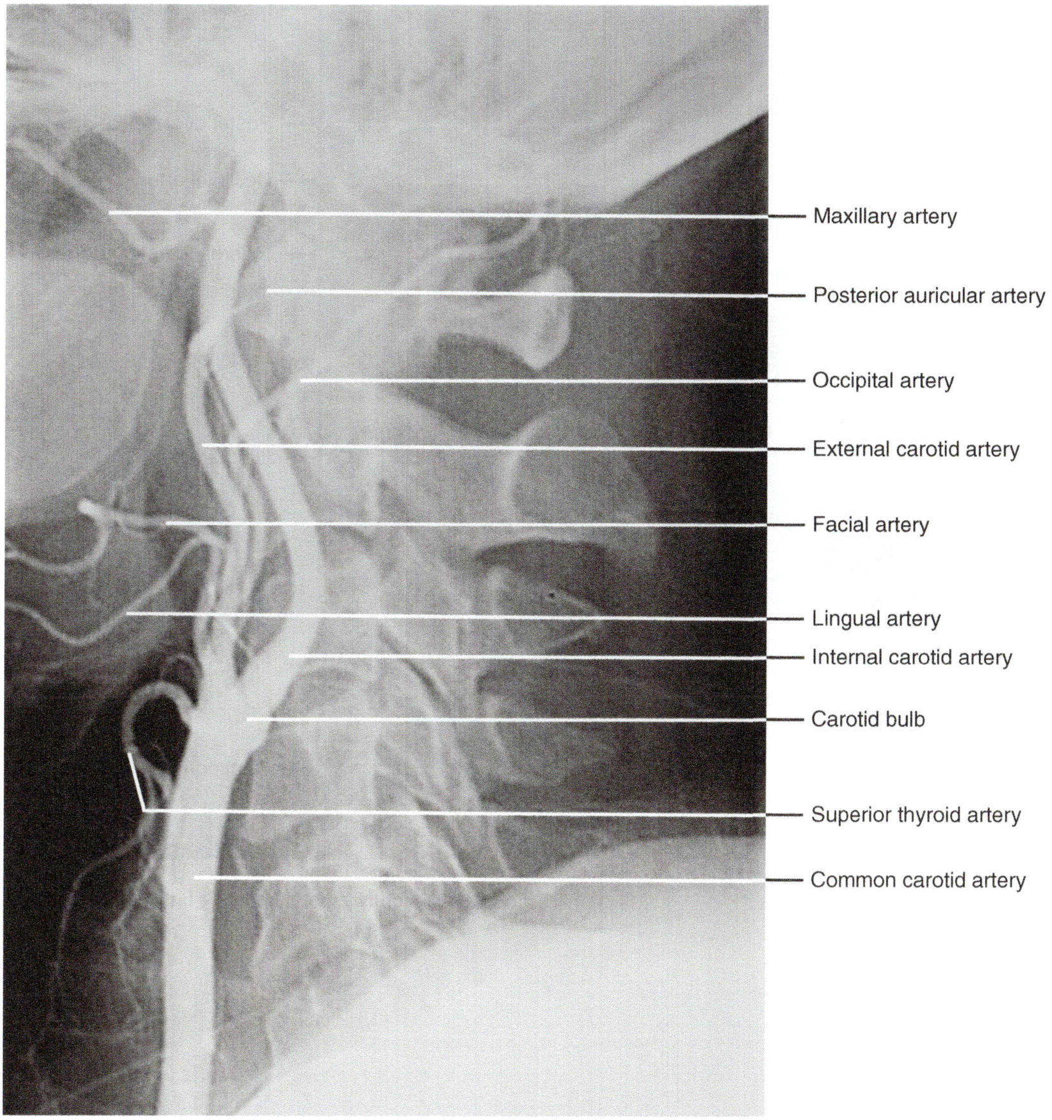

Fig. 13.45 Lateral common carotid artery with bifurcation (without subtraction)

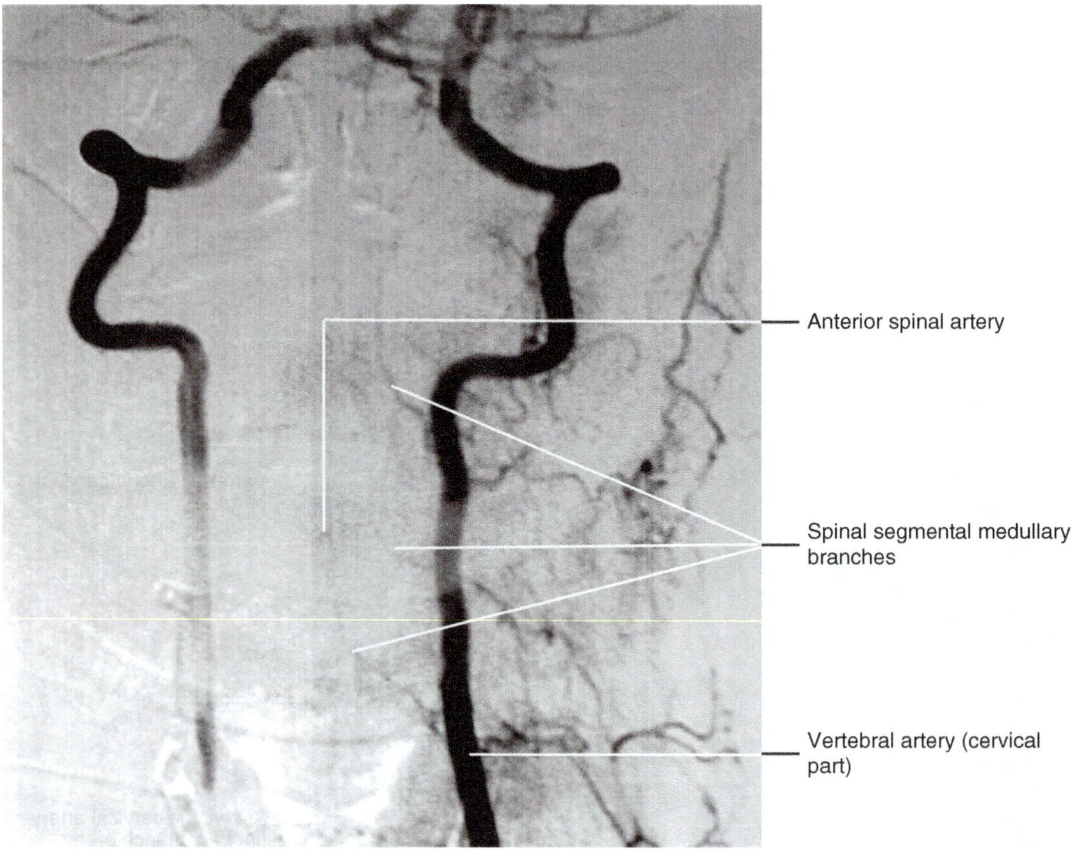

Fig. 13.48 Anteroposterior view of vertebral arteries (with subtraction)

Magnetic Resonance Angiography

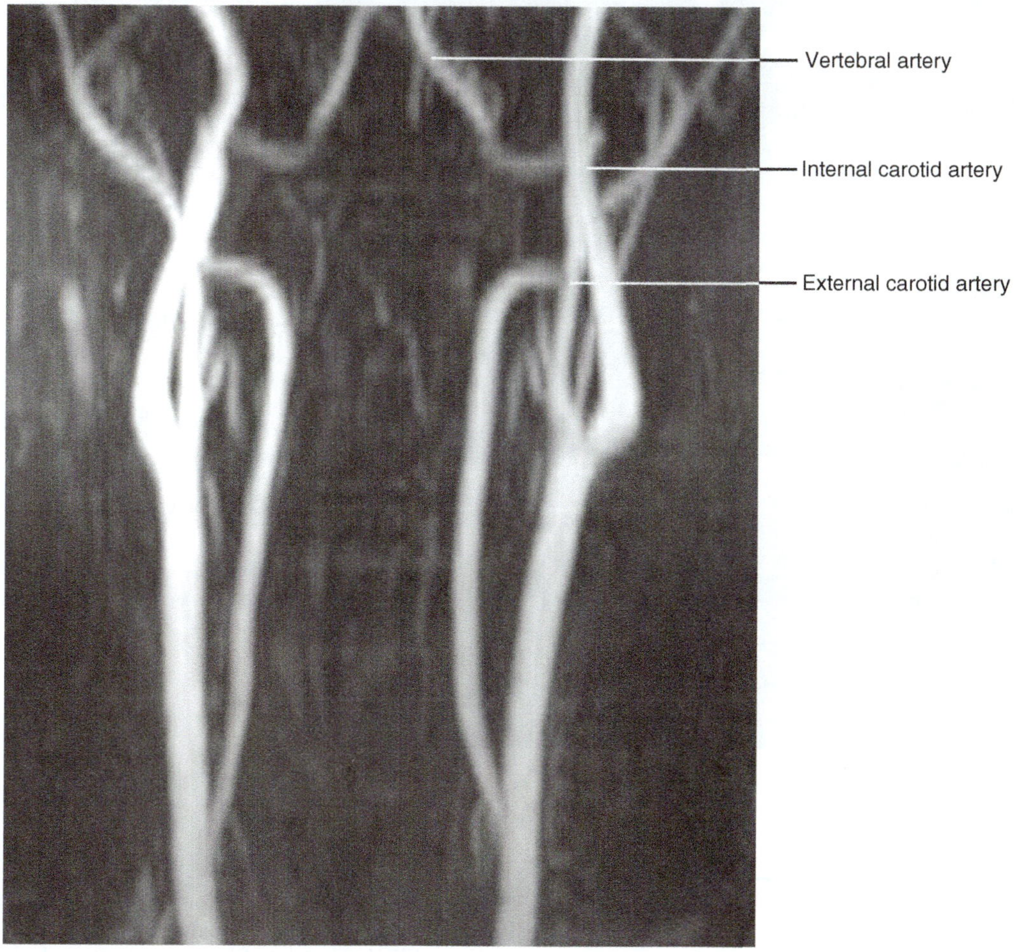

Fig. 13.49 Anteroposterior view of neck vessels

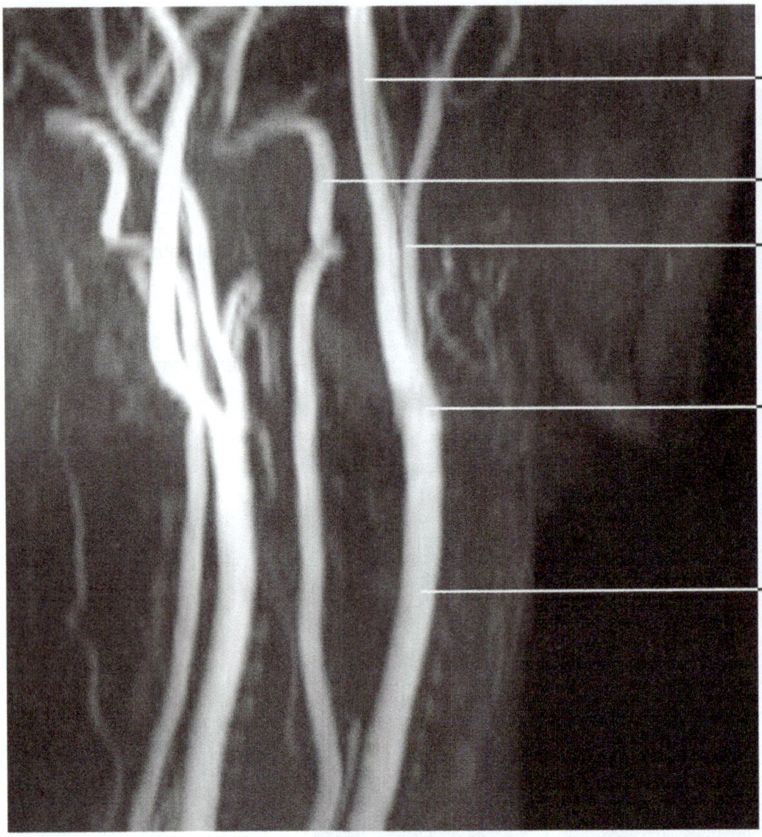

Fig. 13.50 Lateral view of neck vessels

Sources

1. Carter BL, Morehead J, Wolpert SM et al (1977) Cross-sectional anatomy computed tomography and ultrasound correlation, 1st edn. Appleton-Century-Croft, New York
2. Dean D, Herbener TE (2000) Cross-sectional human anatomy, 1st edn. Lippincott Williams and Wilkins, Philadelphia
3. Harnsberger HR (2004) Diagnostic imaging head and neck, 1st edn. Elsevier Saunders, Salt Lake City
4. Meschan I (1975) An atlas of anatomy basic to radiology, 1st edn. WB Saunders, Philadelphia
5. Moore KL (1985) Clinically oriented anatomy, 2nd edn. Williams and Wilkins, Baltimore
6. Netter FH (1997) Atlas of human anatomy, 2nd edn. Novartis, East Hanover
7. Som PM, Curin HD (1996) Head and neck imaging, 3rd edn. Mosby, St Louis
8. Wegener OH (1993) Whole body computed tomography, 2nd edn. Blackwell Scientific Publications, Boston
9. Weir J, Abrahams PH (2000) Imaging atlas of human anatomy, 2nd edn. Mosby, London

Endocrine System

14

A. Pituitary gland	B. Thyroid and parathyroid glands	C. Adrenal glands
Computed tomography	Ultrasound	Ultrasound
Magnetic resonance imaging	Computed tomography	Computed tomography
	Magnetic resonance imaging	Magnetic resonance imaging
	Nuclear medicine	

Pituitary Gland

The pituitary gland is composed of three lobes:

Adenohypophysis (anterior lobe)

Wraps anterolaterally around neurohypophysis.

Pars intermedia

Lies between adeno and neurohypophysis.

Neurohypophysis (posterior lobe)

Diencephalon.

S. Andronikou (ed.), *See Right Through Me*,
DOI 10.1007/978-3-642-23893-2_14,

Computed Tomography

Useful to show calcification. Also useful to assess for bony septae in sphenoid sinus for trans-sphenoidal surgical approach.

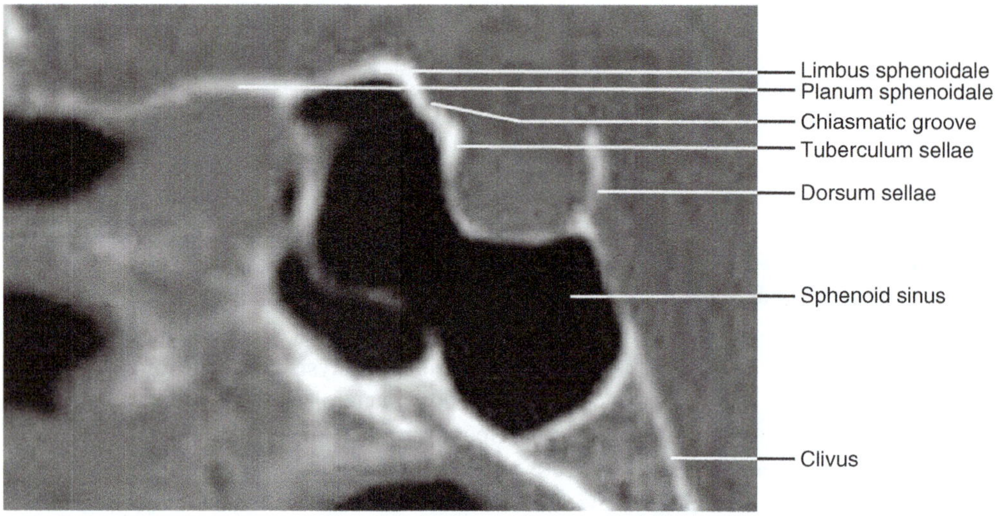

Fig. 14.1 Sagittal CT scan through the sella turcica

Pearl

Sella turcica bound anteriorly by chiasmatic groove, tuberculum sellae, anterior clinoid process (onto which tentorium cerebelli attaches).

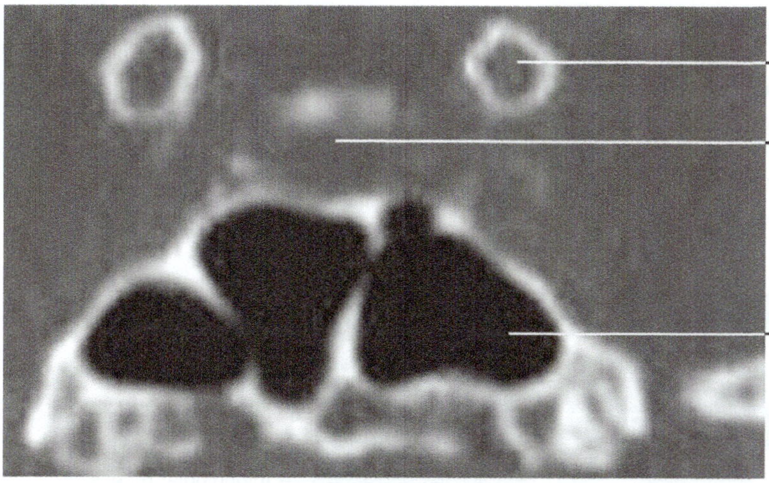

Fig. 14.2 Coronal CT scan (bone windows) to show sphenoid sinus in relation to the pituitary fossa

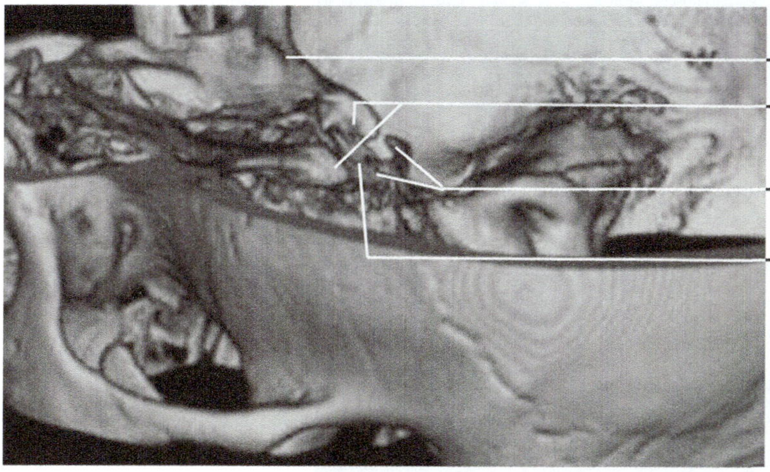

Fig. 14.3 3D Volume Rendered CT reconstruction: view is from above with the patient facing away to the left at an angle

Magnetic Resonance Imaging

MRI is the modality of choice with the coronal plane being the best imaging plane for the pituitary gland. Precontrast T1, T2, and dynamic MR with contrast are most useful. Anterior lobe is isointense to white matter on T1 and T2. In a child younger than 2 months old, the pituitary has higher signal intensity than it does in the rest of infancy. The posterior pituitary has high-signal intensity on T1 and is isointense on T2 in comparison to the anterior lobe. The clivus has a fatty marrow and is bright on T1.

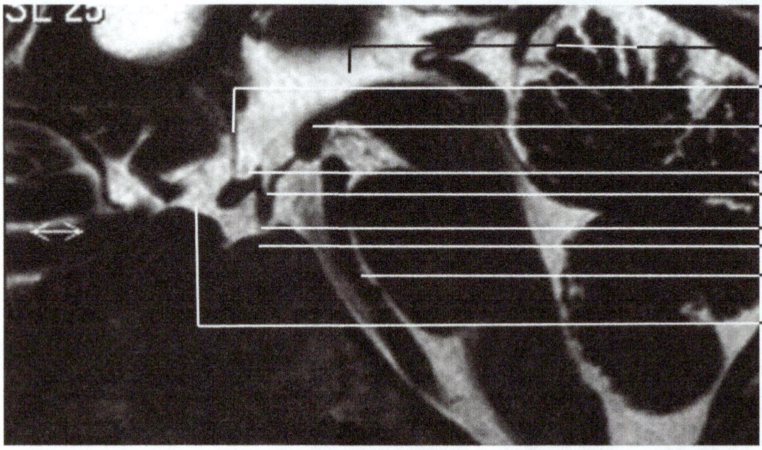

Fig. 14.4 Regional anatomy of pituitary fossa showing the cerebrospinal fluid spaces. Sagittal T2

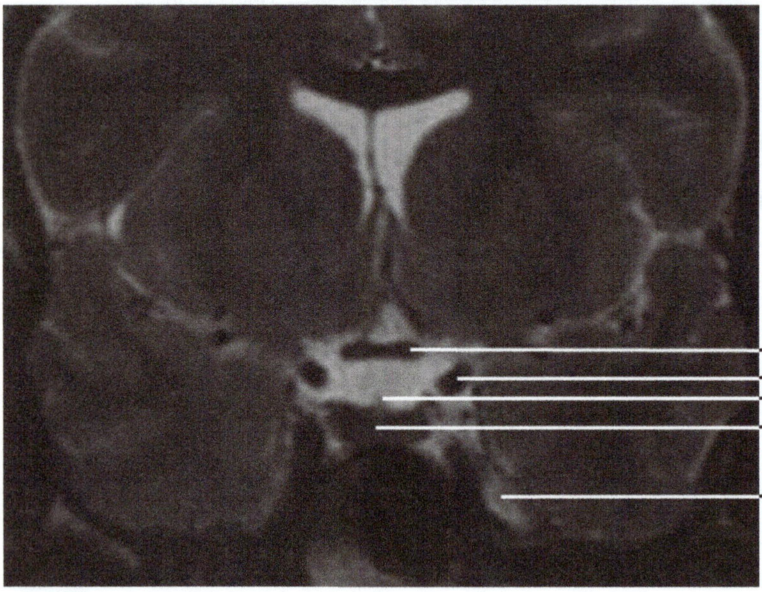

Fig. 14.5 Coronal T2: pituitary gland

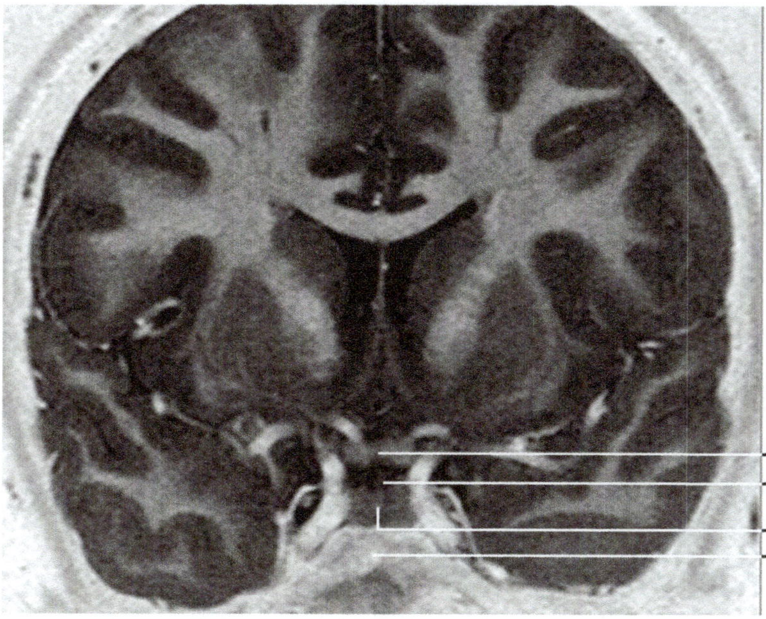

Fig. 14.6 Coronal gradient recalled echo: pituitary gland

Pearl

From the internal carotid and anterior cerebral arteries, the posterior pituitary and the infundibulum have a direct blood supply (compared to the portal system of the anterior pituitary). This accounts for the immediate enhancement on dynamic scanning of the posterior pituitary and the infundibulum.

Venous drainage: via the cavernous sinus.

Pearl: Height of the pituitary gland according to age and gender

Children: 6 mm

Men and postmenopausal women: 8 mm

Young menstruating women: 10 mm (can bulge upwards)

Pregnant lactating women: 12 mm

Pearl: Thickness of the pituitary stalk

At the level of the insertion: 2.5 mm

At the level of the optic chiasm: 3.5 mm

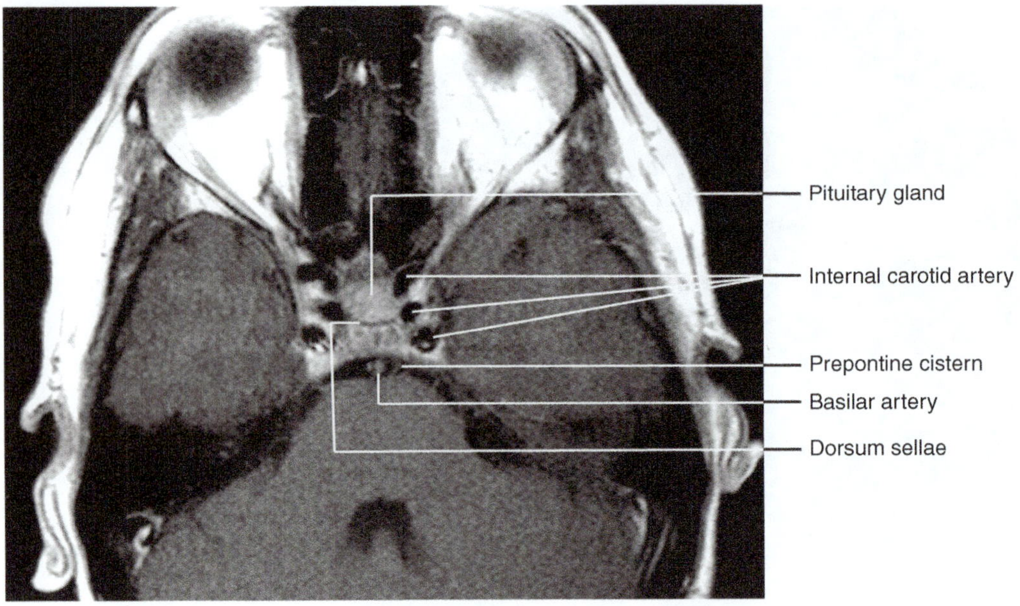

Fig. 14.7 Axial T1

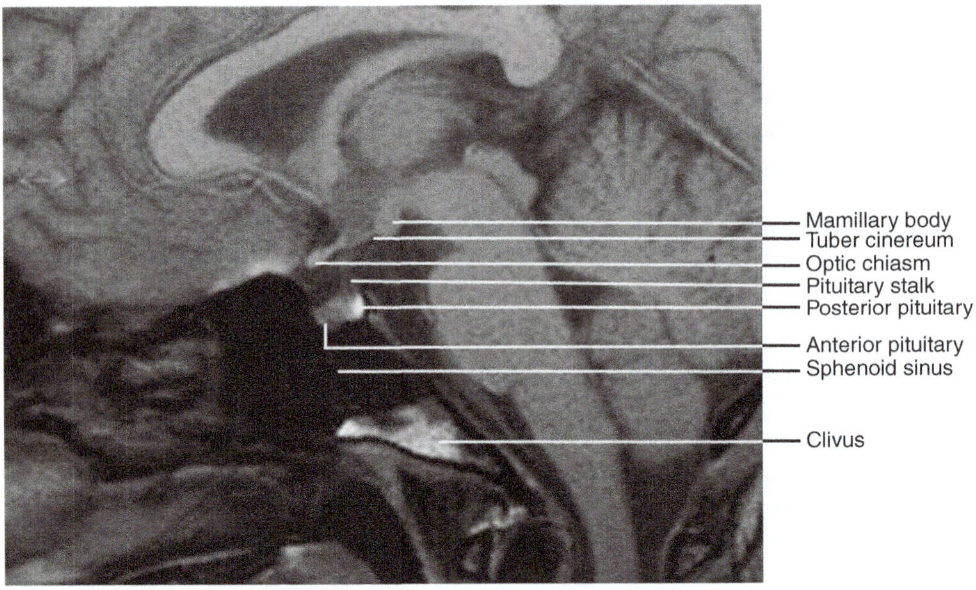

Fig. 14.8 Sagittal T1

Pearl

The signal of the posterior pituitary or neurohypophysis on T1-weighted sequences is characteristically high. This is commonly termed the posterior pituitary bright spot and is visible in up to 95% of normal individuals. The anterior pituitary may also show a high T1 signal in newborns.

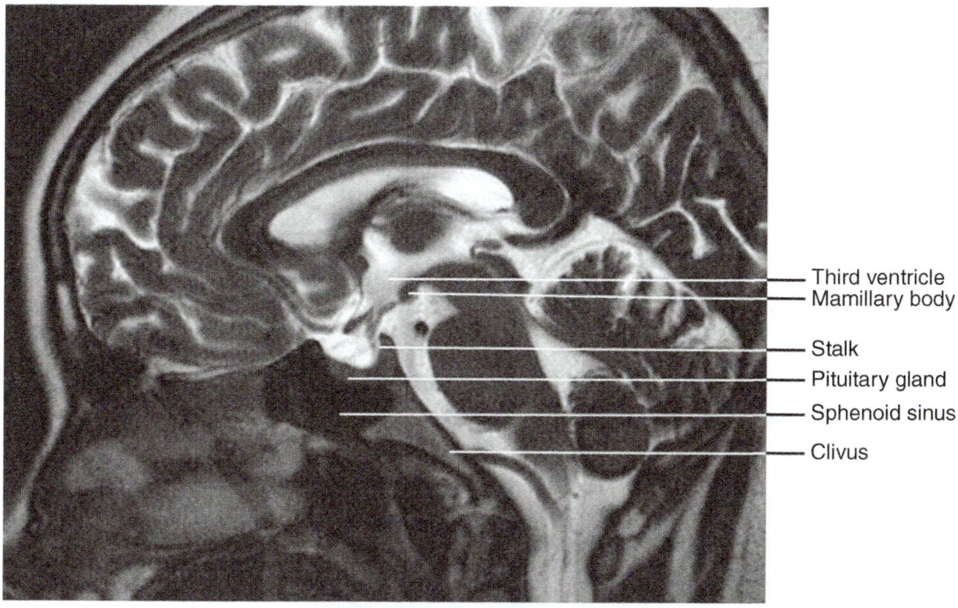

Fig. 14.9 Sagittal T2

Thyroid and Parathyroid Glands

The thyroid gland is an important endocrine organ regulating the body's metabolism. It is situated in the midline, anterior triangle of the neck. It may lie anywhere from the base of the tongue along the line of the thyroglossal duct. Size varies with bodily habitus and transverse diameter of 2 cm is the upper limit of normal.

The parathyroid glands play an important role in calcium and bone metabolism. There are four glands that are situated in close relationship to the thyroid gland, in the anterior neck.

Ultrasound

High frequency (> 10 MHz) linear ultrasound is the most useful imaging tool. The thyroid consists of two lateral lobes connected with a central isthmus. The upper pole is elongated and the lower poles are rounded.

The parenchyma is echogenic and homogeneous. A thin hyperechoic capsule surrounds the gland.

A pyramidal lobe may arise from the left lobe as a normal variant. Vessels are situated peripherally at the poles.

Four parathyroid glands are situated at the poles of the thyroid. The upper parathyroids are posterior to the upper poles and the lower parathyroids are lateral and inferior to the lower poles (35% ectopic). They are isoechoic to thyroid parenchyma.

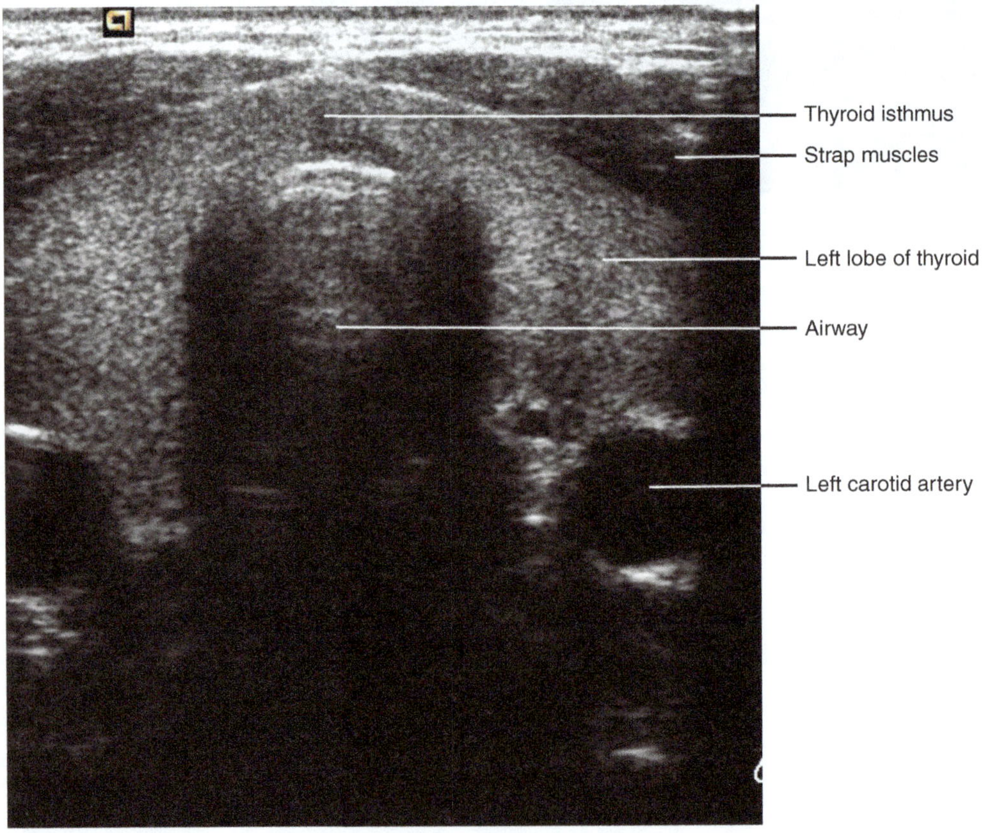

Fig. 14.10 Transverse ultrasound of the thyroid gland

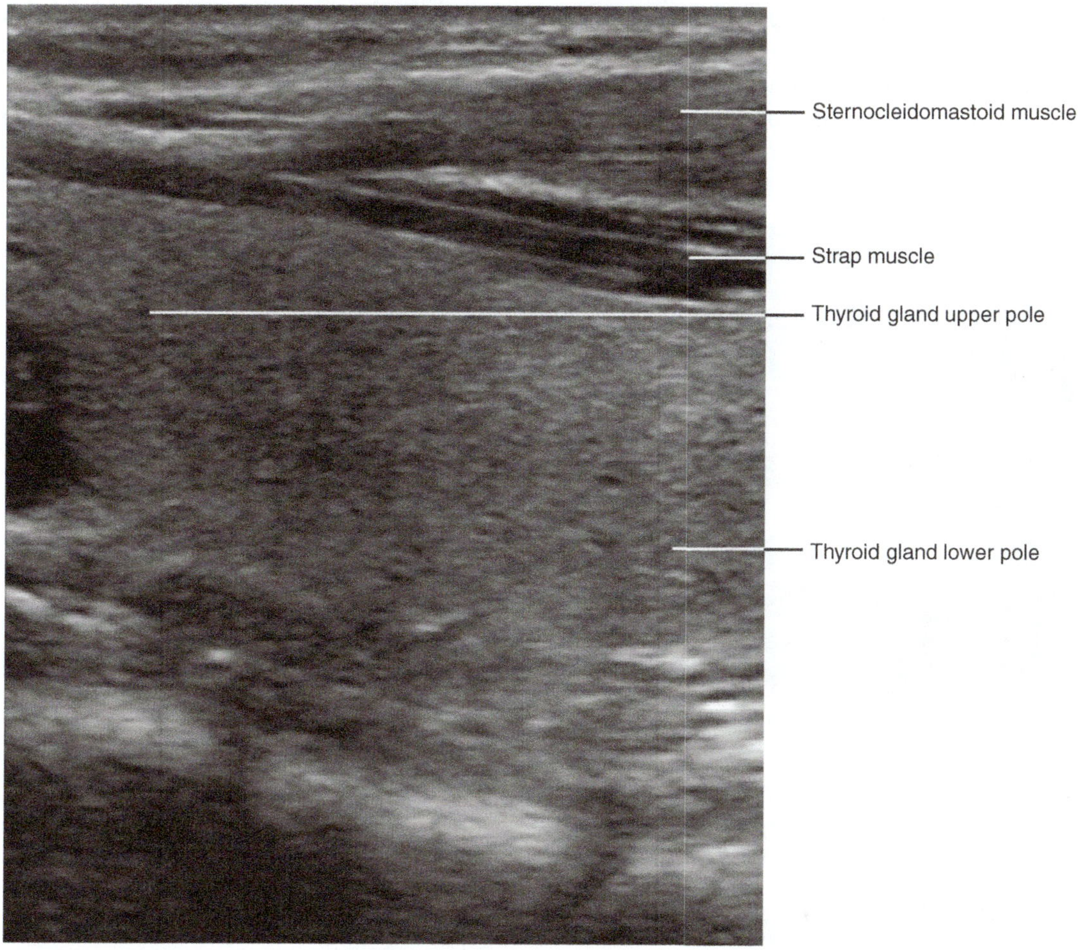

Fig. 14.11 Longitudinal ultrasound of the thyroid gland

Computed Tomography

The **thyroid** is a midline structure situated in the visceral space, extending from the mediastinum to the hyoid bone. Thyroid attenuation on nonenhanced CT is increased due to inherent iodine content. The gland enhances homogeneneously post contrast.

Parathyroids are not routinely visible during CT imaging.

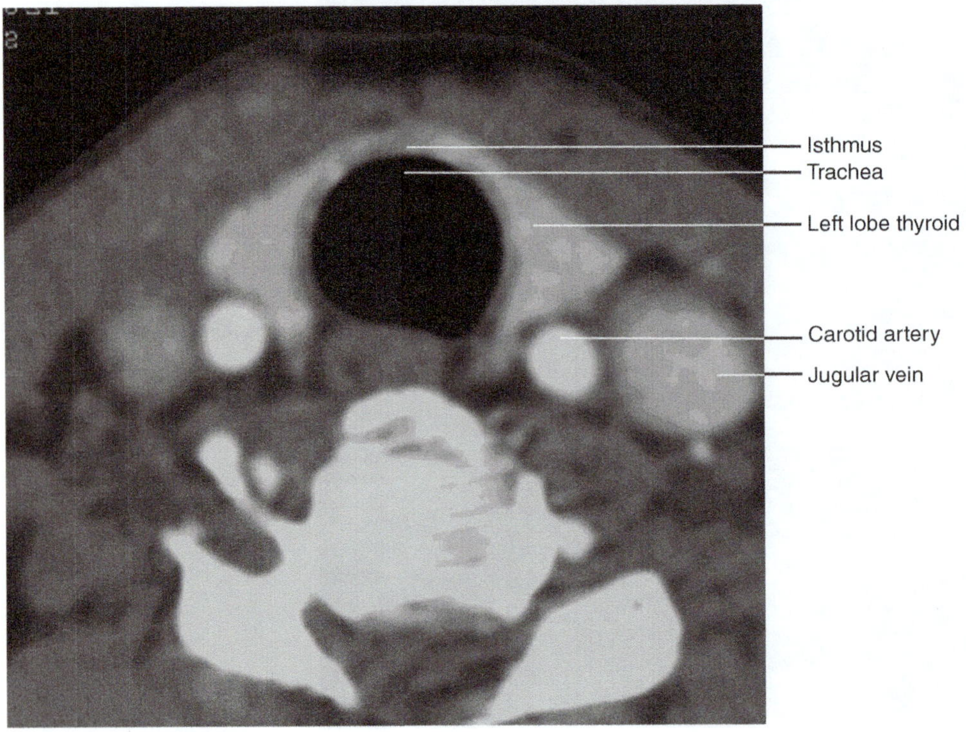

Fig. 14.12 Axial contrast-enhanced CT of the thyroid gland

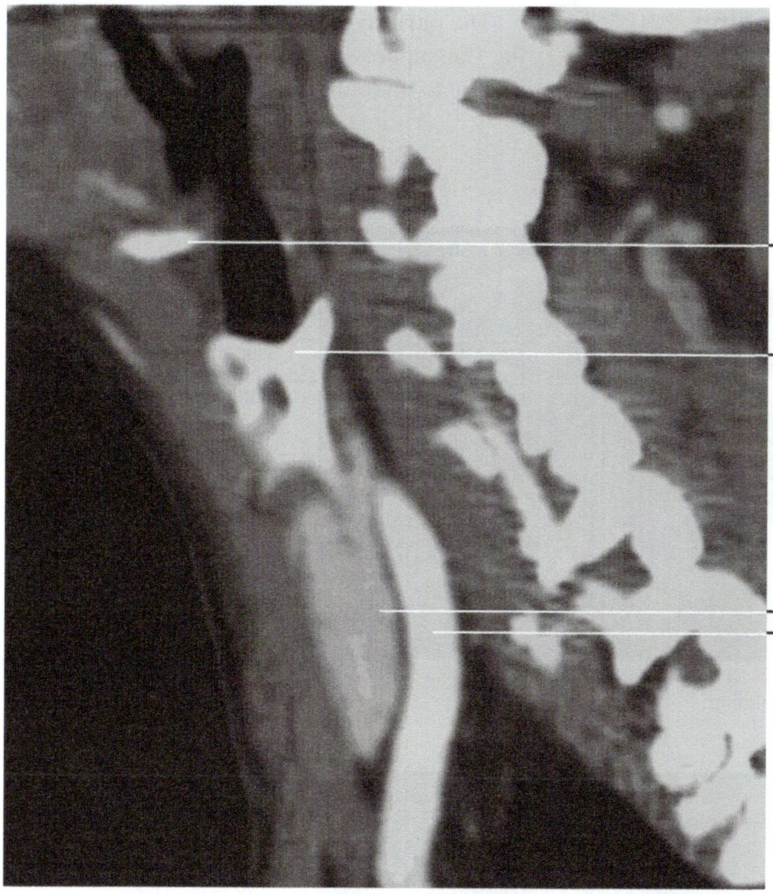

Fig. 14.13 Sagittal reconstructed contrast-enhanced CT of the thyroid gland

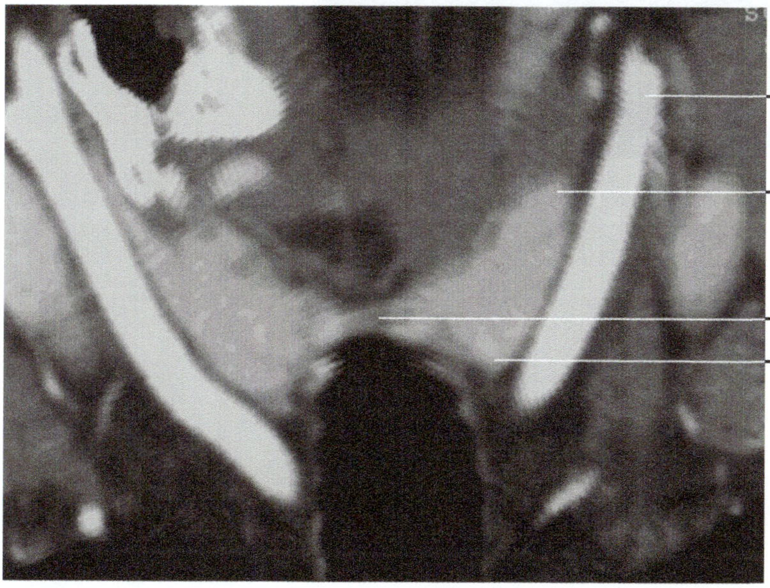

Fig. 14.14 Coronal reconstructed contrast-enhanced CT of the thyroid gland

Magnetic Resonance Imaging

The **thyroid gland** has homogeneneously textured signal iso-tense to musculature on T1- and hyperintense to musculature on T2.

The **parathyroid glands** are best viewed during fat-suppression sequences. They are isointense to thyroid parenchyma.

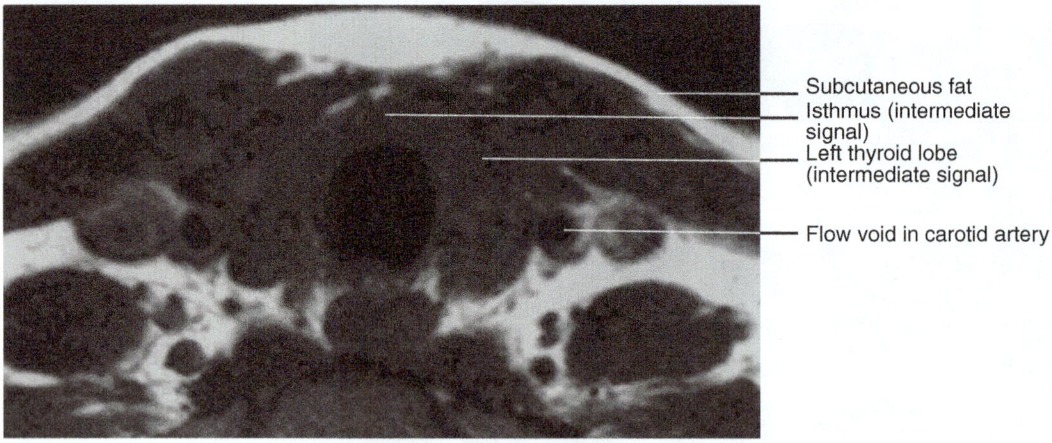

Fig. 14.15 T1 axial without gadolinium

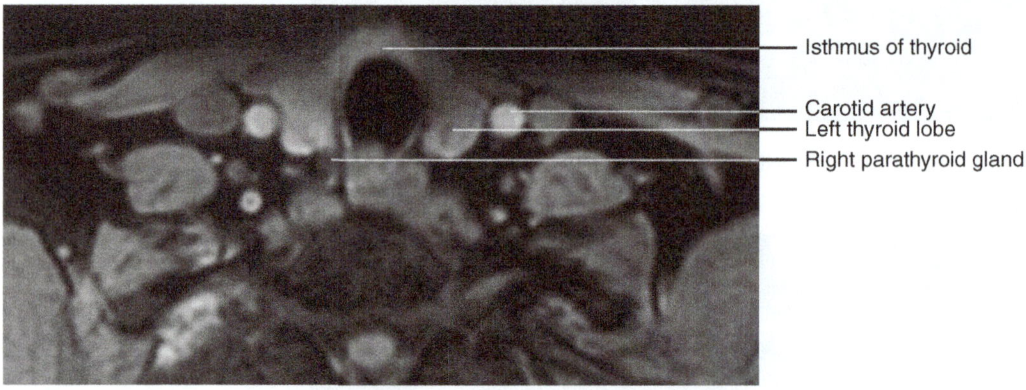

Fig. 14.16 T1 axial fat saturated post-gadolinium

Nuclear Medicine

^{123}I is the most frequently used isotope. Normal thyroid lobes linked with isthmus are shown. Parathyroid glands are not usually visible. Thallium subtraction may be used to demonstrate parathyroid activity.

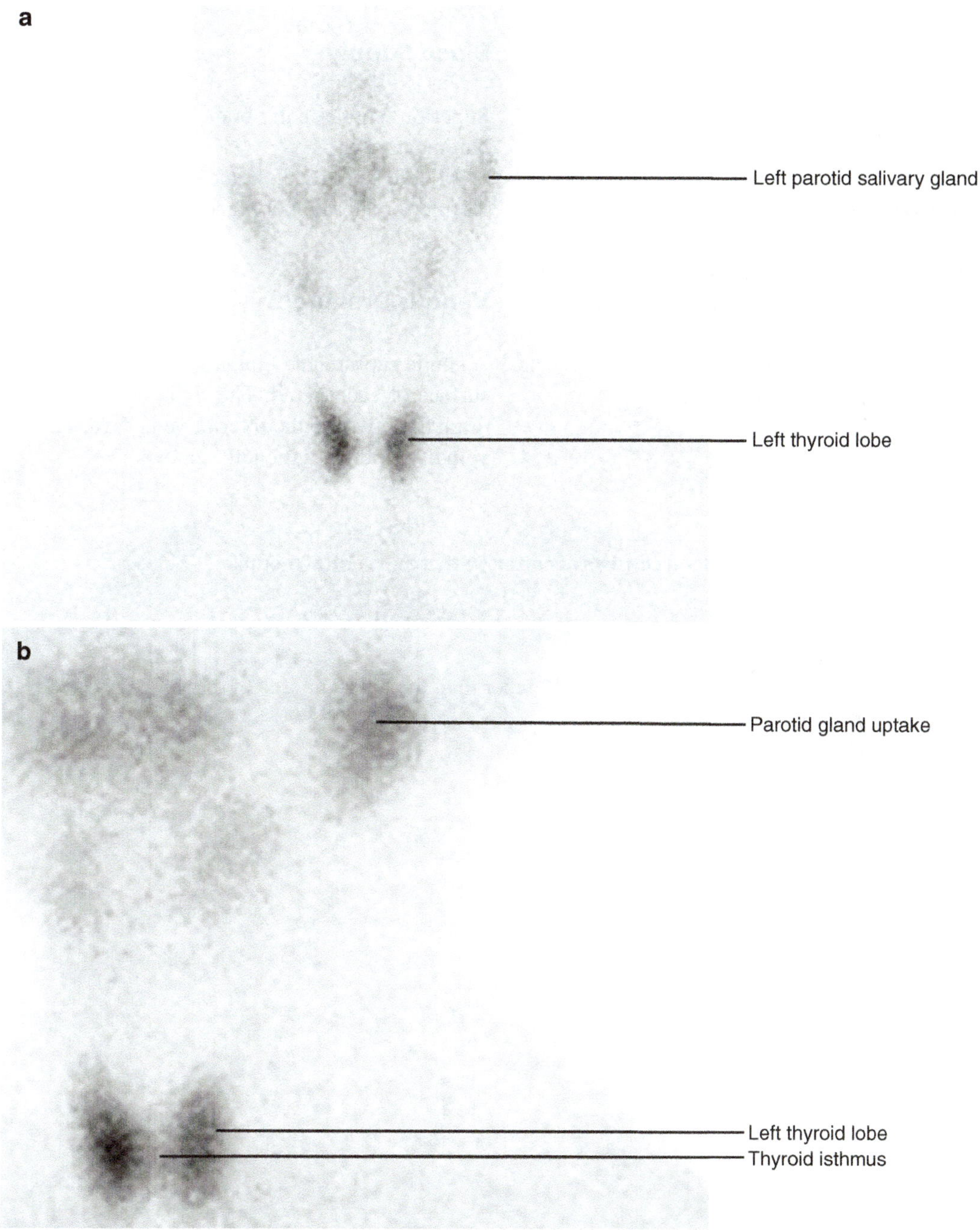

Fig. 14.17 (**a**, **b**) Normal isotope study demonstrating the thyroid in two views

Adrenal Glands

The adrenal glands are paired retroperitoneal structures. They overlap the superomedial poles of the corresponding kidneys. Each gland consists of an anteromedial ridge from which two thin wings extend posteriorly. They are the lateral and medial wings, which straddle the anteromedial aspect of the upper pole of the corresponding kidney.

Right Adrenal

The right adrenal is pyramidal in shape. It resembles a "party hat" and is perched on the upper pole of the kidney. The medial limb is largest superiorly and may be absent inferiorly. The opposite is true of the lateral limb.

Left Adrenal

The left adrenal is crescent shaped. It resembles a hat worn at a tilt, and encroaches more on the medial border of its kidney.

Blood Supply

Superior suprarenal: branch of the inferior phrenic artery.

Middle suprarenal: branch of the aorta.

Inferior suprarenal: branch of the renal artery.

Venous Drainage

A single suprarenal vein passes from the anterior surface of each gland. The right vein ascends to drain directly into the inferior vena cava. The left vein descends into the left renal vein.

Pearl: Size: the bulk of the gland consists of cortex, with a thin central medulla

3–6 cm in overall length

2–3 cm in width

2–6 mm in thickness

A limb of the adrenal gland should not be thicker than the adjacent crus of diaphragm.

Ultrasound

The left adrenal is seen in 45% of patients and the right in 80% of patients. The adrenals can be seen using an intercostal approach in a supine patient. Use is made of the acoustic windows of the liver and the spleen.

On the right, view in the ninth or tenth intercostal space in the mid or anterior axillary line. Scan transversely from the renal hilum upwards. Concentrate on the area posterior to the inferior vena cava. Then rotate the probe through 90° to obtain coronal or longitudinal scans.

The left adrenal is more difficult to detect. Scan in the eighth or ninth intercostal space in the posterior axillary line. Scan coronally through the upper pole of the kidney. The probe is then rotated slightly to locate the adrenal between the spleen, kidney, and the left crus of the diaphragm. A right lateral decubitus position and an erect position may help in difficult cases.

The cortex is echo-poor, whereas the medulla is a thin reflective structure.

Pearl
The cortex is straight or concave in outline. Convexity should be regarded with suspicion. Masses may only affect a part of the gland and therefore the entire gland should be visualized.

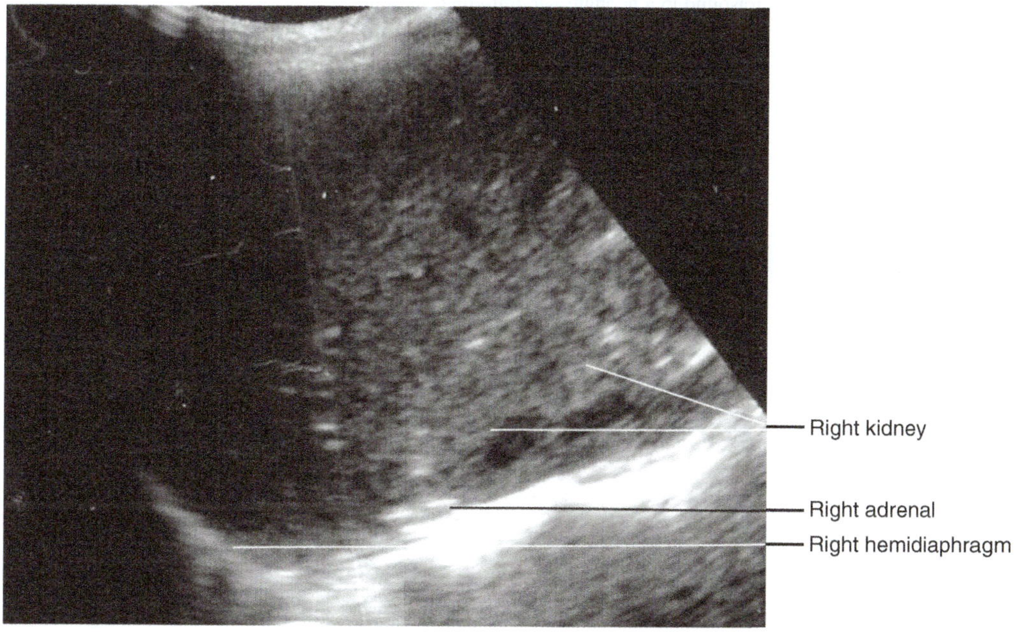

Fig. 14.18 Right adrenal gland: coronal section

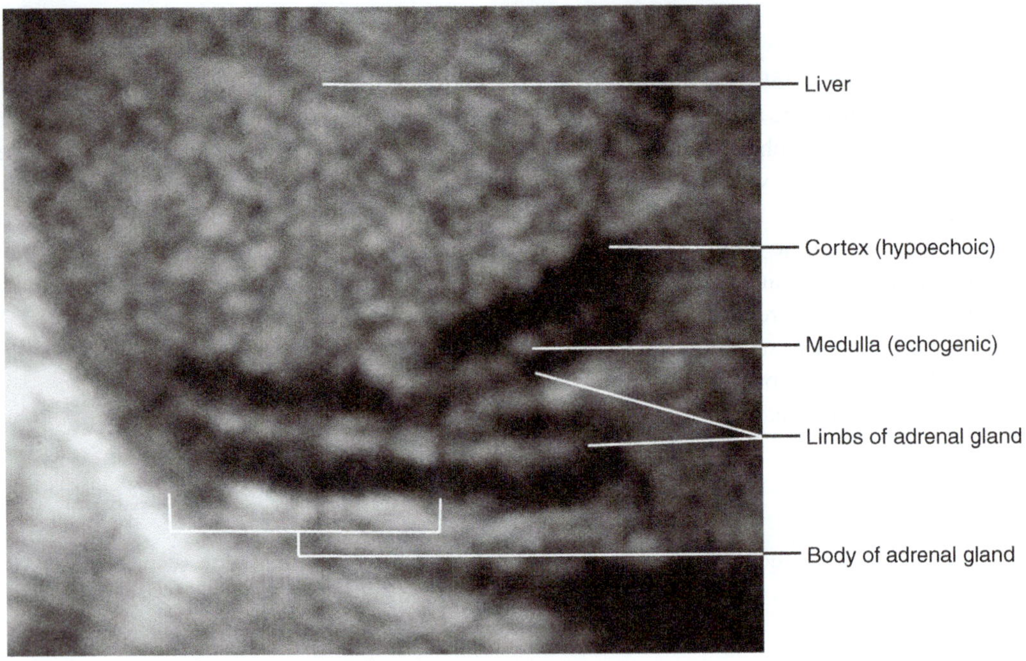

Fig. 14.19 Right adrenal gland ultrasound in a neonate—longitudinal section

Computed Tomography

Routine dynamic abdominal scanning will detect the left adrenal in 100% of scans and the right adrenal in 99% of scans. The presence of retroperitoneal fat aids demonstration of the gland.

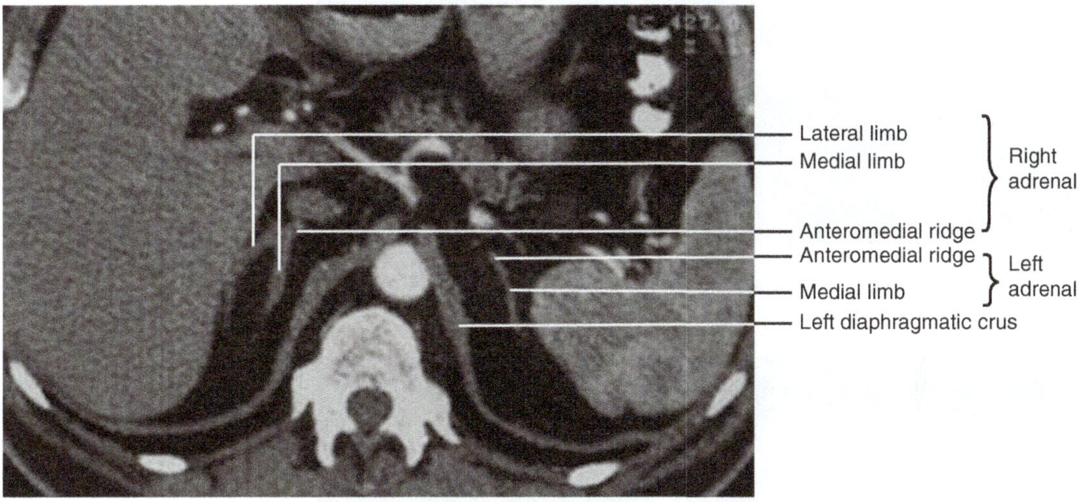

Fig. 14.20 Axial CT post contrast (arterial phase) through the adrenals

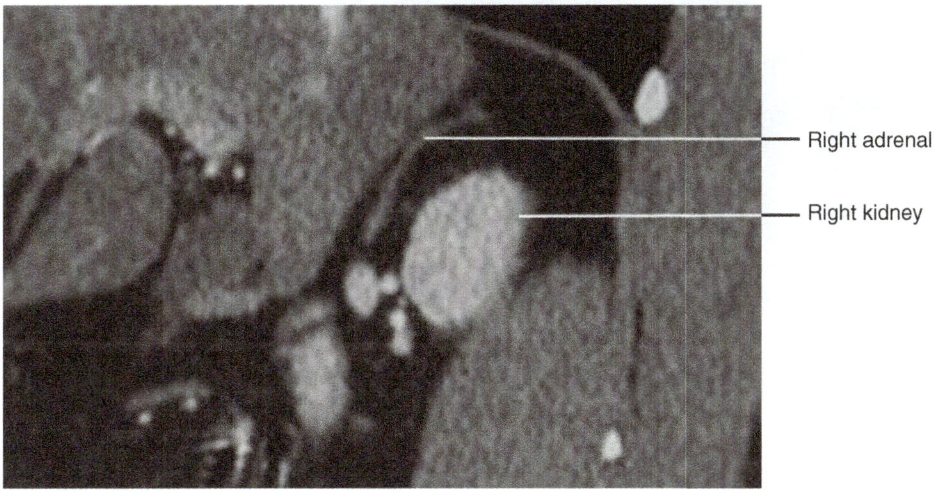

Fig. 14.21 Sagittal CT post contrast through right adrenal gland

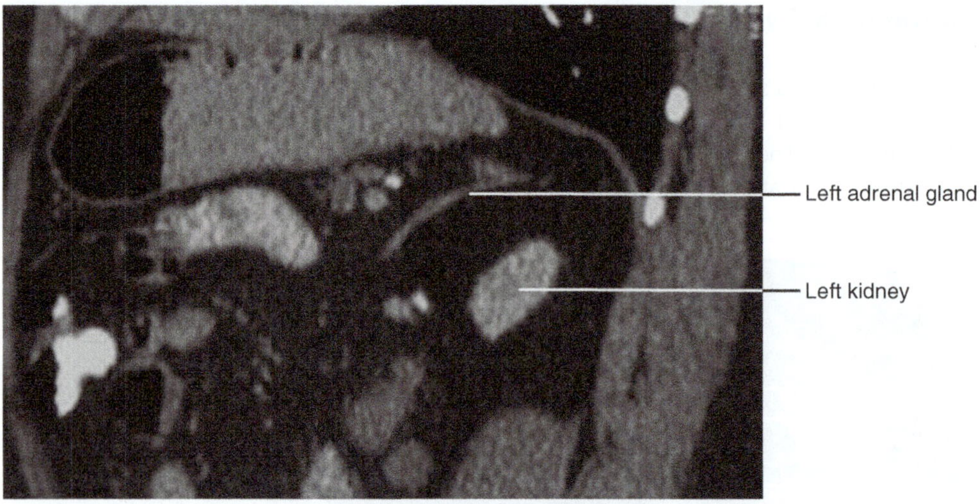

Fig. 14.22 Sagittal CT post contrast through left adrenal gland

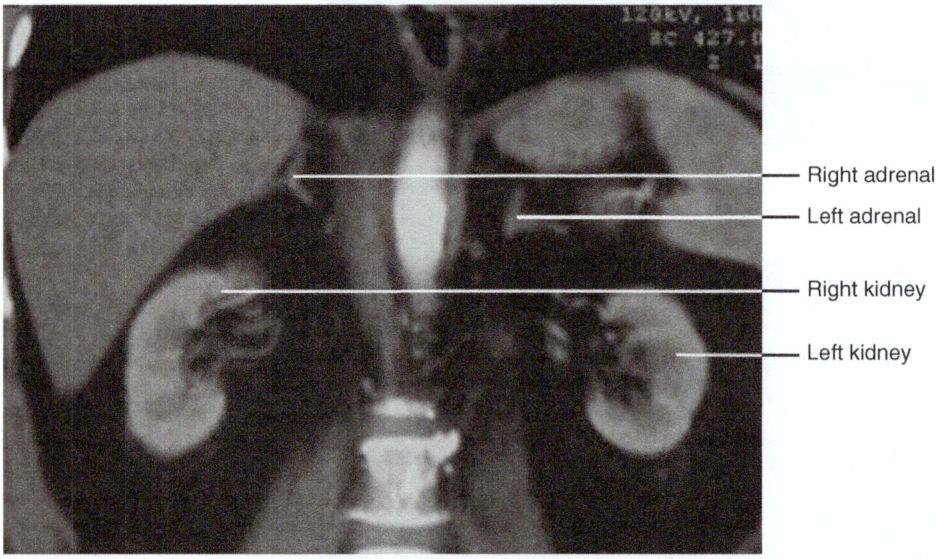

Fig. 14.23 Coronal reformat of CT abdomen at level of adrenal glands

Magnetic Resonance Imaging

The signal intensity of the adrenals is as follows:

Spin echo: hypointense on T1 and isointense to liver on T2.

Fat suppression: hyperintense compared to surrounding fat on T1 and T2.

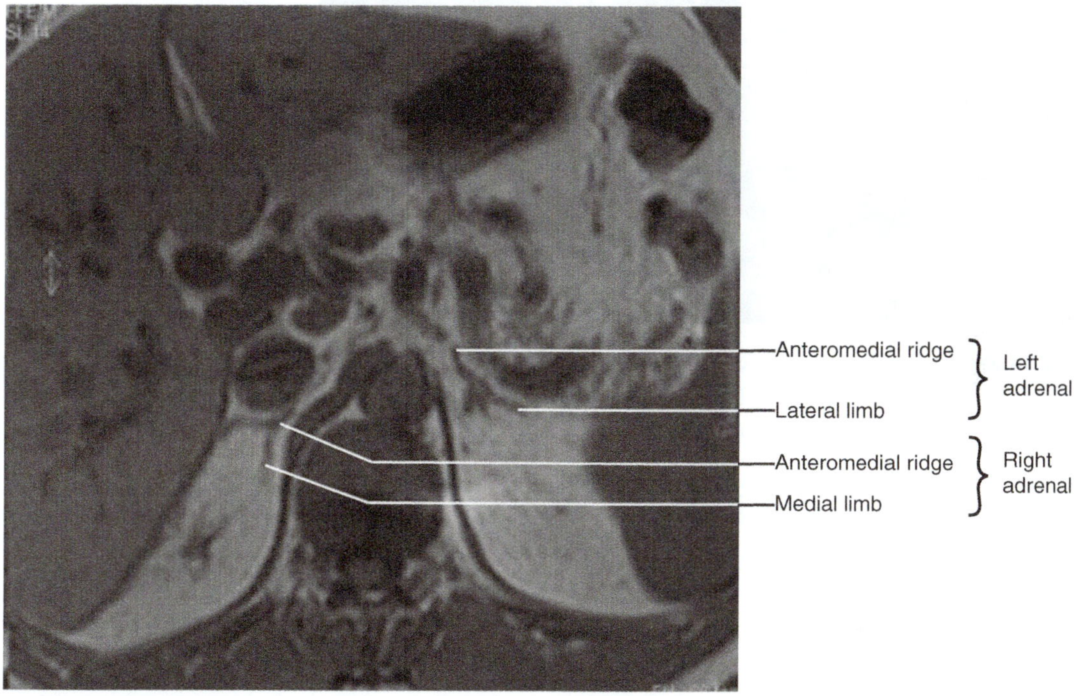

Fig. 14.24 T1 axial

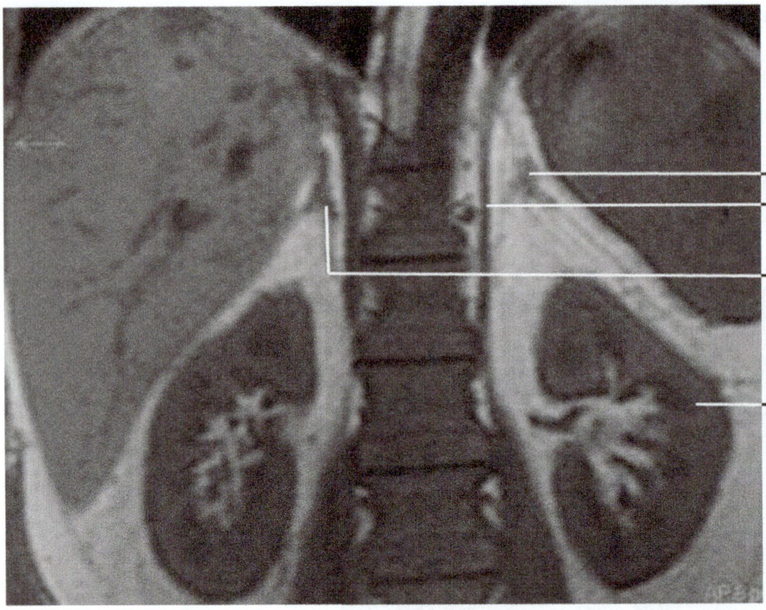

Fig. 14.25 T1 coronal

Acknowledgements Milpark Radiology Department; Donald Gordon Radiology Department; Milpark Nuclear Medicine Department.

Dr. T. Nel Inc. Radiology Practice (Donald Gordon Medical Centre). Special thanks to Dr. M. Haagensen, Dr. S. Brachmeyer, and Ms. C. Gibbs.

Drs. Haagensen and Lurie Inc. Radiology Practice (Milpark Hospital). Thank you to all the radiographers in scanner.

Dr. T. Jenkins, Nuclear Medicine Department, Milpark Hospital.

Sources

1. Dahnert W (1996) Radiology review manual, 3rd edn. Williams and Wilkins, Baltimore, pp 181–182, 658
2. Grossman R, Yousem D (2003) The requisite neuroradiology, 2nd edn. Mosby, Philadelphia, pp 529–531
3. Harnsberger HR, Hudgins P, Michel M et al (2004) Diagnostic imaging of the head and neck. Amirsys, Salt Lake City
4. Meire H, Cosgrove D, Dewbury K, Farrant P (2001) Clinical ultrasound a comprehensive text: abdominal and general ultrasound, 2nd edn. Churchill Livingstone, London, pp 481–495
5. Moeller T, Reif E (2000) Pocket atlas of sectional anatomy CT and MRI, vol 1, 2nd edn, Head, neck, spine, and joints. Thieme, Stuttgart
6. Osborn A, Blaser S, Salzman K et al (2004) Diagnostic imaging brain, vol II, 1st edn. Amirsys, Canada, pp 24–8
7. Solbiati L, Cioffi V, Ballarati E (1992) Ultrasonography of the neck. Radiol Clin North Am 30:941–954
8. Sutton D (ed) (1998) Textbook of radiology and imaging, vol 1 and 2, 6th edn. Churchill Livingstone, Edinburgh
9. Tobias A (1985) Man's anatomy, 3rd edn. University of the Witwatersrand Press, Johannesburg, pp 284–288

Part III

The Thorax

Our definition of thoracic imaging includes the lungs and thoracic cage, as well as the the cardiovascular system of the chest. This chapter demonstratesn detailed imaging of the lung using multiple modalities. The imaging of the central vessels is presented in the 'vascular system' chapter.

Cardiac imaging, including coronary imaging has re-entered the domain of the radiologist, and detailed knoweldge of anatomy using multiple modalities and sequences is essential.

Breast imaging has been included here for convenience, as it is also thoracic imaging, but in practice, breast specialists are usually different from thoracic specilists in the imaging department. Because breast imaging involves screening programs, recognition of normal appearances is important over a wide variety of ages and using multiple imaging modalities.

15 The Breast

Overview of Breast Imaging Modalities

Primary Imaging Modalities

- Mammography (screen-film and digital mammography).
- Ultrasound (US), including three-dimensional (3D) imaging.
- Magnetic resonance imaging (MRI) breast (MR mammography).
- Tomosynthesis: acquisition of 3D mammographic image.
- Experimental modalities: these are not routinely used in clinical practice (e.g., thermography).

Secondary Imaging Modalities

- Computed tomography (CT): CT breast/chest.
- Plain film: chest radiograph (CXR).
- Nuclear imaging; e.g., position emission tomography (PET) and technetium scanning for sentinel nodes.

Investigations

- Galactogram (ductogram).
- US or stereotactic-guided biopsy (MR-guided biopsy).

Intervention and Procedures

- With appropriate modality guidance: drainage, hookwire placement, injection of methylene blue.

Table 15.1 Breast imaging in radiological practice

Older than age 40:	**Younger than age 35–40:**
↓	↓
Mammogram (standard view)	Ultrasound (US)
↓	↓
± Supplemental views	± Mammogram (indicated if breast tissue is fatty; for evaluation of specific clinical problems; or as control for galactogram)
↓	± Supplemental views
± Ultrasound (US)	
± Further investigations and procedures	
Galactogram; Biopsy; Hookwire placement; Drainage (*eg*, cysts, seroma; abscess)	

S. Andronikou (ed.), *See Right Through Me*,
DOI 10.1007/978-3-642-23893-2_15, © Springer-Verlag Berlin Heidelberg 2012

Mammography

The technical aspects of a "good" mammogram include:

1. **Standard images of two-view screening mammography**
   - Mediolateral oblique (MLO) and cranio-caudal (CC) views of left and right breast.
   - The nipples should be in profile in at least one of the views on each side.

The MLO view

- The pectoral muscle is seen postero-obliquely extending to nipple level or below.
- No mammary parenchyma should project posteriorly and off the film.
- A convex anterior margin of the pectoral muscle indicates relaxed positioning versus tense and improper positioning, in which case the muscle margin will appear straight or concave.
- The inframammary fold is seen on the film.

The CC view

- The pectoral muscle is seen posteriorly (not always possible).
- A medial skin fold is visible (not always possible).

2. **Supplemental views ("problem-solving" views)**
   - Exaggerated cranio-caudal (XCC) view.
   - Spot compression view, requires special compression device.
   - Magnification view, with or without spot compression.
   - 90° lateral view.
   - Eklund views: modified compression views for the augmented breast (together with standard views these will yield four views of each side).
   - Less commonly done: axillary view; cleavage view; Cleopatra view (no longer necessary with modern equipment); tangential view; others (see sources for further information).

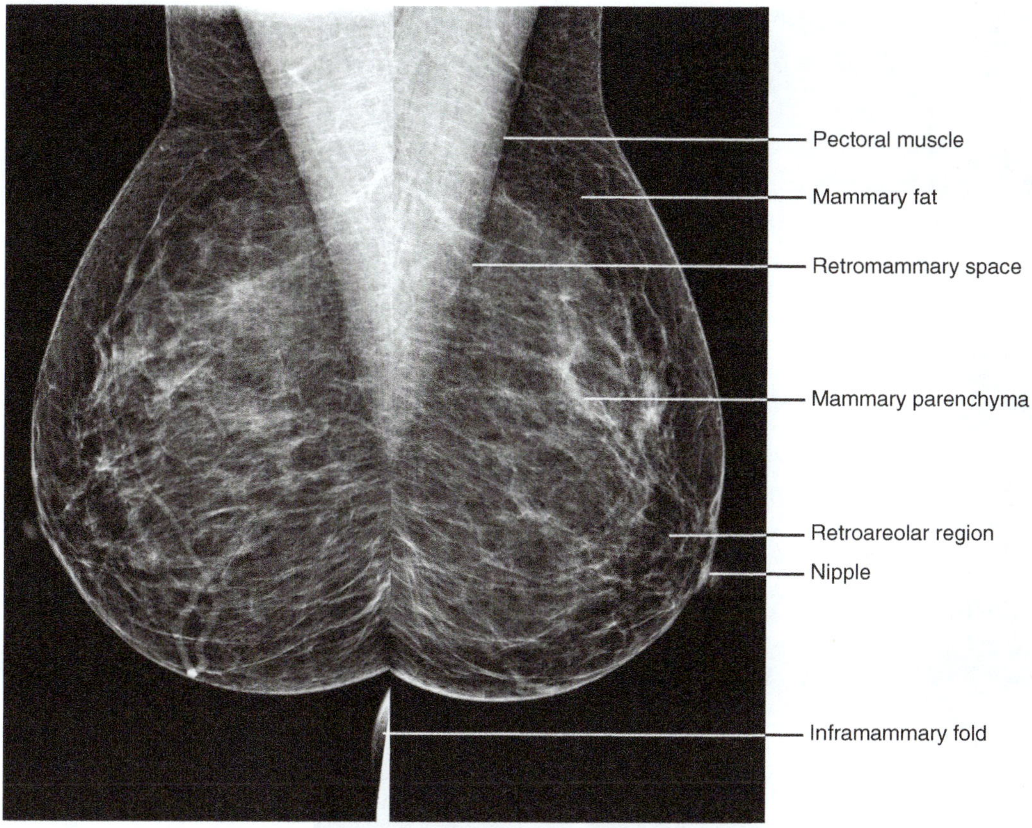

Fig. 15.1 Mediolateral oblique views

Pearl: To view a mammogram correctly
Position films as mirror images with the chest wall side of the breasts against one another.

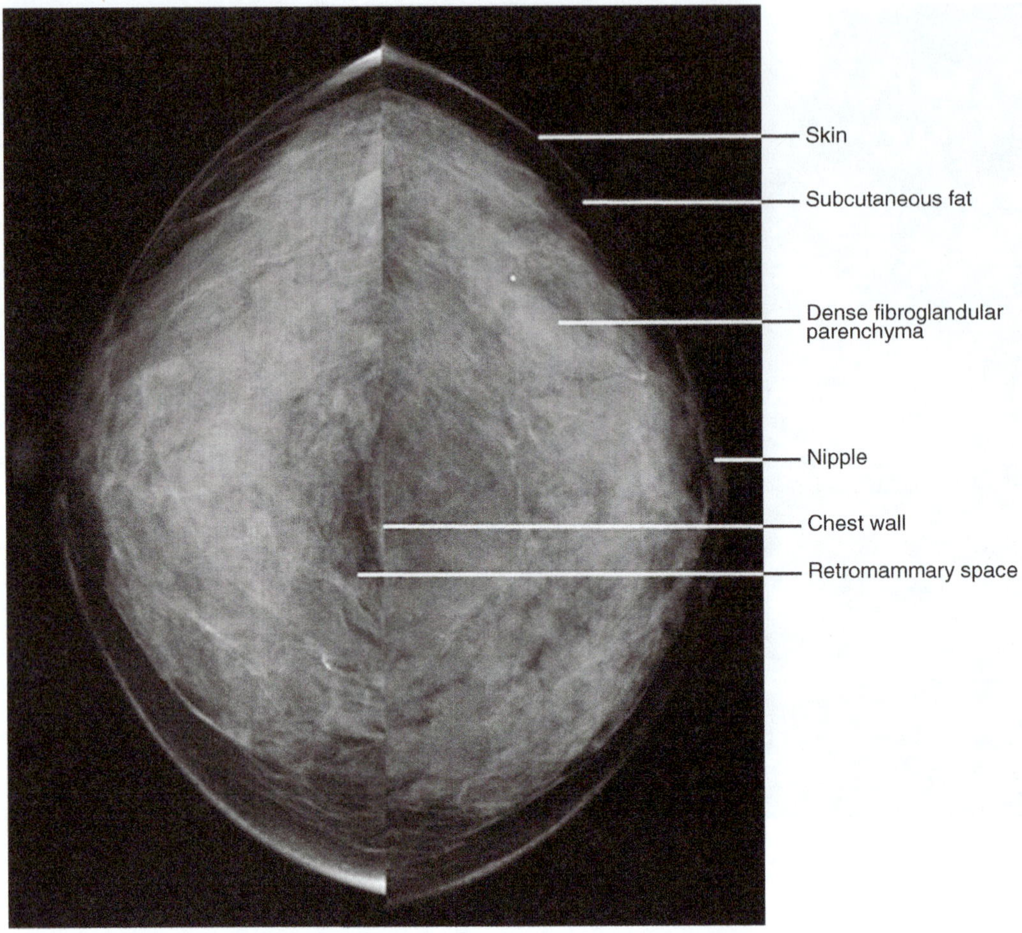

Fig. 15.2 Dense breast: CC views (minimal fatty tissue)

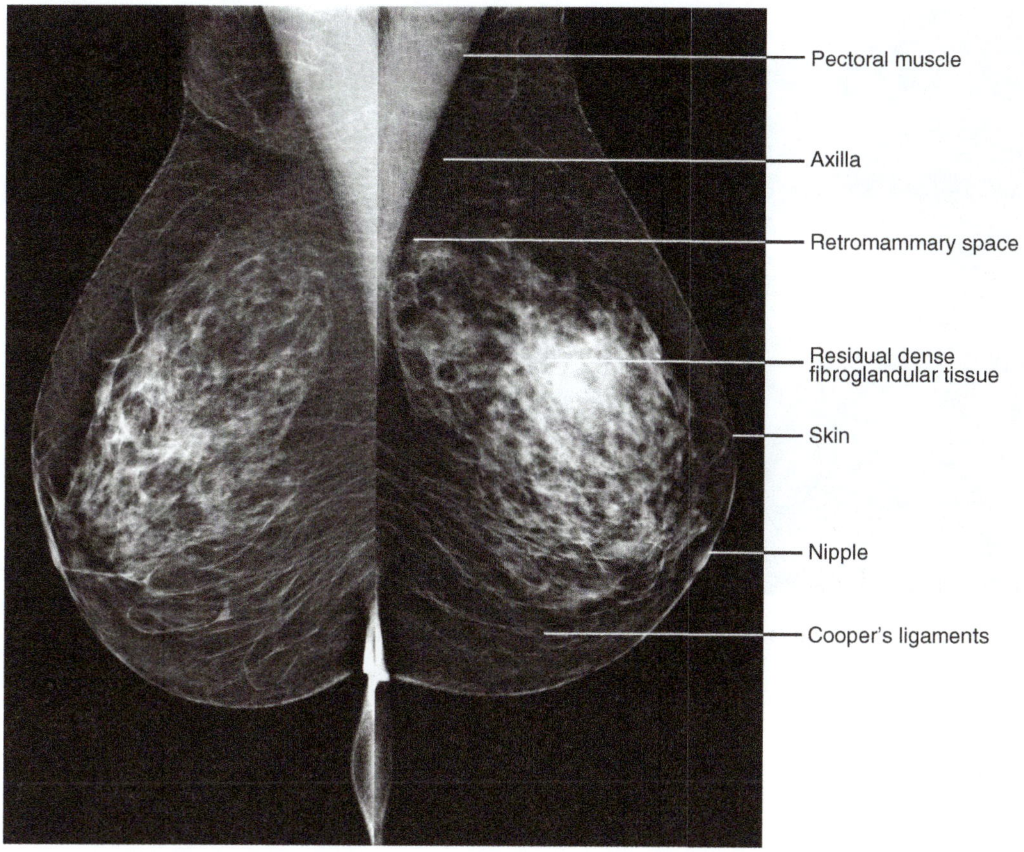

Fig. 15.3 Moderately dense breast: MLO views

Pearl

The upper half of the MLO view represents the upper (superior) portion of the breast. The lower half of the MLO view represents the lower (inferior) portion. The upper half of the CC view is the outer (lateral) portion, the lower half equals the inner (medial) portion. Thus, the MLO view together with the CC view will enable localisation of pathology to a specific quadrant.

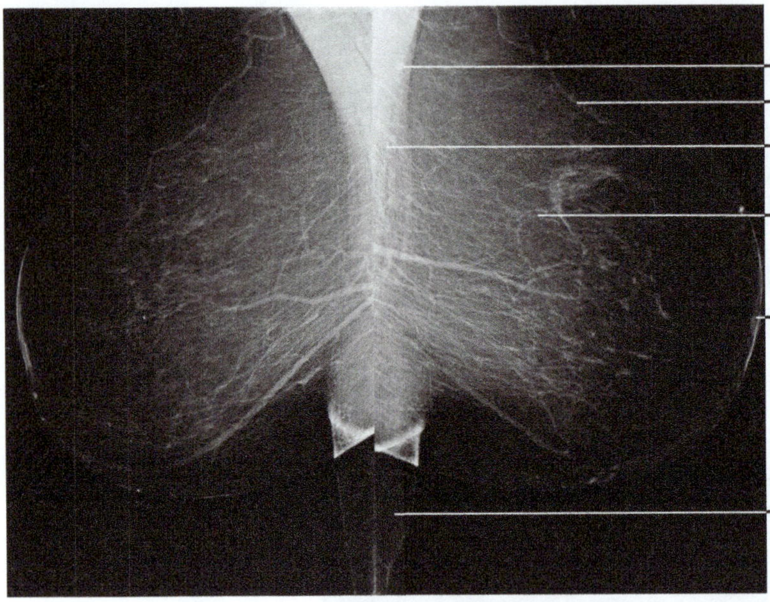

Fig. 15.4 Involuted breast: MLO views. Note fatty replacement of fibroglandular tissue

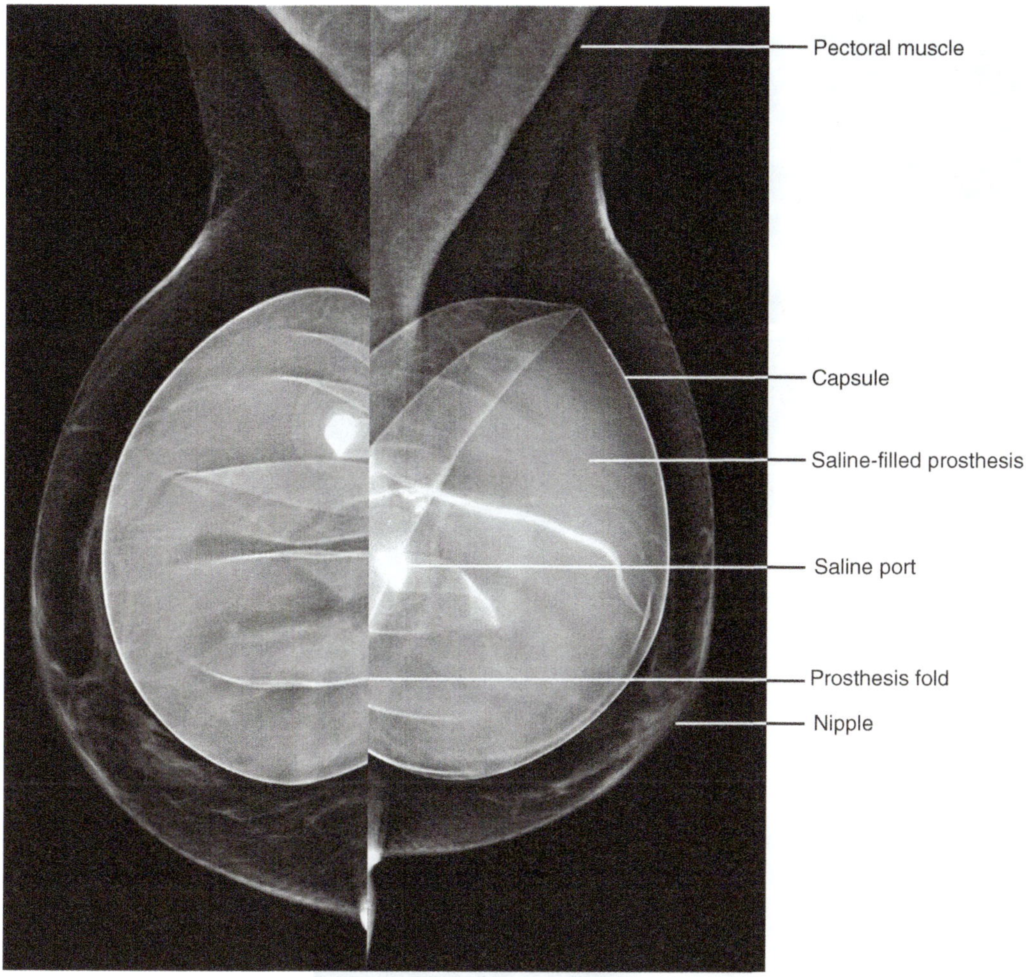

Fig. 15.5 Prepectoral saline breast implants: MLO views

Ultrasound

Routine Technique

Breast US is performed with a high-frequency transducer (5–10 MHz). The patient is asked to lie in a supine position with arms placed behind the head. Whether the sonar is accomplished in a radial, circular, or straight fashion depends on the preference of the operator. Routinely, the entire breast, including the subareolar area and axillary tail are to be imaged, as well as the axilla. Any lesions or questionable features must be viewed in both the sagittal and transverse planes.

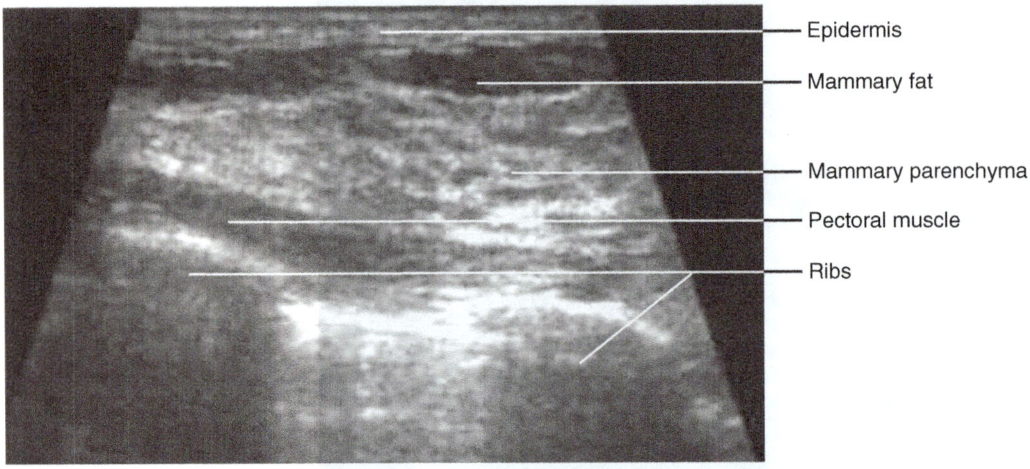

Fig. 15.8 Moderately dense breast

Pearl

Fibroglandular mammary parenchyma is hyperechoic, mammary fat is hypo-echoic, lymph nodes are hypo-echoic and may have a fatty centre which will be hyperechoic. There is little to be gained from an ultrasound of the involuted breast, unless indicated for specific clinical reasons. Most pathological lesions are hypo-echoic as is involuted parenchyma, thus, the pick-up rate in a fatty breast is low.

Pearl

Patients with dense breasts may benefit from US as mammographic sensitivity in these cases may be reduced.

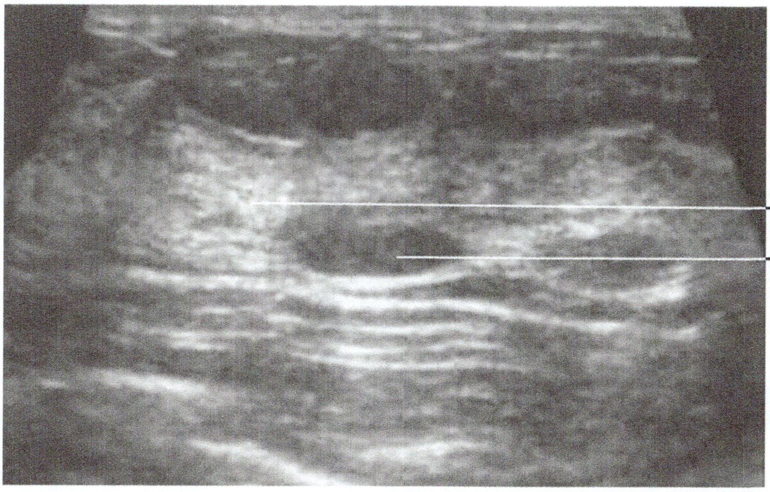

Fig. 15.9 Fatty lobule within parenchyma mimicking the appearance of a focal lesion

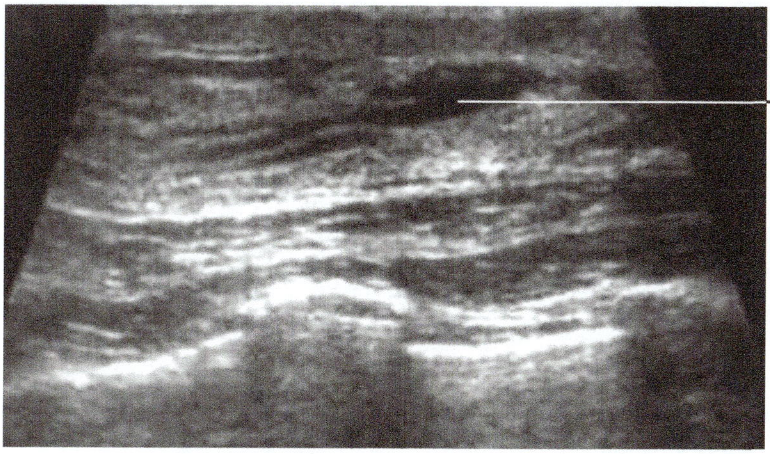

Fig. 15.10 Branching lactiferous ducts

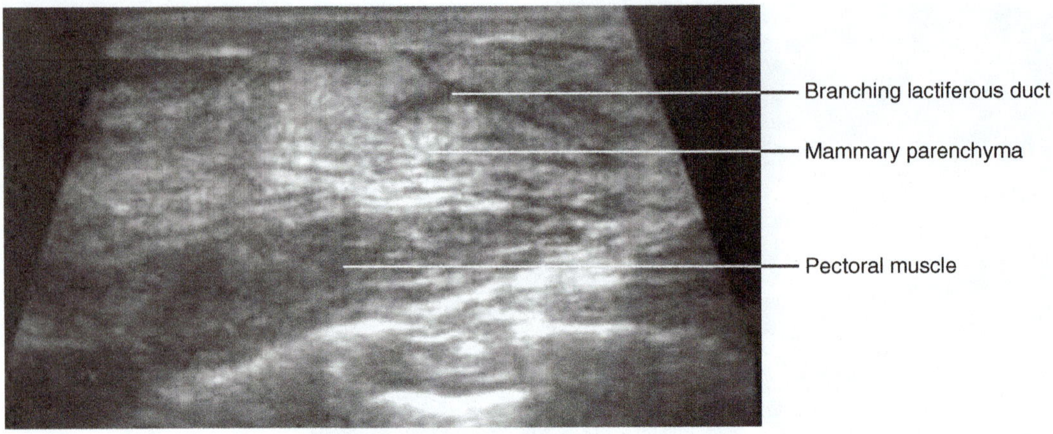

Fig. 15.11 Branching duct within dense parenchyma

Pearl

Colour Doppler may identify a lymph node which will respond with colour as a result of vessels entering at the hilum (versus any other lesions).

Pearl

A "stand-off" devise (e.g., a small saline bag) may be placed between the transducer and the skin for easier identification of superficial lesions as it avoids displacement of the lesion by the probe.

Pearl

For small, short transducers, the field of view can be broadened by making use of the "tangential" option available on some US machines.

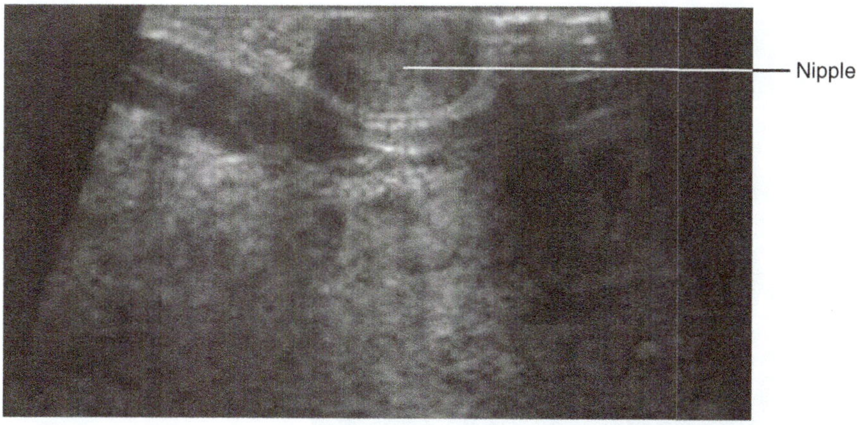

Fig. 15.12 Nipple

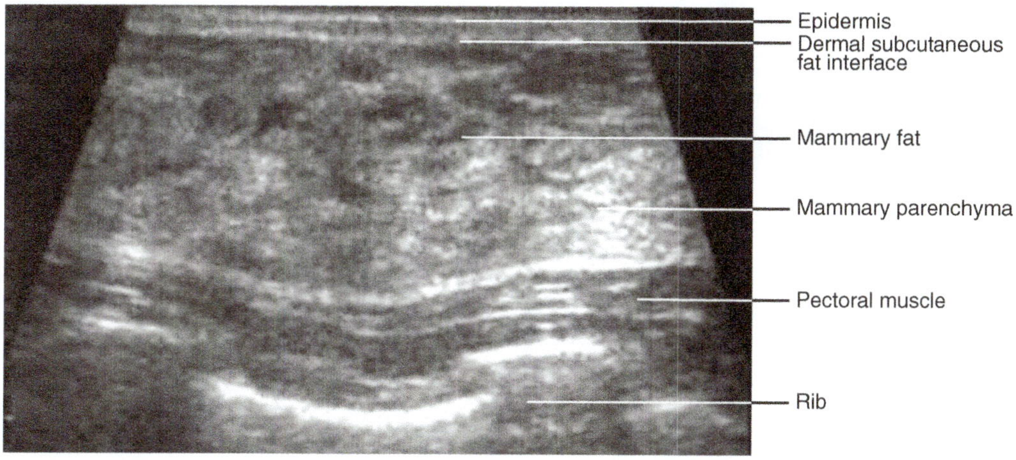

Fig. 15.13 Dense breast

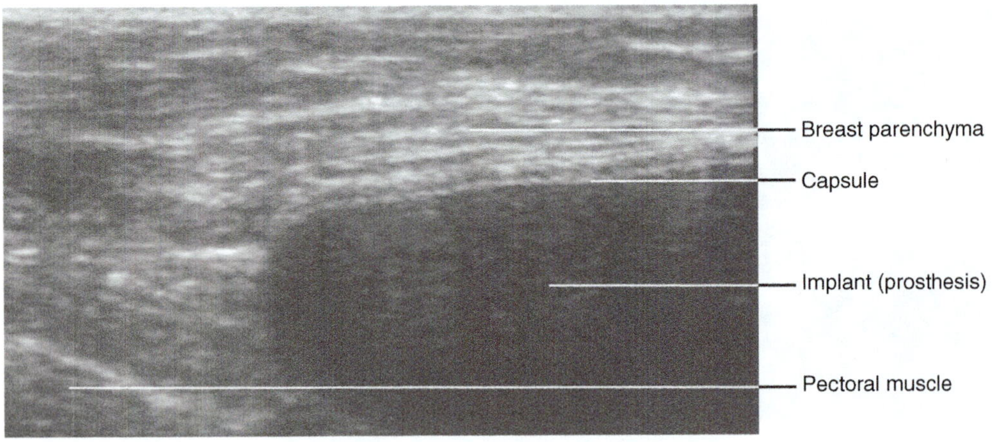

Fig. 15.14 Breast implant (the anechoic appearance of the implant indicates integrity)

Pearl
Information loss occurs near the back edge of the scan field (inferior aspect of screen); this region, therefore, requires careful review.

Magnetic Resonance Imaging

Technique

The patient is placed prone and a surface coil is used.

Routine Sequences

- T1 fat suppression.
- T2 fat suppression.
- T1 high-resolution fat suppression with contrast for enhancement pattern analysis.
- Dynamic enhancement.

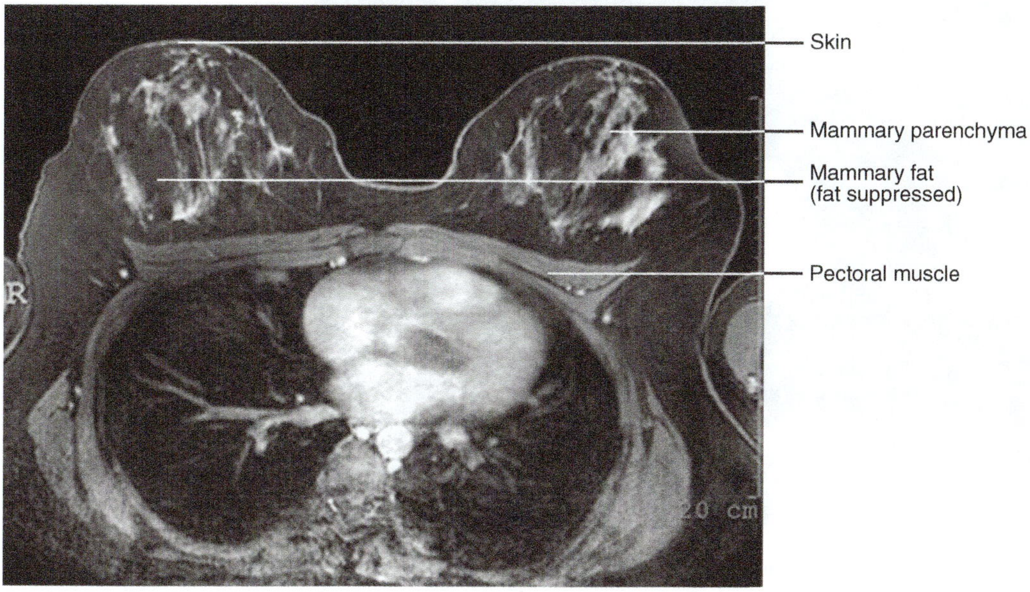

Fig. 15.15 T1 axial high-resolution fat-suppression sequence with contrast

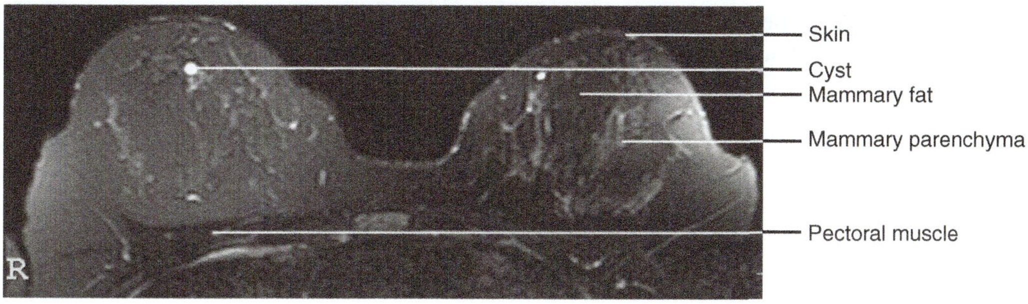

Fig. 15.16 T2 axial fat-suppression sequence

Computed Tomography

Although not commonly used as a primary imaging modality, a breast CT may occasionally be used to guide a biopsy needle or insert a hookwire into a breast lesion. Breast pathology may be noted incidentally on a CT chest scan in the form of a soft-tissue mass, calcification, gas locules, and rib erosions. A common indication for a CT in a patient with breast cancer is the evaluation of local and distant spread.

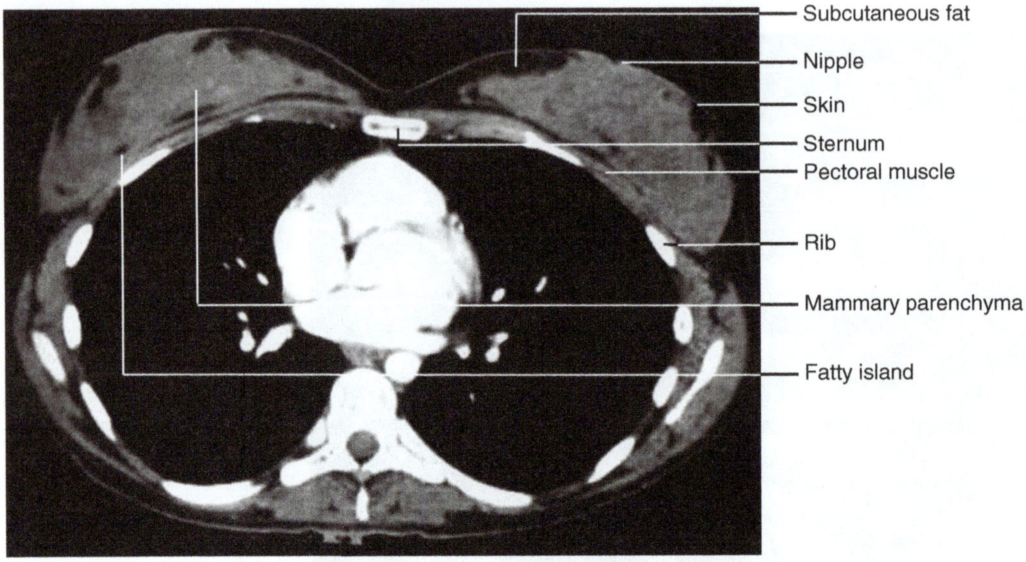

Fig. 15.17 CT chest (contrast enhanced)

Pearl
CT imaging of every chest will yield images of the breast. Always look at breast tissue for asymmetry and focal abnormalities.

Plain Film

As with a CT, breast pathology may also be incidentally noted on a CXR. Also, the CXR is a routine investigation in the work-up of a breast cancer patient.

Nuclear Medicine

Used to identify the sentinel node preoperatively in breast cancer patients. Instruments and techniques include:

1. Gamma camera
2. Injection of Tc-99m nanocolloid, which preferentially travels in the lymphatic system

Acknowledgements Many thanks to Heather Oates for her endless time and help. Image capture was also assisted by Terry Borrain at the medical graphics unit, University of the Witwatersrand. Dr. Mark Haagensen kindly provided the MR images and patiently answered many questions.

Suggested Reading

Andersson I (1986) Mammography in clinical practice. Med Radiogr Photogr 62:1–42

Eklund GW, Cardenosa G (1992) The art of mammographic positioning. Radiol Clin North Am 30:20–53

Feig SA, Piccoli CW, Roebuck EJ (1997) The breast. In: Grainger RG, Allison DJ (eds) Diagnostic radiology. Churchill Livingstone, New York, pp 1993–2036

Kopans DB (1989) Breast imaging. JB Lippincott, Philadelphia, pp 34–115; 227–242

Stensson WE (1996) Breast examination techniques and ultrasonic anatomy. Pretoria, South Africa: mammography workshop and ultrasound workshop, CSIR Conference Centre, pp 62–65

16 The Heart

Plain Film

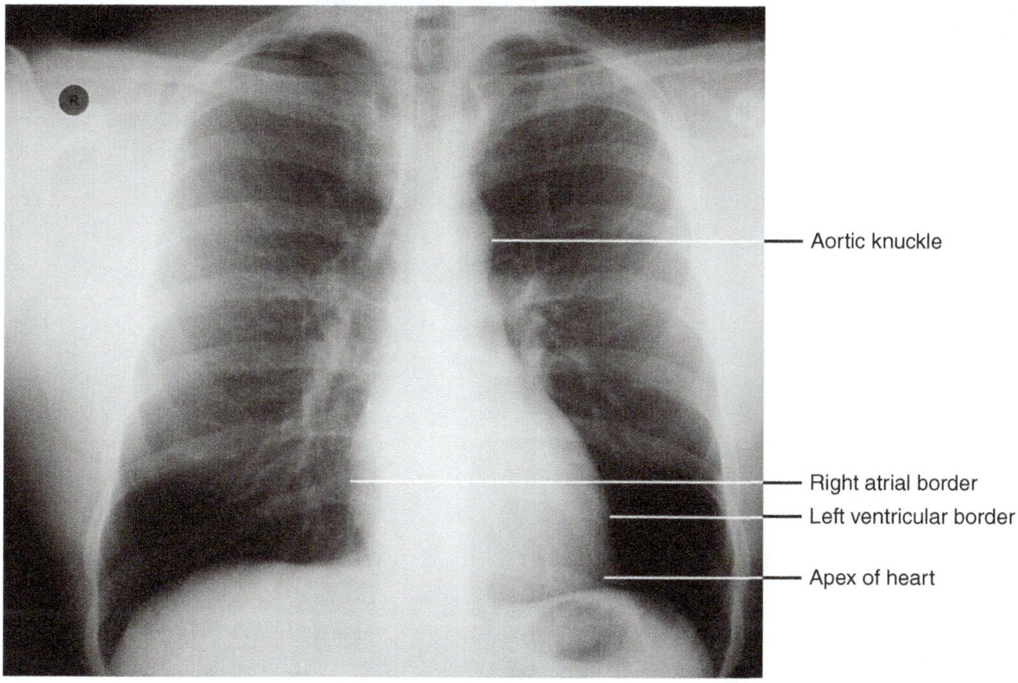

Fig. 16.1 Posterior-anterior view of chest

S. Andronikou (ed.), *See Right Through Me*,
DOI 10.1007/978-3-642-23893-2_16, © Springer-Verlag Berlin Heidelberg 2012

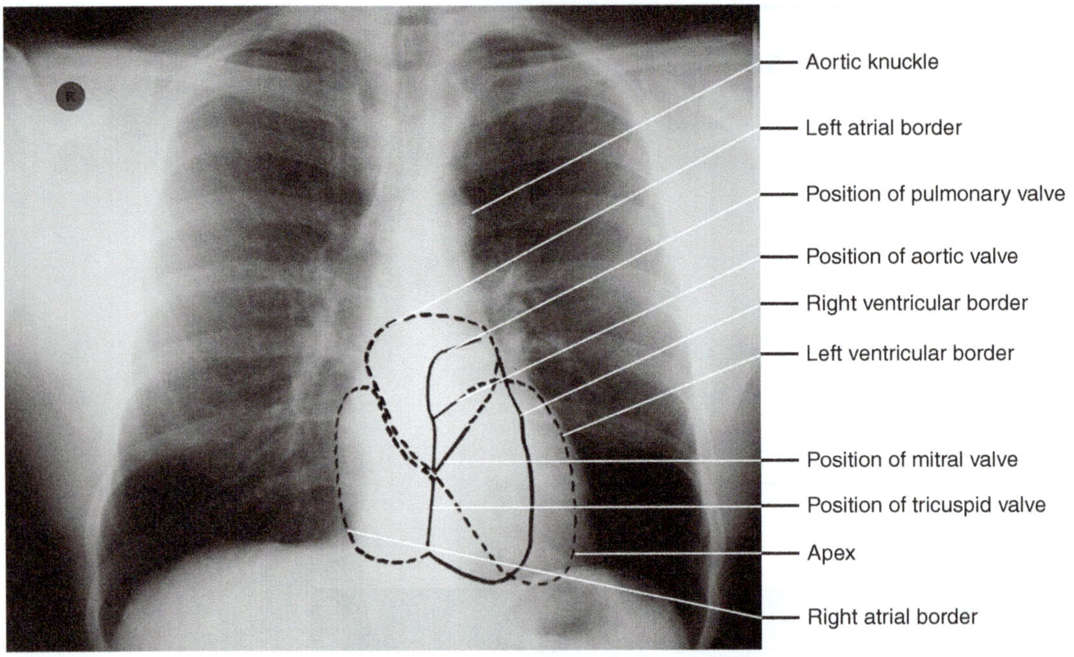

Fig. 16.2 Posterior-anterior view of chest

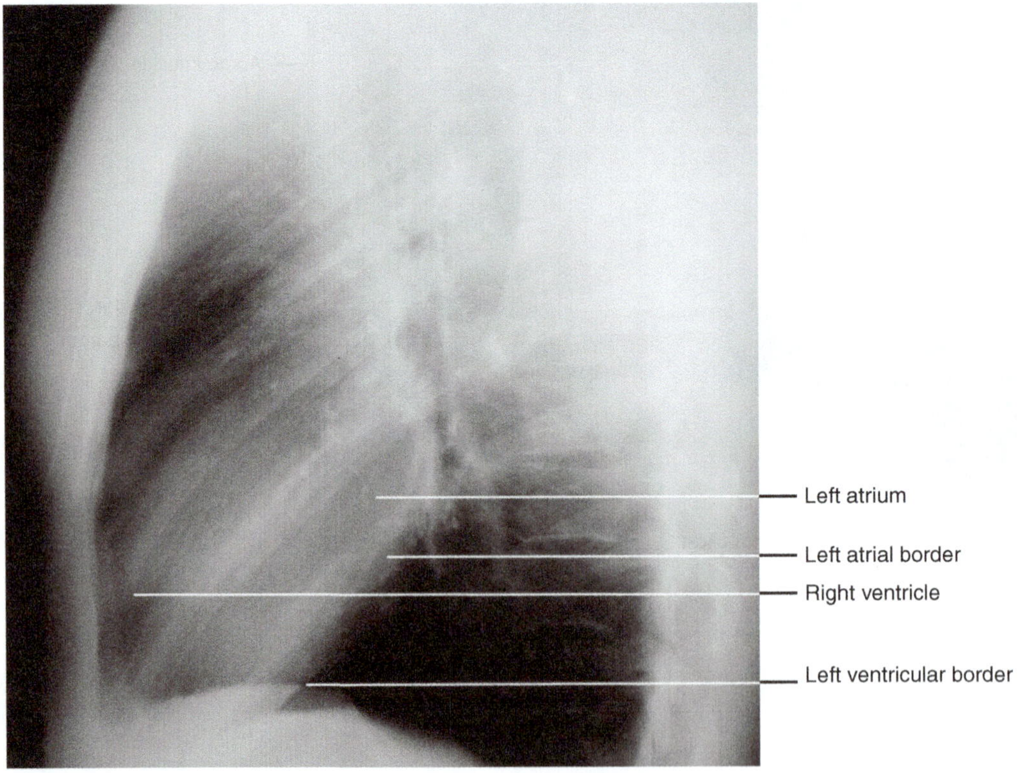

Fig. 16.3 Lateral view of chest

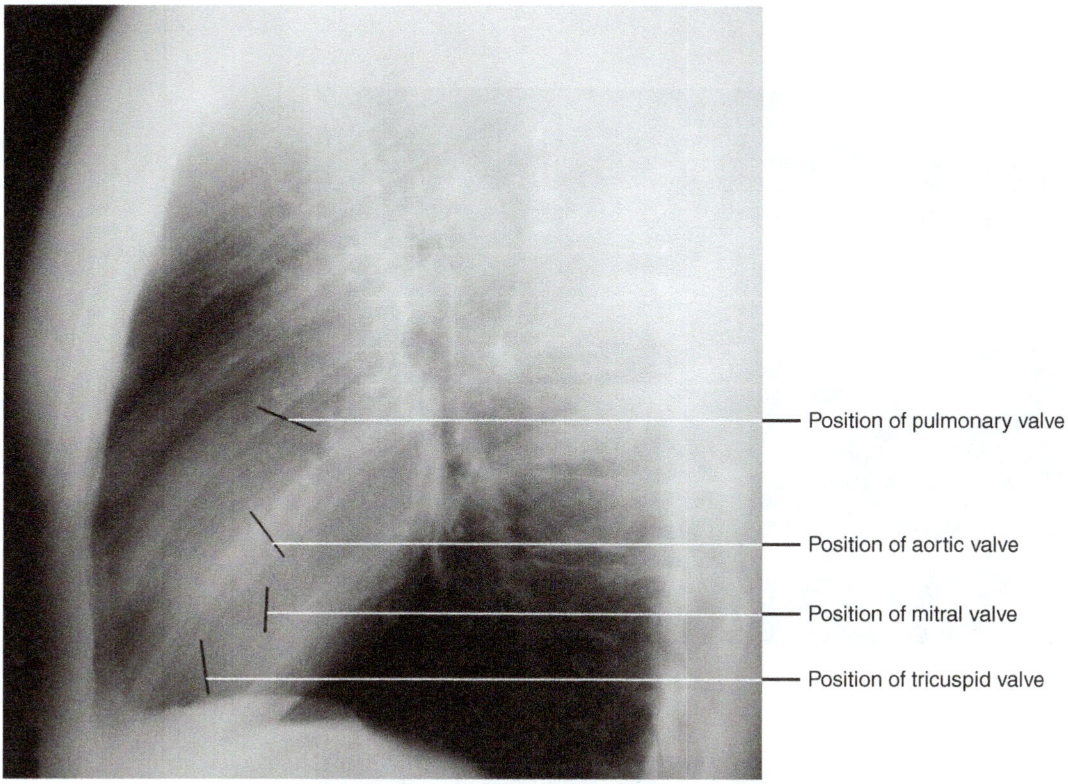

Fig. 16.4 Lateral view of chest

Echocardiogram

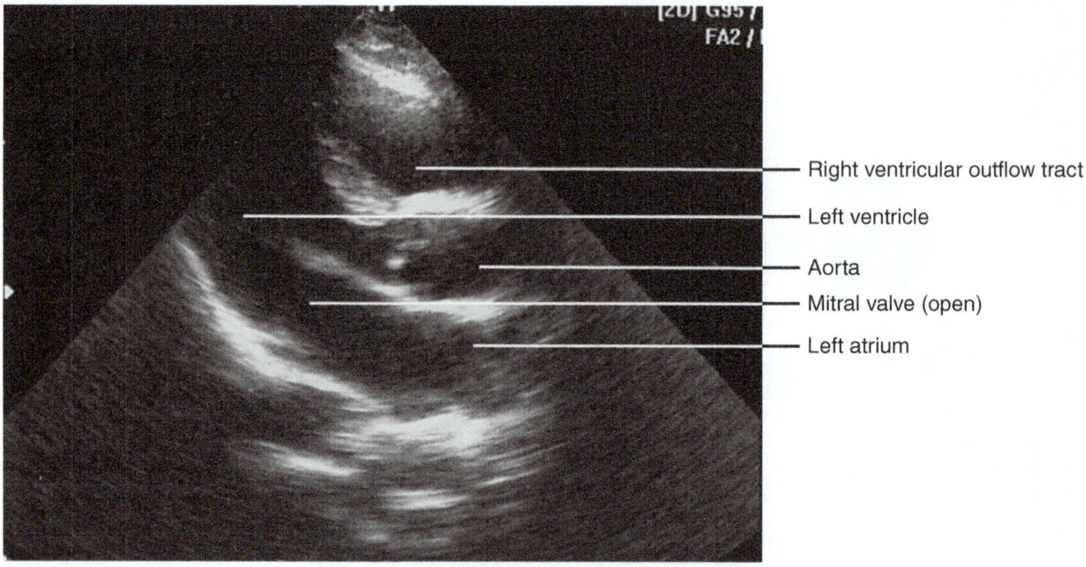

Fig. 16.5 Parasternal long-axis view, end-diastole

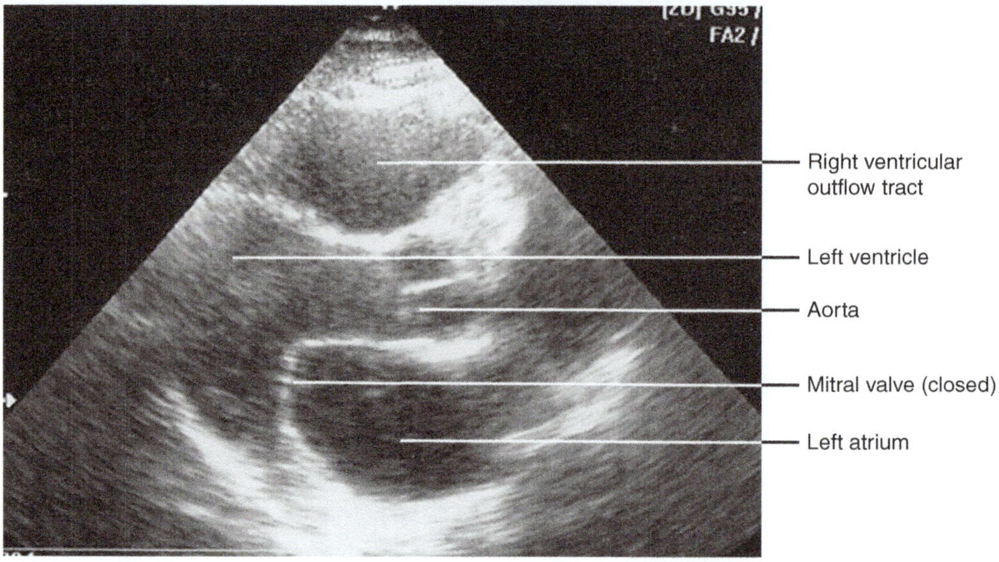

Fig. 16.6 Parasternal long-axis view, end-systole

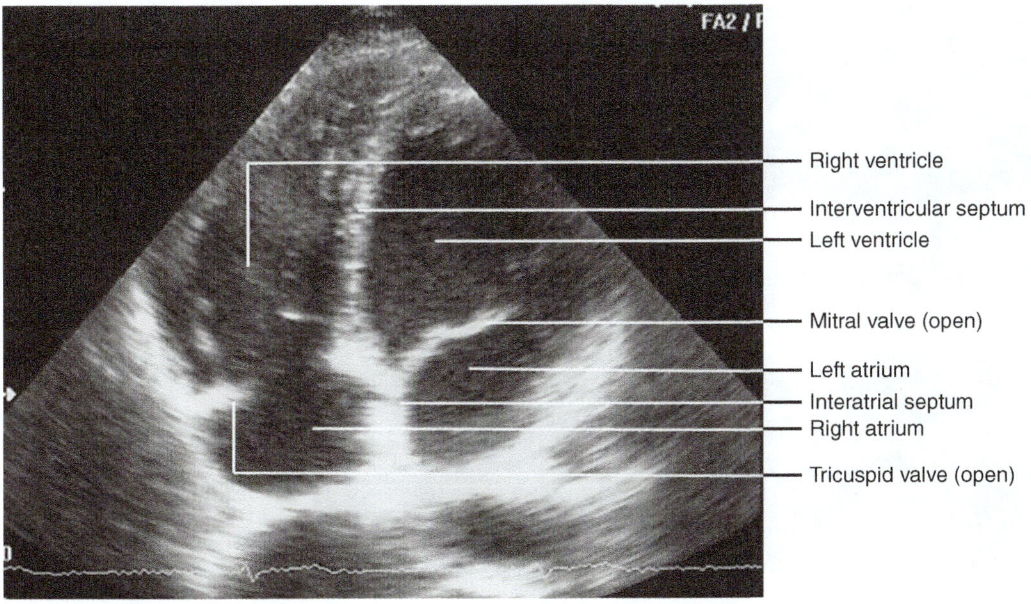

Fig. 16.7 Two-dimensional (2D) apical four-chamber view, end-diastole

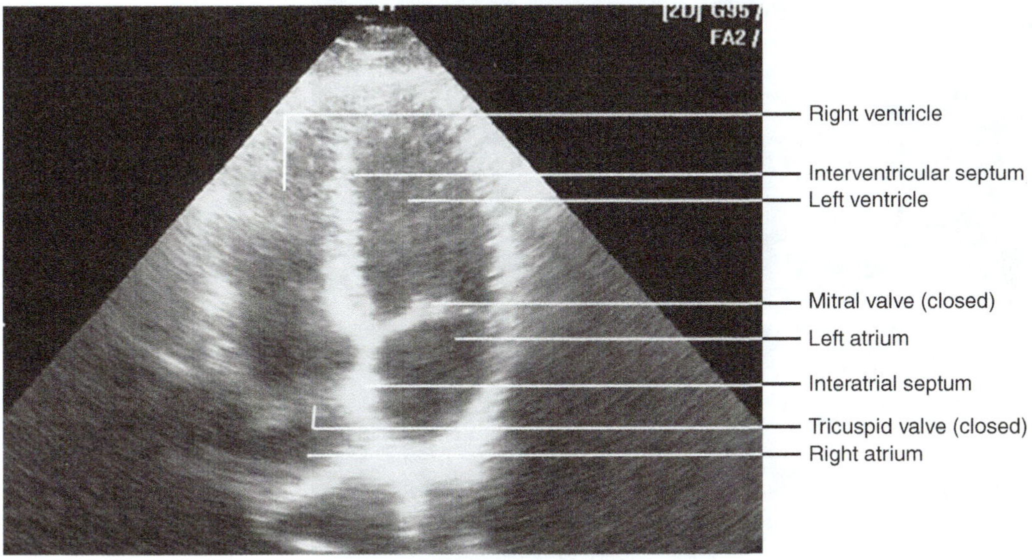

Fig. 16.8 2D apical four-chamber view, end-systole

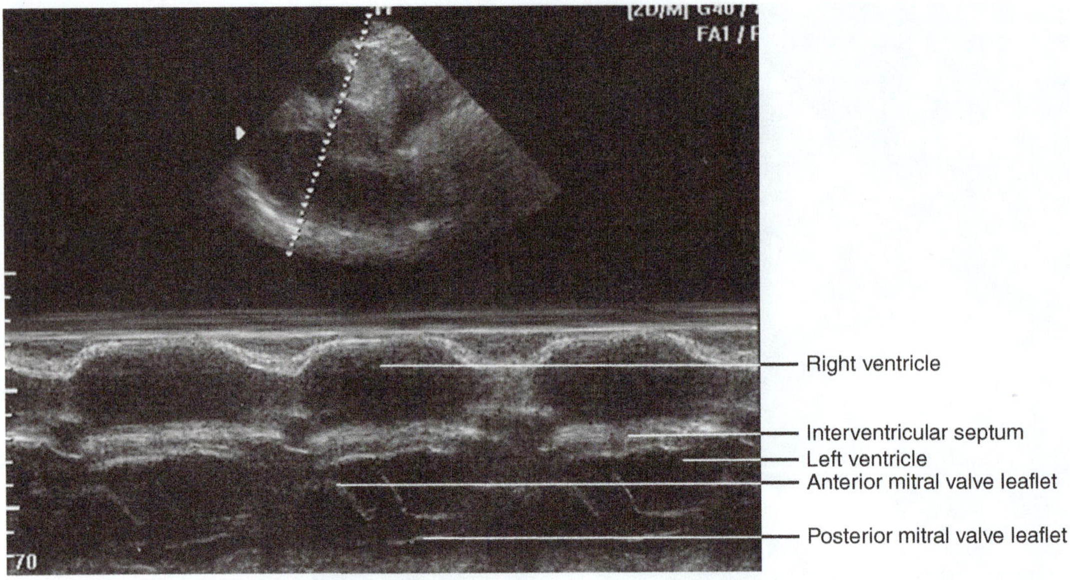

Fig. 16.9 M-mode tracing of a normal mitral valve

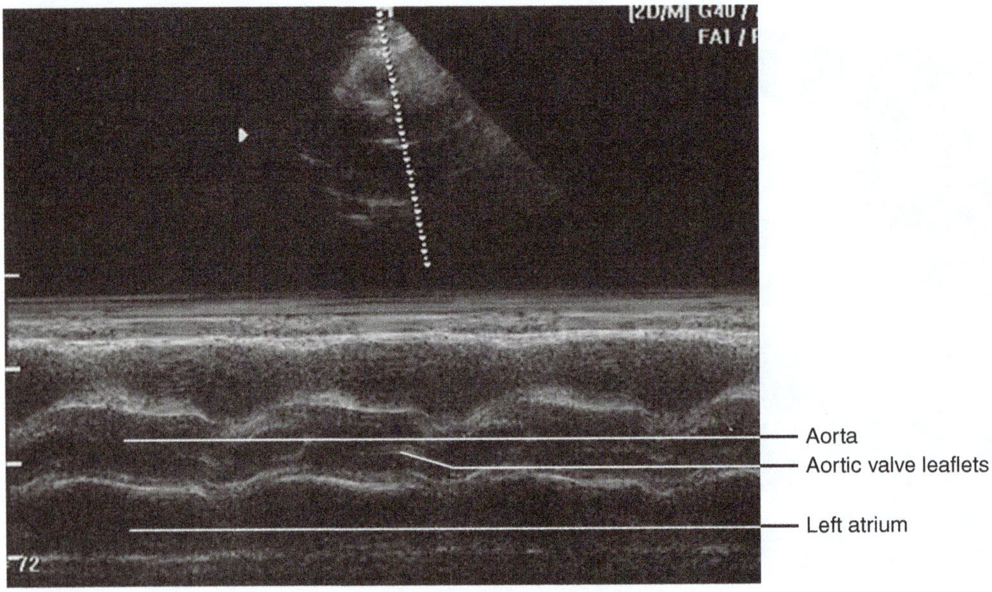

Fig. 16.10 M-mode tracing of a normal aortic valve and left atrium

Angiography

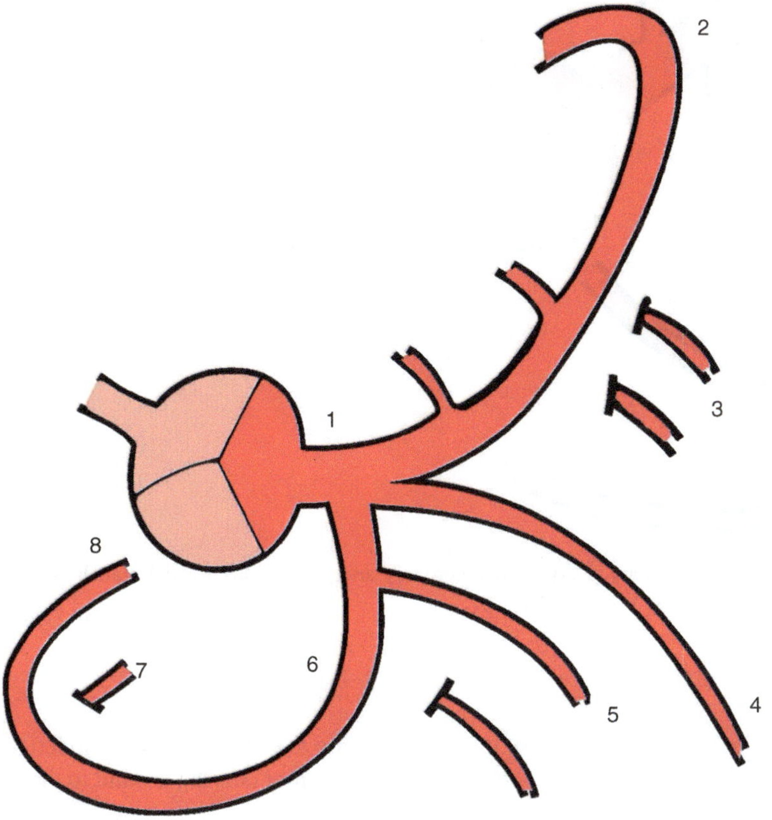

Fig. 16.11 Axial schematic of the left coronary artery. *1* left coronary artery, *2* left anterior descending artery, *3* diagonal branches, *4* ramus intermedius, *5* obtuse marginal arteries, *6* circumflex artery, *7* left posterior lateral branch, *8* posterior descending artery

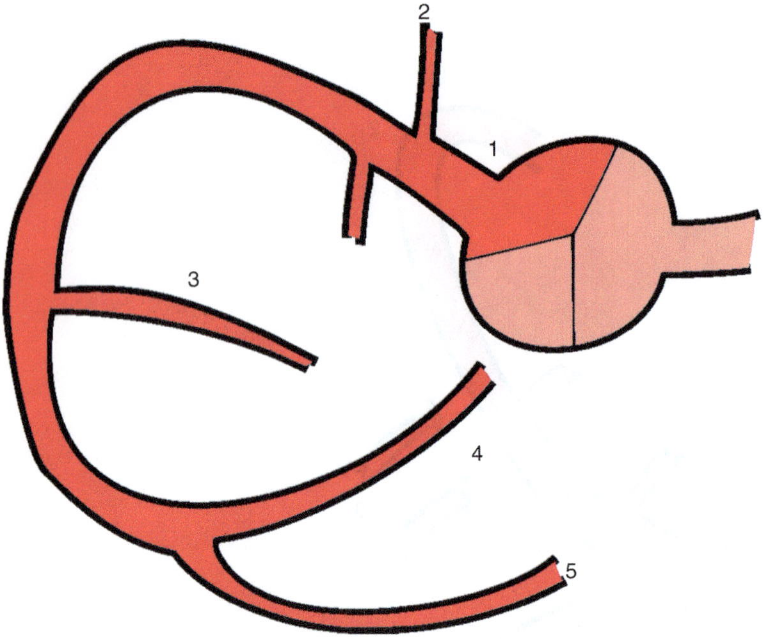

Fig. 16.12 Axial schematic of the right coronary artery. *1* right coronary artery, *2* sinoatrial nodal branch, *3* acute marginal branch, *4* posterior descending artery, *5* right posterior

Pearl

The left coronary artery can be recognised because it splits early into the left circumflex and left anterior descending arteries. The right coronary artery splits distally after curving around to the posterior aspect of the heart.

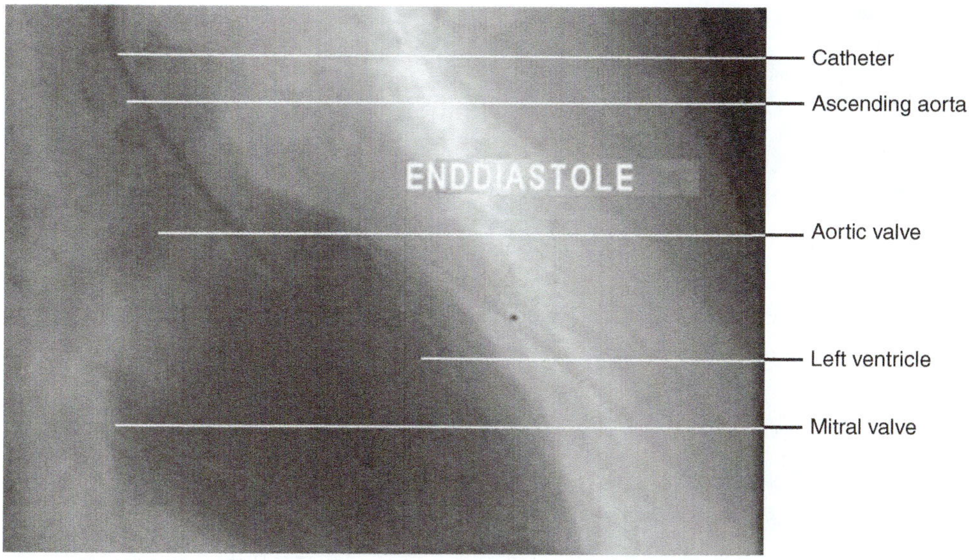

Fig. 16.13 Left ventricular cineangiogram, end-diastole: right anterior oblique (RAO) 60°

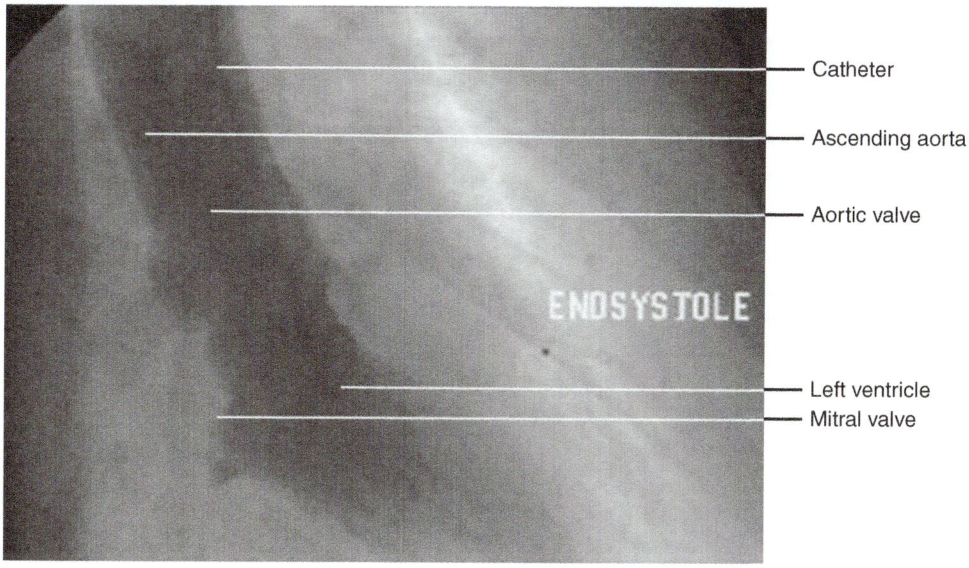

Fig. 16.14 Left ventricular cineangiogram, end-systole

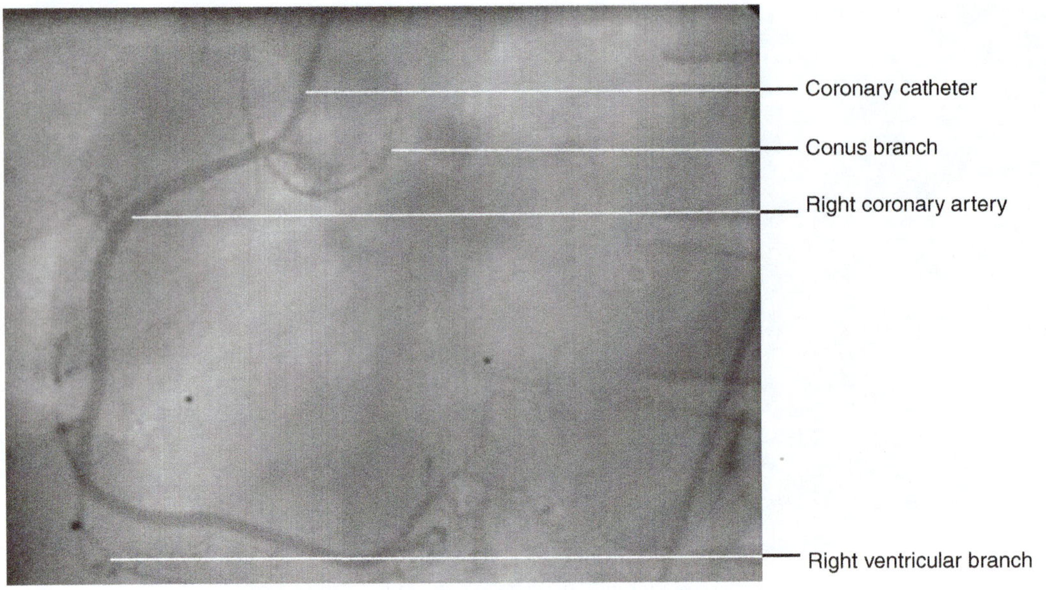

Fig. 16.15 Right coronary artery (dominant), left anterior oblique (LAO) 45°

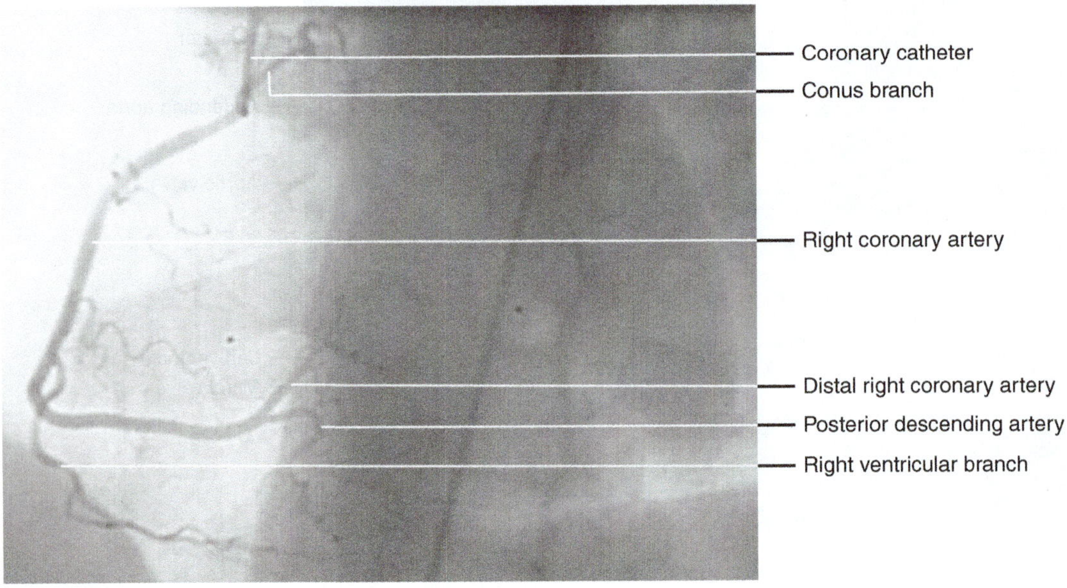

Fig. 16.16 Right coronary artery (dominant), LAO 20°, cranial angulation 20°

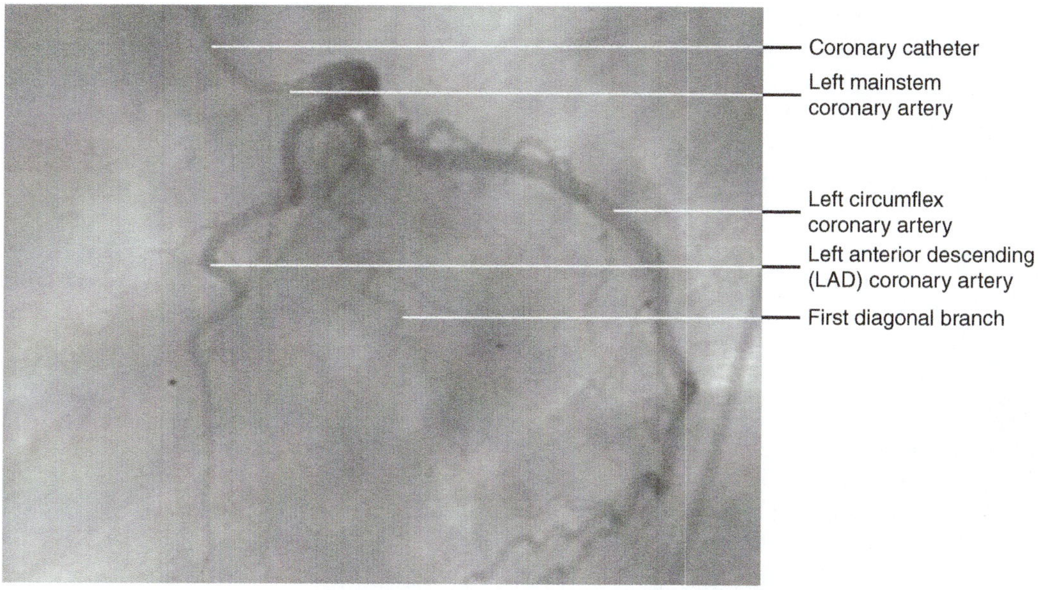

Fig. 16.17 Left coronary artery (nondominant), LAO 60°, caudal angulation 20°

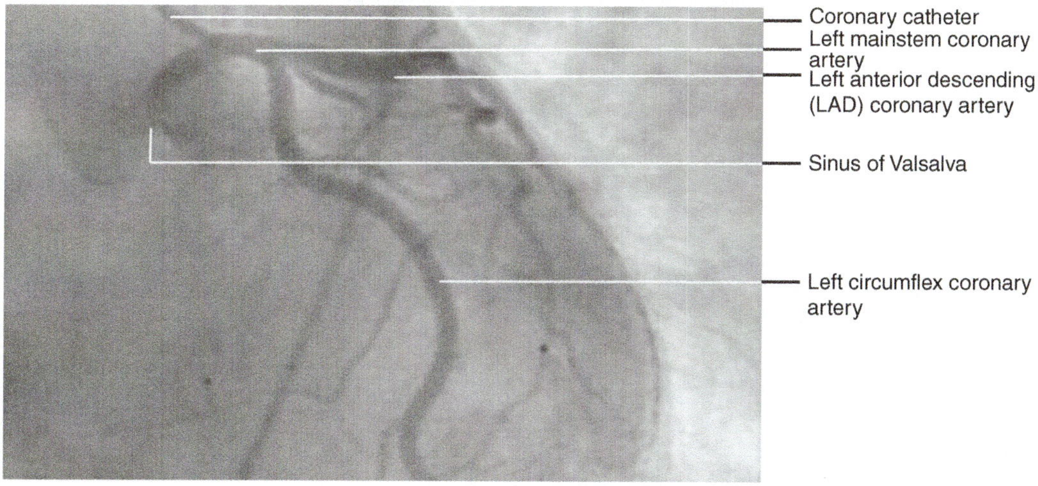

Fig. 16.18 Left coronary artery (nondominant), anteroposterior (AP), caudal angulation 30°

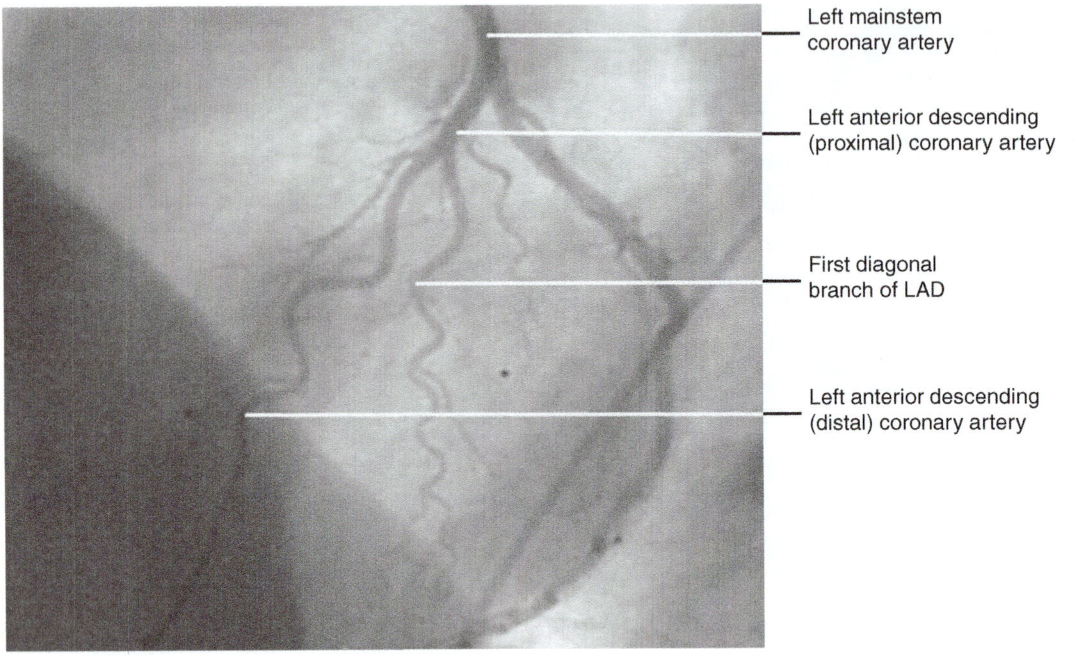

Fig. 16.19 Left coronary artery (nondominant), LAO 60°, cranial angulation 20°

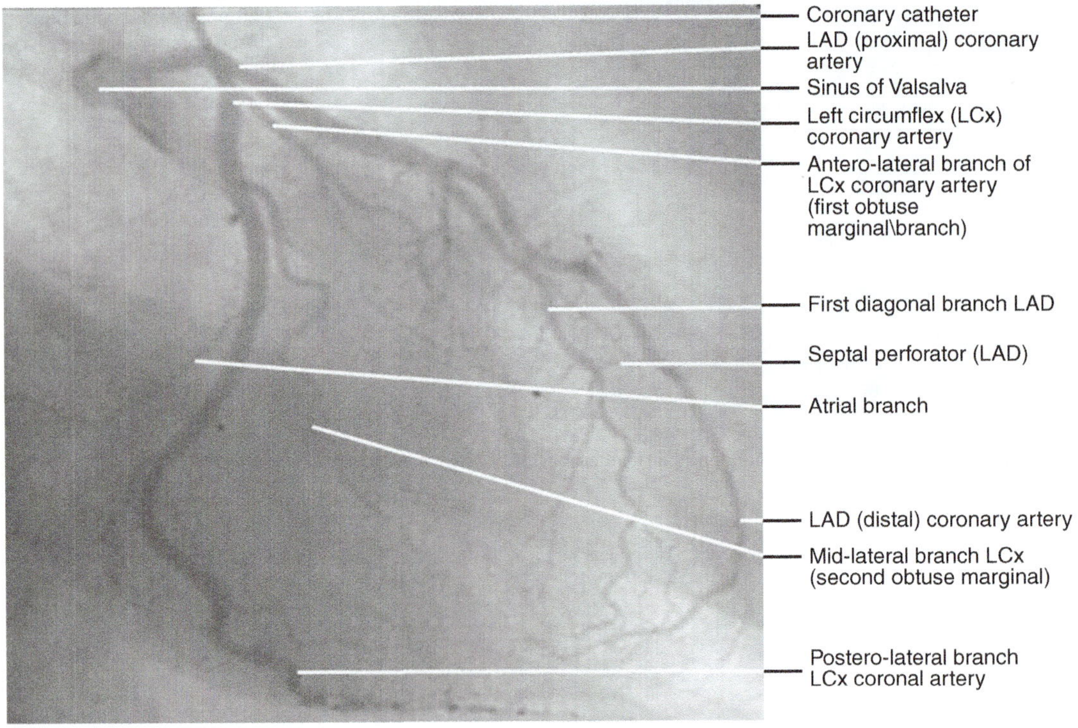

Fig. 16.20 Left coronary artery (nondominant), RAO 44°, caudal angulation 26°

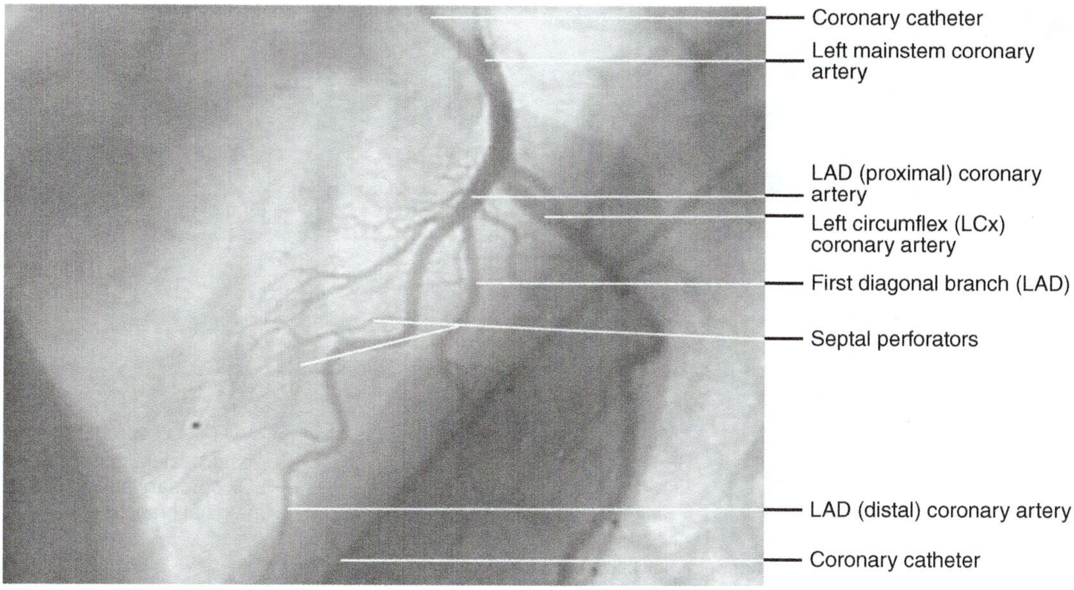

Fig. 16.21 Left coronary artery (nondominant), RAO 45°, cranial angulation 30°

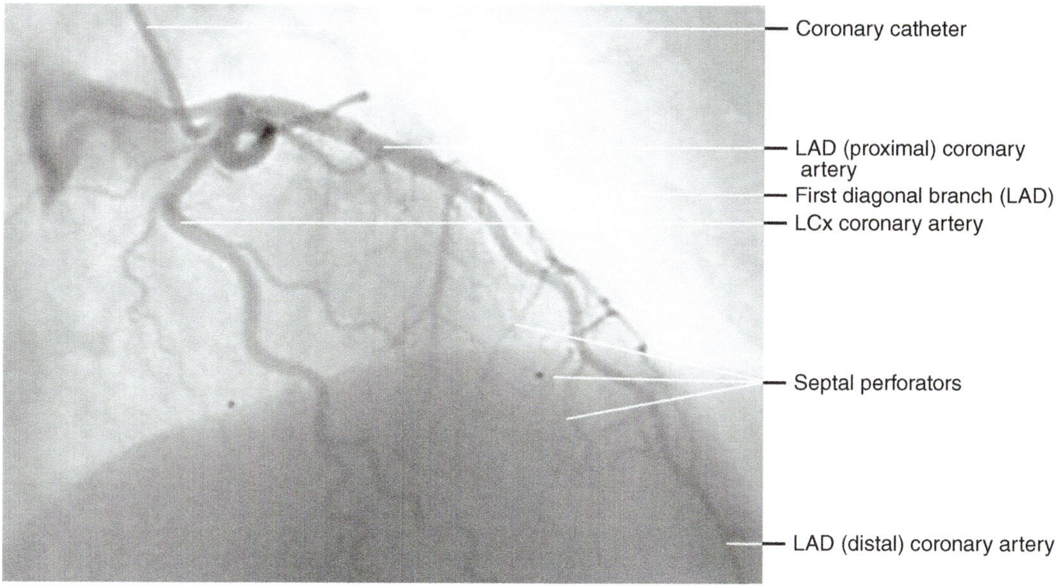

Fig. 16.22 Left coronary artery (nondominant), RAO 40°, cranial angulation 30°

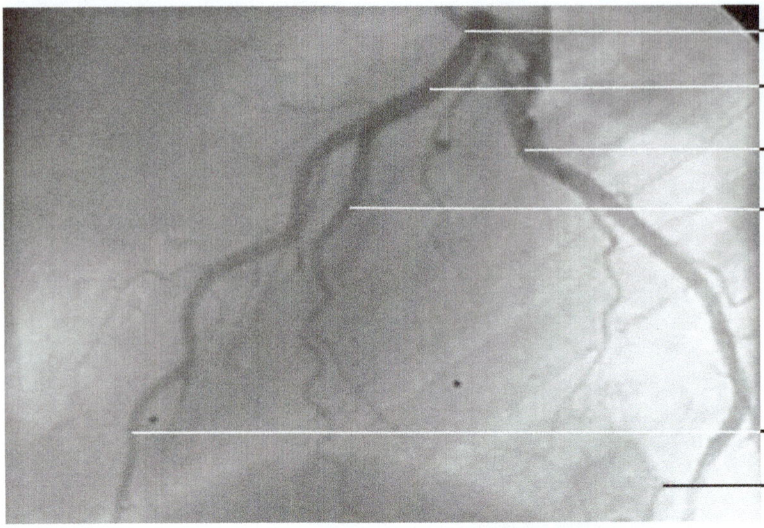

Fig. 16.23 Left coronary artery (nondominant), left lateral projection

Computed Tomography

Coronary Angiography

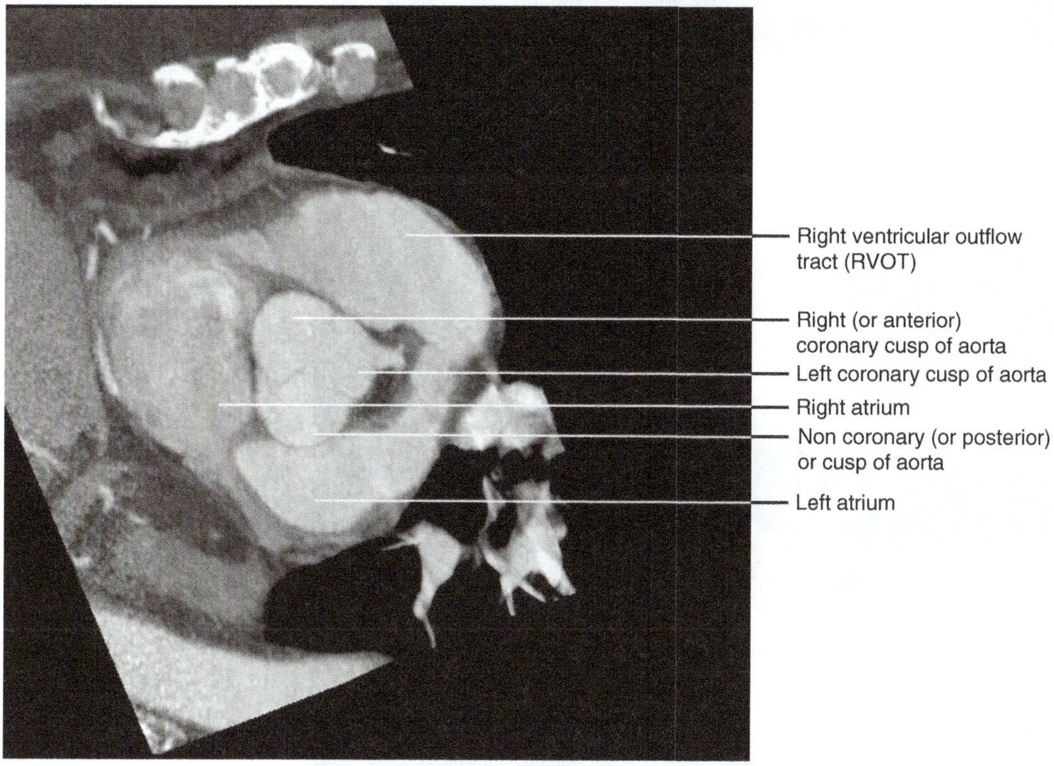

Fig. 16.24 Short-axis view

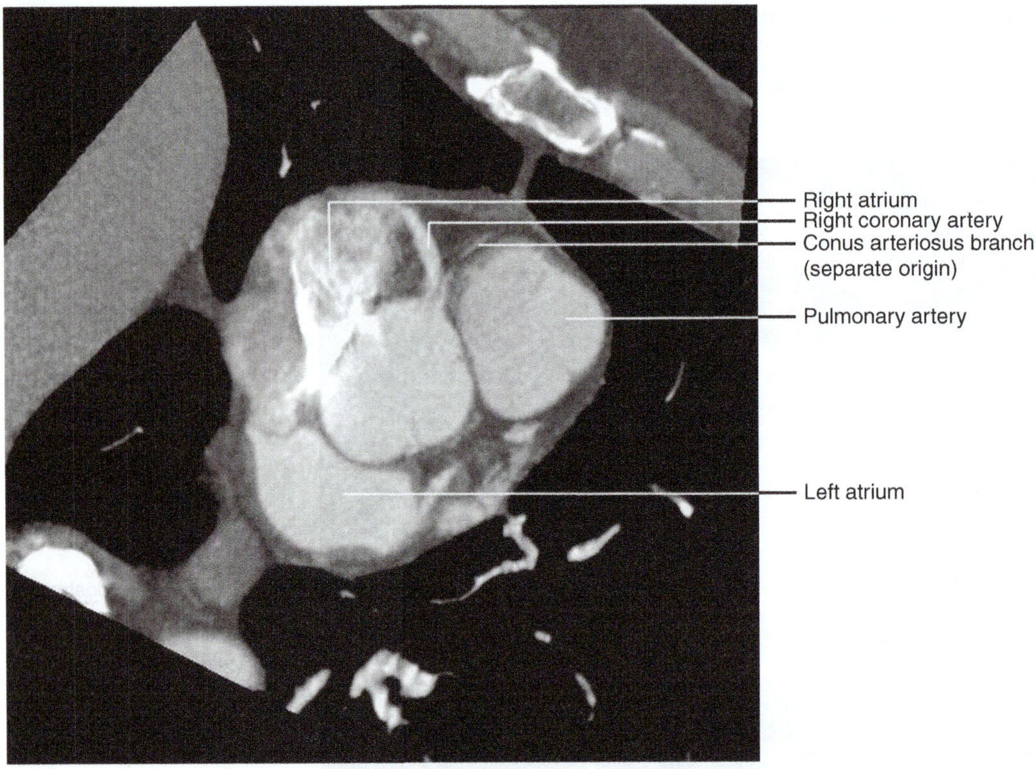

Fig. 16.25 Short-axis view (maximum-intensity projection [MIP])

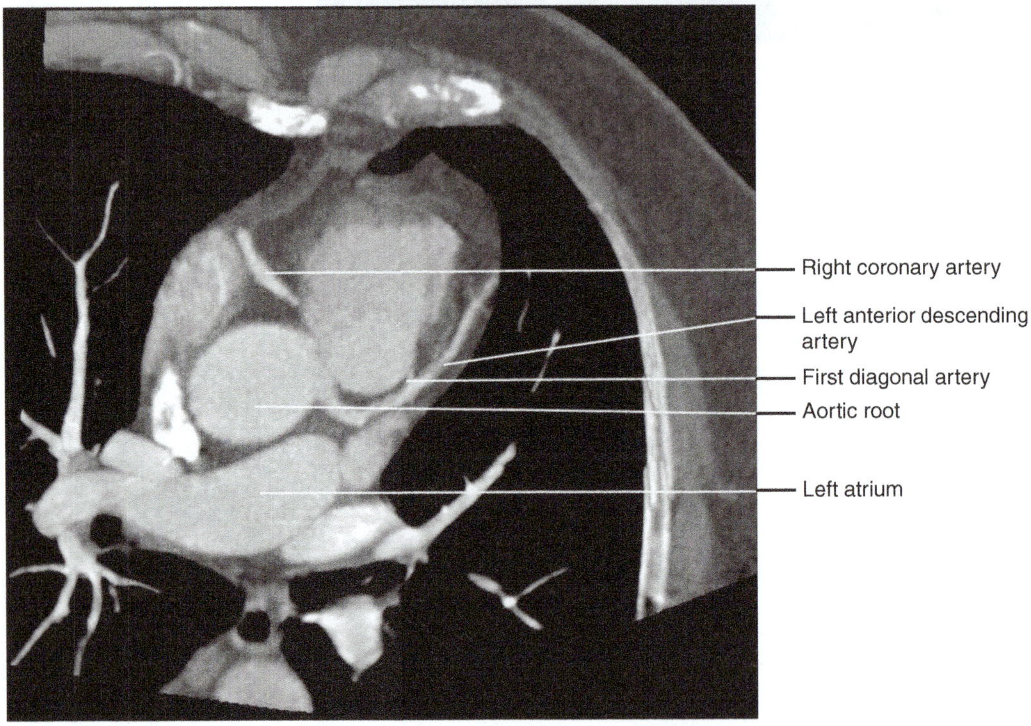

Fig. 16.26 Four chamber orientation (MIP)

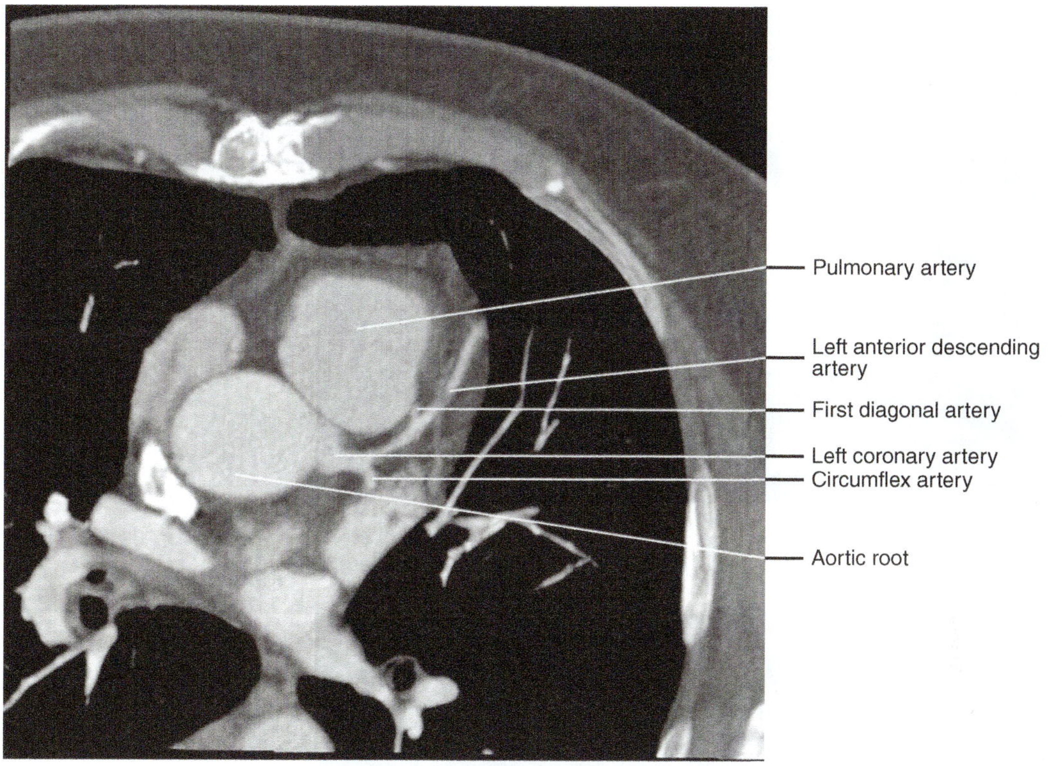

Fig. 16.27 Four-chamber orientation (MIP)

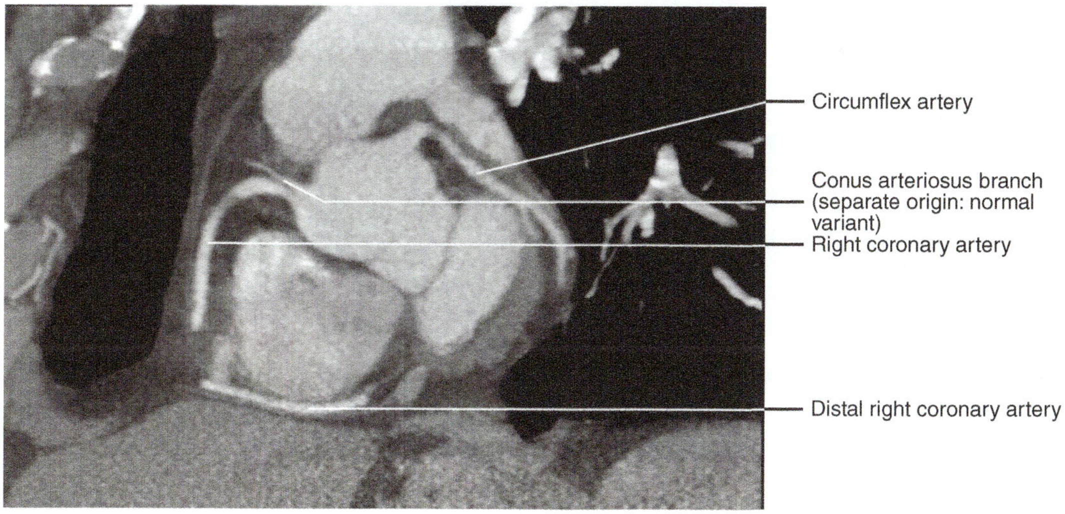

Fig. 16.28 Short-axis view (MIP)

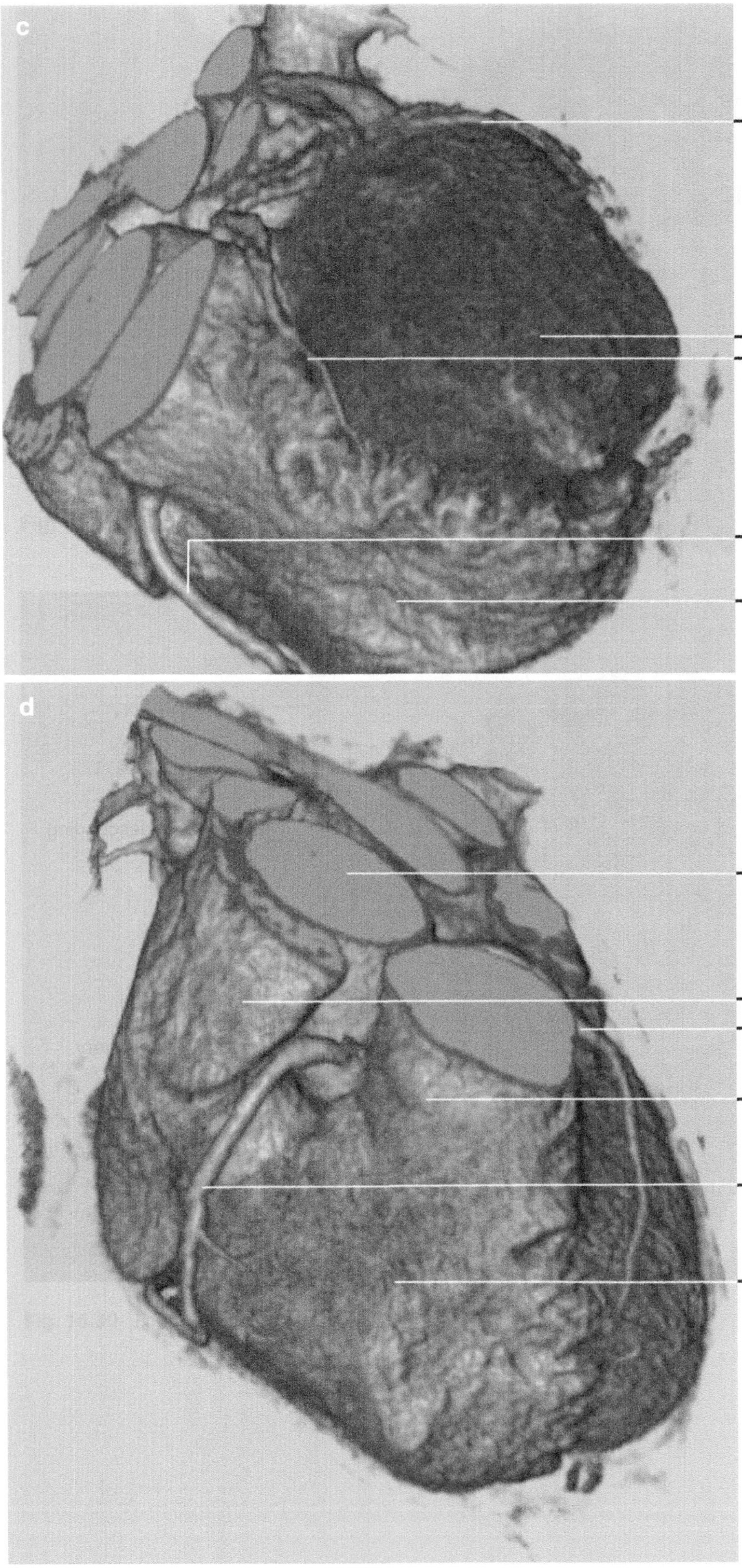

Fig. 16.31 (continued)

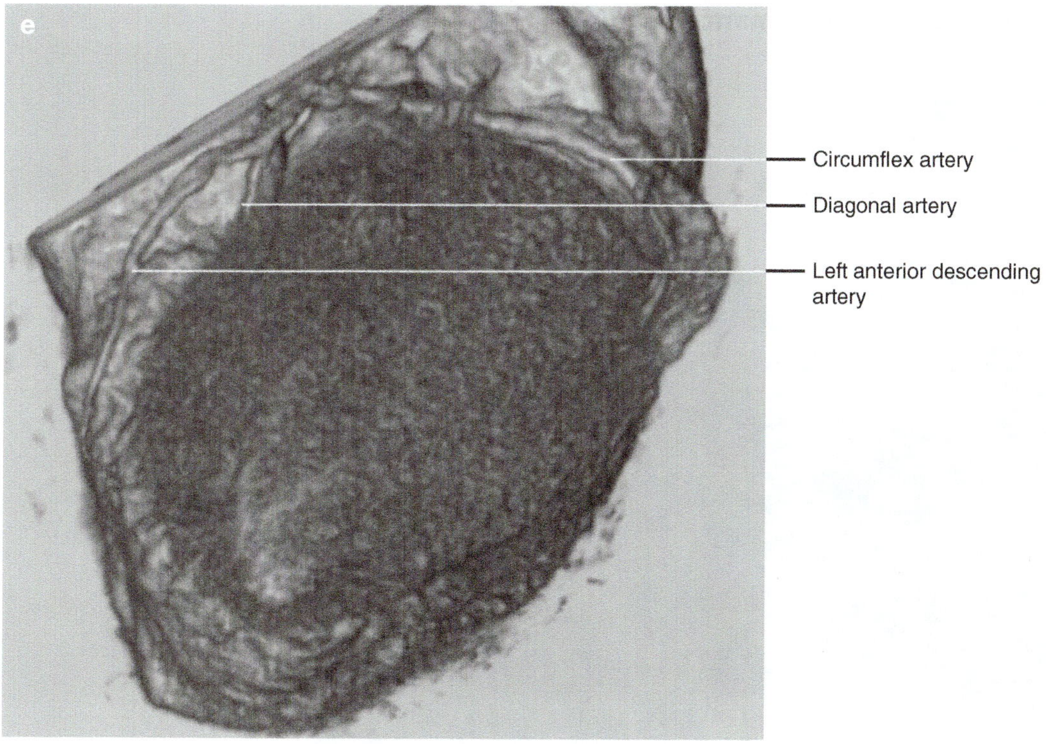

Fig. 16.31 (continued)

Cardiac Magnetic Resonance Imaging

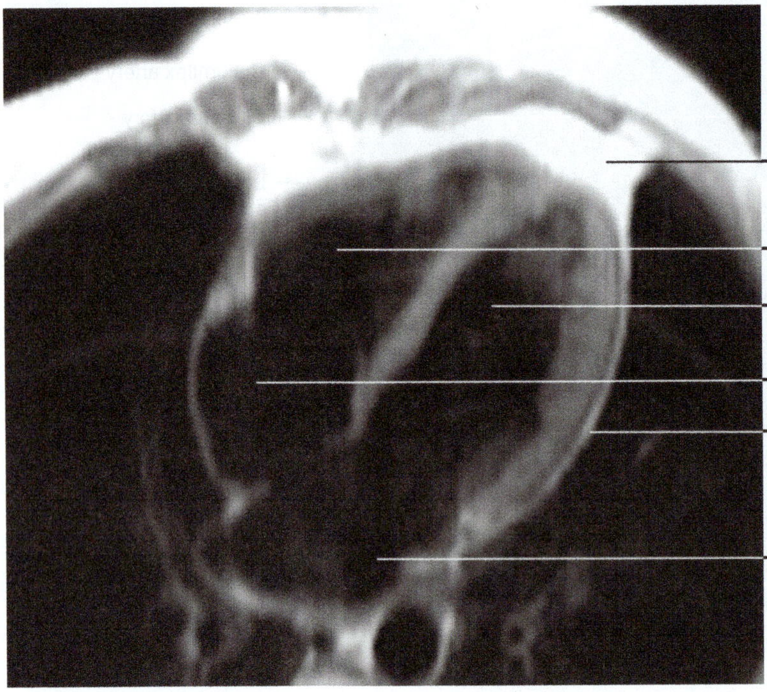

Fig. 16.32 T1 turbo spin echo (TSE) image in a four-chamber orientation

Pearl
The normal pericardium is smooth and <2 mm thick.

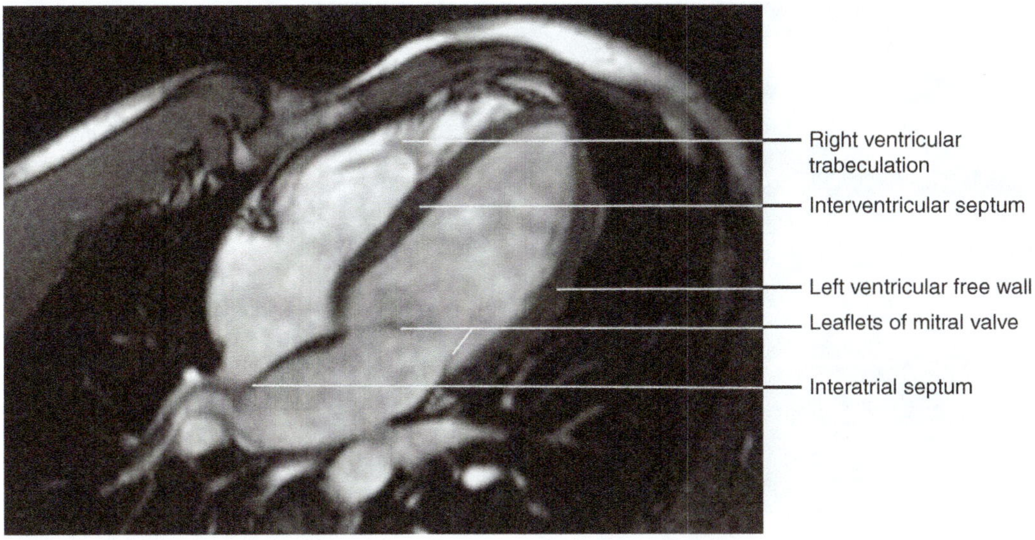

Fig. 16.33 True fast imaging with steady-state precession (FISP) image in a four-chamber orientation

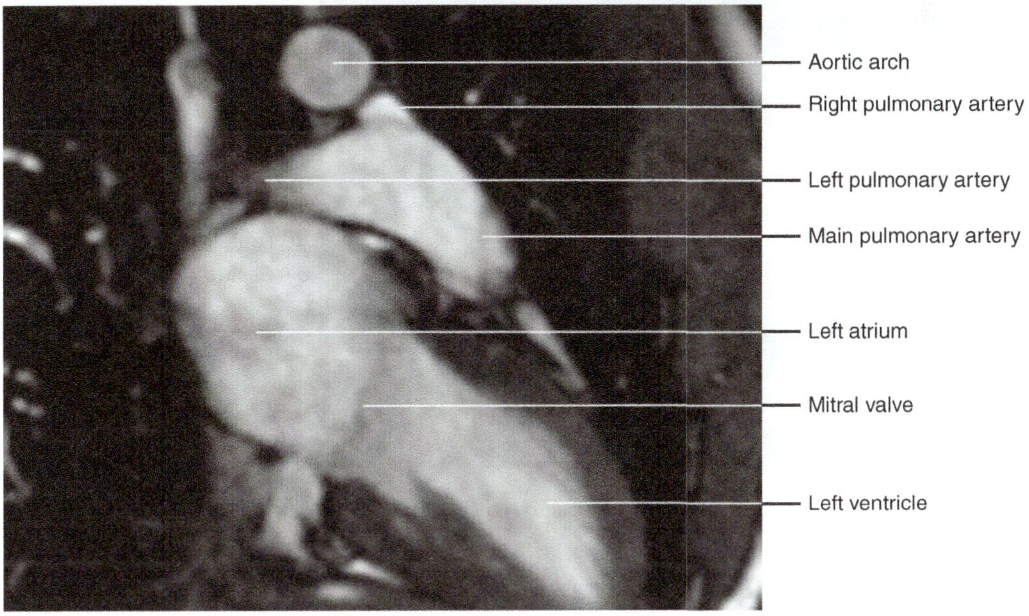

Fig. 16.34 True FISP image in a vertical long axis orientation through the left heart chambers

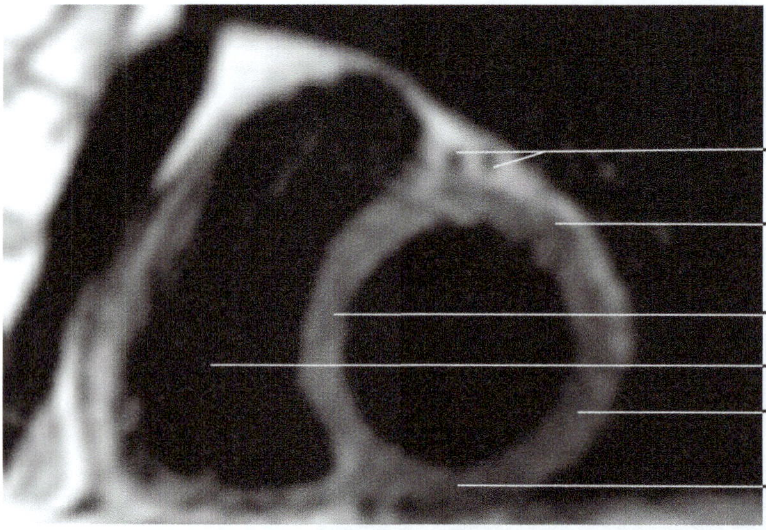

Fig. 16.35 T1 TSE image in short-axis orientation, mid-ventricular level

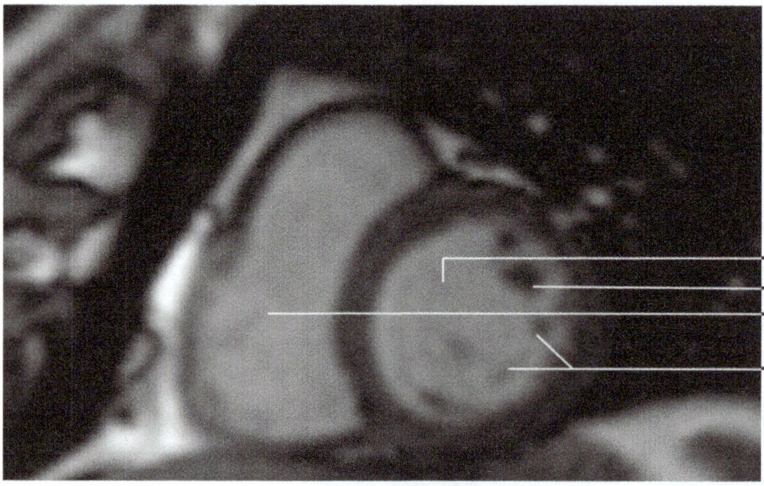

Fig. 16.36 True FISP image in short-axis orientation, mid-ventricular level, diastole

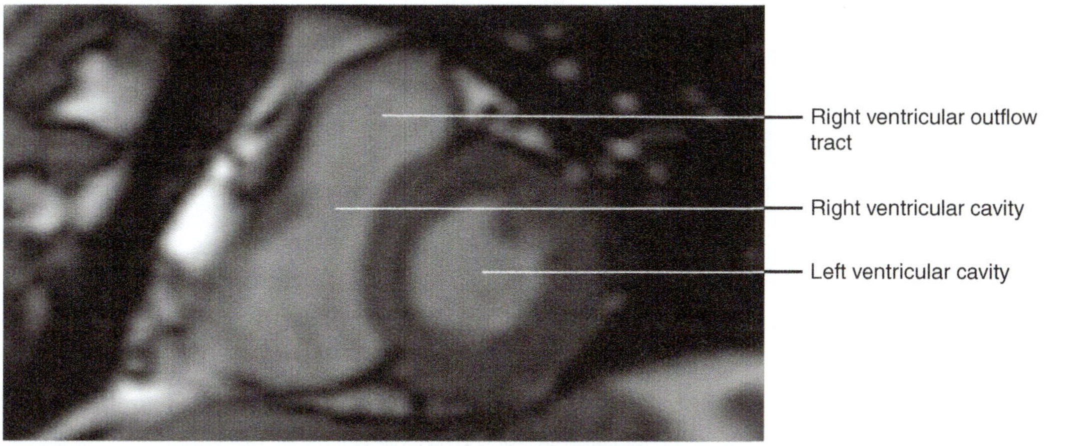

Fig. 16.37 True FISP image in short-axis orientation, mid-ventricular level, systole. Note the uniform left ventricular wall-thickening

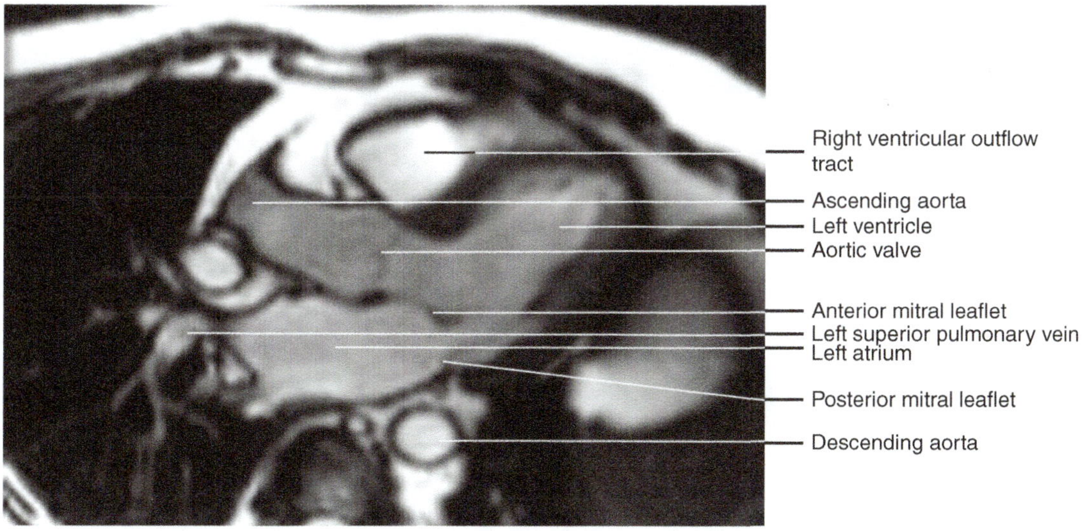

Fig. 16.38 True FISP image in a valve-plane orientation showing the mitral and aortic valves

Pearl

On the cine true FISP images, turbulent blood flow through the valves due to either regurgitation or stenosis results in spin dephasing. This appears as a dark jet in the blood adjacent to the valve. Objective valvar assessment can be made using phase contrast images to measure blood flow and velocity.

The Chest

17

- The standard chest radiograph views are the posteroanterior and lateral projections. The other radiographic views are requested less frequently as cross-sectional imaging becomes more available.
- CT is the primary cross-sectional imaging technique for most chest conditions, with MRI preferred in selective situations such as for the chest wall. CT has better spatial resolution and better depicts the lung parenchyma and cortical bone. Multiplanar reconstruction with the newer multidetector scanners has revolutionized conventional CT scanning.
- CT pulmonary angiography in the chest has replaced conventional invasive aortography, bronchial arteriography, and pulmonary angiography to a large extent. Evaluation of vascular conditions in the chest has also been surpassed by MR angiography and transoesophageal echocardiography.
- High-resolution CT (HRCT) improves visualization of the fissures and depicts the bronchi and vessels out to the peripheral quarter of the lung, thus allowing earlier detection of disease.
- Evaluation of the chest can be supplemented with ultrasound, fluoroscopy, barium/contrast studies, angiography, venography, and nuclear medicine, depending on the primary pathology.

Technique

- High kilovoltage (120–140 kvp) radiographs are used routinely. Low kilovoltage (70 kvp) films better demonstrate pleural pathology (e.g., pleural plaques) and small pulmonary nodules.
- A systemic approach is required to read the chest radiograph. Always review the "blind" areas: (1) apices; (2) hila; (3) outer ends of the clavicles; (4) supraclavicular region; (5) retrocardiac region; (6) lung below the dome of the diaphragm; and (7) strip inside the chest wall.
- The mediastinal and connective tissue compartments in the mediastinum are divided into different spaces: (1) pretracheal space; (2) aorticopulmonary window; (3) subcarinal space; and (4) right paratracheal space.
- On the lateral view, the thoracic spine should appear progressively darker as one looks more inferiorly, and the retrocardiac space and retrosternal space should be relatively dark, and of similar density.
- The diameter of the blood vessels beyond the hilum varies according to the patient's position. On plain films taken in the upright position, the diameter of the arteries and the veins increases gradually from the apex to the base. For comparisons to be valid, the measurements must be equidistant from the hila. The difference is less marked in the supine position.

S. Andronikou (ed.), *See Right Through Me*,
DOI 10.1007/978-3-642-23893-2_17, © Springer-Verlag Berlin Heidelberg 2012

Plain Film

Posteroanterior Chest

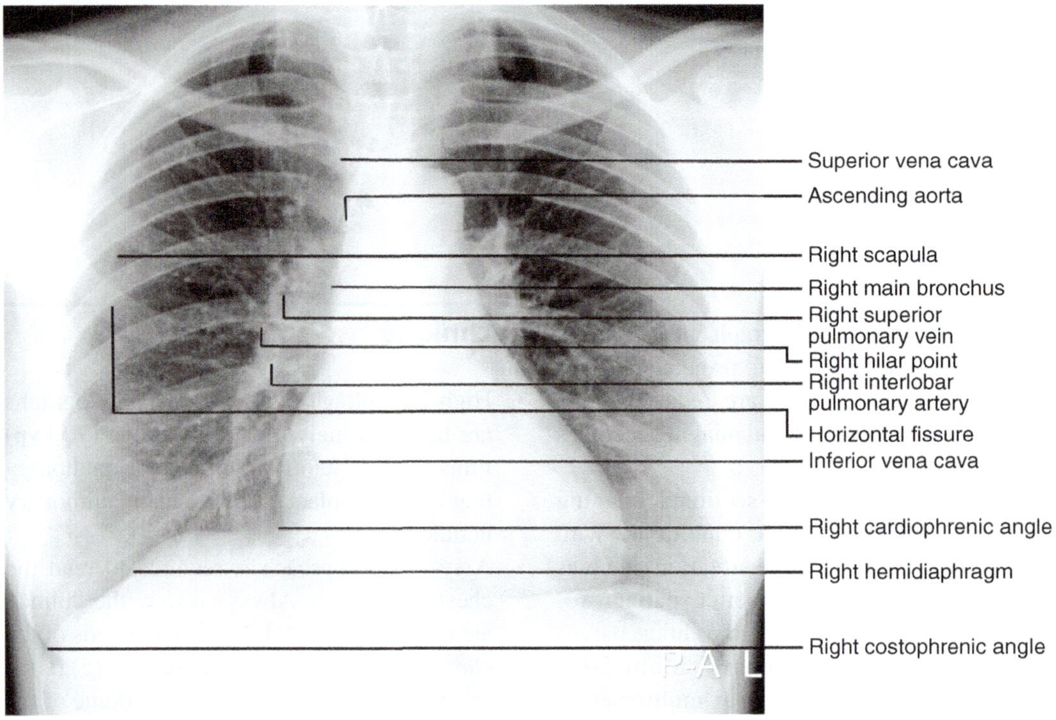

Fig. 17.1 Posteroanterior view

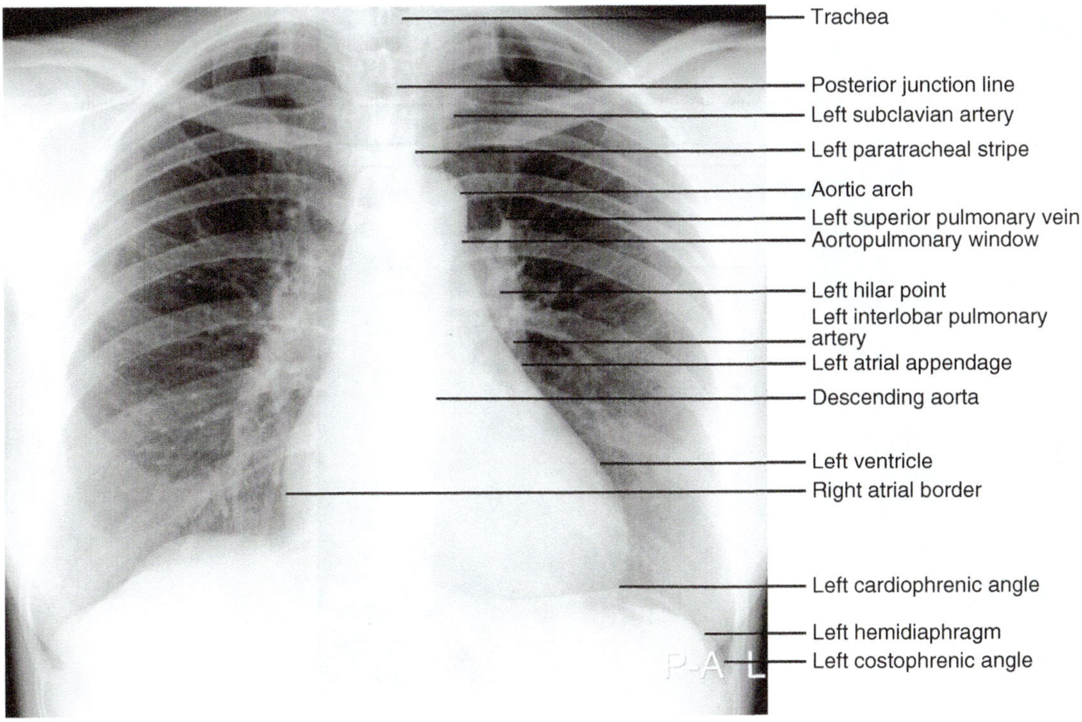

Fig. 17.2 Posteroanterior view

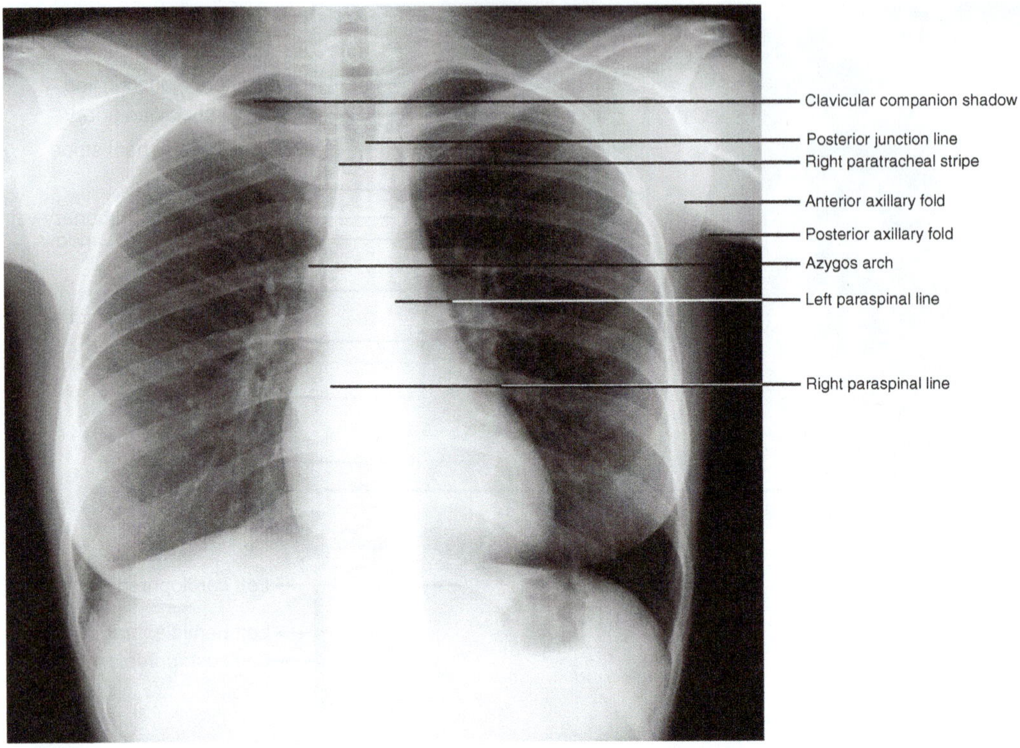

Fig. 17.3 Posteroanterior view of chest demonstrating lines and folds

Pearl

The anterior junction line should not be visible above the clavicles as these represent the upper borders of the anterior mediastinum and can therefore be distinguished from the posterior junction line which can extend above this.

The azygos arch is best identified by its position at the level of the right tracheobronchial angle.

Lateral Chest

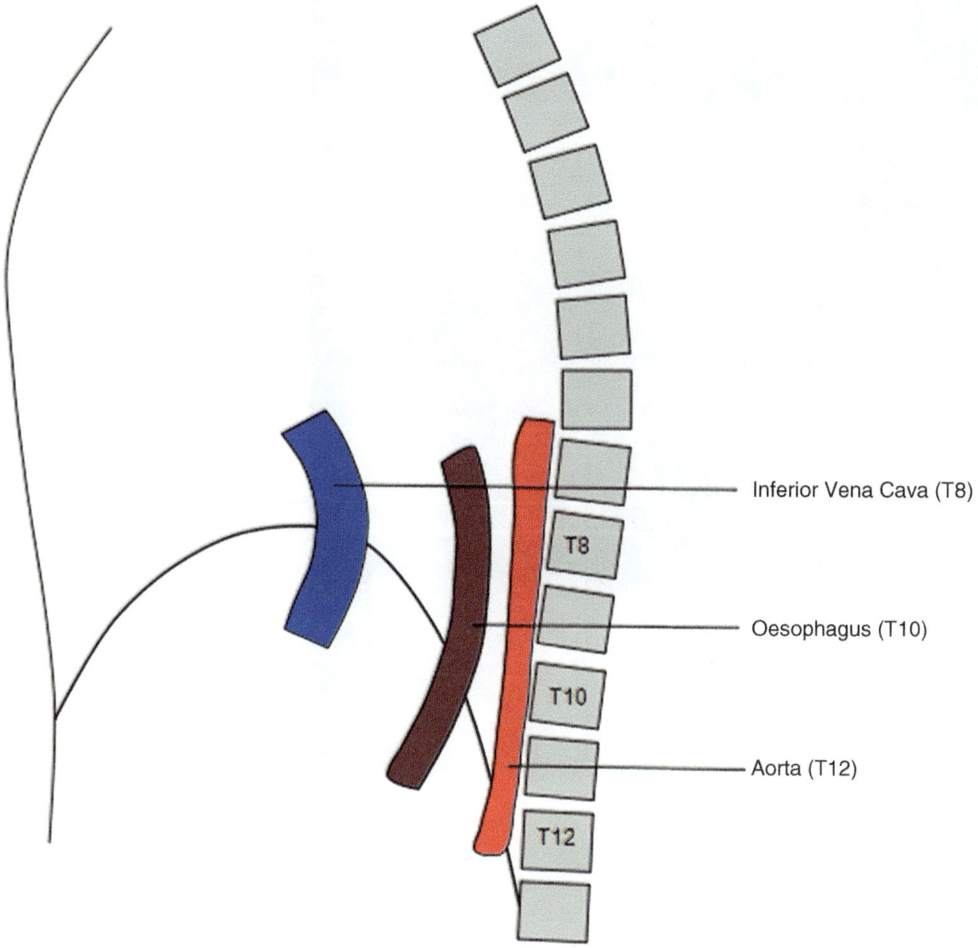

Fig. 17.4 Sagittal representation of the level at which structures traverse the diaphragm

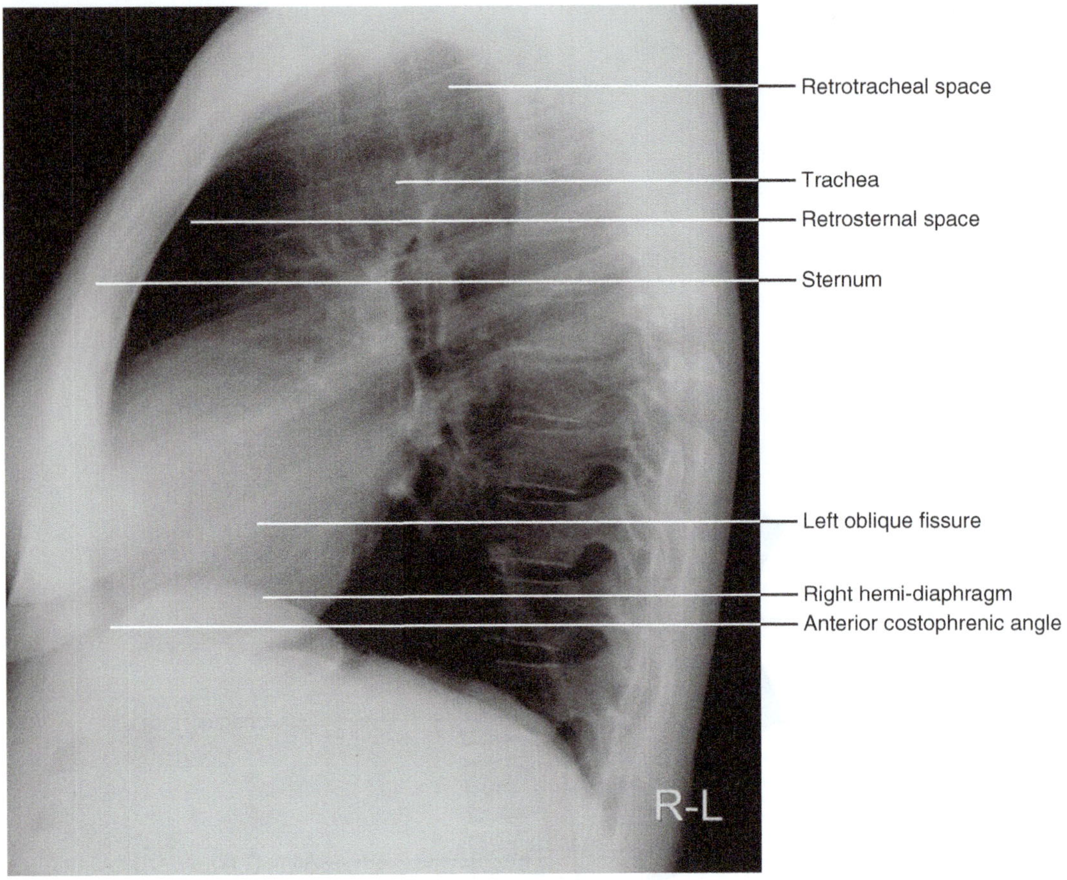

Fig. 17.5 Lateral view of chest

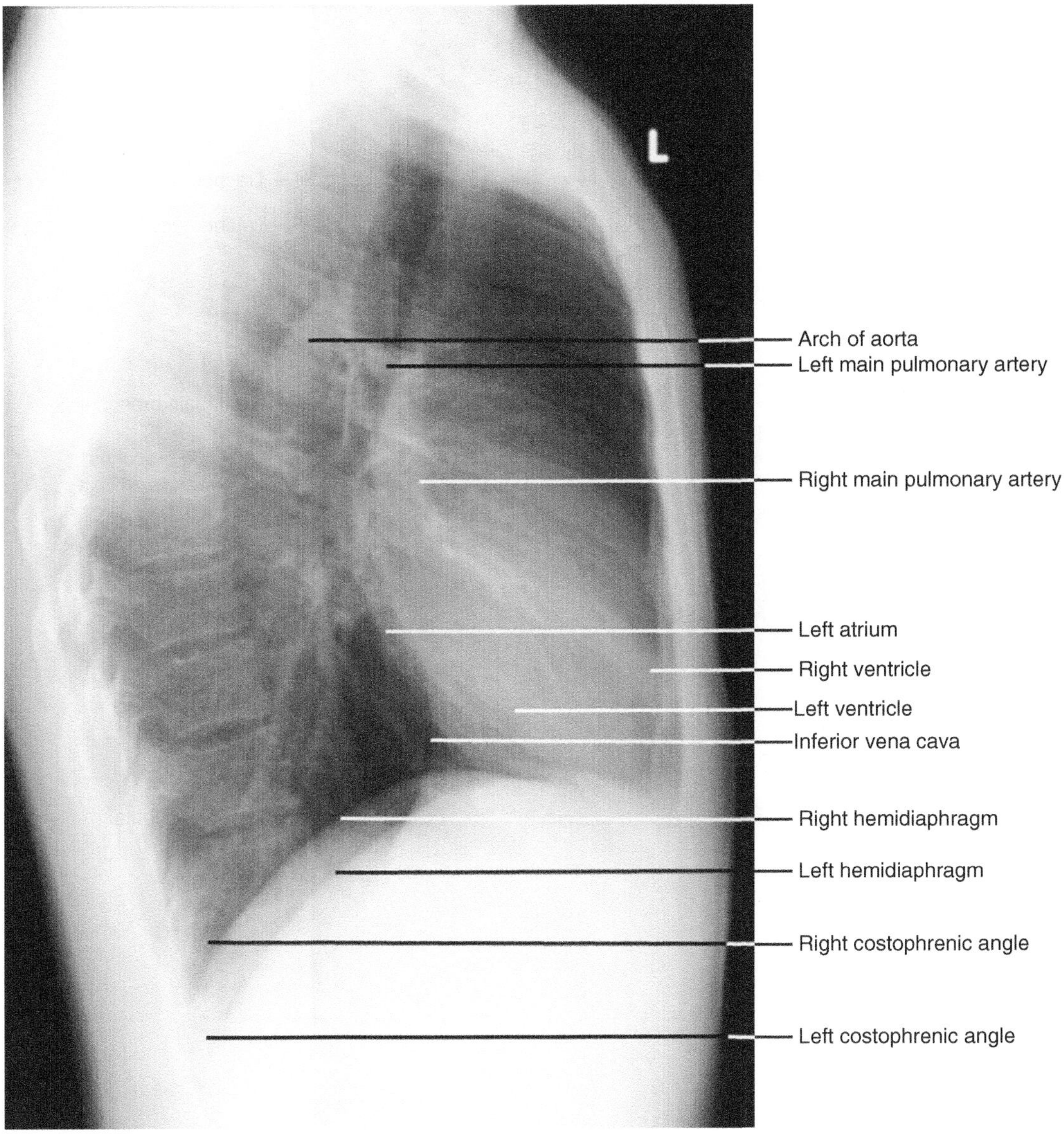

Fig. 17.6 Lateral view of the chest

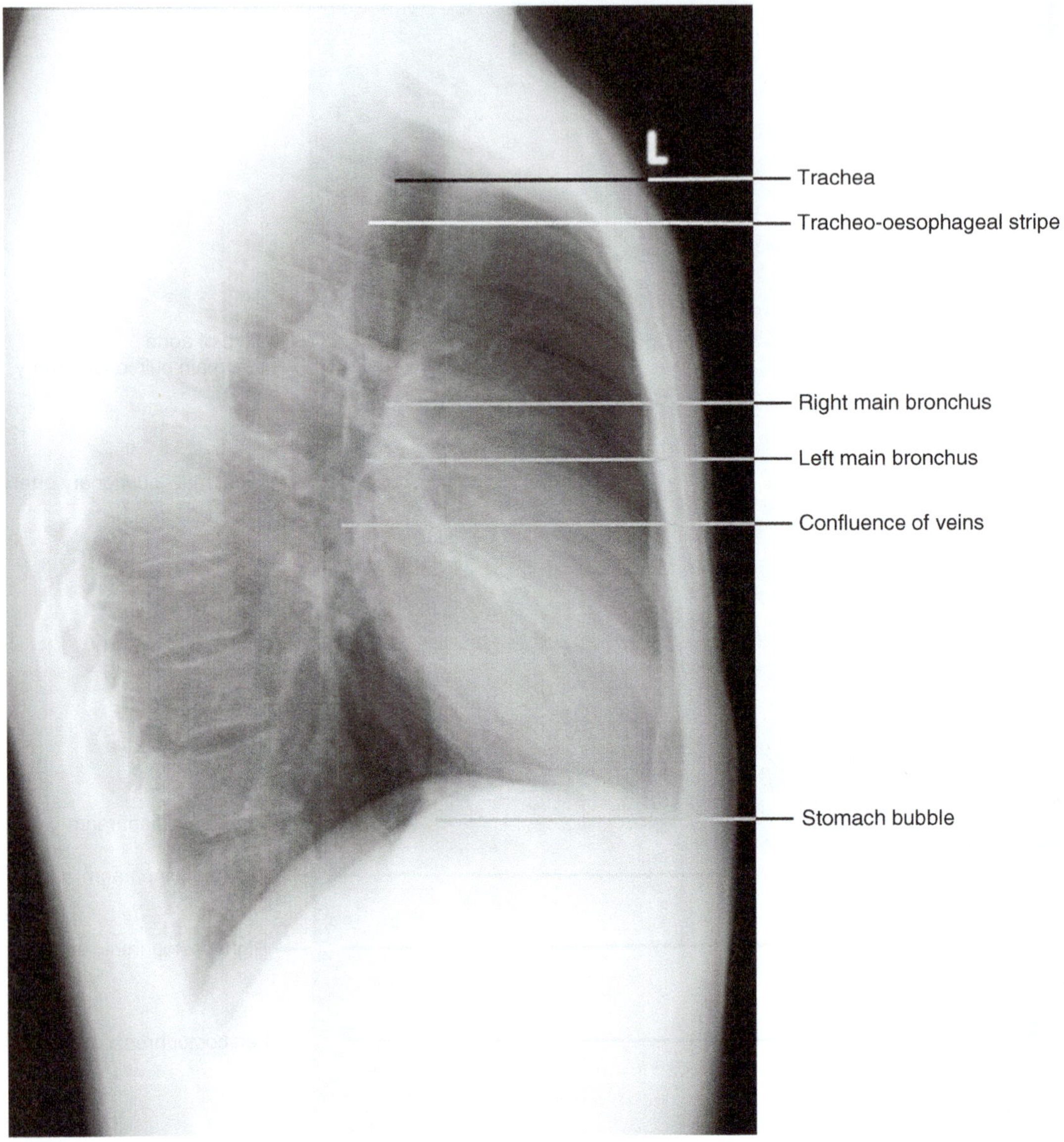

Fig. 17.7 Lateral view of the chest

Additional Views

- **Decubitus view** used to demonstrate pleural effusions, air-fluid levels, and intra-cavitatory bodies.
- **Apical view** for improved visualization of the lung apices.
- **Lordotic view** for demonstration of right middle lobe pathology, including collapse and interlobular effusion.
- **Oblique view** for the retrocardiac area, ribs, and chest wall, including pleural plaques.
- **Paired inspiratory and expiratory views** to demonstrate air trapping and diaphragmatic movement.

Pearl: Distinguishing diaphragm and major fissures

Right diaphragm
Reaches anterior chest
Stomach may be above or below it
Left diaphragm
Falls short of anterior chest and merges with the cardiac density
Stomach always below it
Right major fissure
Reaches right diaphragm near anterior chest
Left major fissure
Falls short of anterior chest and joins left diaphragm

Pearl: Normal measurements

Anterior junction line
Up to 1 mm (seen only below the clavicles)
Azygos arch
Less than 1 cm in the erect view
Decreases in size with inspiration and Valsalva
Carina
60°–75°
Horizontal fissure
Runs from hilum to sixth rib in anterior axillary line
Left hilar point
Usually 2.5 cm higher than right
Left oblique fissure
More vertical than right
Ends 5 cm behind anterior costophrenic angle
Left paraspinal line
Less than 10 mm
Posterior junction line
Up to 2 mm (extends above level of the clavicles)
Right descending pulmonary artery
10–16 mm in men
9–15 mm in women

Ultrasound

Chest Wall

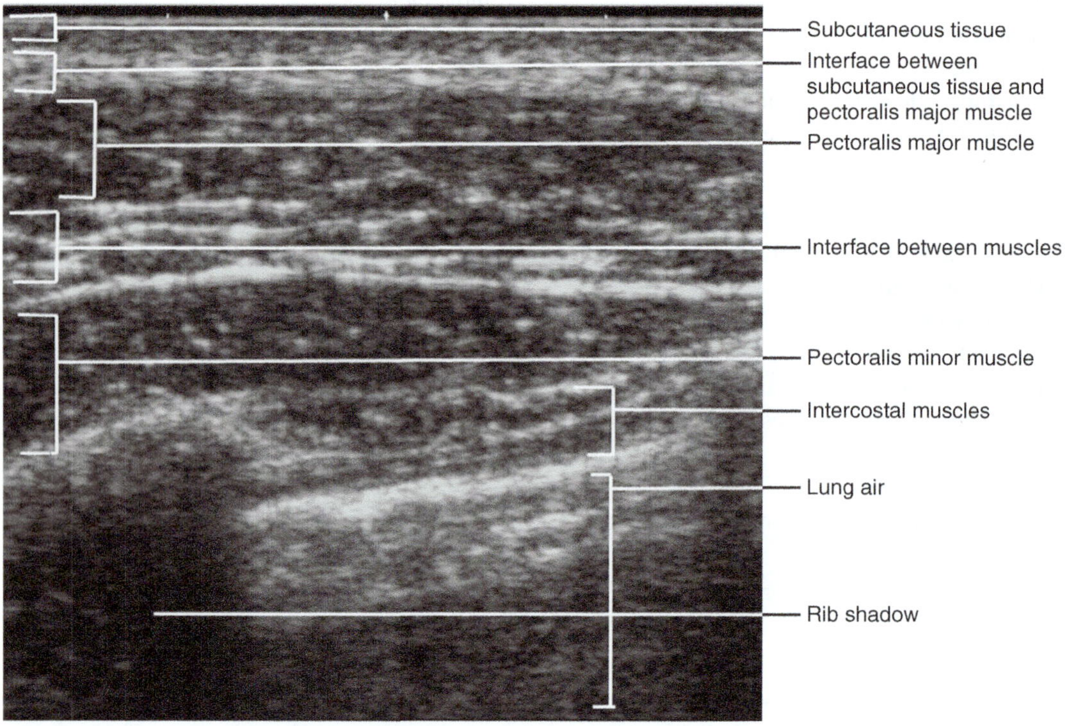

Fig. 17.8 Transverse section through the chest wall

Diaphragm

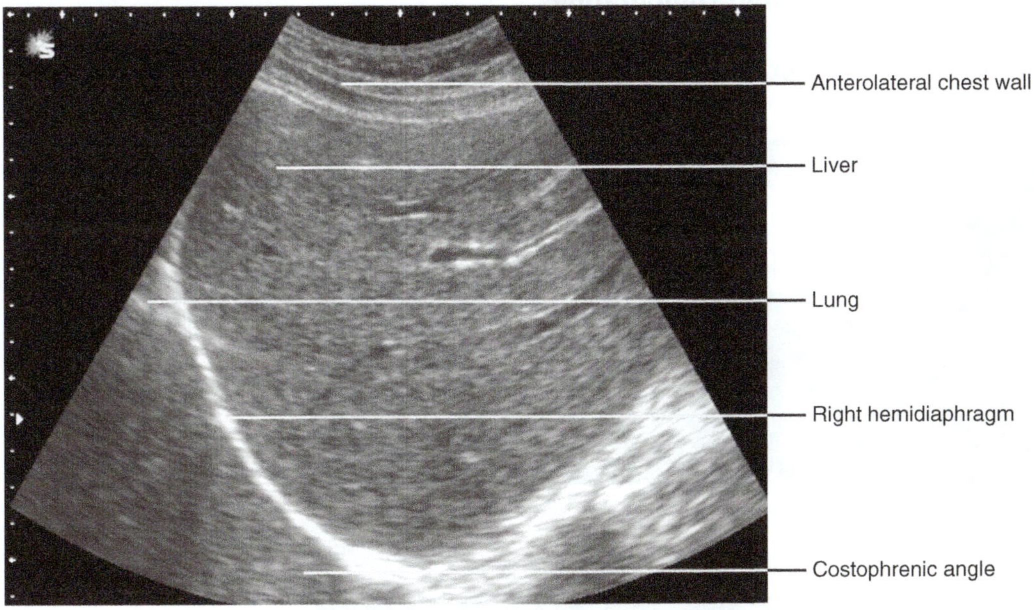

Fig. 17.9 Longitudinal section through the right hemidiaphragm

Computed Tomography

Cross-Sectional Computed Tomography

Axial Computed Tomography Mediastinal Window from Superior to Inferior

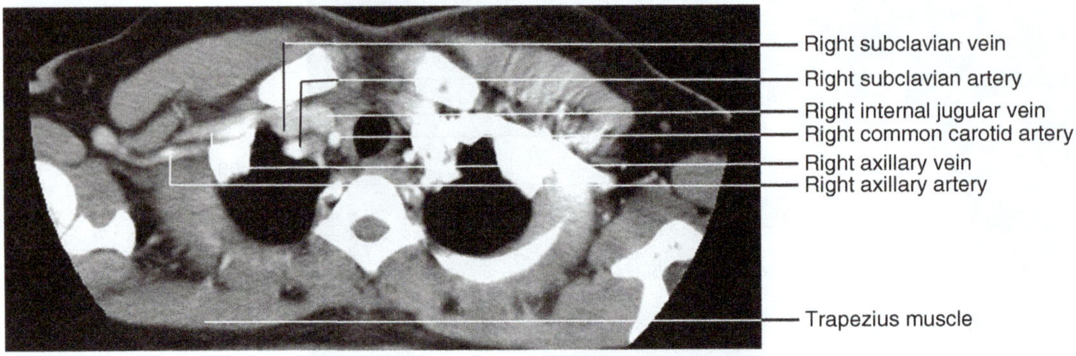

Fig. 17.10 Axial post contrast

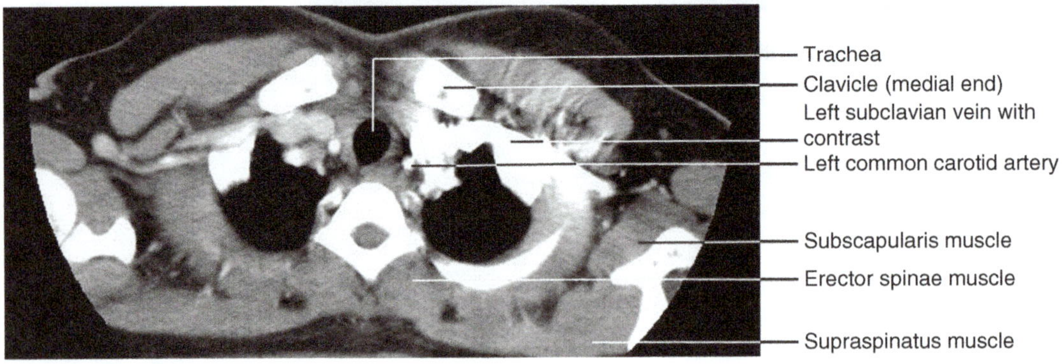

Fig. 17.11 Axial post contrast

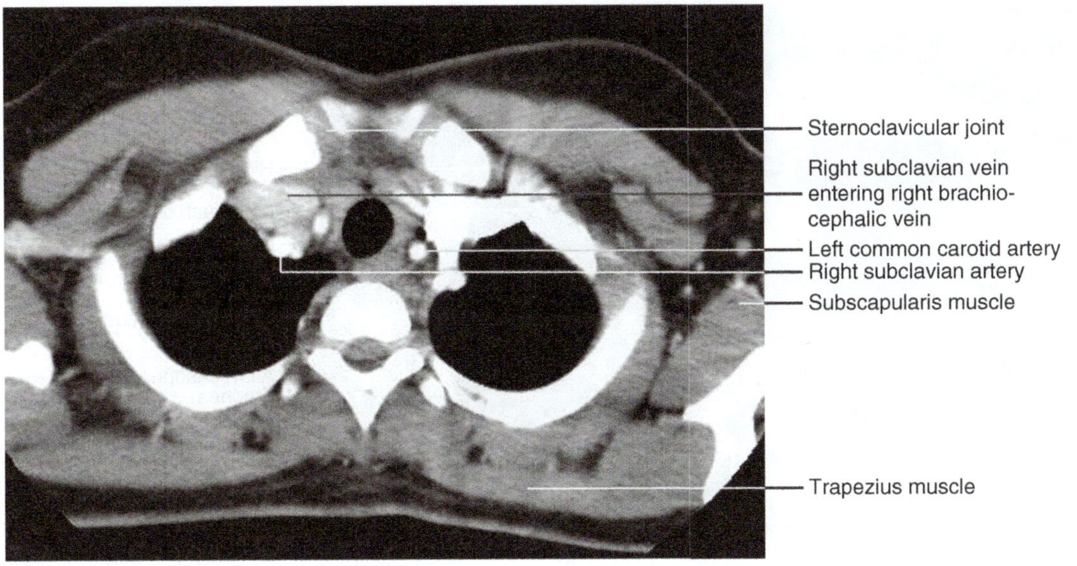

Fig. 17.12 Axial post contrast

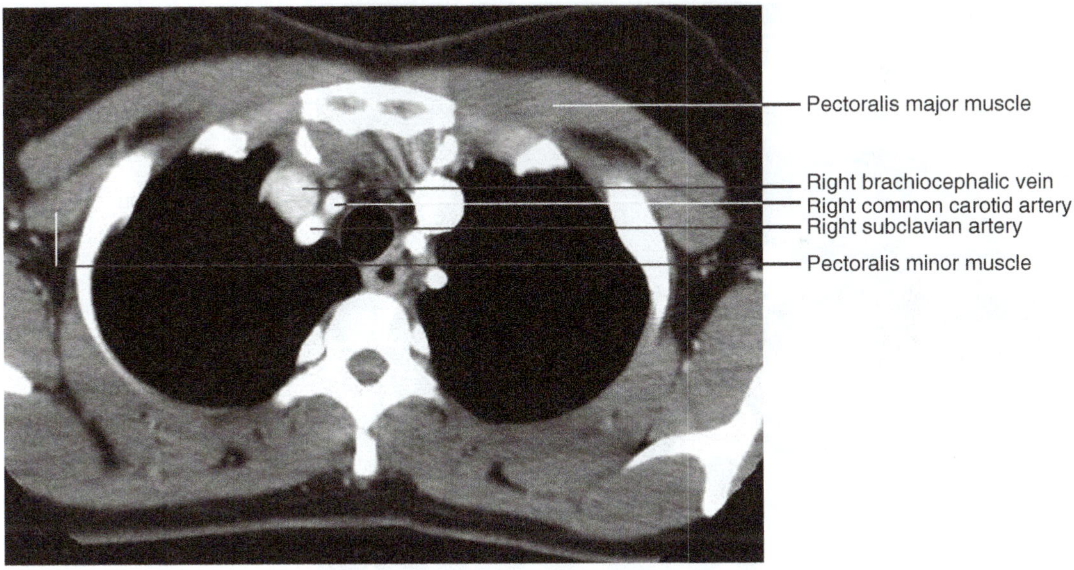

Fig. 17.13 Axial post contrast

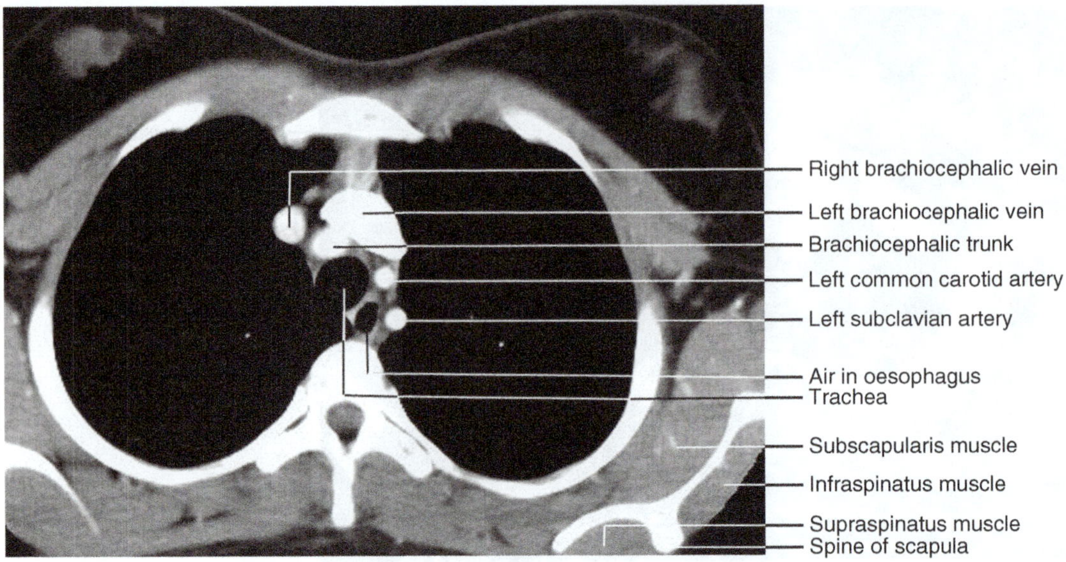

Fig. 17.14 Axial post contrast

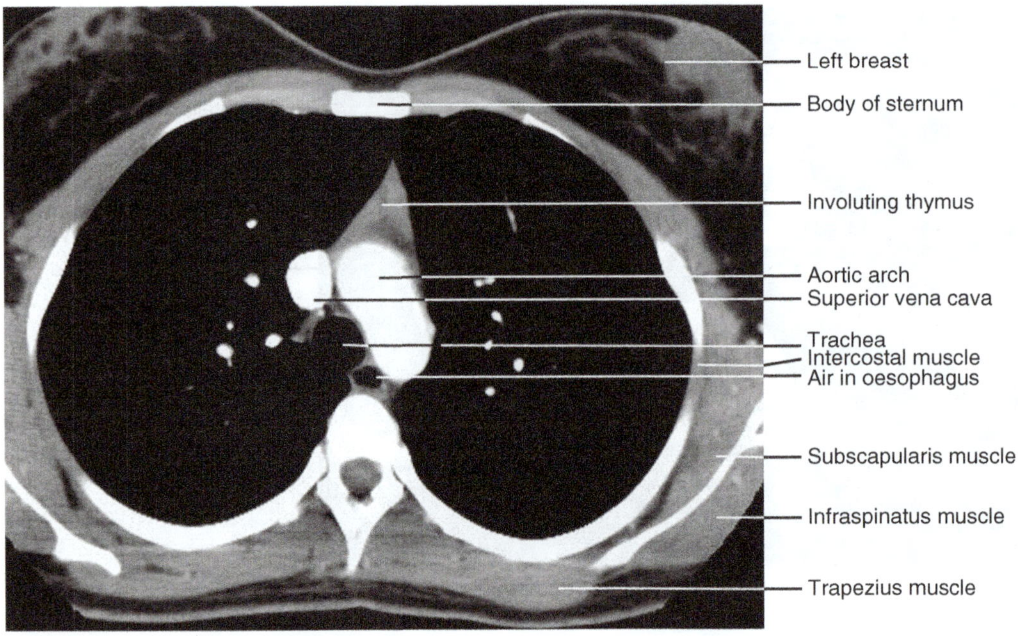

Fig. 17.15 Axial post contrast

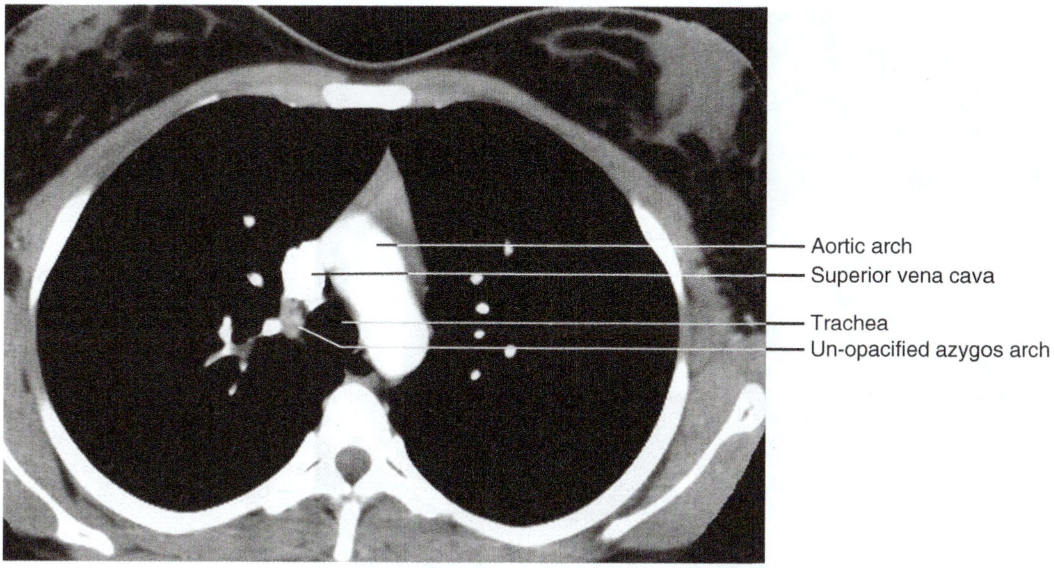

Fig. 17.16 Axial post contrast

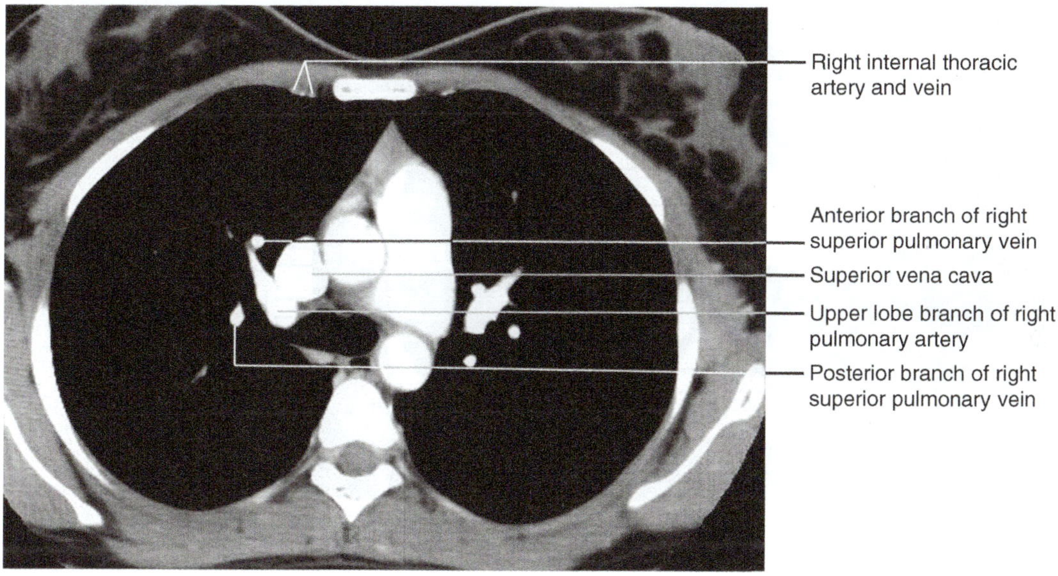

Fig. 17.17 Axial post contrast

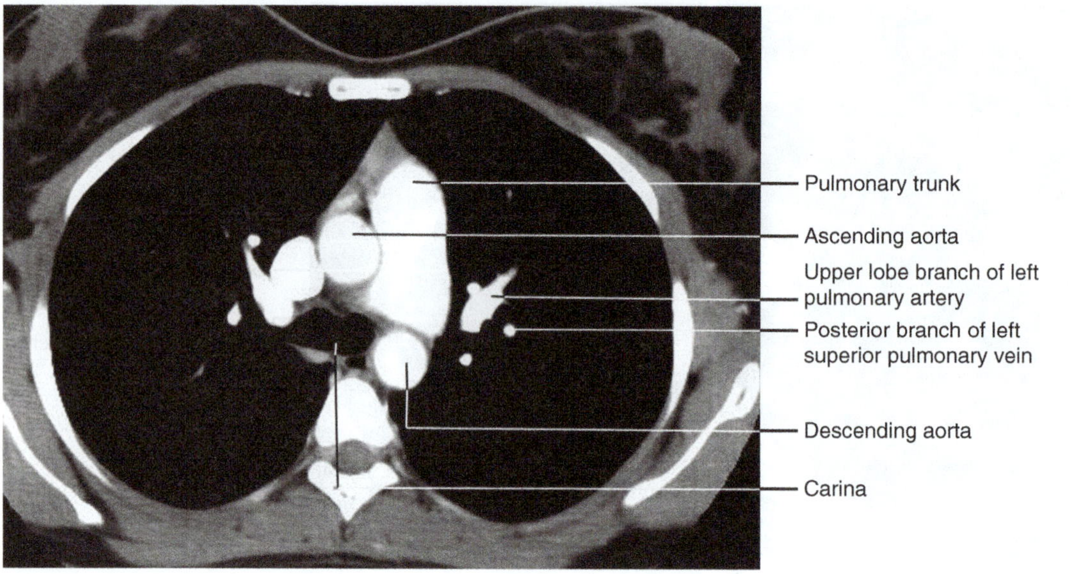

Fig. 17.18 Axial post contrast

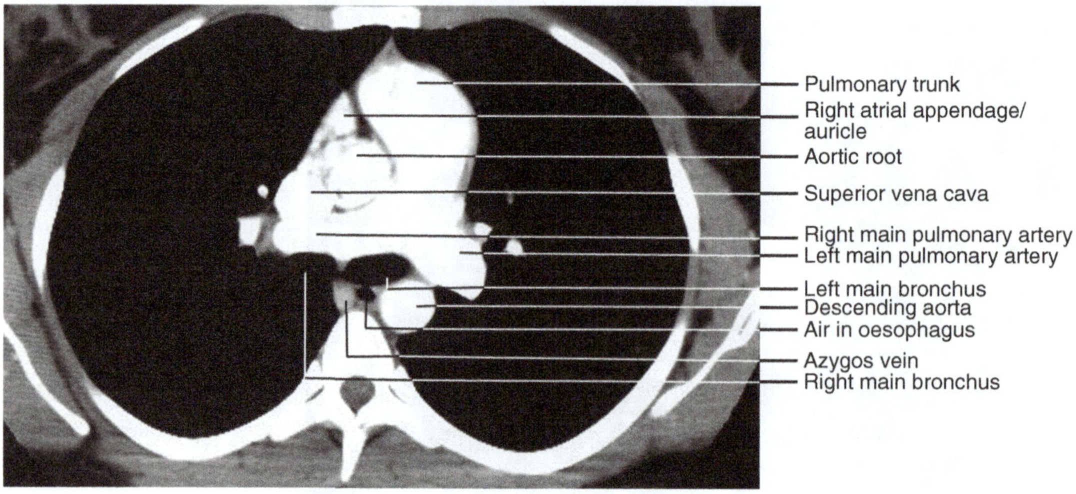

Fig. 17.19 Axial post contrast

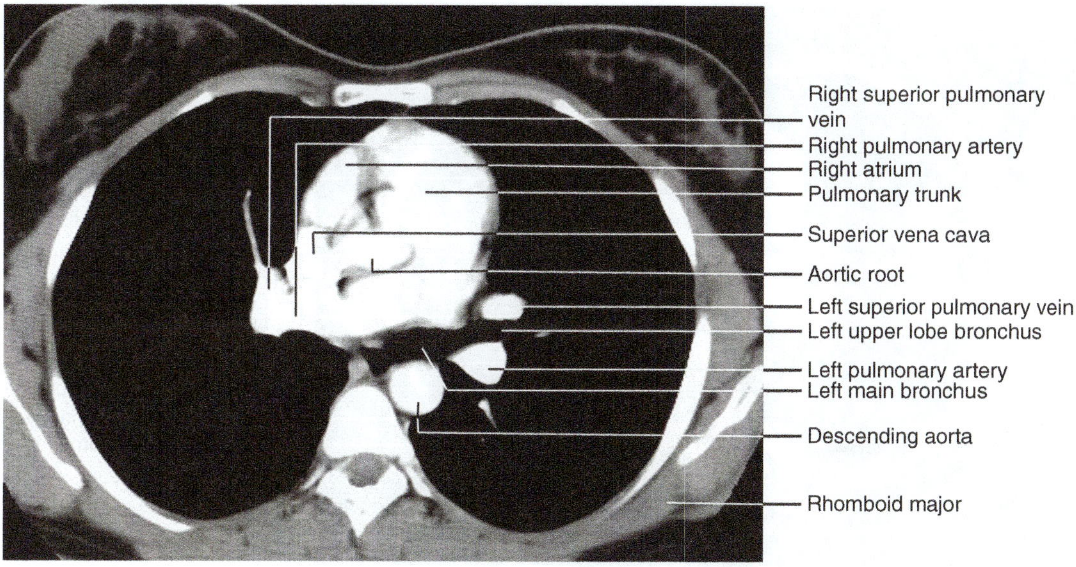

Fig. 17.20 Axial post contrast

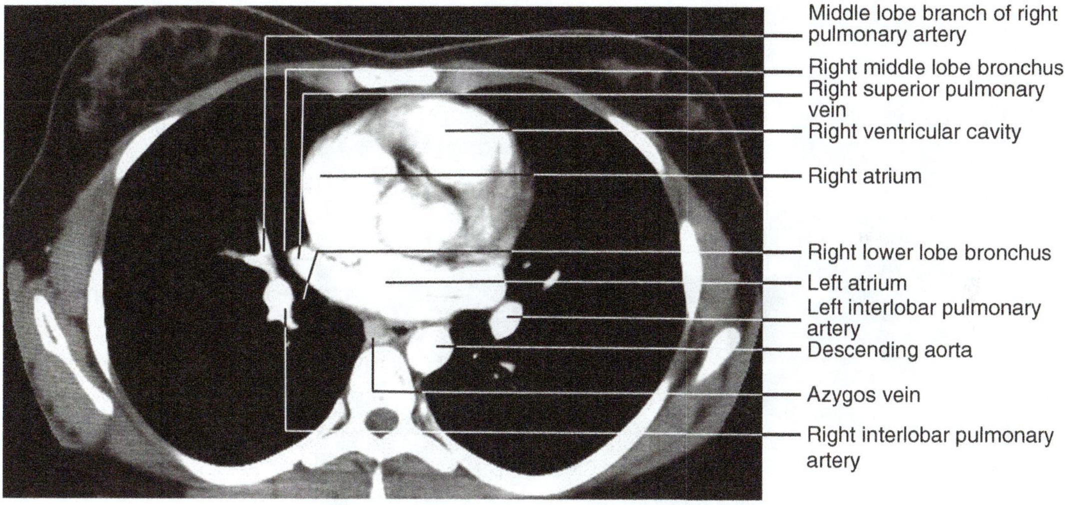

Fig. 17.21 Axial post contrast

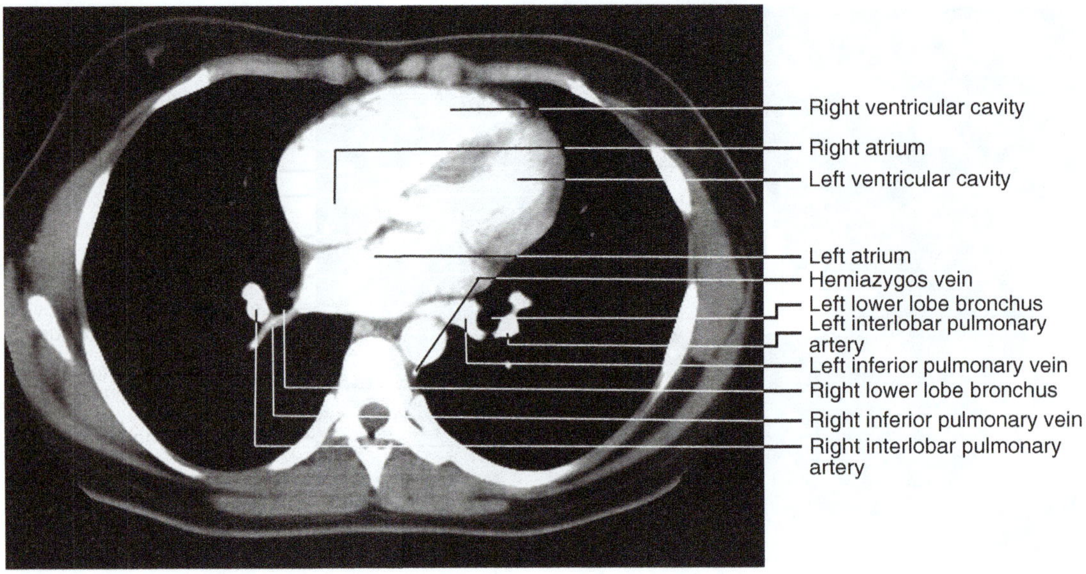

Fig. 17.22 Axial post contrast

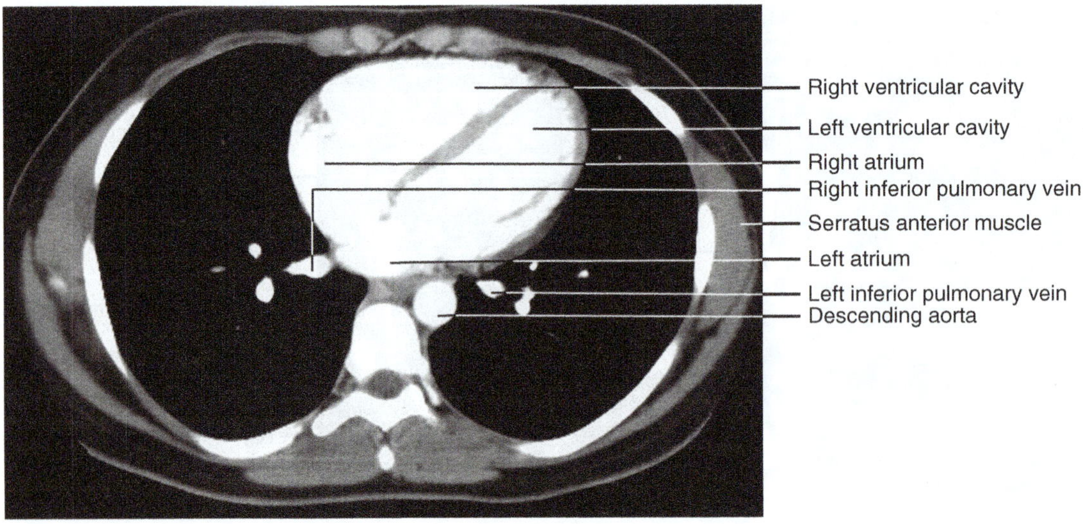

Fig. 17.23 Axial post contrast

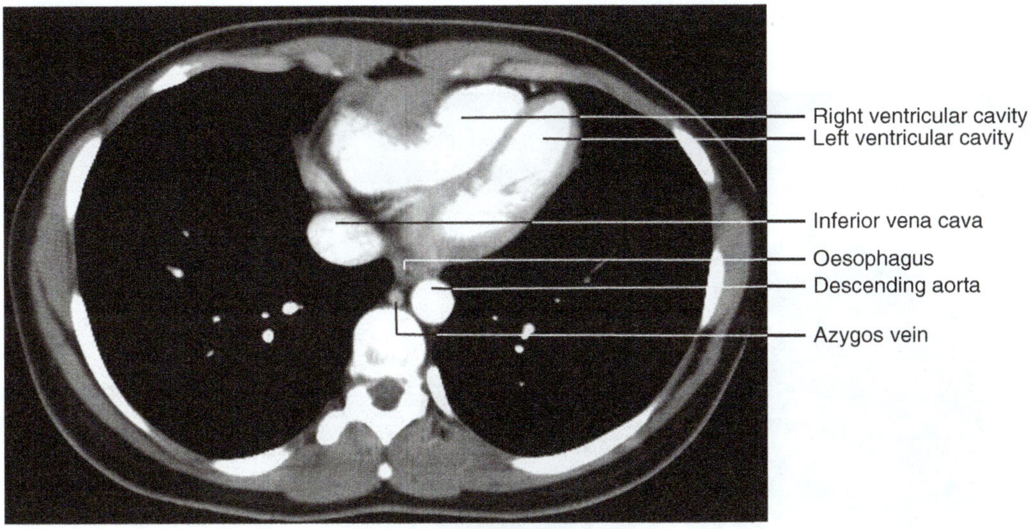

Fig. 17.24 Axial post contrast

Axial Computed Tomography Lung Window

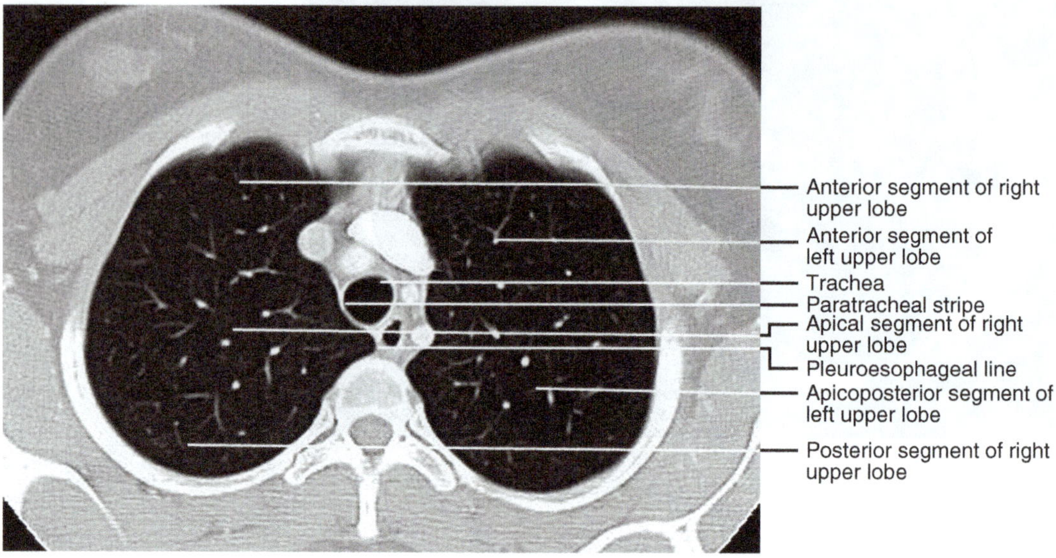

Fig. 17.25 Axial post contrast

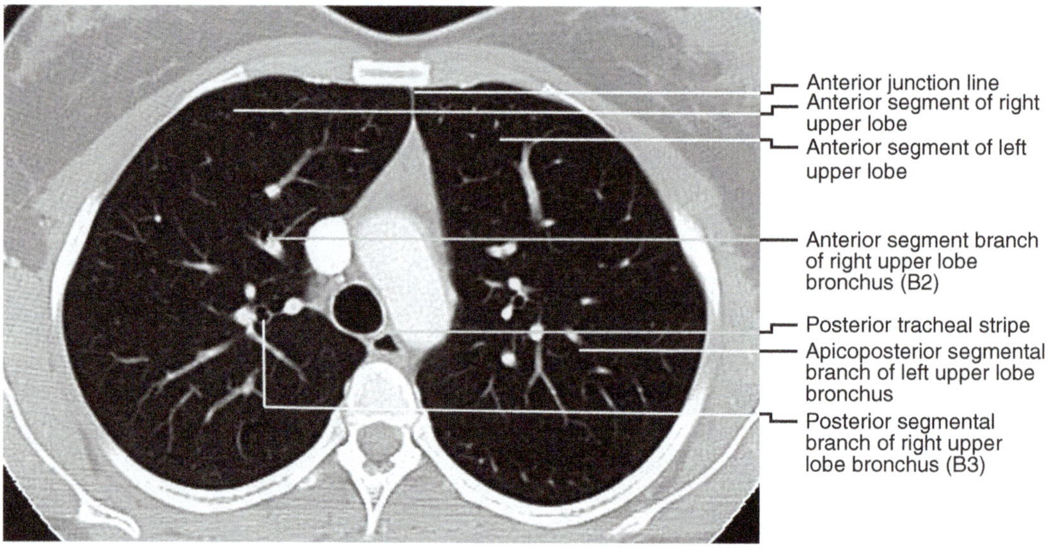

Fig. 17.26 Axial post contrast

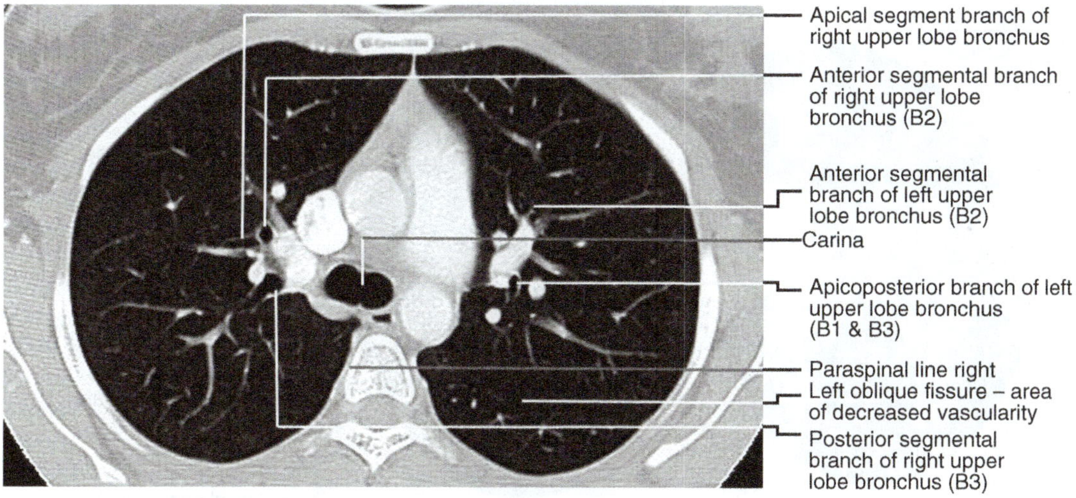

Fig. 17.27 Axial post contrast

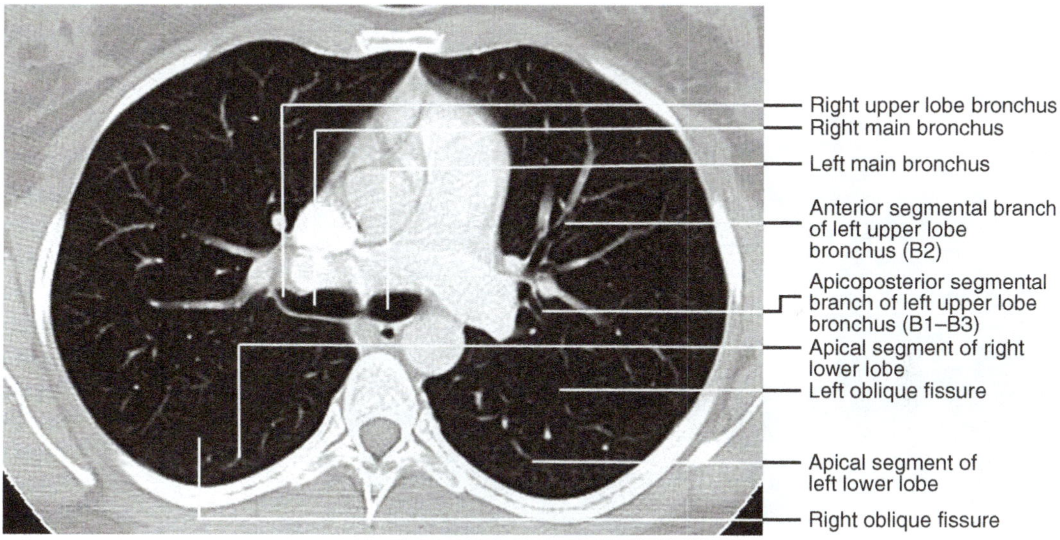

Fig. 17.28 Axial post contrast

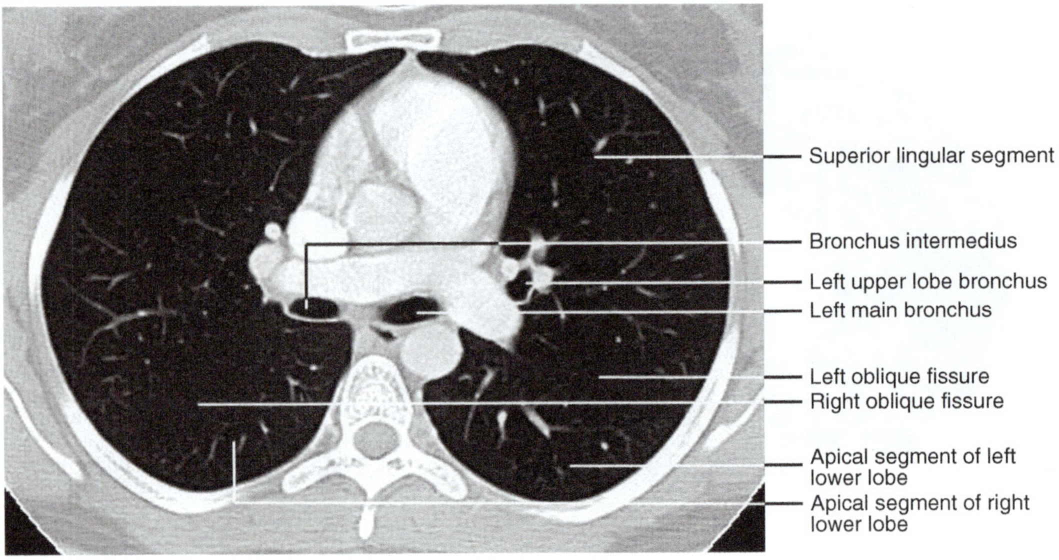

Fig. 17.29 Axial post contrast

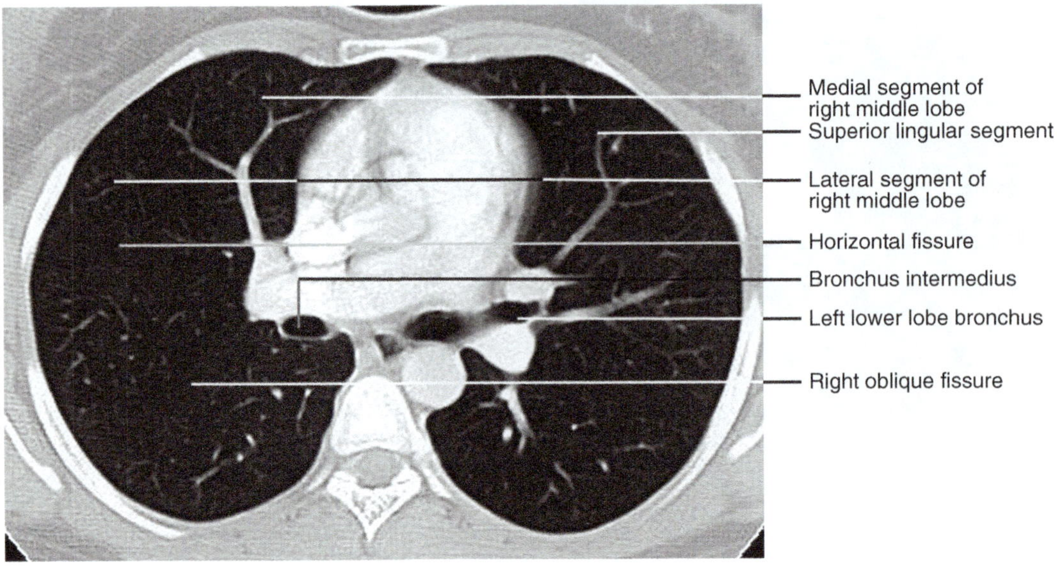

Fig. 17.30 Axial post contrast

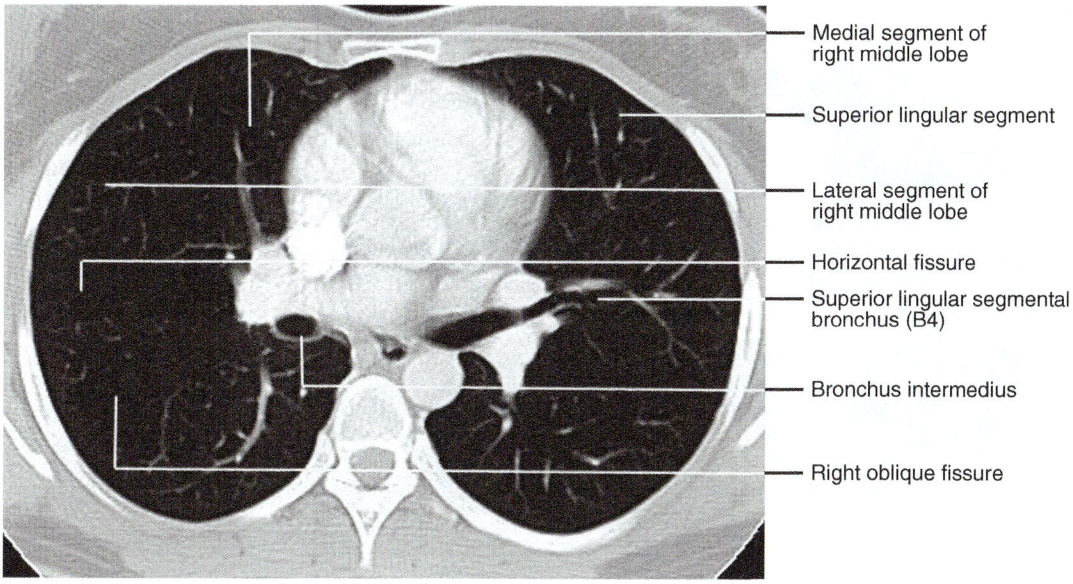

Fig. 17.31 Axial post contrast

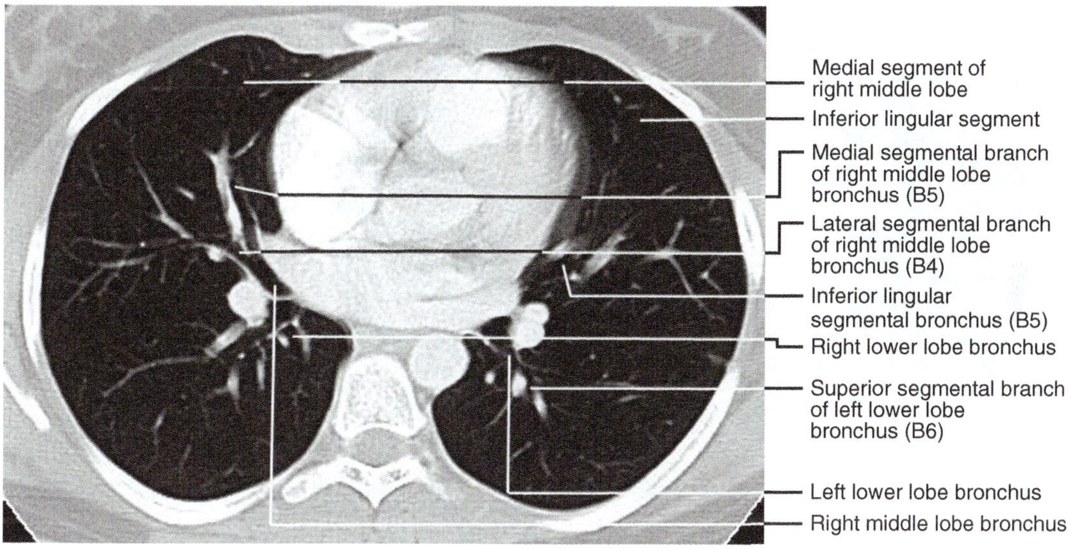

Fig. 17.32 Axial post contrast

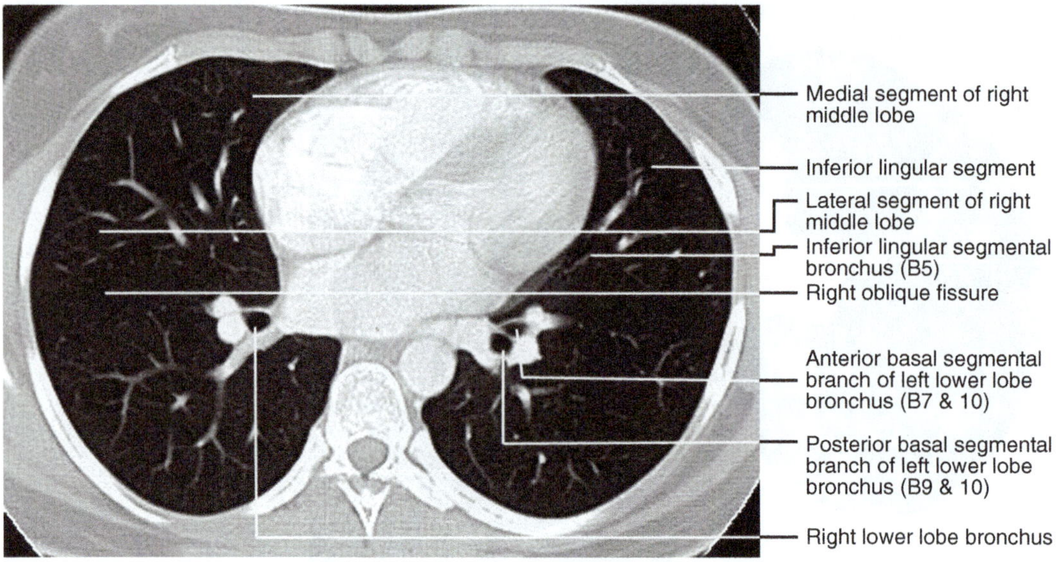

Fig. 17.33 Axial post contrast

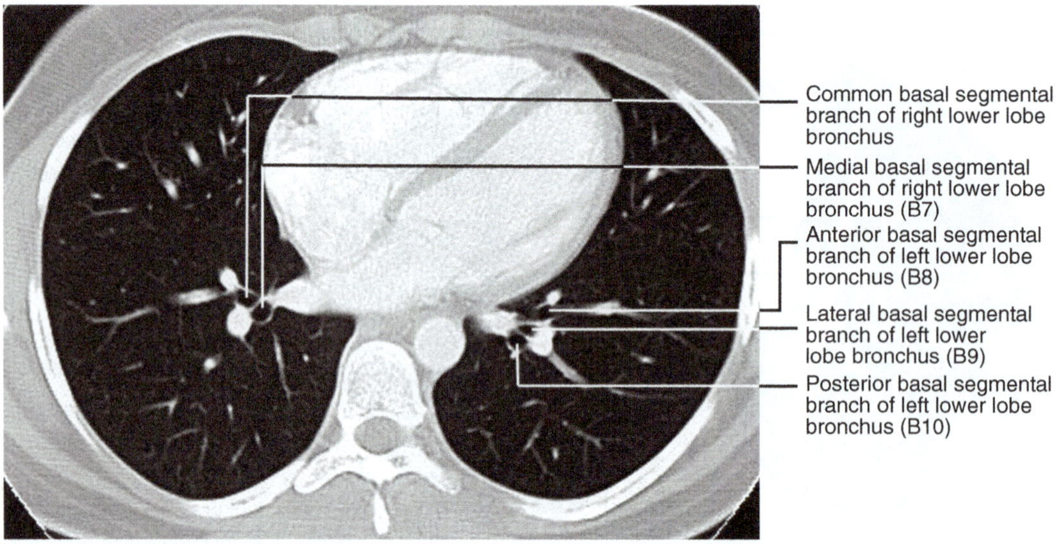

Fig. 17.34 Axial post contrast

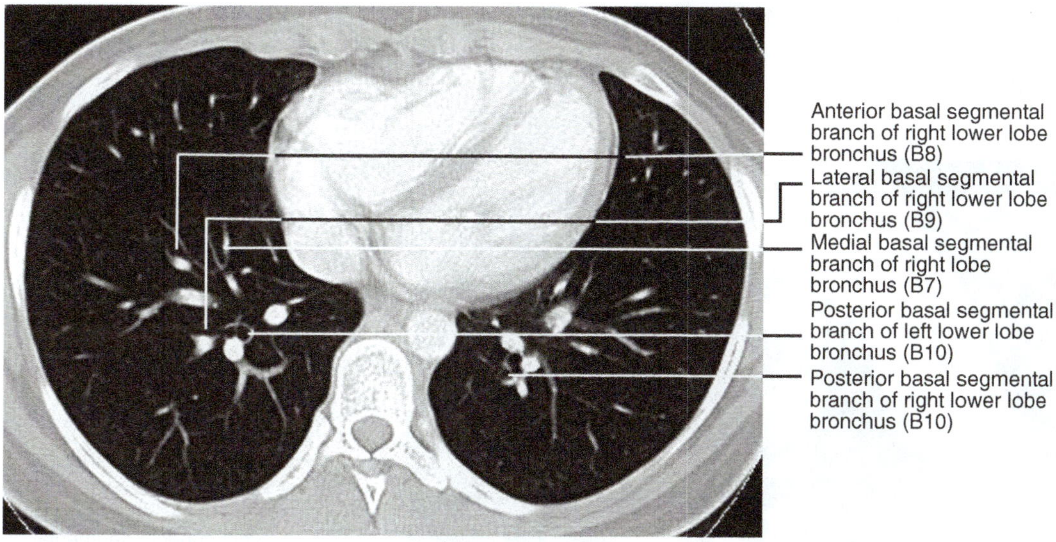

Fig. 17.35 Axial post contrast

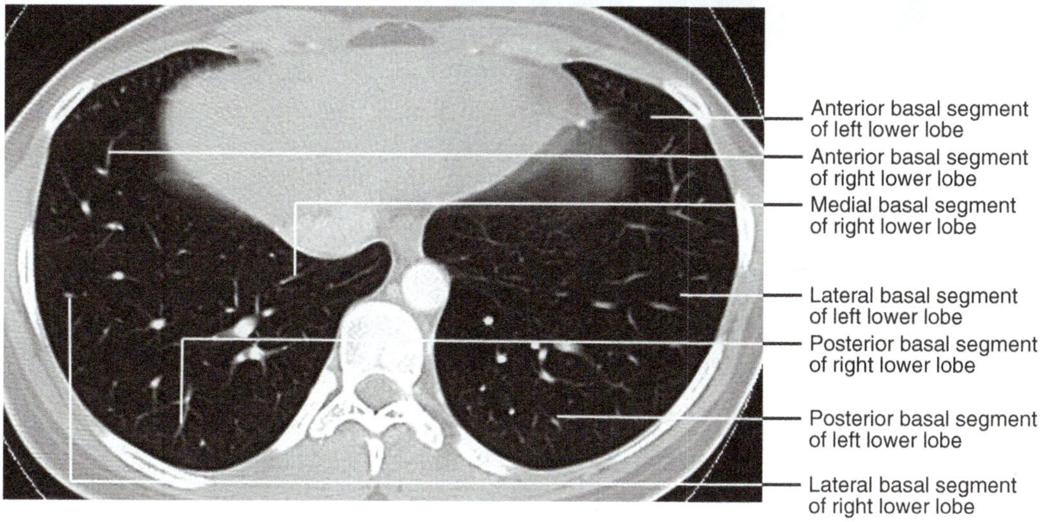

Fig. 17.36 Axial post contrast

Multiplanar Reconstruction Computed Tomography

Sagittal

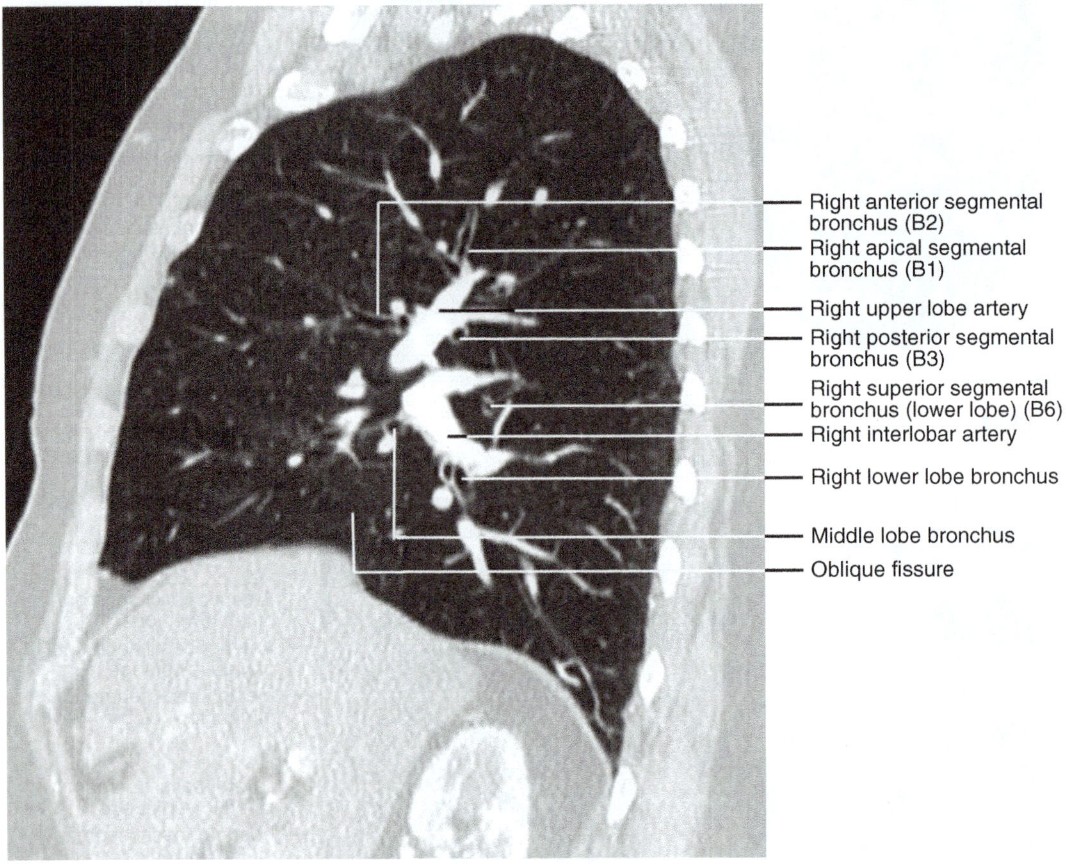

Fig. 17.37 Parasagittal MPR of right lung

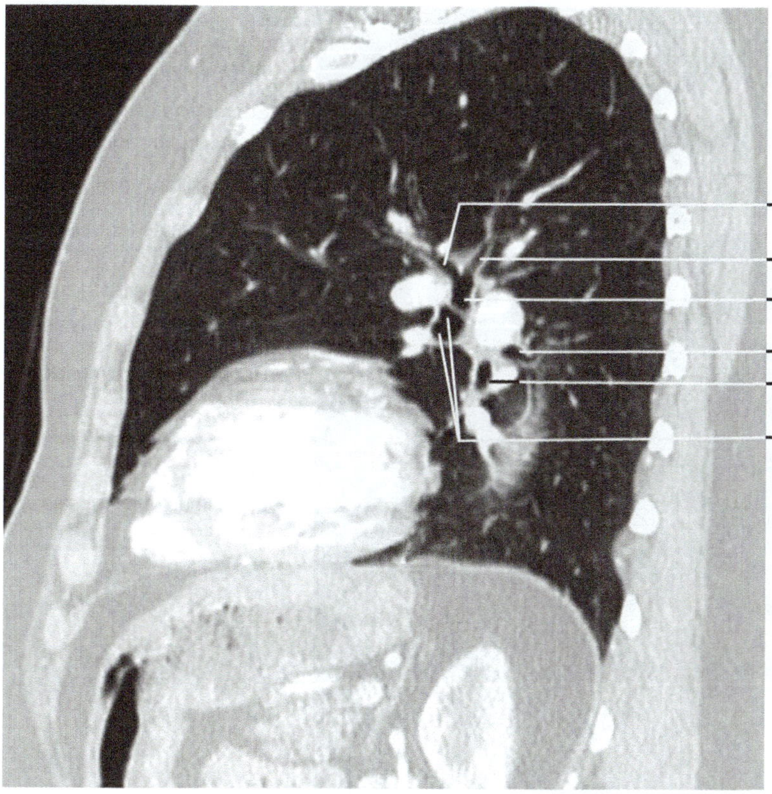

Fig. 17.38 Parasagittal MPR of left lung

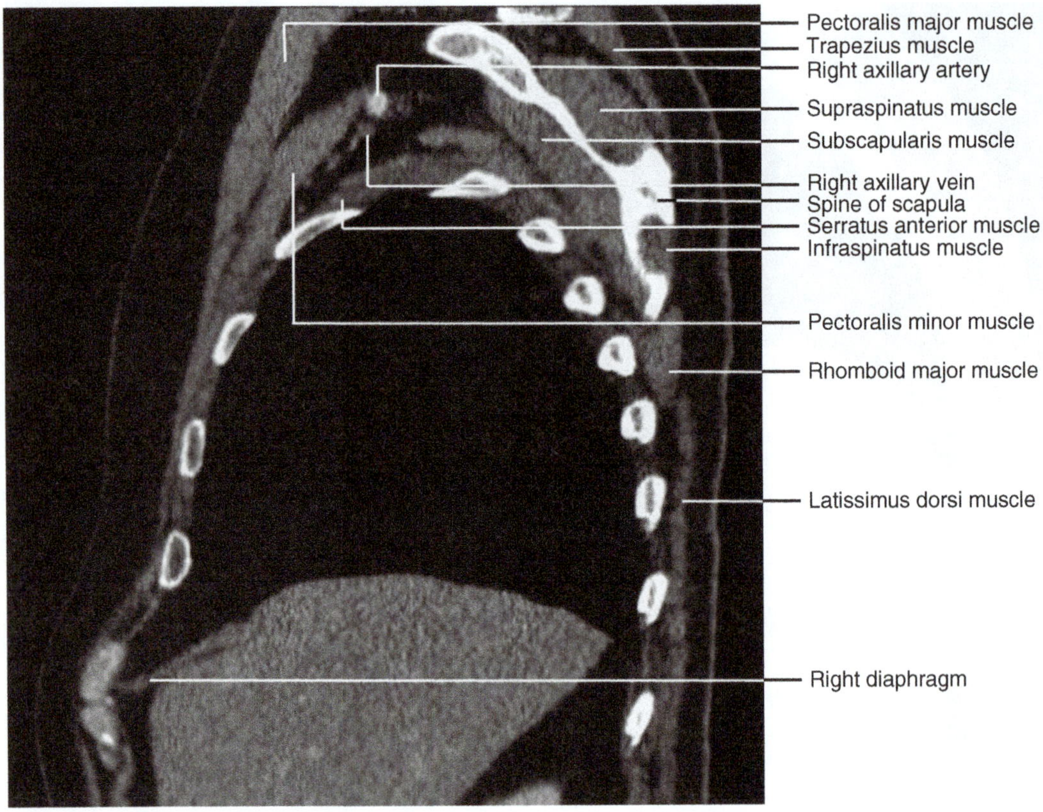

Fig. 17.39 Sagittal MPR: mediastinal window

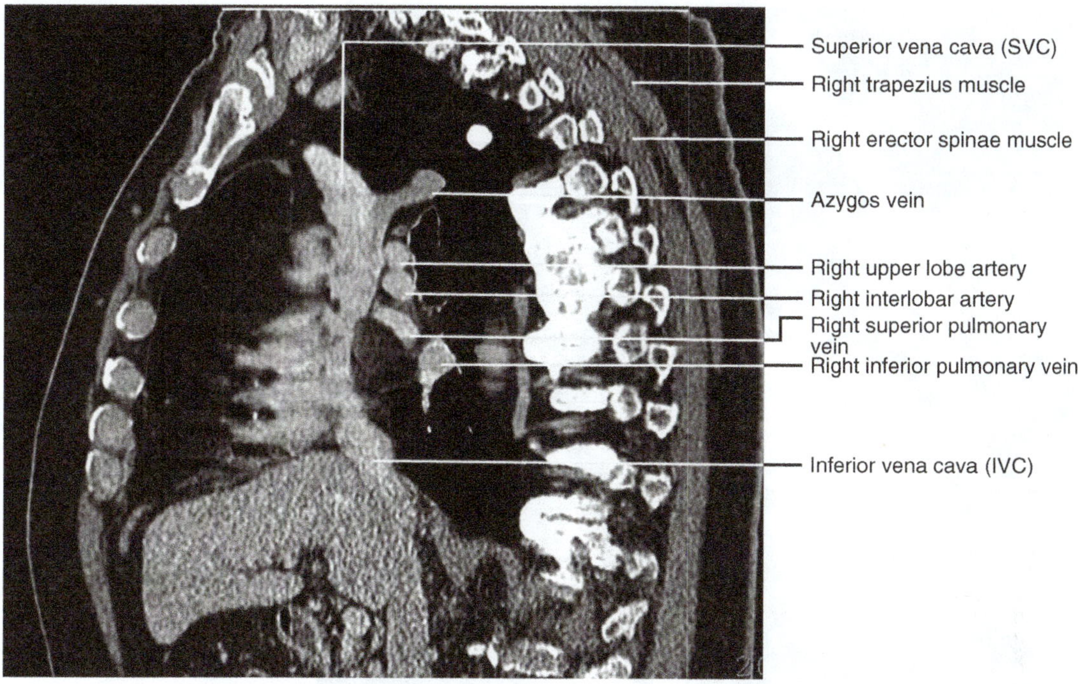

Fig. 17.40 Sagittal MPR: mediastinal window

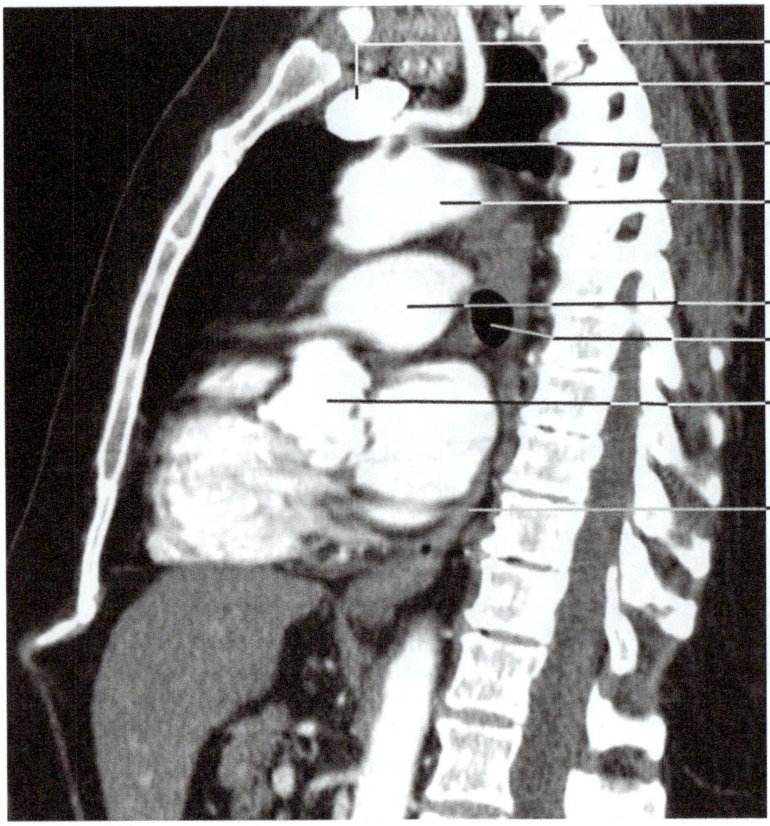

Fig. 17.41 Sagittal MPR: mediastinal window

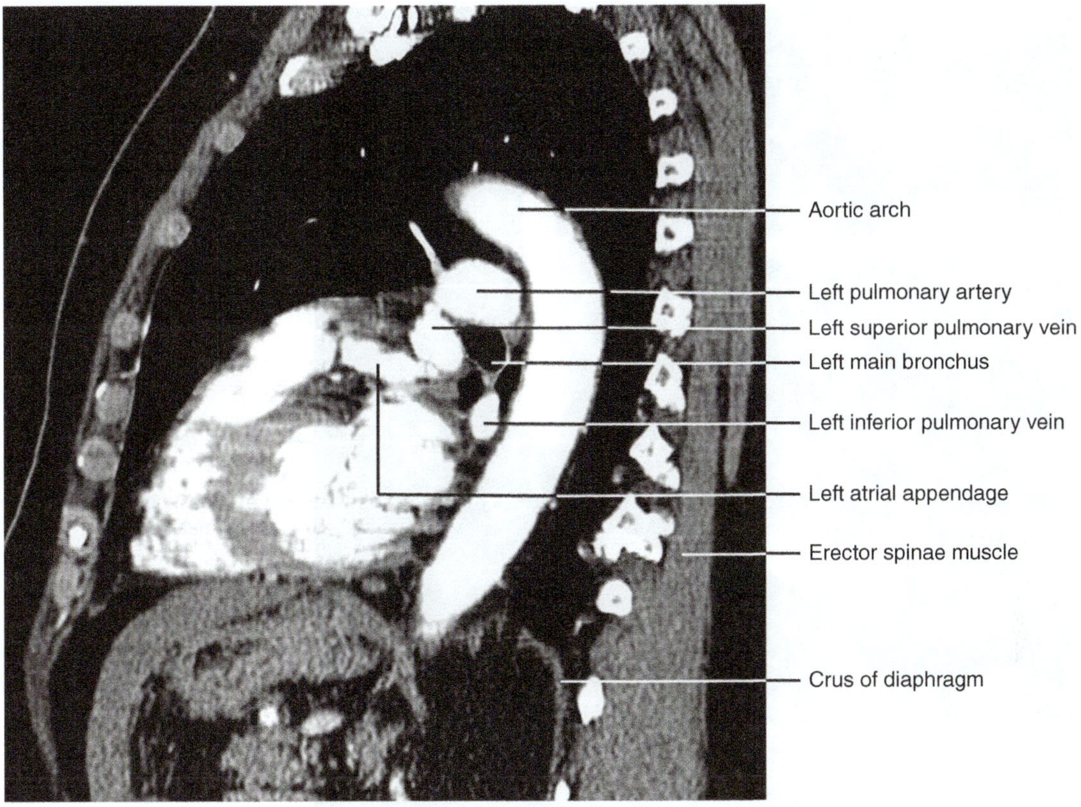

Fig. 17.42 Sagittal MPR: mediastinal window

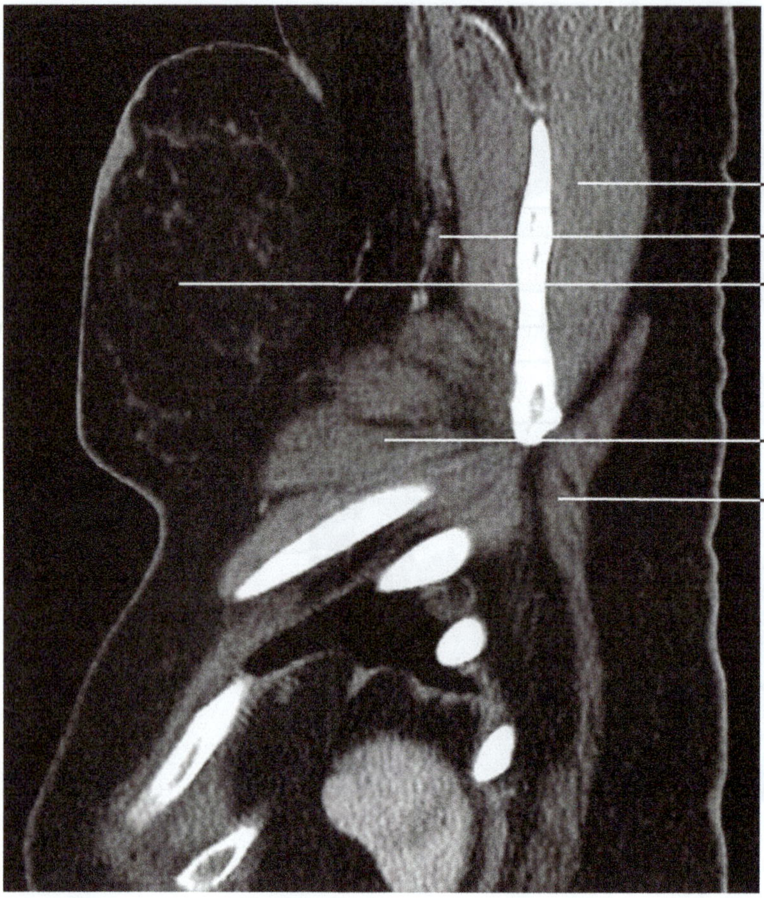

Fig. 17.43 Sagittal MPR: mediastinal window

Coronal

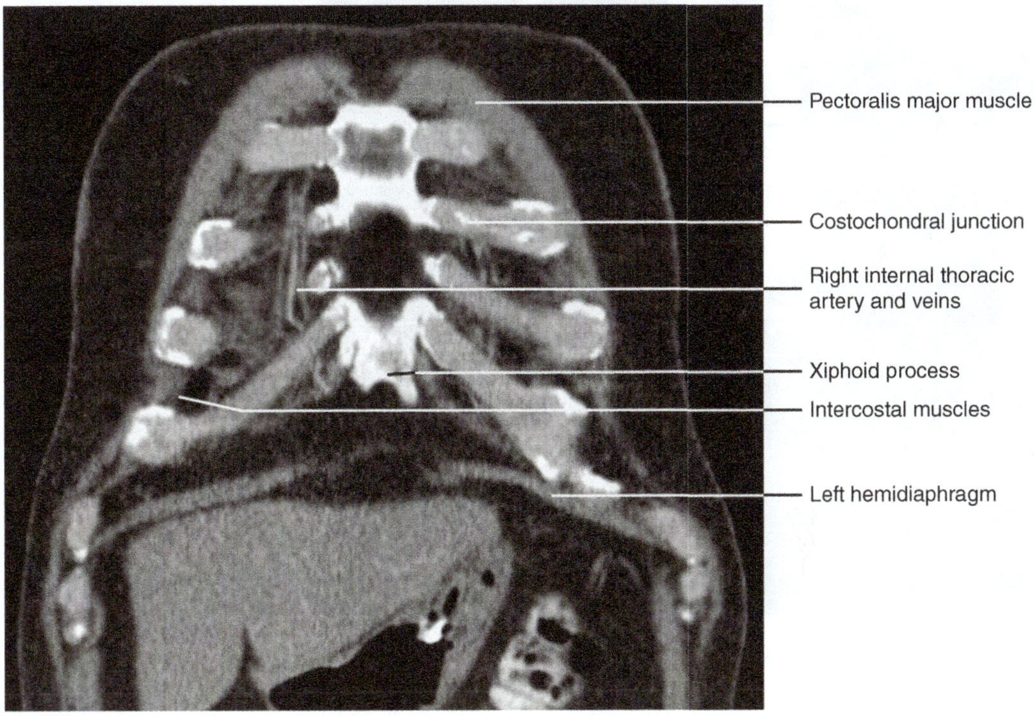

Fig. 17.44 Coronal MPR: mediastinal window

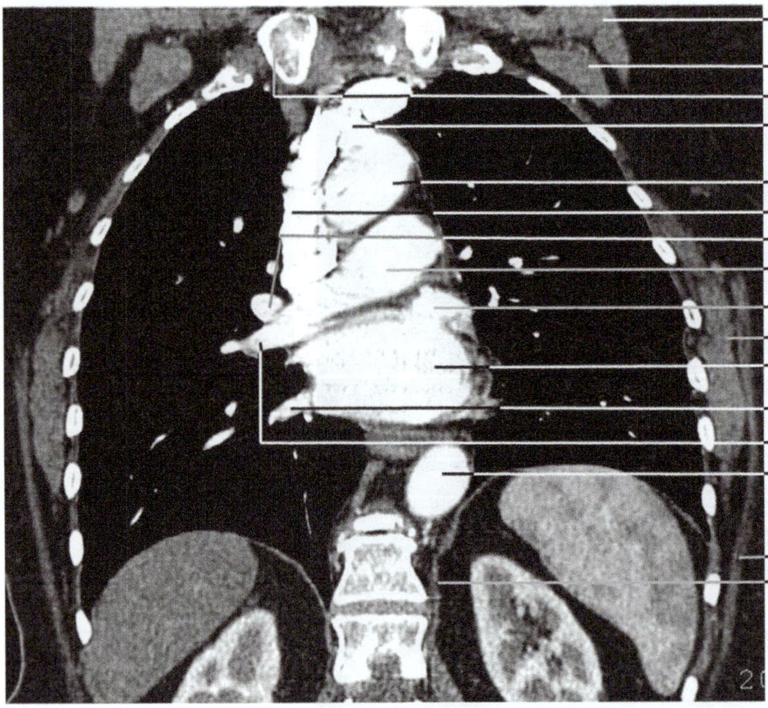

Fig. 17.45 Coronal MPR: mediastinal window

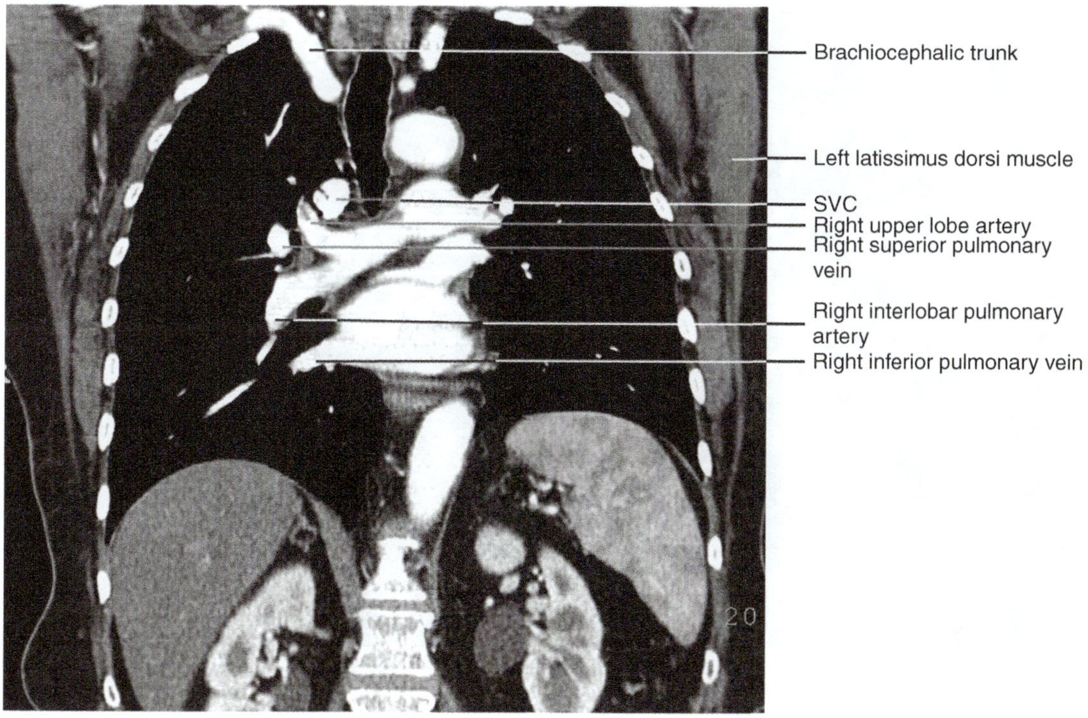

Fig. 17.46 Coronal MPR: mediastinal window

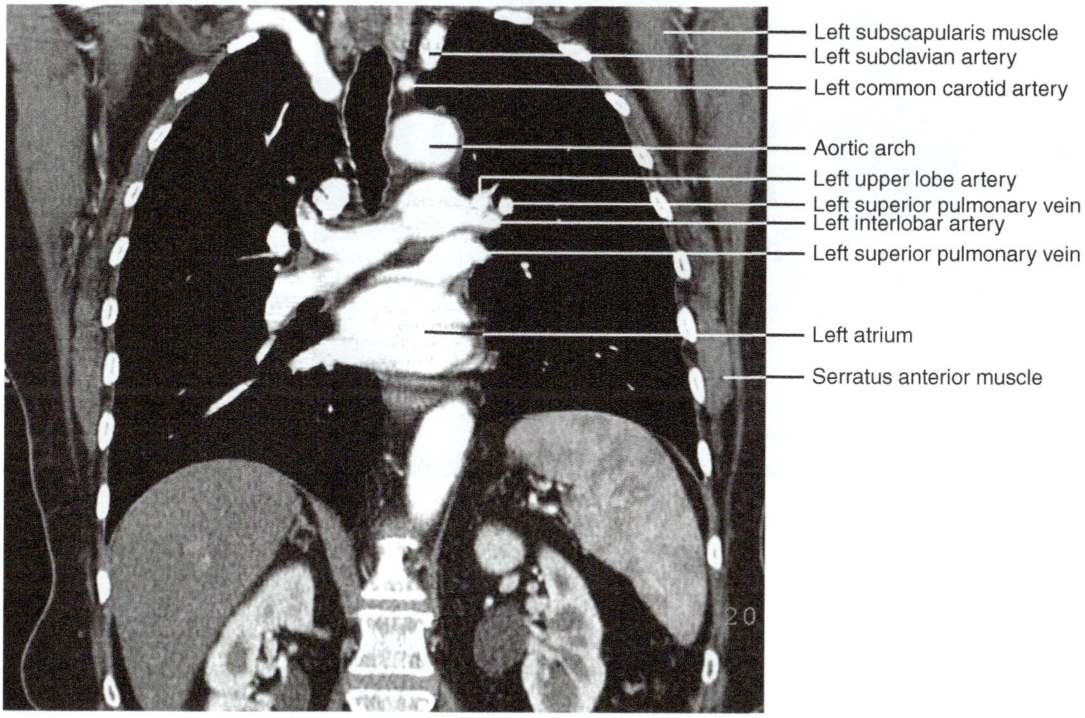

Fig. 17.47 Coronal MPR: mediastinal window

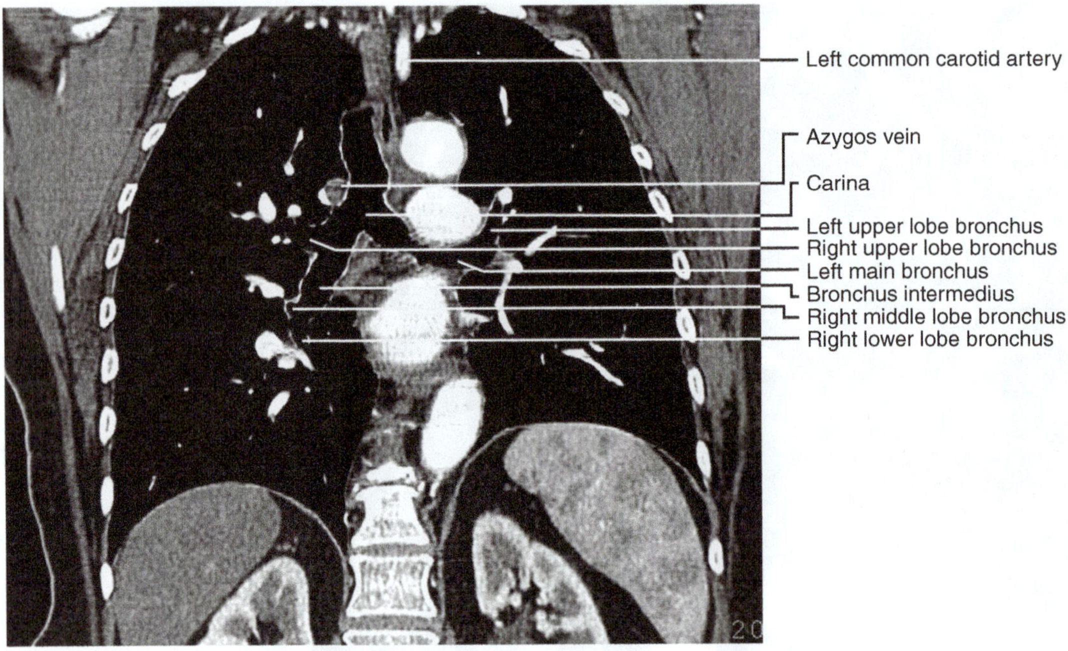

Fig. 17.48 Coronal MPR: mediastinal window

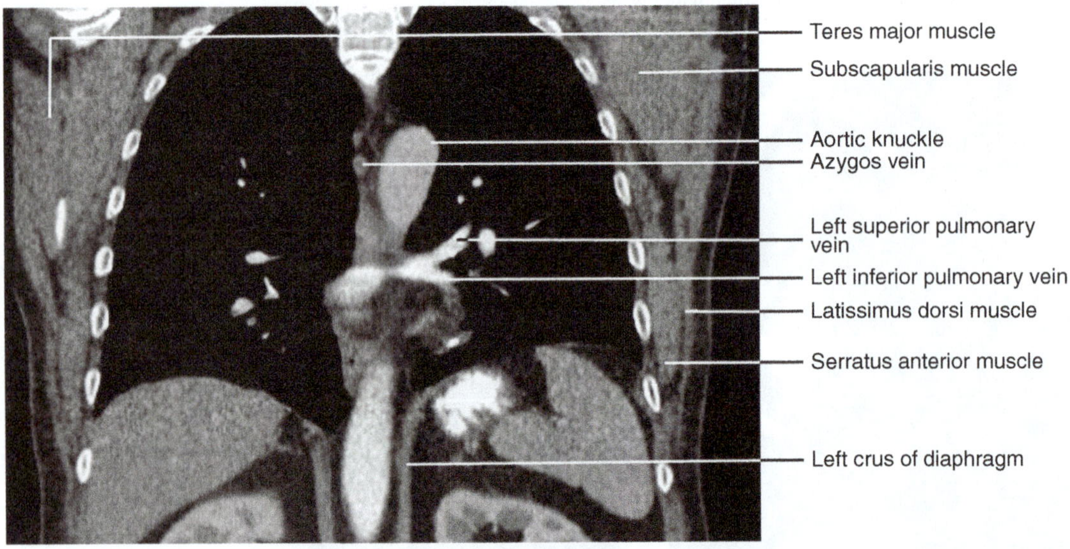

Fig. 17.49 Coronal MPR: mediastinal window

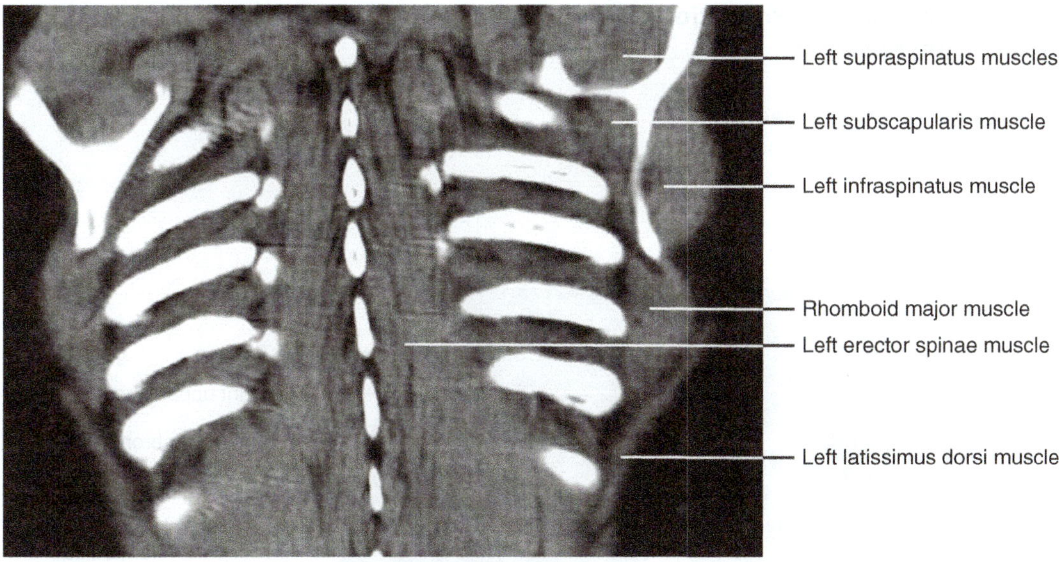

Fig. 17.50 Coronal MPR: mediastinal window

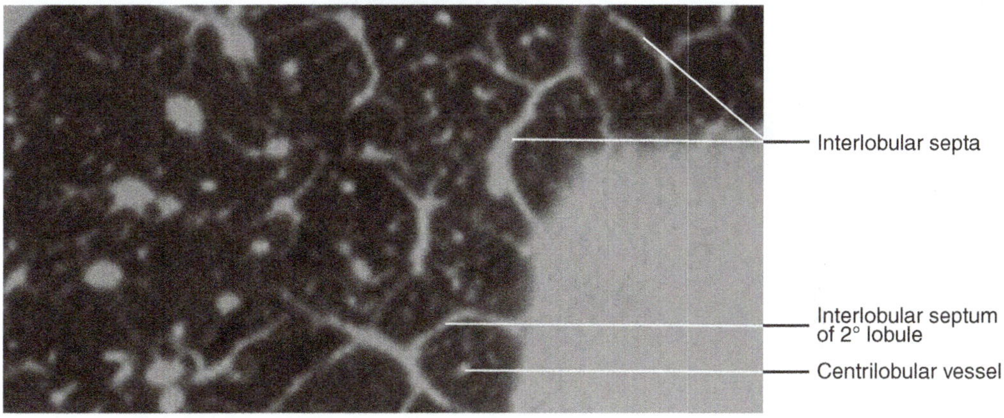

Fig. 17.51 Septal lines

Pearl

Only when pathology affects the interstitium can the interlobular septa and 2° lobule be defined.

CT: Minimum Intensity Projection (MinIP) and 3D Volume-Rendered (VR) Virtual Bronchoscopy

Minimum Intensity Projection

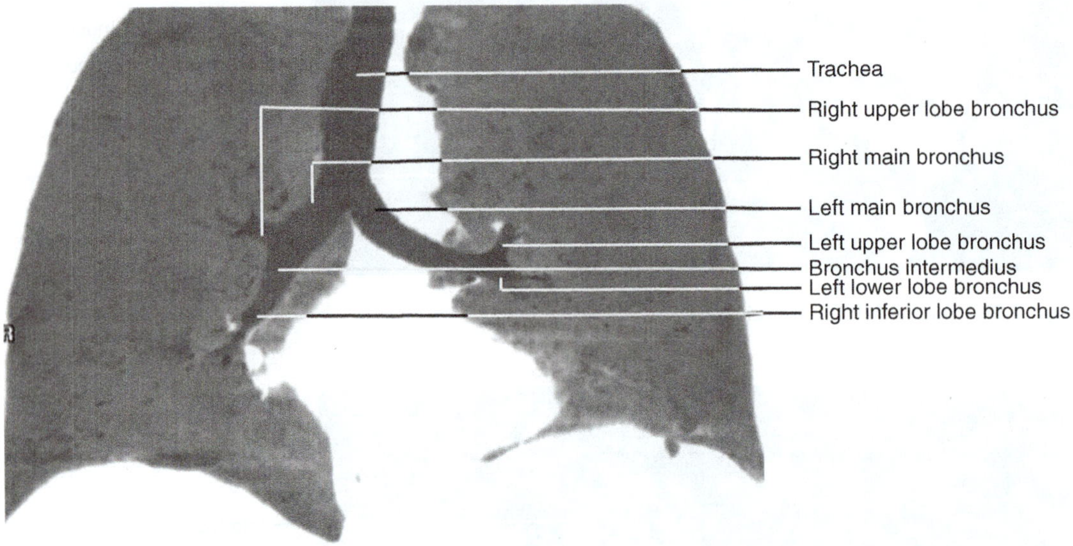

Fig. 17.52 Coronal MinIP

Minimum Intensity Projection

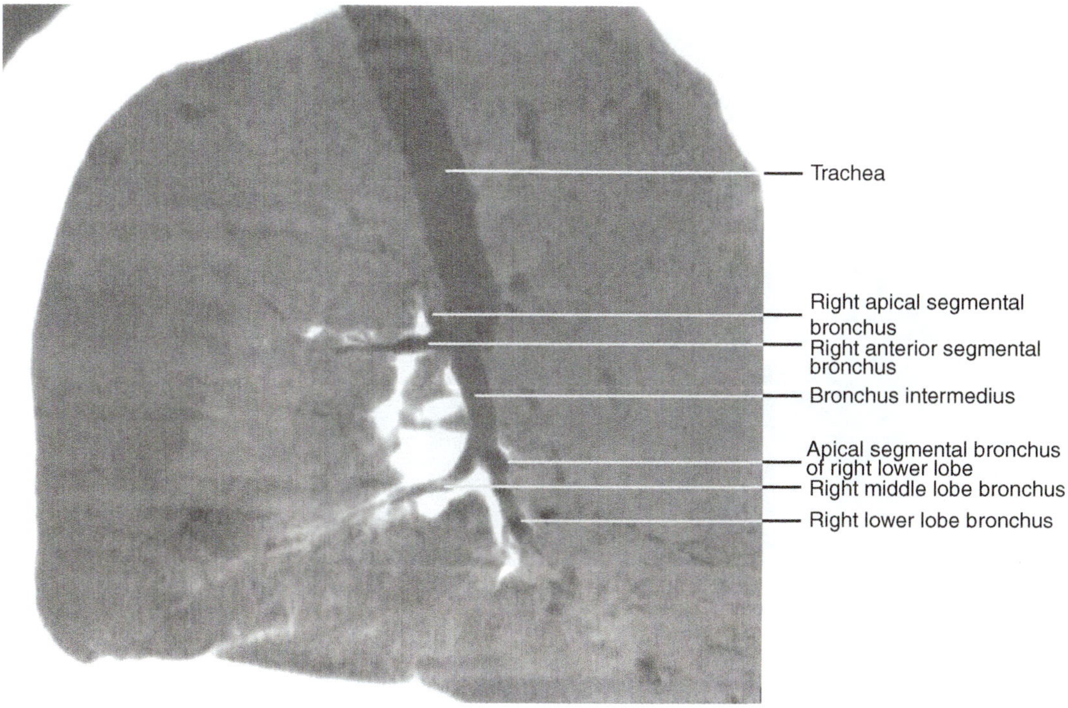

Fig. 17.53 Sagittal MIP at right hilar level

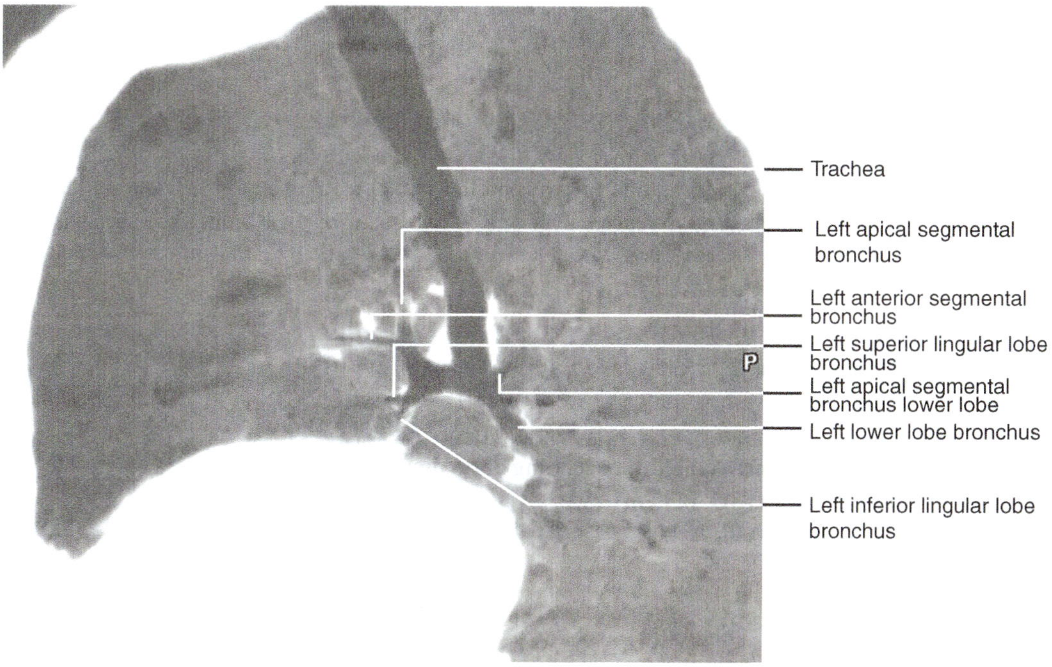

Fig. 17.54 Sagittal MIP at left hilar level

Magnetic Resonance Imaging

Axial MRI: Superior to Inferior

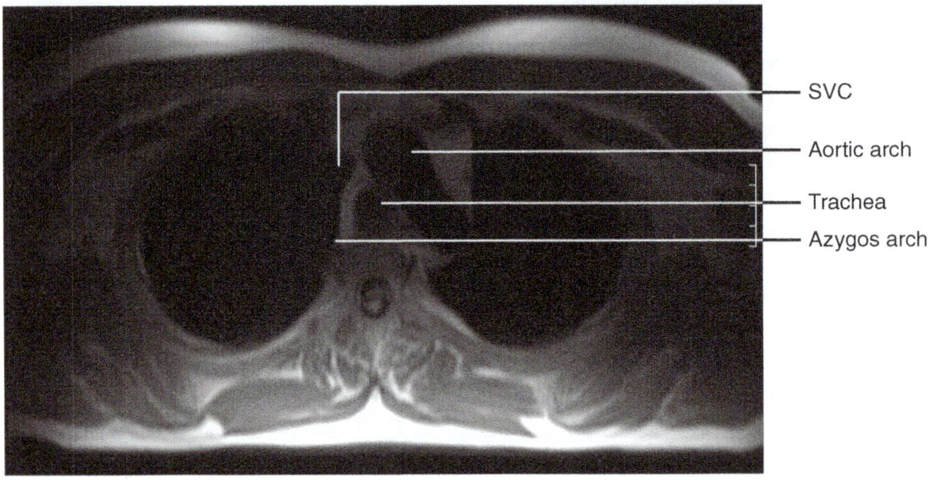

Fig. 17.56 Dark blood T1 axial

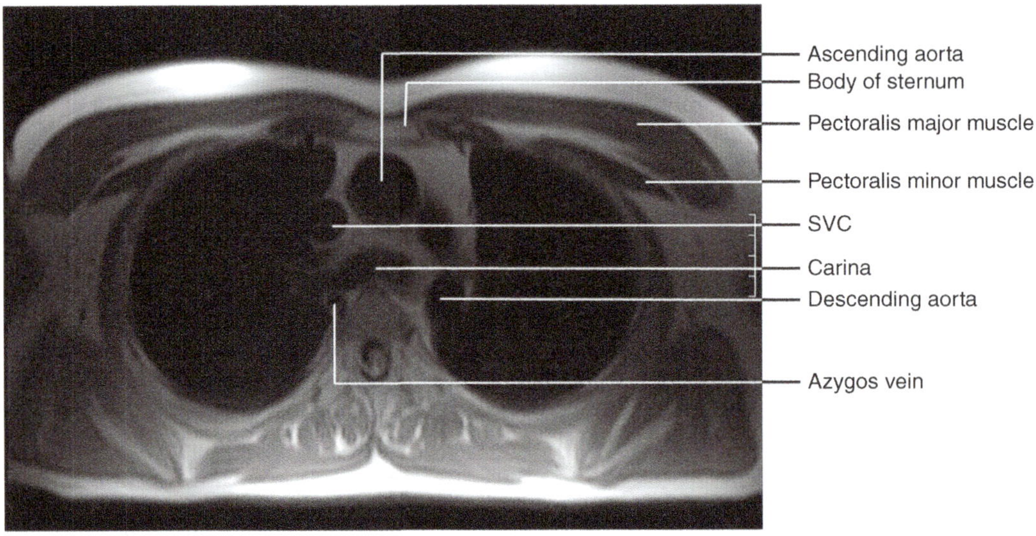

Fig. 17.57 Dark blood T1 axial

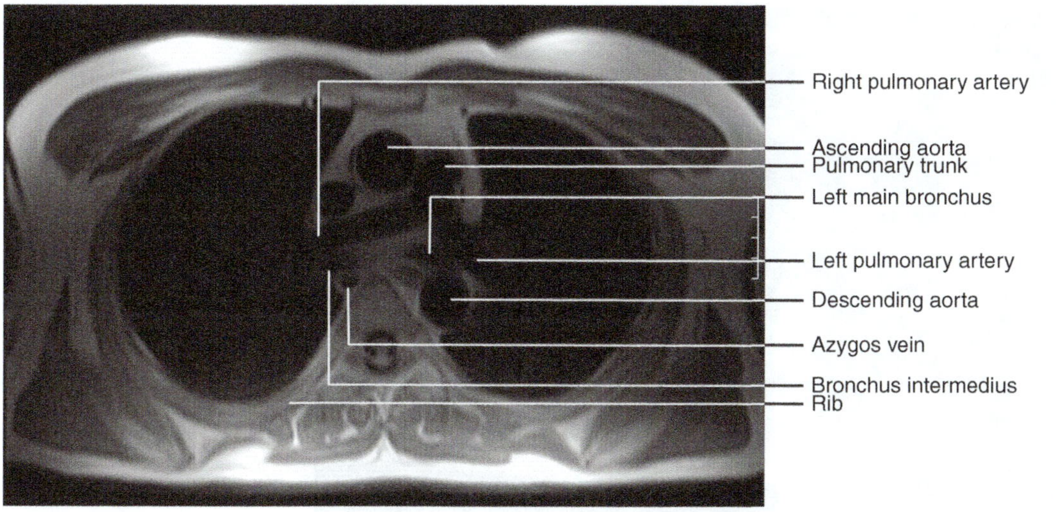

Fig. 17.58 Dark blood T1 axial

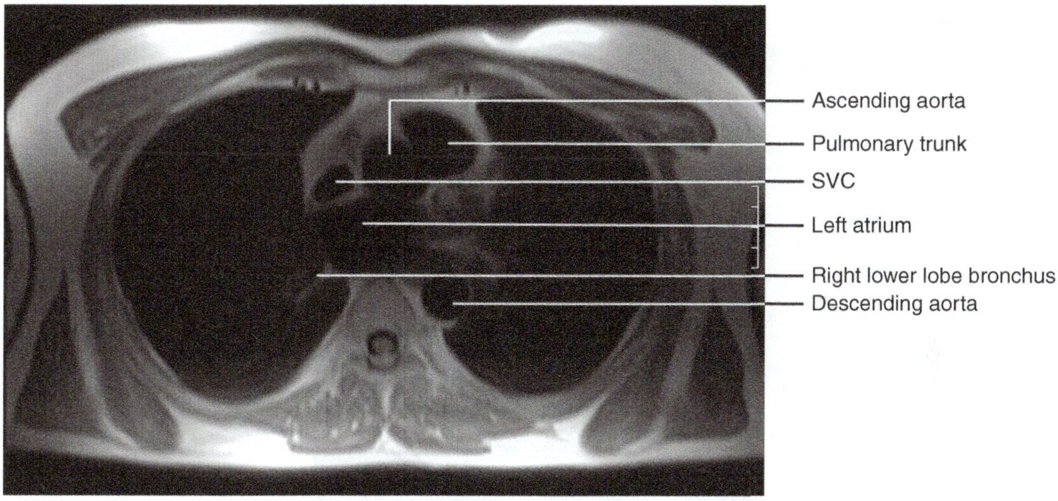

Fig. 17.59 Dark blood T1 axial

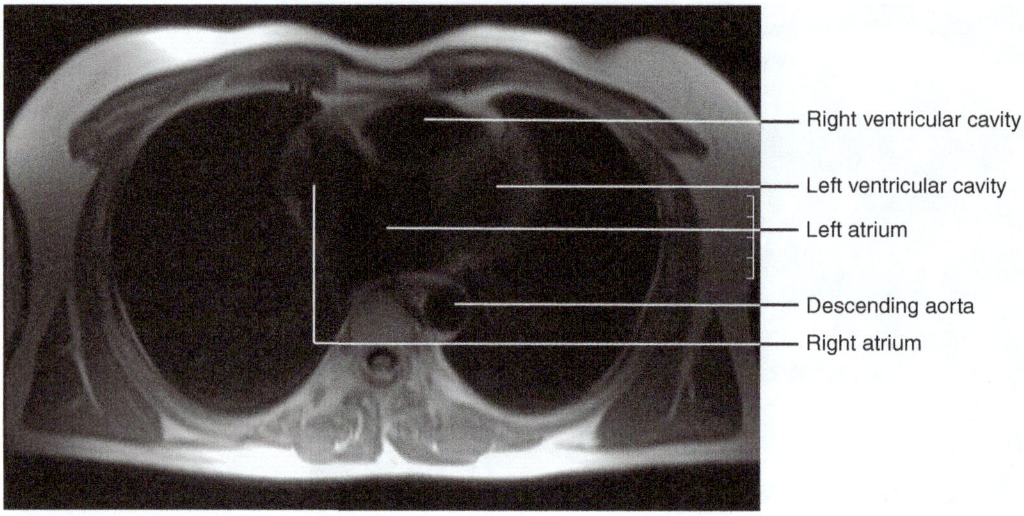

Fig. 17.60 Dark blood T1 axial

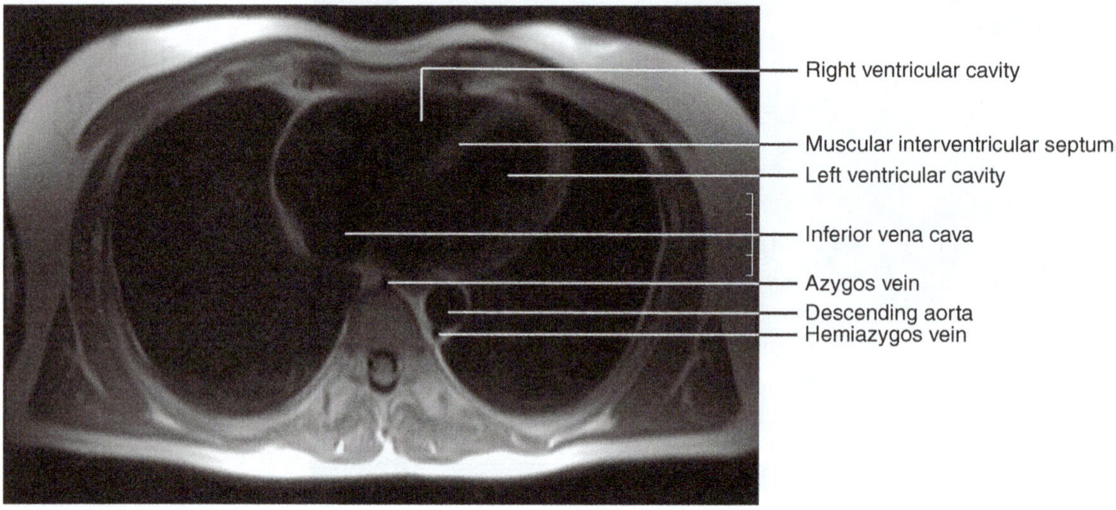

Fig. 17.61 Dark blood T1 axial

Coronal MRI: Anterior to Posterior

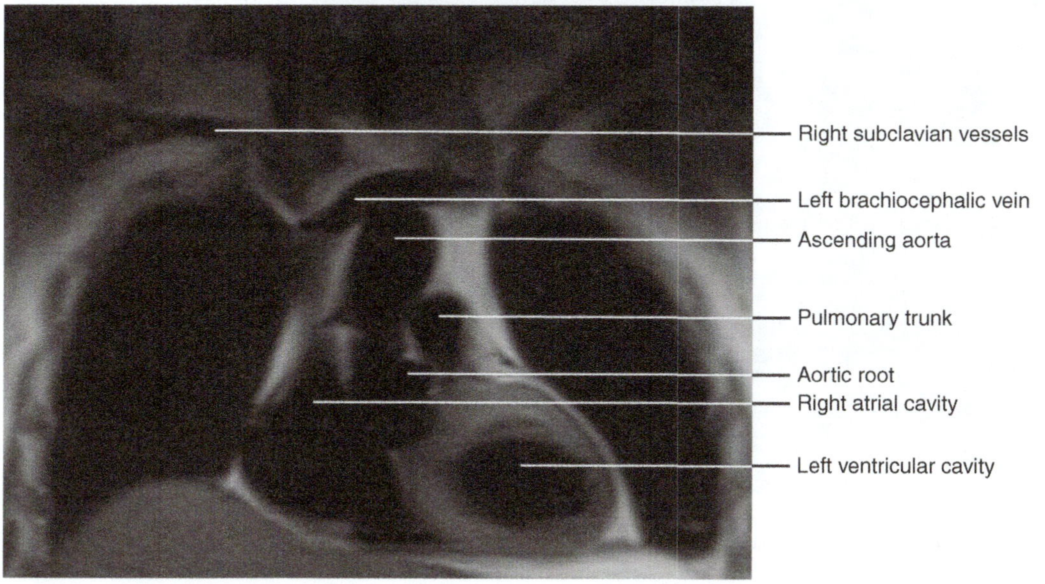

Fig. 17.62 Dark blood T1 coronal

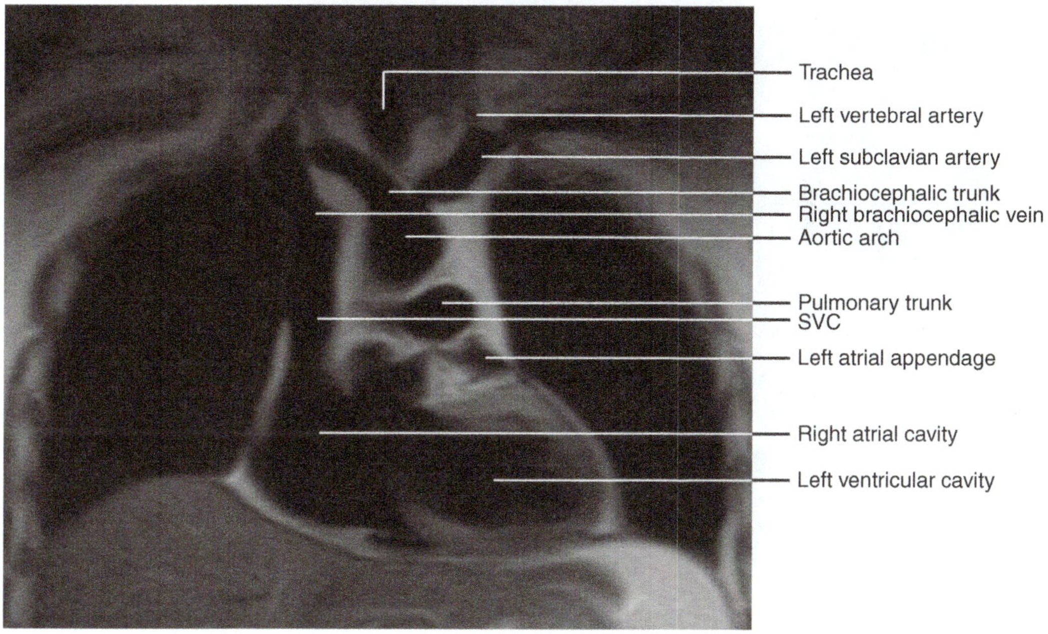

Fig. 17.63 Dark blood T1 coronal

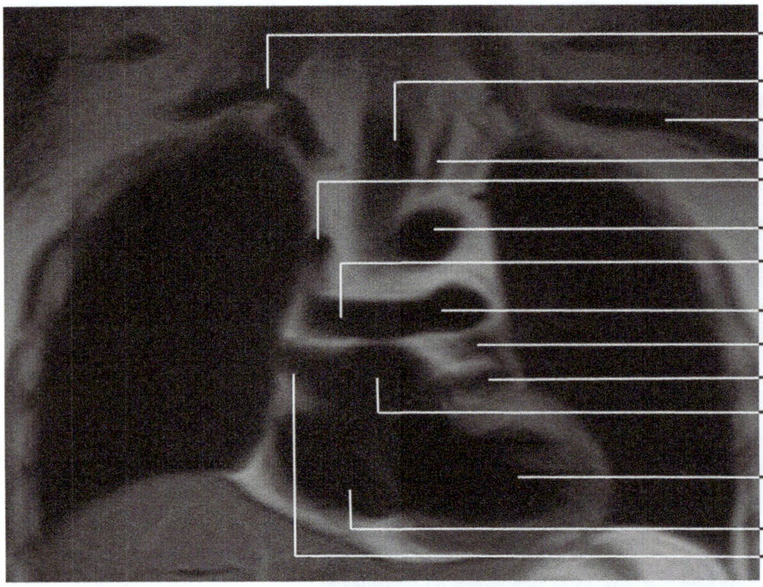

Fig. 17.64 Dark blood T1 coronal

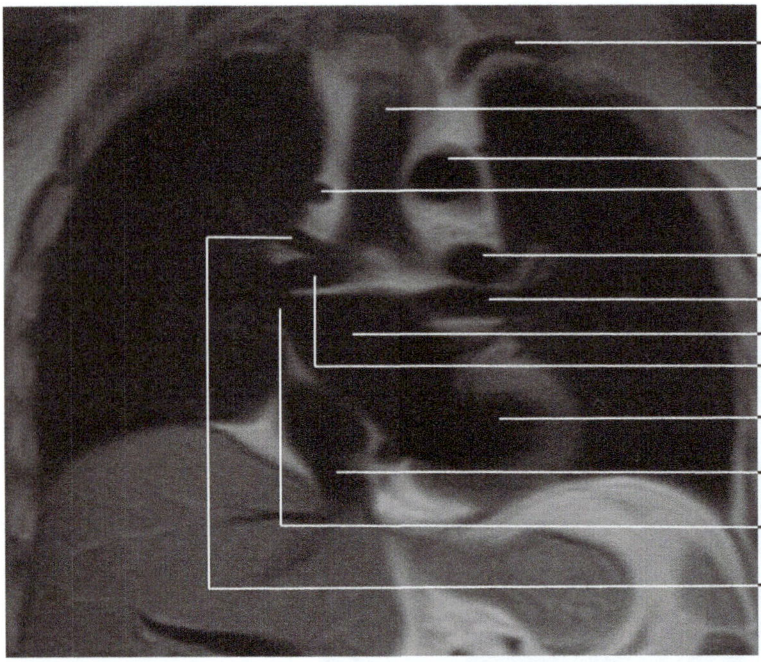

Fig. 17.65 Dark blood T1 coronal

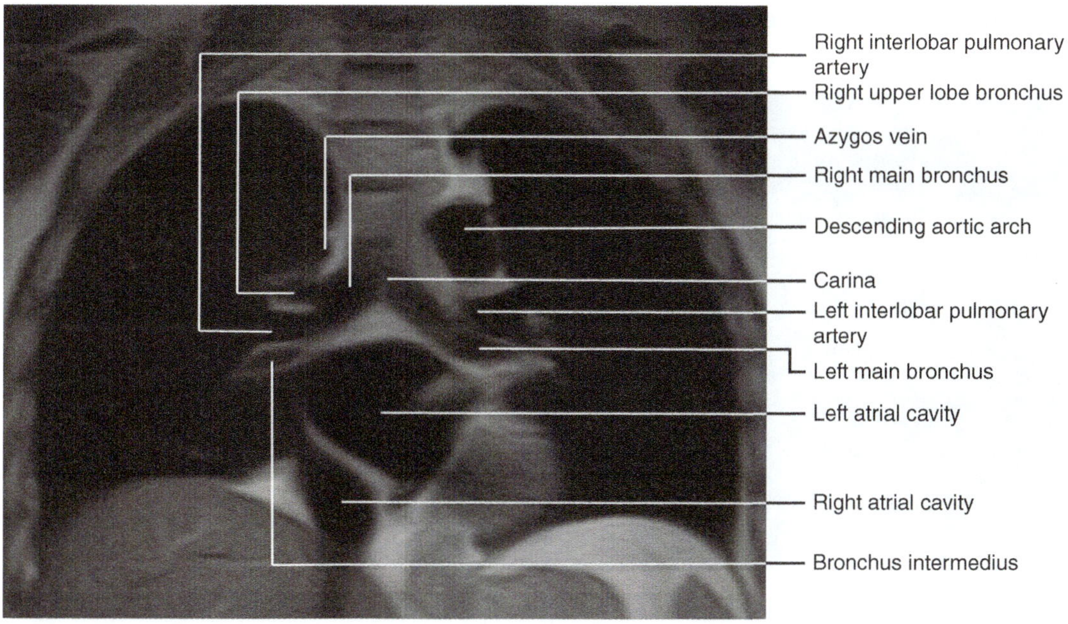

Fig. 17.66 Dark blood T1 coronal

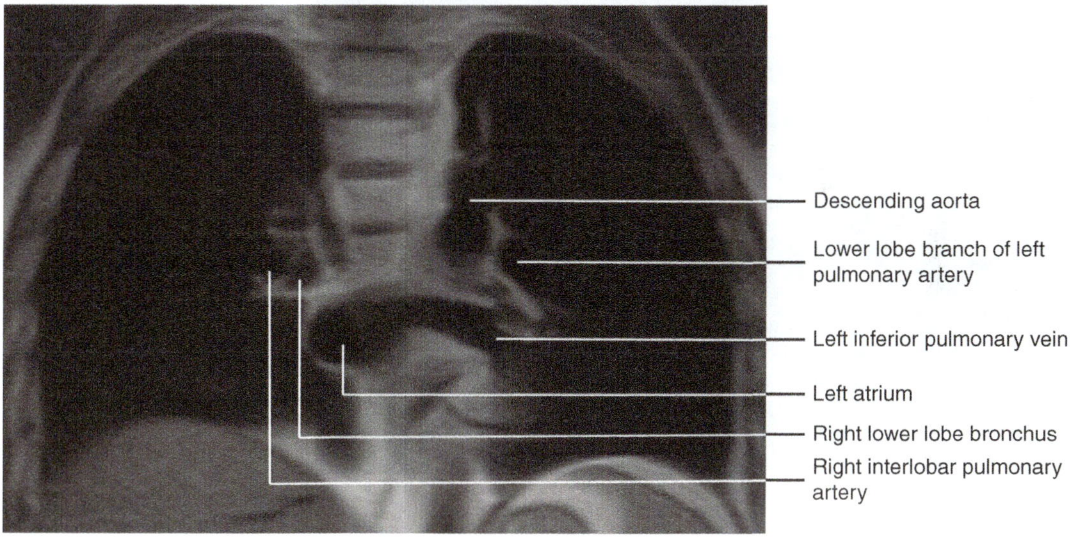

Fig. 17.67 Dark blood T1 coronal

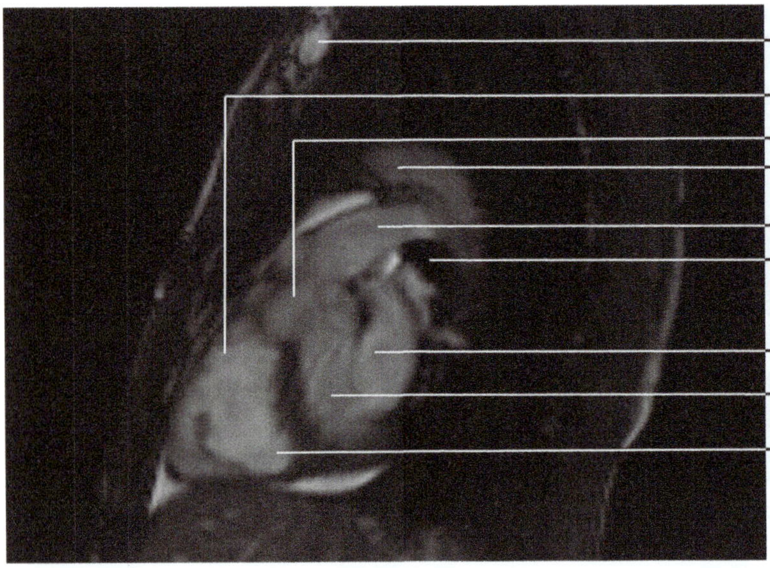

Fig. 17.72 Bright blood T2 sagittal

Nuclear Medicine

Ventilation/Perfusion Scan of the Lung

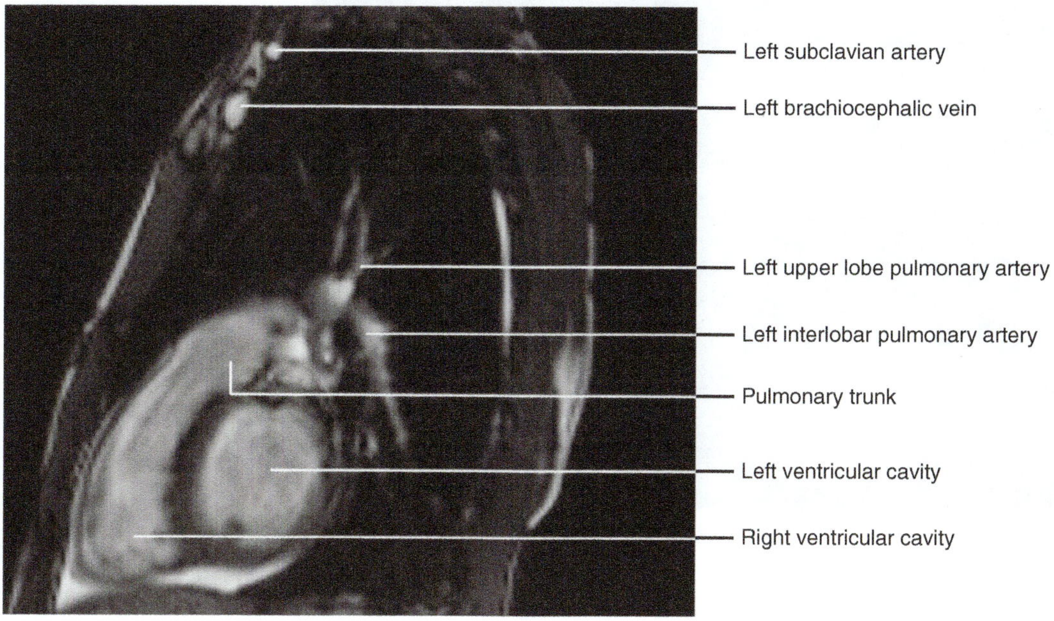

Fig. 17.73 Bright blood T2 sagittal

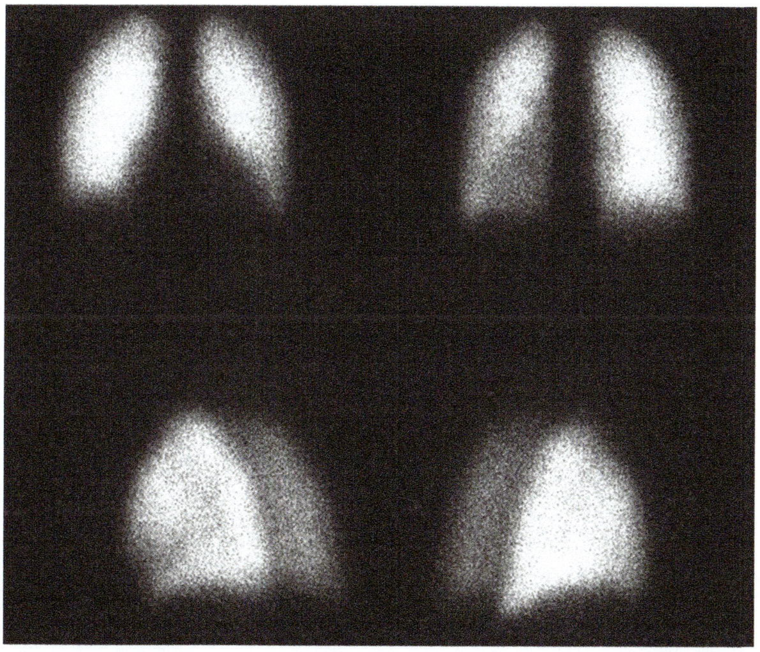

Fig. 17.74 Perfusion scan (P)

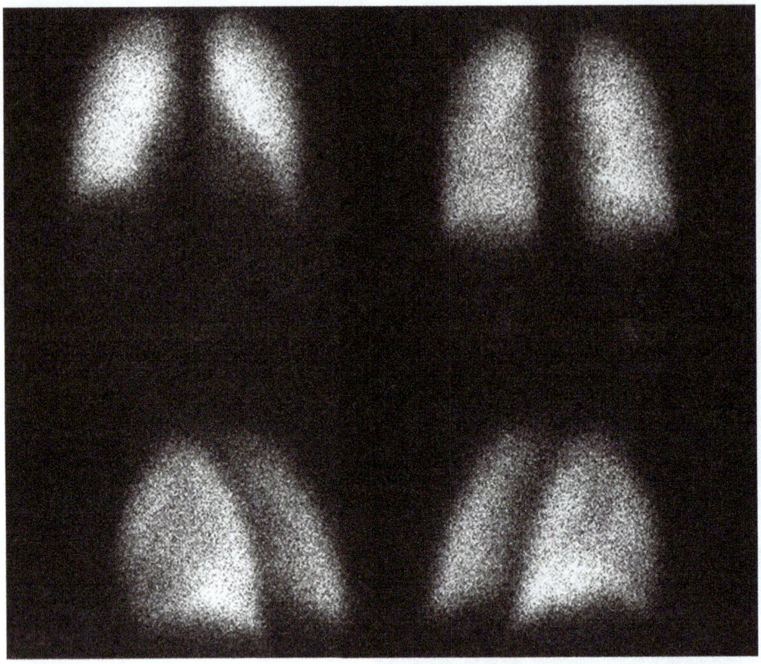

Fig. 17.75 Ventilation scan (V)

Pearl

The perfusion scan is usually performed first. The perfusion agents are Tc-99m macro-aggregated Albumin (MAA) and Tc-99m human albumin microspheres.

The ventilation scan is performed immediately after the perfusion study. Ventilation agents are xenon-133, xenon-127, krypton-81, Tc-99m DTPA aerosol, and carbon dioxide tracer.

Barium Studies

Barium Swallow

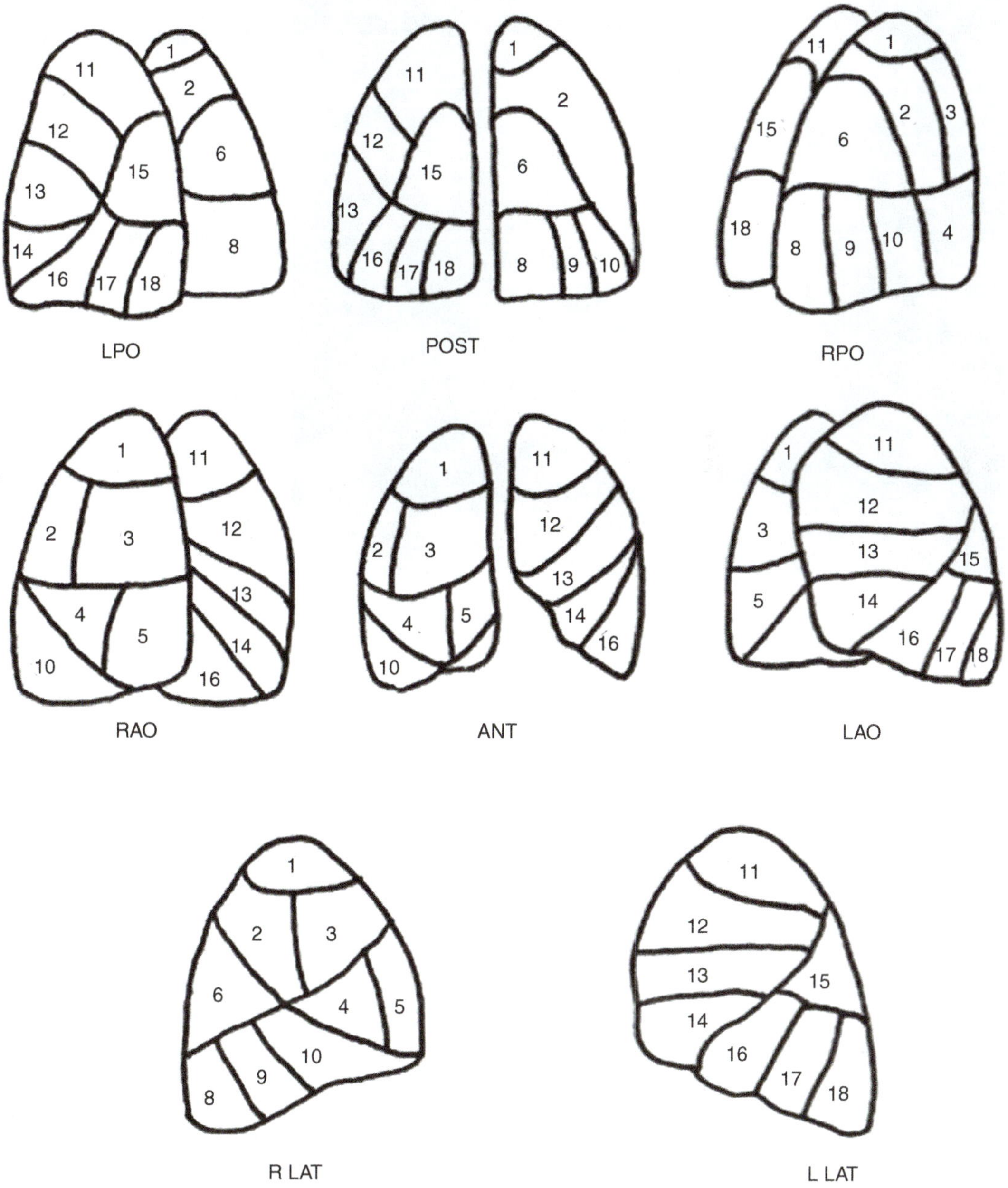

Fig. 17.76 Segmental labeling of Ventilation/Perfusion Scans of the Lung. **Right upper lobe**: (*1*) apical; (*2*) posterior (*POST*); (*3*) anterior (*ANT*). Right (R) **middle lobe**: (*4*) lateral (*LAT*); (*5*) medial. **Right lower lobe**: (*6*) superior; (*7*) mediobasal; (*8*) posterobasal; (*9*) laterobasal; (*10*) anterobasal. Left (L) **upper lobe**: (*11*) apicoposterior; (*12*) anterior; (*13*) superior lingula; (*14*) inferior lingula. **Left lower lobe**: (*15*) superior; (*16*) anteromedial wall; (*17*) laterobasal; (*18*) posterobasal. *LAO* left anterior oblique, *LPO* left posterior oblique, *RPO* right posterior oblique, *RAO* right anterior oblique

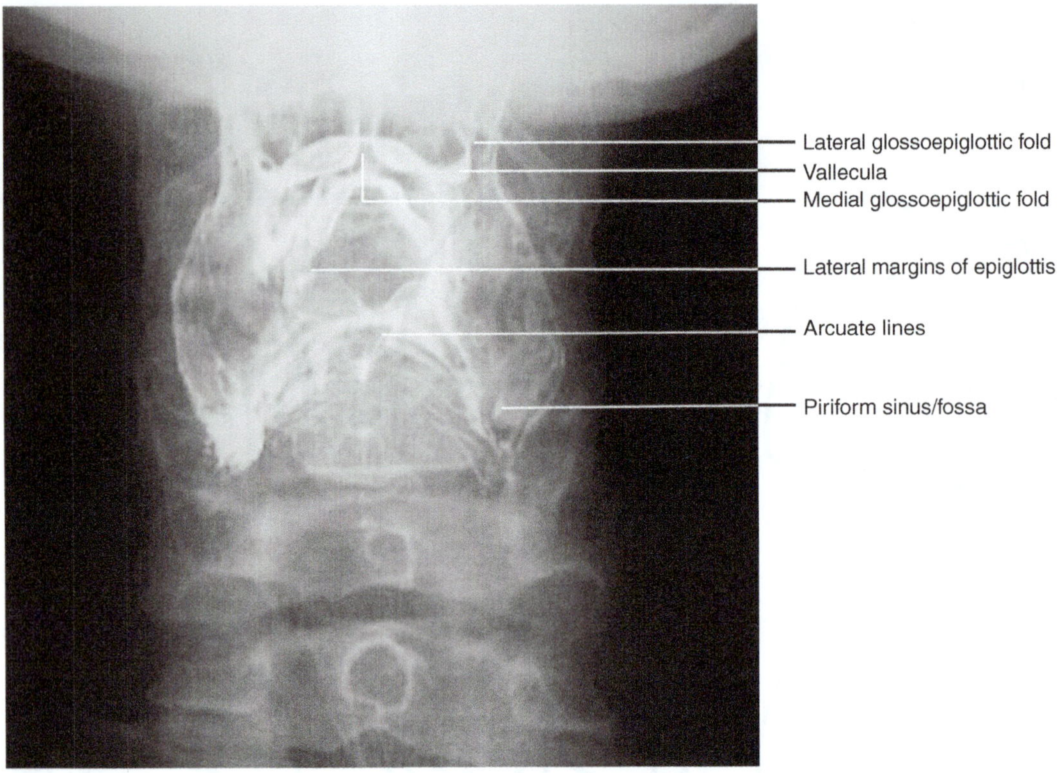

Fig. 17.77 Pharynx and cervical oesophagus (frontal AP view): double contrast

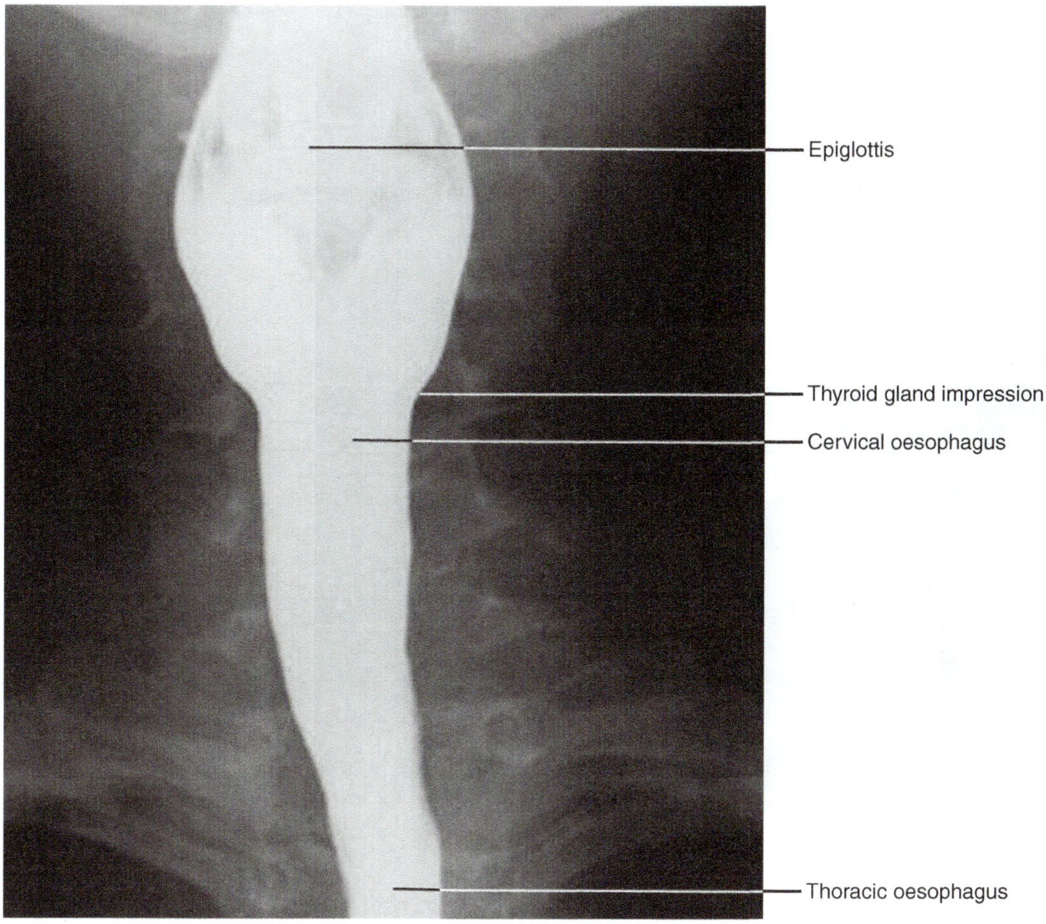

Fig. 17.78 Pharynx and cervical oesophagus (frontal AP view): single contrast

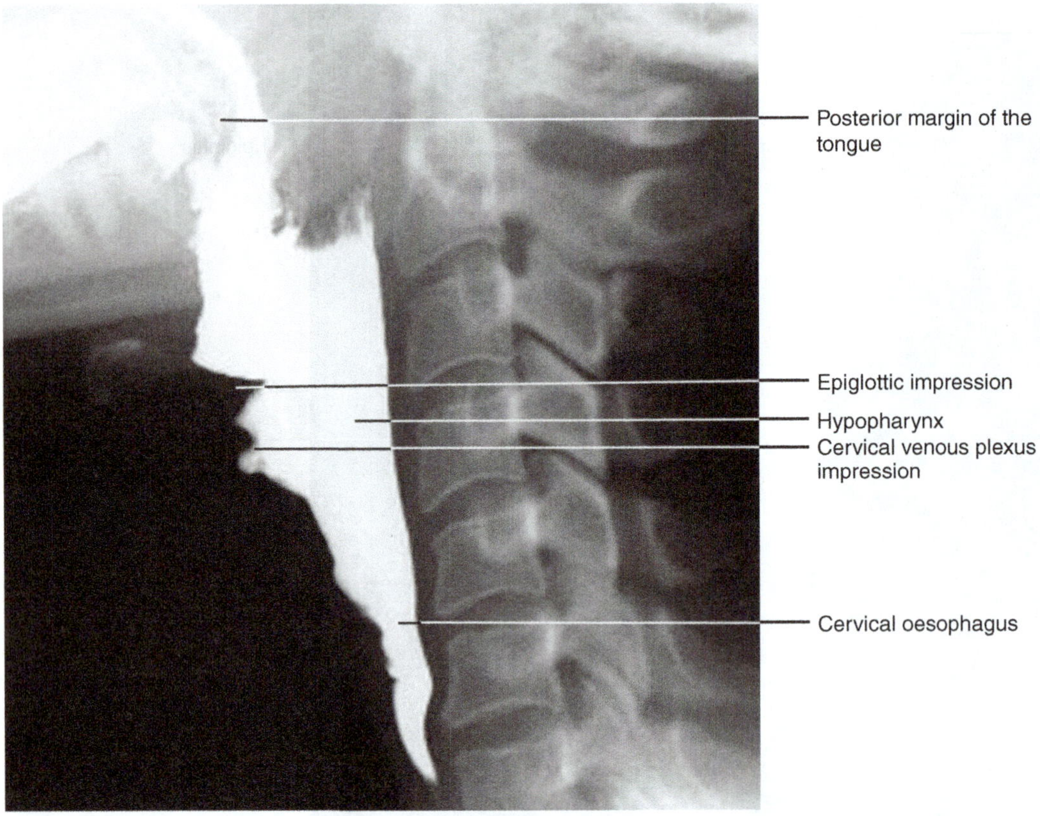

Fig. 17.79 Pharynx and cervical oesophagus: right lateral view

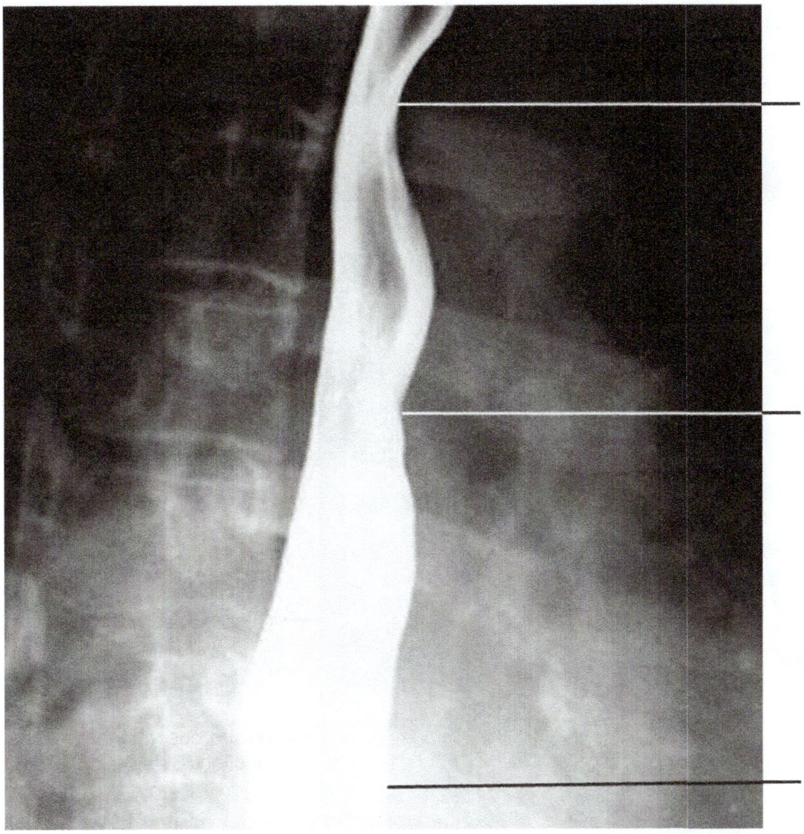

Fig. 17.80 Thoracic oesophagus: left oblique/lateral view

Impressions on the Barium-Filled Oesophagus

> **Pearl: Common anterior impressions on the lateral view**
> Aortic arch
> Left main bronchus
> Left atrium

> **Pearl: Other, less often observed impressions**
> Azygos imprint (posterior aspect of the azygos vein)
> Medial right lower lung (crista pulmonalis)
> Confluence of the pulmonary veins (usually the inferior confluence)

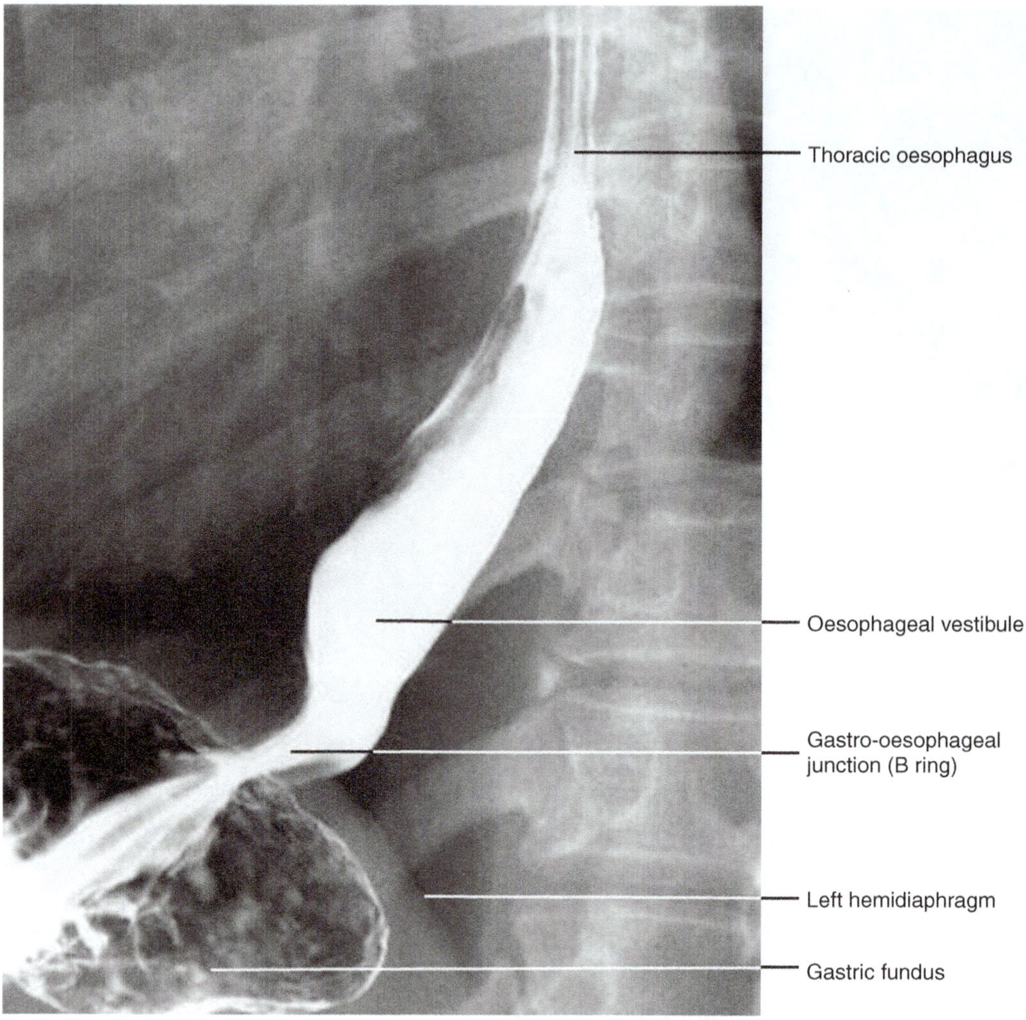

Fig. 17.81 Pharynx and cervical oesophagus (prone PA view)

Suggested Reading

Armstrong P, Wilson A, Dee P, Hansell DM (2000) Imaging of diseases of the chest, 3rd edn. Mosby, London

Chapman S, Nakienly R (1995) A guide to radiological procedures. Respiratory system, 3rd edn. Bailliere Tindall, London, pp 194–197

Chasen C, Rugh KS, Shelton DK (1984) Mediastinal impressions on the dilated oesophagus. Radiol Clin North Am 22:591–605

Chiles C (1990) CT of the pulmonary hilium. Radiol Clin North Am 28:539–544

Kauczor H, Kreitner K (2000) Contrast-enhanced MRI of the lung. Eur J Radiol 34:196–207

Mettler FA, Guiberteau MJ (1998) Essentials of nuclear medicine imaging. Respiratory system, 4th edn. Bailliere Tindall, London, pp 191–235

Müller FA, Guiberteau MJ (2002) Chronic obstructive pulmonary disease. 4: imaging the lungs in patients with chronic obstructive pulmonary disease. Thorax 57:982–989

Proto A, Speckman J (1979) The left lateral radiograph of the chest. Med Radiogr Photogr 55:30–74

Slone RM, Gutierrez FR, Fisher AJ (1999) A practical approach. Thoracic imaging. McGraw-Hill, New York

Specht NT, Russo RD (1998) Practical guide to diagnostic imaging. Nuclear medicine. Mosby, St. Louis, pp 338–345

Sutton D (1998) Textbook of radiology and imaging, 6th edn. Churchill Livingstone, Edinburgh

Weir J, Abrahams PH (1997) Imaging atlas of human anatomy, 2nd edn. Mosby, London, pp 88–90

Part IV

Abdomen and Pelvis

Abdominal imaging includes imaging of the peritoneal and retroperitoneal cavities. These spaces contain organs with their duct and drainage systems, as well as the gastrointestinal tract. The abdominal spaces run into the pelvis but the pelvic organs are more easily appreciated seperately for males and females as complete genital tracts, with regard to their imaging anatomy. Imaging techniques involve plain radiographs with the addition of contrast for hollow vicuses and duct systems as well as cross sectional imaging using ultrasound, CT and MRI, of which each has specific advantages for specific anatomic demonstration.

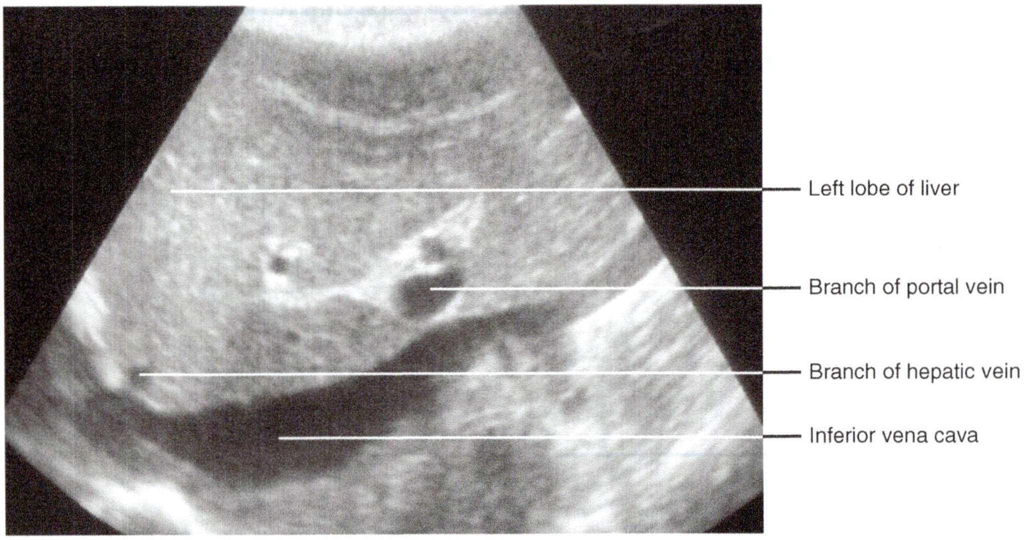

Fig. 18.2 Longitudinal view left lobe/IVC

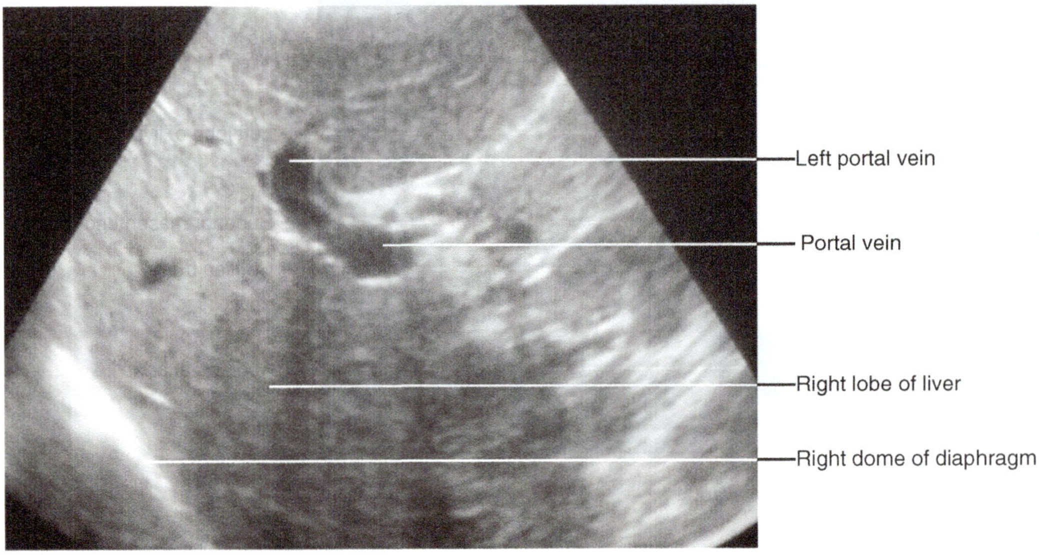

Fig. 18.3 Longitudinal view right lobe

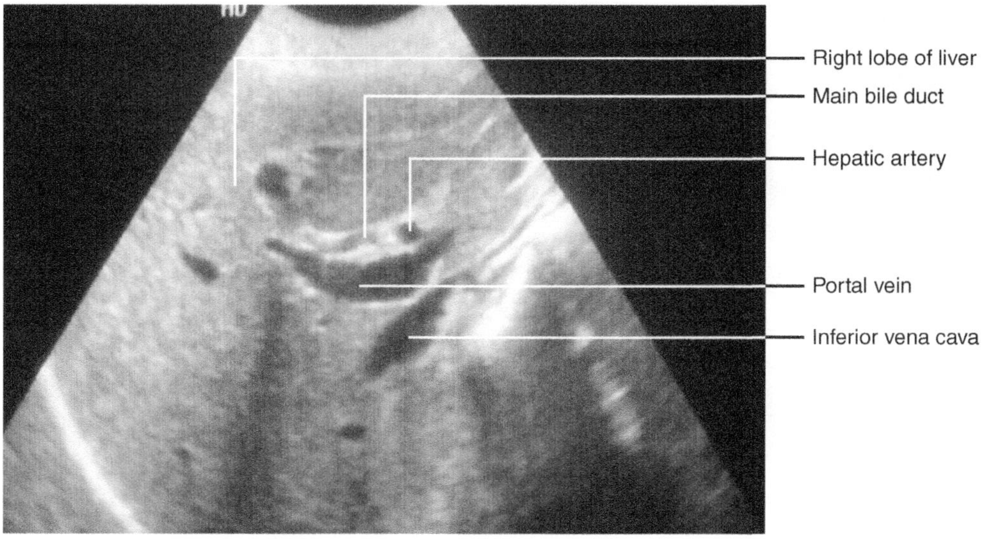

Fig. 18.4 Oblique view through porta hepatis

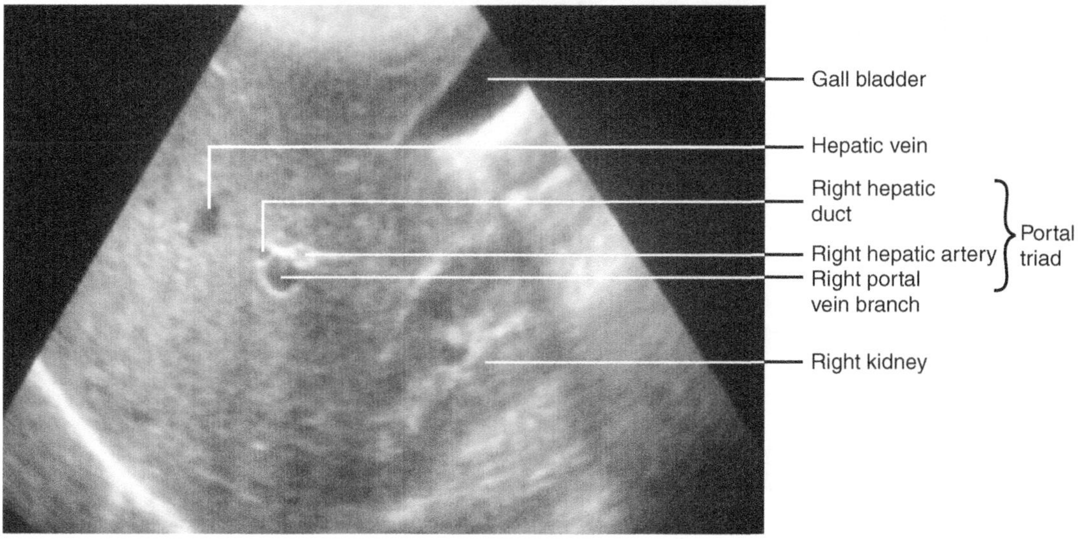

Fig. 18.5 Longitudinal view right lobe

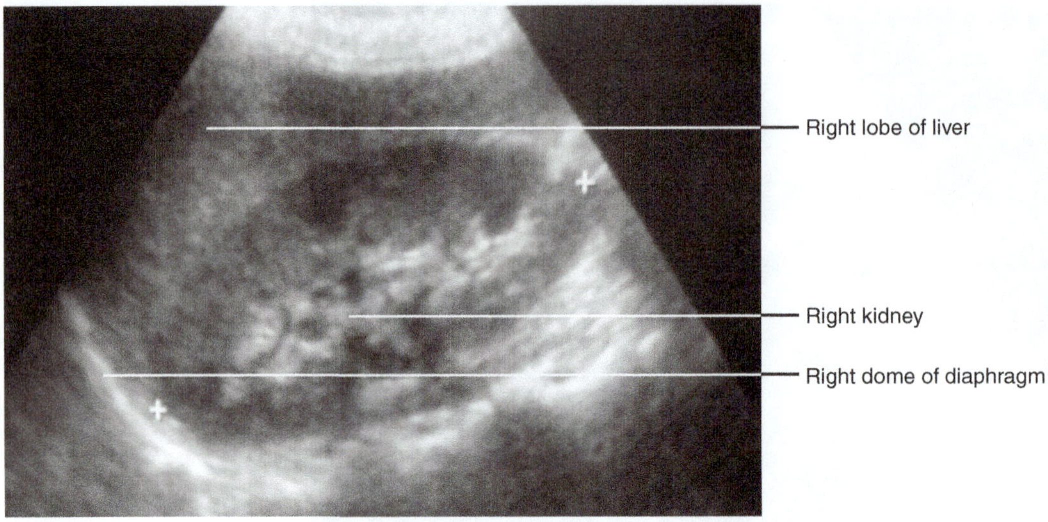

Fig. 18.6 Longitudinal view right lobe

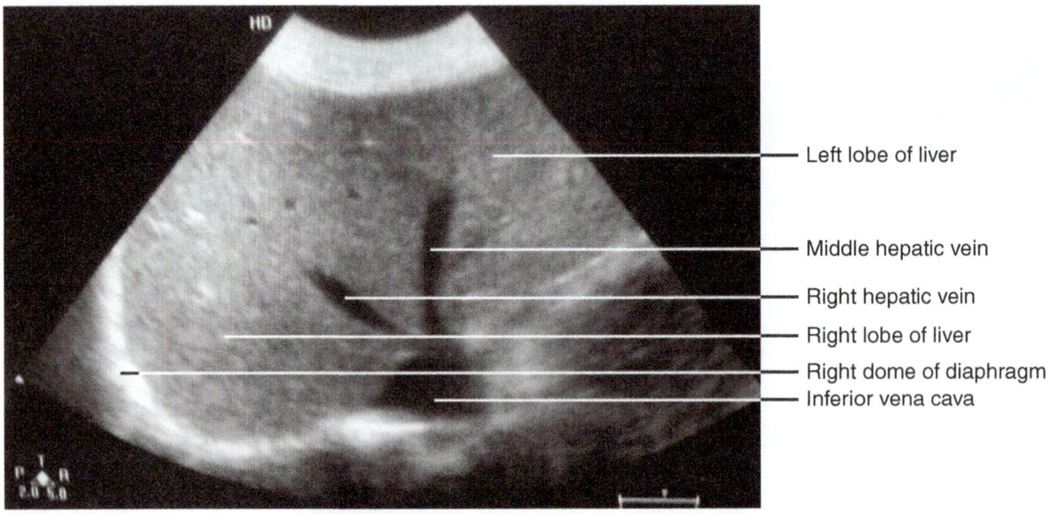

Fig. 18.7 Transverse view at level of hepatic veins

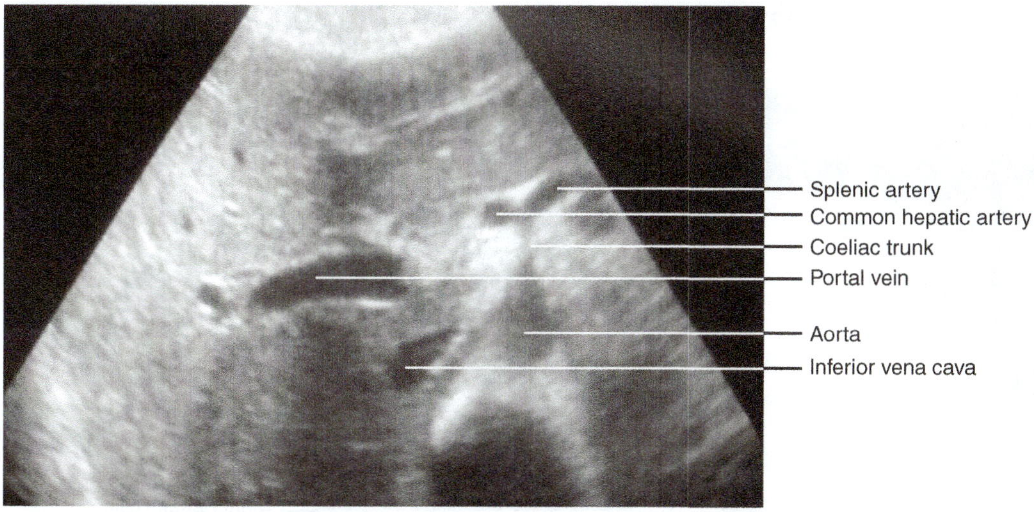

Fig. 18.8 Transverse view at level of hepatic artery

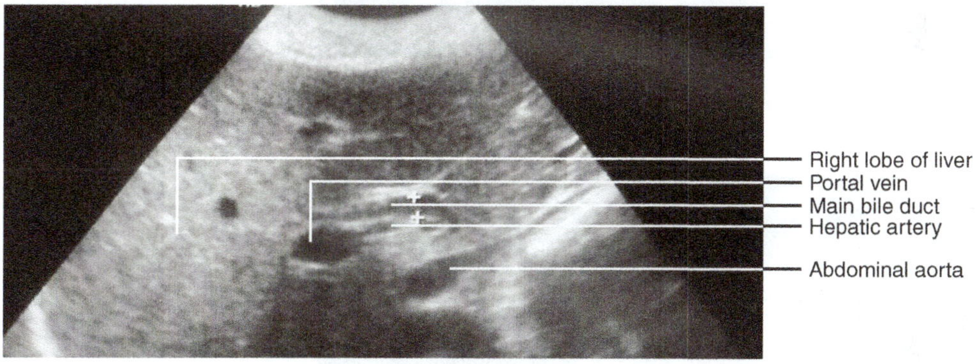

Fig. 18.9 Transverse view porta

Computed Tomography

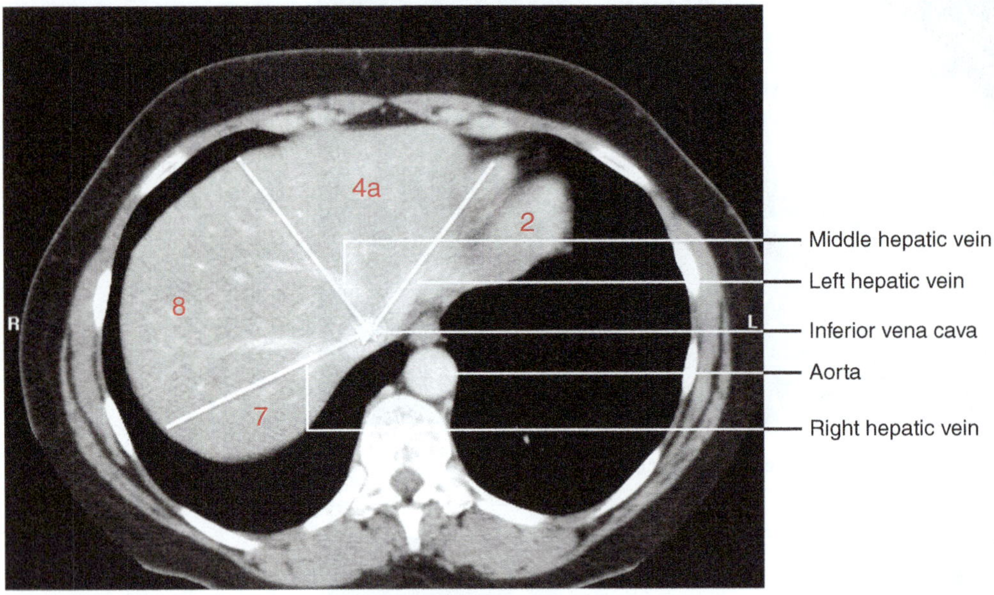

Fig. 18.10 Axial contrasted CT showing segmental anatomy, level of hepatic veins

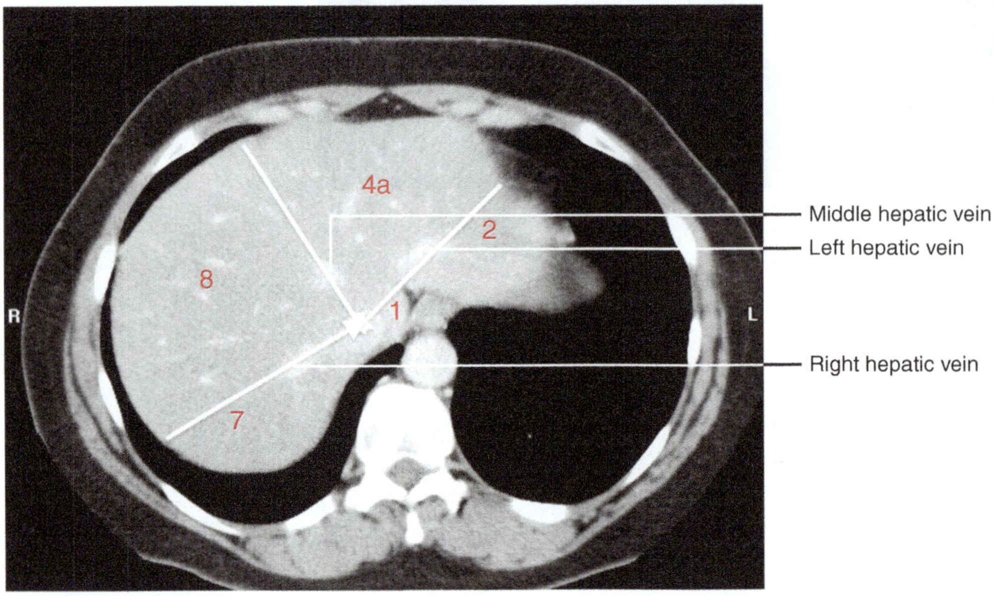

Fig. 18.11 Axial contrasted CT showing segmental anatomy

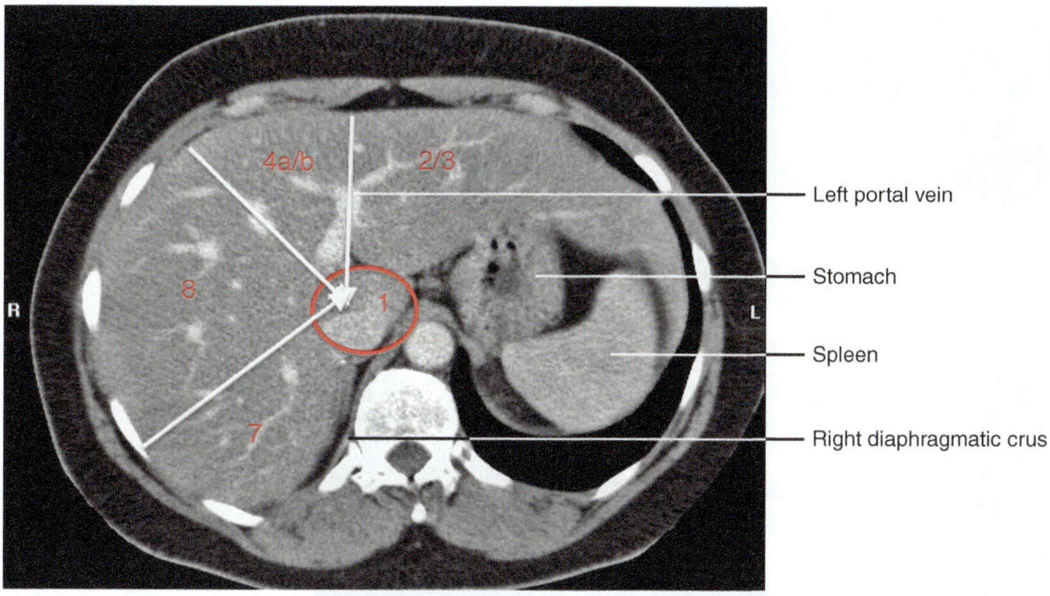

Fig. 18.12 Axial contrasted CT showing segmental anatomy, level of left portal vein

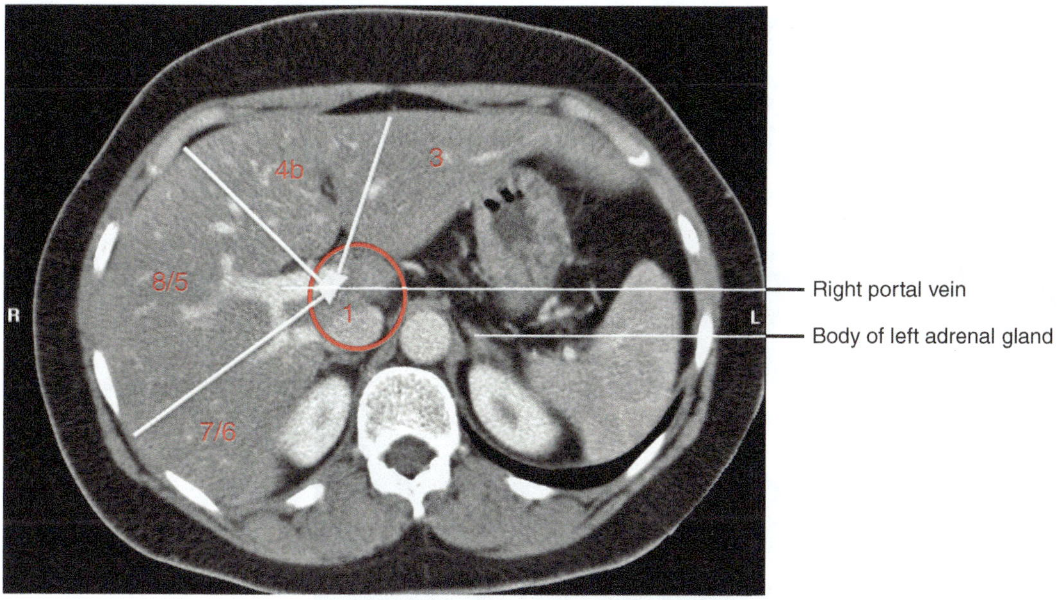

Fig. 18.13 Axial contrasted CT showing segmental anatomy, level of right portal vein

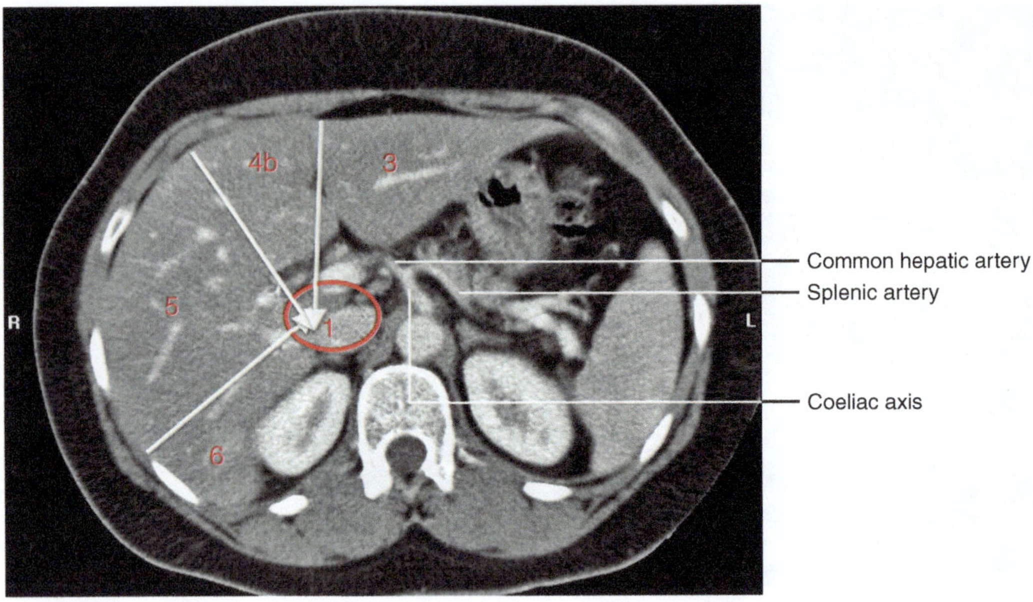

Fig. 18.14 Axial contrasted CT showing segmental anatomy, level of coeliac axis

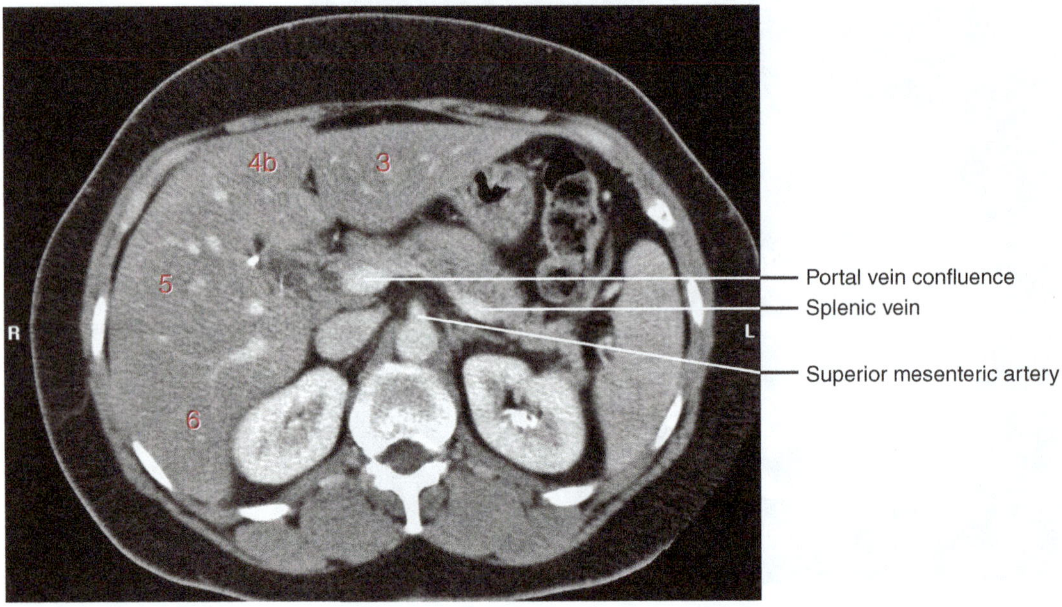

Fig. 18.15 Axial contrasted CT showing segmental anatomy, level of splenic vein

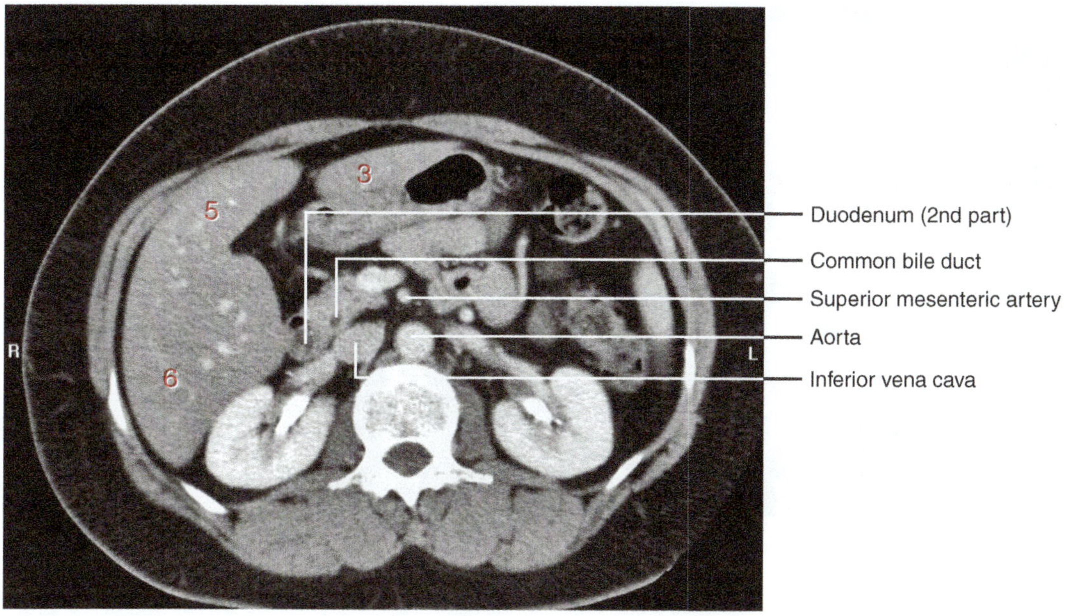

Fig. 18.16 Axial contrasted CT showing segmental anatomy, level of renal hila

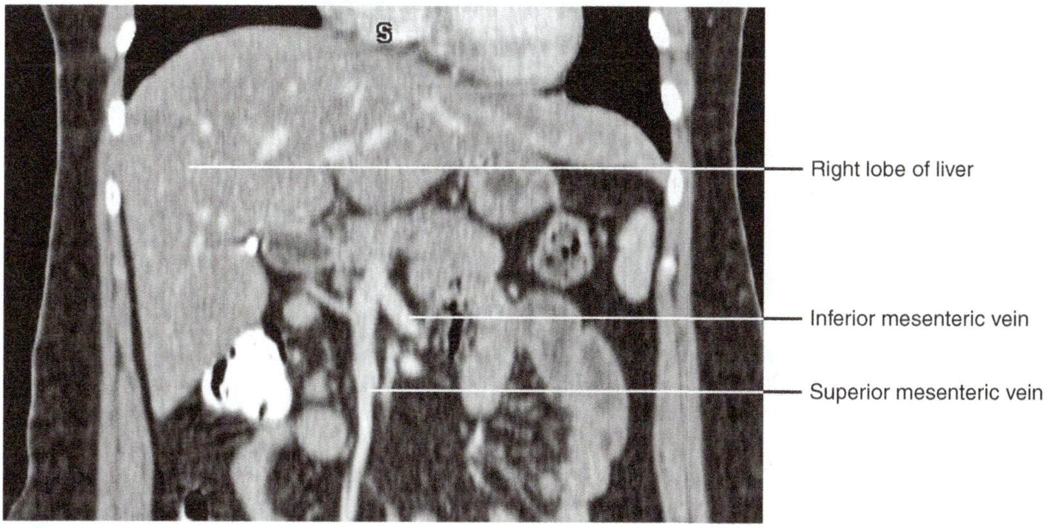

Fig. 18.17 Coronal reformat CT

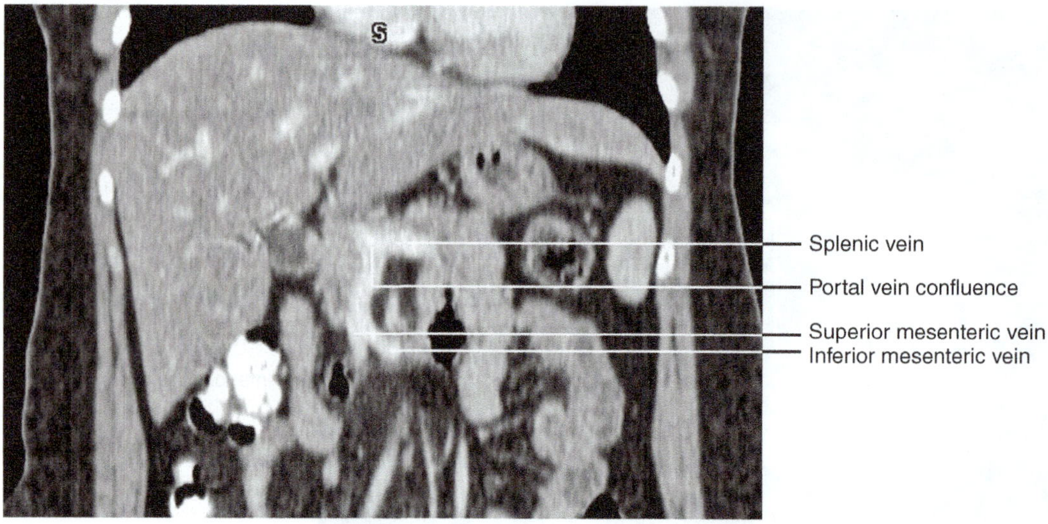

Fig. 18.18 Coronal reformat CT

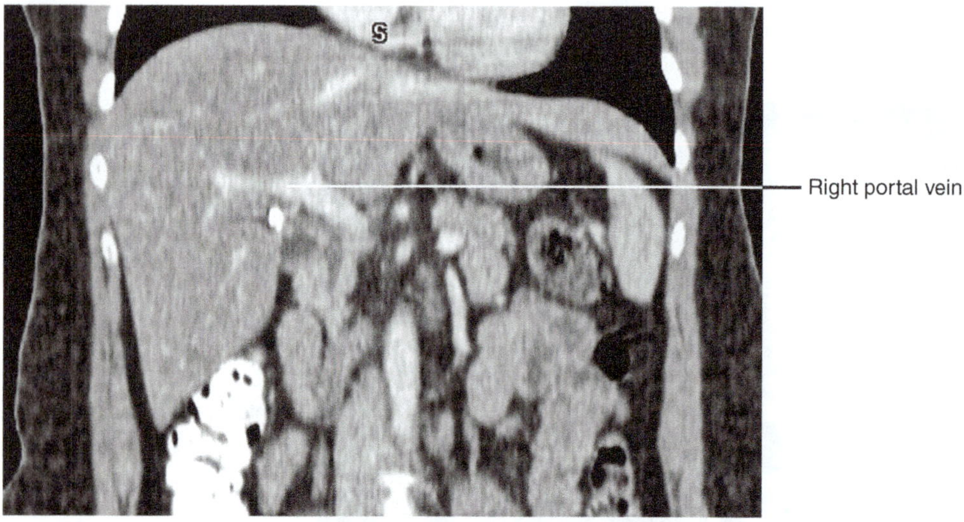

Fig. 18.19 Coronal reformat CT

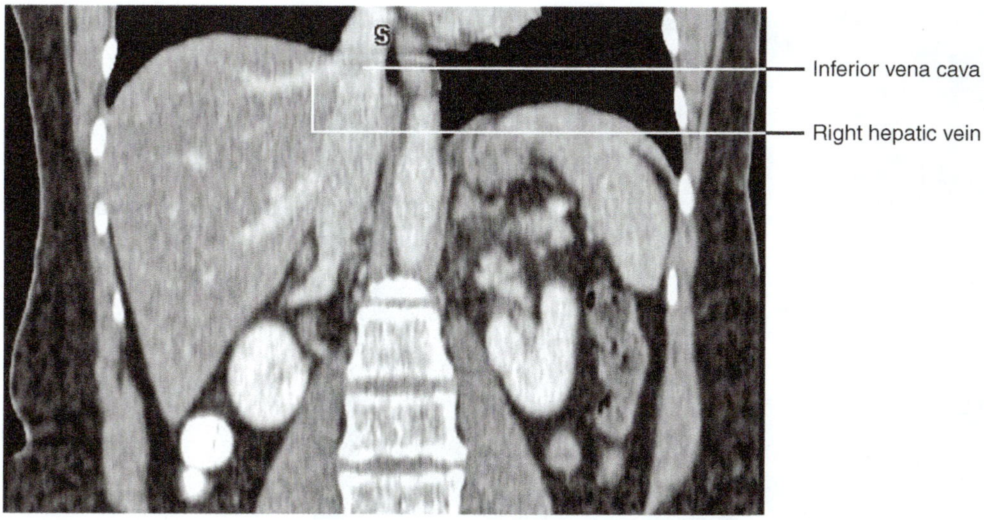

Fig. 18.20 Coronal reformat CT

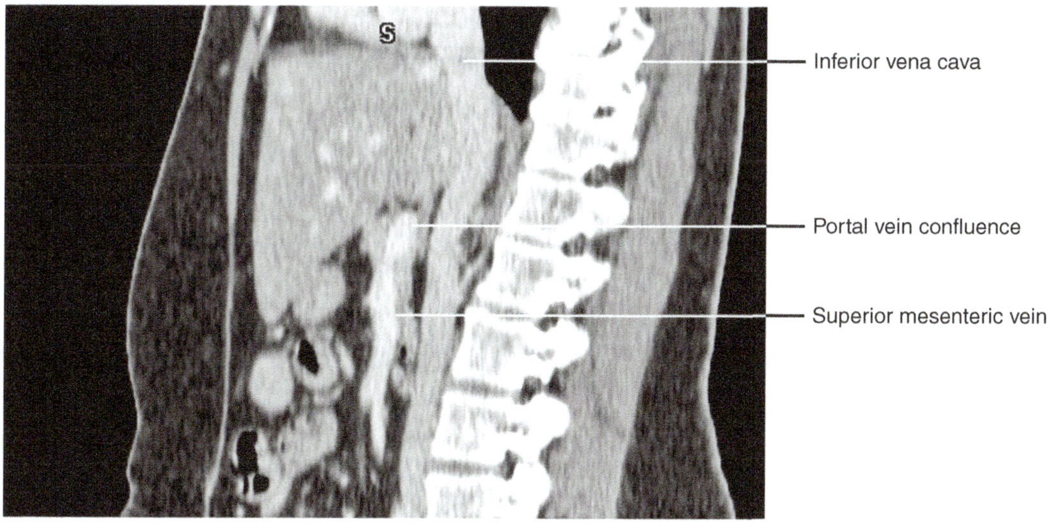

Fig. 18.21 Sagittal reformat CT

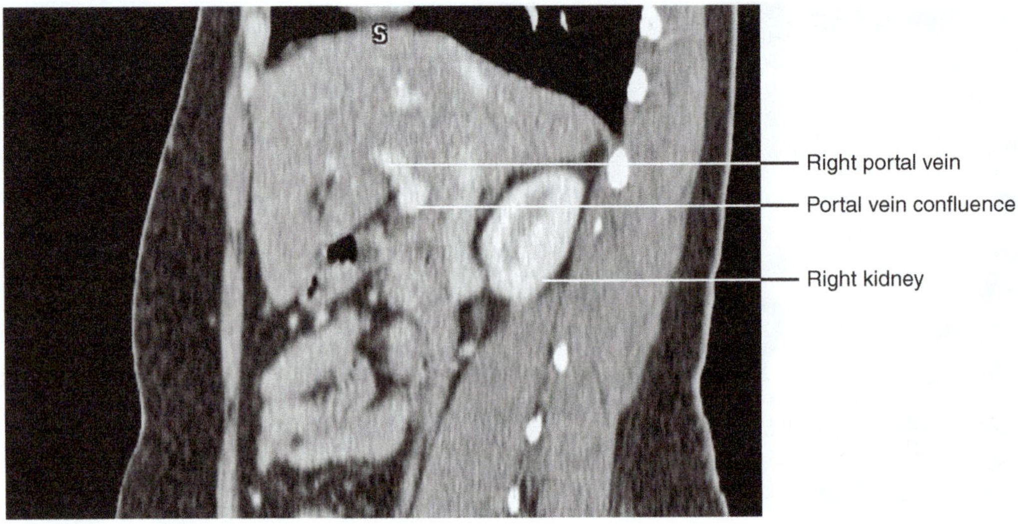

Fig. 18.22 Sagittal reformat CT

Magnetic Resonance Imaging

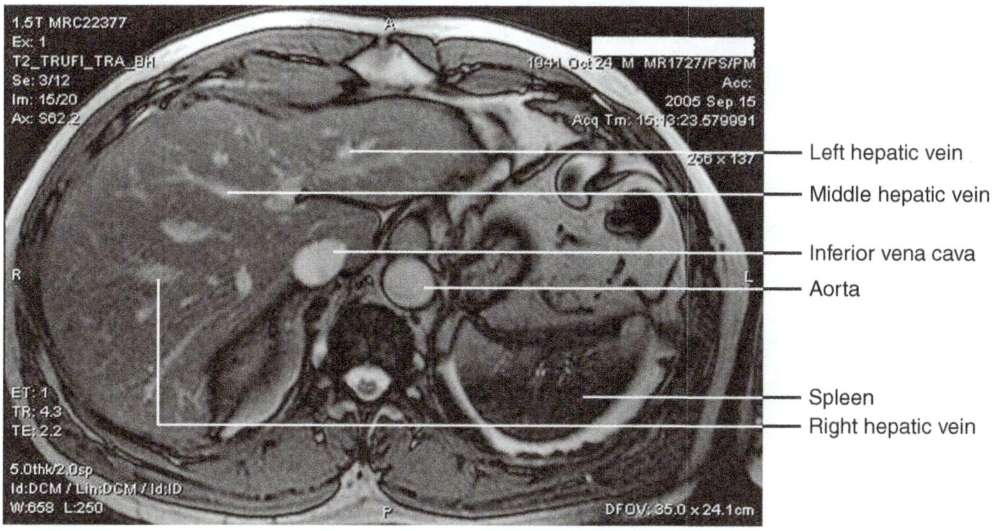

Fig. 18.23 Axial T2 TRUFI liver at the level of hepatic veins

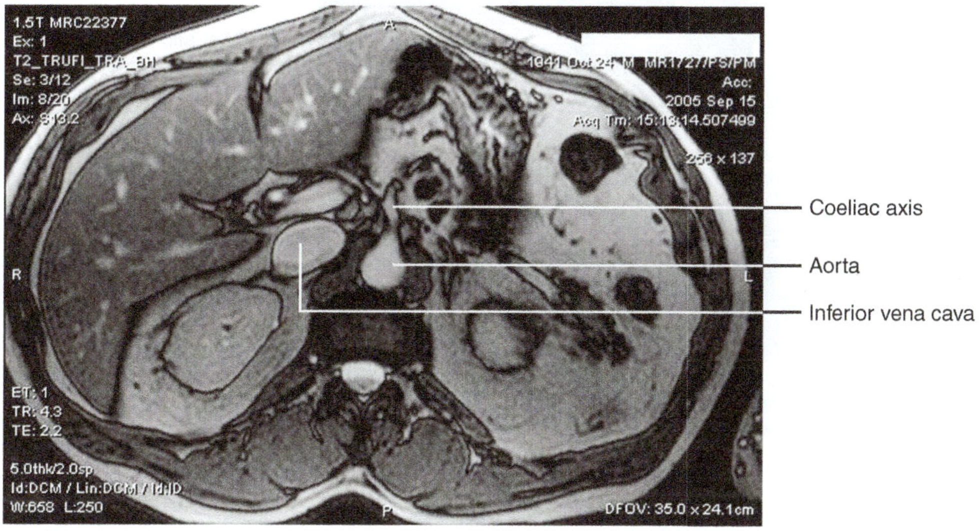

Fig. 18.24 Axial T2 TRUFI liver at the level of coeliac axis

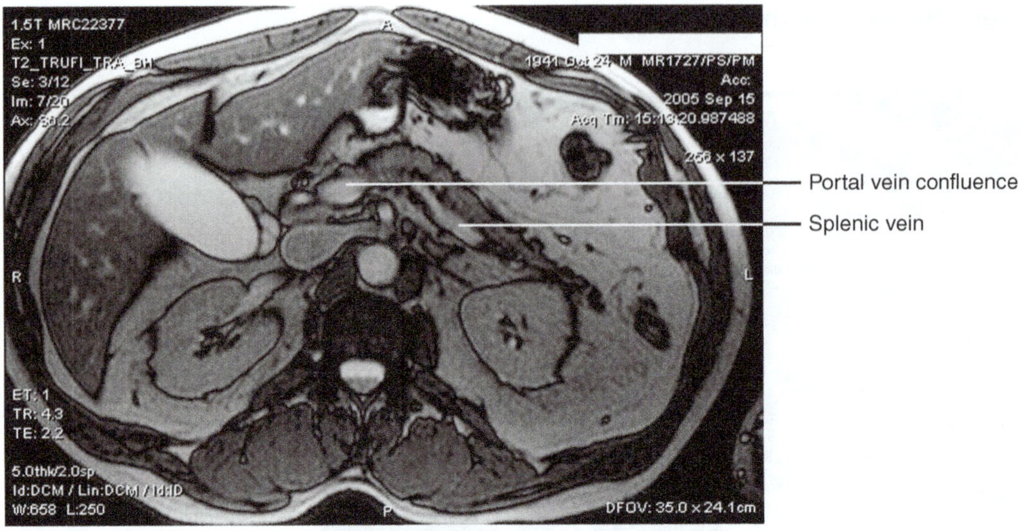

Fig. 18.25 Axial T2 TRUFI liver at the level of splenic vein

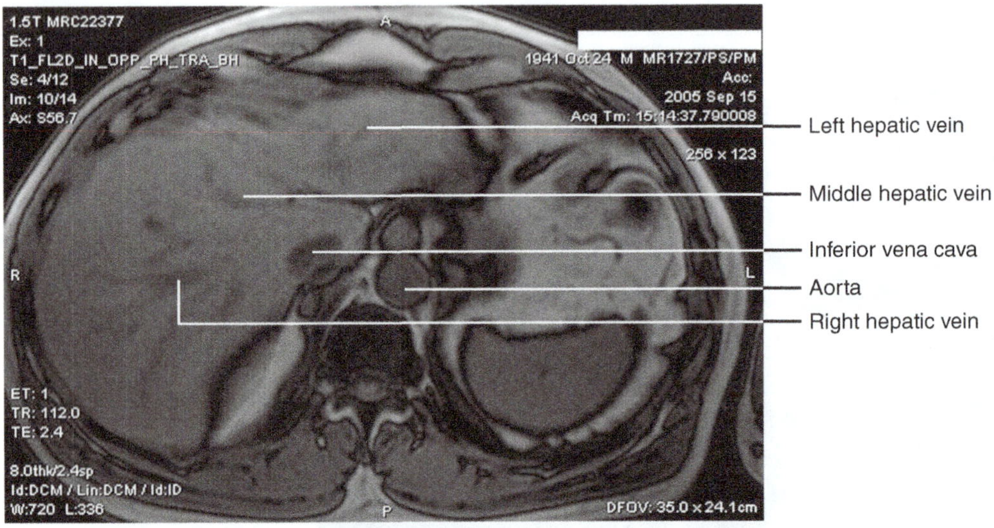

Fig. 18.26 Axial T1 (opposed phase) at the level of hepatic veins

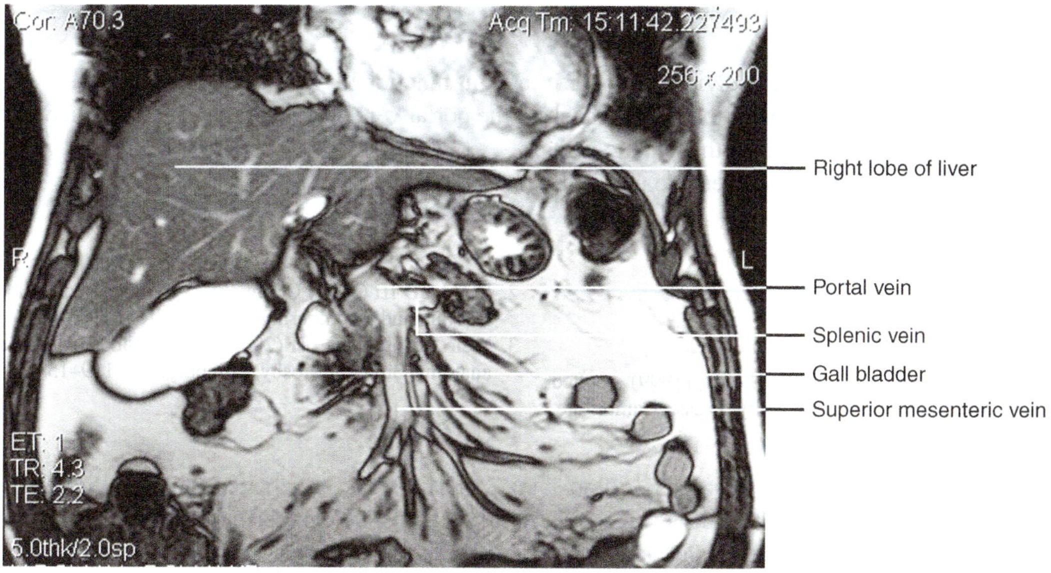

Fig. 18.27 Coronal T2 TRUFI

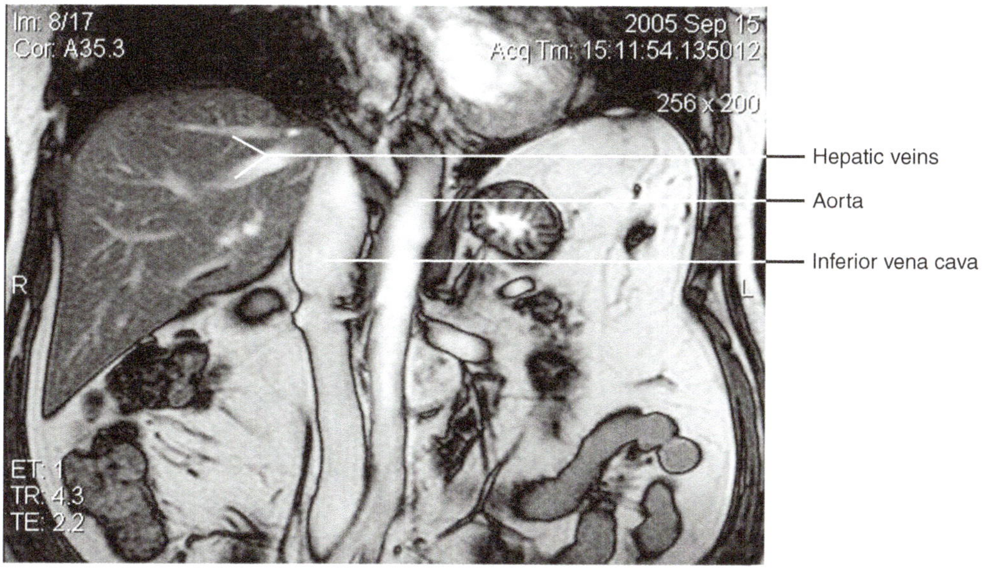

Fig. 18.28 Coronal T2 TRUFI

Endoscopic Retrograde Cholangiopancreatography (ERCP)

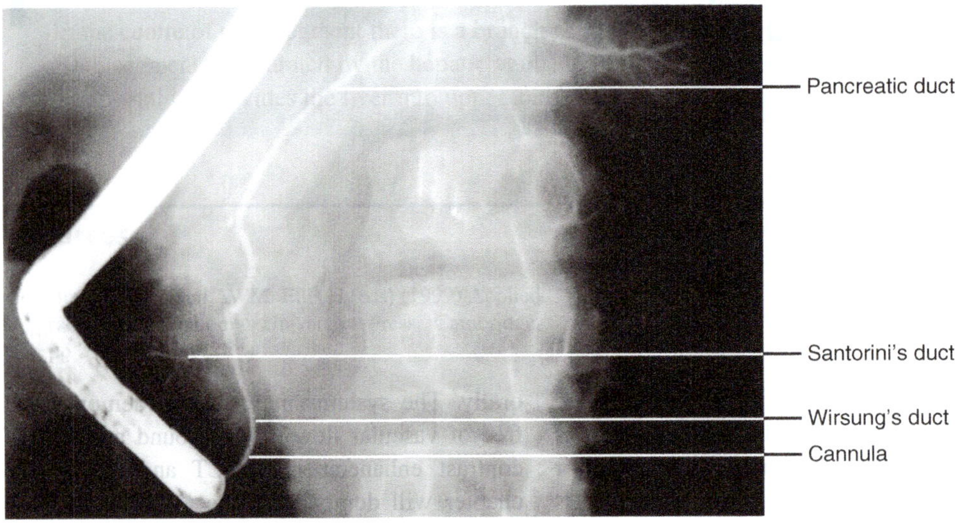

Fig. 19.1 ERCP showing pancreatic duct

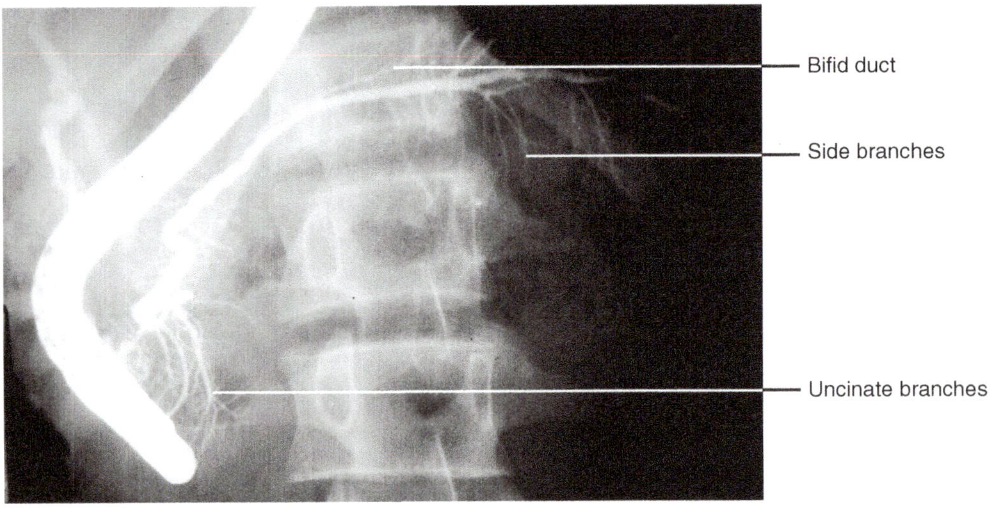

Fig. 19.2 ERCP showing side branches and bifid duct

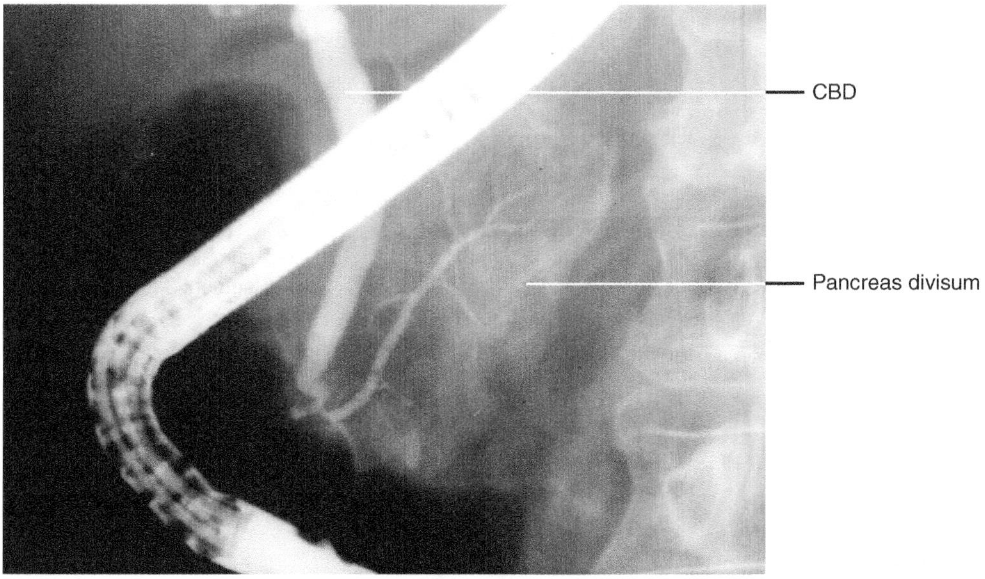

Fig. 19.3 ERCP showing ductus divisum sharing outlet with CBD

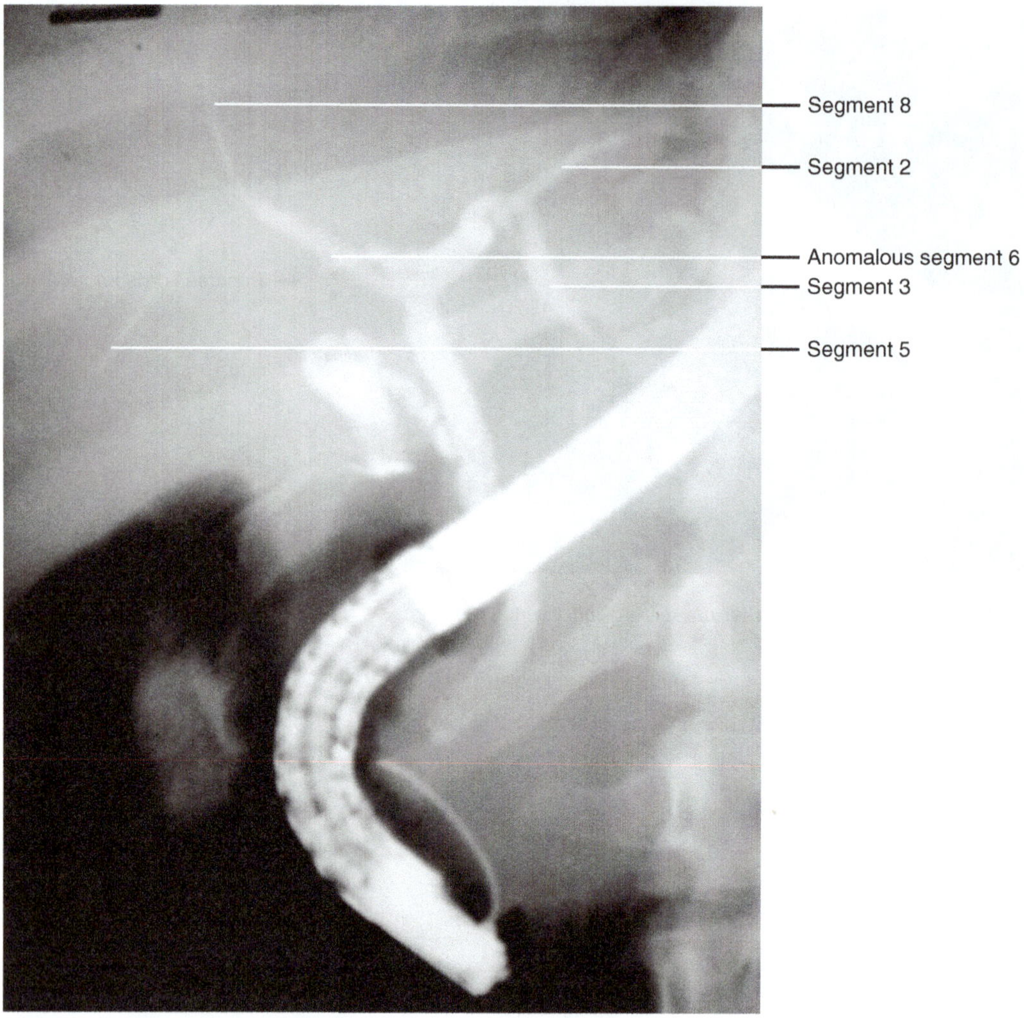

Fig. 19.4 ERCP demonstration of biliary tree – left ducts tend to fill better in the prone position used for ERCP

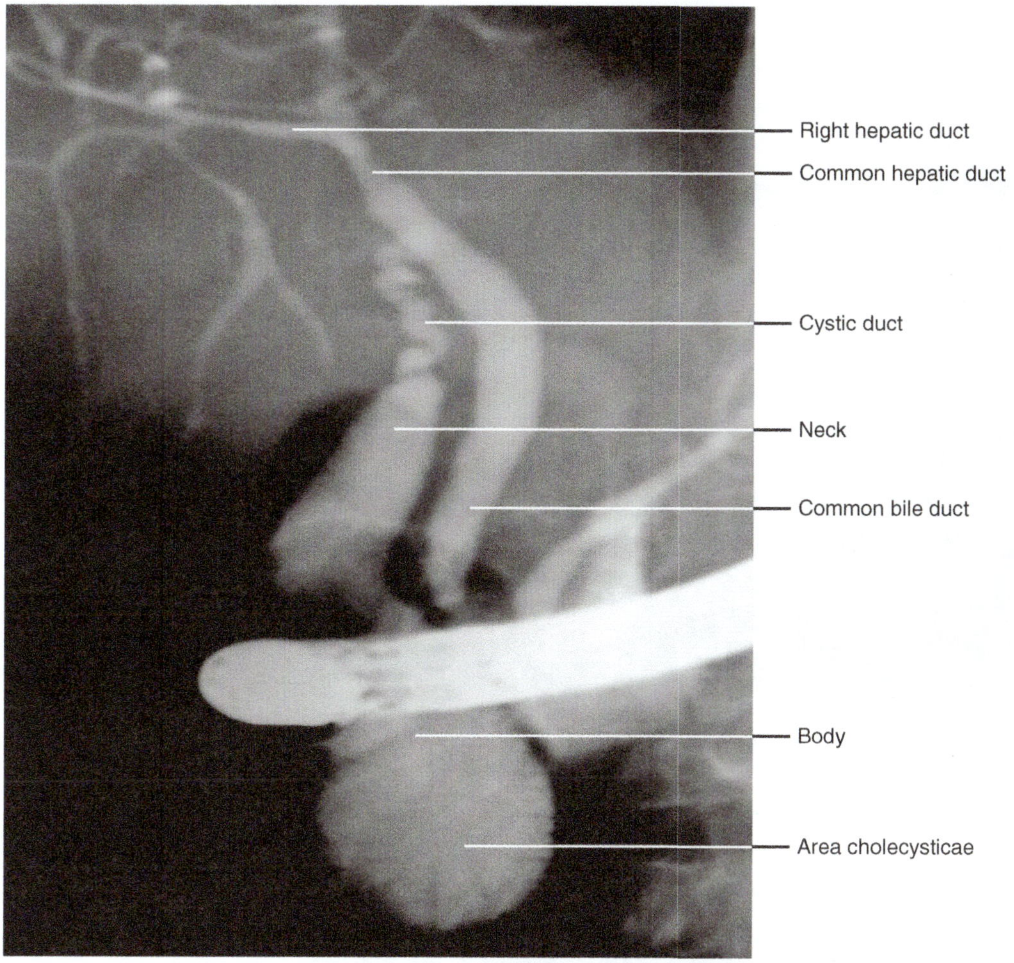

Fig. 19.5 ERCP showing parts of gallbladder

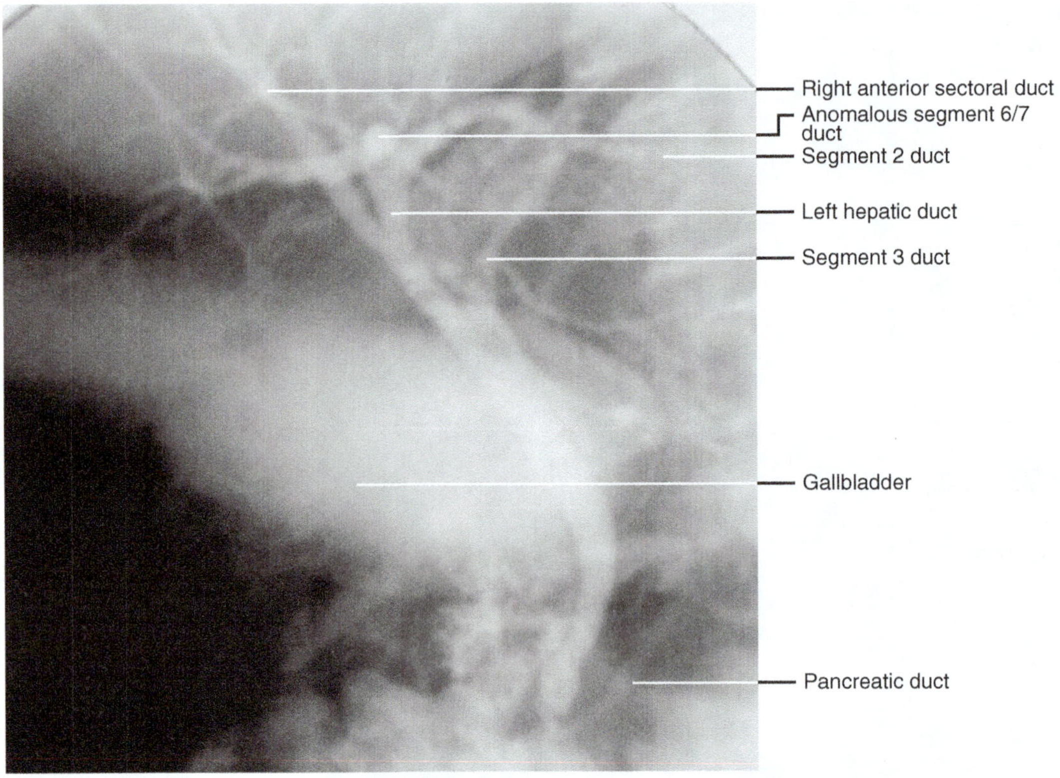

Fig. 19.6 Post-ERCP view with multiple overlapping structures

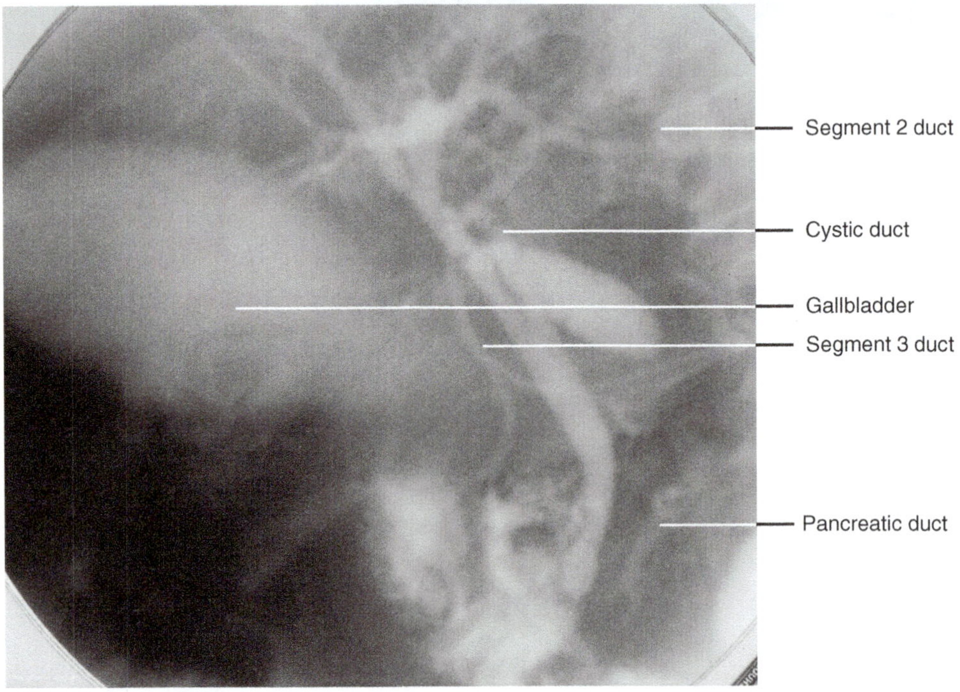

Fig. 19.7 Post-ERCP view multiple overlapping structures

Percutaneous Transhepatic Cholangiopancreatography (PTC)

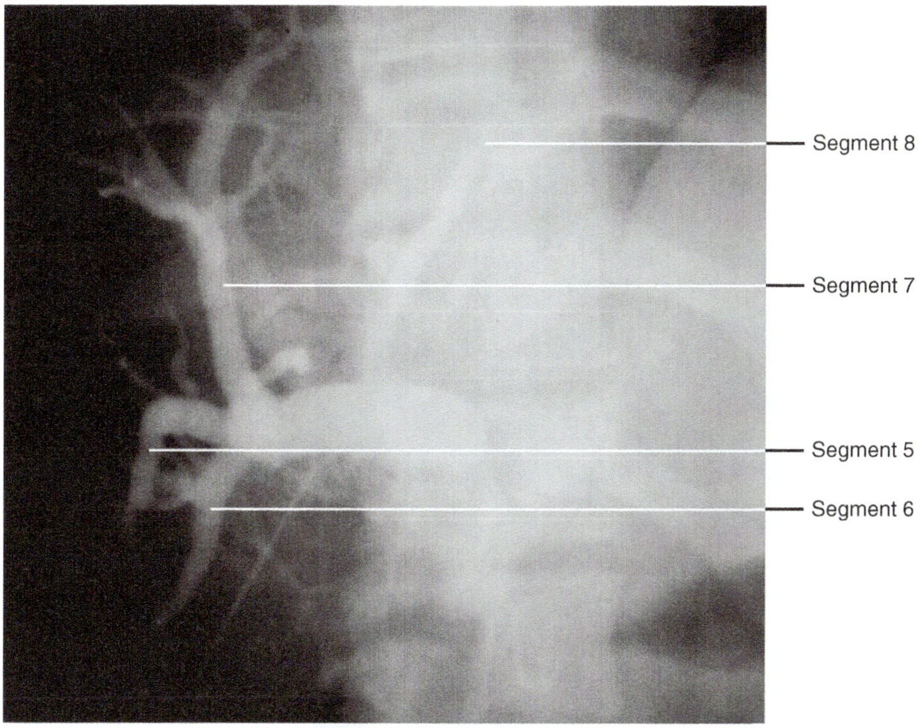

Fig. 19.8 PTC – the RAO allows the segments to be clarified

Magnetic Resonance Imaging

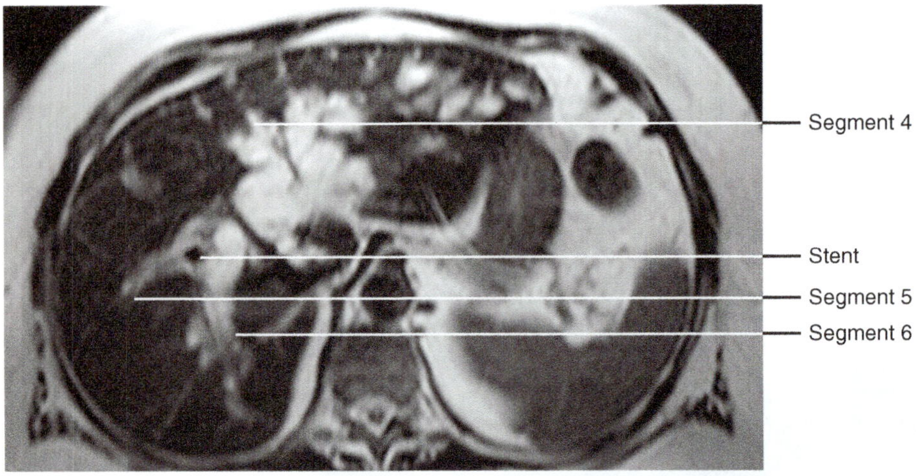

Fig. 19.13 Axial T2 of patient with hilar cholangiocarcinoma and stent in segment 8 shows intrahepatic right ducts well

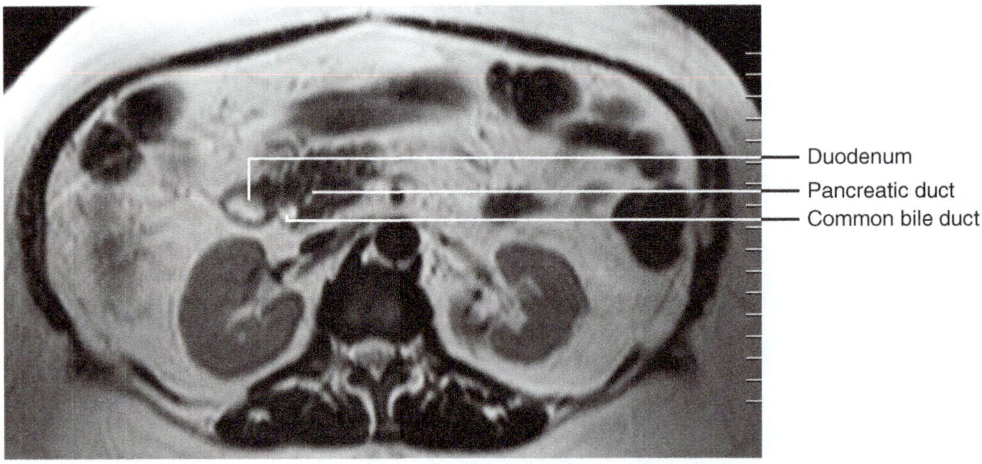

Fig. 19.14 Duodenal, CBD, and pancreatic duct fluid seen as high signal intensity on axial T2 image

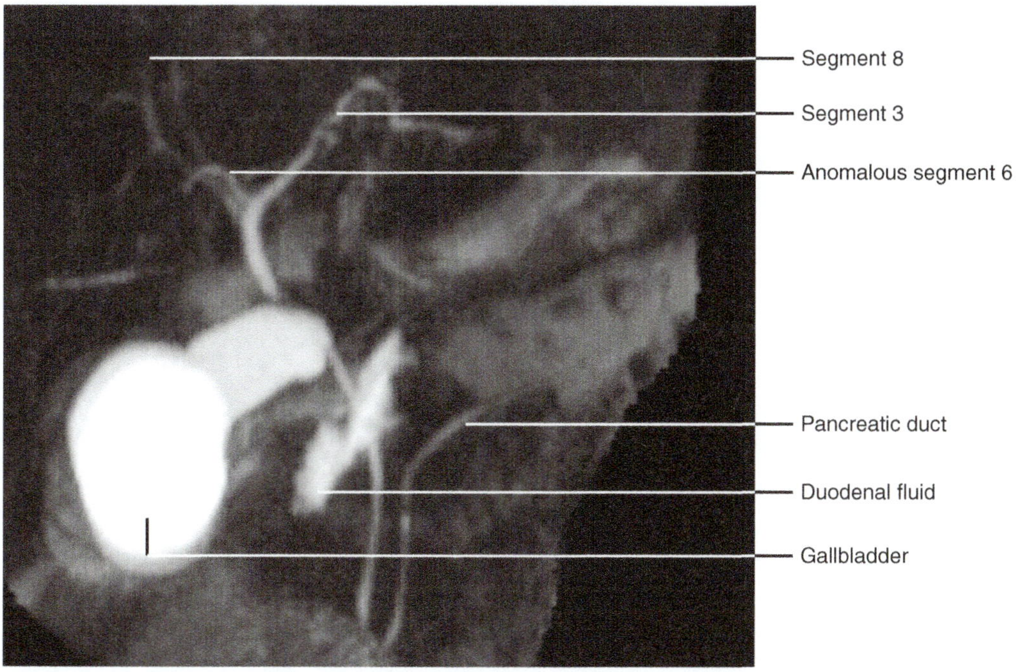

Fig. 19.15 Extracted MRCP data showing ductal systems

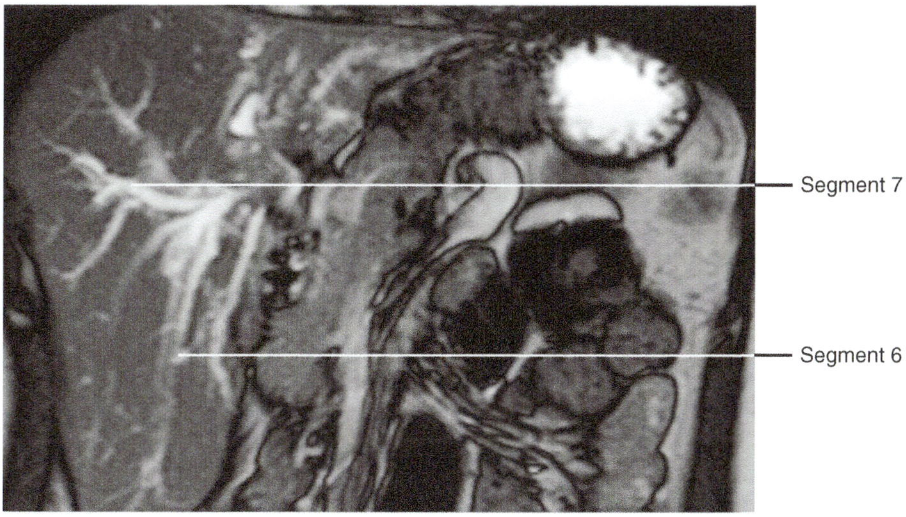

Fig. 19.16 Coronal MRI showing posterior sectoral ducts well

20 Pancreas and Spleen

The pancreas and spleen are well imaged using ultrasound, CT scan and MRI. CT scan is the modality of choice when visualization on ultrasound is suboptimal and where it is available. MRI is fast gaining popularity in imaging both the spleen and pancreas.

S. Andronikou (ed.), *See Right Through Me*,
DOI 10.1007/978-3-642-23893-2_20, © Springer-Verlag Berlin Heidelberg 2012

Ultrasound

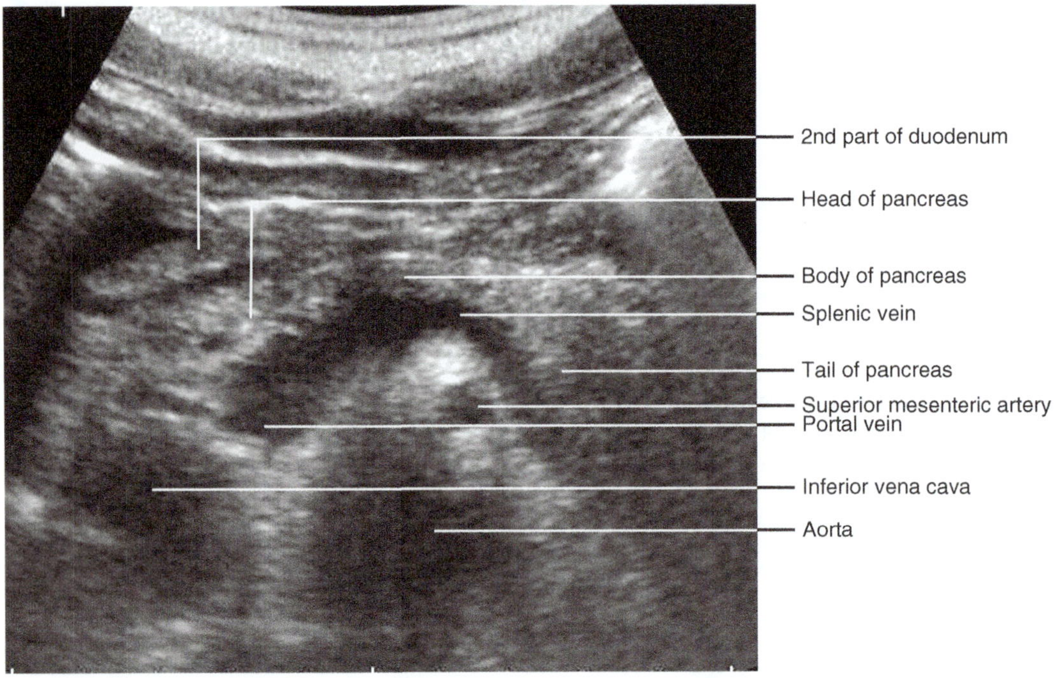

Fig. 20.1 Ultrasound of the pancreas: transverse oblique view

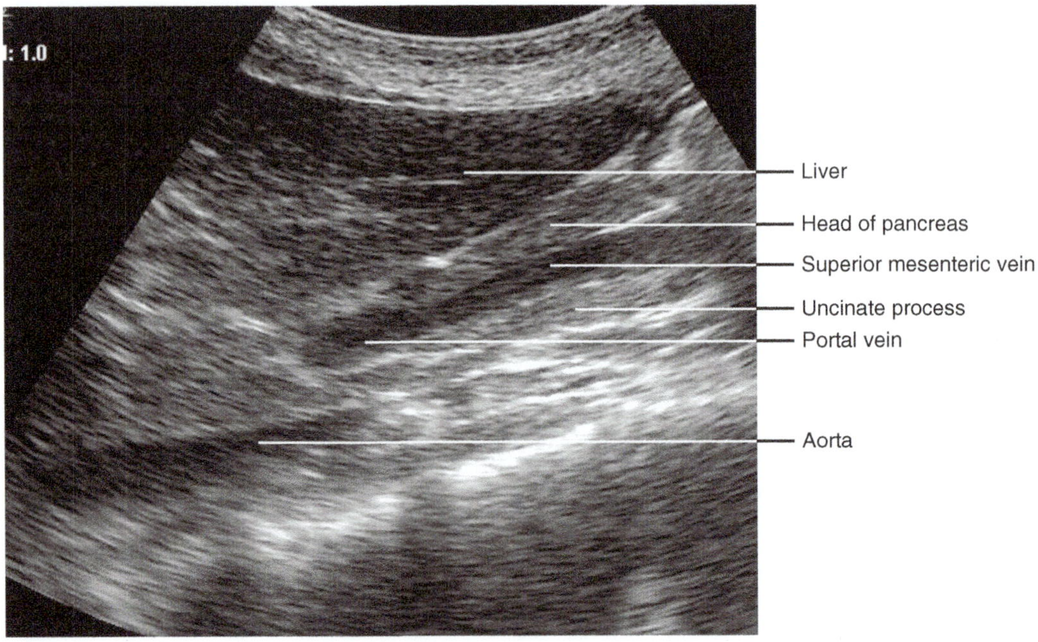

Fig. 20.2 Ultrasound of the pancreas: longitudinal view

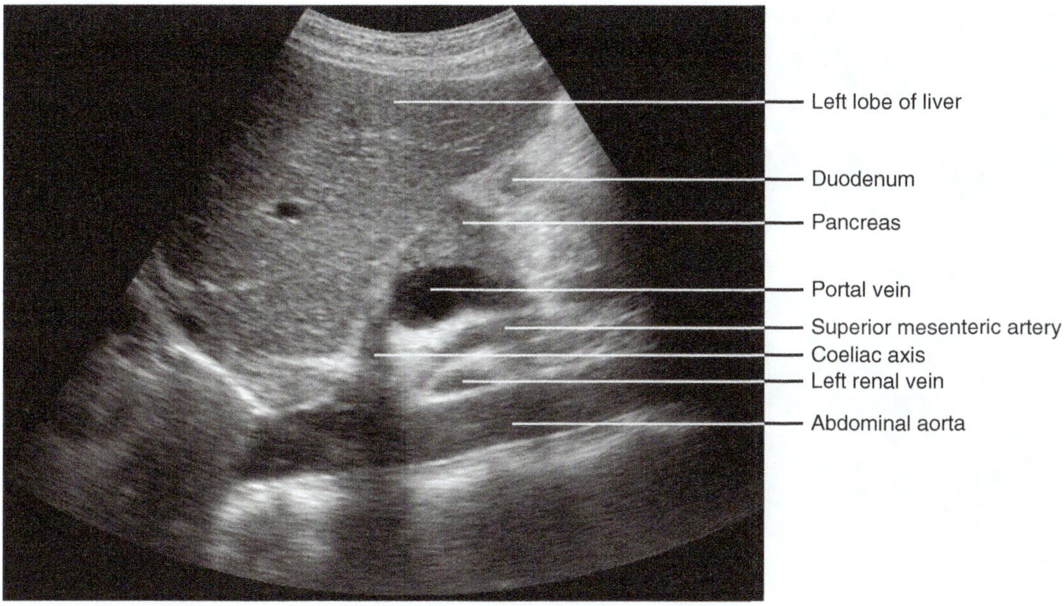

Fig. 20.3 Ultrasound at the level of the aorta: longitudinal view

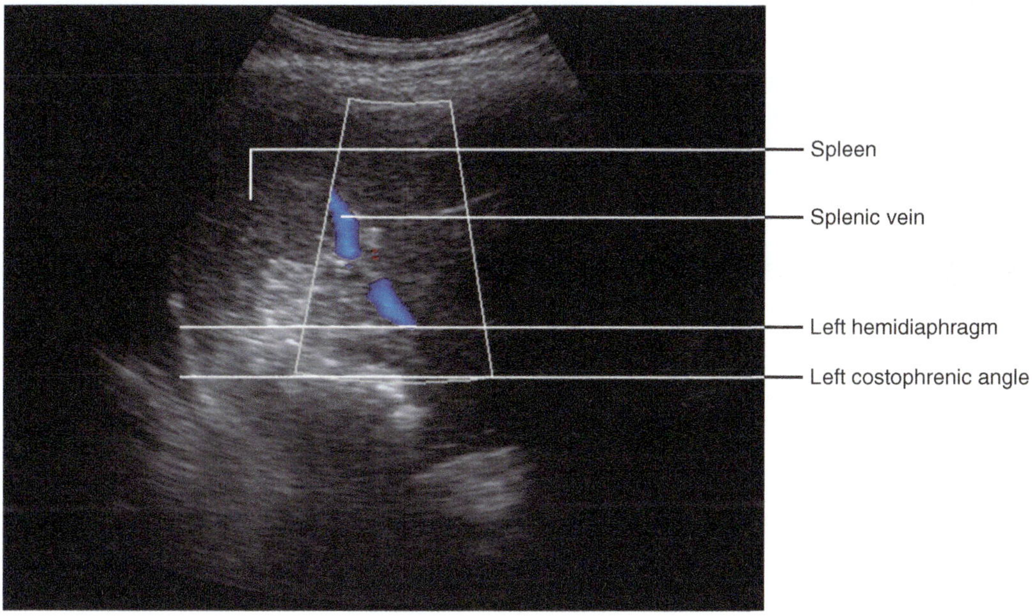

Fig. 20.4 Ultrasound at the spleen: longitudinal view with colour Doppler

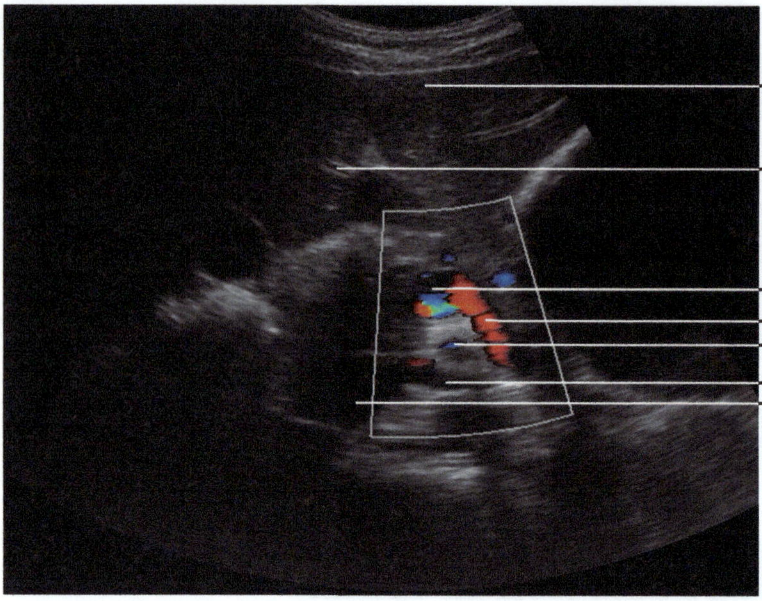

Fig. 20.5 Ultrasound at splenoportal confluence transverse view with: colour Doppler

Pearl

Splenic venous flow should be towards the transducer.

This is often indicated in red but may be either blue or red. Note the colour bar on the screen with colour closest to the probe representing flow towards the transducer.

Computed Tomography

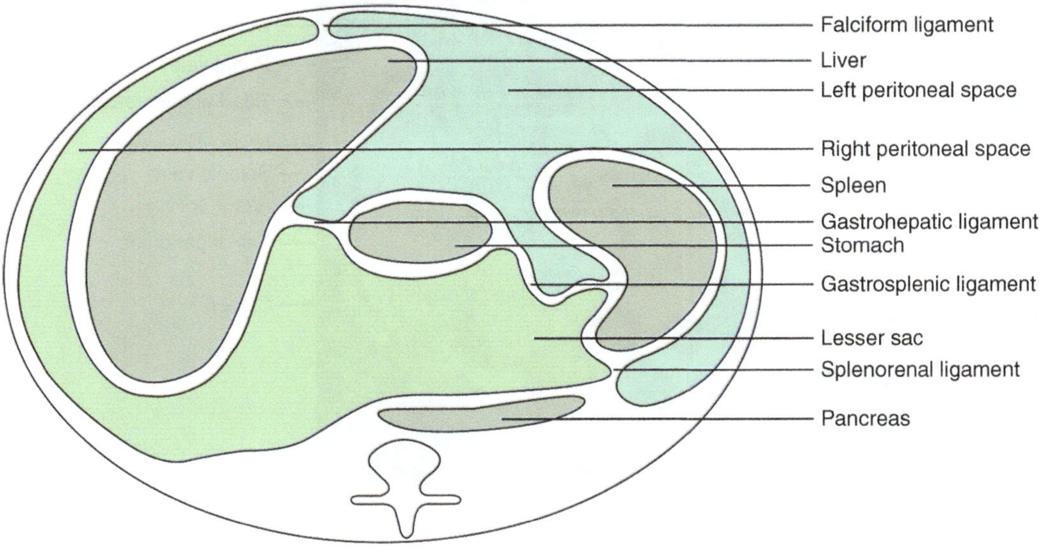

Fig. 20.6 Schematic of peritoneal spaces and ligaments

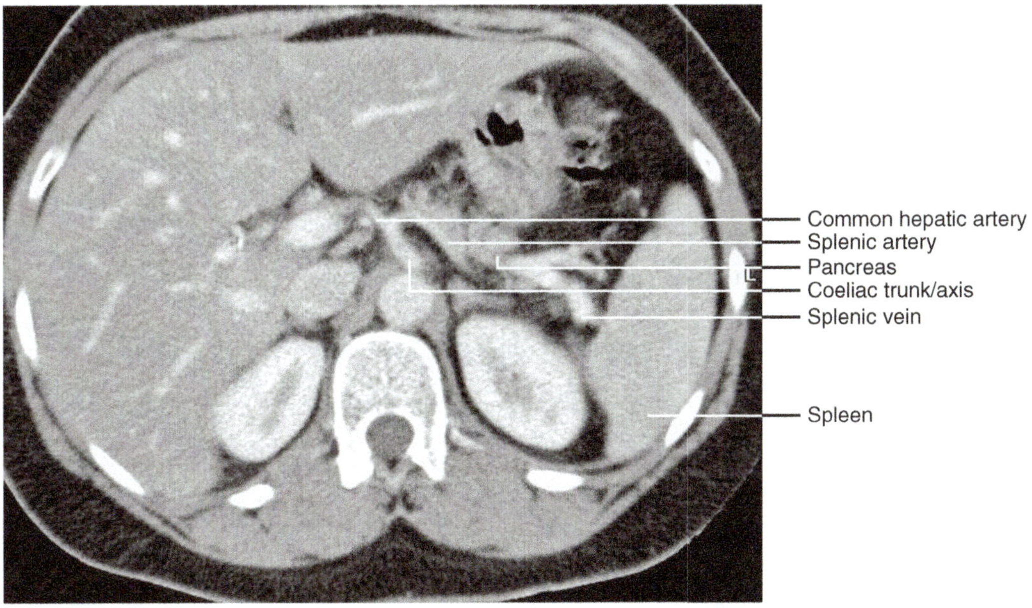

Fig. 20.7 Axial CT – portal venous phase at the level of the coeliac trunk

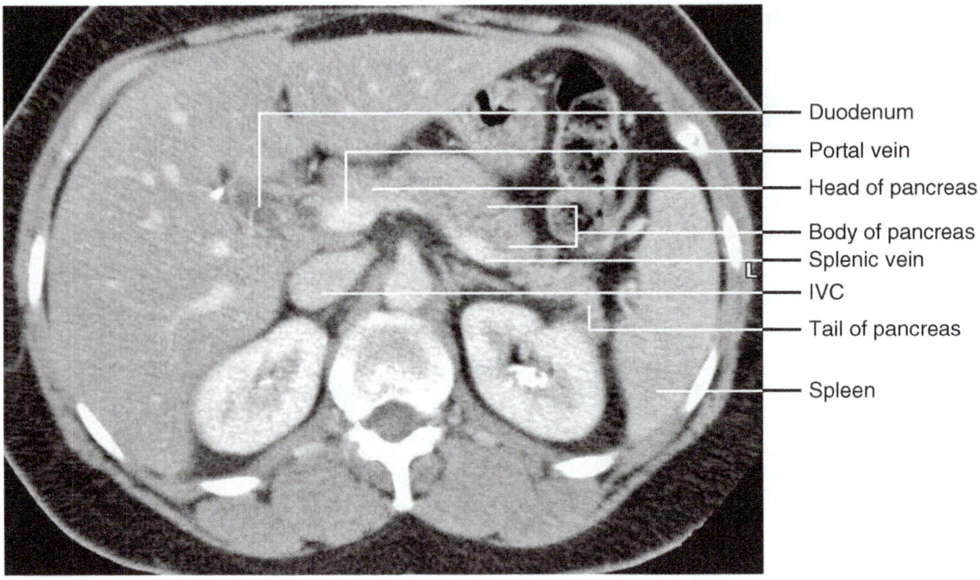

Fig. 20.8 Axial CT – portal venous phase at the level of the pancreas

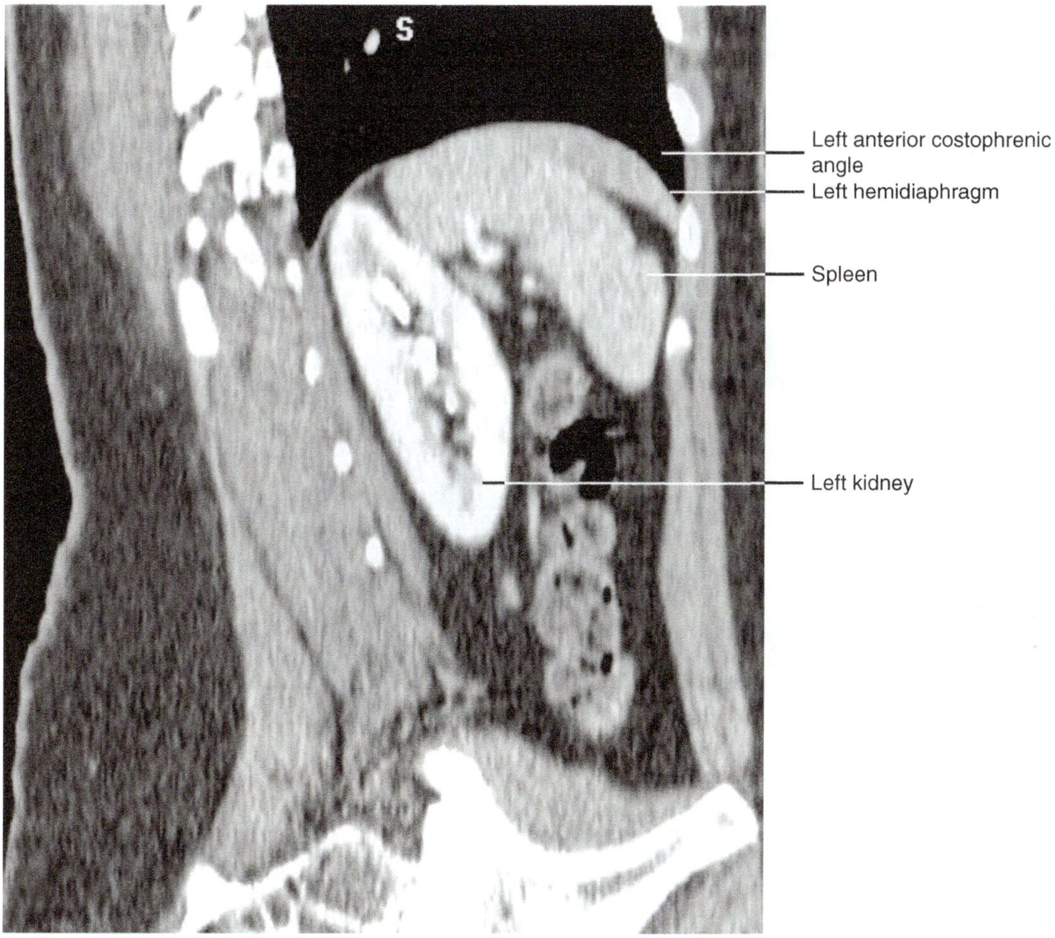

Fig. 20.9 CT scan – sagittal reconstruction

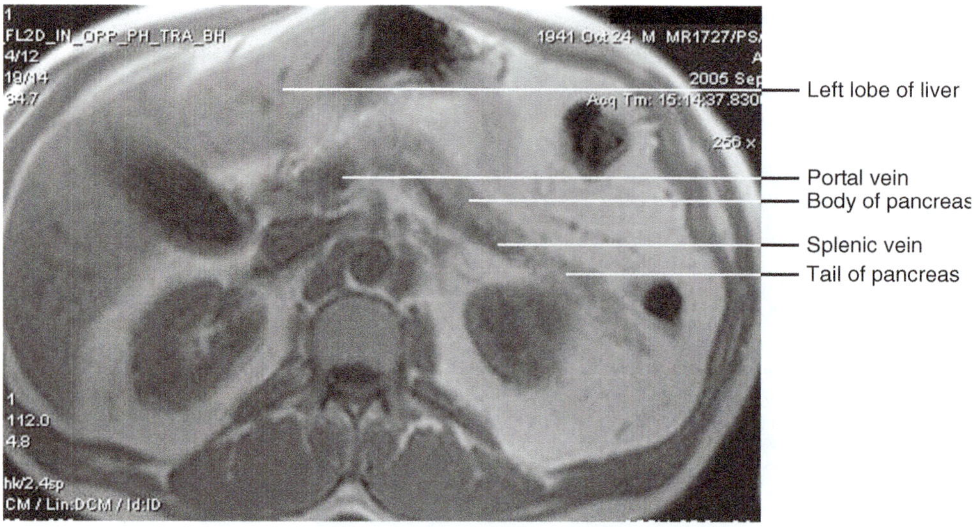

Fig. 20.13 In phase axial (T1)

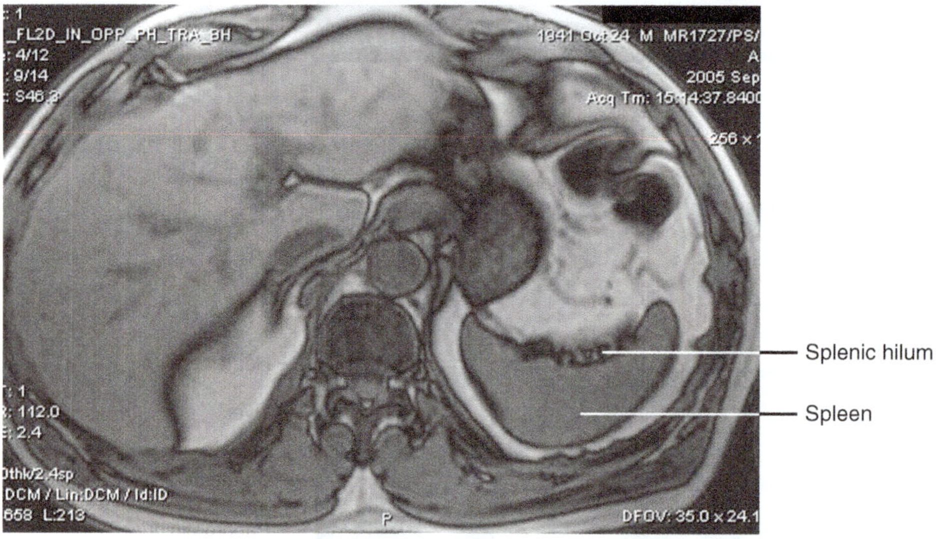

Fig. 20.14 Opposed phase axial (T1)

Pearl: Measurements of the pancreas

Age	Head size in cm ± 1 SD	Body size in cm ± 1 SD	Tail size in cm ± 1 SD
<1 month	1.0 ± 0.4	0.6 ± 0.2	1.0 ± 0.4
1 month–1 years	1.5 ± 0.5	0.8 ± 0.3	1.2 ± 0.4
1–5 years	1.7 ± 0.3	1.0 ± 0.2	1.8 ± 0.4
5–10 years	1.6 ± 0.4	1.0 ± 0.3	1.8 ± 0.4
10–19 years	2.0 ± 0.5	1.0 ± 0.3	2.0 ± 0.4

Pearl

Normal pancreatic duct measures less than 2 mm.

Pearl: Splenic measurements

Age	Range in cm
0–6 years	6–9.5
6–12 years	10–11.5
Above 12 years	Upper limit male = 13; female = 12

Sources

1. Rosenburg HK, Markowitz RL, Kolberg H, Park C, Hubbard A, Bellah RD (1991) Normal splenic size in infants and children: sonographic measurements. AJR Am J Roentgenol 157:119–121
2. Siegel MJ, Martin KW, Wortington JL (1987) Normal and abnormal pancreas in children: ultrasound studies. Radiology 165:15–18
3. Sutton D (2002) Textbook of radiology and imaging, vol 1, 7th edn. Churchill Livingston, London

21 Gastrointestinal Tract

Plain Film

Plain films obtained in the supine position demonstrate the bowel gas pattern.

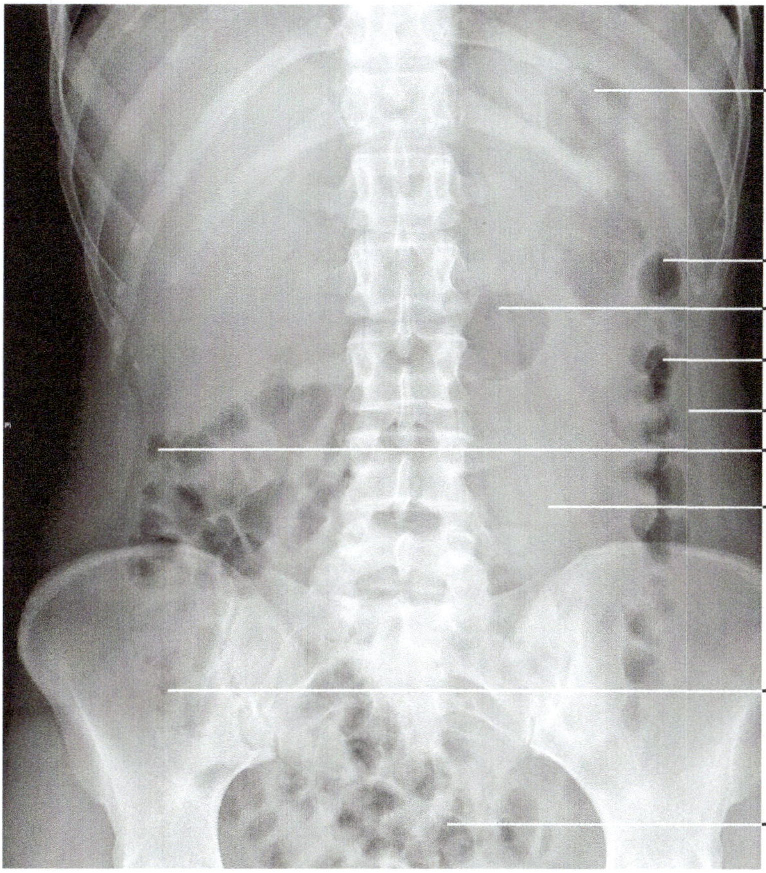

Fig. 21.1 Supine abdominal radiograph

S. Andronikou (ed.), *See Right Through Me*,
DOI 10.1007/978-3-642-23893-2_21,

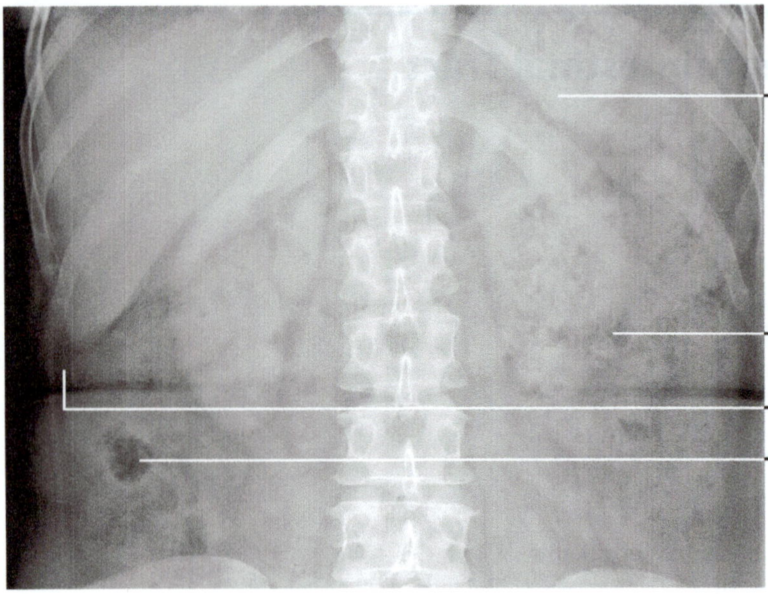

Fig. 21.2 Supine abdominal radiograph after IV contrast during an IVU

Pearl

Prone position for at least 5 min demonstrates the presence of rectal air.
The stomach is normally contracted or distended with food.
The duodenum is not visible.
Small bowel loops <3 cm in caliber are normal.
Maximum allowed caecal diameter is 9 and 5.5 cm for the rest of the colon.
An appendicolith may be visible in 7–15 % of the normal population.

Ultrasound

- Endoscopic ultrasound is used to assess mural anatomy.
- Transabdominal ultrasound may be used to assess gastro-esophageal reflux in children.
- The appendix is seen as a blind-ending tubular structure <6 mm in caliber.
- Anal sphincters can be optimally assessed using an endorectal probe.

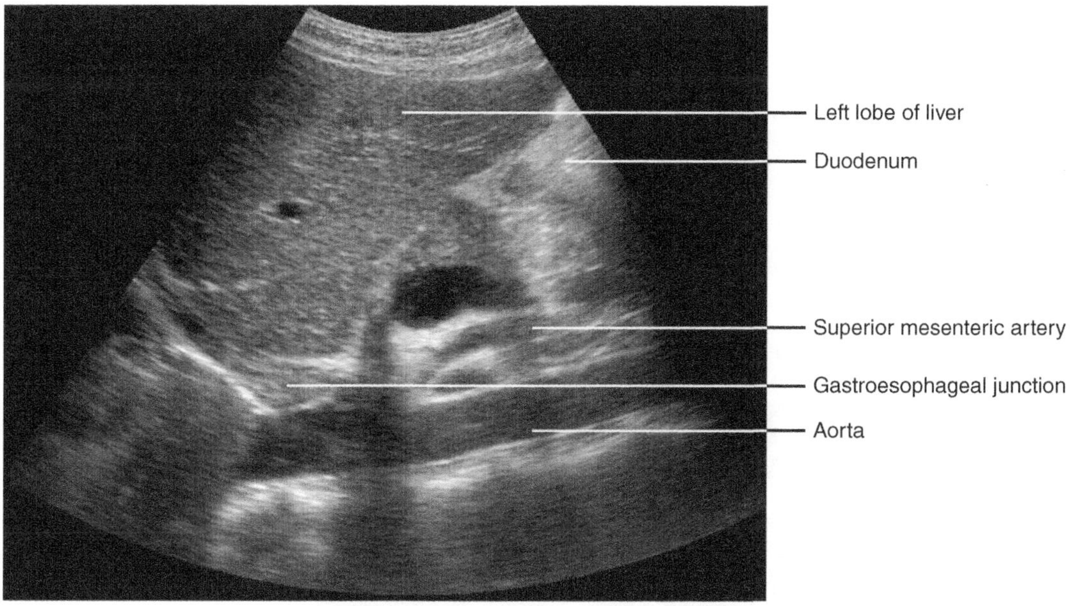

Fig. 21.3 Longitudinal ultrasound through the left lobe of the liver showing the gastroesophageal junction

Contrast Studies

Barium Meal

Assesses morphology and mucosal detail of stomach and duodenum. Usually performed with barium and air, hence called a double contrast barium meal (DCBM).

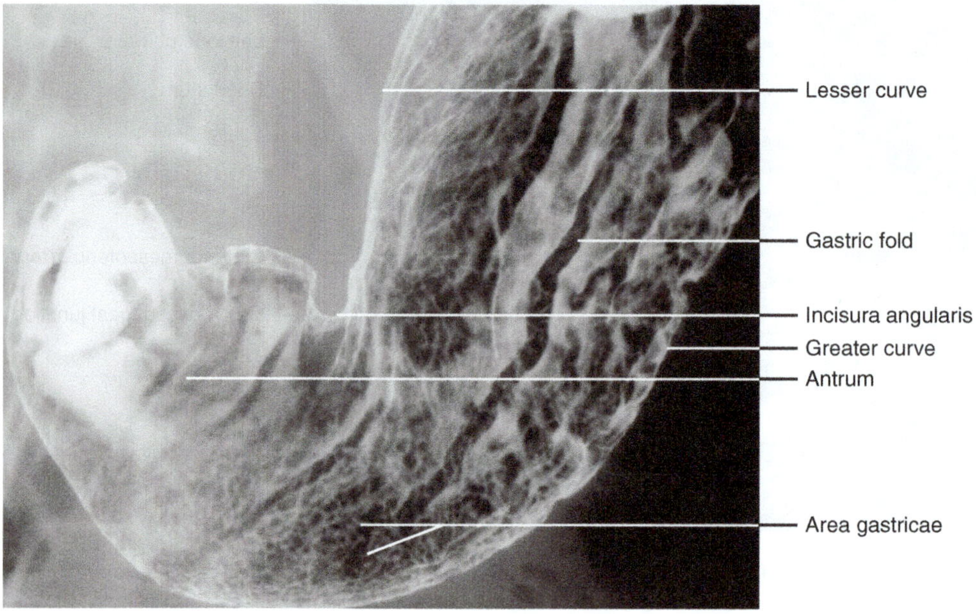

Fig. 21.4 DCBM of stomach

Pearl
Good coating and distension allows visualisation of the area gastricae and "see through" effect.

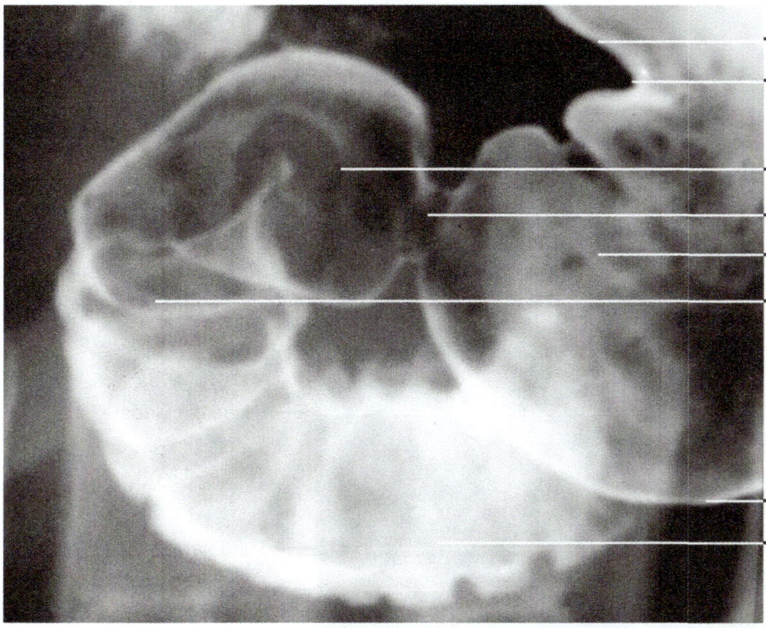

Fig. 21.5 DCBM of the duodenal cap

Pearl

The duodenal cap normally has the shape of a spade.

The duodenal cap may be indented by a normal gall bladder.

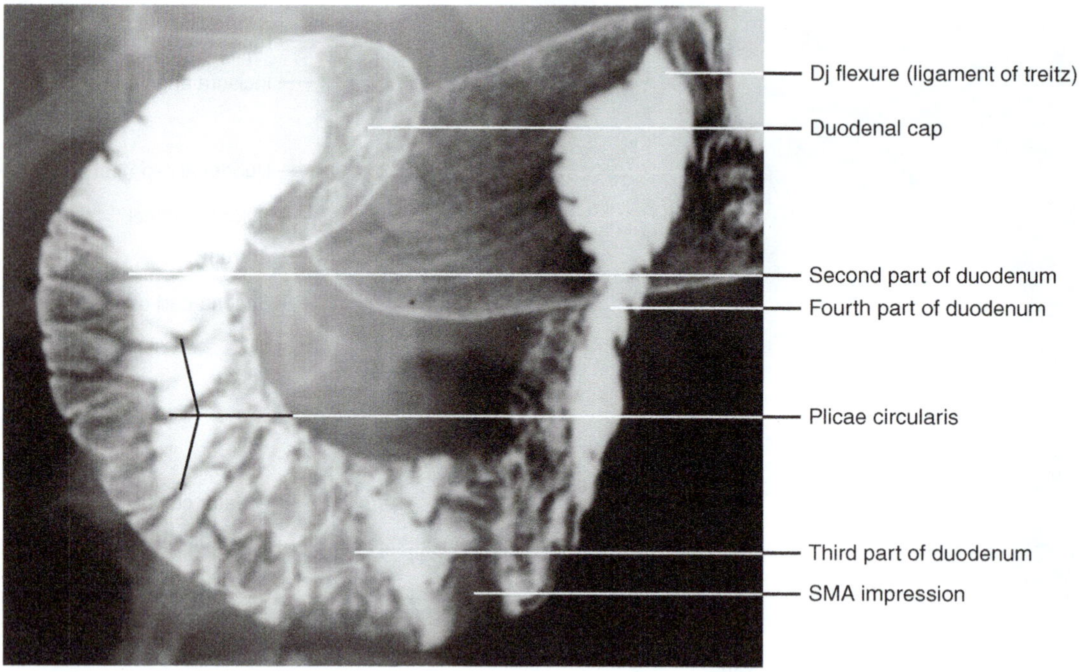

Fig. 21.6 DCBM of duodenum

Pearl

On the AP projection, D1 is located to the right of the spine and the duodeno-jejunal flexure to the left of the left pedicle at the same level or higher.

The major papilla is visualised in 66 % of cases.

The superior mesenteric artery may indent D3 (this may be pathological).

Small Bowel Enteroclysis

Demonstrates small bowel disposition, valvulae conniventes, and wall thickness.

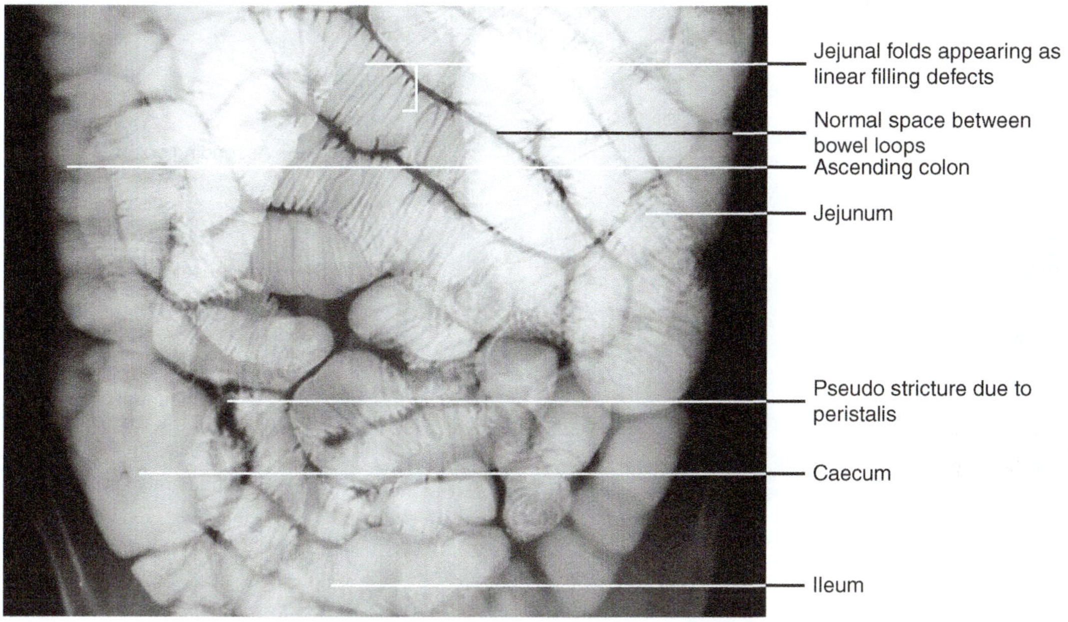

Fig. 21.7 Small bowel enteroclysis (SBE)

Pearl

Prominence of the valvulae conniventes at the jejunum.
Folds are smooth and <3 mm in thickness.
Bowel loops are evenly spaced.
Good contrast and exposure factors results in a "see-through" effect.

Barium Enema

Allows assessment of large bowel morphology and mucosal detail. Usually performed with barium and air and hence called a double contrast barium enema (DCBE).

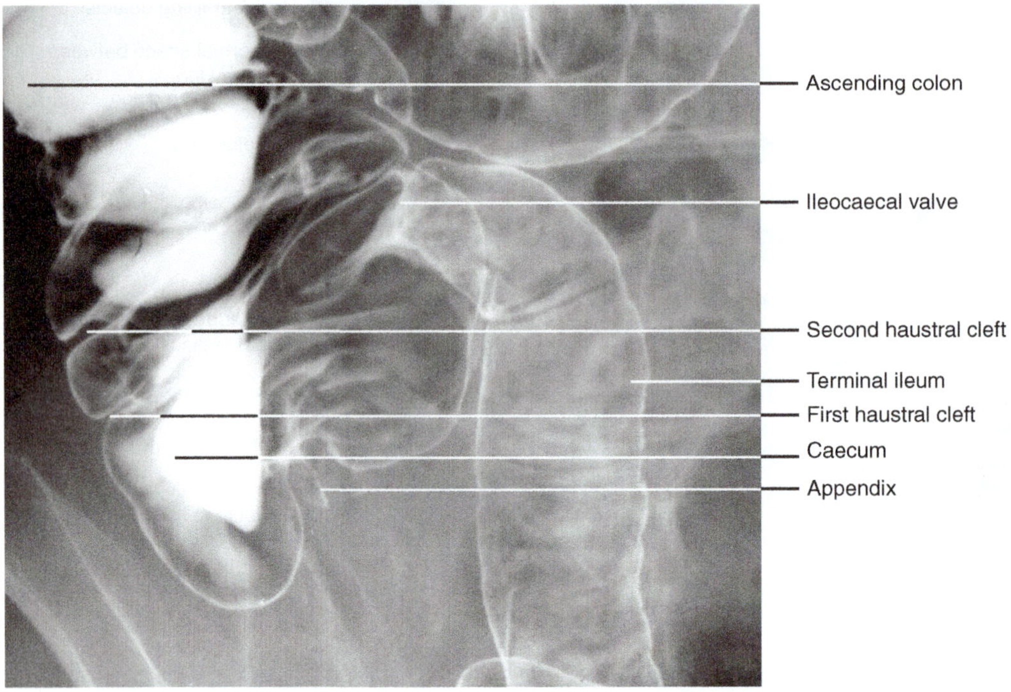

Fig. 21.8 DCBE of the caecum and terminal ileum with reflux into small bowel

> **Pearl**
> The normal terminal ileum resembles a "birds beak" at the ileocaecal valve.

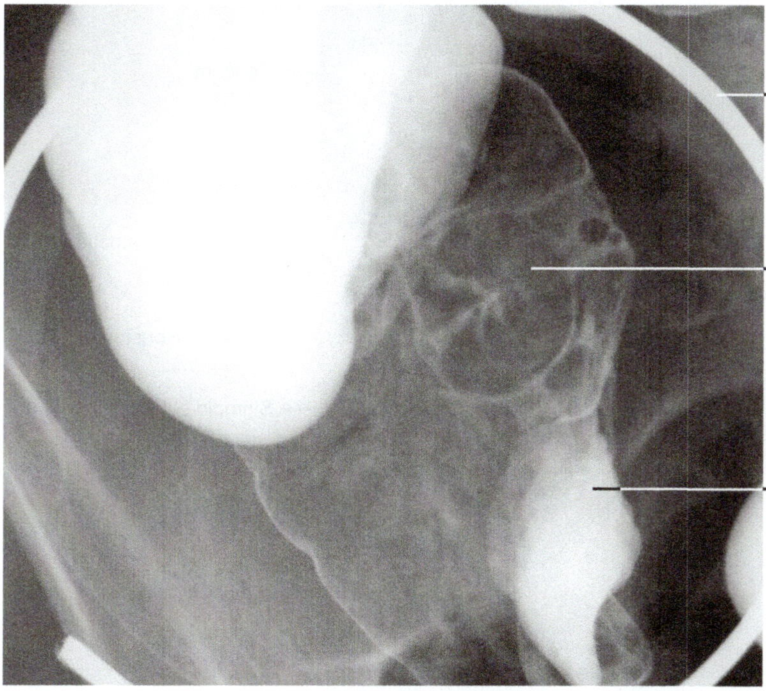

Fig. 21.9 DCBE with compression of the caecum

Pearl
A rounded filling defect in the caecum is caused by the ileocaecal valve.

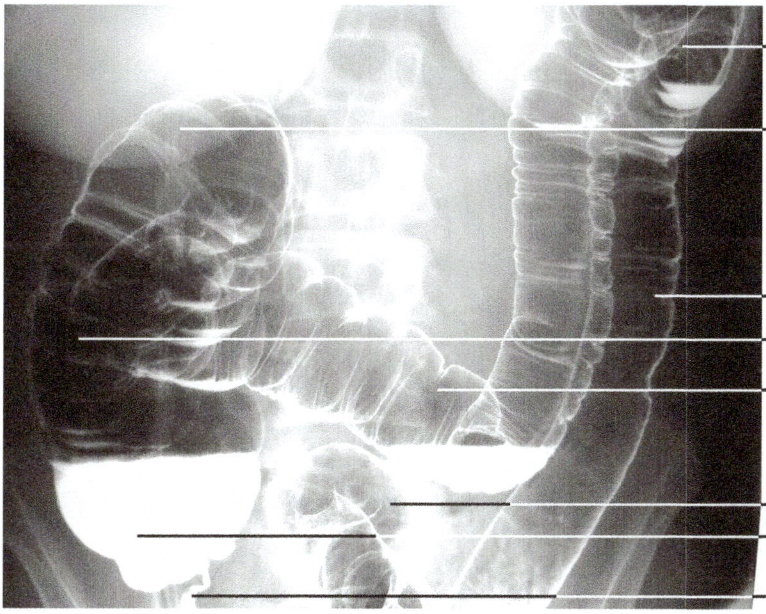

Fig. 21.10 DCBE of the large bowel, erect view

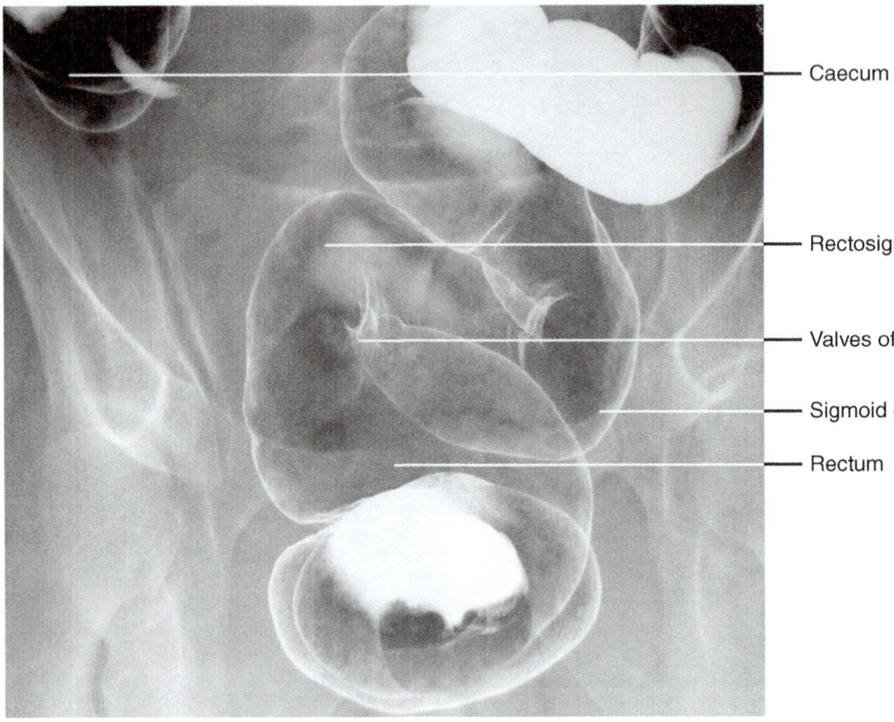

Fig. 21.11 DCBE of the rectosigmoid, AP view

Pearl
This view is performed prone "30° up" to elongate the bowel loops and the enema tube is removed to prevent obscuring of low lesions.

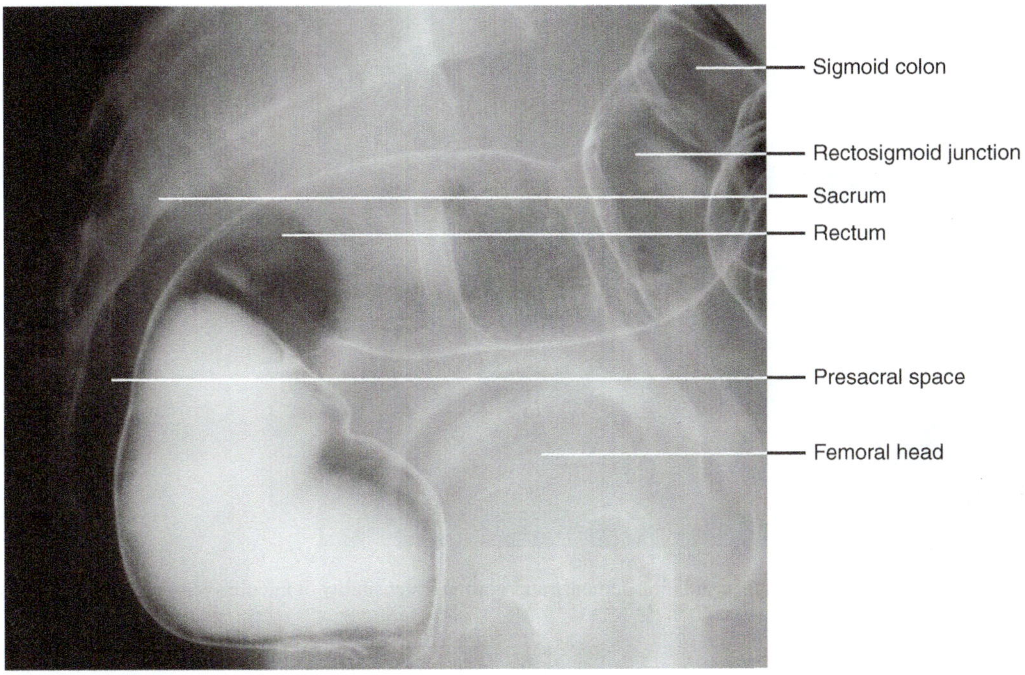

Fig. 21.12 DCBE of the rectosigmoid, AP view

Pearl
The normal presacral width is <15 mm at the fourth sacral element. This increases to 20 mm in obese and elderly patients.

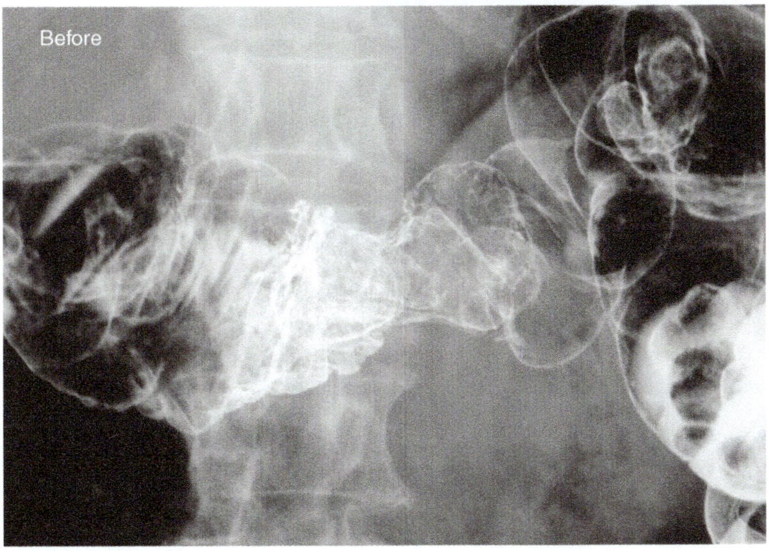

Fig. 21.13 Cannon's ring simulating pathology at the transverse colon during DCBE

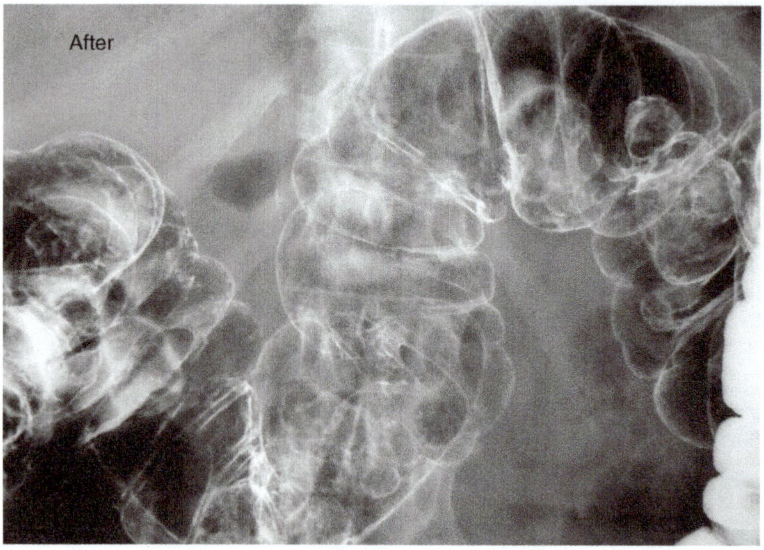

Fig. 21.14 IV muscle relaxant with air inflation demonstrates normal distensibility of the transverse colon at Cannon's ring

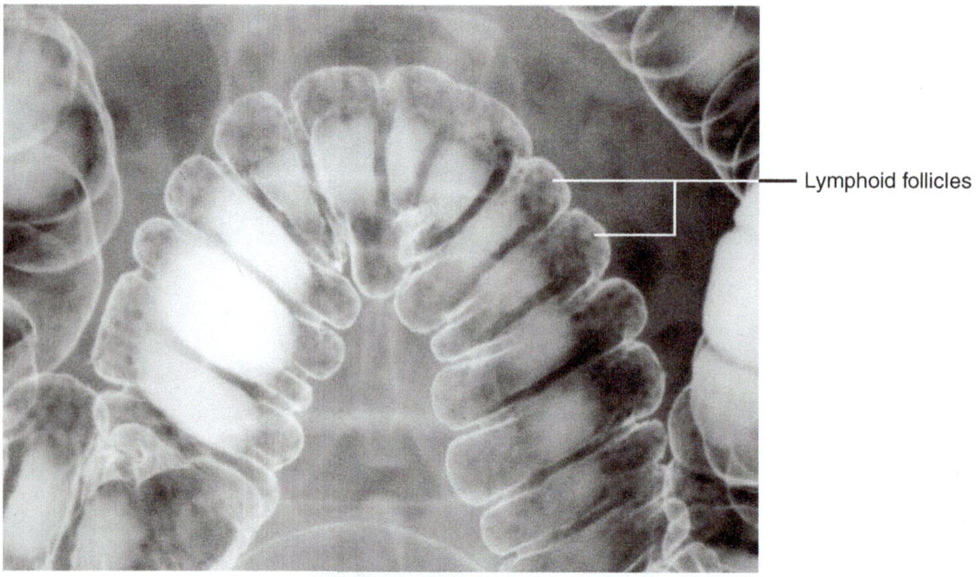

Fig. 21.15 DCBE demonstrating lymphoid follicles

Pearl

Lymphoid tissue aggregates are a normal variant and should measure <4 mm each.

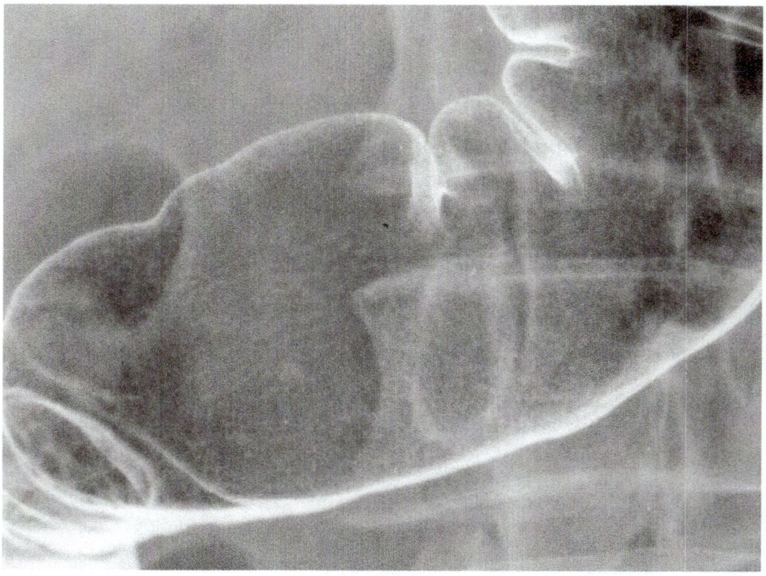

Fig. 21.16 DCBE demonstrating innominate lines

Pearl

These are superficial indentations of the mucosa due to the Crypts of Lieberkühn.

Computed Tomography

Optimal delineation of luminal, mural, and extramural anatomy.

Performed with positive/negative intraluminal contrast (using air, water or oral contrast agent).

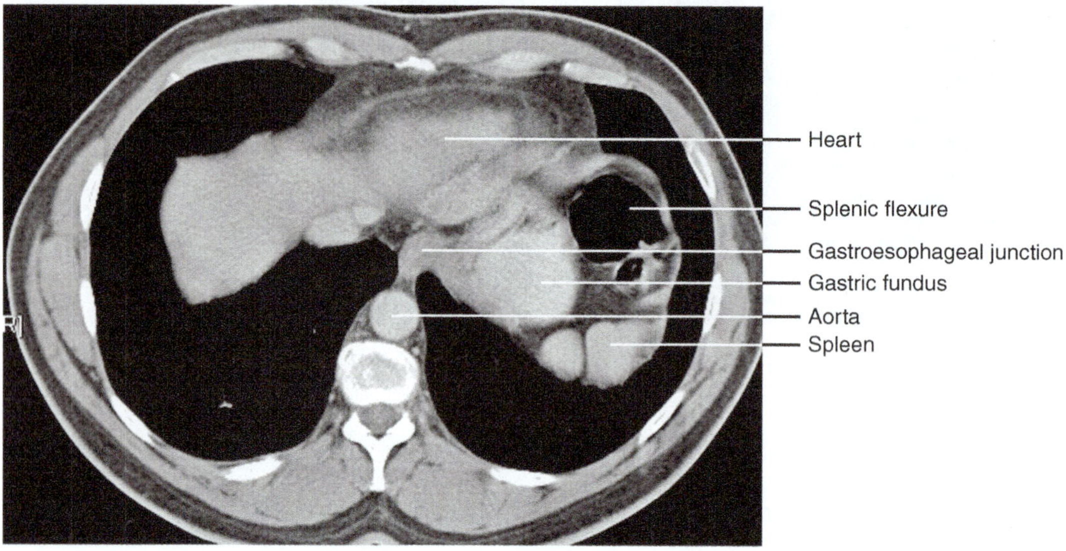

Fig. 21.17 Axial post contrast CT demonstrating the splenic flexure of the colon

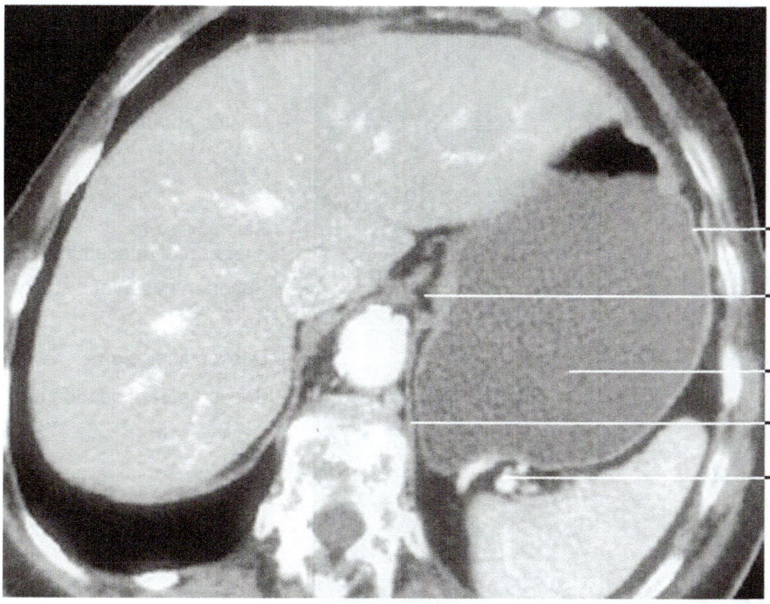

Fig. 21.18 Axial post contrast CT demonstrating the stomach fundus

Pearl

Thickness of the gastric wall is even except at the gastroesophageal junction.
Gastric wall is accentuated by negative intraluminal contrast and perigastric fat.
A fat plane separates the diaphragmatic crura from the gastric wall.

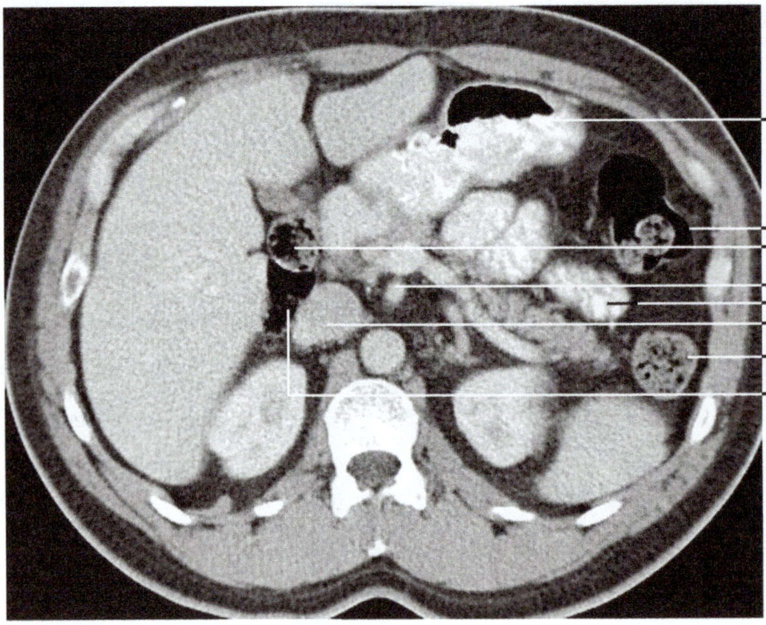

Fig. 21.19 Axial post contrast CT demonstrating the second part of the duodenum

Pearl

The superior mesenteric artery should always be surrounded by a cuff of fat.

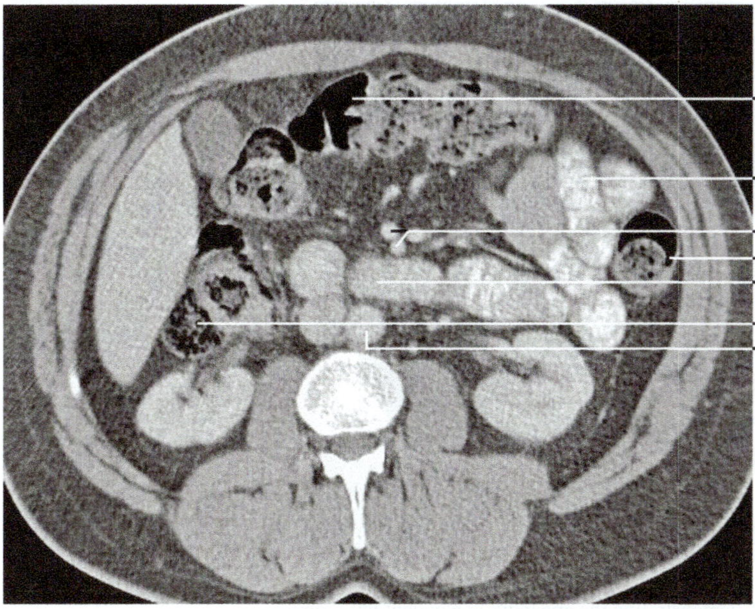

Fig. 21.20 Axial post contrast CT demonstrating the third part of the duodenum

Pearl

D3 is the last of the three structures passing between the SMA and the aorta. The other two are the uncinate process of the pancreas and the left renal vein.

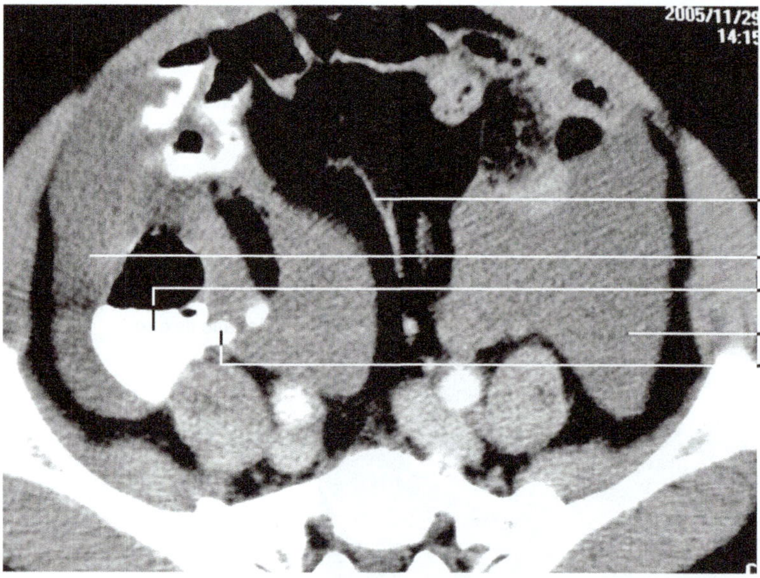

Fig. 21.21 Axial post contrast CT demonstrating the appendix in a patient with ascites

Pearl
The ileocaecal valve can cause a significant filling defect mimicking a soft tissue mass.

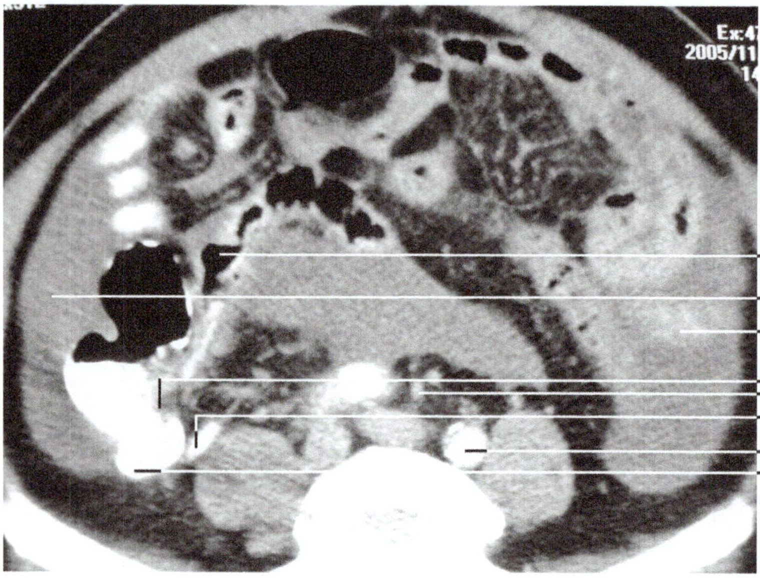

Fig. 21.22 Axial post contrast CT demonstrating the terminal ileum in a patient with ascites

Virtual Colonoscopy

Pearl

Fluid and faecal matter may mimic pathology – hence correlate supine and prone axial CT reference images, with CT virtual colonoscopy images.

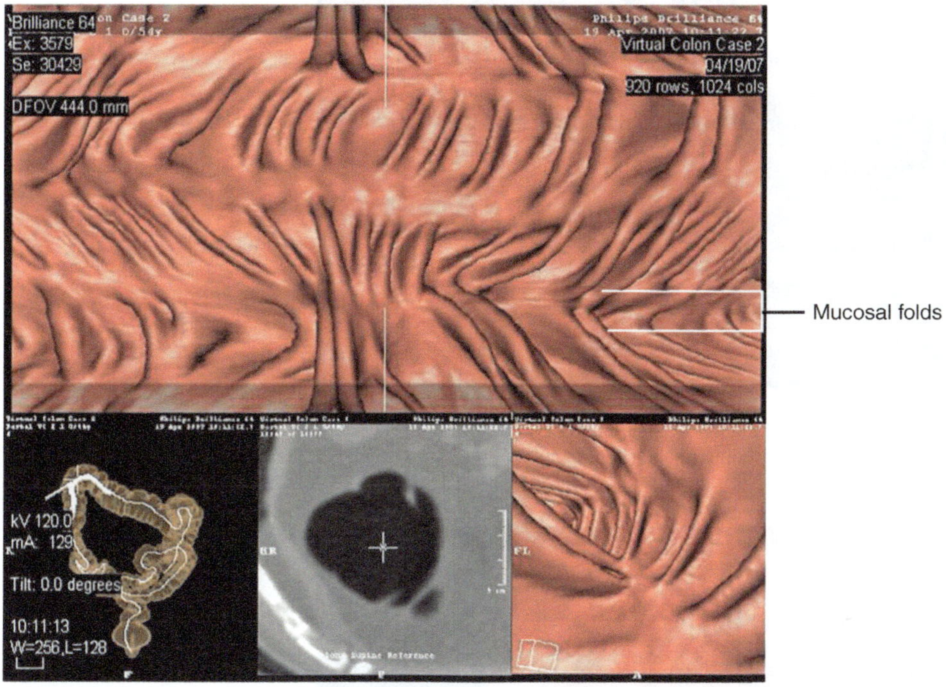

Fig. 21.23 Normal CT virtual colonoscopy. Fillet post processing elongated flattened view of the dissected colon (*top*). Volume-rendered image of the colon in which an automatically central colonic path is created prior to colon virtual dissection (*bottom left*). Axial CT image for reference (*bottom middle*). Virtual endoscopic image (*bottom right*)

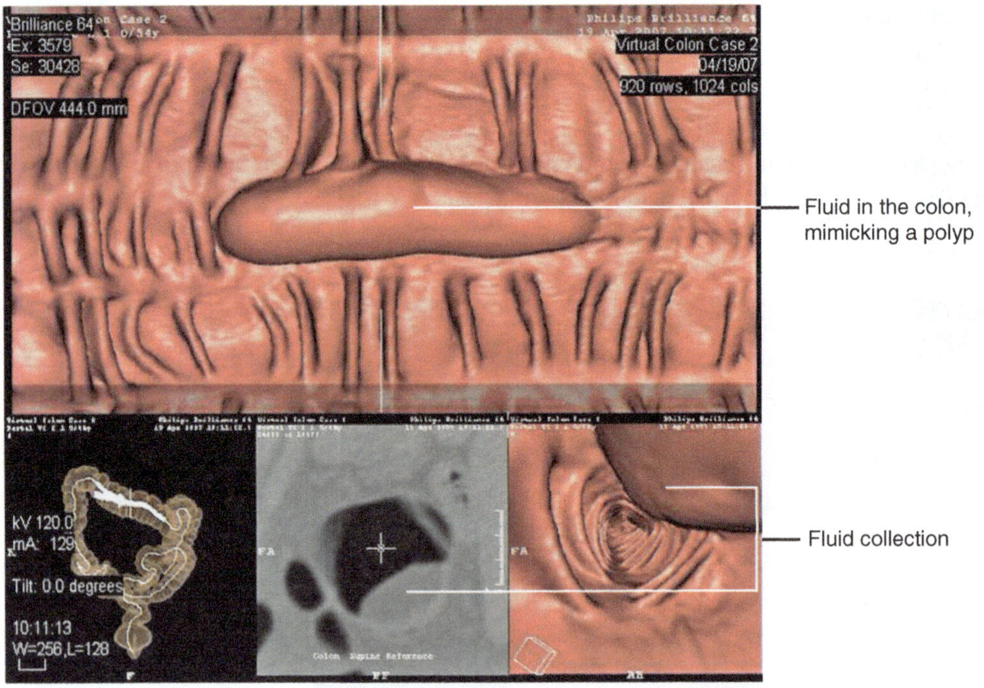

Fig. 21.24 Fluid in colon mimicking a polyp. Fillet view (*top*). Volume-rendered image (*bottom left*). Axial CT reference image (*bottom middle*). Endoscopic image (*bottom right*)

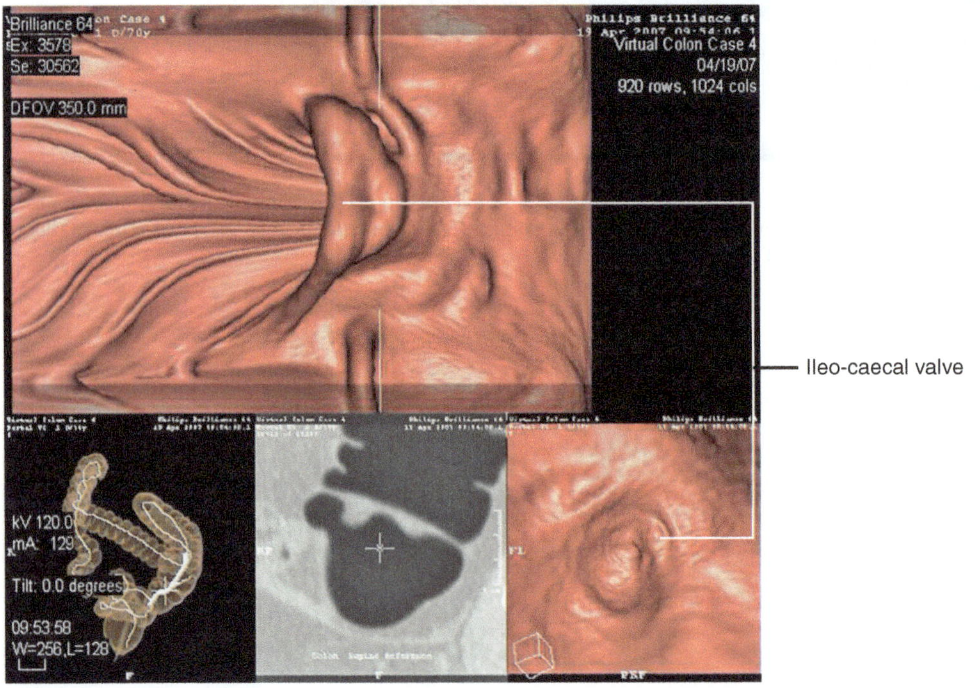

Fig. 21.25 Valve of Bauhini (ileo-caecal valve). CT colonoscopy (*top* and *bottom right*). Axial CT image for reference (*middle*). Prone volume-rendered image of the colon (*bottom left*)

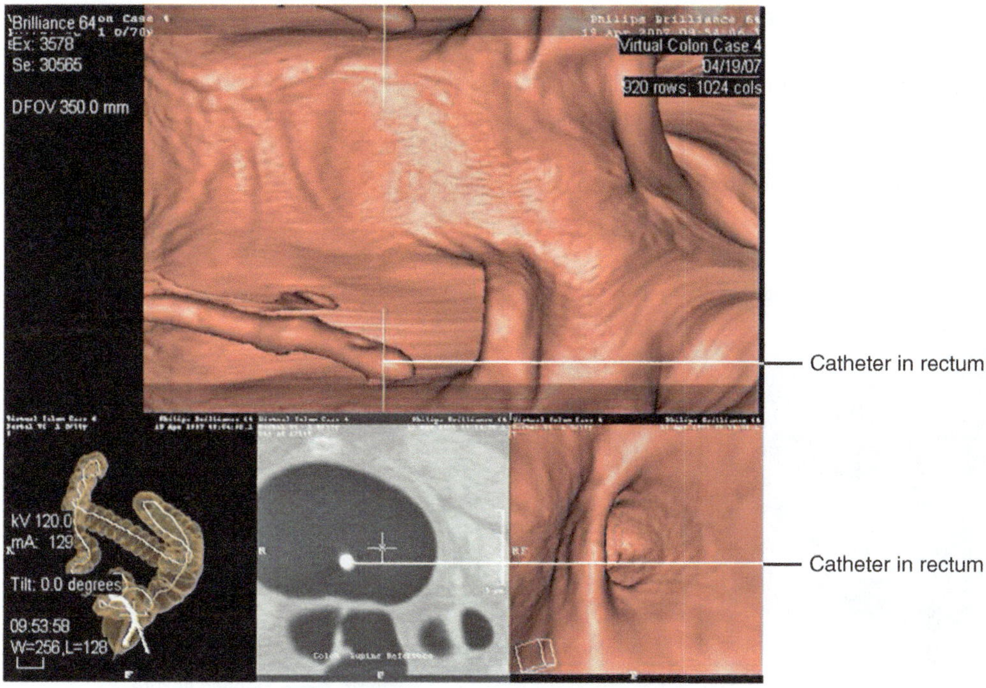

Fig. 21.26 Catheter in the rectum. Fillet view (*top*). Volume-rendered image of the colon in which an automatically central colonic path is created prior to colon virtual dissection (*bottom left*). Axial CT image for reference (*bottom middle*). Virtual endoscopic image (*bottom right*)

Magnetic Resonance Imaging

Demonstrates bowel with similar detail as CT but with additional benefits of better mural and biliary tree assessment.

Optimally demonstrates pelvic floor and anal sphincters.

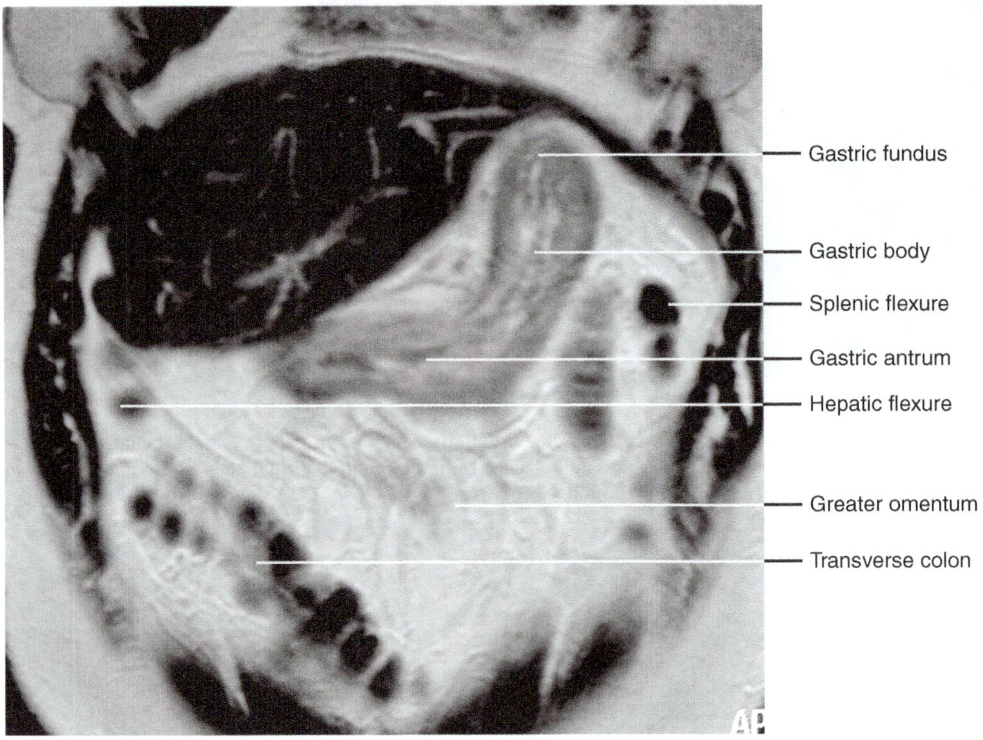

Fig. 21.27 T2 coronal demonstrating the stomach

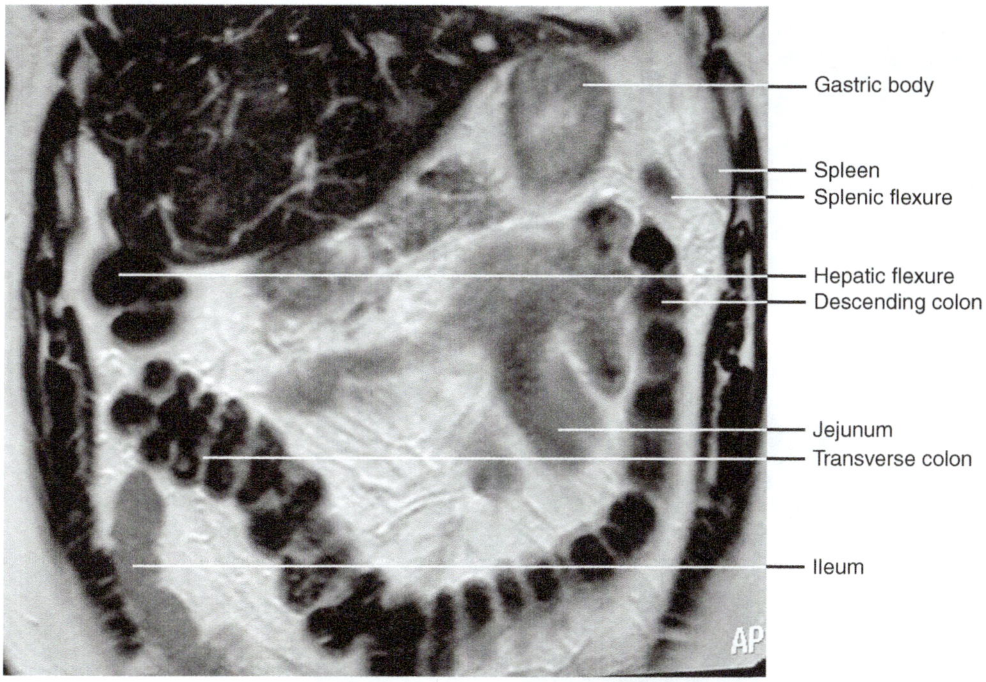

Fig. 21.28 T2 coronal demonstrating the transverse colon

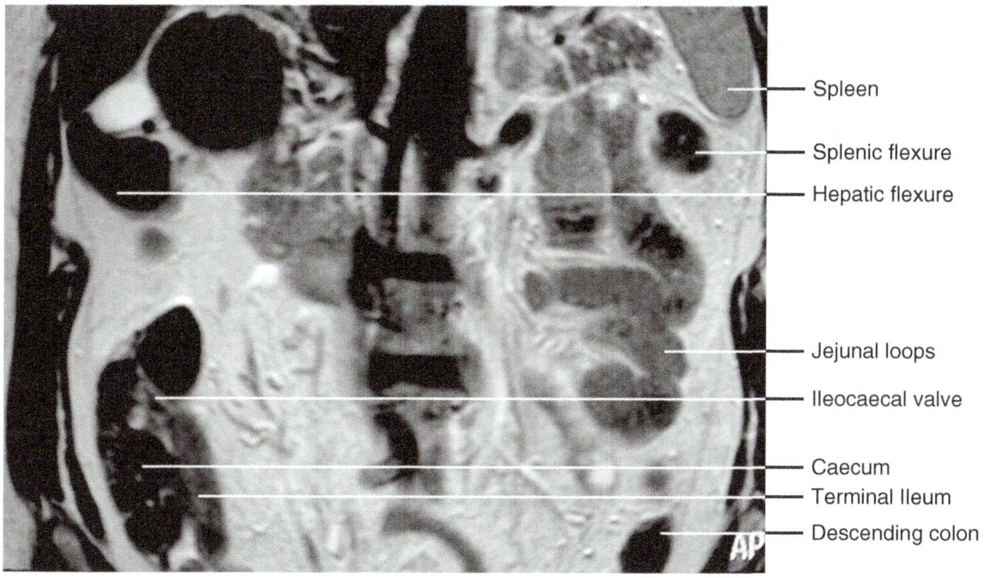

Fig. 21.29 T2 coronal demonstrating the caecum

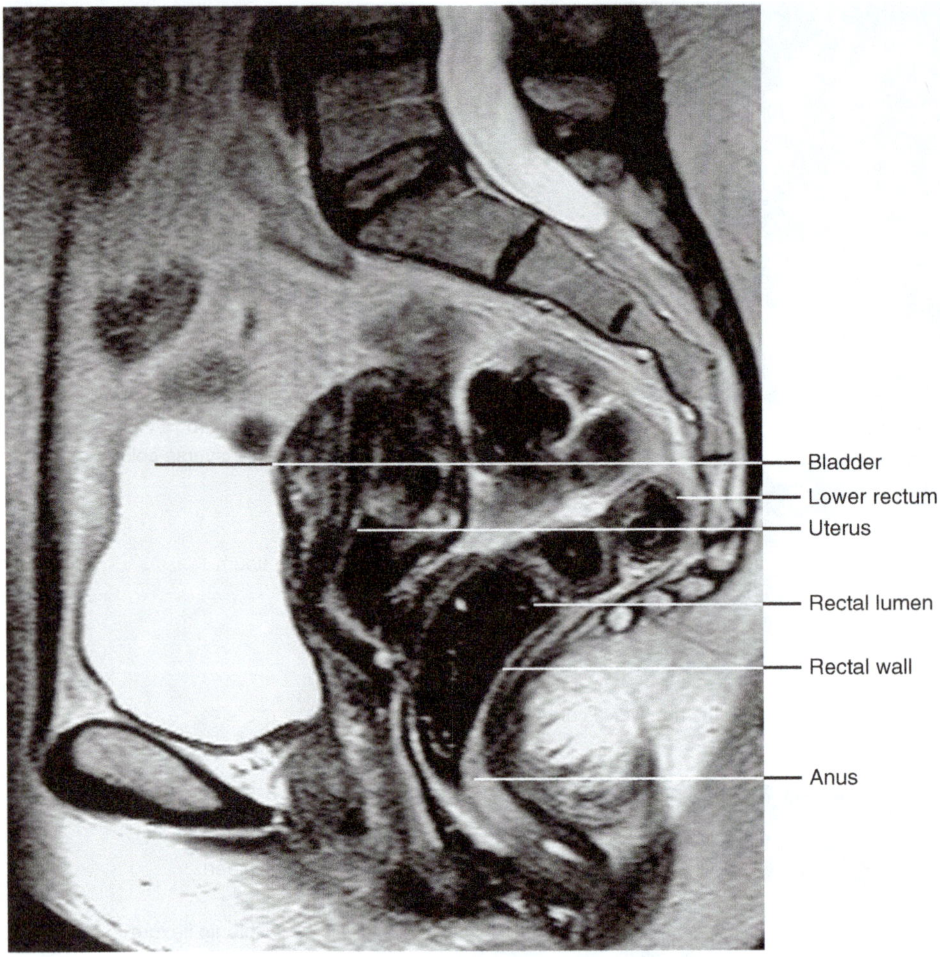

Fig. 21.30 T2 sagittal demonstrating the rectum in a female

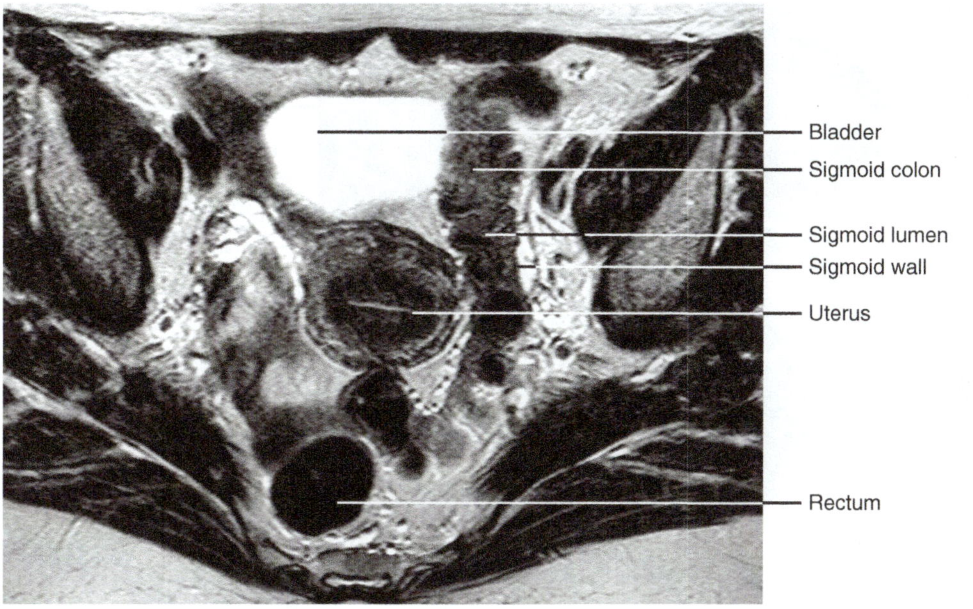

Fig. 21.31 T2 axial demonstrating the rectum in a female

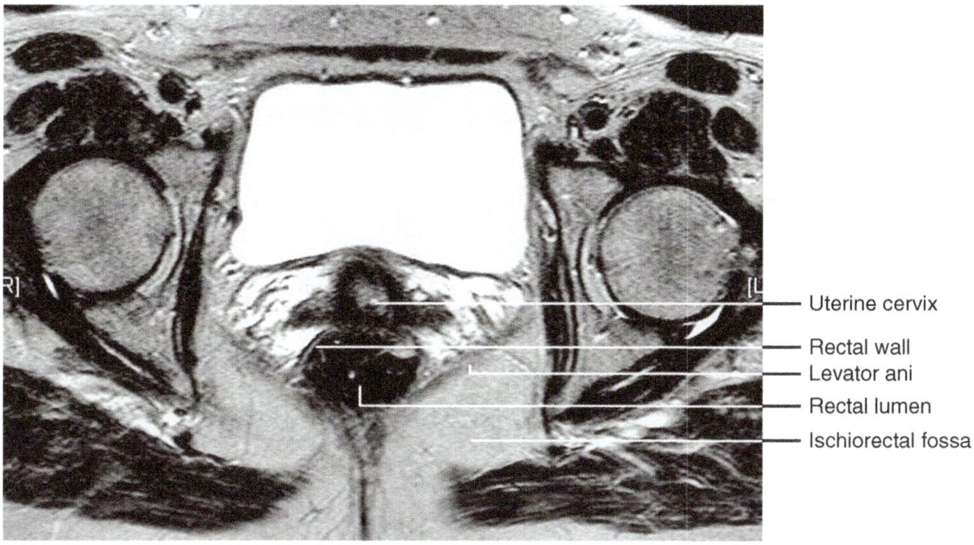

Fig. 21.32 T2 axial demonstrating the rectum in a female

Pearl

The rectal wall consists of:

Mucosa and submucosa: T2 high

Musculari propria: T2 intermediate/high

Perirectal fat: T2 high

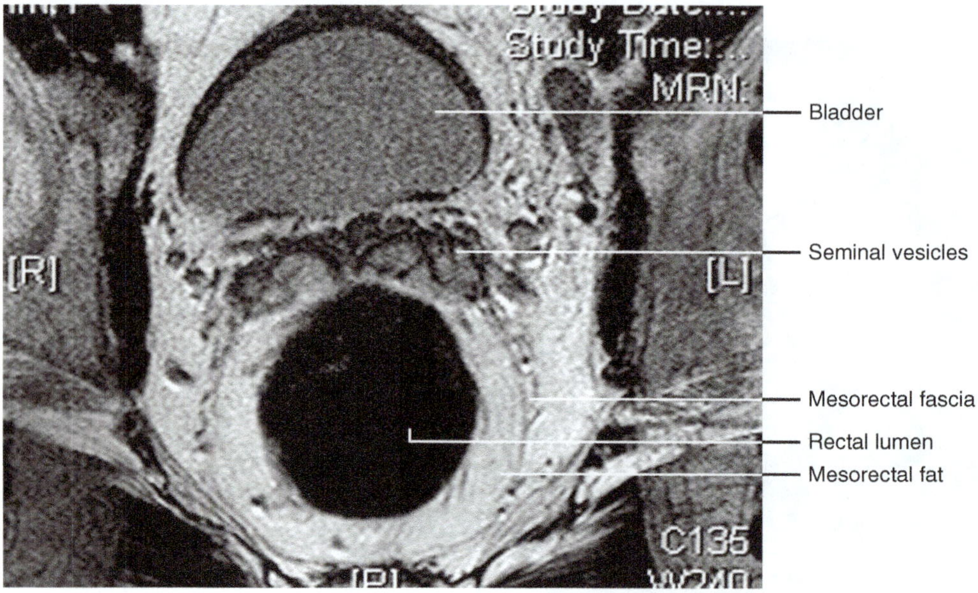

Fig. 21.33 T1 axial demonstrating the rectum in a male

Pearl

The mesorectal space is bounded by the mesorectal fascia surrounding the rectum. It contains fat, lymph nodes, and vessels.

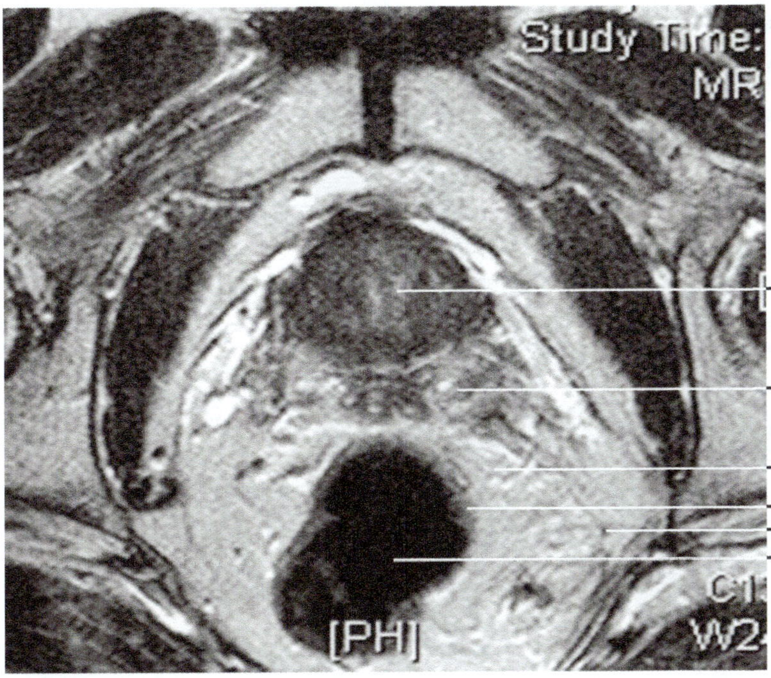

Fig. 21.34 T1 axial demonstrating the rectum in a male

Pearl
The mesorectal fascia blends anteriorly with Devonvillier's fascia of the prostate.

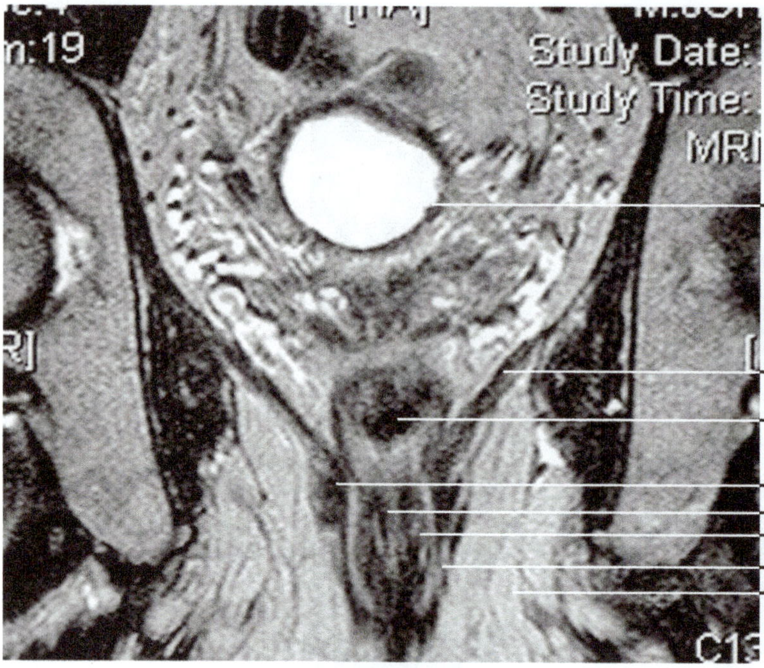

Fig. 21.35 T2 coronal demonstrating the rectum in a male

Pearl

The levator ani muscle is the ceiling of the ischiorectal fossa. Note the rectal wall thickening at the anorectal junction.

Angiography

CT, MR, and catheter angiography are used to assess bowel and visceral blood supply.

Catheter angiography is the gold standard because of better spatial resolution, ability to assess flow dynamics, and intervention in the same setting.

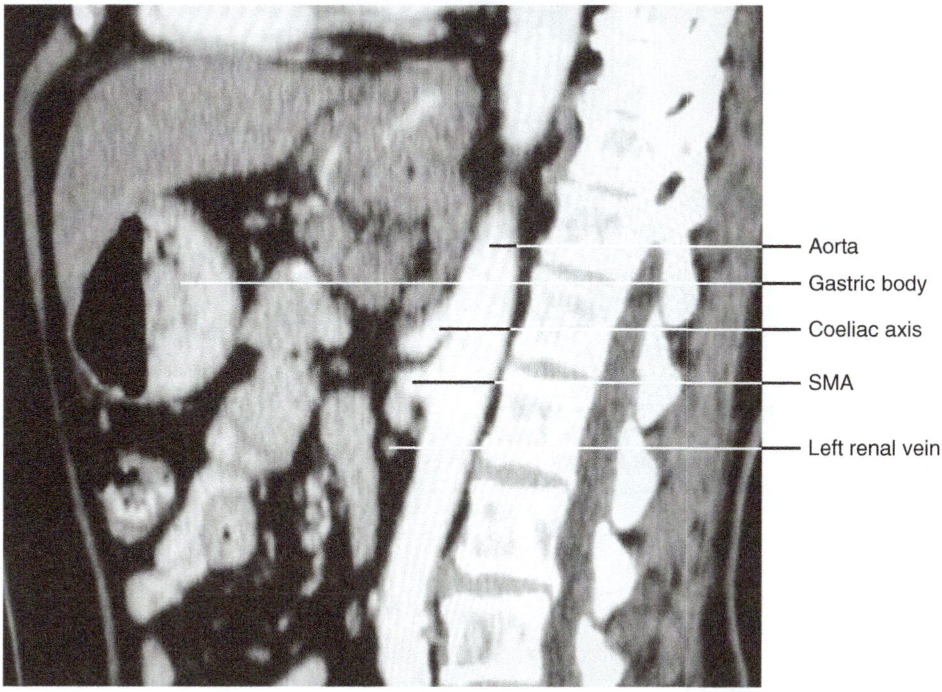

Fig. 21.36 Sagittal CT reconstruction after IV and oral contrast

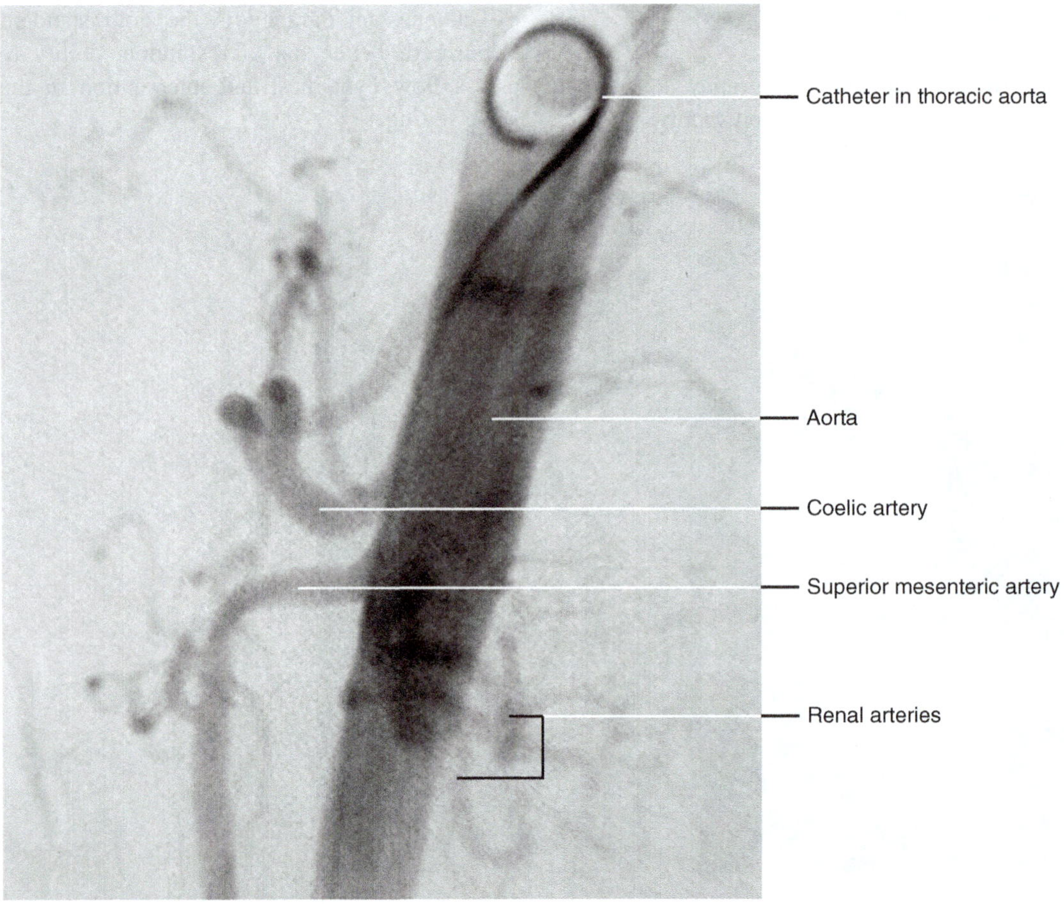

Fig. 21.37 Lateral angiogram of the abdominal aorta

Pearl

The normal angle between the aorta and SMA is 45°–65°.

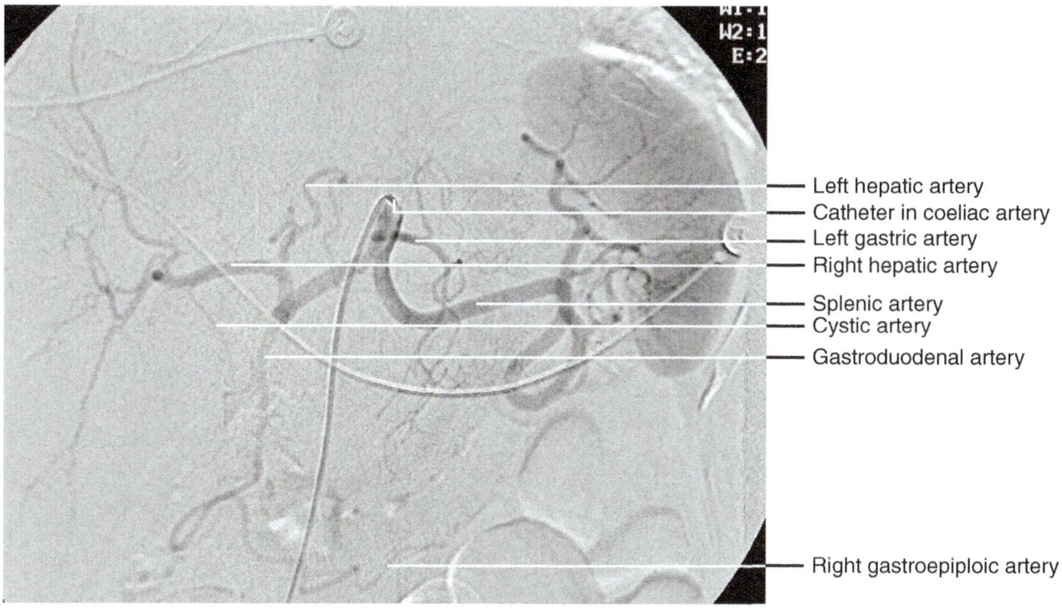

Fig. 21.38 Selective coeliac artery angiogram

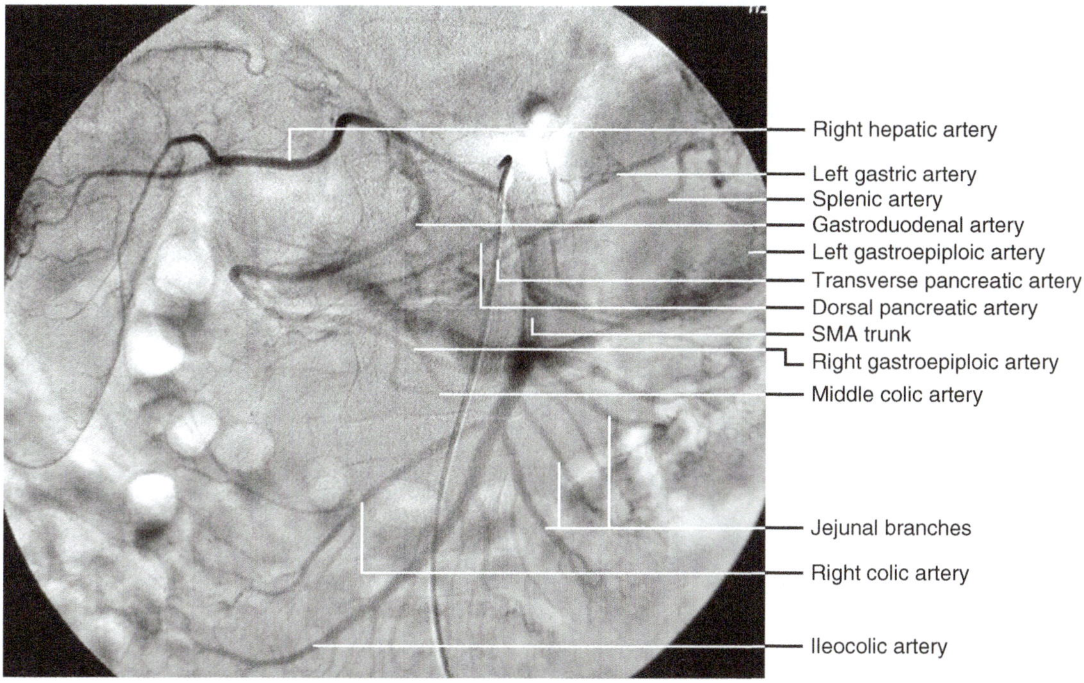

Fig. 21.39 Coeliac and mesenteric artery angiogram

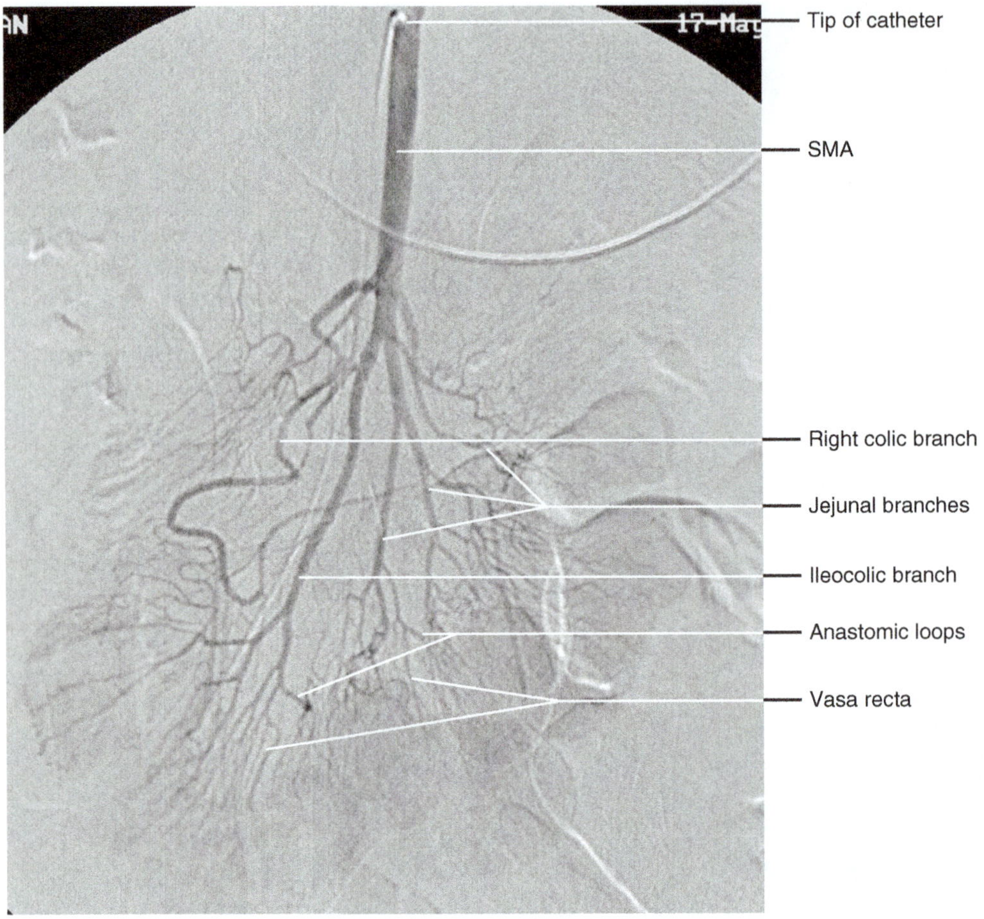

Fig. 21.40 Selective superior mesenteric artery angiogram

22 Renal Tract

Intravenous Pyelogram

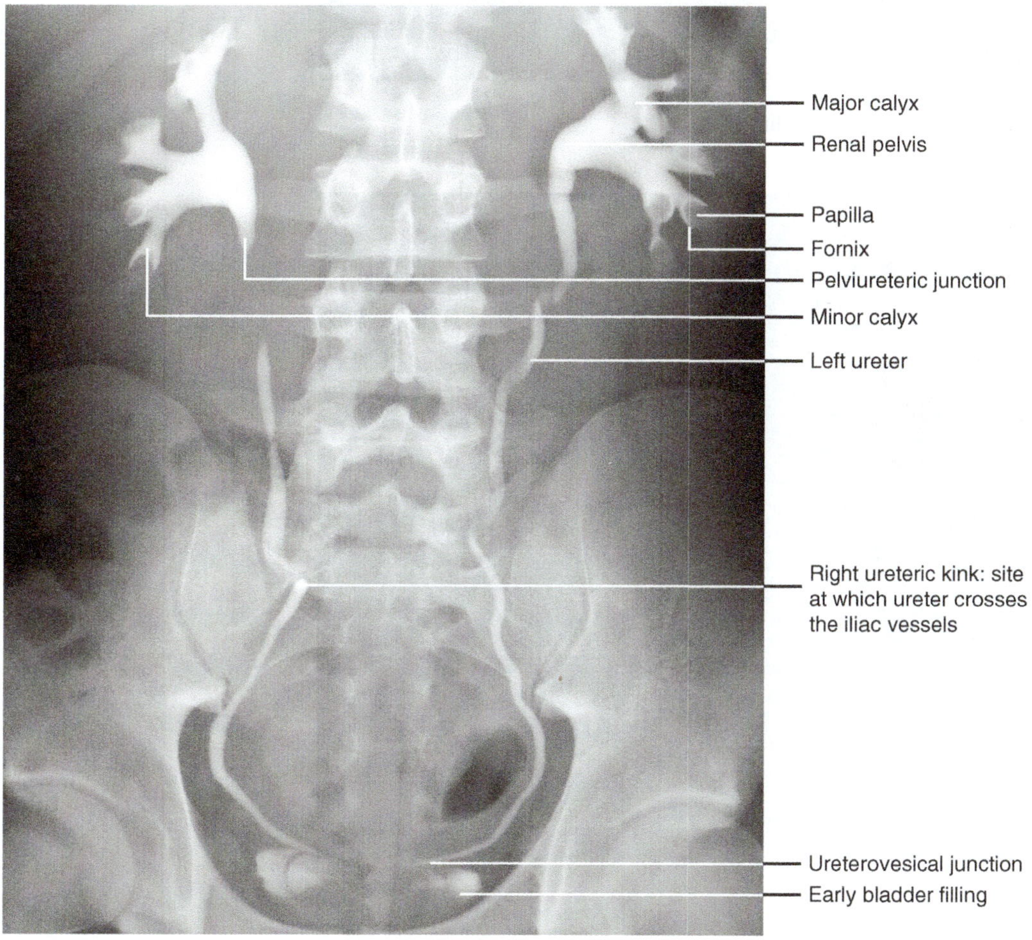

Fig. 22.1 Excretory phase of an intravenous pyelogram (IVP)

DOI 10.1007/978-3-642-23893-2_22,

Ultrasound

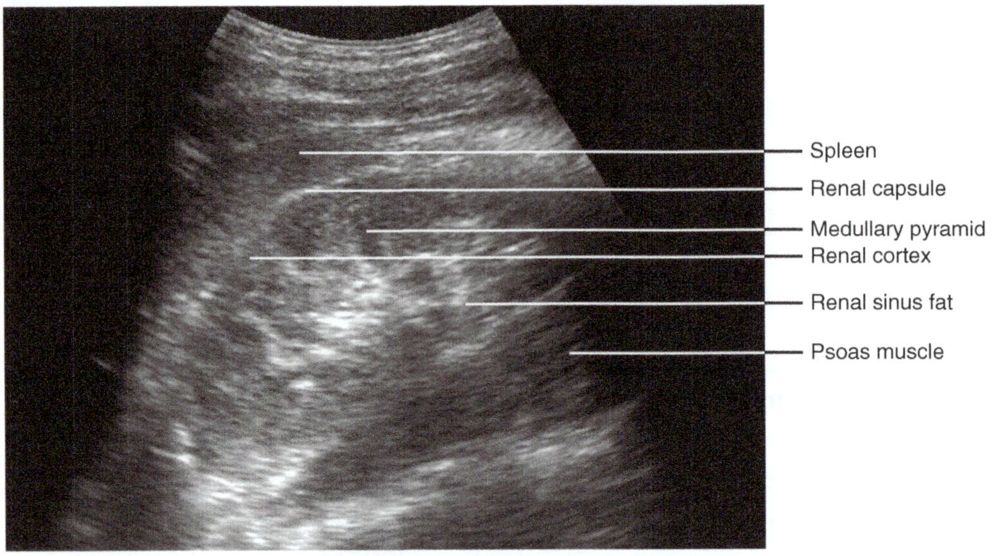

Fig. 22.2 Left kidney – longitudinal

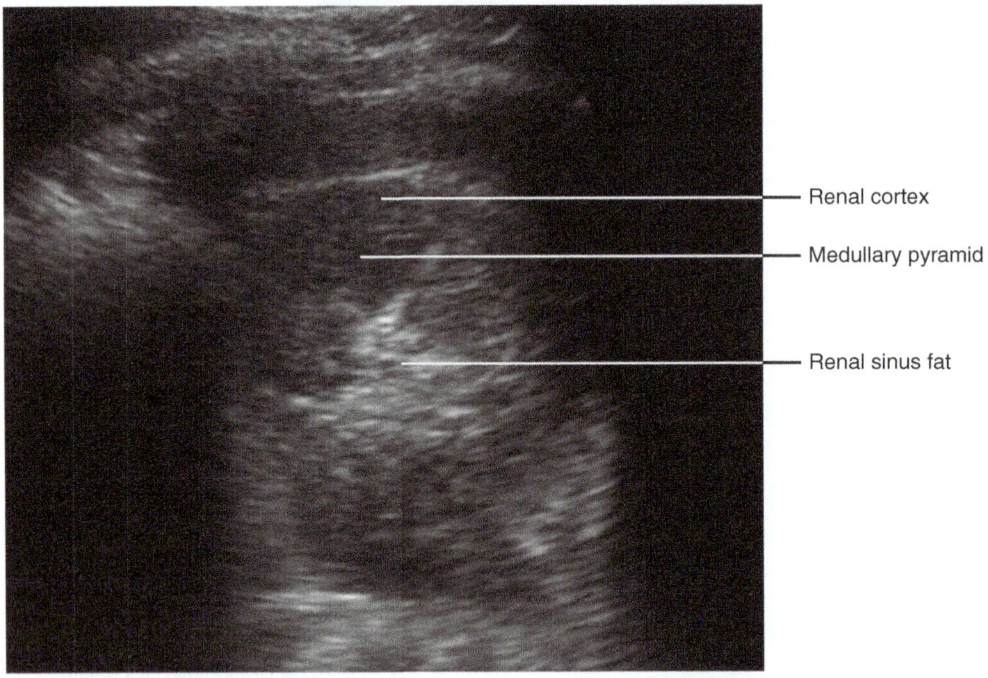

Fig. 22.3 Kidney: transverse

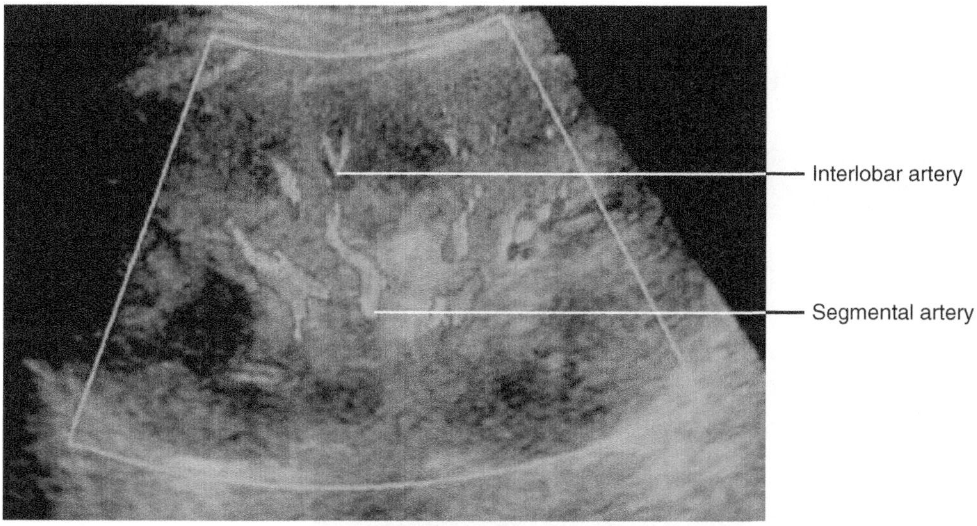

Fig. 22.4 Kidney: longitudinal, with colour Doppler

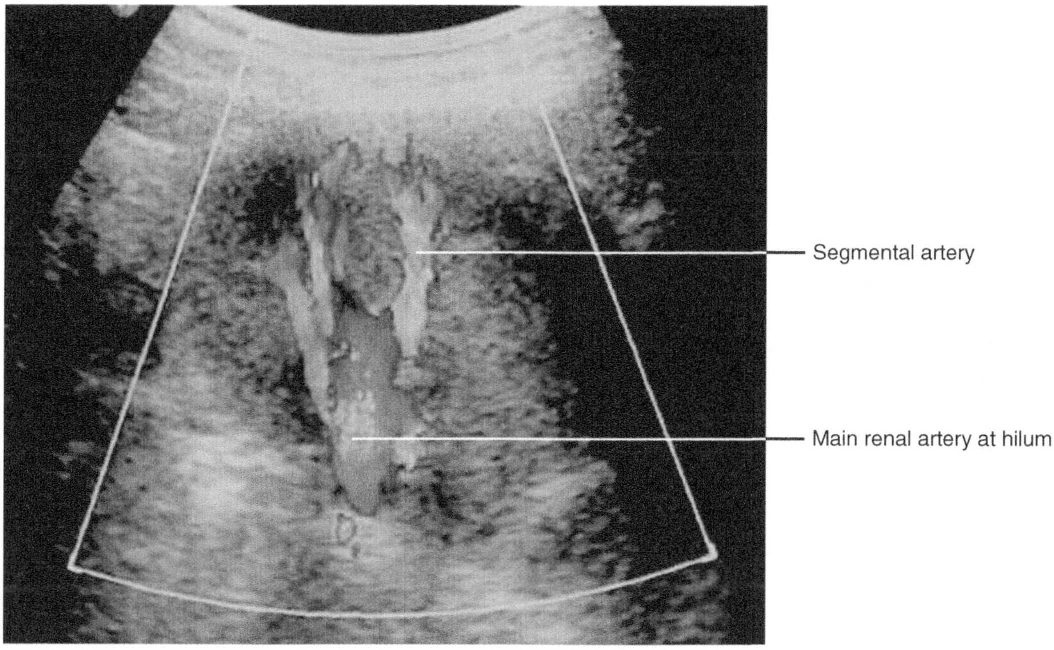

Fig. 22.5 Kidney: transverse, with colour Doppler

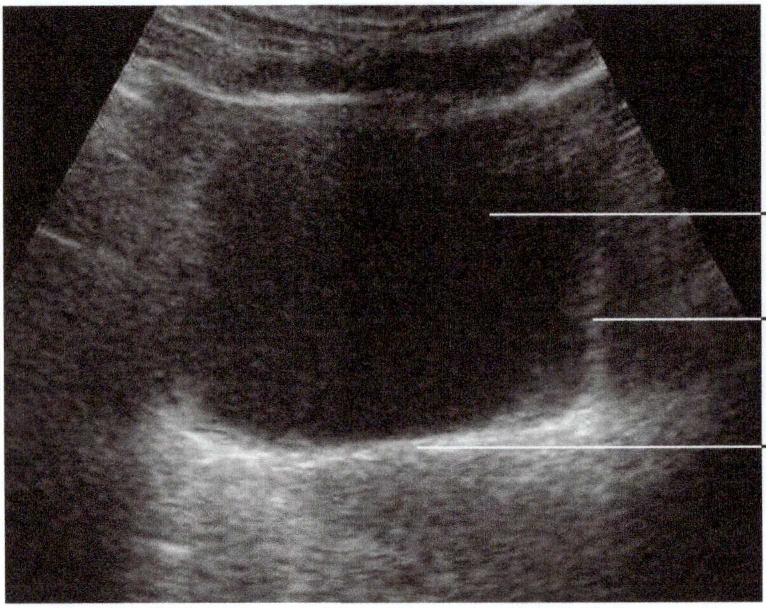

Fig. 22.6 Urinary bladder transverse

Micturating Cystourethrogram (MCUG)

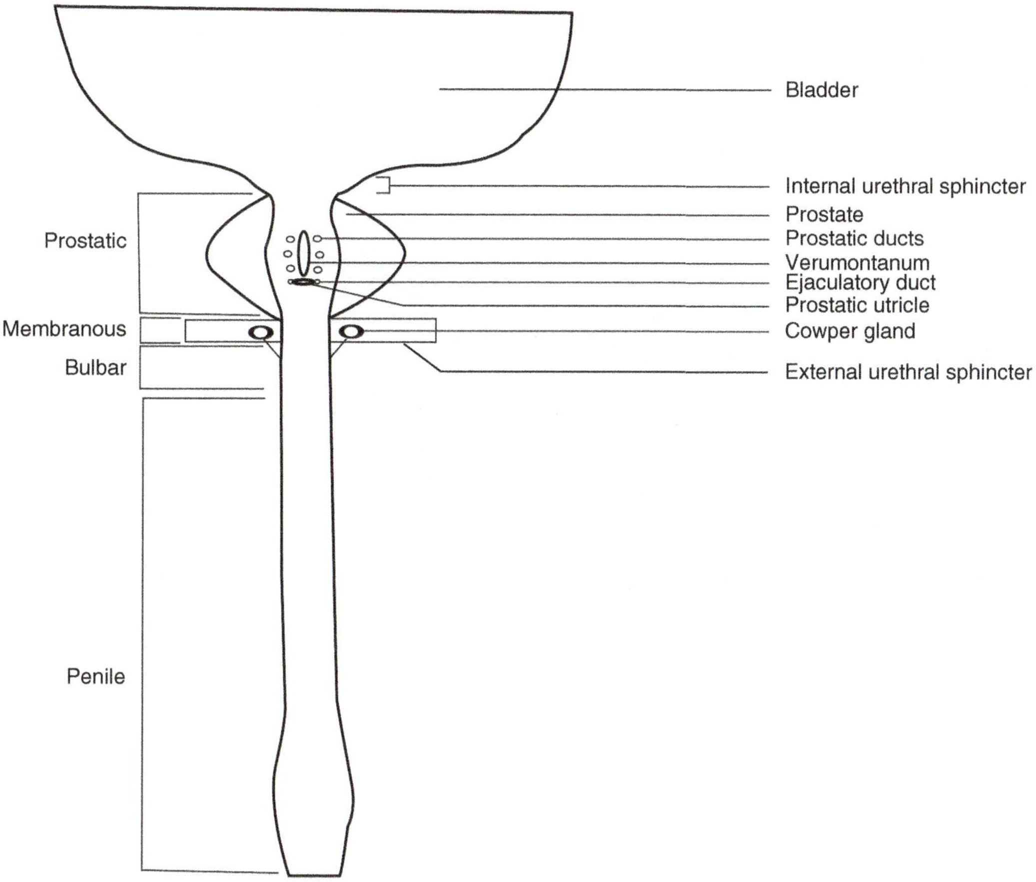

Fig. 22.7 Male urethral schematic anterior view

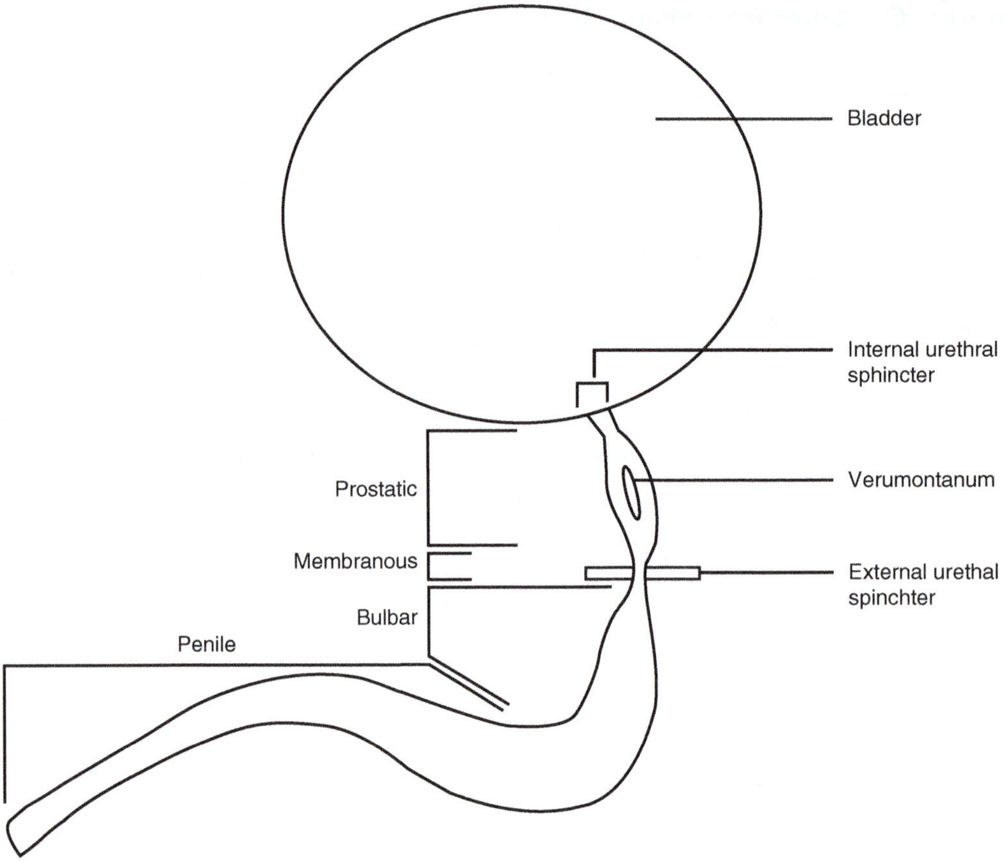

Fig. 22.8 Male urethral schematic lateral oblique view

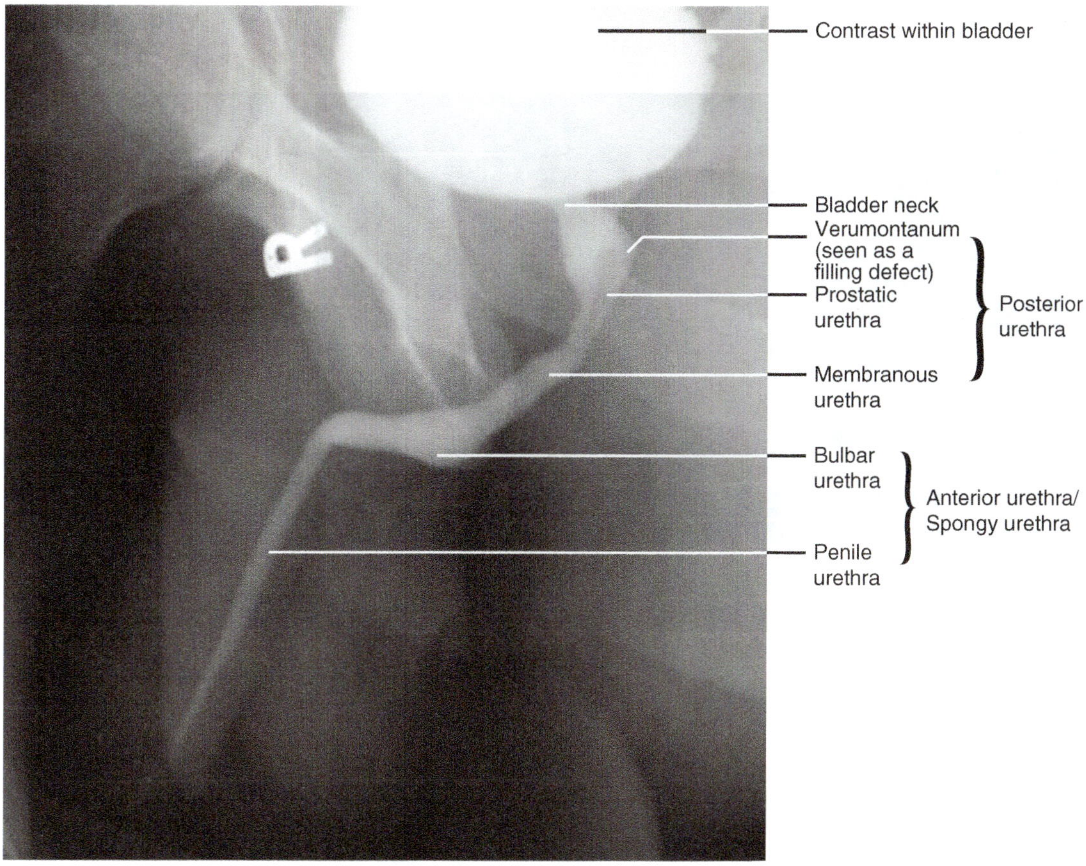

Fig. 22.9 Lateral oblique micturating cystourethrogram in a male

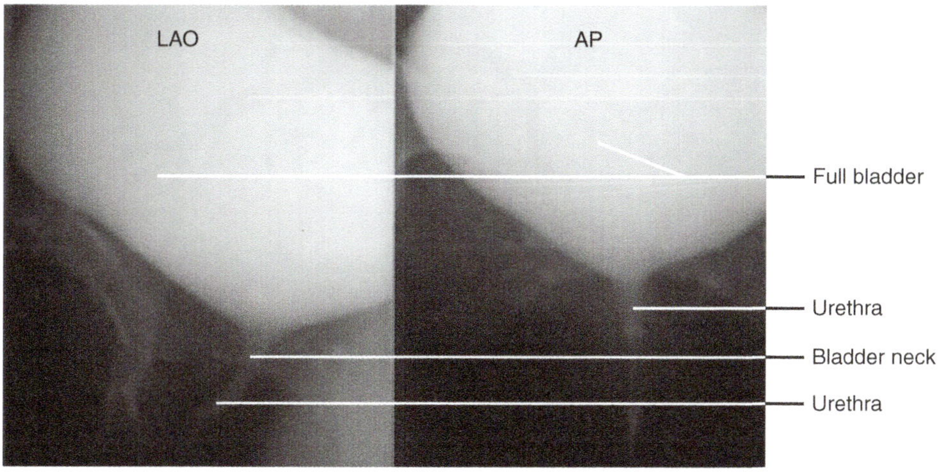

Fig. 22.10 Oblique and AP views of a micturating cystourethrogram in female

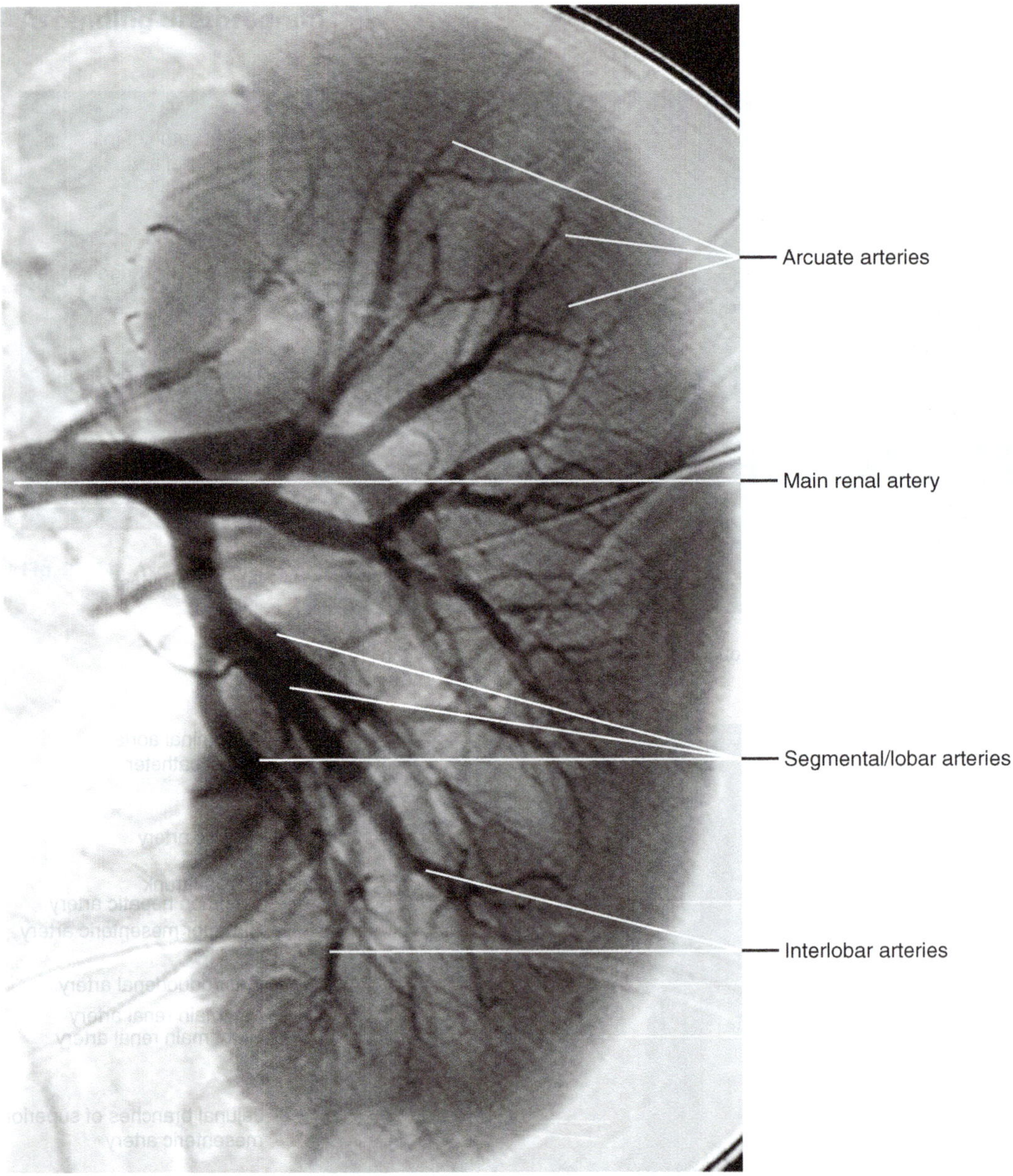

Fig. 22.14 Selective left renal arteriogram

Renal CT Arteriography (CTA)

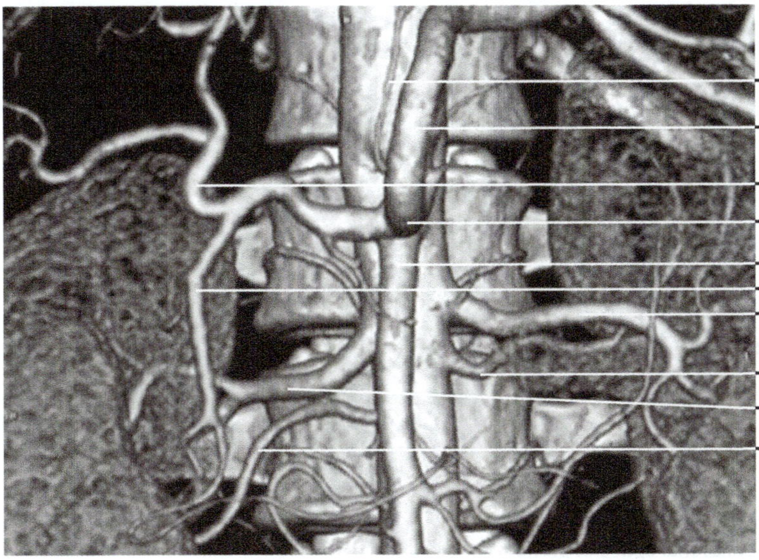

Fig. 22.15 Volume-rendered MDCT after intravenous contrast

Computed Tomography

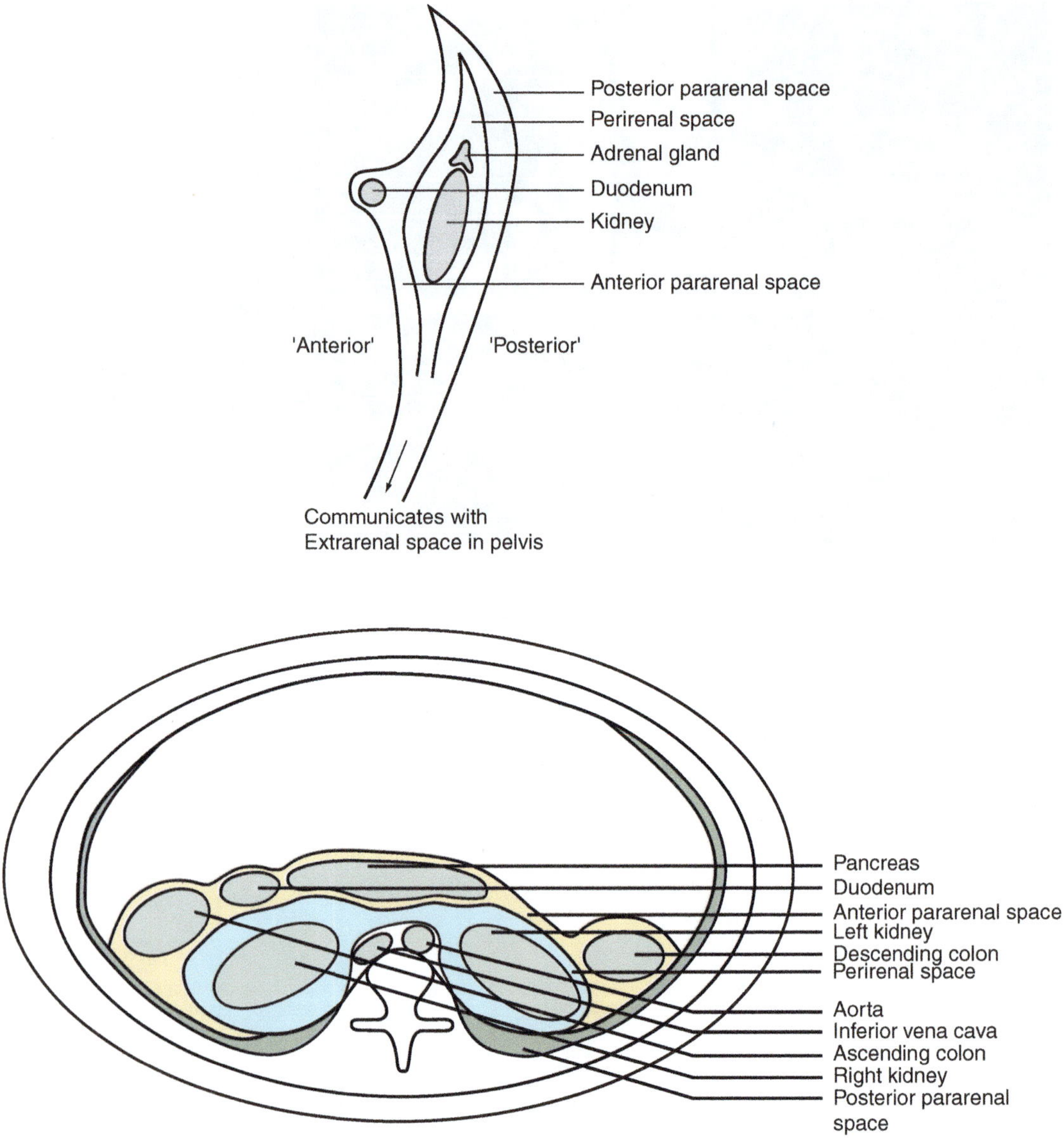

Fig. 22.16 Schematic representation of the retroperitoneal spaces

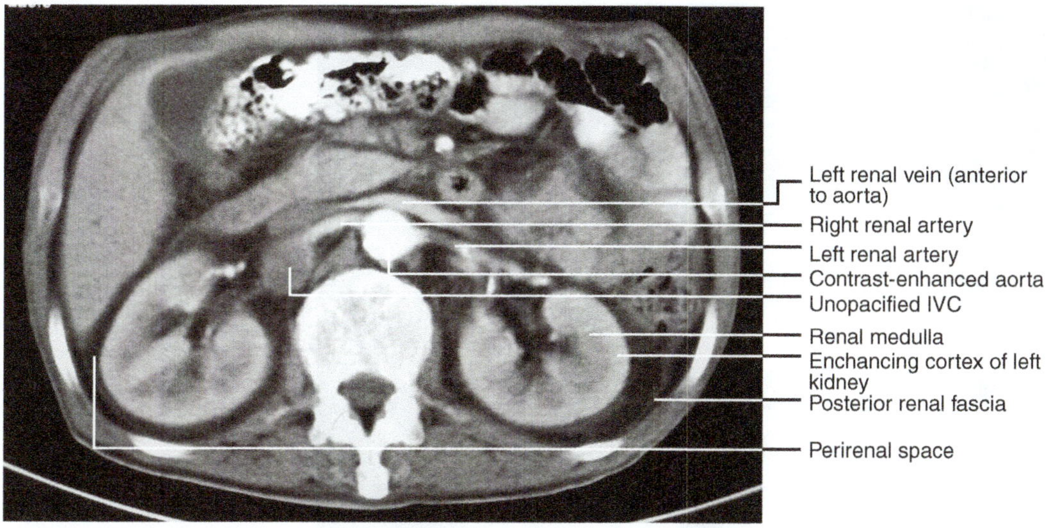

Fig. 22.17 Arterial phase: level of the renal arteries

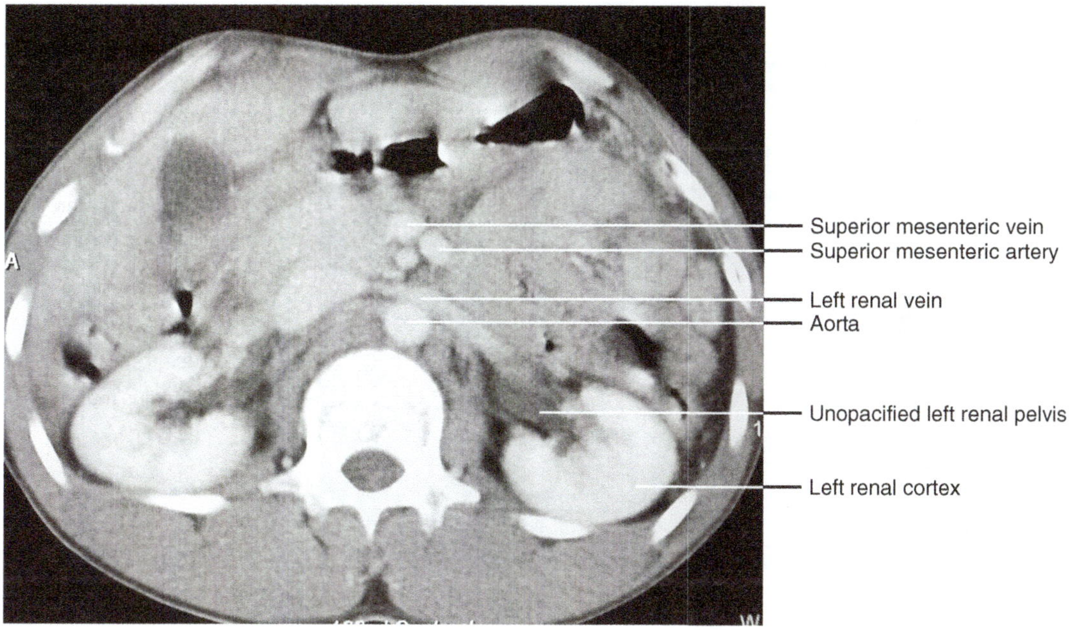

Fig. 22.18 Portal venous phase: level of the renal veins

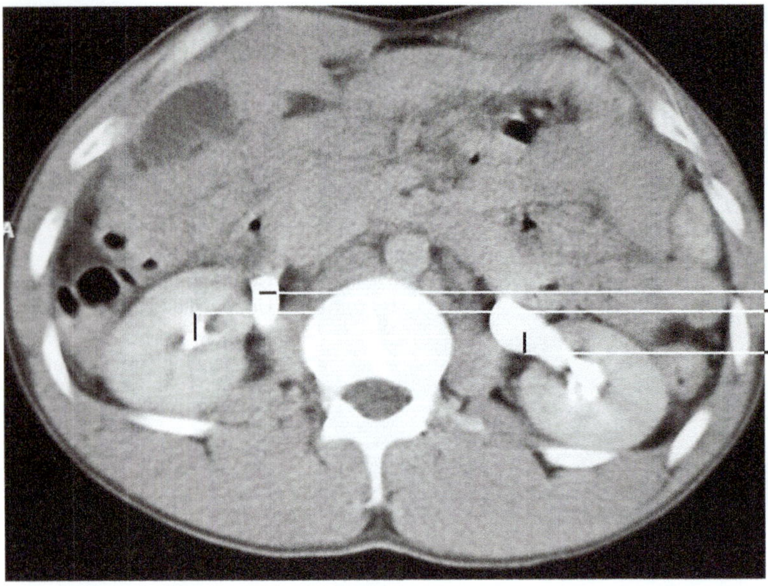

Fig. 22.19 Delayed phase: upper abdomen

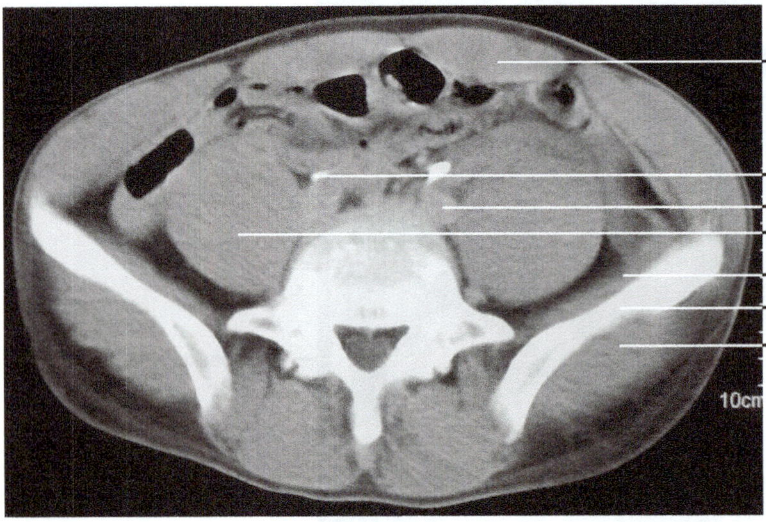

Fig. 22.20 Delayed phase: pelvis

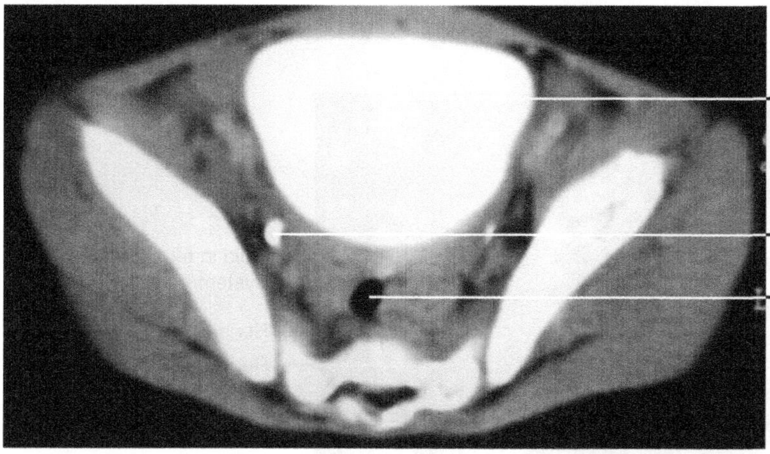

Fig. 22.21 CT cystogram

Magnetic Resonance Imaging

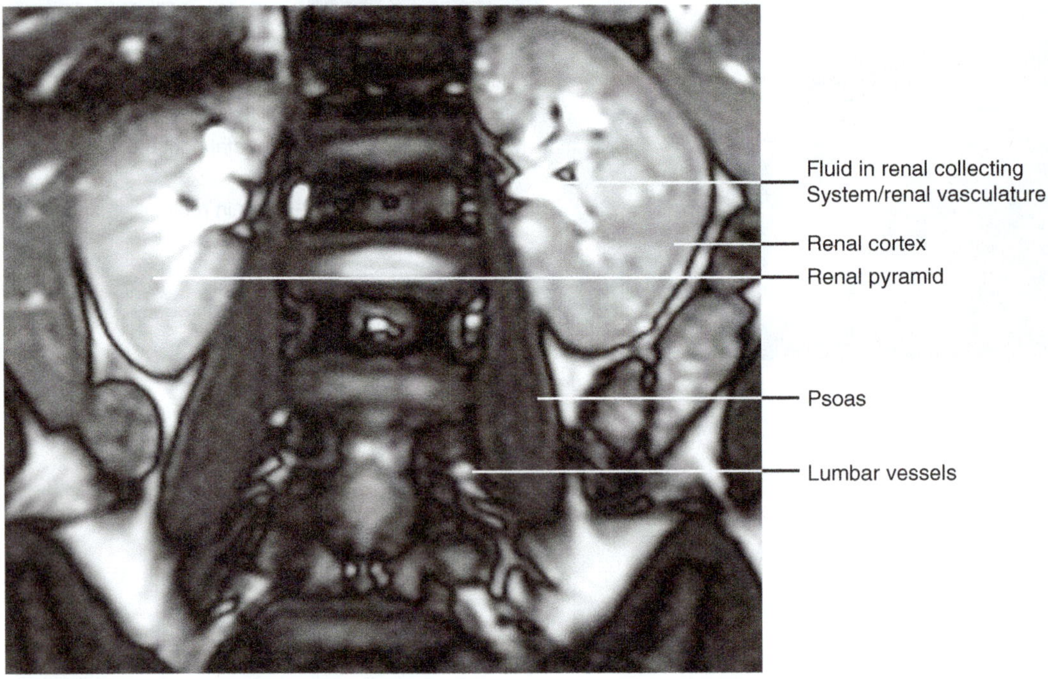

Fig. 22.22 Coronal FISP sequence of renal tract

Pearl

FISP sequences give high spatial resolution with T2 weighting. Both stationary and moving fluid will exhibit high signal.

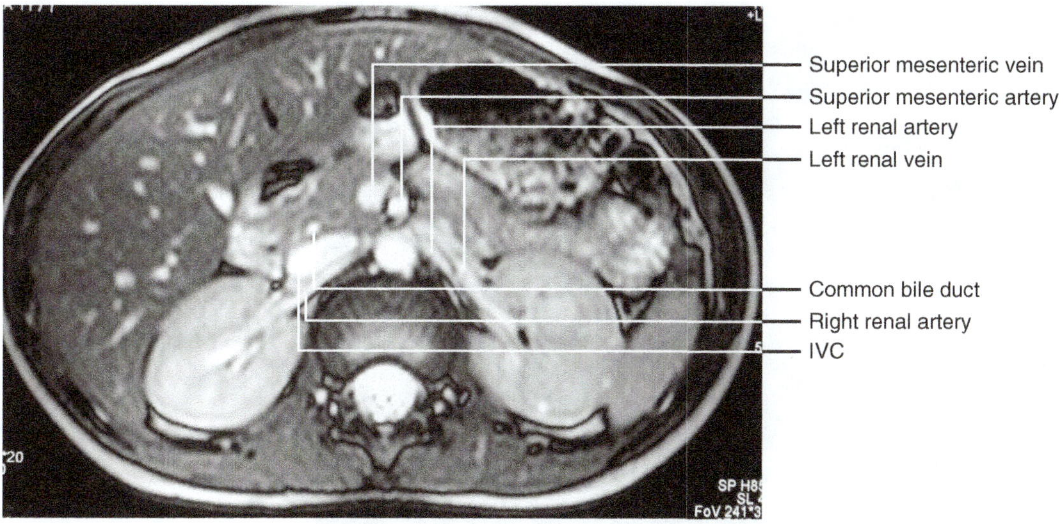

Fig. 22.23 Axial FISP sequence: level of renal arteries

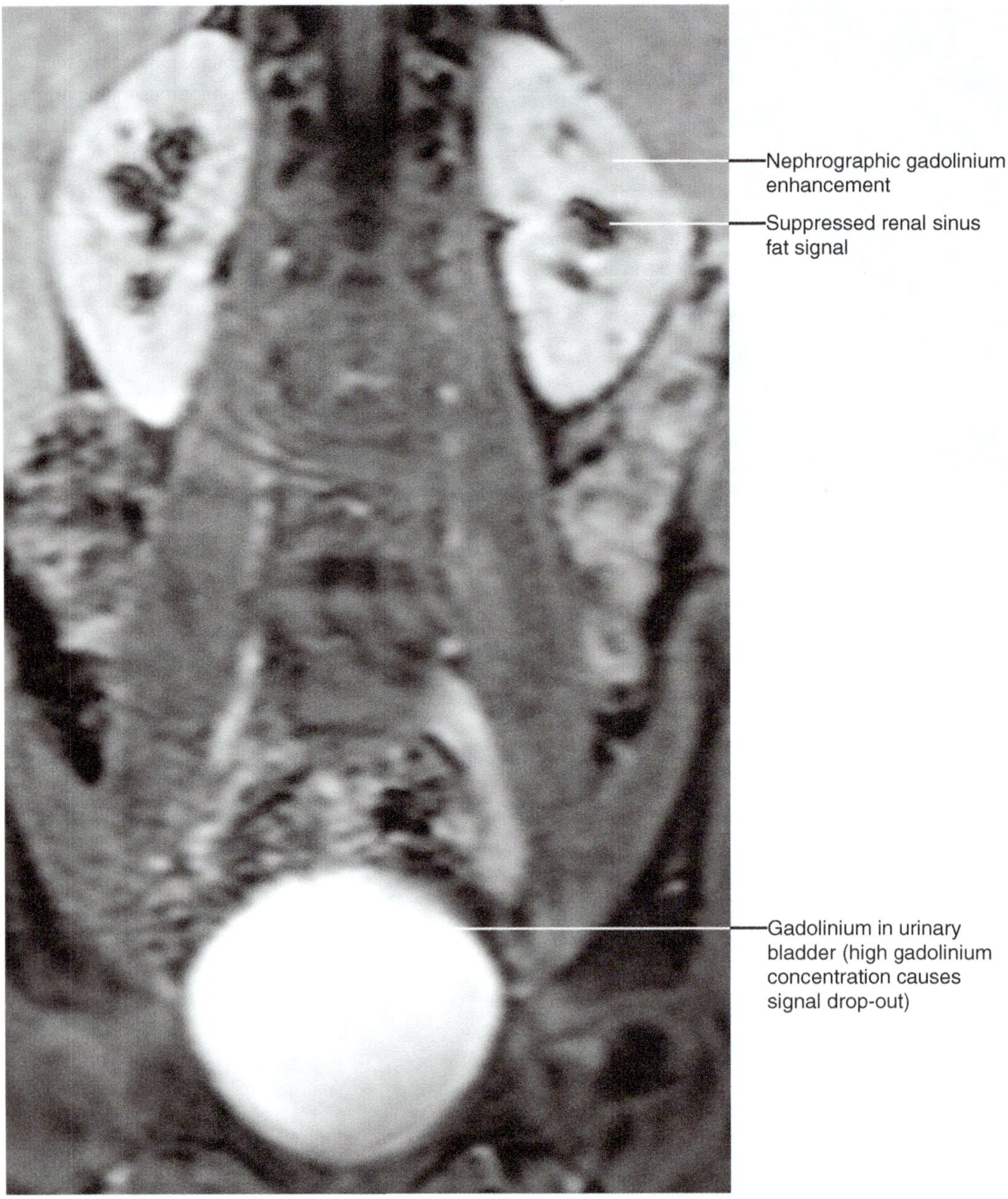

Fig. 22.24 Coronal T1 fat saturated gadolinium enhanced sequence of renal tract – delayed phase with full bladder

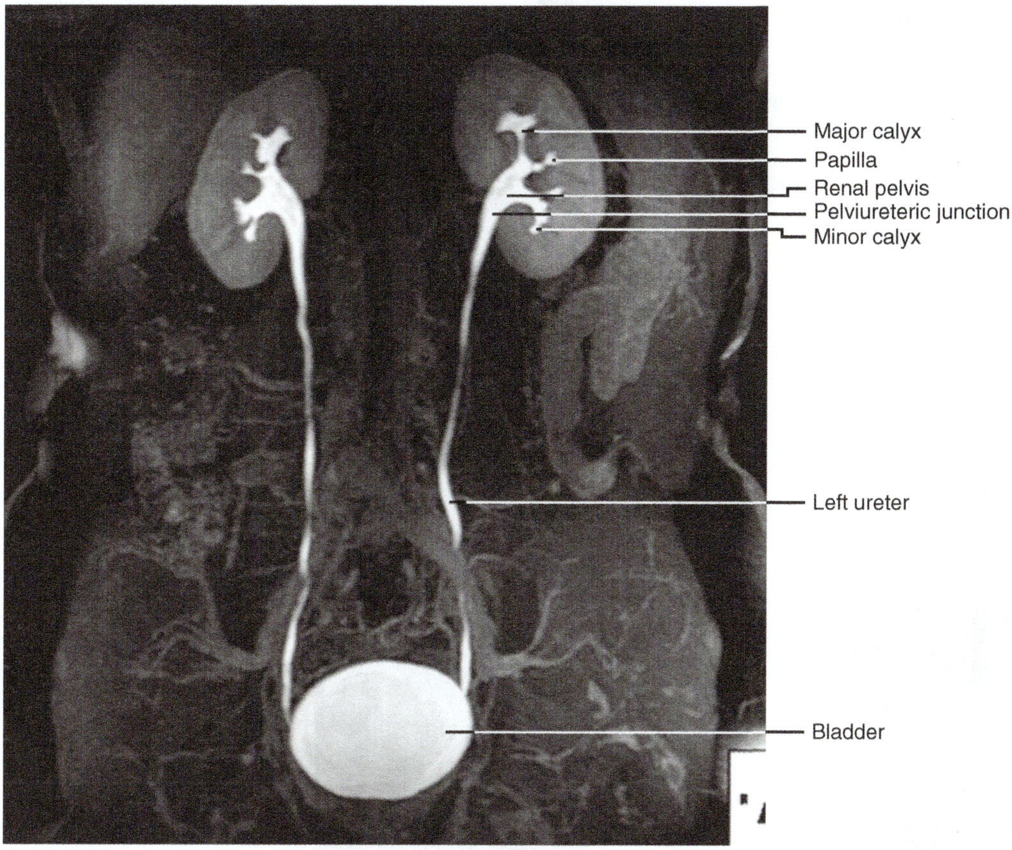

Fig. 22.25 Contrast MRU excretory phase

Nuclear Medicine

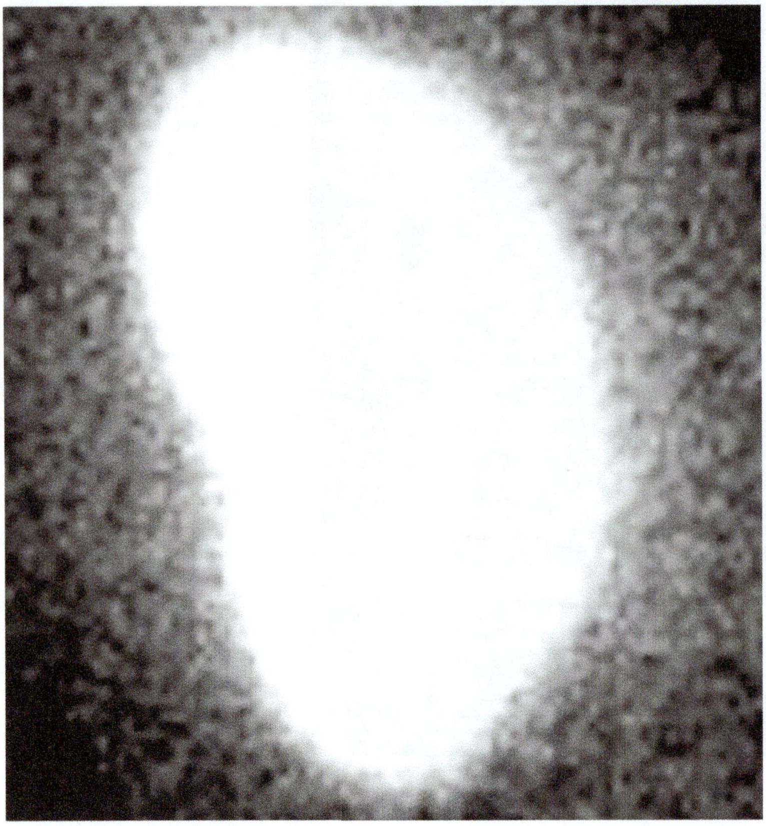

Fig. 22.26 Tc 99 labelled DMSA AP view of kidney

Pearl

DSMA scan provides information on renal morphology and size, and differential function.

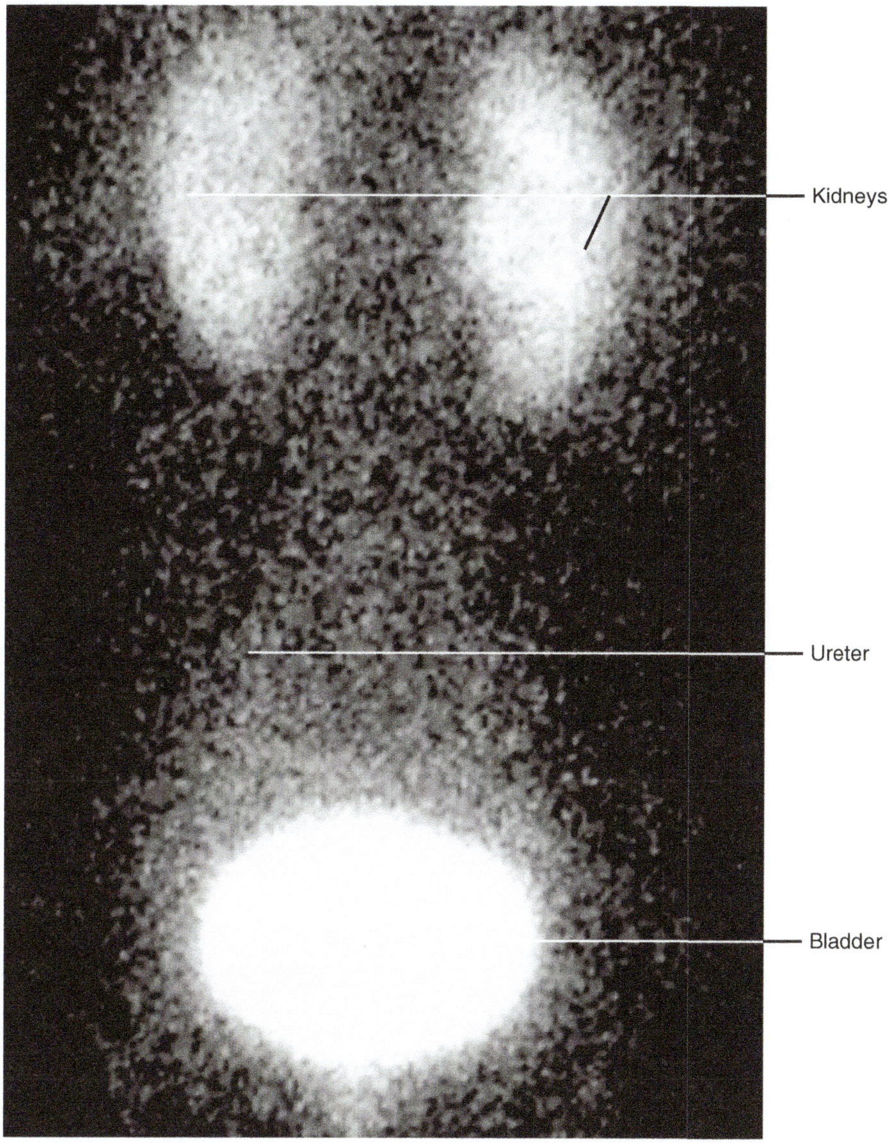

Fig. 22.27 Post intravenous Tc 99 labelled MAG3 AP view of renal tract

Pearl

Intravenous MAG3 studies results in a nephrogram and tracer in the collecting system much like and IVP.

MAG3 scans are also used for cystography to evaluate vesico-ureteric reflux in a co-operative child.

23 Female and Male Genital Tract

Both female and male genital tracts are adequately demonstrated using ultrasound techniques either transabdominally, via the vagina/rectum, or directly as is the case of the testes. Additional information is provided by MRI, which demonstrates the anatomy in more detail. Specific contrast investigations such as the hysterosalpingogram are still in use as is the cavernosogram. Vascular anatomy is demonstrated with ultrasound, angiography, and multidetector CT. The genital system is also imaged incidentally during CT examinations but CT is not primarily used for this purpose.

S. Andronikou (ed.), *See Right Through Me*,
DOI 10.1007/978-3-642-23893-2_23, © Springer-Verlag Berlin Heidelberg 2012

Female Genital Tract

Ultrasound

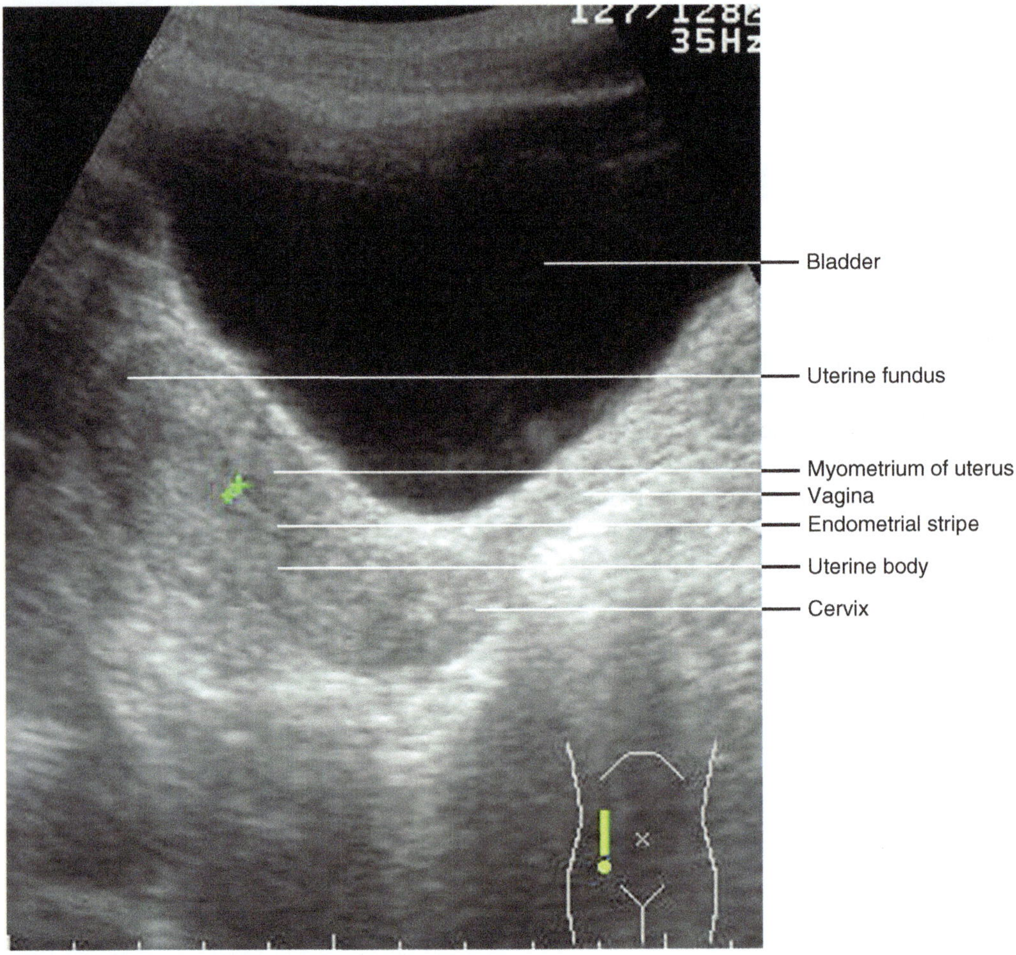

Fig. 23.1 Longitudinal ultrasound of the uterus

Pearl

Normal thickness of the endometrium <5 mm post menopausal, <14 mm premenopausal or postmenopausal on Tamoxifen. Abnormal if >15 mm.

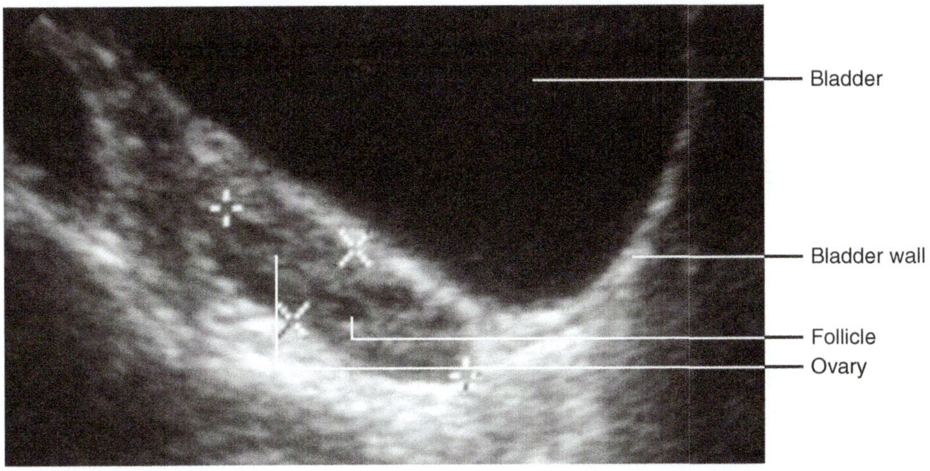

Fig. 23.2 Longitudinal ultrasound of the ovary

Pearl

Normal ovarian dimensions: length 3–5 cm; width 2–3 cm.
Normal diameter of a follicle <25 mm.

Hysterosalpingogram

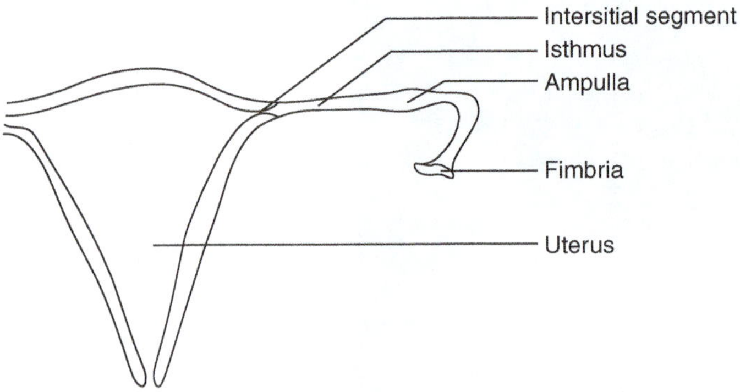

Fig. 23.3 Schematic of the fallopian tube

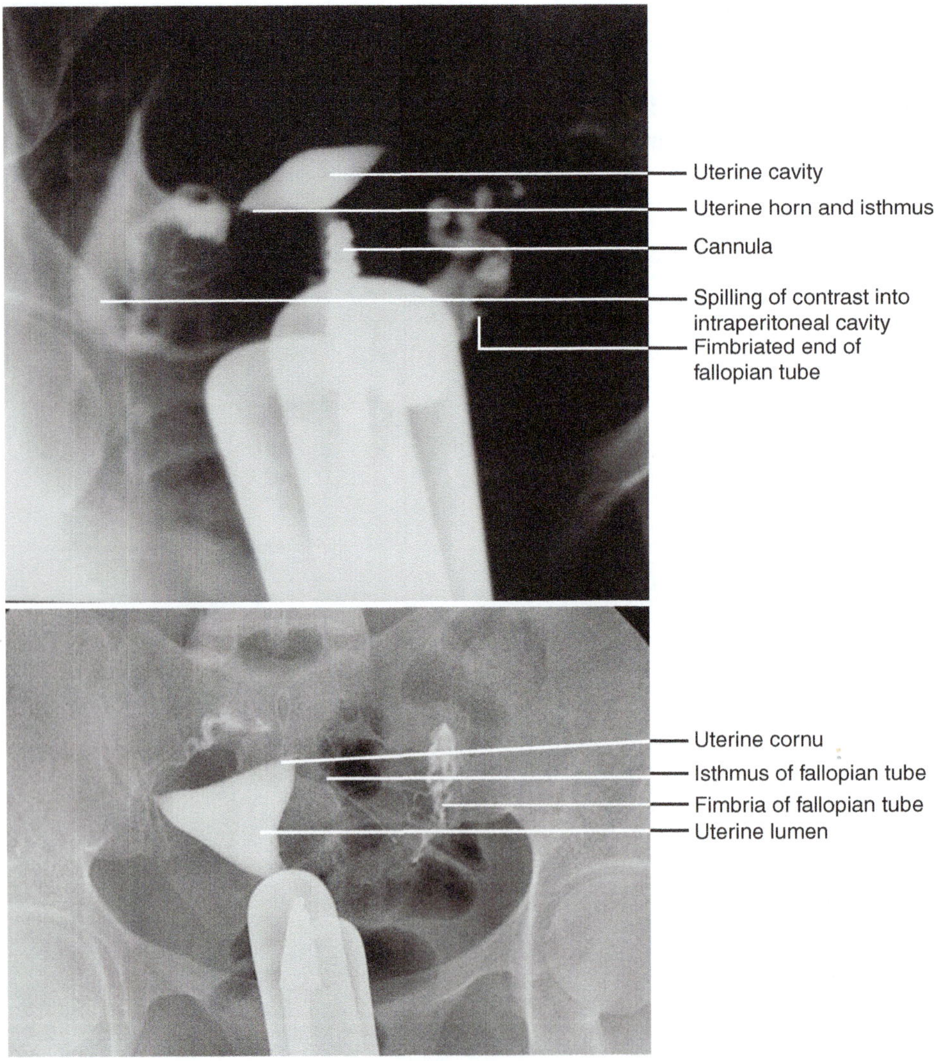

Fig. 23.4 AP radiographs during hysterosalpingography demonstrating the unterine lumen and fallopian tubes

Magnetic Resonance Imaging

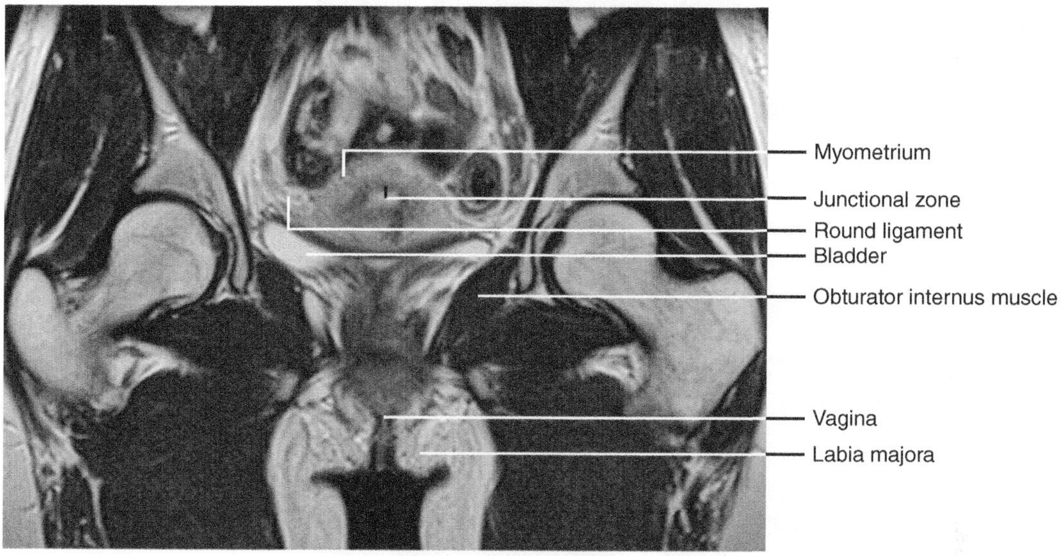

Fig. 23.5 Coronal T2 of the uterus and ovaries

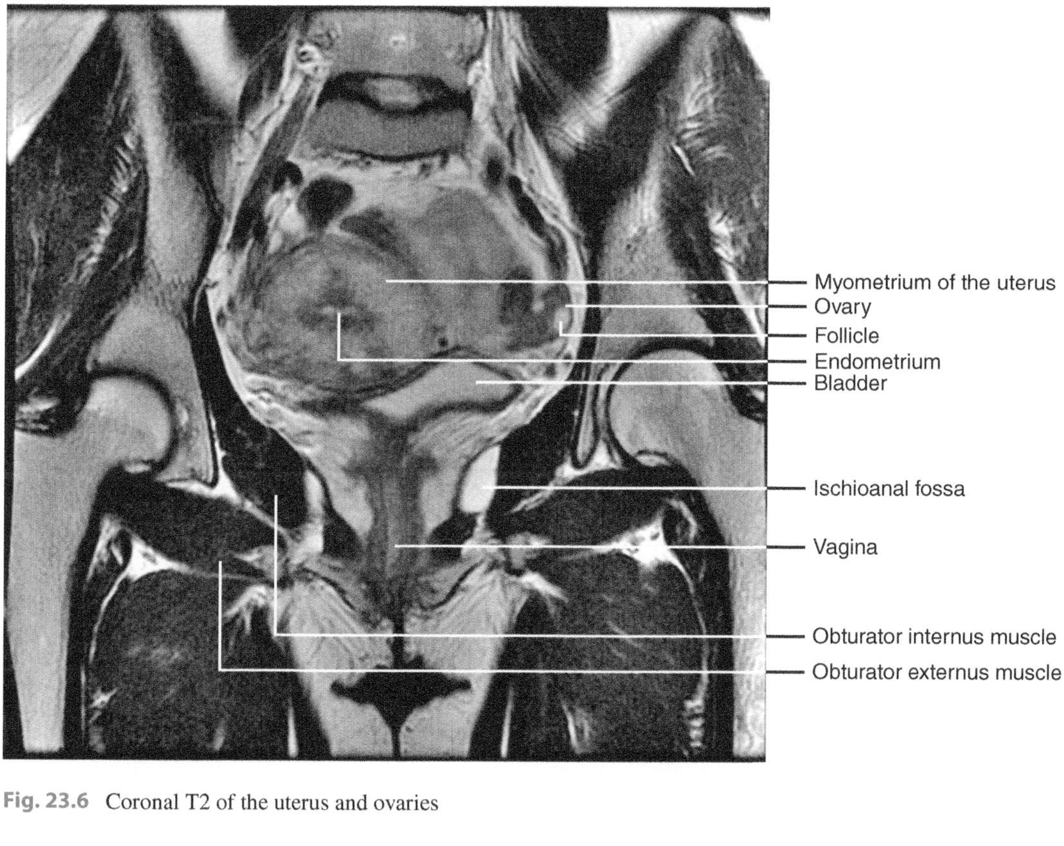

Fig. 23.6 Coronal T2 of the uterus and ovaries

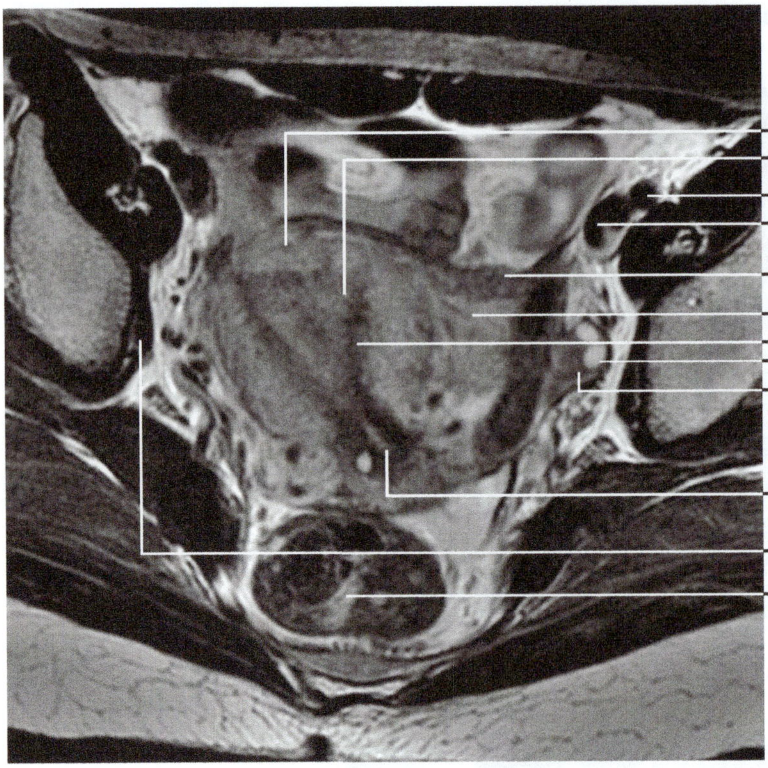

Fig. 23.7 Axial T2 of the uterus and ovaries

Computed Tomography

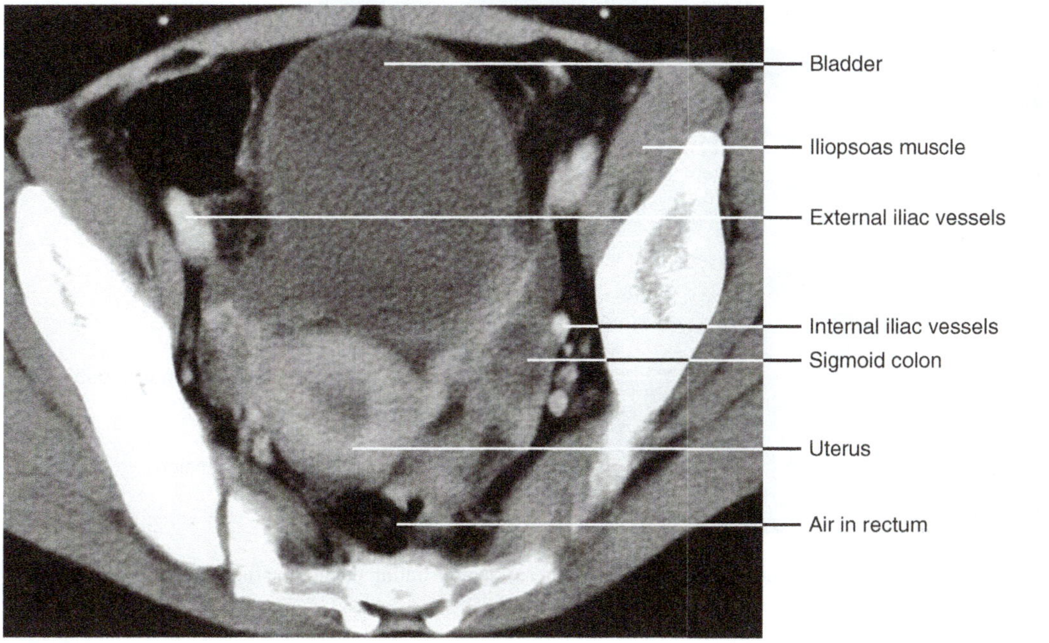

Fig. 23.8 Axial post contrast CT of the pelvis showing the enhancing uterus and ovaries

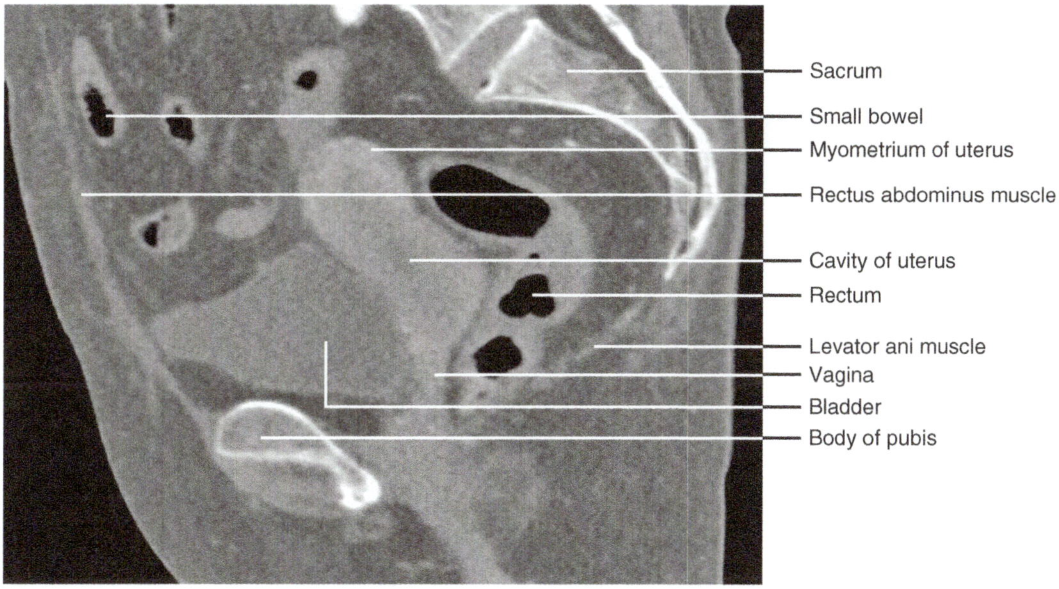

Fig. 23.9 Sagittal reconstruction of the female pelvis

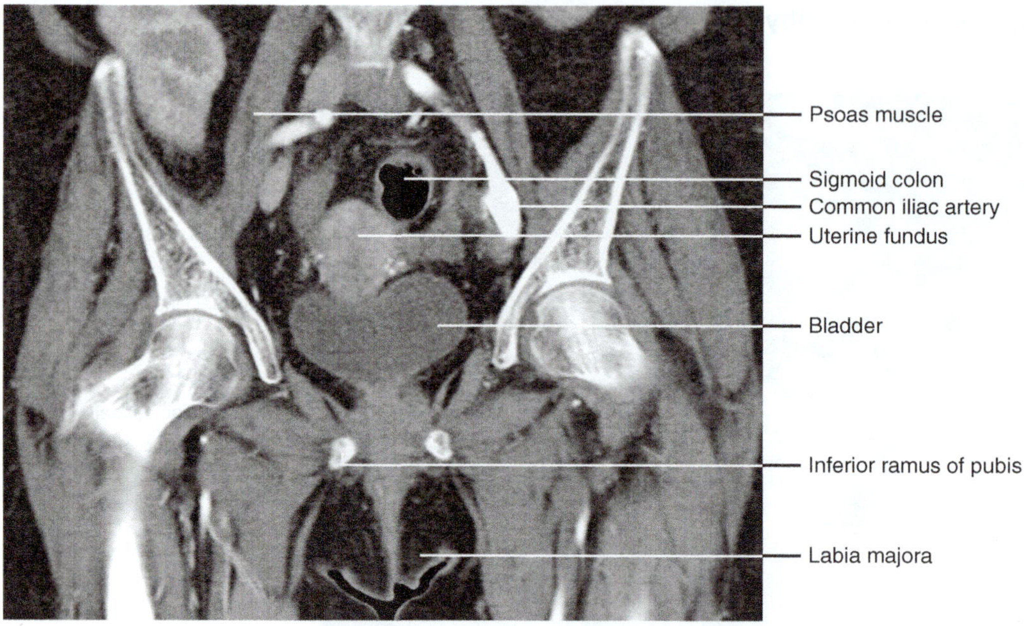

Fig. 23.10 Coronal CT reconstruction of an anteverted uterus demonstrating the uterine fundus above the bladder

Male Genital Tract

Ultrasound

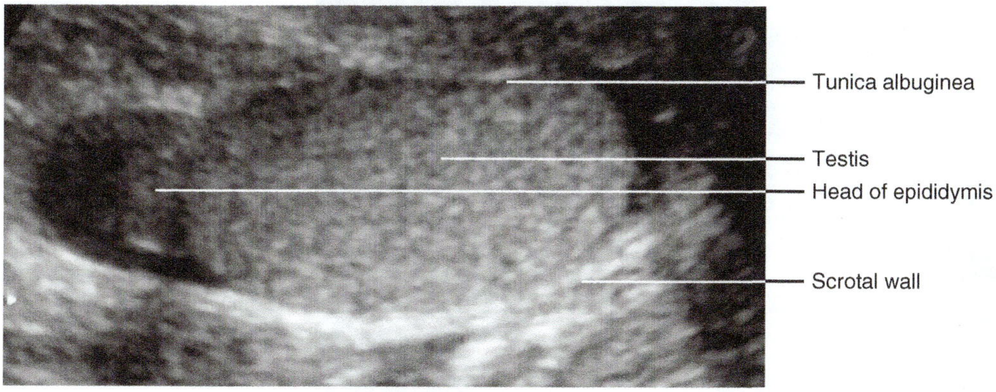

Fig. 23.11 Longitudinal ultrasound of the testis

Pearl

Normal testicular dimensions.

- Length 4–5 cm
- Width 2–4 cm
- AP diameter 3 cm

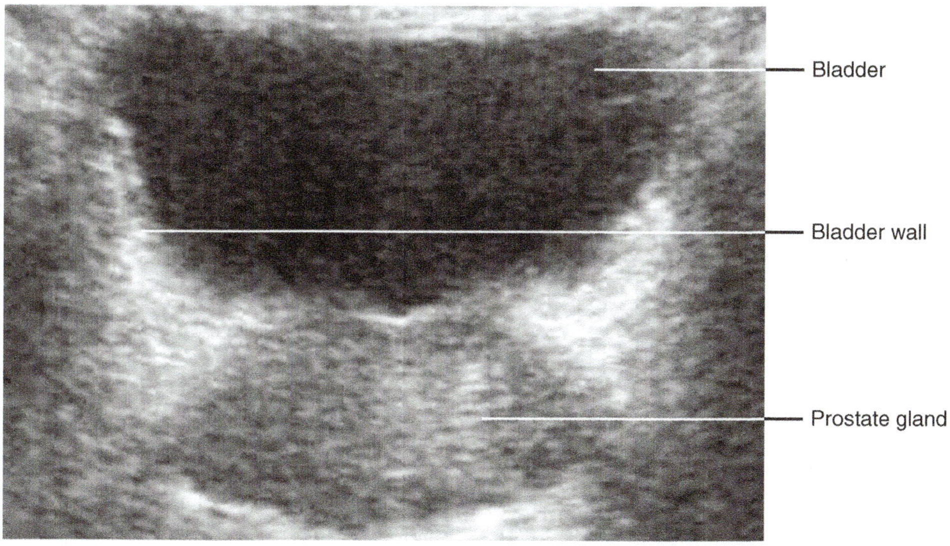

Fig. 23.12 Transverse transabdominal ultrasound of the prostate

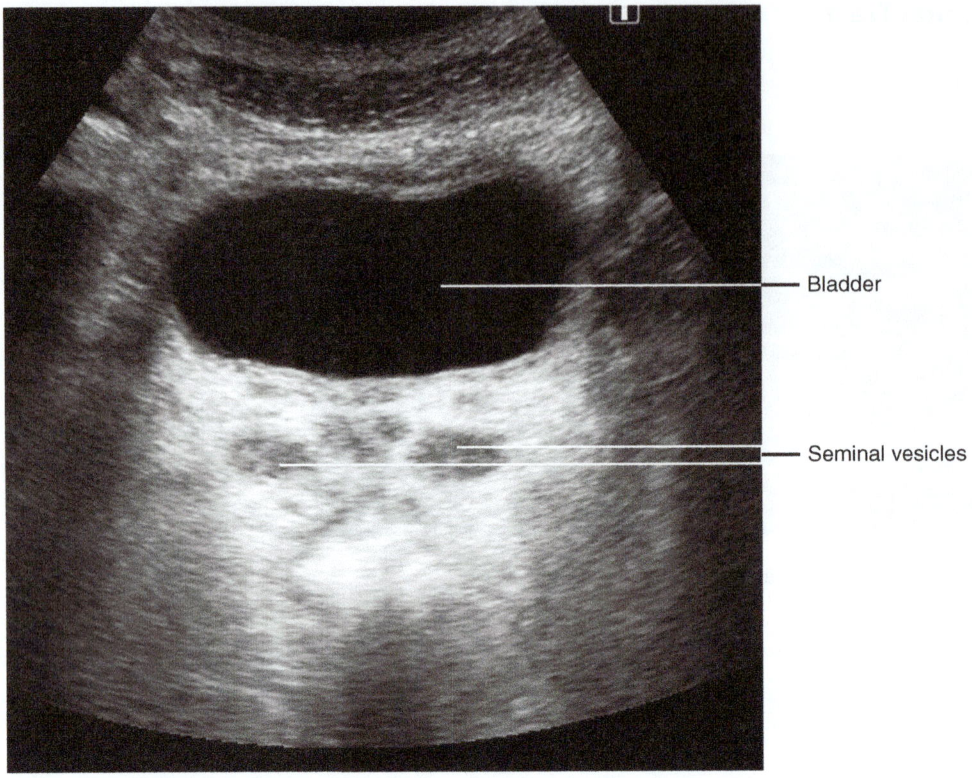

Fig. 23.13 Transverse ultrasound of the seminal vesicles

Magnetic Resonance Imaging

Axial images from inferior to superior demonstrating the testes, penis, prostate, and seminal vesicles.

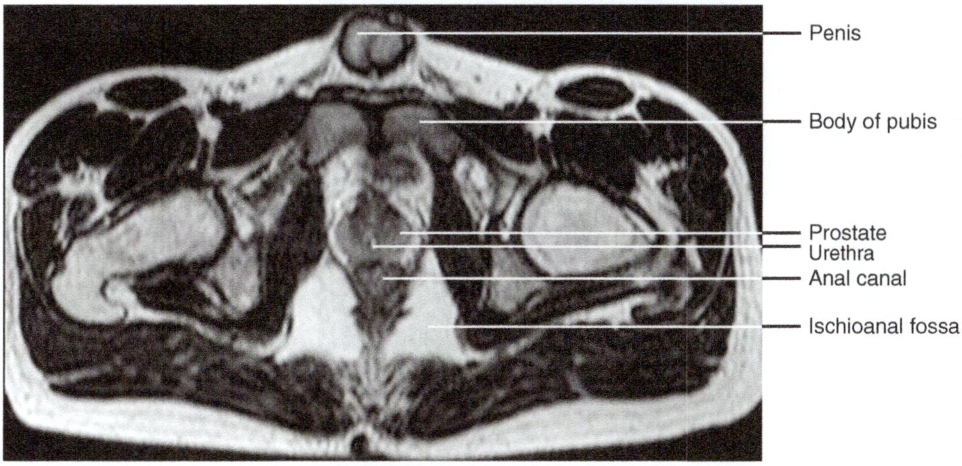

Fig. 23.14 Axial T1 at the level of the pubic symphysis

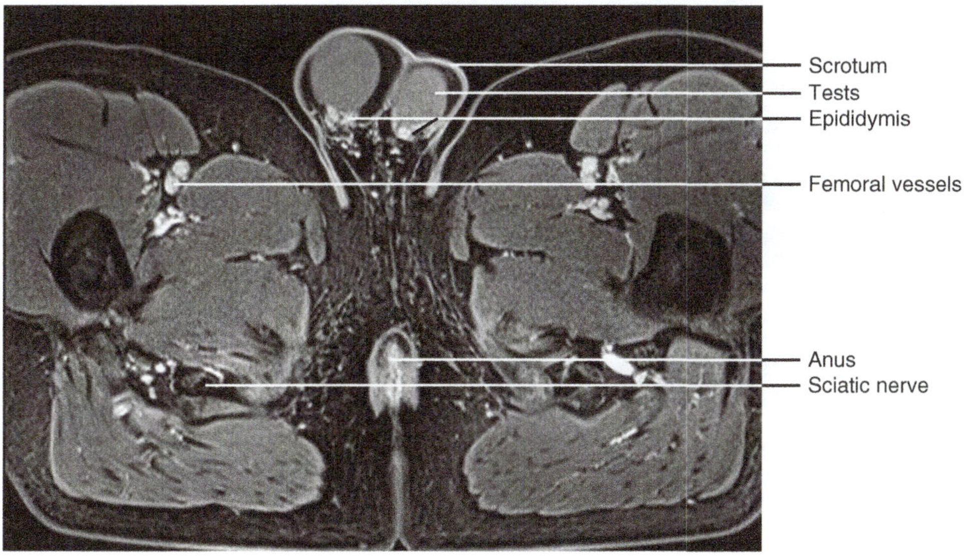

Fig. 23.15 Axial T1 gadolinium-enhanced fat saturated image at level of the scrotum

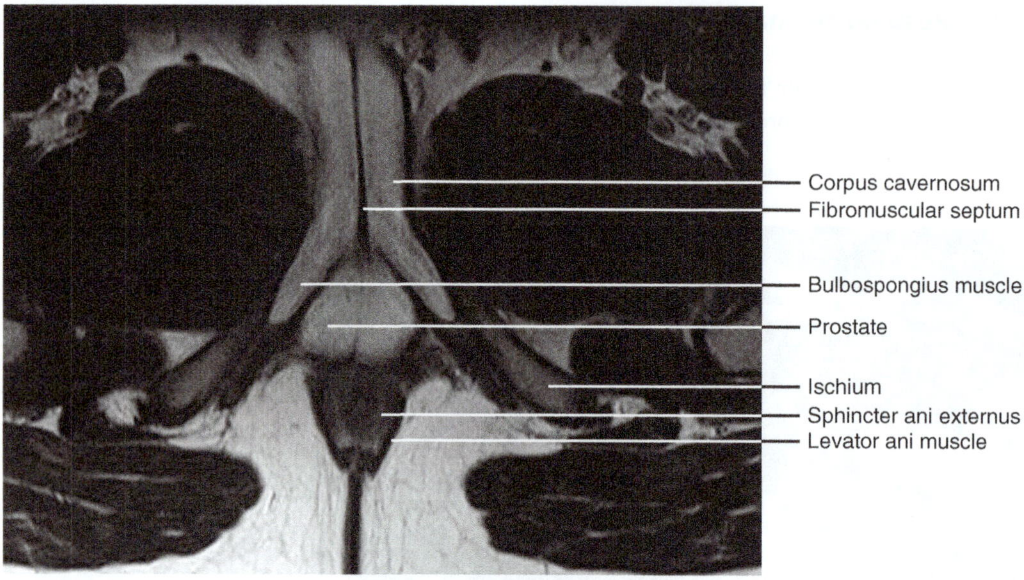

Fig. 23.16 Axial T1 gadolinium enhanced image at the level of the penis

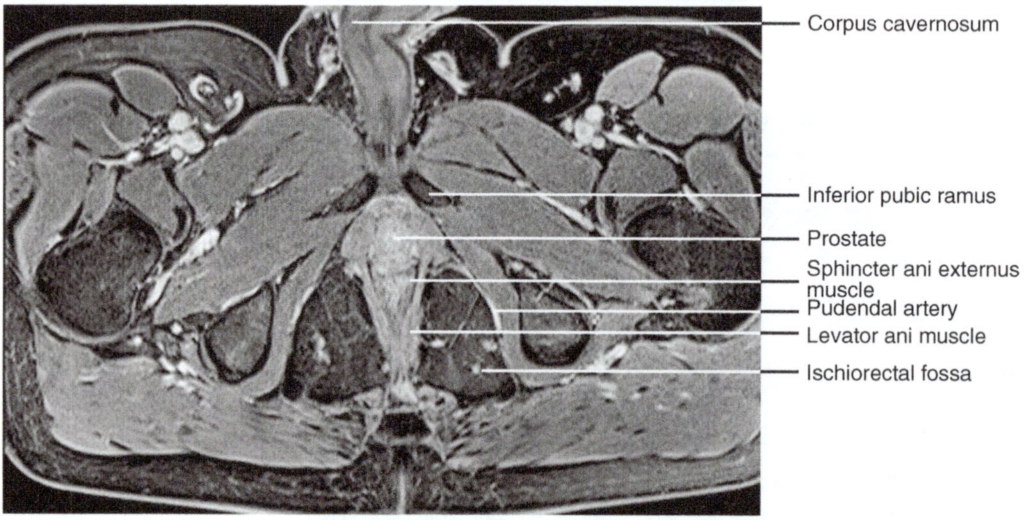

Fig. 23.17 Axial STIR at the level of the penis

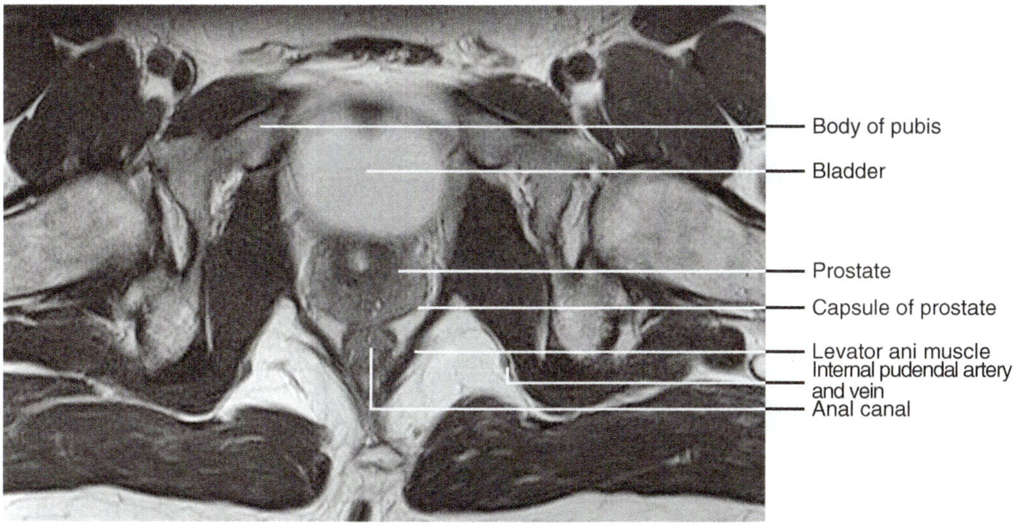

Fig. 23.18 Axial T2 of the prostate

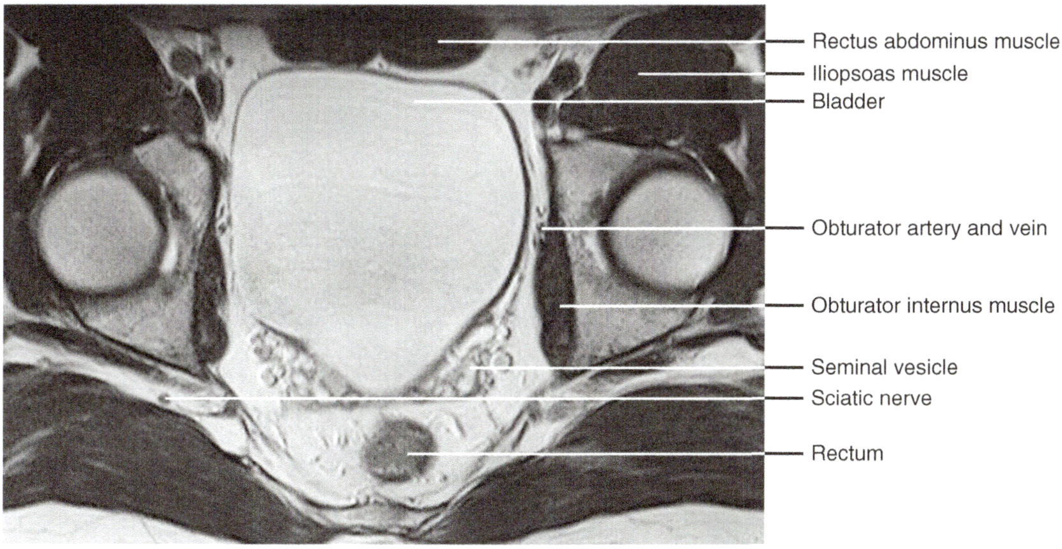

Fig. 23.19 Axial T2 at the level of the seminal vesicles

Spermatic Venogram

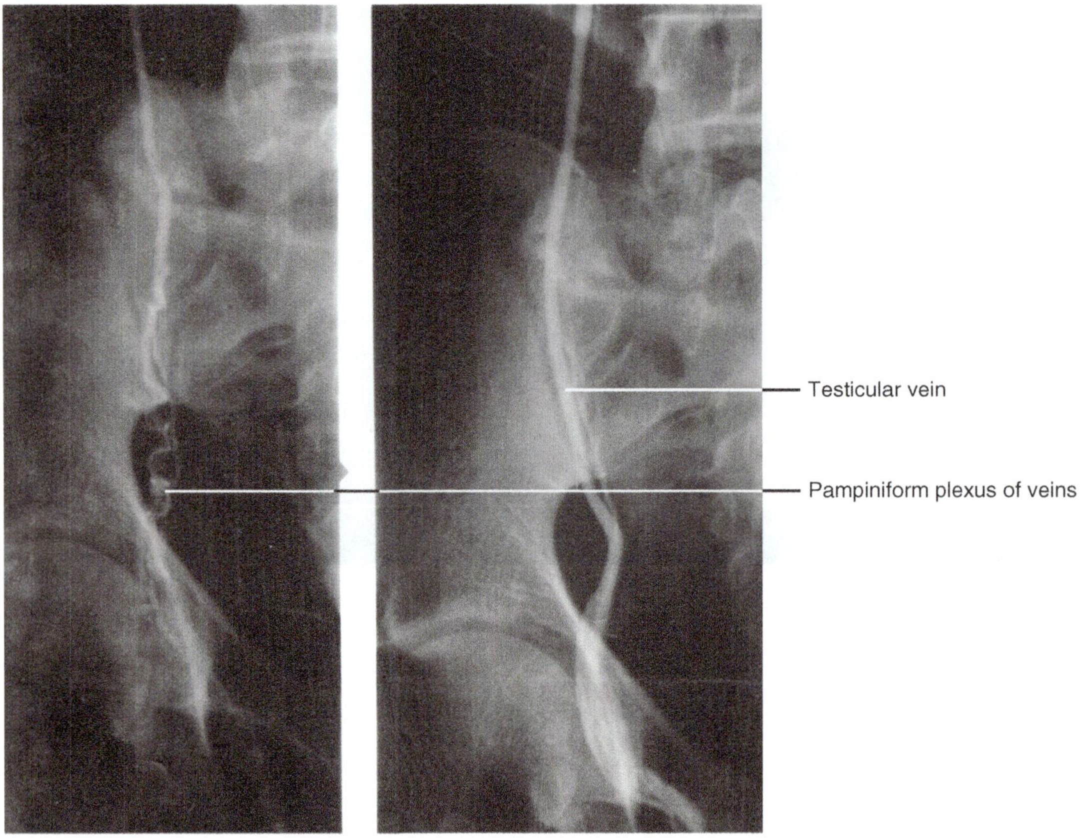

Fig. 23.22 Venogram of pampiniform plexus and testicular vein

Seminal Vesiculogram

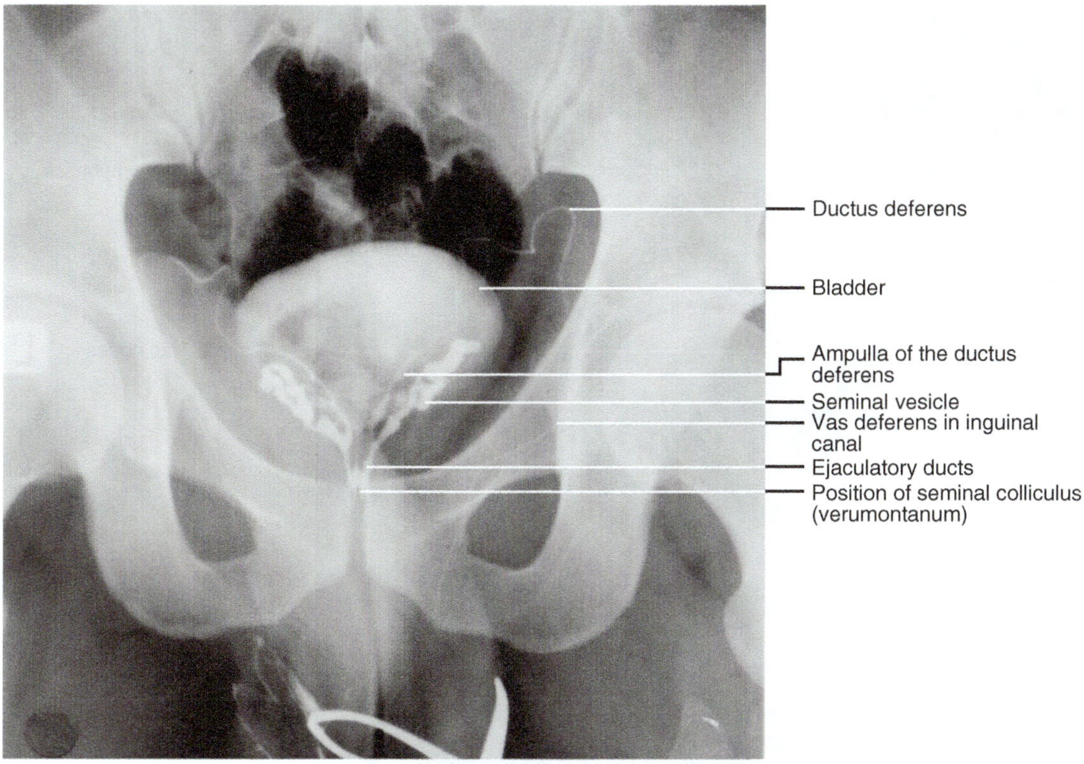

Fig. 23.23 Seminal vesiculogram

Lymphangiogram

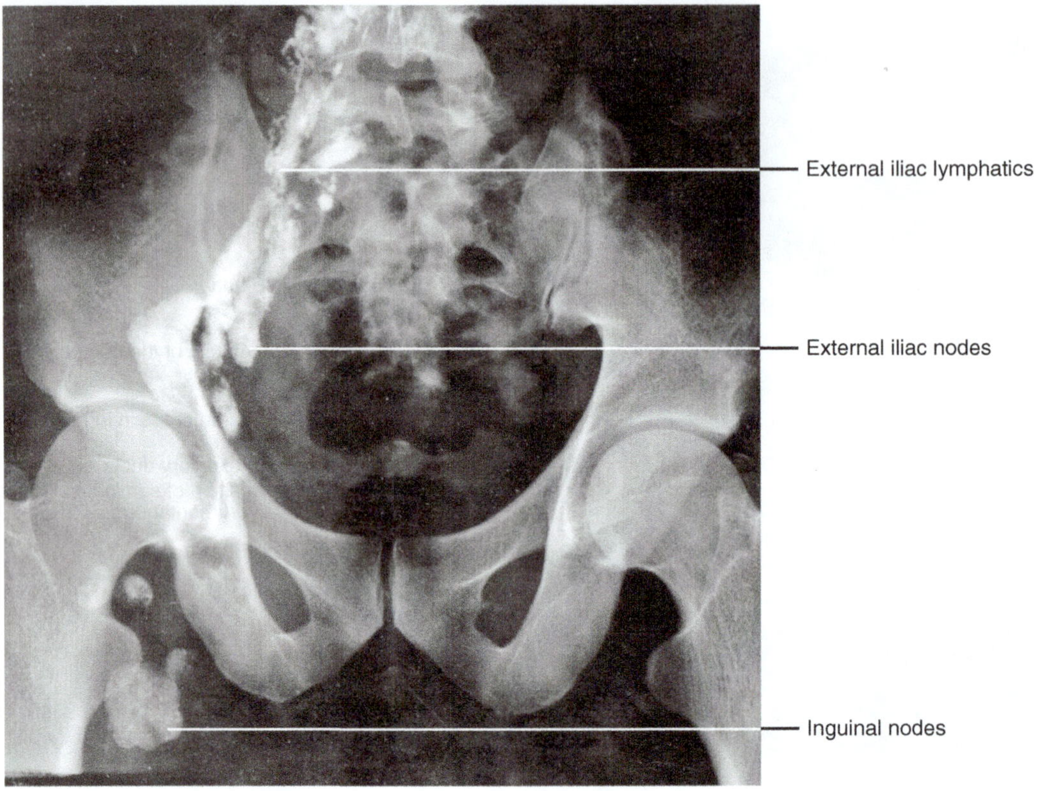

Fig. 23.24 Lymphangiogram

Part V

Musculoskeletal System

The axial skeleton includes the vertebral column and pelvis (and skull). The appendicular system includes the upper and lower limbs. The upper limb extends from the shoulder joint to the hand and includes the elbow joint and wrist. The lower limb extends from the hip to the foot and includes the knee and ankle joints. Imaging is possible using plain film, ultrasound, CT scanning and MRI, which are used to demonstrate the bony components, muscles, ligaments, cartilaginous structures and blood vessels.

Upper Limb

24

Shoulder

Plain Film

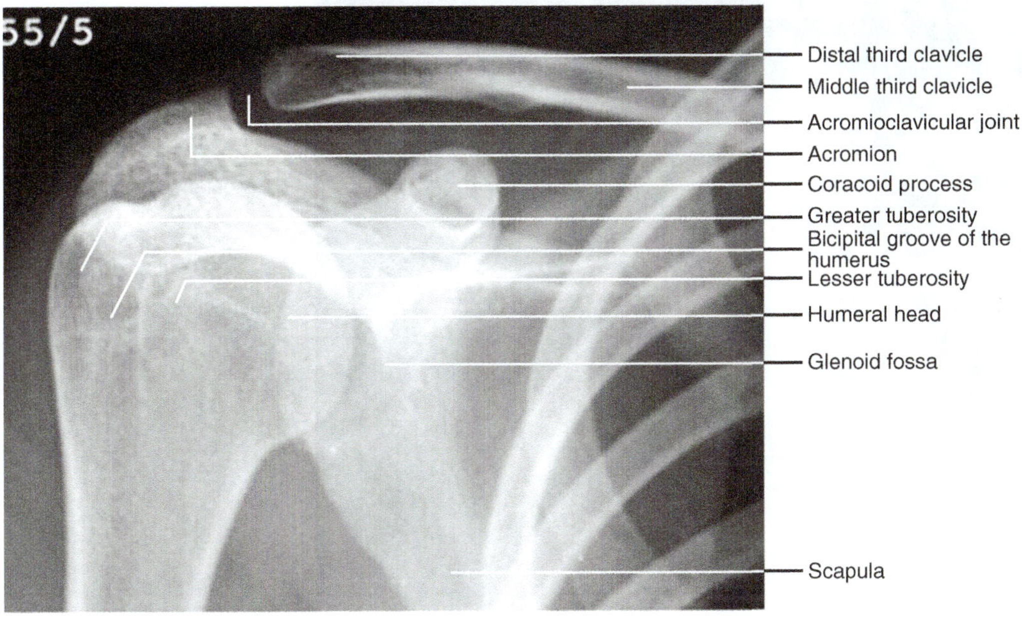

Fig. 24.1 AP view of the shoulder joint

S. Andronikou (ed.), *See Right Through Me*,
DOI 10.1007/978-3-642-23893-2_24, © Springer-Verlag Berlin Heidelberg 2012

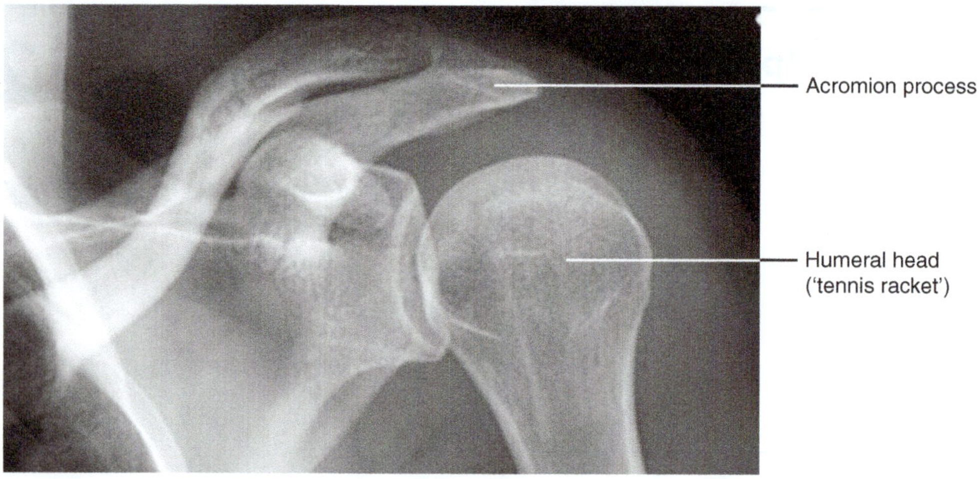

Fig. 24.2 AP view – internal rotation

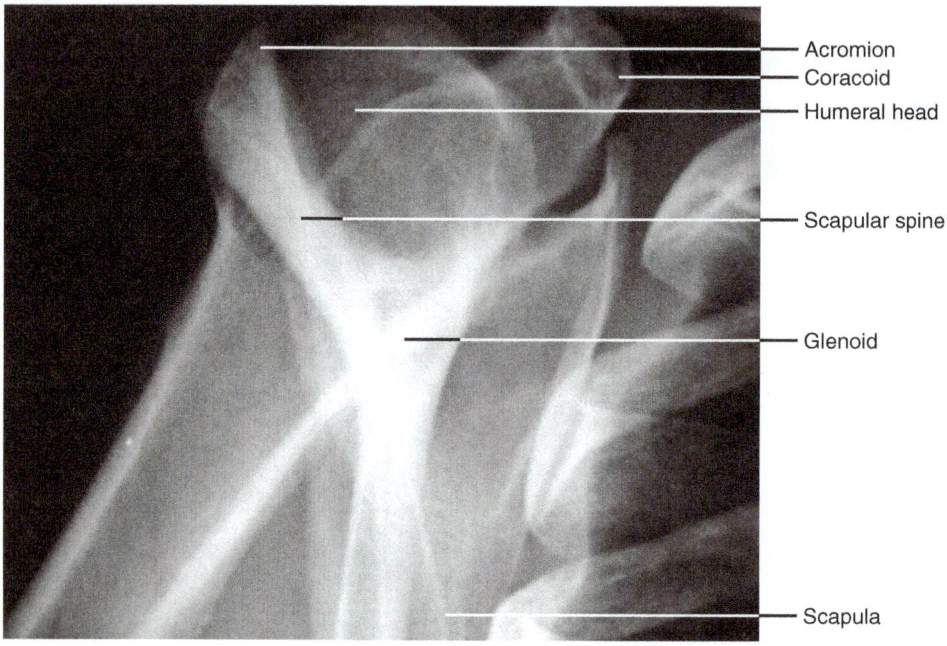

Fig. 24.3 'Mercedes Benz' view of the shoulder

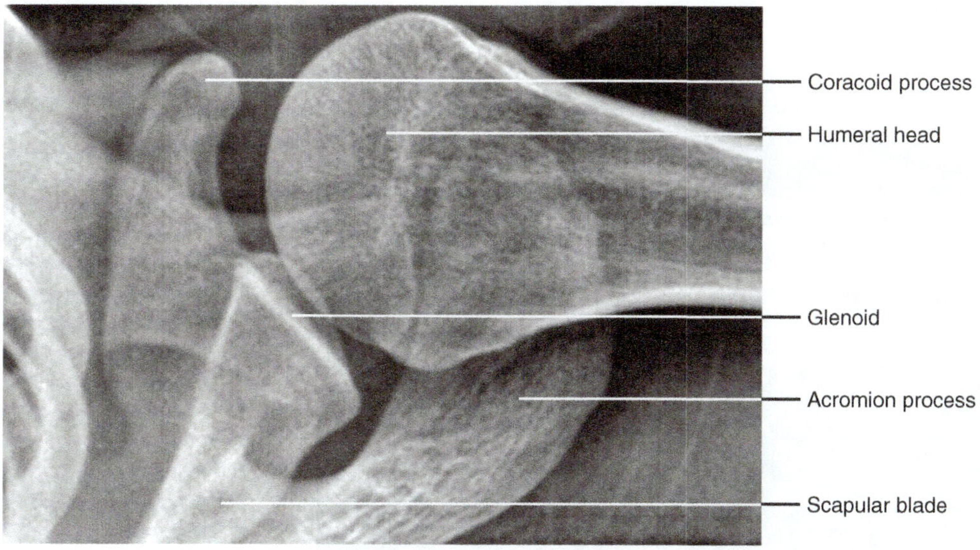

Fig. 24.4 Axial view of the shoulder joint

Ultrasound

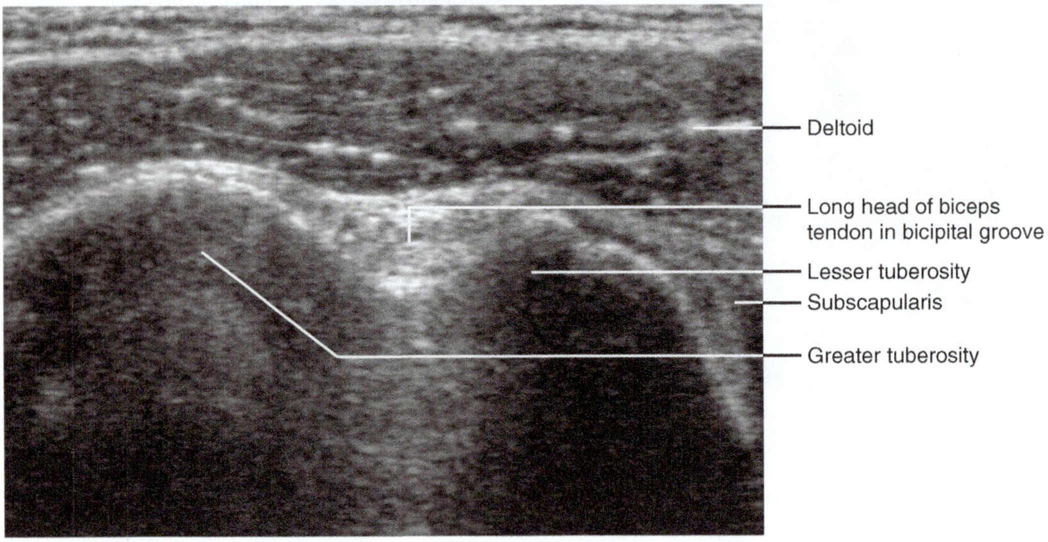

Fig. 24.5 Transverse US of the biceps tendon

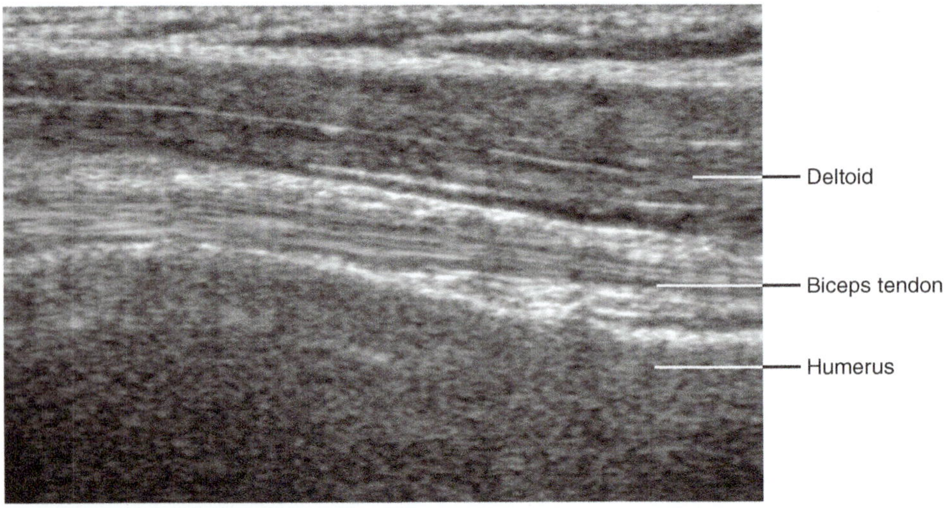

Fig. 24.6 Longitudinal US of the biceps tendon

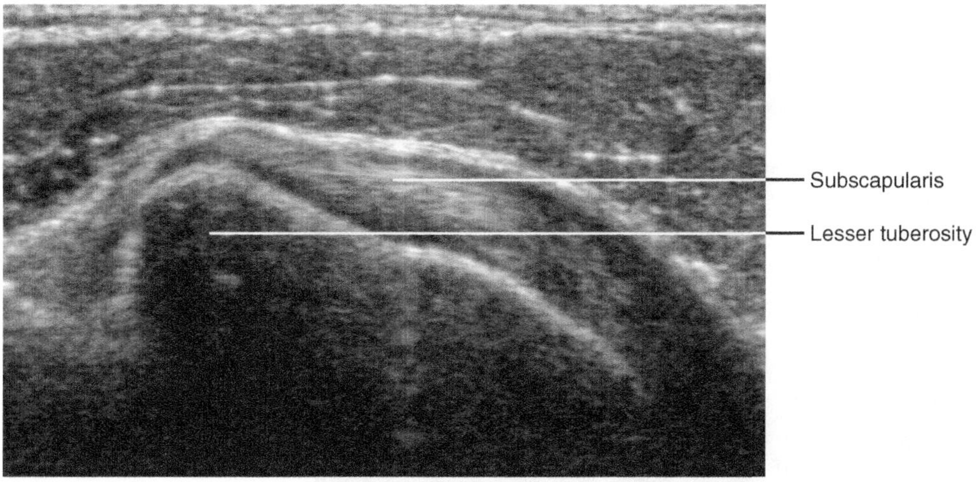

Fig. 24.7 Longitudinal US of subscapularis tendon in internal rotation

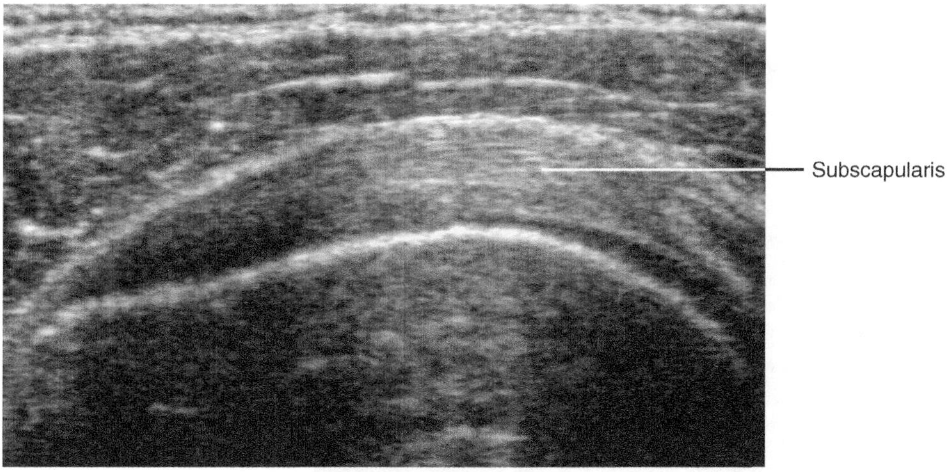

Fig. 24.8 Longitudinal US of subscapularis tendon in external rotation

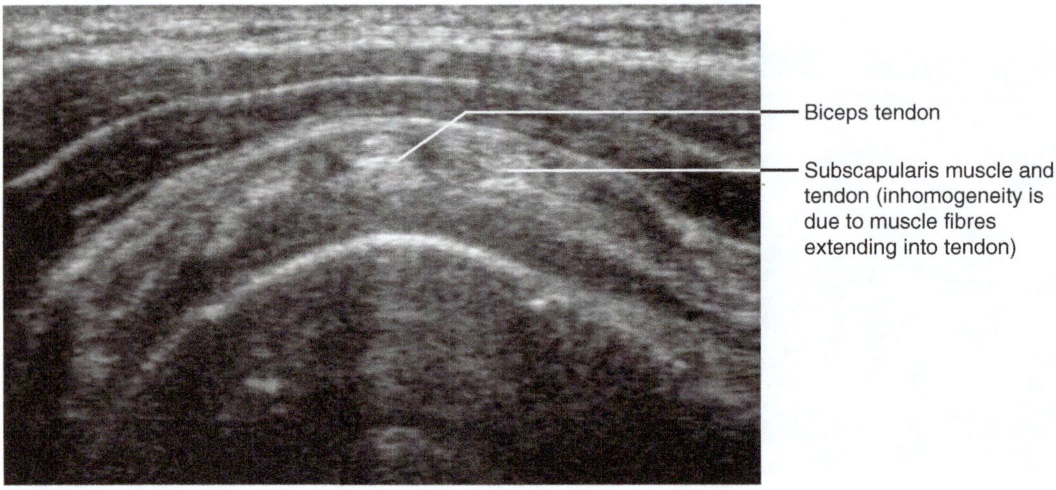

Fig. 24.9 Transverse US of subscapularis

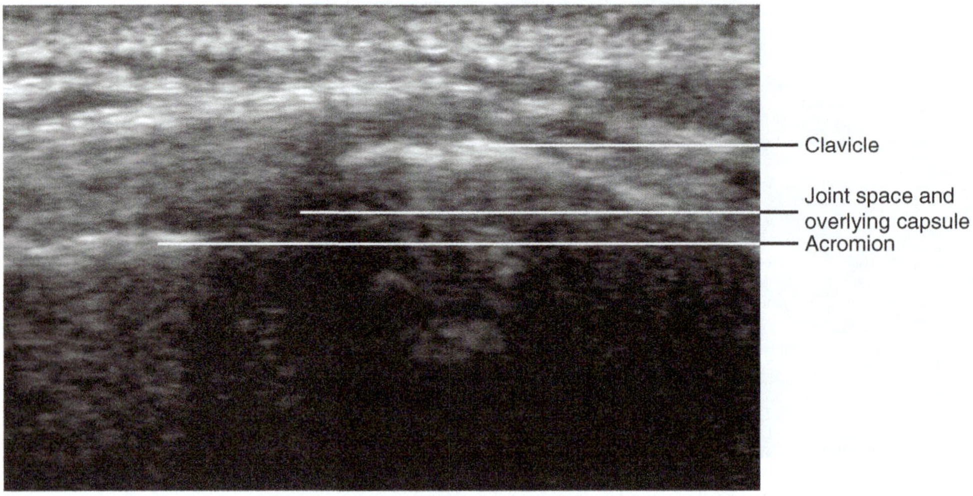

Fig. 24.10 Longitudinal US of the acromioclavicular joint

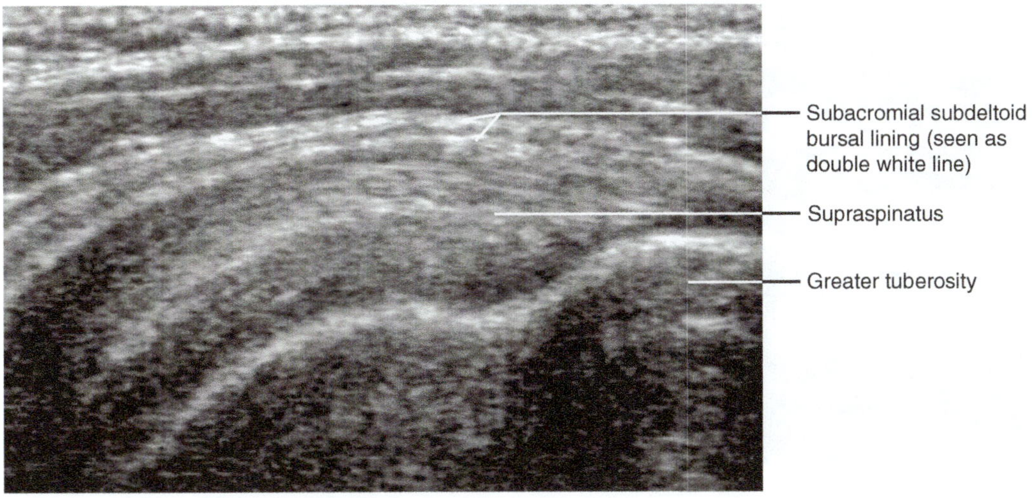

Fig. 24.11 Longitudinal US of supraspinatus tendon

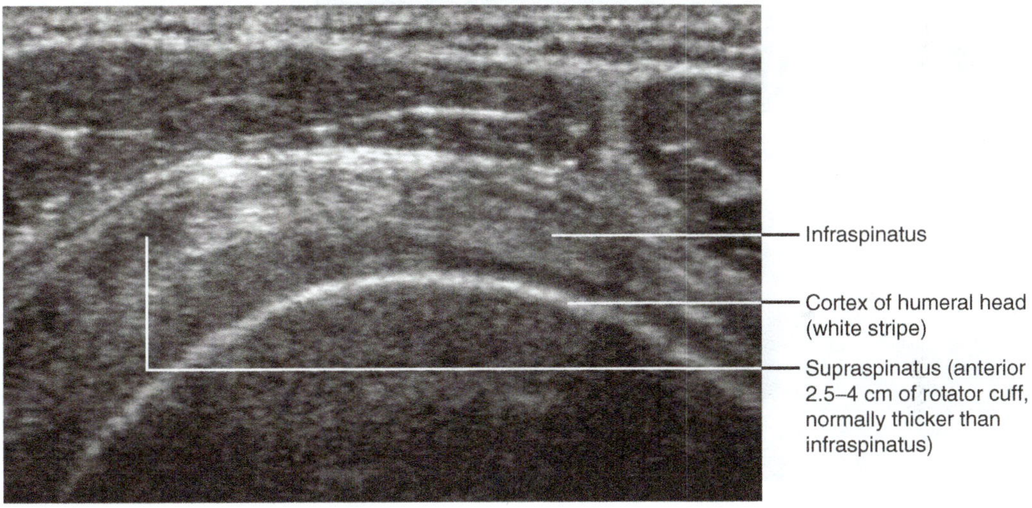

Fig. 24.12 Transverse US of supraspinatus tendon

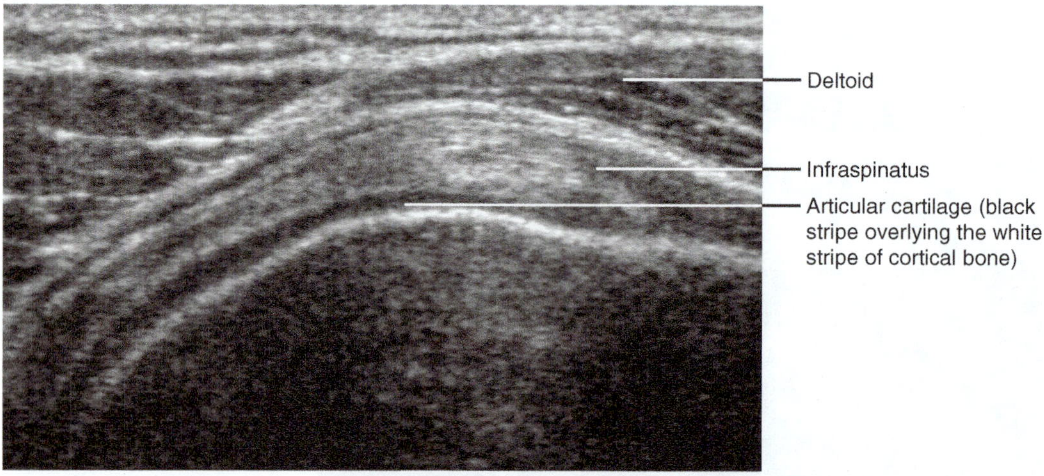

Fig. 24.13 Longitudinal US of infraspinatus tendon

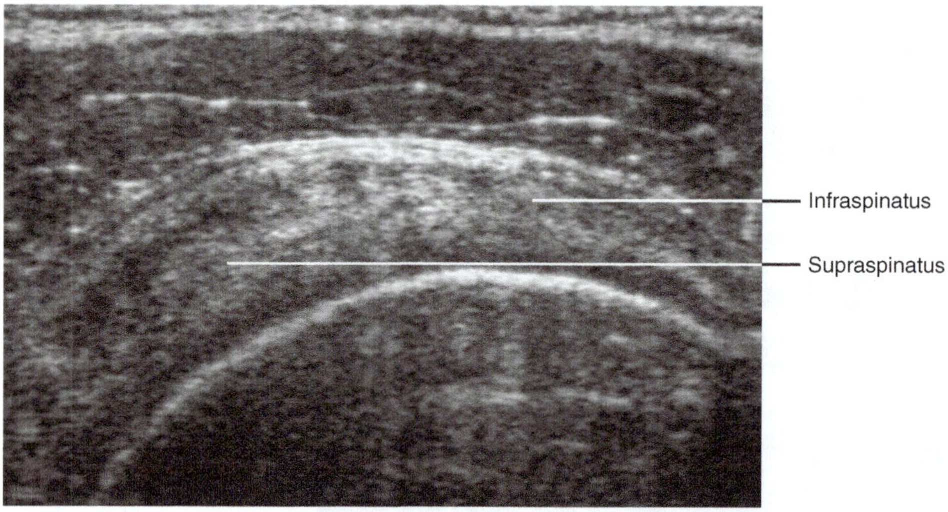

Fig. 24.14 Transverse US of infraspinatus tendon

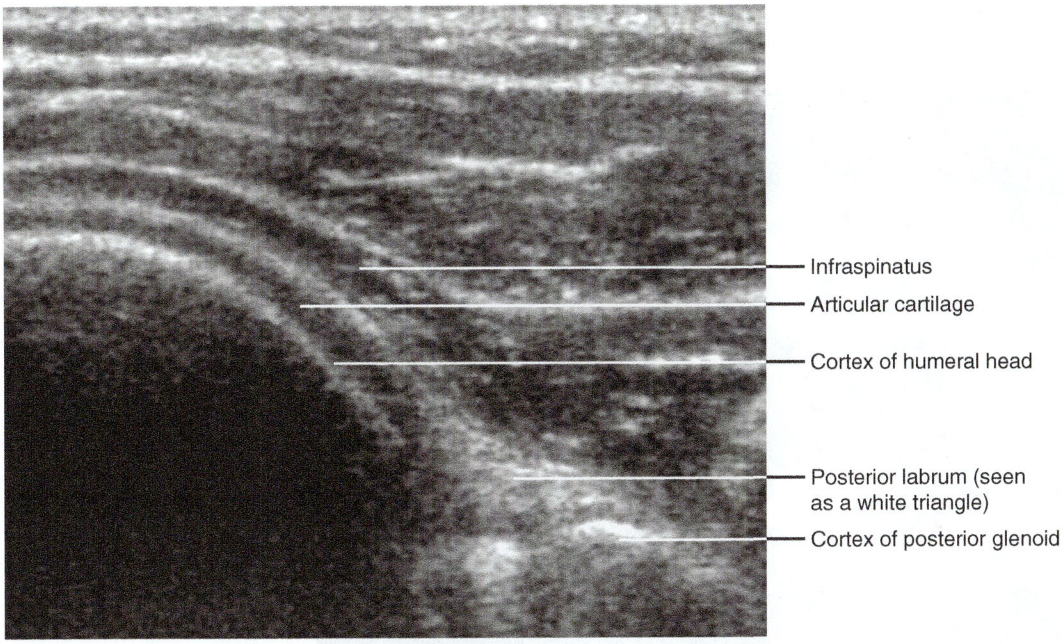

Fig. 24.15 Transverse US of posterior glenohumeral joint

MRI Arthrography

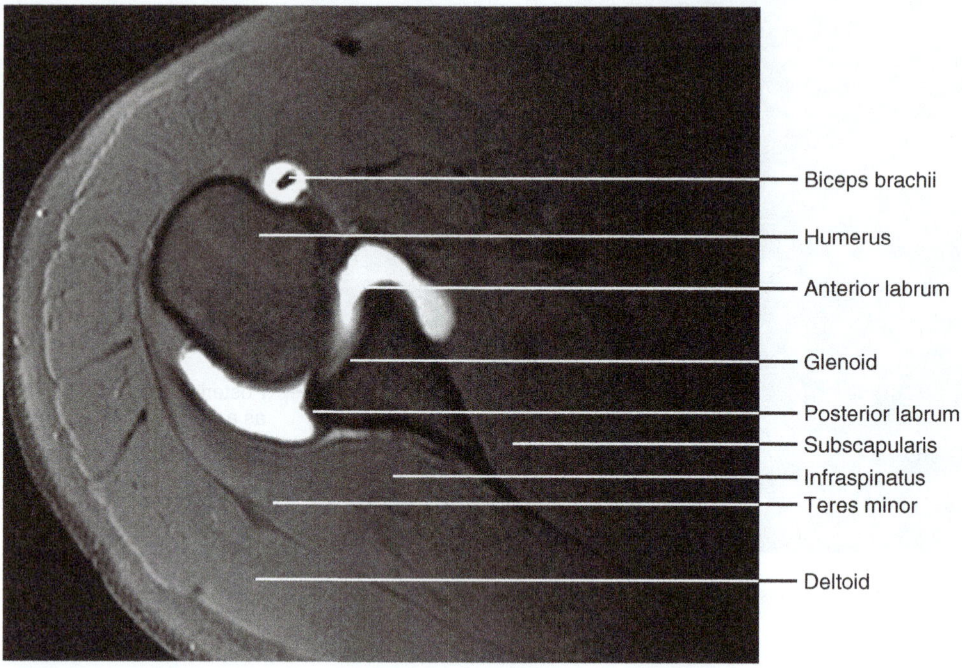

Fig. 24.16 Axial image demonstrating the biceps tendon

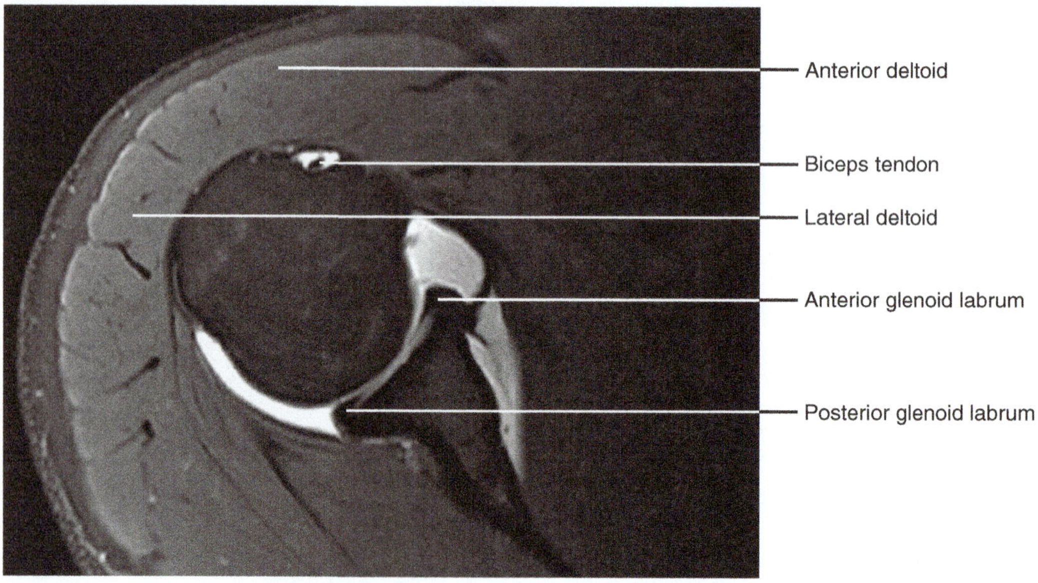

Fig. 24.17 Axial image demonstrating the glenoid labrum

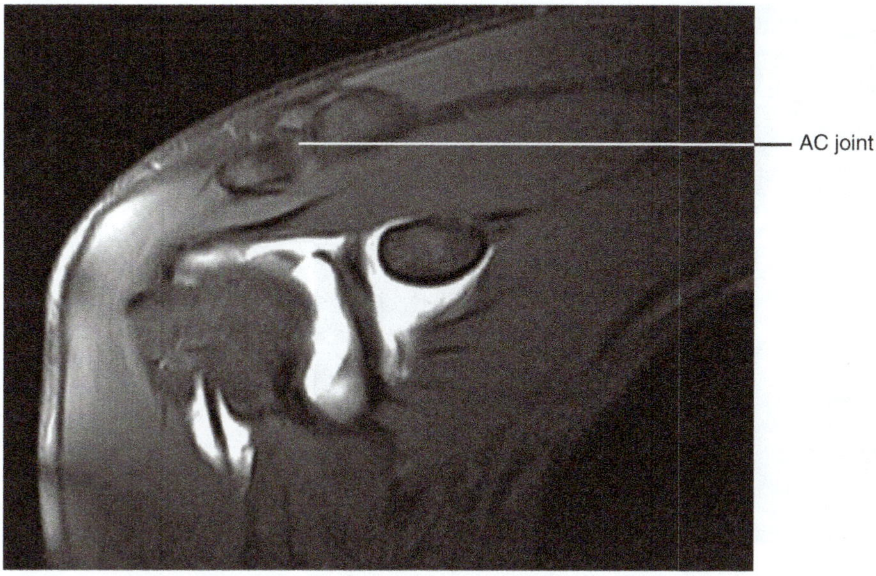

Fig. 24.18 Coronal image demonstrating the acromio-clavicular joint

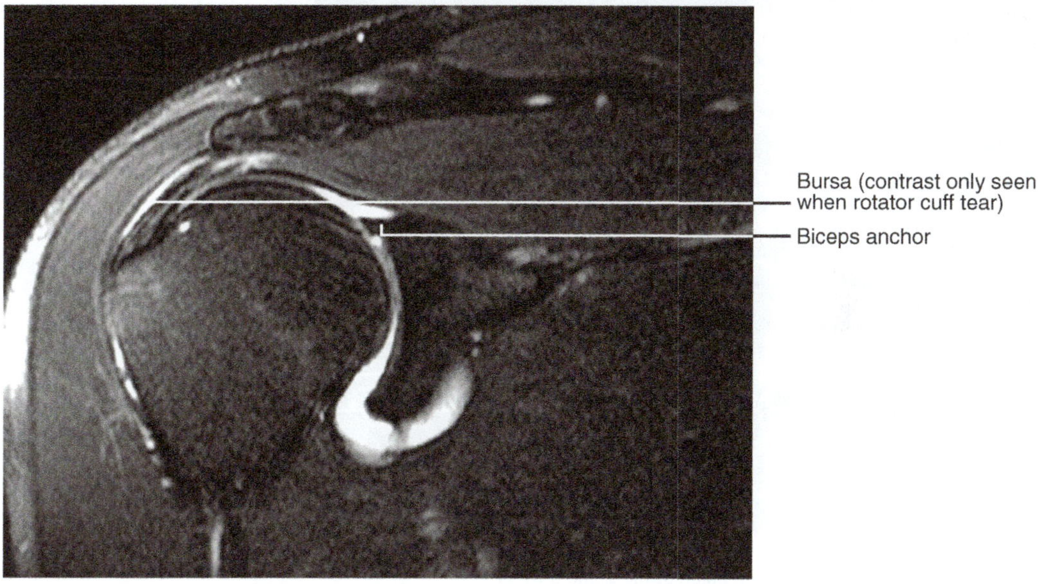

Fig. 24.19 Coronal image of the biceps anchor

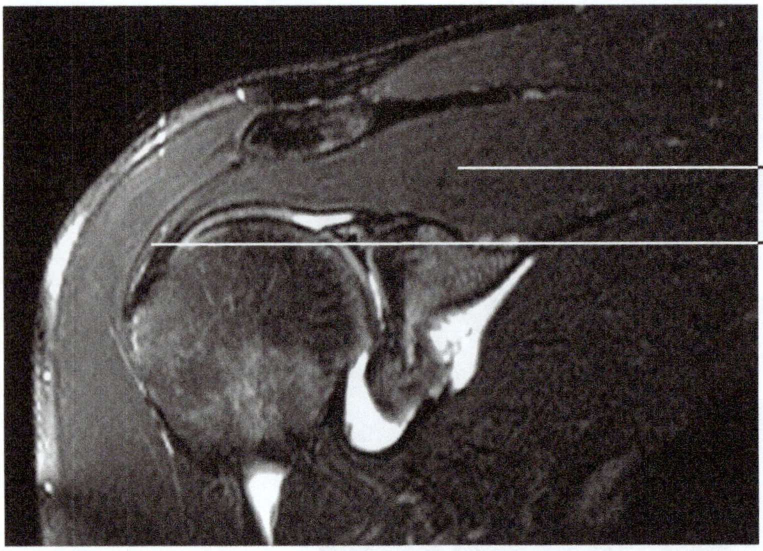

Fig. 24.20 Coronal image of the supraspinatus tendon

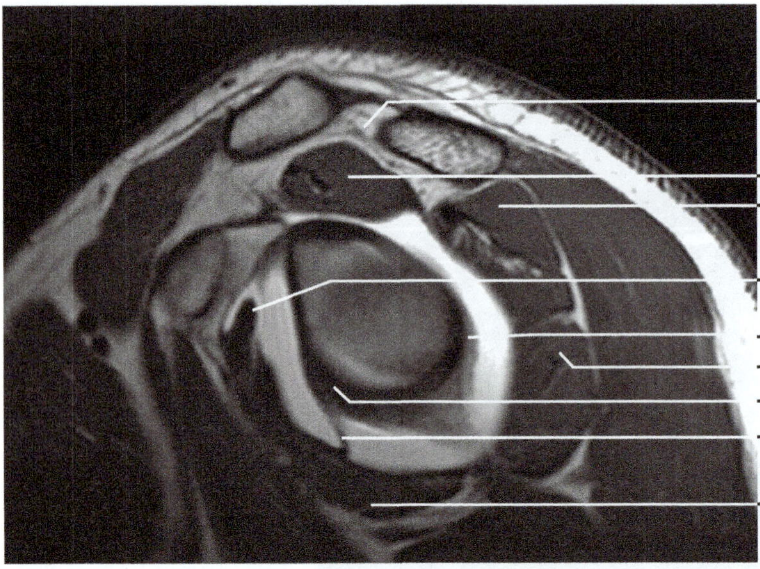

Fig. 24.21 Sagittal T1 image in the plane of the glenoid fossa

Humerus

Plain Film

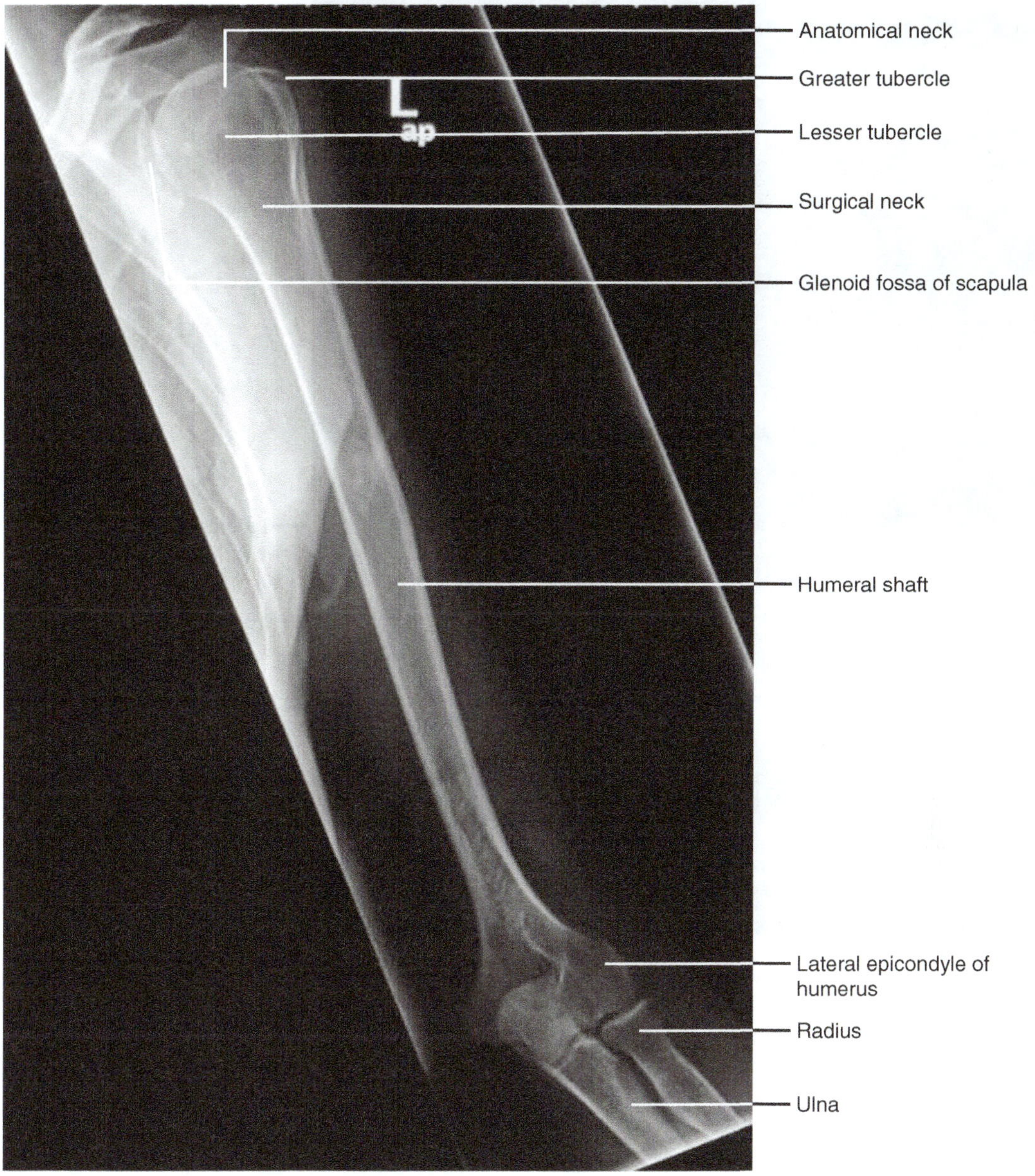

Fig. 24.22 Antero-posterior view of the humerus

Magnetic Resonance Imaging

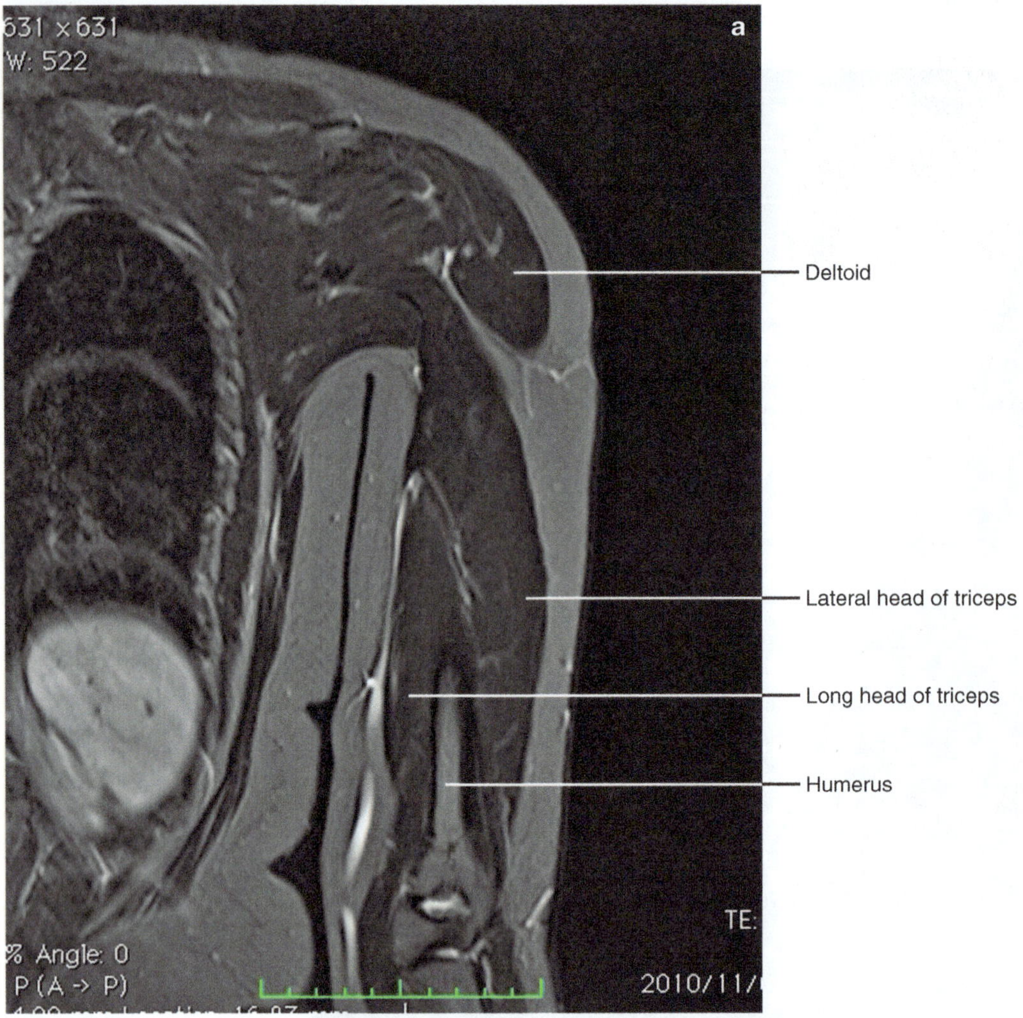

Fig. 24.23 (**a**, **b**) T1 coronal of the humerus

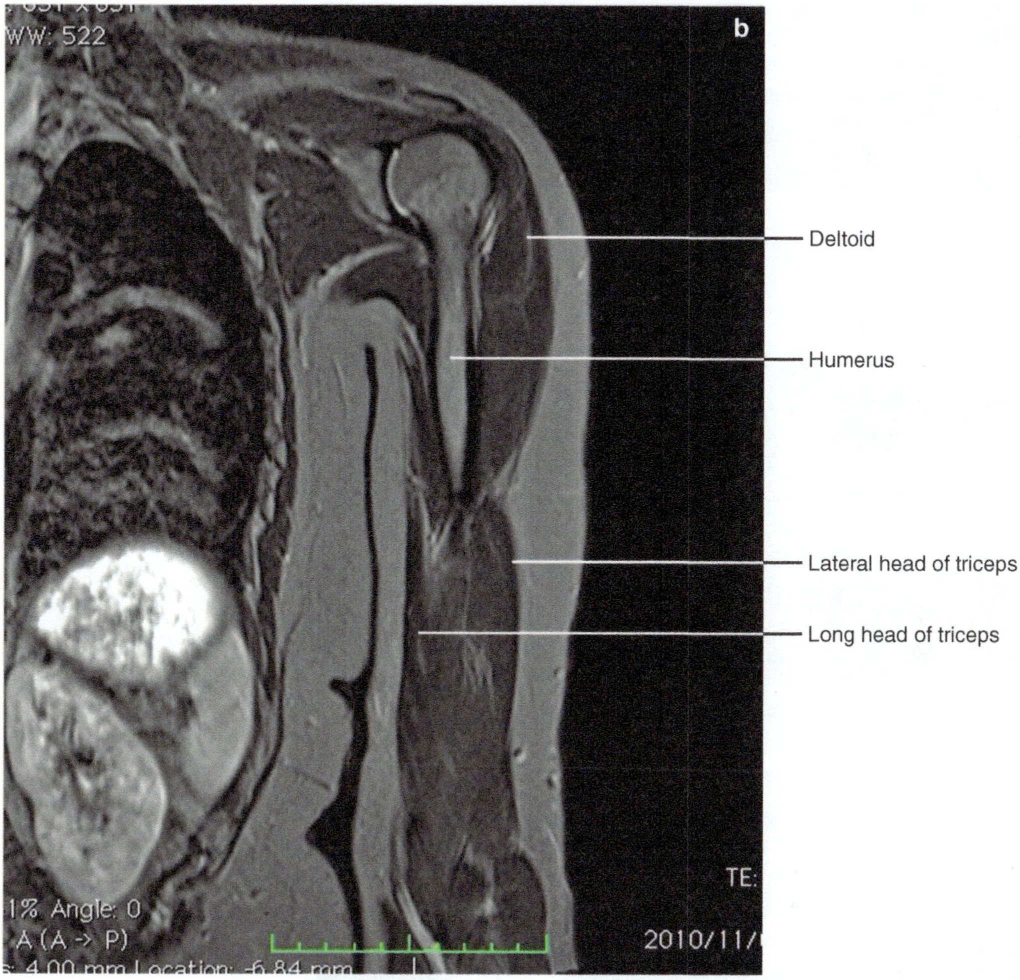

Fig. 24.23 (continued)

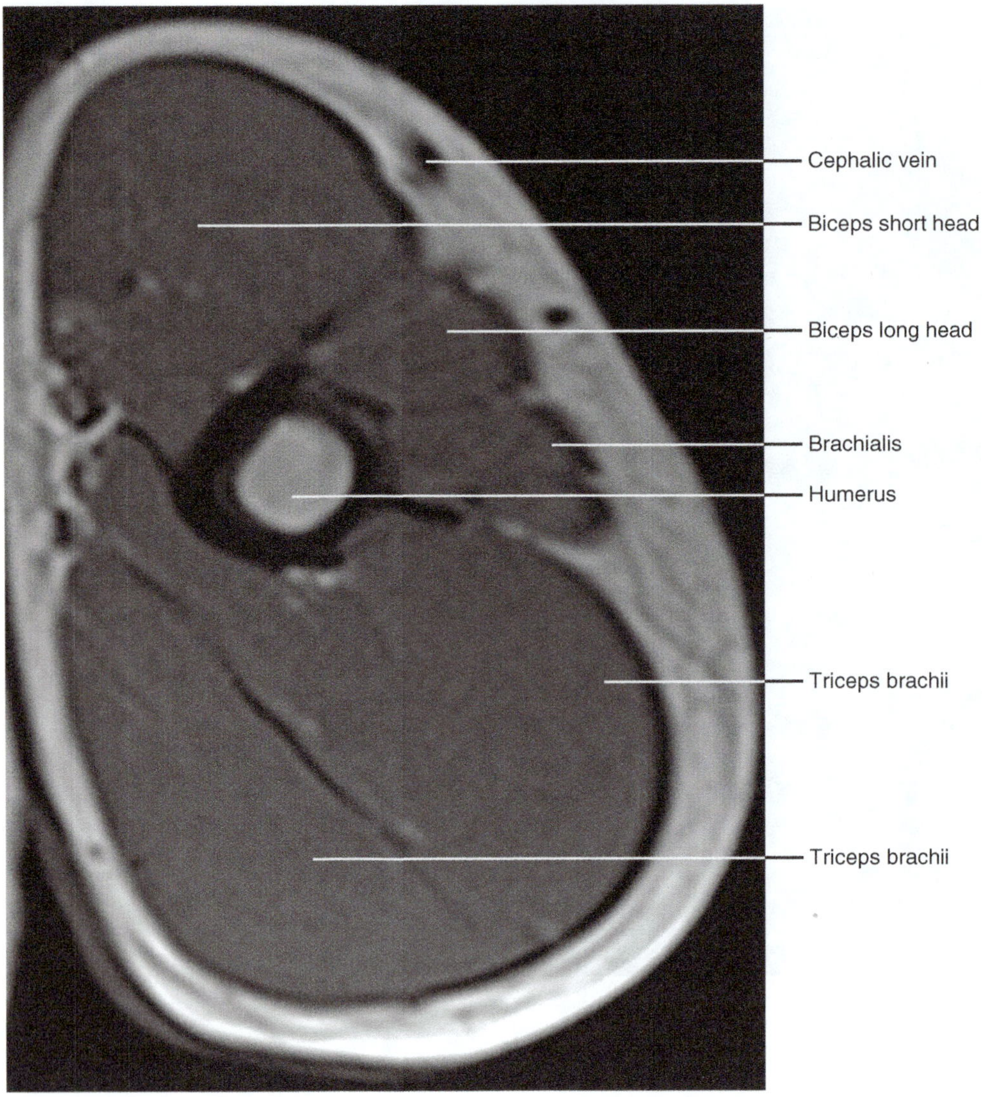

Fig. 24.24 T1 axial at the mid humeral diaphysis

Elbow

Plain Film

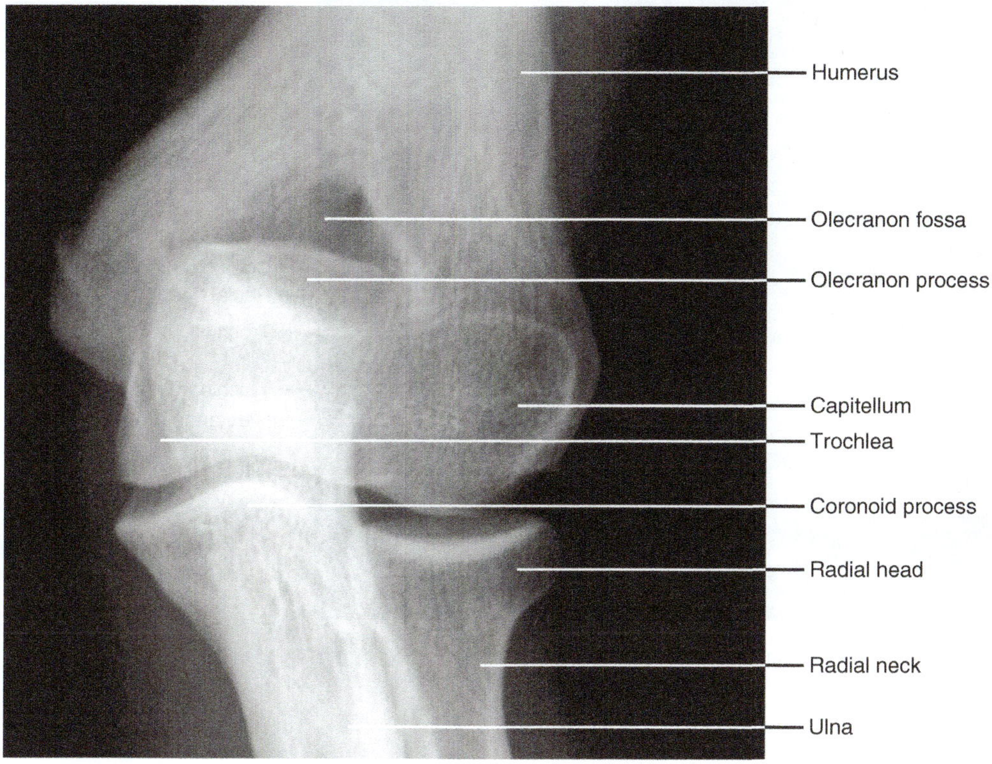

Fig. 24.25 Antero-posterior view of the left elbow

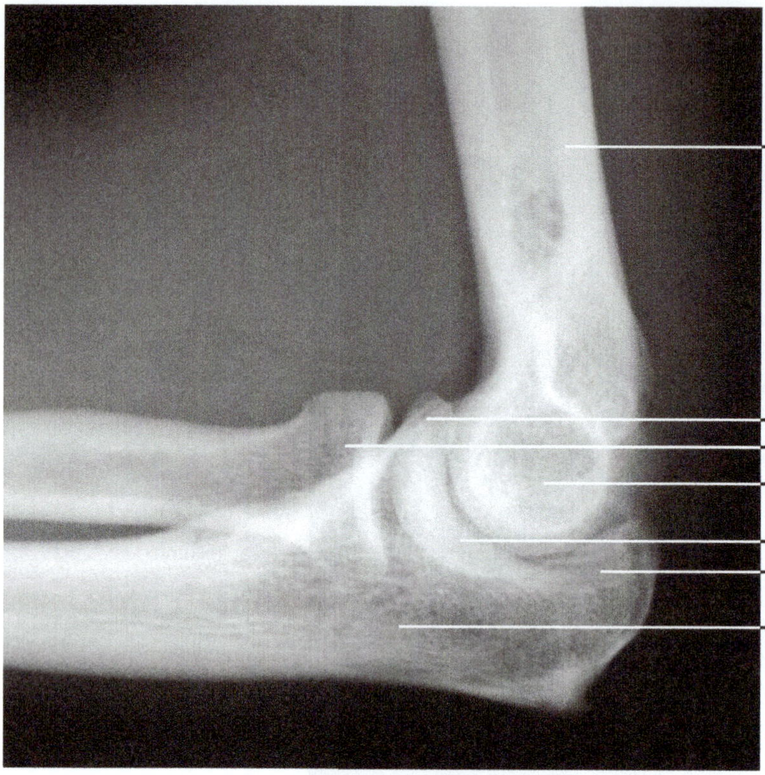

Fig. 24.26 Lateral view of the elbow

Magnetic Resonance Imaging

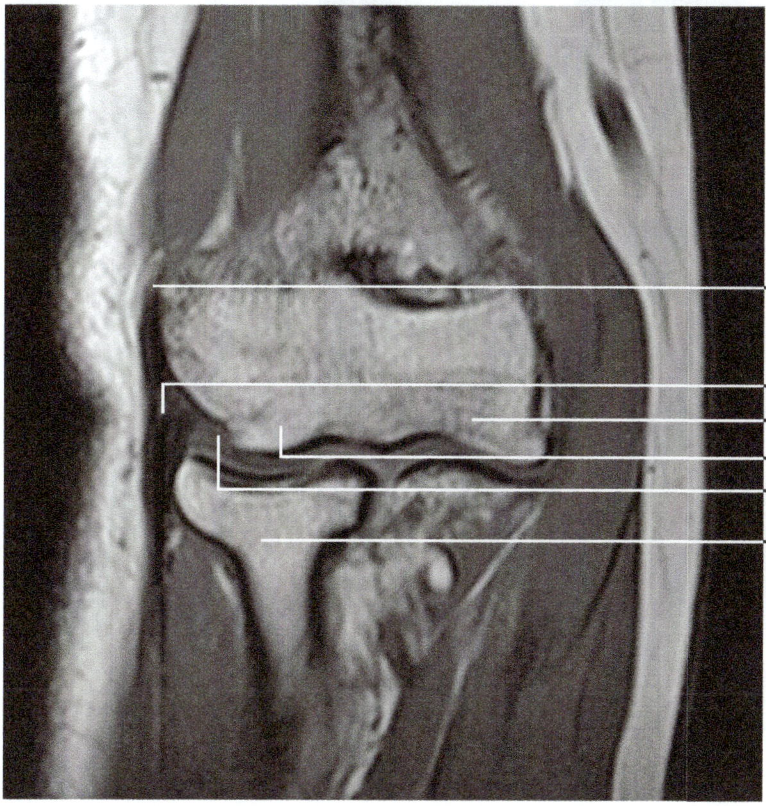

Fig. 24.27 T1 coronal of the elbow at the radial head

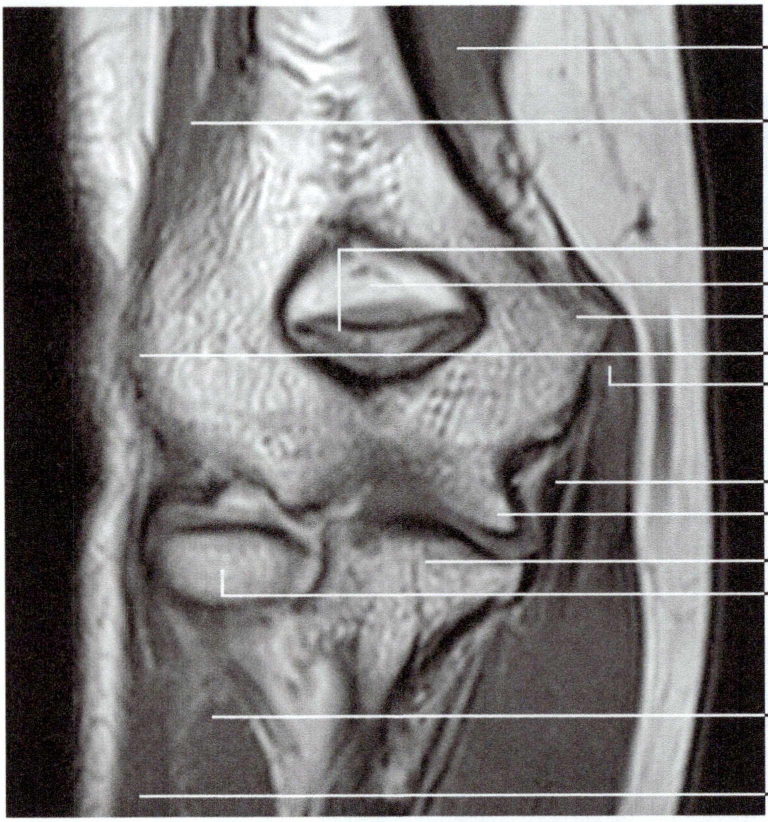

Fig. 24.28 T1 coronal of the elbow at the olecranon fossa

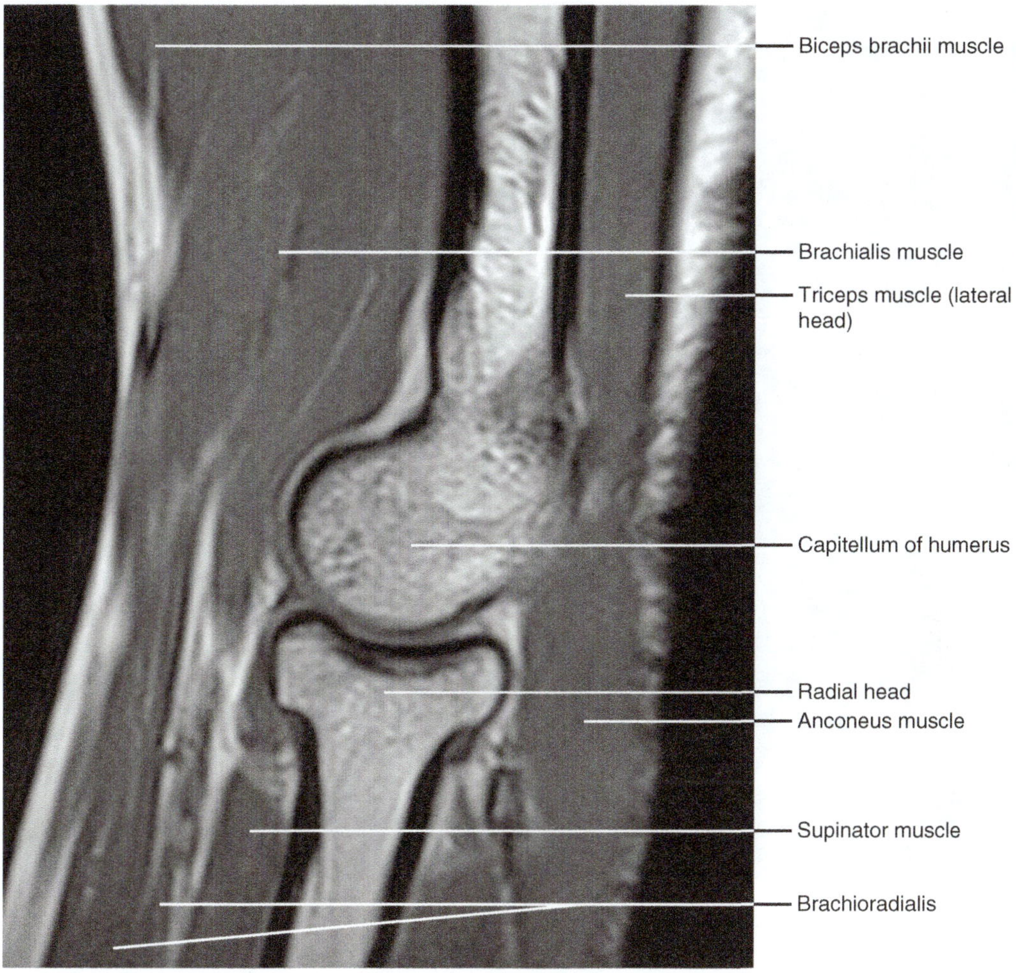

Fig. 24.29 T1 sagittal at the radial head (lateral)

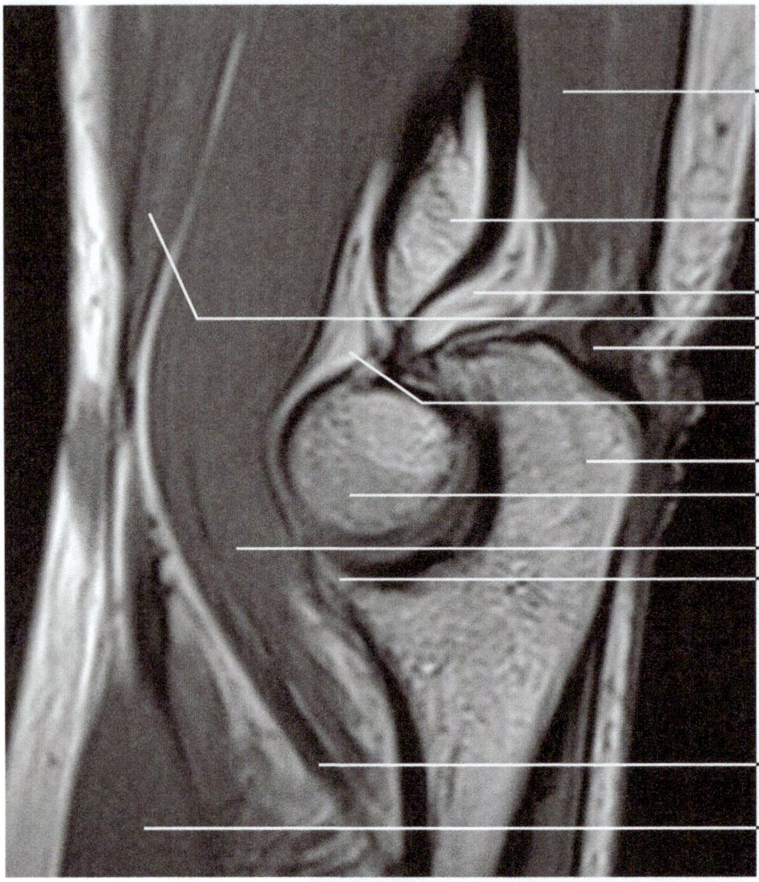

Fig. 24.30 T1 sagittal at the level of the ulna (medial)

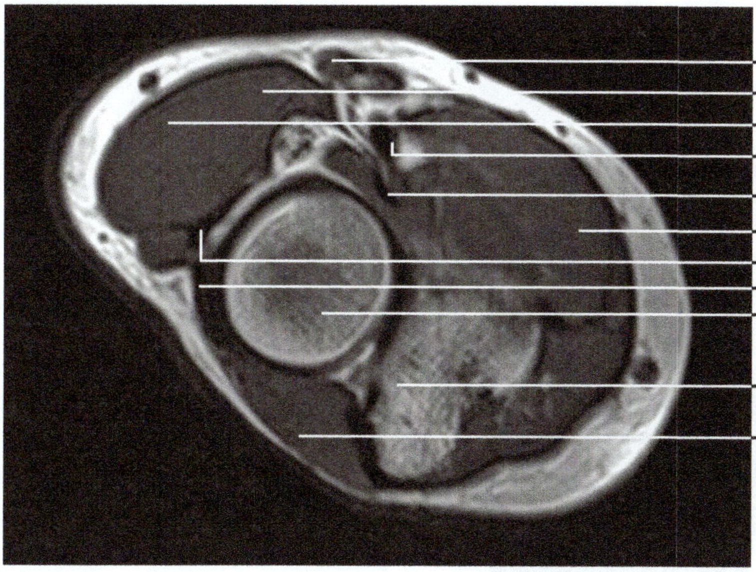

Fig. 24.31 T1 axial at the radial head at the right elbow

Pearl

The posterior fat pad is not seen on a plain radiograph lateral view in flexion, but can be seen on a lateral view in extension.

The radius should be aligned with the capitellum on all views.

The capitellar pseudo-defect is an area of the distal humerus between the inferior capitellum and the lateral epicondyle which mimics an osteochondral injury.

The trochlear ridge is a transversely oriented osseous ridge in the trochlea fossa between the coronoid and olecranon process which is devoid of cartilage.

Radius and Ulna

Plain Film

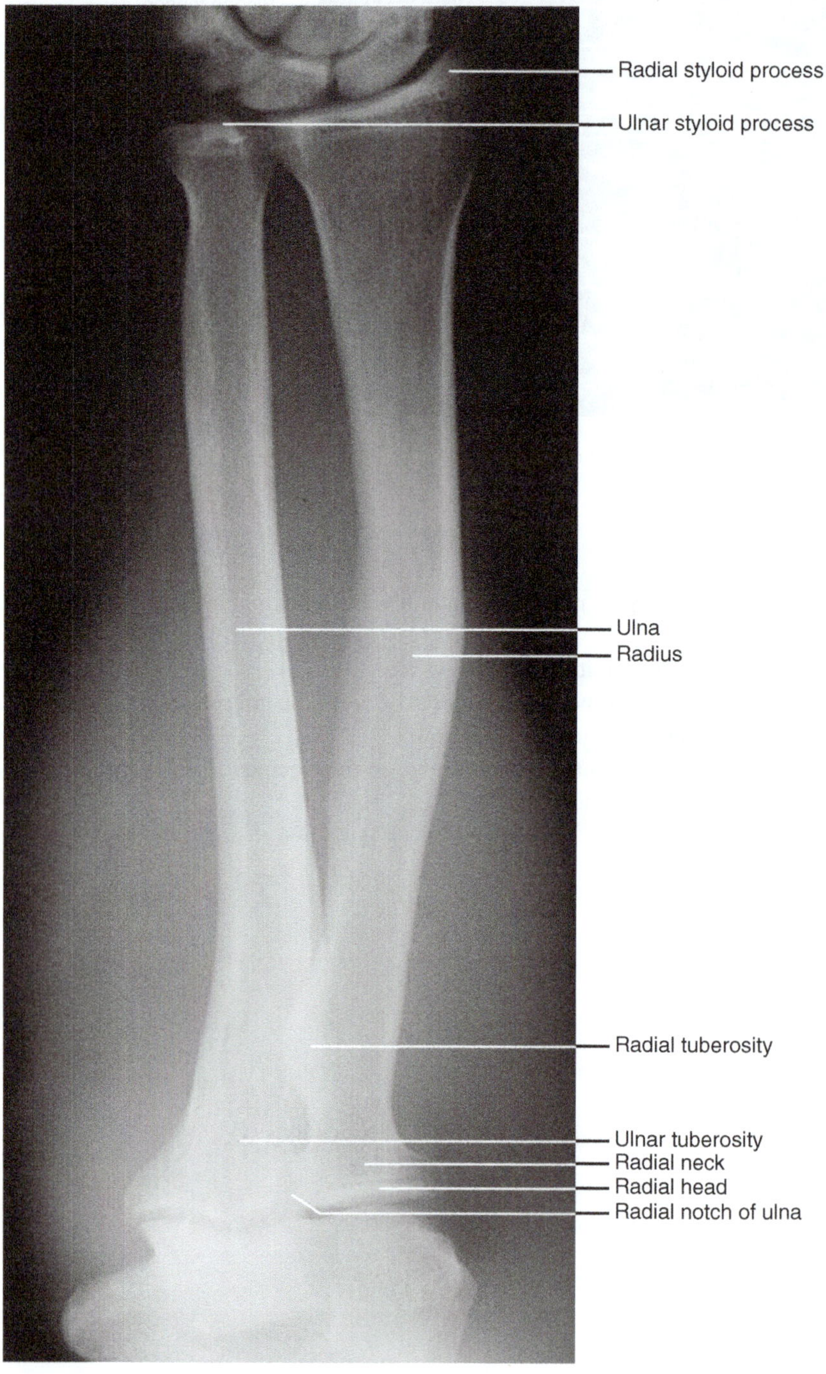

Fig. 24.32 Antero-posterior view of the radius and ulna

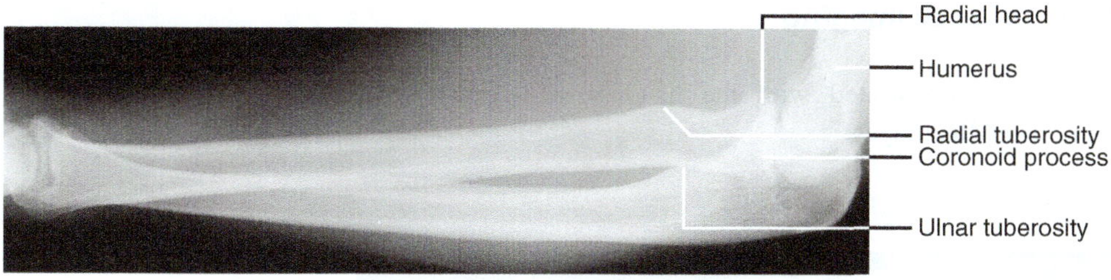

Fig. 24.33 Lateral view of the radius and ulna

Magnetic Resonance Imaging

The left arm is demonstrated with the hand supinated in these MR images.

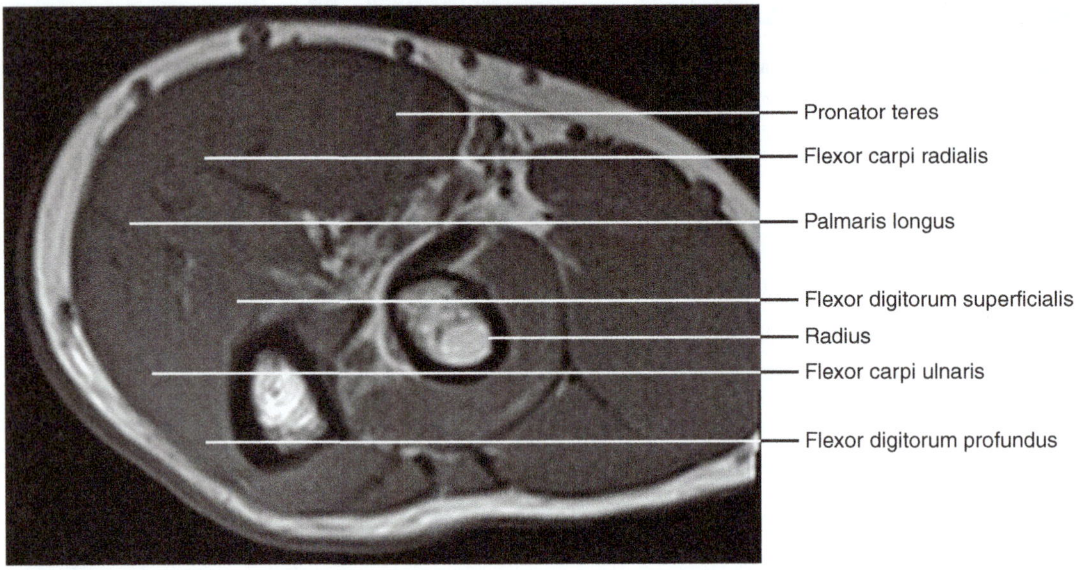

Fig. 24.34 T1 axial of the mid radius and ulna

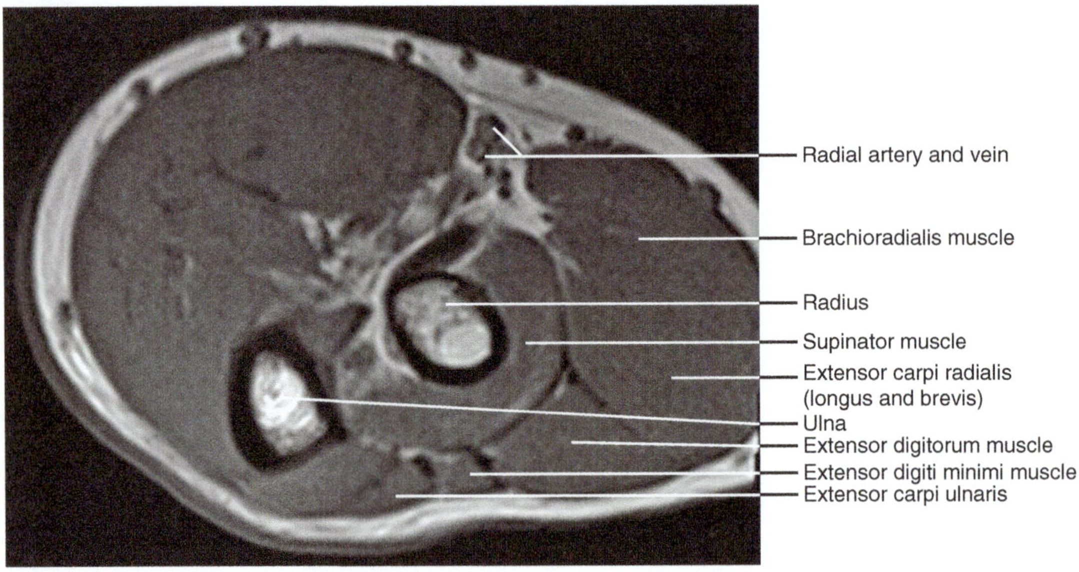

Fig. 24.35 T1 axial of the mid radius and ulna

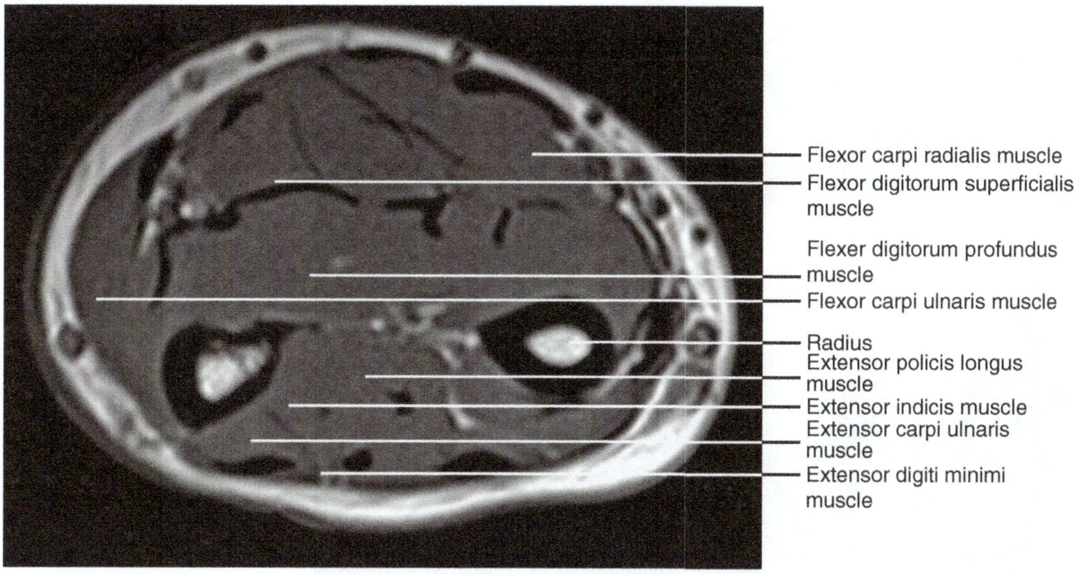

Fig. 24.36 T1 axial of the lower radius and ulna

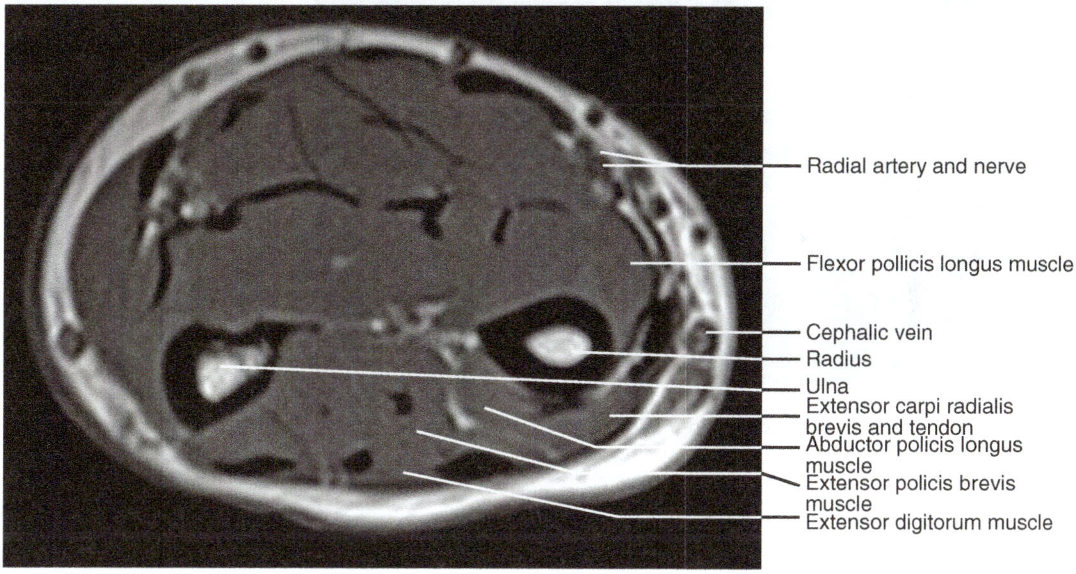

Fig. 24.37 T1 axial of the lower radius and ulna

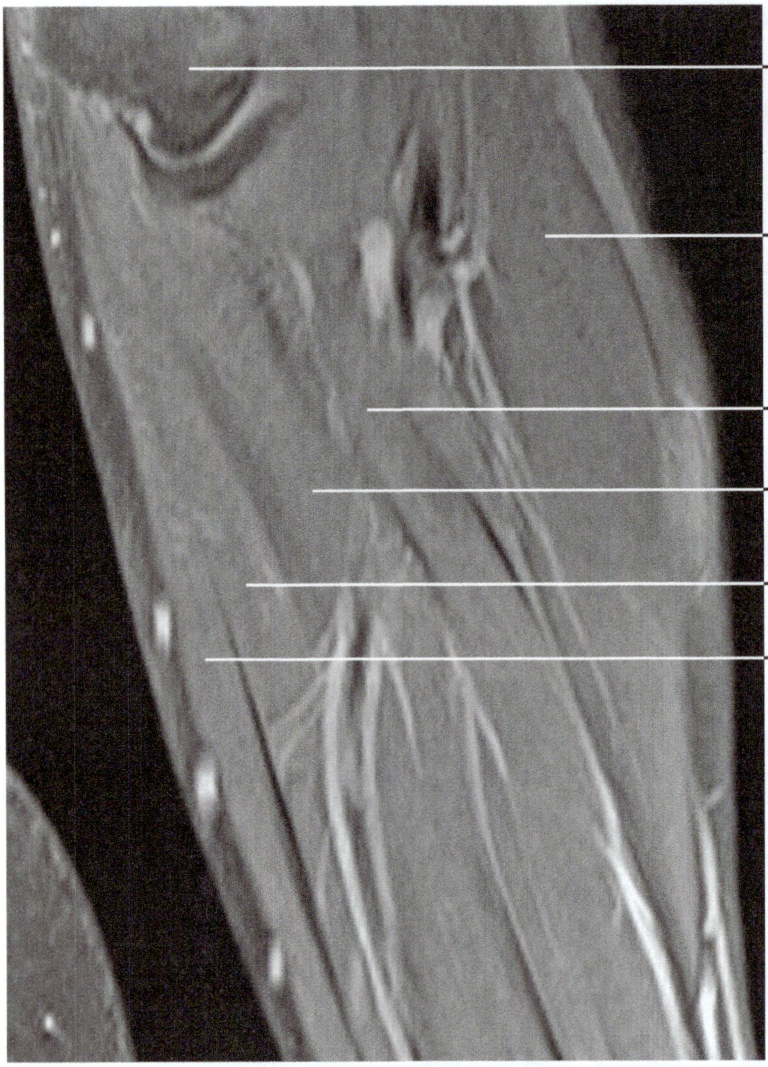

Fig. 24.38 T1 coronal of the anterior part of the forearm

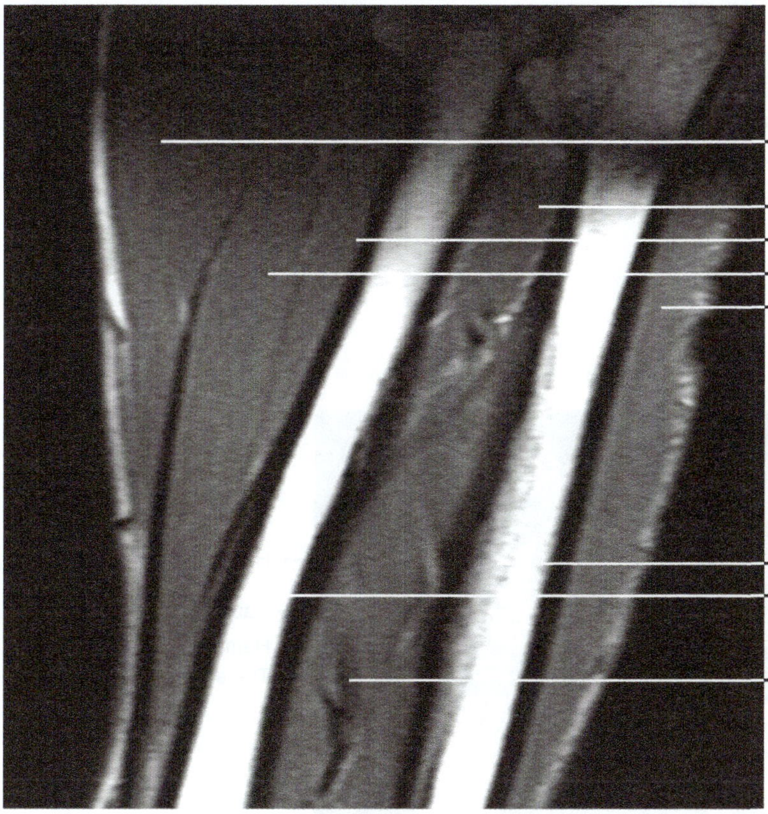

Fig. 24.39 T1 coronal of the radius and ulna

Wrist

Routine plain radiographic views include AP and lateral. Instability may be assessed with ulnar and radial deviation, flexion, extension and grip views. Additional carpal views include the magnified scaphoid view, carpal tunnel view, hook of hamate view and pisiform view.

Plain Film

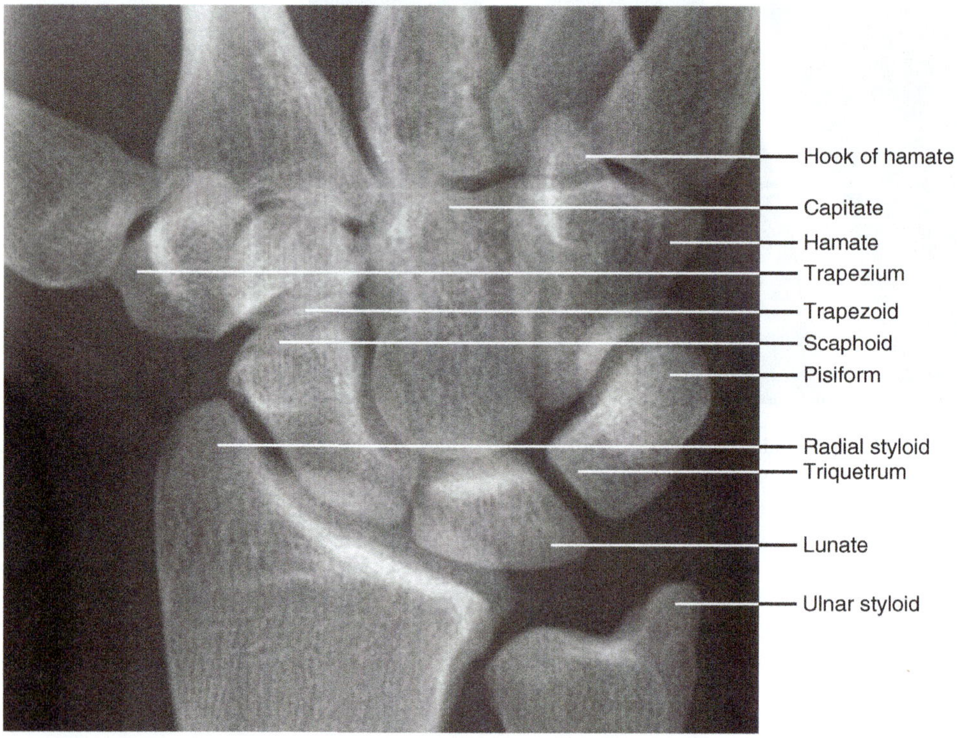

Fig. 24.40 Antero-posterior view of the wrist

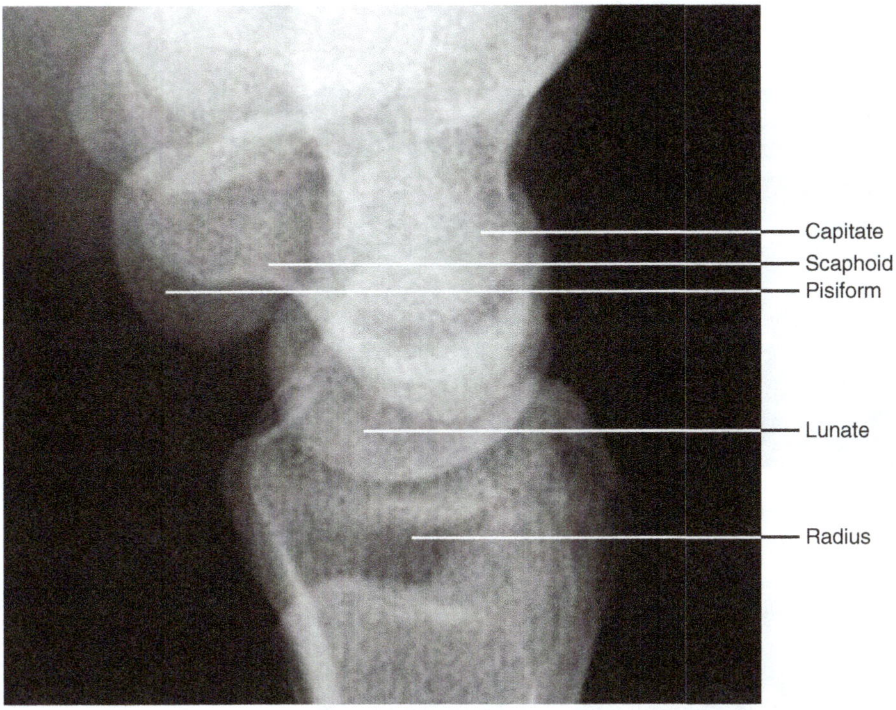

Fig. 24.41 Lateral view of the wrist

Ultrasound

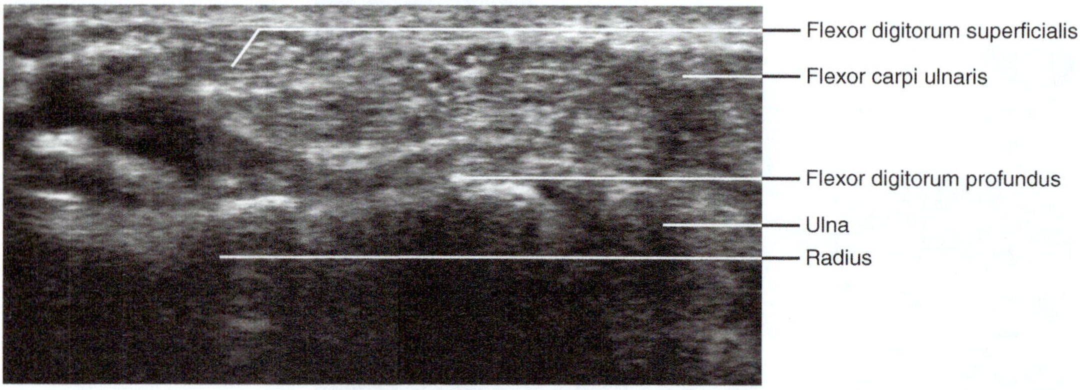

Fig. 24.42 Transverse US of the distal radius

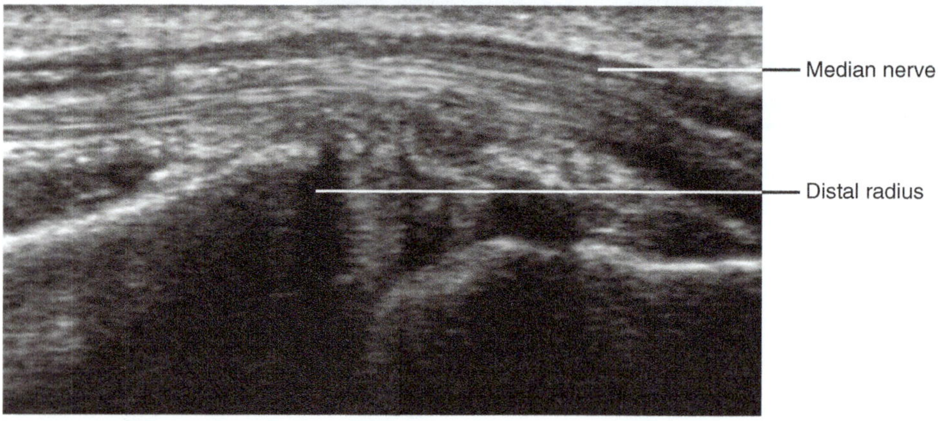

Fig. 24.43 Longitudinal US of the median nerve

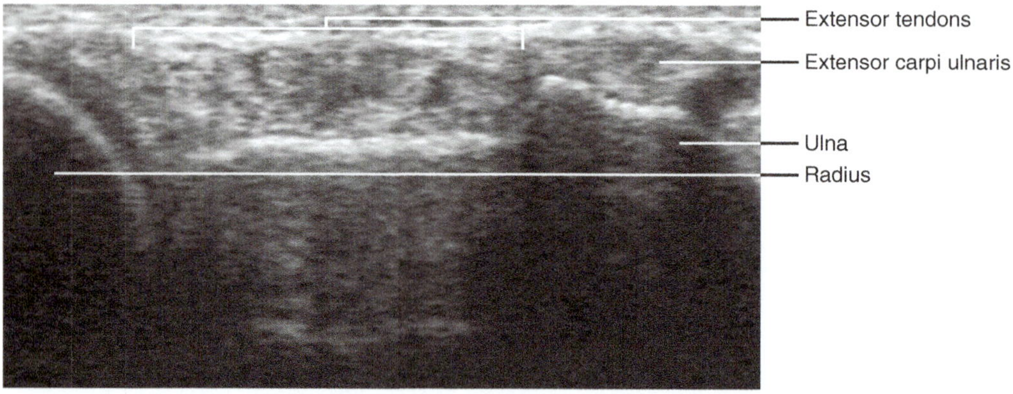

Fig. 24.44 Transverse US of the dorsum of the wrist

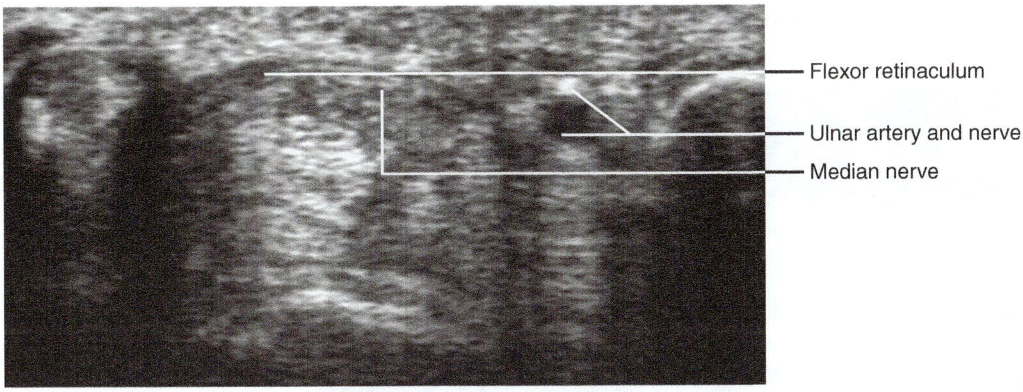

Fig. 24.45 Transverse US of the carpal tunnel

Magnetic Resonance Imaging

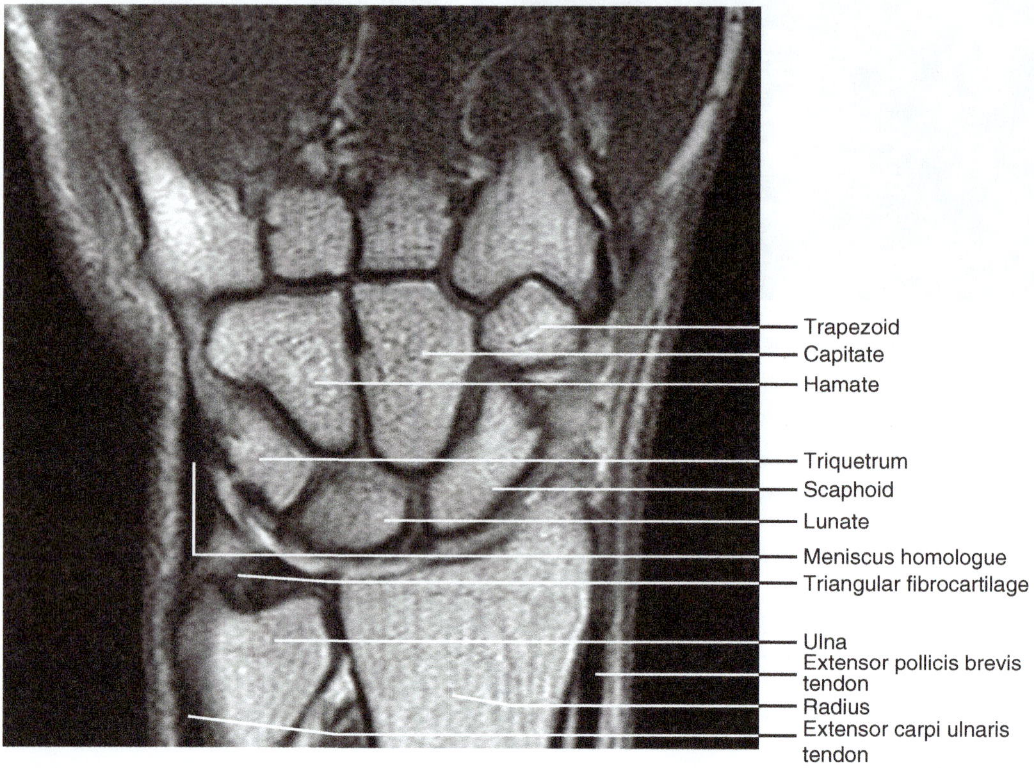

Fig. 24.46 T1 coronal of the wrist

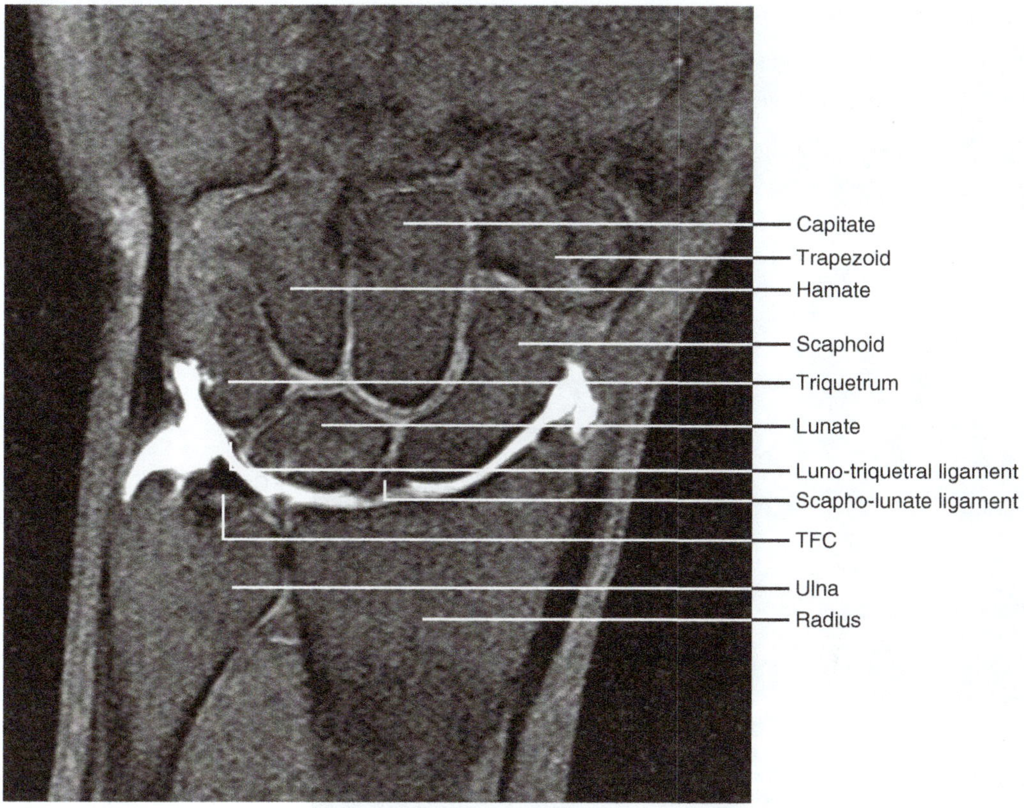

Fig. 24.47 T1 coronal fat saturated image after gadolinium injection into the joint

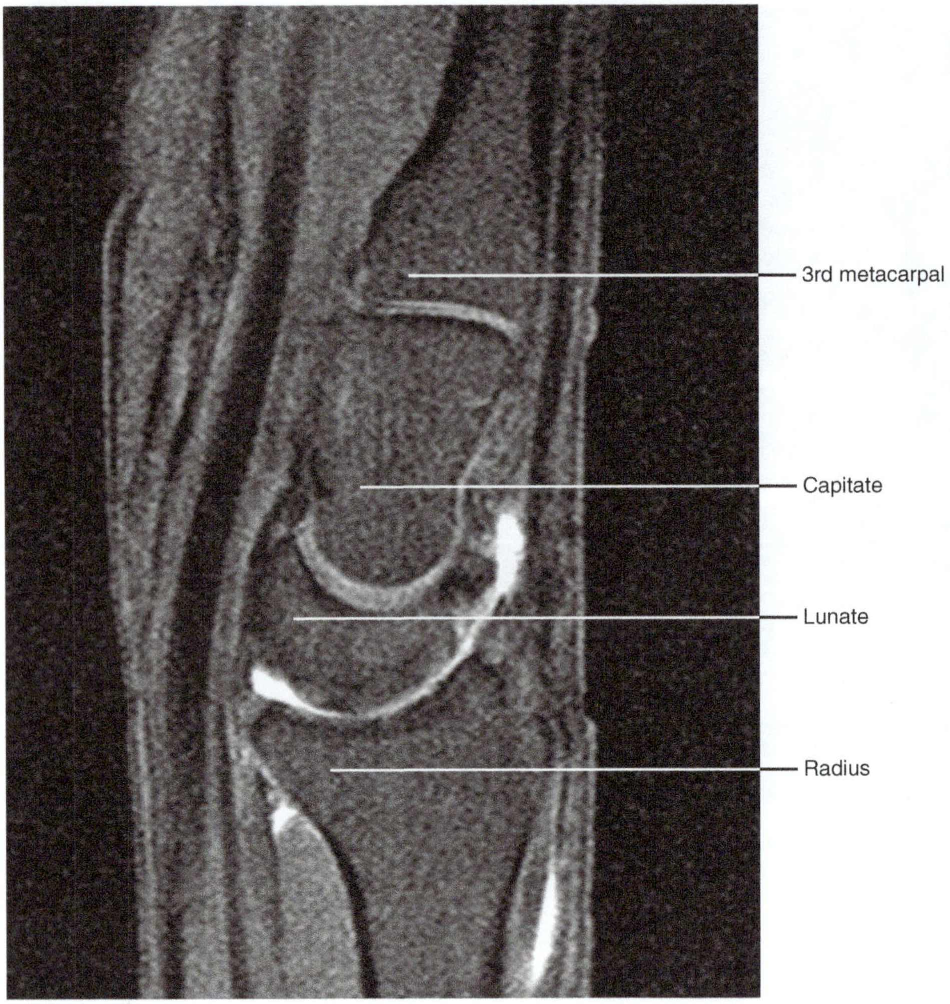

Fig. 24.48 T1 sagittal fat saturated image after gadolinium injection into the joint

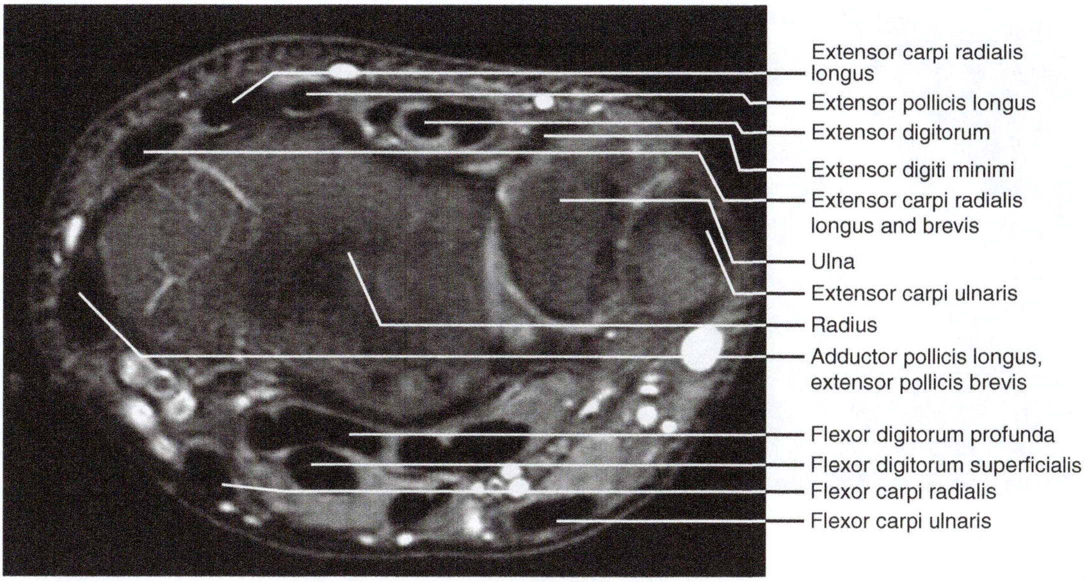

Fig. 24.49 Axial gradient echo image just above the wrist joint

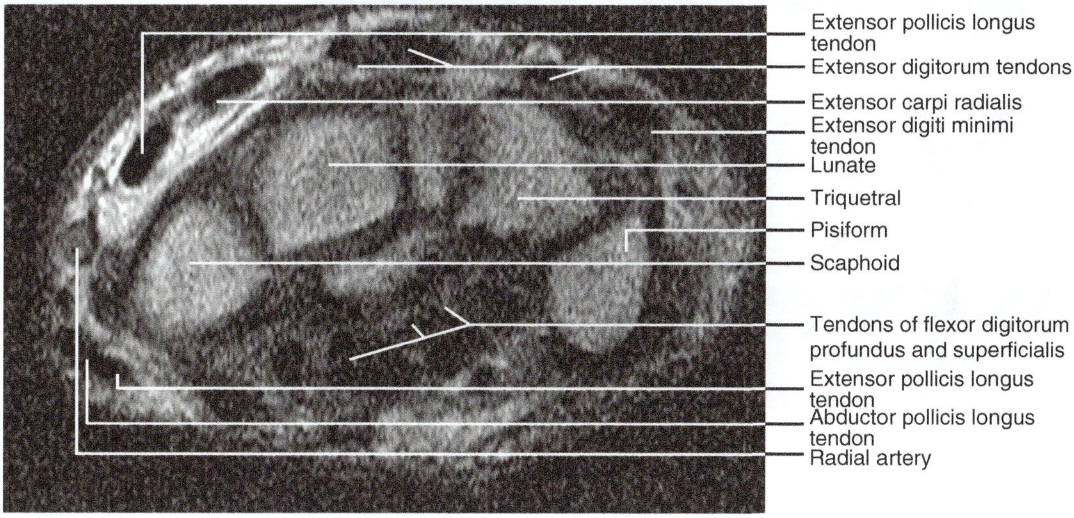

Fig. 24.50 T1 axial at the proximal carpal row

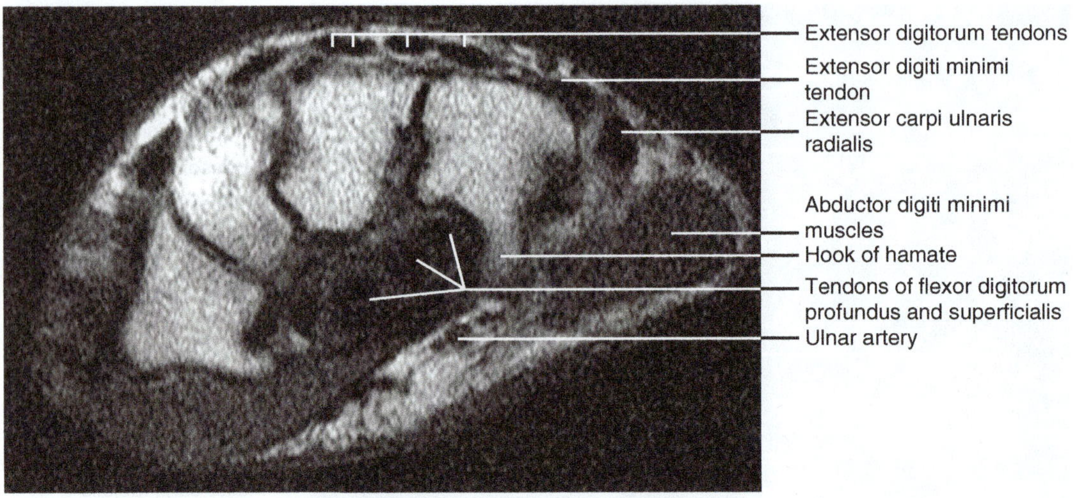

Fig. 24.51 T1 axial at the distal carpal row

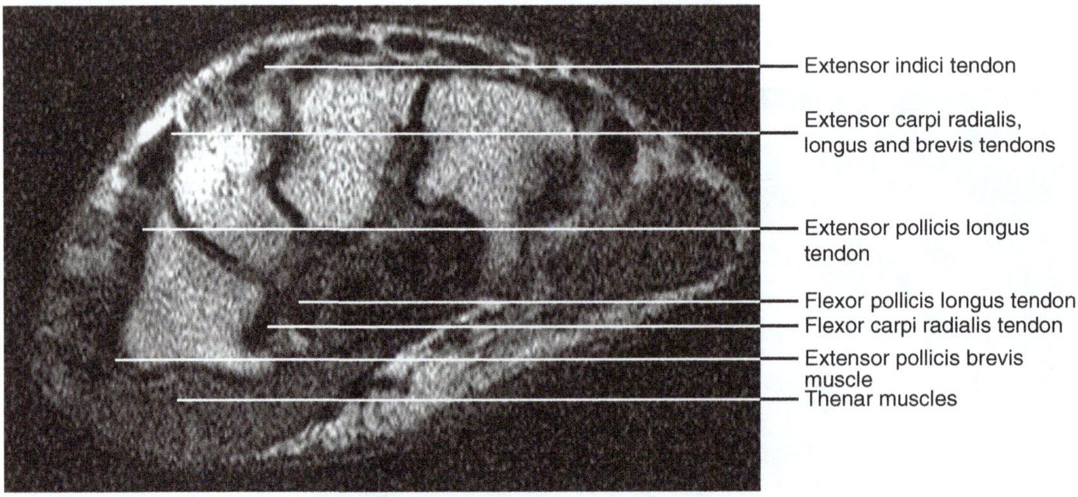

Fig. 24.52 T1 axial at the distal carpal row

Pearl

With intact anatomy, gadolinium injected into the radiocarpal joint space must not flow into the distal radio-ulnar joint or into the midcarpal row. There may however be a normal anatomical communication between the radiocarpal joint and the extensor tendon sheaths.

The meniscus homologue is usually of high signal on fluid sensitive MRI sequences.

On sagittal images, because the medial and lateral aspects of the lunate lie in different planes, the lunate will, on selected images, appear dorsally rotated.

Hand

Plain Film

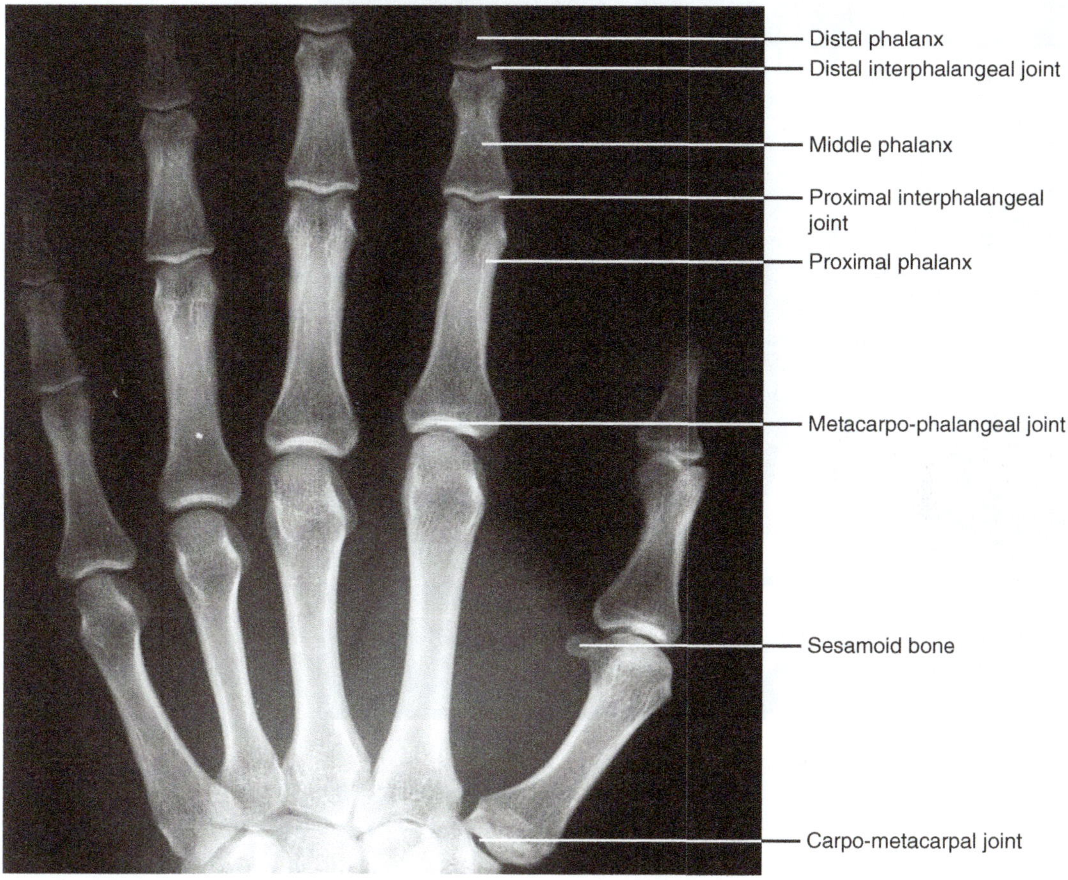

Fig. 24.53 Antero-posterior view of the hand

Magnetic Resonance Imaging

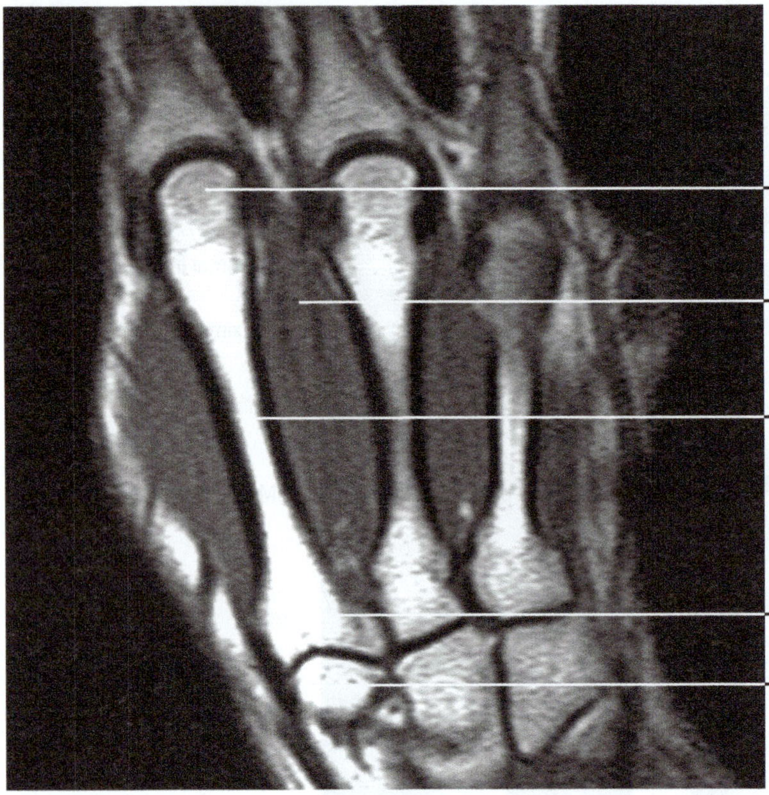

Fig. 24.54 T1 coronal of the dorsum of the hand

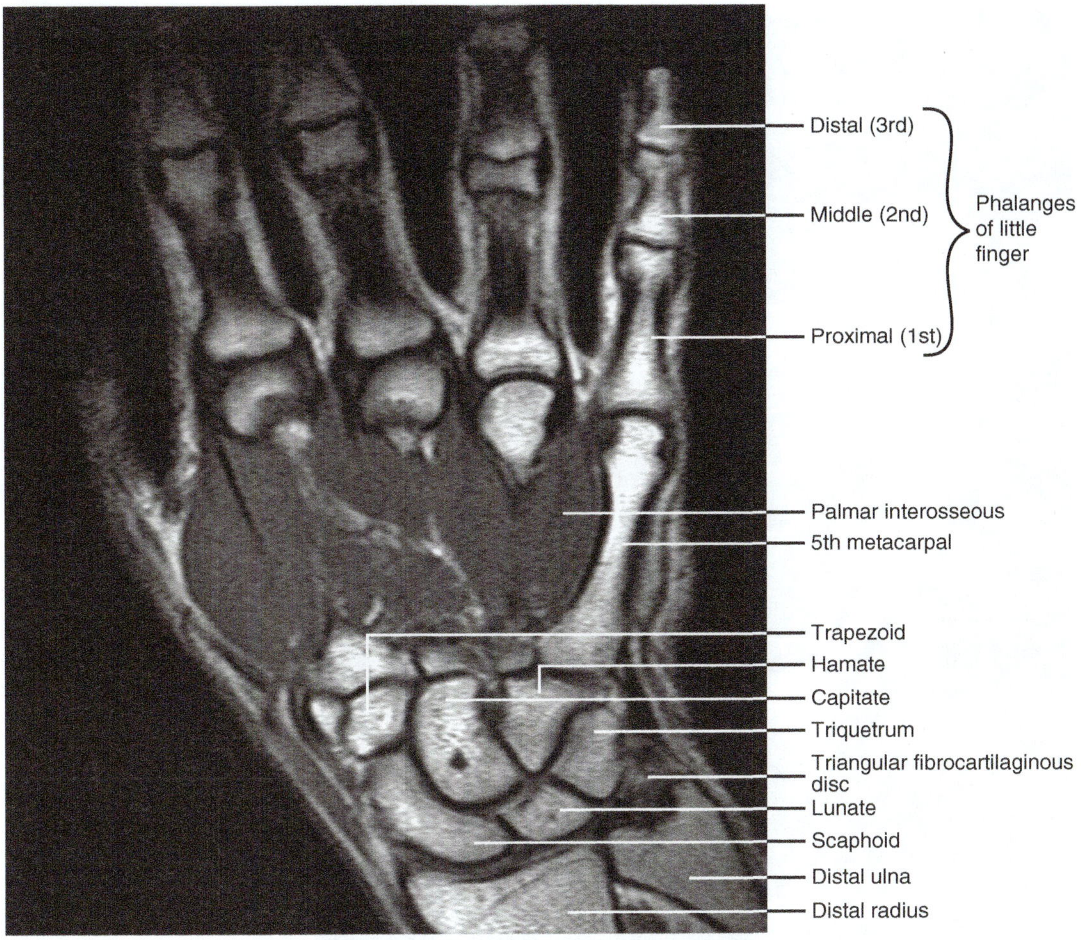

Fig. 24.55 T1 coronal of the mid portion of the hand

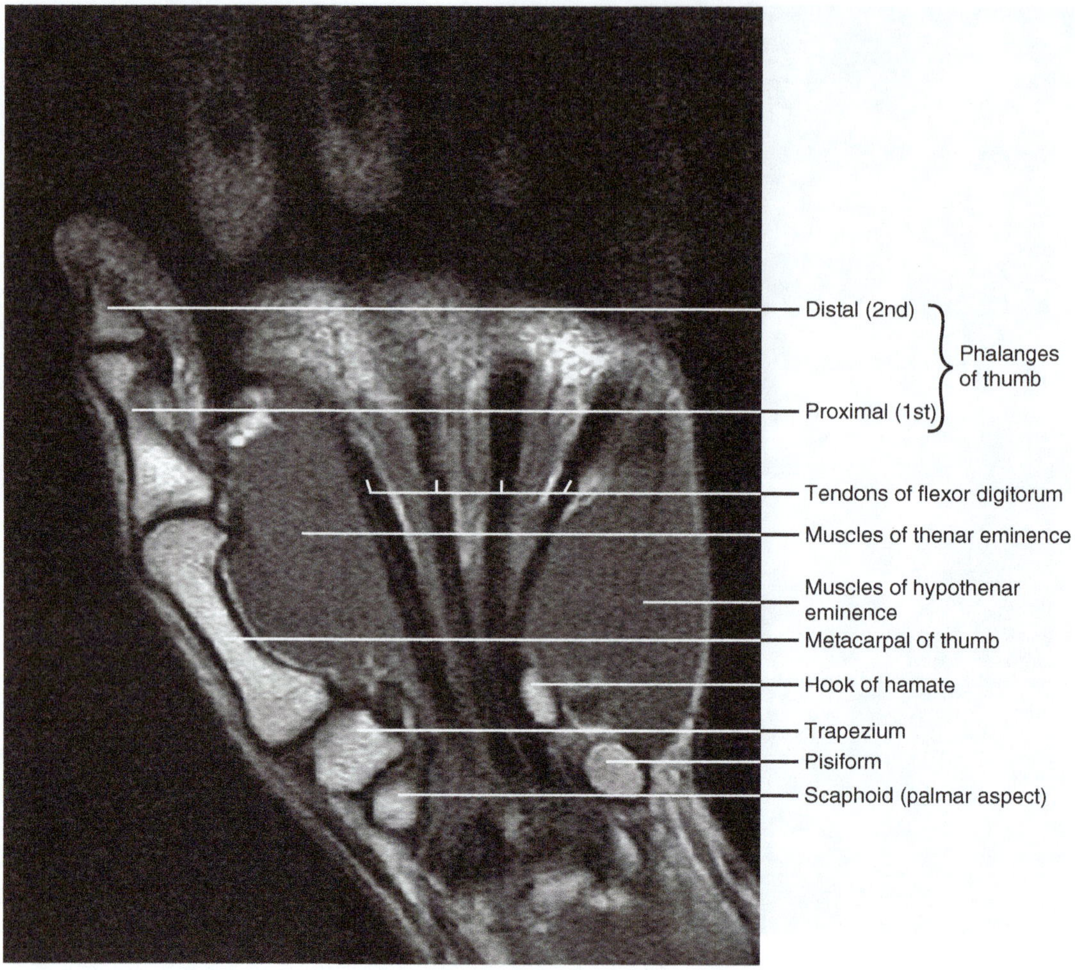

Fig. 24.56 T1 coronal of the palmar aspect of the hand

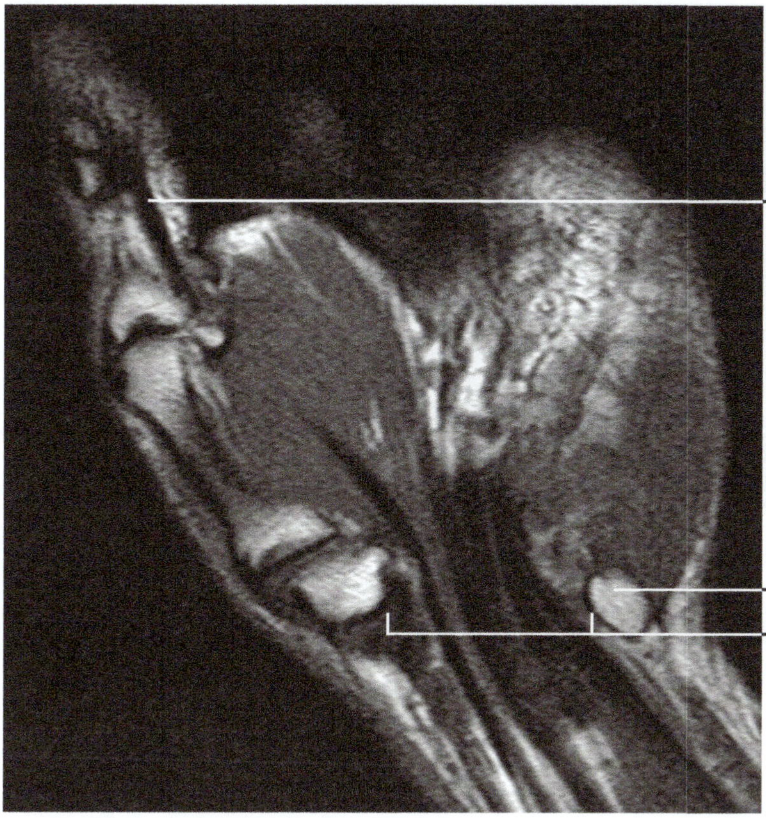

Fig. 24.57 T1 coronal at the thenar and hypo-thenar eminence

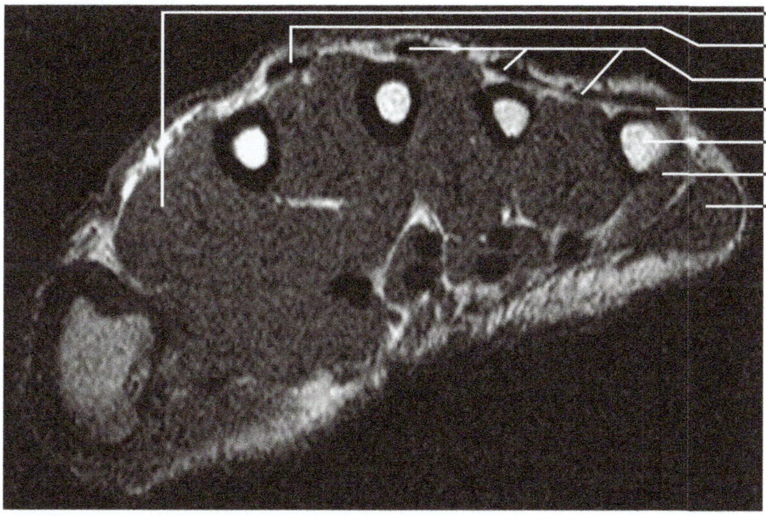

Fig. 24.58 T1 axial at the mid metacarpal level: hand pronated

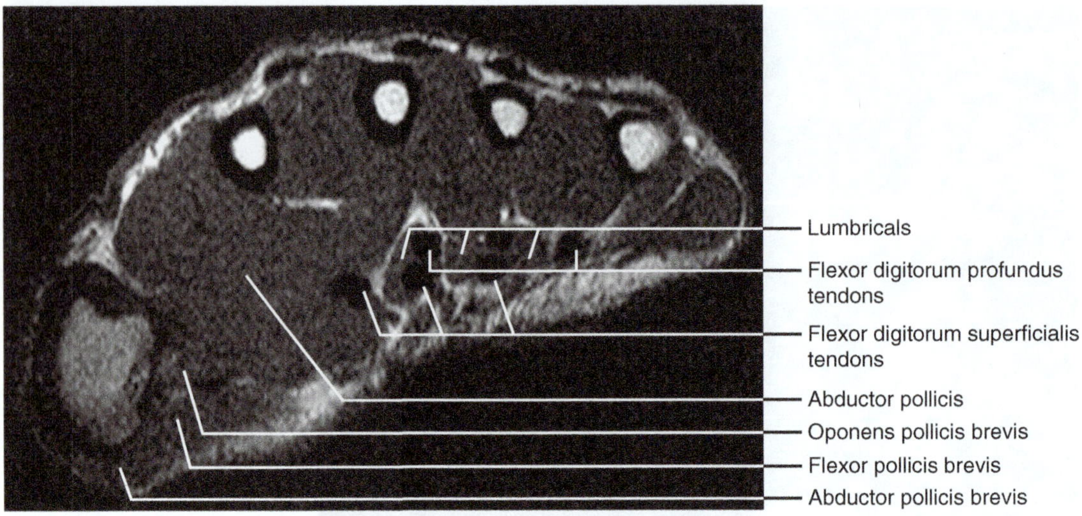

Fig. 24.59 T1 axial at the mid metacarpal level: hand pronated

Lower Limb

25

Hip

Magnetic Resonance Imaging

Periarticular structures and bone or marrow pathology can be well visualised with uncontrasted MRI. Intra-articular pathology, particularly lateral injury, is only assessed accurately with direct MR arthrography.

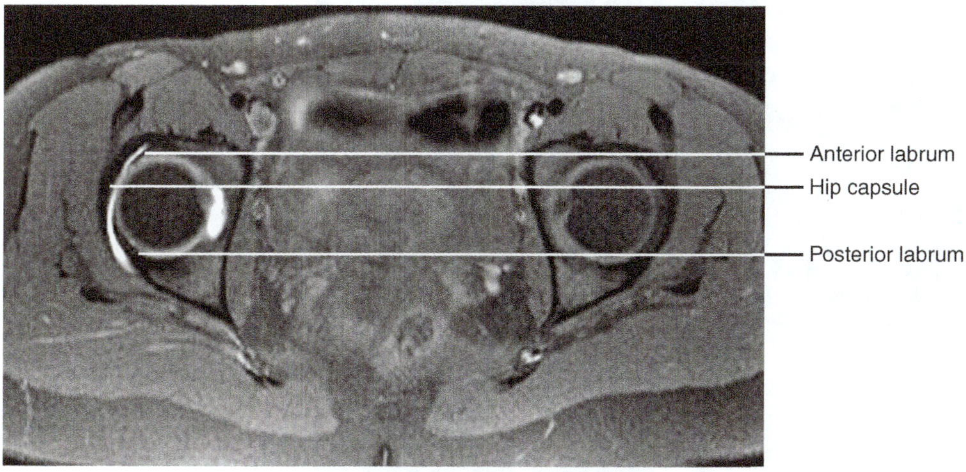

Fig. 25.1 T1 fat saturated axial of the hips with gadolinium injection of the right joint just below the acetabular roof

S. Andronikou (ed.), *See Right Through Me*,
DOI 10.1007/978-3-642-23893-2_25, © Springer-Verlag Berlin Heidelberg 2012

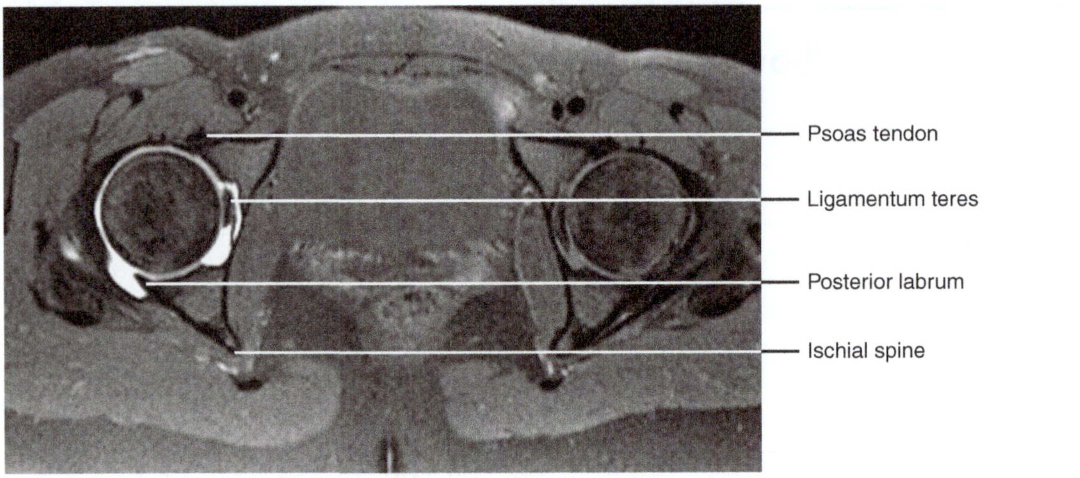

Fig. 25.2 T1 fat saturated axial of the hips with gadolinium injection of the level of the ligamentum teres

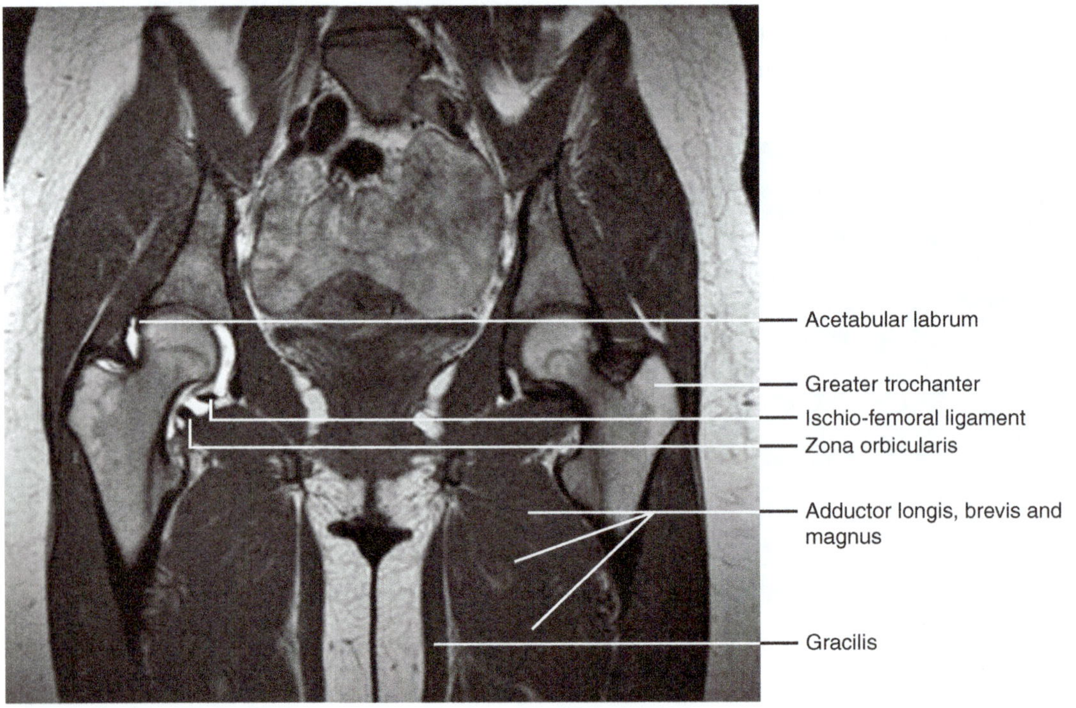

Fig. 25.3 T1 coronal of the hips with gadolinium injection of the joint on the right

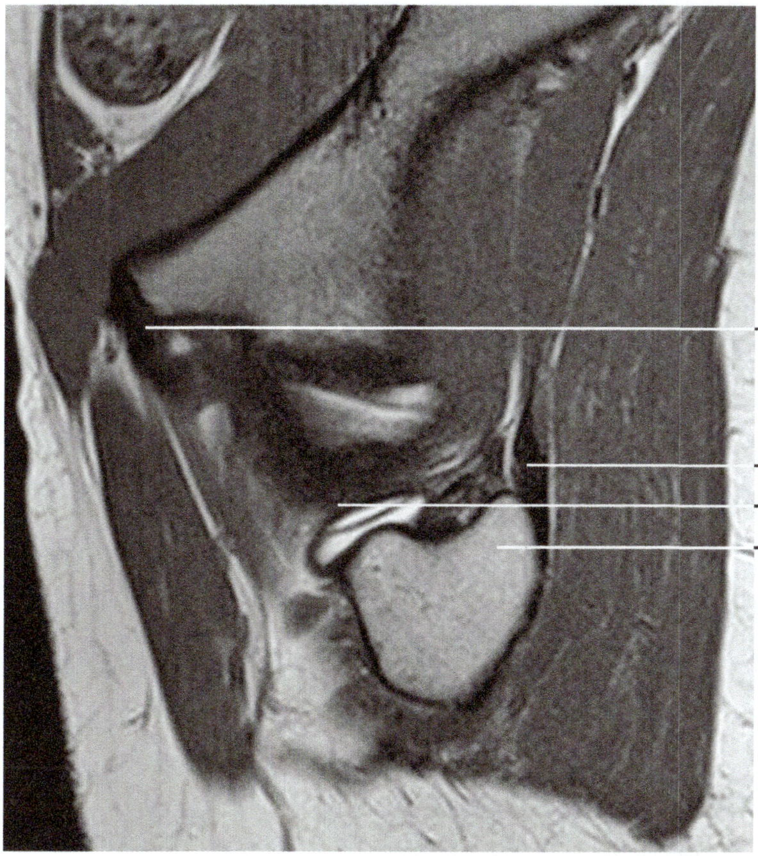

Fig. 25.4 T1 sagittal of the hip with gadolinium injection of the joint at the greater trochanter

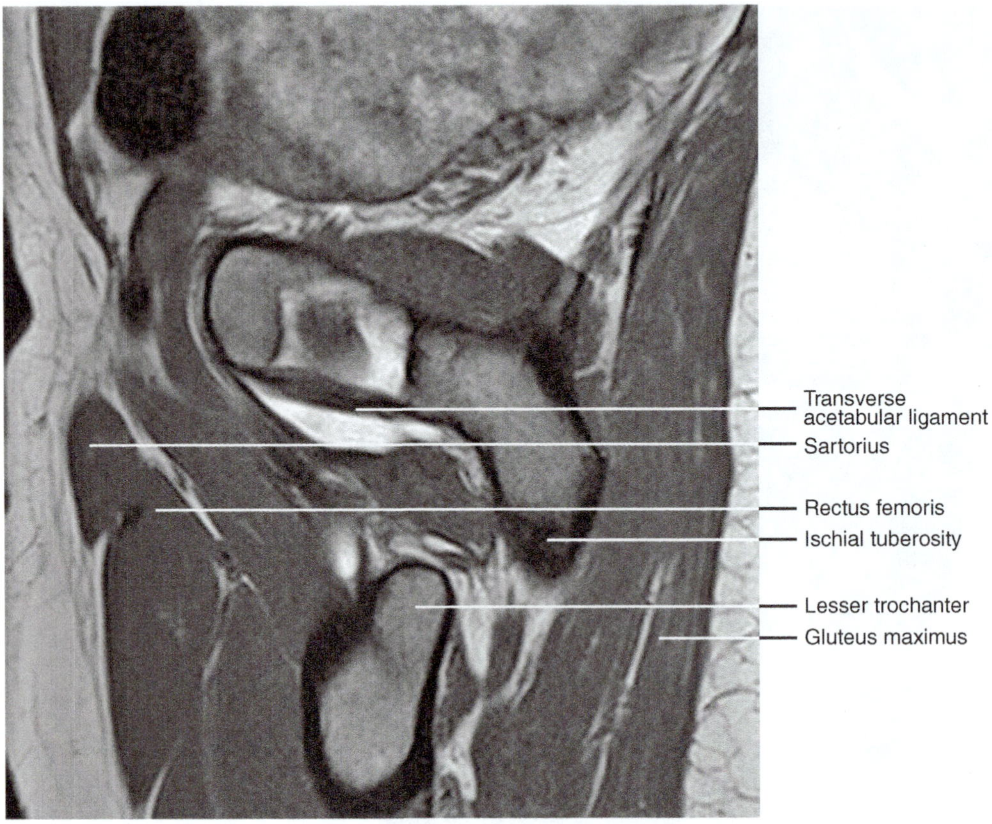

Fig. 25.5 T1 sagittal of the hip with gadolinium injection of the joint at the lesser trochanter

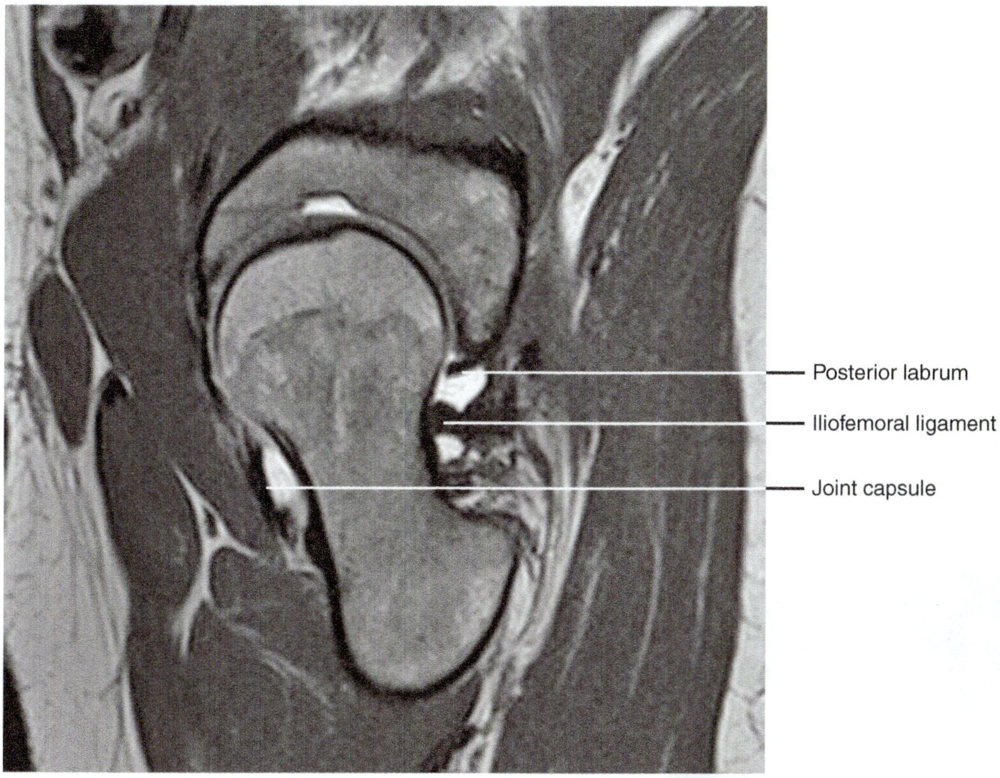

Fig. 25.6 T1 sagittal of the hip with gadolinium injection of the joint at the joint level

Pearl

Beware of inhomogeneous marrow signal on MRI due to the presence of normal red (haemato-poeitic) marrow and yellow (fatty) marrow.

Femur

Plain Film

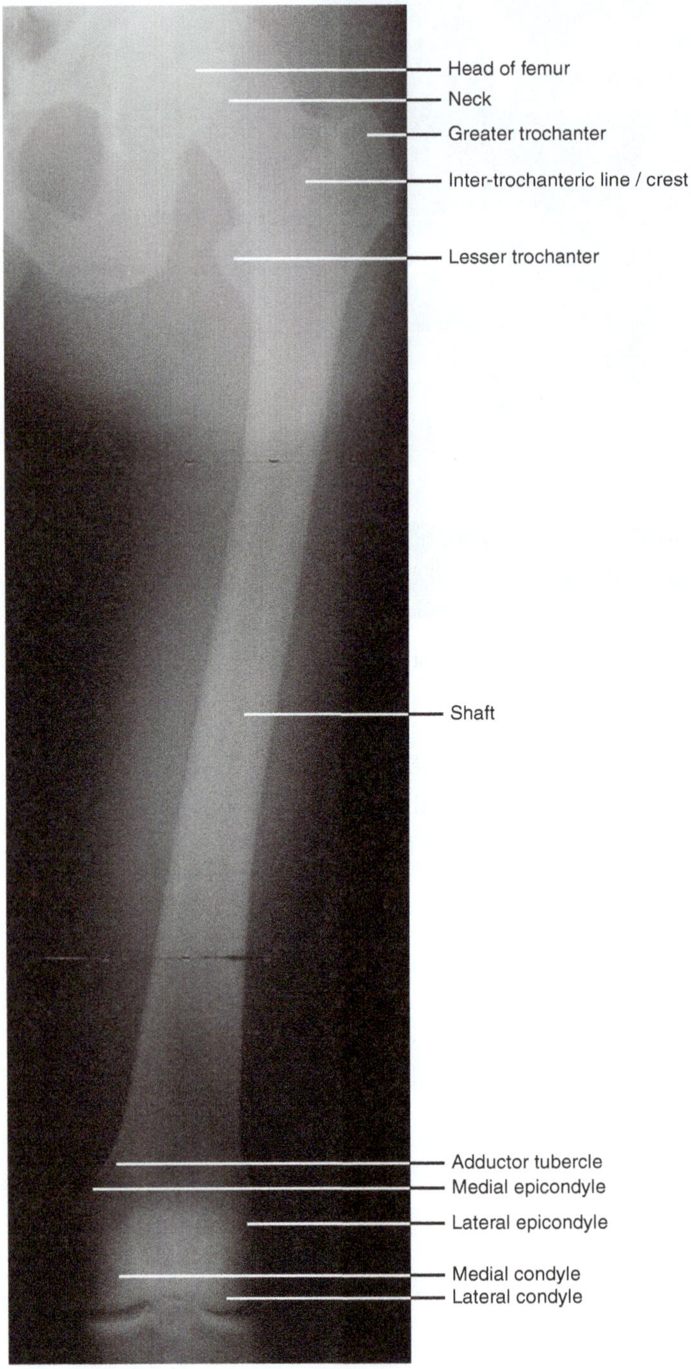

Fig. 25.7 Antero-posterior view of the left femur

Magnetic Resonance Imaging

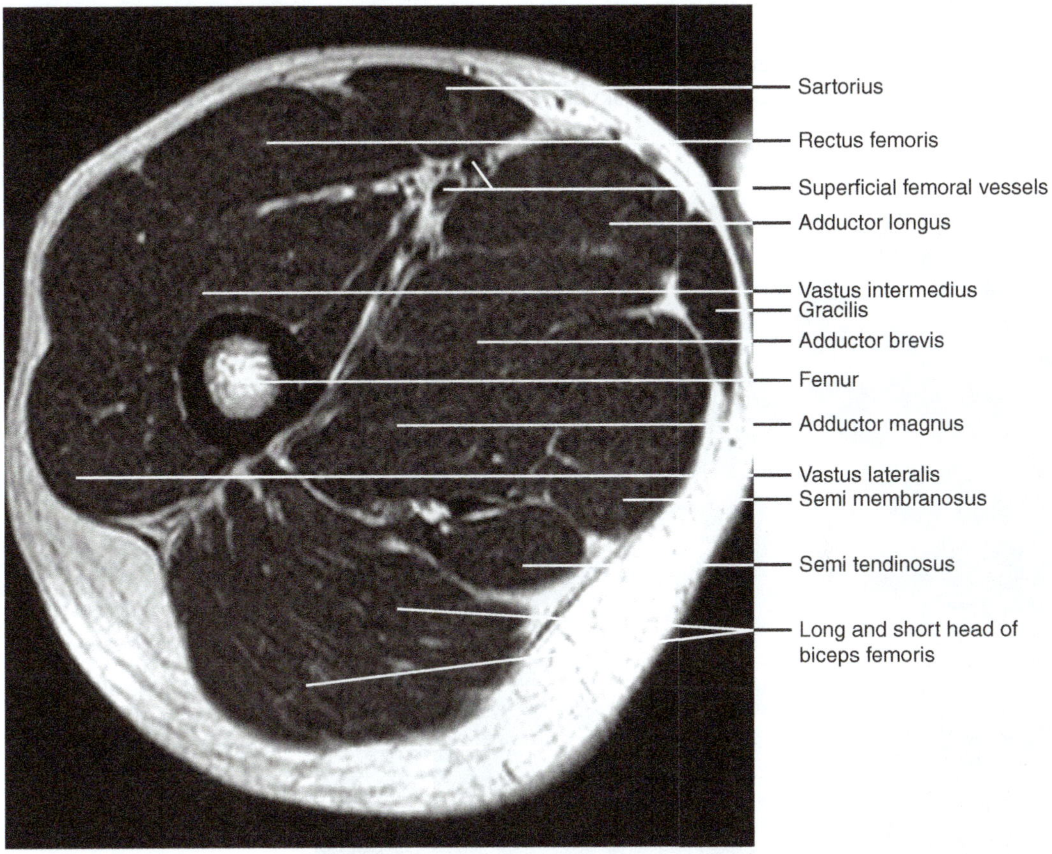

Fig. 25.8 Axial T2 MRI mid-thigh right

Knee

Plain Film

Bony anatomy is well seen with plain film. Standard views include AP, lateral and skyline patellar view.

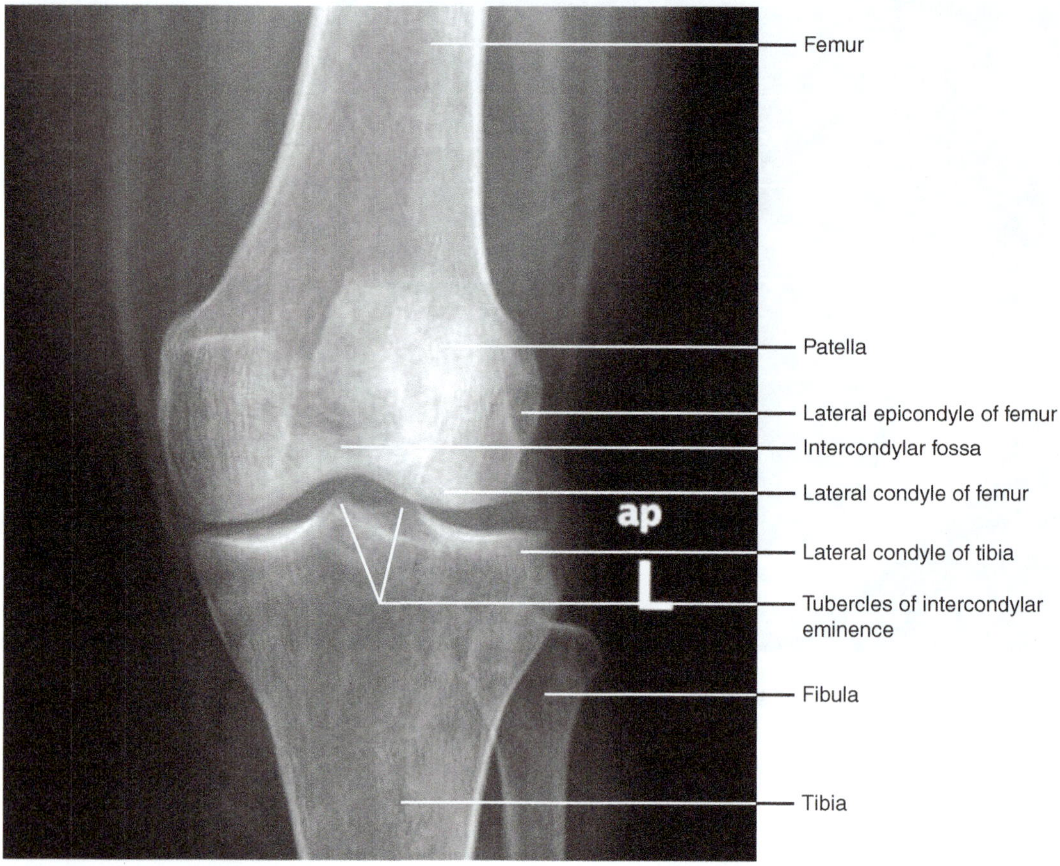

Fig. 25.9 Antero-posterior view of the knee

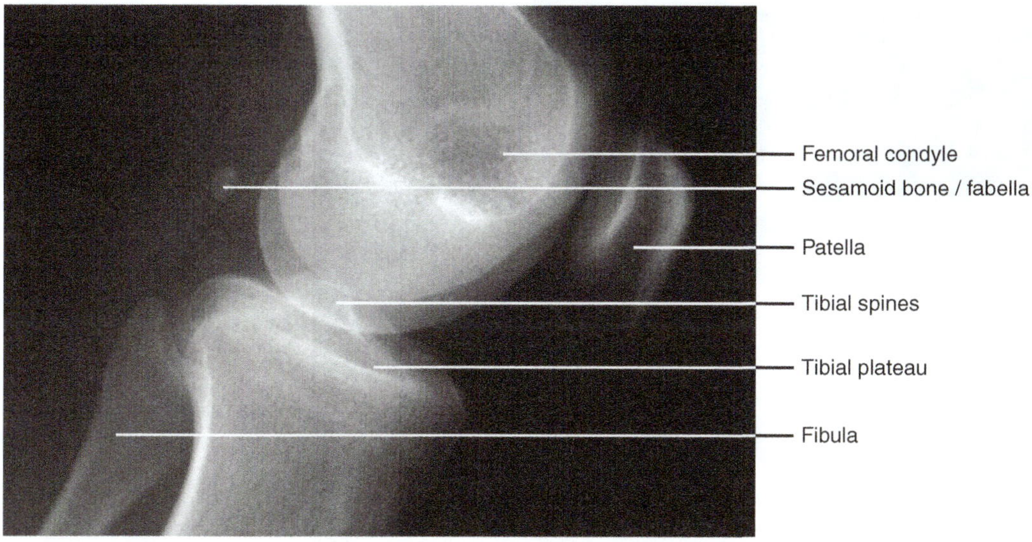

Fig. 25.10 Lateral view of the knee

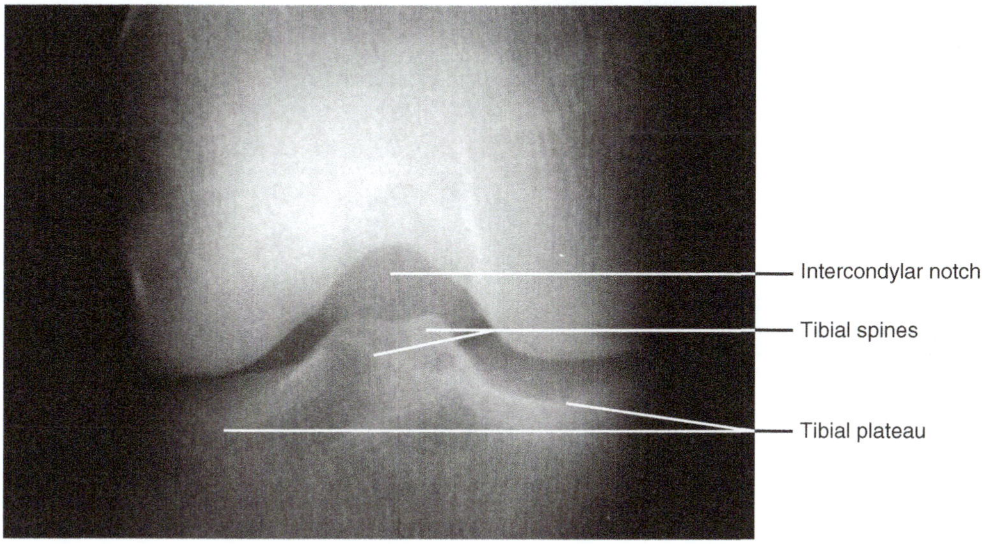

Fig. 25.11 Tunnel view of the knee

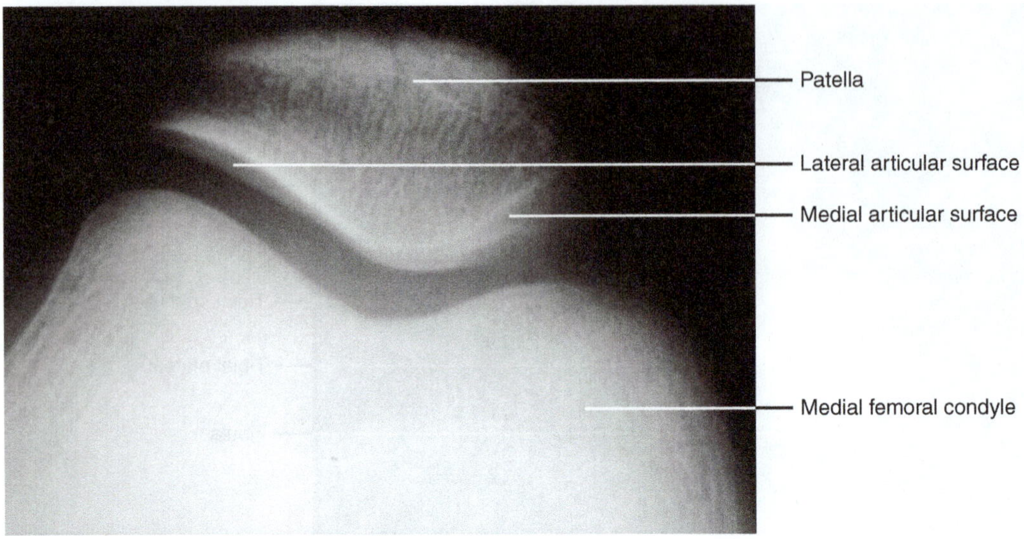

Fig. 25.12 Skyline view of the patella

Pearl

The **Longer** aspect of the patellar articular surface is the **Lateral** aspect.

Ultrasound

There is a limited role for ultrasound due to the predominance of intra-articular disease, but the patellar tendon, quadriceps tendon, collateral ligaments, the popliteal fossa and the suprapatellar pouch are well seen.

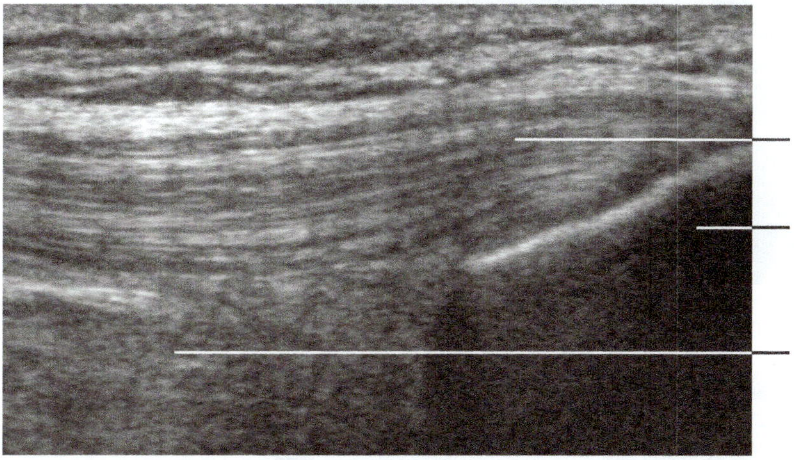

Fig. 25.13 Longitudinal image of the quadriceps tendon

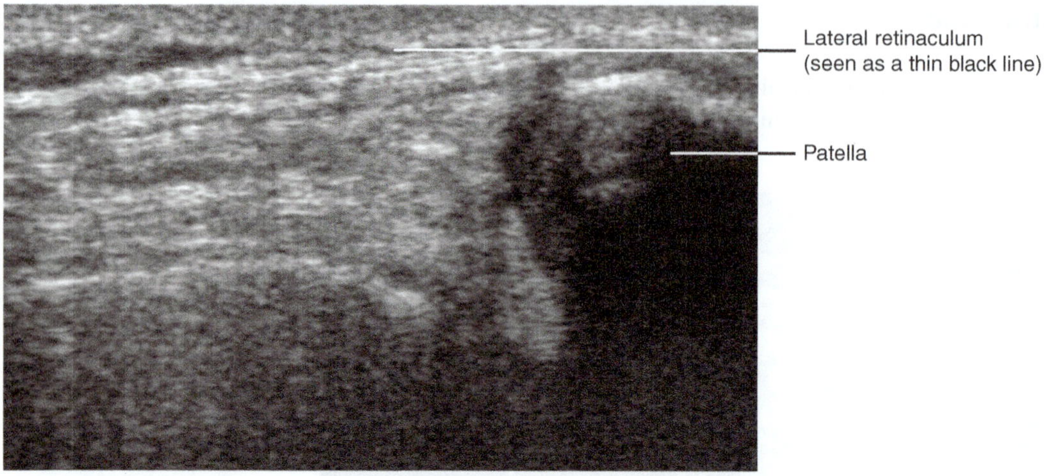

Fig. 25.14 Longitudinal Lateral patellar retinaculum

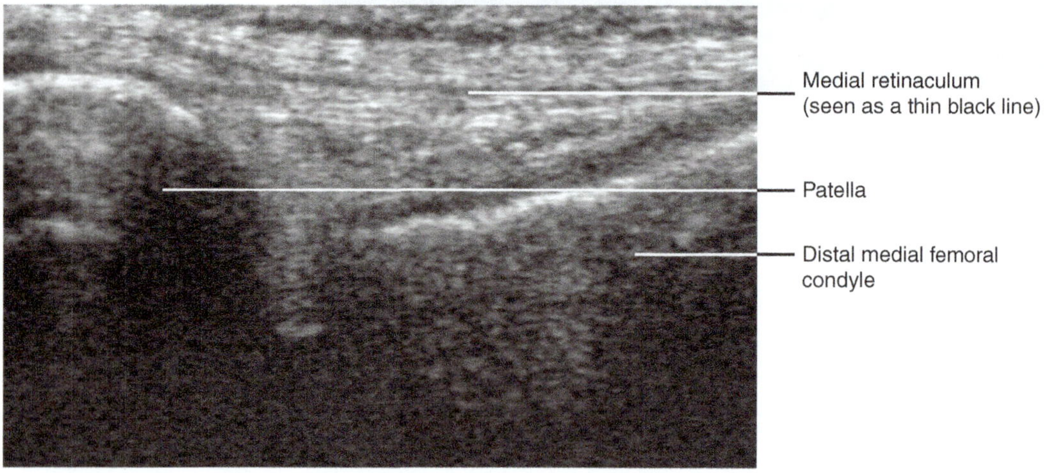

Fig. 25.15 Longitudinal Medial patellar retinaculum

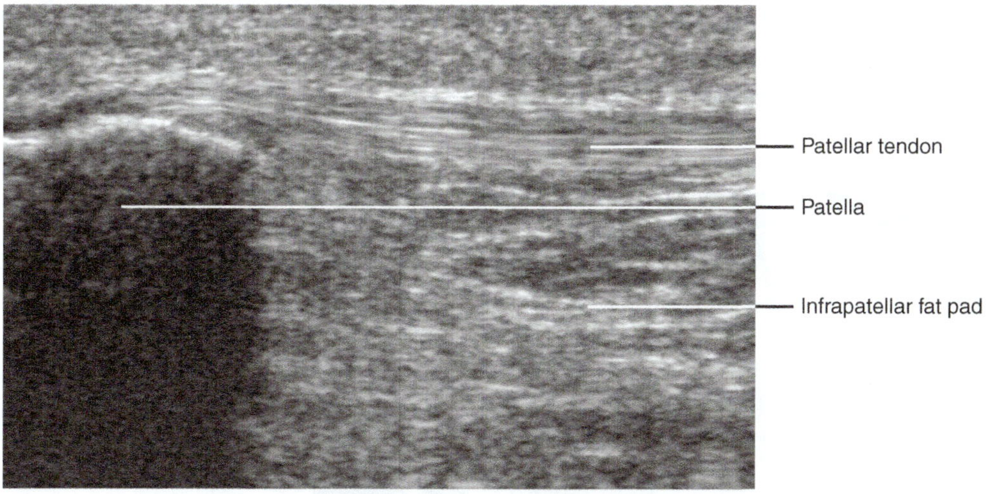

Fig. 25.16 Longitudinal proximal patellar tendon

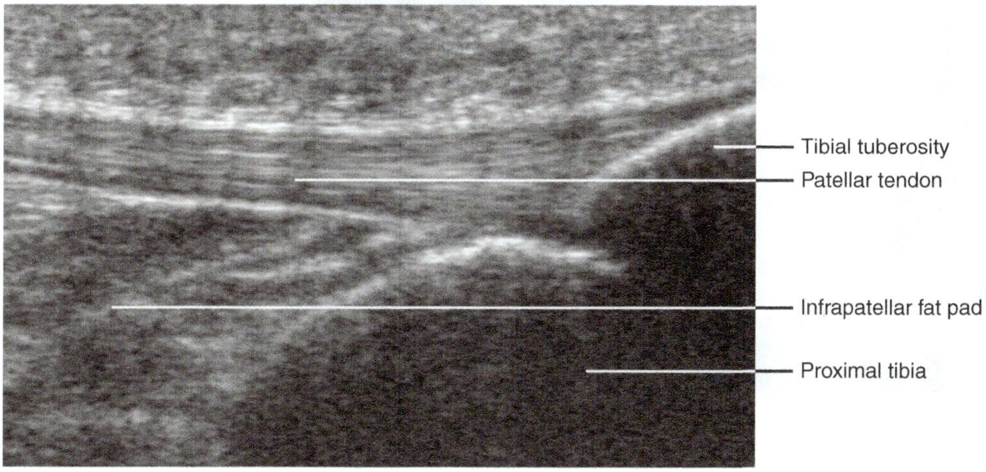

Fig. 25.17 Longitudinal distal patellar tendon

Magnetic Resonance Imaging

MRI is the investigation of choice. Arthrography is only very rarely necessary, predominantly to assess subtle interfaces, like an osteochondral defect or a recurrent meniscal tear.

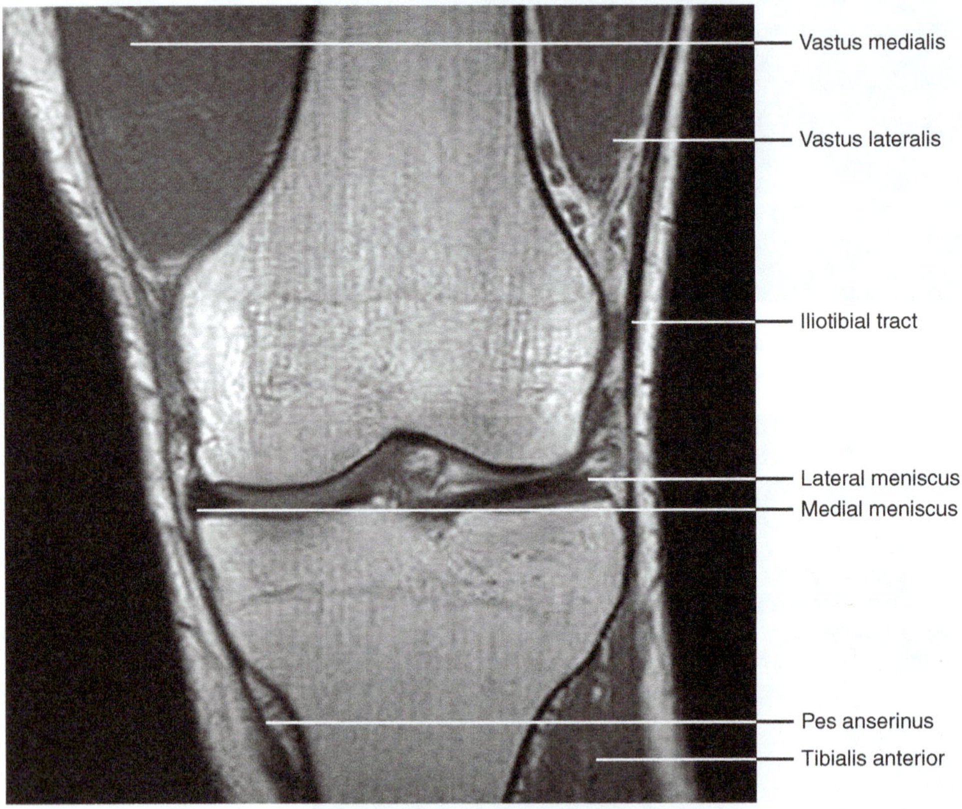

Fig. 25.18 T1 coronal anteriorly at the knee joint

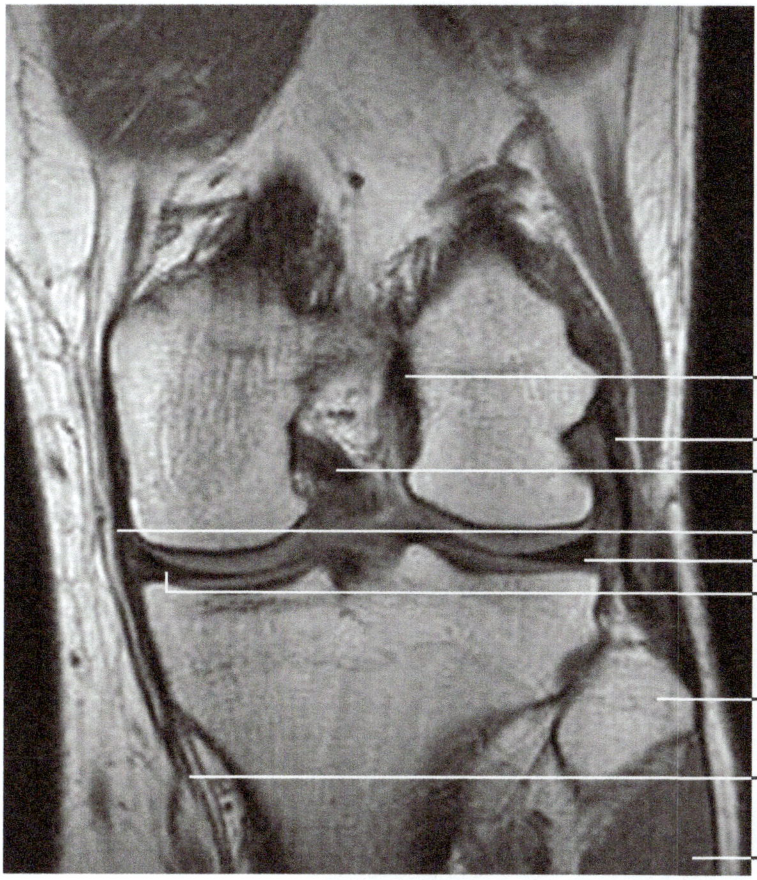

Fig. 25.19 T1 coronal in the mid portion of the knee joint

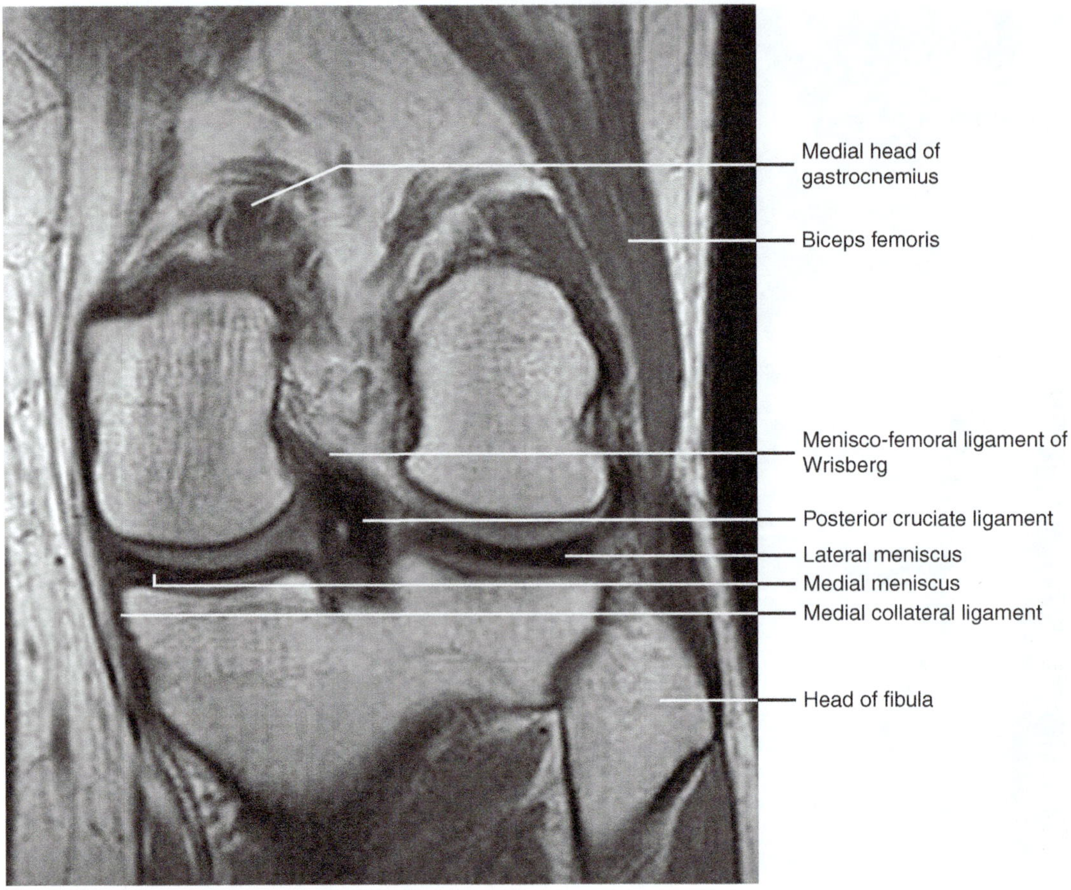

Fig. 25.20 T1 coronal at the posterior aspect of the knee joint

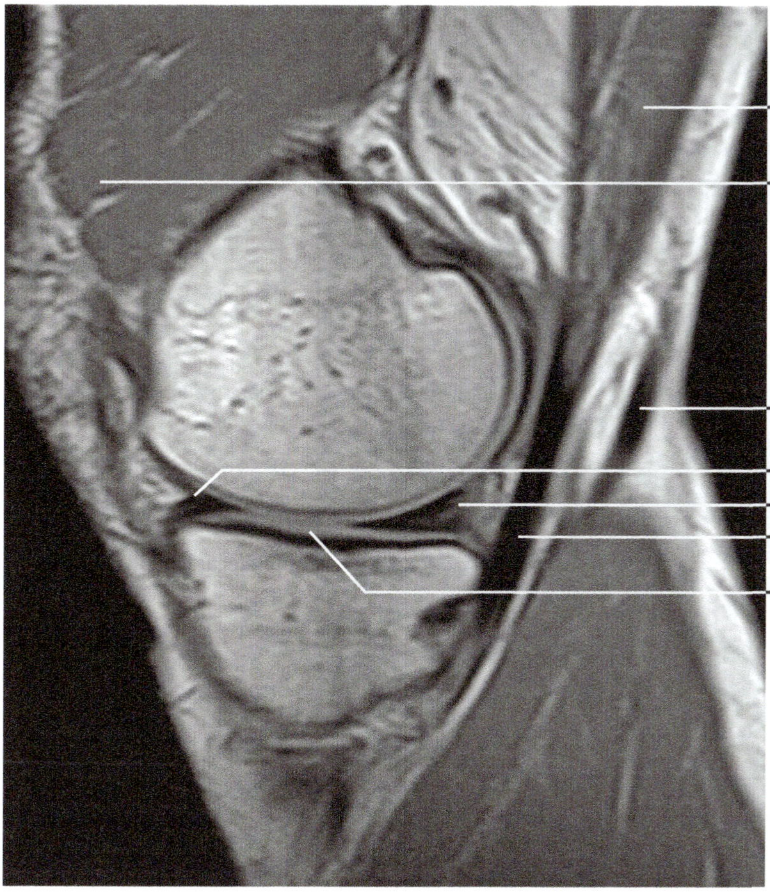

Fig. 25.21 T1 sagittal of the far medial aspect of the knee

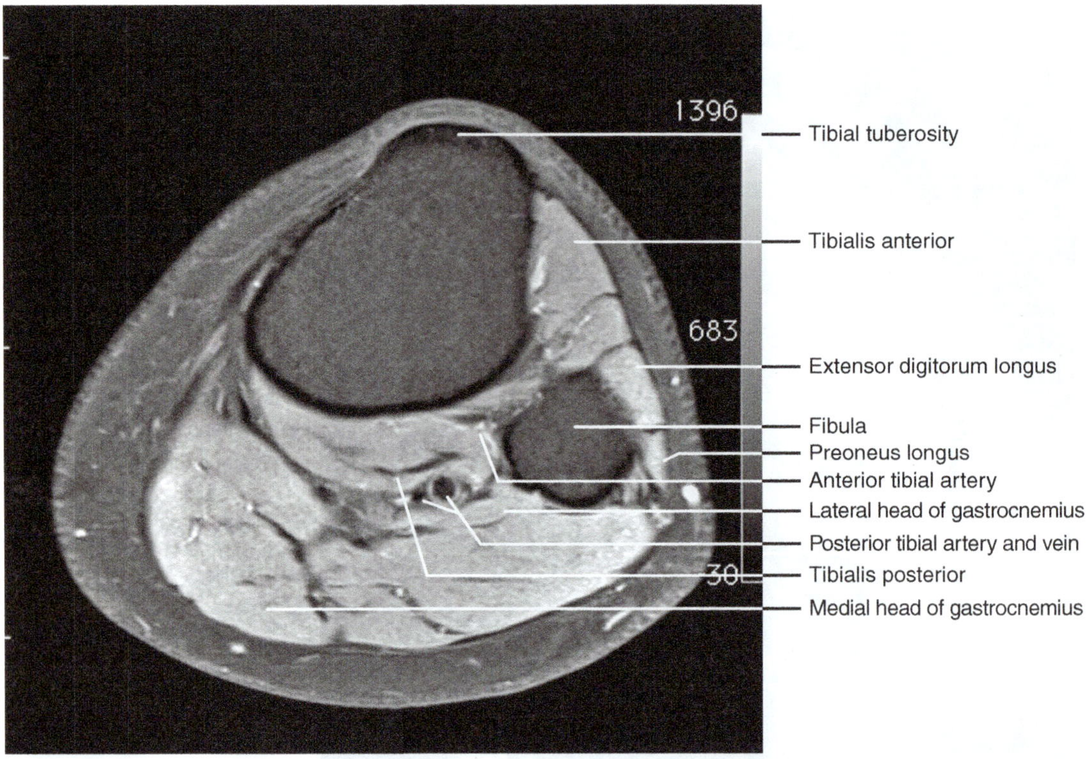

Fig. 25.28 T1 axial with fat saturation above the level of the knee joint at the proximal tibia and fibula

Pearl

The fibula is a good indicator of the lateral aspect of the knee.

The anterior cruciate ligament is lateral and the posterior cruciate ligament is medial.

The lateral collateral ligament is separated from the lateral meniscus by loose areolar tissue which is bright on fluid sensitive sequences.

Intrameniscal high signal is normal as long as is does not contact the articular margins of the meniscus.

The popliteus tendon crosses obliquely past the posterior horn of the lateral meniscus in its intra-articular course, and this interface may simulate a tear.

Movement artifact from the popliteal artery may cross the knee depending on the phase//frequency orientation.

The intermeniscal transverse ligament connects the anterior horns of the menisci and may simulate an oblique tear of the anterior horn of the lateral meniscus on sagittal images.

The MCL bursa may create fluid signal between the MCL and the medial meniscus and simulate a meniscocapsular separation.

The presence of nitrogen in the joint may simulate articular cartilage lesions (vacuum phenomenon).

The anterior aspect of the ACL often appears striated (due to the posterolateral and anteromedial bundles separated by fat).

The menisco-femoral ligaments of Humphry (anterior) and Wrisberg (posterior) cross the PCL and look like loose bodies on the sagittal images.

Fluid in the suprapatellar pouch extending around the knee may look like a pathological focal fluid collection on the sagittal images.

Tibia and Fibula

Plain Film

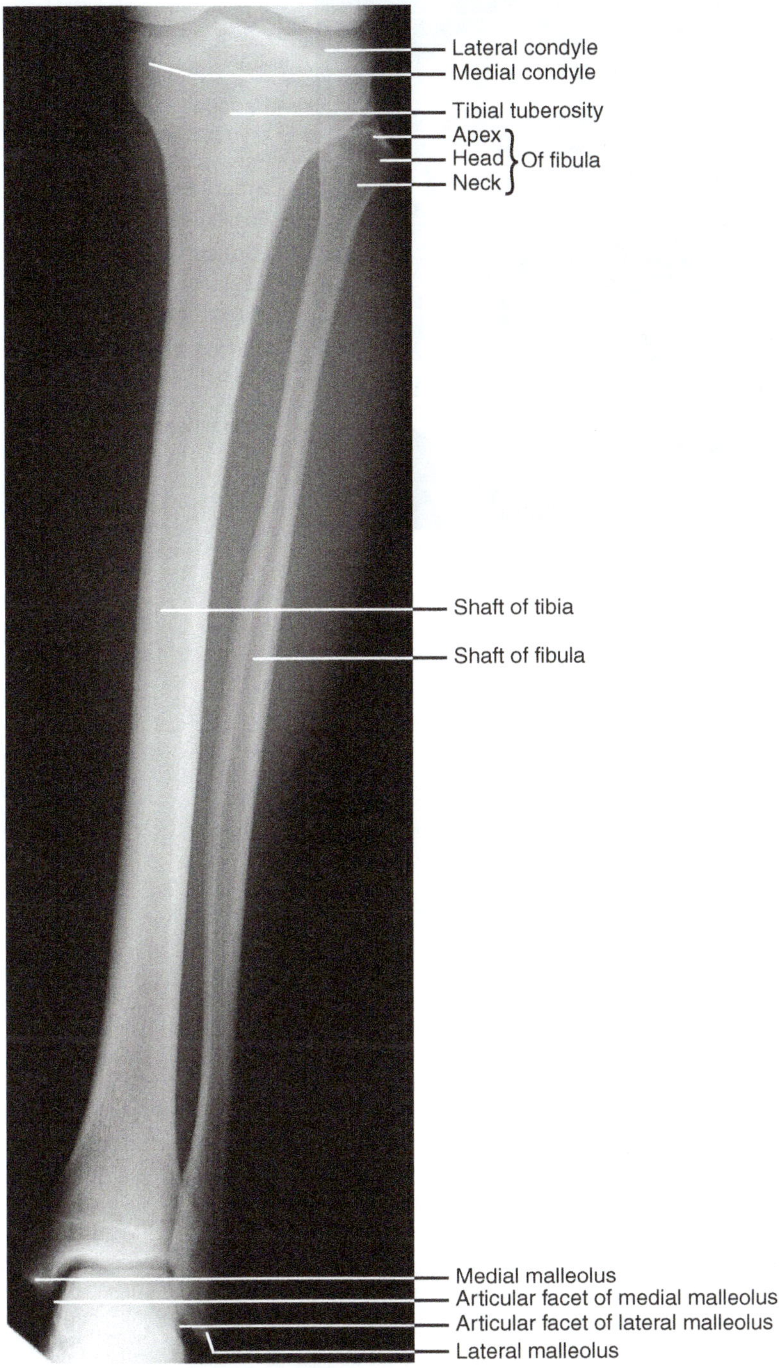

Fig. 25.29 Antero-posterior view of the tibia and fibula

Magnetic Resonance Imaging

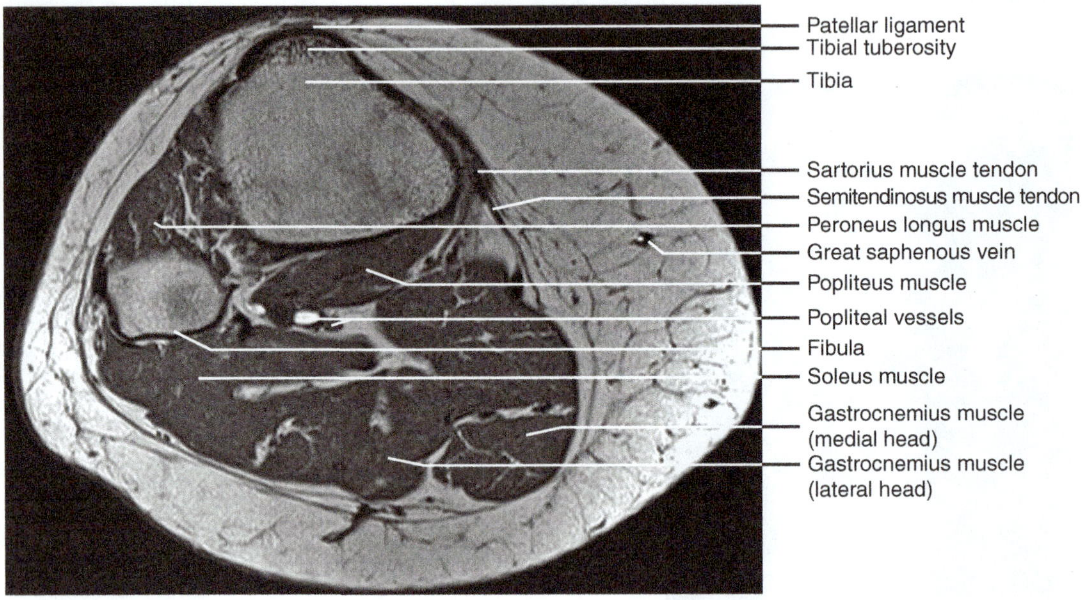

Fig. 25.30 T1 axial proximal mid tibia and fibula

Ankle

Plain Film

Bony anatomy can be assessed accurately with plain films. Routine views include AP, lateral and oblique.

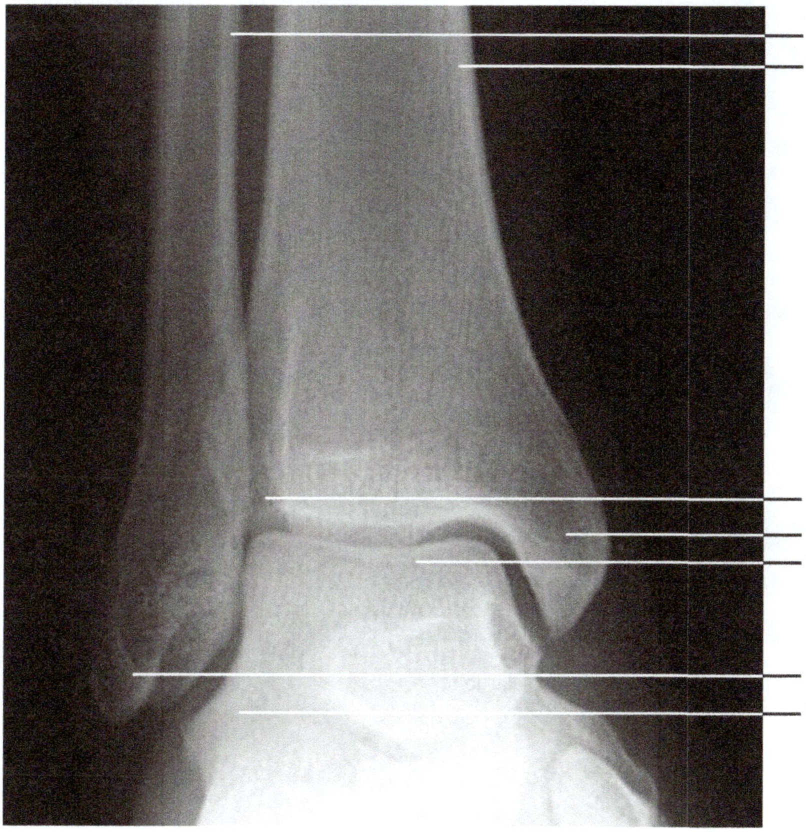

Fig. 25.31 Oblique view of the ankle demonstrating the inferior tibio-fibular joint

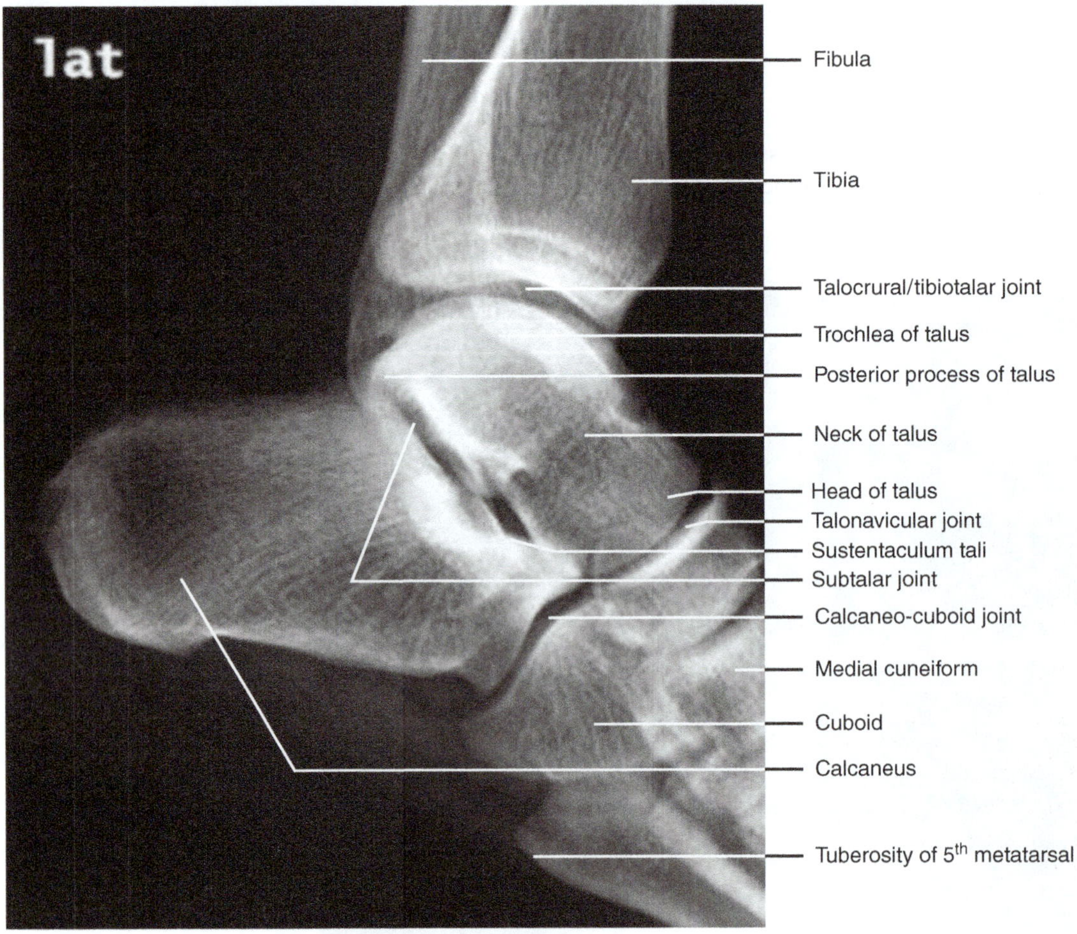

Fig. 25.32 Lateral view of the ankle demonstrating the subtalar joint

Ultrasound

Ultrasound is a very useful technique for assessing tendons around the ankle. Ligaments are also demonstrated with ultrasound, but intra-articular structures are not well shown.

Anterior

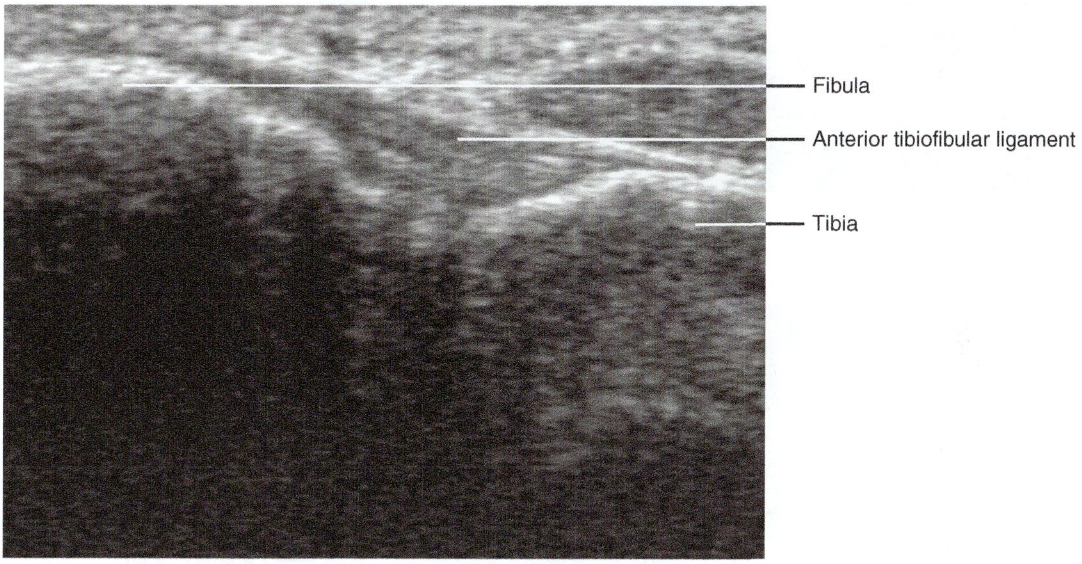

Fig. 25.33 Longitudinal US of the anterior tibio-fibular ligament

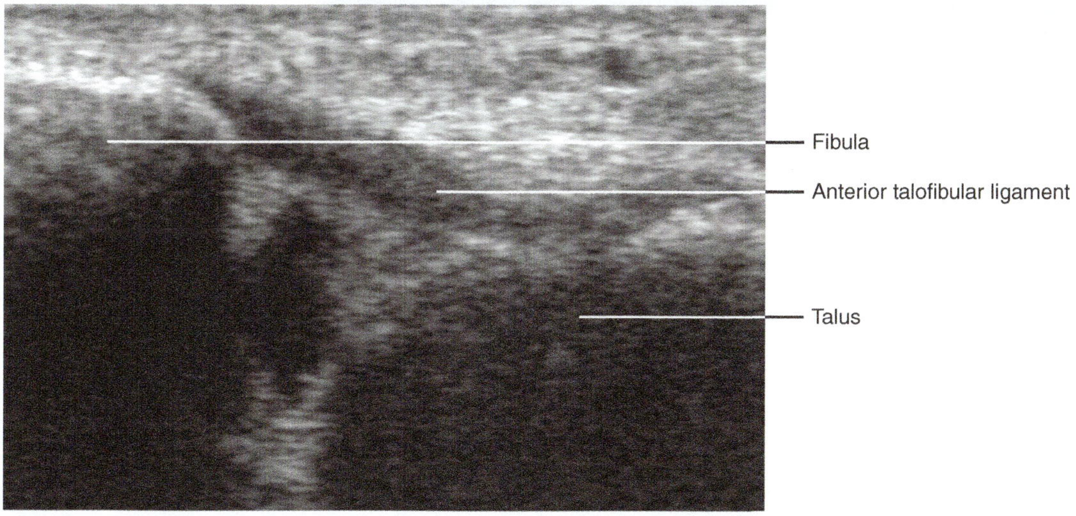

Fig. 25.34 Longitudinal US anterior talo-fibular ligament

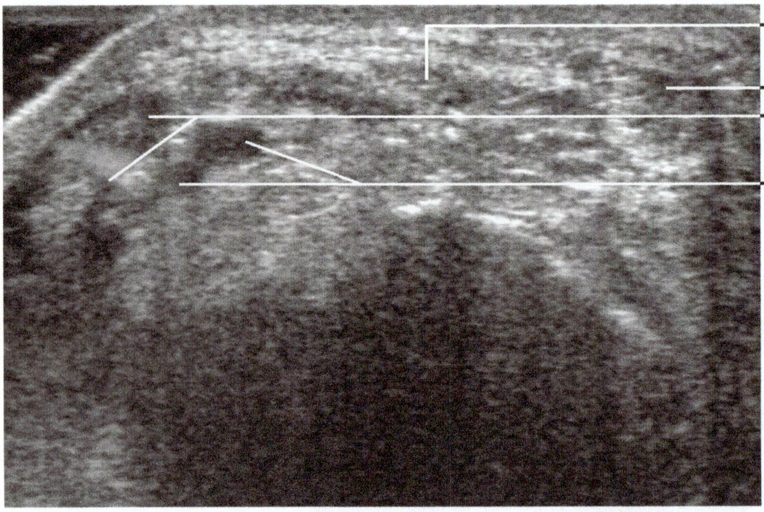

Fig. 25.35 Transverse US of the anterior ankle

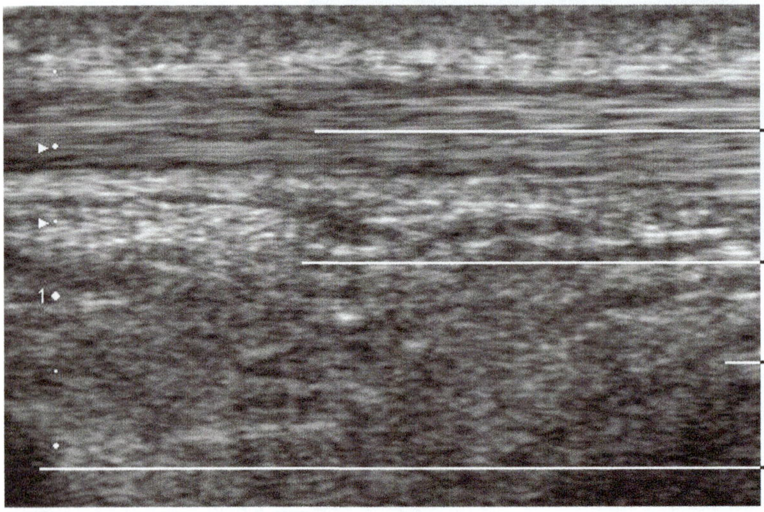

Fig. 25.36 Longitudinal US of the tibialis anterior tendon

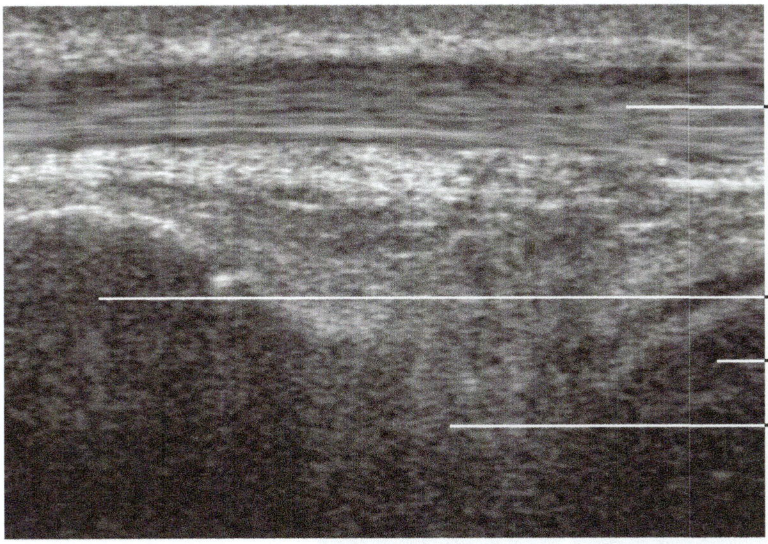

Fig. 25.37 Longitudinal US of the extensor hallucis longus

Posterior

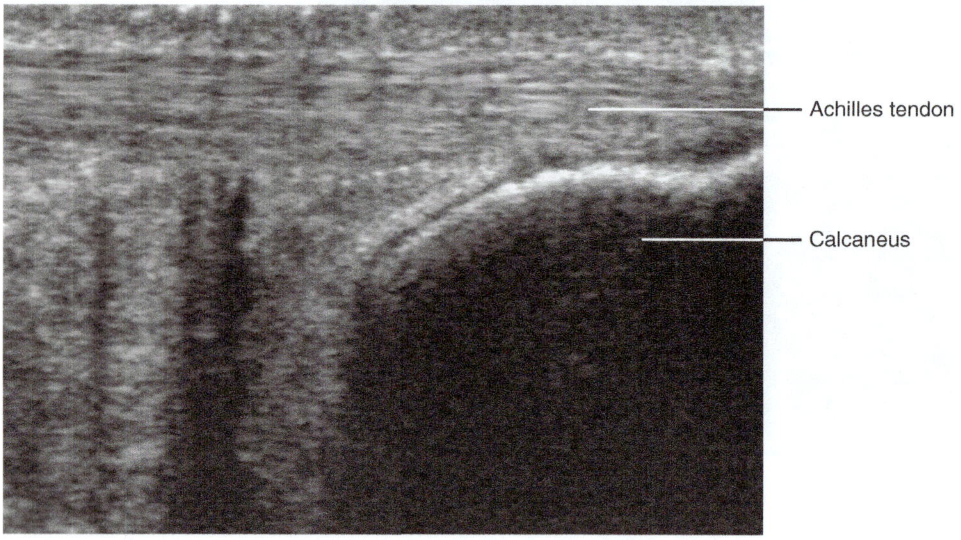

Fig. 25.38 Longitudinal US of the distal Achilles tendon

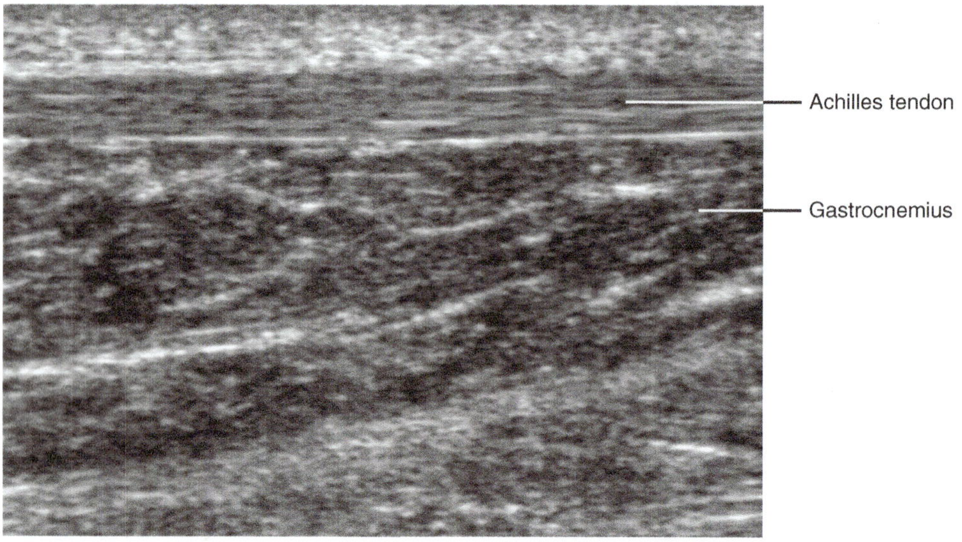

Fig. 25.39 Longitudinal US of the more proximal Achilles tendon and gastrocnemius muscle

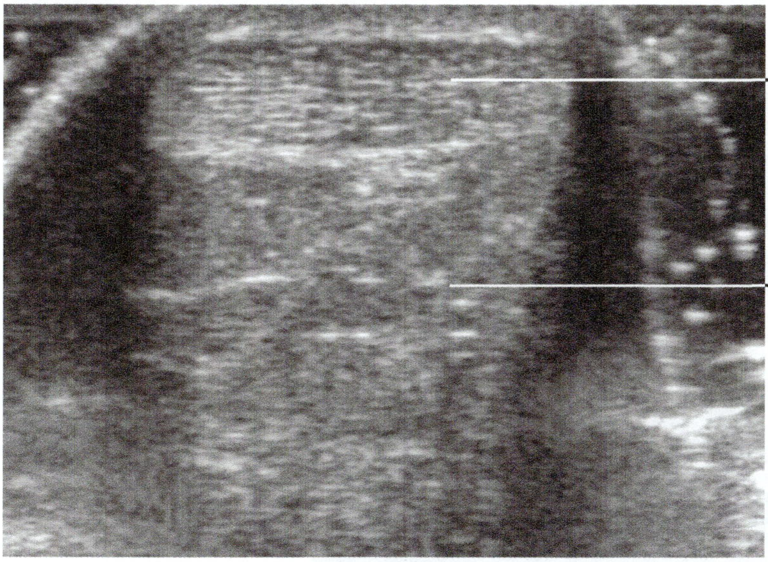

Fig. 25.40 Transverse US of the Achilles tendon

Lateral

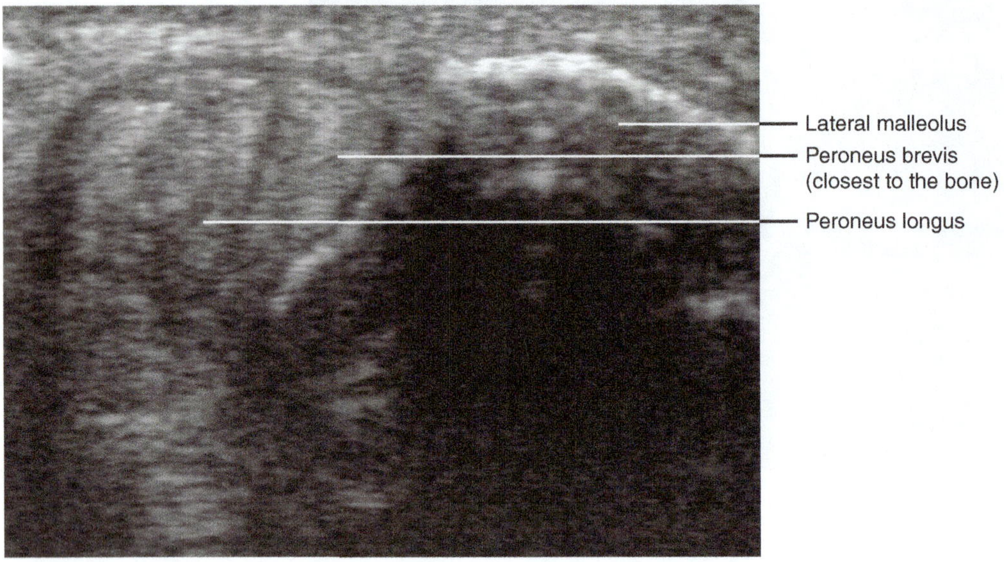

Fig. 25.41 Transverse US of the lateral malleolus

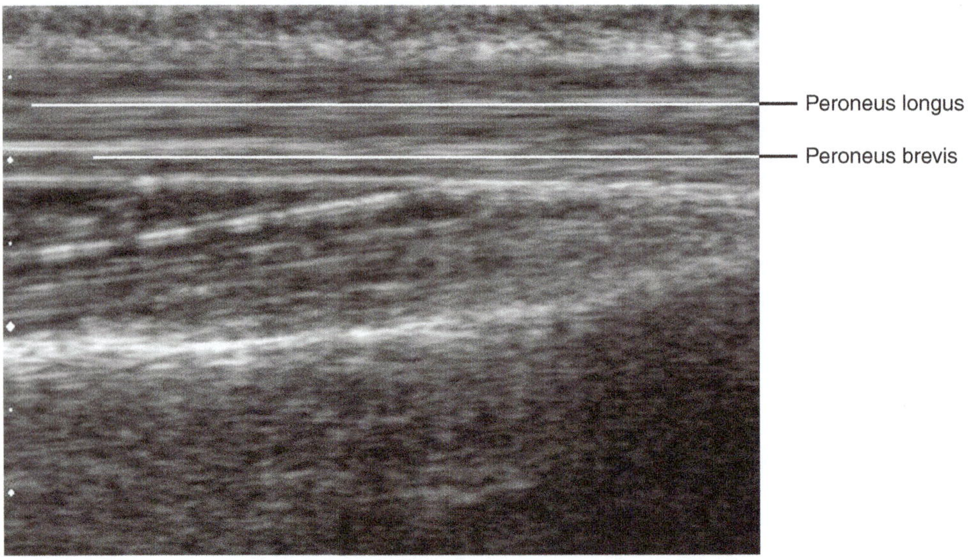

Fig. 25.42 Longitudinal US of the peroneal tendons above the malleolus

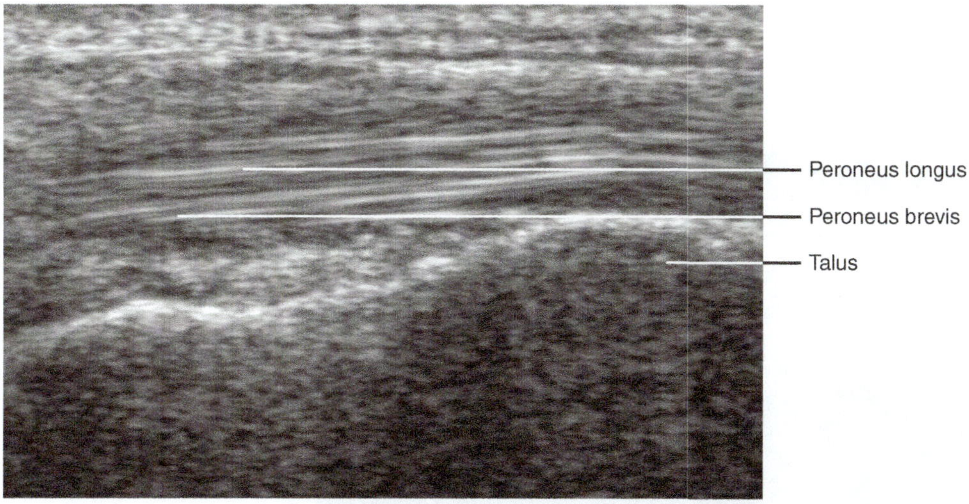

Fig. 25.43 Longitudinal US of the peroneal tendons below the lateral malleolus

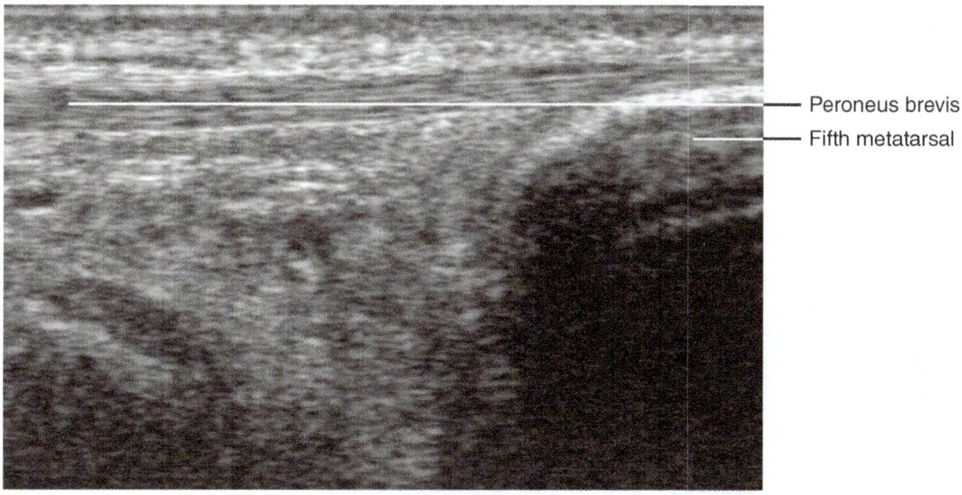

Fig. 25.44 Longitudinal US of the distal peroneus brevis insertion

Medial

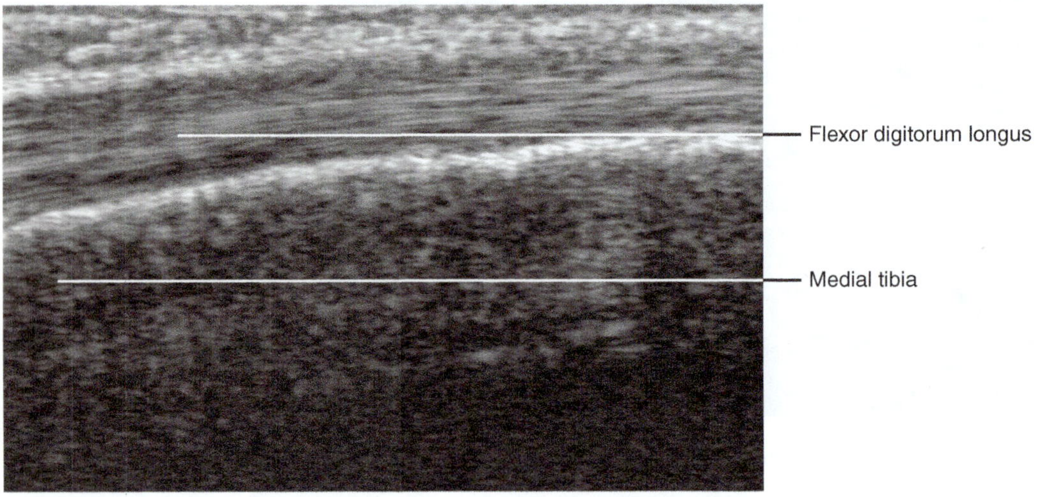

Fig. 25.45 Longitudinal US of the flexor digitorum longus tendon

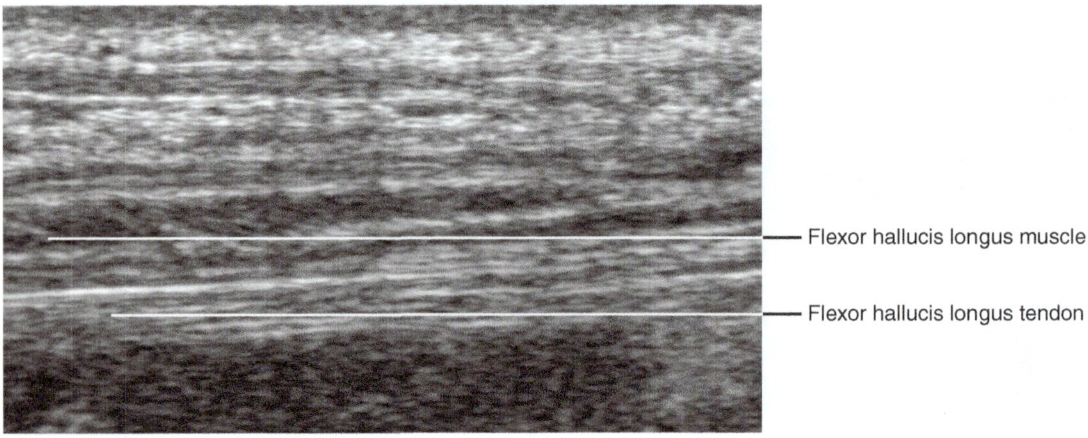

Fig. 25.46 Longitudinal US of the flexor hallucis longus tendon

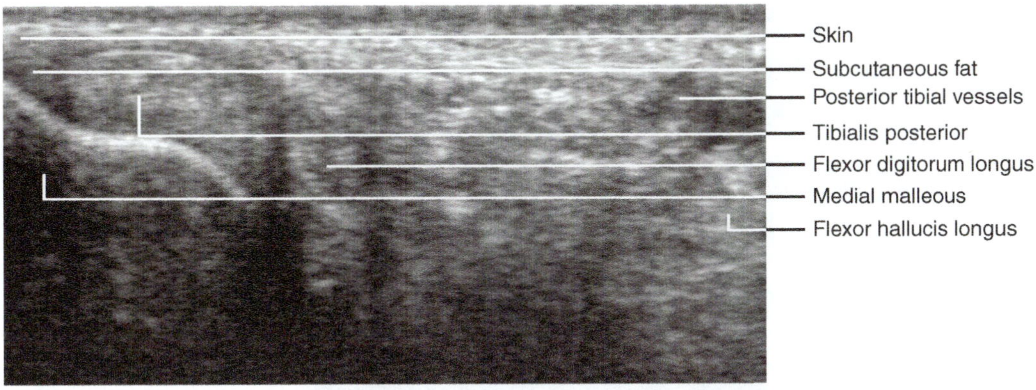

Fig. 25.47 Longitudinal US of the medial malleolus

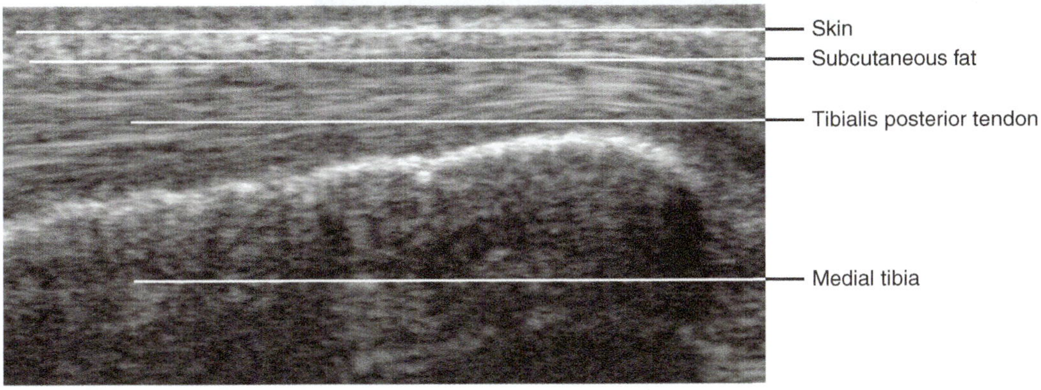

Fig. 25.48 Longitudinal US of the tibialis posterior tendon

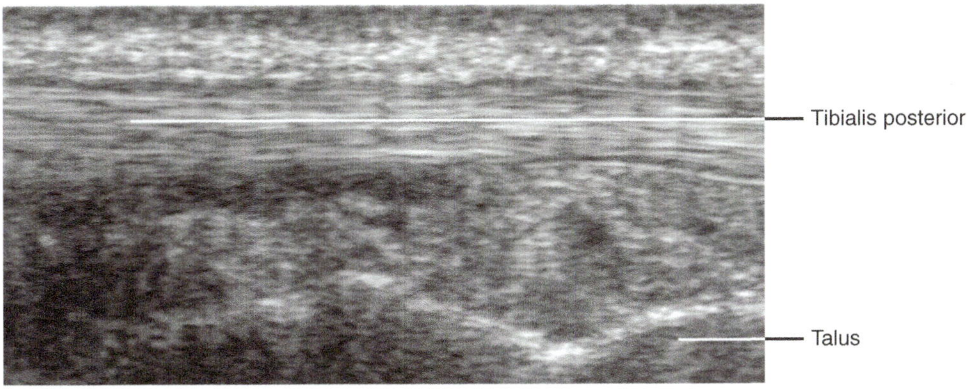

Fig. 25.49 Longitudinal US of the mid tibialis posterior below the medial malleolus

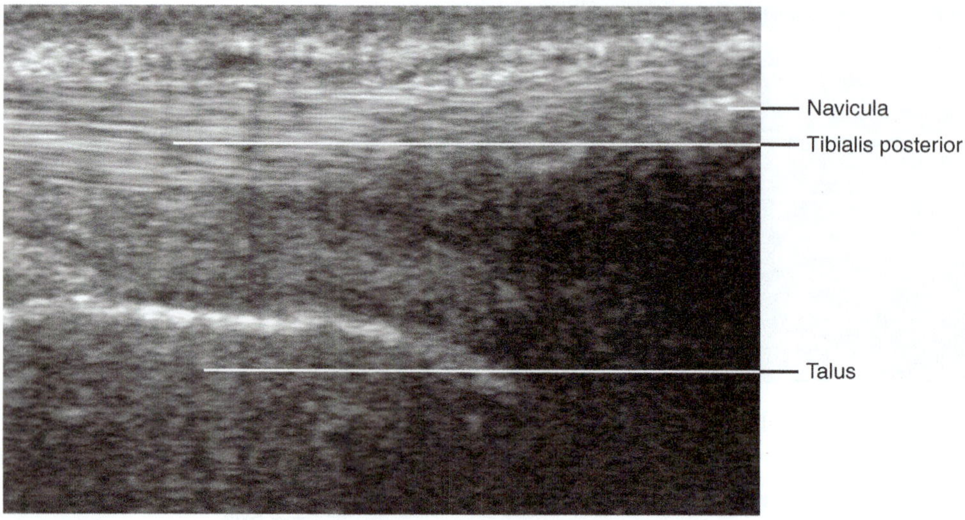

Fig. 25.50 Longitudinal US of the distal tibialis posterior at its insertion

Magnetic Resonance Imaging

MRI demonstrates the articulating bones and ligaments.

Axial Anatomy from Superior to Inferior

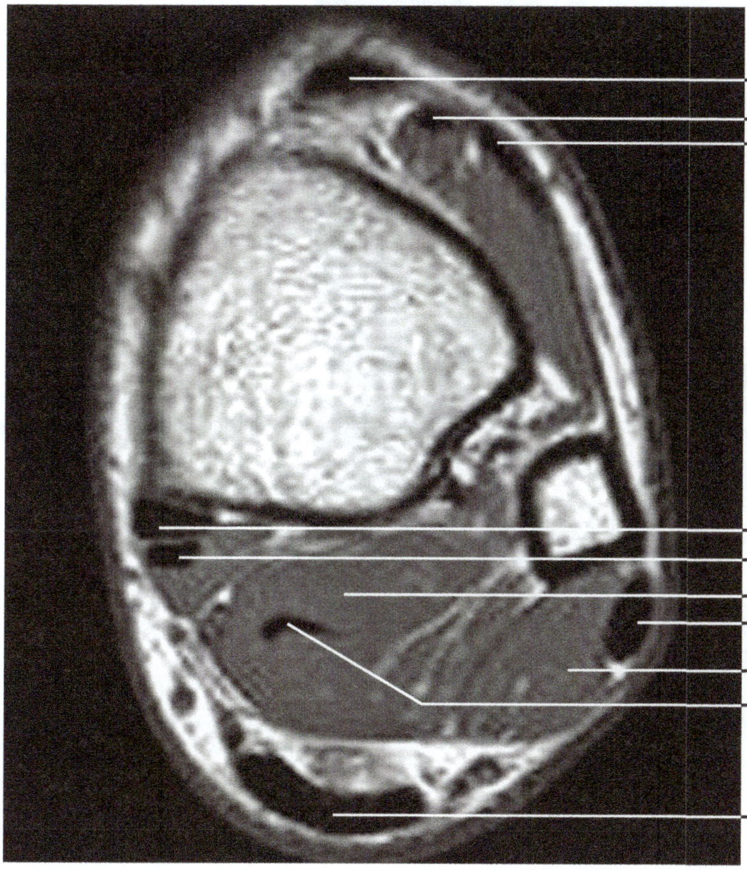

Fig. 25.51 T1 axial at the distal tibia and fibula

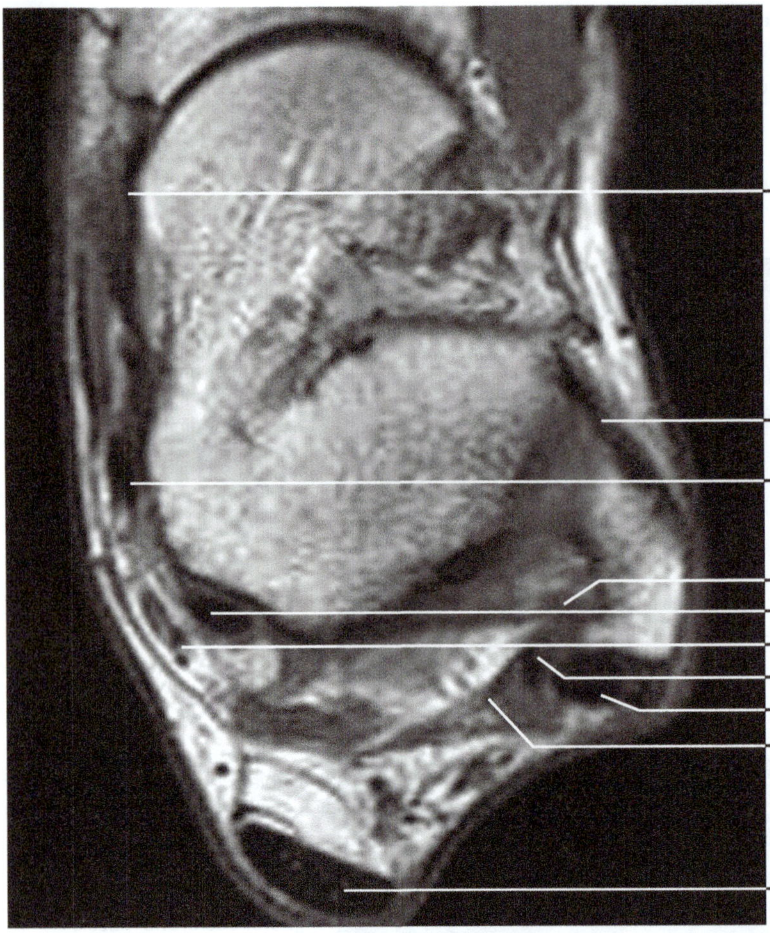

Fig. 25.52 T1 axial at the level of the talo-navicular joint

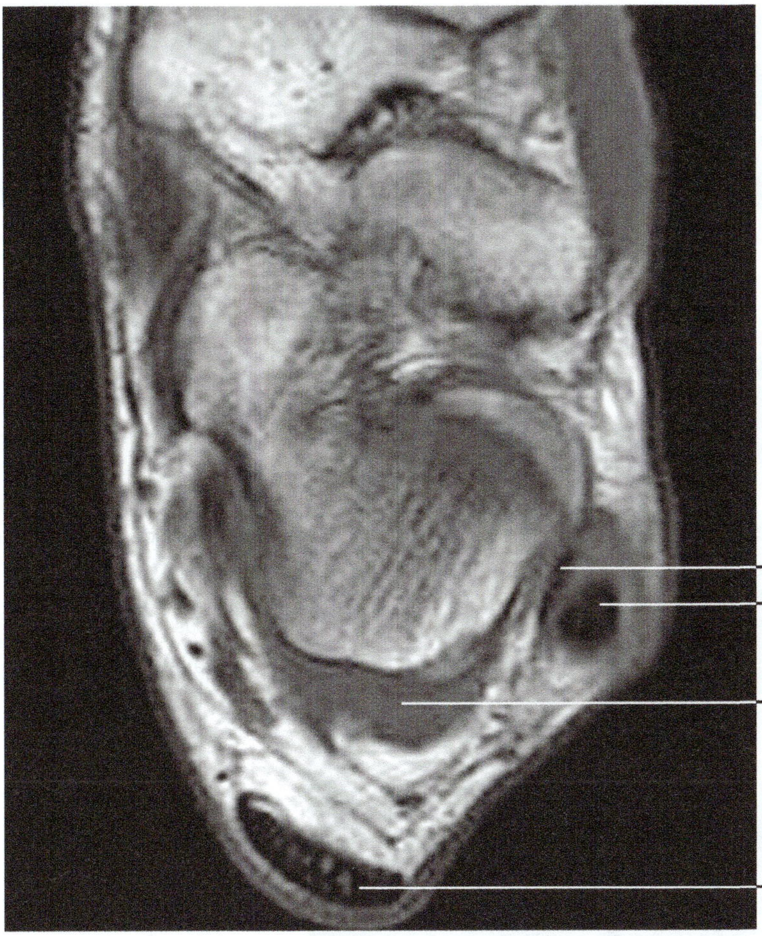

Fig. 25.53 T1 axial at the level of the cuneiform bones

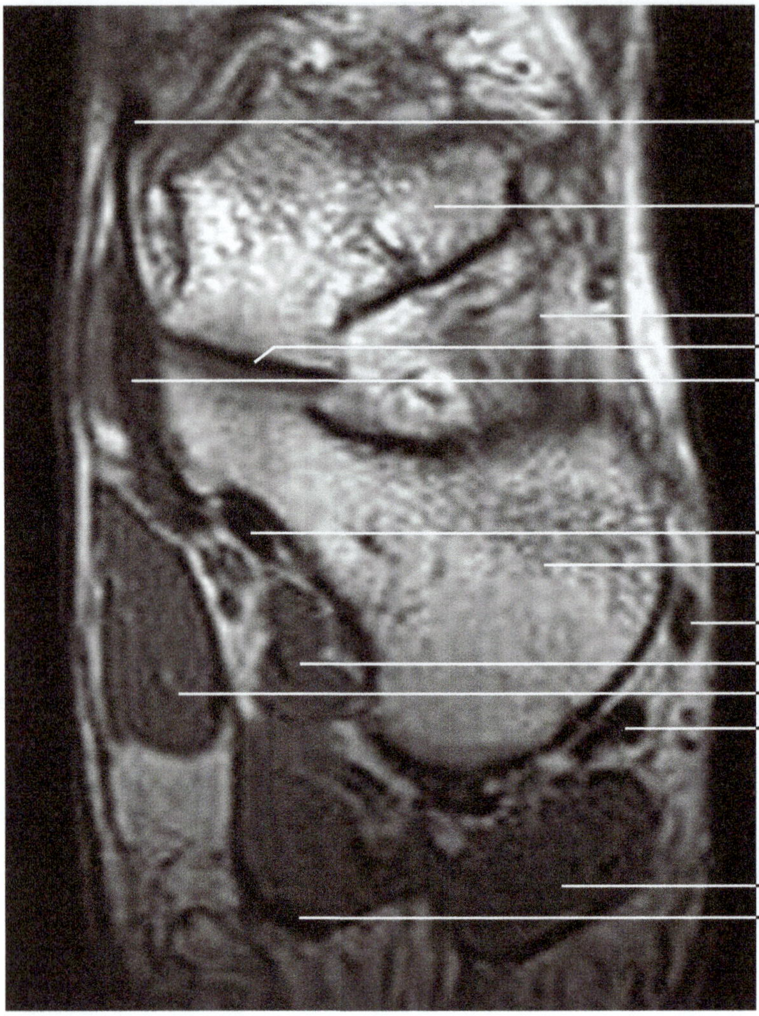

Fig. 25.54 T1 axial at the level of the sinus tarsi

Coronal Anatomy from Anterior to Posterior

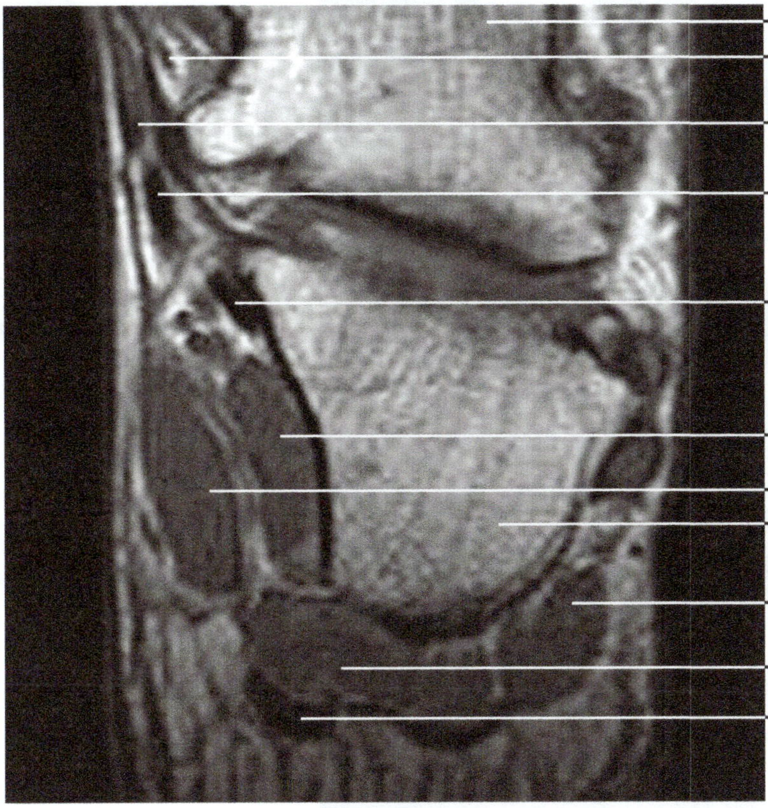

Fig. 25.55 T1 coronal at the level of the tibio-talar joint

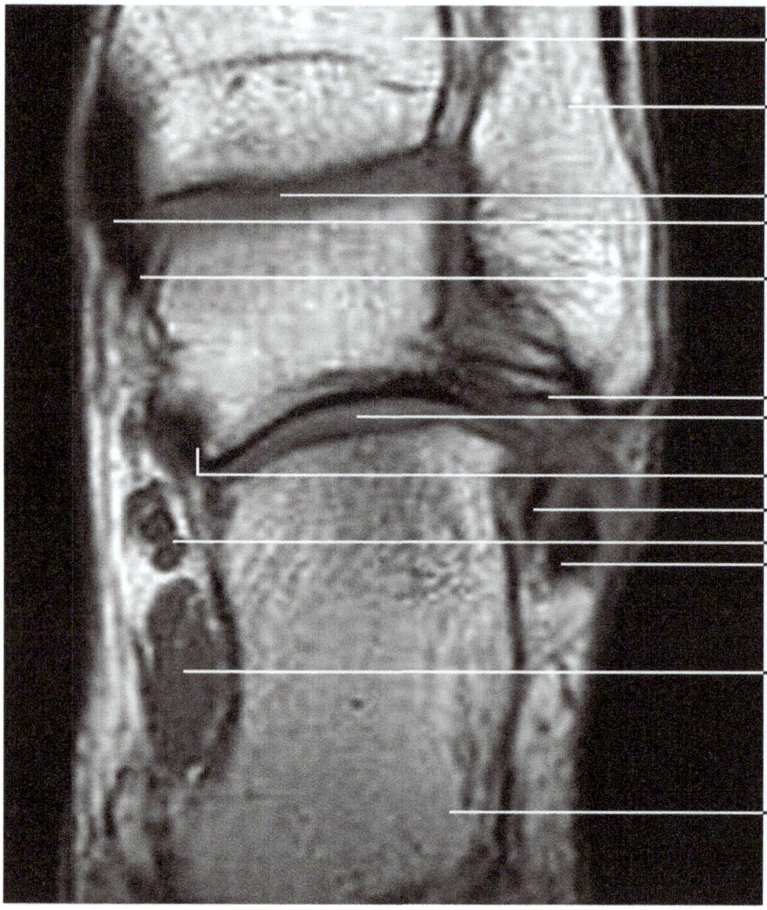

Fig. 25.56 T1 coronal at the level of the posterior subtalar joint

Sagittal Anatomy from Medial to Lateral

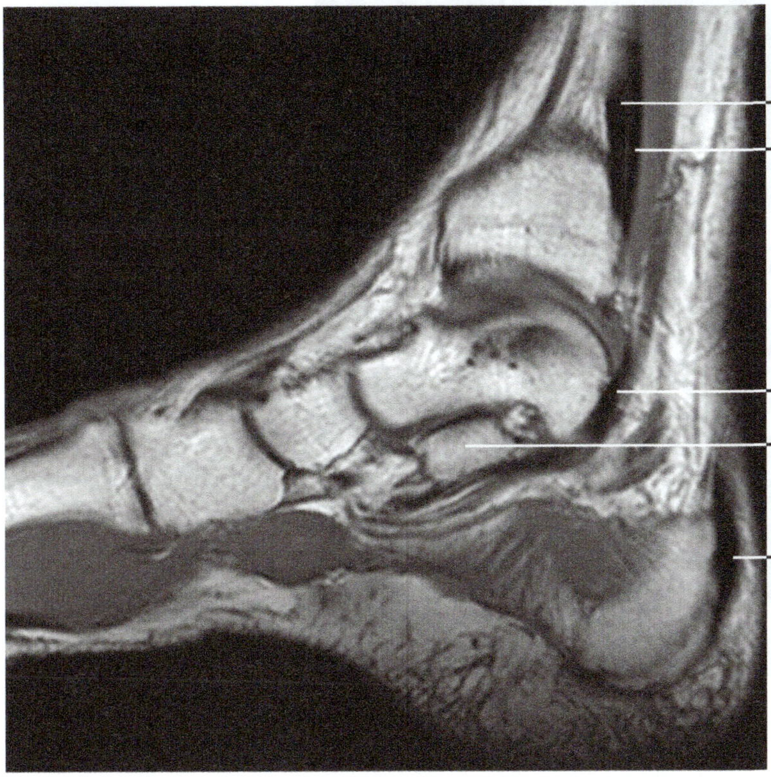

Fig. 25.57 T1 sagittal at the level of the medial aspect of the ankle joint

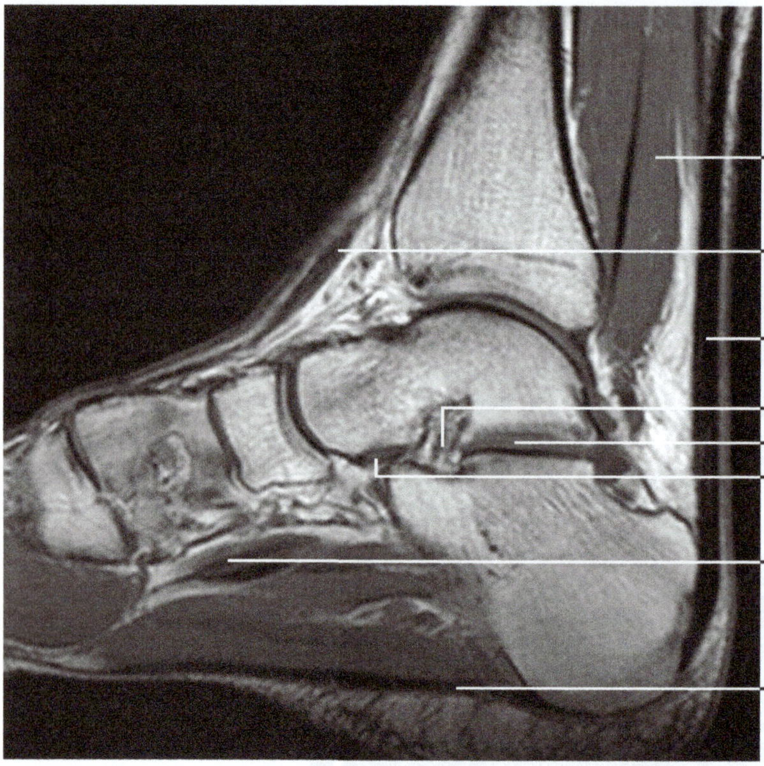

Fig. 25.58 T1 sagittal at the level of the middle facet of the subtalar joint and sinus tarsi

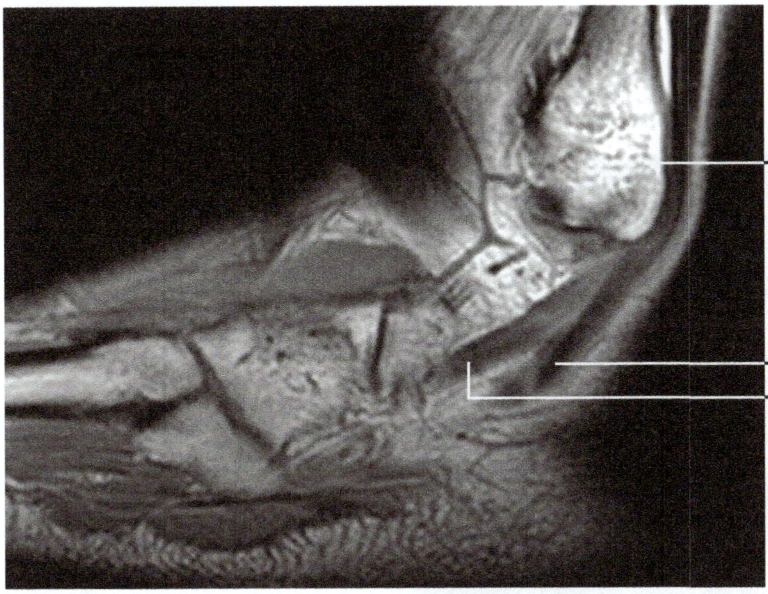

Fig. 25.59 T1 sagittal at the level of the lateral malleolus (fibula)

Pearl

The medial ankle tendons at the level of the malleoli can be remembered in order from superior to inferior by the mnemonic Tom, Dick and Harry. This stands for **T**ibialis posterior [**T**om], flexor **D**igitorum longus [**D**ick] and flexor **H**allucis longus [**H**arry], with the tibialis posterior being closest to the medial malleolus.

The tibialis posterior tendon has a flared, wide insertion at the navicula.

For the anterior tendons the order should be re-arranged and remembered as Tom, Harry and Dick.

The flexor digitorum longus tendon sheath can communicate with the tibiotalar joint and may demonstrate fluid signal intensity.

Foot

Plain Film

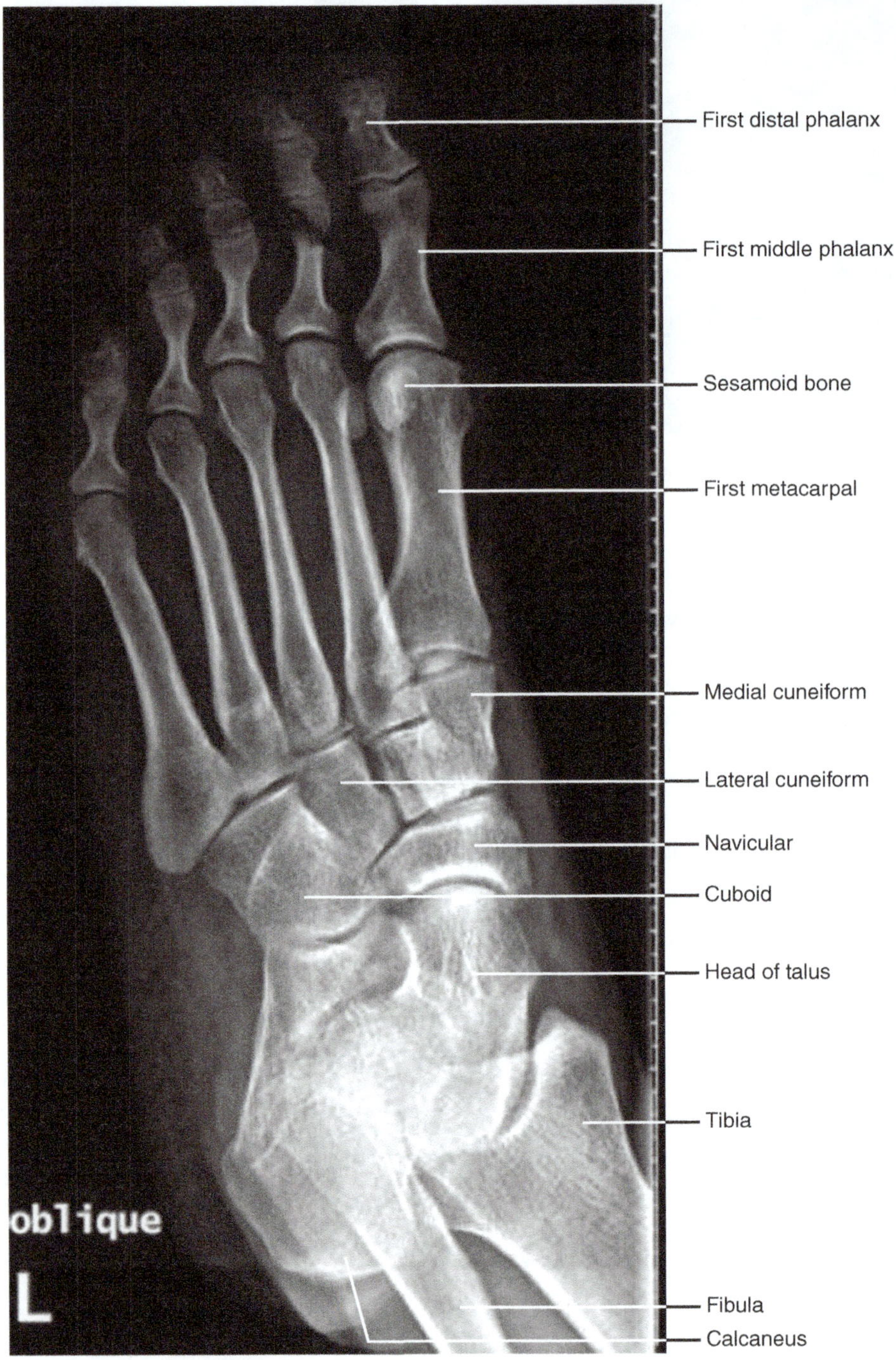

Fig. 25.60 Oblique view of the foot

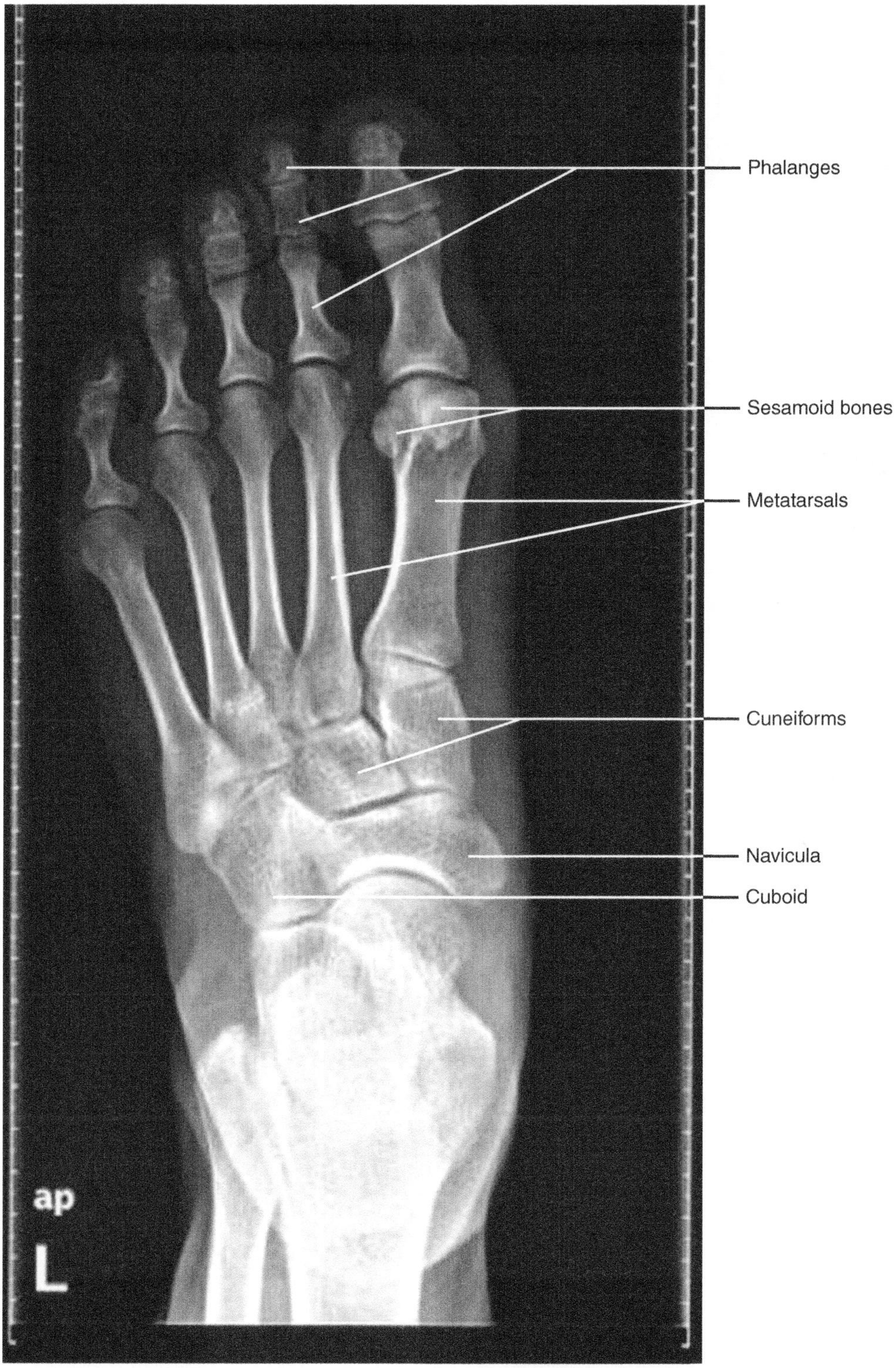

Fig. 25.61 Antero-posterior view of the foot demonstrating the metatarsals and phalanges

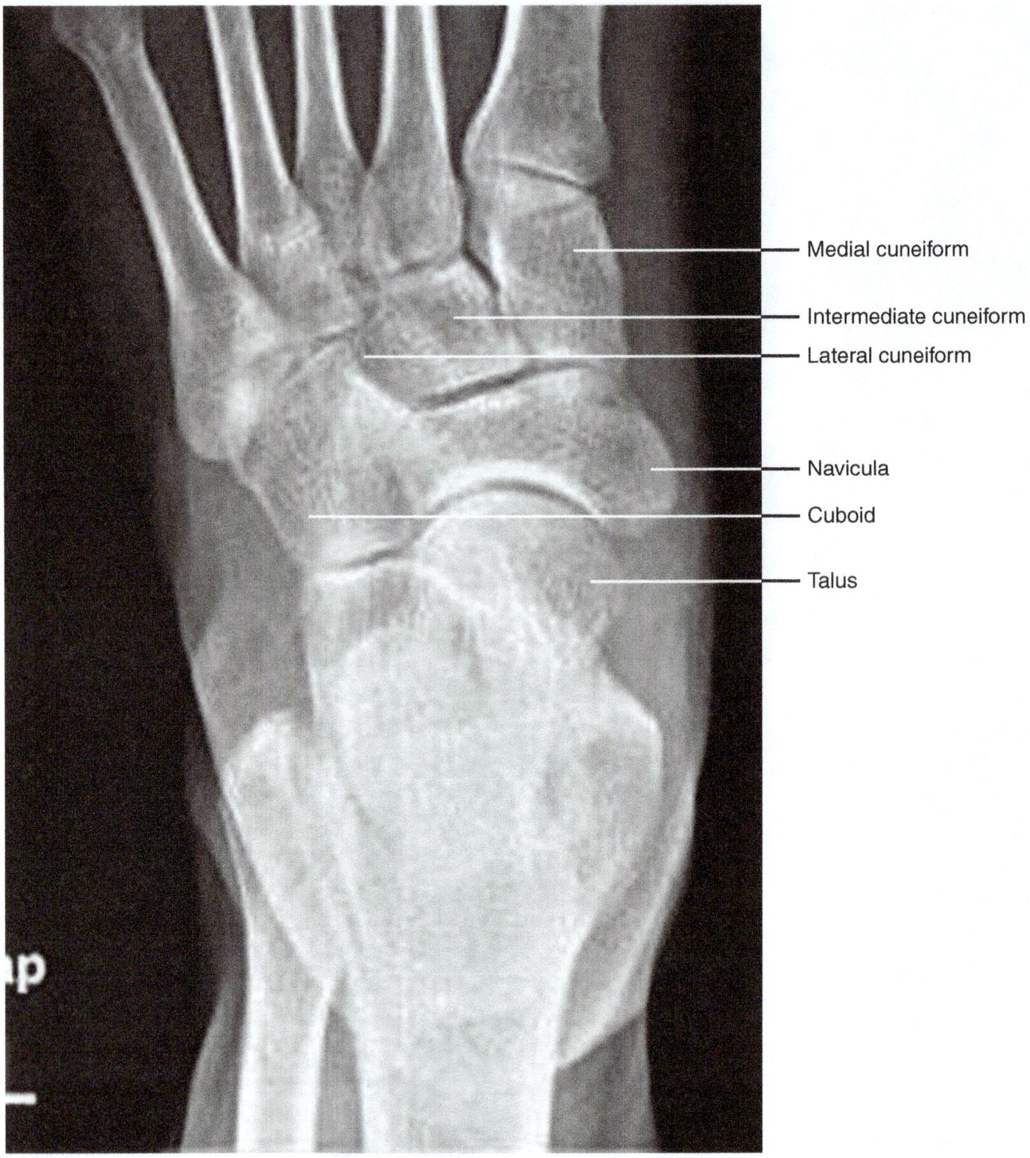

Fig. 25.62 Antero-posterior view of the foot demonstrating the tarsal bones

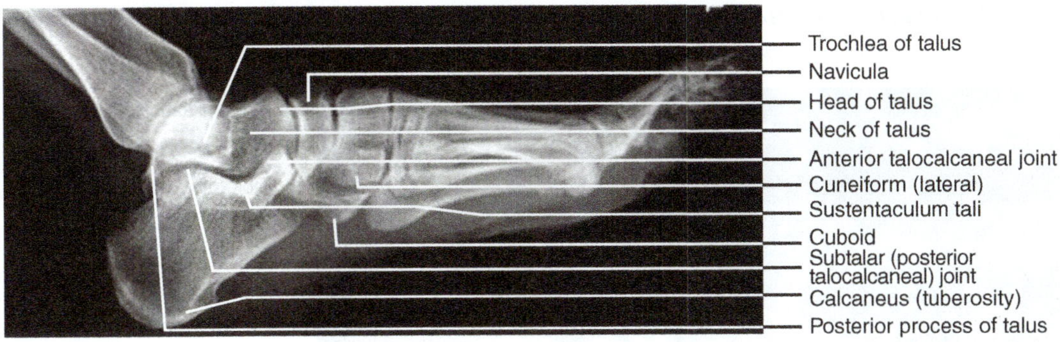

Fig. 25.63 Lateral view of the foot

Magnetic Resonance Imaging

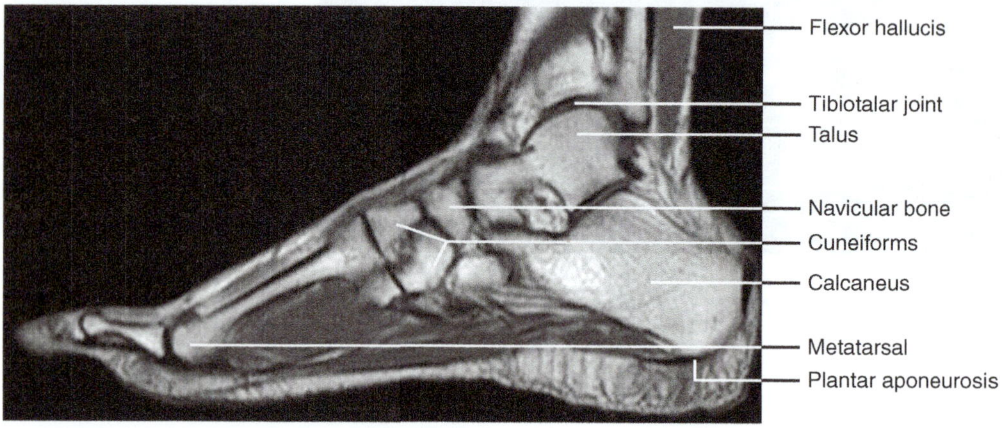

Fig. 25.64 T1 sagittal at the lateral aspect of the foot

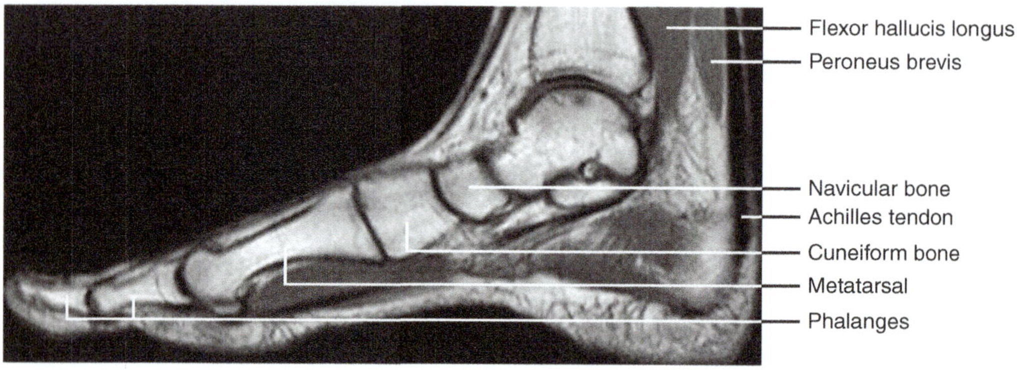

Fig. 25.65 T1 sagittal at the level of the medial aspect of the foot

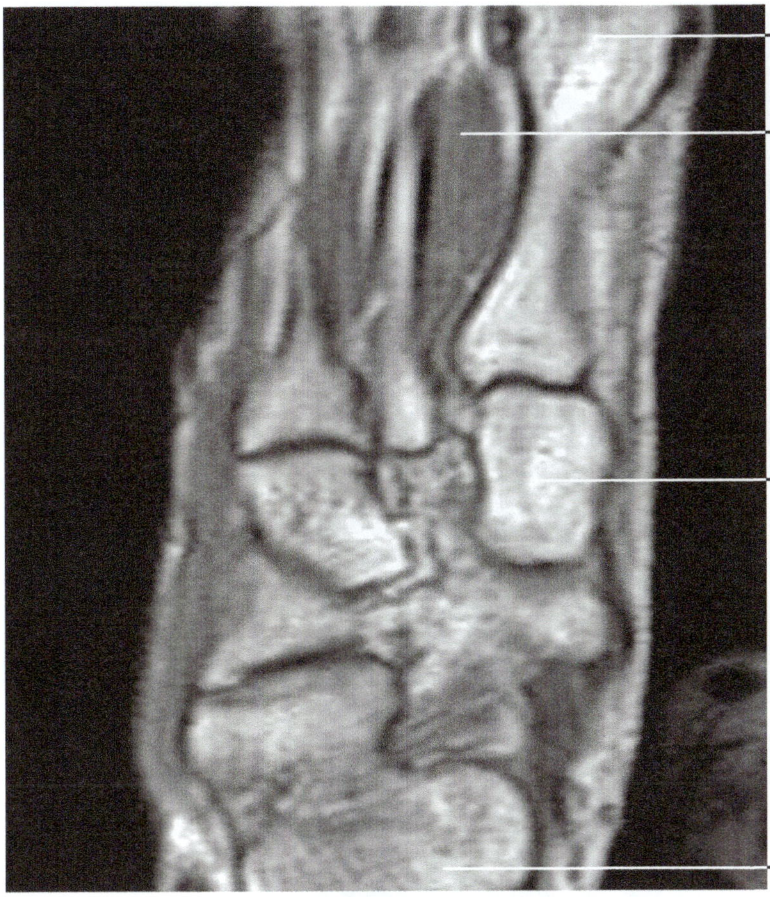

Fig. 25.66 T1 axial at the superior part of the foot showing the cuneiforms

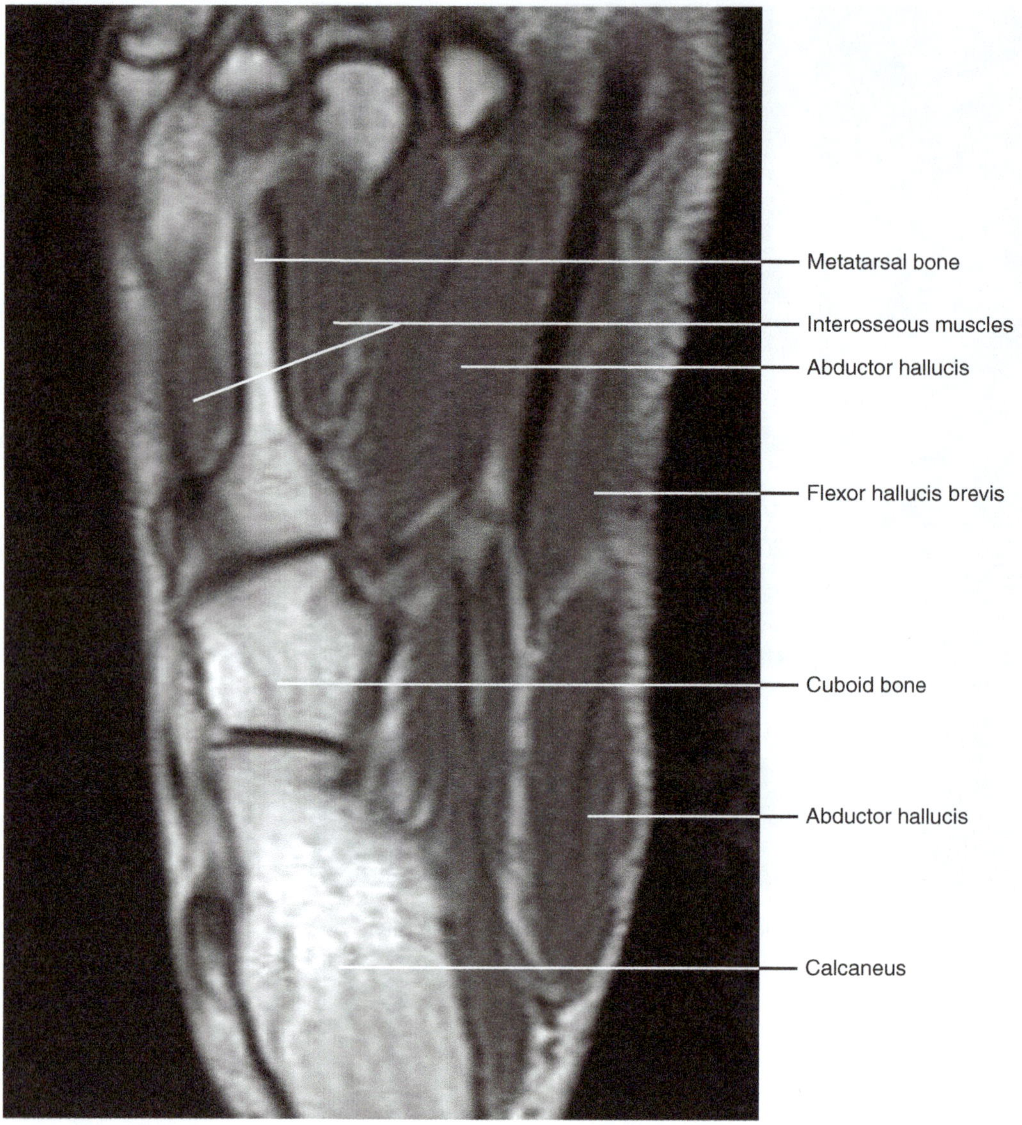

Fig. 25.67 T1 axial at the inferior part of the foot showing the cuboid

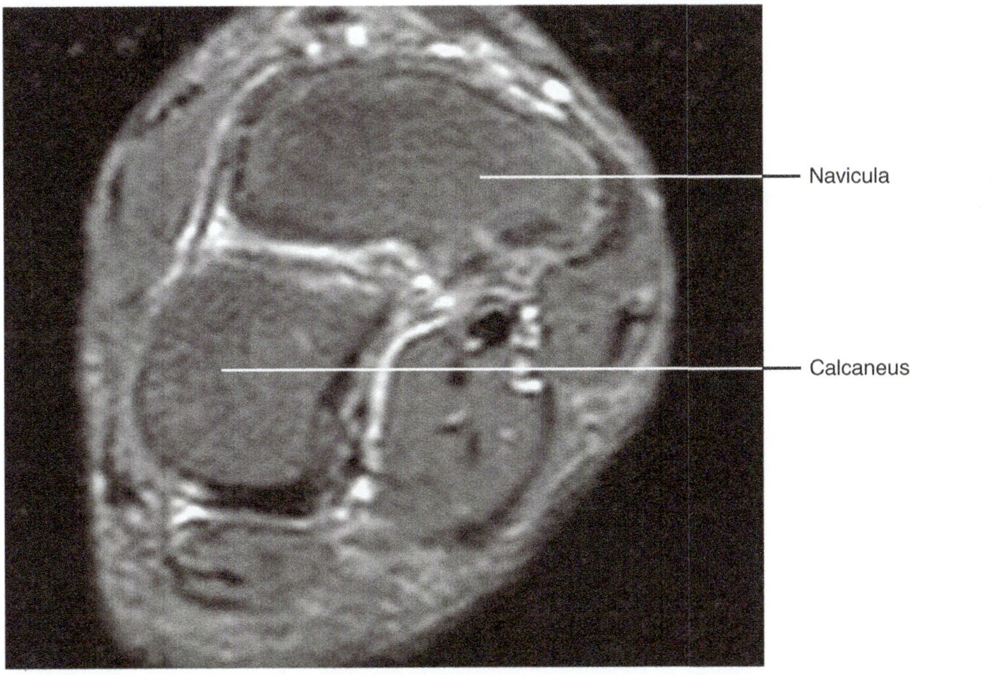

Fig. 25.68 Gradient echo coronal image proximally in the foot at the level of the navicula

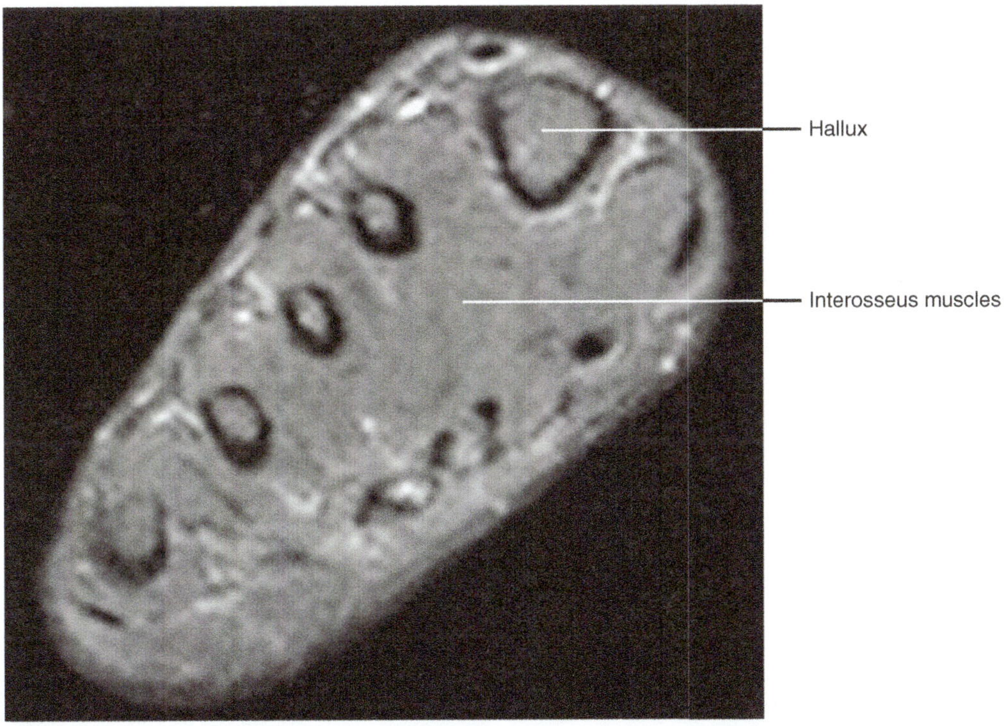

Fig. 25.69 Gradient echo coronal image of the distal foot at the level of the metacarpals

26 The Pelvis

Introduction

Greater part of the pelvis formed by two innominate bones each made up of ilium, ischium and pubis.

The pelvis is divided into true and false pelvis by iliopectineal line.

Acetabulum is formed by union of iliac, pubic and ischial bones

Pubic Symphysis

Fibrocartilaginous non-synovial joint.

Articular surface covered by thin hyaline cartilage.

Pubic bones also joined by superior and inferior pubic ligaments.

Sacrum

Made up of five fused vertebrae.

On either side of S1 are the sacral alae articulating with iliac bones to form SI joint.

Four sets of sacral foramina are present anteriorly and posteriorly.

Anterior sacral foramina (containing ventral divisions of sacral nerves) are more easily visualised on sacral radiographs.

Coccyx is made of four to five fused vertebral bodies to which are attached gluteus maximus, coccygeus, sphincter ani and levator ani muscles.

SI Joint

Anterior inferior part is a true joint.

Cartilaginous surface of true joint is thicker on sacral than iliac side.

Posterior part is a syndesmosis.

The following imaging modalities may be used to examine the pelvic girdle and proximal femur.

1. Plain film
2. CT scan
3. Magnetic resonance imaging (MRI)
4. Ultrasound
5. Arthrography
6. Radionuclide studies
7. Angiography

S. Andronikou (ed.), *See Right Through Me*,
DOI 10.1007/978-3-642-23893-2_26, © Springer-Verlag Berlin Heidelberg 2012

Plain Film

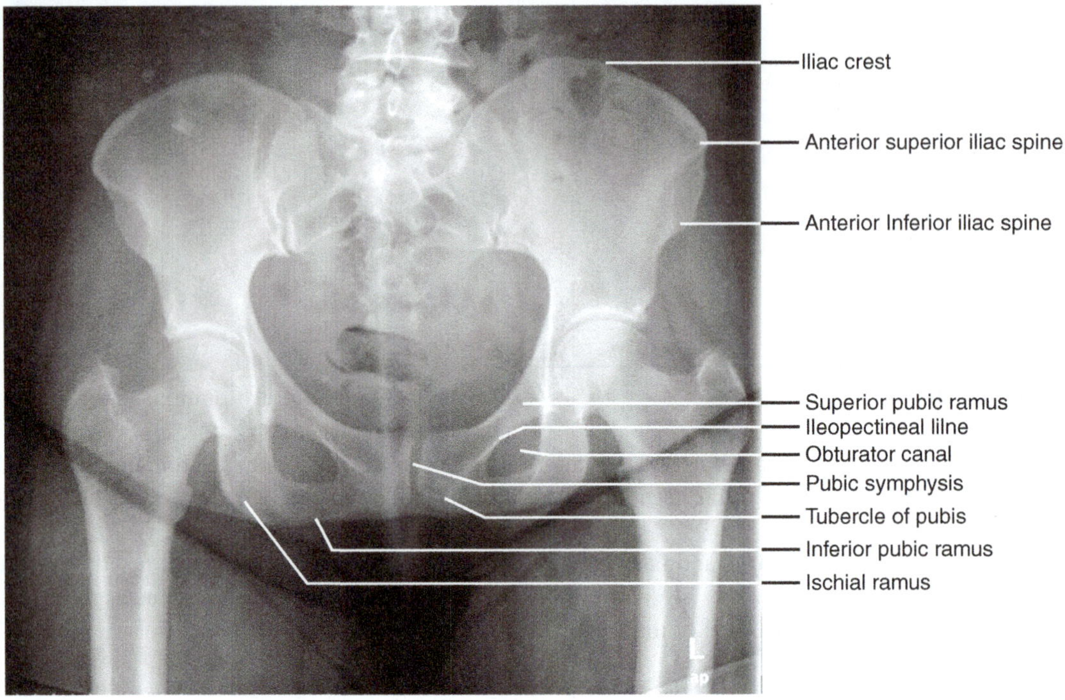

Fig. 26.1 AP pelvis – routine view

Pearl

Anterior sacral foramina have well defined superior margins are horizontally oriented and are more lateral then posterior sacral foramina which are more round and medially positioned.

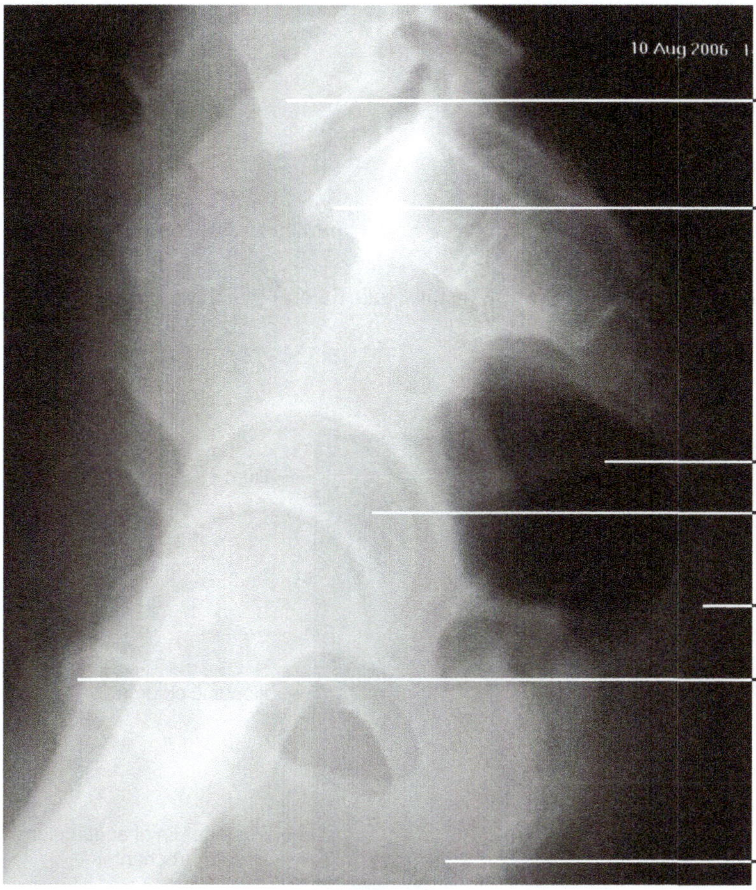

Fig. 26.2 Lateral pelvis – routine view

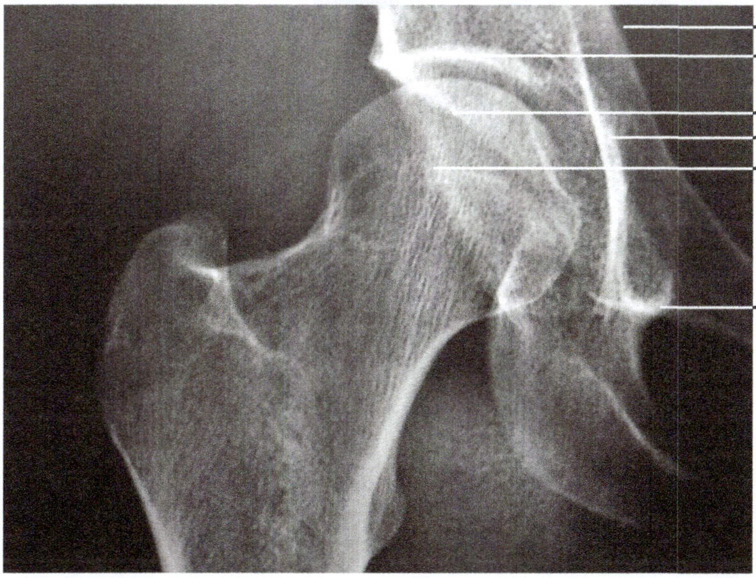

Fig. 26.3 Radiographic landmarks around hip joints

Special Views

Judet Views

Pearl

Teardrop is formed by medial acetabular wall, acetabular notch and anterior portion of quadrilateral plate.

Ilioischial line is formed by the posterior portion of the quadrilateral plate (surface) of iliac bone.

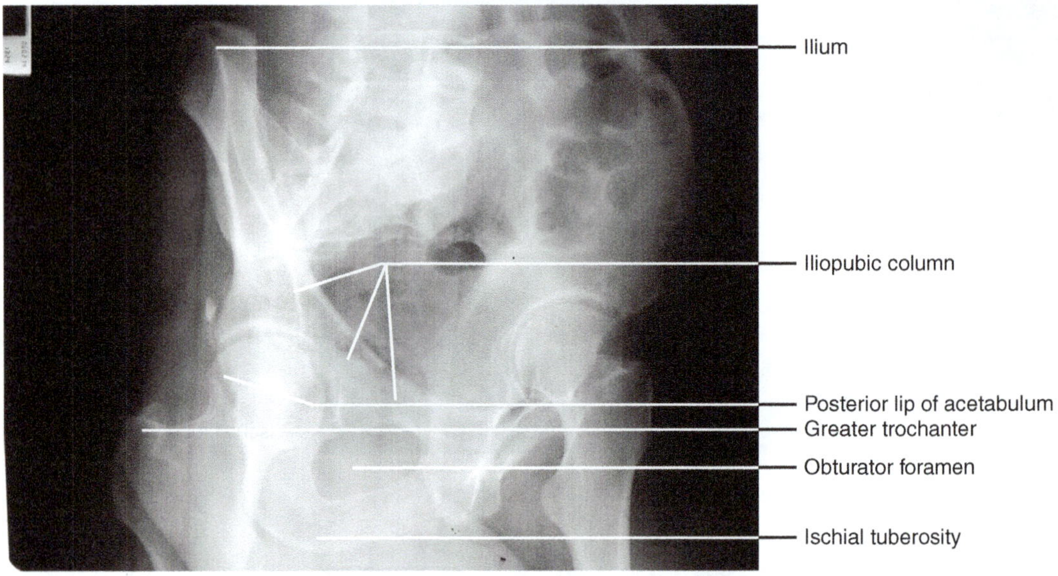

Fig. 26.4 Judet view, anterior oblique

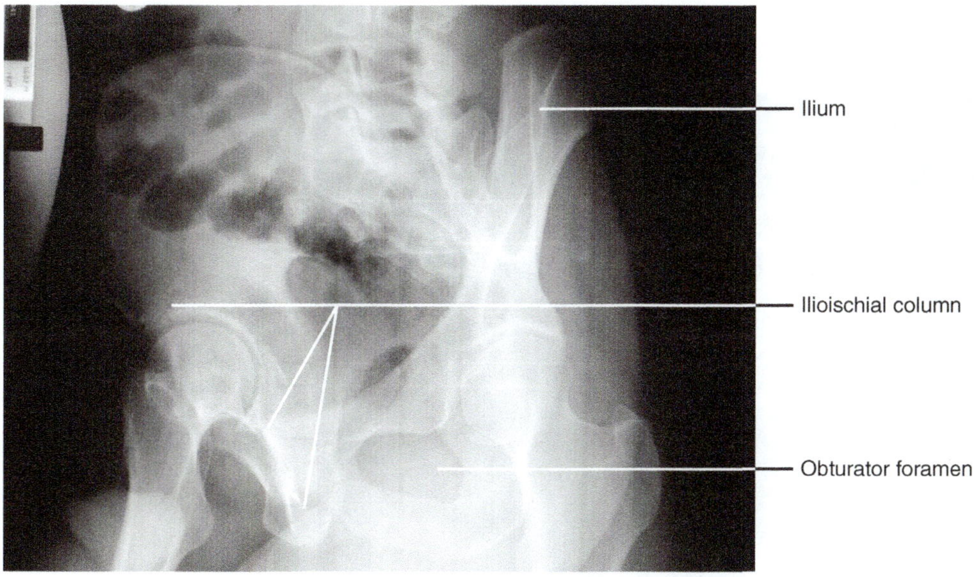

Fig. 26.5 Judet view, posterior oblique

Pearl

Anterior oblique – patient rotated 45 degree away form injured side. Allows for good visualisation of iliopubic column and posterior lip of acetabulum.

Posterior oblique – 45 degree rotation towards involved side. Good for evaluation of ilioischial column and anterior acetabular rim.

Inlet and Outlet Views

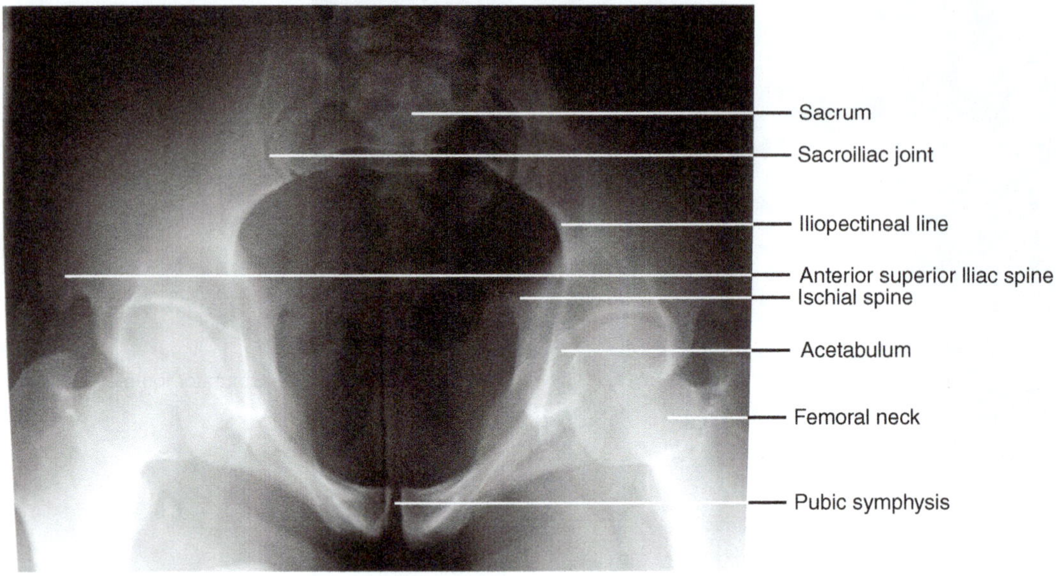

Fig. 26.6 Inlet view of pelvis

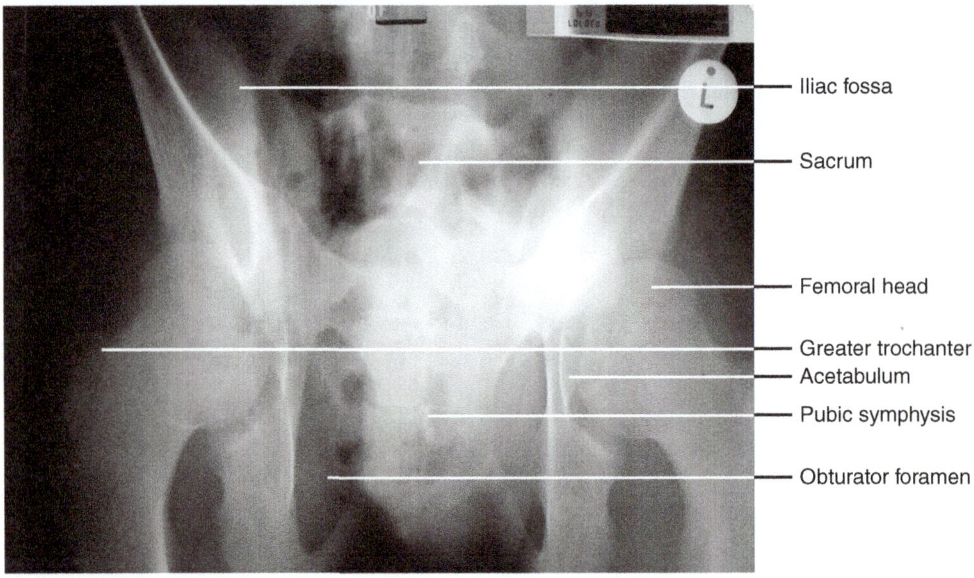

Fig. 26.7 Outlet view of pelvis

Pearl

Looks down lumen of true pelvis.

Better than AP view for showing orientation of pubic rami fractures.

Computed Tomography

Axial: Cranial to Caudal

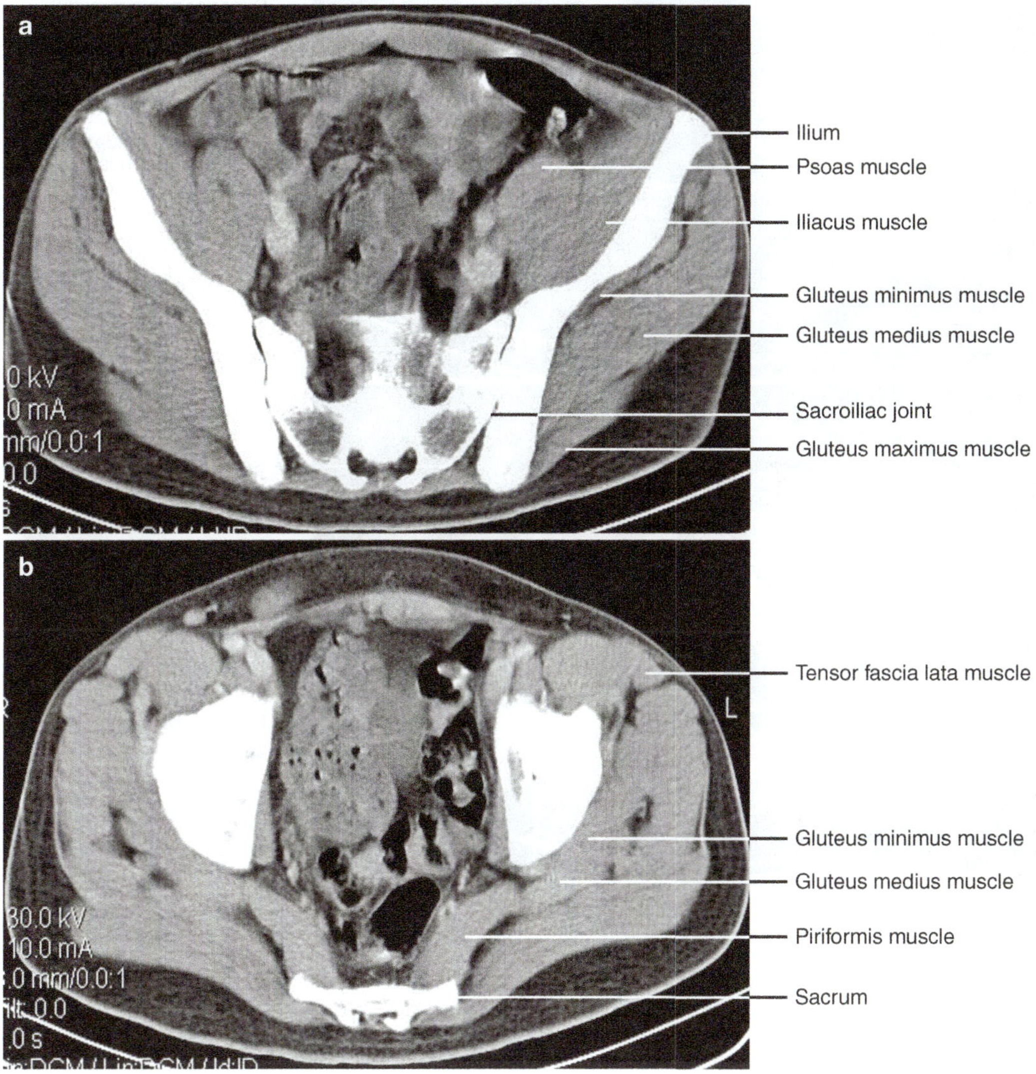

Fig. 26.8 (**a–d**) Axial CT: superior to inferior

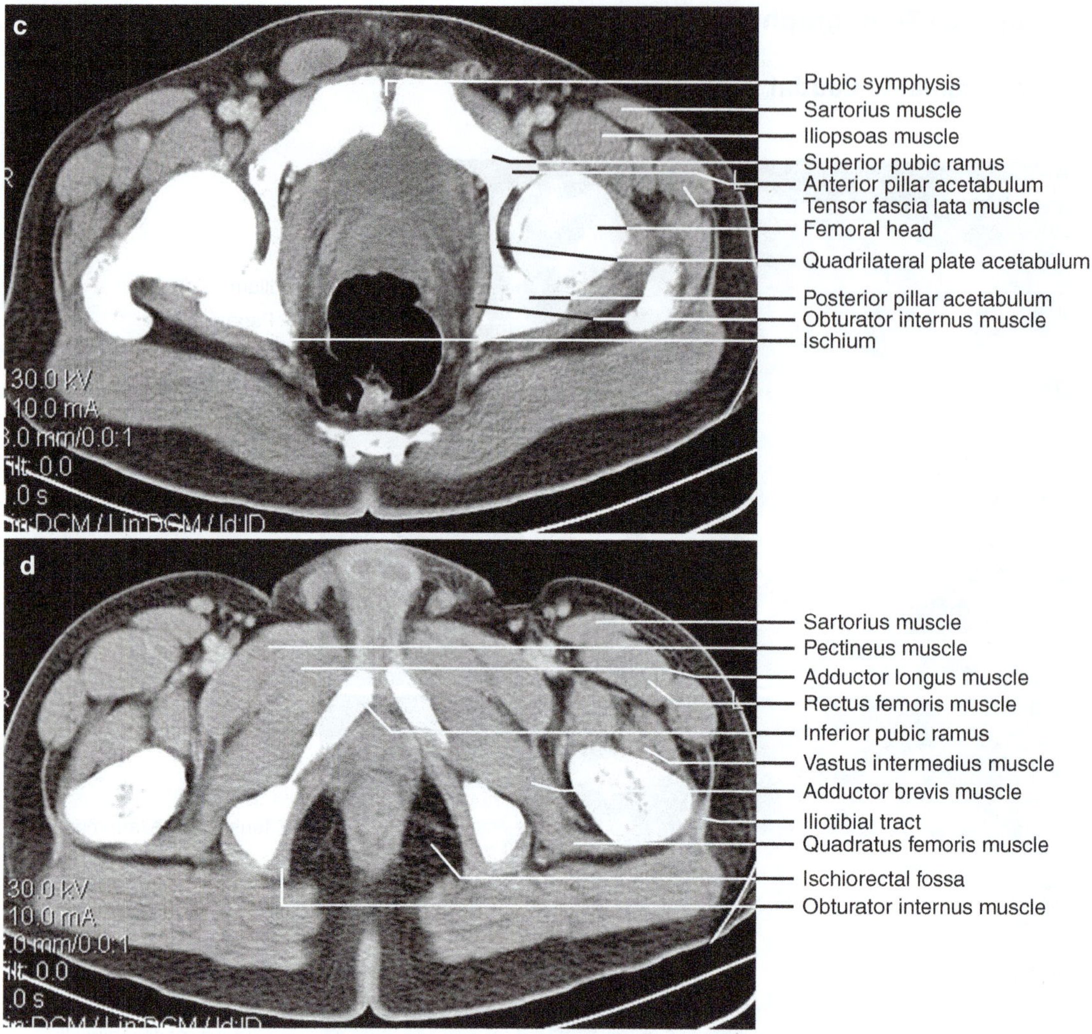

Fig. 26.8 (continued)

Magnetic Resonance Imaging

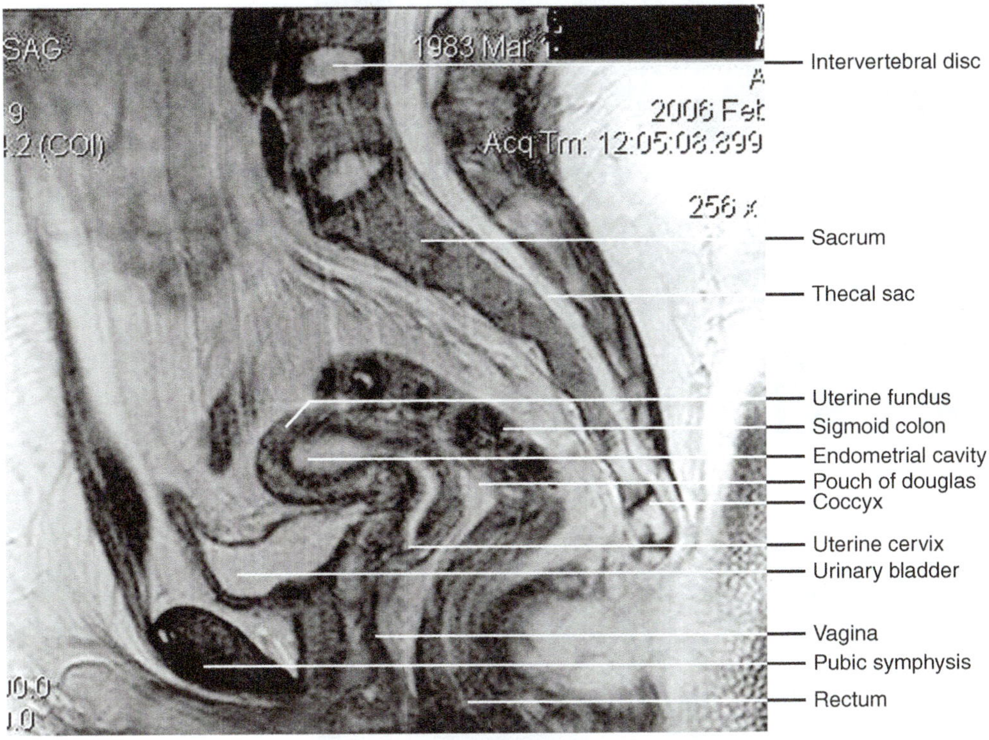

Fig. 26.9 Sagittal T2

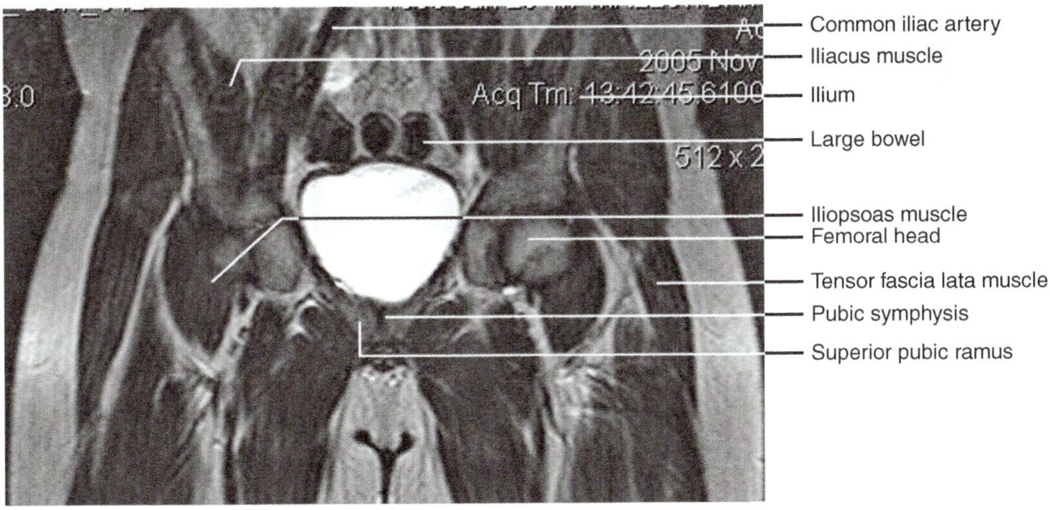

Fig. 26.10 Coronal T2

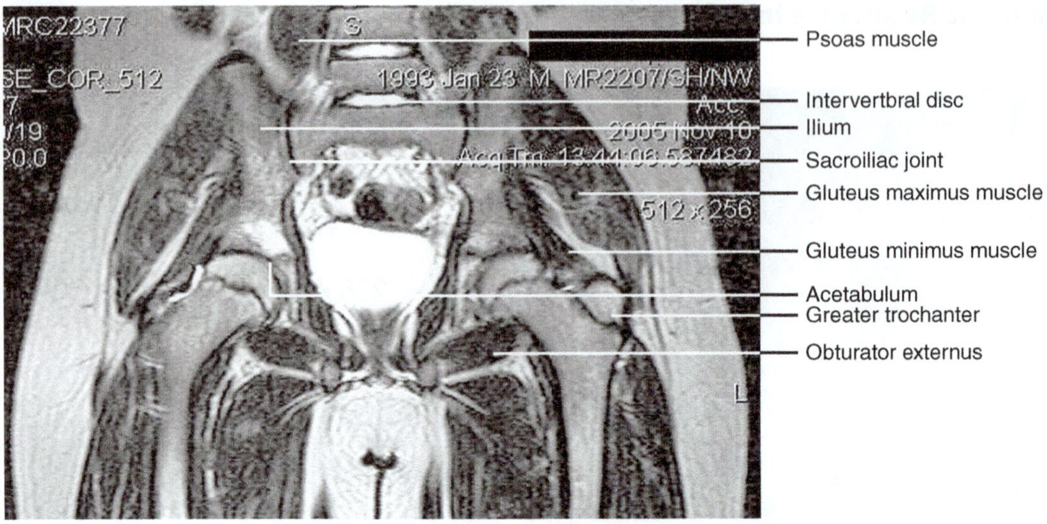

Fig. 26.11 Coronal T2

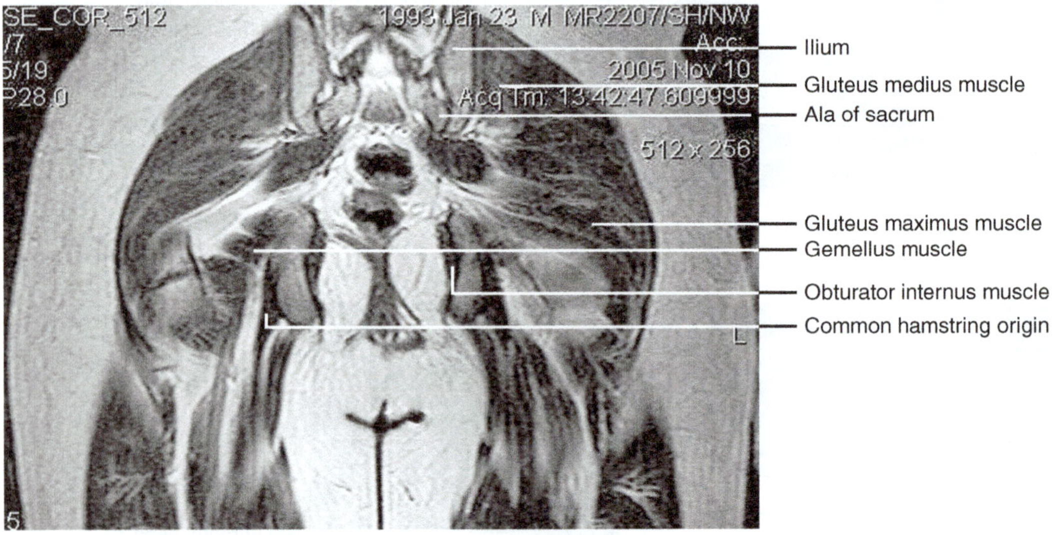

Fig. 26.12 Coronal T2

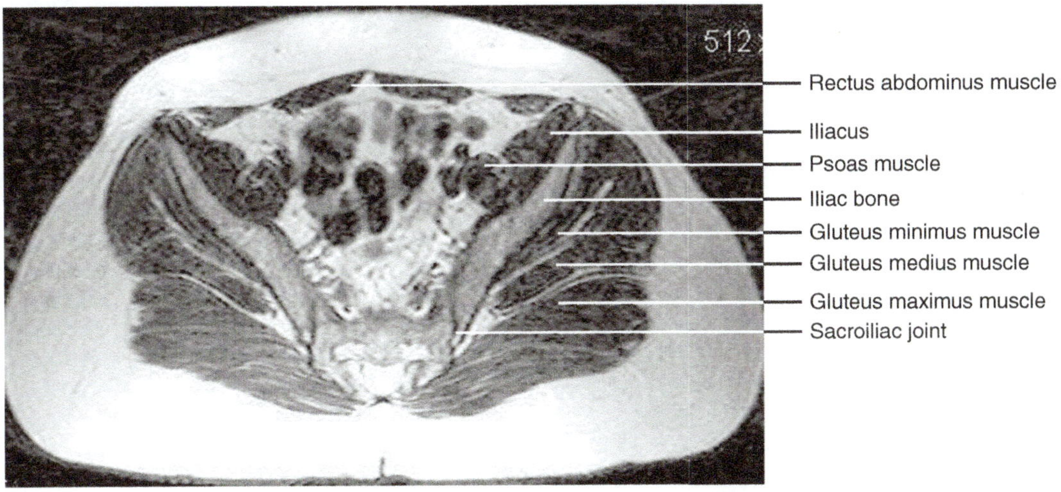

Fig. 26.13 Axial T2

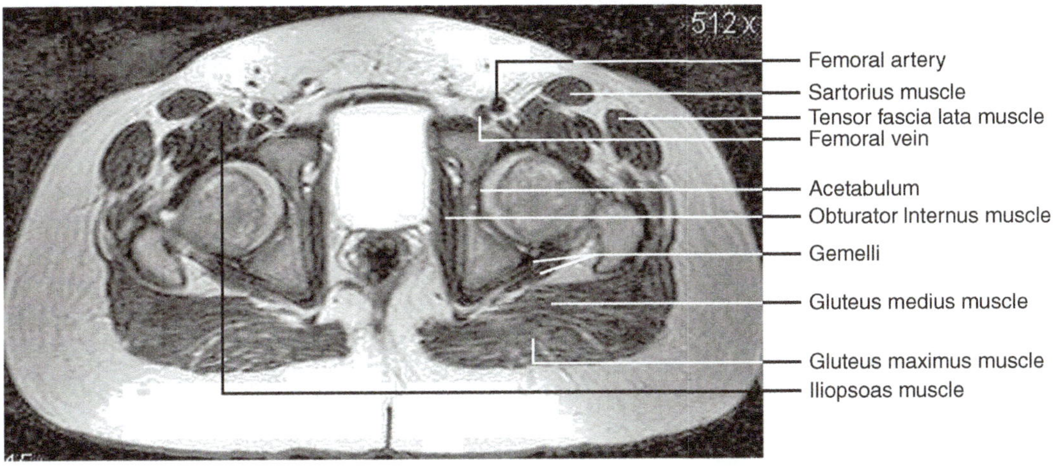

Fig. 26.14 Axial T2

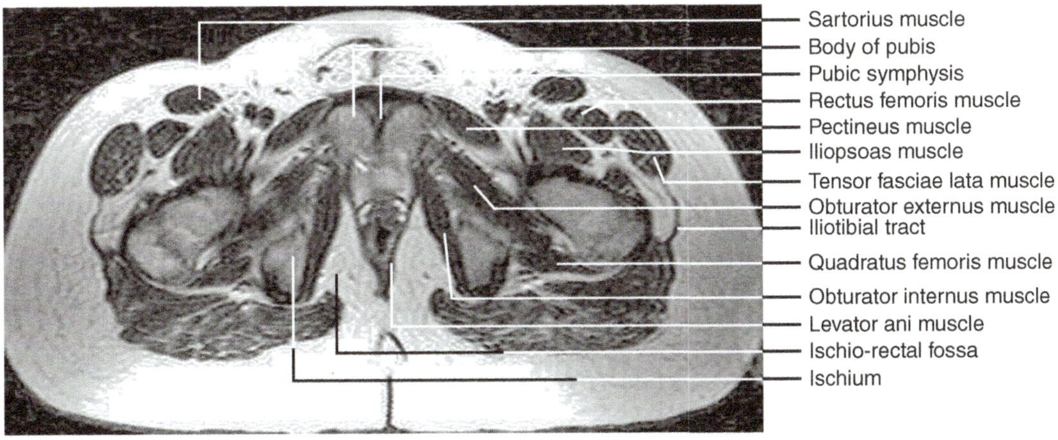

Fig. 26.15 Axial T2

27 Vertebral Column

The vertebral column is composed of cervical, thoracic and lumbo-sacral components. The atlas (C1), axis (C2) and sacrum have specific appearances but other vertebrae have features common to them. Evaluation should include the components of the vertebrae (body, end-plates, pedicles, laminae, spinous processes and marrow), the inter-vertebral disc components (nucleus pulposus and annulus fibrosus) and facet joint components, as well as the ligaments and muscles. This chapter does not include the contents of the spinal canal and neural foramina, which are dealt with in the chapter on the Spinal Cord. Plain radiographs and CT are best suited to evaluating the bony components while MRI is best suited to demonstrating the inter-vertebral discs, ligaments, muscles and marrow.

S. Andronikou (ed.), *See Right Through Me*,
DOI 10.1007/978-3-642-23893-2_27, © Springer-Verlag Berlin Heidelberg 2012

Cervical Spine

Plain Film

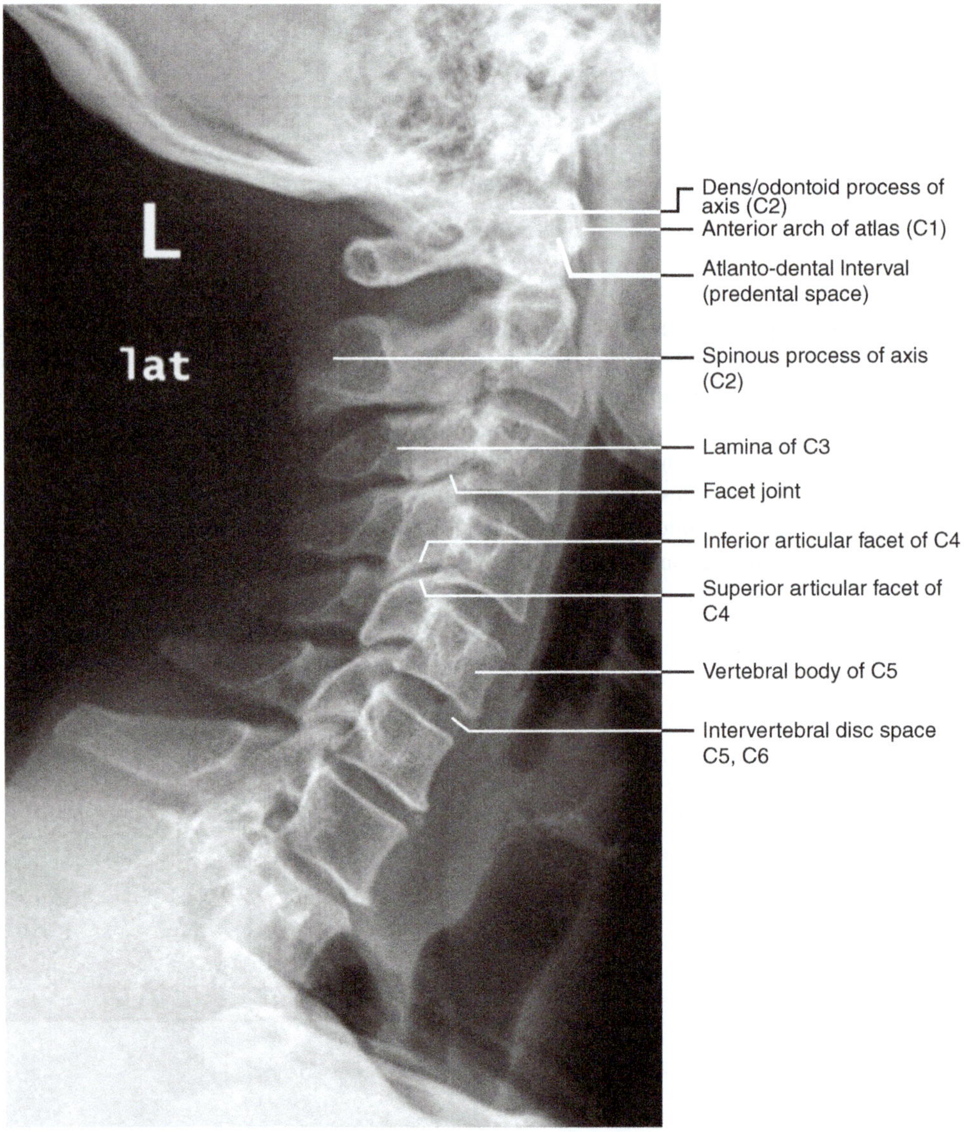

Fig. 27.1 Lateral view of the cervical spine

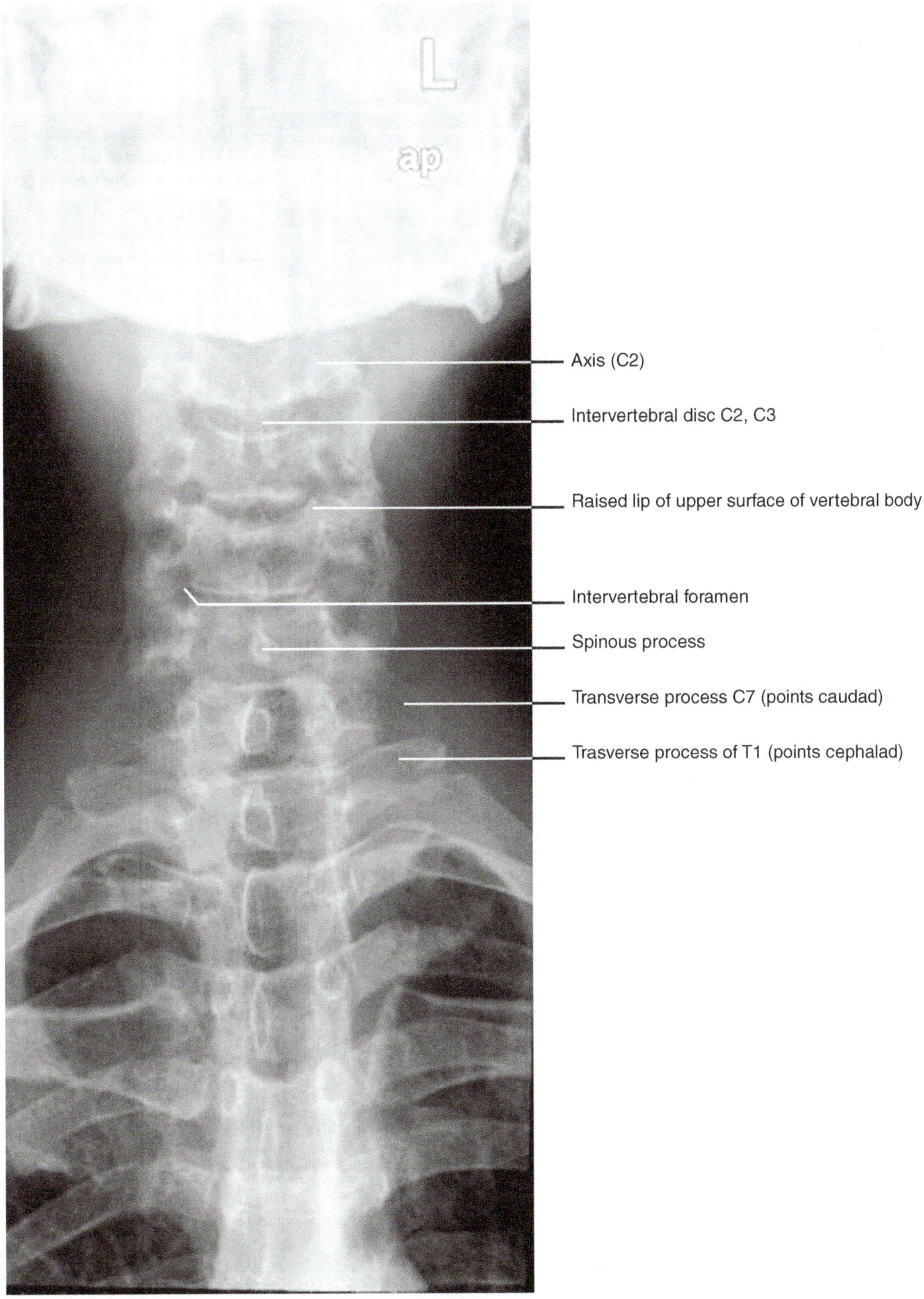

Fig. 27.2 Antero-posterior view of the cervical spine

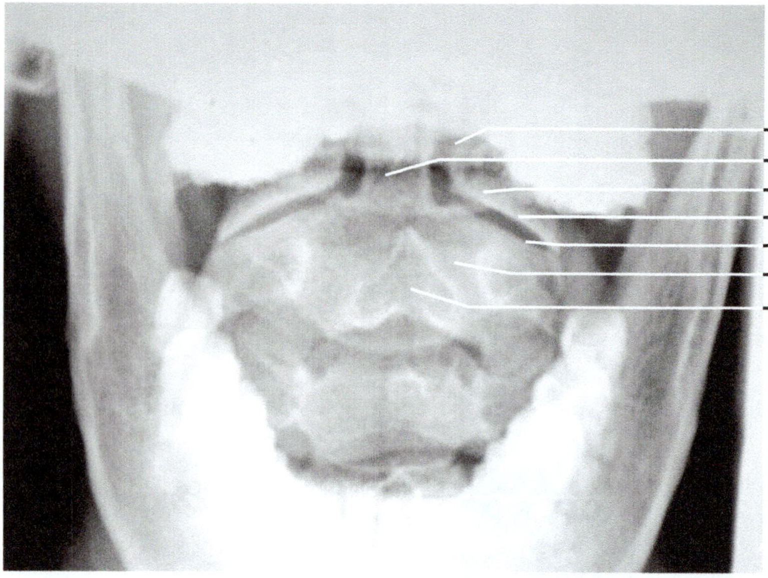

Fig. 27.3 Open mouth view demonstrating the odontoid process

Pearl

Lateral view:

The pre-dental space (atlanto-dental interval) in an adult ≤3 mm.

Pre-vertebral soft tissue in an adult should measure ≤5 mm at C2 and ≤22 mm at C6.

Open mouth view:

There should be ≤2 mm between the lateral margins of the dens and medial margins of the lateral masses of C1.

There should be ≤2 mm unilateral or bilateral atlanto-axial overlap on the open mouth view.

Lateral view:

Power's ratio = BA (basion-atlas)/OA (opisthion-atlas) should be less than 1.

BD (basion-dens) interval should be ≤12 mm.

AP view:

The first thoracic vertebra may be identified by its cephalad-pointing transverse processes on the antero-posterior view as opposed to the caudally inclined cervical transverse processes.

Computed Tomography

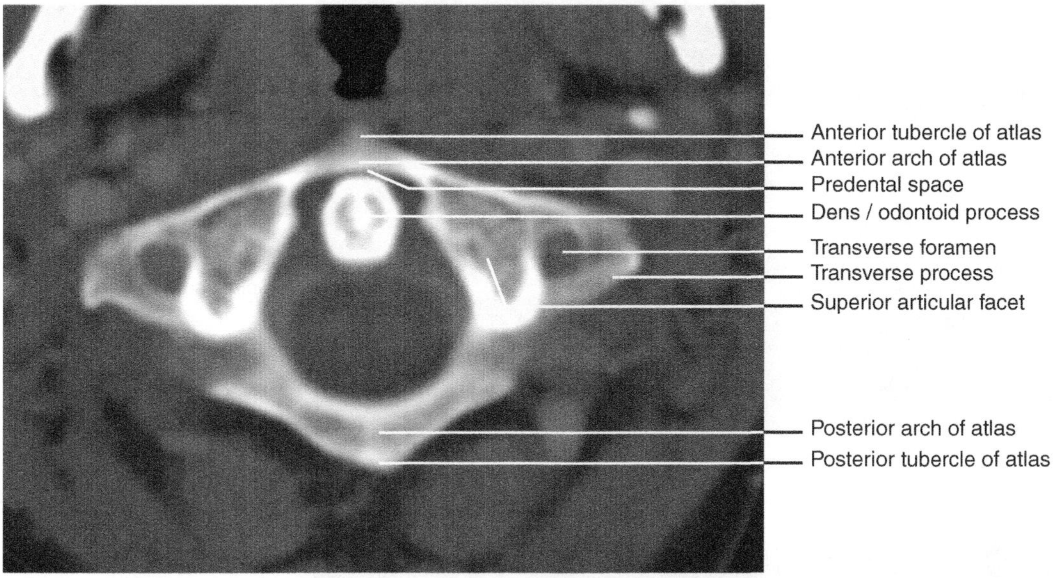

Fig. 27.4 Axial CT scan of the atlas

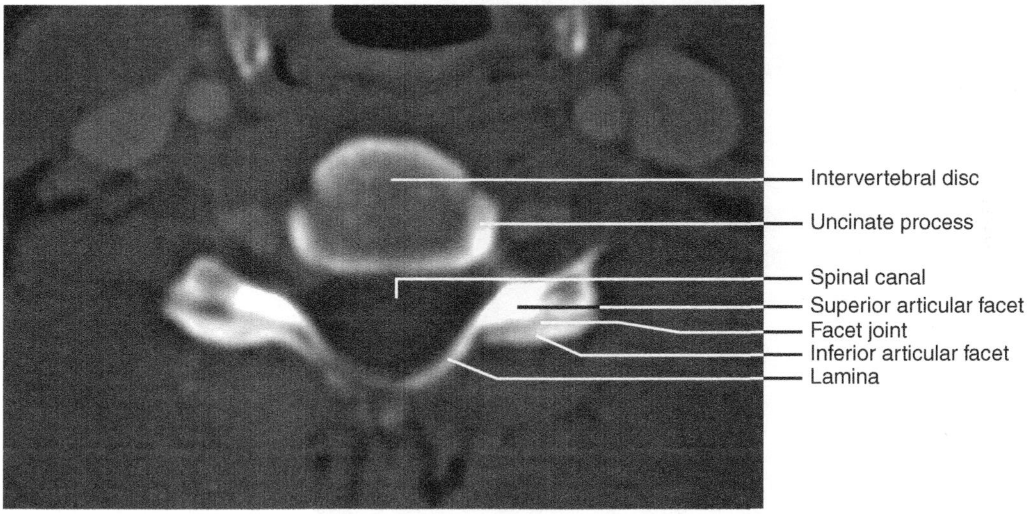

Fig. 27.5 Axial CT scan of a typical cervical vertebra at the intervertebral disc level

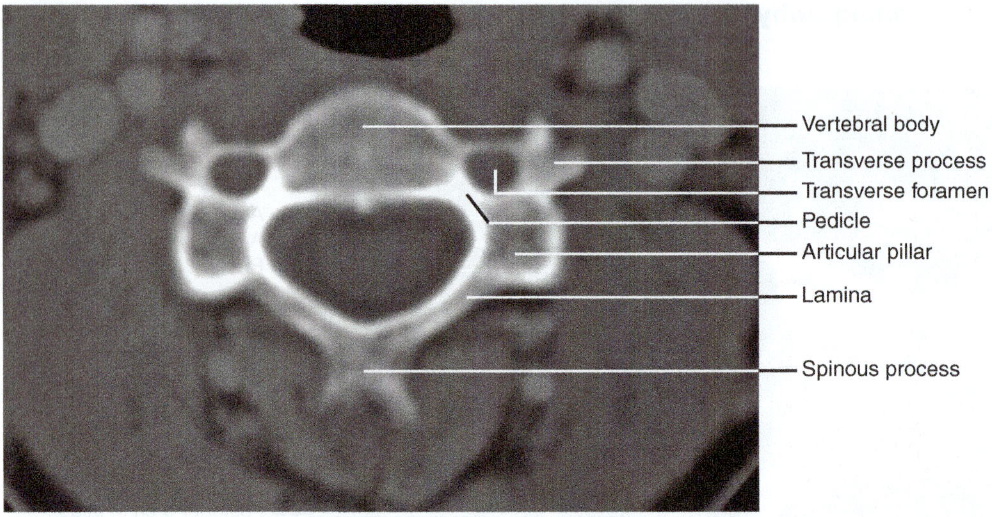

Fig. 27.6 Axial CT scan of a typical cervical vertebra at the level of the vertebral body

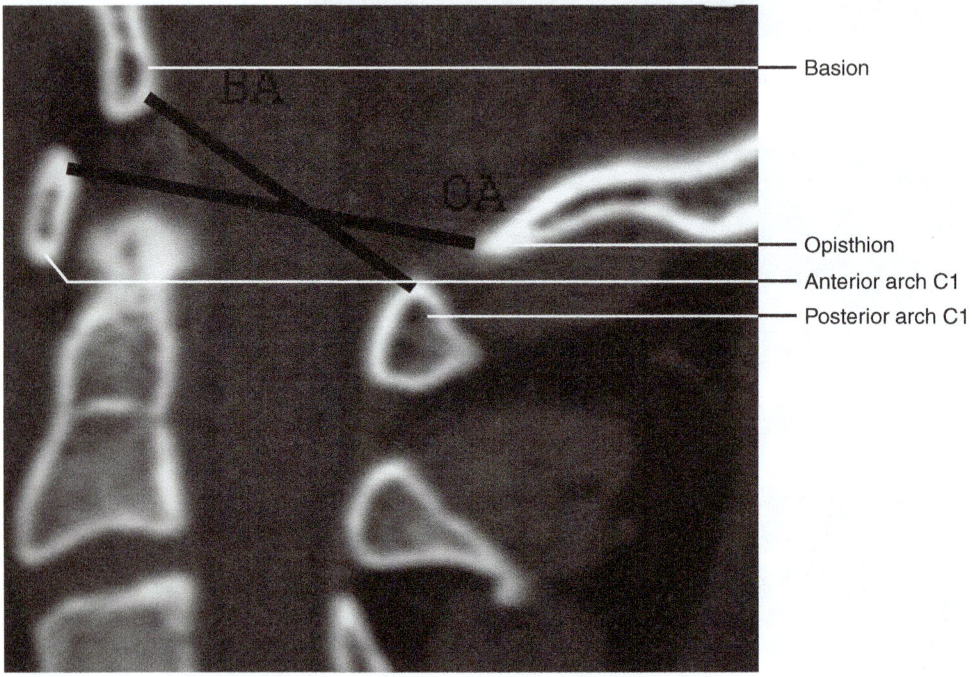

Fig. 27.7 Power's ratio

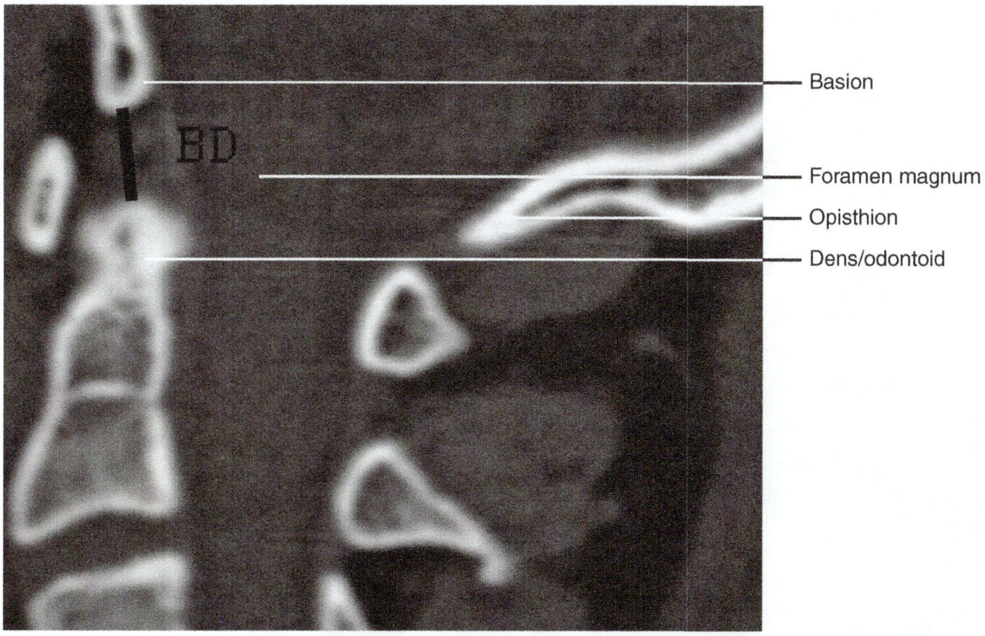

Fig. 27.8 BD interval

Magnetic Resonance Imaging

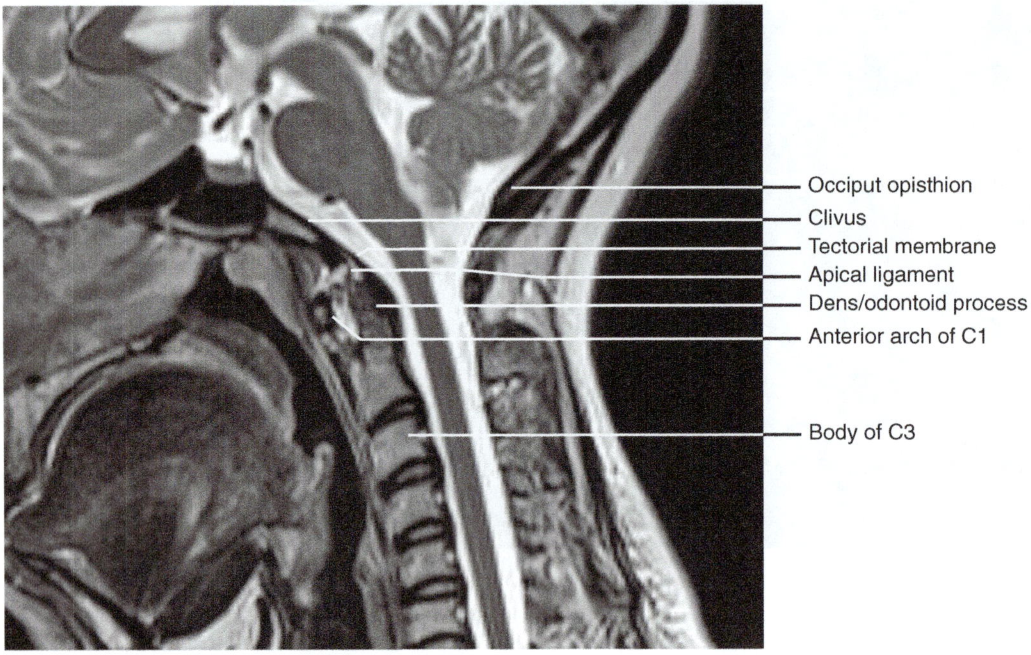

Fig. 27.9 T2 sagittal view of the cranio-cervical junction

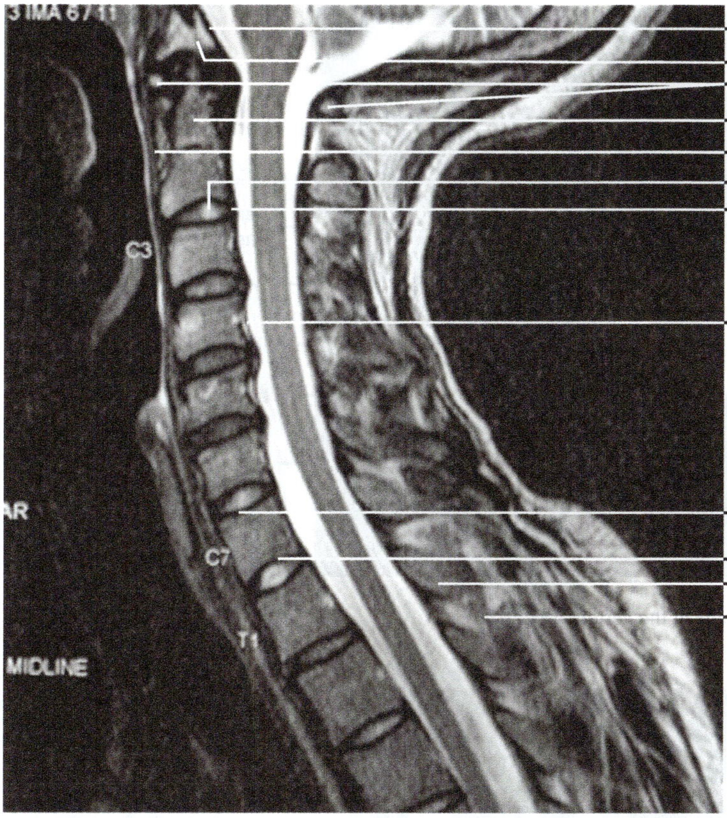

Fig. 27.10 T2 sagittal of the cervical spine

Pearl

The posterior longitudinal ligament continues cranially as the tectorial membrane to attach to the clivus.

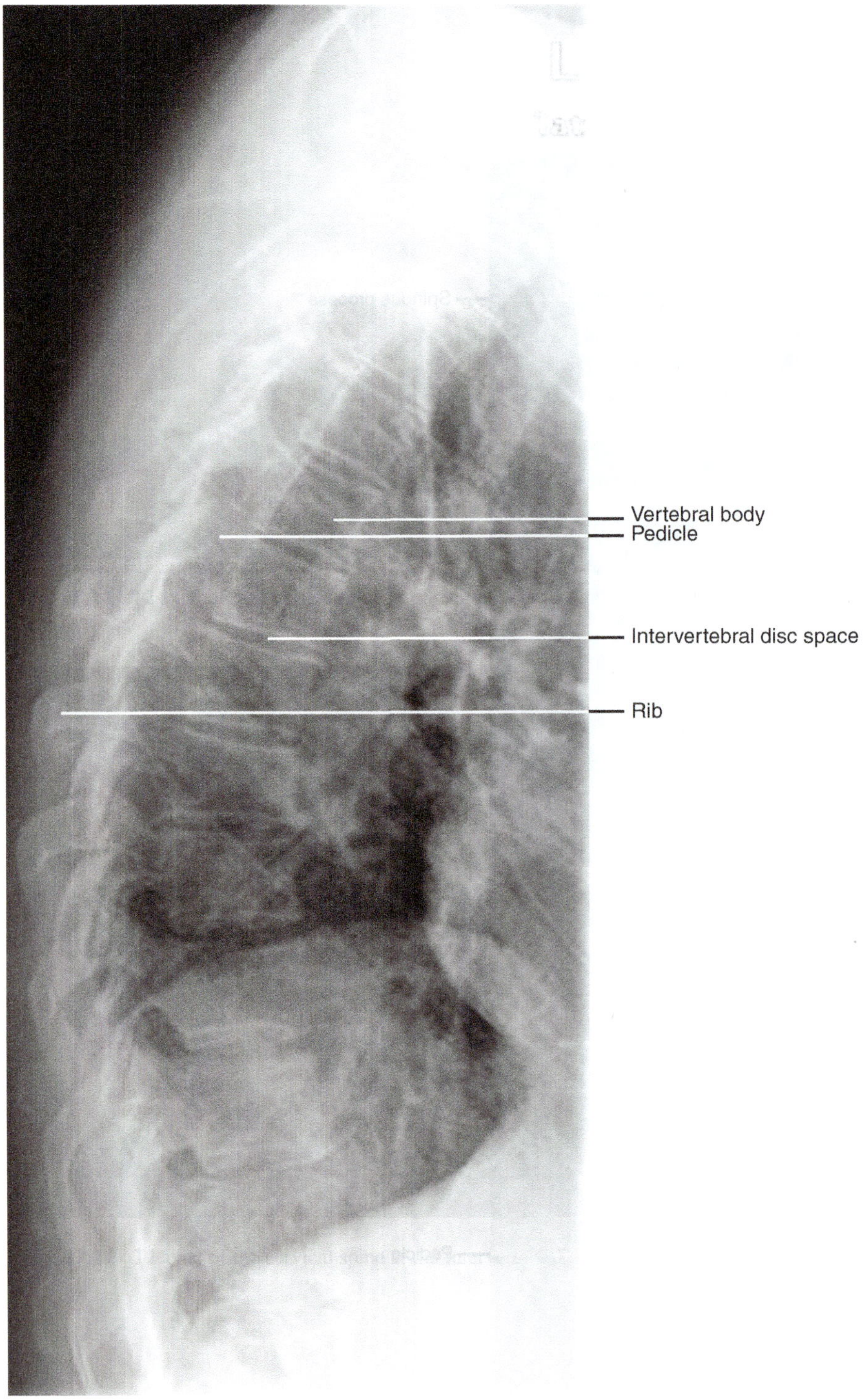

Fig. 27.14 Lateral view of the thoracic spine

Computed Tomography

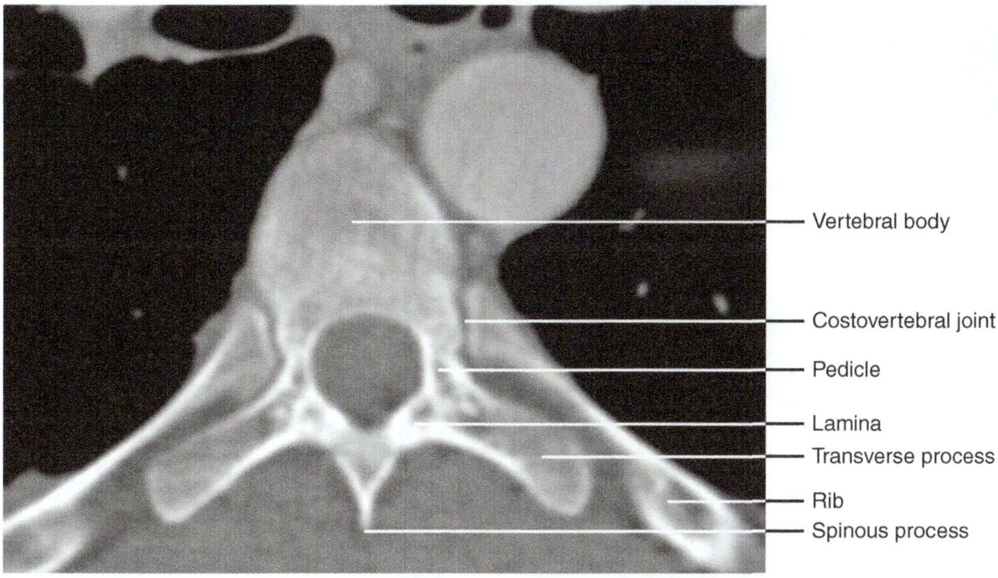

Fig. 27.15 Axial CT scan of a typical thoracic vertebra at the vertebral body level

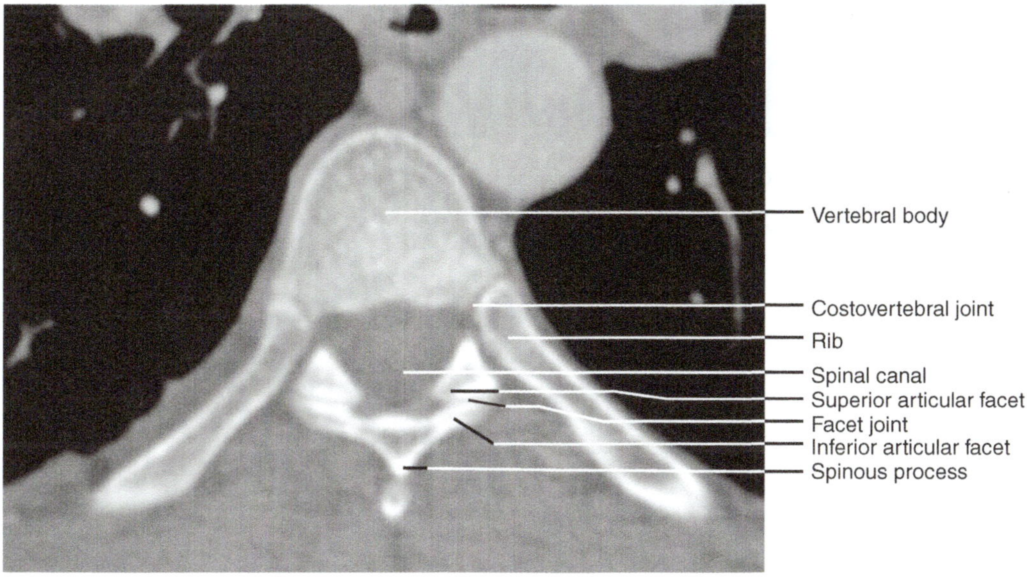

Fig. 27.16 Axial CT scan of a typical thoracic vertebra at the level of the facet joint

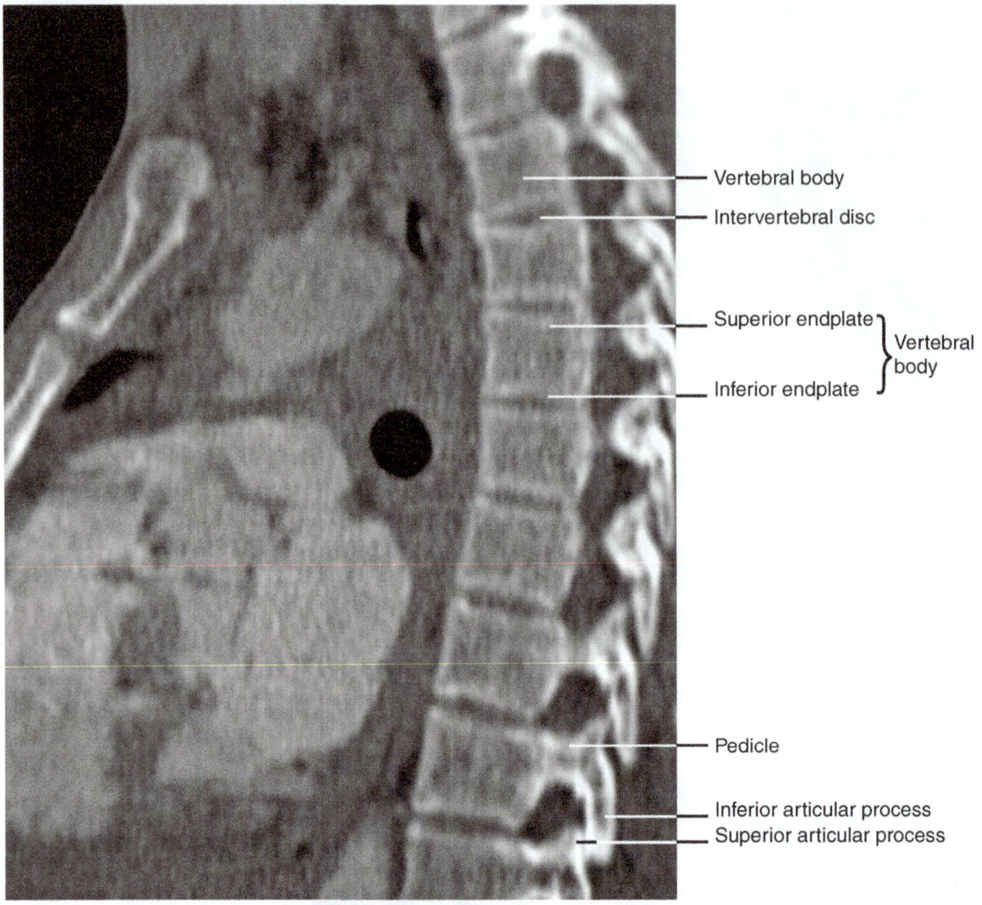

Fig. 27.17 Sagittal reformatted CT scan of the thoracic spine

Lumbar Spine

Plain Film

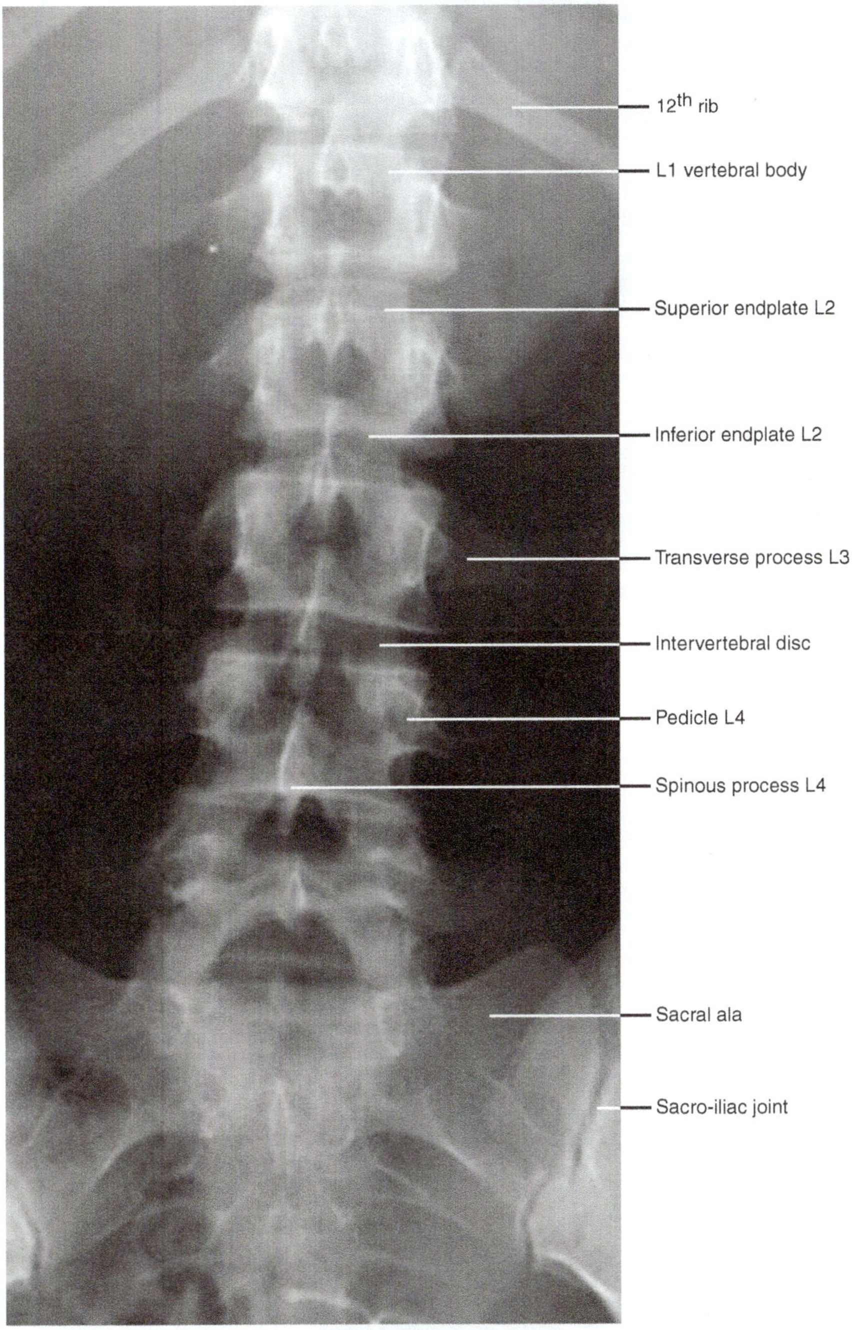

Fig. 27.18 Antero-posterior view of the lumbar spine

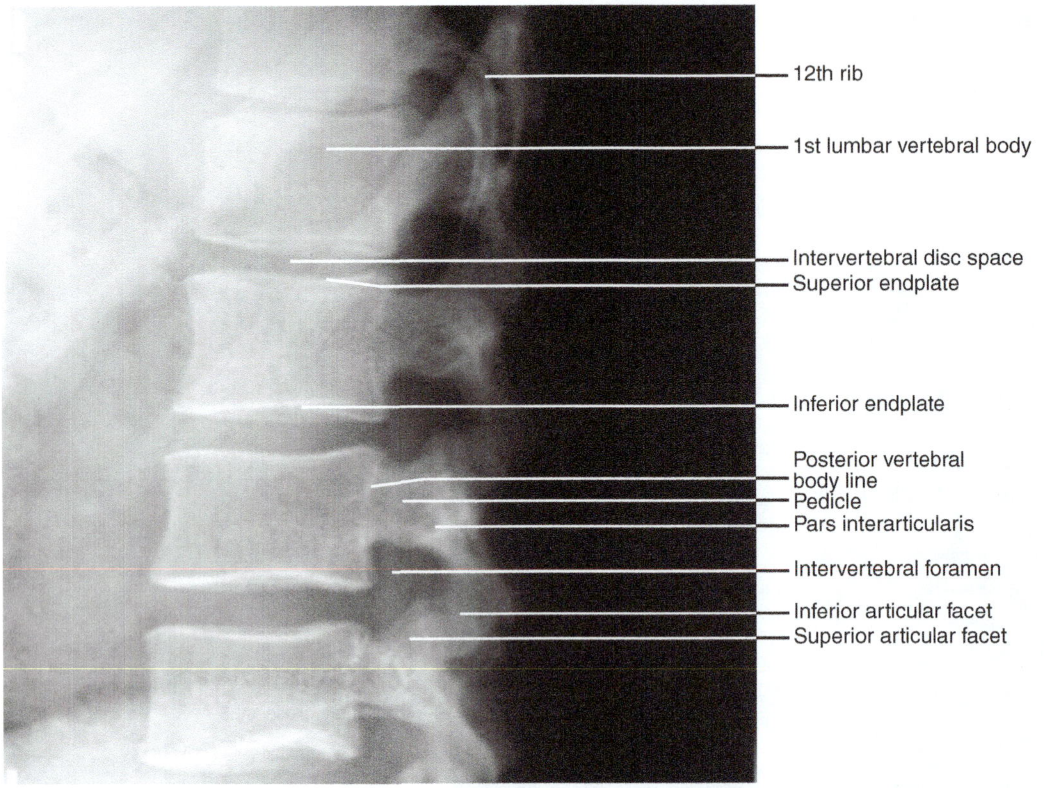

Fig. 27.19 Lateral view of the lumbar spine

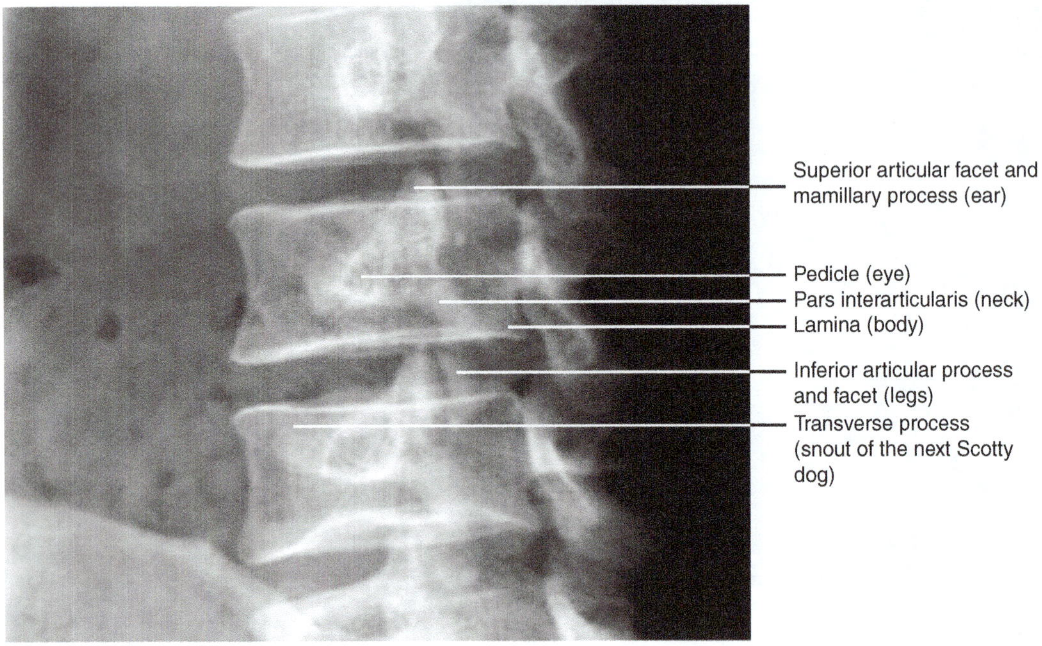

Fig. 27.20 Oblique ('Scotty Dog') view of the lumbar spine

Computed Tomography

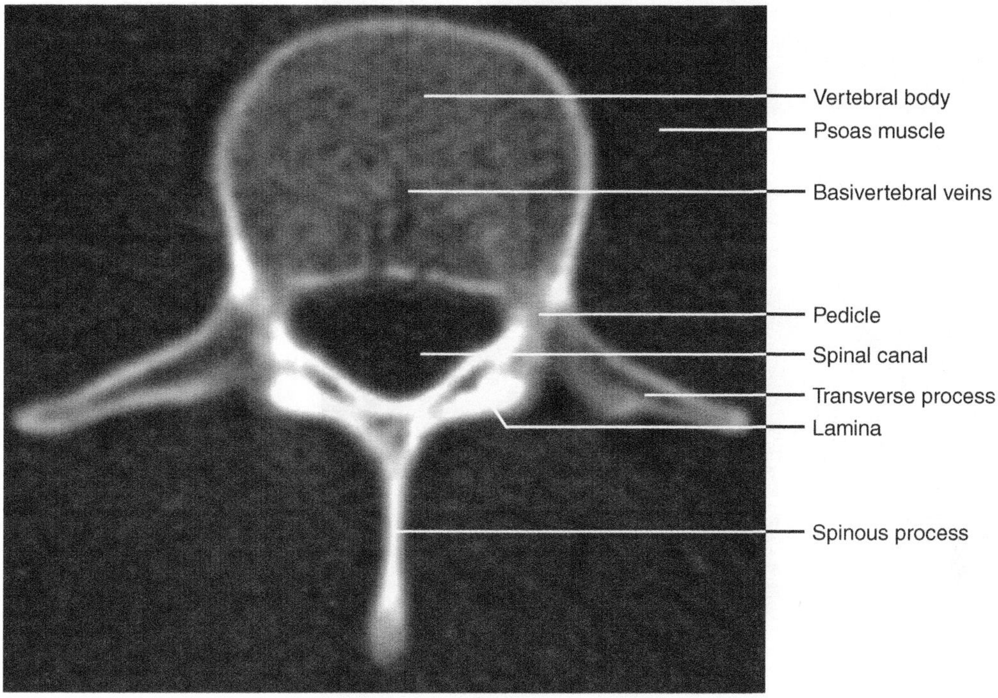

Fig. 27.21 Axial CT scan of a typical lumbar vertebra at the level of the vertebral body

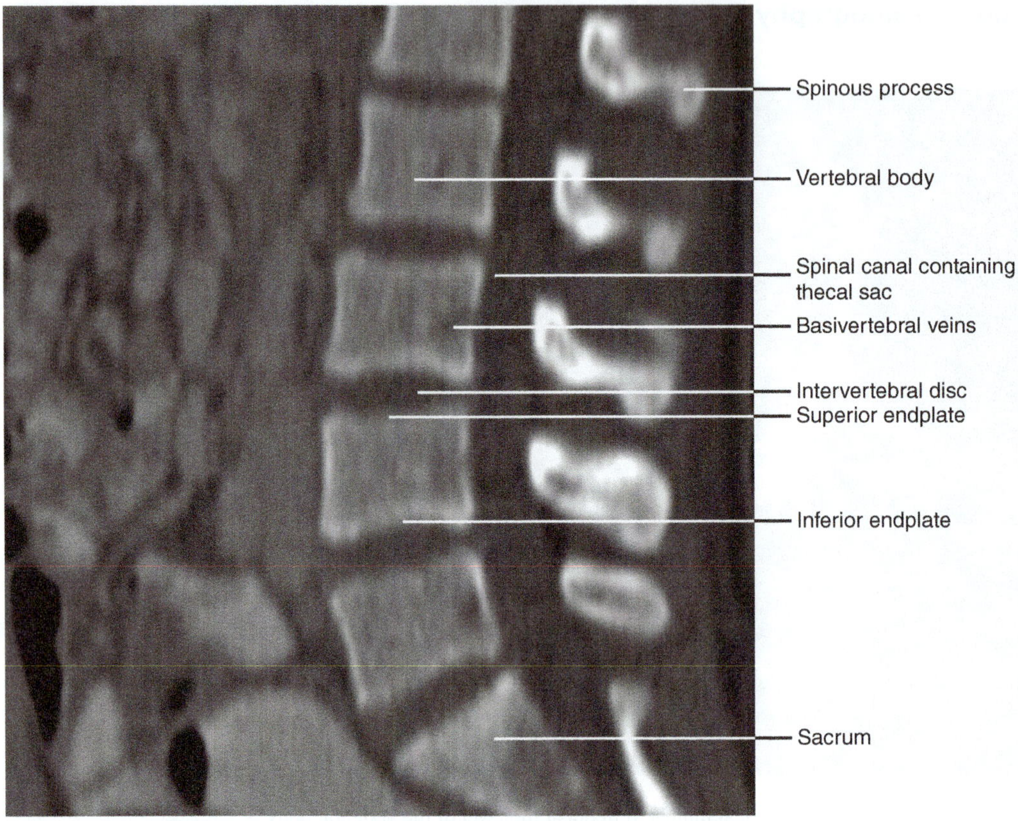

Fig. 27.22 Sagittal reformatted CT scan of the lumbar spine

Magnetic Resonance Imaging

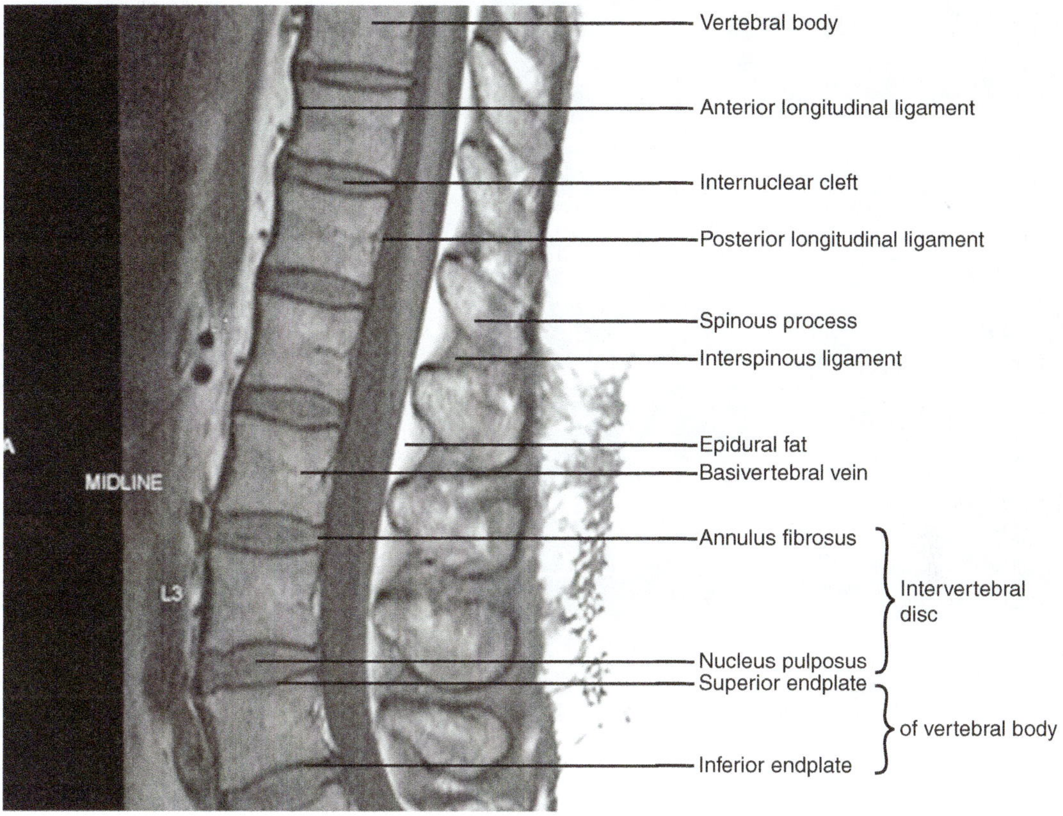

Fig. 27.23 T1 sagittal of the lumbar spine

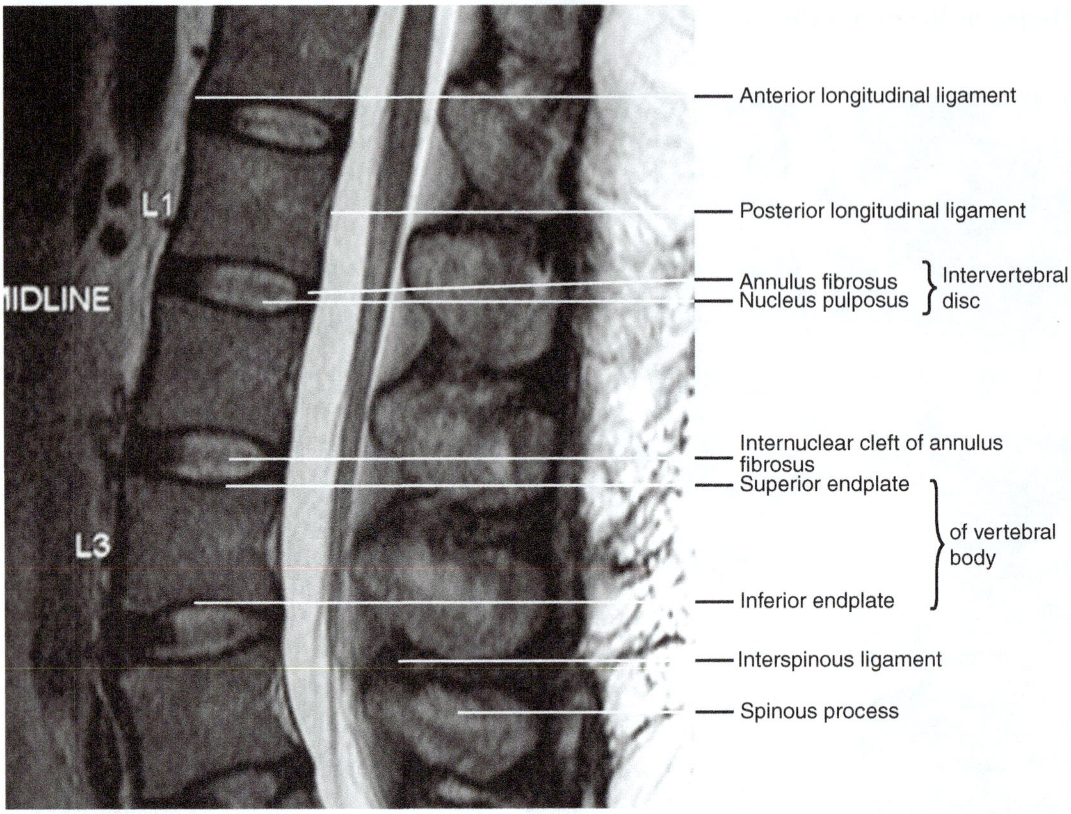

Fig. 27.24 T2 sagittal of the lumbar spine

Pearl

Intervertebral hydration is best assessed on T2 sequences given the water content of the gelatinous nucleus pulposus.

A superior articular facet has a posteriorly oriented articular surface while an inferior articular facet has an articular surface oriented anteriorly.

Normal fatty marrow is best evaluated on T1 sequences in which the marrow is of high signal intensity. On T2 weighted sequences the marrow is also of high signal but less so than on T1.

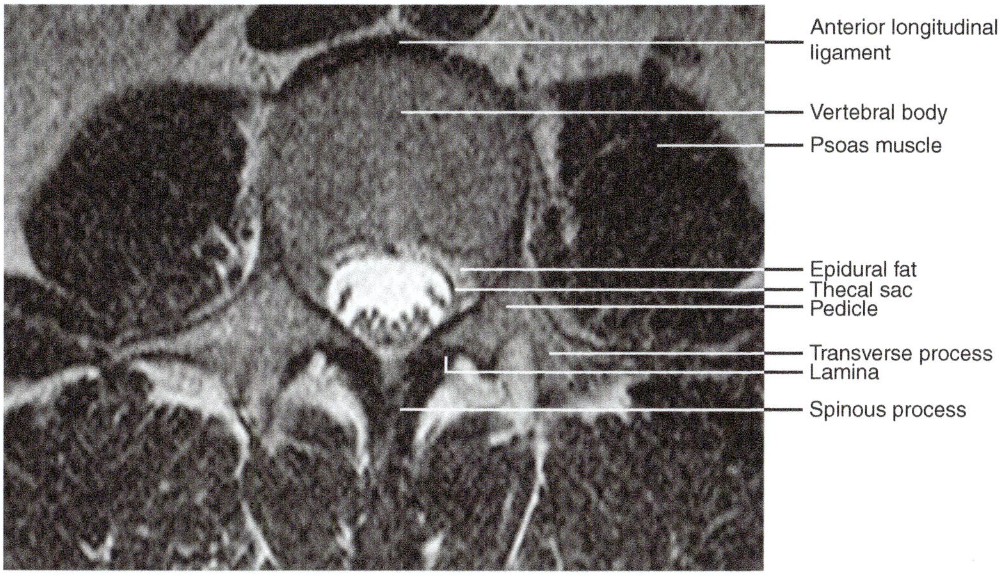

Fig. 27.25 T2 axial at the level of the L3 vertebral body

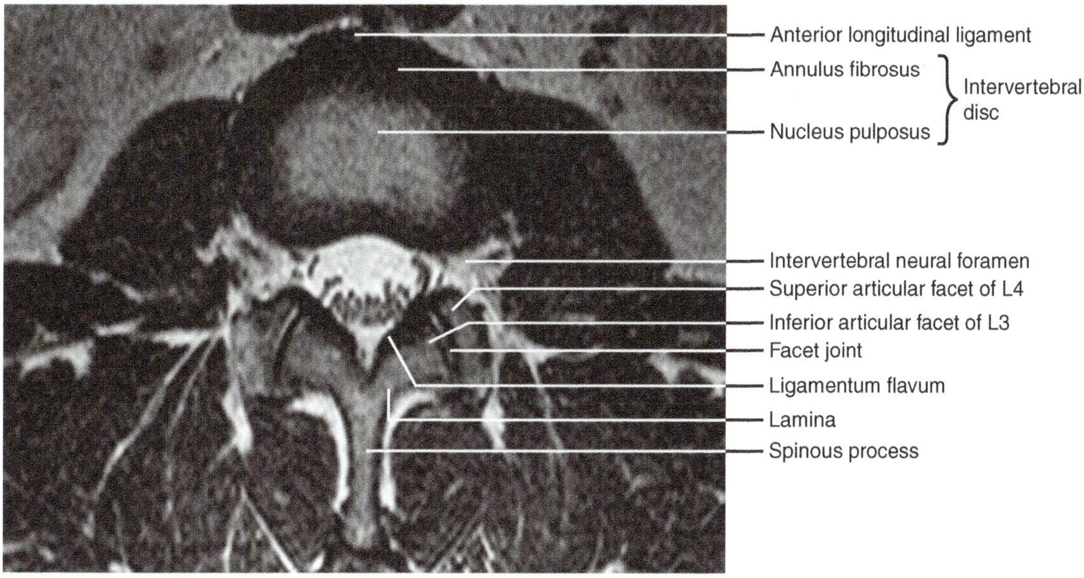

Fig. 27.26 T2 axial at the level of the L3/L4 intervertebral disc/facet joint

Part VI

Vascular System

This section demonstrates 'great vessel' anatomy distinctly from detailed peripheral vascular anatomy, to highlight the appearances, main branches and sizes of the components of the aorta, pulmonary vessels and large veins returning blood to the heart. The section on 'peripheral vessels' demonstrates detailed anatomy of the vessels of the limbs. Vessels of the brain are demonstrated in the chapter on the 'central nervous system' and neck vessels are demonstrated in the 'head and neck chapter'. Angiography includes conventional catheter angiography with and without digital subtraction, CT angiography and MR angiography. There are specific advantages to each with the more invasive techniques demonstrating the most detail and allowing intervention to take place, and the least invasive resulting in lower quality imaging without any risk to the patient.

Great Vessels

28

Conventional Angiography

Digital Subtraction Angiography (DSA)

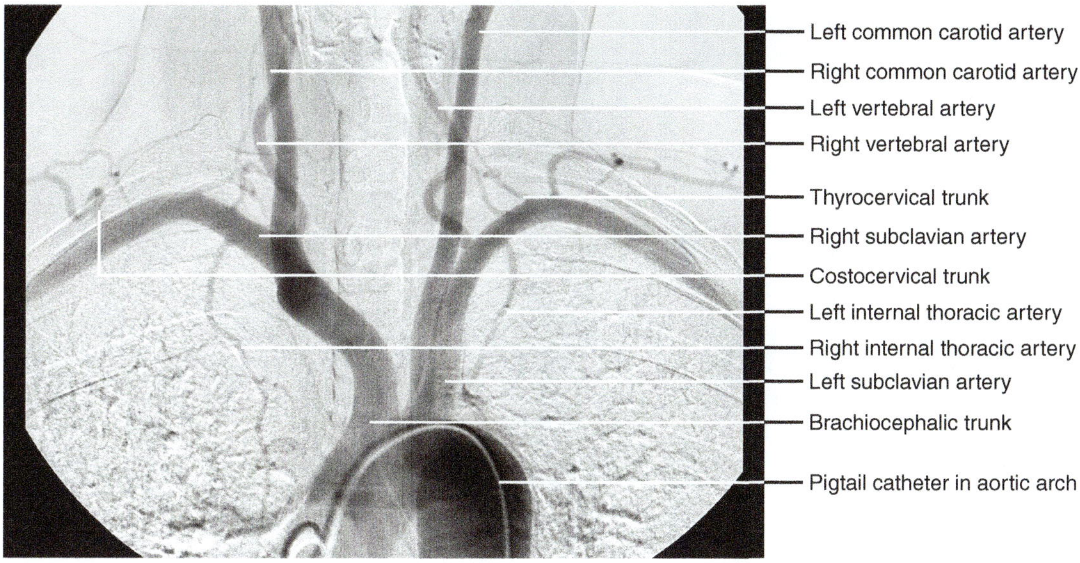

Fig. 28.1 Arch angiogram

S. Andronikou (ed.), *See Right Through Me*,
DOI 10.1007/978-3-642-23893-2_28,

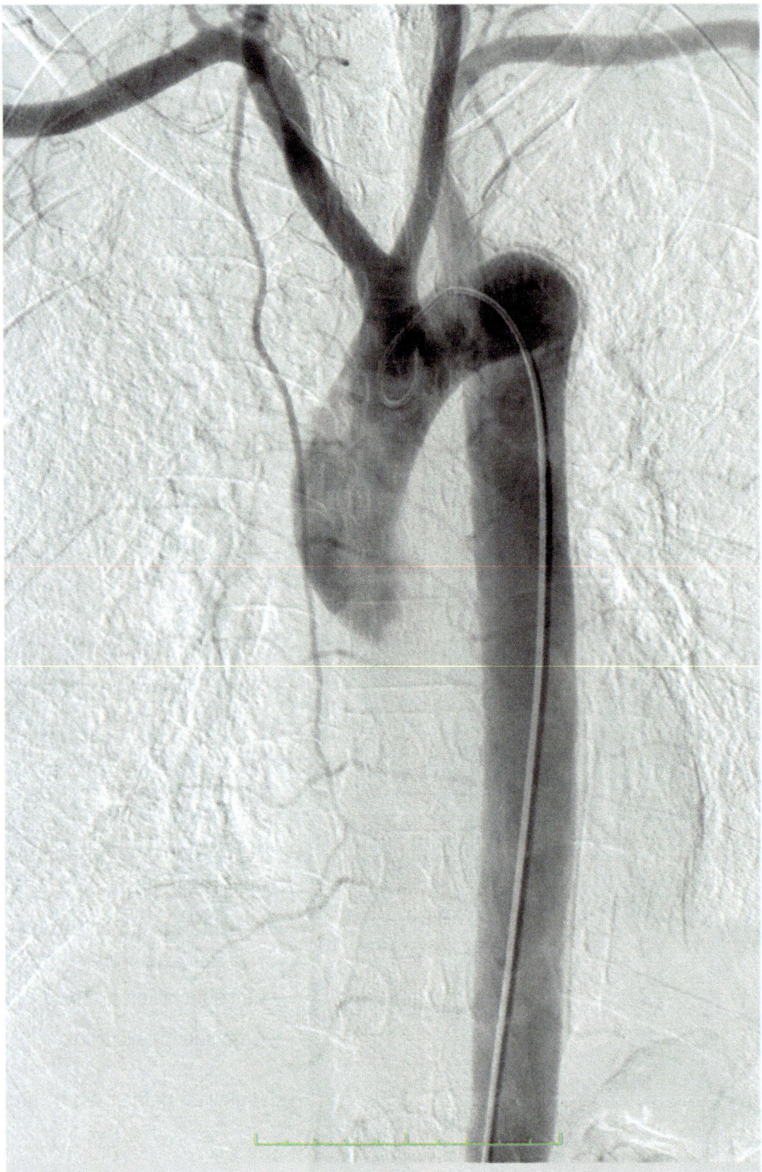

Fig. 28.2 Digital subtraction angiography demonstrating the most common aortic arch variant: the bovine arch

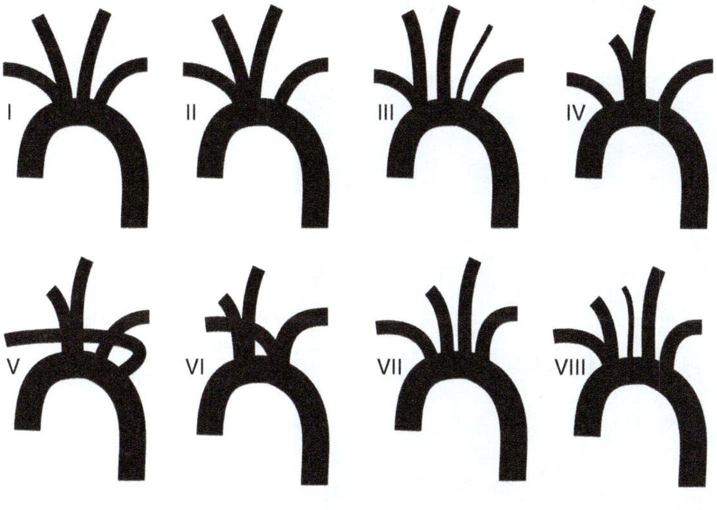

Fig. 28.3 Schematic representation of the aortic arch variants

Vascular Pearls

Common anatomical variants of the aortic arch and branches:

- **Type I:** The "normal" aortic arch giving rise to three branches: (1) the brachiocephalic, which then branches into the R subclavian and R common carotid, (2) the L common carotid and (3) the L subclavian arteries **(65–94 %)**
- **Type II (bovine arch):** The second most common pattern, including two branches of the aortic arch: (1) a common stem which is divided into the R subclavian, R common carotid and L common carotid and (2) the L subclavian **(15 %)**
- **Type III:** Left vertebral arising directly from the arch and not from the left subclavian, providing an aortic arch pattern with four branches: (1) brachiocephalic trunk, (2) L common carotid, (3) L vertebral and (4) L subclavian **(2.5–8 %)**
- **Type IV:** Common origin of the common carotids (arch with three branches): (1) R subclavian, (2) common stem for the R common carotid and the L common carotid (bicarotid trunk) and (3) L subclavian **(<1 %)**
- **Type V:** Common origin of the common carotids and aberrant R subclavian (arch with three branches): (1) common stem for the R common carotid and the L common carotid (bicarotid trunk), (2) L subclavian and (3) R subclavian as the last branch of the arch **(<1 %)**
- **Type VI:** Common origin of the common carotids and common origin of the subclavians (arch with two branches-avian form): (1) common stem for the R common carotid and the L common carotid and (2) common stem for the R subclavian and the L subclavian **(<1 %)**
- **Type VII:** Absence of the brachiocephalic trunk (arch with four branches): (1) R subclavian, (2) R common carotid, (3) L common carotid and (4) L subclavian **(<1 %)**
- **Type VIII:** An additional branch from the aortic arch—the thyroidea ima artery (arch with four branches): (1) brachiocephalic trunk, (2) thyroidea ima artery (3) L common carotid and (4) L subclavian **(<1 %)**

Digital Subtraction Angiogram Demonstrating the Bronchial Arteries

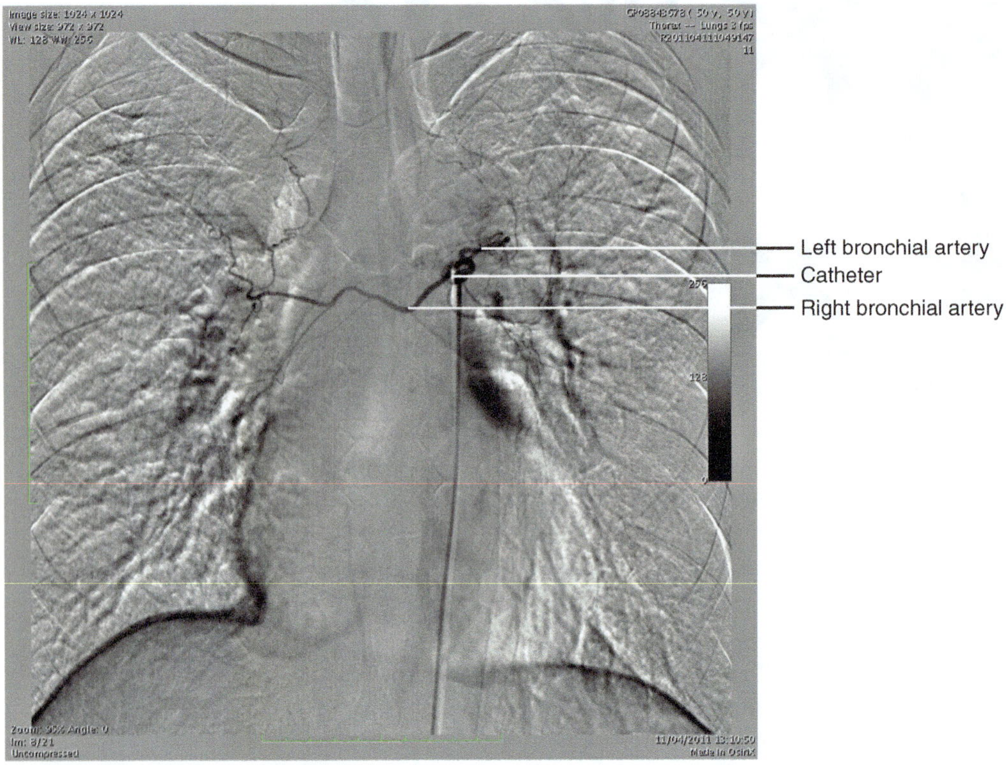

Fig. 28.4 Bronchial arteries-branches of the aorta

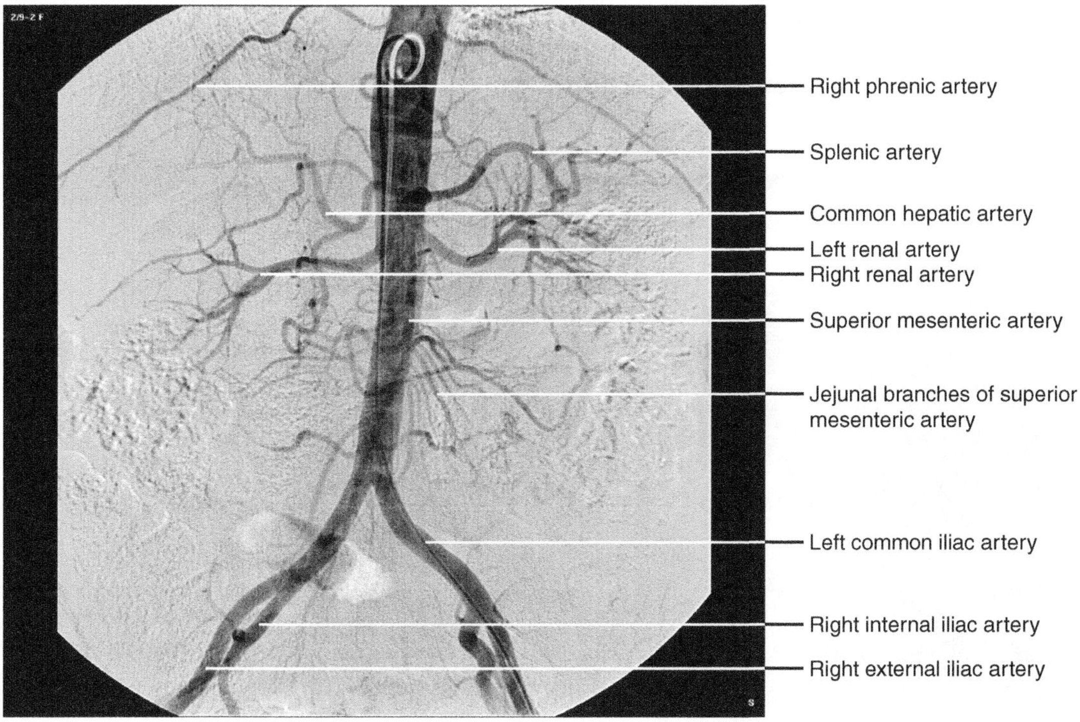

Fig. 28.5 Abdominal aortic angiogram

Multidetector CT Angiography

Coronal Maximum Intensity Projection

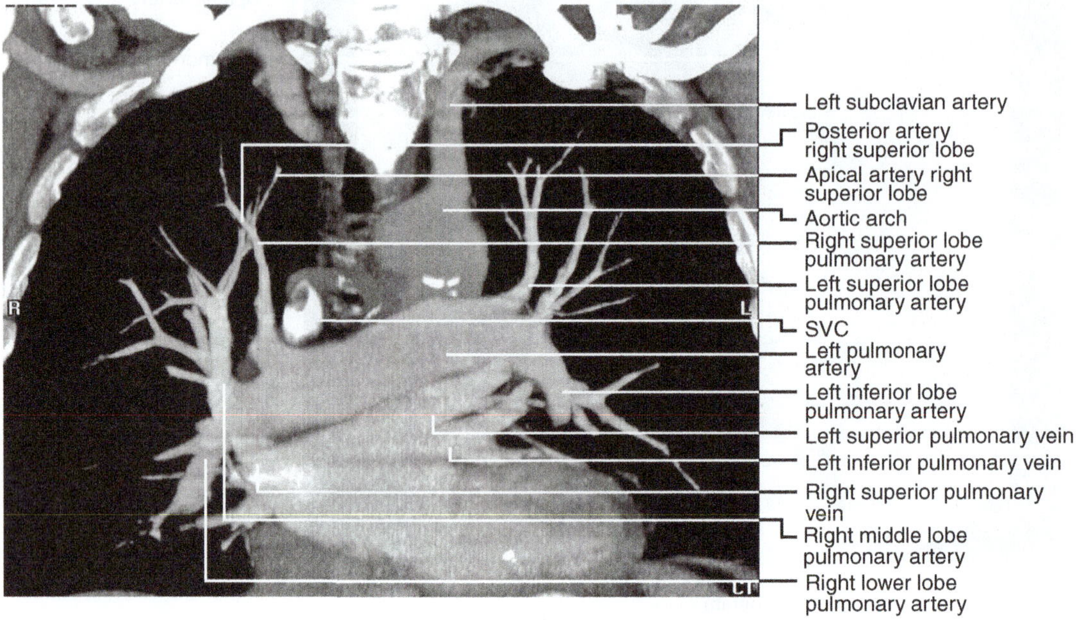

Fig. 28.6 Coronal CT reconstruction in the plane of pulmonary vessels

3D Volume-Rendered CT

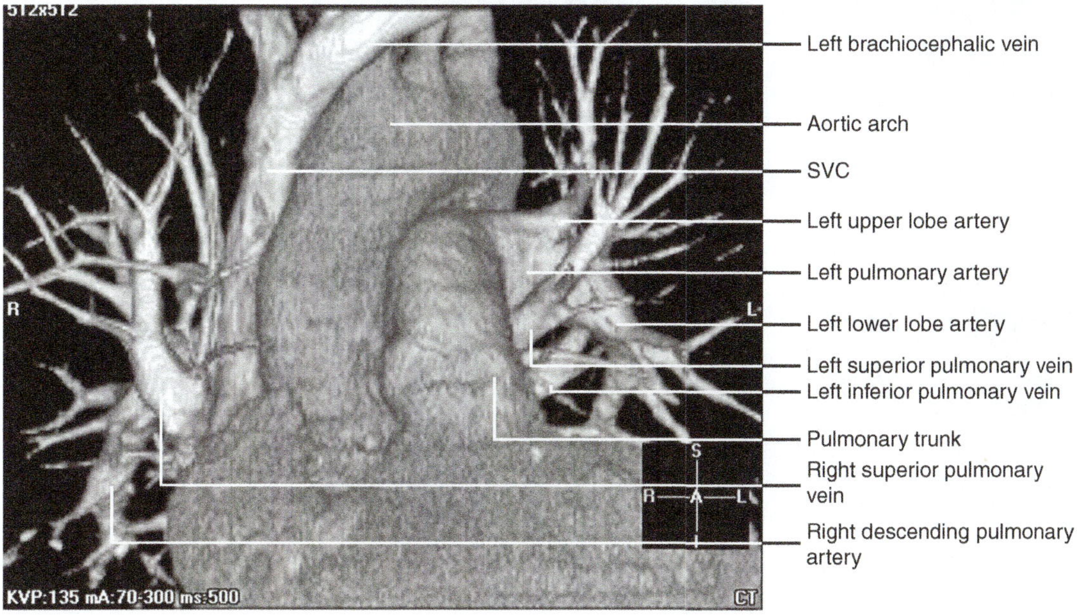

Fig. 28.7 Anterior projection

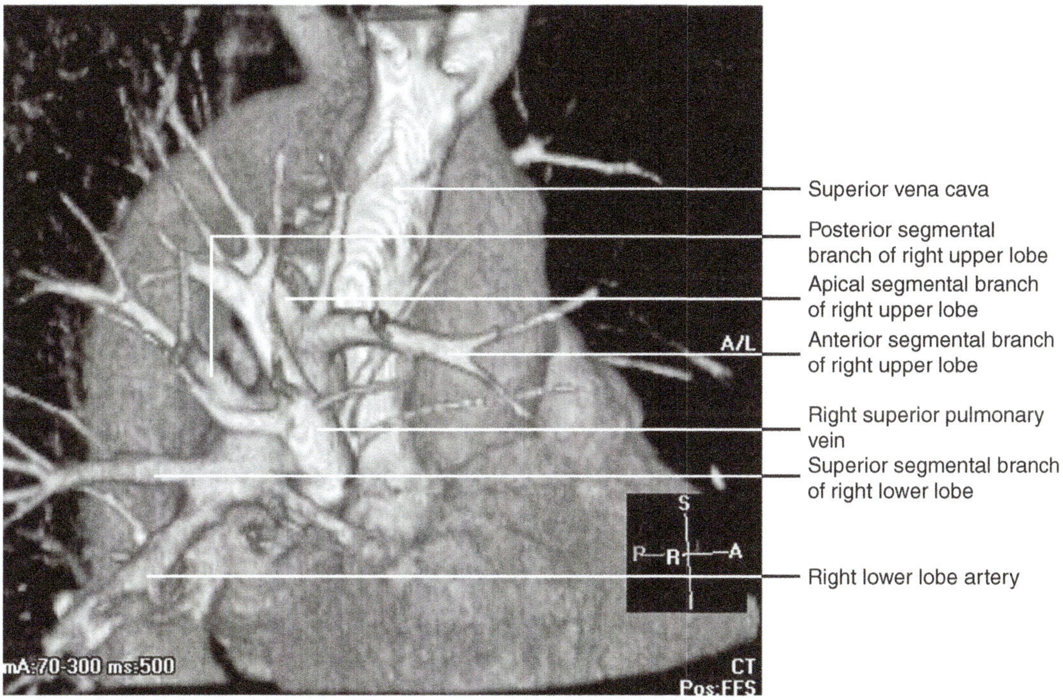

Fig. 28.8 Right anterior oblique projection

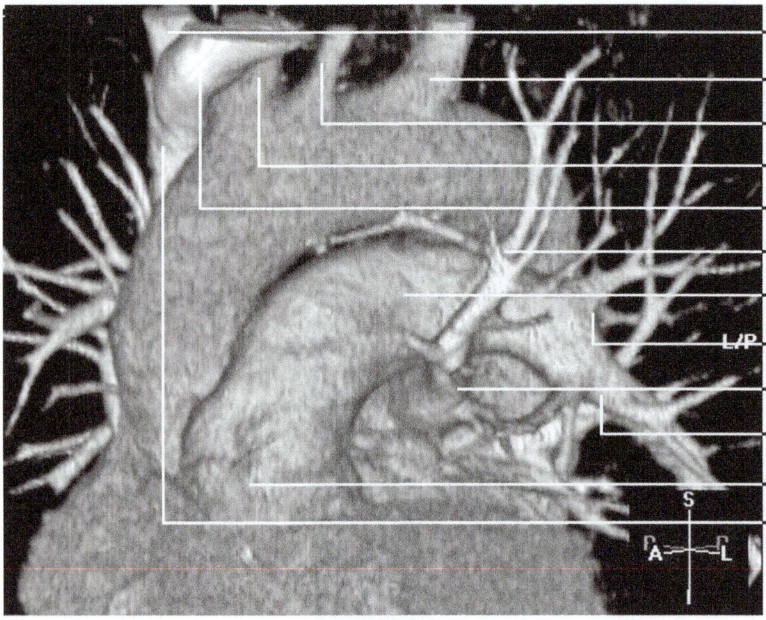

Fig. 28.9 Left anterior oblique projection

Magnetic Resonance Angiography

Contrast-Enhanced MRA

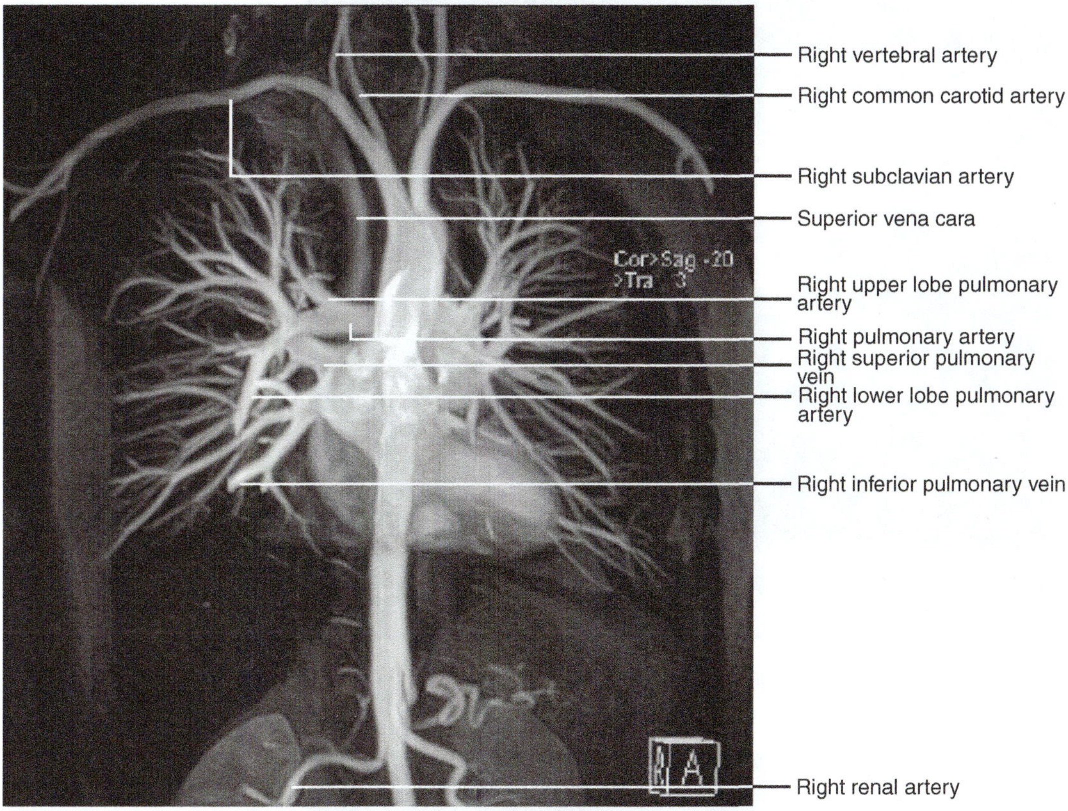

Fig. 28.10 Chest and upper abdomen

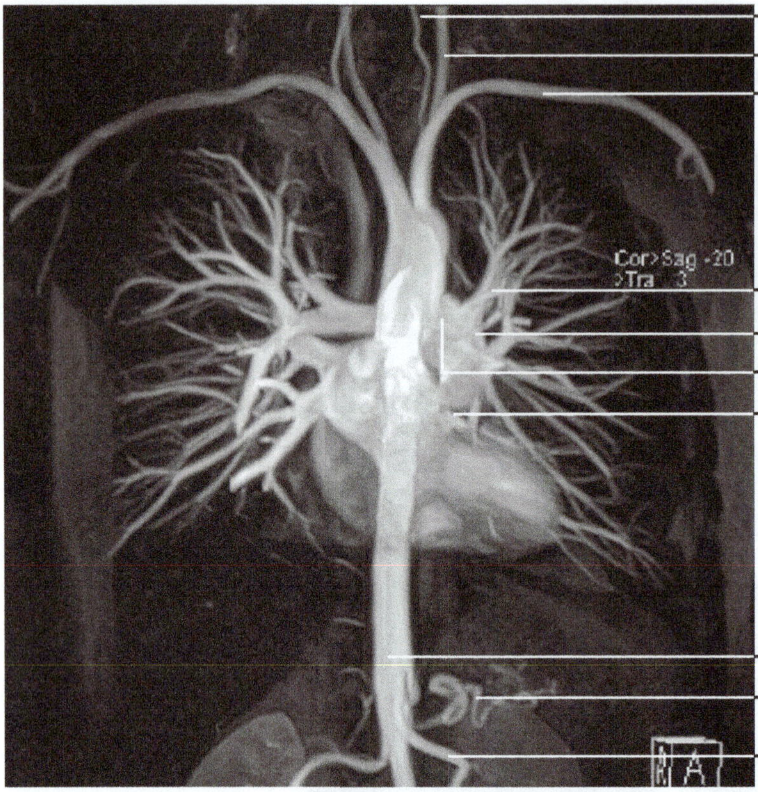

Fig. 28.11 Chest and upper abdomen

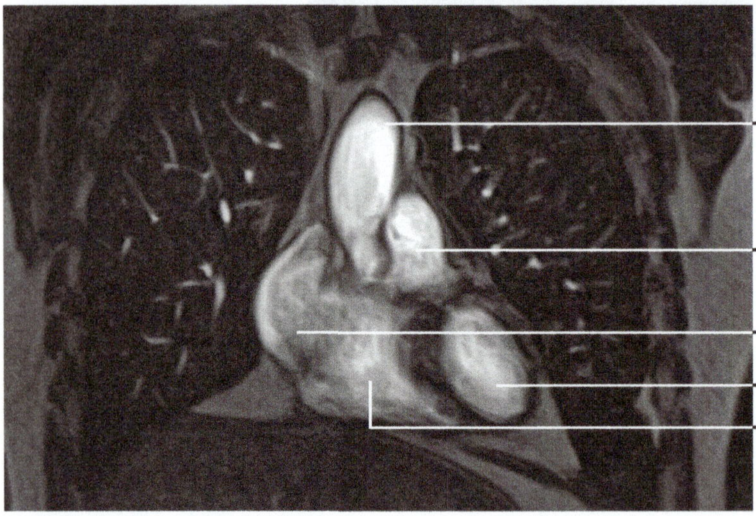

Fig. 28.12 Thoracic source image

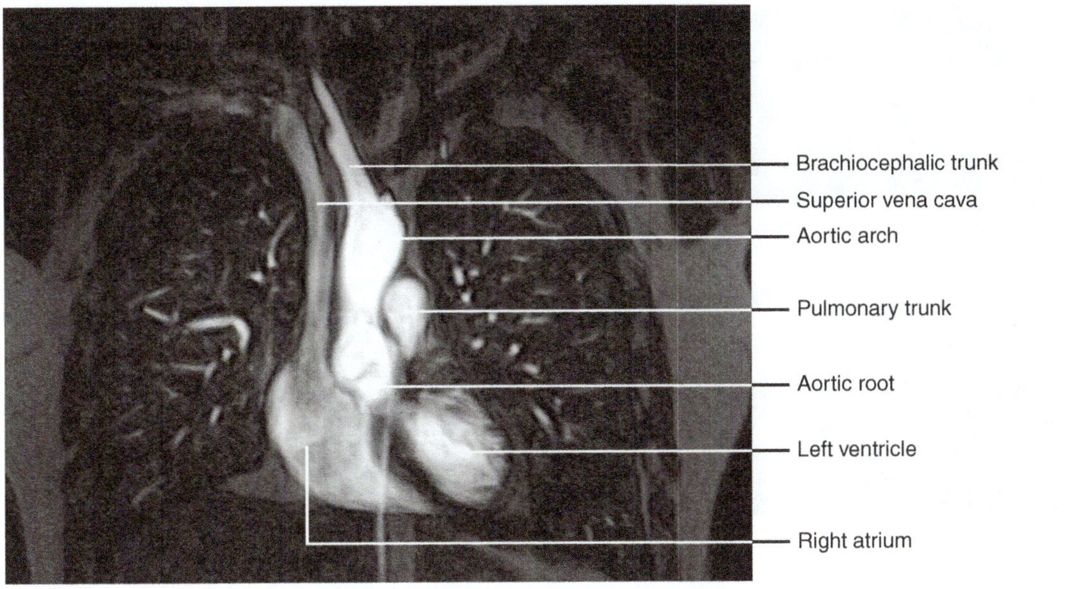

Fig. 28.13 Thoracic source image

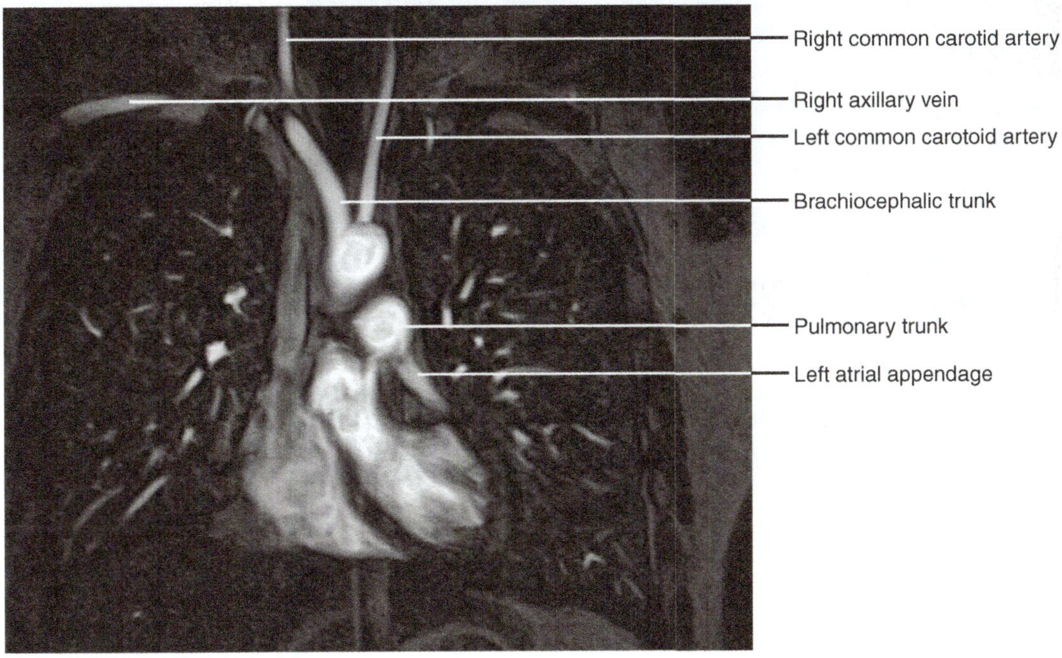

Fig. 28.14 Thoracic source image

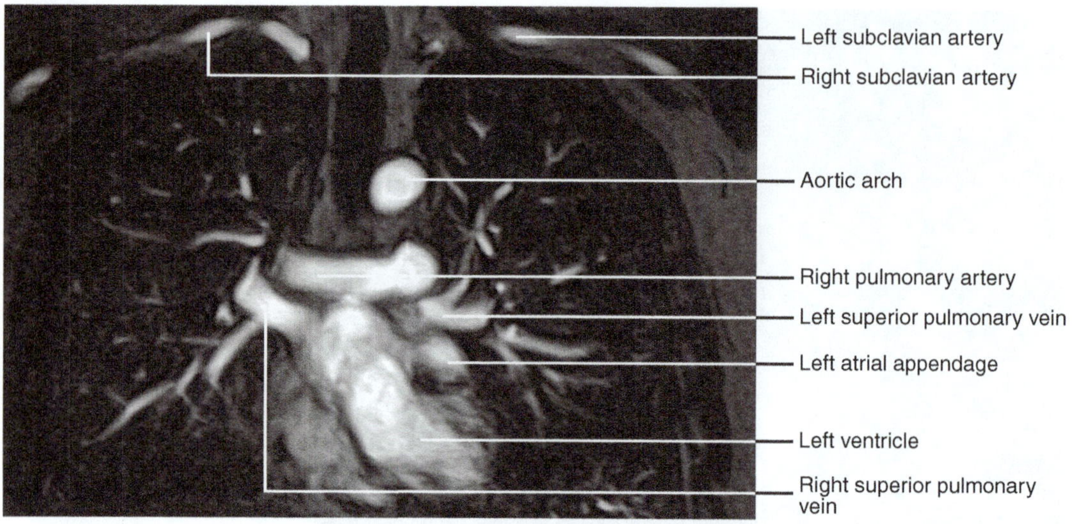

Fig. 28.15 Thoracic source image

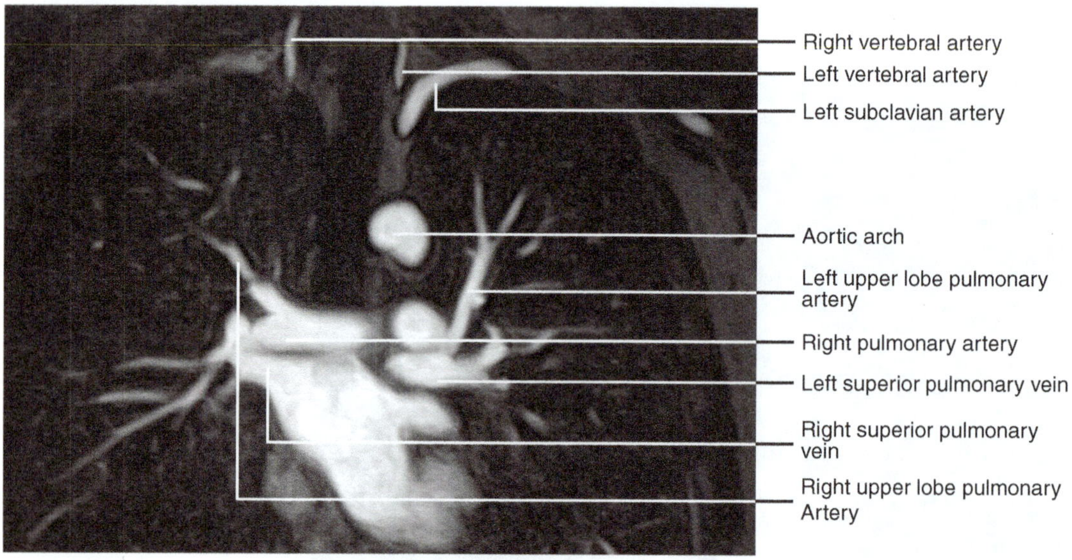

Fig. 28.16 Thoracic source image

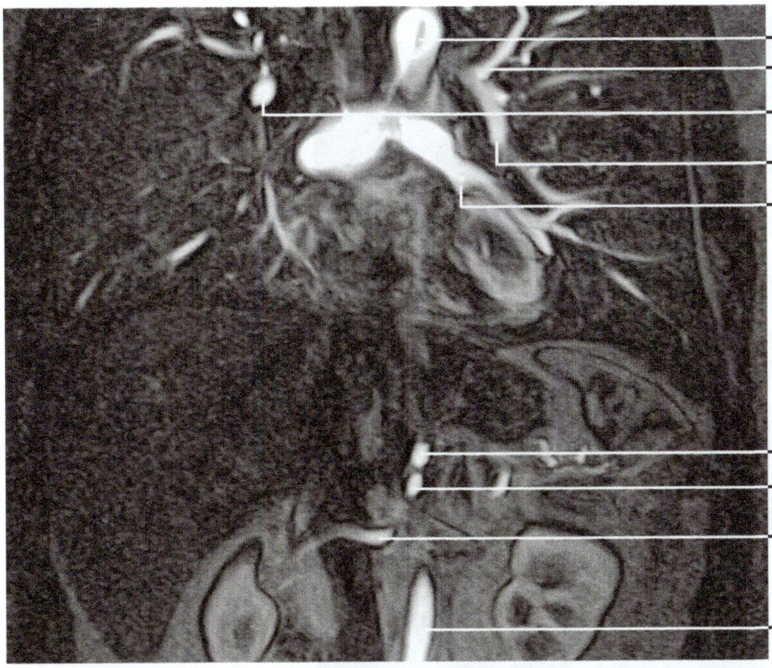

Fig. 28.17 Thoracic source image

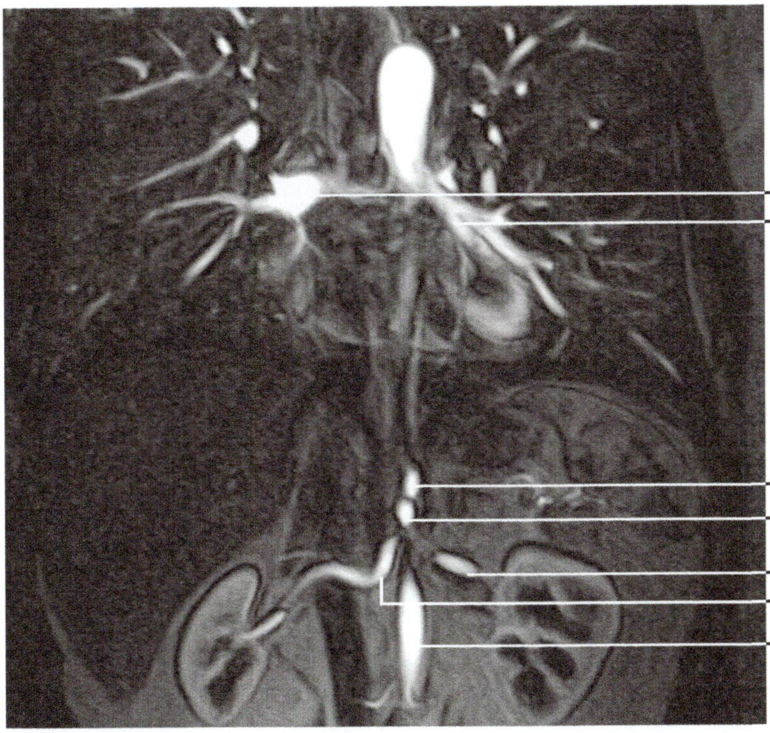

Fig. 28.18 Thoracic source image

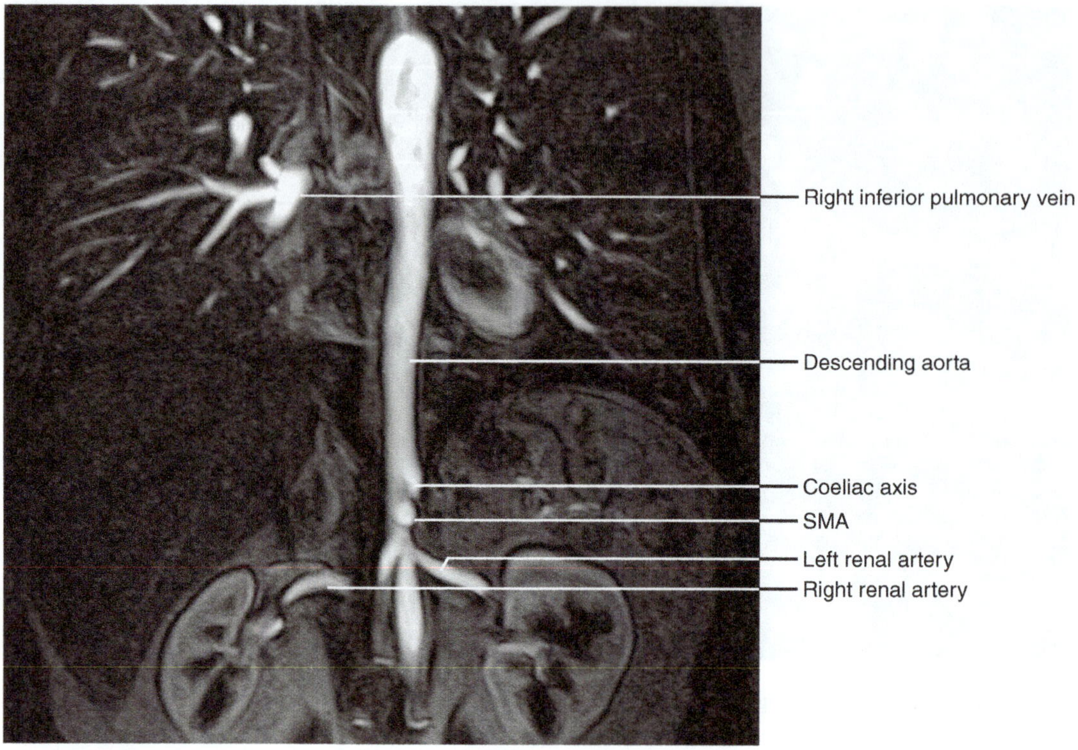

Fig. 28.19 Thoracic source image

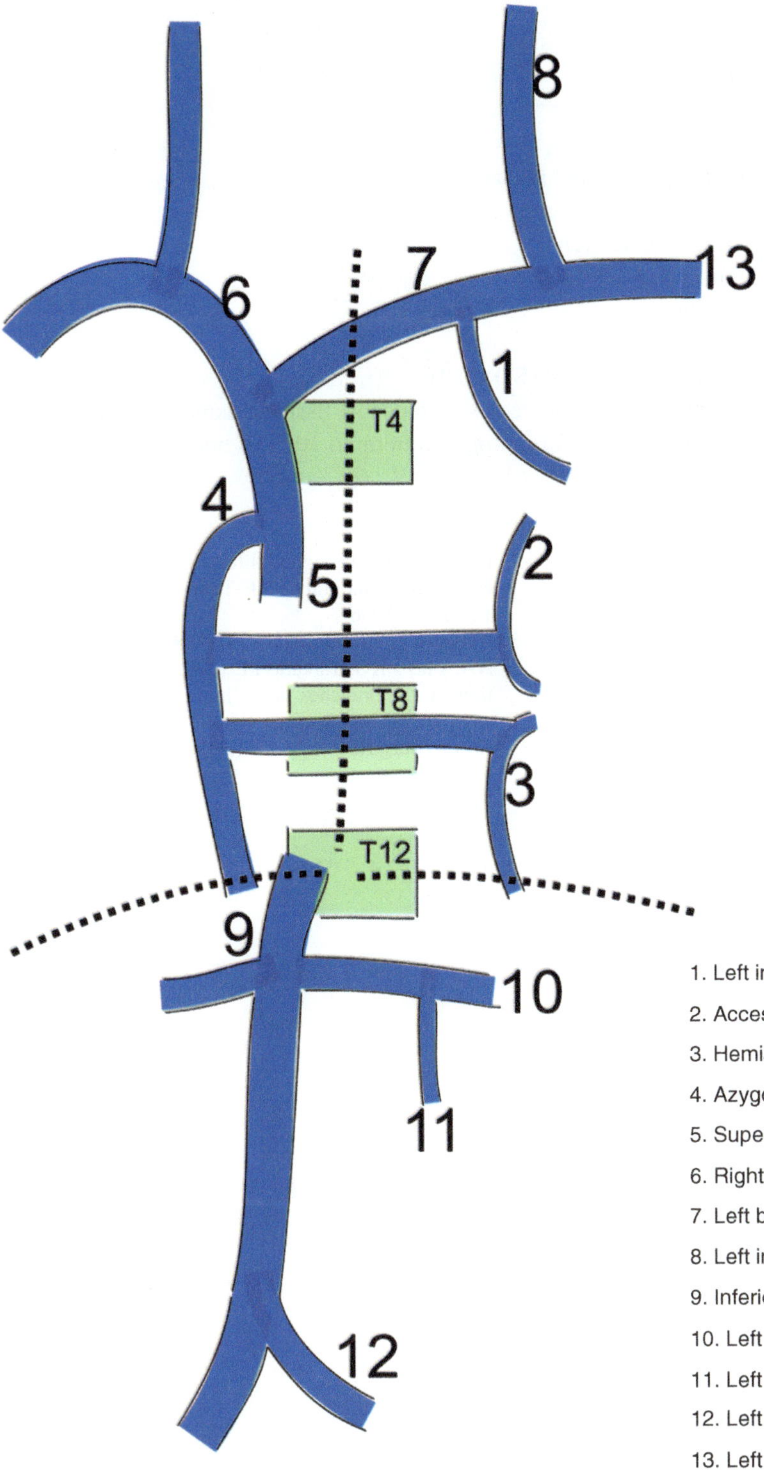

1. Left intercostal vein
2. Accessory hemiazygos vein
3. Hemiazygos vein
4. Azygos vein
5. Superior vena cava
6. Right brachiocephalic vein
7. Left brachiocephalic vein
8. Left internal jugular vein
9. Inferior vena cava
10. Left renal vein
11. Left spermatic / testicular vein
12. Left common iliac vein
13. Left subclavian vein

Fig. 28.20 Schematic superior/inferior vena cava and of Azygos venous drainage system

Pearls: Dimensions and segments

Aortic diameters (mean ± SD)

- 2.98 ± 0.46 cm at the aortic valve sinus
- 3.09 ± 0.41 cm at the ascending aorta
- 2.94 ± 0.42 cm proximal to the brachiocephalic artery
- 2.77 ± 0.37 cm at the proximal transverse arch
- 2.61 ± 0.41 cm at the distal transverse arch
- 2.47 ± 0.40 cm at the isthmus
- 2.43 ± 0.35 cm at the diaphragm

Pulmonary artery

- Main pulmonary artery diameter is 2.72 cm (SD = 0.3)
- Aortic nipple = left superior intercostal vein seen as a small soft-tissue density adjacent to the lateral border of the aortic knob on a frontal radiograph in up to 10 % of normal patients

Sources

1. Edwards PD, Bull RK, Coulden R (1998) CT measurement of main pulmonary artery diameter. Br J Radiol 71:1018–1020
2. Hager A, Kaemmerer H, Rapp-Bernhardt U, Blucher S, Rapp K, Bernhardt TM et al (2002) Diameters of the thoracic aorta throughout life as measured with helical computed tomography. J Thorac Cardiovasc Surg 123:1060–1066
3. Natsis KI, Tsitouridis IA, Didagelos MV, Fillipidis AA, Vlasis KG, Tsikaras PD (2009) Anatomical variations in the branches of the human aortic arch in 633 angiographies: clinical significance and literature review. Surg Radiol Anat 31:319–323

Peripheral Vessels 29

The purpose of the chapter is for the reader to be able to identify and name vessels when reporting vascular studies.

Lower Limb

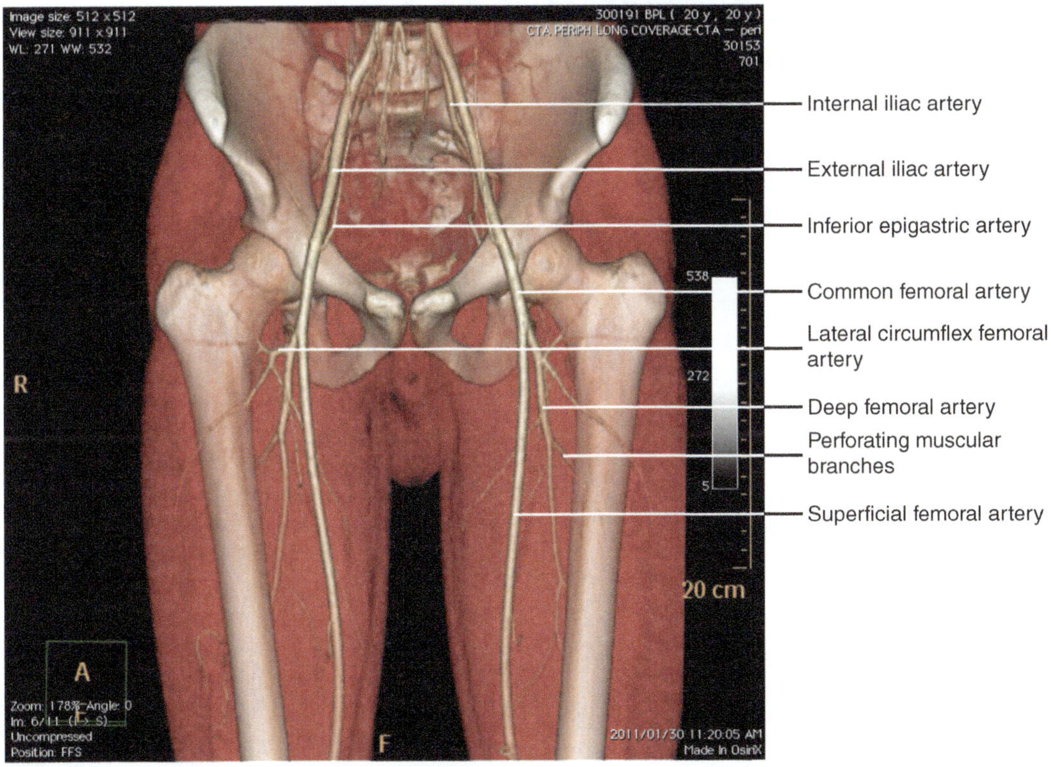

Fig. 29.1 CT angiogram – 3D volume reconstruction

S. Andronikou (ed.), *See Right Through Me*,
DOI 10.1007/978-3-642-23893-2_29, © Springer-Verlag Berlin Heidelberg 2012

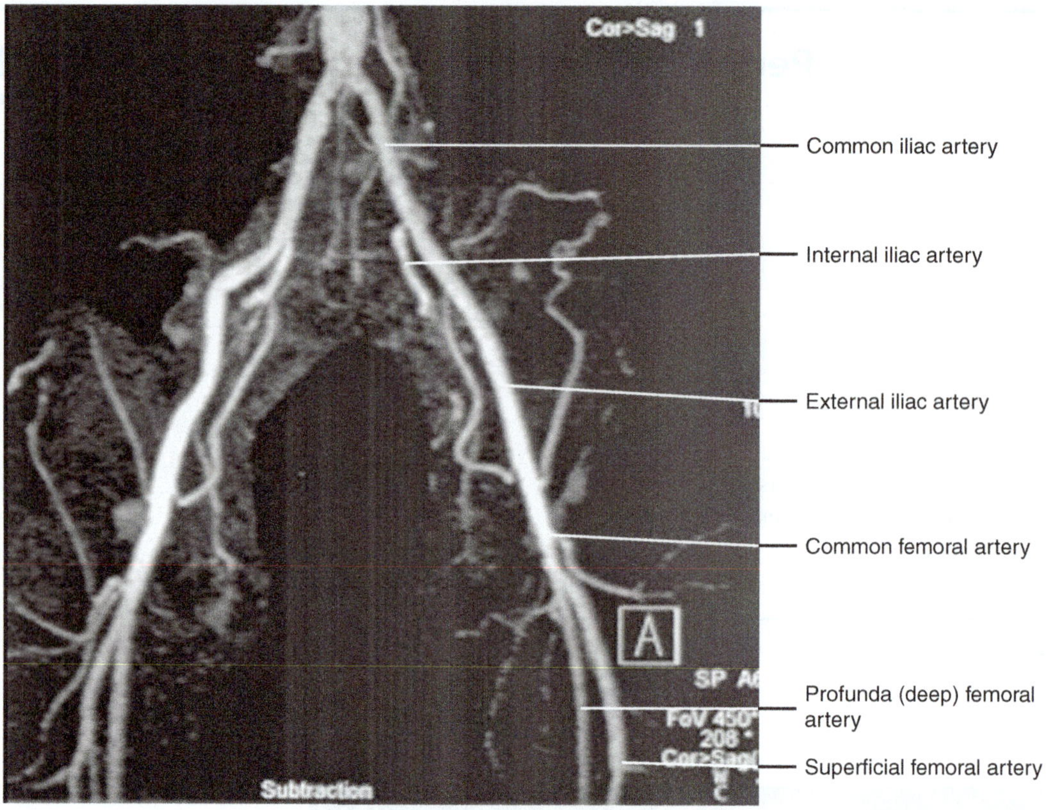

Fig. 29.2 MRI angiography of the pelvis with intravenous gadolinium

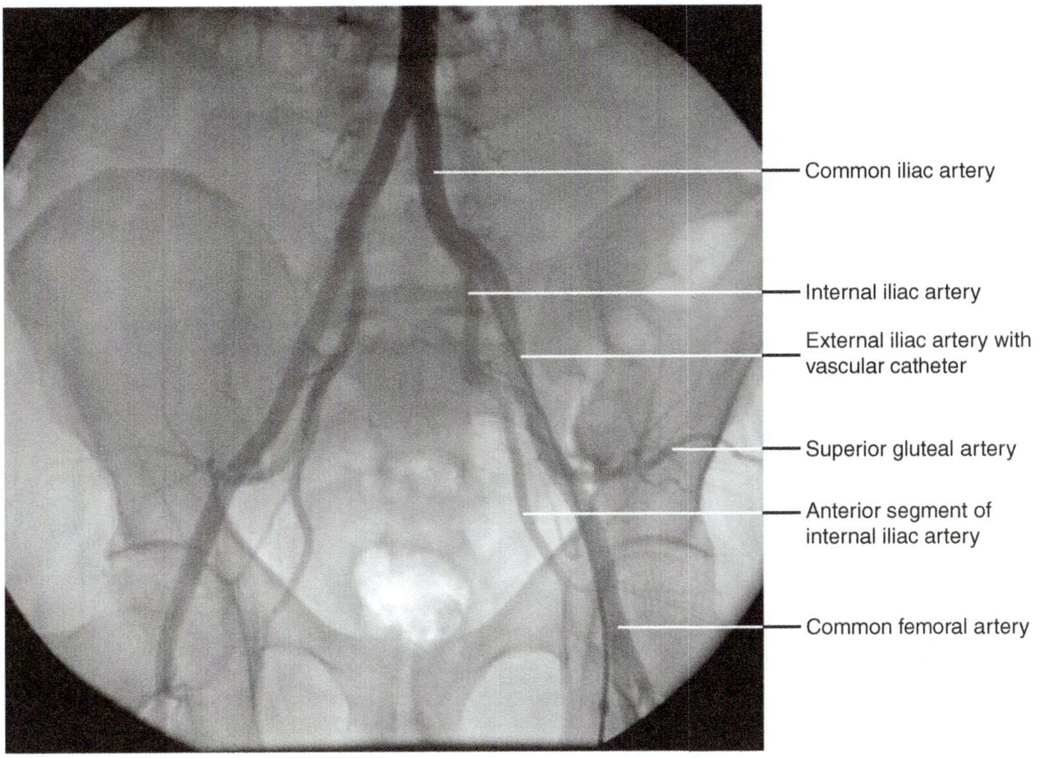

Fig. 29.3 Conventional angiography

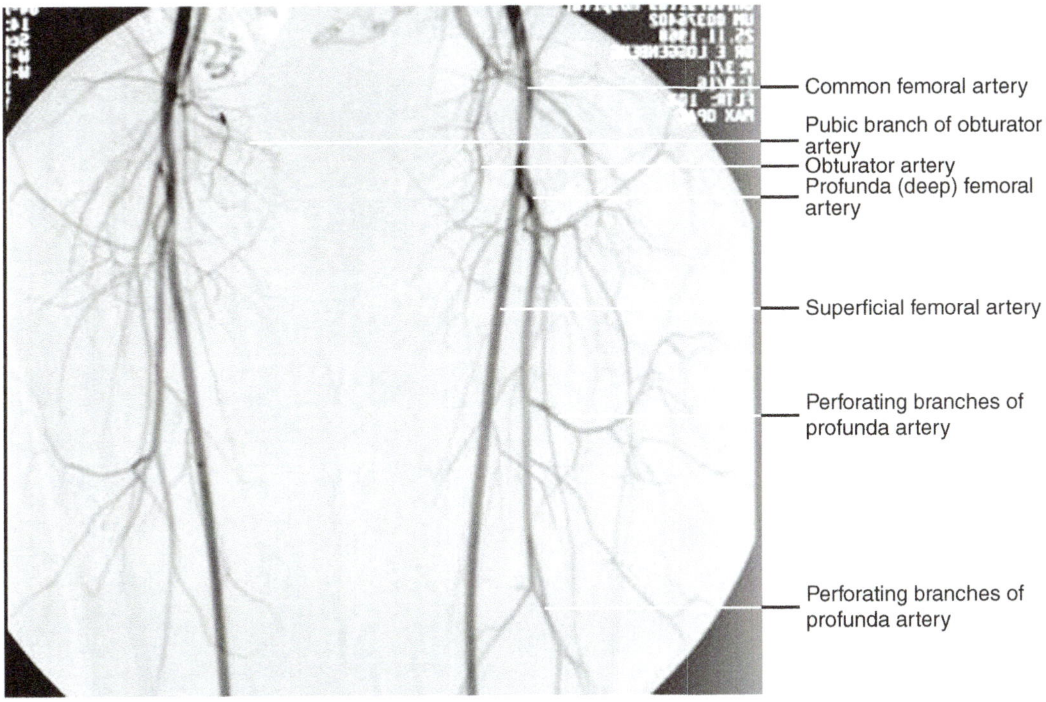

Fig. 29.4 Digital subtraction angiography

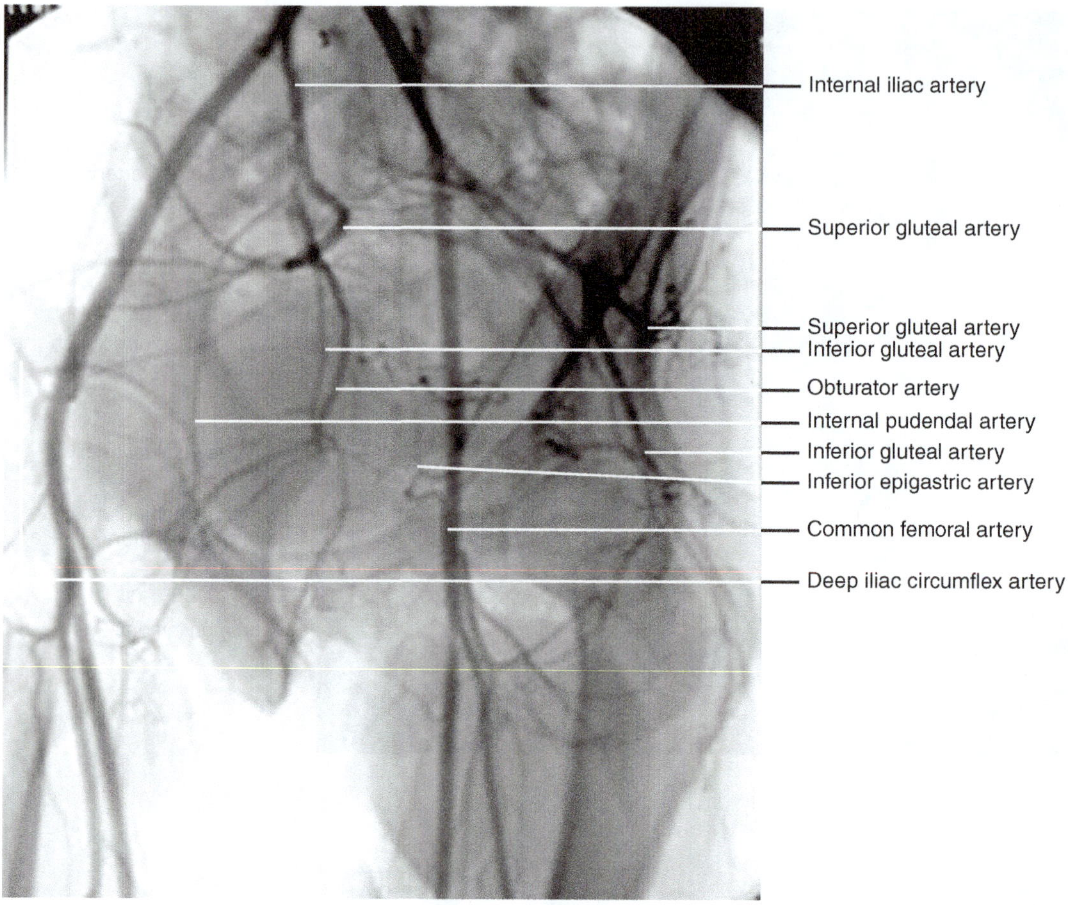

Fig. 29.5 Conventional angiography – oblique view

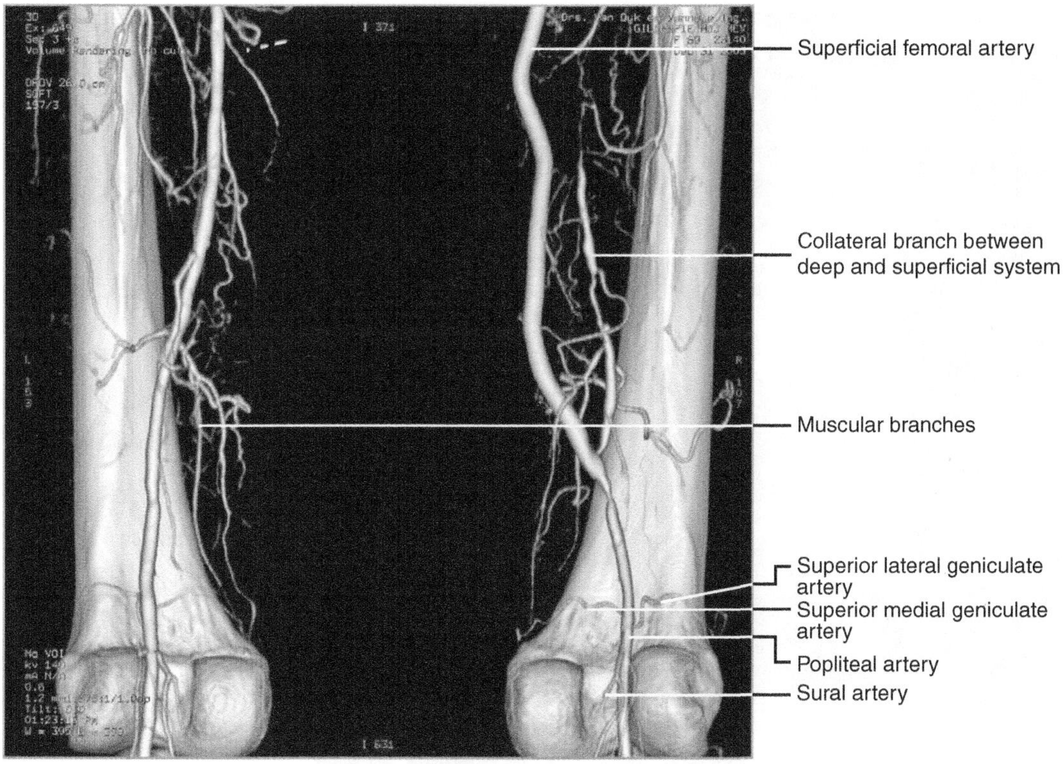

Fig. 29.6 CT angiography with intravenous iodine contrast (posterior view)

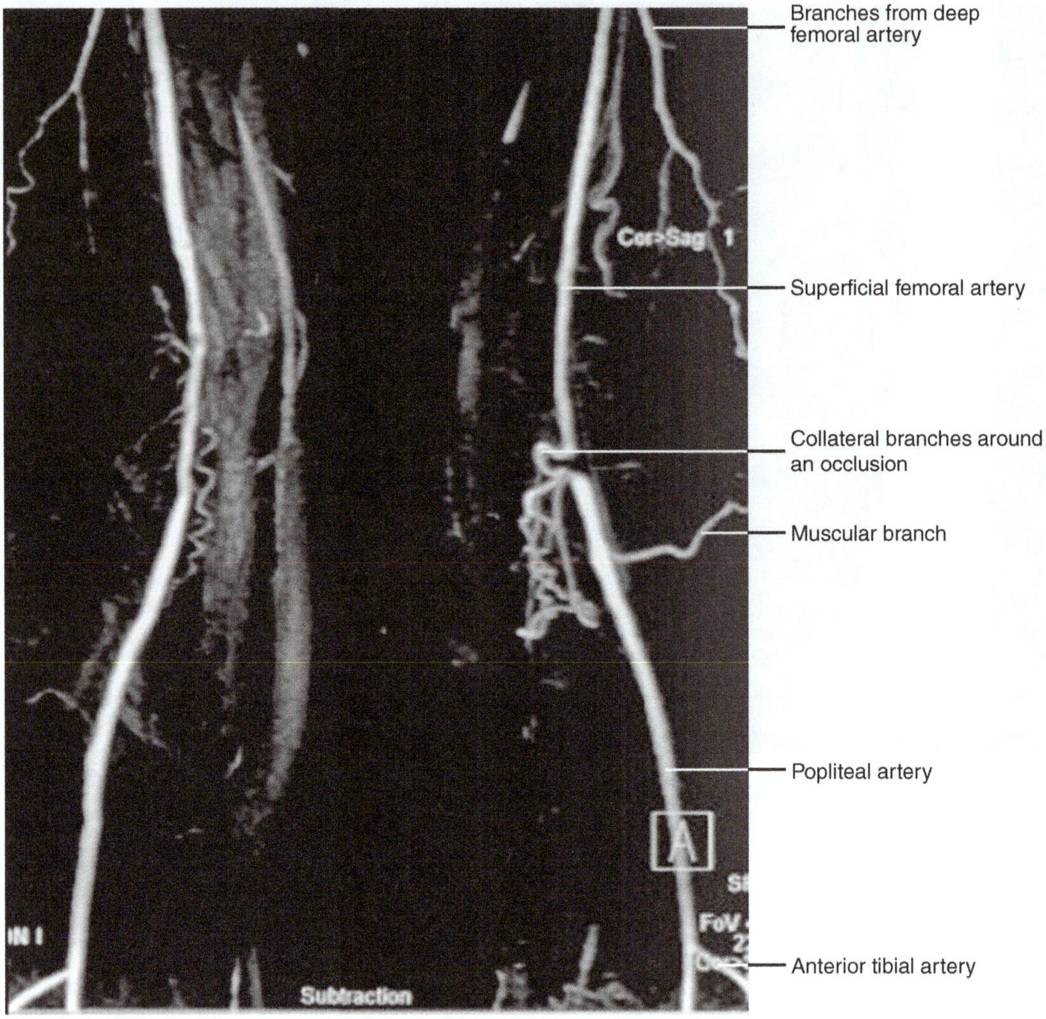

Fig. 29.7 MRI angiography with intravenous gadolinium

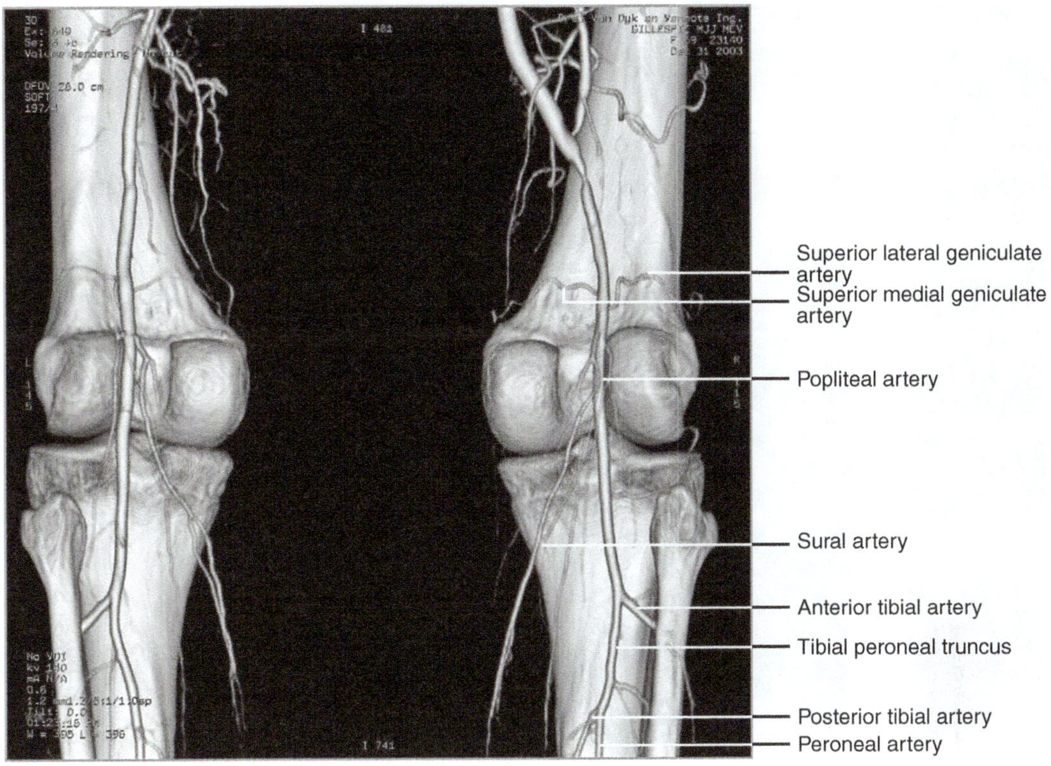

Fig. 29.8 CT angiography with intravenous iodine contrast (posterior view)

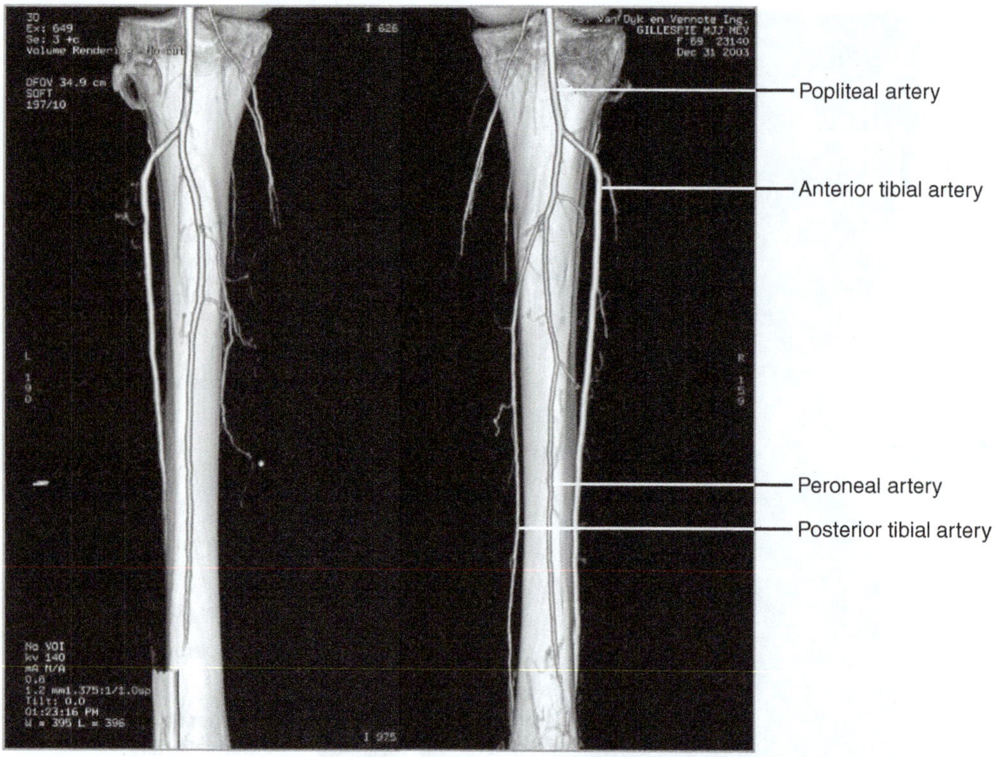

Fig. 29.9 CT angiography with intravenous iodine contrast: fibula removed (posterior view)

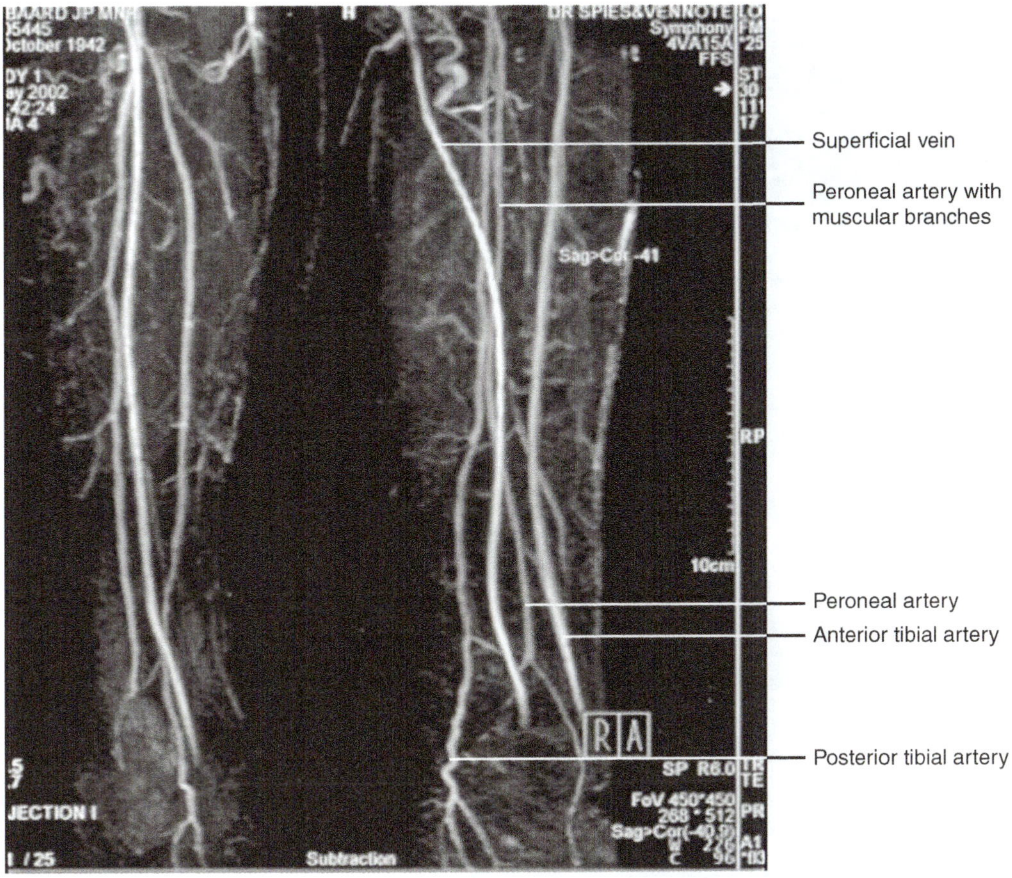

Fig. 29.10 MRI angiography with intravenous gadolinium (image rotated)

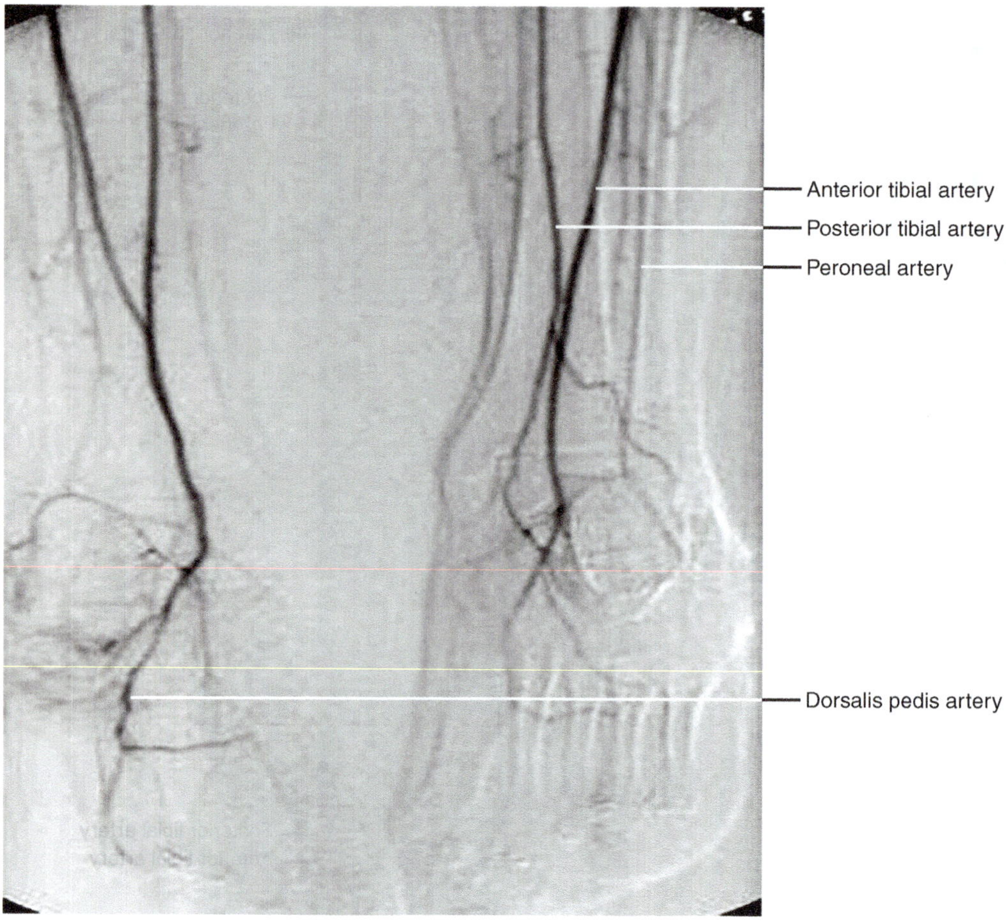

Fig. 29.13 Digital subtraction angiography

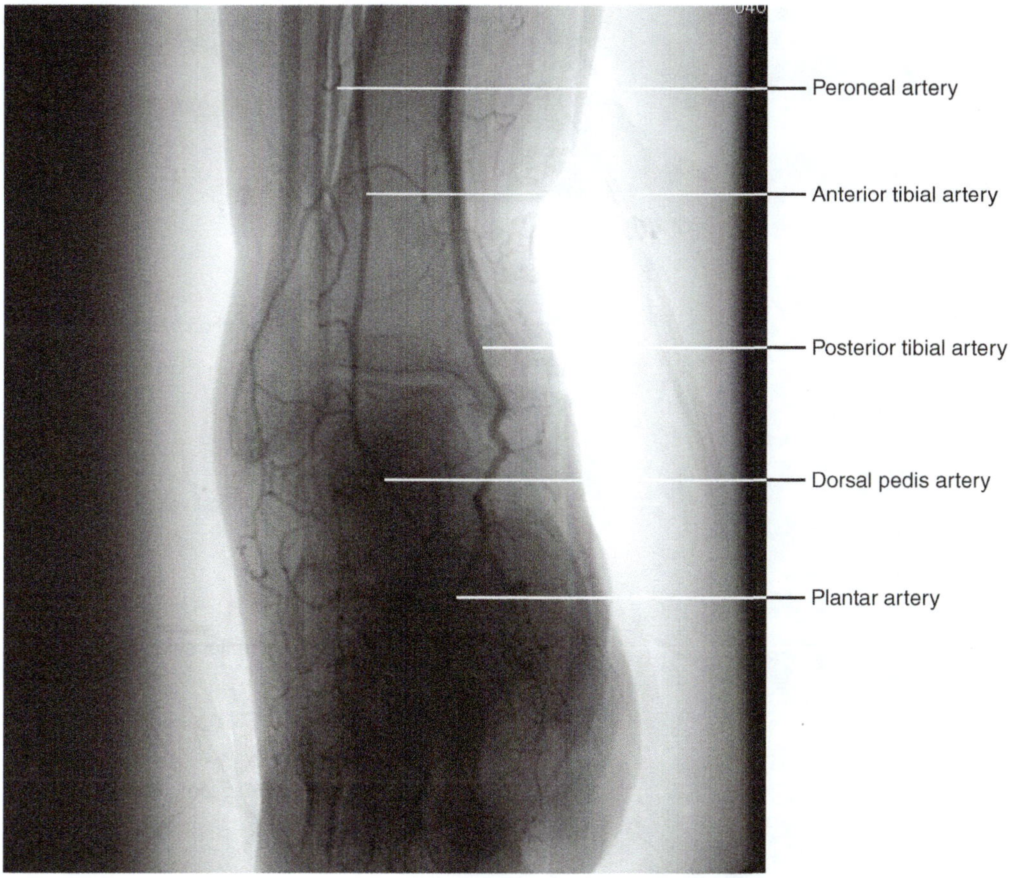

Fig. 29.14 Foot arteriogram

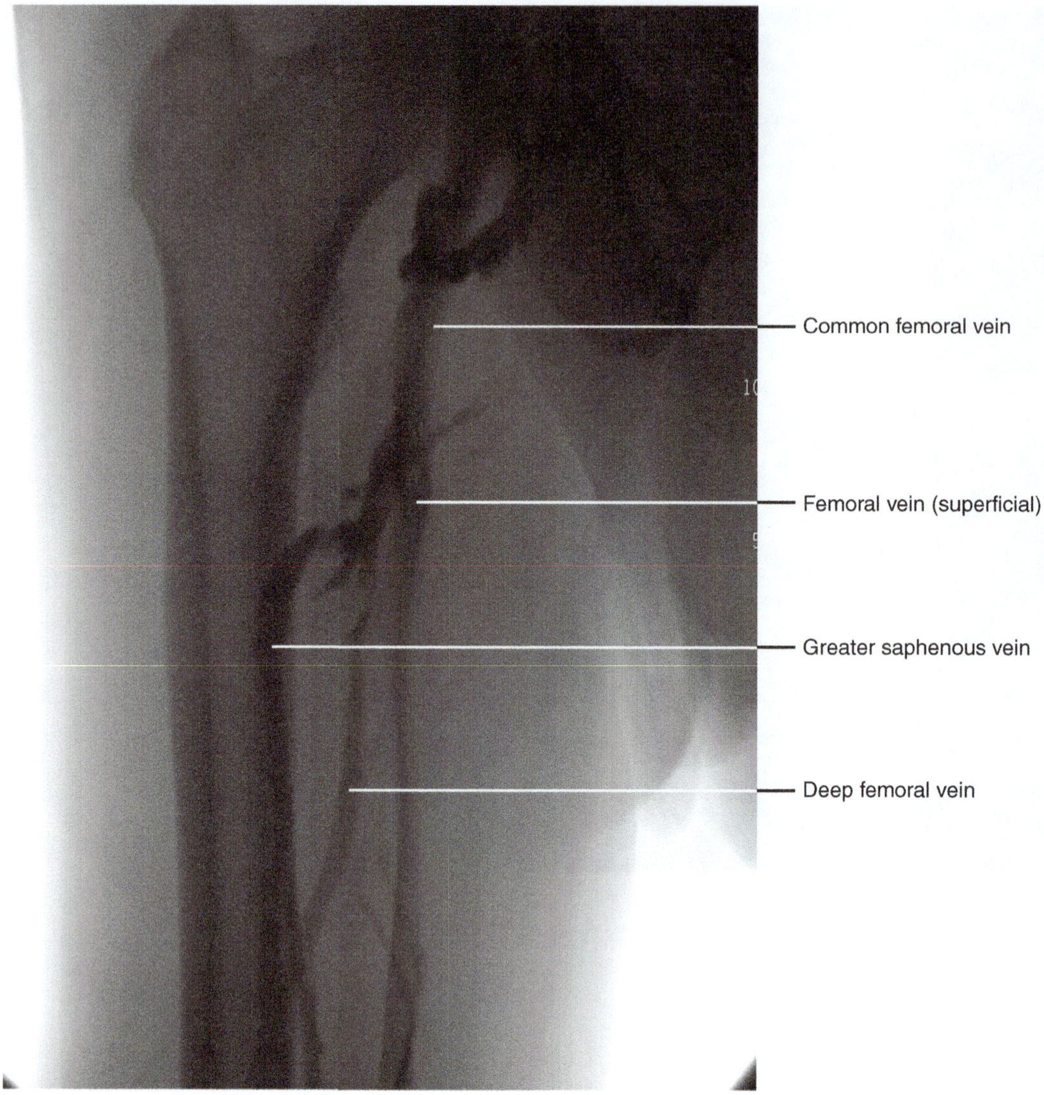

Fig. 29.15 Venogram of the common femoral vein

Upper Limb

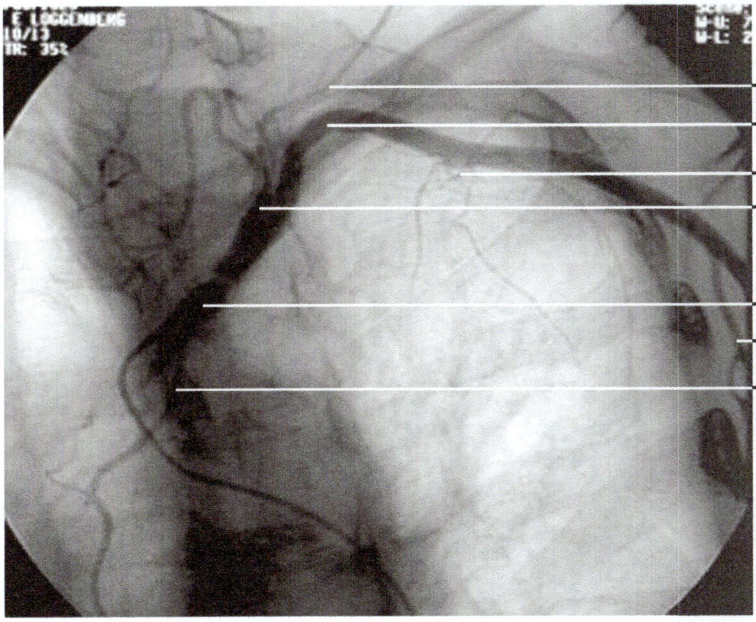

Fig. 29.16 Conventional angiography (the c-arm is angled and rotated)

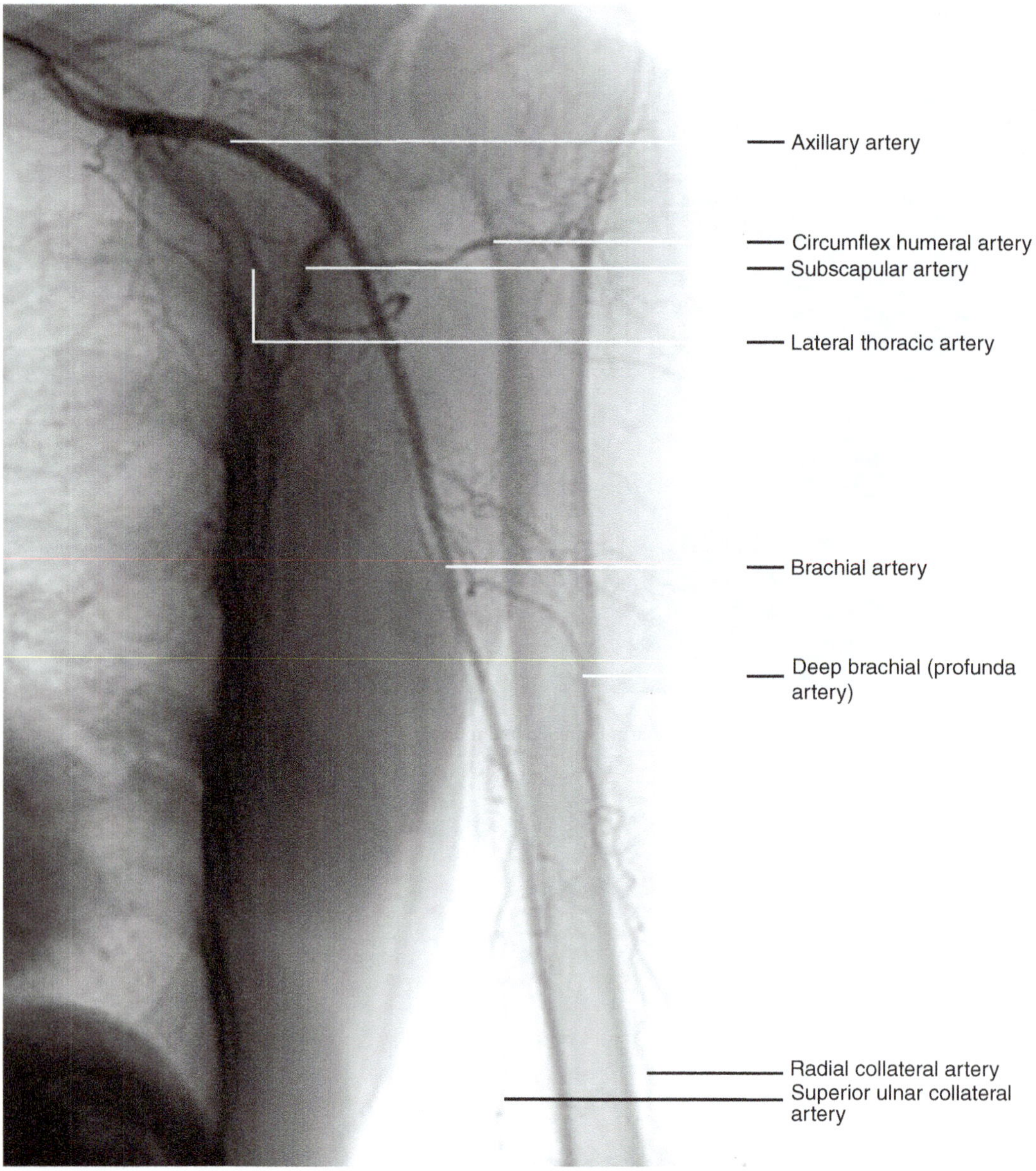

Fig. 29.17 Conventional angiography

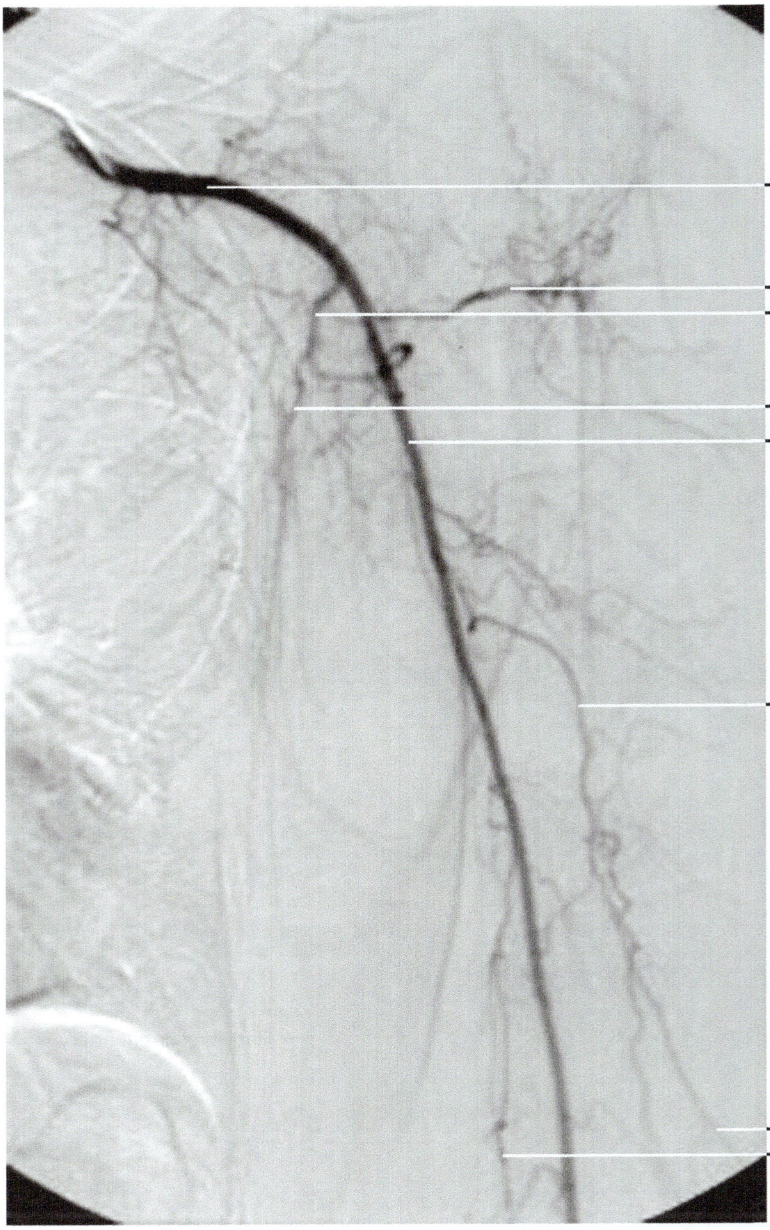

Fig. 29.18 Digital subtraction angiography

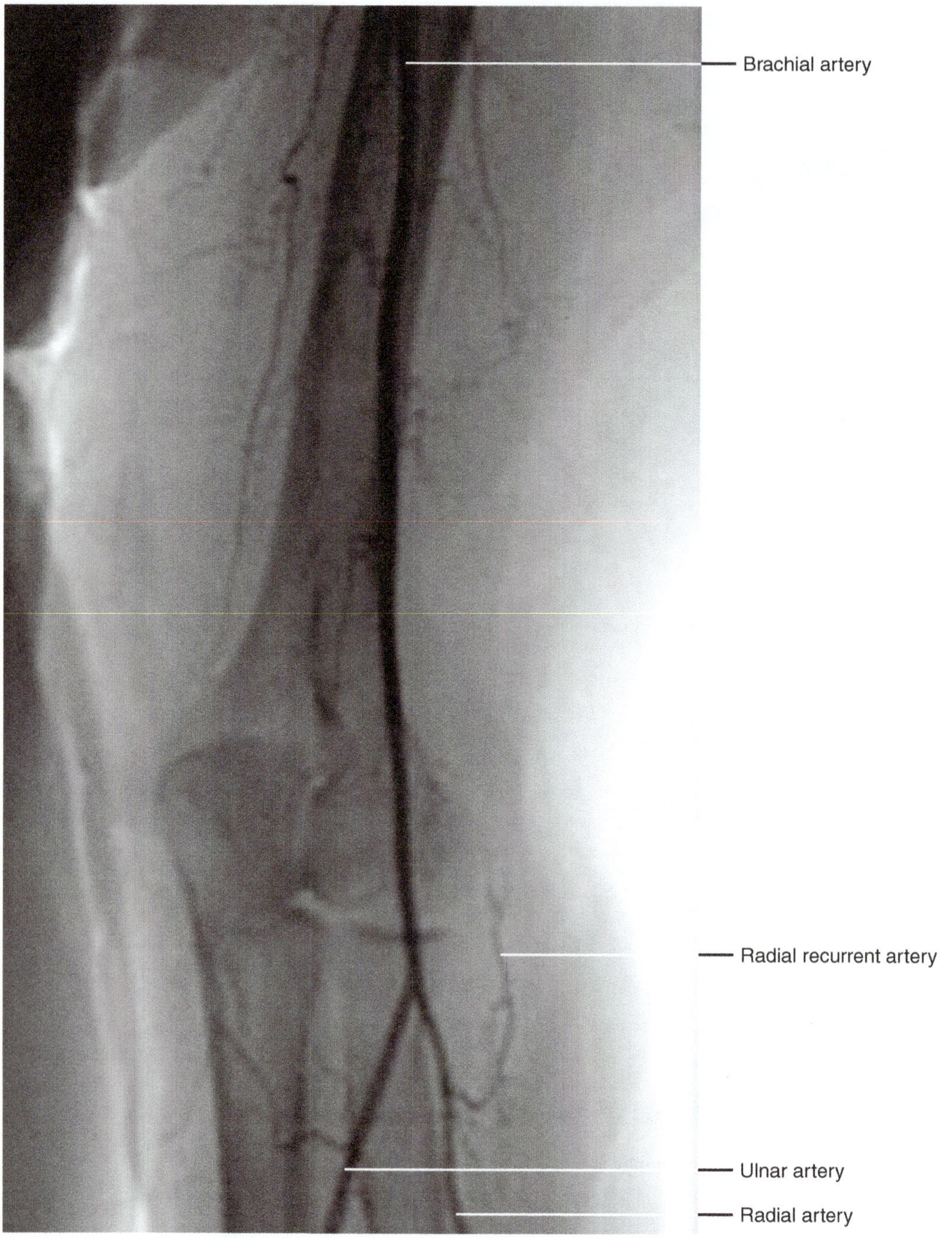

Fig. 29.19 Conventional angiography

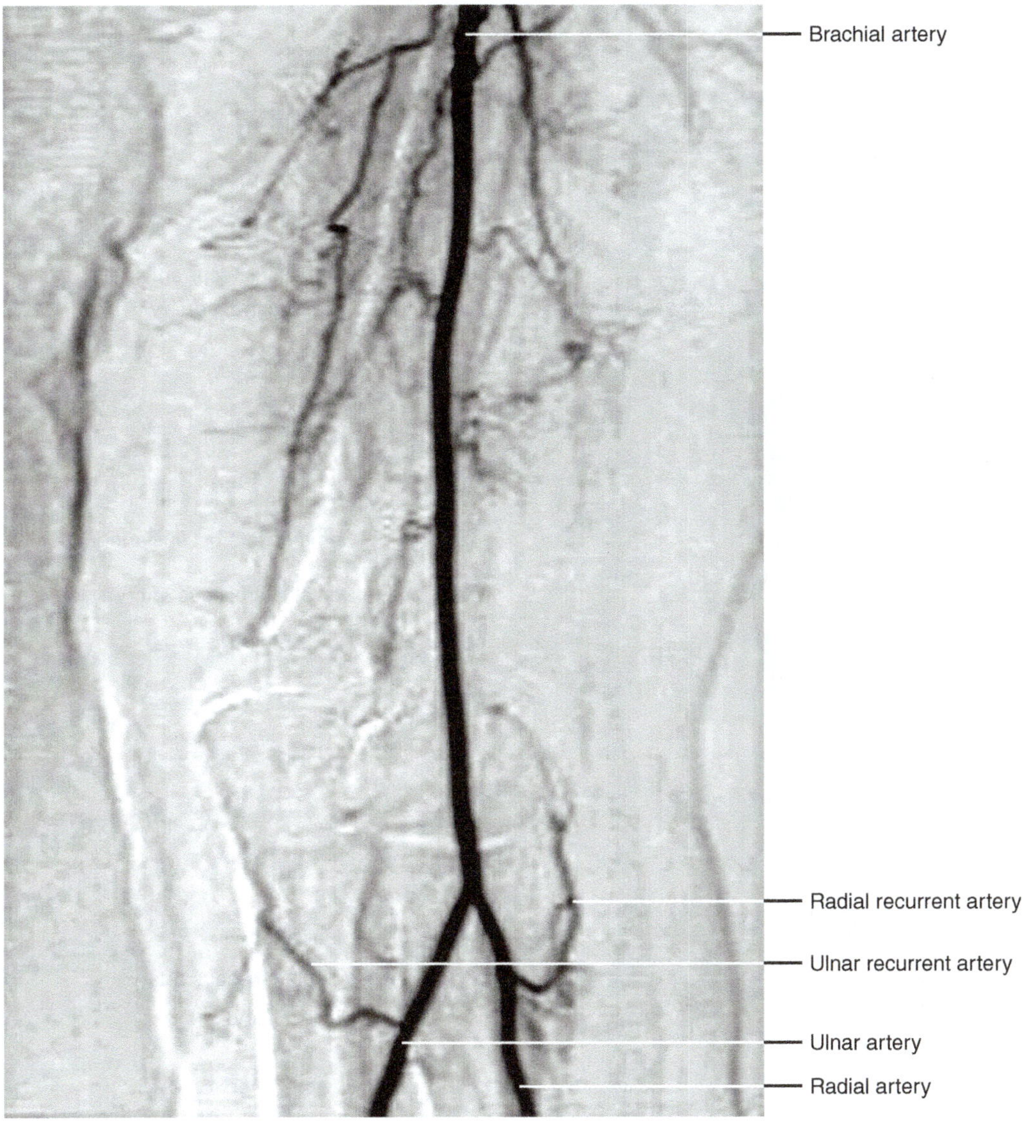

Fig. 29.20 Digital subtraction angiography

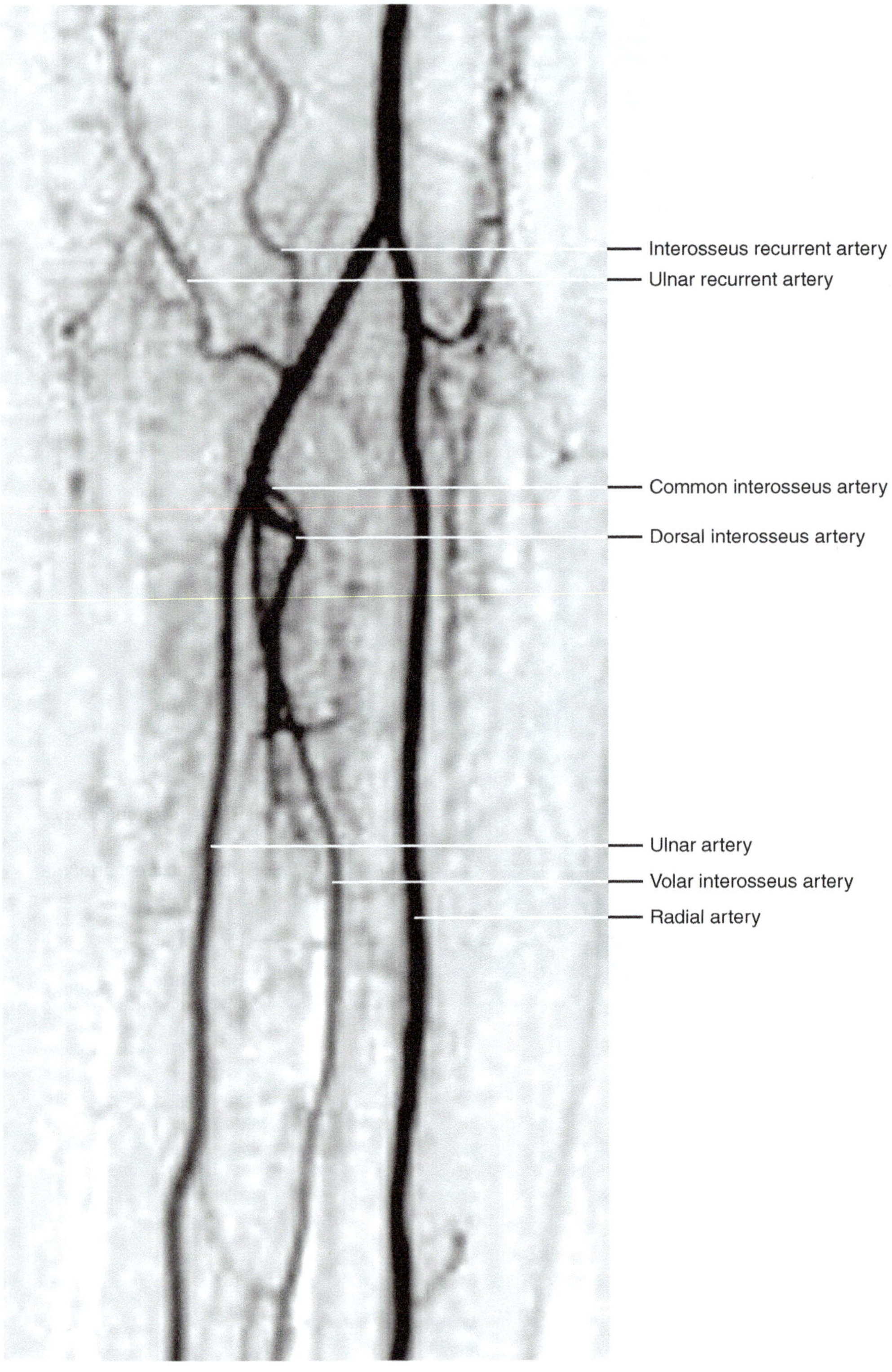

Fig. 29.21 Digital subtraction angiography

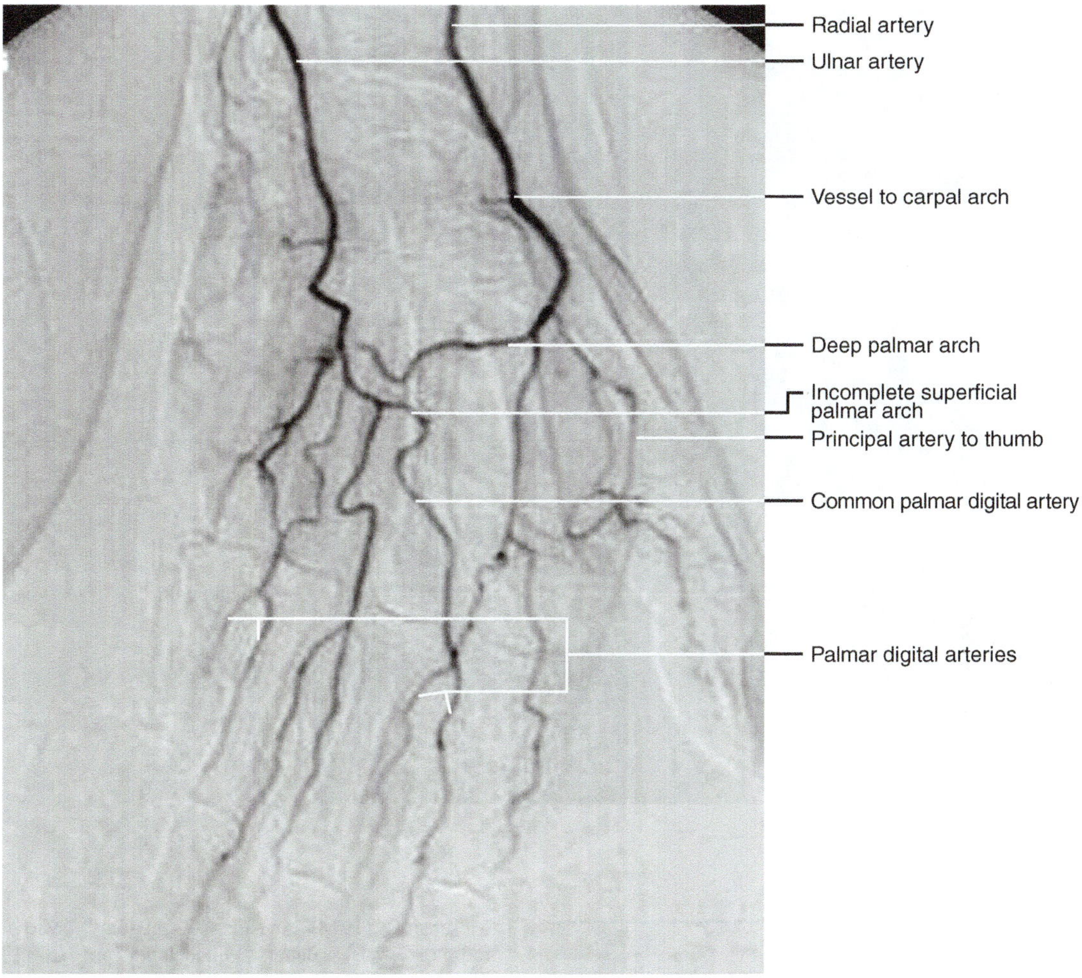

Fig. 29.22 Digital subtraction angiography

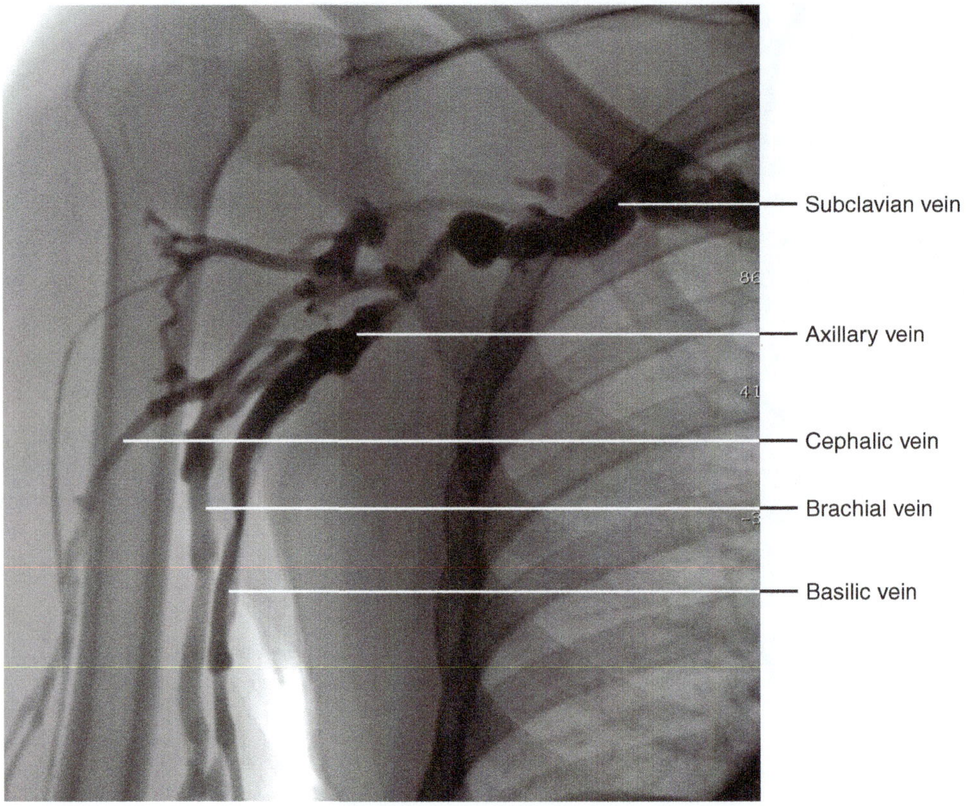

Fig. 29.23 Upper limb venogram

Sources

1. Butler P, Mitchell AWM, Ellis H (1999) Applied radiological anatomy. Cambridge University Press, Cambridge, pp 383–393
2. Last RJ (1978) Anatomy regional and applied, 6th edn. Churchilll Livingston, London, pp 83–84, 155–156
3. Meschan I (1974) An atlas of anatomy basic to radiology. WB Saunders, Philadelphia, pp 116–119, 191–196
4. Weir J, Abrahams PH (1997) Imaging atlas of human anatomy, 2nd edn. Mosby, London, pp 71–73, 181–184

Part VII

Paediatrics

There are many issues unique to the imaging of children as compared with adults which relate to normal developmental during childhood. This part will highlight important normal anatomical differences that may be confused with pathology.

30 Paediatric Central Nervous System

Ultrasound Head

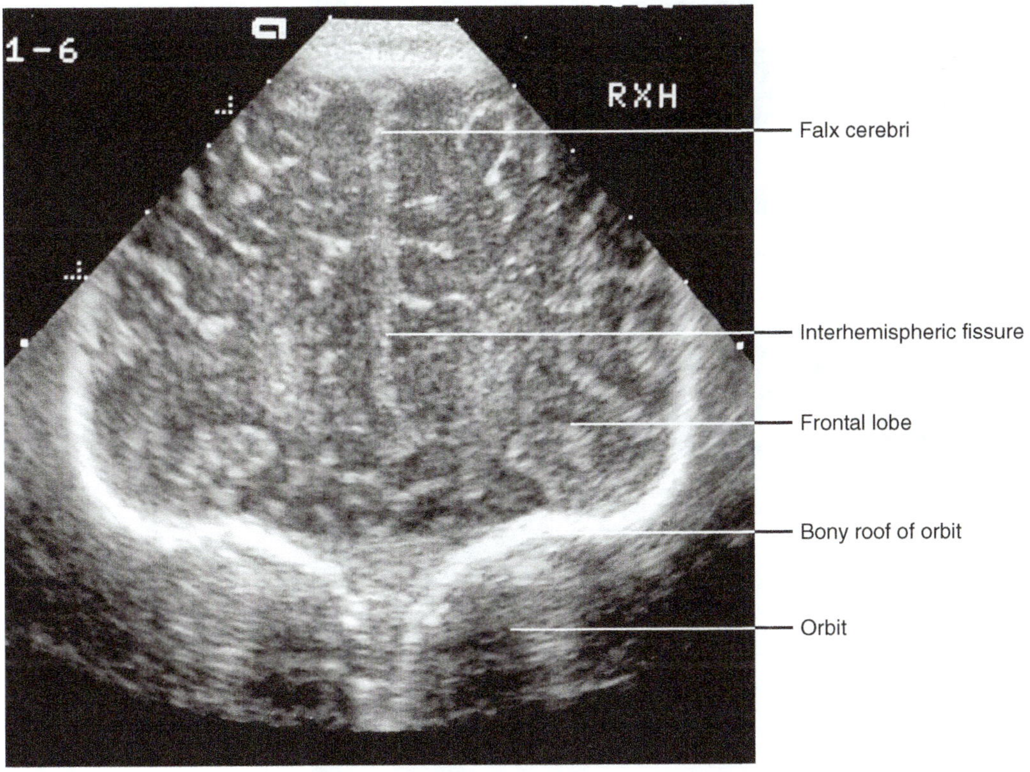

Fig. 30.1 Coronal: anterior to posterior

S. Andronikou (ed.), *See Right Through Me*,
DOI 10.1007/978-3-642-23893-2_30, © Springer-Verlag Berlin Heidelberg 2012

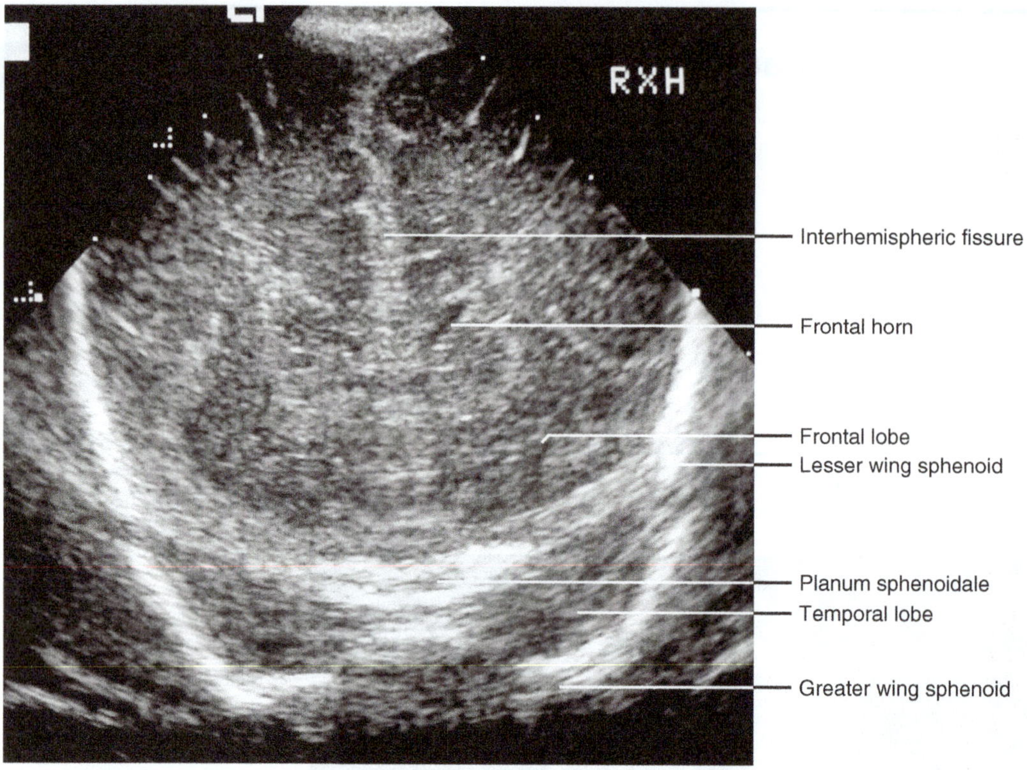

Fig. 30.2 Coronal: anterior to posterior

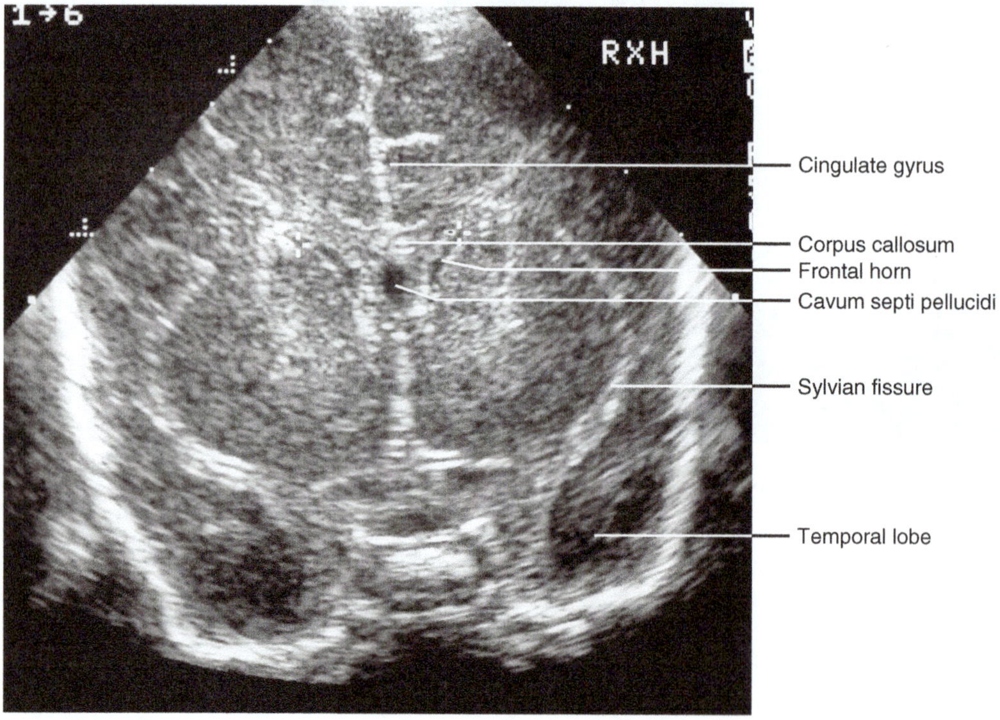

Fig. 30.3 Coronal: anterior to posterior

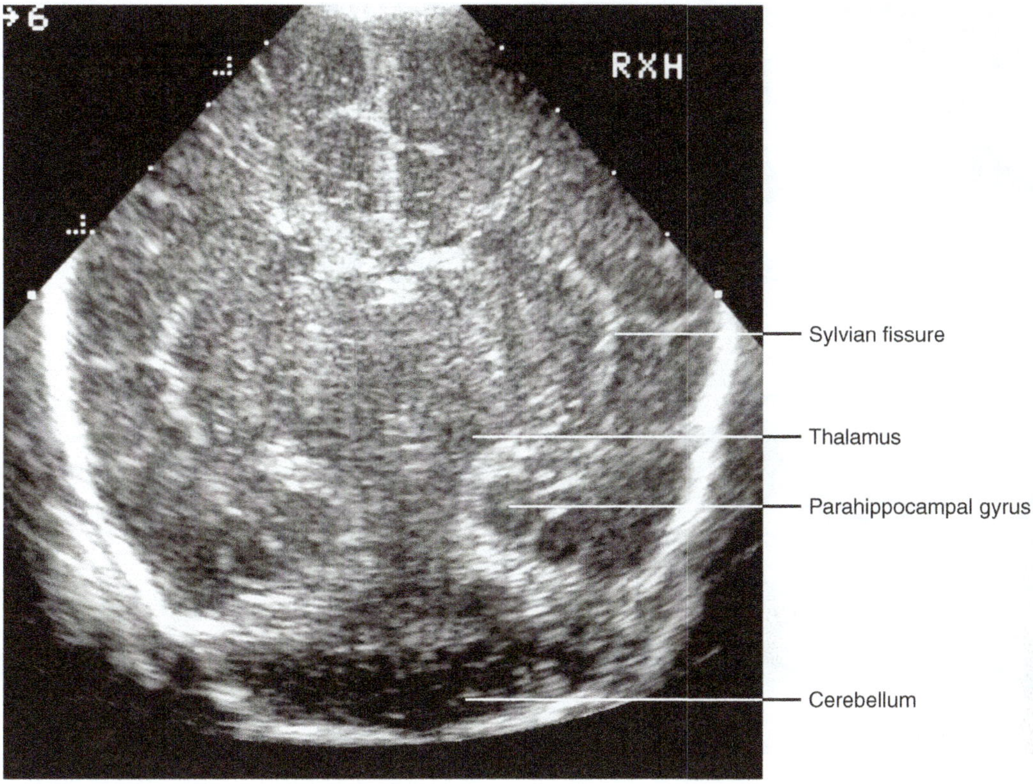

Fig. 30.4 Coronal: anterior to posterior

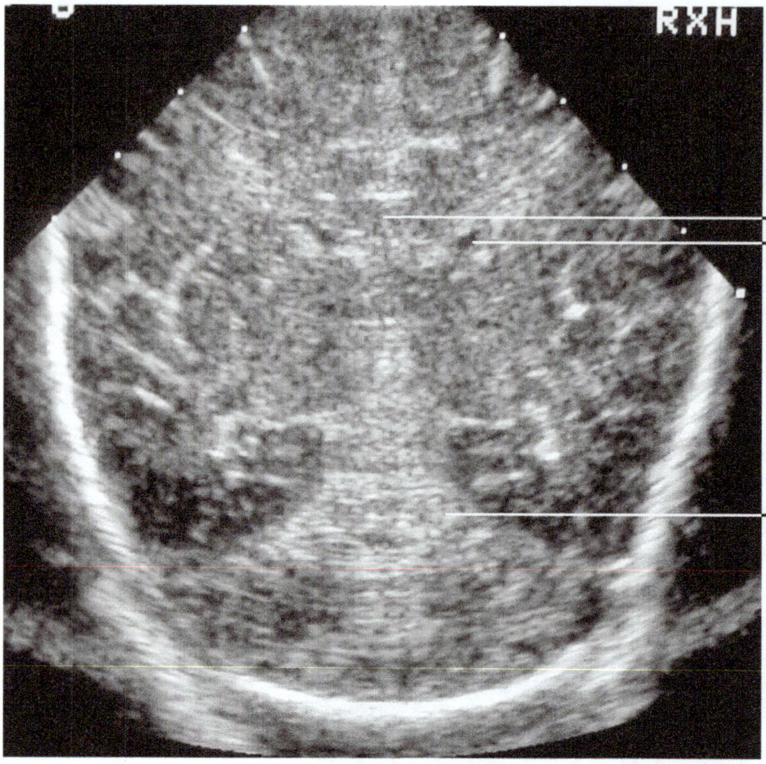

Fig. 30.5 Coronal: anterior to posterior

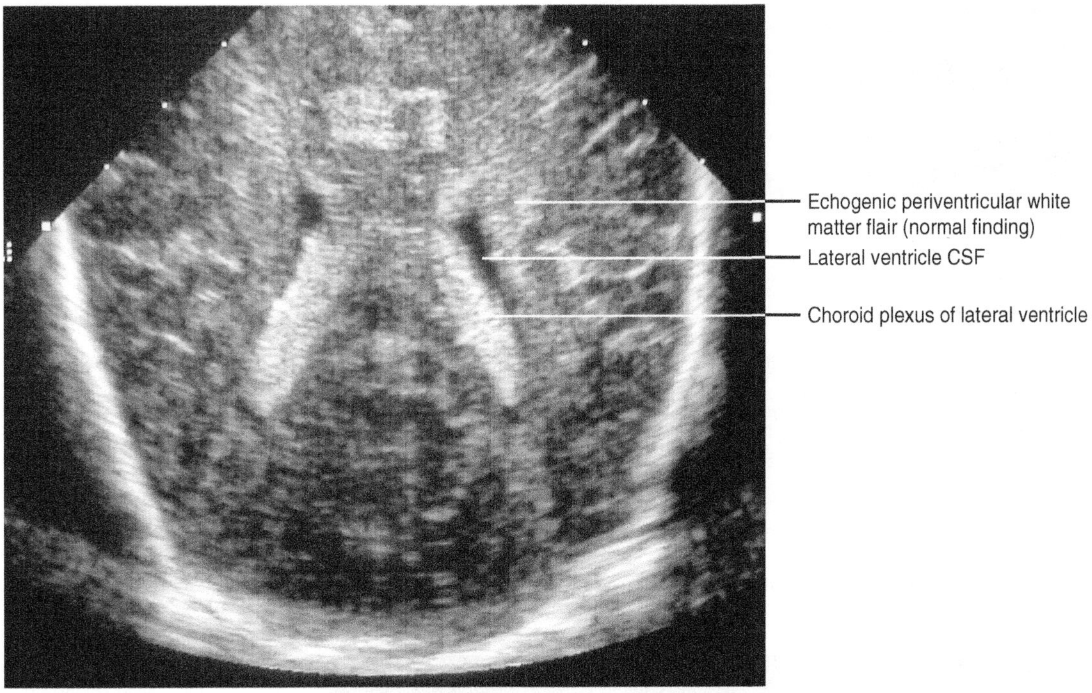

Fig. 30.6 Coronal: anterior to posterior

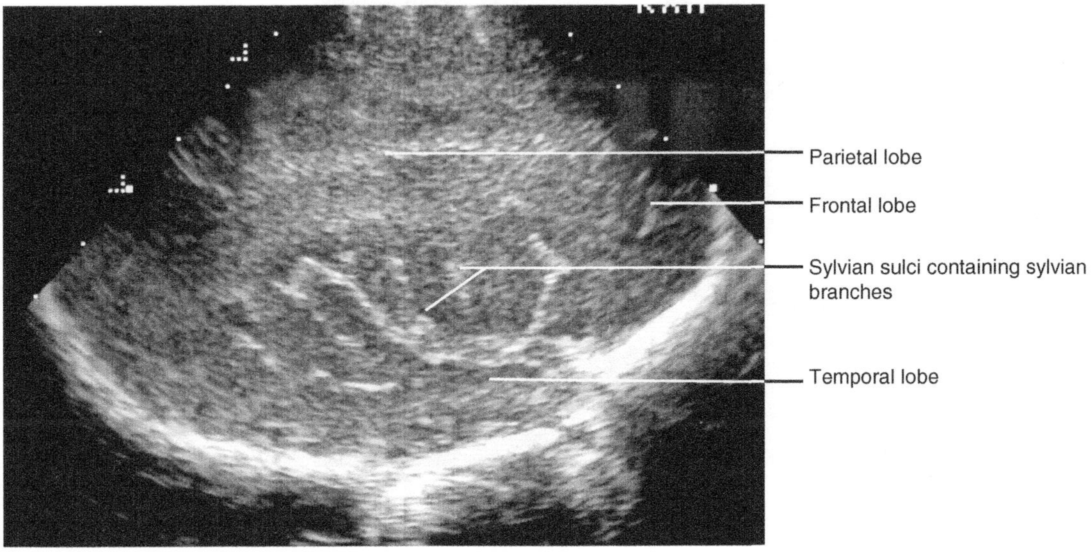

Fig. 30.7 Sagittal: lateral to midline

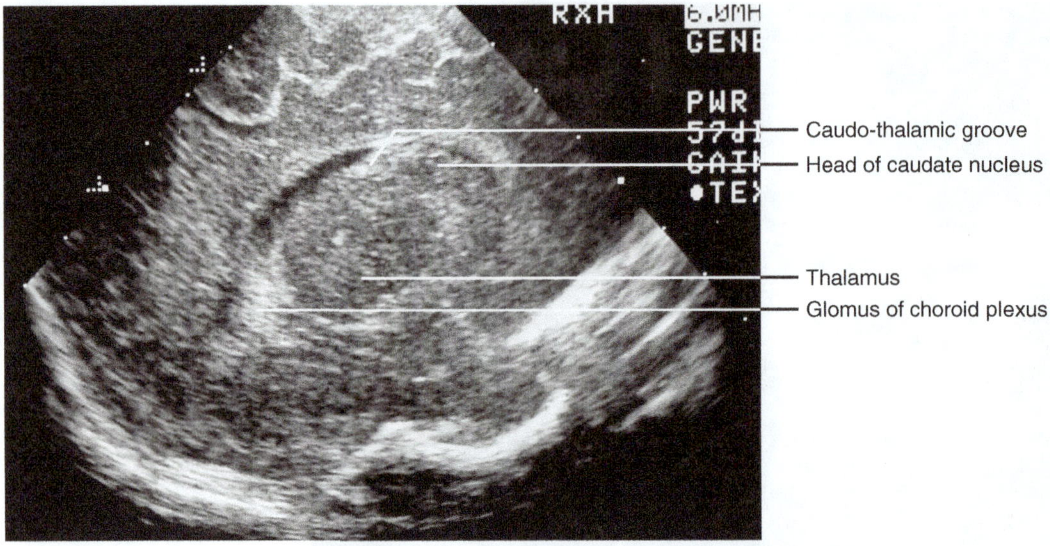

Fig. 30.8 Sagittal: lateral to midline

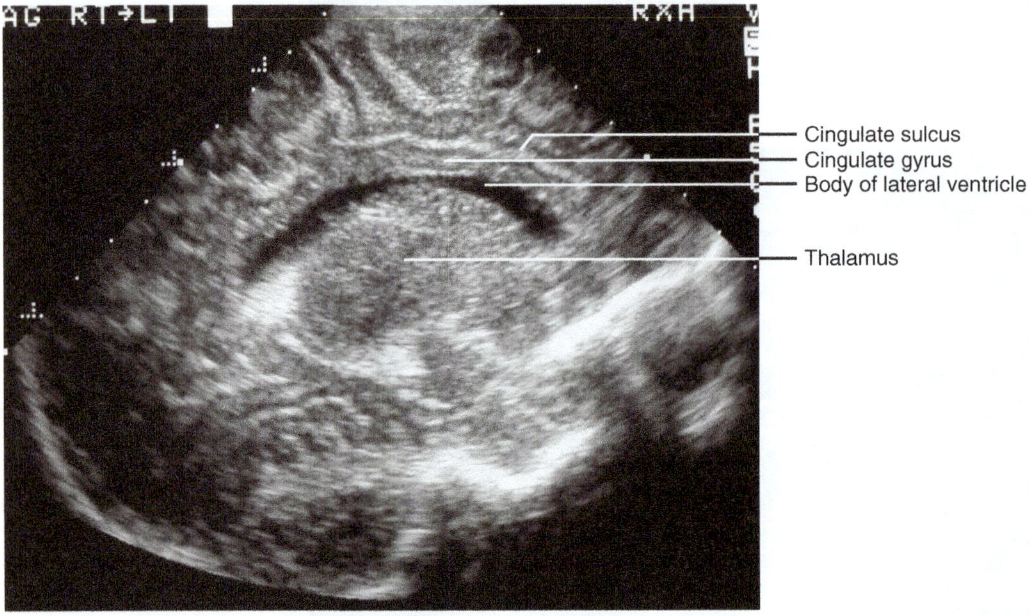

Fig. 30.9 Sagittal: lateral to midline

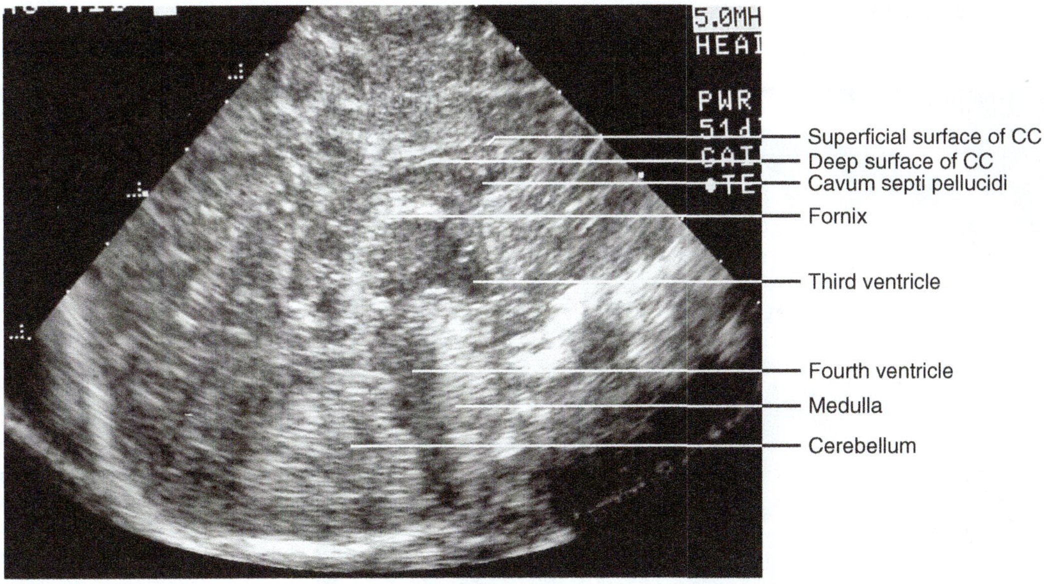

Fig. 30.10 Sagittal: lateral to midline

Pearl
Normal periventricular white matte echodensity should not exceed the echogenicity of the choroid plexus.

Subarachnoid Spaces

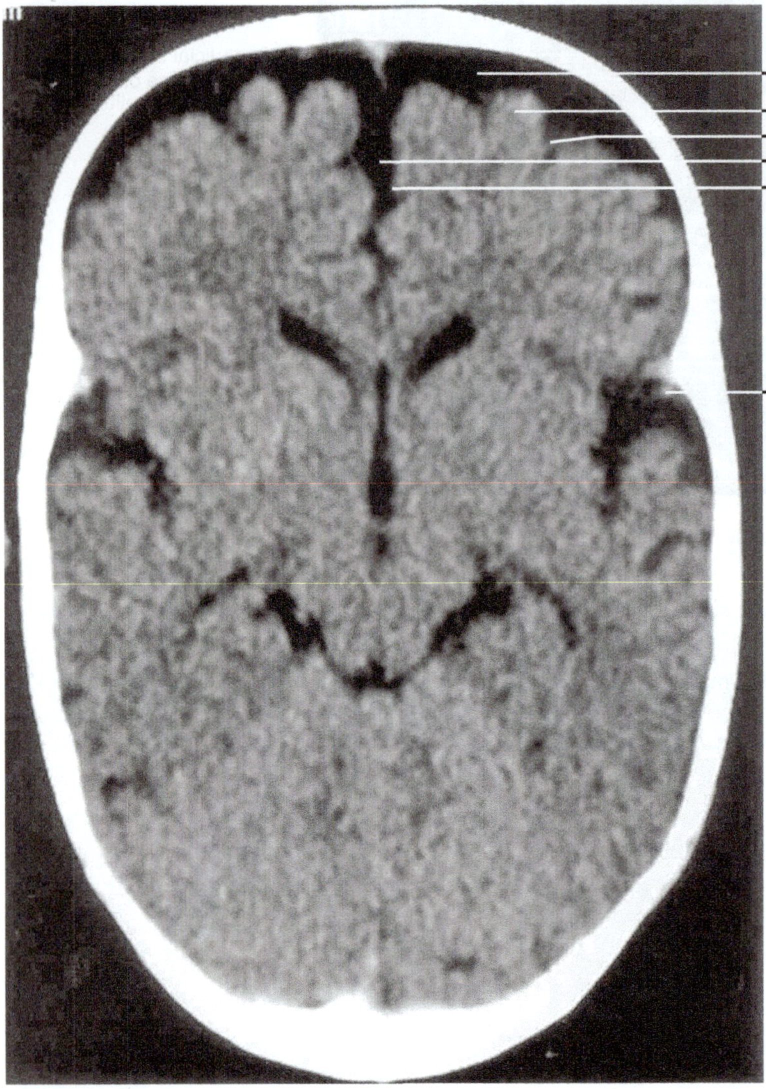

Fig. 30.11 Axial CT

Pearl

Vessels that are seen against the brain surface, displaced away from the calvarium with a large CSF space indicate subdural collections which are pathological.

The subarachnoid space often appears large anteriorly, because patients are scanned supine and the brain settles dependently in the calvarium.

The subarachnoid space is prominent in infants up to 1 year of age (it may extend to 18 months). Thereafter rapid brain growth fills the skull and causes effacement of gyral and sulcal markings.

Myelination

Myelination is best evaluated on T1 in children younger than 8 months of age while it is best evaluated on T2 and FLAIR in older children. On T1 myelin has high signal while on T2 myelin has a low signal.

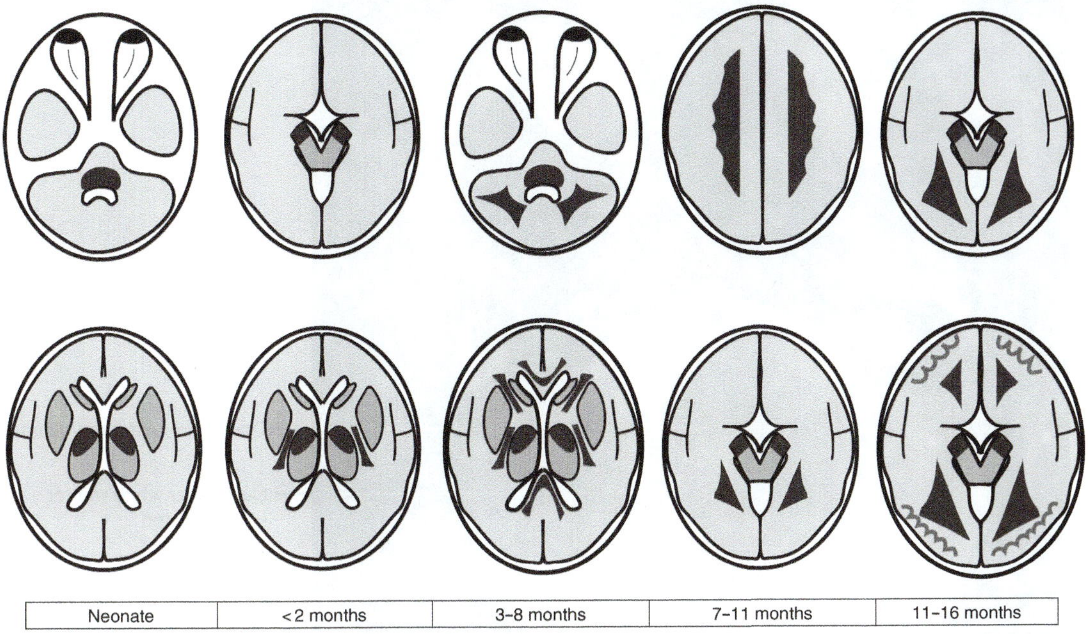

Fig. 30.12 Myelin maturation. Progressive myelination is indicated in here black. It begins with myelination of the dorsal brainstem and ventro-lateral thalami and then involves the internal capsules progressively from posterior to anterior from 2 to 8 months. The corpus callosum begins to myelinate and thicken from 3 to 8 months, while the deep white matter myelinates before the subcortical white matter and progresses from below upwards and from posterior to anterior

Magnetic Resonance Imaging

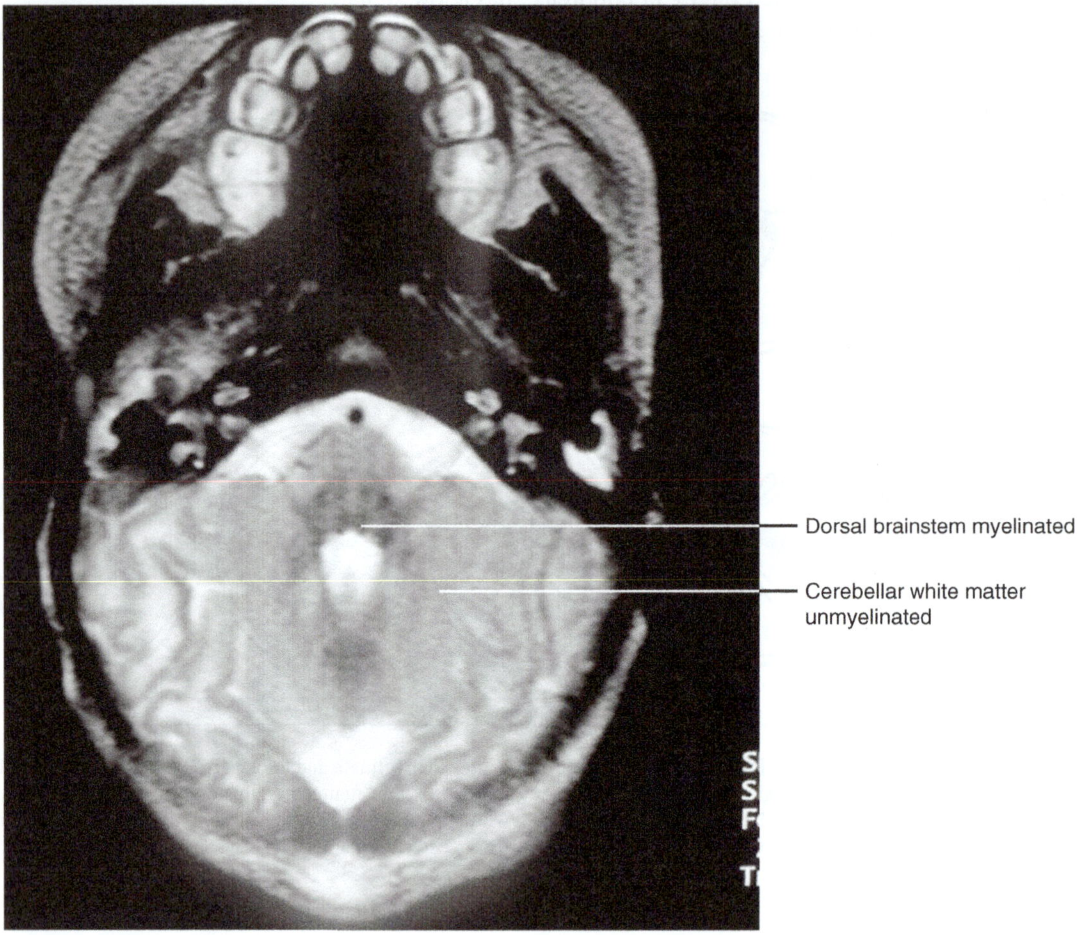

Fig. 30.13 T2 axial: birth

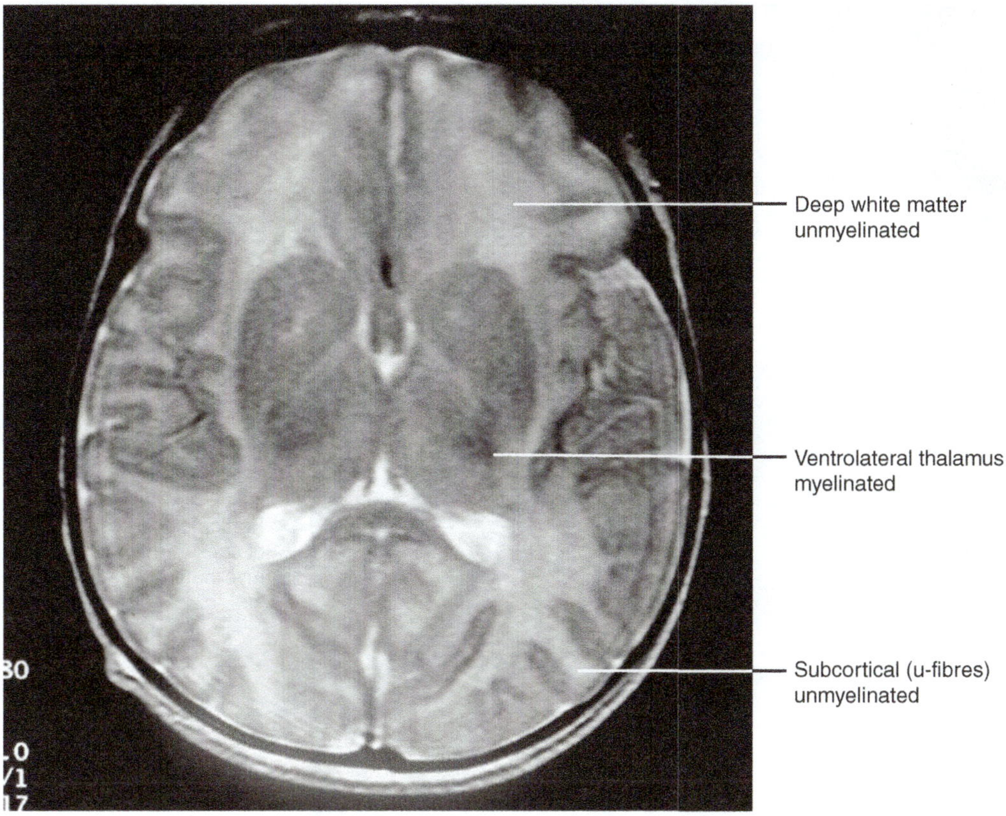

Fig. 30.14 T2 axial: birth

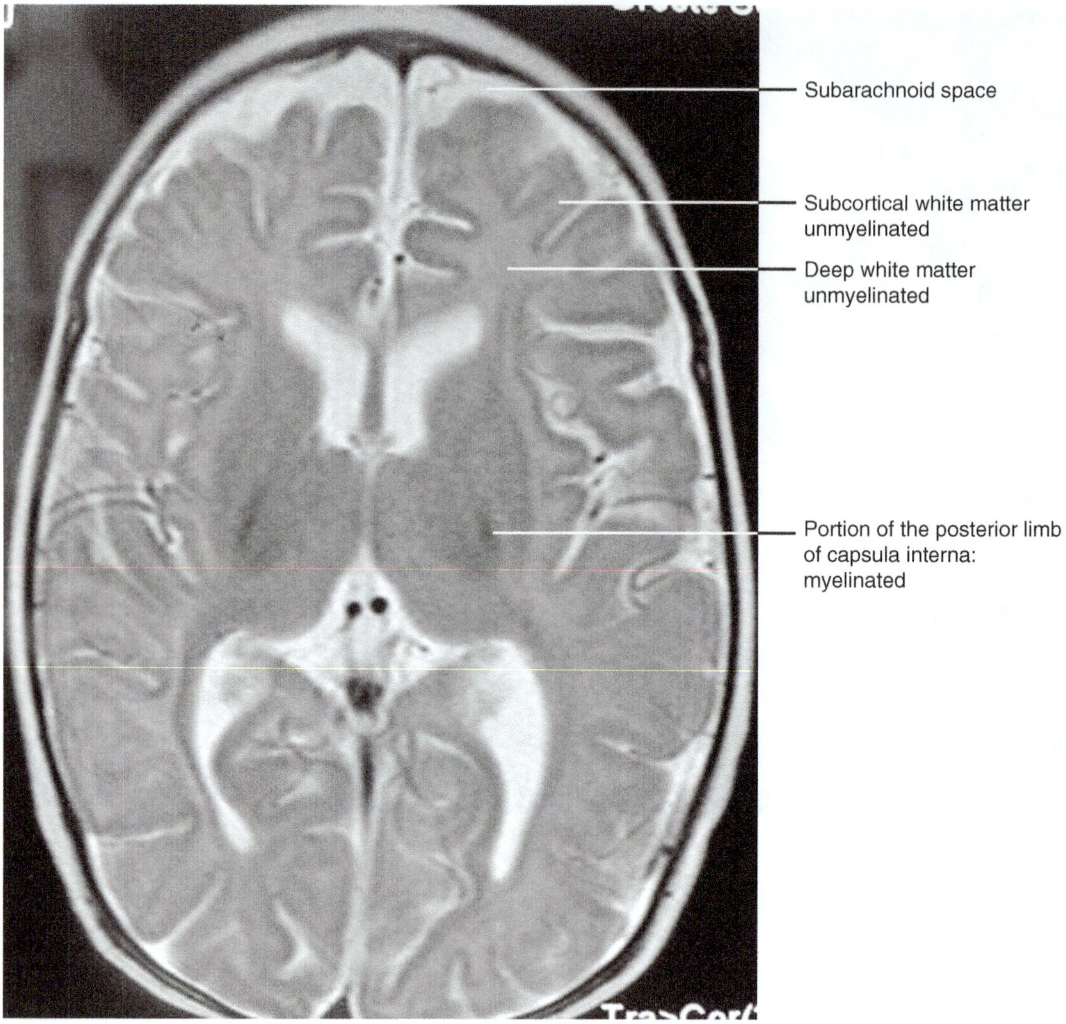

Fig. 30.15 T2 axial: 2 month old

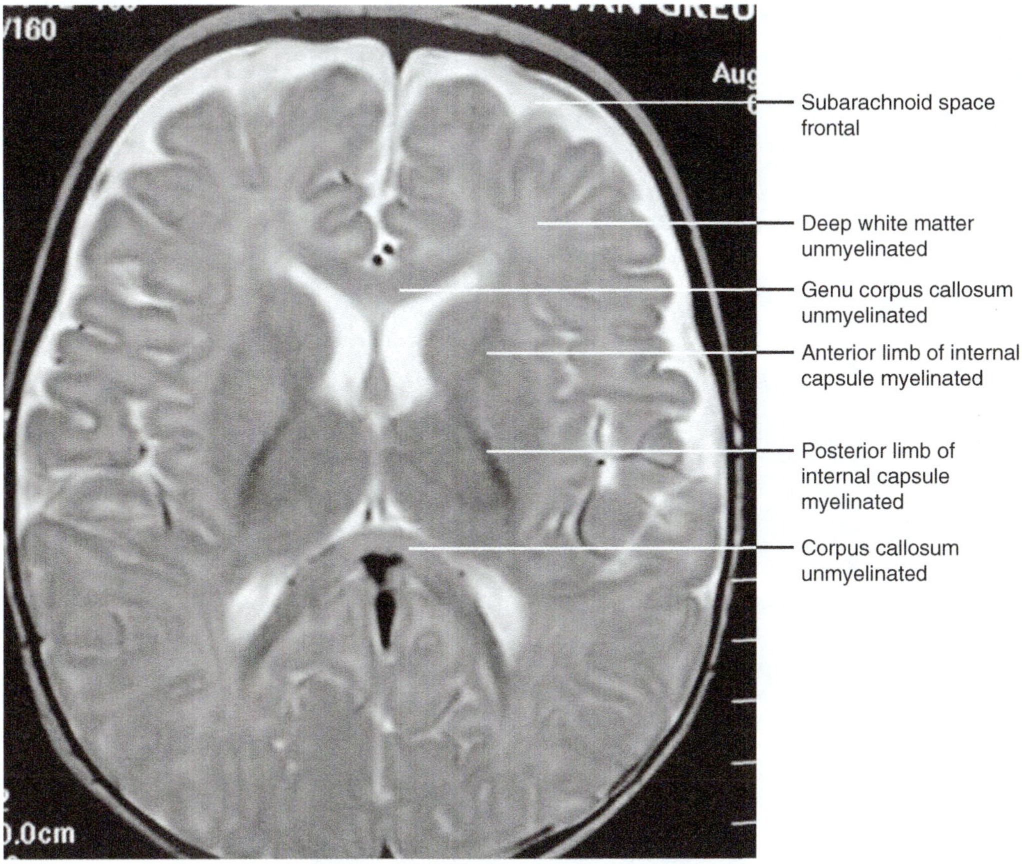

Fig. 30.16 T2 axial: 2 month old

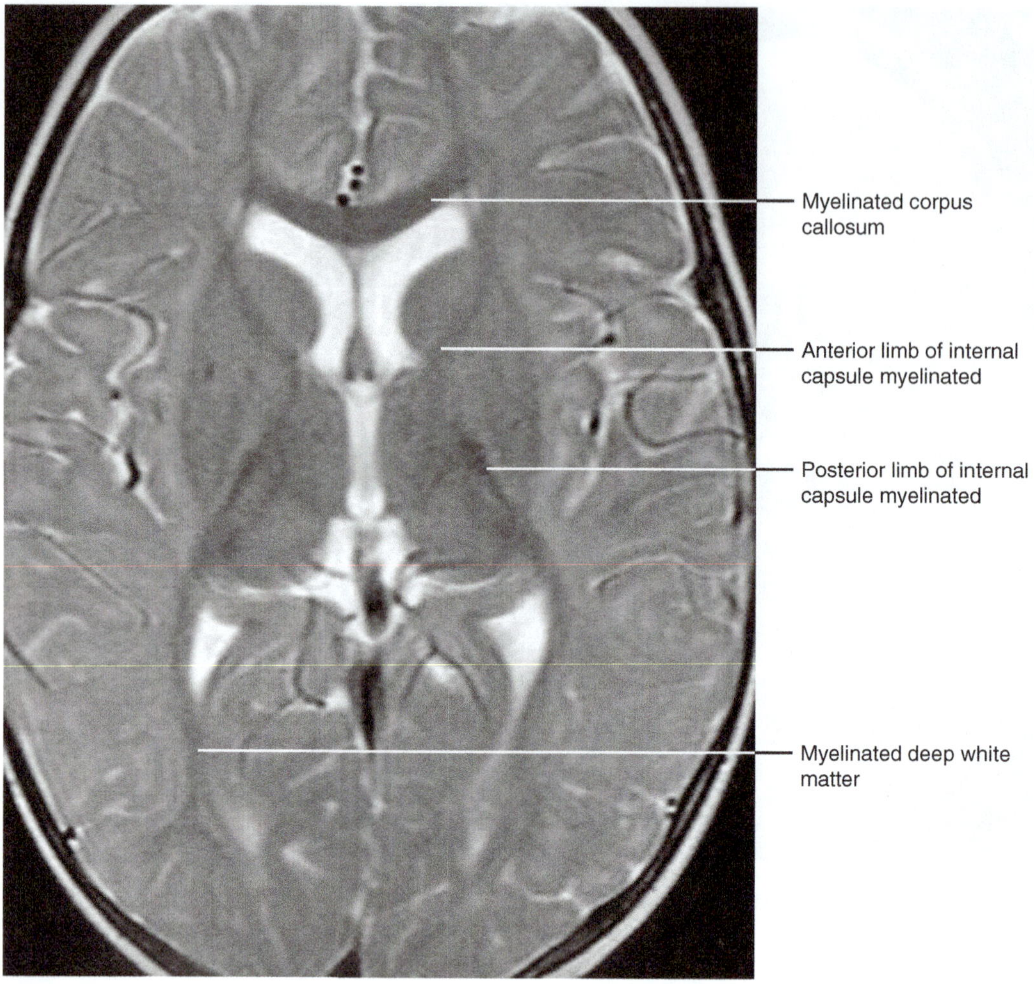

Fig. 30.17 T2 axial: 6 month old

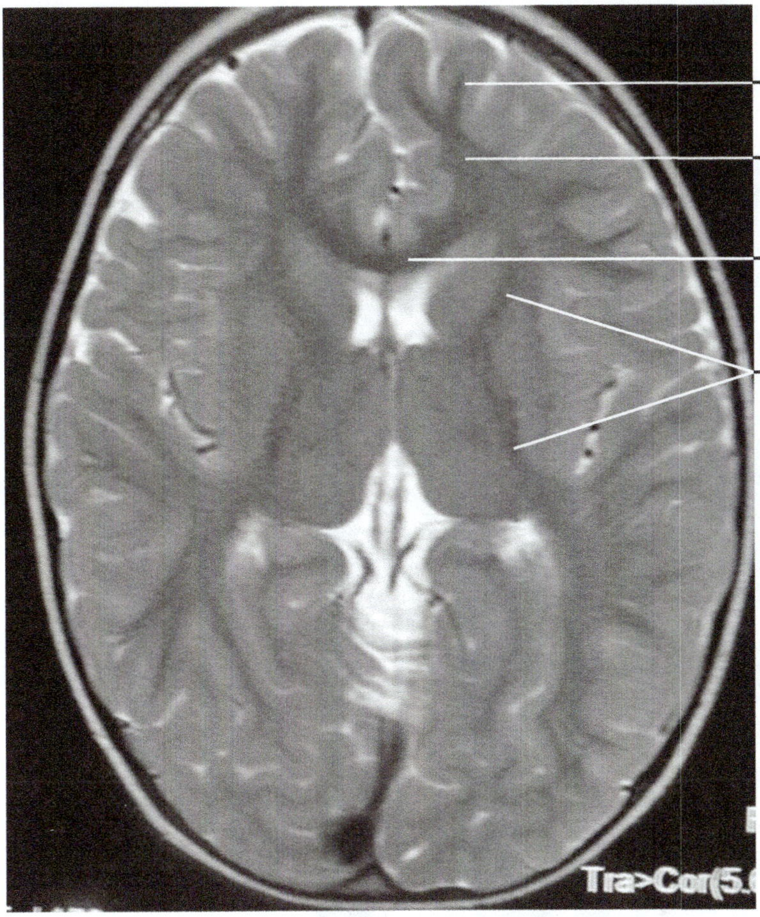

Fig. 30.18 T2 axial: 18 month old

Pearl
Myelination should be complete by age 2 years except for terminal areas.
Terminal areas include peritrigonal, temporal and frontal areas. These areas may persist into adulthood as high signal on T2-weighted images and should not be confused with pathology.

Table 30.1 T2 myelin milestones

Neonate	<2 months	3–8 months	7–11 months	11–16 months
Dorsal brainstem	Cerebral peduncle	Cerebellar white matter	Centrum semiovale	Peripheral white matter, occipital lobe
Ventrolateral thalami	Posterior limb of internal capsule	Anterior limb of internal capsule Corpus callosum Increased thickness	Central white matter, occipital lobe	Central white matter, frontal lobe

Corpus Callosum Development

Magnetic Resonance Imaging

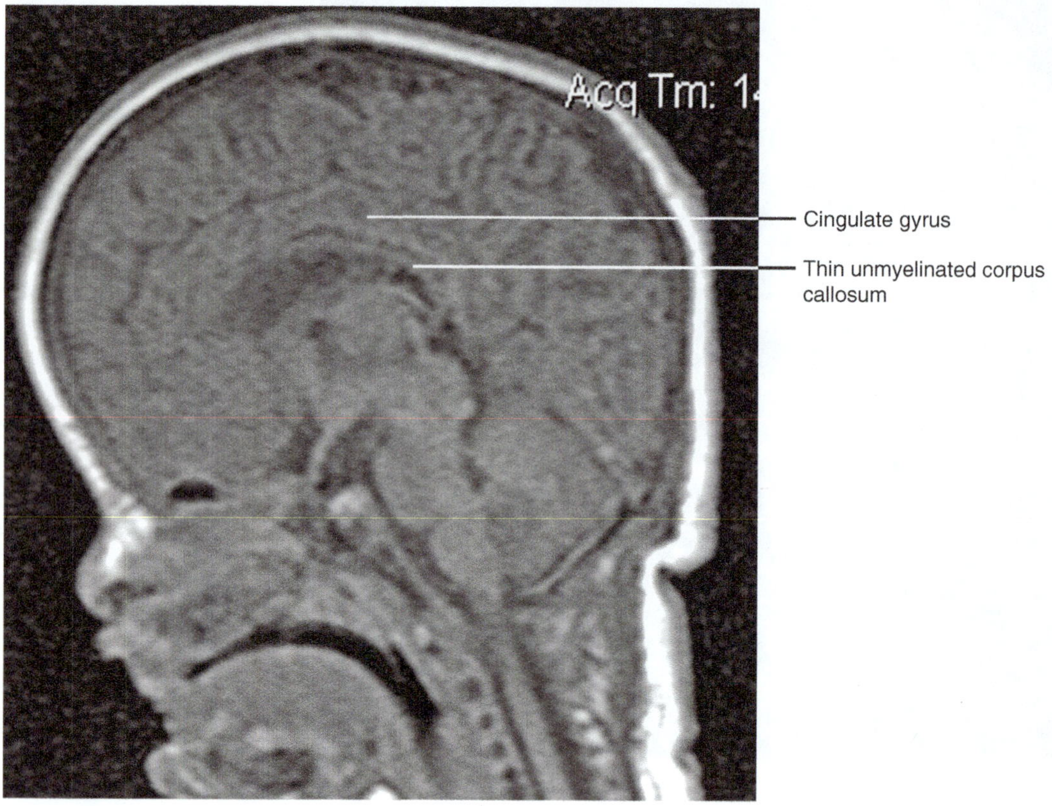

Fig. 30.19 T1 sagittal: 2 days old

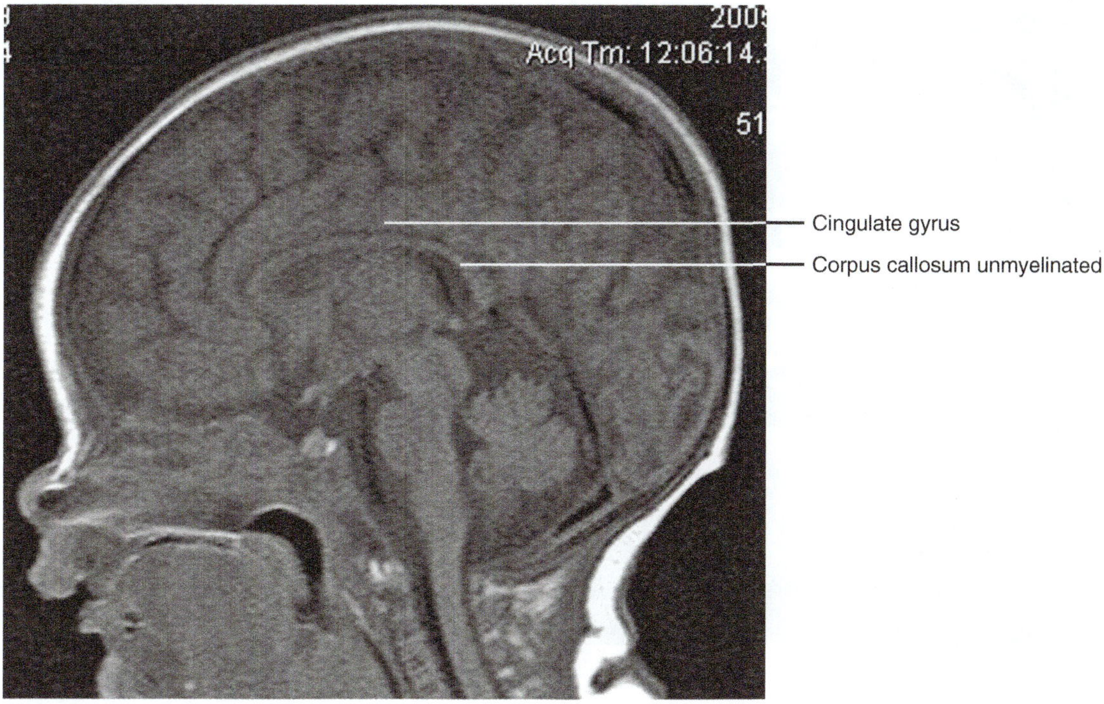

Fig. 30.20 T1 sagittal: 2 month old

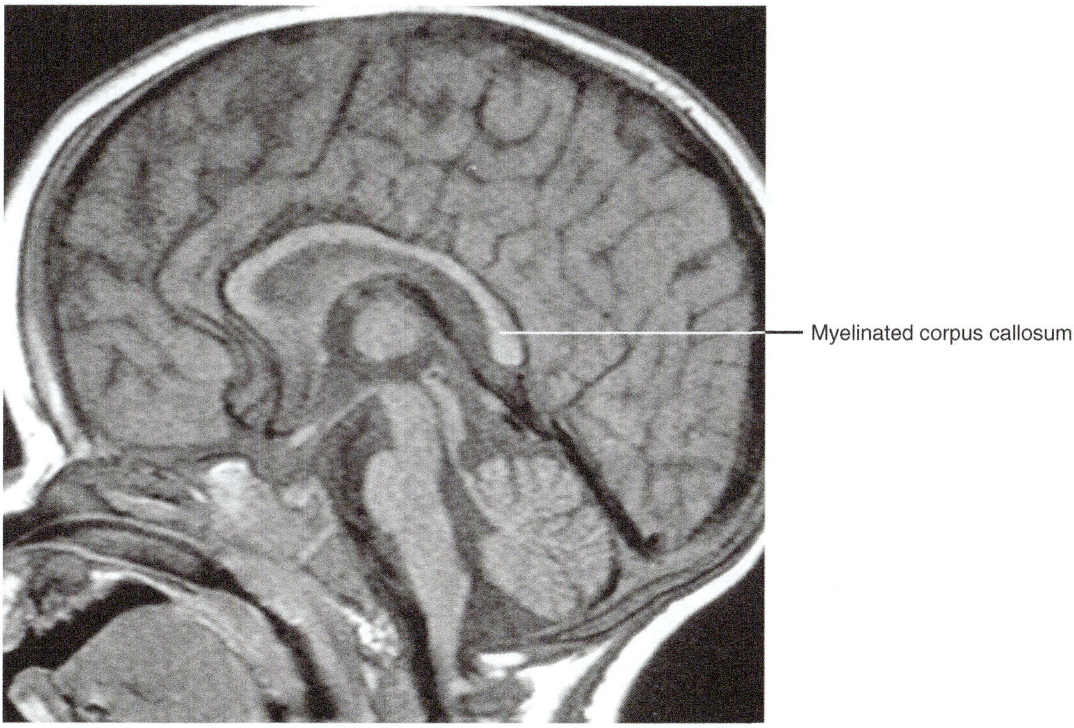

Fig. 30.21 T1 sagittal: 4 months old

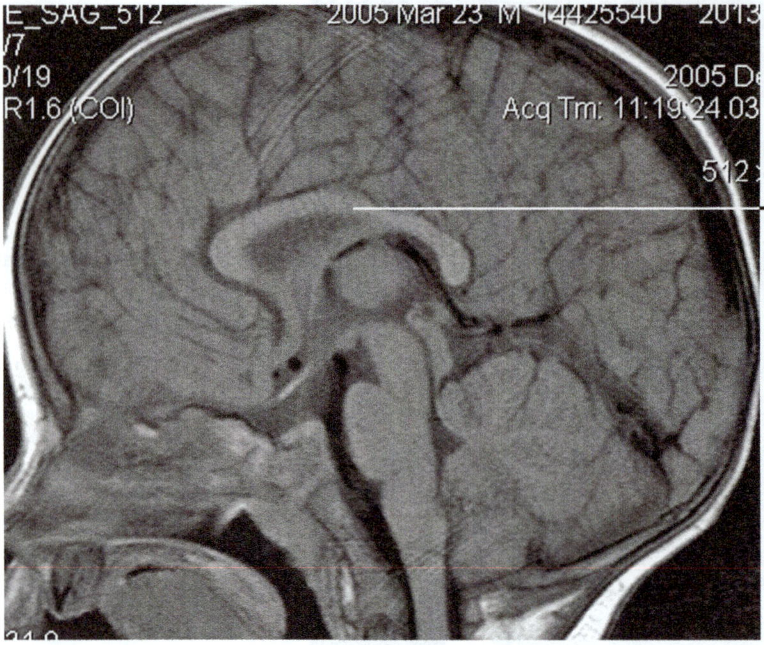

Fig. 30.22 T1 sagittal: 6 months old

Pearl

During the neonatal period the corpus callosum is uniformly thin and of the same signal intensity as white matter throughout the brain.

High signal intensity on T1-weighted images related to the myelination process begins to appear in the splenium at approximately 4 months and in the genu at approximately 6 months.

The corpus callosum shows rapid growth until 3 years of age.

After this there is stabilisation of the Isthmus-splenial thickness ratio (considered indicative of CC modeling).

The maximum CC size is gained at between 6 – 15 years of age.

There is however signficant individual variation in CC morphology.

The anterior CC reaches its maximum size earlier than the middle and posterior portions.

Most variability in size and shape is found posteriorely at the splenium and isthmus.

In addition, most maturation changes across the age spectrum involve this posterior portion of the CC.

Changes occur via myelination, re-organisation and pruning of fibres.

There is also progressive increase in length of the CC throughout childhood.

Cavum Septi Pellucidi and Cavum Vergae

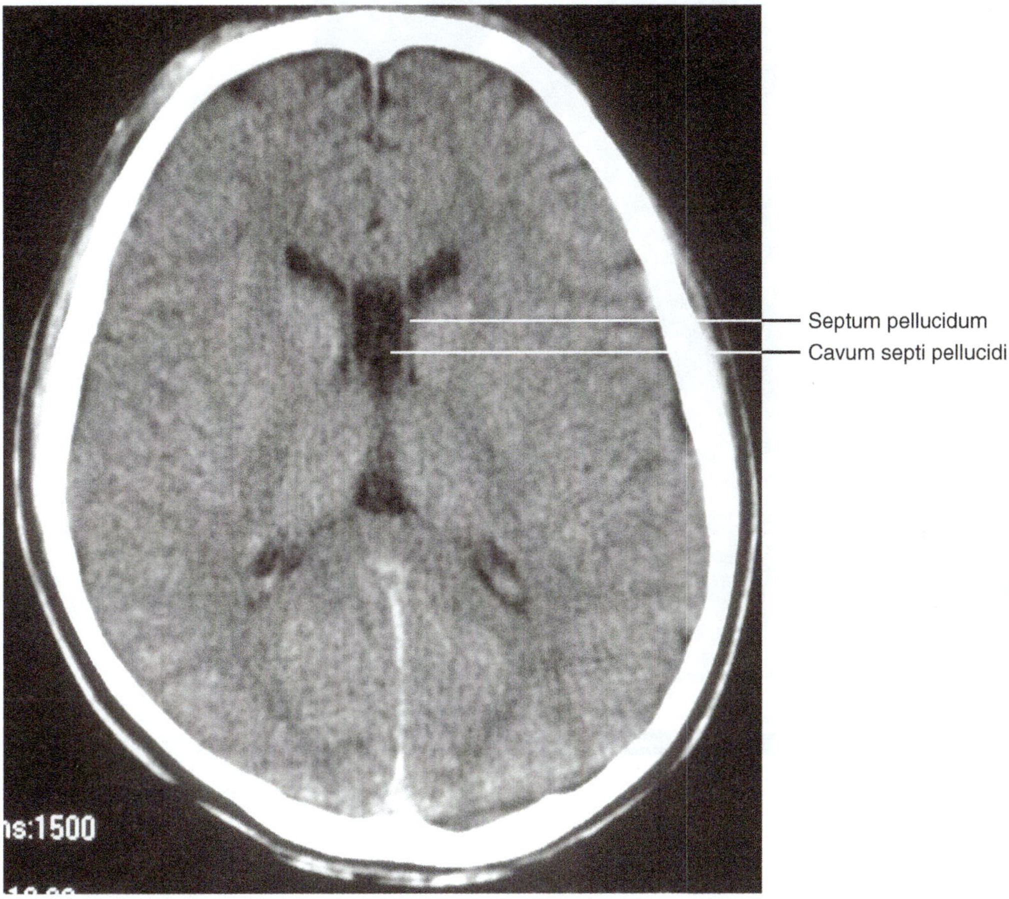

Fig. 30.23 Axial CT

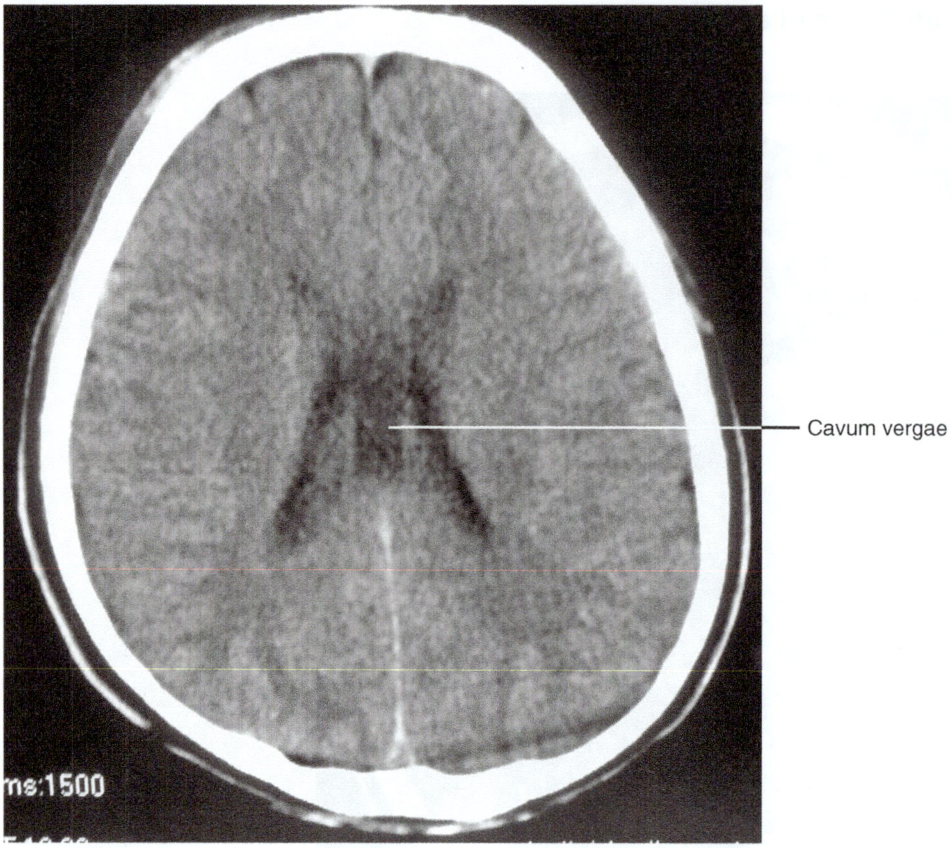

Fig. 30.24 Axial CT

Spinal Cord/Conus Medullaris

Ultrasound

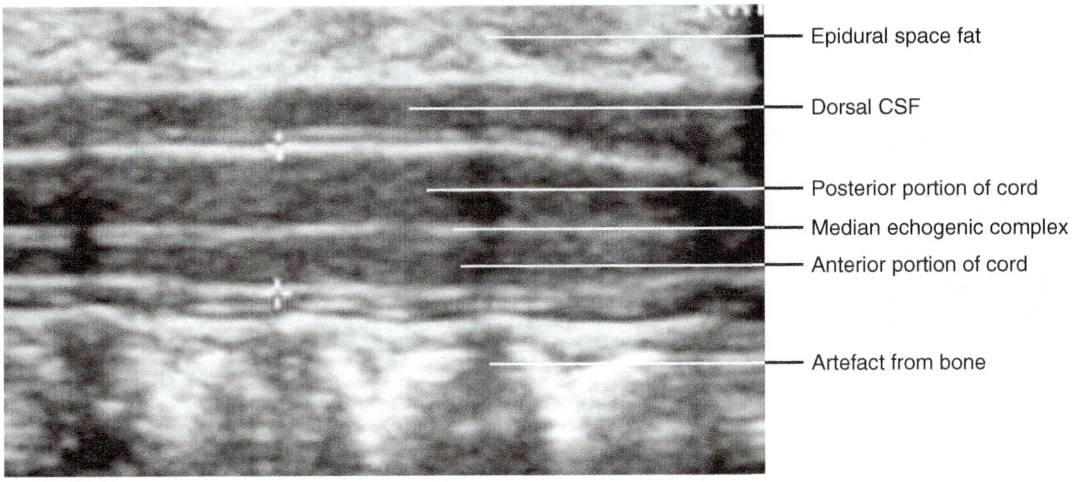

Fig. 30.25 Sagittal US spine

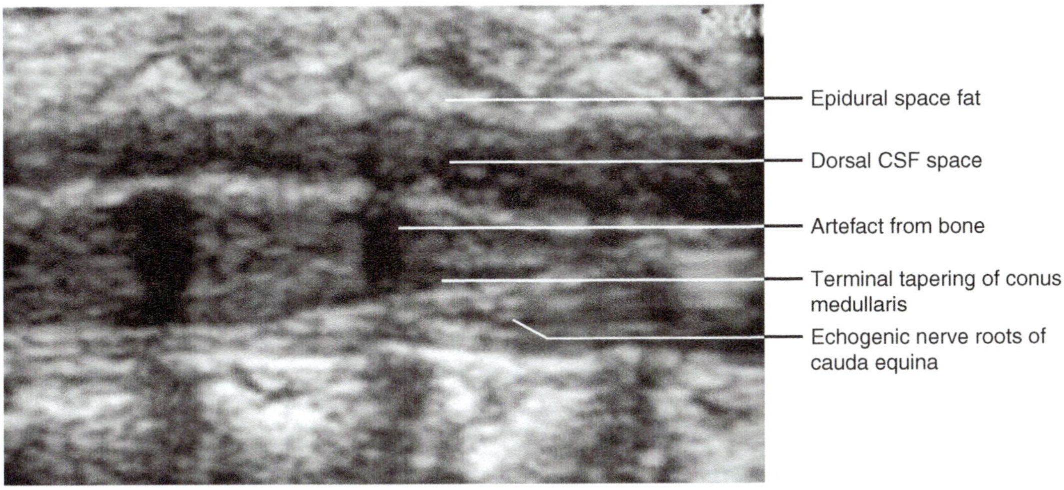

Fig. 30.26 Sagittal US spine

Pearl
Cord – hypoechoic; cauda equina and nerve roots – hyperechoic.

Pearl
The dorsal CSF space is more prominent than the ventral CSF space because the patient is scanned prone and the cord assumes a dependent position.

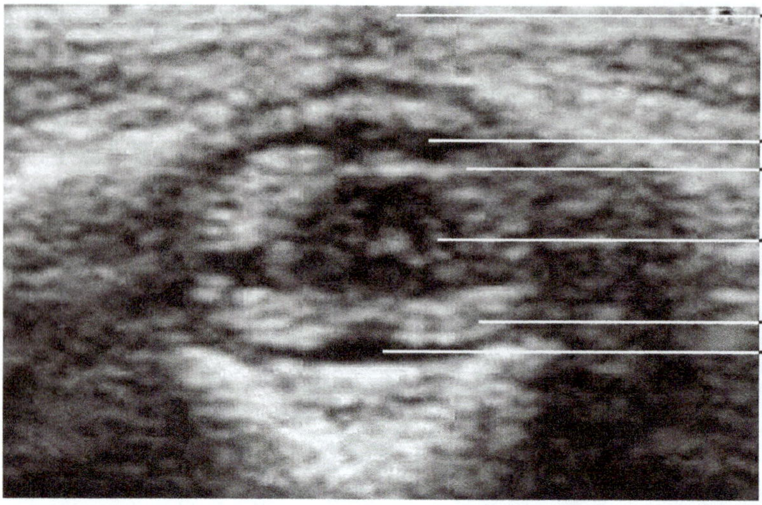

Fig. 30.27 Transverse US spine

Magnetic Resonance Imaging

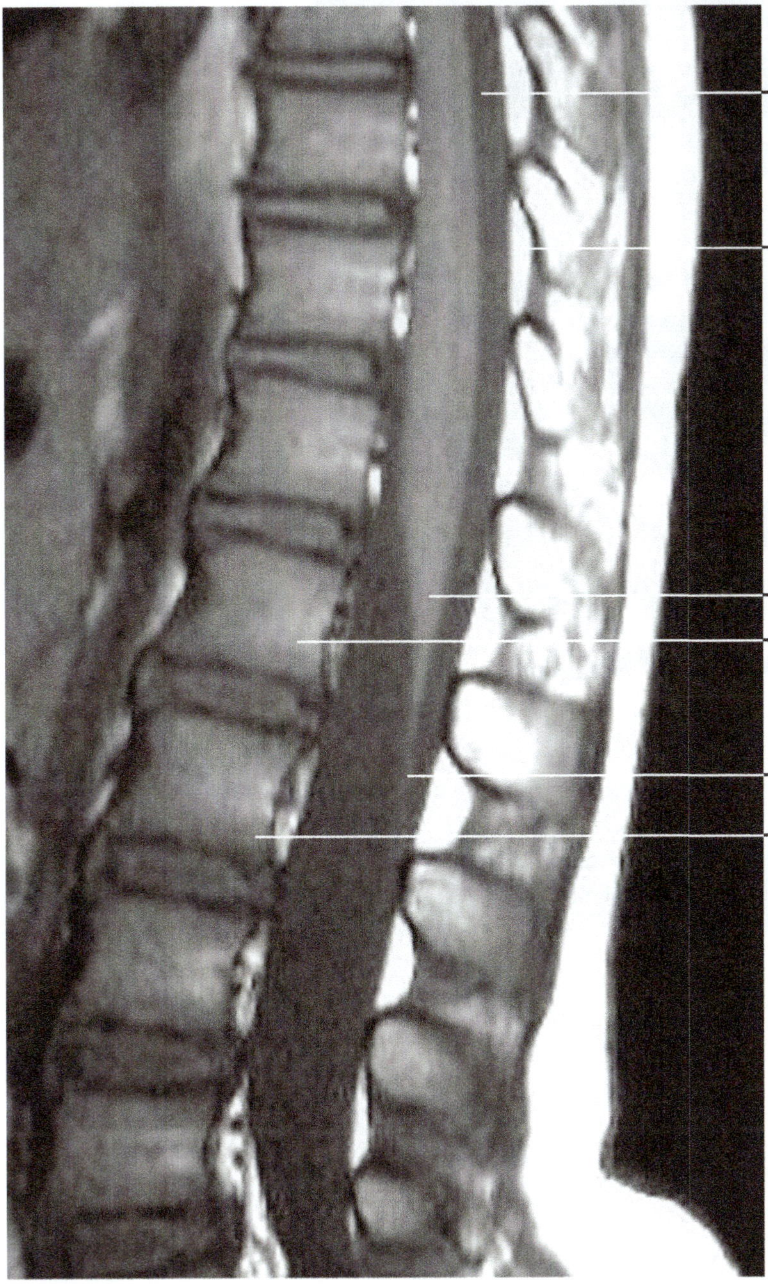

Fig. 30.28 T1 sagittal

Pearl
The conus should taper and terminate at the lower end plate of L1, but may be normal even at L2. In adults the conus ascends relative to the vertebral column up to T12 – use T1 axial to assess.

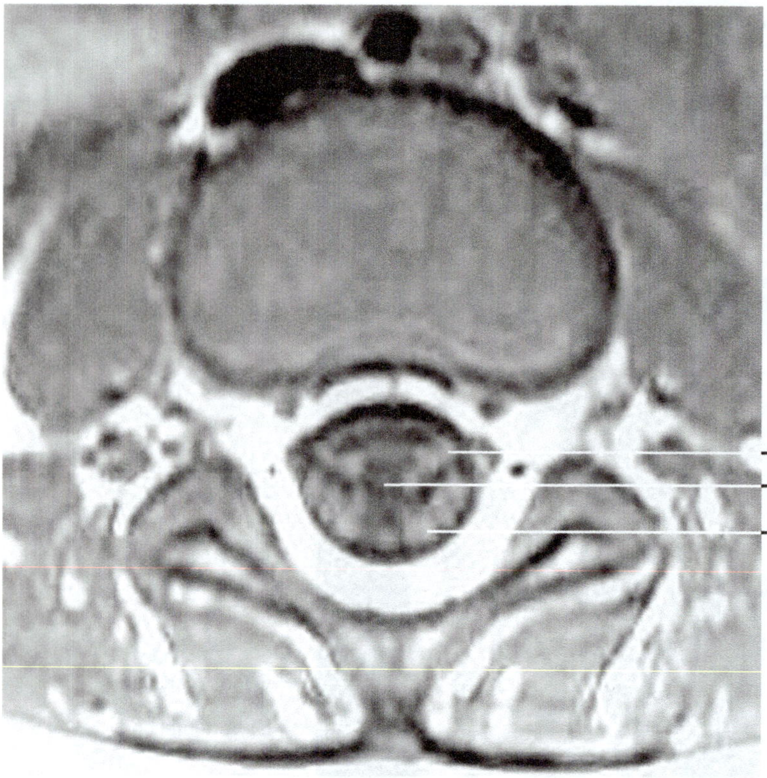

Fig. 30.29 T1 axial

Pearl

The filum terminate should be 2 mm or less in diameter measured on T1 axial. The signal is iso-intense to cord and must not contain fat (high signal on T1).

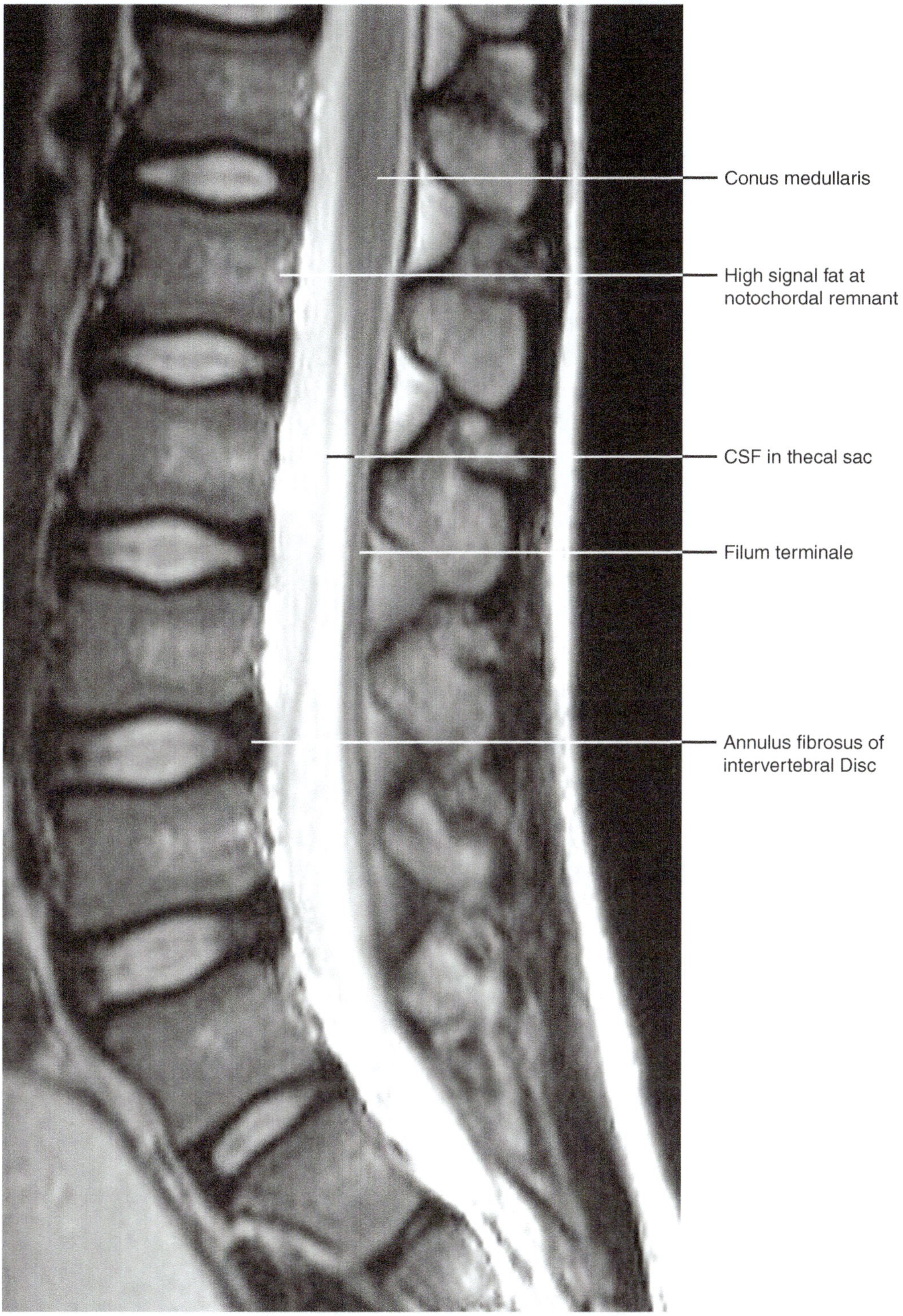

Fig. 30.30 T2 sagittal

Pearl

The intervertebral discs have a higher T2 signal in children than adults.

Paediatric Head and Neck

31

Sutures

Plain Film

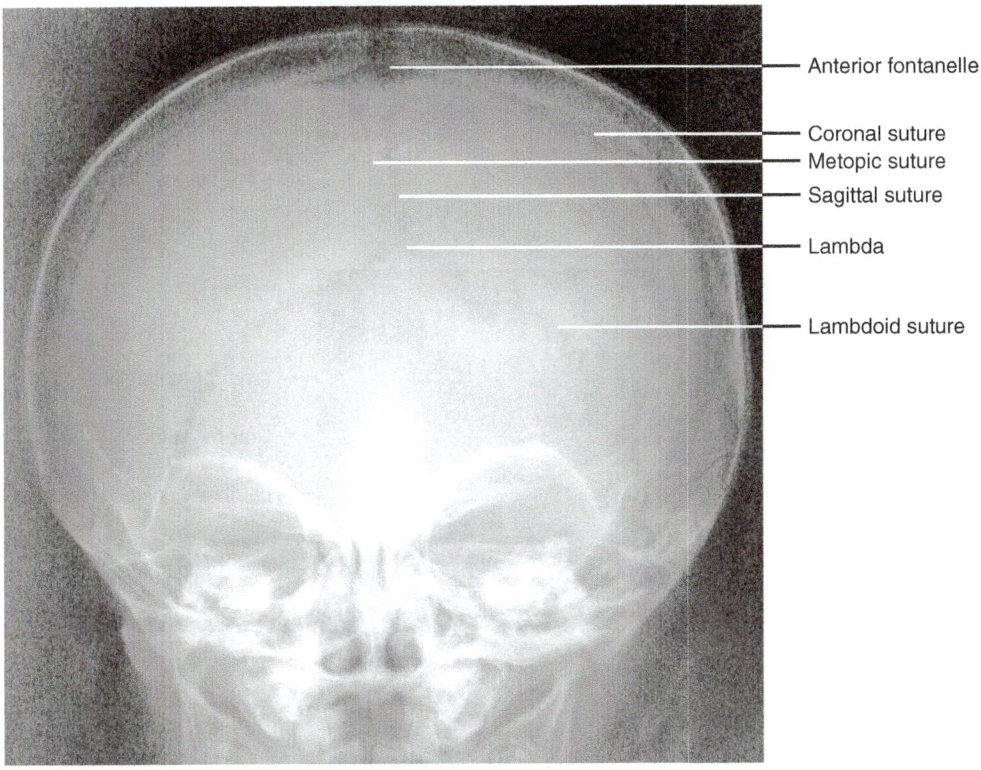

Fig. 31.1 Frontal view

S. Andronikou (ed.), *See Right Through Me*,
DOI 10.1007/978-3-642-23893-2_31, © Springer-Verlag Berlin Heidelberg 2012

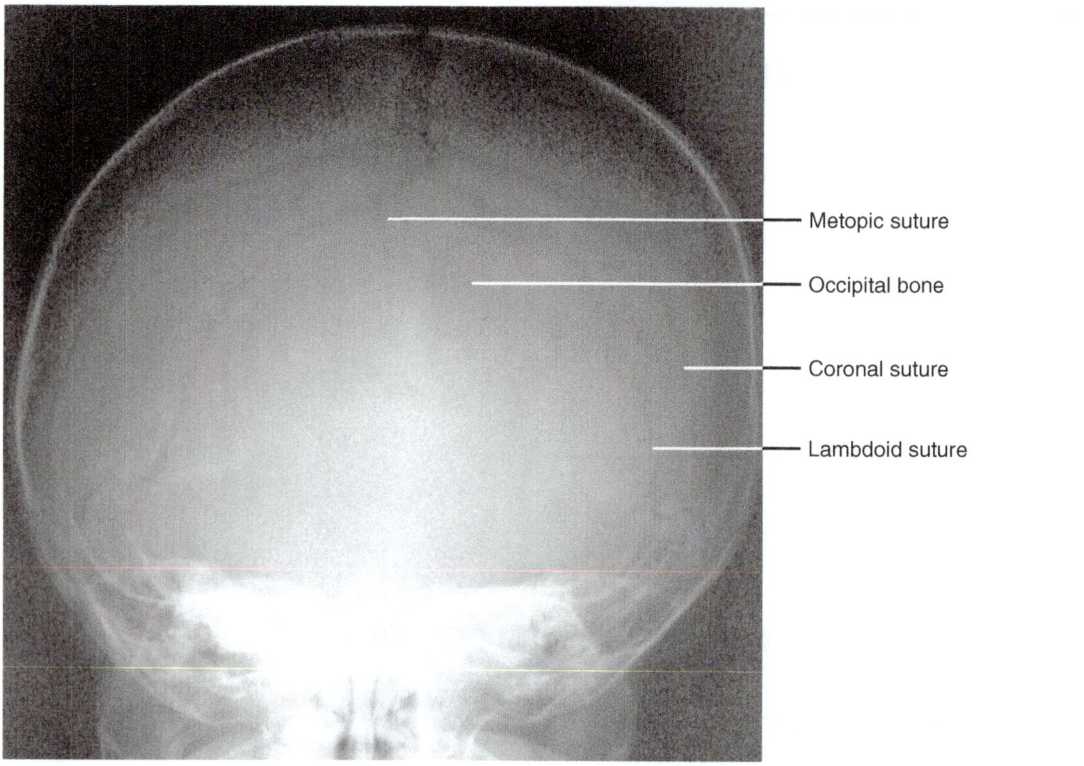

Fig. 31.2 Towne's view

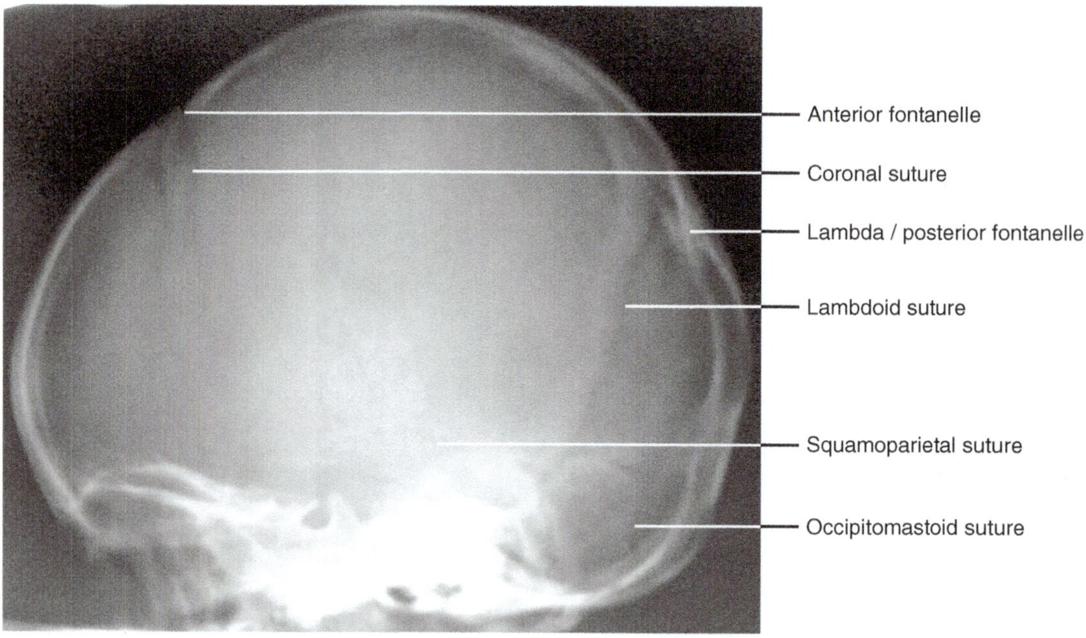

Fig. 31.3 Lateral view

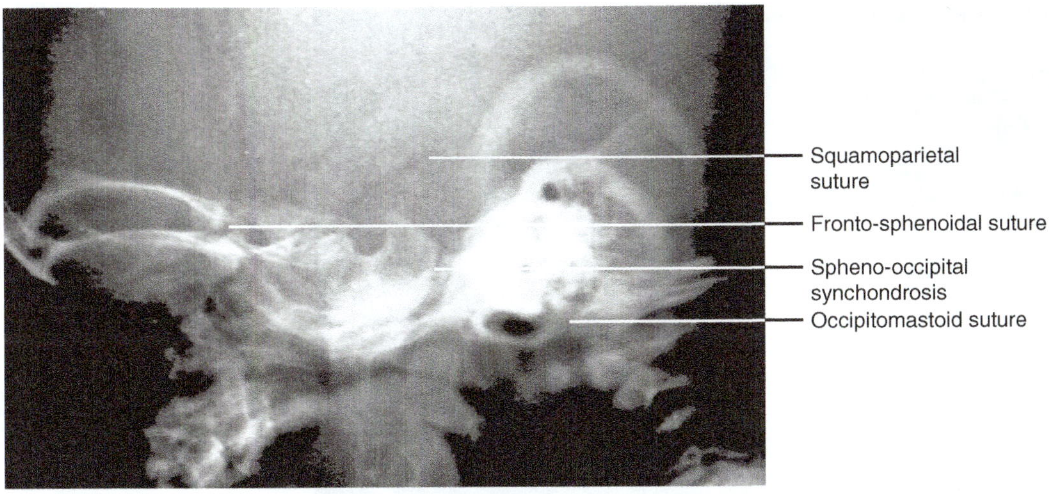

Fig. 31.4 Zoomed lateral view of base of skull

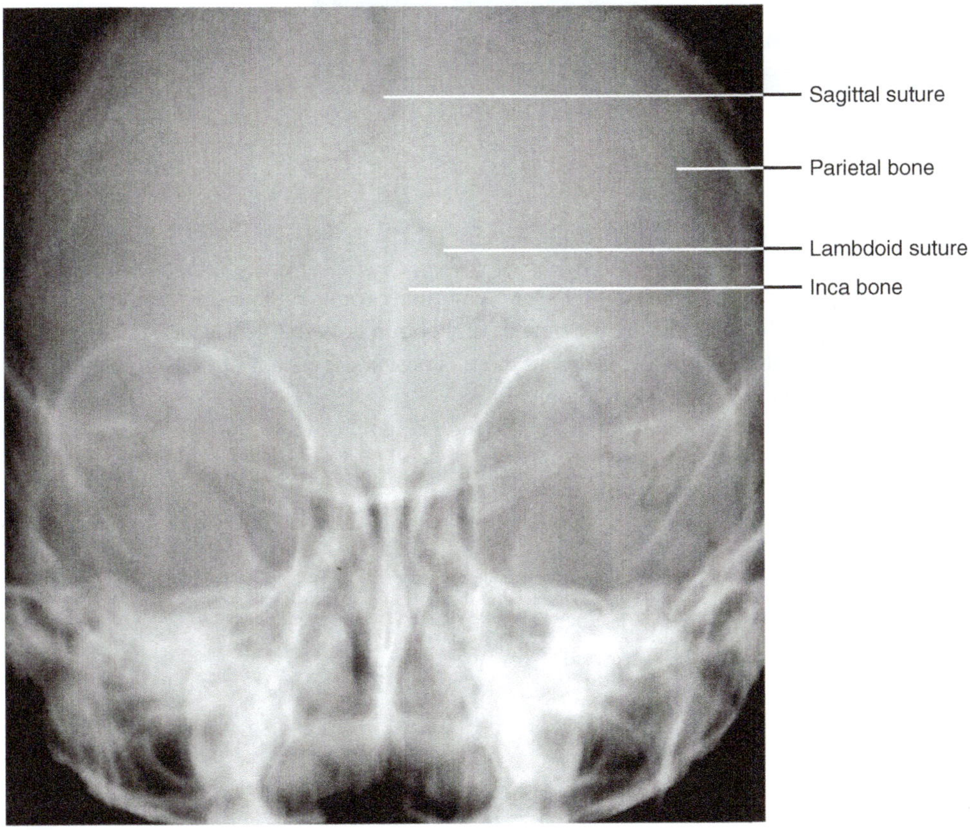

Fig. 31.5 Frontal view showing 'INCA' bone

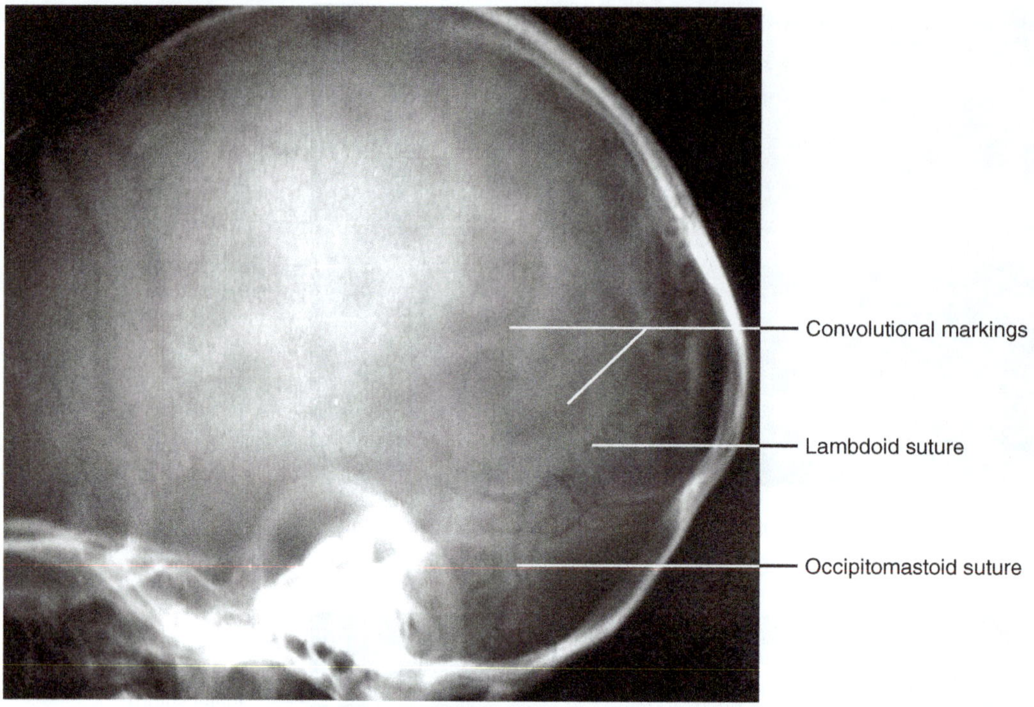

Fig. 31.6 Lateral view showing convolutional markings

Pearl

Convolutional markings = digital markings. These are normal prominent markings at the base of the skull and are thought to occur during the time of rapid brain growth (4–11 years) due to CSF pulsation. When the whole skull is involved and there is premature fusion of the sutures this is termed a 'copper beaten skull' and is pathological.

Computed Tomography

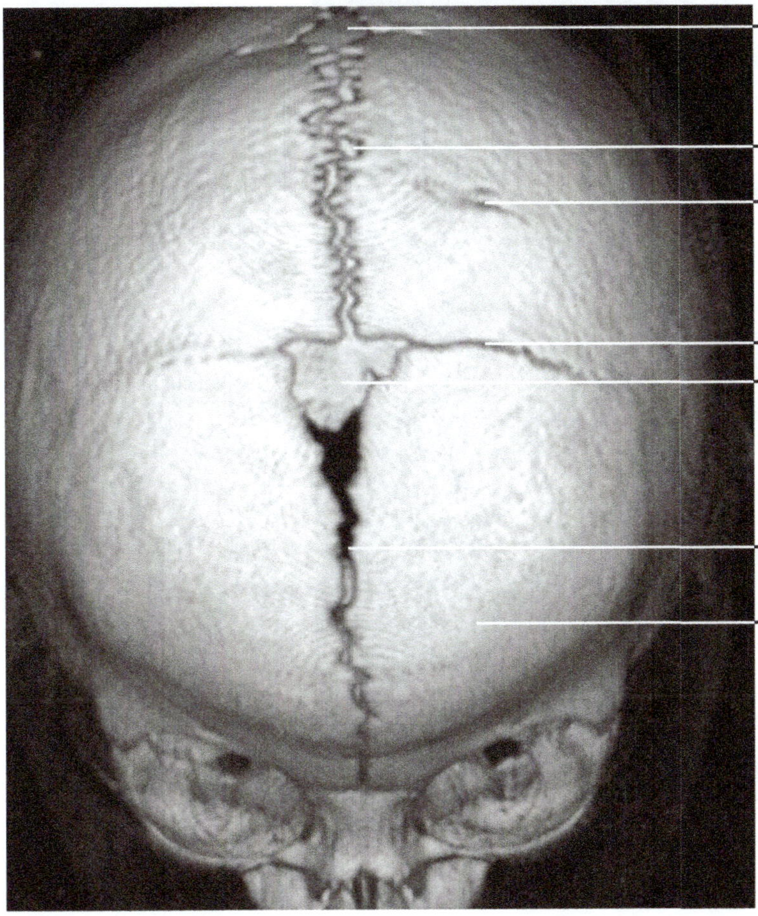

Fig. 31.7 3D reconstruction skull: frontal superior view

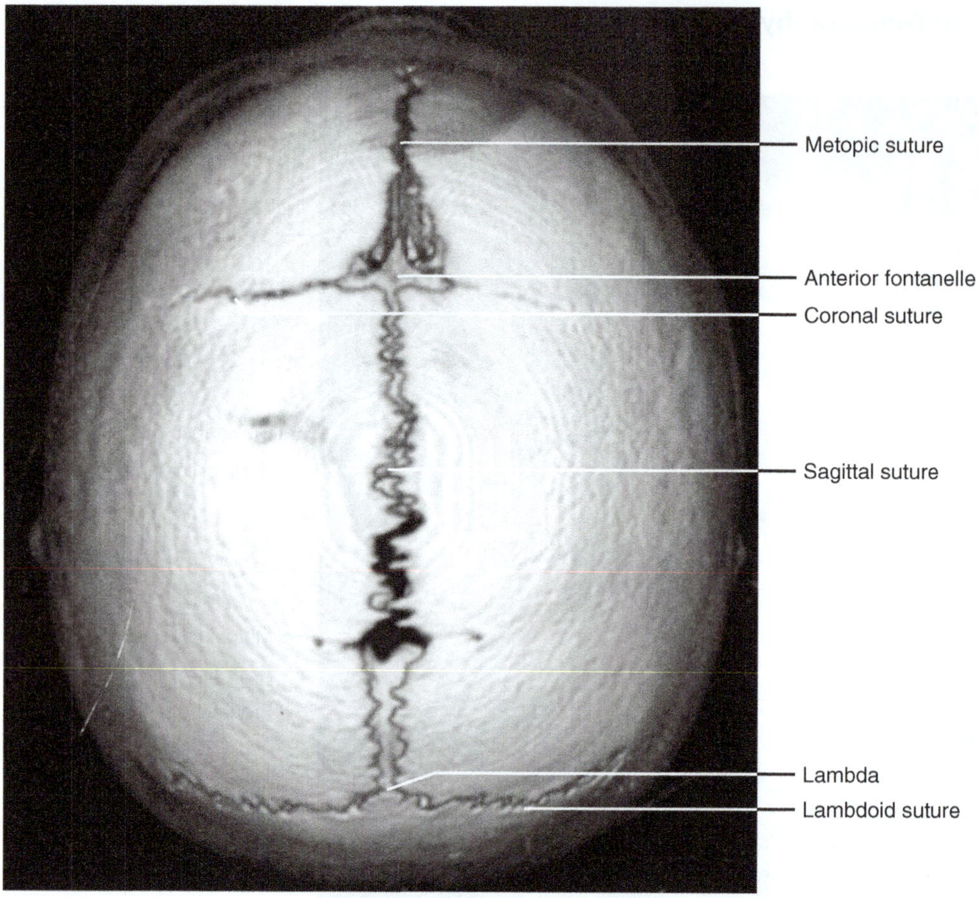

Fig. 31.8 3D reconstruction skull: frontal superior view

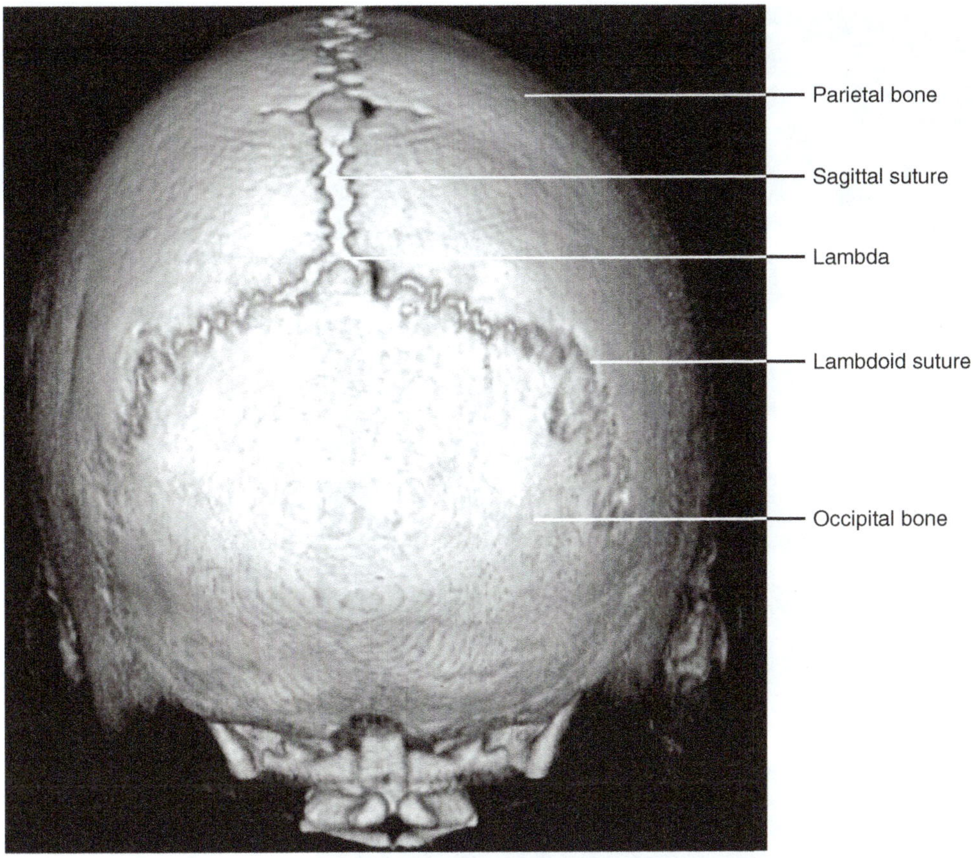

Fig. 31.9 3D reconstruction skull: posterior view

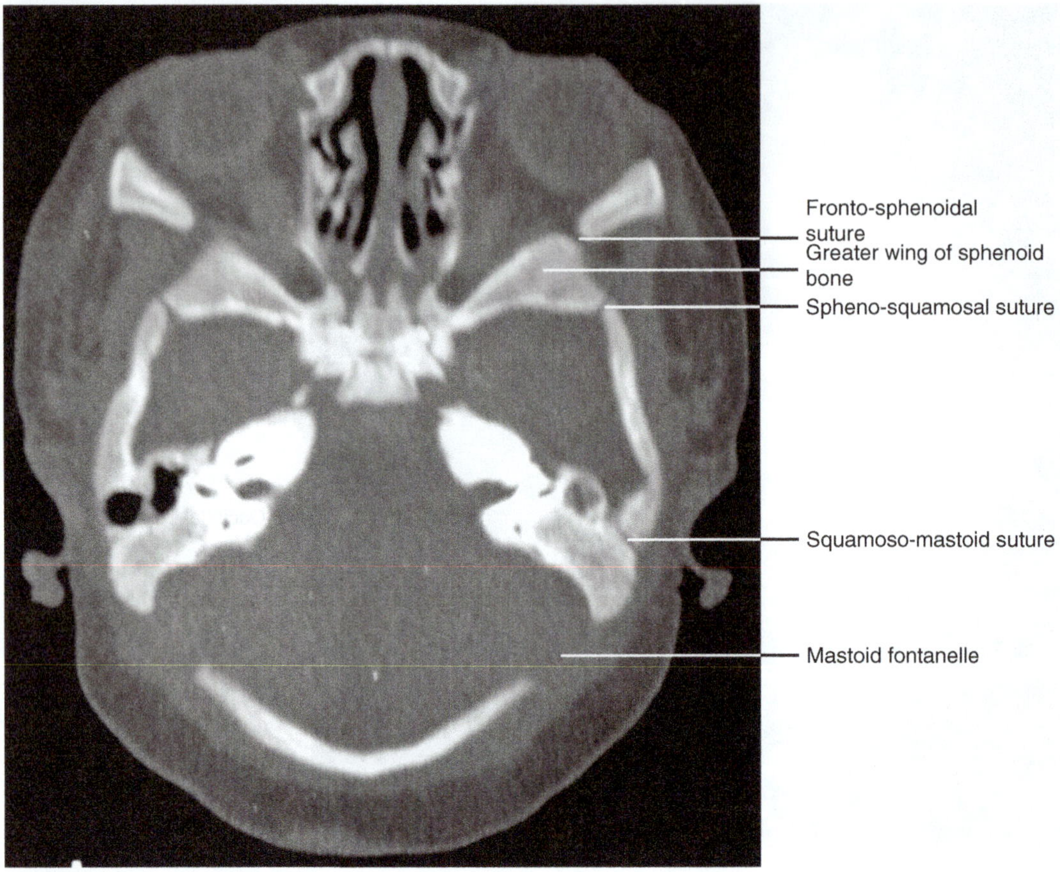

Fig. 31.10 Axial CT base of skull on bone window

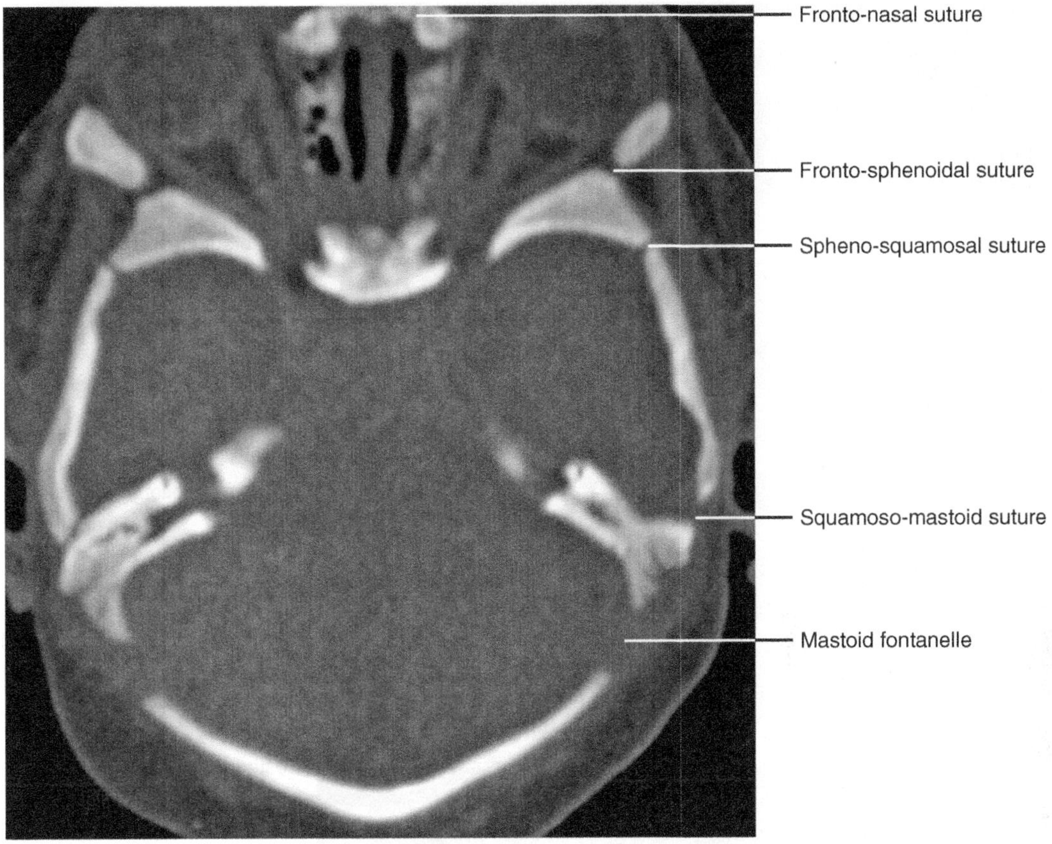

Fig. 31.11 Axial CT base of skull on bone window

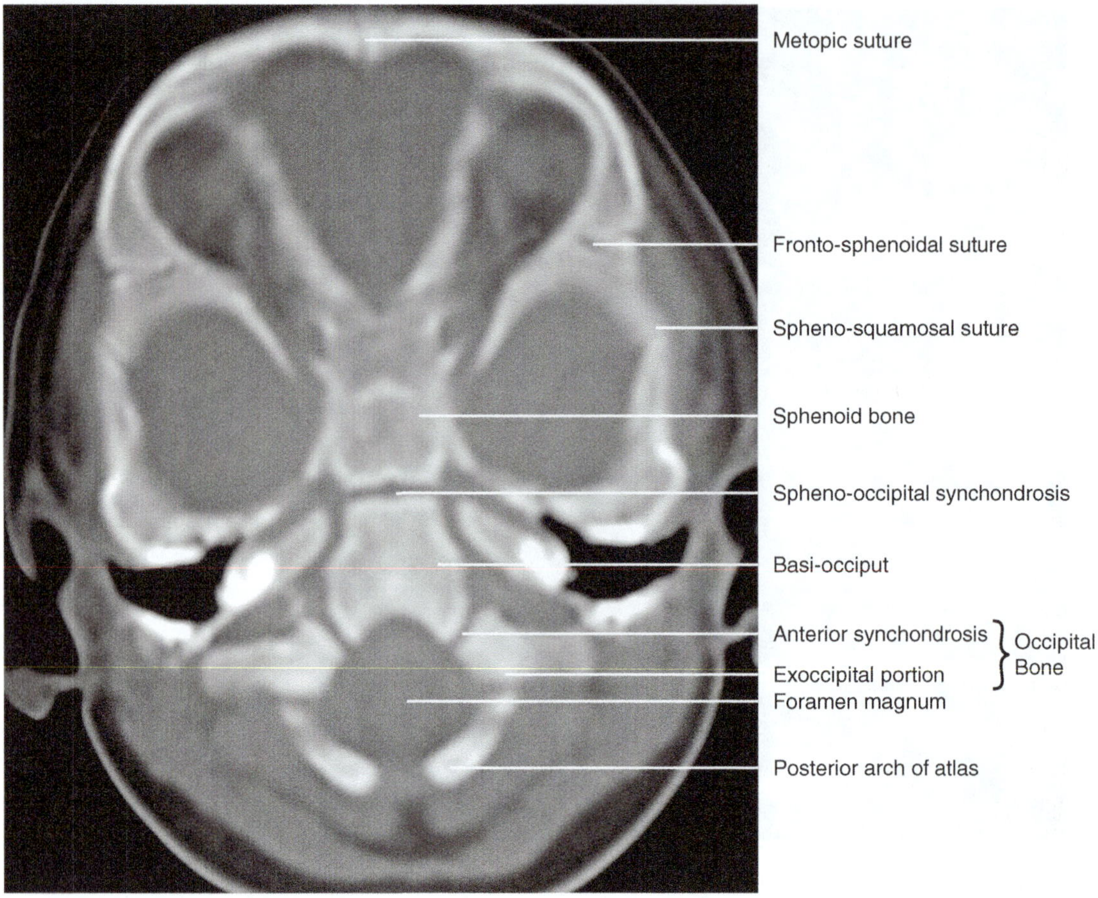

Fig. 31.12 Axial CT base of skull on bone window

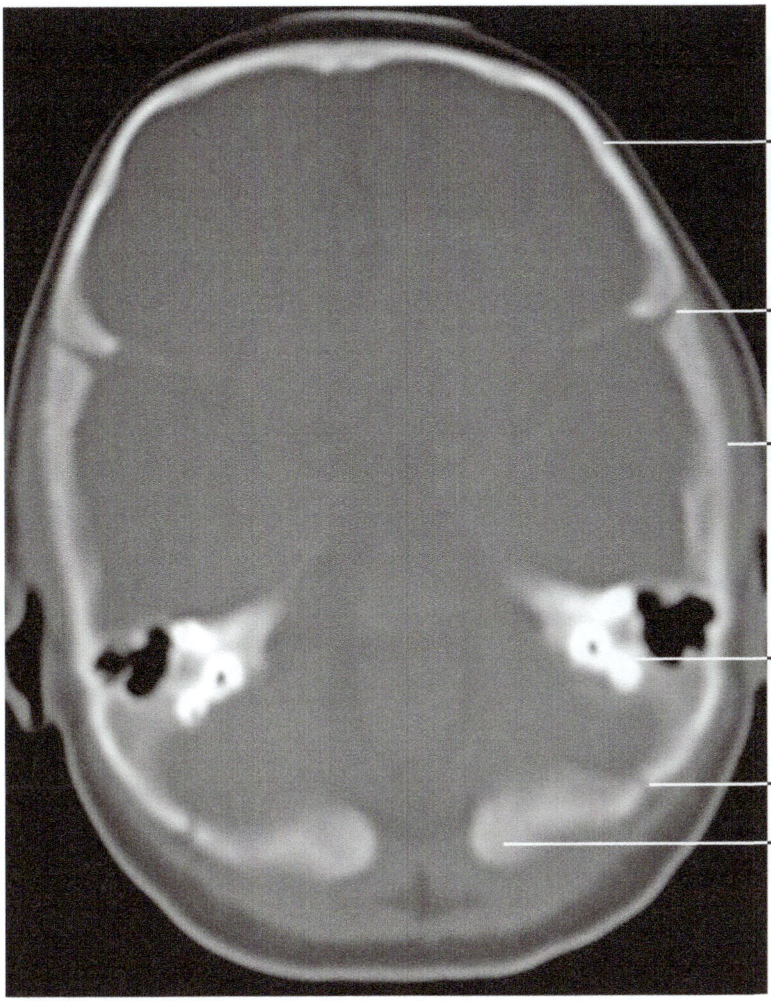

Fig. 31.13 Axial CT base of skull on bone window

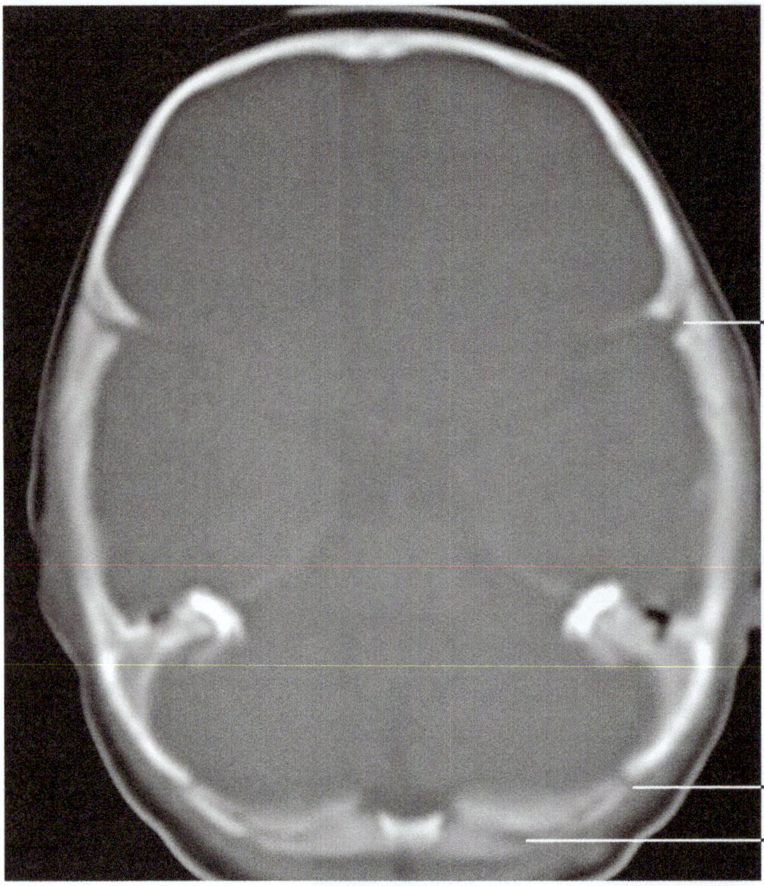

Fig. 31.14 Axial CT base of skull on bone window

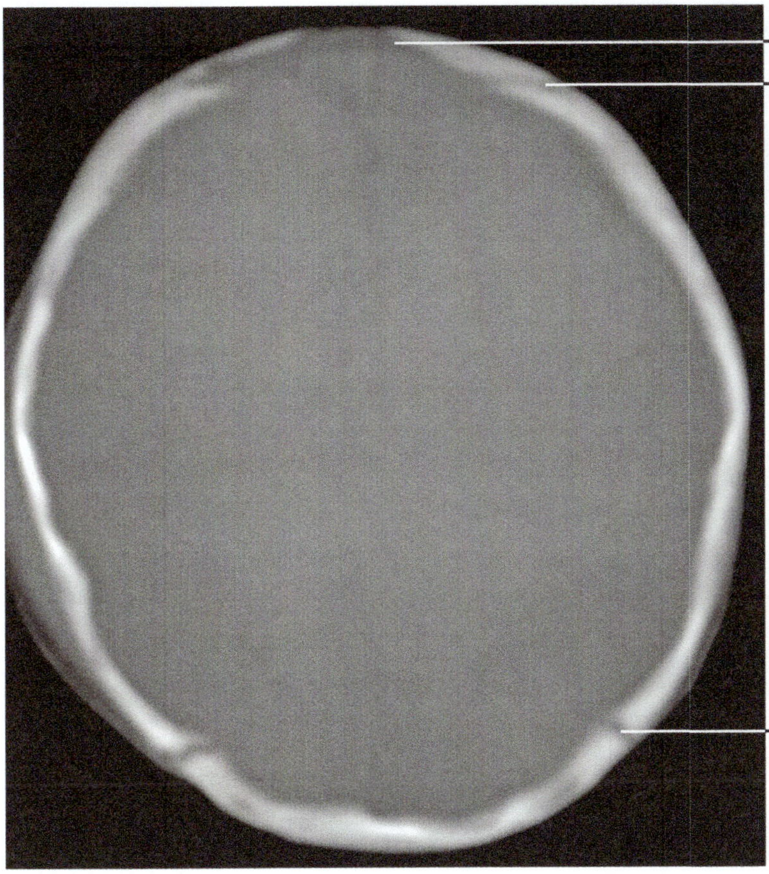

Fig. 31.15 Axial CT base of skull on bone window

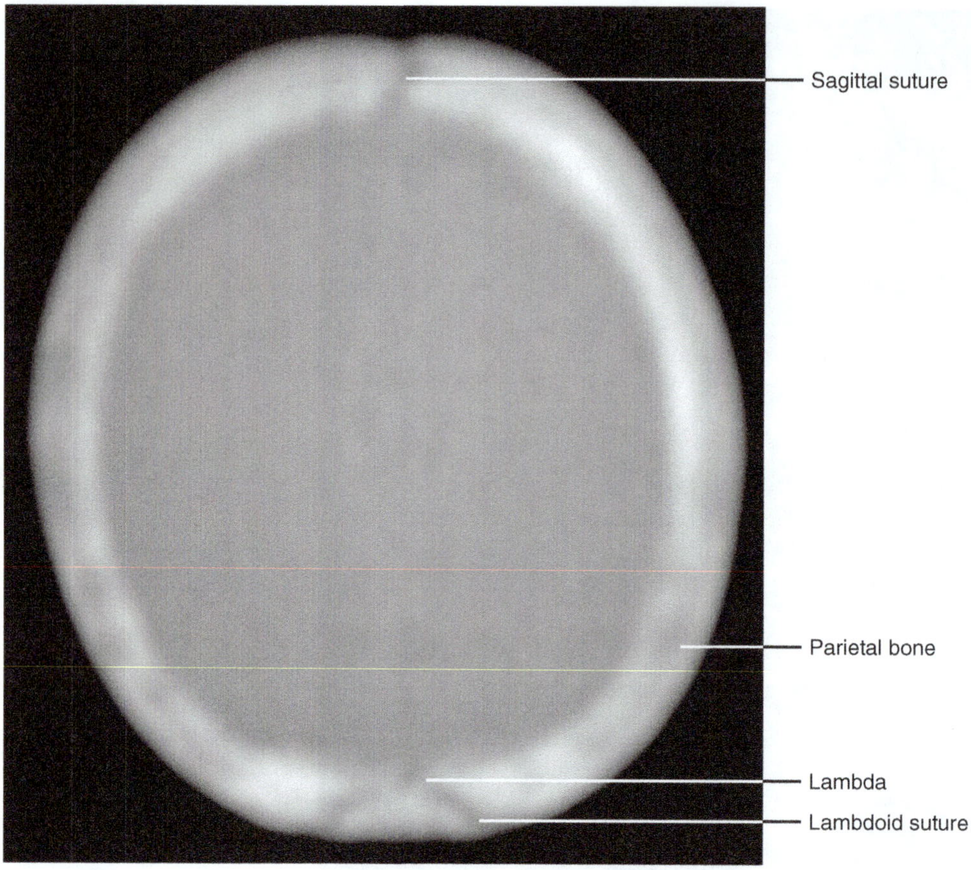

Fig. 31.16 Axial CT base of skull on bone window

Sinuses

Plain Film

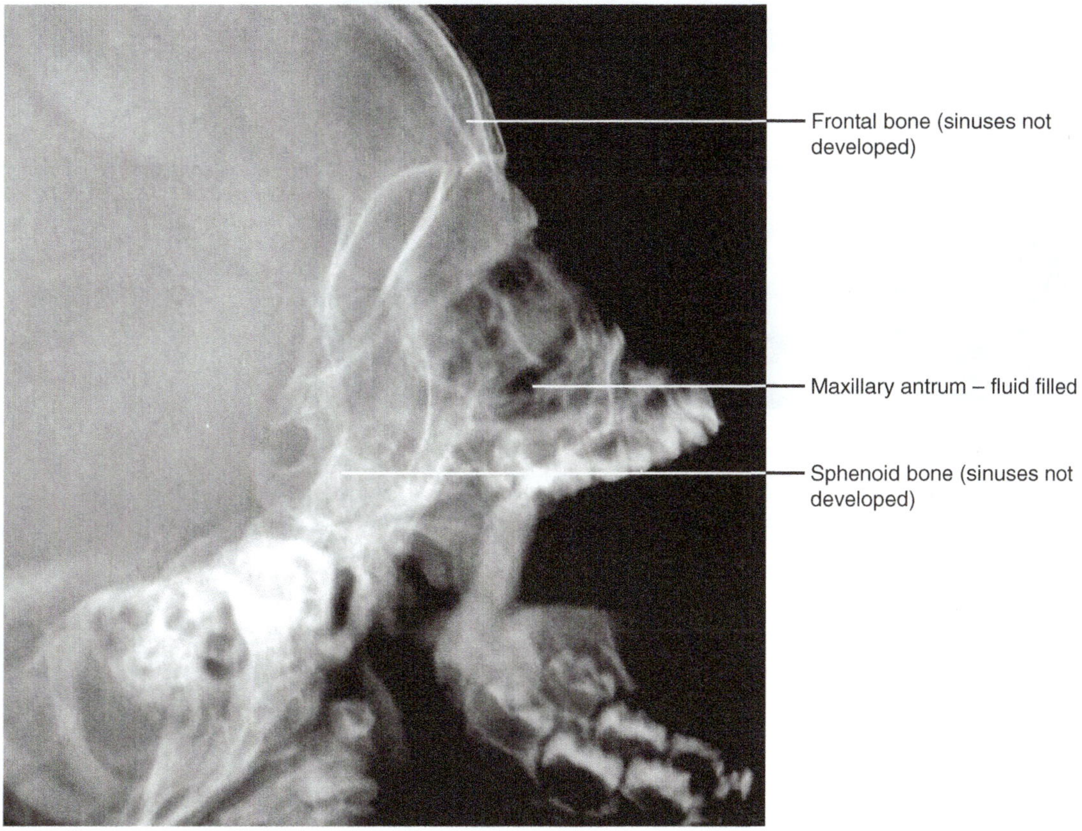

Fig. 31.17 Lateral view: newborn

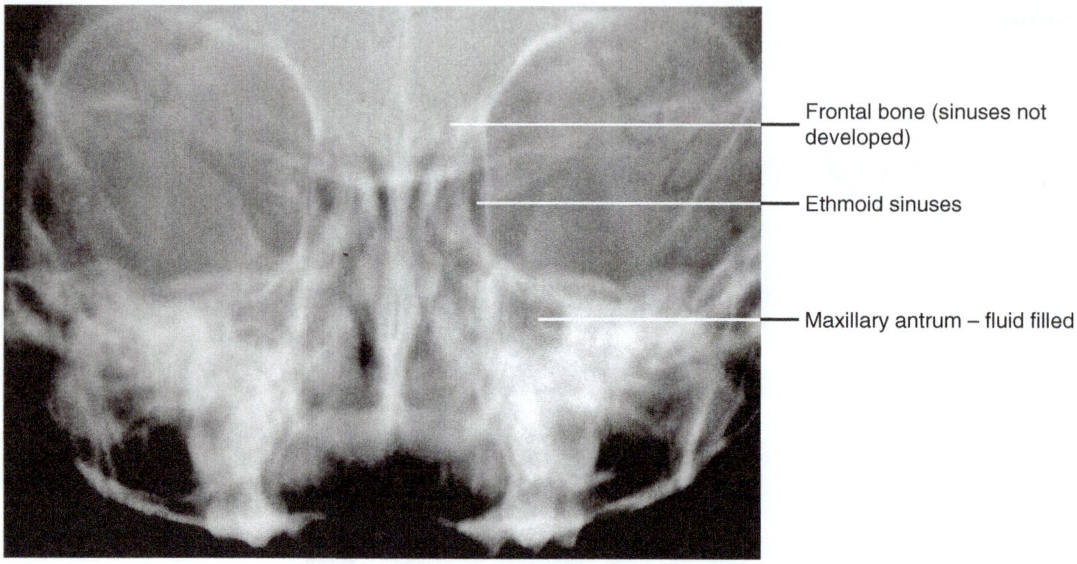

Fig. 31.18 AP view: newborn

Computed Tomography

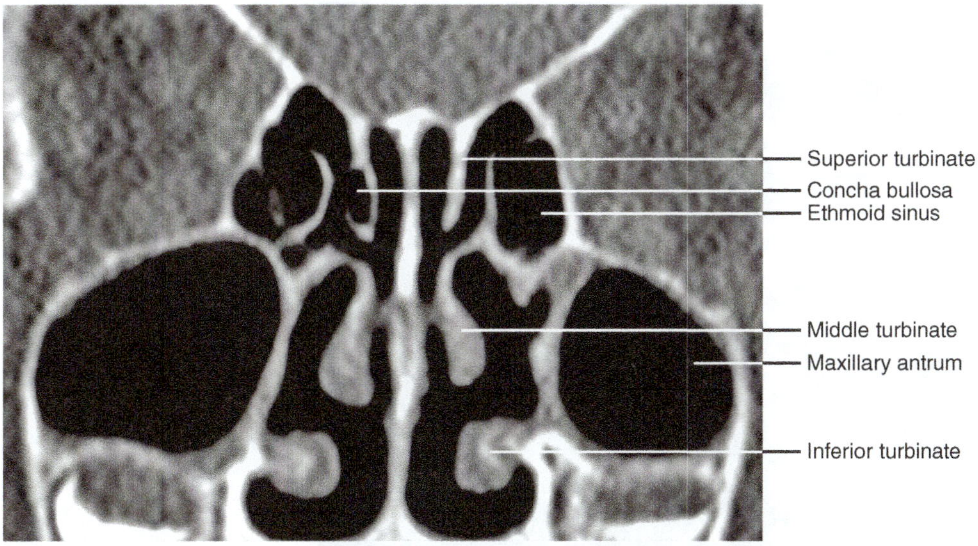

Fig. 31.19 Coronal CT

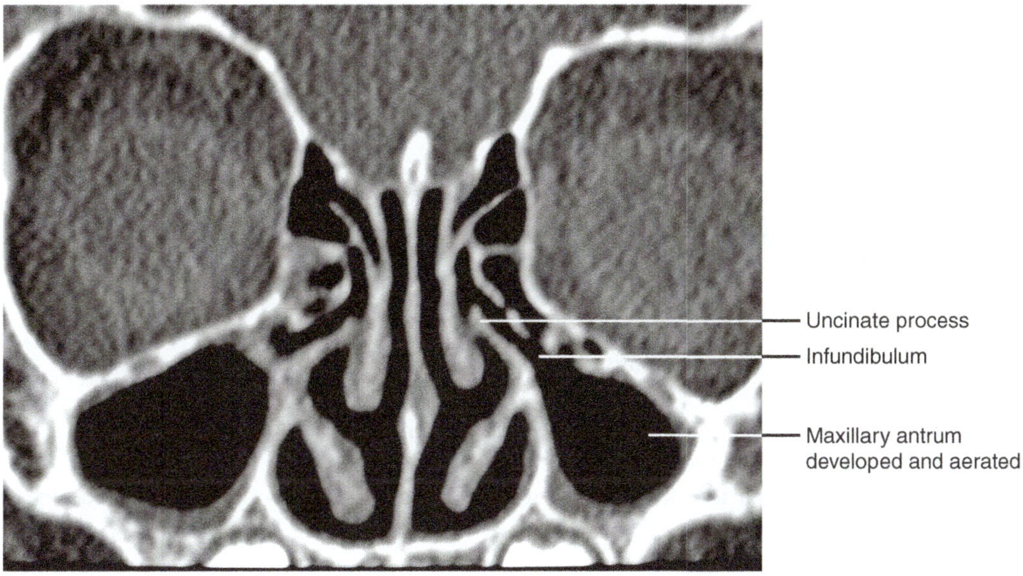

Fig. 31.20 Coronal CT

Magnetic Resonance Imaging

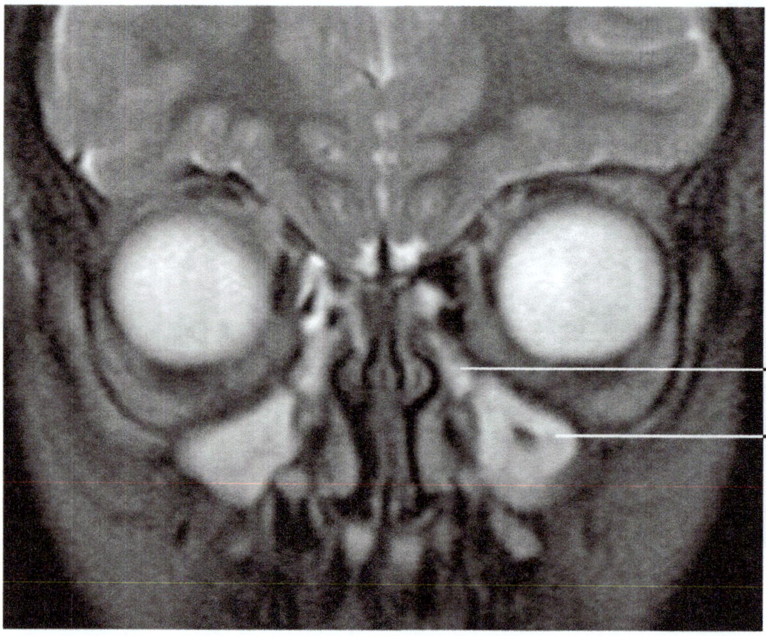

Fig. 31.21 T2 coronal – neonate

Sinus Pneumatisation and Development

Pearl:

Maxillary sinuses are present at birth and are the first sinuses to develop.

At birth they are partly or completely opacified.

They should not be labeled pathological if opacified in the first 2 years of life.

Pearl:

Ethmoid sinuses developed anteriorly at birth.

Posterior ethmoids usually not pneumatised until 6 years.

Pearl:

Sphenoid sinuses are tiny cavities at birth that contain red marrow.

Marrow first become fatty and then progressively pneumatises.

Pneumatisation can be seen as early as 2 years, but generally has an adult appearance by puberty.

Pearl:

Frontal sinuses are not present at birth.

They develop from 2 years onwards.

Seen in the vertical portion of frontal bone by 10 years.

Choanae

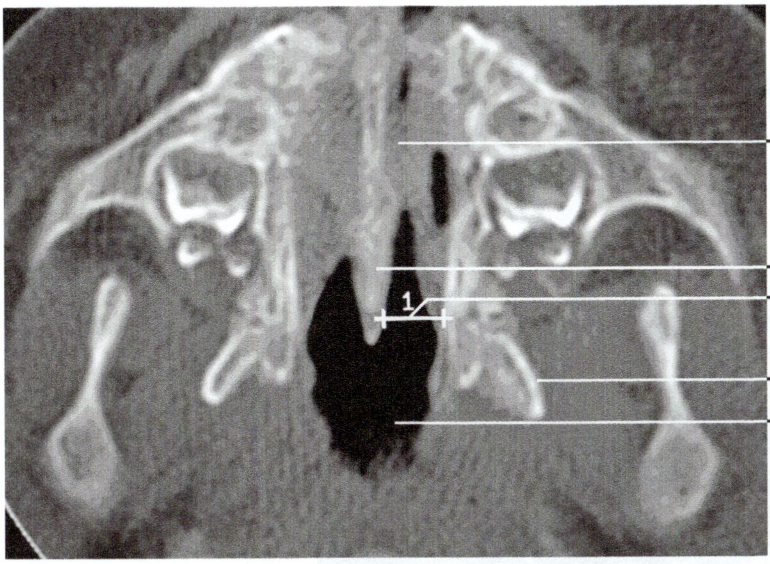

Fig. 31.22 Axial CT nasopharynx

Pearl

The posterior choanal mean width: neonate 6.7 mm; at 6 years 8.6 mm; at 16 years 11.3 mm.
The vomer width in patients less than 8 years : 5.5 mm.
Imaging is achieved parallel to be hard palate in fine slices over 0.5 cm of scan distance.

Lymph Tissue

Plain Film

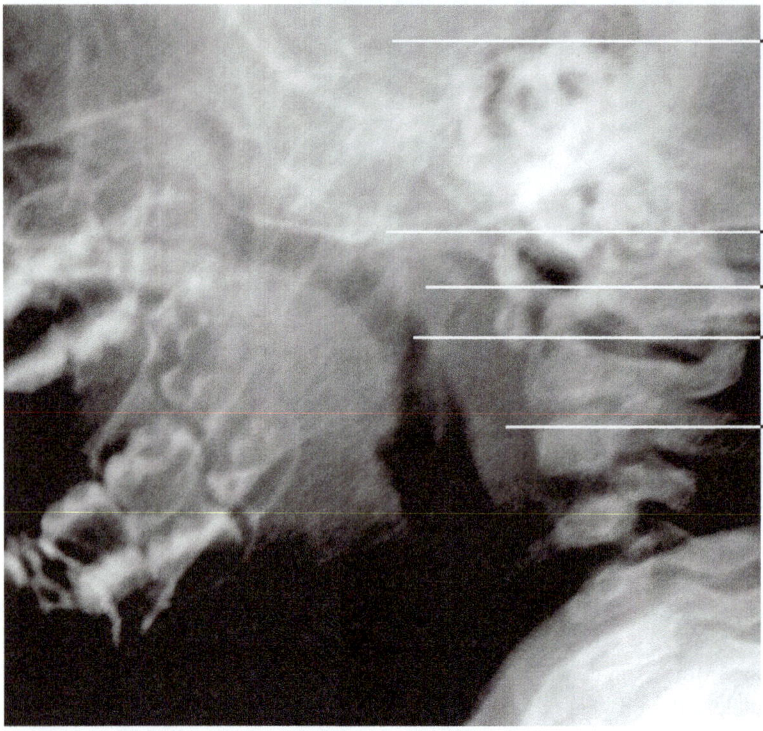

Fig. 31.23 Lateral post-nasal space

Magnetic Resonance Imaging

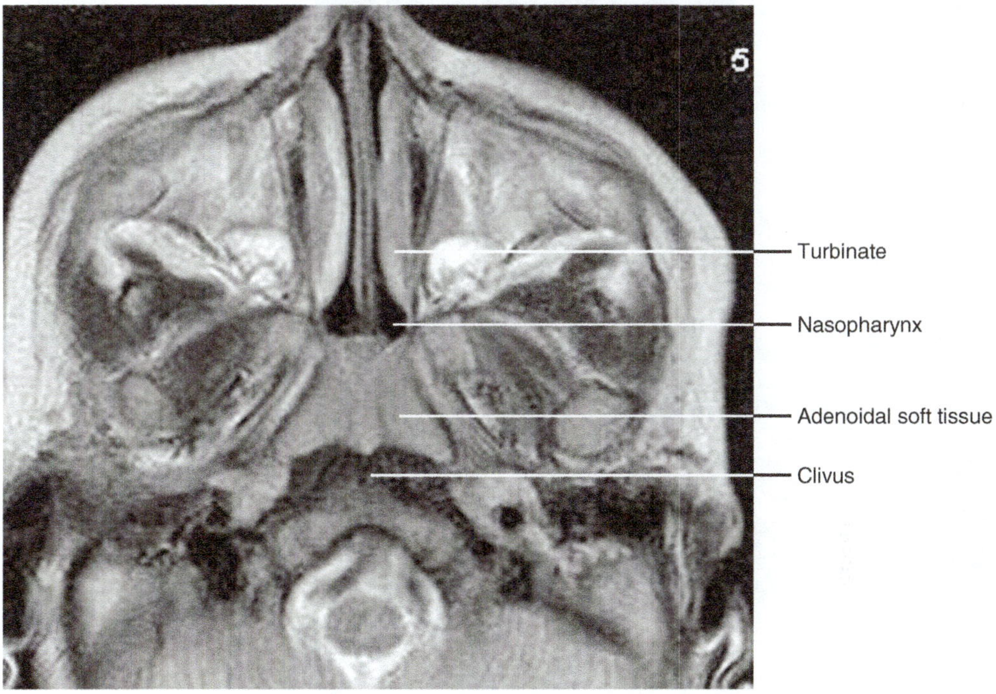

Fig. 31.24 T2 axial nasopharynx (adenoids)

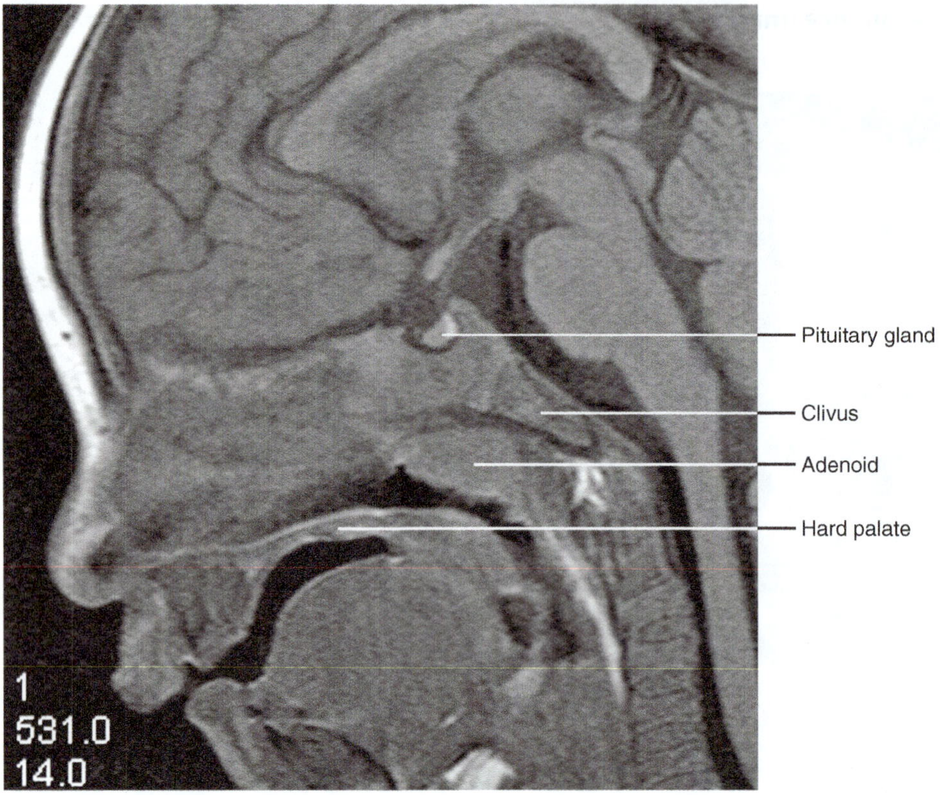

Fig. 31.25 T1 sagittal nasopharynx (adenoids)

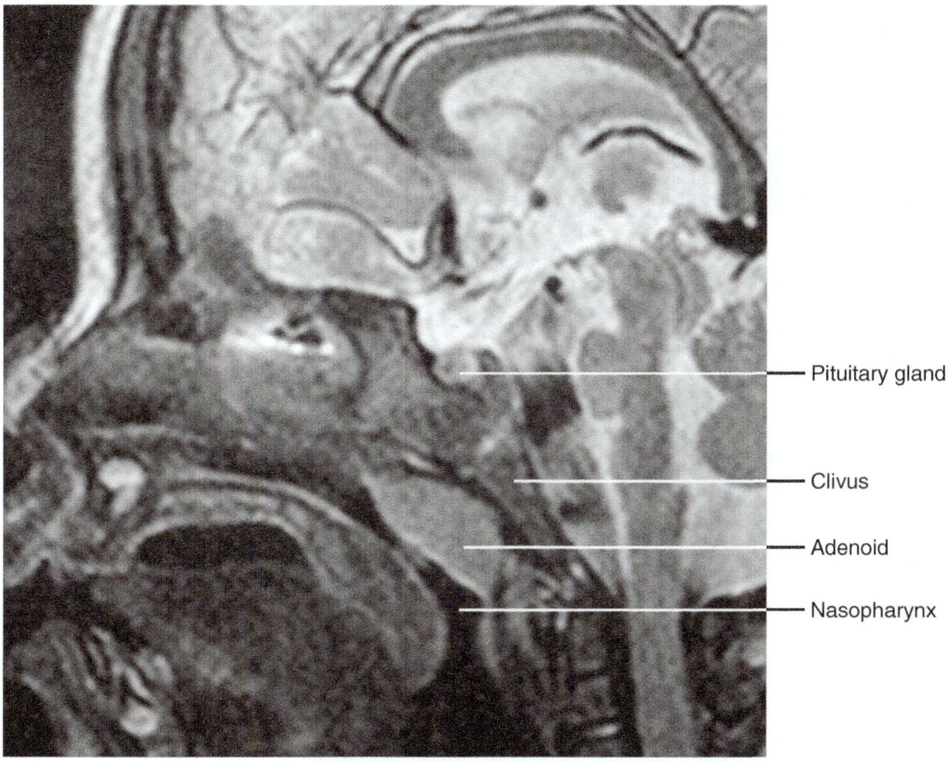

Fig. 31.26 T2 sagittal nasopharynx (adenoids)

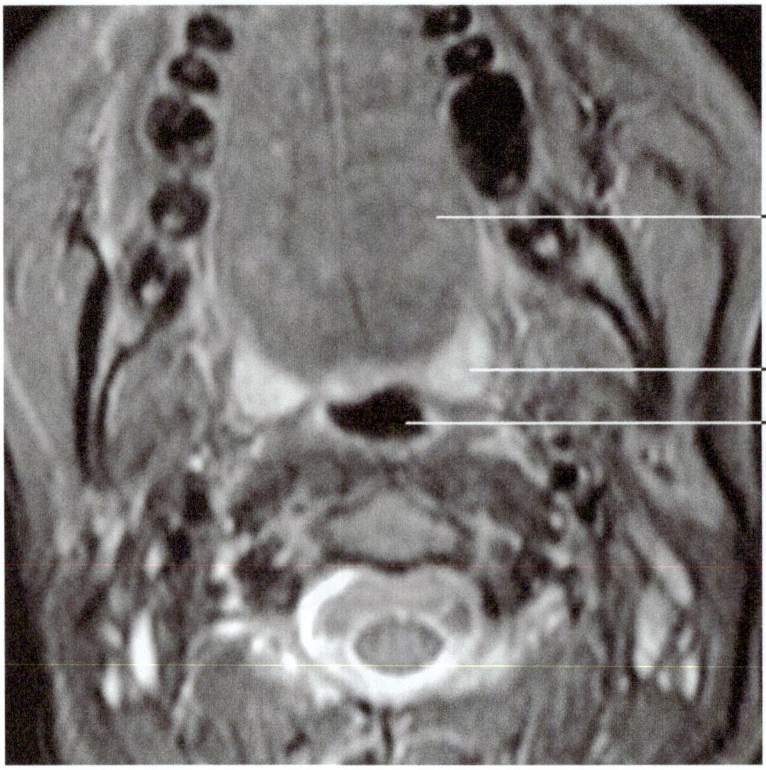

Fig. 31.27 STIR axial (palatine tonsils)

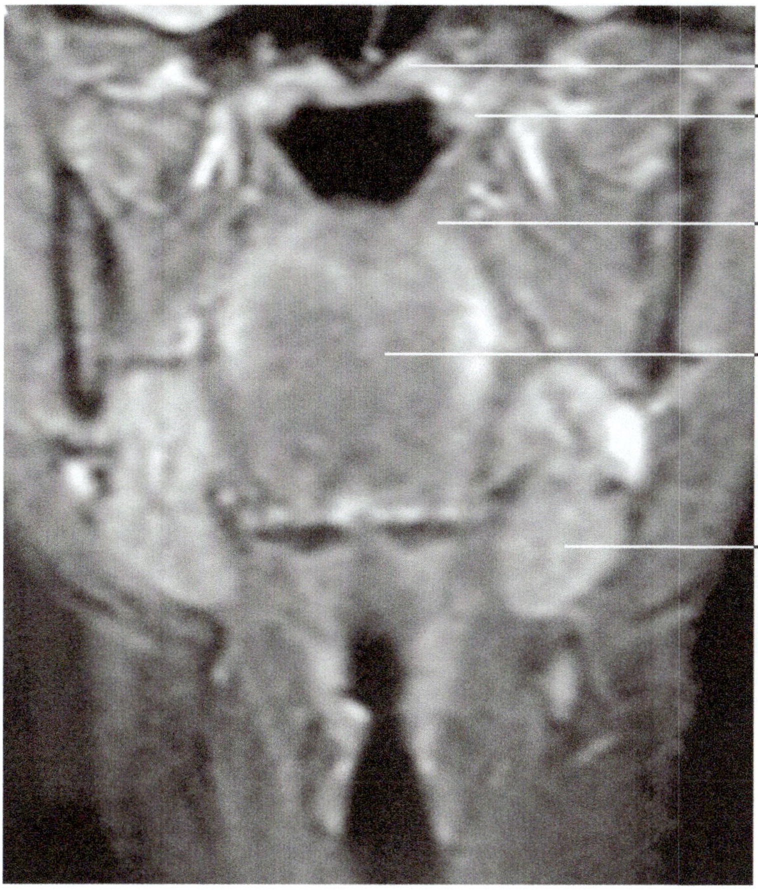

Fig. 31.28 STIR coronal oro/hypopharynx (lingual tonsil/submandibular gland)

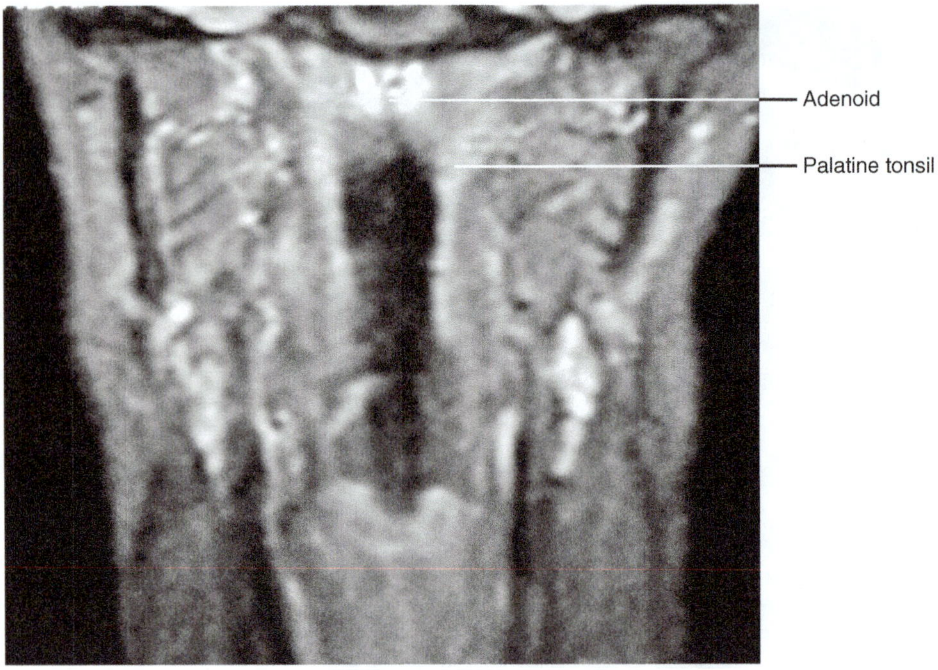

Fig. 31.29 T1 coronal nasopharynx (adenoids)

Larynx

Magnetic Resonance Imaging

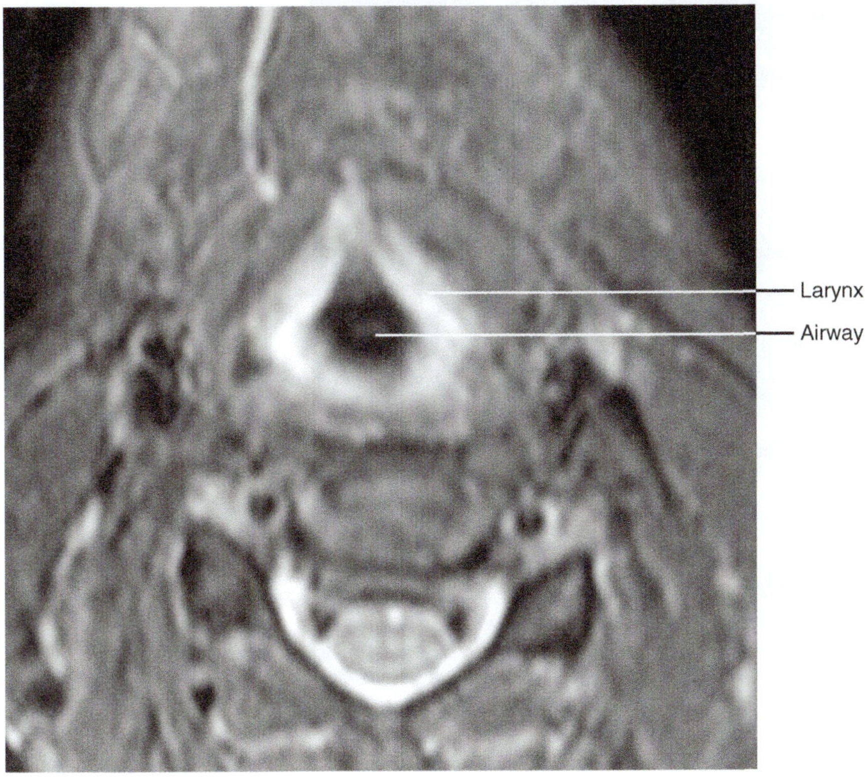

Fig. 31.30 STIR axial

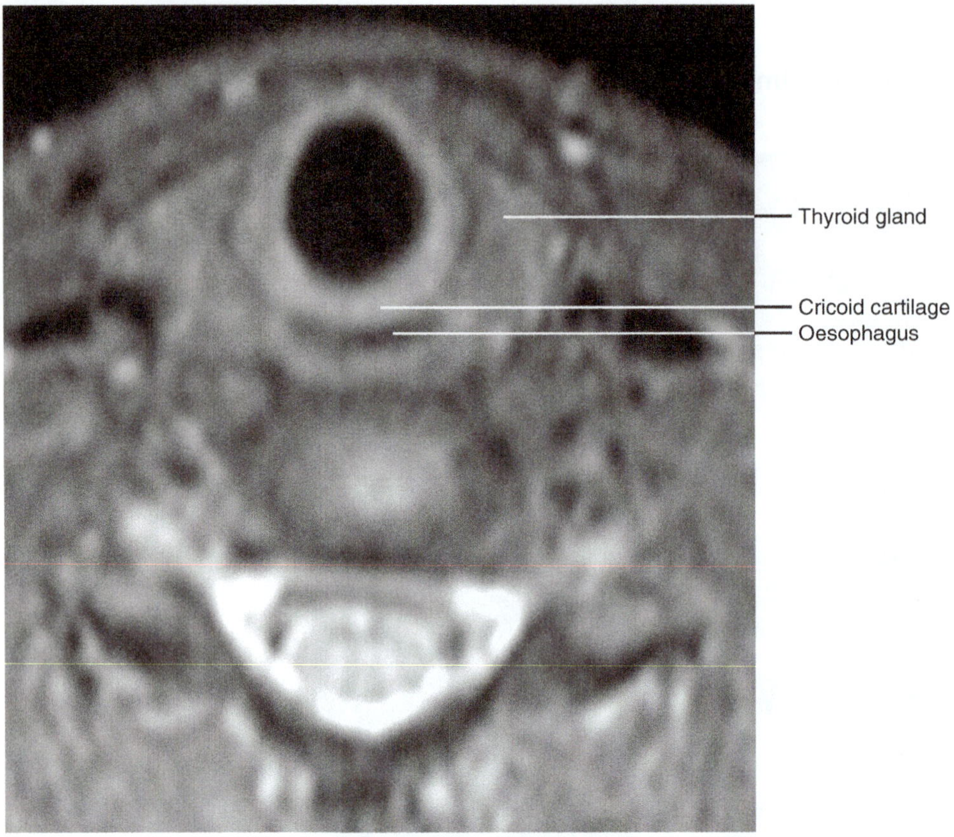

Fig. 31.31 STIR axial

Pearl

The structures of the larynx in children are cartilaginous and not calcified. STIR imaging demonstrates these very well as high signal structures.

Paediatric Thorax

32

Trachea

Pearl

On a well-inspired CXR the trachea will be straight or show a gentle curve away from the aortic arch (usually towards the right).

On a poorly inspired film there is buckling of the trachea away from the arch (usually towards the right) as a result of tracheal redundancy.

Displacement or buckling of the trachea towards the left implies a right-sided aortic arch or paratracheal pathology.

S. Andronikou (ed.), *See Right Through Me*,
DOI 10.1007/978-3-642-23893-2_32, © Springer-Verlag Berlin Heidelberg 2012

Plain Film

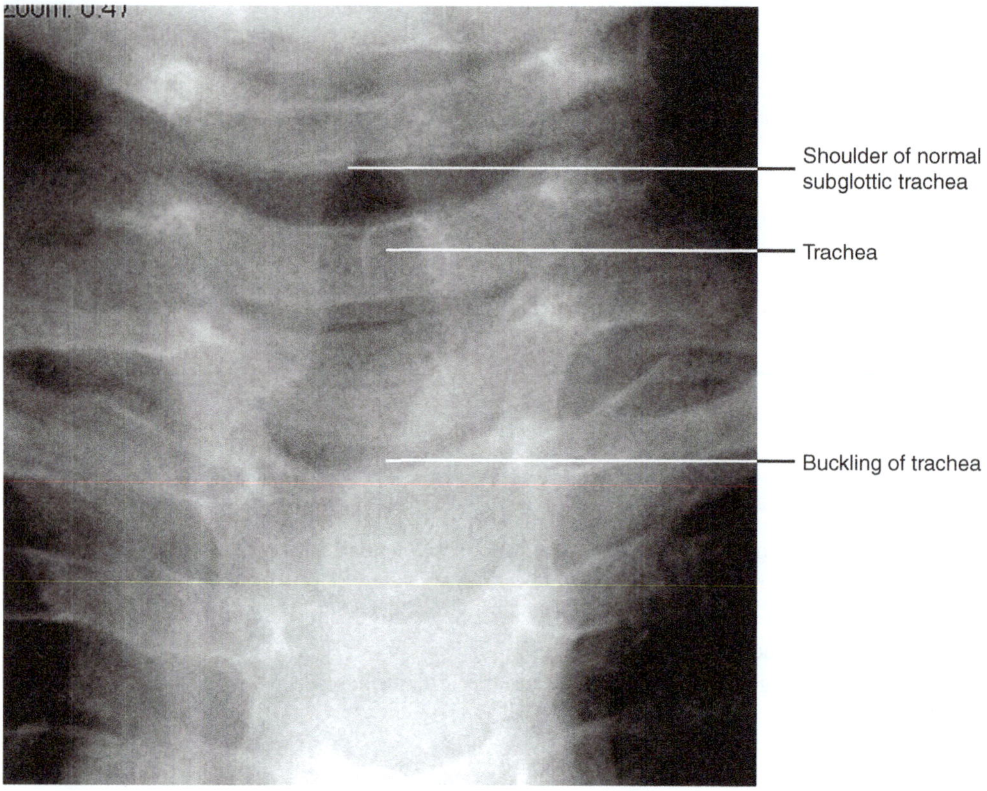

Fig. 32.1 AP view of trachea demonstrating shouldering and buckling to the right

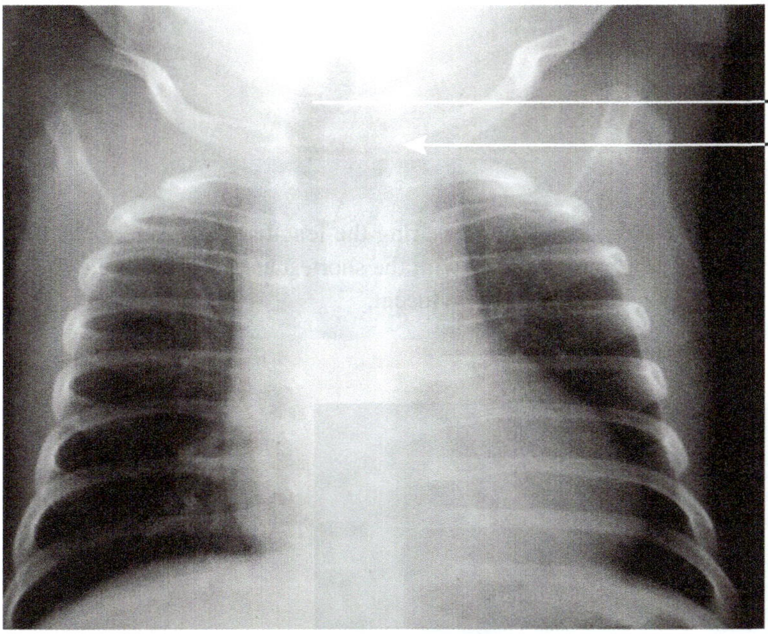

Fig. 32.2 AP CXR demonstrating buckling of trachea towards the right

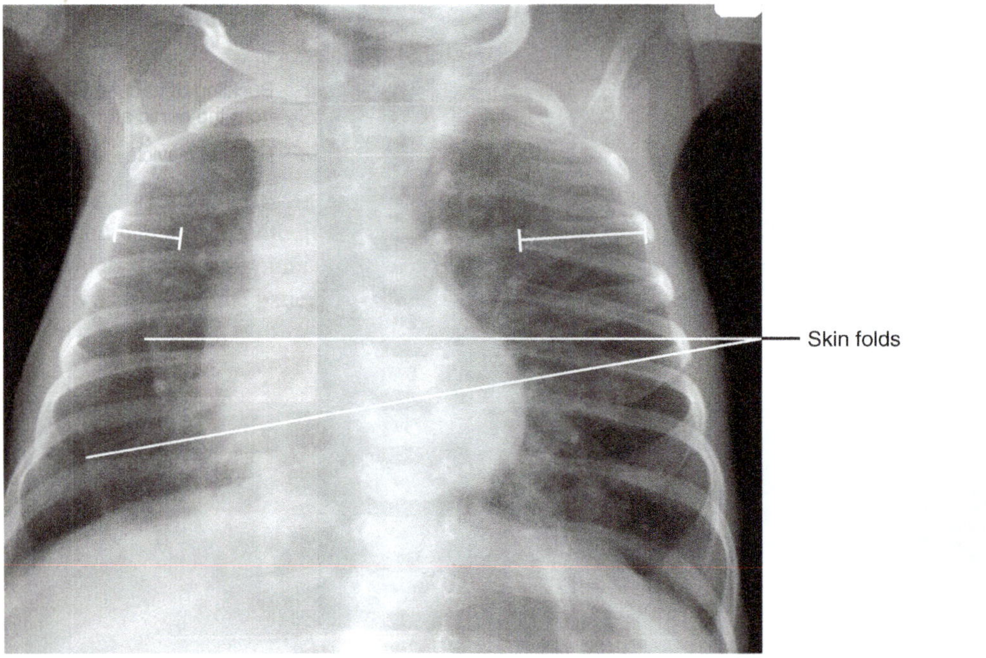

Fig. 32.5 CXR demonstrating skin folds on rotation towards the right

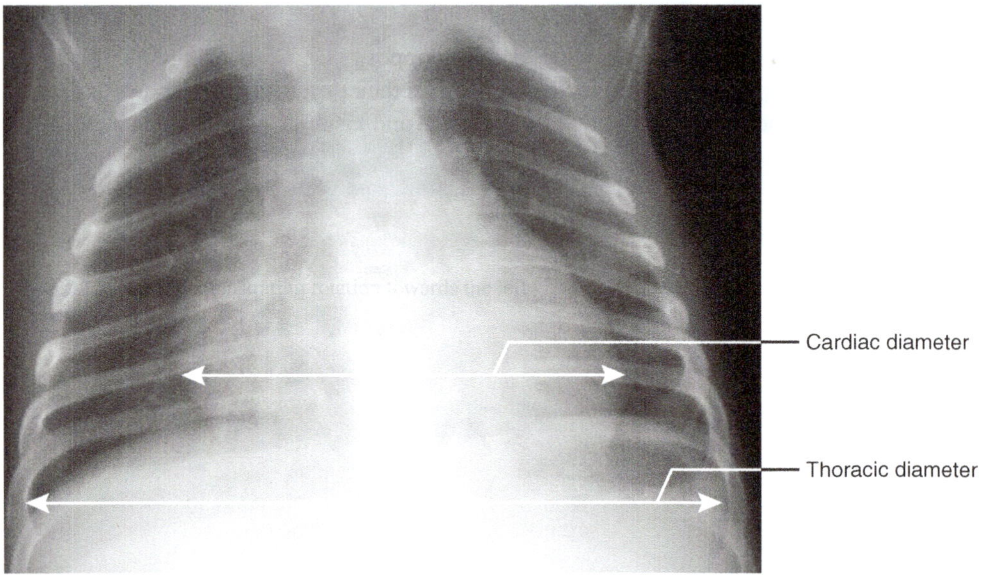

Fig. 32.6 CXR of neonate with CTR of 60%

Pearl

Normal cardiothoracic ratio (CRT) in a child <2 years is up to 60%, 2–4 years is 50–60% and >4 years is up to 50%.

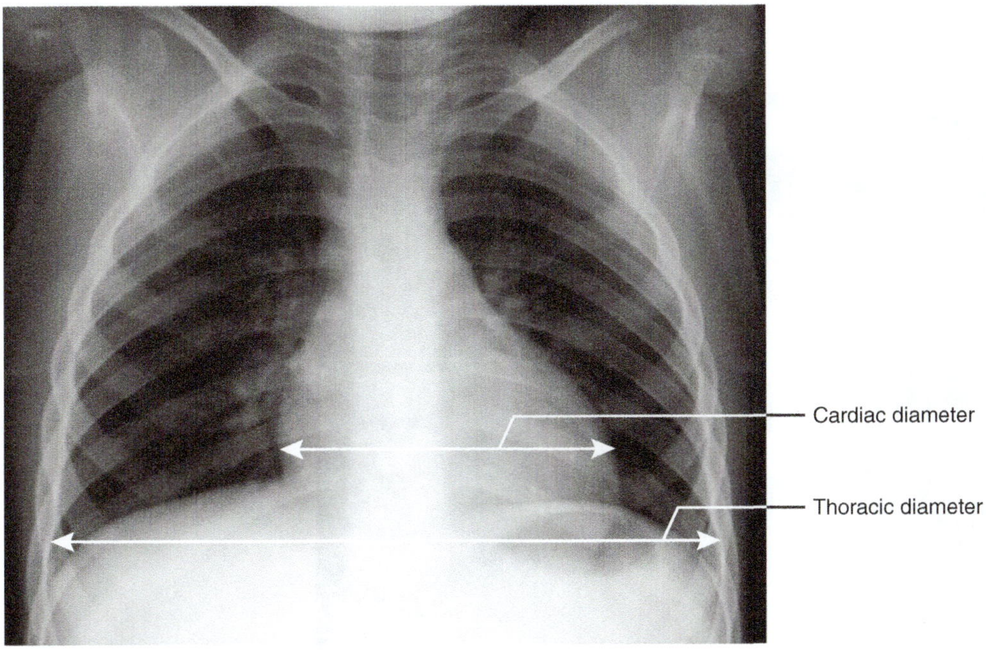

Fig. 32.7 CXR of a child aged 4 years with normal CTR of 50%

Thymus

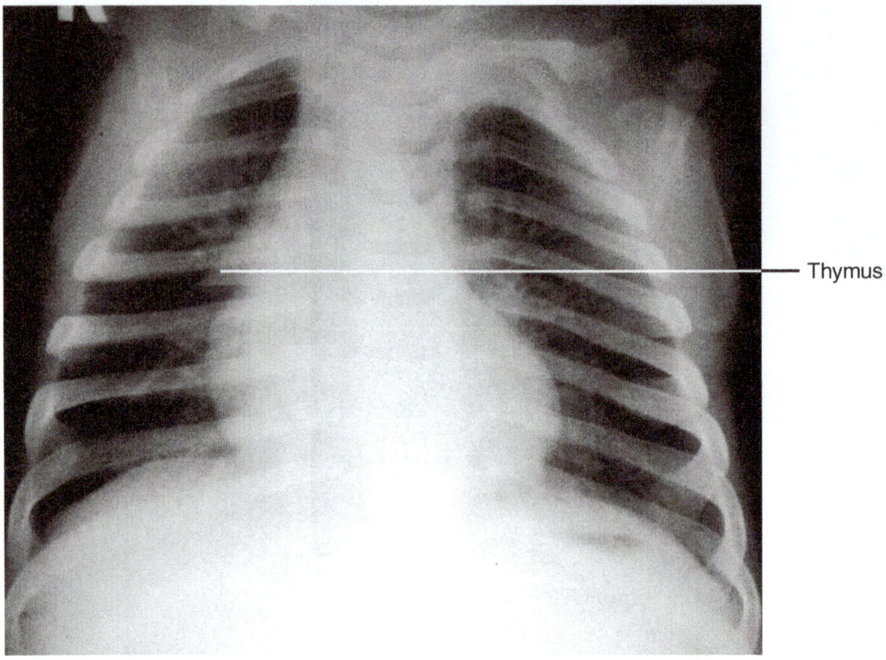

Fig. 32.8 CXR demonstrating the spinnaker-sail sign of the thymus

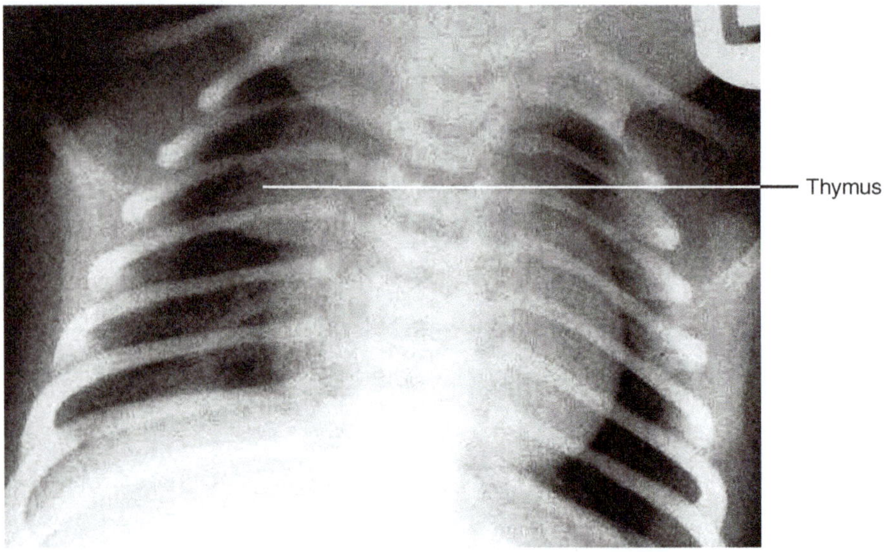

Fig. 32.9 CXR demonstrating the spinnaker-sail sign of the thymus

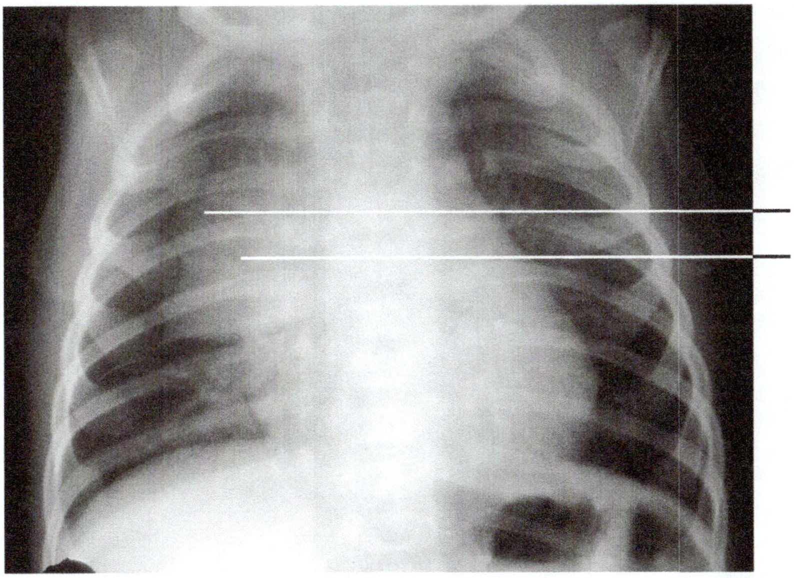

Fig. 32.10 CXR demonstrating wave-like thymus

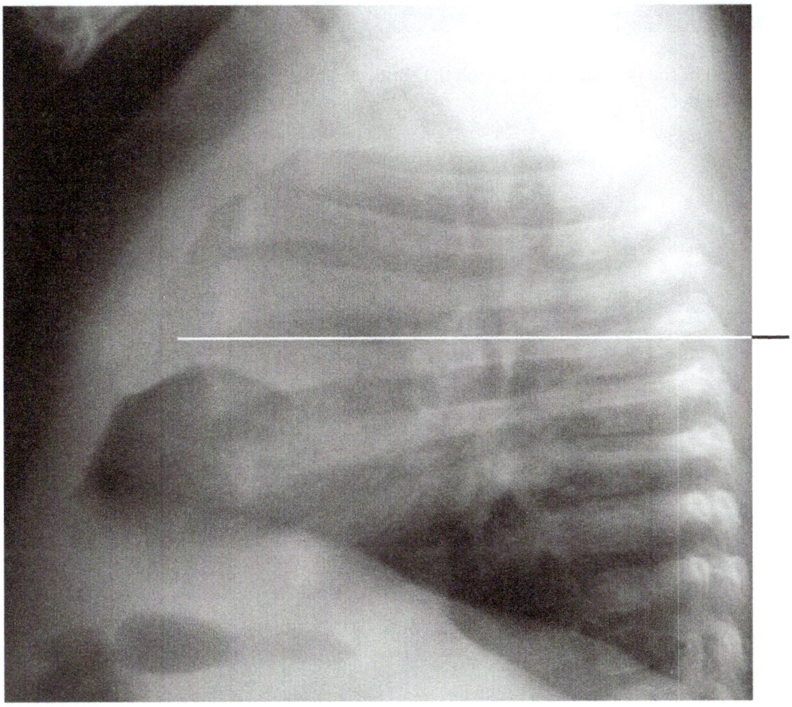

Fig. 32.11 Lateral CXR demonstrating anterior position of thymus

Pearl

The normal thymus never causes displacement of structures. If there is concern about a mediastinal mass, ultrasound using a suprasternal approach will show the normal homogeneous gland draped around normal mediastinal structures.

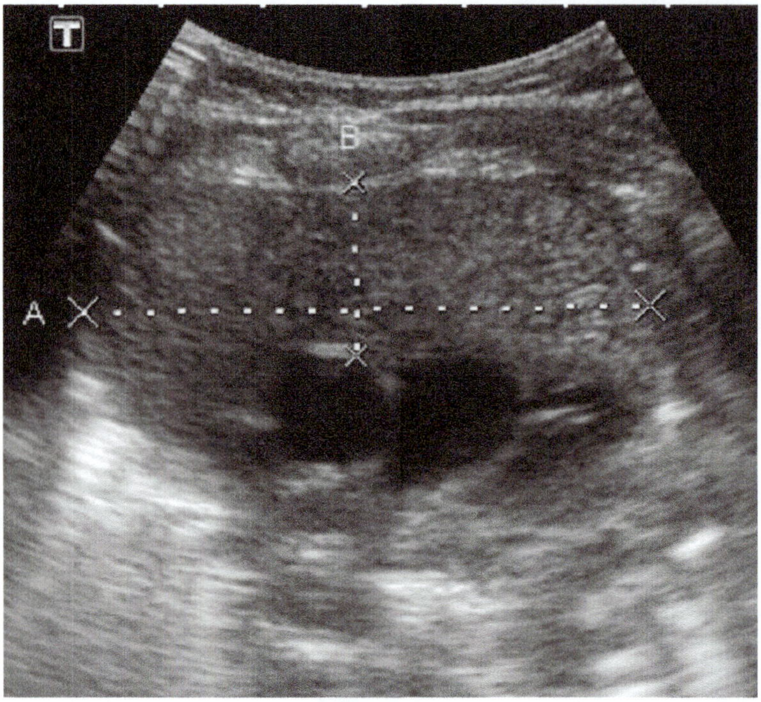

Fig. 32.12 Coronal oblique ultrasound of the thymus seen through the suprasternal approach demonstrating a homogenous organ and the hypoechoic vessels deep to it

Computed Tomography

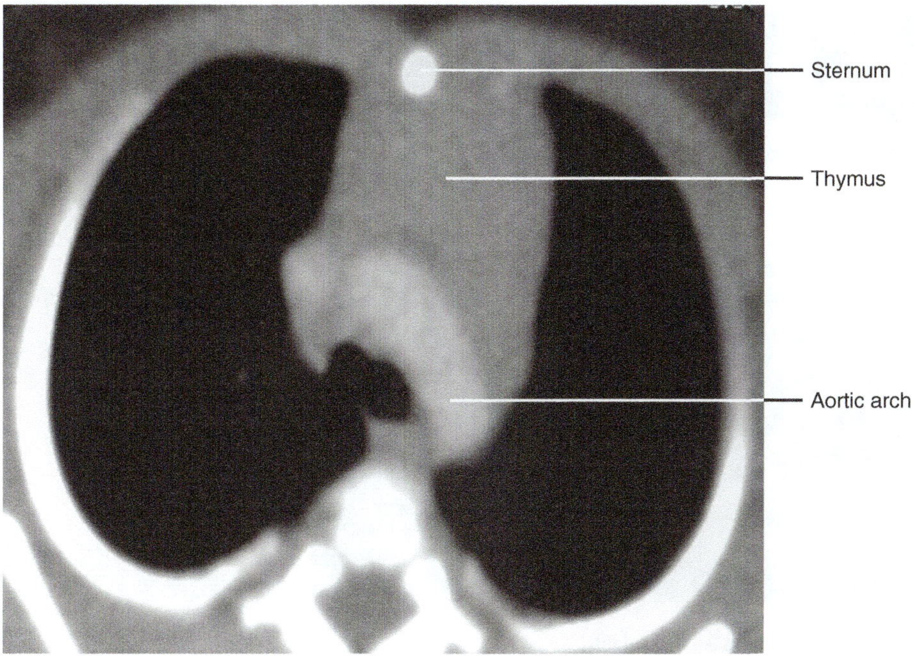

Fig. 32.13 Axial CT at level of sternal notch demonstrating the thymus

Magnetic Resonance Imaging

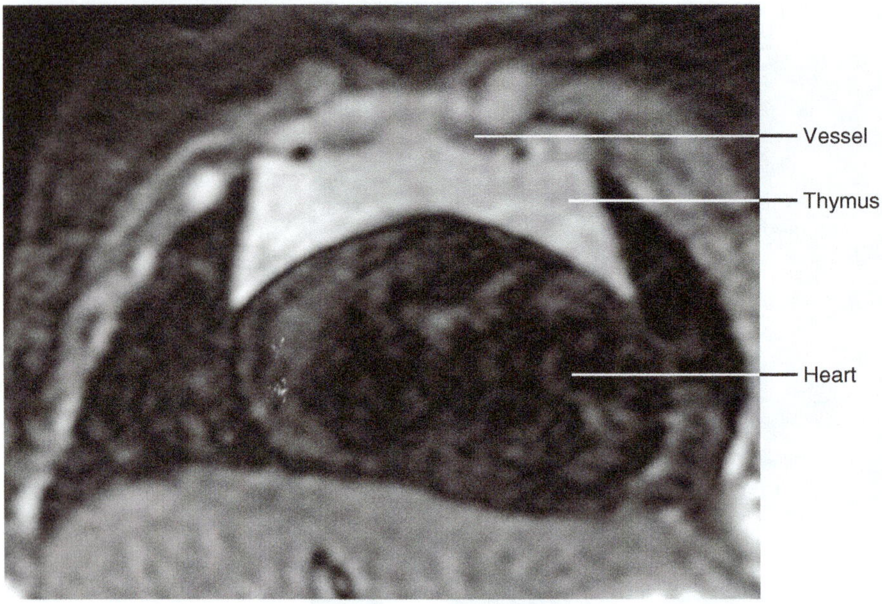

Fig. 32.14 STIR coronal of the thymus in an infant

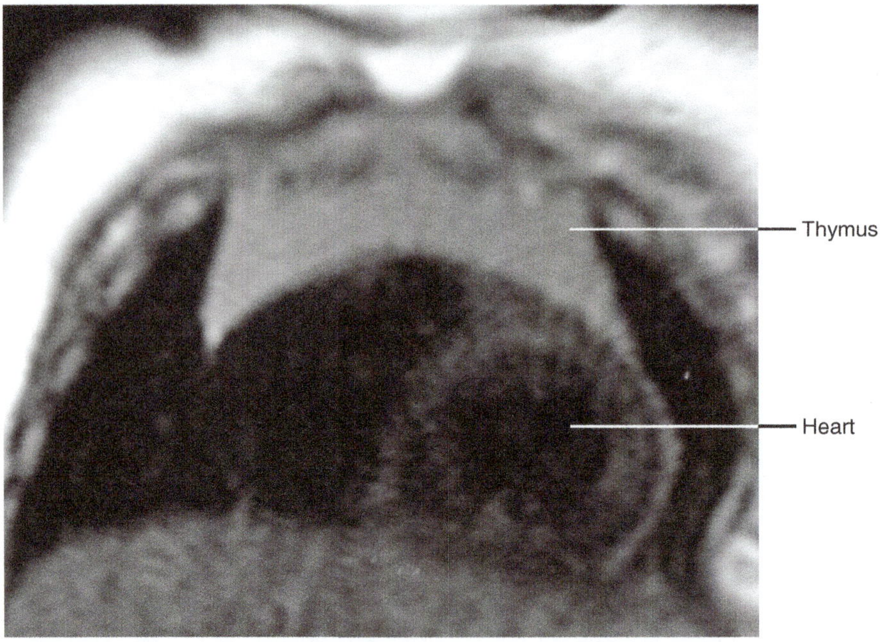

Fig. 32.15 T1 coronal of the thymus in an infant

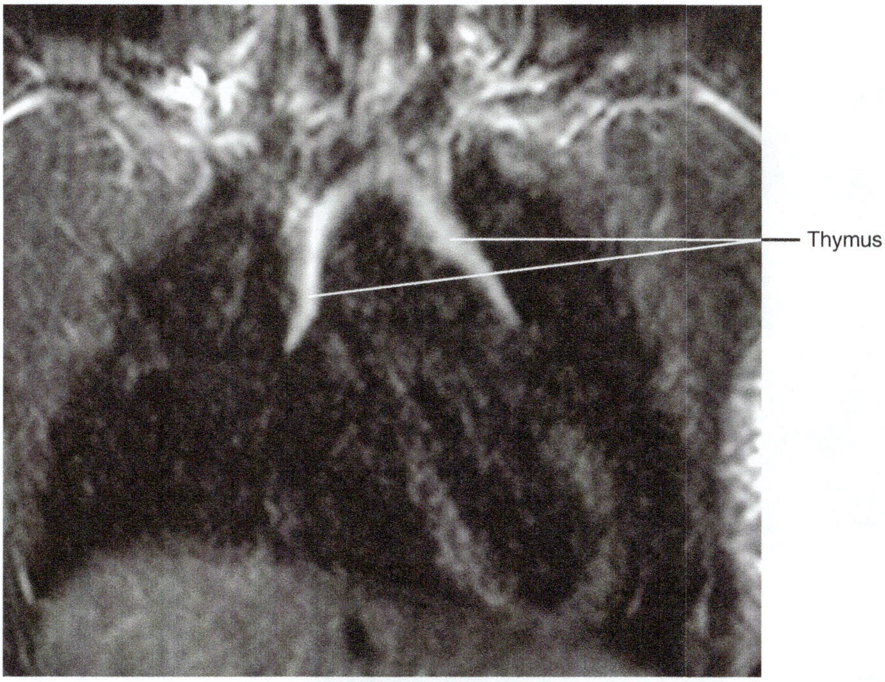

Fig. 32.16 STIR coronal of the thymus in an older child

Breast

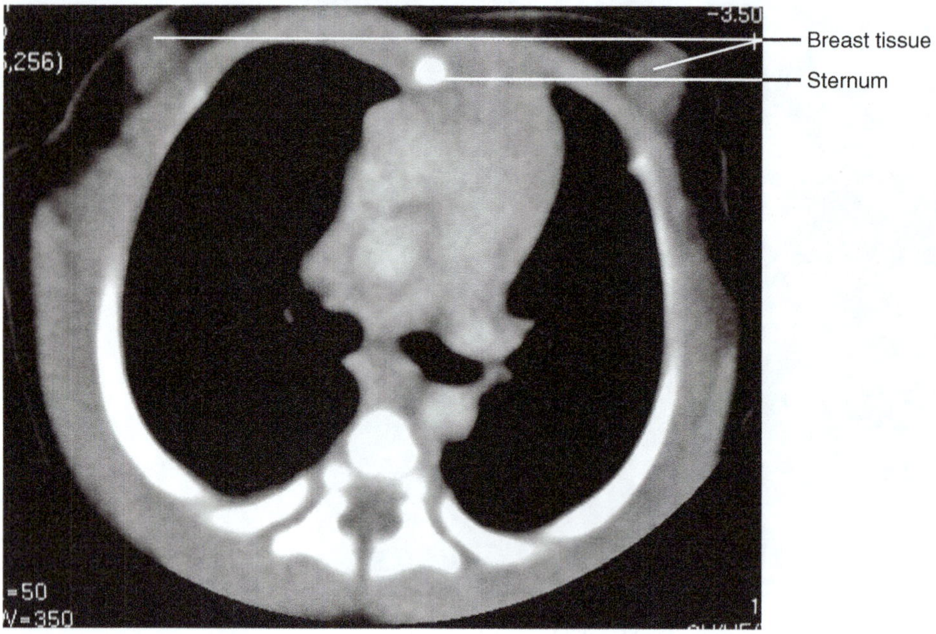

Fig. 32.17 Axial CT showing normal breast tissue in a neonate

> **Pearl**
> Breast tissue may be a normal finding in neonates on CT of the chest as a result of maternal hormones.

33 Paediatric Abdomen and Pelvis

Umbilical Artery and Vein

Plain Film

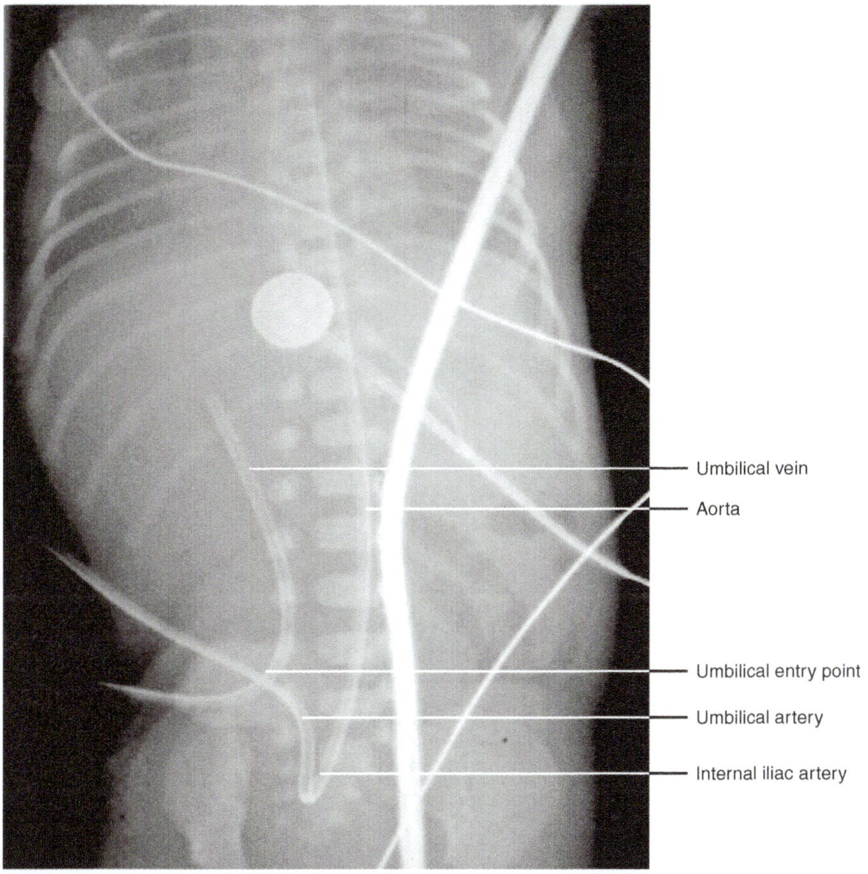

Fig. 33.1 AP Plain radiograph demonstrating the course of the umbilical artery and vein by catheter positions from the umbilicus

S. Andronikou (ed.), *See Right Through Me*,
DOI 10.1007/978-3-642-23893-2_33, © Springer-Verlag Berlin Heidelberg 2012

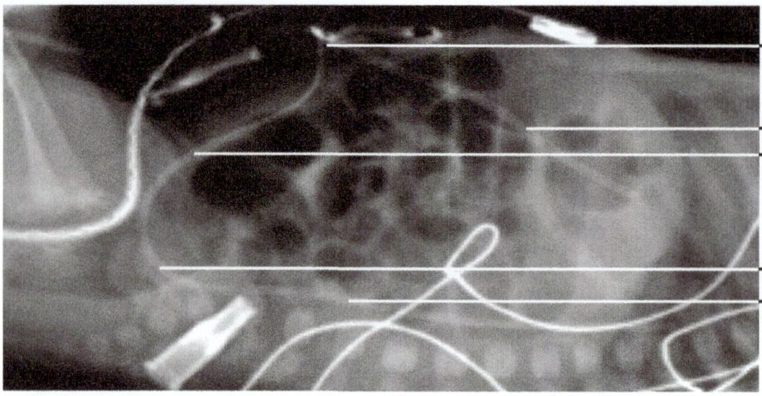

Fig. 33.2 Lateral plain radiograph demonstrating the course of the umbilical artery and vein by catheter positions from the umbilicus

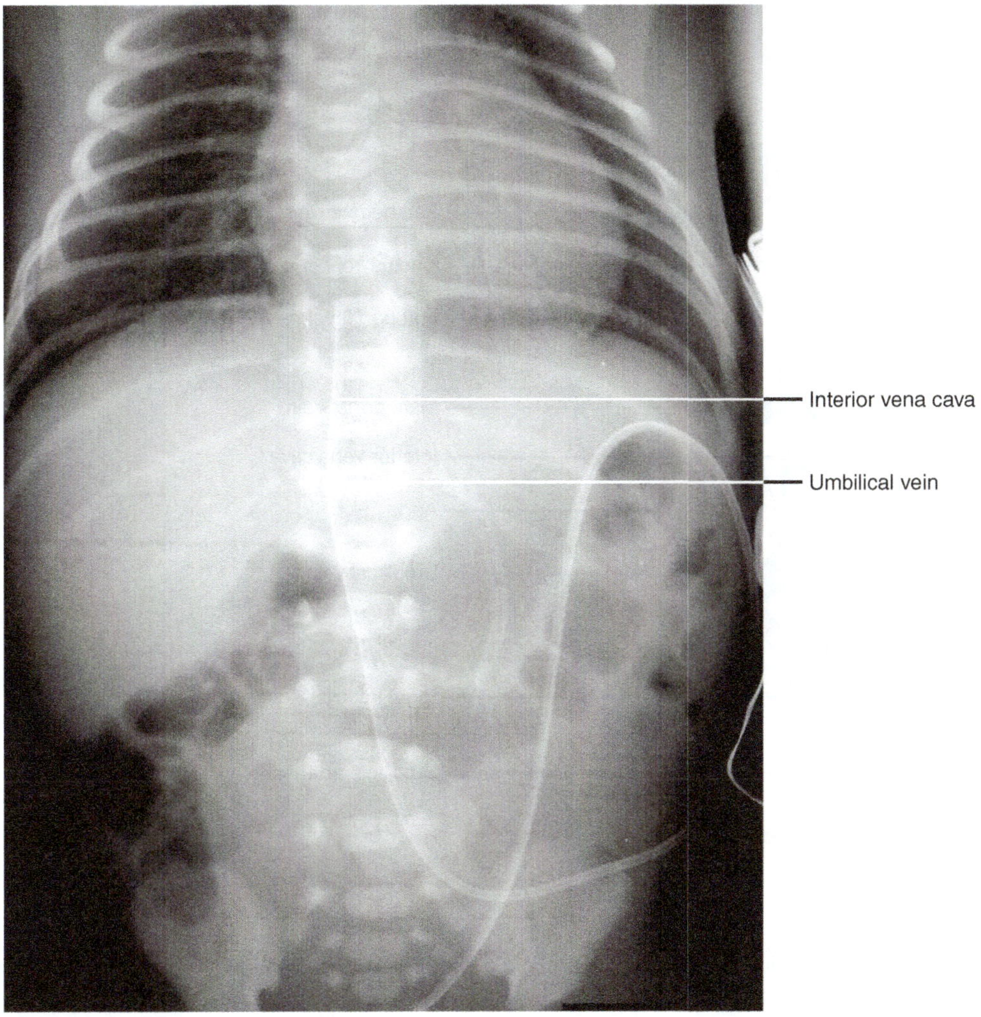

Fig. 33.3 AP plain radiograph demonstrating the position of the umbilical vein (by the course of the umbilical vein catheter) to the right of the midline

Pearl

The UAC passes from the umbilical artery inferiorly into the pelvis to the internal iliac artery then cephalad into the aorta.

The UVC passes from the umbilical vein to the left portal vein, ductus venousus, hepatic vein and IVC. Its course is superior from the umbilicus.

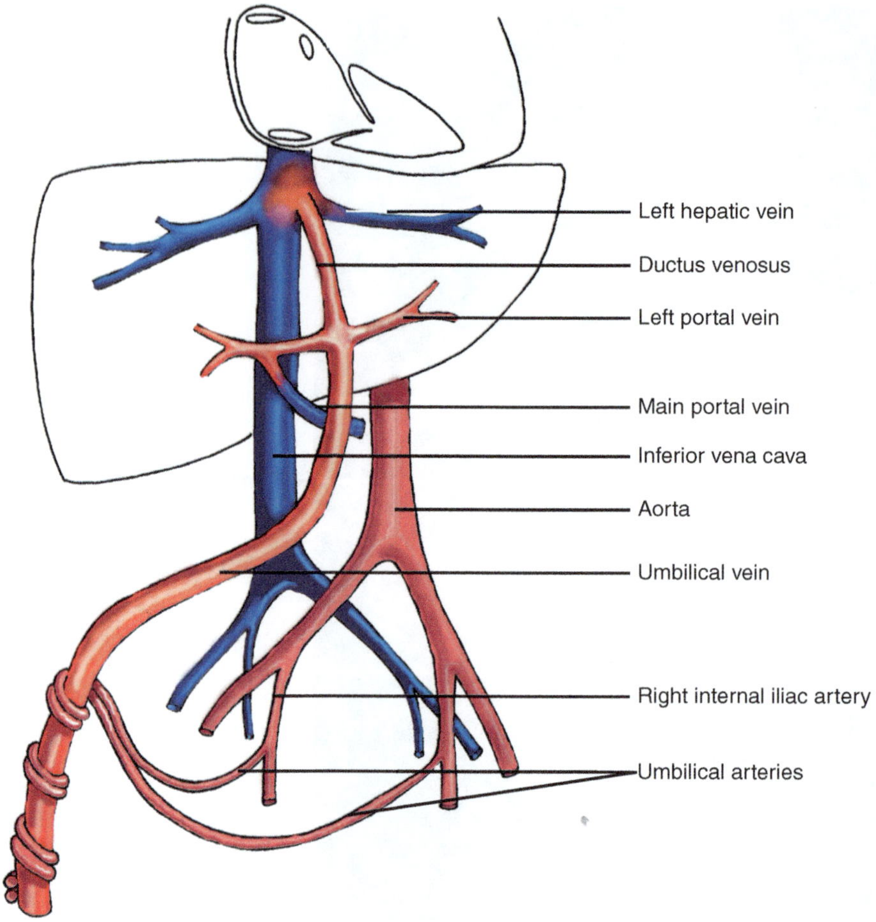

Fig. 33.4 Schematic representation of the course and continuations of the umbilical vein and the umbilical artery in a neonate

Upper Gastrointestinal Tract

Contrast Swallow/Meal

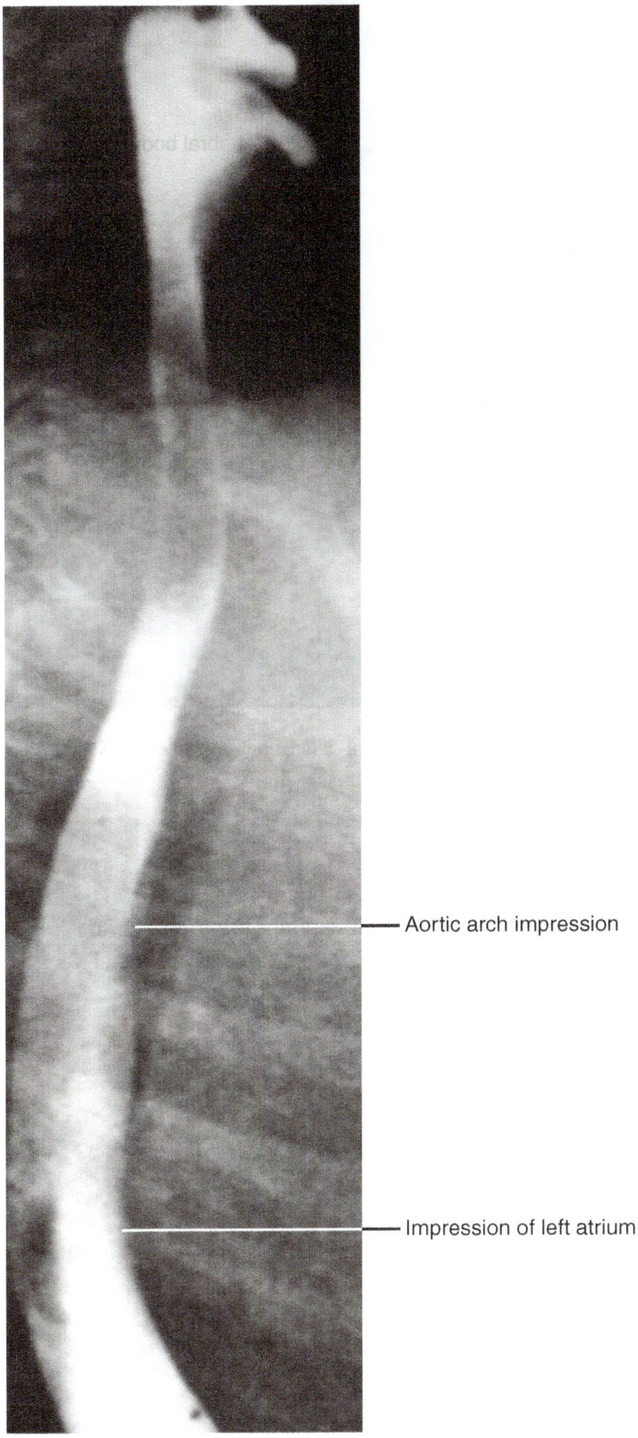

Fig. 33.5 Lateral view of the oesophagus

Pylorus

Ultrasound

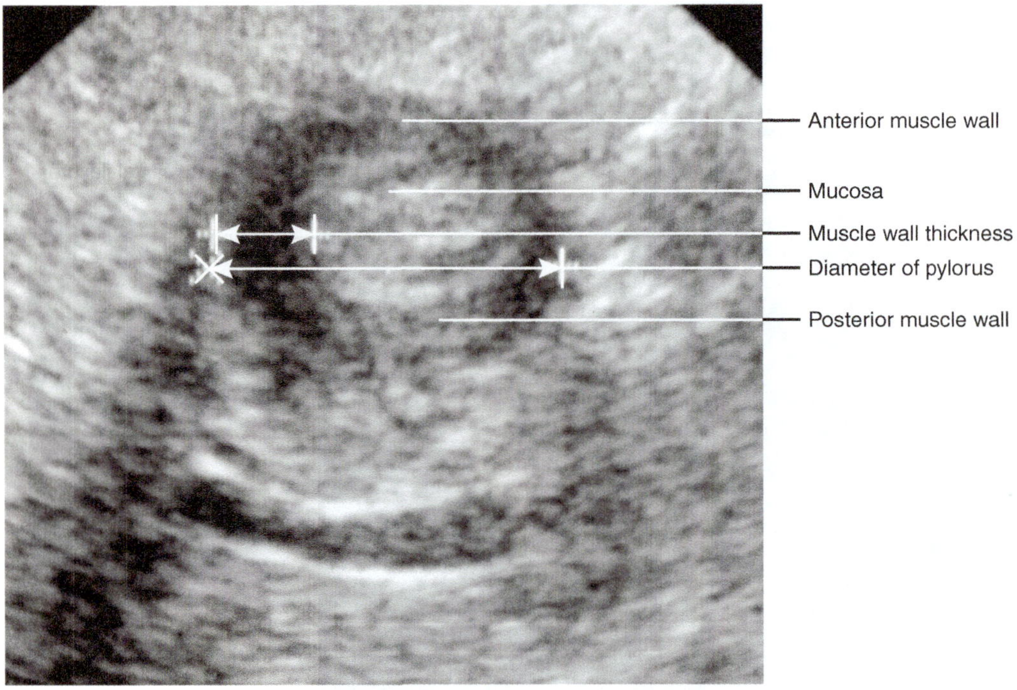

Fig. 33.8 Transverse pylorus (longitudinal abdominal scan)

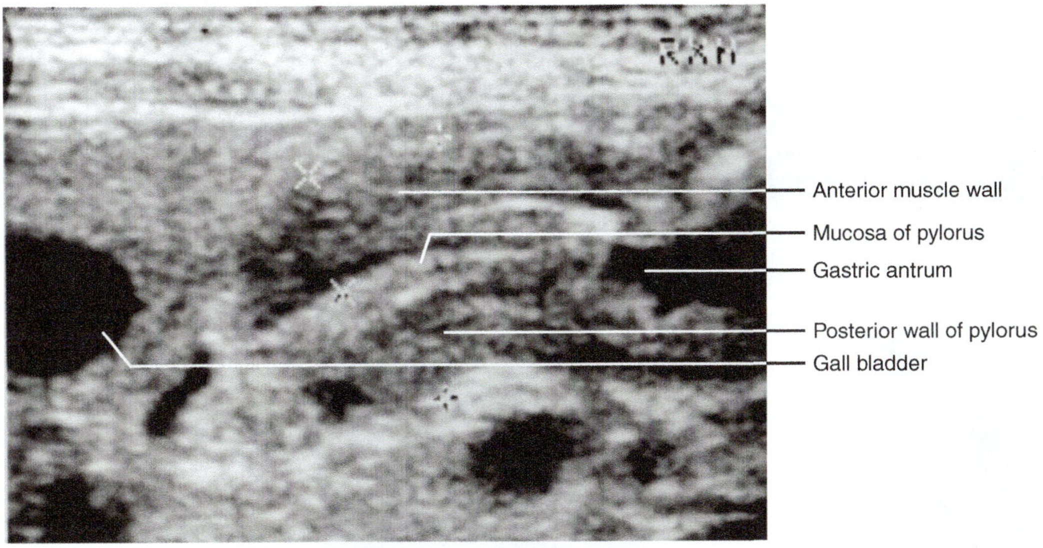

Fig. 33.9 Longitudinal view of an abnormally thickened pylorus to demonstrate the muscle and mucosa

Pearl

Rule of thumb: muscle wall thickness ≤5 mm, diameter pylorus ≤15 mm, length pylorus ≤20 mm.

Pearl

The pylorus is found on ultrasound by identifying the stomach and gallbladder as lateral boundaries.

Rectum and Colon

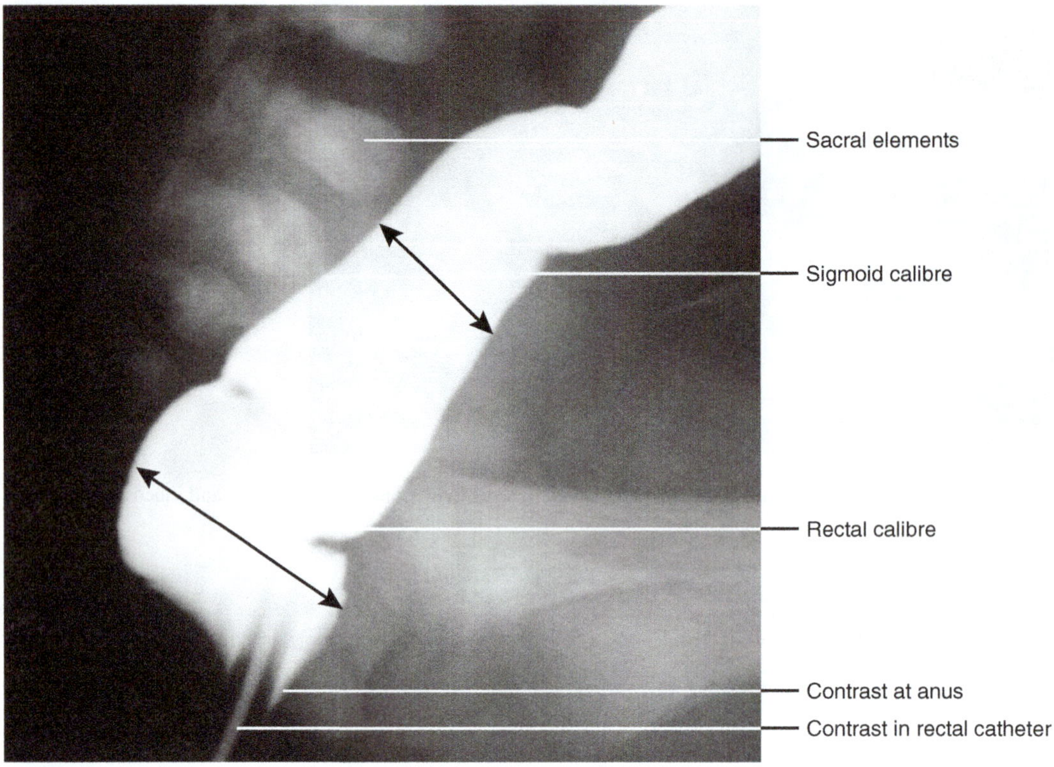

Fig. 33.10 True lateral of the rectum and sigmoid during a contrast enema

Pearl
Rectal caliber should be greater than sigmoid caliber on lateral film.

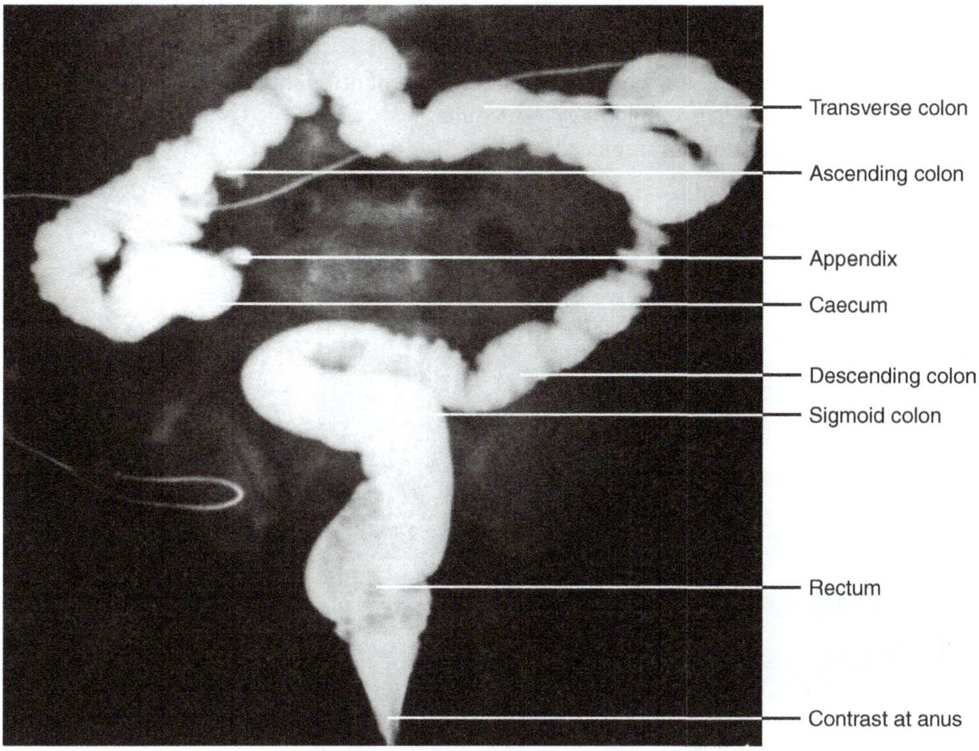

Fig. 33.11 AP of the colon during a contrast enema

Adrenal Glands

Appearances vary with age: in a newborn the medulla is hyperechoic and surrounded by hypoechoic cortex. As the foetal cortex is replaced the echogenicity increases and by 6 months of age the gland is hyperechoic and smaller with poor or absent differentiation between cortex and medulla.

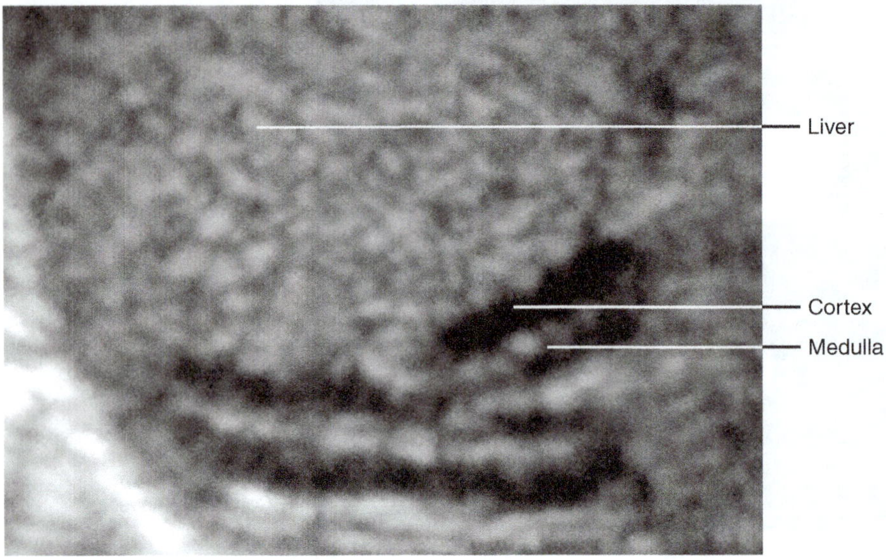

Fig. 33.12 Longitudinal ultrasound of neonatal adrenal gland

Urogenital Tract

In neonates the echogenicity of the renal cortex is increased and is equal to, or slightly greater than that of the liver and spleen. The medullary pyramids are hypoechoic and strikingly prominent. By 12 months of age they are minimally hypoechoic. The renal sinus is not as echogenic as in older children due to the paucity of fat.

Ultrasound

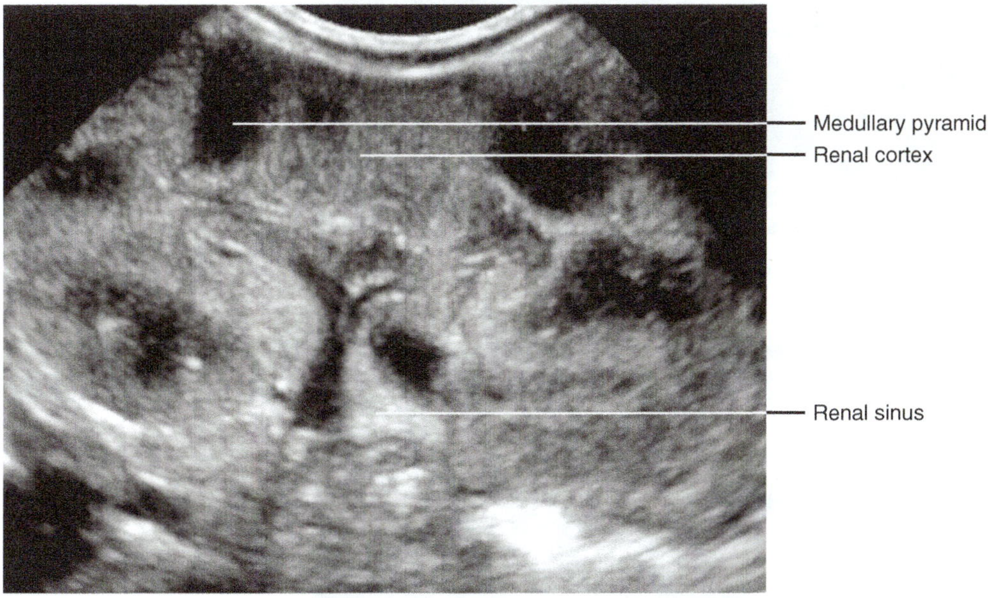

Fig. 33.13 Longitudinal US of neonatal kidney

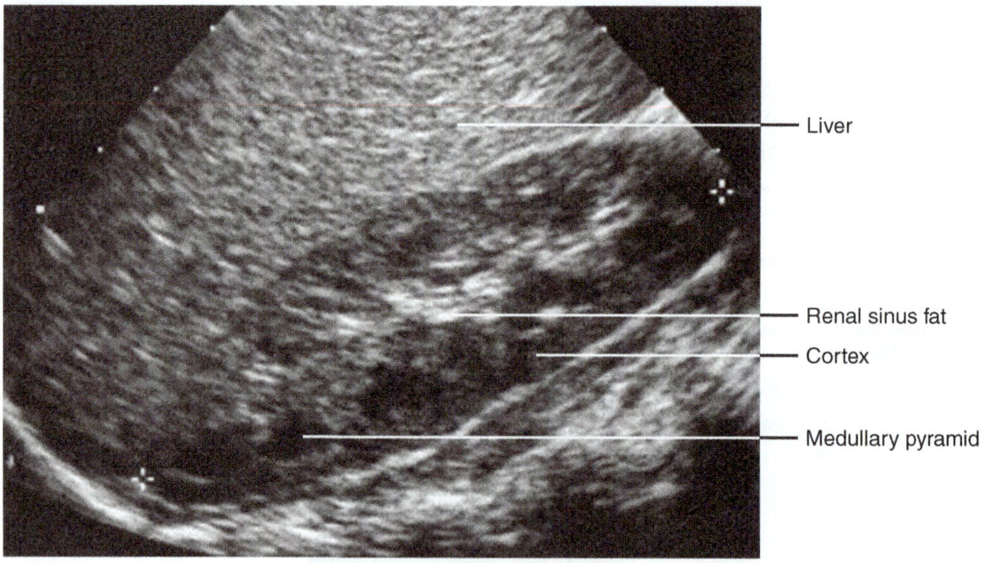

Fig. 33.14 Longitudinal US of the kidney in an older child

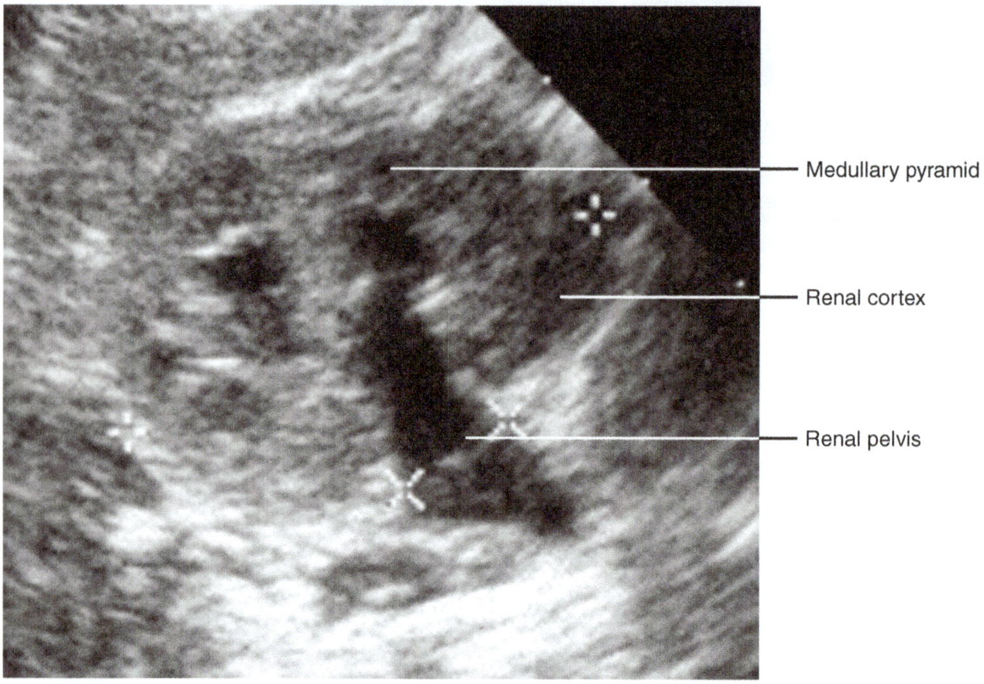

Fig. 33.15 Transverse US of the kidney demonstrating the AP dimension of the renal pelvis within the normal range

Pearl

An AP pelvis diameter < than 10 mm is considered normal.

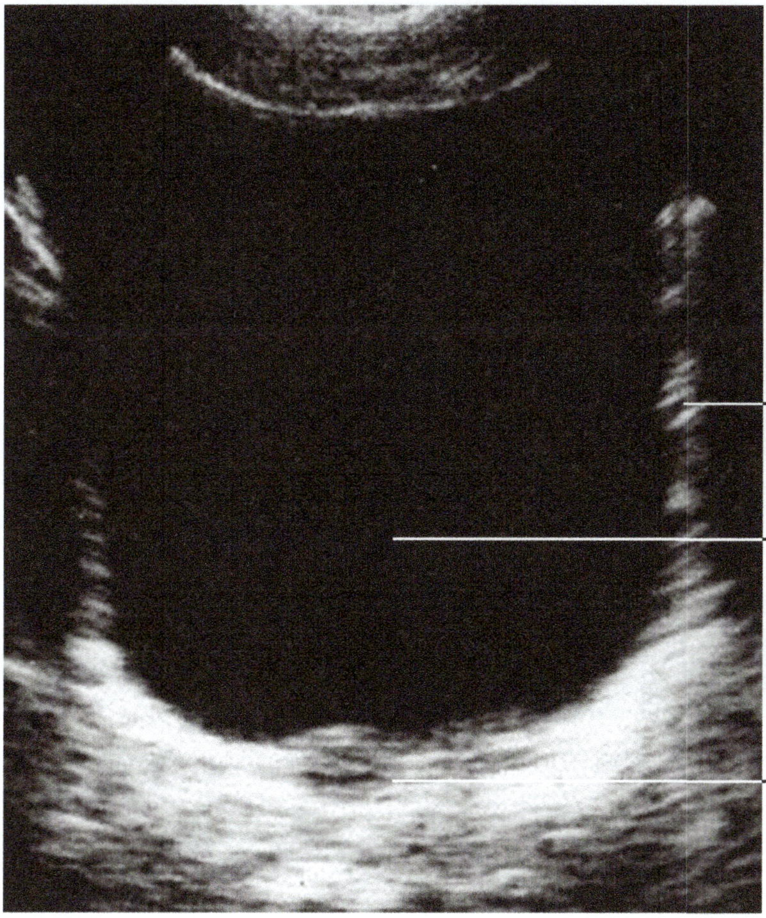

Fig. 33.16 Transverse US of a full bladder

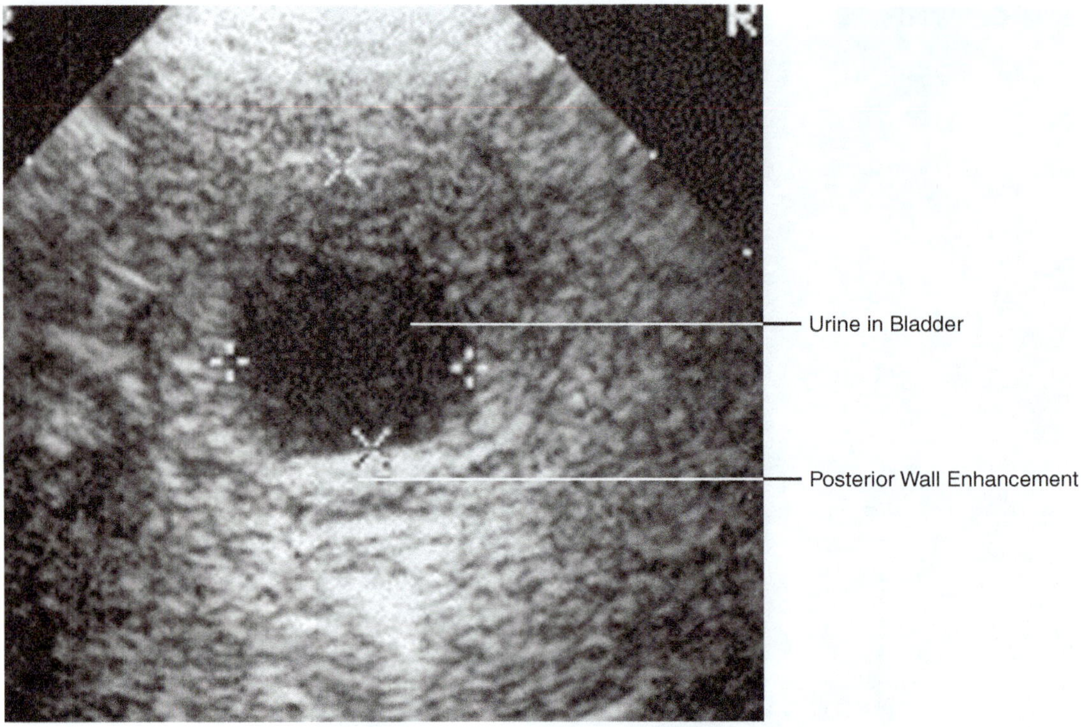

Fig. 33.17 Transverse US of a near-empty bladder post micturition with an acceptable residual volume for a neonate

Female Pelvis

Micturating Cysto-Urethrogram

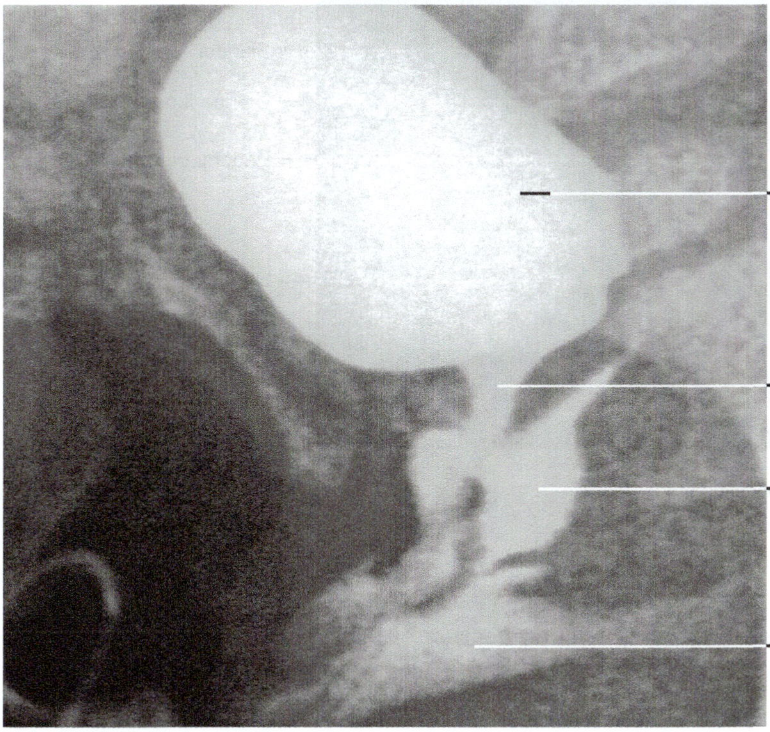

Fig. 33.18 MCUG: right oblique view during voiding in female

Pearl
Vaginal reflux is often seen in females during voiding.

Ultrasound

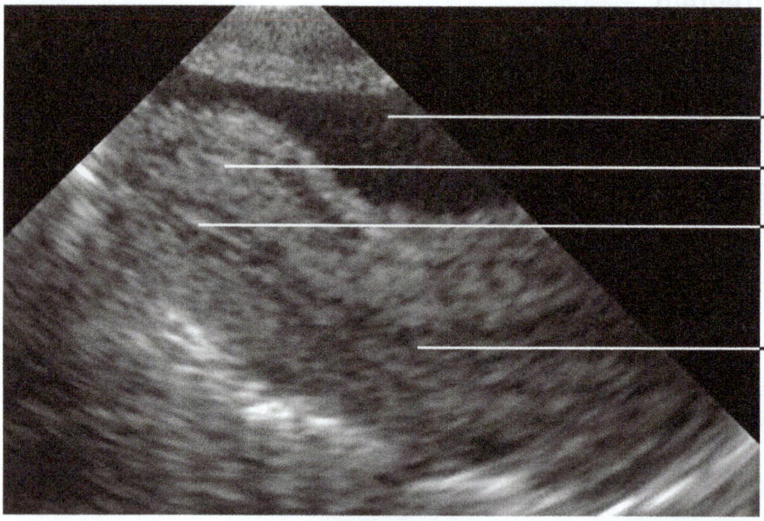

Fig. 33.19 Longitudinal US of neonatal uterus

Pearl
In a normal neonatal uterus the anteroposterior diameter of the cervix is equal to or slightly greater than that of the fundus (inverted pear shape). The Endometrial cavity is visible as a result of maternal hormonal stimulation.

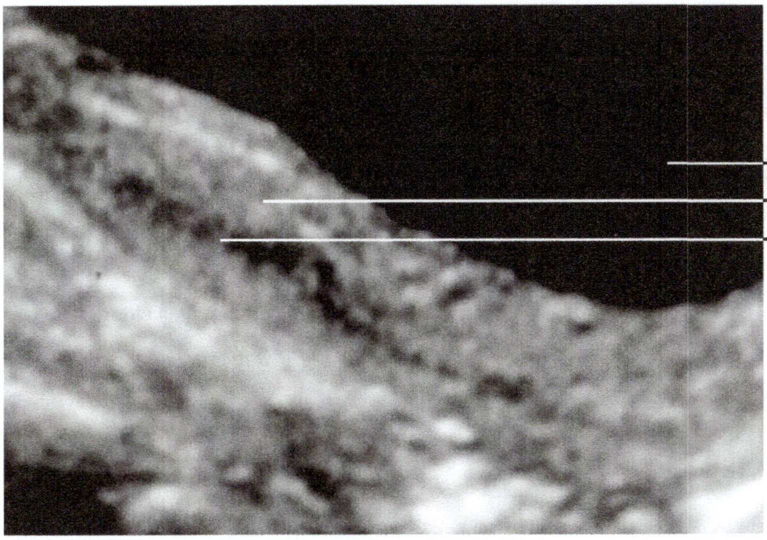

Fig. 33.20 Longitudinal US of uterus (18 months)

Pearl

Immediately after the neonatal period the uterus regresses with the diameter of the cervix being equal to or smaller than that of the uterus giving a tubular shape. The Endometrium is not seen.

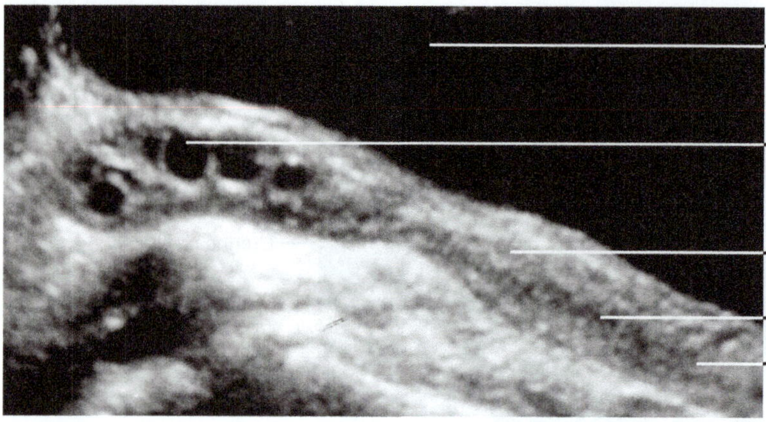

Fig. 33.21 Longitudinal US of uterus (10 years)

Pearl
From 7 years the uterus undergoes an increase in length and width and the endometrial stripe becomes visible. At puberty is assumes an adult configuration.

Pearl
Small follicles or cysts have been reported in the ovaries of 84 % of infants younger than 2 years (usually less than 9 mm). They can exceed this size in neonates still under maternal hormonal stimulation.

Male Pelvis

Micturating Cysto-Urethrogram

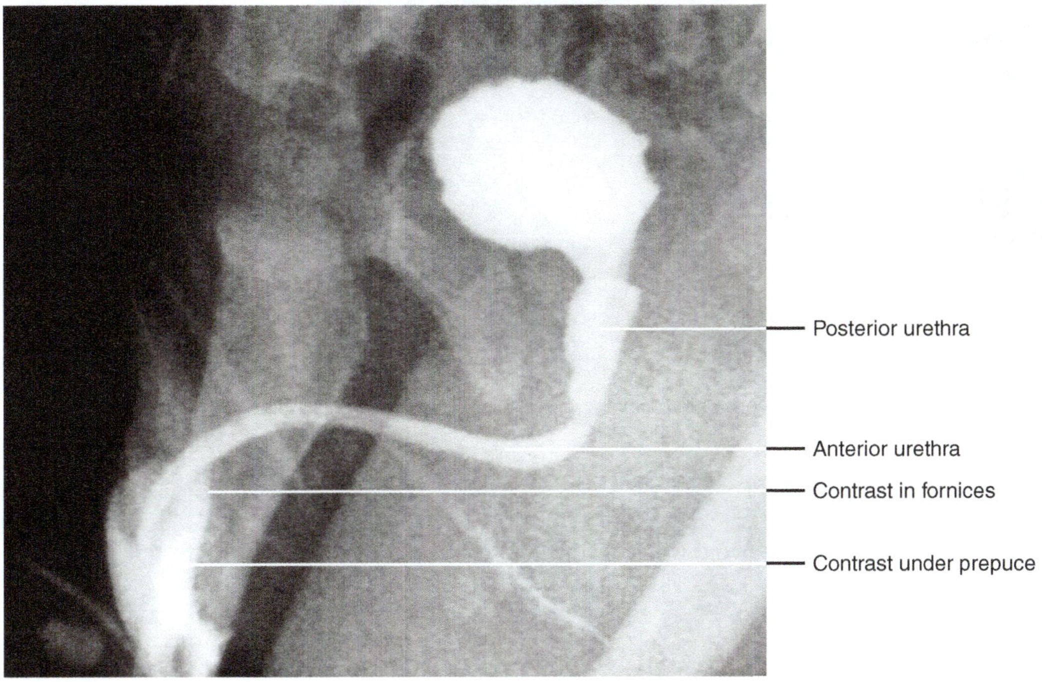

Fig. 33.22 MCUG: right oblique view during voiding in male

Pearl
Contrast filling the fornices of the prepuce during voiding should not be mistaken for pathology.

Pearl
The ratio of the caliber of the posterior urethra to the anterior urethra should remain 3:1 or 2:1 and abnormality should be suspected if this increases to 5:1.

Ultrasound

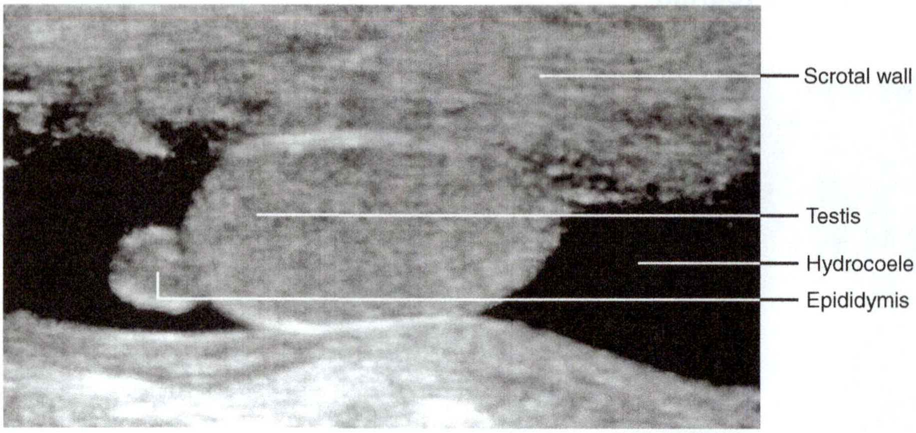

Fig. 33.23 Longitudinal US of neonatal scrotum

> **Pearl**
> The normal infant testis is ovoid with low/medium echogenicity that increases from 8 years to puberty.

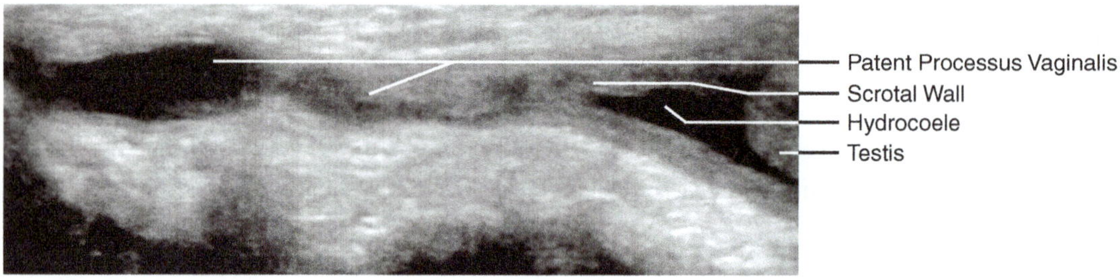

Fig. 33.24 Longitudinal US of inguinal region and scrotum

> **Pearl**
> In neonates and infants virtually all hydrocoeles are congenital and associated with a patent processus vaginalis which allows peritoneal fluid to enter the scrotal sac.

Magnetic Resonance Imaging

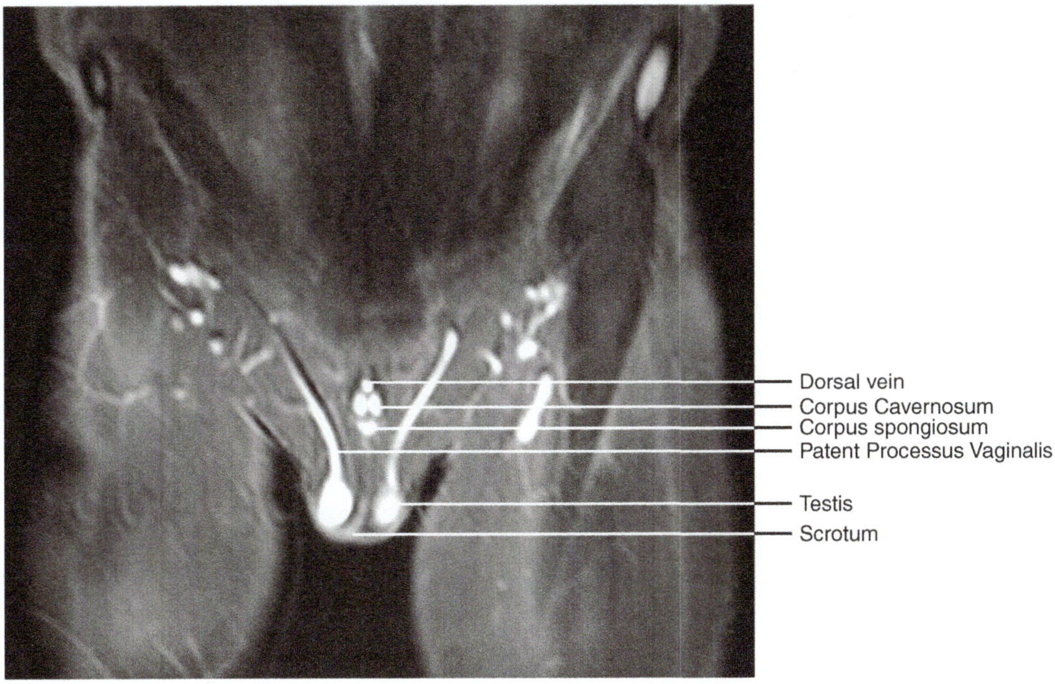

Fig. 33.25 STIR coronal of the scrotum and penis

Pearl
The testes, patent processus vaginalis, corpora cavernosa and corpus spongiosum are all high signal intensity on STIR imaging.

34 Paediatric Musculoskeletal System

Axial Skeleton

Axis

The axis has multiple ossification centers. The summit or tip of the dens ossification centre appears between 2 and 6 years and fuses with the main odontoid mass between 11 and 12 years. The synchondrosis between the dens and body disappears between 3 and 7 years.

S. Andronikou (ed.), *See Right Through Me*,
DOI 10.1007/978-3-642-23893-2_34,

Plain Film

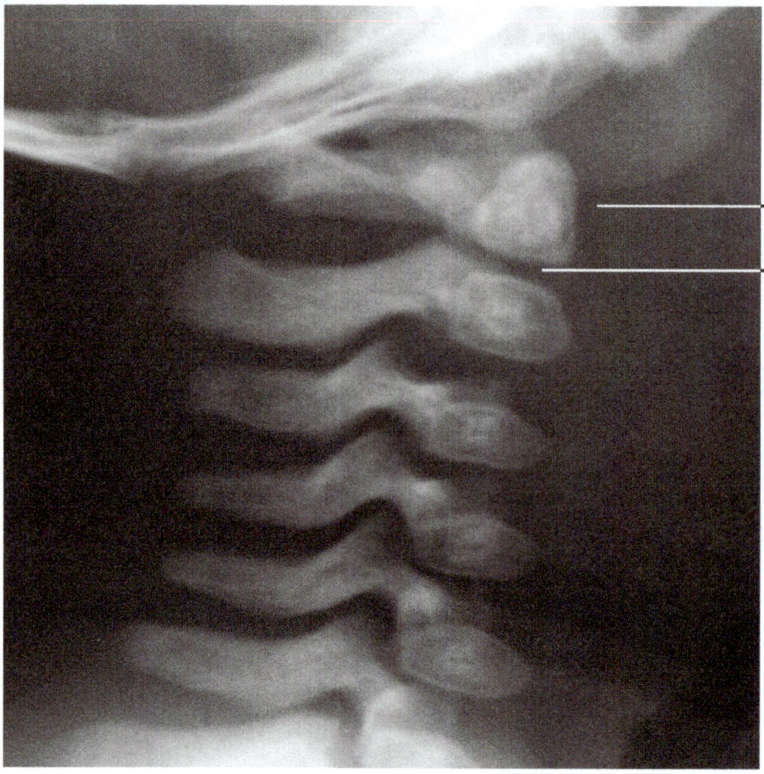

Fig. 34.1 Lateral cervical spine: neonate

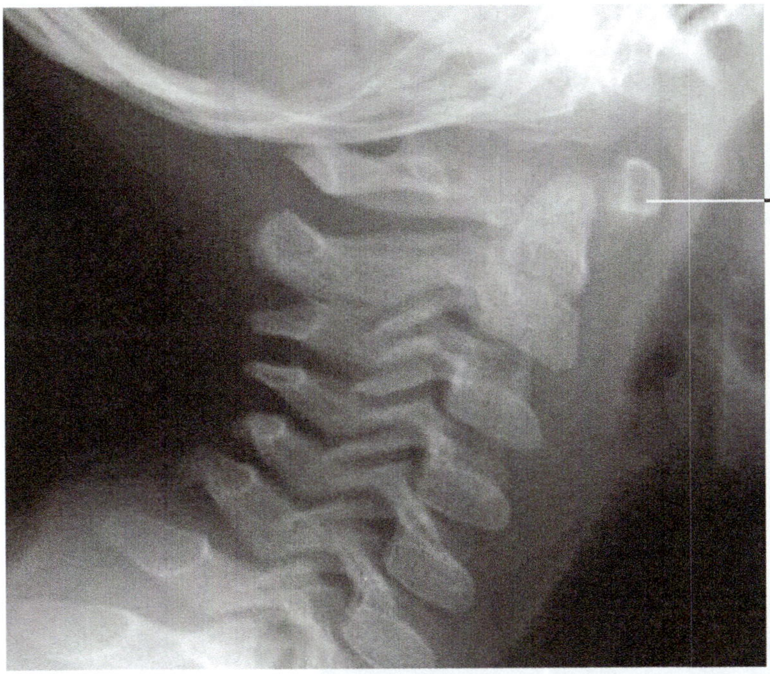

Fig. 34.2 Lateral cervical spine: 2 years old

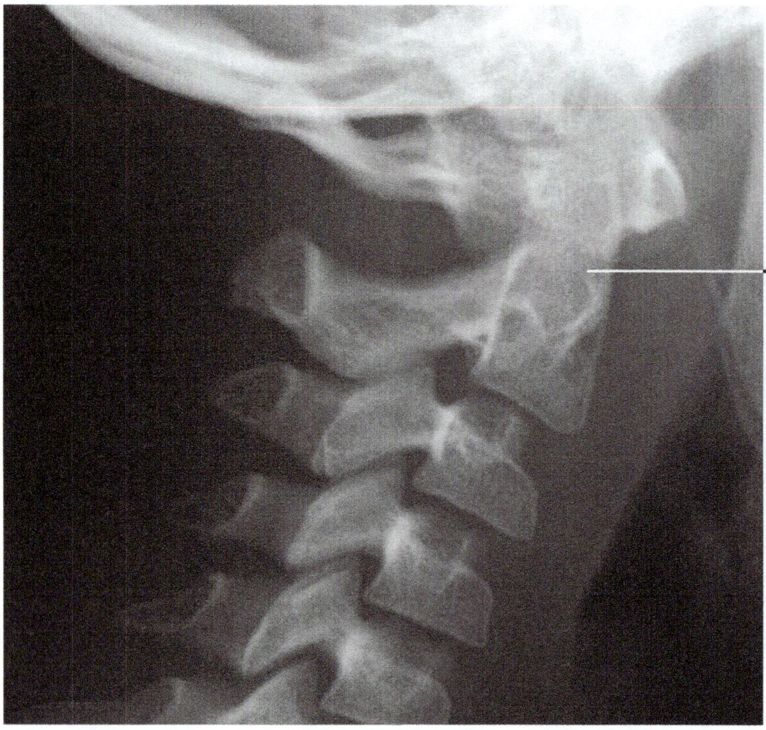

Fig. 34.3 Lateral cervical spine: 9 years old

Pearl

The anterior arch of C1 is not ossified and therefore not visible in an infant.

Magnetic Resonance Imaging

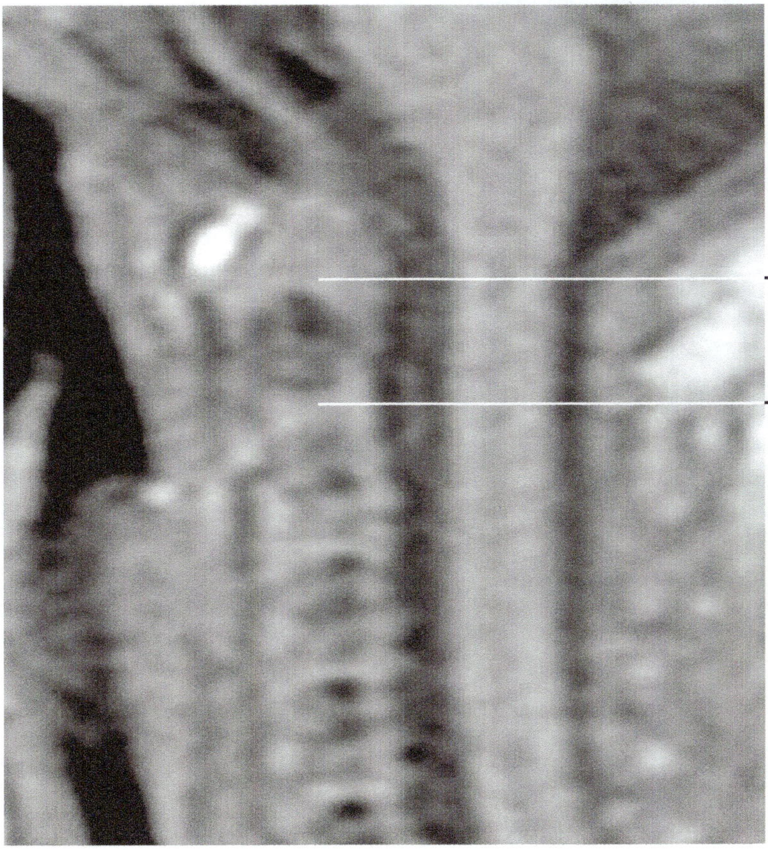

Fig. 34.4 T1 sagittal odontoid: neonate

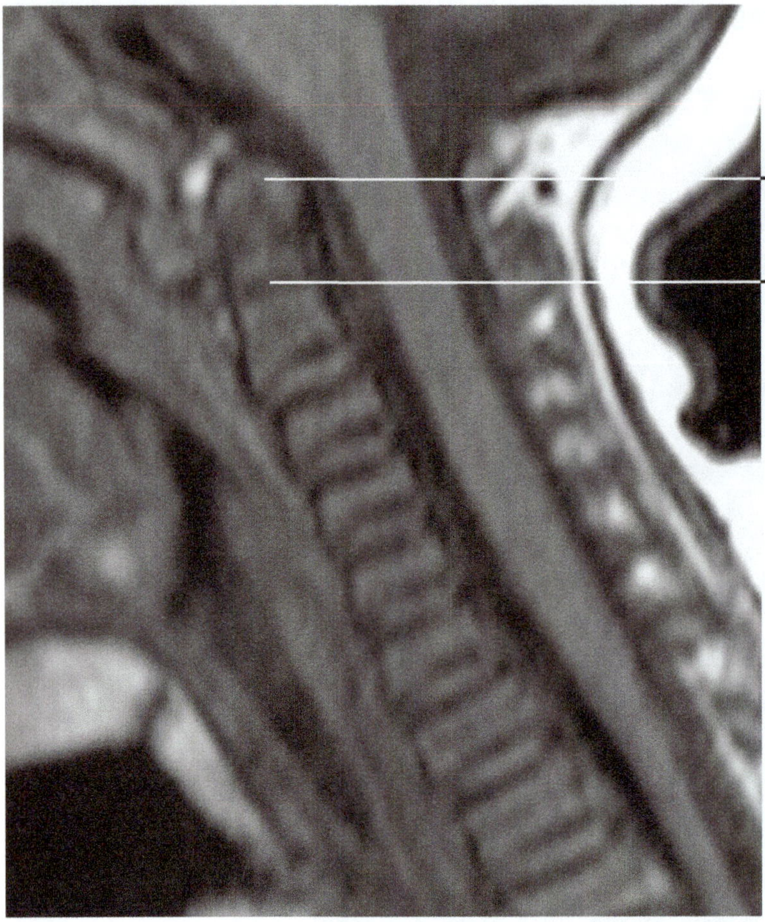

Fig. 34.5 T1 sagittal odontoid: 9 months old

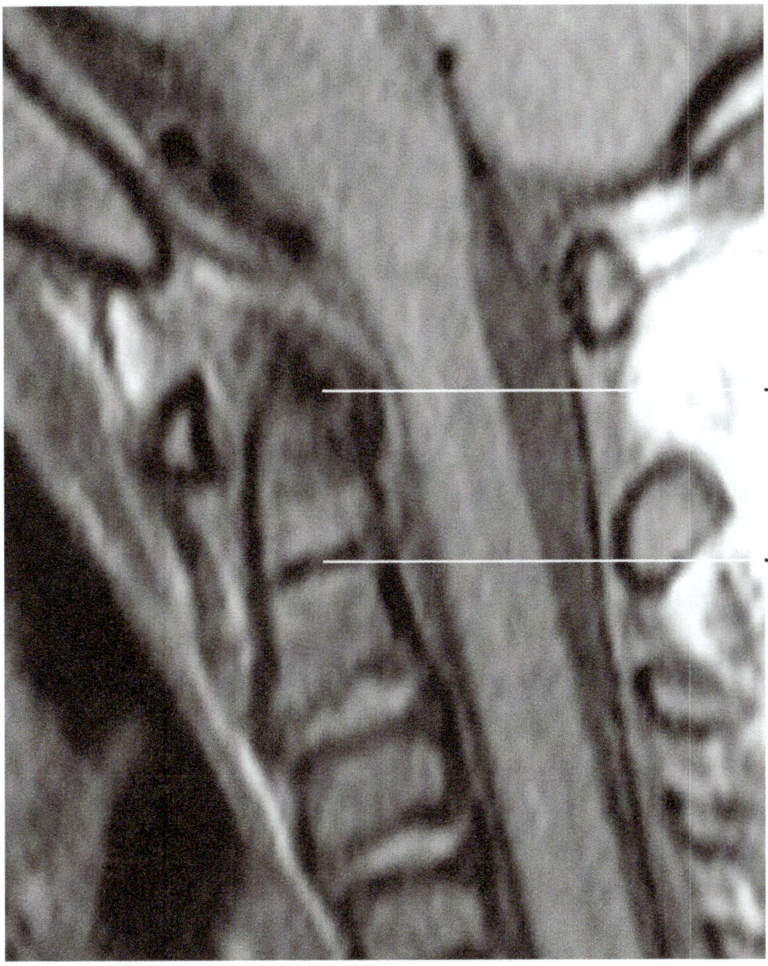

Fig. 34.6 T1 sagittal odontoid: 5 years old

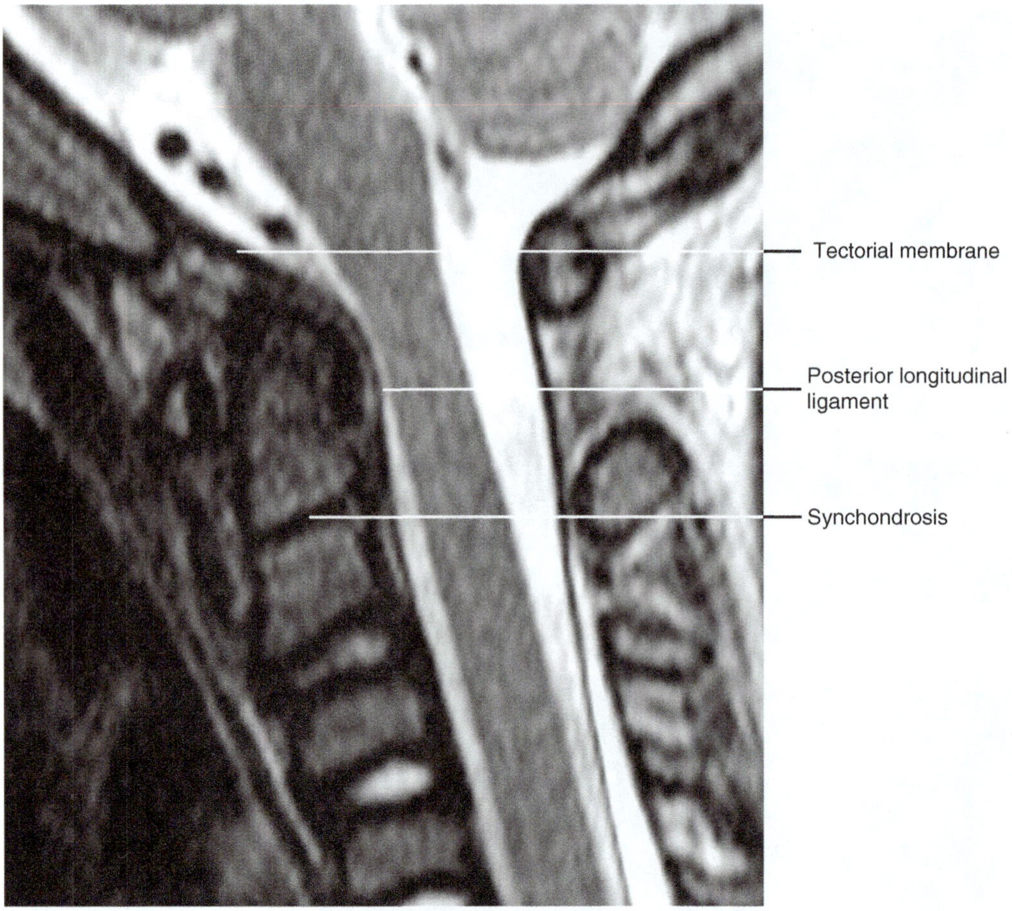

Fig. 34.7 T2 sagittal odontoid: 5 years old

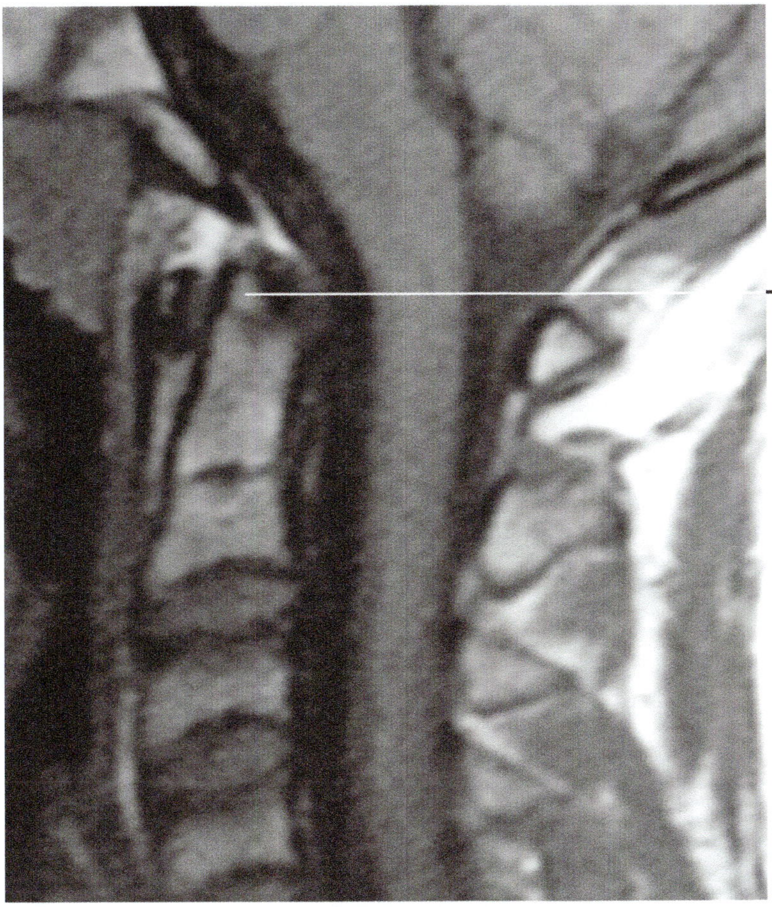

Fig. 34.8 T1 sagittal odontoid: 12 years old

Thoracic Spine

The bony neural arch is ossified by 10 years. Secondary ossification centres appear at puberty and fuse by 25 years.

In infancy and childhood, the disc spaces on plain radiographs appear thicker and the vertebral bodies smaller due to the radiolucent cartilaginous zones in the upper and lower surfaces of vertebral bodies. The neonate has three primary ossification centres separated by radiolucent cartilaginous bridges.

Plain Film

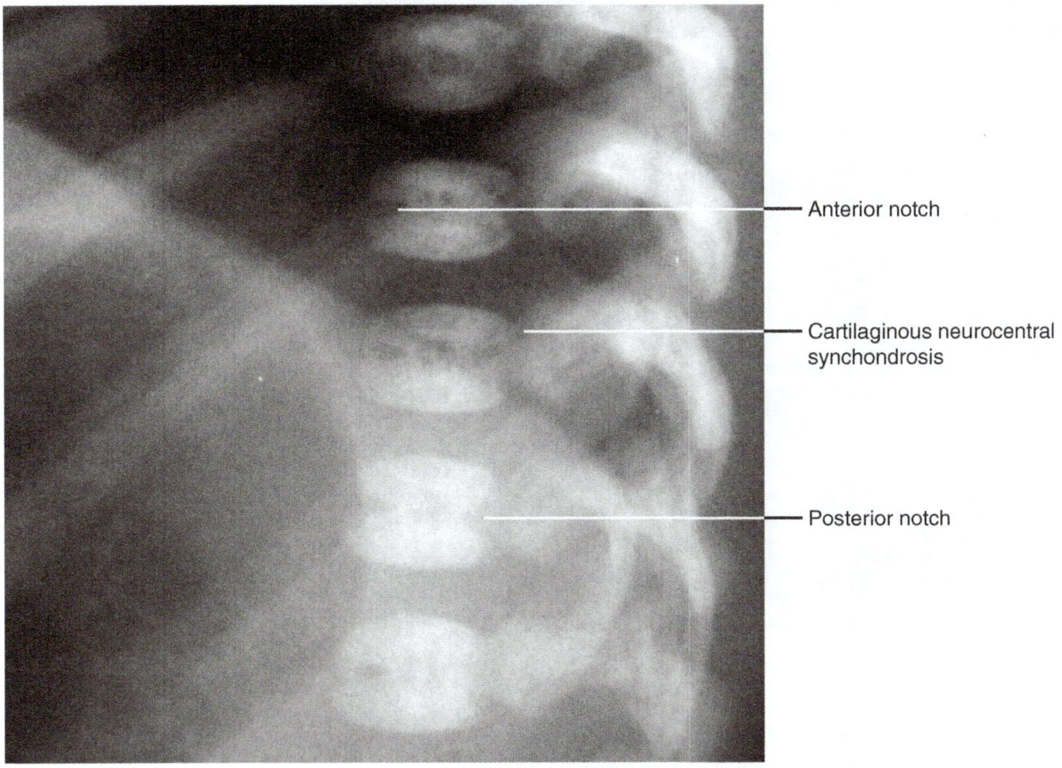

Fig. 34.9 Lateral lower thoracic spine: neonate

Pearl
The anterior notch is due to a large sinusoidal blood space within the ossification centre. It disappears in late childhood. The posterior notch contains nutrient arteries and vertebral veins.

Pearl
As the age increases, the vertebral body becomes more rectangular and less oval.

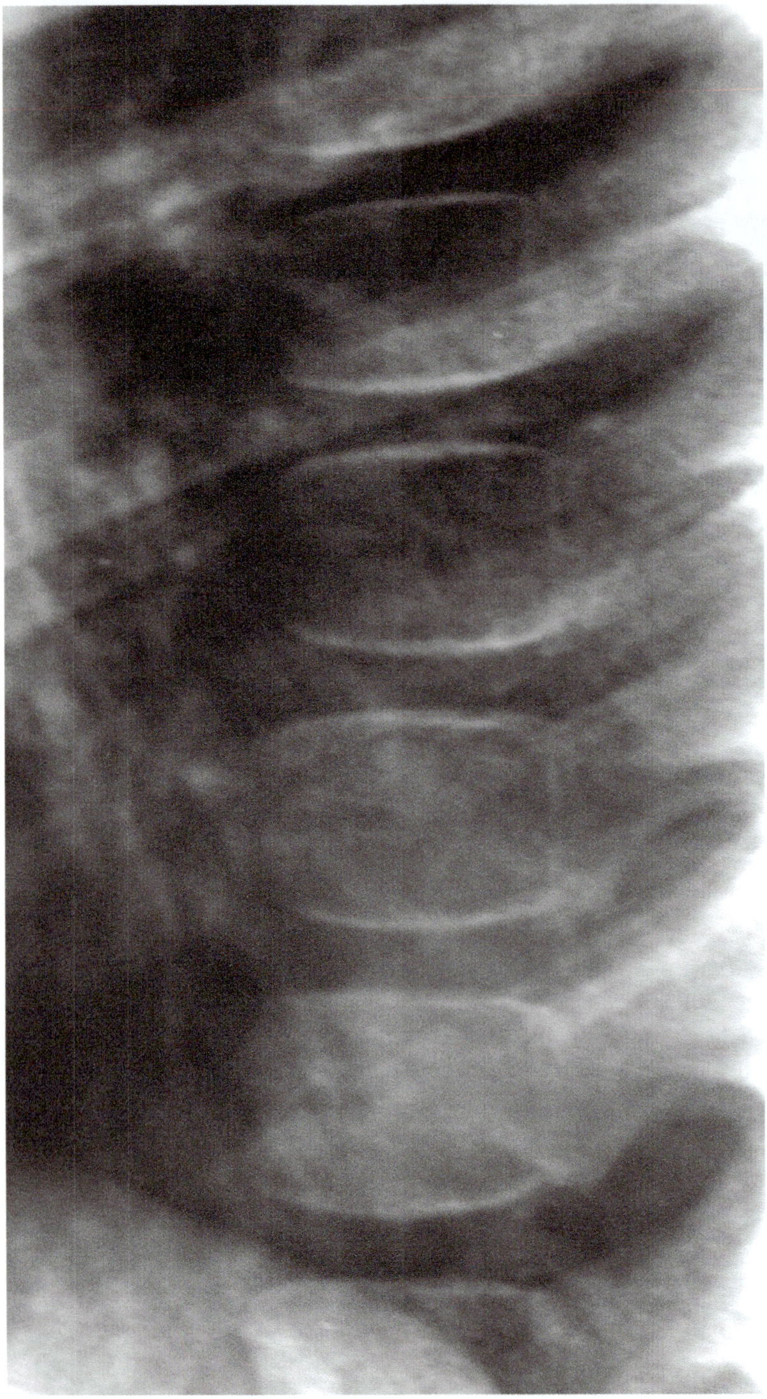

Fig. 34.10 Lateral thoracic spine: mid-childhood

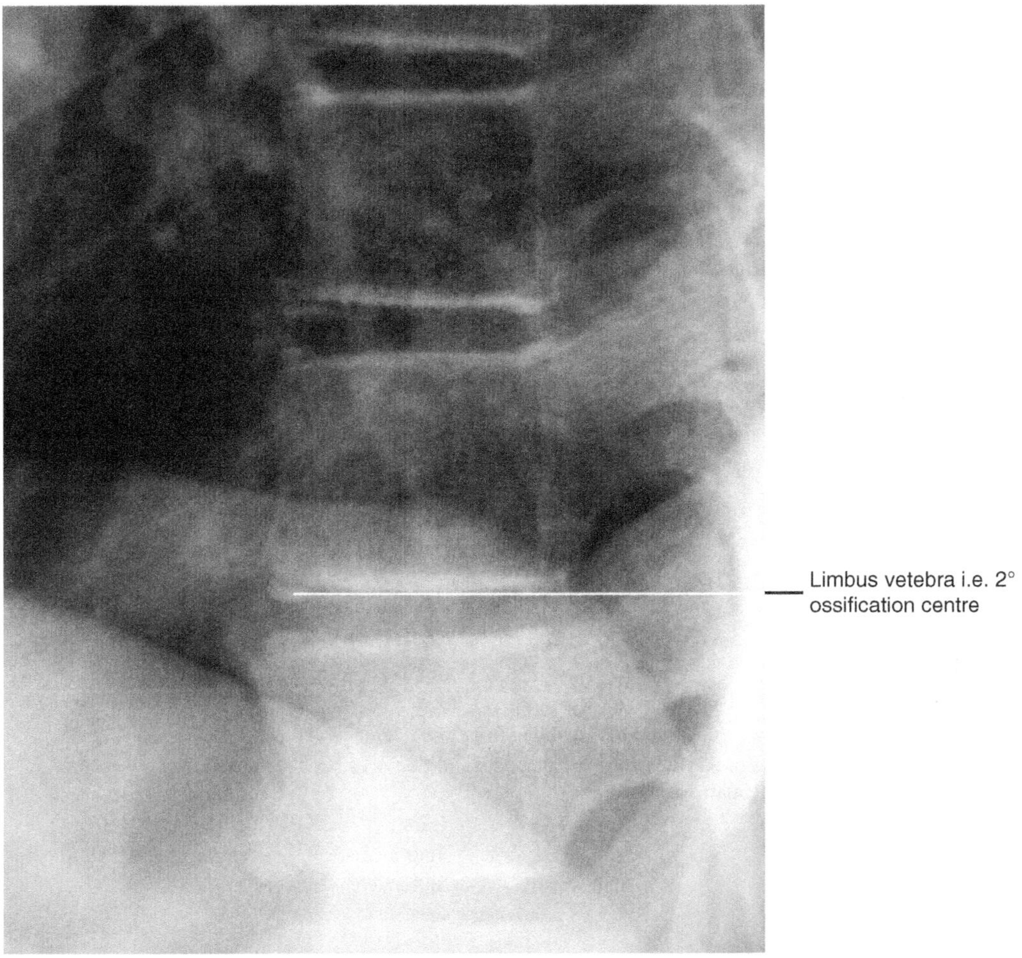

Fig. 34.11 Lateral thoracic spine: puberty/early adolescence

Bone Marrow

Pearl

Two main types of bone marrow – hemapoietically active (red and cellular) and inactive (yellow or fatty) – are readily distinguishable on MRI.
On T1-weighted images, hematopoietic marrow usually shows a signal intensity equal to or slightly higher than that of muscle.
Bone marrow is converted from haemopoetic to fatty marrow from infancy to childhood and occurs in a predictable sequence.
In the spine, the conversion of marrow is much slower than in the appendicular skeleton and occurs into adult life.

Pearl

T1 is the best sequence to assess bone marrow:			
T1 weighted spin-echo sequences allow good differentiation between red and yellow marrow.			
Fat suppressed T2W sequences or STIR can accentuate contrast differences between red and yellow or between normal and pathologic marrow.			
	T1	T2	STIR
Haemopoetic marrow	Low signal	Intermediate	Intermediate / high
Fatty marrow	High signal	Intermediate	Very low

Magnetic Resonance Imaging

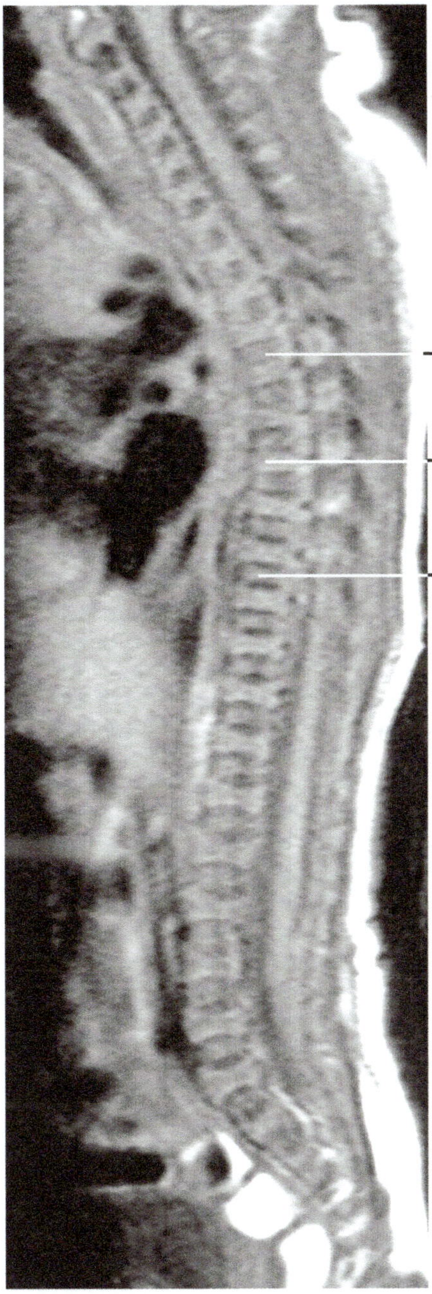

Fig. 34.12 T1 sagittal: neonate

Pearl
On T1, the bodies are of low signal intensity but contain a thin hyperintense band. This represents the basivertebral plexus. The bodies are less intense than the discs.

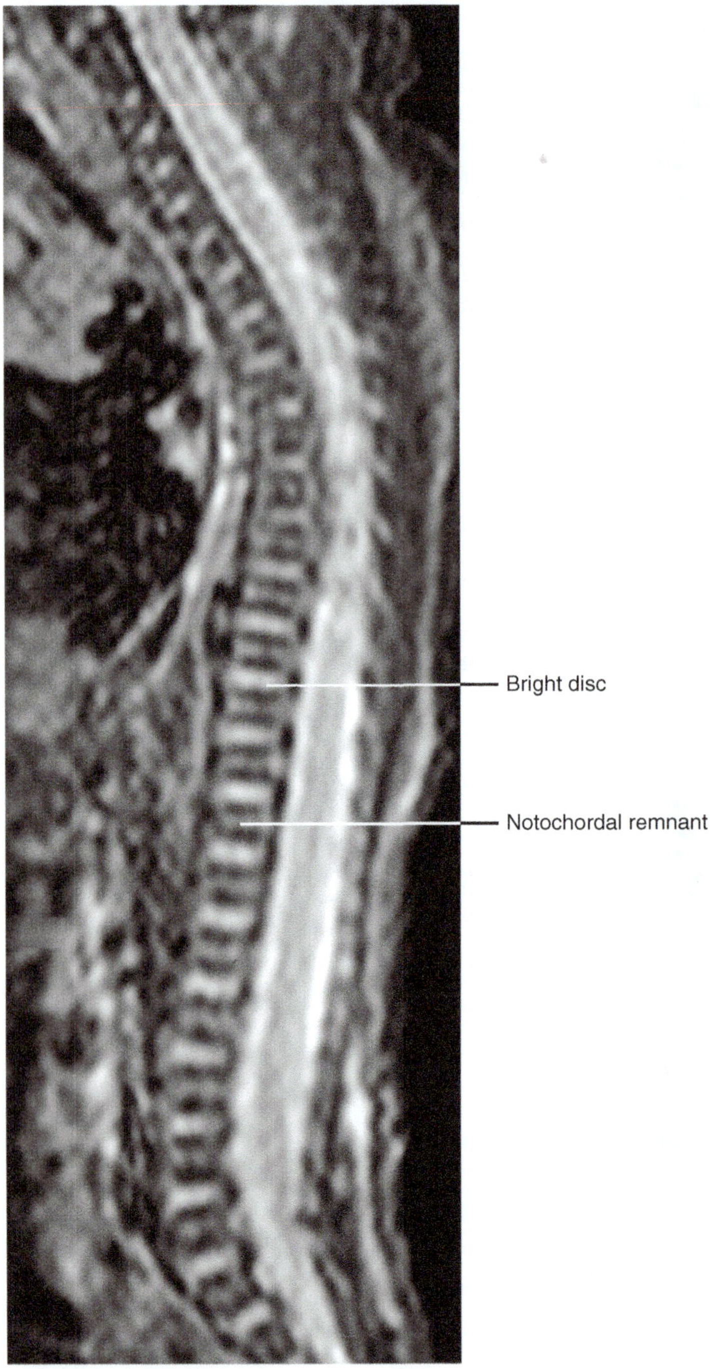

Fig. 34.13 T2 sagittal: neonate

Pearl

The discs have higher water content and are of high signal intensity on T2. Note the thin bright central band in the vertebral body representing the notochordal remnant.

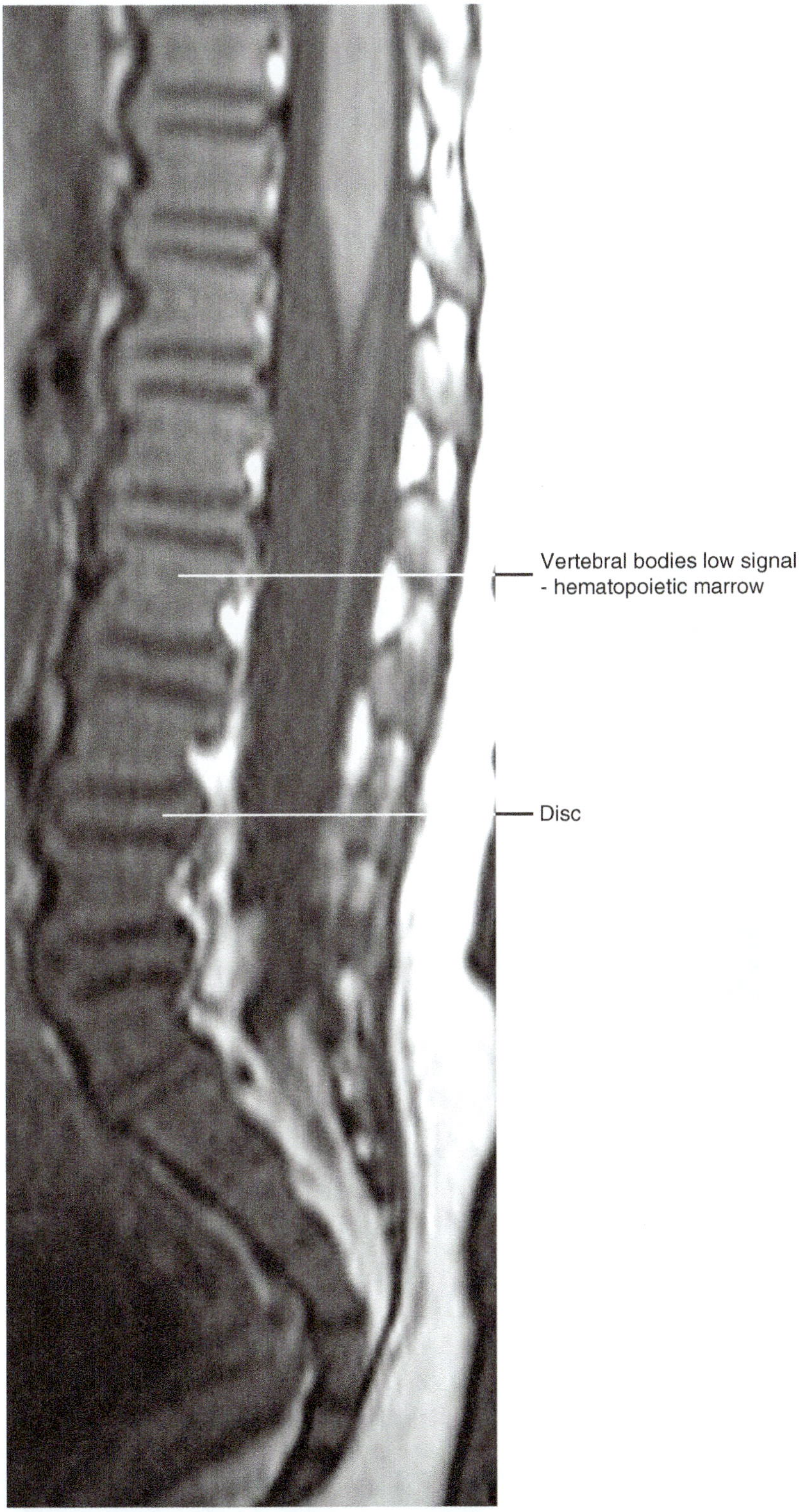

Fig. 34.14 T1 sagittal: 1 year old

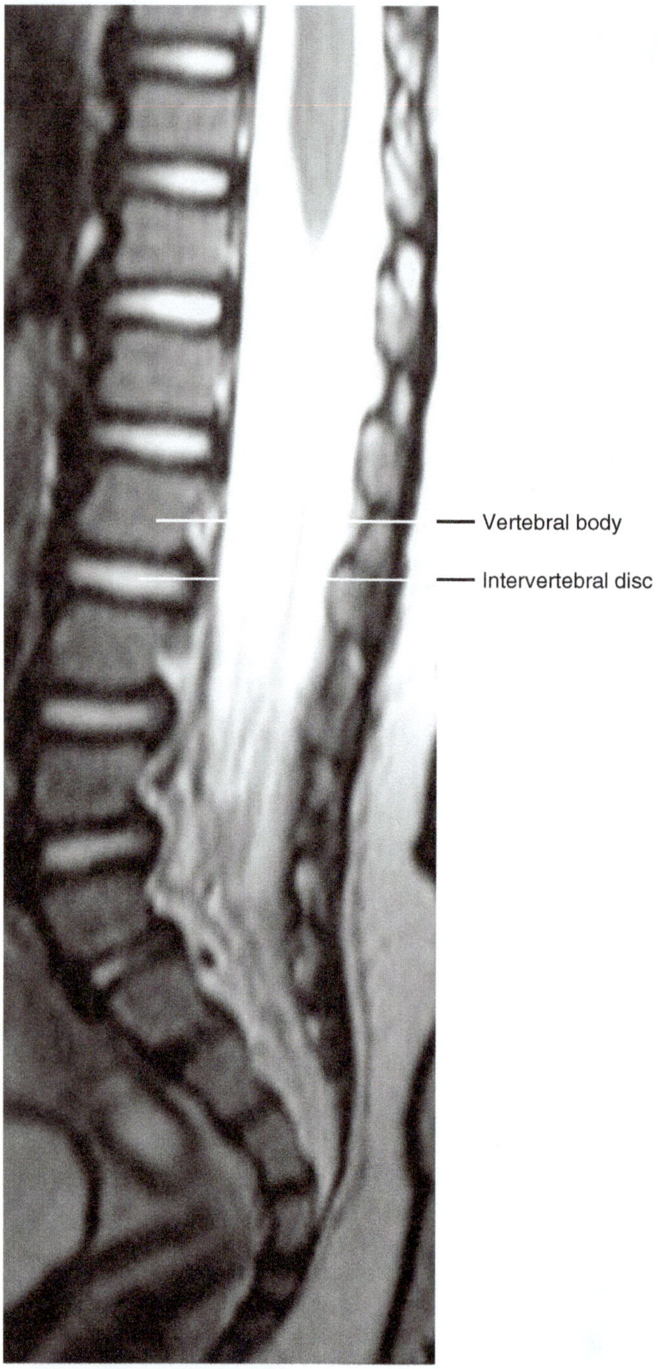

Fig. 34.15 T2 sagittal: 1 year old

Pearl
Between 1 and 5 years the vertebral bodies are isointense with the discs on T1.

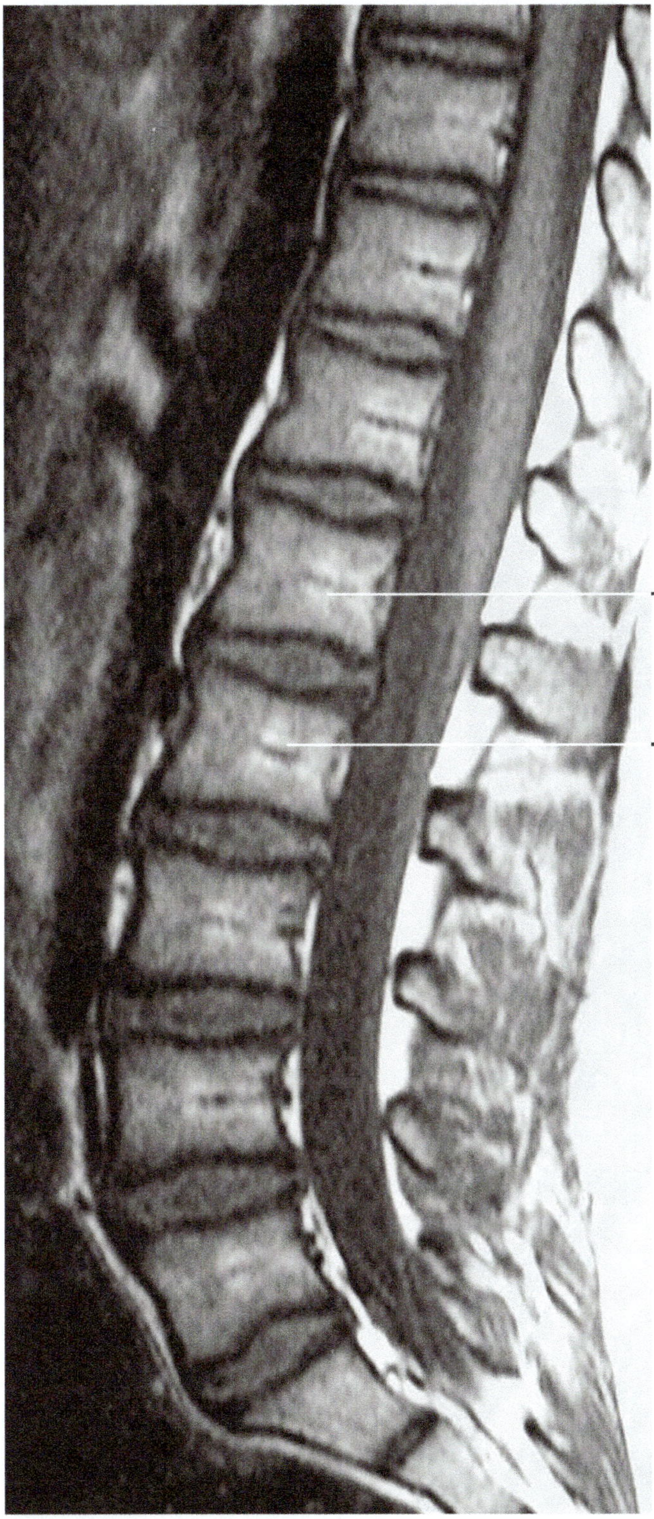

Fig. 34.16 T1 sagittal: 10 years old

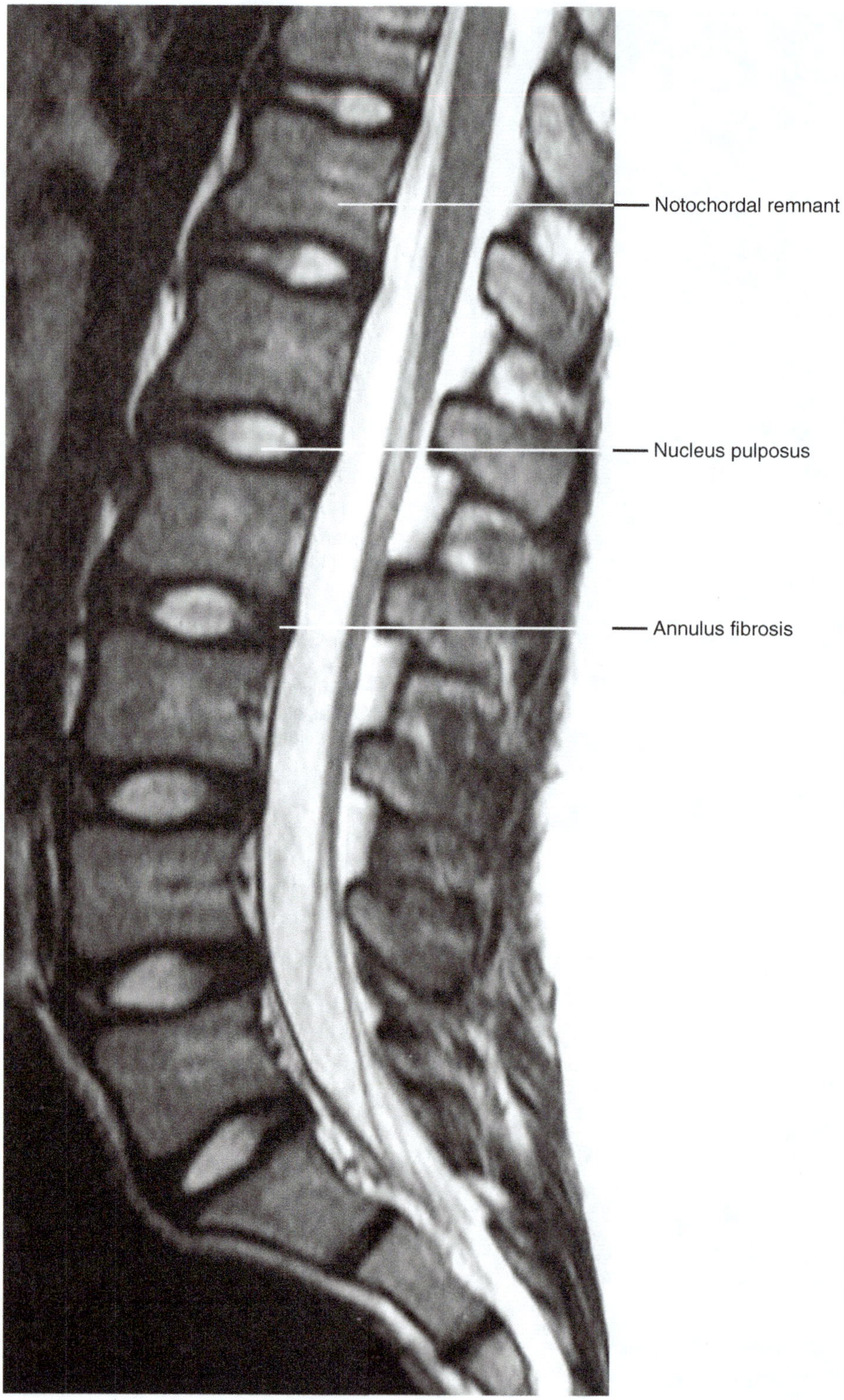

Fig. 34.17 T2 sagittal: 10 years old

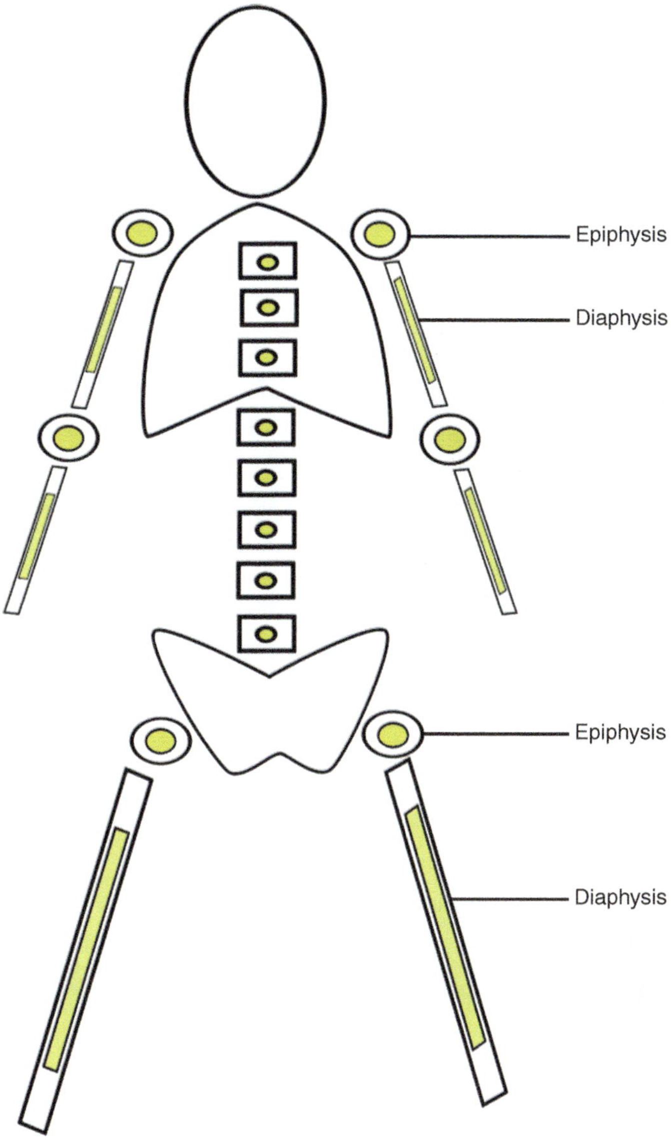

Fig. 34.18 Schematic representations of sequence of marrow maturation

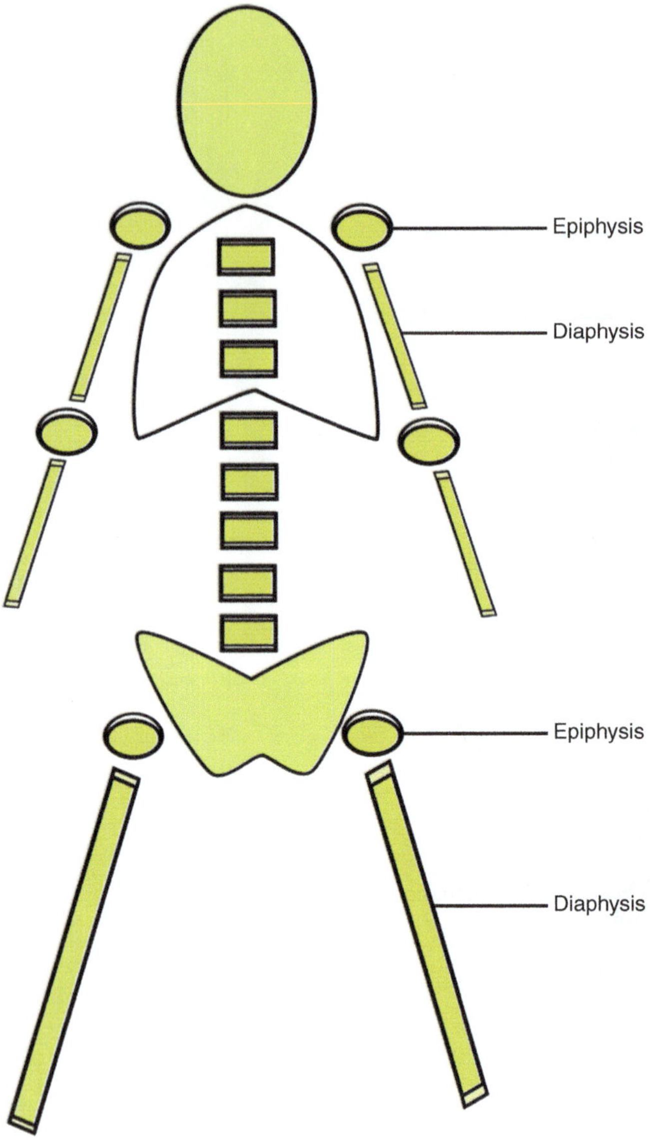

Fig. 34.18 (continued)

Table 34.1 Bone marrow is a dynamic organ that changes composition throughout life

Neonates	Bone marrow contains a predominance of hematopoietically active cells (low SI on T1W sequences)
Shortly after birth	A conversion of hematopoietic to fatty marrow begins in the appendicular or peripheral skeleton and progresses to the axial or central skeleton.
Long bones	Marrow conversion occurs first in the mid **diaphysis** and **epiphysis,** then in the **distal metaphyses,** and finally in the **proximal metaphyses**
Vertebral marrow	***Age 0–10 years:*** the vertebral marrow is predominantly hematopoietic with SI **lower** than that of the adjacent disc space, except for high SI around the basivertebral vein. ***After 10 years:*** the fatty marrow increases and vertebral bodies show a SI **higher** than that of the adjacent disk, but even in adults the vertebral bodies contain red marrow.
Pelvic marrow	Bone marrow is predominantly hematopoietic in the first 20 years of life and it becomes mainly fatty by the third decade.
Adult distribution of bone marrow (after 35 years of age)	Bones show predominantly high T1W SI due to fatty conversion. Hematopoietic/red marrow remains in the axial skeleton (skull, vertebral bodies, flat bones, proximal femoral and humeral memtaphyses).

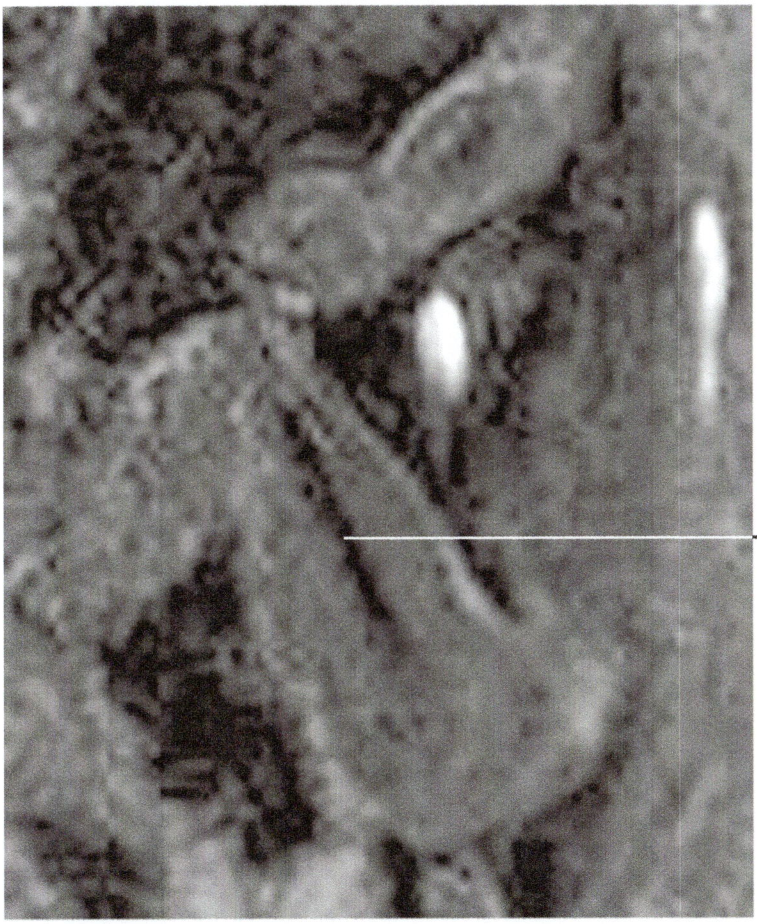

Fig. 34.19 Sagittal T1W image of a 32-week-old fetus: thick cortices and small marrow cavity

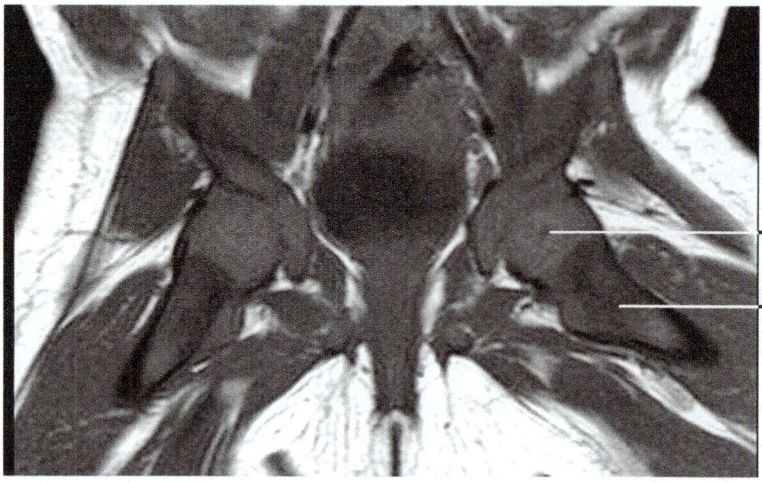

Fig. 34.20 Coronal T1W image of proximal femurs in a 6 months old

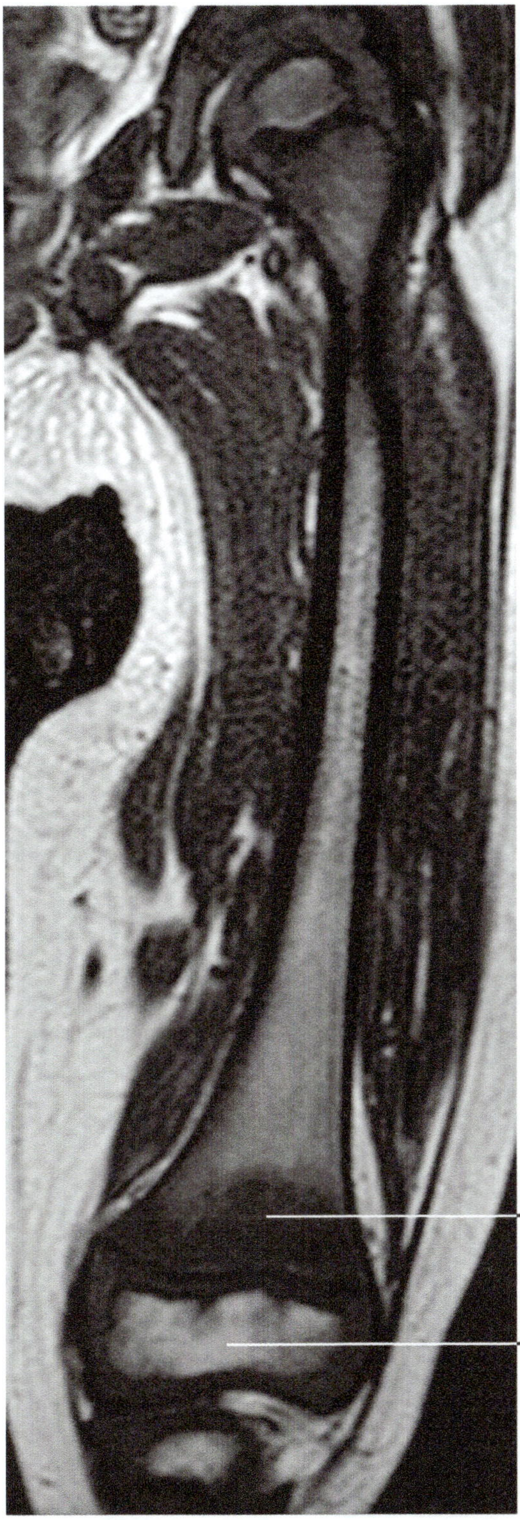

Fig. 34.21 Coronal T1W image of the left femur in a 3 year old

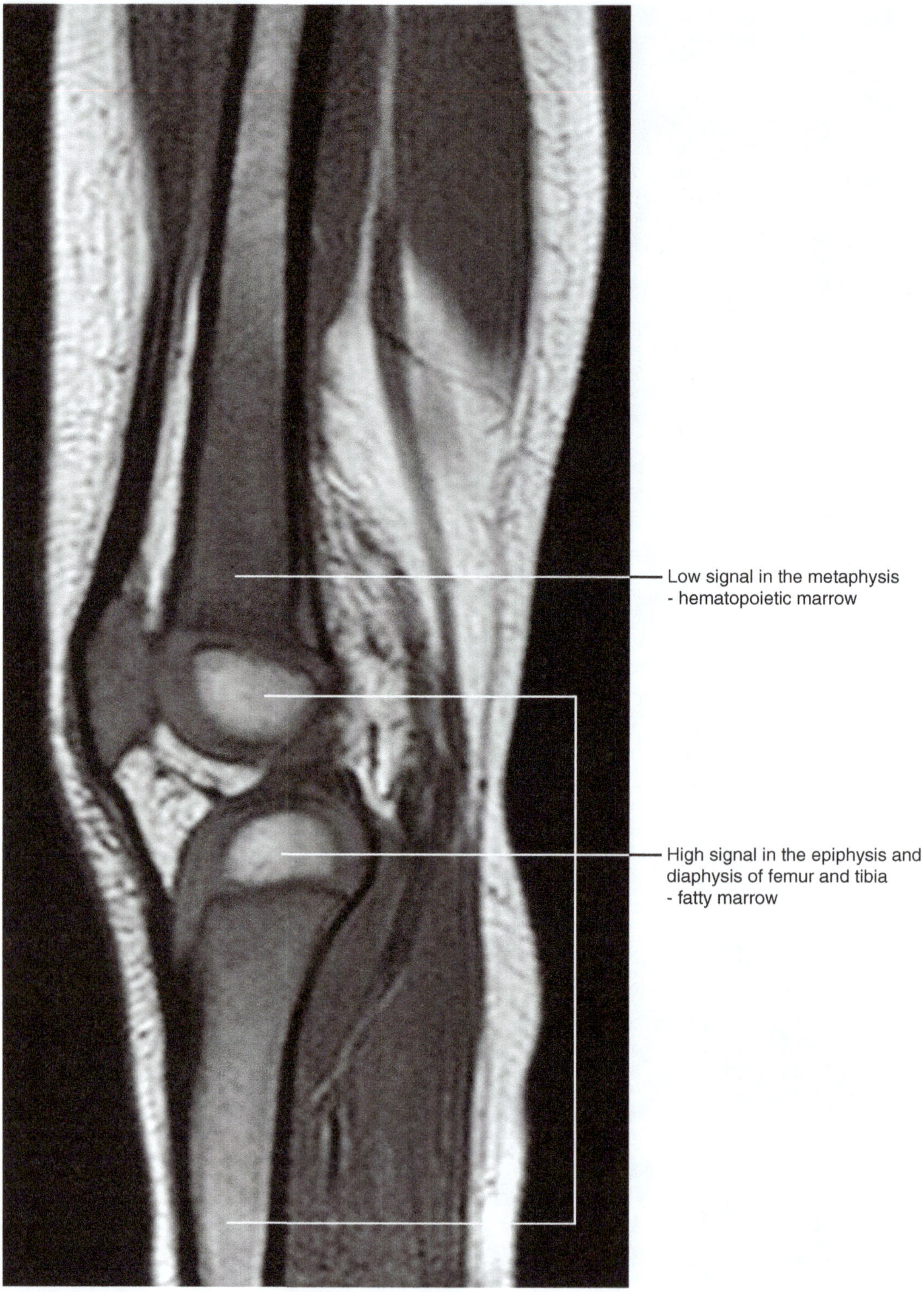

Fig. 34.22 Sagittal T1W image of the knee in a 4-year-old

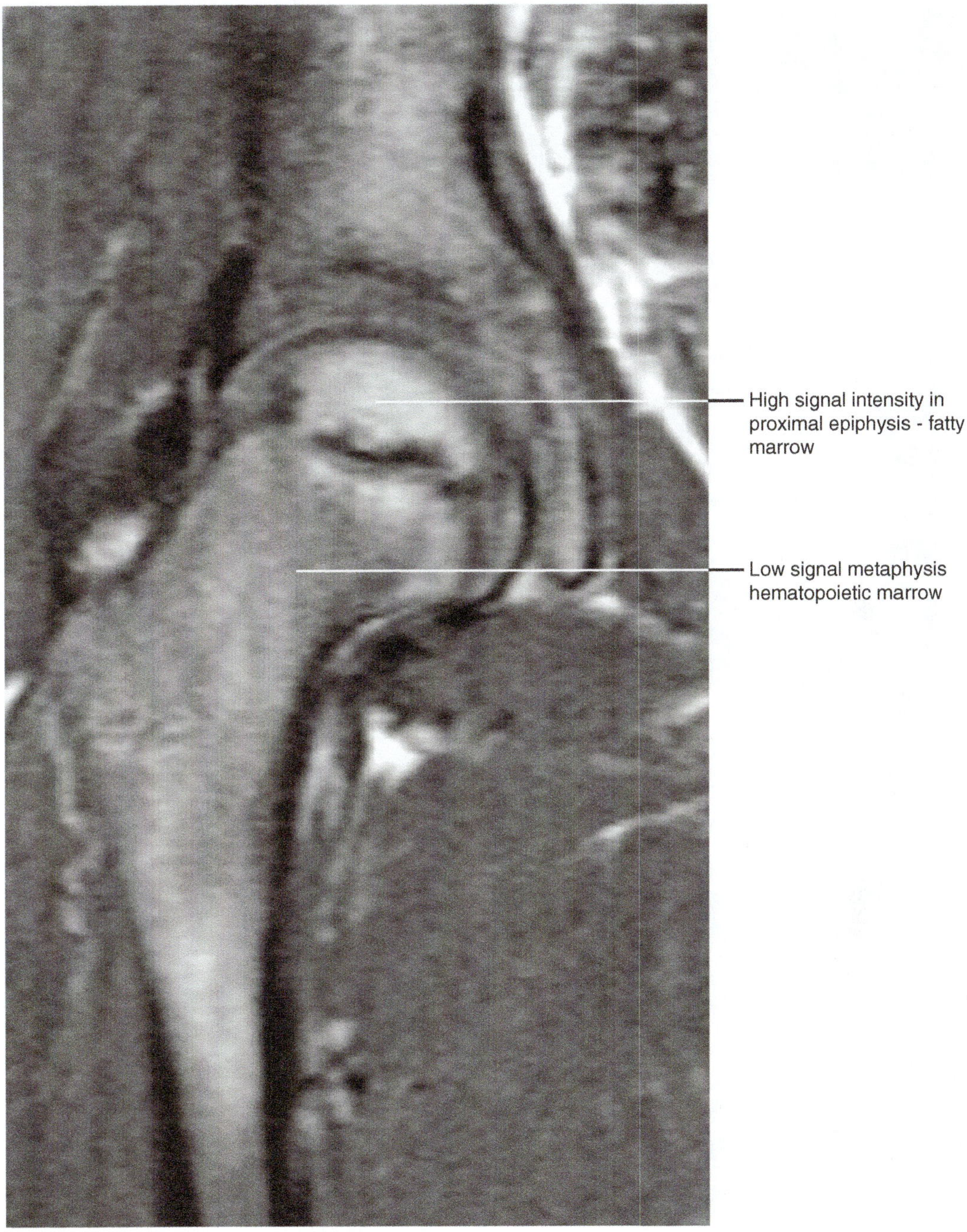

Fig. 34.23 Coronal T1W image of the right femur in a 7-year-old: epiphyseal conversion to fatty marrow

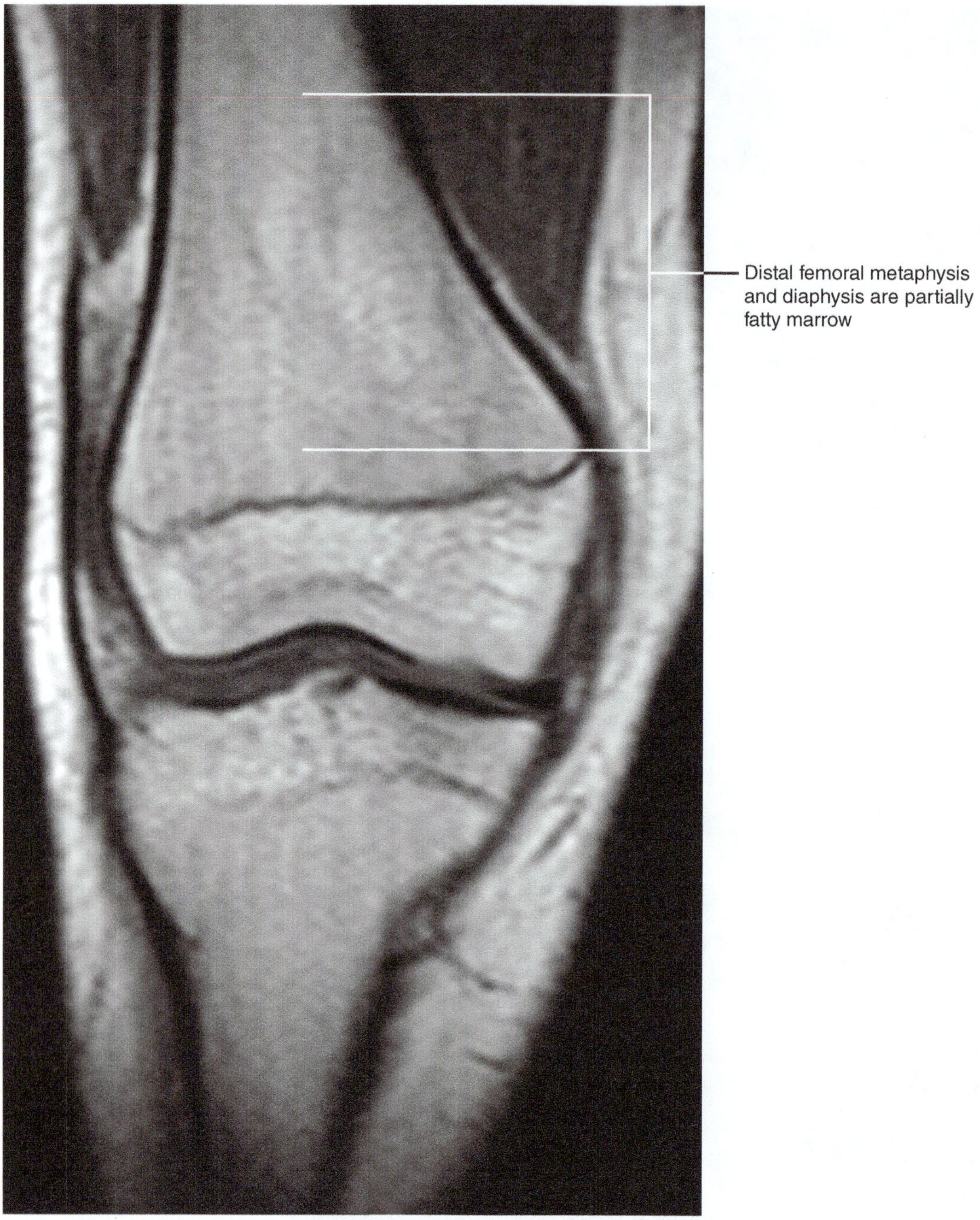

Fig. 34.24 Coronal and sagittal T1W image of the knee in a 13 year old

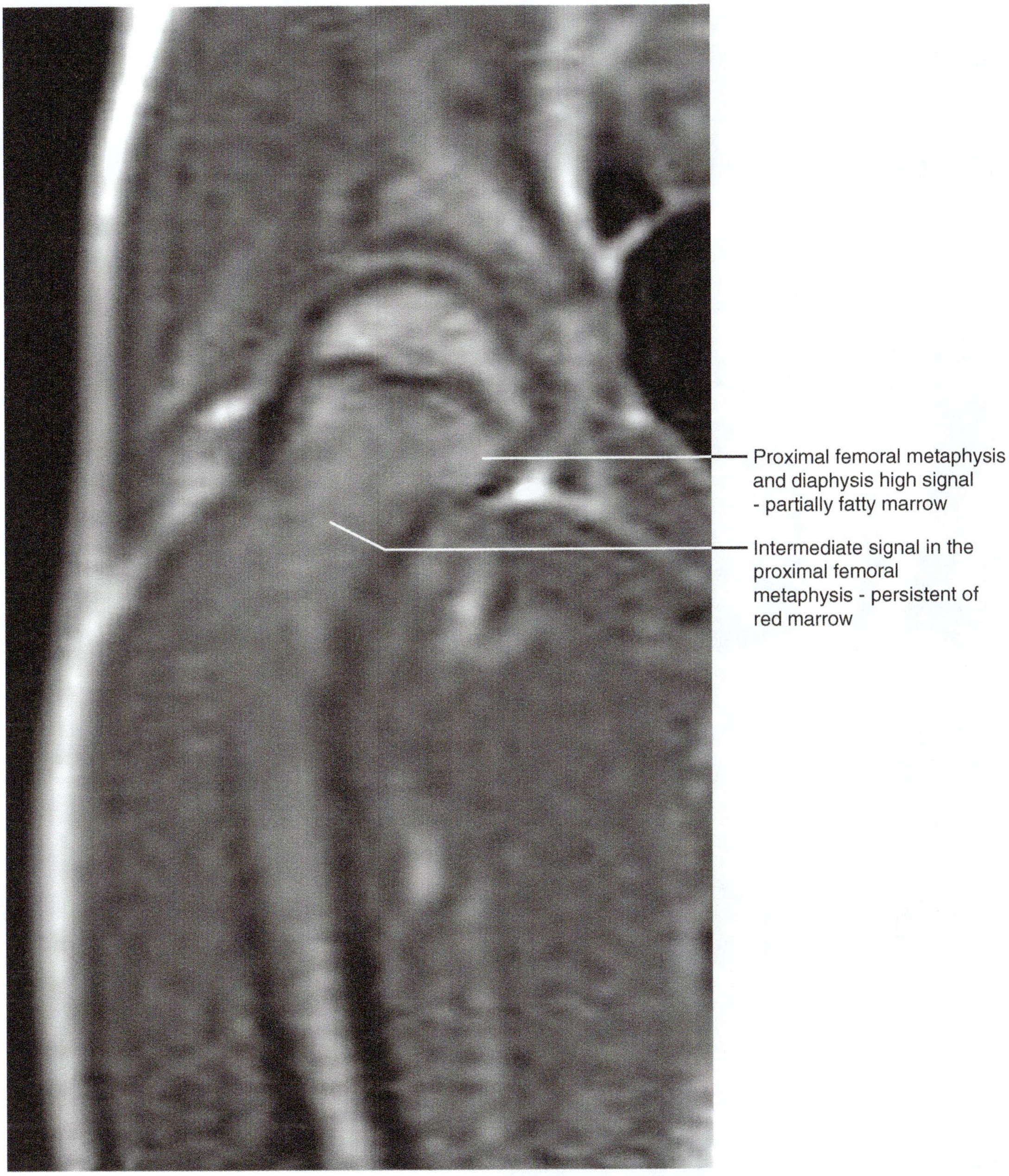

Fig. 34.25 Coronal T1W image of the proximal femur in a 15 year old

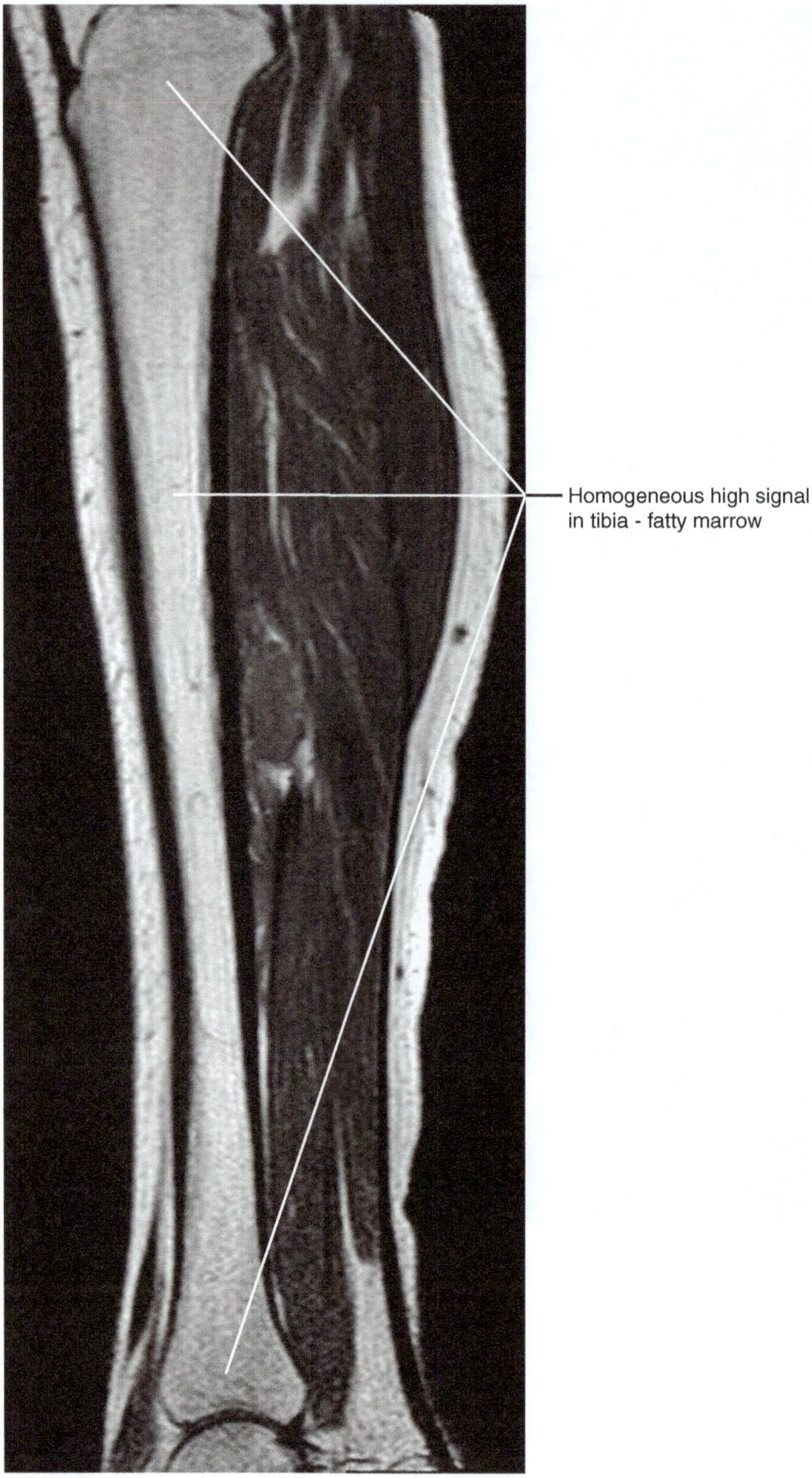

Fig. 34.26 Sagittal T1W image of the tibia in a 17 year old

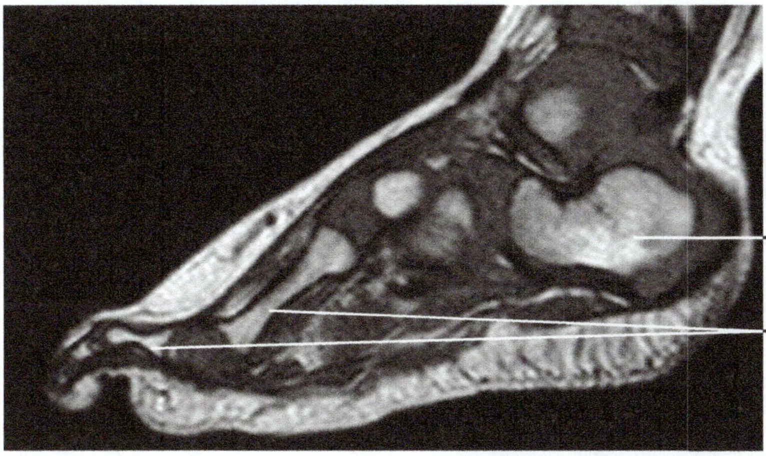

Fig. 34.27 Sagittal T1W image of the foot of a 1 year old

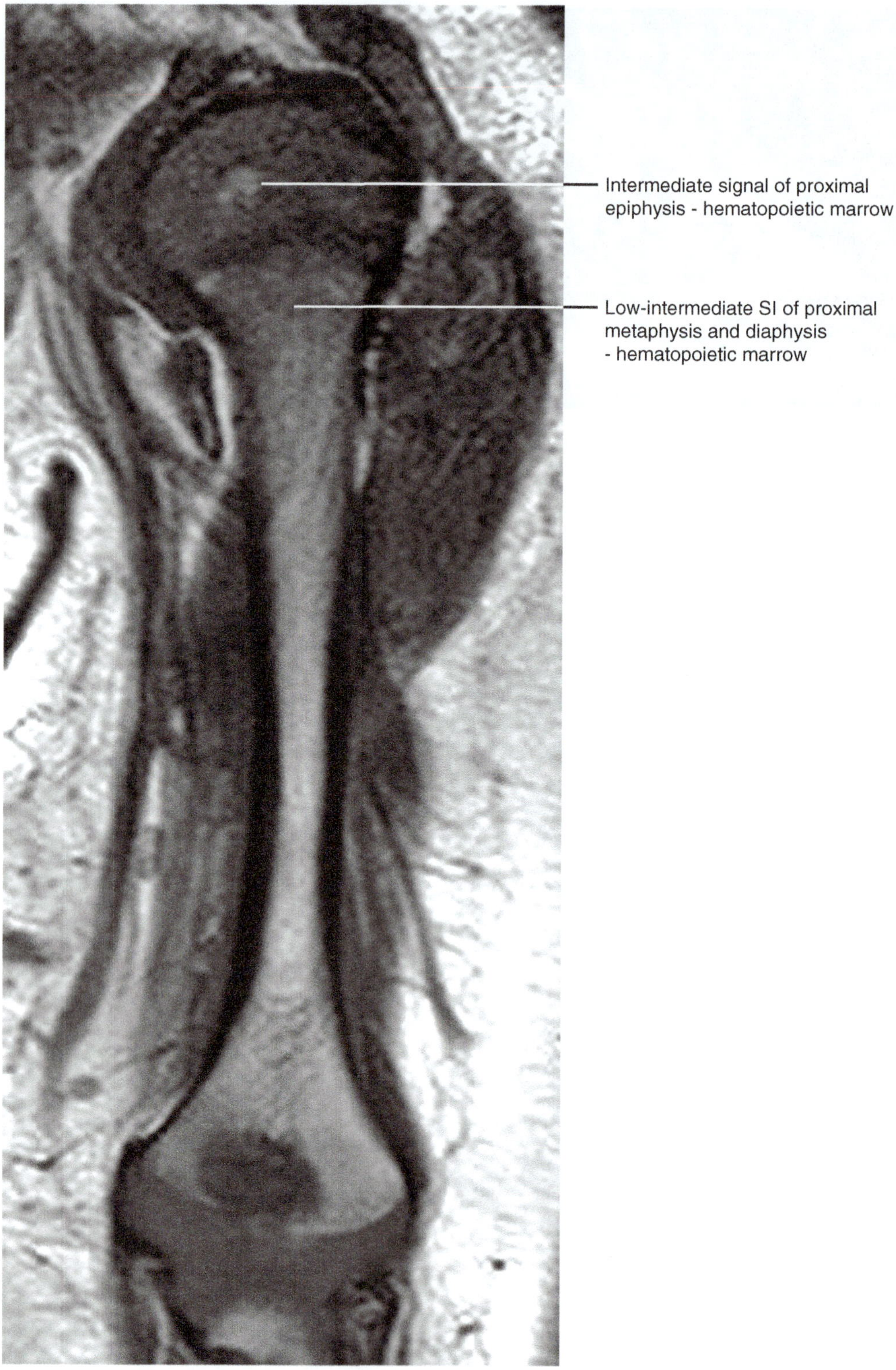

Fig. 34.28 Coronal T1W image of the humerus in a 6 month old

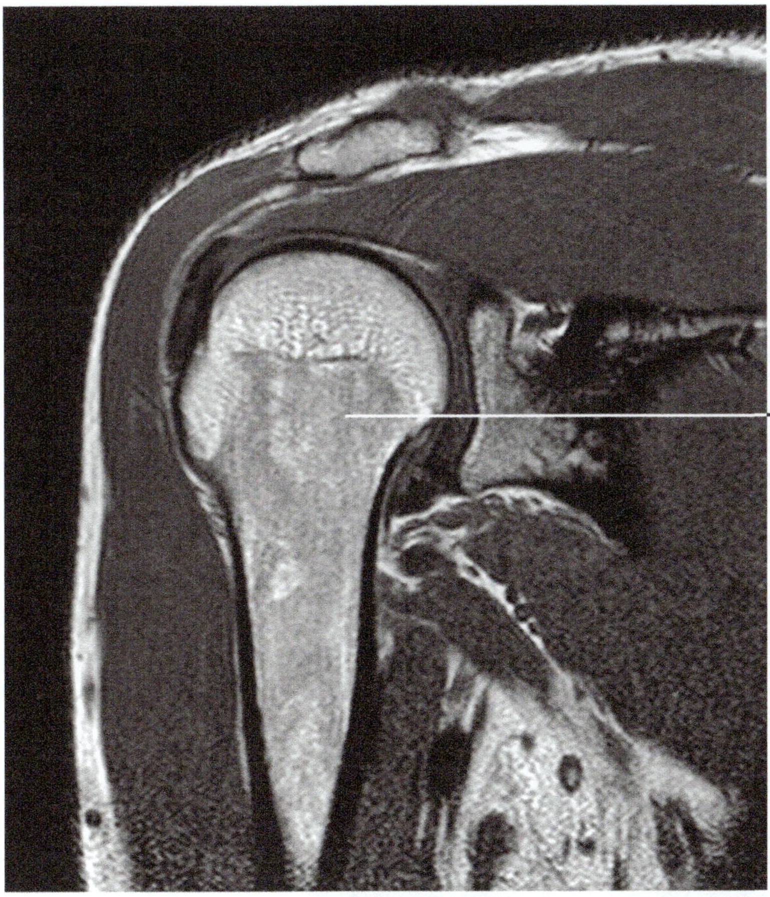

Fig. 34.29 Coronal T1W image of the humeral head in a 25 year old male

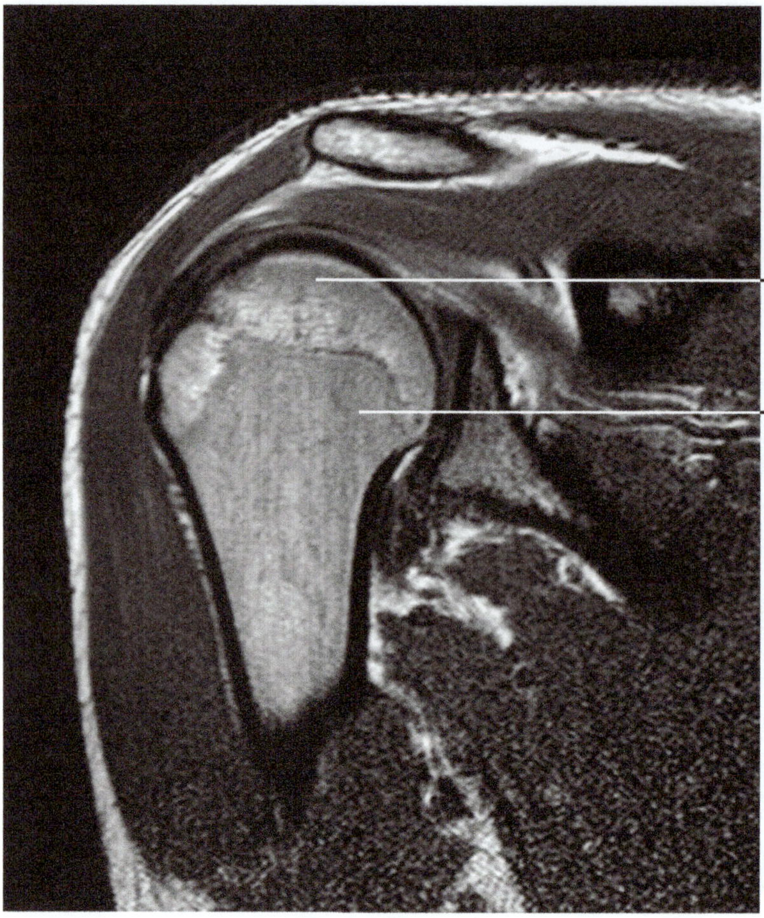

Fig. 34.30 Coronal T1W image of the proximal humerus of a 35 year old

Pelvis/Hip

Table 34.2 Ossification centres

Pelvis	Appears ♀	Appears ♂	Range	Fuses to 'shaft' ♀	Fuses to 'shaft' ♂
Iliac crest	Puberty	Puberty		20 Y	20 Y
Ilium	Birth	Birth		Puberty	Puberty
Ischium	Birth	Birth	Fuse 4–8 Y	Puberty	Puberty
Pubis	Birth	Birth	Fuse 4–8 Y	Puberty	Puberty
Femoral head	4 M	4 M	♀ 6 W –6 M ♂ 2 M – 8 M	16–17 Y	17–18 Y
Greater trochanter	3 Y	3 Y		16–17 Y	16–17 Y
Lesser trochaner	11 Y	12 Y	9–13 Y	16–17 Y	16–17 Y

M months, *Y* years

Plain Film

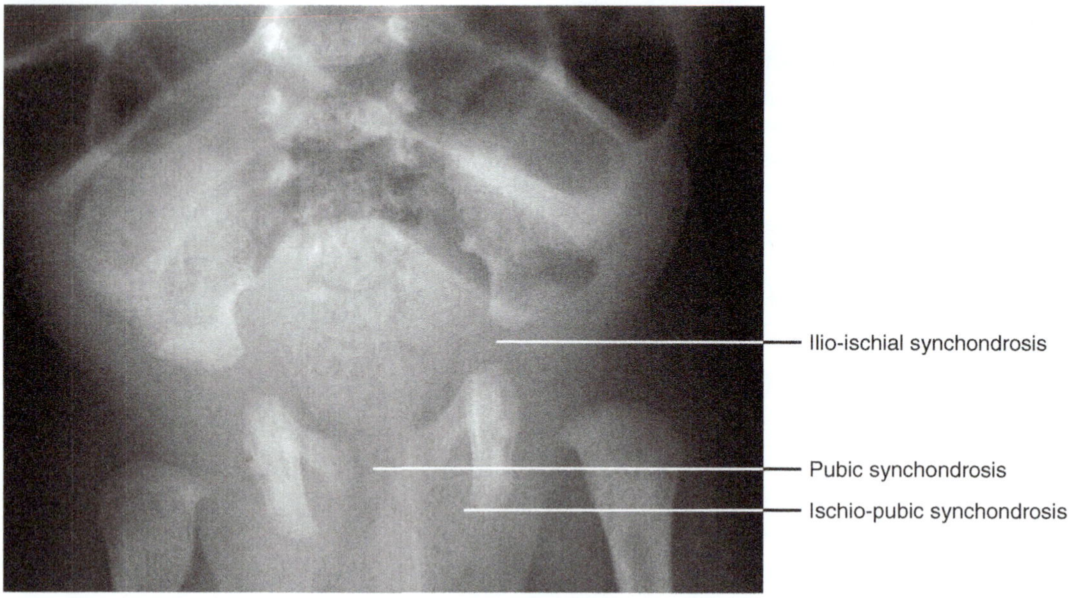

Fig. 34.31 Neonate

Pearl

The proximal femoral epiphysis appears between 4 months and 8 months.
Note the wide open ischiopubic and pubic synchondroses.

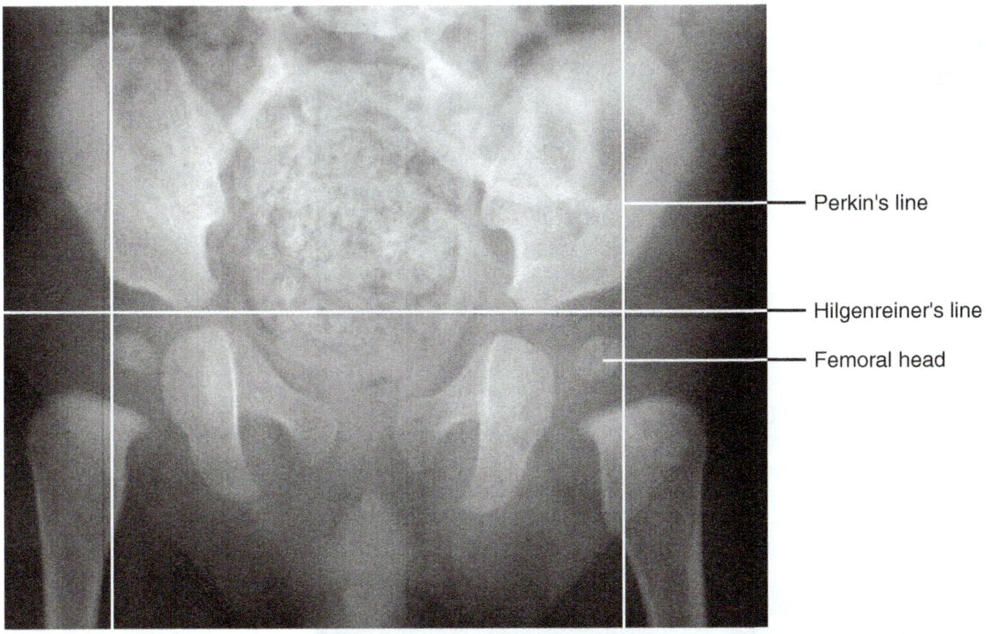

Fig. 34.32 1 year old

Pearl
The femoral head must be in the medial inferior quadrant defined by the crossing Hilgenreiner's and Perkin's lines.

Pearl
Hilgenreiner's line is drawn through the tri-radiate cartilage and Perkin's line is drawn perpendicular to this at the lateral acetabular margin.

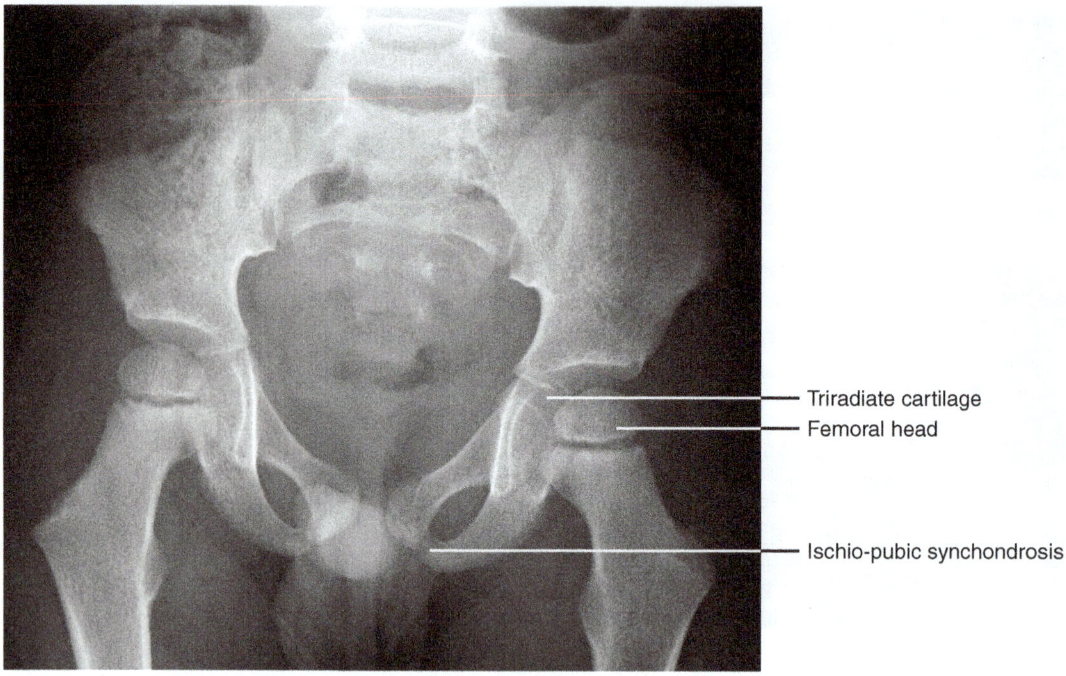

Fig. 34.33 5 years old

Pearl
The ilia and ischia still appear separated by cartilage. The ischiopubic synchondroses are almost completely fused.

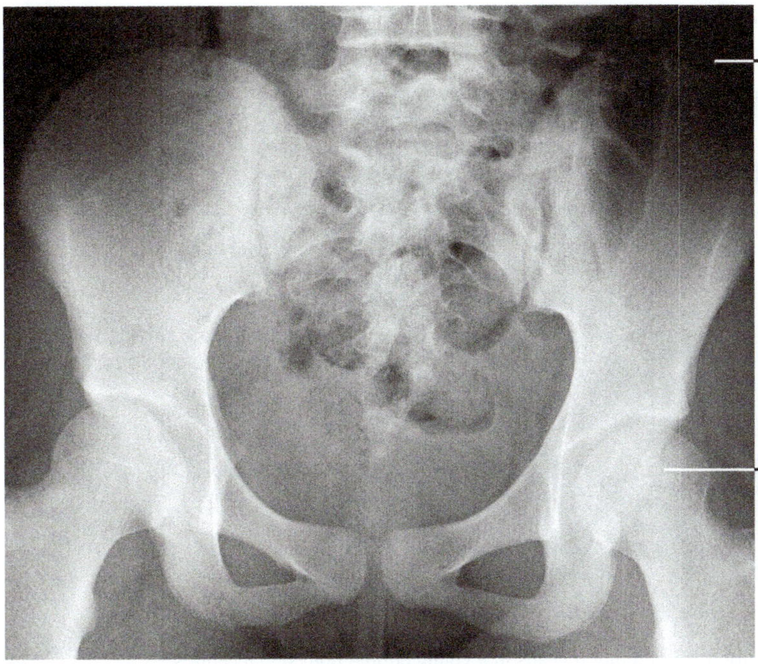

Fig. 34.34 12 years old

Pearl
Secondary ossification centres form at the iliac crests and inferior margins of the ischia between 12 and 14 years and fuse in the late teens.

Pearl
Irregularity along the acetabular roof is a normal phenomenon of growth.

Magnetic Resonance Imaging

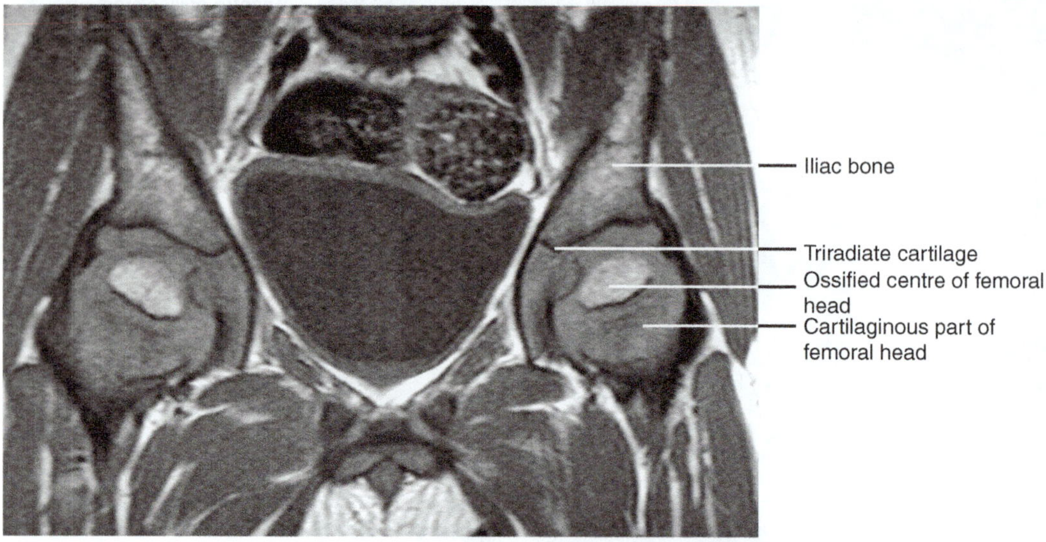

Fig. 34.35 T1 coronal hips: partially ossified femoral heads in a young child – high signal centres

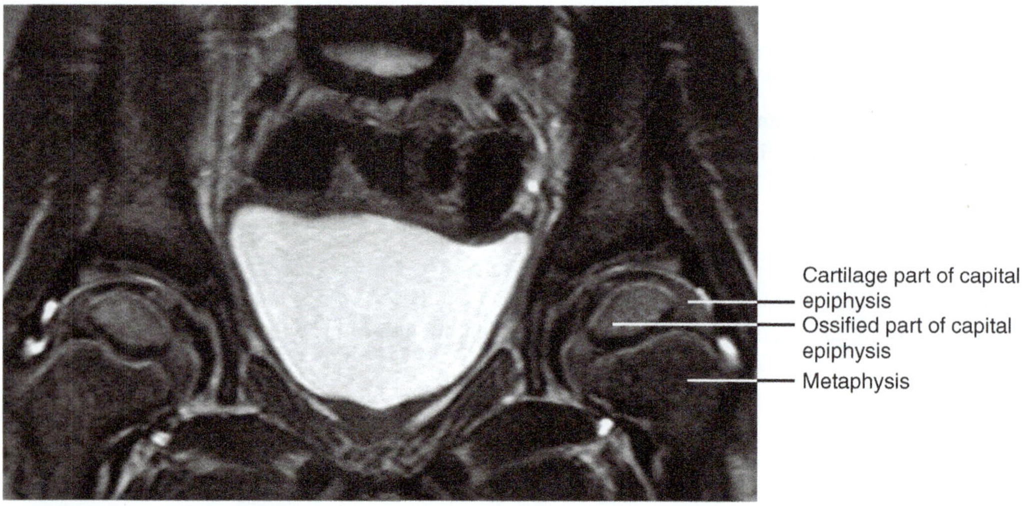

Fig. 34.36 STIR coronal hips: partially ossified femoral heads in a young child – low signal centres

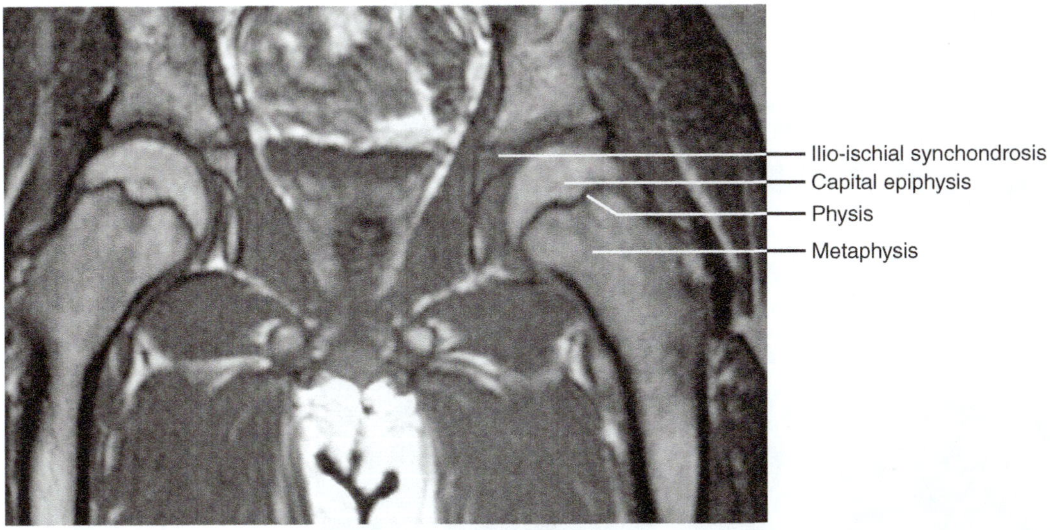

Fig. 34.37 T1 coronal hip: partially ossified femoral heads in an older child – high signal throughout

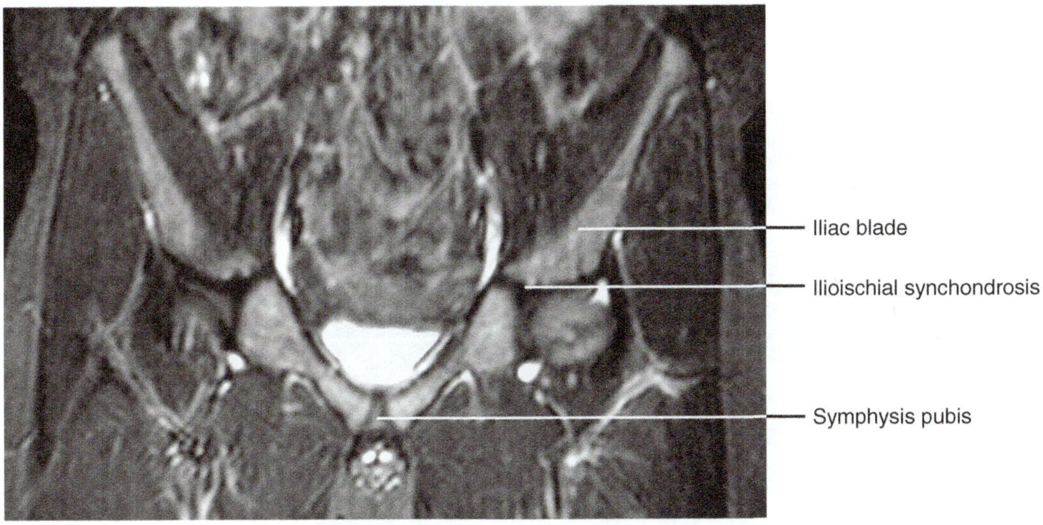

Fig. 34.38 STIR coronal hips: tri-radiate cartilage and symphysis pubis in an older child

Ultrasound

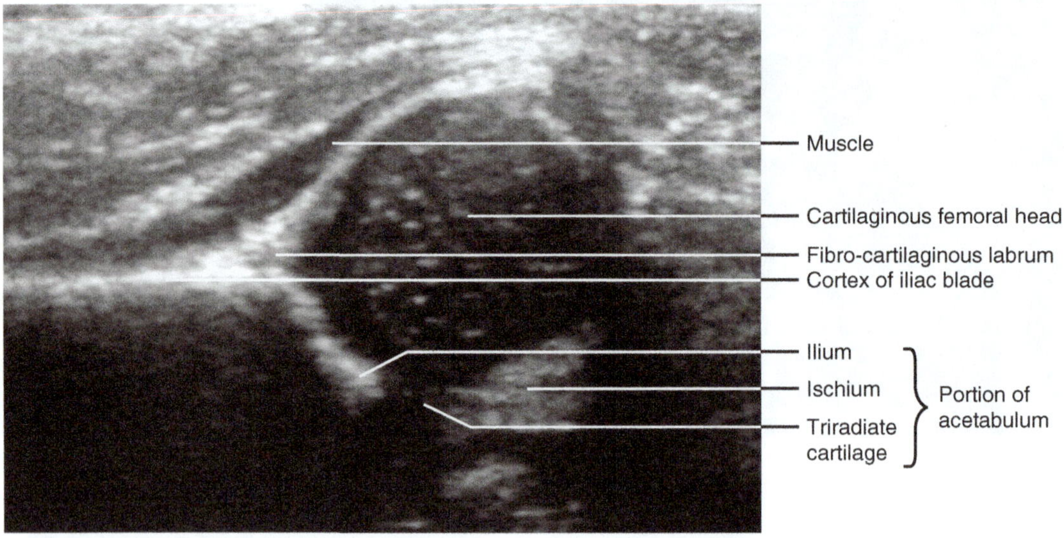

Fig. 34.39 Coronal hips

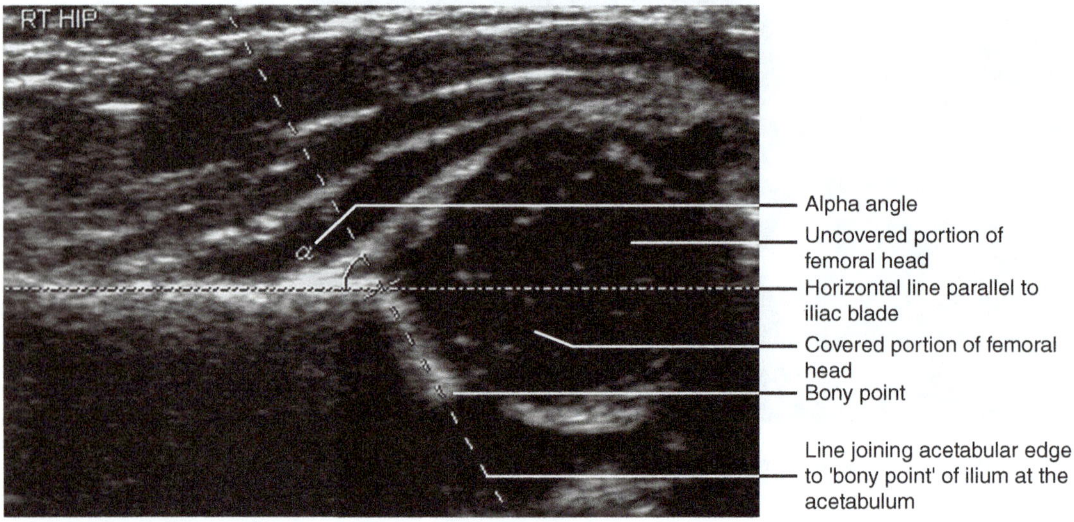

Fig. 34.40 Coronal hips with measurement of the alpha angle representing the adequacy of the acetabular roof

Pearl

The alpha angle on ultrasound measures the acetabular depth and the normal values are between 55 and 56°.

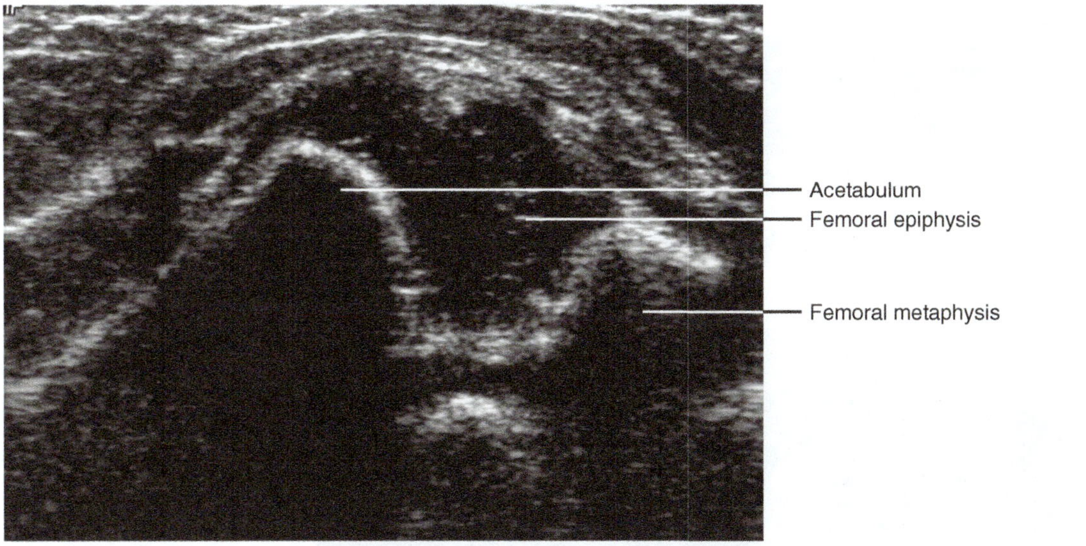

Fig. 34.41 Axial or 'sunset' view at the hip

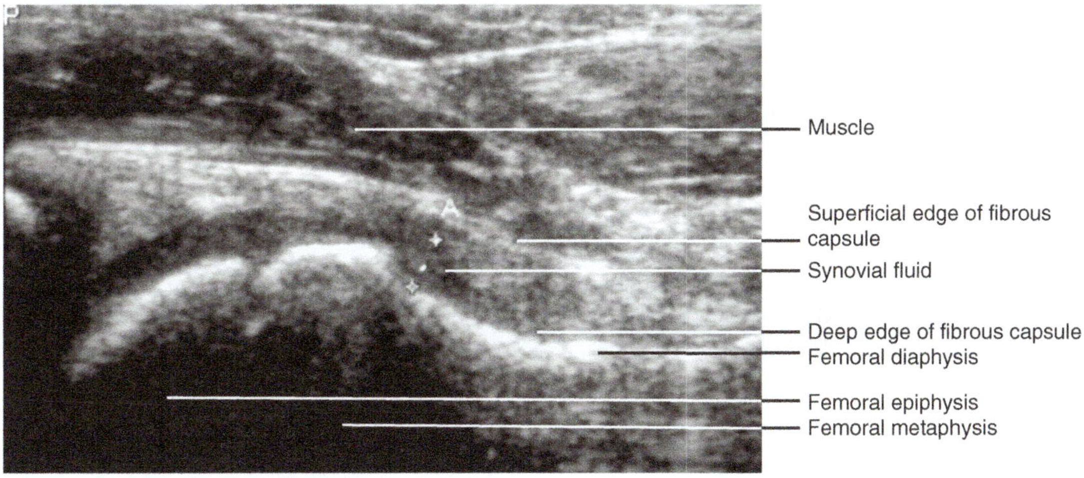

Fig. 34.42 Longitudinal US of capsule

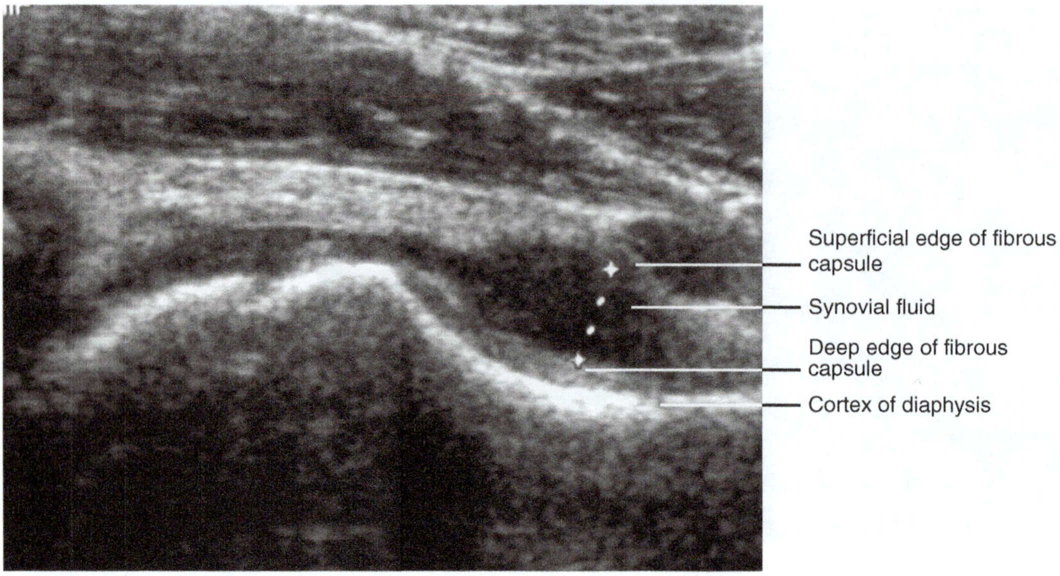

Fig. 34.43 Longitudinal US of the capsule with pathological fluid

Pearl
Capsular fluid (without including the capsule) should measure <2 mm.

Pearl
A difference of >2 mm including fibrous capsule, between left and right joint capsules, is significant.

Pearl
Capsular fluid is accentuated when the leg is externally rotated.

Appendicular Skeleton

Growing tubular bones consist of four portions: epiphysis, physis, metaphysis and diaphysis. Long bones have epiphyses at both ends. Short tubular bones only have an epiphysis at one end. Ossification centres develop within the epiphysis. Growth stops when these ossification centres fuse through the physis with the metaphysis. Extremities show marrow conversion from distal to proximal. Within bone this begins at the epiphysis then diaphysis and lastly the metaphysis. Between 15 and 20 years, the residual haemopoetic tissue in the spine and proximal metaphysis of the shoulders and femurs becomes fatty.

Pearl
The last portions of the appendicular skeleton to convert from haemopoetic to fatty marrow are the humeral and femoral metaphyses.

Knee

Ossification Centres

Pelvis	Appears ♀	Appears ♂	Range	Fuses to 'shaft' ♀	Fuses to 'shaft' ♂
Patella	3 Y	4–5 Y			
Distal femoral epiphysis	36 FW	36 FW	6–10 FM	17 Y	18–19 Y
Proximal tibial epiphysis	40 FW	40 FW	8 FM–1 M	16–17 Y	18–19 Y
Tibial tuberosity			7–15 Y	19 Y	19 Y
Proximal fibula epiphysis	3 Y	4 Y	2–5½ Y	16–18 Y	18–20 Y

FW foetal weeks, *FM* foetal months, *M* months, *Y* years

Plain Film

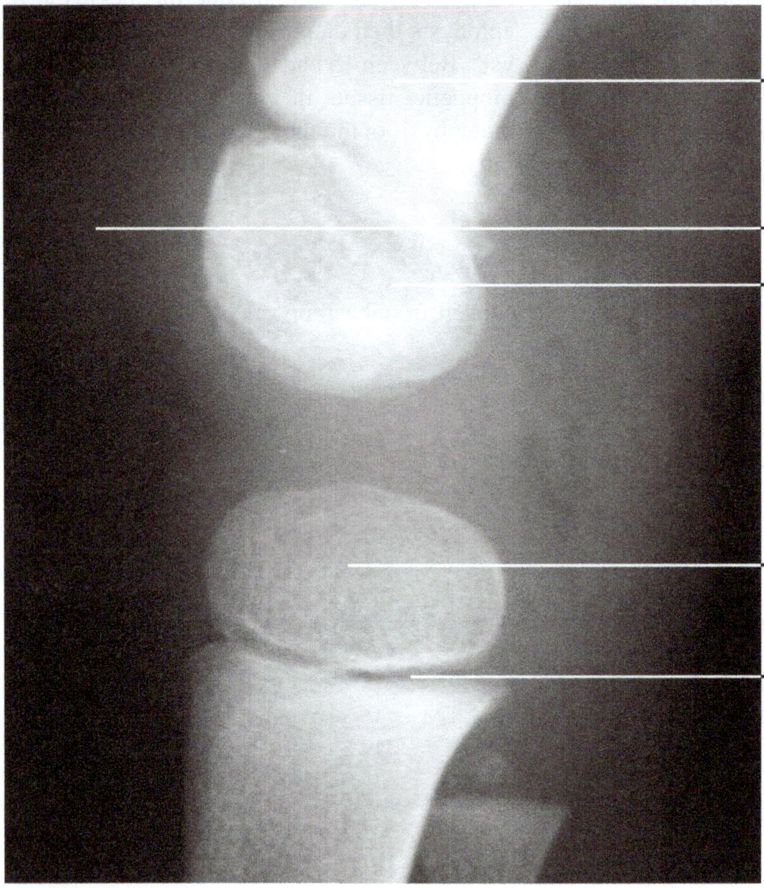

Fig. 34.44 Lateral knee: <6 years old

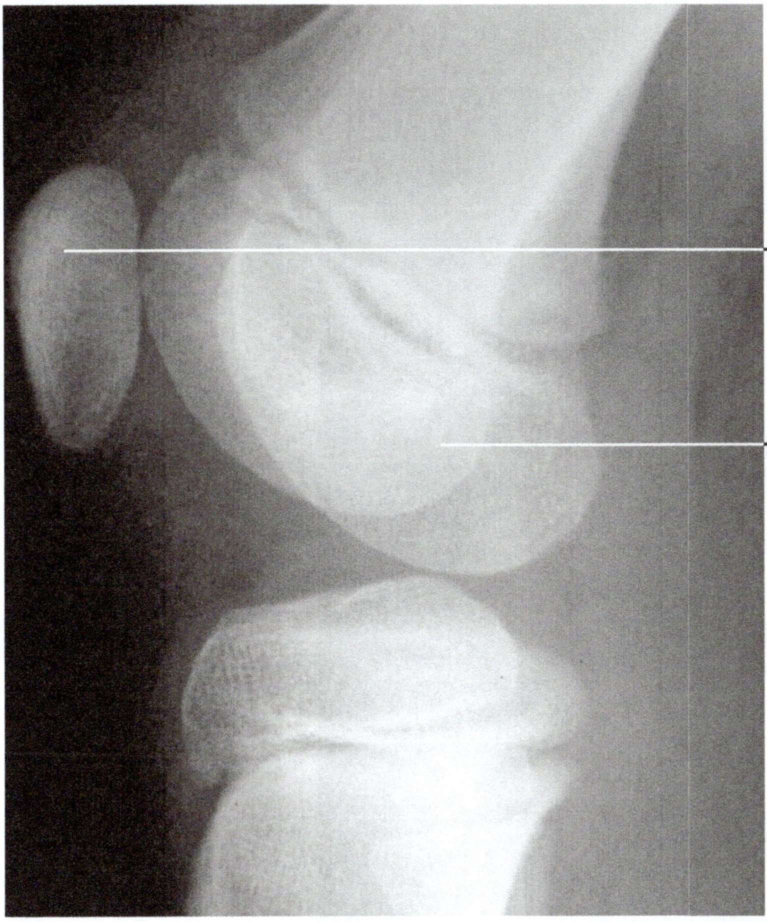

Fig. 34.45 Lateral knee: >6 years old

Magnetic Resonance Imaging

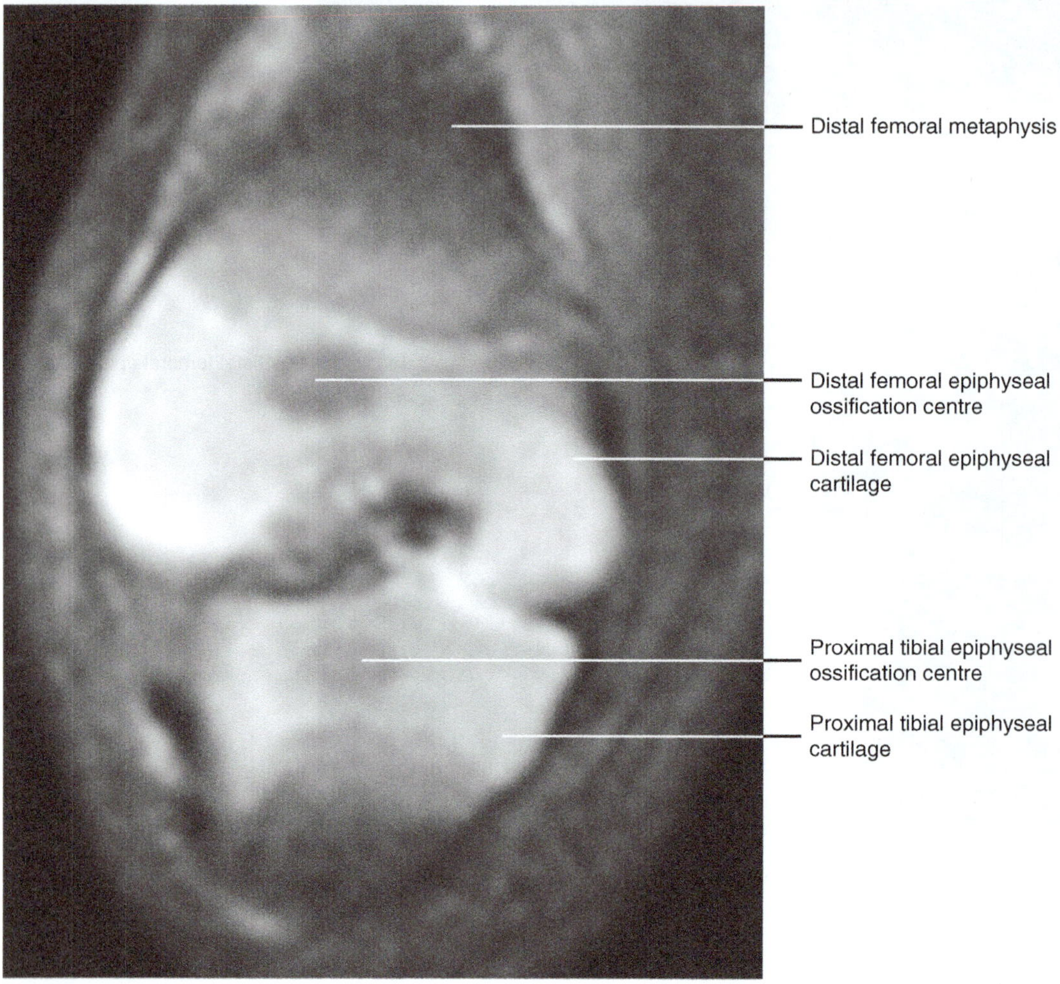

Fig. 34.46 Gradient echo coronal: neonate

Pearl
Central ossification of distal femur and proximal tibia is present in a full term fetus.

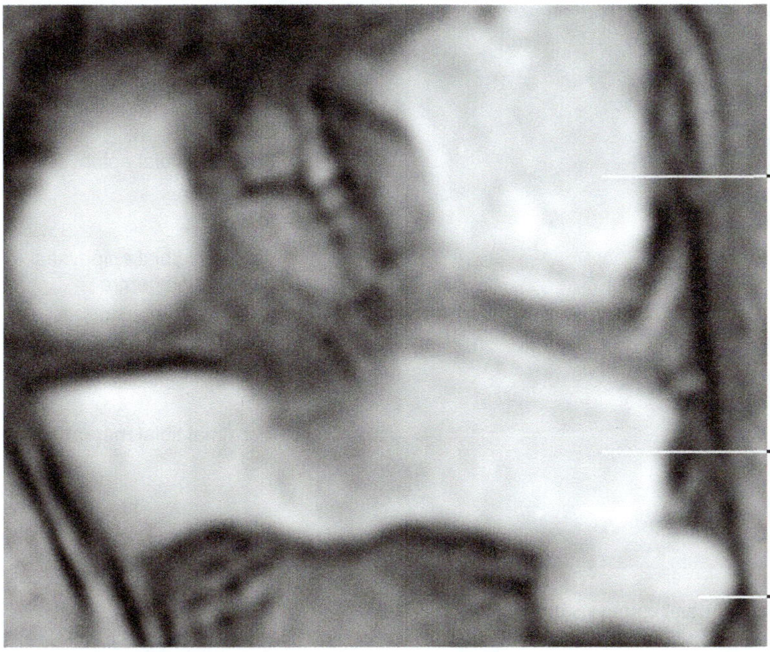

Fig. 34.47 Gradient Echo coronal: infant

Pearl
The epiphyseal cartilage is visible on MRI and has a high signal on gradient echo.

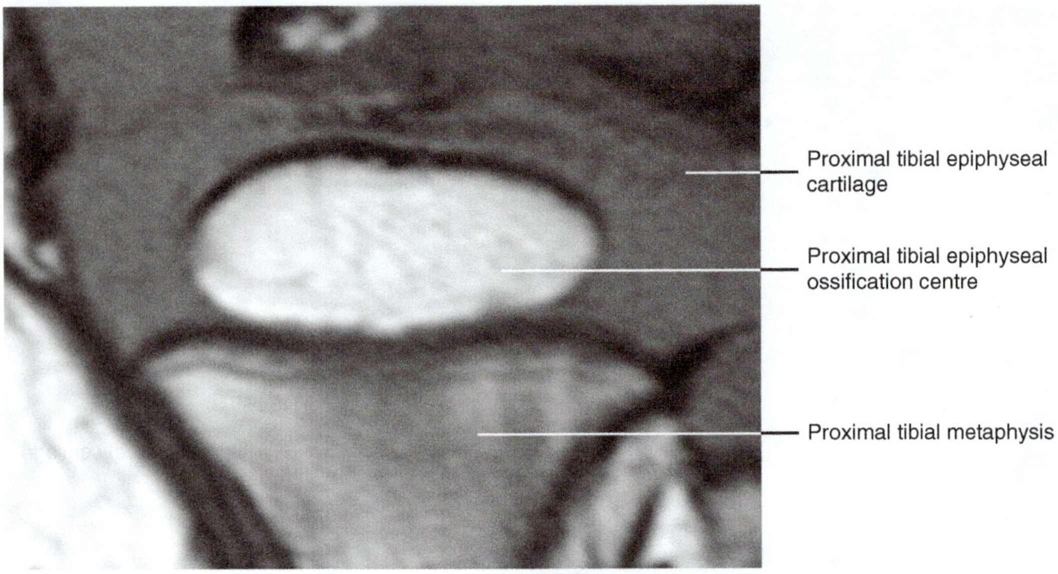

Fig. 34.48 T1 coronal of the proximal tibial epiphysis in a young child

Pearl

Ossified components show high signal marrow on T1 and surrounding cartilage is intermediate signal.

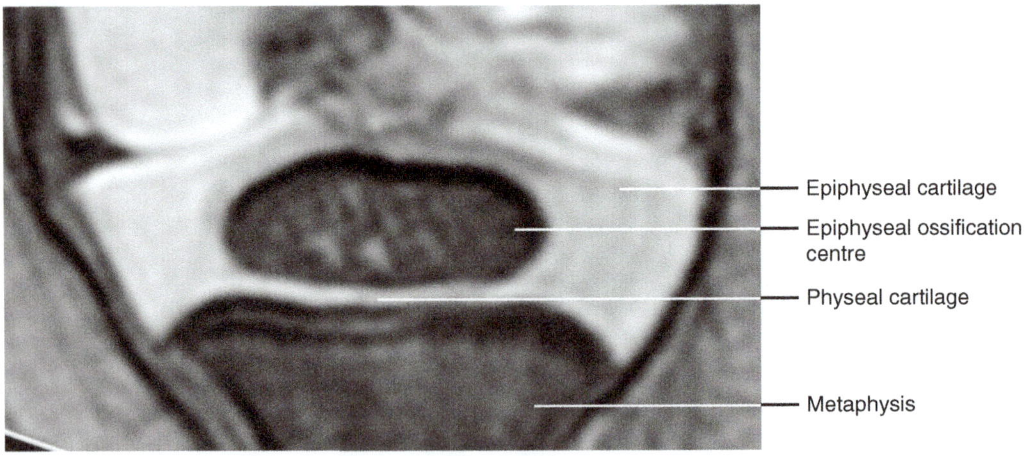

Fig. 34.49 Gradient echo coronal of the proximal tibia in a young child

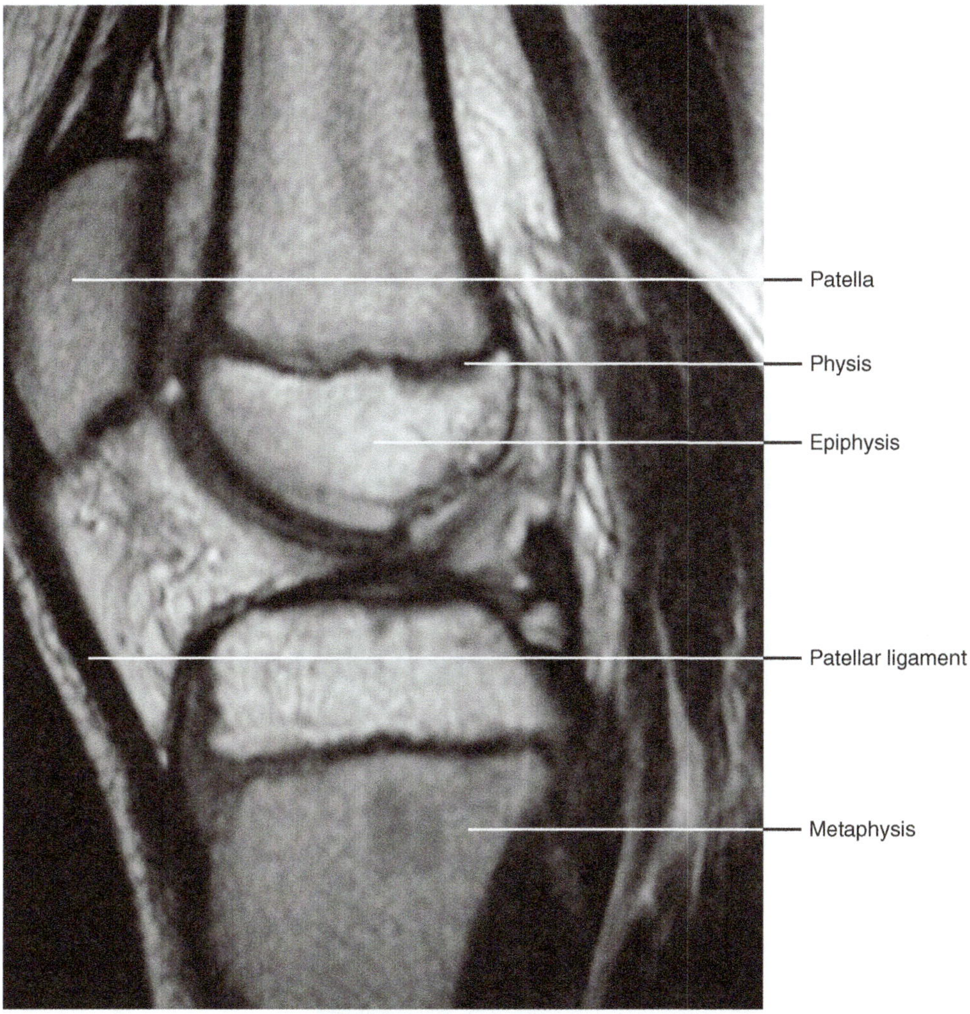

Fig. 34.50 T1 sagittal: adolescent

Pearl
In adolescents, the epiphyses are ossified but not united with the metaphyses. The physis remains hypointense on T1 and hyperintense on gradient echo until fusion.

Ankle and Foot

Accessory ossification centres are common around the ankle and foot and may be mistaken for fractures.

Plain Film

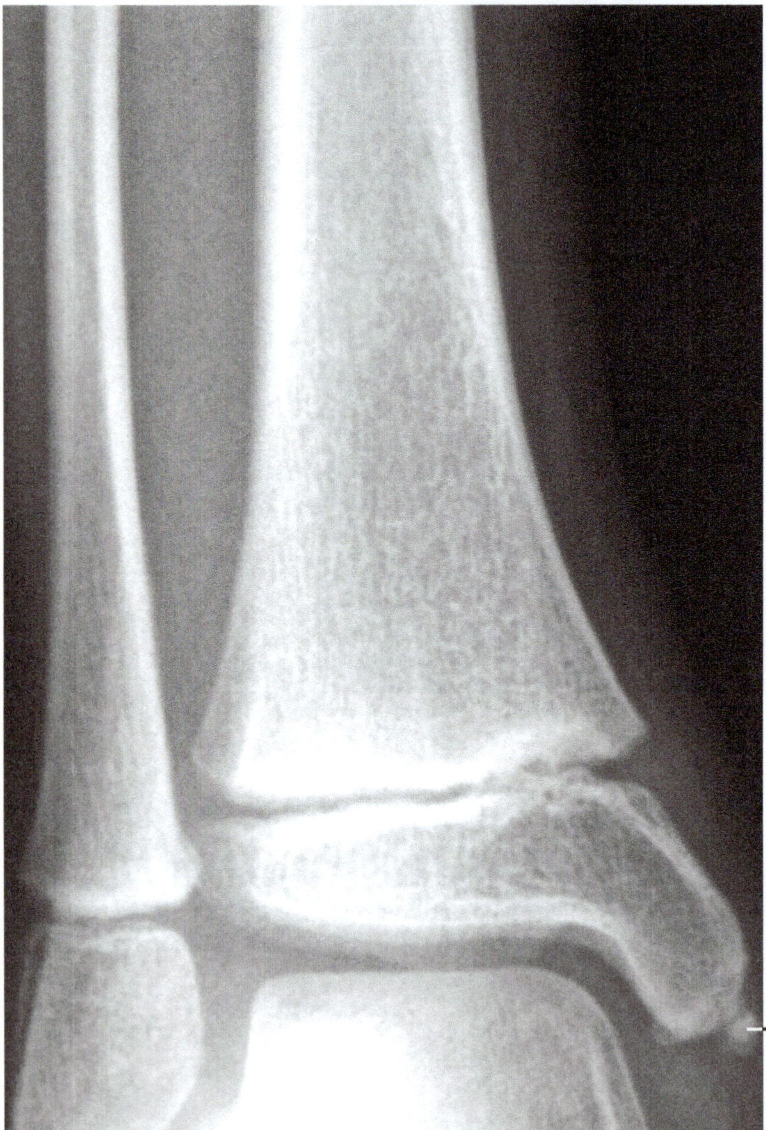

Fig. 34.51 Oblique ankle: accessory ossification centre of tibia

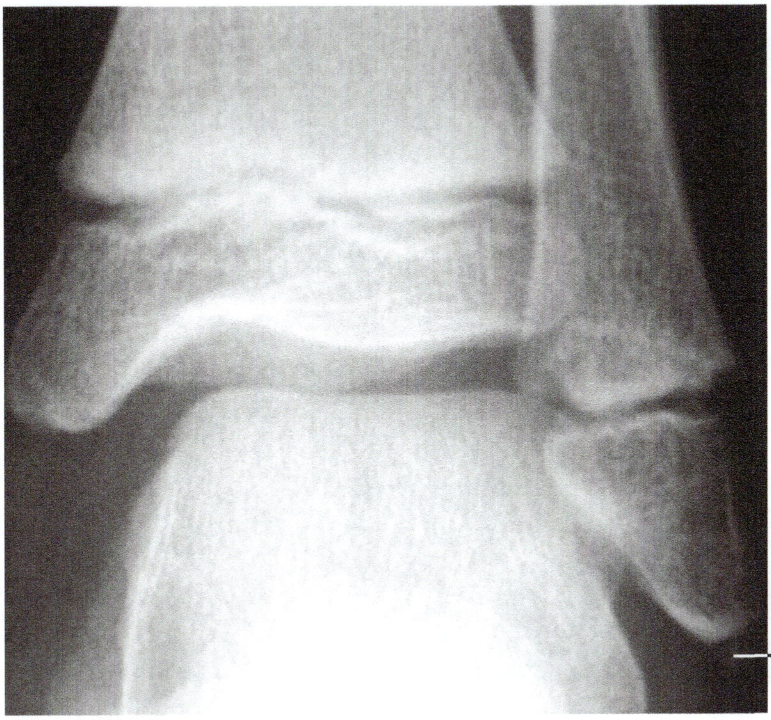

Fig. 34.52 AP ankle: accessory ossification centre of fibula

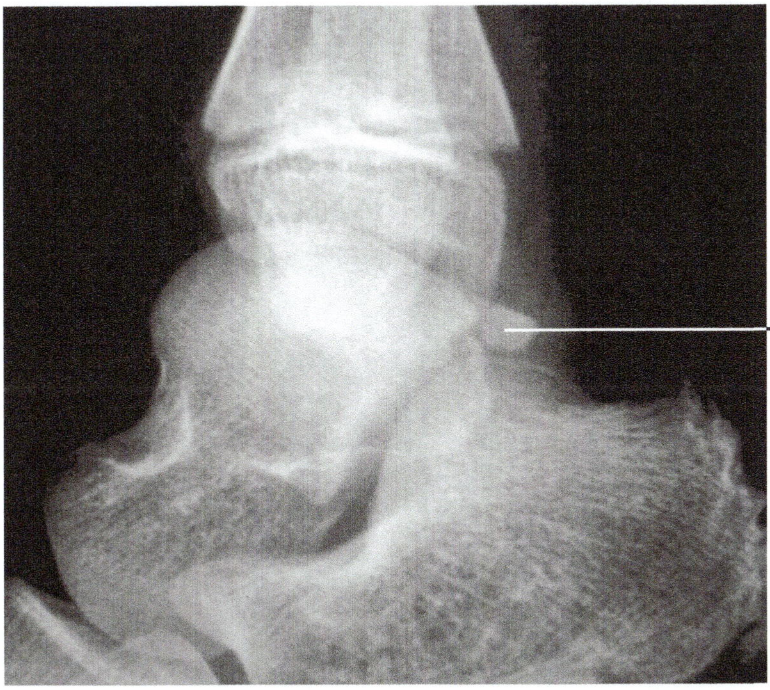

Fig. 34.53 AP ankle: accessory ossification centre of talus

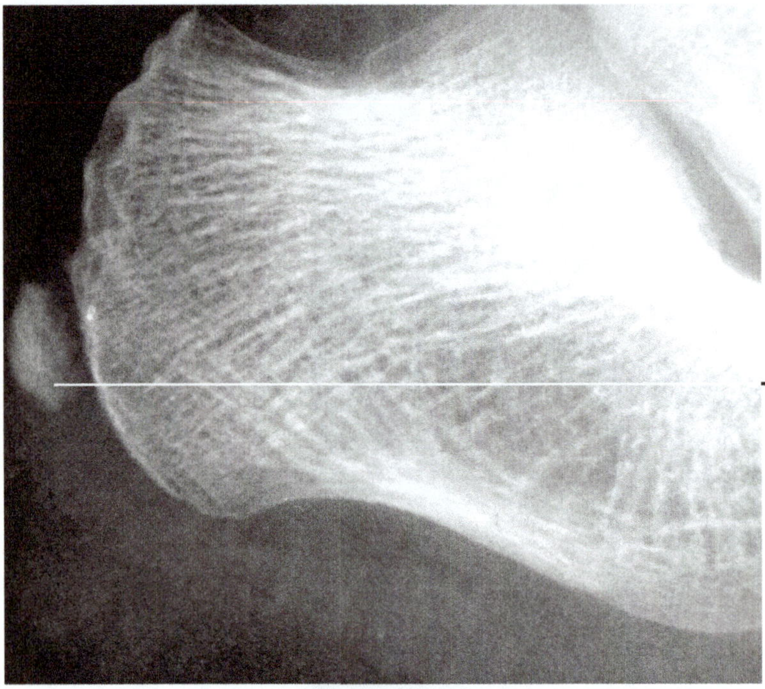

Fig. 34.54 Lateral calcaneus: ossification centre of calcaneus in a child

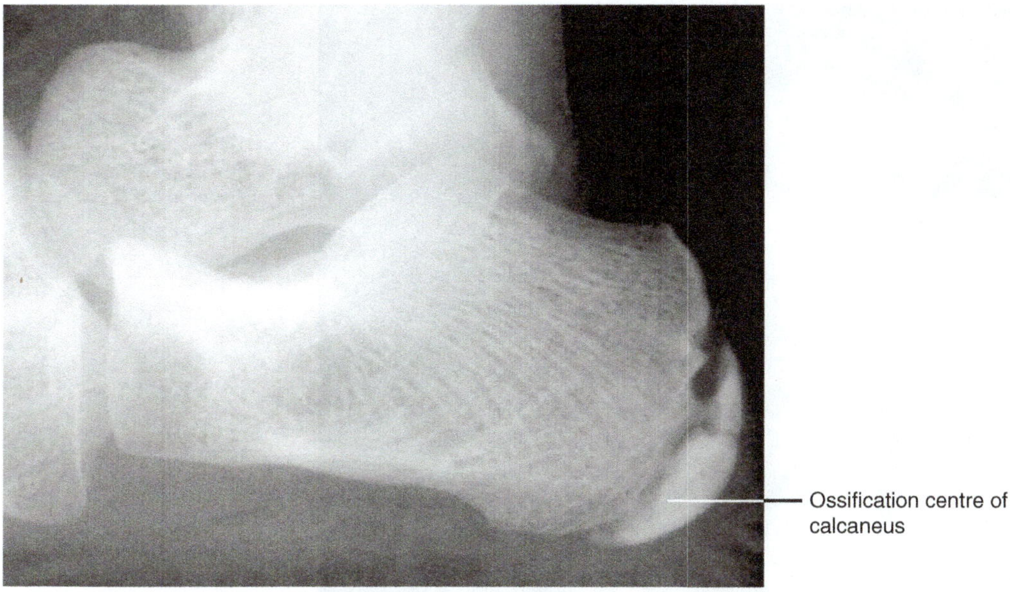

Fig. 34.55 Lateral calcaneus: ossification centre of calcaneus in an adolescent

Pearl

The ossification centre for the calcaneus overlies the irregular posterior margin of the calcaneus. It appears in the middle of the first decade and maintains its fragmented appearance into the second decade.

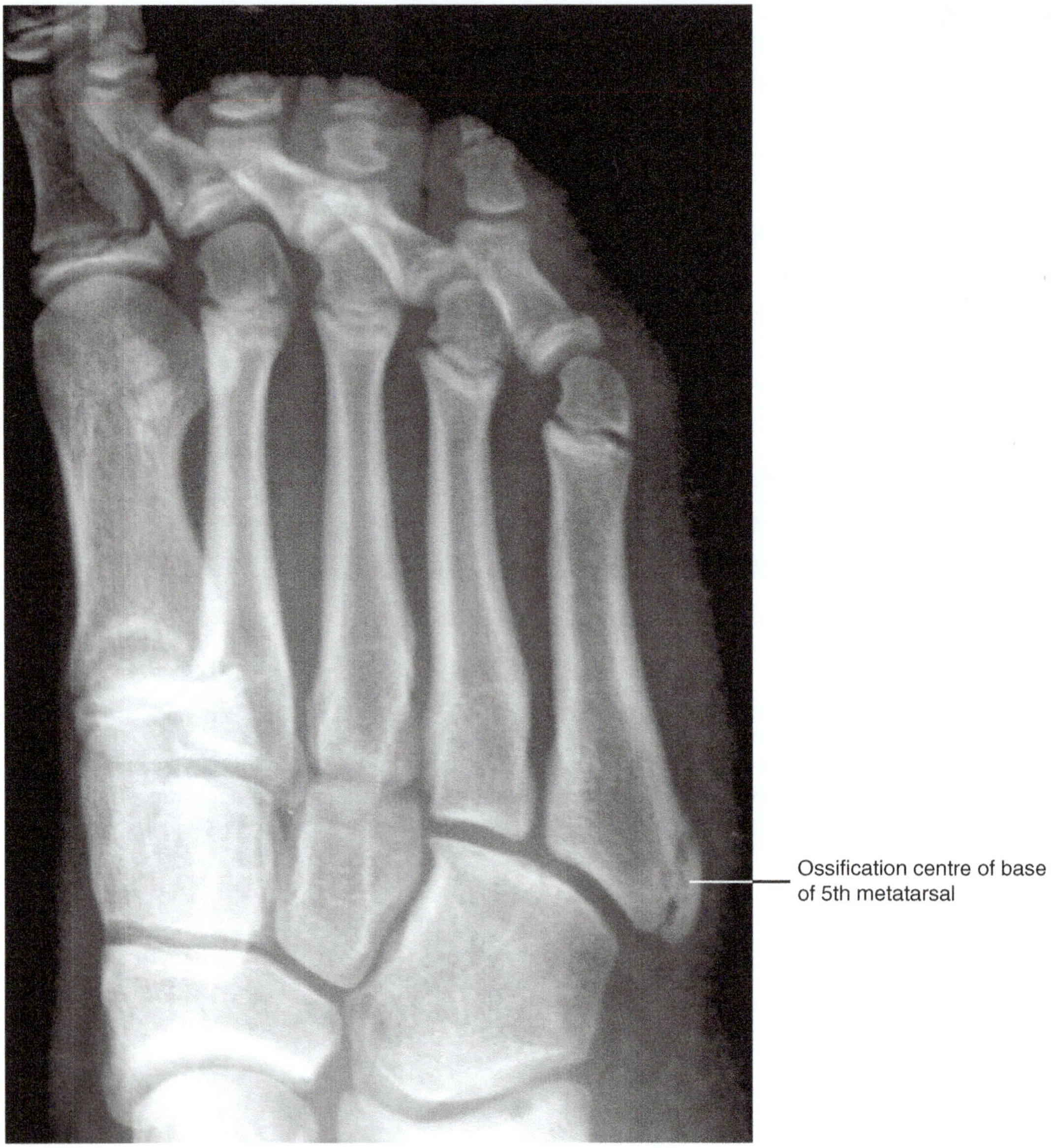

Fig. 34.56 Oblique foot: ossification centre of base of fifth metatarsal

Pearl

Usually appears during puberty and may persist throughout life as an accessory ossicle. Lies parallel to the shaft of the metatarsal and can be differentiated from fractures that are transversely oriented.

Elbow

Ossification Centres

Pelvis	Appears ♀	Appears ♂	Range	Fuses to 'shaft' ♀	Fuses to 'shaft' ♂
Lateral epicondyle	11 Y	12 Y	11–14 y	14 Y	17 Y
Medial epicondyle	5 Y	7 Y	♂ 5–7 Y ♀ 3–6 Y	15 Y	18 Y
Capitellum	4 M	5 M	♂ 6 W–8 M ♀ 1 M–6 M	14 Y	17 Y
Trochlea	8 Y	9 Y	7–10 Y	14 Y	17 Y
Radial head	4 Y	5 Y	3–6 Y	14–15 Y	15–17 Y
Ulna olecranon process	8 Y	10 Y		4–15 Y	15–17 Y

W weeks, *M* months, *Y* years

Pearl

The capitellum appears as early as a month of age whereas the remaining ossification centers only appear from 3 years onwards.

This makes the capitellum significant for evaluation normal alignment around the elbow as it is the only landmark at the elbow during the first 3–4 years of life.

Plain Film

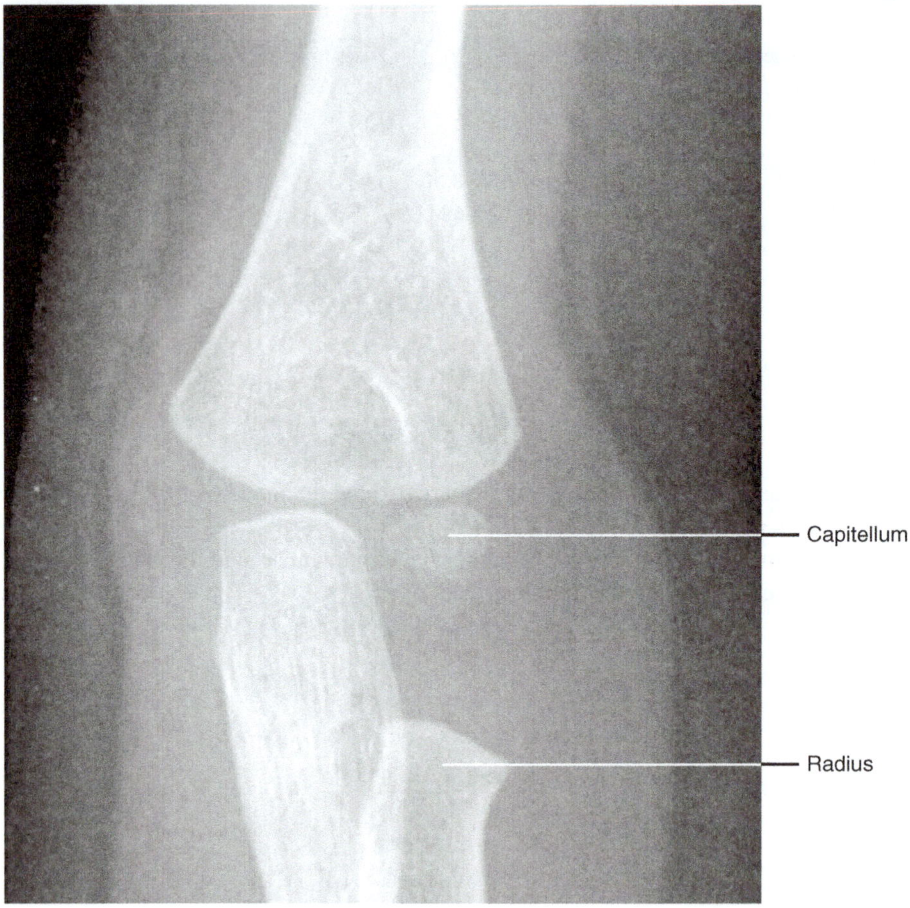

Fig. 34.57 AP elbow: 6 months old

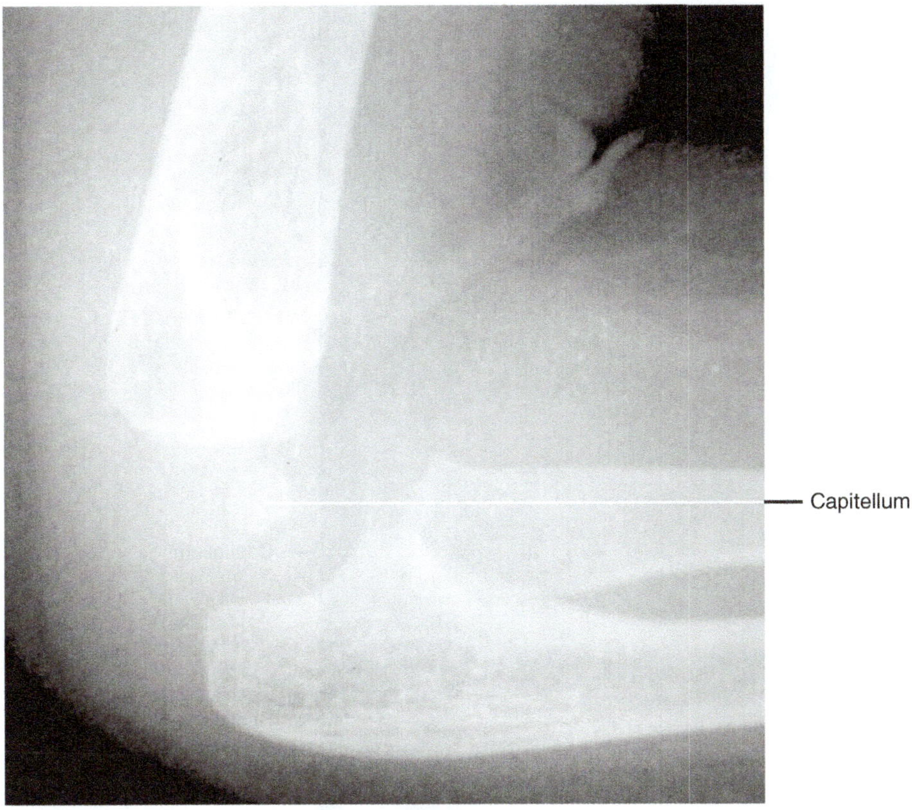

Fig. 34.58 Lateral elbow: 6 months old

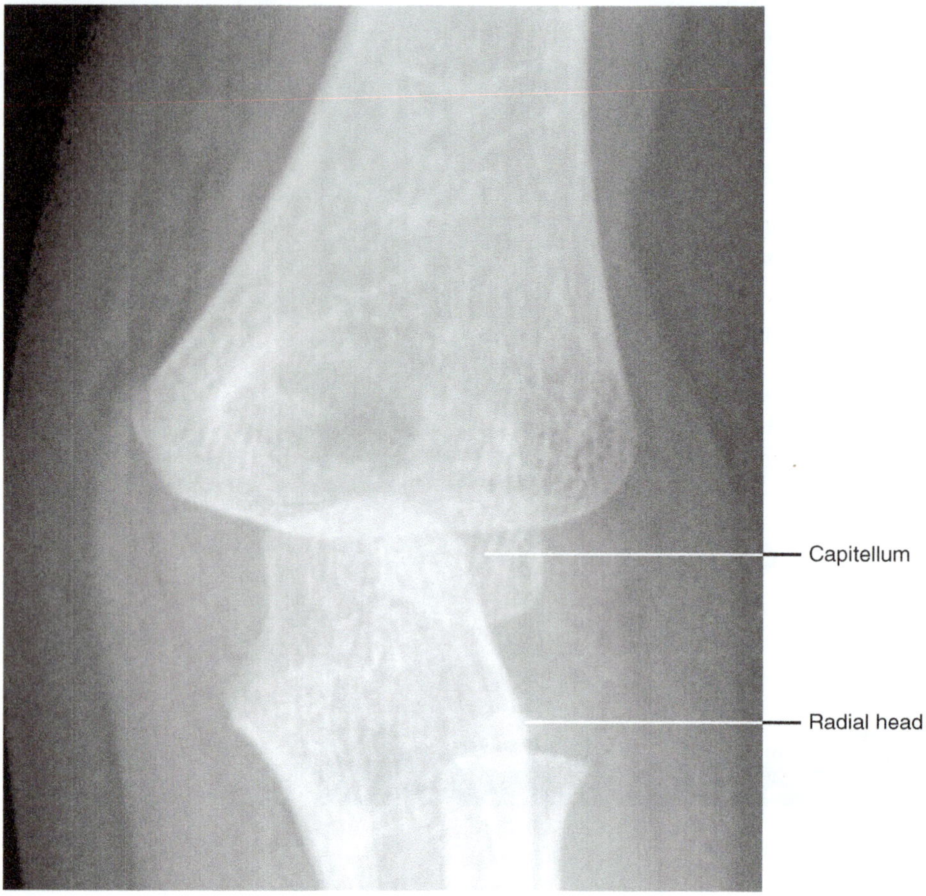

Fig. 34.59 AP elbow: 5 years old

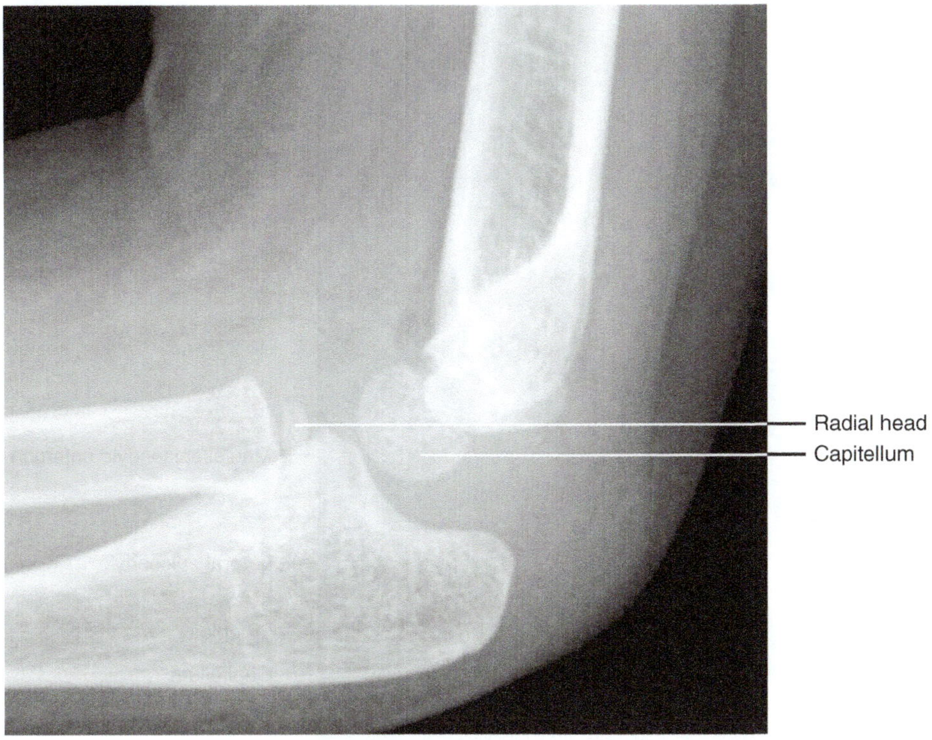

Fig. 34.60 Lateral elbow: 5 years old

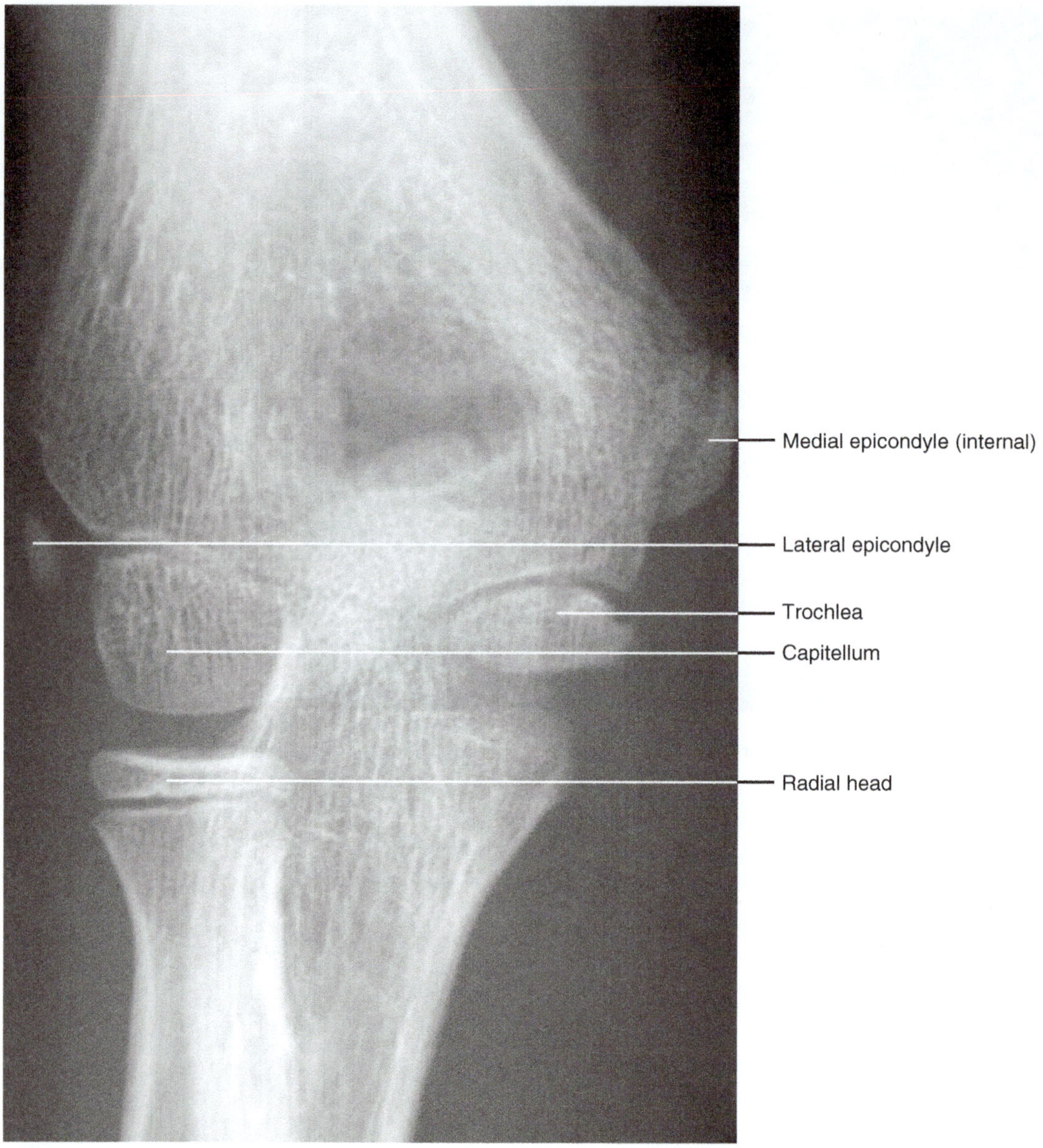

Fig. 34.61 AP elbow: 9 years old

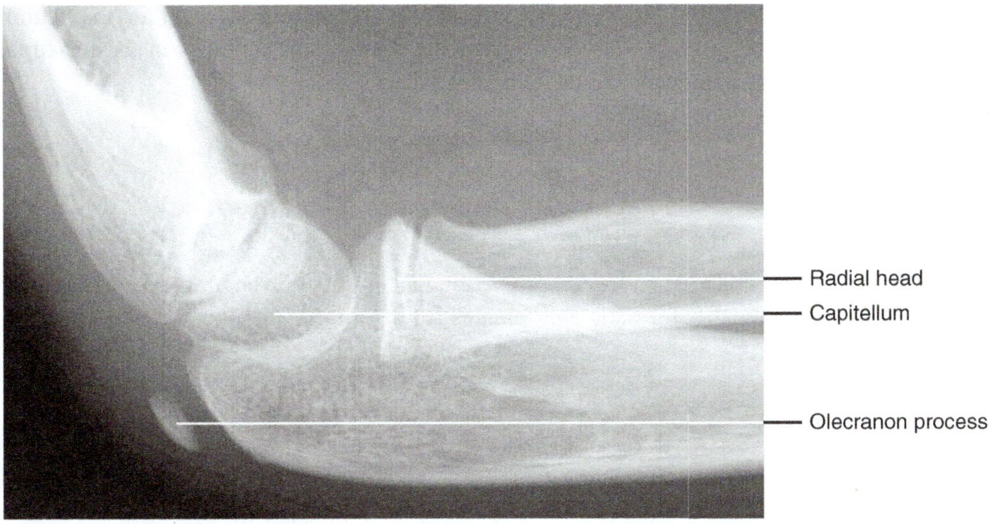

Fig. 34.62 Lateral elbow: 9 years old

Pearl

The mnemonic CRITOL is useful for remembering the order of ossification at the elbow: C=capitellum, R=radial head, I=internal epicondyle, T=trochlea, O=olecranon, L=lateral epicondyle.

Hand

Secondary ossification centres appear in the proximal ends of the phalanges and the distal ends of the metacarpals, except in the thumb where the ossification centre is in the proximal metacarpal.

Pearl
The left hand and wrist are imaged in a frontal projection and ossification centres are compared to a textbook of normal standards for male and females for determining 'Bone Age' as a measure of skeletal maturation.

Pearl
As children get older it is more reliable to use the width of the phalangeal or metacarpal ossification centres as an indicator of bone age rather than the carpal bones.

Plain Film

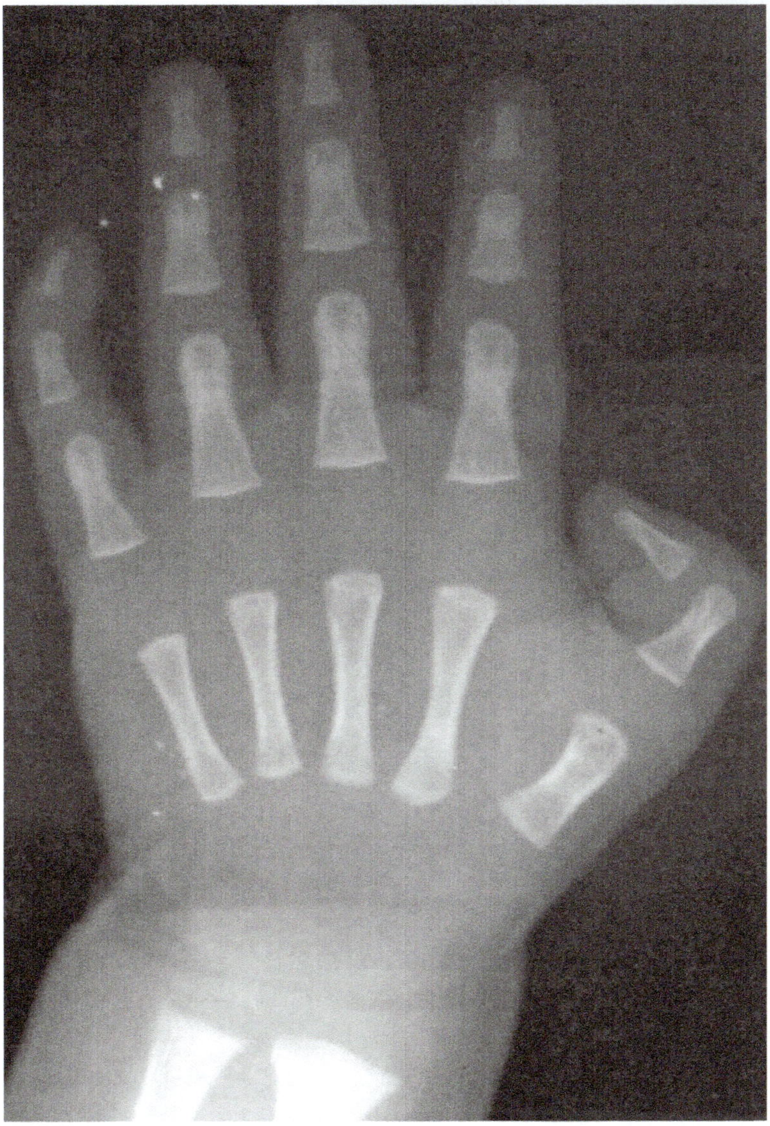

Fig. 34.63 6 months old

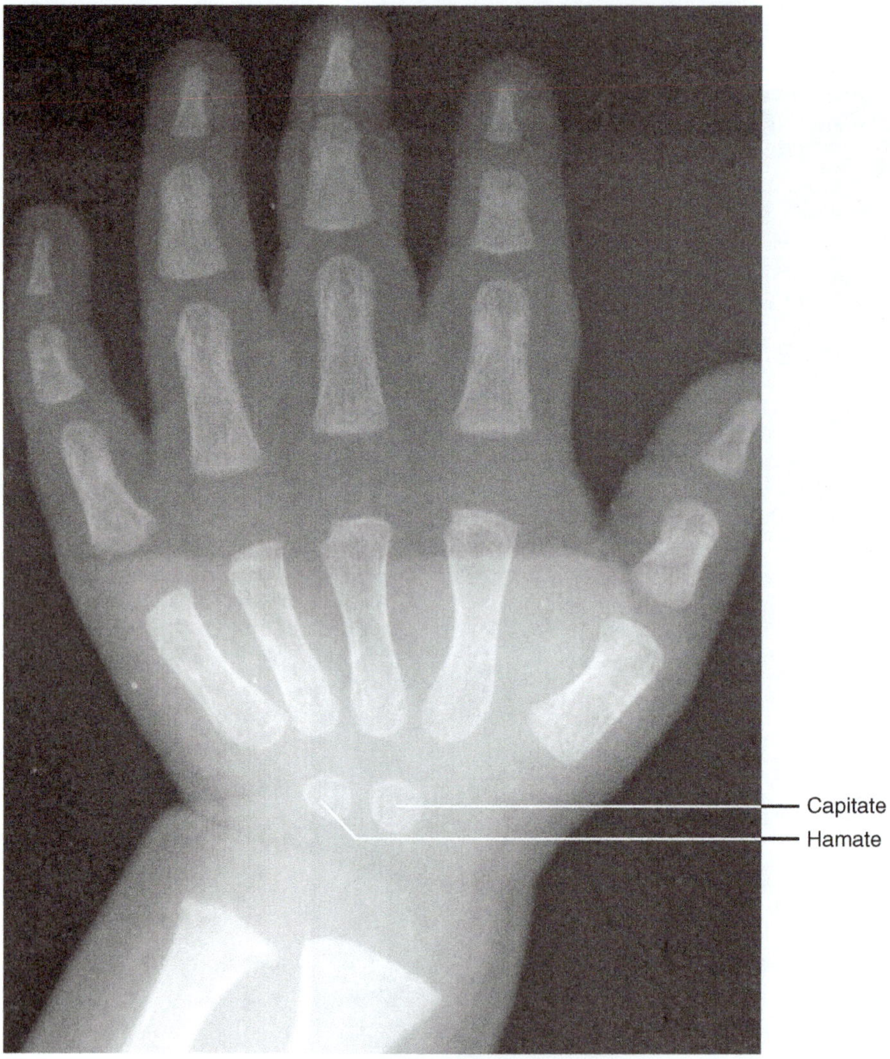

Fig. 34.64 Two years old

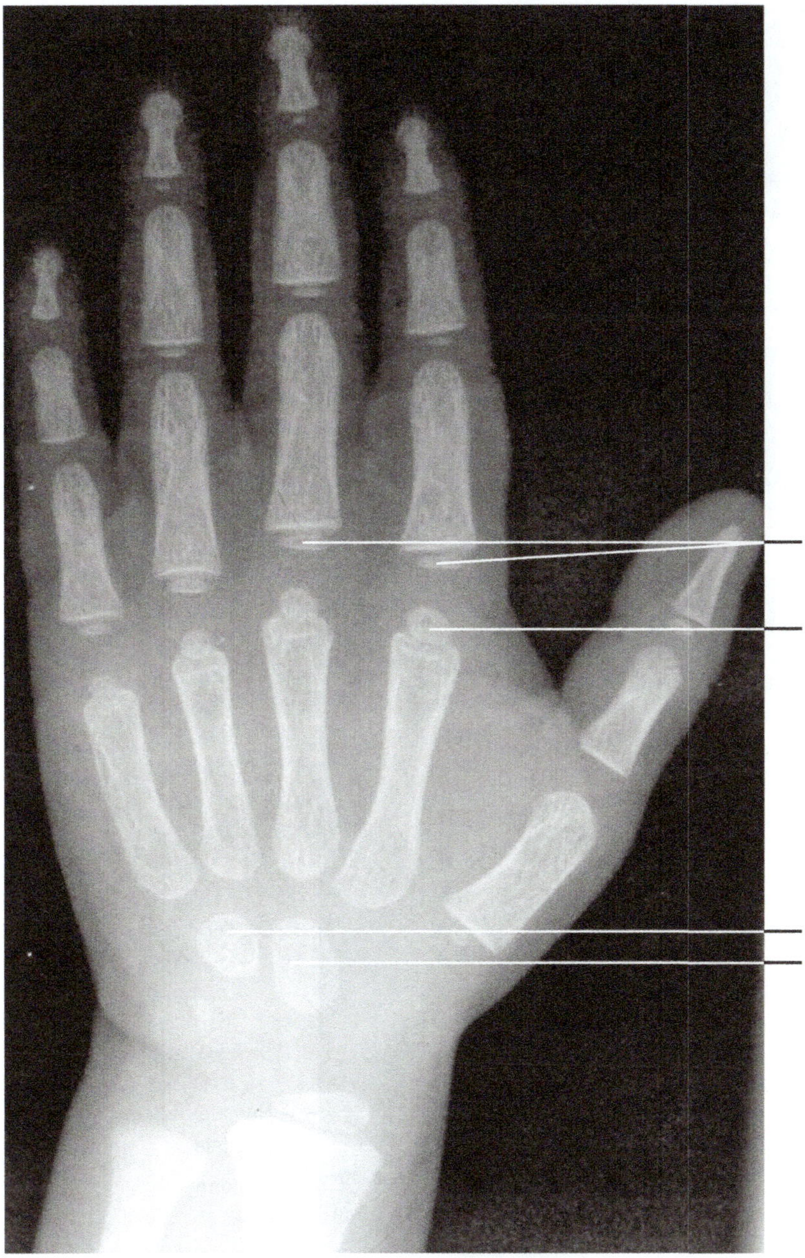

Fig. 34.65 Four years old

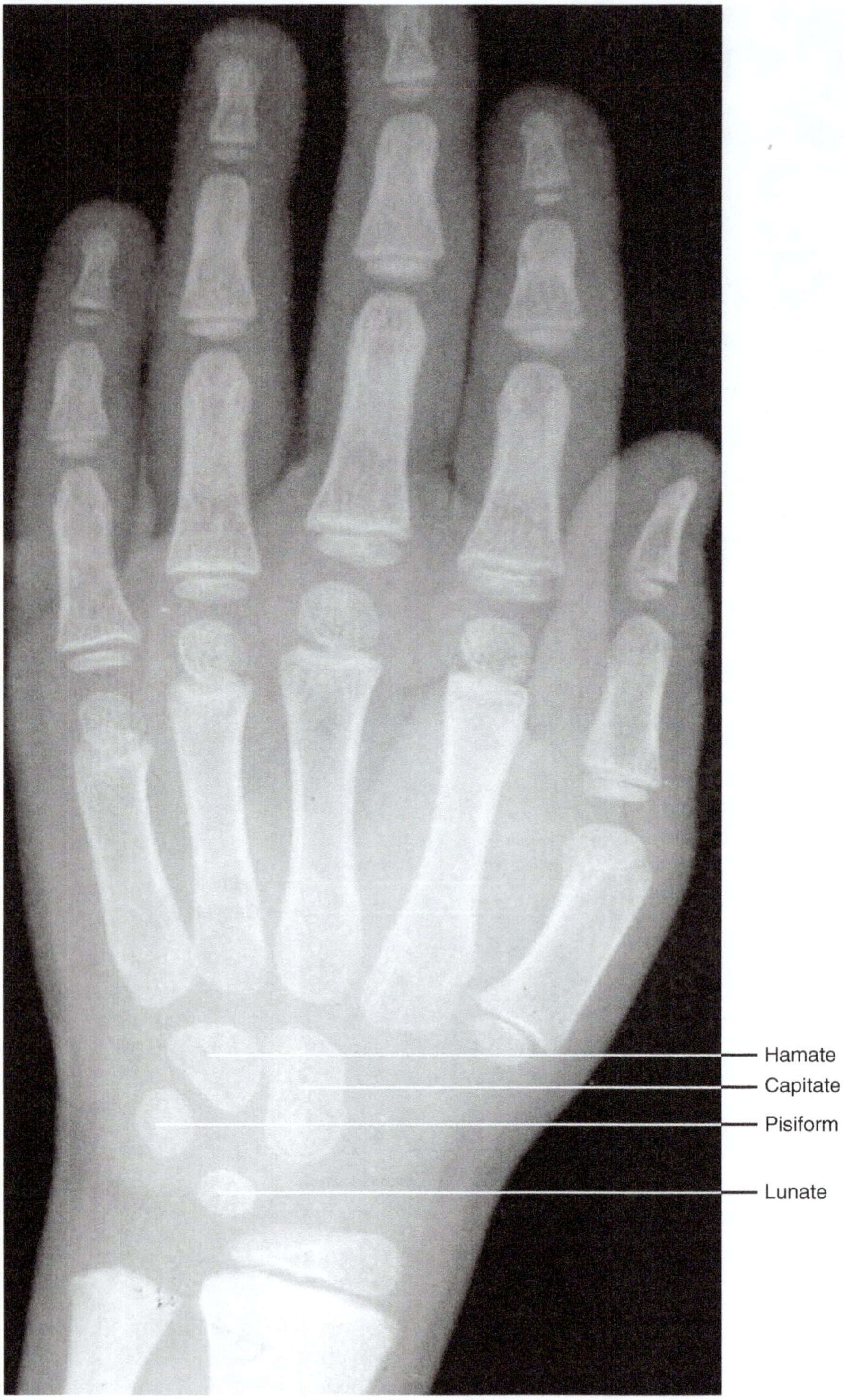

Fig. 34.66 Six years old

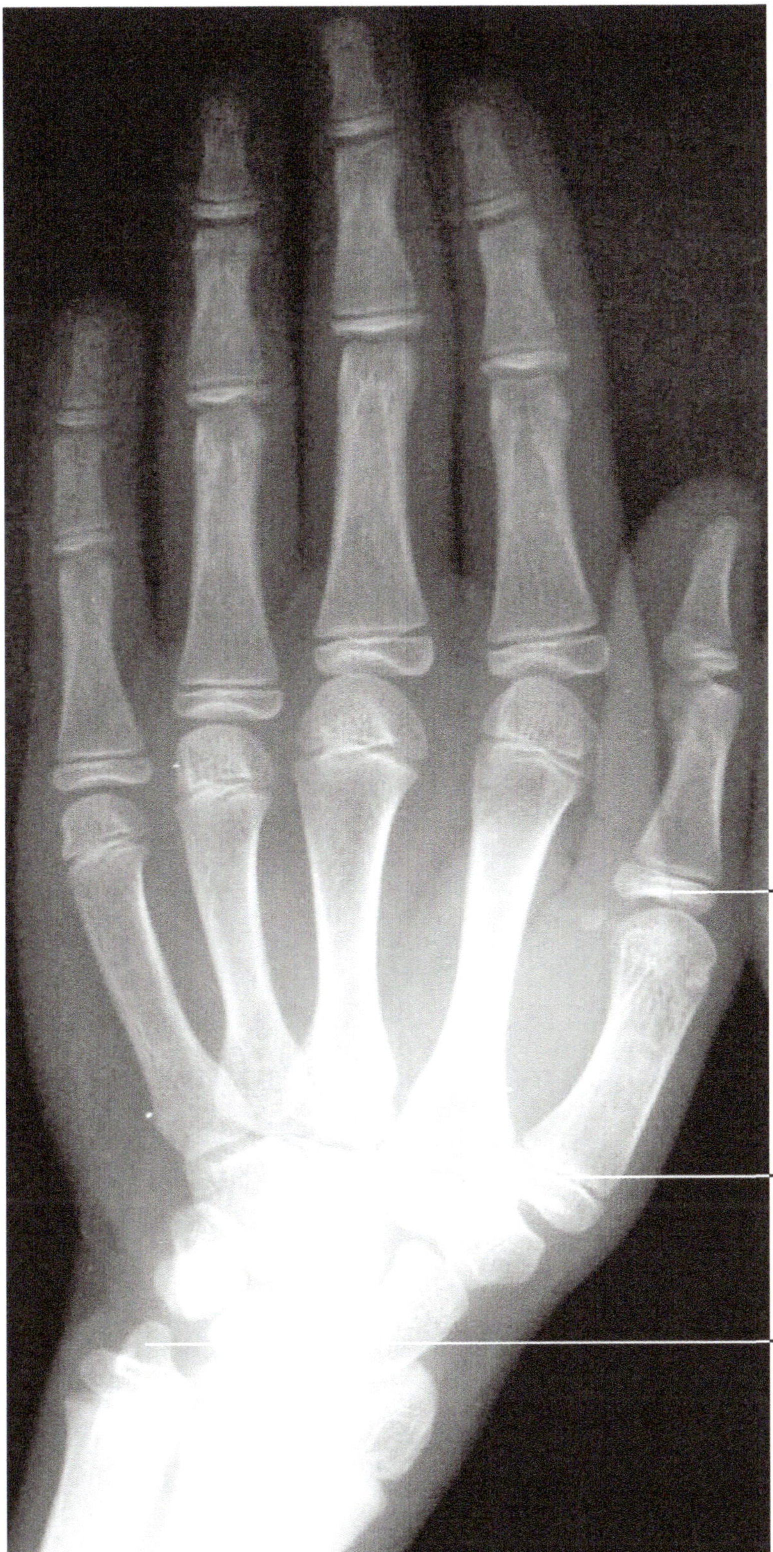

Fig. 34.67 Thirteen years old

GPSR Compliance

The European Union's (EU) General Product Safety Regulation (GPSR) is a set of rules that requires consumer products to be safe and our obligations to ensure this.

If you have any concerns about our products, you can contact us on ProductSafety@springernature.com

In case Publisher is established outside the EU, the EU authorized representative is:

Springer Nature Customer Service Center GmbH
Europaplatz 3
69115 Heidelberg, Germany

Batch number: 10755835

Printed by Printforce, the Netherlands